AF812971

Volume 5 · 1985 · Number 4

Acta Biotechnologica

Journal of microbial, biochemical and bioanalogous technology

Akademie-Verlag
Berlin

ISSN 0138-4988
Acta Biotechnol., Berlin 5 (1985) 4, 319–382
EVP 30,– M

Instructions to Authors

1. Only original papers that have not been published previously will be accepted.
   Manuscripts may be submitted in English, German or Russian (in duplicate). The name of the institute (with the full address) from which the manuscript originates should be stated below the name(s) of the author(s). The authors are responsible for the content of their contributions.

2. Original papers should not exceed 20 typewritten pages (double spaced), including references, tables and figures; short original communications may contain a maximum of six typewritten pages.

3. Each paper should be preceded by a summary in English and the title in English.

4. Latin names of species as well as passages to be printed in italics for greater emphasis should be marked by a waving line. Please use only units and symbols of the SI-system.

5. Tables may be used to shorten the text or to make it more comprehensible. They should be numbered consecutively throughout the text and be supplied with a brief heading. They should not appear in the text, but should be written on separate sheets.

6. The numbers and sizes of illustrations should be limited to a minimum, they should be numbered consecutively and be quoted on separate sheets.
   Line drawings, including graphs and diagrams, should be drawn in black ink. Half-tone illustrations should be presented as white glossy prints. Figure legends are to be typed in sequence on a separate sheet.
   The back of each sheet should bear the name(s) of the author(s).

7. References listed at the end of the contribution should contain only works quoted in the text. They should be numbered in the order in which they are first mentioned in the text.
   Please give surnames and initials of all authors, the name of the journal abbreviated according to "Chemical Abstracts — List of Periodicals", volume number, year of publication, issue number or month, first page number. Books are to be cited with full title, edition, volume number, page number, place of publication, publisher and year of publication.

8. Notes to the text may be presented as footnotes on the same page.

9. 50 offprints are free of charge.
   Additional ones may be ordered on payment.

10. The author will receive two galley proofs for correction. They are to be returned to the managing editor (DDR - 7050 Leipzig, Permoserstr. 15, Dr. Dimter) as soon as possible.

Acta Biotechnologica

Journal of microbial, biochemical and bioanalogous technology

Edited at the
Institute of Biotechnology of the Academy of Sciences of the G.D.R.; Leipzig
and by the
Institute of Technical Microbiology; Berlin
by
M. Ringpfeil, Leipzig and G. Vetterlein, Berlin

Editorial Board:

A. A. Bajev, Moscow
M. E. Beker, Riga
H. W. Blanch, Berkeley
S. Fukui, Kyoto
H. G. Gyllenberg, Helsinki
G. Hamer, Zurich
J. Holló, Budapest
M. V. Iwanow, Moscow
F. Jung, Berlin
H. W. D. Katinger, Vienna
K. A. Kalunjanz, Moscow
J. M. Lebeault, Compiegne
D. Meyer, Leipzig
P. Moschinski, Łodz
A. Moser, Graz
M. D. Nicu, Bucharest
Chr. Panajotov, Sofia
L. D. Phai, Hanoi
H. Sahm, Jülich
W. Scheler, Berlin
R. Schulze, Köthen
B. Sikyta, Prague
G. K. Skrjabin, Moscow
M. A. Urrutia, Habana
J. E. Zajic, El Paso

Managing Editor: L. Dimter, Leipzig

Volume 5

1985

Number 4

AKADEMIE-VERLAG · BERLIN

"Acta Biotechnologica" publishes original papers, short communications, reports and reviews from the whole field of biotechnology.
The journal is to promote the establishment of biotechnology as a new and integrated scientific field.
The field of biotechnology covers microbial technology, biochemical technology and the technology of synthesizing and applying bioanalogous reaction systems.
The technological character of the journal is guaranteed by the fact that papers on microbiology, biochemistry, chemistry and physics must clearly have technological relevance.

Terms of subscription for the journal "Acta Biotechnologica"
Orders can be sent

- **in the GDR:** to Postzeitungsvertrieb, or to the Akademie-Verlag Berlin, DDR - 1086 Berlin, Leipziger Str. 3—4; PF-Nr. 1233;
- **in the other socialist countries:** to a book-shop for foreign languages literature or to the competent news-distributing agency;
- **in the FRG and Berlin (West):** to a book-shop or to the wholesale distributing agency Kunst und Wissen, Erich Bieber, Wilhelmstr. 4—6, D-7000 Stuttgart 1;
- **in the other Western European countries:** to Kunst und Wissen, Erich Bieber GmbH, Dufourstr. 51, CH-8008 Zürich;
- **in other countries:** to the international book- and journal-selling trade, to Buchexport, Volkseigener Außenhandelsbetrieb der DDR, DDR-7010 Leipzig, Postfach 160, or to the Akademie-Verlag Berlin, DDR-1086 Berlin, Leipziger Str. 3—4.

Acta Biotechnologica
Herausgeber: Institut für Biotechnologie der AdW
DDR-7050 Leipzig, Permoserstr. 15
(Direktor: Prof. Dr. Manfred Ringpfeil) und
Institut für Technische Mikrobiologie
DDR-1017 Berlin; Alt-Stralau 62
(Direktor: Dipl. Ing. G. Vetterlein)
Verlag: Akademie-Verlag Berlin, DDR-1086 Berlin, Leipziger Straße 3—4;
Fernruf: 2236201 und 2236229; Telex-Nr.: 114420; Bank: Staatsbank der DDR, Berlin, Konto-Nr.: 6836-26-20712.
Redaktion: Dr. Lothar Dimter (Chefredakteur),
Käthe Geyler (Redakteur),
DDR-7050 Leipzig, Permoserstr. 15; Tel.: 2392255.
Veröffentlicht unter der Lizenznummer 1671 des Presseamtes beim Vorsitzenden des Ministerrates der Deutschen Demokratischen Republik.
Gesamtherstellung: VEB Druckhaus „Maxim Gorki", DDR-7400 Altenburg.
Erscheinungsweise: Die Zeitschrift „Acta Biotechnologica" erscheint jährlich in einem Band mit 4 Heften. Bezugspreis eines Bandes 120,— DM zuzüglich Versandspesen; Preis je Heft 30,— DM. Der gültige Jahresbezugspreis für die DDR ist der Postzeitungsliste zu entnehmen.
Bestellnummer dieses Heftes: 1094/5/4.
Urheberrecht: Alle Rechte vorbehalten, insbesondere der Übersetzung. Kein Teil dieser Zeitschrift darf in irgendeiner Form — durch Photokopie, Mikrofilm oder irgendein anderes Verfahren — ohne schriftliche Genehmigung des Verlages reproduziert werden. — All rights reserved (including those of translation into foreign languages). No part of this issue may be reproduced in any form, by photoprint, microfilm or any other means, without written permission from the publishers.
© 1985 by Akademie-Verlag Berlin. Printed in the German Democratic Republic.
AN (EDV) 42133

03000

Acta Biotechnol. 5 (1985) 4, 321–328

The Inhibitory Effects of Pyrocarbonic Acid Diethyl Ester on Lactobacillus casei and Its Phage Jl — A Novel Strategy for Phage Control in Technical Fermentation Processes

MURATA, A.

Department of Agricultural Chemistry
Saga University
Saga 840, Japan

This paper was presented at the Leipzig Biotechnology Symposium 1984, Leipzig, September 12, 1984

Summary

The effects of pyrocarbonic acid diethyl ester (PADE) on *Lactobacillus casei* Sl and its phage Jl was investigated in relation to the control of phages in the dairy industry and other technica fermentation processes.
PADE exhibited a bacteriostatic effect at 0.5 to 8 mM and a bactericidal effect at 10 mM or higher. It inhibited the growth of the phage at its bacteriostatic and bactericidal concentrations. The growth inhibition of the phage was reversible at the bacteriostatic concentrations but complete and irreversible at the bactericidal concentrations. PADE inactivated the free phage within several minutes; 10 and 30 mM of PADE inactivated 90 and 100%, respectively, of the phage. It completely decomposed into ineffective components in several minutes. The bacteria grew almost normally when they were inoculated after the complete decomposition of PADE.
These four characteristics of PADE — its bactericidal effect, its inhibitory effect on phage growth, its phage-inactivating effect and its decomposition — suggest a novel strategy for phage control in technical fermentation processes, including the dairy industry.

Introduction

In the dairy industry chemical substances to be added into the lactic cultures for the control of phages are very restricted bacause the culture products are for direct human consumption. Addition of harmful substances is thus precluded.
In view of this problem studies of the effects of harmless substances on phages of lactic acid bacteria are in progress. We have so far studied the effects of ascorbic acid [1, 2, 3, 4, 5, 6, 7], other water-soluble vitamins [8], amino acids [9, 10, 11], amino acid derivatives [12, 13, 14, 15], thiol compounds [2, 16, 17, 18] and food preservatives [19, 20, 21, 22], using *Lactobacillus casei*-phage Jl system.
Pyrocarbonic acid diethyl ester* (PADE) is a substance which rapidly decomposes in aqueous solution into harmless components; ethyl alcohol and carbon dioxide. It has

$$(C_2H_5O-CO)_2O \xrightarrow{H_2O} 2\,C_2H_5OH + 2\,CO_2$$

* Dicarbonic acid diethyl ester; diethyl pyrocarbonate, DEPC.

a strong bactericidal effect, and had been developed and used as a preservative for beverages, especially wines, soft drinks and fruit juices [23].
FEDORCSÁK and TURTÓCZKY [24] and KONDOROSI et al. [25] have described that PADE is a potential phage-inactivating agent. On the other hand, no paper has been published on the growth inhibition of phages by PADE, although PADE is expected to inhibit phage growth at least under conditions which inhibit bacterial growth. It appears that nobody has ever studied PADE from the aspect of controlling phages in technical fermentation processes, including the dairy industry. The present paper describes the inhibitory effects of PADE on *Lactobacillus casei* and its phage Jl in relation to the control of phage problems in the dairy industry and other technical fermentation processes.

Materials and Methods

Phage and bacterial strain: Phage Jl (ATCC 27139-B) and *Lactobacillus casei* Sl (ATCC 27139) as a host strain were used throughout the experiments. *Lactobacillus casei* is employed in the production of Lactobacillus drinks in Japan and several other countries. Phage Jl was isolated from a culture which was slow in acid production by *Lactobacillus casei* Sl [26].

Medium: MRT medium, pH 6.0, [27] was used throughout the experiments. It had the following composition per liter of deionized water: 10 g glucose, 10 g polypeptone, 10 g sodium acetate, 5 g yeast extract, 3 g beef extract, 1.5 g $CaCl_2 \cdot 2H_2O$, 1 g NaCl, 0.2 g $MgSO_4 \cdot 7H_2O$, 0.01 g $MnSO_4 \cdot 4H_2O$ and 0.001 g $FeSO_4 \cdot 7H_2O$.

Chemicals: Pyrocarbonic acid diethyl ester (PADE) was supplied by MATSUDA (Research Institute, Ueno Fine Chemical Industries). Before each experiment PADE was diluted in ethyl alcohol to 100 times the desired concentration, and 0.1 ml of the 100 × PADE solution was added to 9.9 ml of the medium containing the bacteria and/or the phage at time zero. The medium therefore contained 1% ethyl alcohol, but this concentration of ethyl alcohol had no effect on the bacteria and the phage. Other chemicals were from commercial sources.

Growth conditions: The bacteria and phage were grown without aeration at 37 °C. The bacteria culture of logarithmic phase cells was prepared by addition of an overnight culture to 20 volumes of the medium and subsequent incubation for 3 to 4 hours. The bacterial growth in a Turbid-ell-cell (Fujimoto Rikaki) was measured at 620 nm in a Spectronic 20 spectrophotometer (Bausch and Lomb-Shimadzu).

Assay of viable bacteria and phage: Colony-formable bacteria and plaqeu-formable phage were assayed by the modified double-layer method [27, 28]. Intracellular mature phage was assayed by the method utilizing the premature lysis of infected cells by *n*-butyl-*p*-hydroxybenzoate [19], i. e., the infected culture was added with 2 mM *n*-butyl-*p*-hydroxybenzoate, and this mixture was incubated for about 40 minutes at 37 °C.

One-step growth experiment: The bacteria, 2×10^8 cells/ml, were infected with the phage at a multiplicity of infection of 0.1. After 2 minutes of adsorption and 1 minute of an anti-phage serum treatment, the phage-infected cells were diluted 1 : 2000 in the medium and incubated at 37 °C.

Others: Other experimental methods were the same as those described by ADAMS [29].

Results

Effect of PADE on bacterial growth

Different concentrations of PADE were added to growing cultures of *Lactobacillus casei* Sl, and the bacterial growth was followed by measuring the turbidity and counting the viable cells.

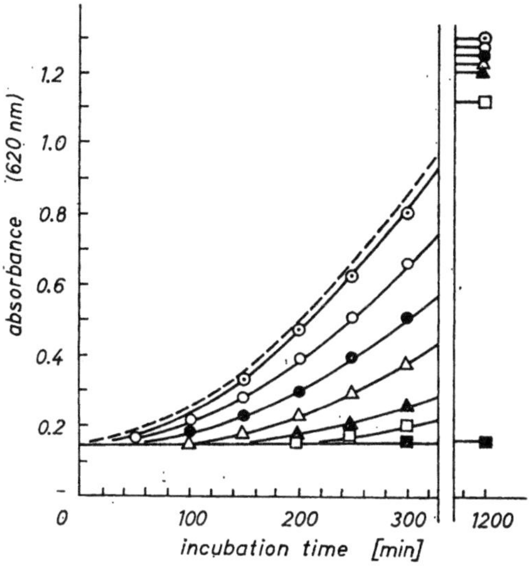

Fig. 1. Effect of PADE on growth curve of *Lactobacillus casei*. Bacterial cells (2×10^8/ml) were incubated with PADE at 37 °C.
Concentration of PADE (mM): ---, 0; -⊙-, 0.5; -○-, 1; -●-, 2; -△-, 4; -▲-, 6; -□-, 8; -■-, 10 and 30

Fig. 1 shows the effect of PADE on the growth curve of the bacteria. PADE at a concentration of 0.5 mM rendered a small effect on the growth. PADE, at higher concentrations up to 8 mM, produced a growth lag, and the lag became longer with the increase in PADE concentrations. After the lag the turbidity increased at a reduced rate, and the ultimate turbidity was nearly equal to that of the PADE-free control culture. On the

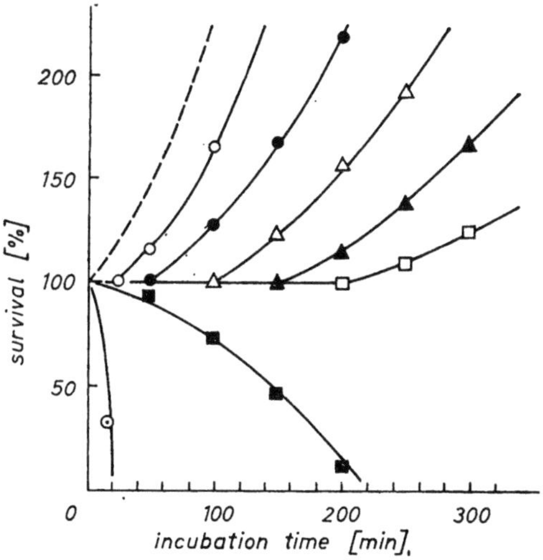

Fig. 2. Effect of PADE on colony-forming ability of *Lactobacillus casei*.
Bacterial cells (2×10^8/ml) were incubated with PADE at 37 °C. The initial number of cells is represented as 100%.
Concentration of PADE (mM): ---, 0; -○-, 1; -●-, 2; -△-, 4; -▲-, 6; -□-, 8; -■-, 10; -⊙-, 30

other hand, complete growth inhibition was achieved at 10 mM PADE; the turbidity failed to increase for a longer period up to 1200 minutes.
Fig. 2 shows the effect of PADE on the viability (colony-forming ability) of bacterial cells. At 8 mM or lower there was neither increase nor decrease in viable cell count during the lag, indicating the bacteriostatic action of PADE in these concentrations. After the lag the viable cell count increased. On the other hand, there was decrease in the viable cell count at 10 mM or higher, indicating the bactericidal action of PADE in these concentrations. Once the viable cell count had reduced to zero, no increase in the viable cell count occurred even after a longer period incubation up to 1200 minutes. This indicates that the killing by PADE is complete. The results coincide with the above-mentioned observation that there was no increase in the turbidity at 10 mM, but the turbidity increased after the lag at 8 mM or lower. In order to measure the time required

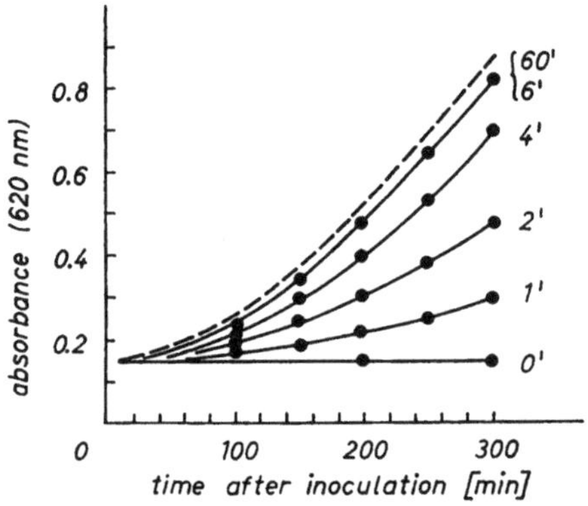

Fig. 3. Effect of preincubation of PADE-added medium on growth of *Lactobacillus casei*.
PADE (10 mM) was added to MRT medium, and the PADE-medium was preincubated for the times indicated in the figure at 37 °C. Then, bacterial cells (2×10^8/ml) were added to the PADE-medium, and this was incubated further at 37 °C.
---: PADE-free control

for the complete decomposition of PADE under the conditions of this experiment, PADE was added to a fresh medium, and the PADE-added medium was preincubated for different times before adding bacteria. Fig. 3 shows the growth curve of the bacteria added to the PADE-medium. The results indicate that PADE decomposes completely in 6 minutes. The bacteria grew almost normally when they were added after PADE had completely decomposed.
Thus, the growth of bacteria after the lag in bacteriostatic concentrations of PADE (Fig. 1 and 2) may be explained by the decomposition of PADE. It should be noted, however, that the bacterial growth did not take place soon after the decomposition of PADE, e. g., there was a lag of about 200 minutes at 8 mM.

Effect of PADE on phage growth

Fig. 4 shows the effect of different concentrations of PADE on the one-step growth curve of phage Jl. PADE at a concentration of 0.5 mM had little effect on the phage growth. Progressively higher concentrations of PADE, up to 8 mM, resulted in progressively longer latent period, e. g., at 8 mM the latent period was about 3 times longer than the PADE-free control. On the other hand, PADE at these concentrations hat little effect on the burst size. It was remarkable that, during the prolonged latent period,

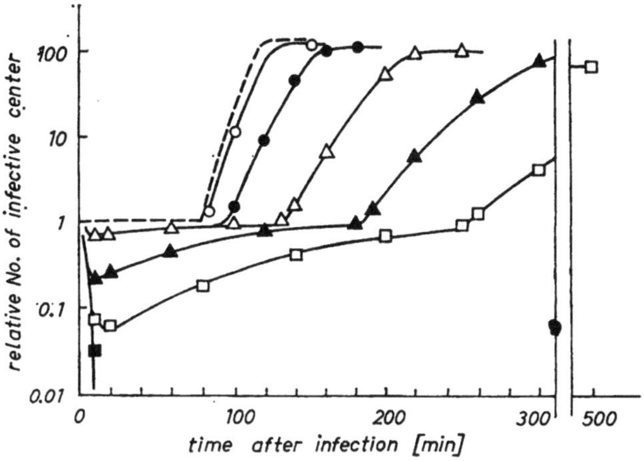

Fig. 4. Effect of PADE on one-step growth curve of phage Jl. PADE was added 3 minutes after infection. The number of initial infected cells is represented as 1. Concentration of PADE (mM): ---, 0; -○-, 0.5; -●-, 2; -△-, 4; -▲-, 6; -□-, 8; -■-, 10

the number of infective center rapidly decreased for the first several minutes and thereafter gradually increased to become nearly equal to the initial number. Since PADE completely decomposes in several minutes, the decrease and increase in the number of infective center may be explained by the inhibition of plaque-forming ability of infected cells by PADE and its recovery after the decomposition of PADE. PADE at 10 mM or higher produced a rapid decrease in the number of infective center, and thereafter no increase in the number of infective center occurred even after a longer period incubation up to 1200 minutes.

It is concluded that the growth inhibition of the phage by bacteriostatic concentrations of PADE is reversible, i. e., the phage burst begins to occur after the decomposition of PADE, and that the growth inhibition of the phage by bactericidal concentrations of PADE is complete and irreversible.

Effect of PADE on intracellular phage growth

Fig. 5 shows the effect of addition of 10 mM PADE at different times during the latent period on the intracellular growth curve of phage Jl. No mature phage particles were formed when PADE was added within 50 minutes after infection, i. e., before the appearance of the first mature phage particle. When PADE was added after the accumulation of mature phage particles had been initiated, the number of infective center rapidly decreased for several minutes after the addition of PADE, and thereafter the decrease in the number of infective center ceased.

The results indicate that the addition of PADE at any time during the second half of the latent period results in an immediate inhibition of the maturation. The reason why the number of infective center decreases after the addition of PADE may be explained by the inhibition of plaque-forming ability of the infected cells by PADE and the inactivation of newly matured phage particles by PADE (cf. Fig. 6).

Effect of PADE on free phage

Fig. 6 shows the effect of different concentrations of PADE on free phage Jl. PADE at a concentration of 3 mM had little effect on the phage. PADE at higher concentrations inactivated the phage; 10, 15 and 20 mM of PADE inactivated 90, 99 and 99.9%,

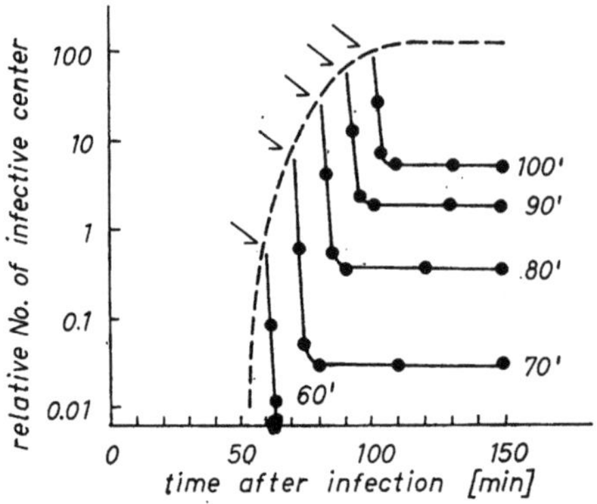

Fig. 5. Effect of delayed addition of PADE on intracellular growth curve of phage Jl. PADE (10 mM) was added to an aliquot from one-step growth tube at the times indicated in the figure, and the culture was incubated further at 37 °C. The number of initial infected cells is represented as 1.

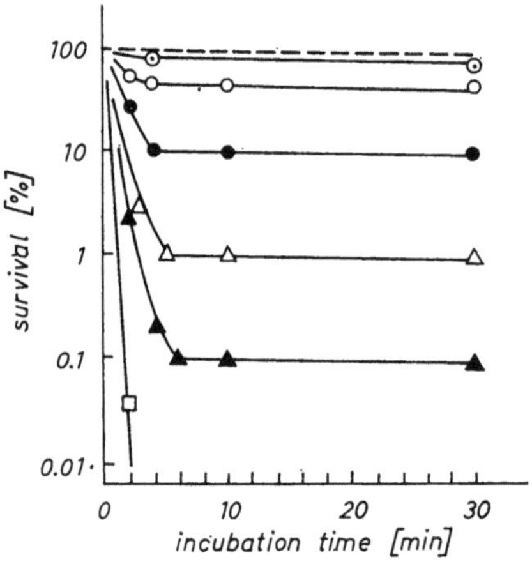

Fig. 6. Effect of PADE on free phage Jl.
Phage (2×10^8 PFU/ml) was incubated with PADE at 37 °C.
Concentration of PADE (mM): ---, 0; -⊙-, 3; -○-, 5; -●-, 10; -△-, 15; -▲-, 20; -□-, 30

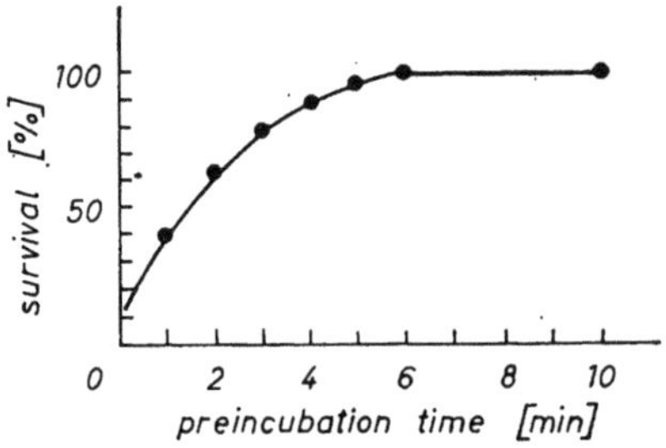

Fig. 7. Effect of preincubation of PADE-added medium on inactivation of phage Jl. PADE (10 mM) was added to MRT medium, and the PADE-medium was preincubated for the indicated times at 37 °C. Then, phage (2×10^8 PFU/ml) was added to the PADE-medium, and this was incubated further for 10 minutes at 37 °C.

respectively, of the phage. The plaque-forming unit rapidly decreased for the first several minutes, and thereafter the decrease in the plaque-forming unit ceased. Complete inactivation was achieved at 30 mM PADE; the phage at the order of 10^8 PFU per ml decreased to zero in a few minutes. In order to confirm that the plateau after the initial decrease is due to the decomposition of PADE into ineffective components, the following experiment was carried out. PADE was added to a fresh medium, and the PADE-added medium was preincubated for different times before adding phage. As shown in Fig. 7, it is evident that PADE decomposes completely within 6 minutes (cf. Fig. 3), and also that the plateau is due to the decomposition of PADE.

Discussion

The purpose of the present work is to see if PADE could be useful for the control of phages in the dairy industry. As to the effects of PADE on *Lactobacillus casei* and its phage Jl, the following results were obtained: (1) PADE inhibits the bacterial growth. (2) It inhibits the phage growth. (3) It inactivates the free phage. Since PADE exhibits no selective action in this system, it cannot be used in the conventional manner when fermenting culture is infected with phage. Advantageously, PADE rapidly decomposes in medium into ineffective components, and the bacteria grow almost normally when they are inoculated after PADE has undergone decomposition.

Taking all of the results into consideration, PADE could be used for the control of phages in two ways. (1) When culture medium (raw milk) is contaminated with phage, PADE should be added before inoculation of starter culture to inactivate the phage (and also bacterial contaminants). A starter culture is then inoculated after PADE has undergone decomposition. (2) When a fermenting culture (an acid-producing culture) is infected with phage, PADE should be added to inactivate free phage and phage-infected bacterial cells completely. The PADE-treated culture must be reinoculated with another starter culture after PADE has undergone decomposition, since uninfected bacterial cells are also completely killed by the PADE treatment. Thus, PADE could be useful for the control of phages in the dairy industry where harmful substances cannot be used. However, it should be noted that the possible formation of urethane (a carcinogenic substance) has been pointed out in beverages treated with PADE [30, 31].

PADE could be rather useful for the control of phage problems in technical fermentation processes where the use of PADE is not restricted. Our preliminary experiments revealed that PADE had similar inhibitory effects on several other bacteria and their phages, including *Escherichia coli* and its T-series phages, and *Bacillus subtilis* and its phages (data not shown).

Four characteristics of PADE — its bactericidal effect, its inhibitory effect on phage growth, its phage-inactivating effect and its decomposition into ineffective components — suggest a novel strategy for phage control in technical fermentation processes, including the dairy industry.

Received November 15, 1984

References

[1] MURATA, A., KITAGAWA, K., SARUNO, R.: Agric. Biol. Chem. **35** (1971), 294.
[2] MURATA, A., KITAGAWA, K., INMARU, H., SARUNO, R.: Agric. Biol. Chem. **36** (1972), 2597.
[3] MURATA, A., KITAGAWA, K.: Agric. Biol. Chem. **37**. (1973), 1145.

[4] MURATA, A.: Proc. 1st Int. Congr. Int. Assoc. Microbiol. Soc. Vol. 3 (ed. by T. HASEGAWA) Science Council of Japan Tokyo 1975, p. 432.
[5] MURATA, A., OYADOMARI, R., OHASHI, T., KITAGAWA, K.: J. Nutr. Sci. Vitaminol. **21** (1975), 261.
[6] MURATA, A., UIKE, M.: J. Nutr. Sci. Vitaminol. **22** (1976), 347.
[7] MURATA, A., SUENAGA, H., INOUE, M., TANAKA, Y., KATO, F.: Bitamin **57** (1983), 515.
[8] MURATA, A., SAKAI, S., ODA, K., NAKATSUMI, K., KATO, F.: J. Nutr. Sci. Vitaminol. **29** (1983), 721.
[9] MURATA, A., TANAKA, K., NAKATSUMI, K., GENNO, S., KATO, F., SARUNO, R., KONDO, M.: Nippon Nogeikagaku Kaishi **52** (1978), 385.
[10] MURATA, A., TANAKA, K., MUKUNO, S., ODAKA, M., GENNO, M., KATO, F., SARUNO, R.: Nippon Nogeikagaku Kaishi **52** (1978), 393.
[11] MURATA, A., TANAKA, K., SHIBUYA, T., IMAZU, K., KATO, F., KONDO, M.: Nippon Nogeikagaku Kaishi **53** (1979), 157.
[12] KONDO, M., SHIMIZU, Y., MURATA, A.: Agric. Biol. Chem. **46** (1982), 913.
[13] KONDO, M., MIYAZAKI, K., YADA, Y., HORIMOTO, H., SAMOTO, M., MURATA, A.: Agric. Biol. Chem. **48** (1984), 1263.
[14] MURATA, A., SAMOTO, M., FUKADA, H., IMAZU, K., NISHI, S., KATO, F., KONDO, M.: Nippon Nogeikagaku Kaishi **58** (1984), 471.
[15] MURATA, A., SAMOTO, M., FUKADA, H., IZUMI, T., YAMAGUCHI, Y., KATO, F., KONDO, M.: Nippon Nogeikagaku Kaishi **58** (1984), 695.
[16] MURATA, A., KITAGAWA, K., INMARU, H., SARUNO, R.: Agric. Biol. Chem. **36** (1972), 1065.
[17] MURATA, A., KITAGAWA, K., OTOKONARI, K., SARUNO, R.: Agric. Biol. Chem. **37** (1973), 1707.
[18] MURATA, A., KITAGAWA, K.: Agric. Biol. Chem. **37** (1973), 2159.
[19] MURATA, A., SHIROURA, Y.: Nippon Nogeikagaku Kaishi **47** (1973), 65.
[20] MURATA, A.: Nippon Nogeikagaku Kaishi **47** (1973), 217.
[21] MURATA, A., IKEDA, M., MITSUTAKE, T., SARUNO, R.: Nippon Nogeikagaku Kaishi **47** (1973) 267.
[22] MURATA, A., MITSUTAKE, T.: Agric. Biol. Chem. **37** (1973), 1763.
[23] FURIA, T. A.: Handbook of Food Additives. Chemical Rubber Cleveland 1968 p. 181.
[24] FEDORCSÁK, I., TURTÓCZKY, I.: Nature **209** (1966), 830.
[25] KONDOROSI, A., SVÁB, Z., SOLYMOSY, F., FEDORCSÁK, I.: J. Gen. Virol. **16** (1972), 373.
[26] HINO, M., IKEBE, N.: Nippon Nogeikagaku Kaishi **39** (1965), 472.
[27] MURATA, A., SOEDA, E., SARUNO, R.: Nippon Nogeikagaku Kaishi **43** (1969), 311.
[28] MURATA, A., SOEDA, E., SARUNO, R.: Nippon Nogeikagaku Kaishi **44** (1970), 262.
[29] ADAMS, M. H.: Bacteriophages. Interscience Publishers New York 1959 p. 443.
[30] LOFROTH, G., GEJVALL, T.: Brauwelt **112** (1972), 371.
[31] SCHMAEHL, D., PORT, R., WAHRENDORF, J.: Int. J. Cancer **19** (1977), 77.

Acta Biotechnol. 5 (1985) 4, 329—332

Isolierung von Bordetella-Phagen aus Bakterienstämmen der Impfstoffproduktion

MEBEL, S.[1], RUSTENBACH, S.[1], LAPAEVA, I. A.[2]

[1] Staatliches Institut für Immunpräparate und Nährmedien
DDR-1120 Berlin, Klement-Gottwald-Allee 317—321
[2] Akademie der Medizinisches Wissenschaften der UdSSR
Gamaleya-Institut für Epidemiologie und Mikrobiologie Moskau, UdSSR

Vortrag auf dem Leipziger Biotechnologiesymposium 1984, 10.—14. 9. 1984

Summary

It has been demonstrated that strains of *Bordetella pertussis* used for vaccine production contain temperate phages. It can be conducted from many experiments performed in our laboratory, that 10—100 phages per 10^{10} bacteria are released. However, the production of bacterial mass is not markedly influenced by lysogeny.
Strains of *Bordetella bronchiseptica* used for production of vaccine against *Rhinitis atrophicans* of pigs have temperate phages too. These phages may cause a complete lysis during a submerse cultivation.
The phages of *Bordetella pertussis* and *Bordetella bronchiseptica* can be propagated on *Bordetella parapertussis*.

Nach wie vor ist der Keuchhusten eine Infektion, die zwar in der Gesamtmorbidität der DDR durch eine langjährige gezielte Prophylaxe keine Bedeutung mehr hat. In den letzten Jahren wurden ca. 170 bis 200 Erkrankungen gemeldet. Dieser Stand ist jedoch nur durch eine konsequent weiterzuführende Immunisierung zu halten. Aus diesem Grund werden in der DDR jährlich über 1 Millionen Impfungen gegen Keuchhusten an Kleinkinder verabfolgt. Unter diesem Aspekt ist jede neue Information, die wir über die Eigenschaften unserer Produktionsstämme erhalten, von Bedeutung.
Bei der Untersuchung einer angeblich homogenen Kultur eines *Bordetella pertussis*-Stammes kann eine große Variabilität innerhalb der Kultur festgestellt werden, die sich sowohl in unterschiedlichen serologischen als auch toxischen und protektiven Eigenschaften ausdrückt [1]. Die Gründe dieser Variabilität der *Pertussis*-Stämme sind vielfältig aber nicht geklärt. Sie haben jedoch bei der Produktion von Impfstoffen eine große Bedeutung, da die Antigene, die für die Schutzeigenschaften der Vakzine verantwortlich sind, in einer von der WHO vorgegebenen Stärke vorhanden sein müssen [2].
Während Phagen bei *Bordetella pertussis* nicht bekannt waren, sind bei der nahe verwandten Species *Bordetella bronchiseptica* schon 1961 durch RAUCH und PICKETT Phagen nachgewiesen worden, die sich auf der 3. Species des Genus — *Bordetella parapertussis* — mittels GRATIA-Test darstellen ließen [3].
Vereinzelt gibt es in der Literatur Hinweise, daß bei nicht näher zu definierenden Umständen [4] eine völlige Lyse submers gezüchteter *Pertussis*-Kulturen stattfindet. Diese Prämissen veranlaßten uns, zumindest hypothetisch, die Existenz von *Bordetella pertussis*-Phagen anzunehmen.

Die größte Schwierigkeit, auf die wir beim Nachweis von *Bordetella pertussis*-Phagen stießen, war, ein System zu finden, um die Phagen darzustellen. *Bordetella pertussis* ist ein Keim mit außerordentlich hohen Ansprüchen an das Medium. Er wächst auf festen Spezialnährböden nur unter Zusatz von mindestens 20% Blut oder Zusatz von Aktivkohle. Auch auf diesen Nährböden bildet *Bordetella pertussis* erst innerhalb von 48—72 h einen sehr feinen Rasen, auf dem sich keine Phagen darstellen ließen.
In Anbetracht der sehr engen Verwandtschaft der 3 Species des Genus *Bordetella*, die für *Bordetella pertussis* und *Bordetella parapertussis* durch japanische Autoren [5] auch genetisch nachgewiesen wurde, versuchten wir, ein System aufzubauen, in dem *Bordetella parapertussis* als Indikatorstamm für *Bordetella pertussis*-Phagen dient. *Bordetella parapertussis* zeigt innerhalb von 24 h auf oben genannten Nährböden ein gutes Wachstum und läßt sich nach mehreren Passagen auf einer Reihe Nährböden adaptieren, z. B. an McConkey-Agar, Caseinhydrolysat-Agar nach Cohen-Wheeler u. a. mehr. Den Caseinhydrolysat-Agar nach Cohen-Wheeler verwendeten wir mit verschiedenen Agarkonzentrationen für beide Schichten im Gratia-Test.

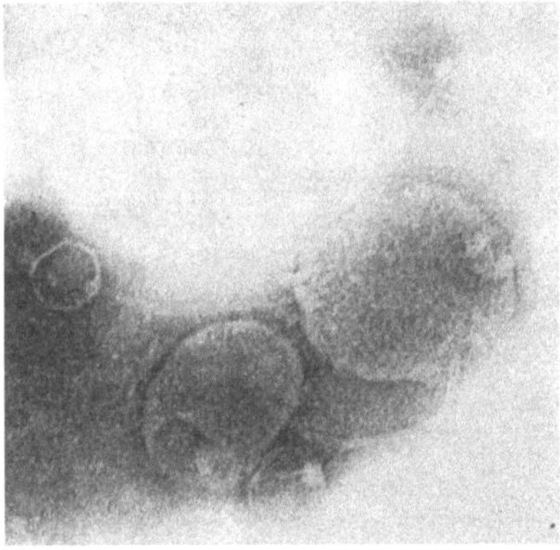

Abb. 1. Bakteriophage aus *Bordetella pertussis* 134,
Negativkontrastierung mit Ammoniummolybdat (× 120000)

Zur Gewinnung der Phagen wurde der lyophilisierte Produktionsstamm auf Medium nach Bordet-Gengou 72 h kultiviert, dann weiter auf festem Medium passagiert, 48 h angezüchtet und davon eine Bakteriensuspension mit 30 IOU (international opacity units) in phosphatgepufferter 0,9%iger NaCl-Lösung angesetzt. Die Suspension wird mit Chloroform versetzt, 2 h bei 20 °C, dann über Nacht im Kühlschrank belassen und anschließend der Überstand nach Gratia titriert.
Mit dieser Methode war es möglich, in allen von uns untersuchten Produktionsstämmen, die ohne Ausnahme auch international für die Impfstoffproduktion verwendet werden, Phagen nachzuweisen und unsere Hypothese vom *Bordetella pertussis*-Phagen zu bestätigen. In einer Bakteriensuspension von 10 IOU, die 10×10^9 Keime *Bordetella pertussis* entsprechen, wurden 10^1 bis 10^2 Plaques gefunden. Insgesamt wurden 8 Produktionsstämme untersucht, wobei Bakteriophagen aus *Bordetella pertussis* 134 und 509 aus Originalampullen, erhalten aus dem Referenzlabor der WHO Moskau und aus dem Rijksinstitut voor Volksgezondheid Bilthoven, Niederlande, gewonnen werden konnten (Abb. 1 und 2).
Wir stellten fest, daß keiner der von uns isolierten Bakteriophagen sich an homologen oder heterologen *Bordetella pertussis*-Stämmen vermehren und darstellen ließ. Jedoch konnten die Phagen in allen in den Versuch genommenen *Bordetella parapertussis*-Stämmen (11 Stämme) und teilweise in *Bordetella bronchiseptica*-Stämmen vermehrt werden.

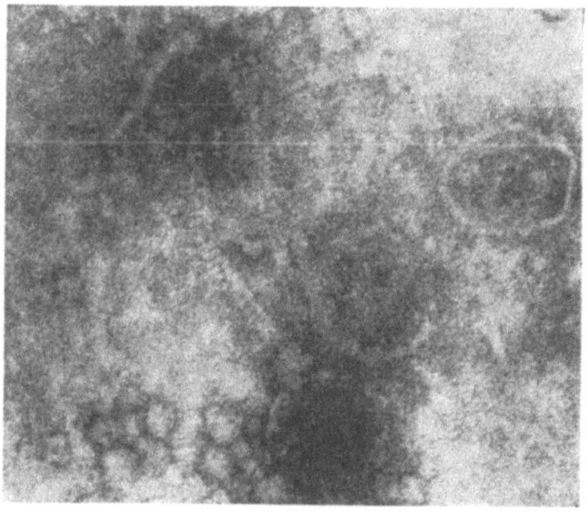

Abb. 2. Bakteriophage aus *Bordetella pertussis* 509, Negativkontrastierung mit Uranylacetat (× 240000)

Am Rande sei erwähnt, daß die Lysogenie offensichtlich bei *Bordetella pertussis* außerordentlich verbreitet ist, und wir auch bei allen von uns untersuchten Patientenstämmen temperierte Phagen isolieren konnten.

Species *Bordetella bronchiseptica* ruft in seltensten Fällen Erkrankungen des Menschen hervor, ist aber die Ursache schwerer Erkrankungen einer ganzen Reihe unserer Nutztiere. Hohe Verluste erleidet die Schweinezucht durch die *Rhinitis atrophicans* der Ferkel, die in einer Reihe Bestände endemisch ist. Da bislang keine Impfstoffe zur Bekämpfung dieser Seuche zur Verfügung standen, wurde uns gemeinsam mit dem Kollektiv des Instituts für Tierseuchenlehre der Humboldt-Universität die Aufgabe gestellt, einen wirksamen Impfstoff zu entwickeln. Die für die Produktion nach bestimmten Kriterien ausgewählten *Bordetella bronchiseptica*-Stämme erwiesen sich alle als lysogen. Bei Anzucht der Stämme auf festen Nährböden konnten die Phagen nach der für *Bordetella pertussis* erwähnten Methode gewonnen und an *Bordetella parapertussis* als Indikatorstamm dargestellt werden (Abb. 3).

Bei der Produktion von Bakterienmasse aus *Bordetella bronchiseptica* für den Impfstoff in Submerskulturen beobachteten wir bei sämtlichen angewandten Produktionsstämmen

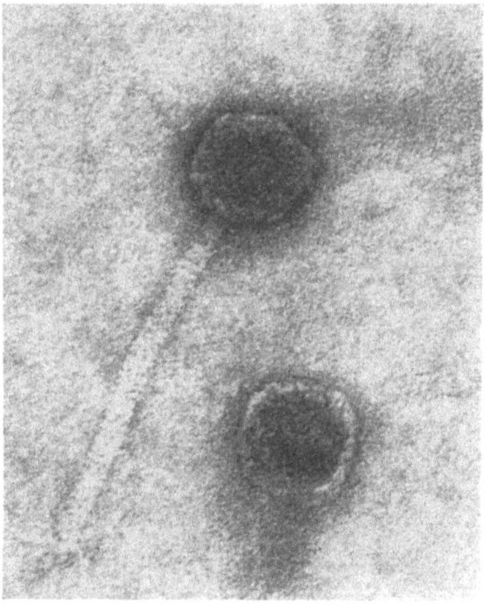

Abb. 3. Bakteriophage aus *Bordetella bronchiseptica*, Negativkontrastierung mit Uranylacetat (× 240000)

spontane Lyse. Ein Teil der Keime ist offensichtlich phagenresistent. Diese Keime vermehren sich bei Fortführung der Submerskultivierung. Die Ursachen der spontanen Infektiosität der Phagen konnten bislang nicht geklärt werden. Diese Fragen bedürfen dringend weiterer Bearbeitung.
Grundsätzlich ist heute nicht geklärt, ob und welche Bedeutung die *Bordetella*-Phagen für die verschiedenen Eigenschaften der 3 Species des Genus haben.
Es ist nicht auszuschließen, daß die Veränderung des Serovars und die Veränderung des Gehaltes an immunologischen Substanzen auf Grund von Phageninfektionen der *Bordetella*-Stämme erfolgt. Konversion verschiedener Eigenschaften von Mikroorganismen im Ergebnis von Phageninfektionen sind auch für andere Genera ausführlich beschrieben.

Die elektronenmikroskopischen Abbildungen wurden von Frau Uerlings in der Abteilung für Elektronenmikroskopie des Pathologischen Instituts der Humboldt-Universität zu Berlin aufgenommen, wofür die Autoren sich bedanken.

Eingegangen: 20. 11. 1984

Literatur

[1] Cameron, J.: Adv. Appl. Microbiol. **20** (1976), 57
[2] WHO Technical Report Series No. 274 (1964).
[3] Rauch, H. C., Pickett, M. J.: Can. J. Microbiol. **7** (1961), 125.
[4] Joó, I. u. a.: Z. Immun.-Forsch. **129** (1965), 244.
[5] Kumazawa, N. H., Yoshikawa, M.: J. Hyg. (Camb.) **81** (1978), 15.

Acta Biotechnol. 5 (1985) 4, 333—338

The Influence of Carbon Catabolism on the Auxiliary Substrate Effect*

BABEL, W., MÜLLER, R. H.

Academy of Sciences of the G.D.R.
Institute of Biotechnology
DDR-7050 Leipzig, Permoserstraße 15

Paper given at the Reinhardsbrunn Symposium
"Physiology of Microbial Growth and Differentiation",
May 20—26, 1984

Summary

Although the carbon/energy ratios of heterotrophic substrates for microbial growth are different this is not reflected in biomass. Nevertheless the macromolecular composition of cells may vary in dependence on growth conditions this does hardly influence the elementary composition and the growth yield. The energy requirement for synthesis of biomass starting from a central precursor, e.g. phosphoglycerate, can be assumed to be constant, hence any differences in carbon conversion efficiency must be attributed to carbon catabolism up to this precursor. This sequence determines if and to what extent an auxiliary substrate effect is possible. However, one has also to consider changes of the P/0 ratio due to simultaneous utilization of substrates which may account for the increase in growth yield with *Hansenula polymorpha* growing on a methanol/glucose mixture.

With chemoorganoheterotrophic growth of microorganisms the carbon substrate is the energy source at the same time. Consequently chemoorganotrophic substrates are all more or less reduced carbon compounds with one or more carbon atoms.
By the oxidation of these compounds their energy is transferred into biologically useful form, where by dehydrogenation (i. e. oxidation of carbon) the generation of ATP via substrate level phosphorylation (ATP_{SLP}) is prepared and by the oxidation of hydrogen in the form of reducing equivalents ATP is synthesized via electron transport phosphorylation (ATP_{ETP}).
The elementary composition of heterotrophic substrates is different. In methane the carbon is most strongly reduced and in oxalate most strongly oxidized. The oxidation number ranges between -4 and $+3$. The oxidation number of the carbon atom, the reduction degree of the substrate and the so-called available electrons reveal the carbon/energy ratio of heterotrophic substrates. Since these parameters are very different it would not be surprising if this constellation were reflected in the cell composition of microorganisms [1].
Depending on growth conditions (i. e. the kind and degree of growth limitation, cf. ref. [2]) and the age of the cells the portion of individual types of macromolecules and polymers varies. To what extent is the elementary composition of the cell influenced by such changes?

* Dedicated to Prof. Dr. F. Mach on occasion of his 60th birthday

Proportion of Cell Constituents and Growth Yield

Proteins are essentially composed of about 20 various amino acids. Assuming these 20 amino acids are represented in equimolecular proportions an elementary composition as shown in Table 1 is obtained. This is compared to those of catalase, methanol oxidase, peroxidase and alcohol oxidase, the former two representing about 20% of the total protein content of methylotrophic yeasts, and to those of the total protein of *Saccharomyces cerevisiae* [3], methanol-grown *Torulopsis* sp. MH 26 and *E. coli* [4]. From these data it is evident that, regardless of considering a specific protein or its mixture in a microorganism, the elementary composition is nearly identical. Furthermore, taking

Table 1. The elementary composition of some proteins

Protein	C	H	O	N
20 amino acids (equimolar, calculated)	4.0	6.2	1.12	1.08
Catalase	4.0	6.12	1.16	1.12
Methanol oxidase	4.0	6.0	1.32	1.02
Peroxidase	4.0	6.35	1.46	0.92
Alcohol dehydrogenase	4.0	6.14	1.05	1.0
S. cerevisiae protein*	4.0	6.33	1.25	1.09
Torulopsis sp. MH 26 protein	4.0	6.32	1.36	1.15
E. coli protein**	4.0	5.75	1.2	1.11

* OURA, 1983 ** STOUTHAMER, 1973

into consideration the macromolecular composition of *Saccharomyces cerevisiae* [3] of 39% protein, 34.1% polysaccharides and 5% trehalose, 10.8% nucleic acids and nucleotides and 7% lipids and 1% sterols the "cell formula" is given by $C_4H_{6.49}O_{2.11}N_{0.61} \cdot P_{0.04}S_{0.006}$. This composition agrees closely with those of yeasts quoted by LEBEAULT [5]. By changing the proportion of these polymers and macromolecular compounds the formulae are obtained (see Table 3). These elementary compositions were calculated on the basis of the molecular formulae of the individual polymers (cf. Table 2) and the *Torulopsis* sp. MH 26 protein. The question arises if these changes have an effect on the growth yield. The growth yields were calculated for methylotrophic yeasts according to STOUTHAMER'S Y_{ATP} concept [2,3] (modified by ANTONY, ref. [6]). From these data

Table 2. The average elementary composition of the main polymers and macromolecules of microbial cells

Cell constituents	Calculated elementary composition				
	C	H	O	N	P
Polysaccharides	4.0	6.67	3.33		
Lipids (Triglycerides)	4.0	7.39	0.45		
DNA	4.0	4.61	2.44	1.5	0.41
RNA	4.0	4.5	2.93	1.61	0.41

summarized in Table 3 it follows, firstly, that relative large differences in the proportion of the individual polymers practically do not influence the elementary composition of the resulting biomass, which allows one to neglect the true cell composition and to use only one "cell formula" with such calculations (e. g. $C_4H_8O_2N_1$, which is often used; cf. Table 3) and, secondly, that if cells of different macromolecular composition are formed in dependence of the growth substrate used, these variations hardly influence the substrate expense for growth. Hence, the question arises why the carbon conversion efficiency of various substrates differ from each other.

Table 3. Dependence of the elementary composition of the cells and the growth yield on the macromolecular composition and the proportion of polymers, respectively

Macromolecular composition, the proportion of polymers				Elementary composition 'cell formula'				Y_{calc}* P/O = 2
Poly-saccharides	Protein	Lipids	RNA + DNA	C	H	O	N	$g \cdot g^{-1}$
15	50	20	15	4.0	6.67	1.47	0.73	0.39
30	50	10	10	4.0	6.59	1.83	0.71	0.41
20	55	10	15	4.0	6.64	1.59	0.76	0.40
20	60	10	10	4.0	6.56	1.65	0.82	0.41
39	40	8	12	4.0	6.64	1.95	0.56	0.41
				4	8	2	1	0.43

* Balance equation of methanol assimilation in yeasts:
3 methanol + 2 ATP → 1 PGA + 1 NADH; methanol → CO_2 + 2 NADH

Causes of Different Carbon Conversion Efficiencies of Various Substrates

In order to clarify theoretically this question, which is relevant for using the auxiliary substrate concept, it is helpful to divide the process of cell substance synthesis into three partial steps starting from a certain carbon and energy source:

1. Precursor synthesis ("carbon-catabolism")

   (i. e. conversion of the substrate molecule into a central metabolite, e. g. 3-phosphoglycerate (PGA) from which all cell components are synthesized)

   substrate → precursor ± ATP_{SLP} ± reducing equivalents for ATP_{ETP} and reductive syntheses

2. "Cell molecule" synthesis

   (i. e. reduction of the precursor to the level of the "cell molecule" and assimilation of nitrogen)

   precursor + nitrogen (e. g. NH_3, NO_3^-, amino acids) + energy (ATP_{SLP} + ATP_{ETP}) + reducing equivalents → biomass $C_aH_bO_cN_d$

3. Dissimilation

   (i. e. oxidation of substrate merely for generation of biologically useful energy)

   substrate + oxygen → energy (ATP_{SLP} + reducing equivalents) + CO_2

This method enables us to take into account different elementary cell compositions, various metabolic pathways for one substrate, the kind of nitrogen source and different P/O quotients.
Since the cell composition is practically independent of the carbon and energy source under optimum growth conditions the energy requirement for synthesis of the cell substance from the central precursor can be assumed to be constant.
In accordance with STOUTHAMER and BETTENHAUSEN [7] the Y_{ATP}^{PGA} value of 10.5 g dry weight per mole of ATP was used. With a "cell formula" of $C_4H_8O_2N_1$, a P/O quotient of 2 for NADH and NH_3 as nitrogen source the overall reaction for the synthesis of cell substance from the precursor reads as:

$$4\,PGA + 3\,NH_3 + 29\,ATP + 5.5\,NAD(P)H \rightarrow (C_4H_8O_2N_1)_3 + 10\,H_2O$$

40 moles of ATP are necessary for this synthesis. Hence, it follows that the reduction degree of the substrate or the oxidation number of the carbon atom, the amount of ATP_{SLP}, the coenzyme dependence of the dehydrogenation reactions on the route to the central precursor and the P/O quotient of NADH determine whether a substrate exhibits an excess of energy or carbon.
Accordingly, with an excess-carbon substrate a part of the total substrate consumed must be oxidized merely for energy production while with excess-energy substrates more energy is provided on the way to the precursor than is needed for the assimilation of the precursor carbon.
The extent of the carbon loss is influenced by the route of complete oxidation of substrate to CO_2 and by the efficiency of the electron transport phosphorylation, which is not constant but depends on several factors such as the oxygen partial pressure and the growth limitation.
Methanol is an excess-energy substrate in a physical sense [8]. Growth yields of 100% are impossible only in so far as decarboxylating reactions are always involved on the route to some monomeric constituents of the cell [9]. The excess in energy might be used to assimilate carbon "foreign" to the substrate proper.
In the light of a biochemical evaluation [10] however, methanol belongs to the energy-deficit substrates which was confirmed by measurements [11].
Assuming that the P/O quotient of bacteria with different assimilation pathways for methanol are uniform then the extent of the energy deficit is only determined by the "carbon-catabolism". Although theoretically the growth yield on methanol with "ribulosebisphosphate pathway bacteria" should be smaller than with "serine pathway bacteria" [12] the experimentally obtained values of both groups are very similar (Table 4). This indicates that the P/O quotients are different.
Since methanol is an energy-deficit substrate like glucose, it is surprising that growth yields with *Hansenula polymorpha* MH 20 on a mixture of both substrates are reached

Table 4. Experimental and theoretical growth yields of "serine-" and "ribulosebisphosphate-pathway bacteria"

Serine pathway* balance equation	3 MeOH	→ 1 PGA + $FADH_2$		
	Y_{theor}		0.5 g/g	62.5% CCE
	Y_{exp}	0.3···0.4 g/g		
RBP pathway balance equation	3 MeOH + 5 ATP	→ 1 PGA + NAD(P)H		
	Y_{theor}		0.42 g/g	53.6% CCE
	Y_{exp}	0.3···0.4 g/g		

* ICL^+ variant via formaldehyde to activated C_1

[11] which exhibit an auxiliary substrate effect (Table 5). If it is established that *H. polymorpha* assimilates glucose via the EMP pathway and a more efficient pathway does not exist to which the assimilation can be switched due to the presence of methanol then this phenomenon could be explained as follows: Because of the presence of glucose the TCA cycle keeps operating in *H. polymorpha* thus methanol can be oxidized to CO_2 via this cycle. If mitochondrial NADH provides 3 moles of ATP per mole (instead of only 2 moles of ATP per mole of cytoplasmic formaldehyde and formate dehydrogenase NADH) the energy yield would increase from 4 moles of ATP to $5^1/_3$ moles of ATP per mole of methanol oxidized. This means that the P/O quotient has changed. The latter has been established with *Paracoccus denitrificans* during growth on methanol plus mannitol [13].

Table 5. Growth yields of yeasts with different pathways for the assimilation of glucose on glucose and substrate mixtures

		Glucose	Methanol	Methanol + Glucose	Formate + Glucose
Hansenula	Y_{exp}	0.52	0.38		0.7
polymorpha	CCE	61.1%	47.7%	59%	82%
EMP path	Y_{theor}	0.53	0.43	(39 : 1)*	0.71
	CCE	61.9%	54.4%	55%	83.3%
				(38.8 : 1)*	
				60.4%	
				(1.1 : 1)*	
Torulopsis	Y_{exp}	0.35			0.5
sp. MH 26	CCE	41.2%			58.8%
HMP path	Y_{theor}	0.42			
	CCE	50%		55%	

* mixing proportion in moles per mole; CCE carbon conversion efficiency Y in g dry; weight per g substrate

With formate as an additional energy donor the carbon conversion efficiency of glucose can be improved up to 83% in the case of *H. polymorpha* [14]. *Torulopsis* sp. MH 26 assimilates glucose via the HMP pathway. If NADPH were taken into consideration as substrate for ATP synthesis glucose would be an excess-energy substrate even at a P/O quotient of only 1.3 [12] and could help to compensate the energy deficit of methanol. The maximum growth yield of 0.42 g/g on glucose might not be exceeded because of the carbon lost on the route to the central precursor. Even with formate as energy donor an increase should be impossible.

The experimental growth yield of *Torulopsis* sp. MH 26 amounts to 0.35 g/g which indicates that NADPH does not provide ATP [15]. The fact that in the presence of formate an improvement is attained even beyond 0.42 g/g shows, firstly, that glucose does not become an excess-energy substrate via the HMP pathway and, secondly, that glucose is assimilated not only via this pathway but the EMP pathway contributes and the immediate incorporation of glucose bypassing PGA has to be considered [15].

With a mixture of methanol/glucose theoretically a 55.2% carbon conversion efficiency should be possible with *Torulopsis* sp. MH 26 (cf. Table 5). This value is markedly greater than the experimental ones on methanol and glucose as sole carbon and energy sources, but 59% which are possible with the EMP pathway and also attainable with *H. polymorpha* cannot be obtained with an exclusive or predominant operation of the HMP pathway.

From this it is evident that the "carbon catabolism" does not only essentially determine the growth yield of single substrates but also the extent of the auxiliary substrate effect. For the methanol assimilation of yeasts this result means furthermore that the growth yield is reducing-equivalent-limited and we are able to predict that substrates yielding reducing equivalents must be suitable as auxiliary substrate in combination with methanol in yeasts.

Received August 9, 1984

References

[1] Babel, W.: In: Abhandlungen der Akademie der Wissenschaften der DDR, N2 Biotechnologie. Ed.: M. Ringpfeil. Berlin: Akademie-Verlag, 1982, pp. 183–188.
[2] Stouthamer, A. H.: In: Internat. Reviews of Biochemistry, vol. 21. Microbial Biochemistry. Ed.: J. R. Quayle. Baltimore: University Park Press, 1979, pp. 1–47.
[3] Oura, E.: In: Biotechnology, Ed.: H. J. Rehm and G. Reed. Vol. 3. Ed.: H. Dellweg, Weinheim, Deerfield/Florida, Basel: Verlag Chemie, 1983, pp. 4–41.
[4] Stouthamer, A. H.: Antonie van Leeuwenhoek **39** (1973), 545.
[5] Lebeault, J. M.: In: Dechema Monographie 83, Nr. 1704–1723. Weinheim: Verlag Chemie, 1979, pp. 135–145.
[6] Anthony, C.: J. Gen. Microbiol. **104** (1978), 91.
[7] Stouthamer, A. H., Bettenhausen, C.: Biochim. Biophys. Acta **301** (1973), 53.
[8] Babel, W.: Z. Allg. Mikrobiol. **19** (1979), 671.
[9] Babel, W., Müller, R. H.: Appl. Microbiol. Biotechnol. **21** (1985), in the press.
[10] Babel, W., Müller, R. H.: J. Gen. Microbiol. **131** (1985), 39.
[11] Müller, R. H., Markuske, K. D., Babel, W.: Z. Allg. Mikrobiol. **23** (1983), 375.
[12] Babel, W.: In: Proceedings of the 3rd Symposium of Socialist Countries on Biotechnology, Bratislava (CSSR), April 25–29, 1983, pp. 169–176.
[13] van Verseveld, H. W., Boon, J. P., Stouthamer, A. H.: Arch. Microbiol. **121** (1979), 213.
[14] Babel, W., Müller, R. H., Markuske, K. D.: Arch. Microbiol. **136** (1983), 203.
[15] Müller, R. H., Babel, W.: Appl. Microbiol. Biotechnol. **20** (1984), 195.

Acta Biotechnol. 5 (1985) 4, 339—345

Zur Kinetik des Wachstums und der Speicherung von Poly-D(—)-3-hydroxybuttersäure bei Alcaligenes latus

Braunegg, G. und Bogensberger, B.

Institut für Biotechnologie, Mikrobiologie und Abfalltechnologie
der Technischen Universität Graz
A-8010 Graz, Schlögelgasse 9, Österreich

Summary

Alcaligenes latus strain DSM 1123 has been tested for the production of Poly-D(—)-3-hydroxybutyric acid using sucrose as the only carbon source. The strain is capable of accumulating the polymer associated to the growth very fast and up to high concentrations in the biomass. *Alcaligenes latus* seems to be very interesting for the production of Poly-D(—)-3-hydroxybutyrate in a cheaper and easier way as with other organisms discussed for the biotechnological process in the past.

Einleitung

Poly-D(—)-3-hydroxybuttersäure (Poly-HB), der optisch aktive lineare Polyester der auch im Stoffwechsel eukaryotischer Organismen vorkommenden D(—)-3-hydroxybuttersäure, ist eines jener biotechnologisch herstellbaren Produkte, welche in den vergangenen Jahren weltweit besonderes Interesse hervorgerufen haben. Das Polymer weist je nach Herkunft und Extraktionsmethode ein Molekulargewicht bis zu $3{,}39 \cdot 10^6$ auf [1] und zeigt thermoplastische Eigenschaften [2]. Auf Grund der zu erwartenden physiologischen Unbedenklichkeit sollte das Material ohne Schwierigkeiten in der Medizin und auch in der pharmazeutischen Industrie einsetzbar sein. So wurde die Verwendung von Poly-HB als chirurgisches Nahtmaterial ebenso in Erwägung gezogen [3] wie sein Einsatz als retardierende Matrix mit kontrolliertem Ausstrom an Pharmaka zum Zwecke der Implantation [4, 5].

Mit der grundlegenden Klärung des Poly-HB-Stoffwechsels und seiner Regulation hatte sich gezeigt, daß dieser nur von Prokaryoten gebildete Speicherstoff dann in hohen Konzentrationen intrazellulär auftritt, wenn das Wachstum von zur Poly-HB-Speicherung befähigten Kulturen durch einen Mangel an Phosphat, assimilierbarem Stickstoff, Sauerstoff, Schwefel oder Magnesium im Nährmedium limitiert wird, gleichzeitig aber noch genügende Mengen an Kohlenstoff vorhanden sind [6, 7, 8].

Untersuchungen zur Klärung der Kinetik der Poly-HB-Speicherung sind an *Alcaligenes eutrophus* und *Mycoplana rubra* [9], aber auch an *Bacillus megaterium* [10] durchgeführt worden. Diese Untersuchungen ergaben für *Alcaligenes eutrophus* und *Bacillus megaterium* übereinstimmend eine sogenannte Speicherphase mit vermehrter Poly-HB-Einlagerung in die Zellen unter den oben genannten das Wachstum limitierenden Bedingungen. Für *Mycoplana rubra* wurde eine deutliche Abhängigkeit der Kinetik der Produktbildung von der Art des limitierenden Substrates in Verbindung mit der eingesetzten Kohlen-

stoffquelle festgestellt. Wurde unter Stickstoff-limitierenden Versuchsbedingungen Methanol als alleinige Kohlenstoffquelle verwendet, so war der Typus der Poly-HB-Synthese deutlich wachstumsassoziiert. Die eigentliche Speicherphase nach der Erschöpfung des limitierenden Substrates konnte nicht gefunden werden. Andererseits konnte, wenn Phosphat-limitierende Bedingungen vorherrschten und Fruktose als alleinige Kohlenstoffquelle eingesetzt wurde, eine mit den biochemischen Erkenntnissen in Einklang stehende Speicherphase nach Beendigung der Vermehrung der Zellzahl festgestellt werden [11].
Alcaligenes latus DSM 1123 [12] zeichnete sich zunächst dadurch aus, daß dieser Stamm eine Vielzahl von verschiedenen Kohlenstoffquellen unter organotrophen, aber auch CO_2 unter chemoautotrophen Wachstumsbedingungen zu verwerten vermag. Unter den verschiedenen verwertbaren Zuckern erscheint insbesondere Saccharose besonders interessant zu sein, weil der Stamm in ersten Vorversuchen mit Saccharose als alleiniger Kohlenstoffquelle neben schnellem Wachstum einen hohen Gehalt an Poly-HB ($< 80\%$ des erreichten Zelltrockengewichtes) zeigte.

Material und Methoden

Organismus und Medien

Alcaligenes latus Stamm DSM 1123 wurde von der Deutschen Sammlung für Mikroorganismen (Göttingen, BRD) bezogen und auf festen Nährböden sowie in der institutseigenen Stammsammlung in flüssigem Stickstoff aufbewahrt. Die Anzucht des Organismus erfolgte in einem Mineralmedium [13] mit Saccharose (10—25 g/l) als alleiniger Kohlenstoffquelle. Das Wachstum wurde durch die Stickstoffquelle $(NH_4)_2SO_4$ (1—3 g/l) begrenzt. Feste Medien wurden durch Zusatz von Agar Agar (15 g/l) erhalten.

Versuchsbedingungen und Inocula

Zur Durchführung der Experimente wurde ein Bioreaktor im Labormaßstab vom Typ L 1523 (Bioengineering AG, Wald, Schweiz) mit einem Arbeitsvolumen von 14 Litern verwendet. Die Mischeinrichtung des Gerätes bestand aus einer Welle mit zwei konventionellen Querblattrührern und vier Schikanen. Die Anzuchttemperatur betrug 33 °C, der pH-Wert der Nährlösung wurde durch ein automatisches pH-Meß- und Regelsystem vom Typ Meredos pH/R (Jungkeit, Göttingen, BRD) durch Zusatz von 10%iger, steriler Natronlauge bei pH $= 7{,}1 \pm 0{,}1$ konstant gehalten. Weiter wurde durch manuelle Regulation der Belüftungsrate aber auch durch Beimischen von Stickstoff zur Zuluft erreicht, daß der in Vorversuchen ermittelte optimale Bereich für den Gelöstsauerstoffpartialdruck mit 20—40% des Sättigungswertes für Luft eingehalten wurde. Zur Messung des Sauerstoffpartialdruckes stand eine nach dem polarographischen Prinzip arbeitende Sauerstoffelektrode nebst Verstärker vom Typ IL 530 (Ingold, Zürich, Schweiz) zur Verfügung. Zum Beimpfen des Bioreaktors wurden Übernachtkulturen von *Alcaligenes latus* Stamm DSM 1123 herangezogen. Diese wurden im gleichen Medium wie anschließend im Bioreaktor in Schüttelkolben mit Schikanen von 1 Liter Gesamtinhalt, befüllt mit 300 ml Nährlösung, bei 30 °C am Kreisschüttler für 10 Stunden inkubiert.

Analytik

Die Bestimmung der Konzentration der Zucker Saccharose, Glukose und Fruktose im Nährmedium erfolgte enzymatisch mit käuflichen Testkombinationen (Boehringer-Mannheim, Wien, Österreich) nach den Bestimmungsvorschriften des Herstellers. Die quantitative Analyse des limitierenden Substrates $(NH_4)_2SO_4$ erfolgte nach einer modifizierten BERTHELOT-Reaktion [9] am Spektralphotometer. Die Bestimmung der Biomasse erfolgte einerseits durch Bestimmung der optischen Dichte der wachsenden Kultur bei 420 nm, andererseits mit Hilfe von Membranfiltern als Zelltrockengewicht [11]. Die Bestimmung der Konzentration der gebildeten Poly-HB in der Biomasse erfolgte mit einem rechnergesteuerten Gaschromatographen vom Typ 5840 A (Hewlett-Packard, Wien, Österreich) nach BRAUNEGG et al. [14].

Versuchsergebnisse

Die jeweiligen zeitlichen Verläufe der Konzentrationen von Zelltrockengewicht, Poly-HB und $(NH_4)_2SO_4$, sowie der prozentuelle Gehalt der Biomasse an Poly-HB während eines diskontinuierlichen Wachstums- und Speicherversuches mit *Alcaligenes latus* DSM 1123 sind in der Abb. 1 dargestellt. Wie ersichtlich, wird das Wachstum der Kultur durch Erschöpfung der Stickstoffquelle in der Nährlösung nach 14,5 Stunden Versuchsdauer begrenzt. Der nach diesem Zeitpunkt feststellbare Zuwachs an Biomasse ist praktisch nur noch auf die Zunahme der Poly-HB in der Biomasse zurückzuführen, was sich auch in der fast sprunghaften Zunahme des prozentuellen Gehaltes des Zelltrockengewichtes an Poly-HB manifestiert.

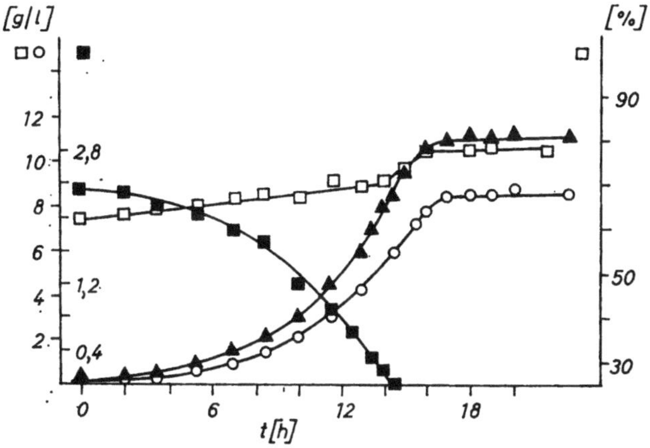

Abb. 1. *Alcaligenes latus* DSM 1123: Zelltrockengewicht (–▲–), Poly-HB (–○–), Ammoniumsulfat (–■–) und prozentueller Gehalt der Biomasse an Poly-HB (–□–) als Funktion der Zeit während eines diskontinuierlichen Wachstums- und Speicherversuches mit Saccharose als Kohlenstoffquelle.
Versuchsbedingungen: Limitiertendes Substrat: $(NH_4)_2SO_4$; Bioreaktor: Bioengineering L 1523, Arbeitsvolumen 14 Liter, 2 Querblattrührer, 4 Schikanen, Drehzahl: 1200 upm. Temperatur: 33 °C, pH-Wert: 7,1 ± 0,1; Gelöstsauerstoff: 20—40% der Luftsättigung. Mineralmedium nach SCHLEGEL et al. [13].

Korrektur: für □ an der Ordinate links lies ▲

Hinsichtlich der Produktion von Poly-HB ist bei diesem Versuch besonders der Zeitraum der ersten 14 Stunden des Experimentes als bedeutend, aber auch als ungewöhnlich zu nennen. Unter den gewählten Versuchsbedingungen steigt der Gehalt der Biomasse an Poly-HB nämlich auch dann stark an, wenn keine wachstumslimitierenden Bedingungen eingestellt wurden. Der Poly-HB-Anteil an der gebildeten Biomasse, gemessen als Zelltrockengewicht, liegt vom Beginn des Versuches an höher als 60% und wird bis zum Einsetzen der Stickstofflimitierung des Wachstums noch auf 68% des Zelltrockengewichtes gesteigert. In der nun anschließenden wachstumslimitierten Phase des Versuches steigt die Poly-HB-Konzentration der Biomasse im Bioreaktor auf ihren Endgehalt von 77% des erreichten Zelltrockengewichtes an. Die Konzentration der Gesamtbiomasse erreicht zu diesem Zeitpunkt ihren Endwert von 11,1 g/l Zelltrockengewicht.
Interessant ist auch der Verlauf der in der Abb. 2 dargestellten Konzentrationen von Saccharose, Glukose und Fruktose während des Experimentes. Der untersuchte Stamm

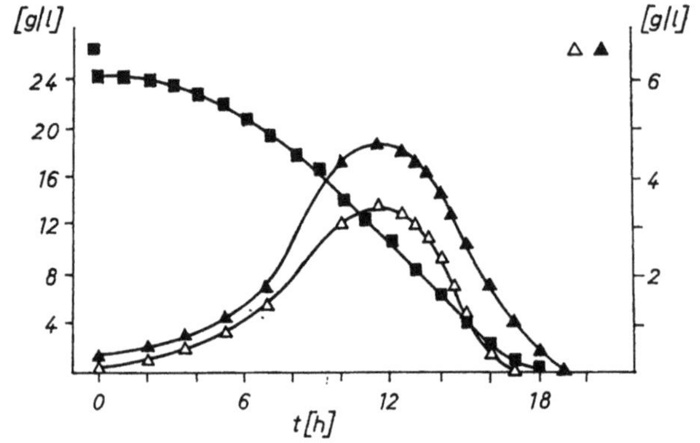

Abb. 2. *Alcaligenes latus* DSM 1123: Verlauf der Konzentrationen von Saccharose (-■-), Glukose (-△-) und Fruktose (-▲)- während eines diskontinuierlichen Wachstums- und Speicherversuches mit Saccharose als Kohlenstoffquelle. Versuchsbedingungen siehe Abb. 1.

verfügt zweifellos über eine kräftige Invertase-Aktivität, auf welche letztlich die in der Nährlösung auftretenden Zucker Glukose und Fruktose zurückzuführen sind. Die Maximalkonzentrationen dieser Zucker in der Nährlösung liegen bei 3,4 g/l für die Glukose und 4,6 g/l für die Fruktose. Weiter fällt auf, daß die gemessenen Konzentrationen für die Glukose durchweg niedriger liegen als die korrespondierenden Konzentrationen für die Fruktose.

Diskussion der Ergebnisse

Bei den bisher im Zusammenhang mit der biotechnologischen Herstellung von Poly-HB genannten Stämmen von Mikroorganismen sind vor allem *Alcaligenes eutrophus* Stamm H 16 [15] und glukoseverwertende Mutanten von *Alcaligenes eutrophus* [16, 17] herauszustellen. Diese Organismen weisen allerdings alle die gleichen Nachteile auf: die Anzahl verwertbarer Kohlenstoffquellen, welche ein noch ansprechendes Wachstum- und Speicherverhalten zeigen, ist letztlich auf Fruktose und Glukose beschränkt, eventuell kommt unter chemolithoautotrophen Wachstumsbedingungen noch Kohlendioxid in Frage. Es war daher bei der Arbeit mit *Alcaligenes latus* von besonderem Interesse, das Disaccharid Saccharose als Kohlenstoffquelle zu testen, weil mit der Verwertbarkeit dieses Zuckers der Zugriff auf Zwischen- oder Abfallprodukte der Zuckerindustrie wie Grünsirup oder Melasse ermöglicht wird. Dies sollte aber die Kosten für die Herstellung des Biopolymers deutlich verringern helfen.
Bei ersten Arbeiten bezüglich der Verwendung von *Alcaligenes latus* Stamm DSM 1123 zur Herstellung von Poly-HB hatte sich gezeigt, daß die höchsten spezifischen Geschwindigkeiten für das Wachstum und die Speicherung von Poly-HB dann auftraten, wenn die Konzentration des zunächst als Kohlenstoffquelle verwendeten Zuckers Glukose kleiner war als 5 g/l. Wurde die Zuckerkonzentration in der Nährlösung hingegen über 10 g/l angehoben, so trat deutliche Substratinhibition auf. Bei der Verwendung von Saccharose als alleiniger Kohlenstoffquelle waren solche Erscheinungen hingegen auch dann nicht feststellbar, wenn die Zuckerkonzentration in der Nährlösung bis auf 35 g/l Saccharose erhöht wurde. Bei diesen Versuchen zeigte sich jedoch, daß das Disaccharid Saccharose in Glukose und Fruktose zerlegt wird, welche dann im Nährmedium frei vorliegen. Es wurde hierbei allerdings niemals beobachtet, daß die Konzentrationen der Einzelzucker

über einen Wert von etwa 5 g/l angestiegen wären. Weiter zeigt sich für die Glukose immer ein geringerer Wert als für die Fruktose zum gleichen Zeitpunkt. Insgesamt läßt dieses Verhalten die Deutung zu, daß die Spaltungsreaktion so reguliert ist, daß der wachsende Organismus sich immer in seinem Optimum hinsichtlich der Monosaccharide befindet.

Wie aus den bisherigen Arbeiten über die Kinetik von Wachstum und Poly-HB-Speicherung bei verschiedenen Organismen hervorgeht, verläuft die Speicherung des Reservepolymers in den meisten der bisher untersuchten Fälle bevorzugt in einer Speicherphase ab, welche nach einer Begrenzung des Wachstums der Kultur durch Nährstofflimitierung bezüglich Stickstoff, Phosphat, Sauerstoff, Magnesium oder Schwefel bei gleichzeitig ausreichendem Angebot an Kohlenstoff auftritt. Die Produktion des Speicherproduktes erfolgt damit vorzugsweise nicht wachstumsassoziiert, was, soll kontinuierlich

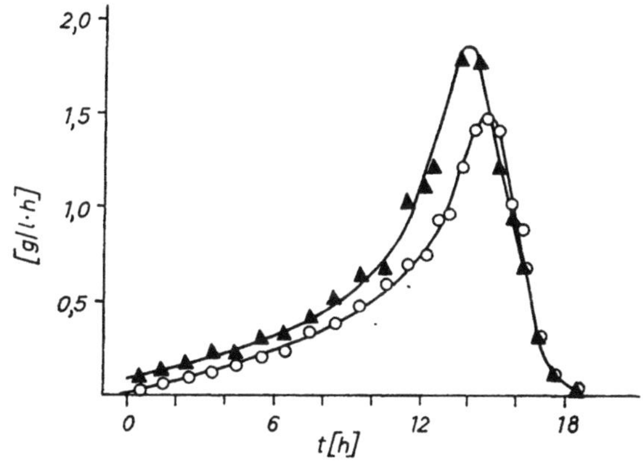

Abb. 3. *Alcaligenes latus* DSM 1123: Die absoluten Geschwindigkeiten der Bildung von Biomasse (Zelltrockengewicht, –▲–) und Poly-HB (–○–) während eines diskontinuierlichen Wachstums- und Speicherversuches mit Saccharose als alleiniger Kohlenstoffquelle. Versuchsbedingungen siehe Abb. 1.

produziert werden, einen zweistufigen Prozess erfordert. Im wesentlichen ergab sich bisher nur für *Mycoplana rubra* Stamm R 14 [11] dann eine Ausnahme, wenn dieser Stamm mit Methanol als alleiniger Kohlenstoffquelle unter ammoniumlimitierenden Bedingungen gezüchtet wurde. Allerdings ist der hierbei rein wachstumsassoziiert gebildete Speicherstoff Poly-HB mit einem maximalen Anteil am Zelltrockengewicht von lediglich 8,8% als sehr gering zu bezeichnen. Mit *Alcaligenes latus* DSM 1123 ließen sich aber unter den hier gewählten Versuchsbedingungen bis zu etwa 70% Poly-HB im Zelltrockengewicht anreichern, ohne daß der Stamm einer Wachstumslimitierung unterworfen wurde. Wie Abb. 3 zeigt, in welcher die absoluten Geschwindigkeiten der Bildung von Biomasse und der Speicherung von Poly-HB bei *Alcaligenes latus* als Funktionen der Zeit dargestellt sind, erfolgt die Produktbildung bei diesem Stamm weitestgehend wachstumsassoziiert. Die geringe zeitliche Verschiebung der Maxima in beiden Kurven ist auf den nach Stickstofflimitierung auftretenden Zuwachs an Poly-HB zurückzuführen, doch ist der Konzentrationszuwachs an Produkt zu diesem Zeitpunkt nicht mehr von wirtschaftlicher Bedeutung. Verbunden mit der hohen erreichbaren Konzentration an Poly-HB in der Biomasse von *Alcaligenes latus* DSM 1123 zeichnet sich damit ein einstufig kontinuierlicher Prozess zur Herstellung von Poly-HB ab.

Abb. 4 zeigt die spezifischen Geschwindigkeiten von Gesamtbiomassezuwachs, Restbiomassesynthese und Poly-HB-Speicherung während des gleichen Versuches mit *Alcaligenes latus* DSM 1123. Die Restbiomasse wird hierbei definiert als biologisch „aktive" Biomasse, was in erster Näherung der erreichten Gesamtbiomasse vermindert um ihren Poly-HB-Gehalt zum jeweiligen Zeitpunkt entspricht. Damit wird dem Umstand Rechnung getragen, daß der Speicherstoff Poly-HB per se keinen Beitrag zu seiner Vermehrung zu leisten vermag [9]. Auch die spezifische Geschwindigkeit für die Speicherung von Poly-HB ist auf dieser Basis berechnet.
Die spezifischen Geschwindigkeiten von Gesamtbiomasse und Restbiomasse zeigen in ihrem Verlaufe ein recht ähnliches Bild. Nach einer ersten Phase rascheren Zuwachses an Biomasse, welcher sich etwa über die ersten 4 Stunden des Versuches erstreckt, werden diese Geschwindigkeiten auf etwas niedrigerem Niveau konstant und fallen schließlich, bedingt durch die einsetzende Stickstofflimitierung auf den Wert Null

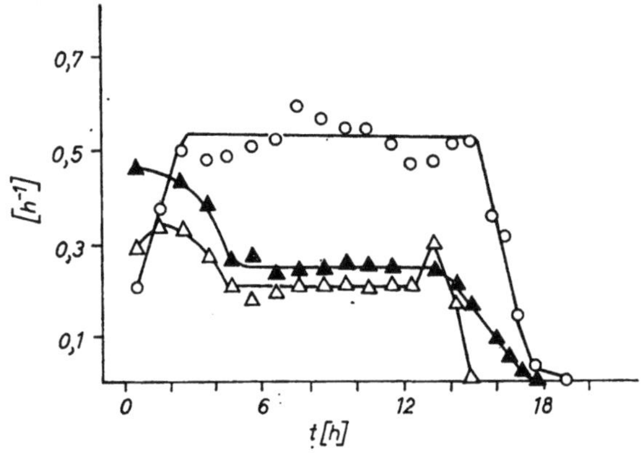

Abb. 4. *Alcaligenes latus* DSM 1123: Die spezifischen Geschwindigkeiten der Zunahme von Gesamtbiomasse (–▲–), Restbiomasse (–△–) und Poly-HB (–○–) während eines diskontinuierlichen Wachstums- und Speicherversuches mit Saccharose als Kohlenstoffquelle. Versuchsbedingungen siehe Abb. 1.

zurück. Hierbei läßt sich aus dem flacheren Abfall der Kurve für die Gesamtbiomasse entnehmen, daß der Stamm nach dem Einsetzen der Wachstumslimitierung noch Poly-HB speichert, bis auch die Kohlenstoffquelle zur Neige geht.
Einen vollkommen anderen Verlauf zeigt hingegen die spezifische Geschwindigkeit des Poly-HB-Zuwachses der Biomasse. Nach eher niedrigen Werten zu Beginn des Experimentes steigt die spezifische Geschwindigkeit der Speicherung auf ihr Maximum in einem Bereich von $q_{Poly\text{-}HB} = 0{,}45-0{,}60\ h^{-1}$ an, welcher für die Dauer des nichtlimitierten exponentiellen Wachstums beibehalten wird. Die bei diesem Experiment gefundenen Werte für $q_{Poly\text{-}HB}$ sind die höchsten, welche bisher überhaupt bei der Untersuchung Poly-HB-speichernder Stämme gefunden wurden. Weiter zeigt sich für den hier untersuchten Stamm nochmals deutlich die Poly-HB-Speicherung während des exponentiellen Wachstums, die deutlich schneller abläuft als die Synthese der Restbiomasse, was letztlich die hohe Poly-HB-Konzentration in der Biomasse von *Alcaligenes latus* DSM 1123 erklärt. Erst knapp vor der Erschöpfung der Stickstoffquelle $(NH_4)_2SO_4$ kommt es zu einem kurzfristigen Unterschreiten dieses Geschwindigkeitsbereiches, mit Eintritt in die Speicherphase steigt die spezifische Produktionsgeschwindigkeit dann wieder etwa auf die vorherigen Werte an, bis sich die ersten Einflüsse der Doppellimitierung, nunmehr auch von der Seite der Kohlenstoffquelle her, auszuwirken beginnen.

Auch aus dieser letzten Auftragung ist damit deutlich zu entnehmen, daß *Alcaligenes latus* Stamm DSM 1123 als für die Poly-HB-Herstellung besonders geeignet zu bezeichnen ist. Es bietet sich der zuvor erwähnte einstufig kontinuierliche Prozeß an, welcher, neben niedrigeren Substratkosten betreffend die Kohlenstoffquelle, auch geringere Investitionskosten für die Produktionsanlage erwarten läßt.

Eingegangen: 24. 10. 1984

Literatur

[1] Akita, S., Einaga, Y., Fujita, H.: Macromolecules **9** (1976), 774.
[2] Baptist, J. N., Werber, F. X.: United States Patent Office (1965) Pat. Nr. 3182036.
[3] Howells, E. R.: Chem. Ind. **8**, (1982), 508.
[4] Korsatko, W., Wabnegg, B., Braunegg, G., Lafferty, R. M., Strempfel, F.: Pharm. Ind **45** (1983), 252.
[5] Korsatko, W., Wabnegg, B., Tillian, H. M., Braunegg, G., Lafferty, R. M.: Pharm. Ind. **45** (1983), 1004.
[6] Dawes, E. A., Senior, P. J.: Adv. Microb. Physiol. **10** (1973), 136.
[7] Oeding, V., Schlegel, H. G.: Biochem. J. **134** (1973) 239.
[8] Tomita, K., Saito, T., Fukui, T.: In: Biochemistry of Metabolic Processes. Elsevier Science Publishing Co., Inc. Edts.: Lennon, D. L. F., Stratman, F. W., Zahlten, R. N., 1983. 353—366.
[9] Sonnleitner, B., Heinzle, E., Braunegg, G., Lafferty, R. M.: Eur. J. Appl. Microbiol. Biotechnol. **7** (1979), 1.
[10] Sonnleitner, B.: Dissertation, Techn. Univ. Graz, Österreich (1979).
[11] Braunegg, G.: Dissertation, Techn. Universität Graz, Österreich (1980).
[12] Palleroni, N. J., Palleroni, A. V.: Int. J. Syst. Bacteriol. **28** (1978), 416.
[13] Schlegel, H. G., Gottschalk, G., von Bartha, R.: Nature **191** (1961), 463.
[14] Braunegg, G., Sonnleitner, B., Lafferty, R. M.: Eur. J. Appl. Microbiol. Biotechnol. **6** (1978), 29.
[15] Lafferty, R. M., Heinzle, E.: Chem. Rundschau **30** (1977), 14.
[16] König, Ch., Sammer, I., Wilde, E., Schlegel, H. G.: Arch. Microbiol. **57** (1969), 51.
[17] Hughes, L., Richardson, K. R.: Europ. Pat. Appl. EP 0046344 A3.

Acta Biotechnol. 5 (1985) 4, 346

Book Review

B. Perbal

A Practical Guide so Molecular Cloning

New York, Chichester, Brisbane, Toronto, Singapore: John Wiley & Sons, 1984, 554 S.

Mit dem Buch von B. Perbal liegt eine weitere umfangreiche Methodensammlung für gentechnische Experimente vor. Besondere Beachtung finden Methoden, die von Firmen unter Verwendung eigener Produkte empfohlen werden. Neben einer genauen Beschreibung des Ablaufs der Experimente werden auch wertvolle Hinweise für die Überwindung technischer Schwierigkeiten gegeben. Zur Ligation von DNA-Molekülen und zum Anlegen von Genbanken sind ausführliche theoretische Abhandlungen enthalten. Zu folgenden Sachgebieten werden Experimente besprochen:

- Reinigung und Charakterisierung von Vektor- und Passagier-DNA
- Restriktionsanalyse
- Gewinnung, Reinigung und Modifikation von DNA-Fragmenten
- Charakterisierung von rekombinanter DNA
- Anlegung von Genbanken
- Reinigung und Charakterisierung von RNA
- Klonierung von cDNA
- DNA-Sequenzanalyse
- Expression klonierter DNA in Pro- und Eukaryoten

Das Buch ist übersichtlich gegliedert. Umfangreiche Tabellen (z. B. zu Eigenschaften von Restriktasen, Restriktionsorten in pBR 322, Charon-Vektoren) und anschauliche Zeichnungen tragen zur Verständlichkeit des Stoffes bei. Ein Nachteil besteht jedoch darin, daß bei verschiedenen Kapiteln auf einführende Erläuterungen verzichtet wurde.
Das vorliegende Buch wird besonders als Labornachschlagewerk einen breiten Leserkreis finden.

J. Engel

Acta Biotechnol. 5 (1985) 4, 347—352

Proteolysis upoń Dehydration of Yeasts Saccharomyces cerevisiae

DAMBERGA, B. E., VILCÁNS, A. P., LAIVENIEKS, M. G., BEKER, M. J.

Latvian SSR Academy of Sciences
August Kirchenstein Institute of Microbiology
226067 Riga, Kleisti, Latvian SSR, USSR

Summary

The effect of dehydration on proteolysis and activity of proteases A, B and C in the cells of baker's yeast *Saccharomyces cerevisiae* was investigated. It can be concluded, that under investigated conditions of yeast *Saccharomyces cerevisiae* drying a decrease of proteases activity takes place. In cells a limited proteolysis takes place which is indicated by an increase in amino nitrogen content and a decrease of tryptophane synthase activity. Adding the protease inhibitor to yeast suspension prevents decrease of tryptophane synthase activity upon dehydration.

Nitrogenous substance content in cells is one of the factors determining baker's yeast xerotolerance. Even upon a sparing convective drying there is a redistribution of nitrogenous compounds in yeast cells, yet data on the character of this redistribution are rather controversive.

Several authors [1, 2] have detected a sharp decrease of protein and a simultaneous increase of amino nitrogen (the content of the latter at times reaching 20% of amino compounds of yeasts) upon drying of baker's yeast. Others [3] observed a decrease of both, protein and amino nitrogen content upon drying of yeast, while still others [4] detected a considerable decrease of protein amount and insignificant changes in amino nitrogen content. It must be noted, that upon yeast dehydration the solubility of cellular proteins decreases: the higher the temperature of yeast drying, the lower protein solubility [5].

At our laboratory, with the help of consecutive extraction and a subsequent electrophoresis of separate protein fractions in polyacrylamide gel, it was demonstrated that changes in the quantitative and qualitative contents of separate protein fractions depend on the applied drying methods [6].

Yeast *Saccharomyces cerevisiae* contains a number of proteolytic enzymes: endoproteinases A and B, carboxypeptidases Y and S, several aminopeptidases and dipeptidase [7]. These enzyme activity depends on cultivation conditions and growth phase of the yeast.

The mode of action and role of proteases A, B and C (carboxypeptidase Y) have been studied in quite great detail. *In vitro* experiments demonstrated that under the effect of proteases A, B and C some cytoplasmatic enzymes quickly lose their activity [8]. Tryptophane synthase, NAD-dependent glutamate dehydrogenase and hexose-dyphosphatase are very sensitive in this respect.

Proteases are responsible also for the regulation of some enzymes in yeast cells [9]. Of late, studies on yeast mutants devoid of some proteases give a slightly different idea about role of proteases A, B and C in cell metabolism.
Protease A activity in yeast cells sharply increases upon yeast cultivation on nitrogen-deficient media. Some authors [10] suppose, that under these conditions supply of cells with the necessary amino acids takes place on account of the fact that protease A degrades proteins that are not needed by the cell.
Protease B takes part in the processes of cell reparation, since its activity sharply increases at the beginning of cultivation of yeasts treated with radiation [11]. Protease C has a definite effect on the growth of yeasts on peptide-containing media: mutants, devoid of protease C, lose their ability to use peptides as a nitrogen source [12].
We failed to find data on the effect of dehydration upon proteolytic enzyme activity in yeasts and their role in proteolysis processes upon yeast drying, with the only exception of a note, that upon drying of *Saccharomyces cerevisiae* their "proteolytic activity" increases [13].
The present work studies the effect of dehydration on proteolysis and some protease activity in the cells of baker's yeast *Saccharomyces cerevisiae.*

Materials and Methods

We used yeast *Saccharomyces cerevisiae* strain L-1 from the culture collection of August Kirchenstein Institute of Microbiology, Latvian SSR Academy of Sciences. Yeast was cultivated on media of the following contents (g/l): $(NH_4)_2SO_4$ — 3.0; $MgSO_4 \cdot 7H_2O$ — 0.7; NaCl — 0.5; KH_2PO_4 — 1.0; K_2HPO_4 — 0.1. Molasses (2% sugar), ethanol (1.5%) or glucose (2%) were used as carbon sources. Yeast was cultivated in flasks (750 ml, 100 ml of medium in each flask) on a shaker (200 rpm) at 30 °C, medium pH — 4.5—5.5.
The biomass was harvested by centrifugation at the stationary growth phase, rinsed with sterile tap water on a vacuum filter and granulated by pressing through a sieve (cell size 0.63 mm) for a subsequent dehydration.

In experiments with the C^{14}-labelled leucine the acid (0.5 μCi) was added to the molasses medium and yeast was grown to the stationary growth phase. Cells were separated by filtration through a "Sinpor" membrane filter and dehydrated on the same filters at 37 °C.
Cycloheximide (50 μg/ml) and iodoacetic acid (0.2 mg/ml) were added 2 hours before the end of the cultivation. In some experiments two hours before the end of yeast cultivation there were added protease inhibitors — pepstatin (0.5 μg/ml) and phenylmethylsulphonyl fluoride (PMSF) (3.5 mg/ml) [14].

Yeast Dehydration

The granulated biomass was dehydrated as follows:

1. In a stationary layer (2—3 mm thick) in a thermal chamber at 37 °C for 24 hours.
2. By aeration in a laboratory-scale equipment. Biomass temperature — 37 °C, duration of drying — 1 hour.

Water content of dried yeasts — 8—10%.

Establishing of the Number of Living Cells

The number of living cells in dehydrated preparations after reactivation in a 10% sucrose solution (37 °C, 10 min) was determined by luminiscent microscopy with primuline [15].

Establishing of Enzyme Activity

Yeast cells were disintegrated on the equipment L-17 with glass beads in a 0.1 M phosphate buffer, pH 7.0. Disintegrate was centrifugated at 15000 g for 30 minutes and the following was determined in the supernatant fluid:

1) Tryptophane synthase (E. C. 4.2.1.20) [16]. Activity unit — the amount of the enzyme converting 1 μM of indole into tryptophane in 1 minute.
2) Protease A (E. C. 34.23.8) [17]. Activity unit — the amount of the enzyme, splitting 1 μg of tryosine from haemoglobin in 1 minute.
3) Protease B (E. C. 34.22.9) [18]. Activity unit — the amount of the enzyme, increasing the optic density of the reacting mixture by 1.0 per hour at 590 nm.
4) Protease C (carboxypeptidase Y) (E. C. 31.4.12.8) [18]. Activity unit — the amount of the enzyme, producing 1 μM of p-nitroaniline from N-benzoyl-L-tyrosine-p-nitroanilide in 1 minute.

Proteases were determined after a preliminary activation, as described earlier [19].

Reagents

Azocoll, haemoglobin, N-benzoyl-L-tyrosine-p-nitroanilide, indole, pyridoxal-5-phosphate, phenylmethylsulphonyl fluoride, cycloheximide, pepstatin and EDTA (ethylene diamine tetraacetic acid), purchased from "Sigma" (USA); L-serine and L-leucine — from "Serva" (FRG); C^{14}-leucine — from "Izotop" (USSR). The other reagents we used were chemically pure.

Analytical Methods

Protein amount in disintegrates was determined as to LOWRY, that in yeast cells — as described earlier [20]. Amino nitrogen content was established as to the reaction with ninhydrine. Free amino acid pool was extracted from yeast cells by 0.5 M trichloracetic acid. The radioactivity of the compounds was determined after the extraction of trichloracetic acid by ether, using "Unisolve" mixture on the counter SL-4000. The obtained experimental data were treated statistically [21].

Results and Discussion

By way of a direct determination of protein and amino nitrogen, as well as by studies on the redistribution of C^{14}-labelled leucine, the present work investigates the effect of dehydration on the protein amount and soluble amino acid pool in yeast *Saccharomyces cerevisiae*. Under sparing conditions of yeast dehydration in our experiments (Tab. 1, 2) the amount of protein in cells tends to decrease, yet this decrease is statistically unreliable. At the same time there is an increase of amino nitrogen content. Thus, amino nitrogen content in yeast cells, grown on a glucose medium, upon a slow drying increases by 42%, while the number of living cells diminishes to 10.9%. Amino nitrogen content sharply increases (by 28%) upon drying of molasses yeast, too, yet in this case cell viability is considerably better preserved. Ethanol yeast proved to be rather resistant to drying, and changes of amino nitrogen content in the cells of this yeast are insignificant. Thus, there is no direct relation between the number of living cells and the increase of amino nitrogen content.

Amino acids, formed in the result of proteolysis may get involved in cell metabolism. Some proteins are synthesized during dehydration, acid phosphatase, for example [22], and there is an increase in trehalose content [23]. As to the data by some authors [23] trehalose biosynthesis takes place on account of amino acids.

We carried out inhibitor analysis of molasses yeast, which, as to their properties are the most close ones to the trade form of baker's yeast, and upon a slow drying showed a marked increase of their amino nitrogen content (28%). Tab. 2 shows that cyclo-

Table 1. Protein and amino nitrogen content in *Saccharomyces cerevisiae* cells upon their dehydration (mg/g dry wt.)

C source in growth medium	Indicator	Variants		
		Initial	Dehydrated at 37 °C for 1 h	Dehydrated at 37 °C for 24 h
Glucose 2%	Living cells, %	95.1	22.3	10.9
	Protein	397.7 ± 7.2	398.5 ± 7.2	378.3 ± 9.4
	Nitrogen — NH_2	4.08 ± 0.07	4.75 ± 0.23	5.81 ± 0.27
Ethanol 1.5%	Living cells, %	100.0	61.3	79.1
	Protein	357.3 ± 10.0	349.2 ± 12.0	341.6 ± 5.6
	Nitrogen — NH_2	4.18 ± 0.12	4.71 ± 0.25	4.65 ± 0.14
Molasses 2%	Living cells, %	100.0	62.6	59.1
	Protein	337.9 ± 7.1	360.8 ± 12,4	363.4 ± 15.6
	Nitrogen — NH_2	7.51 ± 0.21	8.01 ± 0.24	9.61 ± 0.30

heximide does not affect the size of soluble amino acid pool, i. e. — there is no notable synthesis of protein in yeast cells during their dehydration. Addition of monoiodine acetate — a glyceraldehyde phosphate dehydrogenase inhibitor — also does not affect the protein amount and the size of soluble amino acid pool (Tab. 2), that means that amino acids are not used in glyconeogenesis.

Upon an increased temperature and decreased water content in yeast cells there may be formed melanoidines [24]. These compounds are partly soluble in alkaline solutions, yet they do not sediment together with proteins under the effect of trichloracetic acid. Since the sum of the labelled leucine, included in protein and the free amino acid pool, in actual fact is the same in the initial and dehydrated yeast cells, it can be assumed that upon our conditions of yeast drying there is no decrease of amino nitrogen content on the account of sugar-amino compound formation. Consequently, upon our conditions of yeast dehydration no essential disintegration of cellular proteins takes place.

Table 2. Effect of inhibitors on the content of protein and free amino acid pool upon yeast cell dehydration (cpm/mg dry wt.)

Yeast cells	Molasses medium 2%	
	Free amino acid pool	Protein
Initial	6400 ± 200	31500 ± 1000
Dehydrated at 37 °C for 24 h	6800 ± 300	29800 ± 800
Initial	7200 ± 300	32000 ± 1300
Dehydrated at 37 °C for 24 h + cycloheximide	7000 ± 300	29700 ± 1600
Initial	9400 ± 1000	32900 ± 1200
Dehydrated at 37 °C for 24 h + iodoacetic acid	8700 ± 200	39600 ± 900

There are data [13], that upon a long-term drying of yeast *Saccharomyces cerevisiae* their proteolytic activity increases three- to four-fold. Prior to and after yeast drying we have established the activity of proteases A, B, C, i. e. — of enzymes, which under certain conditions affect cellular proteins. In our experiments yeast dehydration caused a decrease of the content of all the three proteases we studied (Tab. 3). This decrease of activity is the most marked in yeast subject to a long-term dehydration process, disregarding the fact, that the number of living cells in several experiments was higher than in yeast, dehydrated for 1 hour.

In the cytoplasm of yeast cells there are polypeptides — protease inhibitors, which upon cell disintegration block protease activity. During activation these inhibitors are splitted. Protease activation in the initial and dehydrated yeasts takes place on an equal level (data are not given), which indicates to the fact that intracellular protease inhibitors are not inactivated during yeast drying.

Table 3. Protease A, B and C activity in yeast *Saccharomyces cerevisiae* cells prior to and after dehydration

C source in growth medium	Indicator	Variants		
		Initial	Dehydrated at 37 °C for 1 h	Dehydrated at 37 °C for 24 h
Glucose 2%	Protease A	4.5 ± 0.8	4.6 ± 0.5	2.8 ± 0.3
	Protease B	8.9 ± 0.7	6.7 ± 0.5	2.5 ± 0.3
	Protease C	2.3 ± 0.2	1.8 ± 0.1	1.7 ± 0.3
Ethanol 1.5%	Protease A	4.7 ± 0.3	4.1 ± 0.4	3.6 ± 0.4
	Protease B	0.7 ± 0.1	0.6 ± 0.1	0.4 ± 0.1
	Protease C	0.2 ± 0.0	0.2 ± 0.0	0.1 ± 0.0
Molasses 2%	Protease A	4.3 ± 0.9	2.9 ± 0.6	3.2 ± 0.3
	Protease B	3.4 ± 0.3	2.6 ± 0.4	2.6 ± 0.2
	Protease C	3.7 ± 0.5	3.5 ± 0.5	2.2 ± 0.4

Note: Activity is expressed: for protease A — units/mg protein per min;
for protease B — units $\times 10^2$/mg protein per min;
for protease C — units $\times 10^4$/mg protein per min.

There are data, that intracellular proteases, the enzymes under study included, directly participate in the processes of cell metabolism regulation, destroying enzymes that are not necessary for the cells. Tryptophane synthase is an enzyme very sensitive to proteases A and B [8]. It was observed that during dehydration its activity decreases by 30—50% (Tab. 4). In our experiments tryptophane synthase inactivation in molasses yeast was more marked than in ethanol yeast, like the increase of amino nitrogen during yeast dehydration.

If two hours before the end of yeast cultivation protease inhibitors pepstatin and phenylmethylsulphonyl fluoride were added into the flasks, the decrease of tryptophane synthase activity in dehydrated yeast was insignificant if compared with the initial one, which confirms the presence of proteases in tryptophane synthase inactivation during yeast dehydration (Tab. 4).

Thus, it can be concluded, that under our conditions of yeast *Saccharomyces cerevisiae* drying, a selective proteolysis takes place in its cells. It is indicated by an increase in amino nitrogen content and a decrease of tryptophane synthase activity upon dehydration. Obviously, the primary role of biosynthesis enzymes is lost during dehydration,

Table 4. Effect of dehydration and protease A and B inhibitors on tryptophane synthase activity in yeast *Saccharomyces cerevisiae* cells (U/mg protein per min)

C source in growth medium	Variants		
	Initial	Dehydrated at 37 °C for 1 h	Dehydrated at 37 °C for 24 h
Glucose 2%	9.9 ± 0.9	5.5 ± 0.5	6.6 ± 0.4
Ethanol 1.5%	24.8 ± 0.5	17.8 ± 0.3	15.4 ± 0.3
Ethanol 1.5% + pepstatin + PMSF	57.0 ± 0.9	55.8 ± 0.7	not determined
Molasses 2%	14.1 ± 0.3	6.8 ± 0.5	6.2 ± 0.8
Molasses 2% + pepstatin + PMSF	15.7 ± 0.4	13.8 ± 0.3	13.7 ± 0.3

and therefore they are subject to proteolysis. The increase of amino nitrogen content in yeast cells during dehydration does not correlate with their viability. Consequently, proteolysis is not the main reason for cell death during drying.

Received July 3, 1984

References

[1] Nechayeva, A. S.: Biokhimiya 5 (1940), 48.
[2] Notkina, L. G.: Trudi Instituta Mikrobiologii AN SSSR 6 (1959), 183.
[3] Medvedeva, J. I., Lukina, G. D., Petrenko, J. B., Belous, V. N.: Prikladnaya biokhimiya i mikrobiologiya 4 (1968), 690.
[4] Vitrinskaya, A. M., Meledina, T. V.: Prikladnaya biokhimiya i mikrobiologiya 15 (1979), 227.
[5] Labuza, T. P., Jones, K. A., Sinskey, A. I., Gomez, R., Wilson, S., Miller, B.: J. Food Sci. 37 (1972), 103.
[6] Beker, M. J., Damberga, B. E., Rapoport, A. I.: Anabiosis of Microorganisms. Riga, Zinatne, 1981, 136.
[7] Beck, J., Fink, G. R., Wolf, D. H.: J. Biol. Chem. 252 (1980), 4821.
[8] Jusic, M., Hinze, H., Holzer, H.: Hoppe-Seyler's J. Physiol. Chem. 357 (1976), 735.
[9] Mazon, M. J., Hemmings, B. A.: J. Bacteriol 139 (1979), 686.
[10] Mechler, B., Wolf, D. A.: Eur. J. Biochem. 121 (1981), 47.
[11] Schwenke, J., Moustacchi, E.: Mol. Gen. Genet. 185 (1982), 290.
[12] Wolf, D. H., Ehmann, C.: J. Bacteriol. 147 (1981), 418.
[13] Bachmann, B., Kosiek, E., Wlodarczyk, C.: Mitteilungen der Versuchsstation für das Gärungsgewerbe in Wien 3 (1973), 45.
[14] Galkin, A. V., Tsoi, T. V., Luzikov, V. N.: FEBS Letters 105 (1979), 373.
[15] Graham, R. K.: J. Inst. Brew. 76 (1970), 16.
[16] Betz, H., Hinze, H., Holzer, H.: J. Biol. Chem. 249 (1974), 4515.
[17] Saheki, T., Holzer, H.: Eur. J. Biochem. 27 (1972), 520.
[18] Jones, E. W.: Genetics 85 (1977), 23.
[19] Lenney, J. F., Matile, P., Wiemken, A., Schellenberg, M., Meyer, J.: Biochem. Biophys. Res. Comm. 60 (1974), 1378.
[20] Termkhitarova, N. G., Schulga, A. V.: Prikladnaya biokhimiya i mikrobiologiya 10 (1974), 928.
[21] Bailey, N. T. I.: Statistical Methods in Biology. The English Universities Press Ltd, 1959.
[22] Laivenieks, M. G., Tsaimenko, A. B., Bobyk, M. A., Kulaev, I. S.: Prikladnaya biokhimiya i mikrobiologiya 14 (1978), 690.
[23] Panek, A. D.: Eur. J. Appl. Microbiol. 2 (1975), 39.
[24] Ellies, G. P.: Adv. Carbohydrate Chem. 14 (1959), 63.

Acta Biotechnol. 5 (1985) 4, 353—361

Autohydrolysis Extraction Process as a Pretreatment of Lignocelluloses for their Enzymatic Hydrolysis

TARGOŃSKY, Z.

Department of Food Technology, Agricultural University
20-934 Lublin, Akademicka 13, Poland

Summary

Autohydrolysis was studied as a pretreatment to enhance sugar yields from enzymatic hydrolysis of wheat and rape straw, beech, birch and poplar sawdust. Reaction temperatures were 185 °C to 212 °C and the reaction time 20 min. The pretreated slurries were hydrolyzed with "Novo" cellulase and *Fusarium* sp. 27 cellulase at 45 °C and pH 4.8 for 24 h with addition of *Fusarium* sp. 27 cell-bound cellobiase. From 85% to 90% sugar content of substrates were converted to reducing sugars after 24 h enzymatic hydrolysis, with exception of poplar wood.
10.8 g biomass was obtained after cultivation of *Fusarium* sp. 27 with water solution hemicellulose fraction from 100 g beech sawdust autohydrolyzed at 200 °C during 20 min.

Introduction

The obtaining ethanol, SCP and other products by means of enzymatic hydrolysis of lignocelluloses is connected with the necessity of pretreatment [1, 2, 3]. In recent years, special attention was paid to the process of autohydrolysis of waste lignocelluloses of agricultural and wood industry origin [4, 5, 6]. The process of autohydrolysis which consists in heating lignocelluloses with water or steam can be operated continuously at 180—200 °C for 5 to 30 min or periodically at 245—260 °C for a dozen or several dozen seconds with the explosive expansion of the mixture [5]. Under these conditions there is a degradation of hemicellulose into water soluble products and partial lignin depolymerization [4]. Lignin, at least partly, can be removed with the help of alkaline solutions [7] or organic solvents [8]. The cellulosic material obtained in this way can successfully undergo enzymatic or acidic hydrolysis obtaining glucose as the main product.

Materials and Methods

Substrates

Wheat and rape straw were crushed in the impact mill. Beech, birch and poplar sawdust were obtained from the local sawmills. All the crushed substrates were shaken through a sieve and the fraction of 0.43—1.5 mm diameter was used in further studies.

Enzymatic Preparations

A cellulolytic preparation made by the firm "Novo" (SP 122) with FPU activity 62 units/g preparation was used in the examinations and a cellulolytic preparation obtained in the submerged culture of *Fusarium* sp. 27 made on the modified SAUNDER's medium with the addition of 1% filter paper and 0.1% malt sprouts [9]. The culture filtrate condensed under reduced pressure at 30 °C had a FPU activity of 1.6 units. The mycelium obtained after the *Fusarium* sp. 27 culture under the conditions mentioned above was an additional source of cellobiase added to the cellulolytic preparations. The cell-bound cellobiase activity was 0.164 nmol/min × cm^3 × g mycelium. The cellulolytic preparation from the firm "Novo" was dissolved in 0.1 M acetate buffer, pH 4.8, containing 1 mM sodium azide in such proportion to obtain the solution activity of 1.6 measured in FPU units. The obtained preparations were mixed in ratio 1 : 1 with the condensed culture filtrate of *Fusarium* sp. 27. The cellulase solution prepared in this way was diluted with 0.1 M acetate buffer at pH 4.8 containing 1 mM sodium azide to obtain the activity of the preparation added to the reaction mixture of 16—64 units of FPU/g of hydrolyzed lignocelluloses.

Autohydrolysis and Extraction of Lignocelluloses

Autohydrolysis of substrates was carried out in the 250 cm^3 autoclave with 20 g dry lignocelluloses and 100 cm^3 distilled water and it was submerged in the oil bath so that the temperature inside the autoclave could reach 185—212 °C. Beginning at moment, the desired temperature was reached, 20 min were allowed for the duration of autohydrolysis and then the reactor was completely vented. The obtained material was placed in the Büchner funnel and washed with hot distilled water and 200 cm^3 filtrate were made. Next, the material was treated with 100 cm^3 1% NaOH solution to extract the part of lignin and left for 16 hours, then filtrated and washed with phosphoric acid and distilled water till pH was 5.0. The wet material was used as a substrate in the enzymatic hydrolysis and a lignin fraction was precipitated from the alkali extract by its acidification up to pH 3.0.

Enzymatic Hydrolysis of the Cellulosic Materials

Approximately 250 mg cellulosic material, converted into dry weight, was put into 25 cm^3 flasks 10 cm^3 cellulase solution was added as above and incubated at 45 °C in the shaker with the water bath, taking samples for analysis from time to time.

The Water Solution Hemicellulose Extract for Production of Single Cell Protein

Phanerochaete chrysosporium from Institute of Wood Technology in Poznan and *Fusarium* sp. 27 and *Candida tropicalis* from our own collection were used to obtain SCP. *Fusarium* sp. 27 was cultivated on SAUNDER's mineral medium [9] mixed with the autohydrolysis water extracts. The ratio of both solutions in the mixture was 2 : 1 and pH was 5.5. *Phanerochaete chrysosporium* and *Candida tropicalis* strains were cultivated on ANDER and ERIKSSON's medium [10] with the extract added as above. The mineral medium and water extract were sterilized separetely for 30 min at 0.5 bar pressure. 60 cm^3 medium were placed in 300 cm^3 ERLENMAYER's flasks and inoculated. After incubation at 28 °C for 4—7 days, the culture has been finished. The biomass was centrifuged at 3000 rpm, washed with distilled water twice and centrifuged again. The biomass was dried at 80 °C to determine the final weight. The course of the culture was controlled by the loss of reducing substances in the culture medium.

Analytical Methods

The content of reducing substances in the water extract obtained after the autohydrolysis of lignocelluloses was determined in the reaction with 3,5-dinitrosalicylic acid (DNS) [11]. The content of reducing sugars produced during the enzymatic hydrolysis of lignocelluloses was determined in reaction with DNS [11] and the content of glucose in the hydrolysate with the enzymatic method with the glucose oxidase and the peroxidase [12]. The extent of carbohydrates saccharification contained in the lignocellulose was calculated from the following formula

$$\% \text{ reducing sugars yield} = \frac{\text{mg/cm}^3 \text{ reducing sugars in solution}}{\text{mg/cm}^3 \text{ potential reducing sugars}}.$$

Efficiency of glucose from the cellulose contained in the substrates examined was calculated from the formula

$$\% \text{ glucose yield} = \frac{\text{mg/cm}^3 \text{ glucose in solution}}{\text{mg/cm}^3 \text{ potential glucose}}.$$

The content of cellulose was determined by the method of KÜRSCHNER and HANAK [13], the content of lignin by JAYME and Knoll's method [14] and that of pentosan by TOLLEN's method [15].
Cellulolytic activities were expressed in FPU units according to the method of MANDEL et al. [16], the activity of cellobiase according to the method of STERNBERG [17] and adjusted by TARGOŃSKY [18] to determine cell-bound cellobiase of the *Fusarium* sp. 27 mycelium. Protein was estimated by KJELDAHL determination of nitrogen ($N \times 6.25$). The total content of acids in the water extract after the autohydrolysis of lignocelluloses was determined by means of 0.1 N NaOH titration in the presence of phenolphthalein and the result was given in terms of acetic acid and expressed in relation to the amount of the substrate used in the autohydrolysis. The content of furfural in the water extract was determined by the reaction with phloroglucin [15].
Presence of monomeric lignin derivatives in the water extracts was detected after etheric extraction by thin-layer chromatography on DC-Alufolien Kieselgel 60, using benzene-methanol-acetic acid 45 : 8 : 4 as a following solvent, and diazotized benzidin was used for detection [19].

Results

Wheat and rape straw as well as beech, birch and poplar sawdust were autohydrolyzed. Changes of the chemical composition of wheat straw and beech sawdust autohydrolyzed at 185 °C, 200 °C, 212 °C and the remaining substrates at 200 °C which were then subjected to water and alkali extraction are presented in Tab. 1. An increase of autohydrolysis temperature from 185 °C to 200 °C favoured the pentosan and lignin degradation, whereas cellulose was degrated only to a very small degree which was manifested by an increased content of cellulose in the lignocellulosic material. A further increase of the autohydrolysis temperature to 212 °C caused unfavourable changes in wheat straw since a smaller amount of cellulose and a greater amount of lignin were found to be in the autohydrolyzed material at 200 °C. In the case of beech sawdust a further increase of the cellulose content in the lignocellulose material and a small decrease of lignin in pentosan was observed. Concerning the remaining lignocellulose, autohydrolyzed at 200 °C for 20 min, the highest mass decrement after the water and alkali extraction (53%) was observed in rape straw and the lowest one (35%) in poplar sawdust. In the case of rape straw the conditions of the autohydrolysis seemed to be too drastic but in the case of poplar sawdust they did not cause any advanced degradation of the substrate, thus a considerable part of lignin was not degraded (Tab. 1).
In view of a considerable degradation of hemicellulose of the lignocelluloses taking place during the autohydrolysis process the water extract was temporarily put in a group of components soluble in hot water (Tab. 2). The results from Table 2 show that the content of reducing substances determined in reaction with 3,5-dinitrosalicylic

Table 1: Chemical composition of autohydrolyzed lignocelluloses after water and alkalic extraction

Substrate	Autohydrolysis temperature	Amount of mass obtained	Substrate composition after autohydrolysis			
			Cellulose	Lignin	Pentosan	Others
	[°C]	[%]	[%]	[%]	[%]	[%]
wheat straw	control	100	43.8	19.0	30.2	7.0
	185	66.2	61.0	21.6	10.5	6.9
	200	55.6	74.2	20.3	2.4	5.1
	212	53.2	71.2	24.2	0.6	4.0
beech sawdust	control	100	45.0	19.3	25.8	9.9
	185	67.4	61.2	19.5	10.8	8.5
	200	55.4	74.8	19.6	2.2	3.2
	212	51.0	78.2	18.2	0.5	3.1
rape straw	control	100	43.2	17.9	12.4	26.5
	200	46.5	60.1	24.2	3.5	12.2
birch sawdust	control	100	45.8	19.0	26.3	8.9
	200	56.8	72.6	19.3	5.3	2.8
poplar sawdust	control	100	46.1	27.1	18.7	8.1
	200	64.6	68.0	29.8	1.2	1.0

Table 2. Characteristics of water extract obtained after autohydrolysis of lignocelluloses

Substrate	Autohydrolysis temperature [°C]	Reducing substances [%]	Acid in terms of acetic acid [%]	Furfural [%]	Qualitative composition of aromatic compounds
wheat straw	185	5.0	1.05	0.32	ferulic acid
	200	11.3	3.0	0.75	
	212	6.8	3.1	1.32	
beech sawdust	185	7.2	1.4	0.62	ferulic acid
	200	12.4	3.8	1.08	syringaldehyde
	212	6.6	4.7	1.94	veratraldehyde
rape straw	200	10.8	4.06	1.12	Ferulic acid synapic acid syringaldehyde
birch sawdust	200	15.1	2.8	1.25	NT
poplar sawdust	200	6.9	2.3	0.69	NT

Content of substances are given in relation to amounts of substrate subjected to autohydrolysis

acid is only from 5.0% to 15.1% in relation to the substrate used in the autohydrolysis whereas after the beech sawdust hydrolysis with 0.1 N HCl solution at 155°C for 30 min the amount of reducing sugars in terms of xylose was approximately 25%. It should be mentioned that the enzymatic hydrolysis of carbohydrates contained in the water extract only in the case of the extract obtained after the autohydrolysis at 185 °C increased the content of reducing substances in reaction with DNS. In the other cases having added the enzymatic preparation and after 1 hour incubation, a decreased amount of reducing substances was observed, as well as formation of sediments (unpublished data). Increased acidity of the extract and increased amount of furfural were observed when the autohydrolysis temperature was also increased which pointed to the progressing process of degradation of hemicellulose and their monomers. In the water extract the presence of lignin monomeric compounds was found, such as ferulic acid, synapic acid, syringaldehyde, veratraldehyde. These compounds may exert a significant effect on utilization of substances contained in the water extract by microorganisms in the biosynthesis and fermentation process.

Table 3. Results of saccharification of autohydrolyzed wheat straw with some cellulolytic preparations

Origin of preparation	Activity FPU μmol/cm³ · min	% saccharification of carbohydrates wheat straw after	
		1 h	24 hs
Fusarium sp. 27	0.4	23	57.5
Novo (SP 122)	0.4	24.1	61.4
Novo (SP 122) + *Fusarium* sp. 27 1 : 1	0.45	27.5	68.7

Autohydrolysis of wheat straw: 200 °C, 20 min
Enzymatic hydrolysis: 2% autohydrolyzed wheat straw after water and alkali extraction, pH 4.8, temperature of hydrolysis 45 °C.

For the enzymatic hydrolysis of lignocelluloses a mixture of cellulolytic preparation of the firm Novo (SP 122) and *Fusarium* sp. 27 culture was used. The results from Tab. 3 point to a more favourable action of the mixed preparations than that of the individual preparations used separately.

Dynamics of saccharification of lignocelluloses before and after the autohydrolysis is presented in Fig. 1. An increase of the autohydrolysis temperature from 185 °C to 200 °C affected favourably the hydrolysis efficiency of both beech sawdust and wheat straw. After 24 hours hydrolysis of autohydrolyzed lignocelluloses at 200 °C the degree of saccharification was 85% to 90% with only one exception when the autohydrolyzed poplar sawdust was used, saccharification was only 53.5%. The degree of cellulose saccharification to glucose ranged within 80–86% in the case of pretreatment of wheat straw and beech sawdust by autohydrolysis-extraction process. Native substrates were saccharified from 7.7% to 25.6% but pentoses were the main product of the hydrolysis.

A high degree of saccharification of the lignocelluloses was obtained when a cellulolytic preparation of high FPU (filter paper units) activity was used (64 units per gram hydrolyzed substrate). An increase of the concentration of the substrate in the reaction mixture or a decrease of the activity of the cellulolytic preparation limited the degree of saccharification in the hydrolyzed substrate (Fig. 2).

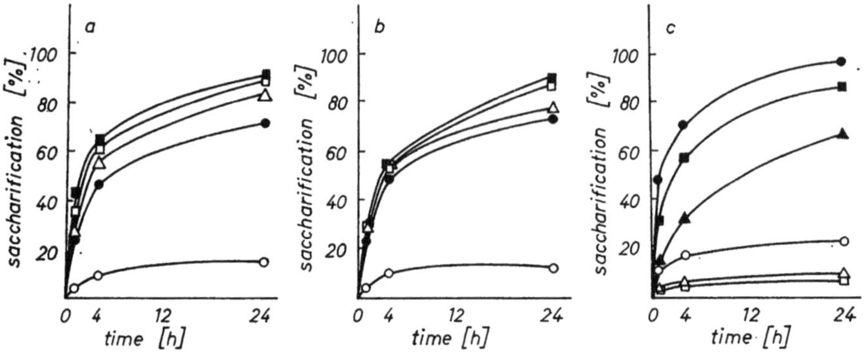

Fig. 1. Sugar yields of enzymatic hydrolysis after pretreatment.
a) wheat straw, b) beech sawdust
○—○ untreated; ●—● 185 °C
□—□ 200 °C; ■—■ 212 °C
△—△ glucose yields; pretreatment at 200 °C
c) rape straw, ○—○ untreated, ●—● 200 °C
birch sawdust □—□ untreated, ■—■ 200 °C
poplar sawdust, △—△ untreated, ▲—▲ 200 °C
Cellulase activity was 64 FPU/g + 200 mg biomass *Fusarium* sp. 27/g substrate

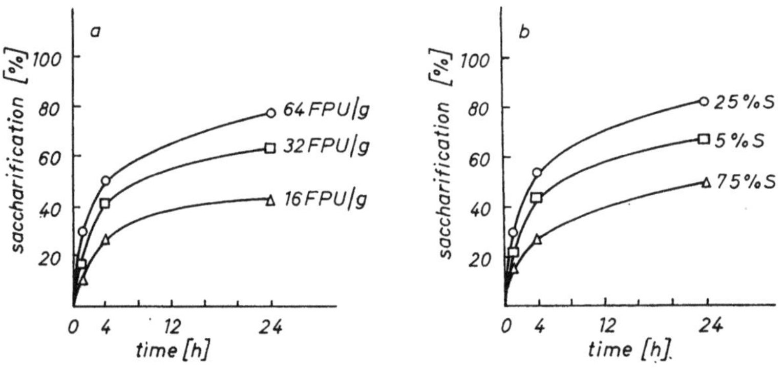

Fig. 2. Sugar yields of enzymatic hydrolysis of pretreated beech.
a) effect of cellulase activity; concentration substrate 2.5%.
b) effect of substrate concentration; cellulase activity 64 FPU/g substrate.

Furtheron, possibilities of using the water extracts obtained after the autohydrolysis of beech sawdust in the production of SCP. Strains *Phanerochaete chrysosporium* F-16 and *Fusarium* sp. 27 appeared to be much better producers of biomass and protein in comparison with the strain *Candida tropicalis*, which follows from the data in Tab. 4. It should be stressed that during 72—96 hours cultivation minimal growth of the examined strains was observed, which was most probably caused by the presence of substances like furfural inhibiting the culture medium. The maximum efficiency of the biomass could be reached when the examined strains were cultivated for 144—168 hs. It was similar in the case of culture carried out in the water extracts obtained after the autohydrolysis of wheat and rape straw as well as birch and poplar sawdust.

Table 4. Biomass efficiency obtained in culture of some strains on water extract from autohydrolyzed beech sawdust

Strain	Biomass after culture on water extract from autohydrolysis at temperature [g/100 g substrate]			Mean content of protein in biomass [%]
	185 °C	200 °C	212 °C	
Fusarium sp. 27	8.1	10.8	4.2	38.7
Phanerochaete chrysosporium	8.2	10.4	4.2	40.7
Candida tropicalis	2.4	2.8	1.3	41.2

Discussion

The process of hydrothermal treatment of lignocelluloses is a well-known and worked out process [15] but as a method of pretreatment of lignocelluloses for acid or enzymatic hydrolysis it became of interest for researchers only a few years ago [4, 5]. Lignocelluloses free from hemicelluloses and native parts of lignin due to the autohydrolysis can be successfully subjected to enzymatic or acid hydrolysis. LINDEN et al. autohydrolyzed wheat straw at 180 °C and then extracted it with water and ethanol. The material which was obtained in such a way was saccharified 90% during 24 hours. DEKKER and WALLIS [21] autohydrolyzed the sunflower-seed hulls at 200 °C for 5 min and after the explosive defibring obtained the material containing 33% cellulose, 45% lignin and 7% heteroxylan. The yield of enzymatic hydrolysis of the remaining residue obtained with the cellulolytic preparations Meicelase and Onozuka 3S and the preparation obtained from *T. reesei* C-30 was 40%. But it was 70% when cellobiase was added. The cellulolytic activity of the preparations was 20 FPU/g substrate. PULS et al. [7] treated the sawdust and straw with saturated steam at 170 °C—200 °C in laboratory defibrator and then defibrinated them. Hemicelluloses were extracted with water and part of lignin with 0.4% solution. The cellulose contained in the material was saccharified to glucose in a degree of 90% during 24 hours, at the same time the cellulase activity in the reaction mixture was 75 FPU/g substrate. The content of carbohydrates in the water extract was 15—23% of the used substrate.

Similar and slightly lower yields of glucose from cellulose were obtained in this study. The content of reducing sugars in the water extract determined in reaction with DNS was considerably lower than those obtained by PULS et al. [7] analysing the sugars content in the water extract with the method of ion-exchange chromatography.

The water extract obtained after the autohydrolysis of lignocelluloses may be used in SCP production, yet the amounts of biomass and protein obtained by means of fungus cultures were generally higher than those by means of the yeast cultures carried out on the same extracts. One of the reasons of the initial growth inhibition of organisms was the presence of various compounds, because according to the studies of BARNET et al. [22] water extracts obtained from the autohydrolysis contained 15 monomers and 8 carbohydrate dimers. Apart from carbohydrates and their products of degradation there also occur monomeric derivatives of lignin in the water extracts which may significantly affect the growth of microorganisms. EKLUND et al. [23] obtained 35—36 g biomass from 100 g reducing sugars originating from the acidic hydrolysis of hemicellulose of sunflower-seed hulls after the culture of *Candida* yeasts, and 63 g mycelium *Paecilomyces variotii* from 100 g reducing sugars.

Production of ethanol by means of yeasts or fungi fermentation may appear to be an important way of using water extracts obtained after the autohydrolysis of lignocellulosic materials [24, 25, 26]. CHAHAL et al. [27] confirmed certain problems to exist in utilizing the water extract as a result of the presence of substances inhibiting the growth of microorganisms. They obtained cellulosic material containing 57% cellulose, 28% lignin and 11% hemicellulose and the water extract treating poplar sawdust with steam pressure of 500—550 psig for 40—50 s. Presence of the extract in the culture medium inhibited the growth of *Trichoderma reesei* C-30 for 96 hours.
Increased autohydrolysis temperature to over 200—212 °C and increased duration to over 20—10 minutes deteriorates the water extract properties but also decreases the glucose yield of the cellulosic materials obtained after the autohydrolysis. MORJANOFF et al. [28] changed the temperature of bagasse autohydrolysis from 200—213 °C to 236—241 °C for 5 min and observed a decreased effect of enzymatic digestion of the lignocellulosic material. The results of this paper confirm it also. It applies to wheat straw autohydrolyzed at 200 °C and 212 °C for 20 min and to the mixture of sawdust treated with water steam at 167—235 °C reported by SCHULTZ et al. [29].
Lignin extraction from lignocelluloses which had been autohydrolyzed with application of NaOH water solution may significantly decrease the amount of the cellulolytic enzymes added [30]. According to WAYMAN [4] an increased concentration of the NaOH solution from 2% to 10% to extract lignin from autohydrolyzed aspen sawdust decreased she duration of the total hydrolysis of cellulose into glucose from 8 to 1.5 days. Due to high costs of this operation and the necessity to neutralize NaOH other authors extract lignin with ethanol. According to MEDNICK et al. [31] the costs of lignin extraction with ethanol are so high that it is more favourable not to extract lignin from lignocellulose subjected to autohydrolysis. Further studies of the increase of susceptibility of lignocelluloses subjected to autohydrolysis to enzymatic degradation should considerably contribute to a decrease of the production costs of hydrolyzates and ethanol, respectively.

Received April 13, 1984

References

[1] RYU, D. D.Y., MANDELS, M.: Enzyme Microb. Technol. 2 (1980), 91.
[2] SADDLER, J. N., HOGAN, C., CHAN, K. H., LOUIS-SEIZE, G.: Can. J. Microbiol. 28 (1982), 1311.
[3] UMER, D. G., TENGERDY, R. P., MURPHY, V. G.: Biotechnol. Bioeng. Symp. 11 (1981), 449.
[4] WAYMAN, M.: IVth International Symposium on Alcohol Fuels Technology. Guaruja, S. P., Brazil, October 5—8, 1980. 1.
[5] LORA, J. H., WAYMAN, M.: Tappi 61 (1978), 47.
[6] MURPHY, V. G., LINDEN, J. G., MOREIRA, A. R., DOCKREY, K.: Second Chemical Congress of the North American Continent, Las Vegas, Nevada. August 1, 1980.
[7] PULS, J., AYLA, C., DIETRICHS, H. H.: IXth Cellulose Conference May 24—27, 1982. State University of New York, College of Environmental Science and Forestry, Syracuse N.
[8] MURPHY, V. G., DOCKREY, K., LINDEN, J. C., MOREIRA, A. R.: AIChE 1982. Annual Meeting, Los Angeles, California. November 14—19, 1982.
[9] TARGOŃSKI, Z.: Acta Microbiol. Pol. 32 (1983), 153.
[10] ANDER, P., ERIKSSON, K.-E.: Arch. Microbiol. 109 (1976), 1.
[11] MILLER, G. L.: Anal. Chem. 31 (1959), 426.
[12] BERGMEYER, H. U.: Methods of Enzymatic Analysis. New York, London: Academic Press, 1963, 123.
[13] KÜRSCHNER, K., HANAK, A.: Z. Lebensm.-Unters. Forsch. 59 (1930), 484.
[14] MODRZEJEWSKI, K., OLSZEWSKI, J., RUTKOWSKI, J.: Metody badań w przemyśle celulozowo-papierniczym. P. Ł. Łodz, 1966.

[15] PROSIŃSKI, S.: Chemia drewna. PWRiL, Warszawa, 1969.
[16] MANDELS, M., ANDREOTTI, R., ROCHE, C.: Biotechnol. Bioeng. Symp. **6** (1976), 21.
[17] STERNBERG, D.: Appl. Environ. Microbiol. **31** (1976), 123.
[18] TARGOŃSKI, Z.: Acta Alimentaria Pol. 1983. In the press.
[19] OPIEŃSKA-BLAUTH, J., KRACZKOWSKI, H., BRZUSZKIEWICZ, H.: Zarys Chromatografii Cienkowarstwowej. PWRiL, Warszawa, 1971.
[20] LINDEN, J. C., MURPHY, V. G., MOREIRA, A. R.: VIth International Fermentation Symposium, London, Ontaria, 1980.
[21] DEKKER, R. F. H., WALLIS, A. F. A.: Biotechnol. Lett. **5** (1983), 311.
[22] BARNET, D., DUPEYRE, D., EXCOFFIER, G., GAGNOIRE, D., NOVA, S. J. E., VIGNON, M. R.: Comm. Eur. Communities [Rep.] EUR 1983, EUR 8245, Energy Biomass, 889.
[23] EKLUND, E., HATAKKA, A., MUSTRANTA, A., NYBERGH, P.: Eur. J. Appl. Microbiol. **2** (1976), 143.
[24] SUIHKO, M.-L., DRAZIC, M.: Biotechnol. Lett. **5** (1983), 107.
[25] SUIHKO, M.-L., ENARI, T. M.: Biotechnol. Lett. **3** (1981), 723.
[26] DETROY, R. W., CUNNINGHAM, R. L., HERMAN, A. I.: Biotechnol. Bioeng. Symp. **12** (1982), 81.
[27] CHAHAL, D. S., MCGUIRE, S., PIKOR, H., NOBLE, G.: Biomass **2** (1982), 127.
[28] MORJANOFF, P., DUNN, N. W., GRAY, P. P.: Biotechnol. Lett. **4** (1982), 187.
[29] SCHULTZ, T. P., BIERMAN, Ch. J., MC GINNIS, G. D.: Ind. Eng. Chem. Prod. Res. Dev. **22** (1983), 344.
[30] SADDLER, J. N., HOGAN, C., CHAN, M. K. H., LOUIS-SEIZE, G., BROWNELL, H. H.: Proceedings of Fourth Bioenenry R D Seminar Winnipeg, Canada. March **4**, 1982. 509.
[31] MEDNICK, R. L., WEISS, L. H., XIPPLITOS, E. G.: AIChe, Engineering Progress, August 1982. 68.

Acta Biotechnol. 5 (1985) 4, 362

Book Reviews

Y. Y. Gleba, K. M. Plants

Protoplast Fusion — Genetic Engineering in Higher Plants

Berlin, Heidelberg, New York, Tokyo: Springer-Verlag, 1984. 220 S., 62 Abb., 148 DM

The problems in genetic engineering in higher plants are more difficult than in microorganisms. For genetic engineering in higher plants there are only few types of plasmids and other methods for genetical manipulations need protoplasts. Two pioneers in the field of protoplast isolation and subsequent fusion are the authors of these important volume 8 in the series "Monographs on Theoretical and Applied Genetics".
This book is a revised translation of the Russian edition. It is written in a clear style and the following topics are mainly discussed:

Techniques of parasexual hybridization.
Protoplast fusion and parasexual hybridization of higher plants.
Transmission genetics of parasexual hybridization in closely related crosses.
Protoplast fusion and hybridization of distantly related plant species.
Use of somatic hybridization.

Some experiments and results are discussed critically and in a concluding chapter the different fields of view are summarized and discussed. A detailed list with references (up to 1984) and an extensive register close this volume.
This excellent volume is necessary for all laboratories interested in this topic. For libraries in this field it will be a precious completion.

H.-P. Schmauder

J. Coombs

The International Biotechnology Directory 1985

Products, Companies, Research and Organizations.
London: Macmillan Publishers Ltd. 1985, pp. 494, £ 65

This Directory covers biotechnology in Western Europe, North America, Brazil, Australasia and Japan. It provides both an overview of the extend of present interest with a summary of activities in the various geographical areas and a catalogue whereby suppliers of materials and services can be identified.
Antibiotics and vitamins as well as biomass, energy and genetic engineering are included in this Directory.
The book comprises the three parts: International Organizations and Information Services; National Profiles; Non-commercial Organizations and Companies.
The Directory is a very useful book for all, who are working in the field of biotechnology, including these who intend to use the results of biotechnological research. A good arrangement makes it easy to use this book.

L. Dimter

Acta Biotechnol. 5 (1985) 4, 363—373

Study and Theoretical Modelling of the Drying of the Aminoacids — The Products of Microbial Synthesis

Saduikov, R. A.[1], Migunov, V. V.[1], Karpov, A. M.[2], Golubev, L. G.[1] and Pobedimsky, D. G.[1]

[1] Kazan Chemical Industry Institute
420015 Kazan, K. Marx street 68, U.S.S.R.
[2] All-Union Research Institute Biotechnika
119034 Moscow, Kropotkin street 38, U.S.S.R.

Summary

Some properties of the pure aminoacids — the products of the microbiological synthesis as an object of drying are established by the experiments. It is shown that the vacuum-oscillating drying method is the effective solution of the technological problem of drying the Leucine, Isoleucine, Triptophan and Threonin. The mathematical model of this process is elaborated, whose adequacy is verified by experiments. This model may be used for technological calculations and for technology optimization.

Microbiological synthesis is the most effective way of producing the L-aminoacids now. The drying is the final stage of this process is playing an important role because of its great influence on the quality of the product. Thus the detailed study of the process fundamentals, theoretical and experimental substantiation of the adopted solutions are required in designing of the stage.

As it is stated [1, 2], the aminoacids are non-thermostable materials; the quantity influence of the heating effect of drying on the aminoacids consistence in mushrooms was studied in [2]. The analysis of the drying means extended in microbiological industry [3—6] shows, that according to a number of considerations the methods with "mild" regimes of heating effect, such as vacuum drying and drying in fluidized layer (DFL), are suitable to aminoacids.

This article deals with the study of the problems of the application of these two drying methods and their combination — vacuum-oscillating method (VOD) [7] for drying of L-leucine (Leu), L-isoleucine (Ile), L-tryptophan (Trp), L-threonin (Thr) and with the elaboration of the physical and mathematical models of these processes.

Experimental Methods

The possibilities of vacuum drying of the aminoacids have been studied in the experimental vacuum dryer with heated rotor-coil pipe and walls [8]. It is stated that these materials are disposed to agglomerate and to stick to the rotor and walls (in particular Thr), which brings to naught the mixing vacuum drying advantages and confirms the remark [5] on the difficulty of the paste product drying.

The investigation of DFL and VOD were carried out on installation [9, 10] Fig. 1. Experiments on VOD were started with the heating of the installation with hot air, after

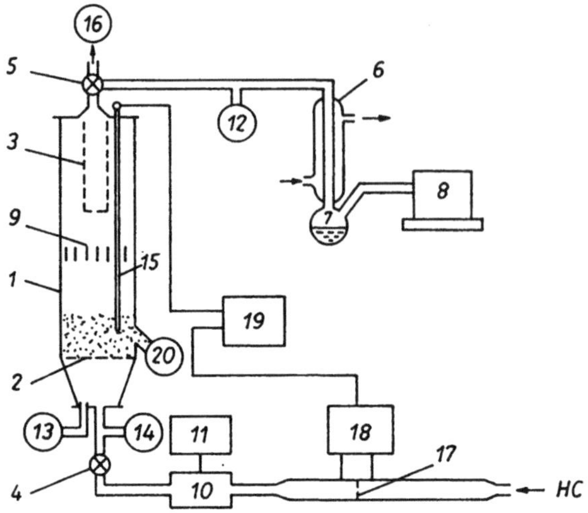

Fig. 1. Experimental installation scheme.
1 — drying camera with lenght 0.7 m and diameter 0.15 m, 2 — net, 3 — filter, 4 and 5 — valves, 6 — condensator, 7 — bulb for the condensate, 8 — vacuum pump, 9 — braking unit, 10 — electric heater, 11 — autoconnected transformer, 12 — vacuum manometer, 13 — manometer, 14 — mercury thermometer, 15 — thermistor, 16 — psychrometer, 17 — calibrated hole, 18 — differential manometer, 19 — registering apparatus, 20 — sampler.

which the moist material was loaded into it and the height of the layer h was measured. At the stage of convective heating-up of the product in fluidized layer the heat-carrier (HC) at the temperature T_{HC} is supplied into the installation through high-speed valve 4, and the output of the dryer is communicated with the atmosphere through the individual filter 3 and valve 5; the temperature of the material T, the dry-bulb T_d and wet-bulb T_w temperatures and HC volumetrical consumption G are measured. At the stage of pumping valve 4 is closed and the output of the dryer through valve 5 is communicated with the vacuum system, in which the vapour-gas mixture is pumped; the pressure in the installation P and T is measured.

After the stage of pumping valve 5 seals the dryer hermetically and the underlayer space is communicated with the atmosphere by means of high-speed valve 4 (the impulsive material throw up is going on, the layer is moving up like a piston). When the layer comes into collision whith braking unit 9, the braking of the agglomerates of the particles takes place. Then the output of the dryer is communicated with the atmosphere by means of valve 5, valve 4 open the HC supplying line and the cycle heating-up-pumping reiterates as many times as it is required. The samples of the aminoacids in the process of drying are taken with the vacuum sampler [11], the specific moisture content U is determined by means of the thermogravimetrical method [12]. DFL is accomplished as a particular case of VOD without the stage of pumping.

The errors in observation are determined by multiply repeating of the measurements and statistic treatment of the results and are characterized by the values of its dispersion: T_{HC} — 0.36 °C, T_d — 0.36 °C, T_w — 1.83 °C, T — 0.5 °C, U — 6.4×10^{-4} kg/kg.

Actually thermistor thermometer 15 measures the average temperature between the T and the temperature of the HC near thermistor during the heating-up stage. The special experiments were made to determine the difference between their values, in which thermometer 15 T-readings before and after the cut of the HC supply were registered. These readings proved to be constant with a mistake lower than the error in observation, which resulted in such conclusions: the difference between T and the temperature of the HC in the fluidized layer is negligible, the fluidized layer is isothermal [3], the heat exchange between the material and the HC is finished in the layer nearest to net [13].

The reproducibility of drying kinetics and thermograms was valued by making three experiments, in which the DFL Leu under the same conditions took place (Fig. 2). It is seen that the reproducibility of U is $\pm$ 0.02 kg/kg and that of T is $\pm$ 5 °C. These values are respectively 8% and 12% of the alteration ranges of U and T and that seems to be sufficient for our purposes. The irregularity of moisture distribution in the material moist-cured in the exsiccator for 12 hours (when the material is loaded into the dryer) is characterized by the magnitude 2.25×10^{-3} kg/kg — the value of the U dispersion in the 9 samples.

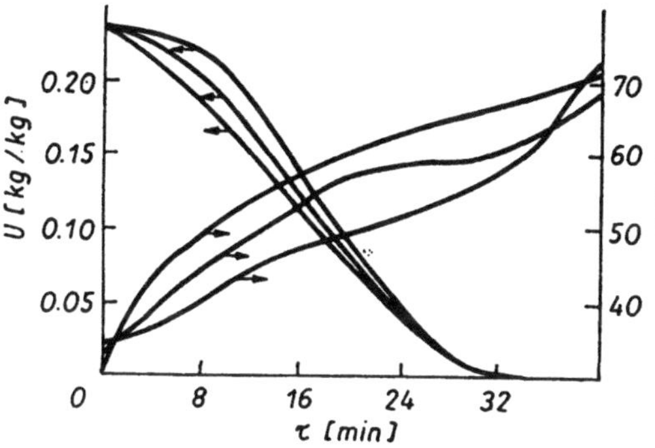

Fig. 2. Experiment results reproducibility.
$G = 7 \times 10^{-3}$ m³/s, $T_{HC} = 80$ °C, $h = 0.13$ m.

In the experiments the Leu, Ile, Thr and Trp were being dried by means of DFL and VOD with some modifications: the HC supplying in under- and in over-layer space, pumping with and without impulsive material throw up. The regime parameters were changed within the ranges: h from 0.03 m to 0.2 m, G from 1.4×10^{-3} m³/s to 8.3×10^{-3} m³/s, T_{HC} from 70 °C to 120 °C, the heating-up duration τ_H from 4 min. to 15 min., the pumping duration τ_P from 3 min. to 10 min., the initial moisture content U_0 from 0.1 kg/kg to 0.55 kg/kg. Different hydrodynamic regimes occured in the heating up stage: the HC flow through the system of canals, gushing and "boiling".

Results and Discussion

The comparison of the modifications has shown, that the drying method with HC supplying in the underlayer space makes it possible to decrease the general drying duration two times as compared to HC supplying in the overlayer space. Thus the most experiments were made with the first method of HC supplying.

During DFL of the Leu with $U_0 < 0.2$ kg/kg the fluidized layer was obtained at the start of the process and intensive drying took place. When $U_0 > 0.2$ kg/kg the agglomerates appeared; there was "boiling" only of the upper part of the layer, where the minute particles were concentrated. In the lower part of the layer there appeared canals and local gushing was observed. Under favourable conditions ($h < 0.07$ m, $G > 5 \times 10^{-3}$ m³/s) the agglomerates reduced to fragments later on, the lower part of the layer became fluidized and fine-dispersive product was received in accordance with the requirements for it (the white crystalline powder; $U < 0.01$ kg/kg). Otherwise the friable clots with diameters up to 0.02 m remained in the material. The typical picture of the alteration of U, T, T_d and T_w during DFL Leu is shown in Fig. 3a, b. As one can see, mainly free moisture is moved off from the Leu (the first period of drying); T changes up to the great values (60 °C and more), particularly when G increases (Fig. 2). T decreasing after achieving the maximum in Fig. 3b can be explained by the agglomerates reducing to fragments and the whole layer fluidizing, because of which the moisture

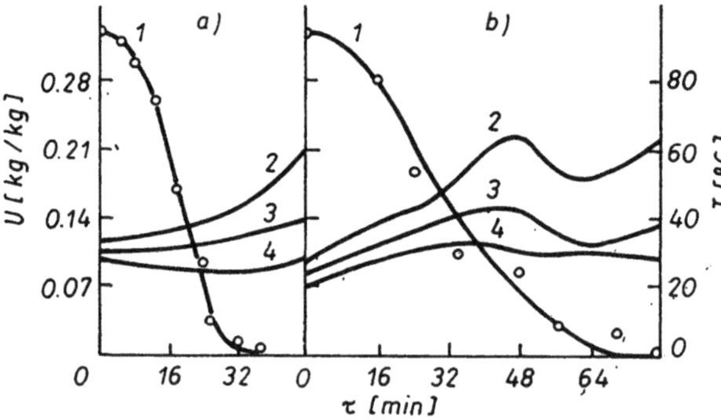

Fig. 3. Leucine drying in fluidized layer.
a — $G = 1.4 \times 10^{-3}$ m³/s, $T_{HC} = 90$ °C, $h = 0.07$ m; b — $G = 1.4 \times 10^{-3}$ m³/s, $T_{HC} = 80$ °C, $h = 0.15$ m; 1 — U, 2 — T, 3 — T_d, 4 — T_w.

exchange is intensified. This explanation has a confirmation in the drawing together T_w and T_d, i.e. in increasing the relative humidity of the HC at the output of the dryer. As experiments show, all the quality and quantity relationships of DFL Leu can be spread on DFL Ile and Trp with little modifications. However, in DFL Thr case the fine-grained product was not received because of the clots' durability — they did not reduce to fragments. The conclusions concerning the T increase and the free state of moisture are also verified for Thr. The most important result of the DFL experiments is the conclusion about the necessity of the effective means to reduce the agglomerates to fragments.

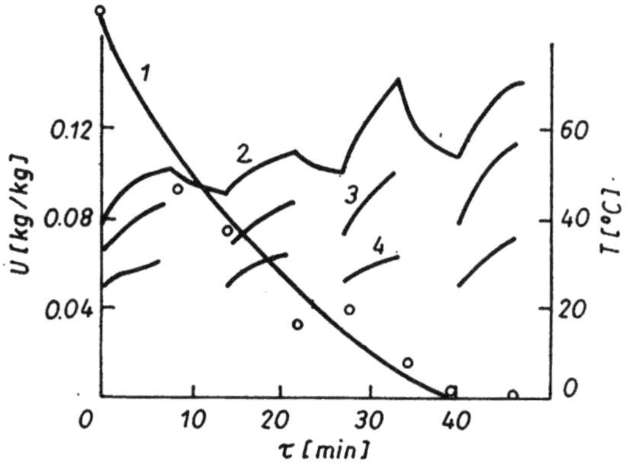

Fig. 4. Threonin vacuum-oscillating drying with impulsive material throw up.
$G = 7 \times 10^{-3}$ m³/s, $T_{HC} = 94$ °C, $h = 0.07$ m, $\tau_H = 8$ min., $\tau_P = 5$ min.; 1 — U, 2 — T, 3 — T_d, 4 — T_w.

The changes of U, T, T_d and T_w during VOD Thr and Leu are represented in Figs. 4, 5, 6. It is clear that the pumping makes it possible to decrease T and to eliminate the aminoacids overheating at the time of drying. The Thr agglomerates can be destroyed and the required quality product can be received by applying the impulsive material throw up. In this case the fluidized layer regime occurs at all the heating-up stages. Therefore VOD with the impulsive material throw up after the pumping stage can solve the technological problem of the aminoacids drying.

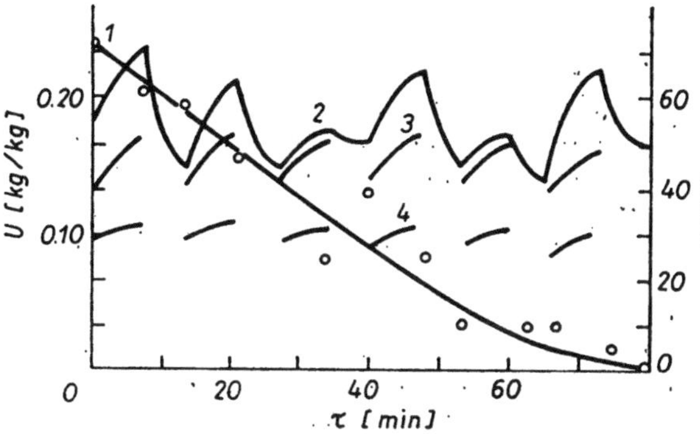

Fig. 5. Threonin vacuum-oscillating drying without impulsive material throw up. $G = 7 \times 10^{-3}$ m³/s, $T_{HC} = 86$ °C, $h = 0.07$ m, $\tau_H = 8$ min., $\tau_P = 5$ min.; 1 — U, 2 — T, 3 — T_d, 4 — T_w.

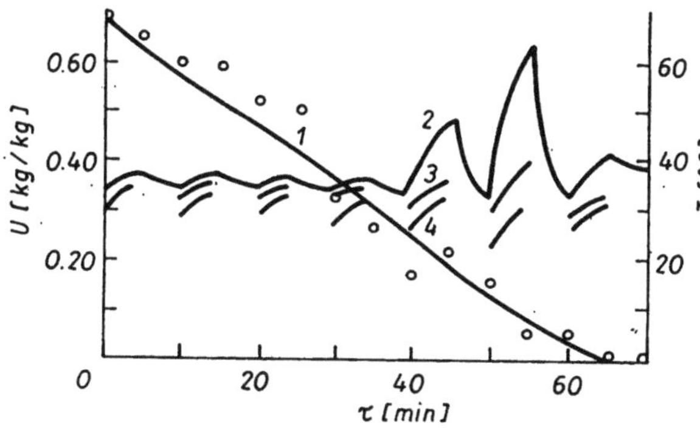

Fig. 6. Leucine vacuum-oscillating drying with impulsive material throw up. $G = 7 \times 10^{-3}$ m³/s, $T_{HC} = 96$ °C, $h = 0.1$ m, $\tau_H = 5$ min., $\tau_P = 5$ min.; 1 — U, 2 — T, 3 — T_d, 4 — T_w.

The VOD Process Mathematical Modelling

The mathematical modelling is aimed at the understanding of the process physical essence and makes clear the determining parameters influence on the process and serves for solving the problem of the technology optimization. This article does not deal with the questions of choosing the optimum criterion and the optimization itself, it is devoted to the elaboration the mathematical model of the VOD.

The regressive formula for the L-isoleucine drying duration was received in [10] by the treatment of the systematic experiments results:

$$\tau_d = -2007 + 8.616\tau_H + 4.5\tau_P + 88056G + 174.6h + 33.28\tau_H h - 0.0075\tau_H\tau_P - 782.76\tau_H G; \quad (1)$$

This expression is well-grounded in definite ranges of the regime parameters in the right part changing. In this formula some factors which had had influence on the process were not taken into consideration (for example, humidity and temperature of HC). It was received under the condition of τ_{H} and τ_P constance for all the cycles and it was heating-up that the first stage had to be; at last, there is no solution of the optimization problem in that region of arguments (1) change where (1) is well-grounded. So it is necessary to elaborate more general mathematical model, however the [10]th methods would require a great number of experiments and it is better to elaborate the physical model of the process (but not the statistic one).
VOD modelling essentially reduces to modelling of the DFL and the drying by pressure decreasing, but it has some peculiarities — the oscillating character of VOD makes the role of the transitive processes greater, the problem is essentially non-stationary. Besides the moisture moving off from aminoacids is the binary mixture of ethanol and water (50 : 50 or 30 : 70 volumetrically), specific heat, heat of vaporization and the boiling temperature of which are greatly influenced by the moisture composition, which, in its turn, changes continuously during VOD. In [14] it the model of the drying by pressure decreasing was proposed, when the individual liquid is moved off; the condition of the start of the intensive vapour pumping was formulated. This condition consists in pressure decreasing down to the value of the saturation pressure. The relationships between U, T and P were investigated in [14] without the analysis of the time development of the process. In [15] the same relationships at the state of the vapour pumping in case of real n-component liquid mixture were investigated; the dependence of the ith component saturation pressure P_i on T was used there in the form

$$\ln P_i(T) = A_i + \frac{B_i}{T} \tag{2}$$

(A_i and B_i are constants), which does not provide a good approximation of the experimental values in all the cases (the other expressions are used in [16] for the ethanol and water in particular). The general form of $P_i(T)$ was used in [17], but the time development of the process was not also investigated there. This aspect was considered for the individual liquid in [18], where (2) and the material balance of the moisture in the installation were used:

$$j - \varrho Q_P = V \frac{d\varrho}{d\tau}. \tag{3}$$

In (3) j is vapourization intensity, ϱ is the vapour density, V is the installation volume with the exception of the material volume, Q_P is the volumetrical vacuum-pump delivery (the vacuum pump pumping rapidness). But there is always a pipe-line between the dryer and the pump, and instead of the Q_P we must use the effective pumping rapidness Q [19], the dependence of which from the P is not so simpler, as the formula used in [18]. In literature there are no works dealing with the modelling of the moving off the mixtures by means of DFL, as well as with VOD physical modelling. In the present article the dynamic model of DFL, drying by pressure decreasing and VOD when the real n-component liquid mixture is moving off are proposed. General forms of the functions $P_i(T)$ and $Q(P)$ are assumed in this models.
Let us consider the periodical VOD process excepting the stages of product loading and unloading. From the point of view of the physical description we divide VOD into 5 stages. Two of them are the regimes switchings: heating-up-pumping and pumping-heating-up do not require the physical description and have the only characteristic — switching time; U and T are constant at these stages. The pumping we consider consists of two stages — the HC pumping and the pumping of the vapour of material moisture. The HC pumping is assumed to be finished when P decreases to the mixture saturation

pressure P_S at the temperature of T. Intensive evaporation and T lowering began from this moment — the vapour-pumping stage. When the HC pumping U and T are supposed to be constant, the heat and mass exchanges between the material and the HC pumping from the underlayer space are left out of account.

The HC pumping thermodynamic process is assumed to be polytropic with polytropic exponent $\varkappa$, which must be determined experimentally for this installation and vacuum system. The material balance of HC is expressed in (3) when $j = 0$

$$V \frac{d\varrho}{d\tau} + \varrho Q(P) = 0, \tag{4}$$

where $Q(P)$ is determined experimentally or by means of calculations. The polytropic process equation

$$\frac{P}{P_{at}} = \left(\frac{\varrho}{\varrho_0}\right)^{\varkappa}$$

(ϱ_0 and P_{at} are the characteristics of any standard HC state) together with (4) determine the HC pumping duration in this way:

$$\tau_{HC} = -\frac{V}{\varkappa} \int\limits_{P_{at}}^{P_S} \frac{dP}{PQ(P)}. \tag{5}$$

There are some general ideas in the modelling of the stages of heating-up and vapour pumping. We assume the installation walls to be adiabatic — this condition was provided by the thermal insulation of the dryer in the experiments. We examine the case, when the moving-off liquid is a real non-azeotropic mixture of n components having unlimited mutual solubility. The additional procedure we use to calculate the specific molar heat of vaporization r, the molar specific heat c, the molecular masses of the vapour μ_v and the liquid μ_l is

$$r = \sum_{i=1}^{n} r_i y_{fi}, \qquad c = c_0 + U_m \sum_{i=1}^{n} c_i x_i,$$

$$\mu_v = \sum_{i=1}^{n} \mu_i y_i, \qquad \mu_{vf} = \sum_{i=1}^{n} \mu_i y_{fi}, \qquad \mu_l = \sum_{i=1}^{n} \mu_i x_i.$$

Here $i = \overline{1, n}$ is the mixture components' numbers, c_0 is the molar specific heat of the moisture-free material, U_m is the molar specific moisture content; x_i, y_i und y_{fi} are the molar fractions of the ith component in the liquid, vapour and moving off from the material phases, respectively. The finestilling material balance [20] is used:

$$U_m \frac{d\bar{x}}{d\tau} = (\bar{y}_f - \bar{x}) \frac{dU_m}{d\tau}, \tag{6}$$

where $\bar{x}$ and $\bar{y}_f$ are the molar fractions n-dimensional vectors.

The fine-grained material structure (the crystalls less than 10^{-3} m) provides the developed surface of exchanges between the phases, which makes it possible to apply the thermodynamic equilibrium relationships to determine the vapour phase composition. The equilibrium idea has been suggested for DFL in [13], for drying by pressure decreasing — in [14] and verified by the experiments. Because of small values of the pressure and temperature in the process the ideal gas physical equation is used:

$$\varrho R T_g = P\mu, \tag{7}$$

where R is the universal gas constant, T_g is the gas temperature. The vapour phase ideality makes it possible to use formula [20]:

$$P_S = P(\bar{x}, T) = \sum_{i=1}^{n} x_i \gamma_i(\bar{x}, T)\, P_i(T), \tag{8}$$

$$y_i = x_i \gamma_i(\bar{x}, T)\, P_i(T)\, P^{-1}(\bar{x}, T) \qquad i = \overline{1, n}, \tag{9}$$

where $\gamma_i(\bar{x}, T)$ is the ith component activity coefficient.

The heating-up stage is characterized by heat and mass balances. As it was pointed in [3, 13, 21], the temperature of HC at the output of the layer may be considered equal to T and the fluidized layer may be considered isothermal; this assumptions are verified by the author's experiments with aminoacids. The heat supplied by HC is expended on moisture vaporization and the material heating-up:

$$cM \frac{dT}{d\tau} - rM \frac{dU_m}{d\tau} + G\varrho C_P(T - T_{HC}) = 0. \tag{10}$$

Here M is the moisture-free material molar number, ϱ is the HC density before the heater, C_P is the HC isobaric specific heat. The mass balance takes such a form:

$$G_1\varrho_1 - G\varrho + \mu_{vf}\, M \frac{dU_m}{d\tau} = 0, \tag{11}$$

where index "1" applies to HC at the output of the material layer. As experiments show, the dryer hydraulic resistance is small and the differential pressure between the HC supplying line and the atmosphere is not higher than 1 kPa. It follows that in the heater the isobaric process takes place and the atmospheric density ϱ_{at} may be used in the formula (10) instead of ϱ (if the HC is air). In order to deduce G_1 we are to apply the condition of constancy of the moisture-free HC mass consumption

$$G\varrho_{mfHC} = G_1\varrho_{mfHC1} \tag{12}$$

and the moisture-free HC physical equation. The result is

$$G_1 = \frac{P_{at} - \varphi P_n(T_{at})}{P_{at} - \varphi_1 P(\bar{x}, T)} \frac{T}{T_{at}} G, \tag{13}$$

here φ is the HC relative humidity, P_n is the water saturation pressure (index "n" here and below means "water"). It is important to emphasize that on account of water vapour presence in HC the presence of water in the material results from the thermodynamic equilibrium condition regardless the water presence in the material at the beginning of the process.

When using (7) for the partial vapour density we can obtain a mass-balance differential equation from (11)

$$\frac{dU_m}{d\tau} = \frac{G}{MRT_{at}\mu_{vf}} \left[\varphi\mu_n P_n(T_{at}) - \varphi_1\mu_v P(\bar{x}, T) \frac{P_{at} - \varphi P_n(T_{at})}{P_{at} - \varphi_1 P(\bar{x}, T)}\right]. \tag{14}$$

The φ in (14) is determined with the help of psychrometer, φ_1 depends on U, T and, because of the assumption concerning the equilibrium, on the way to reaching this equilibrium [3]. The sorption and desorption isotherms, used for drying calculations usually [3, 21], are not fit for researching essentially non-isothermal process, therefore the function $\varphi_1(U, T)$ was calculated from the real drying process experimental data and thus a greater adequacy of the means of equilibrium achievement was reached.

The experimental points were approximated by the expression

$$\varphi_1(U, T) = a_1 \ln U + a_2 \ln (a_3 + T) + a_4, \tag{15}$$

where $a_i = \text{const}\ i = \overline{1, 4}$; the method of the least squares was used to solve the approximation problem. The normal system of nonlinear equations was solved with the help of Newton's method by using scientific subroutines OS ES FORTRAN-4 and the ES-1033 computer.

When calculating the $\bar{y}_f$ one must consider that in the vapour phase there is water vapour, which was brought by the HC:

$$y_{fi} = \frac{y_i}{1 - y_b} \qquad i = \overline{1, n-1}, \qquad y_{fn} = \frac{y_n - y_b}{1 - y_b}. \tag{16}$$

Here y_b is the molar fraction of water brought by the HC in the vapour phase at the output of the layer,

$$y_b = \frac{\varphi P_n(T_{at})}{\varphi_1 P(\bar{x}, T)}. \tag{17}$$

It is seen that the obtained aggregate of the heating-up stage mathematical model equations is the closed system of ordinary differential and transcendental equations with the independent variable τ and when the initial conditions

$$U_m(0), \qquad \bar{x}(0), \qquad T(0) \tag{18}$$

are fixed it can be solved as a KOSHI problem with the help of any crude method after transformation to the convinient form for solving. The correctness of this KOSHI problem in the class of functions having continuous first derivatives in the half-space $\tau > 0$ when $U_m > 0$, $x_i > 0$ $i = \overline{1, n}$ may be shown when the dependents $\gamma_2(\bar{x}, T)$ and $P_i(T)$ are fixed — for example, with the help of [20] equations.

When modelling the vapour-pumping stage the mass balance is written by analogy with (3):

$$V \frac{d\varrho}{d\tau} + \mu_{vf} M \frac{dU_m}{d\tau} + \varrho Q[P(\bar{x}, T)] = 0. \tag{19}$$

The material vapour-cooling is described by means of the formula

$$c \frac{dT}{d\tau} - r \frac{dU_m}{d\tau} = 0, \tag{20}$$

and the equation

$$\bar{y}_f = \bar{y} \tag{21}$$

is true because of the HC's absence in the installation.

Thus, the close system of equations is present to describe the vapour-pumping stage; the digital solving of the KOSHI problem for this system with the (18) conditions offers no difficulties.

It is easier to solve the problems stated in the form, solved in regard to the derivatives. It is achieved in an evident manner for the heating-up stage, but for the vapour pumping stage it is necessary to use the following formula:

$$\frac{dU_m}{d\tau} = -\frac{c\varrho Q[P(\bar{x}, T)}{V(\alpha c - \beta r) + c\mu_v M}, \tag{22}$$

for deducing which the derivative $d\varrho/d\tau$ was calculated. Here

$$\alpha = \frac{1}{RTU_m} \sum_{i=1}^{n} \mu_i P_i(T) \left[(y_i - x_i)\, \gamma_i + x_i \sum_{j=1}^{n} \frac{\partial \gamma_i}{\partial x_j} (y_j - x_j) \right]; \tag{23}$$

$$\beta = \frac{1}{RT} \sum_{i=1}^{n} \mu_i x_i \left[\frac{\partial \gamma_i}{\partial T} P_i(T) + \gamma_i \frac{dP_i(T)}{dT} \right] - \frac{\varrho}{T}. \tag{24}$$

The stated Koshi problem of the vapour pumping stage is correct for the same class of functions and under the same conditions as the Koshi problem of the heating-up stage. It is the argument in favour of the whole mathematical model validity.

For the aims to verify the proposed model adequacy the comparison with the experimental data had been made. All the data required for the calculations were taken from [16] and assumed to be constant. The $Q(P)$ was stated for the experimental installation by means of the constant volume method [19] and then substituted by the approximate expression, approached the experimental points in the segment $[P_{min}, P_{at}]$, where P_{min} is the minimum pressure in the dryer. The subroutine HPCG from the FORTRAN-4 scientific subroutines library was used to integrate the Koshi problems, all the calculations being fulfilled on the computer ES-1033.

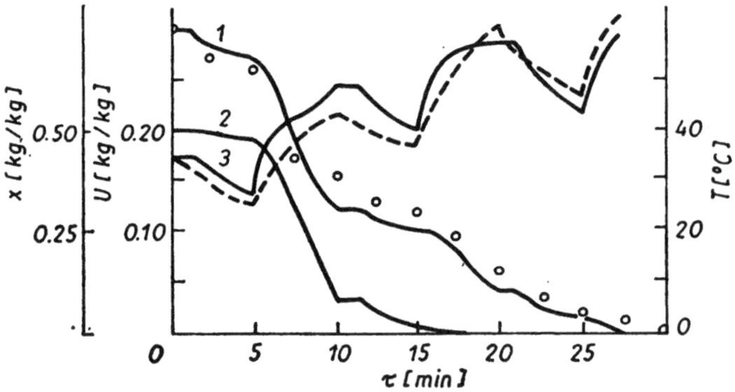

Fig. 7. Leucine vacuum-oscillating drying rated and experimental data comparison $G = 7 \times 10^{-3}$ m³/s, $T_{HC} = 95\,°C$, $h = 0.05$ m, $\tau_H = 5$ min., $\tau_P = 5$ min.; 1, 2 and 3 — computation (U, x_1 and T, respectively); ○ — U, experiment; ---- — T, experiment.

The rated and experimental data comparison is shown in the Fig. 7. It is seen that the mathematical model describes adequately the aminoacids drying process with the use of the vacuum-oscillating regime when the ethanol-water mixture is moved off. This model may be used for variety of technological calculations, for the substitution of the real experiments by the computations — the duration of the computer calculations of the drying process being above 20 seconds.

Conclusions

1. It is shown that the vacuum-oscillating drying method with impulsive material throw up solves the technological problem of drying of the pure aminoacids — leucine, isoleucine, threonin and tryptophan.
2. The physical and mathematical models of the drying in the fluidized layer and drying by pressure decreasing when the real n-component liquid mixture is moved off are proposed.

3. The vacuum-oscillating drying model and the computation programme are elaborated with the help of this models. The computation with the programme used verifies the adequacy of the model to the studying process.

Received July 30, 1984

References

[1] TUTOVA, E. G.: Microbiological and Enzymatic Aminoacids Synthesis. Conference materials, Pushchino, 1980. P. 49.
[2] DUDAREVA, N. T.: Appl. Biochem. Microbiol. **10** (1974) 2, 326.
[3] KUTSZ, P. S., TUTOVA, E. G.: Microbiological Products Drying. Moskva, ONTITEImicrobioprom, 1975. P. 80.
[4] KARPOV, A. M., ULUMIEV, A. A.: Microbiological Synthesis Products Drying. Moskva, Lyogkaya i Pishevaya Prom., 1982. P. 216.
[5] SARUKHANOV, A. V., VERIEVKIN, M. S.: Mikrobiol. Prom. (1983) 2, 17.
[6] SARUKHANOV, A. V., VERIEVKIN, M. S., MOSKEVICH, Yu. R.: Mikrobiol. Prom. (1983) 3, 21.
[7] LABUTIN, V. A., KAPITOV, N. A., GOLUBEV, L. G. et al.: Author's certificate N 954742 USSR, Invention Bulletin, 1982. N. 32.
[8] LABUTIN, V. A., SADUIKOV, R. A., GOLUBEV, L. G. et al.: Author's certificate N 827917 USSR, Invention Bulletin, 1981. N 17.
[9] ENIKEEV, E. Sh., ZOLOTOV, A. V., KARAEVA, G. A., MIGUNOV, V. V., SADUIKOV, R. A., PHARAKHOV, T. I.: Dep. ONTITEImicrobioprom N 167. 12. 10. 1983.
[10] IVANOVA, E. N., KAPITOV, N. A., MIGUNOV, V. V., PEVZNER, I. L., SADUIKOV, R. A., PHARAKHOV, T. I.: Dep. ONTITEImicrobioprom N 168. 12. 10. 1983.
[11] LABUTIN, V. A., SADUIKOV, R. A., GOLUBEV, L. G. et al.: Author s certificate N 720363 USSR. Invention Bulletin, 1980. N 9.
[12] BERLINER, M. A.: Moisture Measurements. Moskva, Energya, 1973. P. 400.
[13] GELPERIN, N. I., AINSHTEIN, V. G.: Fluidization. Moskva, Znanie, 1968. P. 64.
[14] LABUTIN, V. A., SAPHIN, R. G., GOLUBEV, L. G., PANICHKIN, V. I.: Theoretical Bases of the Chemistry Technology **17** (1983) 4, 527.
[15] SAPHIN, R. G., LABUTIN, V. A., GOLUBEV, L. G.: Modern Apparatuses for Heterogenium Mediums Treatment. Leningrad, LTI im. Lensoveta, 1982. P. 41.
[16] STABNIKOV, V. N., ROITER, I. M., PROCYUK, T. B.: Ethyl Alcohol. Moskva, Pishchevaya Prom. 1976. P. 272.
[17] MIGUNOV, V. V., PEVZNER, I. L., SADUIKOV, R. A.: Young Scientists and Specialists Researchs for the Industry. Conference materials, Kasan, 1983. P. 14.
[18] LABUTIN, V. A., SAPHIN, R. G., KAPITOV, N. A., GOLUBEV, L. G.: Mass Exchange Process and Apparatuses of the Chemical Technology. Kazan, Kazan Chemical Industry Institute, 1980. P. 31.
[19] ROZANOV, L. N.: Vacuum Technique. Moskva, Vysshaya Shkola, 1982. P. 207.
[20] KOGAN, V. B.: Theoretical Bases of the Chemical Industry Typical Processes. Leningrad, Khimya, 1977. P. 592.
[21] LUIKOV, A. V.: Drying, Theory. Moskva, Energya, 1972. P. 470.

Acta Biotechnol. 5 (1985) 4, 374

Book Review

A. Pühler and K. N. Timmis (Editors)

Advanced Molecular Genetics

Berlin, Heidelberg, New York, Tokyo: Springer-Verlag 1984
347 S., 98 Abb., 89 DM

In den letzten Jahren sind verschiedene exzellente Laborhandbücher über Methoden der Molekulargenetik geschrieben worden (z. B. die Bücher von J. H. Miller, R. W. Davies et al. und T. Maniatis et al. von Cold Spring Harbor Laboratory). Der vorliegende Band setzt diese Reihe fort, ohne jedoch eine einfache Wiederholung des Vorhandenen zu sein. Vielmehr wird versucht, neue Aspekte aus dem breiten methodischen Gebiet darzustellen. So werden z. B. praktische Hinweise für die Erschließung neuer, bisher wenig untersuchter Bakterienarten als Objekte moderner genetischer Forschung gegeben.
Im einzelnen werden folgende Sachgebiete behandelt:

- grundlegende Methoden (Gentransferprozesse, Isolierung und Charakterisierung von Plasmid-DNA, Hybridisierungsmethoden)
- Mutagenesemethoden (z. B. unter Nutzung von Hydroxylamin bzw. Bal 31, Transposonmutagenese)
- Genkonierung (mit Plasmiden, Phagen und Cosmiden)
- Untersuchungen zur Genexpression
- DNA-Sequenzanalyse
- Elektronenmikroskopie
- Experimente zur Aufklärung von Transkriptions- und Replikationsfunktionen.

Der Inhalt ist sehr übersichtlich gestaltet. Die Darstellung jedes Versuchs beginnt mit einer knappen theoretischen Einführung. Es folgt eine exakte Beschreibung der experimentellen Arbeiten in ihrer zeitlichen Folge. Eine Aufstellung über die benötigten Reagenzien und ein Literaturverzeichnis ergänzen das jeweilige Kapitel. Das Buch eignet sich damit ausgezeichnet als Leitfaden für Praktikumsversuche. Es ist weiterhin geeignet, den mit molekulargenetischen Methoden bisher weniger vertrauten Mikrobiologen, Biochemiker oder Biotechnologen in die Geheimnisse dieses modernen Wissenschaftszweiges einzuweihen. Aber auch dem auf molekulargenetischem Gebiet tätigen Wissenschaftler wird es als Nachschlagewerk für spezielle Methoden nützliche Dienste erweisen.

J. Engel

Acta Biotechnol. 5 (1985) 4, 375—378

Short Communications

Horseradish Peroxidase Based Bienzyme Electrode for Isocitrate

SCHUBERT, F.[1], KIRSTEIN, D.[1], ABRAHAM, M.[2], SCHELLER, F.[1], BOROSS, L.[2]

[1] Academy of Sciences of the G.D.R.
Central Institute of Molecular Biology
Department of Applied Enzymology
DDR-1115 Berlin-Buch, Lindenberger Weg 70
[2] Institute of Biochemistry
Attila Jozsef University Szeged
Szeged, Hungary

Summary

Coimmobilized horseradish peroxidase and D-isocitrate dehydrogenase were fixed to an oxygen electrode to assemble a bienzyme electrode for isocitrate determination. The linear measuring range for isocitrate of the sensor is between 0.1 and 2.0 mmol · l^{-1}, the coefficient of variation (20 measurements) is 3.6%. 8 samples per hour can be assayed. With one sensor preparation 140 measurements can be carried out.

The development of enzyme membrane electrodes for physiologically or industrially important compounds continues to be an expanding area of research. In case of amperometric enzyme electrodes oxidase enzymes are applied whenever available, allowing the rather simple electro-chemical indication of the reactant, O_2, or the product, H_2O_2. However, for many substrates of analytical importance specific oxidases are not available to the research or analytical laboratory. That is why coupling of these substrates to oxidase reactions via additionally immobilized enzymes has been investigated. Research in this direction has led to bi- or multienzyme electrodes. They are based either on the competition of the oxidase with the coimmobilized enzyme [1, 2] or on the sequential action of the enzymes [2—5]. The latter principle can also be applied to dehydrogenase substrates, provided a pyridine nucleotide oxidase is coimmobilized with the respective dehydrogenase. This has been demonstrated in a hybrid sensor using coimmobilized dehydrogenases and liver microsomal NADPH oxidase [6].
The coupling of the horseradish peroxidase (HRP) catalyzed air oxidation of NADH with dehydrogenases for ethanol and lactate determination has been reported by CHENG and CHRISTIAN [7, 8]. However, the enzymes were used in the soluble form.
Recently in the authors' laboratory an enzyme electrode for NADH and NADPH based on the oxidic reaction of HRP coupled to an O_2-sensor was developed [9]. In the present paper the extension of this electrode to the determination of isocitrate (IC) using an enzyme membrane with coimmobilized HRP and isocitric acid dehydrogenase (ICDH) is described.

Experimental

Materials

HRP (RZ = 0.42) and $NADP^+$ were purchased from VEB AWD Dresden. ICDH (14 U/ml) from pig heart and D,L-isocitric acid (sodium salt) were obtained from Reanal Budapest. Gelatin was from VEB ORWO Wolfen. All other substances were reagent grade.

If not otherwise stated the measuring solution consisted of tris-titric acid-sodium citrate buffer, 0.1 mol · l^{-1}, pH 8.0, containing $MgCl_2$ (9.5 mmol · l^{-1}), $MnCl_2$ (90 µmol · l^{-1}) and dichlorophenol (70 µmol · l^{-1}).

Procedures

Immobilization of the enzymes in gelatin membranes was performed as follows: 250 µl ICDH and 10 mg HRP were mixed and diluted with 250 µl doubly distilled water. 20 µl of 5% glutardialdehyde were added and the mixture was allowed to crosslink for 15 min at room temperature. Another 20 µl glutardialdehyde and 250 µl of 7.5% gelatin solution were added and the mixture was cast on a 10 cm^2 glass plate and dried at 4°C. The membranes were stored in dry state at 4°C until use.

Enzyme electrodes were prepared by placing a piece of 4 × 4 mm of the enzyme layer on the polyethylene membrane of a CLARK oxygen electrode (Forschungsinstitut Meinsberg). The immobilized enzyme was held in place with a dialysis membrane and an O-ring. For IC determination the enzyme electrode was horizontally inserted into the measuring cell containing 2 ml of stirred background solution. The temperature was kept at 25 ± 0.2° by thermostating the cell. 100 µl of $NADP^+$, dissolved in water, were added and the system was allowed to equilibrate. After 6 to 15 min a 100 µl sample of IC was injected and the derivative current was registrated with a polarograph GWP 673 (ZWG Berlin) connected to an x-y-recorder. When not in use the electrode was stored in the measuring solution containing $NADP^+$ (0.46 mmol · l^{-1}).

Results and Discussion

For the determination of IC with the bienzyme electrode two reactions proceed in the biocatalyst layer:

$$\text{D-IC} + NADP^+ \rightarrow \text{2-oxoglutarate} + CO_2 + NADPH + H^+ \quad (1)$$

$$2H^+ + 2NADPH + O_2 \rightarrow 2NADP^+ + 2H_2O \quad (2)$$

The application of reaction (2) has been extensively studied in the reported sensor for NAD(P)H [9]. The optimal cocatalyst concentrations for the coupled reactions (1) and (2) have been elucidated in a study using soluble HRP and ICDH [10]. ICDH requires millimolar amounts of Mn^{2+} or Mg^{2+} for activation [11]. The optimum $MnCl_2$ concentration for the oxidic HRP reaction was determined at 175 µmol · l^{-1} [9]. Beyond this value a drastic decrease of activity was observed. A further activation of ICDH was attained by addition of $MgCl_2$ to the measuring solution. 9.5 mmol · l^{-1} $MgCl_2$ caused an increase of the electrode sensitivity to IC of 125% of the value without $MgCl_2$. At higher concentrations the sensitivity decreased.

Fig. 1 shows the dependence of the sensitivity of the enzyme electrode to IC on $NADP^+$ concentration. The electrode signal increased with increasing $NADP^+$ concentration showing a tendency to level off at values beyond 0.4 mmol · l^{-1}. For further IC determinations an $NADP^+$ concentration of 0.46 mmol · l^{-1} was applied, first, because the sensitivity at this value should be suitable for IC measurement in fermentation samples and second, to keep the cofactor expense on a low level.

The calibration curve of the sensor for IC is shown in Fig. 2. The derivative current — concentration dependence is linear up to 2 mmol · l^{-1} with a lower limit of detection of 0.2 mmol · l^{-1}. The response time of the bienzyme electrode was 50 s and the equilibration time between two measurements (including cell rinsing) was 6 min, thus allowing the determination of 8 samples per hour. A coefficient of variation of 3.6% for 20 measurements was calculated. With one electrode preparation at least 140 measurements can be carried out without significant loss of sensitivity. After that the basic current of the sensor dropped to about 60%.

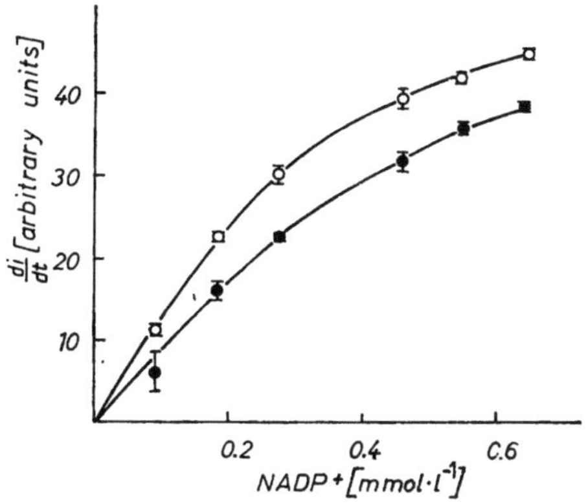

Fig. 1. Dependence of the sensitivity of the bienzyme electrode on NADP+ concentration in the solution for two different electrode preparations. [IC] = 3.1 mmol · l^{-1}. The values are means ± S.E.M. for 3 determinations.

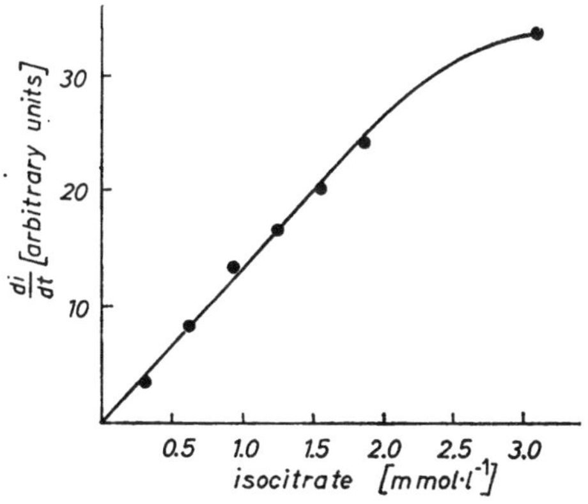

Fig. 2. Calibration curve of the bienzyme electrode. [NADP+] = 0.46 mmol · l^{-1}.

Conclusions

At present the quantitation of isocitric acid in fermentation processes is performed by titrimetric determination of the total acid and subsequent subtraction of the complexi-metrically measured citric acid. This method is time consuming and suffers from a low specificity and accuracy. The use of isocitric acid dehydrogenase, which catalyzes reaction (1), in combination with photometric NADPH determination [10] overcomes these problems. This procedure is expensive since new enzyme solution has to be applied for each measurement. Furthermore, fermentation broths of high turbidity cannot be assayed with reasonable accuracy.

Therefore a rapid, selective, and simple method is required for IC measurement.

Coupling of ICDH with HRP in the present bienzyme electrode combines the selectivity of the enzymatic reactions with the reusability of enzymes and the sensitivity and optical independence of electrochemical indication. With the exception of the paper by NAKAMURA et al. [12] to our knowledge no report on an enzyme electrode for IC has appeared so far. These authors coupled the dehydrogenase reaction to the electrode by means of a chemical mediator. Experimental data and operational parameters have not been communicated. Our ICDH-HRP sensor is characterized by a low response time achieved by use of the kinetic indication mode, and a good precision. The long lifetime of the sensor should allow isocitrate determination at relatively low cost. Therefore the bienzyme electrode seems to be well suited for application in fermentation control.

Received June 4, 1984

References

[1] Scheller, F., Pfeiffer, D.: Anal. Chim. Acta **117** (1980), 383.
[2] Renneberg, R., Pfeiffer, D., Scheller, F., Jänchen, M.: Anal. Chim. Acta **134** (1982), 359.
[3] Jensen, M. A., Rechnitz, G. A.: J. Membr. Sci. **5** (1979), 117.
[4] Pfeiffer, D., Scheller, F., Jänchen, M., Bertermann, K.: Biochimie **62** (1980), 587.
[5] Kulys, J. J., Peslakiene, M. V., Samalius, A. S.: Bioelectrochem. Bioenerg. **8** (1981), 81.
[6] Schubert, F., Scheller, F., Mohr, P., Scheler, W.: Anal. Lett. **15** (1982), 681.
[7] Cheng, F. S., Christian, G. D.: Clin. Chem. **24** (1978), 621.
[8] Cheng, F. S., Christian, G. D.: Clin. Chim. Acta **91** (1979), 295.
[9] Kirstein, D., Schubert, F., Scheller, F.: Anal. Lett. **15** (1982), 1479.
[10] Kirstein, D., Abraham, M., Boross, L., Scheller, F.: FEMS Lett., submitted.
[11] Ochoa, S.: Methods Enzymol. **1** (1955), 699.
[12] Nakamura, K., Nankai, S., Iijima, T.: Natl. Techn. Report **26** (1980), 497.

Book Review

Klaus Müntz

Stickstoffmetabolismus in Pflanzen

Jena: VEB Gustav Fischer Verlag, 1984
331 S., DDR 40 M, Ausland 72 M

Das Buch gibt einen Überblick über die in den letzten Jahrzehnten erzielten Fortschritte bei der Erforschung des Stickstoffmetabolismus der Pflanzen, wobei in konzentrierter Form und auf der Basis einer umfangreichen Literaturauswertung (bis 1983) ein teilweise bis in das Detail gehender Einblick in die Kompliziertheit des N-Stoffwechselgeschehens gegeben wird.
Das Buch ist übersichtlich in 3 Teile mit insgesamt 16 Kapiteln gegliedert.
Im ersten Teil „Assimilation anorganischer und Metabolismus niedermolekularer Stickstoffverbindungen in Pflanzen" werden die biologische Stickstoff-Fixierung, die Nitrat- und Ammoniumassimilation, die Prozesse der Nitrifikation und Denitrifikation, der Aminosäurestoffwechsel und der Stoffwechsel sekundärer Pflanzenstoffe sowie die Biosynthese und der Abbau von Nukleotiden und Ureiden behandelt.
Der Teil „Ferntransport stickstoffhaltiger Verbindungen in Pflanzen" beschreibt den Transport und die Verteilung löslicher N-Verbindungen in Sproßpflanzen.
Im dritten und letzten Teil „Metabolismus hochmolekularer Stickstoffverbindungen in Pflanzen: Nukleinsäuren und Proteine" wird auf der Grundlage des molekularbiologischen Geschehens die Spezifität des pflanzlichen Nukleinsäure- und Proteinstoffwechsels dargestellt, wobei basierend auf einer allgemeinen Einführung die pflanzlichen Organe gesondert behandelt werden. Der Metabolismus der Samenproteine und die Samenkeimung werden umfassend dargestellt.
Als methodische Ergänzung hätte sich eine kurze Darstellung der Einsatzmöglichkeiten der ^{15}N-Tracertechnik für pflanzliche N-Stoffwechseluntersuchungen angeboten.
Das übersichtlich gestaltete und gut gegliederte Buch ist auf einem hohen Niveau gehalten und dürfte sowohl Studenten als auch Fachwissenschaftlern wertvolle Informationen liefern.

K. Jung

Acta Biotechnol. 5 (1985) 4, 379—381

Isolation Method of Lysine Overproducers from Hydrocarbon Utilizers

SEN, S. K.

Department of Botany
Suri Vidyasagar College
Suri 731101, Birbhum, India

Hydrocarbons as the substrates of microbial production of amino acids was first initiated by YAMADA et al. [1]. Ever since, extensive search for amino acid production from hydrocarbons has been undertaken by several workers [2] and by industrial sectors [3]. In the course of a survey of lysine producers from the hydrocarbon utilizing microorganisms, mainly from the soils of Assam and West Bengal has been undertaken. In this communication results of the survey of hydrocarbon utilizing lysine producers has been summarized. Preliminary screening of lysine producing microorganisms from amongst the hydrocarbon utilizers was done by the cross streak assay method. For this purpose isolated cultures were streaked across one side of a petriplate containing the medium of MAKULA and FINNERTY [4]. The plates were incubated at 30 °C in an inverted position with hydrocarbon vapours. When good growth of the streak appeared, the lysine auxotroph of *Candida guilliermondii* was streaked at right angle to the growth of the hydrocarbon utilizer. As the agar medium of the plate did not contain glucose, the suspension of the auxotroph to be streaked was prepared in 1% (w/v) sterile glucose solution. After further incubation for 48 h, the plates were examined for growth of the auxotroph. In order to confirm the nature of the amino acid produced, two dimensional paper chromatography was employed further, with the culture filtrates by growing the pure cultures of the producing strains in the same medium with 2% (v/v) straight run gas oil as the sole carbon source. Identity of the amino acids was confirmed by cochromatography with authentic samples.

During this survey, 263 hydrocarbon utilizers were tested for their lysine production. Among these hydrocarbon utilizers only 24 isolates (9%) produced lysine but most of them as mixtures with other amino acids. Two isolates (8.5%) were found to produce lysine alone (Tab. 1). The discrepancy could be seen (in the number) between the two methods of detection. Where detection by both bioautography and chromatographic methods are possible, it is found that the chromatographic method gave a higher number of amino producers. This indicated a greater sensitivity of the chromatographic method for the detection of producing organisms. On the basis of the production of lysine alone and also the quantity, the isolates Lb and 2 Fa were identified [5] as *Arthrobacter globiformis* and *Micrococcus varians*, respectively.

Present address:

Department of Botany, School of Life Sciences, Visva-Bharati
Santiniketan — 731235, West Bengal, INDIA

Table 1. Screening for lysine producing organisms from hydrocarbon utilizers

Index No.	Amino acid(s) in culture filtrate	Rf value (× 100)	
		Solvent 1	Solvent 2
Cd	Lysine	81	18
	Aspartic acid	20	34
Fa	Lysine	81	18
	Methionine	82	58
	Unidentified spot	14	22
Hb	Lysine	81	18
	α-Alanine	61	40
Jd	Lysine	81	18
	Glutamic acid	32	38
Kc	Lysine	81	18
	Aspartic acid	20	34
	Glutamic acid	32	38
Lb	Lysine	81	18
NC	Lysine	81	18
	Arginine	90	20
	Unidentified spot	30	24
Nd	Lysine	81	18
	Aspartic acid	20	34
	Unidentified spot	15	26
Of	Lysine	81	18
	Glutamic acid	32	38
Pc	Lysine	81	18
	Aspartic acid	20	34
	Glutamic acid	32	38
Ta	Lysine	81	18
	Arginine	90	20
	Aspartic acid	20	34
Ue	Lysine	81	18
	Aspartic acid	20	34
	Methionine	82	58
Va	Lysine	81	18
	Unidentified spot	24	16
We	Lysine	81	18
	Methionine	82	58
	Unidentified spot	14	22
Xb	Lysine	81	18
	α-Alanine	61	40
	Unidentified spot	12	24
2Bb	Lysine	81	18
	Glutamic acid	32	38
2Ca	Lysine	81	18
	Glutamic acid	32	38
	Unidentified spot	14	16
2Df	Lysine	81	18
	Unidentified spot	22	16

Table 1 (Continuation)

Index No.	Amino acid(s) in culture filtrate	Rf value (× 100)	
		Solvent 1	Solvent 2
2Fa	Lysine	81	18
2Ib	Lysine	81	18
	Arginine	90	20
2Kd	Lysine	81	18
	Aspartic acid	20	34
	Unidentified spot	24	16
2Pc	Lysine	81	18
	Methionine	82	58
	Unidentified spot	14	22
2Xa	Lysine	81	18
	α-Alanine	61	40
	Unidentified spot	14	22

Solvent 1. Phenol : water (100 : 20),
Solvent 2. n-Butanol : acetic acid : water (4 : 1 : 1)

Though quite a number of microorganisms produce lysine from hydrocarbon, none of these belong to the genera *Arthrobacter* and *Micrococcus* [2]. In an extensive survey involving screening of microorganisms for amino acid production, KINOSHITA et al. [6] observed that many kinds of microorganisms produced amino acid in the culture broth in small quantities. It was only rarely they found that a single amino acid was accumulated. From a survey of oil bearing areas, KASTURI and TAMAHNE [7] reported that out of 168 isolates 13 were producing different amino acids using hydrocarbon. KVASNYKOV et al. [8] found about 50% of the 89 hydrocarbon utilizers produced several amino acids simultaneously.

The present survey though limited in nature indicated that lysine producing organisms are fairly well spread in nature. Results also supported the view that though small quantities of different amino acids are produced by these lysine producers, the production of lysine alone in considerable quantity from hydrocarbon is a comparatively rare event.

The author is grateful to the University of Burdwan and UGC, India, for financial assistance. The author gratefully acknowledges the constant guidance of Dr. S. P. CHATTERJEE, Reader in Botany, Burdwan University and also the help rendered by Dr. Mrs. M. CHATTERJEE.

Received July 11, 1984

References

[1] YAMADA, K., TAKAHASHI, J., KOBAYASHI, K.: Agric. Biol. Chem. **26** (1962), 636.
[2] ABBOTT, J. B., GLEDHILL, W. E.: Adv. Appl. Microbiol. **14** (1971), 249.
[3] NAKAYAMA, K.: Hakko Kogyo Kaishi **56** (1968), 425.
[4] MAKULA, R., FINNERTY, W. R.: J. Bacteriol **95** (1968), 2108.
[5] SEN, S. K., CHATTERJEE, M., CHATTERJEE, S. P.: Biol Bull. India **4** (1982) 3, 131.
[6] KINOSHITA, S.: Adv. Appl. Microbiol. **1** (1959), 201.
[7] KASTURI, K., TAMAHNE, D. V.: Curr. Sci. **38** (1969), 92.
[8] KVASNYKOV, E. I., BURAKOVA, O. A., NESTERENKO, O. A., DYKAN, S. A.: Microbiol. Zh. **31** (1969), 427.

Acta Biotechnol. 5 (1985) 4, 382

Book Review

H. U. Bergmeyer (Editor-in-Chief)
J. Bergmeyer, M. Grassl (Editors)

Methods of Enzymatic Analysis
Vol. VI. **"Metabolites 1: Carbohydrates"**

3rd Edition. Weinheim; Deerfield Beach, Florida; Basel: Verlag Chemie, 1984
701 (724) S., 295 DM (Serienpreis); 350 DM (Einzelpreis)

Der gesamte 6. Band (4 Abschnitte, 86 Kapitel) ist der enzymatischen Analyse von Kohlehydratmetaboliten vorbehalten; einem Gebiet, auf dem die enzymatische Analyse seit langem mit bestem Erfolg angewendet wird. Es verwundert deshalb nicht, daß neben den neu eingeführten Bestimmungsmethoden auch zahlreiche althergebrachte in diesem Band nach wie vor enthalten sind. Von den neuen Methoden wird vom Autor mit Recht besonders auf die zur Bestimmung von Stärke, Pektinen, Sialinsäure, Amygdalin, Glucosinolate, Glucarat, Ascorbinsäure, Fucose, Oxalsäure, Formaldehyd und Fructose-2,6-biphosphat hingewiesen. Neben der Photo- und Fluorimetrie finden sich mehrere Hinweise auf die Anwendung der Luminometrie, GC und HPLC bei den einzelnen Bestimmungsmethoden.
Im ersten Abschnitt werden insgesamt 21 Methoden zur Bestimmung von 19 Poly-, Oligo- bzw. Disacchariden vorgestellt. Dabei wurden im Vergleich zur 2. Auflage (1974) 5 Metabolite zusätzlich aufgenommen. Die Bestimmung des hier nicht abgehandelten Heparins ist bereits im 5. Band enthalten.
Der zweite Abschnitt bringt insgesamt 61 bzw. 66 Methoden (schließt man Hinweise auf automatische, halbautomatische und semiquantitative Verfahren bei der D-Galactose mit ein) zur Bestimmung von 55 bzw. 58 (altro- und ido-Octulose—1,8-biphosphat, 2-Deoxyglucose- bzw. Fluoroglucose-6-phosphat) Monosacchariden und Monosaccharidderivaten. Das sind im Vergleich zur 2. Auflage 10 bzw. 13 neu aufgenommene Metabolite. Durch Anwendung der GC kann im speziellen Fall der Glucosinolate die Zahl der bestimmbaren Metabolite noch beträchtlich erweitert werden, Die in diesem Band nicht aufgeführten Methoden zur Bestimmung Tartrat sind im folgenden 7. Band enthalten.
17 Methoden zur Bestimmung von 15 C_3-Kohlehydratverbindungen sind im dritten Abschnitt des Bandes enthalten, in dem im Vergleich zur 2. Auflage lediglich auf die Bestimmung des 3-Mercaptopyruvats verzichtet wurde.
Im letzten Abschnitt über C_1- und C_2-Kohlenhydratmetabolite werden 12 Methoden zur Bestimmung von 10 Metaboliten vorgeschlagen; dabei wurden 2 Verbindungen (Ethylenglykol und Formaldehyd) neu aufgenommen, das Forminioglutamat dagegen nicht mehr.
Das sehr positive Urteil über die bisherigen Bände bezüglich der Qualität (und Quantität) des fachlichen Inhalts, der Literaturaktualität sowie des Einbandes und Druckes kann nur wiederholt werden.

R. Berger

Acta Biotechnol. 5 (1985) 4

Book Review

Noel R. KRIEG (Editor)
John G. HOLT (Editor-in-Chief)

Bergey's Manual of Systematic Bacteriology — Volume 1 —

Baltimore, London: Williams & Wilkins, 1984. pp 964, $ 96

Ten years after the publication of the 8th edition a new "BERGEY" (new title!) has come out. The Editors decided in favour of a completely revised and enlarged work, consisting of four subvolumes comprising the Gram-negatives of general, medical or industrial importance; the Gram-positives other than actinomycetes; the archaeobacteria, cyanobacteria and remaining Gram-negatives; the actinomycetes. The scope of the Manual was expanded to give more information of importance for systematic bacteriology: ecology, enrichment and isolation, descriptions of species and their determinative characters, maintenance and preservation. Thus the Manual contains not only determinative material for identification (e. g. diagnostic keys and tables) but also detailed descriptive information (inclusive excellent electron micrographs) and taxonomic comments on the identification of bacteria and the relationships that exist between the various kinds of bacteria. The microbiologists will now have the option of purchasing only that particular subvolume containing the organisms in which the yare interested. The contributions were made by approximately 130 authors, about 90 of them had not been involved in the preparation of the 8th edition of BERGEY's Manual. The articles represent the authors' views and do not contain any "official classification" of bacteria; citation in the scientific literature should refer to the author(s), the chapter title and its inclusive pages, but not to the Editor. It is very advantageous that Vol. 1 was supplemented by chapters concerning modern methods used in bacterial classification and dealing with classification of procaryotic organisms, numerical taxonomy, nucleic acids in bacterial classification, genetic methods, serology, and chemotaxonomy. Other chapters deal with other principles of bacterial identification and taxonomy.

The systematic part of Vol. 1 is subdivided into 11 "sections", which agree for the most part with the corresponding "parts" in BERGEY's 8th edition and are supplemented by new families and genera: the spirochetes; aerobic/microaerophilic, motile, helical/vibrioid Gram-negative bacteria; nonmotile (or rarely motile) Gram-negative curved bacteria; Gram-negative aerobic rods and cocci; facultatively anaerobic Gram-negative rods; anaerobic Gram-negative straight, curved and helical rods; dissimilatory sulfate- or sulfur-reducing bacteria; anaerobic Gram-negative cocci; the rickettsias and chlamydias; the mycoplasmas; endosymbionts (of protozoa, insects, other invertebrates, and fungi). Each article dealing with a bacterial genus is presented wherever possible as follows: name of the genus; name of author(s); synonyms (if necessary after publication of the "Approved Lists"); etymology of the genus name; capsule description; further descriptive information; enrichment and isolation; maintenance procedures; procedures for testing special characters; differentiation of the genus from other genera; taxonomic comments; further reading; differentiation of the species of the genus; list of the species of the genus; species incertae sedis; literature cited. As an interimsolution a few genera difficult to classify are described not only in Vol. 1 but also in an appropriate subsequent volume.

With regard to recent developments in research and handling the field of taxonomy there are differences between the individual authored chapters, of course. For instance the latest developments in the new family of *Acetobacteraceae* (1983) are reported by DE LEY et al., whereas I feel a "hole" in the chapter on the *Methylococcaceae* (WHITTENBURY and KRIEG): genera created in 1970 by WHITTENBURY et al. are not classified taxonomically, and the validly published *Methylobacillus glycogenes* (1977) and *Methylobacterium organophilum* (PATT et al. 1976) are only mentioned as "incertae sedis" without giving details on the species or their type strains. In spite of everything Vol. 1 of the Manual contains so much summarized information on each bacterial group and in an excellent presentation that it should belong to every microbiologist working in the field of Gram-negatives.

K. Karbaum

Acta Biotechnologica

Volume 5 | 1985 | Number 4

Contents

Short Communications

Acta Biotechnologica is indexed or abstractet in Current Contents/ET & AT; Chemical Abstracts; Biotechnology Abstracts; Excerpta Medica

Bei Fragen zur Produktsicherheit wenden Sie sich bitte an:
If you have any questions regarding product safety,
please contact:

Walter de Gruyter GmbH
Genthiner Straße 13
10785 Berlin
productsafety@degruyterbrill.com

Sozialpädiatrie in der Praxis

2. Auflage

Harald Bode, Hans-Michael Straßburg, Helmut Hollmann

AF812969

ELSEVIER
URBAN & FISCHER

URBAN & FISCHER München

Zuschriften an:
Elsevier GmbH, Urban & Fischer Verlag, Hackerbrücke 6, 80335 München

Wichtiger Hinweis für den Benutzer
Die Erkenntnisse in der Medizin unterliegen laufendem Wandel durch Forschung und klinische Erfahrungen. Herausgeber und Autoren dieses Werkes haben große Sorgfalt darauf verwendet, dass die in diesem Werk gemachten therapeutischen Angaben (insbesondere hinsichtlich Indikation, Dosierung und unerwünschter Wirkungen) dem derzeitigen Wissensstand entsprechen. Das entbindet den Nutzer dieses Werkes aber nicht von der Verpflichtung, anhand weiterer schriftlicher Informationsquellen zu überprüfen, ob die dort gemachten Angaben von denen in diesem Werk abweichen und seine Verordnung in eigener Verantwortung zu treffen.
Für die Vollständigkeit und Auswahl der aufgeführten Medikamente übernimmt der Verlag keine Gewähr.
Geschützte Warennamen (Warenzeichen) werden in der Regel besonders kenntlich gemacht (®). Aus dem Fehlen eines solchen Hinweises kann jedoch nicht automatisch geschlossen werden, dass es sich um einen freien Warennamen handelt.

Bibliografische Information der Deutschen Nationalbibliothek
Die Deutsche Nationalbibliothek verzeichnet diese Publikation in der Deutschen Nationalbibliografie; detaillierte bibliografische Daten sind im Internet über http://www.d-nb.de/ abrufbar.

Alle Rechte vorbehalten
2. Auflage 2014
© Elsevier GmbH, München
Der Urban & Fischer Verlag ist ein Imprint der Elsevier GmbH.

13 14 15 16 4 3 2 1

Dieses Buch enthält auch Links auf externe Webseiten Dritter. Auf die Inhalte dieser Webseiten haben wir keinen Einfluss, da es sich nicht um unsere eigenen Inhalte handelt. Für die Richtigkeit der über die Links erreichbaren Inhalte ist der jeweilige Anbieter verantwortlich. Wir übernehmen daher keine Garantie für deren Richtigkeit, Vollständigkeit und Aktualität. Ein Überprüfung der Inhalte der von uns verlinkten externen Seiten ohne tatsächliche und konkrete Anhaltspunkte für einen Rechtsverstoß leisten wir nicht. Falls uns aber entsprechende Hinweise bekannt werden, werden wir unverzüglich eine Überprüfung, soweit möglich, einleiten und die dabei erzielten Ergebnisse bei Neuauflagen berücksichtigen.

Das Werk einschließlich aller seiner Teile ist urheberrechtlich geschützt. Jede Verwertung außerhalb der engen Grenzen des Urheberrechtsgesetzes ist ohne Zustimmung des Verlages unzulässig und strafbar. Das gilt insbesondere für Vervielfältigungen, Übersetzungen, Mikroverfilmungen und die Einspeicherung und Verarbeitung in elektronischen Systemen.

Um den Textfluss nicht zu stören, wurde bei Patienten und Berufsbezeichnungen die grammatikalisch maskuline Form gewählt. Selbstverständlich sind in diesen Fällen immer Frauen und Männer gemeint.

Planung und Lektorat: Inga Schickerling, Dr. med. Constance Spring
Herstellung: Elisabeth Märtz
Satz: abavo GmbH, Buchloe/Deutschland; TnQ, Chennai/Indien
Druck und Bindung: Printforce, Alphen/NL
Umschlaggestaltung: Spiesz Design, Neu-Ulm

ISBN Print 978-3-437-31630-2
ISBN e-Book 978-3-437-31631-9

Aktuelle Informationen finden Sie im Internet unter **www.elsevier.de** und **www.elsevier.com.**

Vorwort

Die „Sozialpädiatrie in der Praxis" hat seit ihrem Erscheinen in der 1. Auflage 2009 eine erfreuliche Akzeptanz und Verbreitung gefunden. Als Herausgeber und Autoren, denen die Sozialpädiatrie seit vielen Jahren Anliegen und Auftrag zugleich ist, haben wir daher gern diese 2. Auflage erstellt. Sie wurde nötig, um die aktuellen Entwicklungen für das Fach Sozialpädiatrie angemessen darzustellen.

Die Sozialpädiatrie hat in den letzten Jahren ihren Stellenwert in der Kinder- und Jugendmedizin in Deutschland nicht nur behauptet, sondern weiter verbreitet und vertieft. Davon zeugen inzwischen über 140 Sozialpädiatrische Zentren und mehrere Lehrstühle und Professuren für Sozialpädiatrie. Sozialpädiatrische Themen nehmen auf zahlreichen Kongressen und Fortbildungsveranstaltungen eine bedeutende Stelle ein. Die Kinder- und jugendärztliche Tätigkeit in der Praxis ist ohne sozialpädiatrische Grundkenntnisse und Fertigkeiten heute nicht mehr vorstellbar. Dies spiegelt sich neuerdings auch in ausgewiesenen sozialpädiatrischen Abrechnungsziffern für die vertragsärztliche Tätigkeit wider.

Die Bedeutung des familiären und sozialen Kontextes und der Lebenswelten für die bestmögliche Entwicklung von allen Kindern und Jugendlichen ist inzwischen unbestritten. Sie spiegelt sich zunehmend in gesellschaftlichen Prioritäten und politischen Entscheidungen. Fachlich kompetente sozialpädiatrische Beratung ist dabei äußerst wertvoll und wird nachgefragt.

Inhaltlich ist die Sozialpädiatrie mit beharrlicher Arbeit vieler engagierter Kolleginnen und Kollegen verschiedener Professionen in ihrem Selbstverständnis und ihrem Qualitätsanspruch weiter vorangeschritten. Dies wird in der aktuell für die ärztliche Weiterbildungsordnung vorgeschlagenen Definition deutlich:

„Die Zusatzweiterbildung Spezielle Sozialpädiatrie umfasst in Ergänzung der Facharztkompetenz in Kinder- und Jugendmedizin und zusätzlich zur sozialpädiatrischen Basisversorgung die Diagnostik und Therapie von komplexen somatischen, seelischen und verhaltensbezogenen sowie sozialen Folgen und Begleiterscheinungen von Krankheiten bei Kindern und Jugendlichen in einem multiprofessionellen Team unter Zuhilfenahme der „Mehrdimensionalen Bereichsdiagnostik der Sozialpädiatrie" (MBS), die zu komplexen Entwicklungs- und Verhaltensstörungen oder Behinderungen führen. Dabei werden Ursachen und Auswirkungen im familiären, vorschulischen, schulischen und beruflichen Umfeld auf der Grundlage der Klassifikation von Funktionsfähigkeit, Behinderung und Gesundheit berücksichtigt und entsprechende Behandlungspläne sowie Förderprogramme entwickelt und koordiniert.

Diagnostik und Therapie erfordern die interdisziplinäre Zusammenarbeit in einem multiprofessionellen Team."

Das vorliegende Buch richtet sich aber keineswegs nur an Spezialisten. Die „Sozialpädiatrie in der Praxis" soll ein praxisnahes und alltagstaugliches Werk sein – sowohl für die in der Praxis tätigen Kinder- und Jugendärzte als auch für nichtärztliche Fachpersonen im medizinischen, psychologischen und pädagogischen Bereich sowie in den Verwaltungsämtern. Das Buch soll die im Berufsalltag notwendigen sozialpädiatrischen Grundkenntnisse vermitteln, die in den gängigen Lehrbüchern nicht oder nur marginal dargestellt sind.

Konsequenterweise wurde daher in dieser 2. Auflage der klinisch-praktische Teil stärker hervorgehoben. So wird die Sozialpädiatrie in der kinder- und jugendärztlichen Grundversorgung von in diesem Bereich tätigen Autoren dargestellt, insbesondere die Vorgehensweise im Rahmen der

Früherkennungsuntersuchungen. Sozialpädiatrisch relevante Krankheitsbilder werden ausführlicher und differenzierter besprochen, z.B. die Entwicklungsbegleitung von Kindern nach sehr und extrem früher Geburt. Neu ist auch das Kapitel über sozialpädiatrische Aspekte häufiger chronischer Erkrankungen und die damit verbundene Herausforderung einer gelingenden Transition. Der Praxisbezug wird durch Fallbeispiele und Merksätze vertieft.

Im Bewusstsein der sich wandelnden Lebenswelten von Kindern und Jugendlichen werden die Themen Kinderschutz, Krippenbetreuung, Inklusion und Medien ausführlich dargestellt. Die Therapie- und Fördermöglichkeiten wurden stärker untergliedert, die Abschnitte zum öffentlichen Gesundheitsdienst zusammengeführt.

Alle Kapitel des vorliegenden Werkes wurden von den jeweiligen Autoren gründlich überarbeitet, ergänzt und aktualisiert.

Herausgeber und Autoren hoffen wiederum auf eine gute Akzeptanz und Verbreitung dieser 2. Auflage der „Sozialpädiatrie in der Praxis", die gedruckt und – dem Zeitgeist entsprechend – auch in elektronischer Form zu beziehen ist. Sie freuen sich über konstruktive Kritik und Ergänzungsvorschläge.

Die Herausgeber danken den Mitautoren für die zeitgerechte und kompetente Erstellung der Manuskripte. Den Mitarbeiterinnen und Mitarbeitern des Verlages Elsevier Urban und Fischer, insbesondere Frau Dr. Constance Spring, danken wir für Ermutigung und Begleitung.

Ulm, Würzburg, Bonn, im Oktober 2013

Harald Bode
Hans-Michael Straßburg
Helmut Hollmann

Herausgeber

Prof. Dr. med. Harald Bode
Universitätsklinik für Kinder-und Jugendmedizin
Sozialpädiatrisches Zentrum und Kinderneurologie
Frauensteige 10
89075 Ulm

Prof. Dr. med. Hans-Michael Straßburg
Emil-von-Behring-Weg 8
97218 Gerbrunn

Dr. Helmut Hollmann
Kinderneurologisches Zentrum
LVR-Klinik Bonn
Waldenburger Ring 46
53119 Bonn

Autoren:

Dr. Peter Borusiak
Zentrum für Kinder- und Jugendmedizin
Heusnerstr. 40
42283 Wuppertal

Dr. med. Folkert Fehr
Gemeinschaftspraxis für Kinder- und Jugendmedizin
Karlsplatz 5
74889 Sinsheim

Dr. med. Christian Fricke
Werner Otto Institut GmbH
Bodelschwinghstr. 23
22337 Hamburg

Dr. Ulrike Horacek
Städtisches Gesundheitsamt
Kurt-Schumacher-Allee 1
45657 Recklinghausen

Dr. Elke Jäger-Roman
Köhlerstr. 23
12205 Berlin

Dr. Renate Klein
Ministerium für Justiz, Gesundheit und Soziales
Franz-Josef-Röder-Str. 23
66119 Saarbrücken

Dr. med. Norbert Knieß
Gemeinschaftspraxis für Kinder- und Jugendmedizin
Schillerstr. 2
85055 Ingolstadt

Prof. Dr. med. Olaf Kraus de Camargo
McMaster University
Evel Building Rm311
1200 Main Street West
Hamilton ON L8N 3Z5
Canada

Dr. med. Johannes Oepen
Viktoriastift Bad Kreuznach
Cecilienhöhe 1-3
55543 Bad Kreuznach

Dipl.-Soz. Robert Schlack
Robert Koch-Institut
Abteilung Epidemiologie und Gesundheitsberichterstattung
General-Pape-Str. 62-64
12101 Berlin

Dr. med. Dirk Schnabel
Humboldt-Universität Berlin
Charité / Campus Virchow-Klinikum
Pädiatrische Endokrinologie
Otto-Heubner-Centrum
Augustenburger Platz 1
13353 Berlin

Dr. Heidrun Thaiss
Kinder- und Jugendgesundheit
Med. Kinderschutz und Frühe Hilfen
Kinder- und Jugendärztlicher Dienst des ÖGD
MSGFG des Landes Schleswig-Holstein
Adolf-Westphal-Str. 2
24143 Kiel

Prof. Dr. med. Bernd Wilken
Klinikum Kassel gGmbH
Abteilung Neuropädiatrie
Sozialpädiatrisches Zentrum
Mönchebergstr. 41-43
34125 Kassel

Inhaltsverzeichnis

I Einleitung

1 Die pädiatrische Grundversorgung von Kindern und Jugendlichen

Elke Jäger-Roman

1.1 Einleitung

Die ambulante medizinische Grundversorgung von Kindern und Jugendlichen ist in Deutschland traditionell Aufgabe von Ärzten/Ärztinnen der Kinder- und Jugendmedizin und der Allgemeinmedizin. Sie ist im SGB V §73 gesetzlich geregelt. Dort wird festgelegt, dass Allgemeinärzte, Kinderärzte, Internisten ohne Schwerpunktbezeichnung und andere Ärzte an der „**hausärztlichen Versorgung**" teilnehmen, und es wird orientierend beschrieben, was eine hausärztliche Versorgung beinhaltet.
Die Einführung einer primären Gesundheitsversorgung zum Schutz und zur Förderung der Gesundheit aller Menschen wurde 1978 von einer Internationalen Konferenz unter Federführung der WHO in der „**Declaration of Alma Ata on Primary Health Care**" gefordert. Die Konferenz hielt darin unter anderem fest, dass Gesundheit ein grundlegendes Menschenrecht ist, und dass das Erreichen eines möglichst guten Gesundheitszustands einer Bevölkerung ein wichtiges soziales Ziel ist, zu dessen Verwirklichung nicht nur die Gesundheitspolitik sondern viele ökonomische und soziale Bereiche beitragen müssen. Diese Feststellung war insbesondere wichtig im Hinblick auf die schwerwiegenden gesundheitlichen Ungleichheiten der Menschen in den Entwicklungsländern, in denen viele keinen Zugang zu den grundlegenden Mitteln und Einrichtungen zum Erhalt ihrer Gesundheit haben. In derselben Zeit begann auch der enorme Wissenszuwachs in allen Fächern der Medizin und damit einhergehend eine überproportionale Zunahme medizinischer Spezialisten, zu denen die arme Bevölkerung erst recht keinen Zugang hat.

Sozialpädiatrie in der Praxis. http://dx.doi.org/10.1016/B978-3-437-31630-3.00001-8
Copyright © 2014 Elsevier GmbH. All rights reserved.

Unter den vielen auch heute noch bedeutenden Aussagen dieser Konferenz zur primären Gesundheitsversorgung wurden folgende **Mindestanforderungen** genannt, die unter den gegebenen ökonomischen Rahmenbedingungen eines Landes und seiner soziokulturellen und politischen Merkmale eingehalten werden sollten:
Impfung gegen die schwersten Infektionskrankheiten;
Prävention und Bekämpfung endemischer Krankheiten; angemessene Behandlung der häufigsten Krankheiten und Verletzungen;
Versorgung mit unentbehrlichen Arzneimitteln".
Die WHO beschrieb **7 Wesensmerkmale der medizinischen Grundversorgung:** sie sollte „**allgemein, kontinuierlich, umfassend, koordinierend, kooperativ, Familien- und Gemeinde-orientiert**" sein.
Der Begriff der **hausärztlichen Versorgung (Primary health care)** wurde in den letzten Jahrzehnten kontinuierlich durch WHO und insbesondere WONCA (World Organisation of National Colleges, Academies and Academic Associations of General Practitioners/Family Physicians) weiter entwickelt.

1.2 Pädiater in der Grundversorgung

Pädiater gehören im Spektrum der Medizin zu den Spezialisten, da sich ihre Weiterbildung nur auf eine bestimmte Altersgruppe (0 bis 18 Jahre) bezieht und nicht die gesamte Krankheitslehre umfasst. Die Zuordnung der Pädiater zu den medizinischen Spezialisten hat dazu geführt, dass ihre Teilhabe an der Grundversorgung von Kindern und Jugendlichen in vielen entwickelten Ländern kontrovers diskutiert wird. Allgemeinmediziner machen geltend, dass nur sie in der Arbeitsweise der ambulanten Hausarzt- und Familienmedizin geschult werden und deshalb die ersten Ansprechpartner für Patienten aller Altersgruppen sein sollten („**reines primärärztliches System**"), Pädiater führen dagegen ihre Expertise für alle medizinischen und psychosozialen Belange von Kindern und Jugendlichen an, die sie in einer mindestens 5-jährigen rein pädiatrischen Weiterbildung erwerben. Beide Argumente beziehen sich auf unterschiedliche Kompetenzen, nämlich einmal auf die Wesensmerkmale bzw. auf die Arbeitsweise in der Grundversorgung, zum anderen auf das ärztliche Expertenwissens. Idealerweise sollte ein guter Arzt/Ärztin der Grundversorgung beide Kompetenzen haben.
In den Disput um die Grundversorgung der Bevölkerung einschließlich der Kinder schalten sich auch Gesundheitspolitiker und Ökonomen ein, die angesichts der exorbitant steigenden Kosten im Gesundheits-Sektor auf eine reine primärärztliche Versorgung setzen unter der Annahme, dass „**Spezialistenmedizin**" im ambulanten Sektor wesentlich für die steigenden Kosten verantwortlich ist. Bezogen auf die Grundversorgung von Kindern und Jugendlichen gibt es jedoch inzwischen eine zunehmende und gute internationale Datenlage, die belegt, dass die Grundversorgung von Kindern und Jugendlichen in entwickelten Ländern durch Pädiater qualitativ besser und letztendlich kostengünstiger ist, weil sich Pädiater unter anderem besser an klinische Leitlinien halten, weniger Medikamente wie Antibiotika verschreiben, wenn diese nicht indiziert sind, höhere Durchimpfungsraten erreichen und vor allem seltener in Krankenhäuser einweisen.

1.3 Die ambulante Grundversorgung von Kindern und Jugendlichen

1.3.1 Definition der Grundversorgung

Angelehnt an die internationale Definition der medizinischen „Primary Health Care" (Grundversorgung) hat die **Deutsche Gesellschaft für Ambulante Allgemeine Pädiatrie (DGAAP)** zusammengefasst, welche Versorgungsaufgaben ein Pädiater in der Grundversorgung hat (➤ Tab. 1.1).

Grundsätzlich sollte die Gesundheits-Versorgung von Säuglingen, Kindern und Jugendlichen auf den UN-Kinderrechten beruhen, und Gesundheitsförderung, Prävention, Kinderschutz und die bestmögliche medizinische Behandlung beinhalten. „**Bestmögliche medizinische Behandlung**" bedeutet sowohl das rechtzeitige Erkennen des abwendbaren schweren Verlaufs einer Erkrankung als auch die seelische Gesundheit von Kindern und Jugendlichen in alle diagnostischen und therapeutischen Überlegungen mit einzubeziehen. Damit eine solche Grundversorgung gelingt, bedarf es neben medizinisch-pädiatrischer Expertise einer ganzen Reihe zusätzlicher Schlüsselkompetenzen wie guter Kommunikationsfähigkeit, Förderung aller Aspekte von Gesundheit, Qualitäts- und Fehlermanagement, inter-professionelle Kooperation, Lehrtätigkeit (für Studenten und Weiterbildungsassistenten), professionelle Vorbildfunktion, Fürsprache für die Belange der Kinder im Sinne einer politischen Kindermedizin und einer guten kommunalen Netzwerkarbeit mit den Mitarbeitern des medizinischen-, sozialen- und des Bildungs-Systems (➤ Tab. 1.2).

Tab. 1.1 Versorgungsaufgaben der ambulant tätigen Allgemeinpädiater (AAP) (erarbeitet von einer Arbeitsgruppe der Deutschen Gesellschaft für Ambulante Allgemeine Pädiatrie – DGAAP)

- AAP sind die ersten Ansprechpartner im Gesundheitssystem für die von ihnen betreuten Kinder, Jugendlichen und ihre Familien bei allen medizinischen und psychosozialen Problemen und Fragen zu Wachstum und Entwicklung
- Sie versorgen Kinder und Jugendliche kontinuierlich und umfassend von der Geburt bis zum 18. Geburtstag
- Sie arbeiten primär kindzentriert (# nicht organspezifisch).
  Dazu gehört eine effektive und altersgerechte Kommunikation mit Kindern und Jugendlichen unter Einbeziehung ihrer Familien und anderer Bezugspersonen in ihrem jeweiligen kulturellen, sozialen und religiösem Kontext
- Sie behandeln akute sowie angeborene und erworbene chronische Erkrankungen
- Sie führen alle Präventionen durch (Früherkennungsuntersuchungen, Impfungen, Gesundheitserziehung und vorausschauende Beratung)
- Sie sind Lotsen im Gesundheits- und Sozialsystem
- Sie arbeiten auf kommunaler Ebene in Netzwerken mit Sozial- und Bildungs-Behörden/Institutionen zusammen (wie: KJGD, Jugendamt, Kindergärten und Schulen). Im medizinischen Kontext kooperieren sie mit Kinder-Krankenhäusern, pädiatrischen Subspezialisten und mit Einrichtungen, die chronisch kranke und behinderte Kinder und Jugendliche multiprofessionell versorgen (SPZ), außerdem mit ärztlichen Kolleginnen und Kollegen anderer Fachdisziplinen, die auf ihrem Gebiet eine besondere pädiatrische Kompetenz erworben haben.

Tab. 1.2 Kompetenzen, die ambulant tätige Allgemeinpädiater benötigen (erarbeitet von einer Arbeitsgruppe der DGAAP; siehe Fehr 2012)

Als medizinischer Experten
- Hält optimale und patientenzentrierte Versorgung vor
- Erwirbt und erhält anwendungsorientiertes Wissen, Fertigkeiten und Haltungen
- Erhebt Anamnese und Befund und führt die klinische Beurteilung durch
- Nutzt effektiv präventive und therapeutische Interventionen
- Wendet Fähigkeiten und Fertigkeiten kompetent an

Als Kommunikatoren
- Fördert Beziehung, Vertrauen, Verständnis mit Kindern und Jugendlichen durch alters entsprechende Kommunikation
- Arbeitet ganzheitlich und kindzentriert im familiären, sozialen und kulturellen Kontext
- Erkennt Krankheitszusammenhänge im individuellen und sozialen Kontext
- Fördert ein gemeinsames Verständnis aller Beteiligten, um einen gemeinsamen Behandlungsplan zu entwickeln („shared decision making")
- Respektiert die Schweigepflicht

Als Verantwortungsträger und Manager
- Nimmt an Aktivitäten zur Förderung der Effizienz, Sicherheit und Qualität in seiner Praxis teil
- Setzt die begrenzten Ressourcen des Gesundheitssystems effektiv ein
- Übernimmt Aufgaben und Funktionen in der ärztlichen Selbstverwaltung und im öffentlichen Gesundheitsbereich

Als Gesundheitsberater und -fürsprecher
- Reagiert auf die individuellen Bedürfnisse der Kinder und Jugendlichen
- Reagiert auf die gesundheitlichen Bedürfnisse und Themen im Sozialraum
- Fördert die gesunde Entwicklung der Kinder und Jugendlichen
- Identifiziert die Determinanten für gesunde Entwicklung und wirkt entsprechend gesundheits- und gesellschaftspolitisch ein

Als Interprofessionelle Partner
- Teilt Verantwortung mit anderen Mitgliedern des Gesundheits- und Sozialwesens und erkennt deren Kompetenzen an
- Arbeitet mit pädiatrischen Subspezialisten, Spezialeinrichtungen (wie SPZ's) und Selbsthilfeorganisationen zusammen
- Pflegt gute Zusammenarbeit und loyale Haltung zu Mitarbeitern
- Erkennt die eigenen Grenzen und lernt, Rat und Hilfe einzuholen

Als Wissenschaftler und Lehrer
- Erhält und verbessert seine Kenntnisse, Fertigkeiten und Haltung durch fortgesetztes Lernen
- Erlernt, medizinische Sachverhalte klar und verständlich darzustellen
- Unterstützt gesundheitsbezogenes Lernen von Kindern, Jugendlichen, ihren Bezugspersonen und der Öffentlichkeit
- Beteiligt sich an Aus-, Weiter- und Fortbildung
- Evaluiert kritisch Informationen und Quellen; trägt zur Erzeugung, Verteilung, Anwendung und Übersetzung medizinischen Wissens aus seinem Fachbereich bei

Als professionelles Vorbild
- Zeigt seinen Einsatz für Patienten, Beruf und Gesellschaft durch vorbildliche Haltung
- Zeigt seinen Einsatz für Patienten, Beruf und Gesellschaft durch Mitwirkung in Selbstverwaltung und politischer Pädiatrie
- Achtet auf die eigene Gesundheit und gesunde Arbeitsbedingungen

1.3.2 Versorgungs- und Patientenstruktur

In Deutschland werden derzeit über 90 % aller Kinder im Vorschulalter zumindest anlässlich der Vorsorge-Untersuchungen von einem Kinder- und Jugendarzt gesehen. Je älter Kinder und insbesondere junge Menschen werden, desto häufiger wechseln die Familien mit Kindern im Schulalter und mit Jugendlichen insbesondere im ländlichen Raum vom Pädiater als Arzt der Grundversorgung zum Allgemeinmediziner/Familienarzt (➤ Abb. 1.1).
Dies wurde sowohl in der Studie zur Gesundheit von Kindern und Jugendlichen in Deutschland (KGGS, 2007) als auch in der **Versorgungsstudie der Deutschen Akademie für Kinder- und Jugendmedizin (DAKJ-Studie, 2011)** gesehen. Eine Ausnahme davon machen chronisch kranke Jugendliche, die mehrheitlich bei den ambulant tätigen Pädiatern bleiben (➤ Abb. 1.2).
In der DAKJ-Studie hatten 25 % der Jugendlichen in der Altergruppe 10 bis 18 Jahren eine chronische Grundkrankheit (vs. 16 % der Jugendlichen in der Allgemeinbevölkerung). Da die chronisch kranken Jugendlichen häufiger mit Erkrankungen, die in der Grundversorgung behandelt werden, vorgestellt

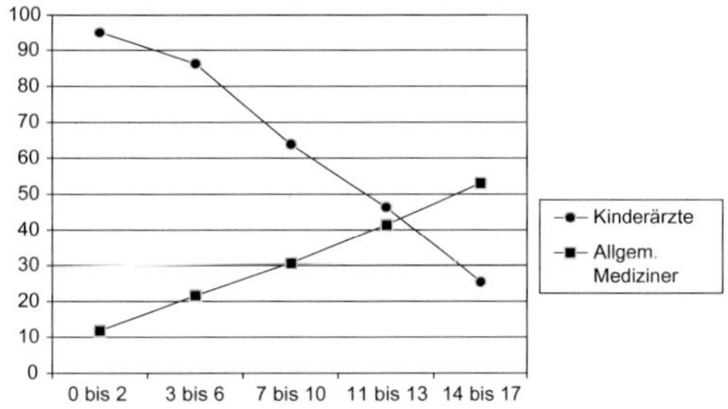

Abb. 1.1 Inanspruchnahme niedergelassener Kinderärzte und Allgemeinmediziner/praktischer Ärzte in den letzten 12 Monaten vor Befragung (Abbildung erstellt nach Daten des „Kinder- und Jugendgesundheitssurveys: KiGGS"; Kamtsiuris P, Bergmann E, Rattay P, Schlaud M [2007] Inanspruchnahme medizinischer Leistungen. In: Gesundheitsbl-Gesundheitsforsch-Gesundheitsschutz 50: 836–850)

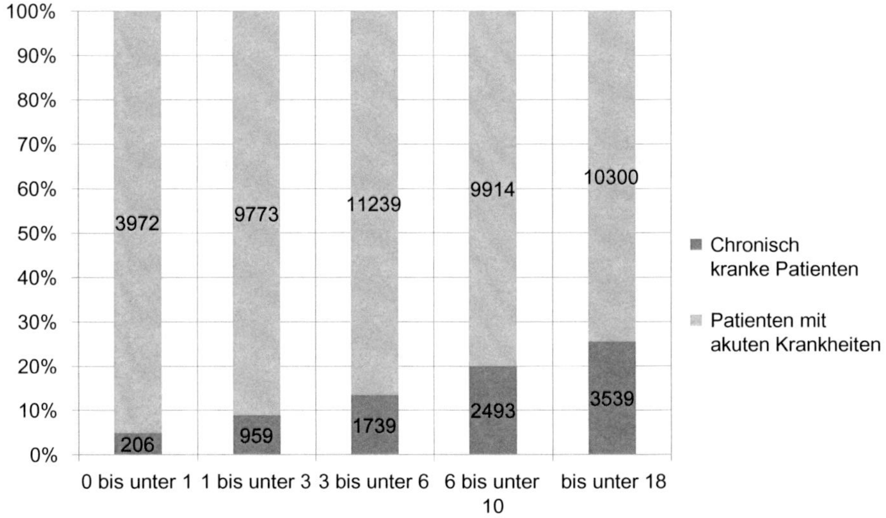

Abb. 1.2 Anteil chronisch kranker Patienten in den verschiedenen Altersgruppen (DAKJ-Versorgungsstudie)

wurden als anderweitig gesunde Kinder und Jugendliche (5,3 vs. 3,2 mal pro Jahr), betrafen 40 % aller Vorstellungen im Jugendalter junge Leute, die eine chronische Grunderkrankung hatten.

1.3.3 Vorstellungsanlässe

Die meisten Erst-Vorstellungen von Kindern und Jugendlichen in pädiatrischen Praxen erfolgen laut der DAKJ-Studie wegen einer akuten Erkrankung/Störung oder einem anderweitig sehr vielfältigen Beratungsbedarf der Familien (➤ Tab. 1.3). Am häufigsten wollen Eltern das Symptom „Husten" abgeklärt wissen, danach folgen Krankheiten und Störungen, die mit Schmerzen oder Fieber verbunden sind. Unerwartet häufig geben Haut-Erscheinungen jedweder Art Anlass zur Konsultation.

Tab. 1.3 Vorstellungsanlässe (n = 96.036) in 30 deutschen Kinder- und Jugendarzt-Praxen im Jahr 2010. Aus der Studie der Deutschen Akademie für Kinder- und Jugendmedizin (DAKJ) und der Dresden International University; Ulrich Fegeler, Elke Jäger-Roman, Ronny Martin, Hans-Jürgen Nentwich

Erstvorstellungs-/Beratungsanlässe n=51.188 Husten, Fieber, Hauterscheinungen, Durchfall u/o Erbrechen, Unfälle, akute Schmerzen, Störungen/Erkrankungen in den Organsystemen außer obere und untere Atemwege, Beratungsbedarf im Bereich „Neue Morbidität"	53,3 %
Prävention (Früherkennungsuntersuchungen und Impfungen)	30,3 %
Kontrollen (nach Erstvorstellung aus akutem Anlass)	10,2 %
Andere (Zuweisungen z. B. zum Ultraschall; Therapien wie Verbandswechsel, Inhalationen; Wiederholungsrezepte etc.)	6,2 %

Aus den entsprechenden zugehörigen Diagnosen geht hervor, dass der größte Anteil (67 %) aller akuten Erkrankungen immer noch infektiöser (viraler und bakterieller) Genese ist, obwohl die klassischen schwerwiegenden Infektionskrankheiten durch Impfungen (Beispiel: Polio, Masern) oder rechtzeitige antibiotische Therapie (Beispiel: Scharlach) beherrscht sind. Nur wenige der infektiösen Erkrankungen, die heutzutage in der ambulanten Praxis gesehen werden, sind ernsthafter Natur, dies bedeutet aber auch, dass es immer einer erhöhten Aufmerksamkeit bedarf, um die seltenen abwendbar gefährlichen Verläufe rechtzeitig zu erkennen (➤ Tab. 1.4).

Tab. 1.4 Die 10 häufigsten akuten Erst-Vorstellungs-(Beratungs)anlässe (n = 51.188; DAKJ Versorgungsstudie)

Husten mit/ohne Atemnot	30,2 %
Akute Schmerzen	15,4 %
davon:	
Hals-, Mundschmerzen	5,7 %
Ohrenschmerzen	4,1 %
Kopf-, Bauch-, Muskel-/Skelettschmerzen	5,6 %
Fieber	13,3 %
Haut-, Haar-, Nagel-Probleme	13,1 %
Durchfall u/o Erbrechen	8,2 %
Störungen aus dem Bereich „Neue Morbidität"	7,2 %
Rote u/o verklebte Augen	4,3 %
Unfälle	2,3 %
Probleme/Schmerzen beim Wasserlassen	1,6 %
Erkrankungen u/o Störungen in den einzelnen Organsystemen außer den oben bereits aufgeführten Vorstellungsanlässen	1,2 %

Derzeit erfolgen 30 % aller Vorstellungen von Kindern und Jugendlichen zum Zweck der Durchführung einer **Präventions-Maßnahme** (Früherkennungsuntersuchungen und Impfungen). Es ist absehbar, dass sich dieser Anteil der Arbeit an der Grundversorgung erhöhen wird, wenn sich in Zukunft die Anzahl der Früherkennungsuntersuchungen erhöht und auch Information und Beratung zur primären Prävention durch das neue Präventionsgesetz in den Aufgabenbereich der ärztlichen Grundversorger fallen. Der Tätigkeitsbereich „Prävention“ nimmt in der Arbeit des niedergelassenen Kinder- und Jugendarztes schon jetzt mehr Arbeitszeit in Anspruch als der Bereich „akute Erstvorstellungen“.

1.3.4 Pädiatrische Grundversorgung mit Beratungsbedarf zu Problemen aus dem Bereich „Neue Morbidität"

Eine besondere Anforderung an die pädiatrische Versorgungsqualität stellt der Beratungsbedarf der Familien zu Problemen aus dem Bereich „**Neue Morbidität**“ (➤ Kap. 18; ➤ Kap. 20; ➤ Kap. 22; ➤ Kap. 25; ➤ Kap. 35) dar (➤ Tab. 1.5).

Tab. 1.5 Beratungsanlässe mit Problemen aus dem Bereich „Neue Morbidität" (DAKJ-Versorgungsstudie; Bereich akute Erstvorstellungen: n = 51.188)

	Anzahl	**% Gesamt**
Entwicklungsprobleme (motorisch, sprachlich, kognitiv)	1130	7,2
Verhaltensprobleme; Konzentrationsprobleme	622	
Rezidivierende oder chronische Schmerzen	534	
Schreien, Schlafproblem, Essstörungen im Säuglingsalter im Sinne der frühkindlichen Regulationsstörung	503	
Schwierigkeiten mit Sauberkeitstraining	252	
Schulleistungsprobleme	228	
Störung des Appetits	155	
Müdigkeit, Abgeschlagenheit	75	
Einnässen; Einkoten	33	
Ängste, Depressionen, Tics	25	

Diese Vorstellungsanlässe machten in der DAKJ-Studie 7,2 % der akuten Erstvorstellungen aus (Tab. 4). Der gesamte Beratungsbedarf ist jedoch deutlich höher, da viele Fragen und Beratungen anlässlich von Vorsorgeuntersuchungen (➤ Kap. 14) behandelt werden. Häufig werden entsprechende Fragen und Themen auch während anderer akuter Vorstellungs-Anlässe „nebenbei“ und scheinbar nachrangig vorgebracht („hidden agenda“/„verdeckte Frage“/„verborgenes Motiv“). Dieser quasi Zweit-Beratungsbedarf konnte aus Gründen der Dokumentation und Analyse der Daten in der DAKJ-Studie nicht dargestellt werden.
Die Art und Häufigkeit des Beratungsbedarfs ist abhängig vom Alter der Kinder und Jugendlichen (Tab. 1.6): im Säuglingsalter überwiegen elterliche Fragen zu Schrei-, Fütter- und Schlafproble-

Tab. 1.6 Beratungshäufigkeit zu den „neuen Morbiditäten" bei Kinder und Jugendlichen

Altersgruppe (Jahre)	0 bis < 1	1 bis < 3	3 bis < 6	6 bis < 10	10 bis < 18
Prozentualer Anteil	3.9	2.1	6	11.2	9.7

men und zur kindlichen Entwicklung. Die kindliche Entwicklung bleibt ein Hauptanliegen der Eltern bis zum Schulalter.
In dieser Zeit spielt in Deutschland die **Beurteilung der kindlichen Sprachentwicklung** eine große Rolle („Sprache als Schlüssel-Kompetenz für den schulischen Erfolg“), da von vielen Seiten (Eltern, Pädagogen in Kindergärten und Schule, Logopäden) bei Verdacht auf das Vorliegen einer **Sprachentwicklungsstörung (SES)** (Kap. 20) auf frühe therapeutische Intervention gedrängt wird. Die Prävalenz einer SES liegt bei 4–8 %, eine SES kann aber erst ab ca. vier Jahren sicher diagnostiziert werden. Ein später Sprachbeginn („late talker“), der bei bis zu 20 % der Zwei- bis Dreijährigen Kinder zum Zeitpunkt der Vorsorgeuntersuchung U7 und U7a mittels Screeningverfahren (z. B. FRAKIS K – Elternfragebogen) festgestellt wird, lässt keine Vorhersage bezüglich der Entwicklung einer SES zu. Diese diagnostische Unsicherheit hat dazu geführt, dass im Jahr 2010 19.8 % der Jungen und 14,2 % der Mädchen im Alter von 5 Jahren eine Logopädie-Verordnung erhielten (Barmer GEK Arztreport, 2012; aus dem wissenschaftlichen Institut der AOK liegen vergleichbare Daten vor). Ein Großteil dieser medizinischen Logopädie-Verordnungen gehen auf Sprachauffälligkeiten oder Sprachschwächen zurück, die soziogen bedingt sind, d. h. aufgrund mangelnder familiärer Anregung und Förderung. In großstädtischen Ballungsgebieten, in denen es eine ausgeprägte Segregation bestimmter Bevölkerungsgruppen gibt, kommen mangelnde Deutsch-Kenntnisse von Kindern aus **Migrantenfamilien** hinzu, die ebenfalls zur Verordnung von Logopädie führen. Für alle diese Kinder wäre die erforderliche Intervention nicht Logopädie, sondern Sprachförderung im Rahmen einer die gesamte kognitive Entwicklung fördernden sozialen Umgebung. Die Beratung der Eltern über sprachförderndes Verhalten durch den Kinder- und Jugendarzt/-ärztin kann deshalb nicht früh genug beginnen. Leider gibt es keine hinreichenden kommunalen Angebote zur frühzeitigen Sprachförderung oder zum Eltern-Training außerhalb des Medizinsystems.
Zwischen dem 4. (5.) und 10. Lebensjahr werden zunehmend Verhaltens- und schulische Probleme in der Praxis thematisiert: Störungen des Sozialverhaltens, emotionale Störungen, Schulleistungsprobleme und oft Probleme, die eine Abklärung auf ein ADHS erforderlich machen (➤ Kap. 22). 9,7 % aller Studienkinder waren im DAKJ-Studienzeitraum mit der Dauerdiagnose „ADHS“ versehen.
Im Jugendalter (10. bis 18. Lebensjahr) überwiegen Vorstellungen wegen psychosomatischer Störungen wie rezidivierende oder chronische Kopf-, Bauch-, Skelett- Schmerzen und Schwindel (➤ Kap. 25), aber auch Störungen, die dem psychiatrischen Spektrum zuzuordnen sind wie Depressionen, Ängste, Tics, Phobien, Schlafstörungen, Essstörungen, Adoleszentenkrisen, Ausscheidungs-Störungen, Schulleistungsabfall, Verhaltensprobleme, Suchtverhalten (➤ Kap. 22) Die zuletzt genannten Diagnosen stammen aus einer Diagnosenstatistik der eigenen Gemeinschaftspraxis im Jahr 2009.
Selbst-initiierte Vorstellungen von Jugendlichen oder Eltern mit ihren Kindern zu **Übergewicht und Adipositas** spielen im Verhältnis zur Prävalenz des Übergewichts in der Bevölkerung und den verschiedenen Altersgruppen eine eher geringe Rolle (Kap. 35). Entsprechende Beratungen müssen in der Regel seitens der Praxis angestoßen und kontinuierlich angeboten werden.
Kinder und Jugendliche, die mit einem Problem aus dem Bereich „Neue Morbidität“ vorgestellt werden, brauchen einen notwendigen Zeitrahmen zur Evaluation des Problems. In einer vollen Akutsprechstunde kann es geboten sein, zuerst eine Symptom-bezogene Anamnese z. B. bei chronischen Kopfschmerzen zu erheben und eine neurologische Untersuchung durchzuführen, um eine organische Erkrankung auszuschließen. Sodann sollte ein zeitnaher längeren Termin vereinbart werden. An diesem wird die erweiterte Anamnese mit Familien- und Sozialanamnese erho-

ben, soweit diese nicht schon durch die langzeitige Betreuung bekannt ist. Das Ausmaß der Beeinträchtigung des Alltags, der Lebensqualität und eventuell der Schulfehlzeiten ist abzuklären.
Je nach Vorstellungsanlass (wie umschriebene Entwicklungsstörung, Verhaltens-, emotionale-, Sozialstörung) sollte in der Praxis eine Basis-MBS (mehrdimensionale Bereichsdiagnostik Sozialpädiatrie) durchgeführt werden. Für eine solche **Basisdiagnostik** stehen eine Fülle von strukturierten und evaluierten Interviewleitfäden und Fragebögen zur Verfügung wie der MEF = Mannheimer Elternfragebogen zur Erfassung psychischer Auffälligkeiten oder der SDQ = Strengths and Difficulties Questionnaire und auch Vorlagen für strukturierte Tagebuchaufzeichnungen, die z. B. bei der Abklärung von Spannungs- und Migränekopfschmerzen oder Bauchschmerzen hilfreich sein können. Entwicklungsdiagnostische Testmaterialien ermöglichen vorab ein schnelles Screening (z. B. Denver-Screening bis zum 6. Lebensjahr), dem eine weitergehende Basis-Diagnostik angeschlossen wird z. B. mit der MFED (Münchner Funktionelle Entwicklungsdiagnostik) 1. Lebensjahr; MFED 2.–4. Lebensjahr; MFED 3.–6. Lebensjahr; dem Entwicklungstest ET 6–6 (6 Monate bis 6 Jahre); der BUEVA (Basisdiagnostik umschriebener Entwicklungs-störungen) im Vorschulalter 4–6 Jahre und der BUEGA 6 Jahre im Grundschulalter (➤ Kap. 17).
Auch für die häufigste bereits erwähnte umschriebene Entwicklungsstörung, die Sprach-Entwicklungsstörung stehen Screening-Tests zur Basis-Bewertung der Sprachproduktion (nicht des Sprachverständnisses) zur Verfügung wie der SBE-2-KT, SBE-3-KT, ELFRA 2, SETK-3–5 u. a. (➤ Kap. 17)
Eine gute Basisdiagnostik erlaubt dem ambulant tätigen Kinder- und Jugendarzt/-ärztin in der Regel eine Einschätzung des Schweregrades und der Prognose einer Störung und ist damit Grundlage für das weitere Vorgehen. Bei Vorliegen von schweren oder komplexen Störungen sollte der Patient zur differenzierten diagnostischen Abklärung und Therapieplanung an eine **Einrichtung der tertiären Versorgung** mit interdisziplinärem Team weitergeleitet werden. Je nach örtlicher Gegebenheit kann dies ein **Sozialpädiatrisches Zentrum** sein, eine **neuropädiatrische Abteilung,** ein **niedergelassener Neuropädiater** mit interprofessionellem Team oder eine **sozial-psychiatrische Praxis eines Kinder- und Jugendpsychiaters.**
Die meisten der vorgestellten Störungen sind kurz- oder mittelfristig durch gute ärztliche und/ oder psychologische Beratung und ggf. eine Unterstützung durch eine zeitlich begrenzte **Verordnung von Heilmitteln** (Logo-, Ergo-, Physiotherapie) im Rahmen der Möglichkeiten einer primären Gesundheitsversorgung gut zu versorgen (➤ Kap. 37; ➤ Kap. 38).
Gemeinsam mit dem Kind, dem Jugendlichen und den Angehörigen muss ein Therapieplan des ambulanten Vorgehens erarbeitet werden, der die Bedürfnisse, Vorlieben, Stärken und Ressourcen des Kindes/Jugendlichen und seiner Familie mit einbezieht.

1.4 Sozialraumvernetzung

Ein wichtiges Merkmal der ambulanten Grundversorgung ist ihre gemeindenahe Ausrichtung und die Notwendigkeit einer guten **Sozialraumvernetzung** zwischen den **Mitarbeitern des Gesundheitssystems** (Ärzte, nichtärztliche Heilmittelerbringer, SPZ's, Entbindungsklinik, Kinderkliniken mit sekundärer und tertiärer Versorgung, Kinder- und Jugendpsychiatrie, Kinderhospiz, KJGD etc.), **dem sozialen Hilfesystem** (Jugendamt, Frühförderstelle, Schulpsychologischer Dienst, Beratungsstellen staatlicher, freier und kirchlicher Träger, Kinder- und Jugendnotdienst, Kinderschutzzentrum etc.), **dem Bildungssystem** (Kindertagesstätten, Schulen, Sportvereine, Jugendfreizeitzentren) und **den Freien Organisationen** (Elternselbsthilfegruppen, Berufsständige örtliche Verbände, „Stammtische" etc.). Da eine ärztliche Mitwirkung an der Hilfeplanung im § 36 SGB VIII nicht vorgeschrieben ist, sollte der Kinder- und Jugendarzt den Mitarbeitern des Jugendamtes grundsätzlich seine Bereitschaft signalisieren, an fallbezogenen Hilfeplanungen für die eigenen Patienten teilzunehmen.

1.5 Weiter- und Fortbildung für die pädiatrische ambulante Grundversorgung

Die allgemeine pädiatrische Weiterbildung in Deutschland beträgt 5 Jahre, wovon mindestens 3 Jahre im stationären Bereich verbracht werden müssen und maximal weitere 2 Jahre in einer zur Weiterbildung zugelassenen ambulanten pädiatrischen Praxis abgeleistet werden können. Die Weiterbildung soll kompetenzbasiert sein und den Arzt am Ende befähigen, selbstständig alle professionellen Tätigkeiten auszuführen die notwendig sind für den allgemein stationären als auch für den allgemein ambulanten pädiatrischen Bereich.

De facto werden die meisten Pädiater aber noch ausschließlich d. h. 5 Jahre im stationären Bereich weitergebildet, da es bislang keine Finanzierungsmöglichkeit für eine Weiterbildung in ambulanten pädiatrischen Praxen gibt. Dies bedeutet, dass sie bei Beginn ihrer ambulanten Tätigkeit oft nicht ausreichend vorbereitet sind für die Aufgaben in der pädiatrischen Grundversorgung, in der sie andere Versorgungsaufgaben haben als Allgemeinpädiater im Krankenhaus oder als Pädiater in der tertiären ambulanten und stationären Versorgung. Zudem werden sie in der ambulanten Grundversorgung mit einem anderen Spektrum von Beratungsanlässen, akuten Beschwerden und Krankheiten konfrontiert als während ihrer Weiterbildung im Krankenhaus.

Die DGAAP erarbeitet eine Orientierungshilfe für einen Weiterbildungs-Abschnitt in der ambulanten pädiatrischen Praxis, in dem neben dem Erwerb medizinisch-pädiatrischer Kompetenz für die Erfordernisse der Grundversorgung weitere zu vermittelnde Schlüsselkompetenzen dargestellt werden (➤ Tab. 1.2).

Der Berufsverband der Kinder- und Jugendärzte (BVKJ) organisiert seit vielen Jahren Fortbildungskongresse, die sich insbesondere der Themen annehmen, die für die Arbeit in einer ambulanten pädiatrischen Praxis wichtig sind.

1.6 Die Zukunft der Kindergrundversorgung – ein Ausblick

Kinder, Jugendliche und ihre Familien brauchen auch in Zukunft eine gut erreichbare, umfassende, kontinuierliche, Kind- und Familien-zentrierte, empathische und kulturell kompetente **Gesundheitsgrundversorgung**. In einigen ländlichen Regionen sind bereits jetzt zu wenige oder gar keine Ärzte in einer zumutbaren Entfernung erreichbar. Angesichts der Feminisierung der Ärzteschaft (80 bis 90 % aller Ärzte/Ärztinnen in allgemeinmedizinischer und pädiatrischer Weiterbildung sind derzeit Frauen) und damit der Reduzierung der Lebensarbeitszeit der Ärzte und Ärztinnen in der medizinischen Grundversorgung müssen grundsätzlich neue Wege für eine flächendeckenden Grundversorgung überlegt werden. Konkurrierende Einzelpraxen, die lange von der Sozialgesetzgebung begünstigt wurden, werden wahrscheinlich in regionale medizinische Familien-Versorgungszentren überführt werden, in denen Ärzte für Allgemeinmedizin und für

Kinder- und Jugendmedizin zusammen mit nichtärztlichen Therapeuten vernetzt arbeiten. Wünschenswert wäre, wenn diese **Familien-Versorgungszentren** auch Raum böten für eine Sozialraumvernetzung mit dem sozialen Hilfesystem und dem Bildungssystems im Sinne einer gemeindenahen Solidargemeinschaft zum Wohle der Kinder und Familien.

Literatur

Fegeler U, Jäger-Roman E, Martin R, Nentwich HJ (2011/2012) Studie der Deutschen Akademie für Kinder- und Jugendmedizin (DAKJ) und der Dresden International University;

Fehr F (2012) Kompetenzen und Lernziele in der pädiatrischen Weiterbildung. Kinder- und Jugendarzt 43/11: 666–668

Kamtsiuris P, Bergmann E, Rattay P, Schlaud M (2007) Inanspruchnahme medizinischer Leistungen. In: Bundesgesundheitsbl-Gesundheitsforsch-Gesundheitsschutz 50: 836–850

WHO (1978) Declaration of Alma-Ata. International Conference on Primary Health

Care, Alma-Ata, USSR. http://www.who.int/hpr/NPH/docs/declaration_almaata.pdf

Wolfe I, Thompson M, Gill P, Tamburlini G, Blair M, Bruel A, Ehrich J, Pettoello-Mantovani M, Janson S, Karanikolos M, McKee M (2011) Health Services for Children in Western Europe BMJ 342:d1277

www.thelancet.com Published online March 27, 2013 http://dx.doi.org/10.1016/S0140-6736(12)62085-6 11

WONCA Europe (2005) The European Definition of General Practice/Family Medicine. http://www.euract.eu

2 Was ist Sozialpädiatrie?

Harald Bode

2.1 Einleitung

Soziale Ursachen und Folgen von Krankheiten bei Kindern und Jugendlichen spielen seit jeher eine herausragende Rolle: Infektionen, Fehl- und Mangelernährung, Unfälle, frühe Sterblichkeit und Verwahrlosung waren schon immer eng mit der sozialen Situation der Familien verknüpft. Der hohe Stellenwert der Hygiene resultierte schon früh aus der Erkenntnis, dass Umweltfaktoren eine große Bedeutung für Gesundheit und Entwicklung von Kindern und Jugendlichen hatten. So wurde die Prävention zu einem Grundprinzip ärztlichen und speziell kinderärztlichen Denkens und Handelns.
Seit etwa drei Jahrzehnten stehen wir vor **neuen Morbiditäten,** die z. T. als neue Epidemien angesehen werden müssen: chronische Erkrankungen, Entwicklungs- und Verhaltensstörungen sowie Behinderungen unterschiedlicher Art sind so häufig geworden, dass die Kinder mit diesen Problemen und ihren besonderen Bedürfnissen eine Herausforderung im beruflichen Alltag eines jeden Kinder- und Jugendarztes, von Psychologen, Therapeuten, Pädagogen oder Erziehern darstellen. Ein großer Teil dieser Erkrankungen lässt sich nur mit Einbeziehung des familiären, sozialen und gesellschaftlichen Kontextes befriedigend und wirkungsvoll behandeln. Für die angemessene Behandlung sind darüber hinaus spezifische, meist **interdisziplinäre Konzepte** erforderlich.
Weitere öffentlich stark diskutierte Aufgabenfelder sind frühe Hilfen für belastete Familien, Kinderschutz, die vielfältigen Probleme und Gefährdungen von Kindern und Jugendlichen mit Migrationshintergrund und von sozial benachteiligten oder marginalisierten Gruppen, die oft verbunden sind mit mangelhafter Sprachkompetenz und anderem soziokulturellen Hintergrund. Strukturen des Gesundheitssystems, die aktive Inanspruchnahme der Betroffenen voraussetzen, erreichen diese Bevölkerungsgruppen häufig nicht. Präventive Ansätze sind hier besonders gefragt.
Der **soziale Gradient** der Kindergesundheit bei Seh-/Hörstörungen, Sprachstörungen, motorischen Störungen, Verhaltensstörungen, Störungen der Aufmerksamkeit, Unfällen, Verbrennungen, Karies und Adipositas ist erwiesen. Gewandelte, wechselnde, unvollständige und z. T. sich

Sozialpädiatrie in der Praxis. http://dx.doi.org/10.1016/B978-3-437-31630-3.00002-X
Copyright © 2014 Elsevier GmbH. All rights reserved.

auflösende Familienstrukturen, hohe Mobilitätsanforderungen, eine zunehmende Häufigkeit mütterlicher Berufstätigkeit, gewandelte Wertvorstellungen, Verunsicherungen von Eltern und anderen Bezugspersonen und die neue Armut erfordern die Entwicklung und den Ausbau neuer bzw. komplementärer **Betreuungsstrukturen** für Kinder und Jugendliche. Die Bedeutung der genannten Probleme reicht in alle gesellschaftlichen Bereiche hinein (z. B. Krippen, Kindergärten, Schulen, Kindertagesbetreuung, berufliche Bildung, Arbeitsmarkt, soziale Sicherung etc.).
Kinder- und Jugendärzte, Psychologen, Therapeuten, Pädagogen und Erzieher erwerben im Rahmen ihrer Weiterbildung für diese Themen und Aufgabenbereiche nur sehr eingeschränkte Kenntnisse und Fähigkeiten. Sie sind andererseits zu Recht für Familien oft Ansprechpartner bei Problemen ihrer Kinder, die sich aus den vielfältigen Entwicklungs- und Integrationsaufgaben in unserer heutigen Gesellschaft ergeben.
Die Sozialpädiatrie entwickelt und bietet spezifische Konzepte für diese Aufgaben an oder macht Konzepte aus anderen Disziplinen für die Diagnostik, Beratung und Behandlung von Kindern und Jugendlichen und ihren Familien bekannt und verfügbar. Sozialpädiatrische Grundkenntnisse sollte heute jeder besitzen, der mit Kindern und Jugendlichen mit besonderen Bedürfnissen beruflich im Gesundheits-, Förder- oder Erziehungsbereich zu tun hat. Andererseits sind sozialpädiatrische Qualifikationen heute so differenziert und spezifisch, dass vertiefte Kenntnisse, Fähigkeiten und Fertigkeiten nur in geeigneten Institutionen und Tätigkeitsfeldern erlernt werden können.

2.2 Definition von Sozialpädiatrie

Die Internationale Gesellschaft für Sozialpädiatrie (ISSOP) definiert „**Social Pediatrics**“ wie folgt: **„Social paediatrics is an approach to child health that focuses on the child, in illness and in health, within the context of their society, environment, school, and family.“** (Sozialpädiatrie ist ein Zugang zur Kindergesundheit, mit Fokus auf dem Kind in Krankheit und Gesundheit im Kontext seiner Gesellschaft, des Umfeldes, der Schule und Familie).
In Deutschland wird **Sozialpädiatrie** als eine Querschnittswissenschaft in der Kinder- und Jugendmedizin verstanden. Sie befasst sich mit den Bedingungen von Gesundheit und Entwicklung sowie deren Störungen und Auswirkungen. Dabei wird besonderer Bezug auf das Gemeinwesen genommen. In ihrer interdisziplinären Arbeitsweise ist sie mit natur-, geistes- und gesellschaftswissenschaftlichen Disziplinen vernetzt. Das Spektrum der Sozialpädiatrie schließt neben den Methoden der Pädiatrie solche der Epidemiologie, der Psychologie und Entwicklungspsychologie, der Entwicklungsneurologie und der Gesundheitsökonomie ein.
Sozialpädiatrie untersucht empirisch-wissenschaftlich die Häufigkeit und Verteilung von Gesundheits- und Entwicklungsstörungen im Zusammenhang mit der sozialen, insbesondere familiären und der natürlichen Umwelt. Sie befasst sich mit der Organisation des Gesundheitswesens, der vorschulischen und schulischen Förderung sowie der Einrichtungen der sozialen Sicherung. Sie beteiligt sich an der gesundheitsfördernden Gestaltung der kindlichen Lebensräume im Gemeinwesen.
Im Bereich der Versorgung befasst sie sich mit der Habilitation, der ambulanten und stationären Rehabilitation und der Integration von Kindern und Jugendlichen mit langfristigen Beeinträchtigungen der gesundheitlichen, psychischen, geistigen, seelischen und sozialen Entwicklung im Rahmen eines interdisziplinären Behandlungskonzepts unter Einschluss der notwendigen Heil- und Hilfsmittelversorgung. Dabei werden immer die Eltern bzw. Bezugspersonen und das jeweilige Lebensumfeld einbezogen. Übergeordnetes Therapieziel ist die bestmögliche gesellschaftliche Teilhabe.

Aktuell wird in Deutschland eine **sozialpädiatrische Grundversorgung** von jedem Kinder- und Jugendarzt erwartet. Für viele Kinder und Jugendliche mit komplexen Problemen sind darüberhinausgehende Kenntnisse und Fähigkeiten („**Spezielle Sozialpädiatrie**") erforderlich.

2.3 Aufgabenfelder und Zielgruppen

2.3.1 Individualmedizinische Aufgaben

Sozialpädiatrie betrachtet als eine ihrer Hauptaufgaben die kinder- und jugendärztlich geleitete und verantwortete, medizinisch-therapeutisch-psychologisch-(heil)pädagogisch-soziale Behandlung, d.h. Diagnostik, Beratung und Therapie von **Kindern und Jugendlichen mit besonderen Bedürfnissen und ihren Familien.** Traditionell handelt es sich dabei besonders um Säuglinge und Kleinkinder mit Regulationsstörungen (Ess-, Schlaf-/Schreistörungen), Kinder und Jugendliche mit umschriebenen, kombinierten und tiefgreifenden Entwicklungsstörungen, mit emotionalen und Verhaltensstörungen (z. B. hyperkinetische Störungen, Störungen des Sozialkontaktes), neurologischen und psychosomatischen Erkrankungen, Tic-Störungen, Enuresis, Enkopresis, genetischen Syndromen, körperlichen, seelischen und/oder geistigen Behinderungen bzw. Intelligenzminderung, vernachlässigte, misshandelte und sexuell missbrauchte Kinder sowie bestimmte Risikopopulationen (z. B. ehemalige sehr kleine Frühgeborene, Kinder mit schwerwiegenden psychosozialen Risikofaktoren).

Bei diesen traditionellen Zielgruppen sozialpädiatrischer individualmedizinischer Betreuung bestehen Überschneidungen mit den Aufgabengebieten der Neuropädiatrie bzw. Kinder- und Jugendpsychiatrie.

In den letzten Jahren wird immer deutlicher, dass darüber hinaus bei vielen Kindern und Jugendlichen mit chronischen Erkrankungen (z. B. Adipositas, Mukoviszidose, Diabetes, onkologischen oder Herzerkrankungen, Stoffwechselstörungen u. a.) über die organmedizinischen Aspekte ihrer Erkrankung hinaus akute und/oder langfristige psychosoziale Probleme entstehen. Für diese Kinder und ihre Familien haben sozialpädiatrische Behandlungskonzepte eine wachsende Bedeutung erlangt, damit die notwendigen medizinischen Behandlungen zum größtmöglichen Erfolg führen. Sozialpädiatrie beansprucht dabei nicht die fachspezifische Kompetenz, z. B. die eines pädiatrischen Onkologen, Kardiologen, Diabetologen oder eines spezialisierten Neuropädiaters. Vielmehr bringt die Sozialpädiatrie Kompetenz ein für den Umgang mit den besonderen psychosozialen Bedürfnissen der Kinder und Jugendlichen und ihren Familien. Sehr häufig ist dafür ein multiprofessionelles Team erforderlich, in dem ein „Case-Manager" die erforderlichen Kontakte im Netzwerk des Helfer- und Sozialsystems herstellt und pflegt.

Neben den klassischen kurativen Behandlungsansatz tritt in der Sozialpädiatrie für jedes individuelle Kind gleichwertig eine gesundheitsfördernde und/oder krankheitspräventive Perspektive. Dabei spielen niederschwellige sozialraumbezogene und eltern-/familienorienierte Maßnahmen, z. B. in Form von Aufklärung und Beratung eine wichtige Rolle.

Oberstes Ziel aller sozialpädiatrischen Arbeit muss es dabei sein, die Kinder/Jugendlichen und ihre Eltern im Umgang mit und in der Bewältigung der Probleme des jeweiligen Lebens- oder Krankheitsabschnitts kompetent und autonom zu machen. Oft ist hierfür eine langjährige sozialpädiatrische Begleitung erforderlich.

Ein wesentlicher sozialpädiatrischer Aufgabenbereich in der individuellen Betreuung von Kindern, Jugendlichen und ihren Familien ist heute die fachliche Beratung und die Begutachtung bei Fragen der Integration/Inklusion in Kindertagesstätten und Schulen, Pflegebedürftigkeit, Heil- und Hilfsmittelversorgung, Erziehungshilfe, Fremdunterbringung, Berufsförderung, Rehabilitation und in versorgungsrechtlichen Fragen.
Zu den bekannten Aufgaben jeder pädiatrischen Diagnostik und Therapie sind die spezifischen Aufgaben einer sozialpädiatrischen Behandlung komplementär.
Spezifische Aufgaben und Themen sozialpädiatrischer Behandlung sind:

- Sensibilisierung der Eltern für spezifische Bedürfnisse des Kindes
- Vermittlung einer realistischen Perspektive für die Zukunft des Kindes
- Bewältigung von Krankheit/Behinderung
- Bewältigung des Alltags/Entlastung
- Häusliche Pflege
- Individuelle und spezifische Erziehungs-/Fördermaßnahmen
- Wahl von Kindergarten/Schule
- Förderung der Integration/Inklusion
- Berufsfindung
- Vermittlung an andere fachspezifische Institutionen/Dienste
- Gesetzliche/rechtliche Rahmenbedingungen.

Je nach Aufgabenschwerpunkt und Qualifikation können diese Aufgaben durch verschiedene Berufsgruppen wahrgenommen werden, z. B. Kinder- und Jugendarzt, Kinder- und Jugendlichenpsychotherapeut, klinischer Psychologe, medizinischer Therapeut (Heilmittelerbringer), Sozialpädagoge, Heilpädagoge, Kinderkrankenschwester.
Die individualmedizinische sozialpädiatrische Versorgung findet in Sozialpädiatrischen Zentren und Kliniken, Rehabilitationskliniken statt, aber auch in großem Umfang in kinder-/jugendärztlichen Praxen und Kliniken, Spezialambulanzen, bei medizinischen und psychologischen Therapeuten, in Zusammenarbeit mit verschiedenen anderen Einrichtungen und Institutionen (z. B. Frühförderstellen, Kindergärten, Schulen, psychologische Dienste, Sozial- und Jugendämter, Sozialversicherungen, Berufsförderungswerke, Arbeitsagenturen, sozialmedizinische Nachsorgeeinrichtungen, Familiengerichte).
In der sozialpädiatrischen Behandlung der Kinder und Jugendlichen und ihrer Familien finden die Prinzipien der evidenzbasierten Medizin und die Leitlinien der Fachgesellschaften Anwendung. Der Qualitätssicherung wird eine große Bedeutung zugemessen.

2.3.2 Gemeinwesen – bezogene Aufgaben

Diese sind insbesondere das Arbeitsfeld des Öffentlichen Kinder- und Jugendgesundheitsdienstes (KJGD). Dieser ist an den jeweiligen Gesundheitsämtern bzw. Gesundheitsfachverwaltungen der Gemeinden und Kreise organisiert. Wesentliche Aufgaben des KJGD sind der unten stehenden Aufzählung sowie Kapitel 7(➤ Kap. 7)und Kapitel 9(➤ Kap. 9) zu entnehmen.
In der Regel werden je nach regionaler Bedarfslage Schwerpunkte gesetzt, die sich auch an der je nach Bundesland und Gemeinde recht unterschiedlichen personellen und sachlichen Ausstattung des KJGD orientieren. Zur Erfüllung seiner Aufgaben arbeitet der KJGD mit zahlreichen, insbesondere regionalen Institutionen und Einrichtungen zusammen. Intensität, Vernetzung und Inhalte der Zusammenarbeit sind regional verschieden.

Bevölkerungsbezogene Sozialpädiatrie (z. B. Epidemiologie, Gesundheitsberichterstattung zur Lage von Kindern und Jugendlichen und Gesundheitsökonomie) wird außerdem in verschiedenen wissenschaftlichen Einrichtungen mit unterschiedlichen fachlichen Schwerpunkten, kommunalen sowie Landes- und Bundesbehörden realisiert.

Aufgaben des öffentlichen Kinder- und Jugendgesundheitsdienstes:

- Beobachtung des Gesundheitszustands von Kindern und Jugendlichen
- Identifizierung gesundheitsabträglicher Lebensumstände für Kinder und Jugendliche
- Koordination und Implementation gesundheitsfördernder Aktivitäten für Kinder und Jugendliche
- Stärkung der elterlichen Versorgungs- und Erziehungskompetenz in gesundheitsrelevanten Bereichen
- Heimärztliche Betreuung von Kindern in vorschulischen Einrichtungen
- Schulgesundheitspflege, u. a. Untersuchung aller Schulanfänger
- Begutachtung von Kindern und Jugendlichen bei Fragestellungen aus dem Sozial-, Jugendhilfe- und Bildungsbereich
- Maßnahmen für Kinder und Jugendliche, die infolge sozialer Benachteiligung unzureichend von den Regelleistungen des Gesundheitssystems profitieren
- Koordination der Förderung behinderter Kinder und Jugendlicher auf Gemeindeebene
- Steigerung der Durchimpfungsraten von Kindern und Jugendlichen
- System- sowie Politikberatung zu regionalen Vorhaben, soweit diese Rückwirkung auf Gesundheits- und Entwicklungsbedingungen für Kinder und Jugendliche haben.

2.4 Sozialpädiatrische Methodik

Diese soll beispielhaft am Konzept der Diagnostik und Therapie in Sozialpädiatrischen Zentren verdeutlicht werden. Mit entsprechender Qualifikation und Zeit sowie persönlichem Interesse und Engagement lassen sich diese Prinzipien jedoch auch in kinder- und jugendärztlichen Praxen, Spezialambulanzen, auf kinder-/jugendärztlichen Stationen, bei Heilmittelerbringern und in pädagogischen und therapeutischen Einrichtungen realisieren. Auch in den Bereich der stationären Kinder-Jugend-Rehabilitation haben diese Prinzipien Eingang gefunden.

Kern der sozialpädiatrischen Methodik ist die interdisziplinäre Zusammenarbeit in einem multiprofessionellen Team bei Diagnostik, Beratung und Therapie. Das Kind/der Jugendliche und seine Familie sowie sein soziales Umfeld werden dabei umfassend einbezogen und gewürdigt. Besonderes Augenmerk wird auf Kompetenzen und Autonomie des Kindes/Jugendlichen und seiner Familie gelegt. In diesem Kontext wird auch die Arbeit von Elternselbsthilfegruppen geschätzt und unterstützt.

2.4.1 Sozialpädiatrische Diagnostik

Anamnese und Untersuchung sind die Grundlagen der sozialpädiatrischen Diagnostik und des Behandlungsplans. Dabei wird in der Anamnese neben dem bisherigen Krankheitsverlauf auch eine biografische Anamnese erhoben.

Die sozialpädiatrische Untersuchung berücksichtigt in Anlehnung an das multiaxiale Klassifikationssystem folgende Bereiche: Entwicklungsstand/Intelligenz, körperlich-neurologischer Befund, psychischer Befund und Verhalten, soziale Kontextfaktoren, die Abklärung der Ätiologie und die Erhebung von Kriterien der Teilhabe an der Gemeinschaft nach ICF-CY-Kriterien **(EKPSA+T)**. Hierbei finden – je nach Problemstellung oft kombiniert oder nacheinander – verschiedene diagnostische Techniken Anwendung (s. u.).

Mehrdimensionale Bereichsdiagnostik in der Sozialpädiatrie (MBS):

- Entwicklungsstand/Intelligenz
- Körperlich-neurologischer Befund
- Psychischer Befund
- Sozialer Hintergrund
- Aetiologische Abklärung
- Teilhabe/Partizipation

= **EKPSA+T**

Diagnostische Techniken in der Sozialpädiatrie:

- Biographische Anamnese
- Pädiatrische, neuropädiatrische und entwicklungsneurologische Untersuchung
- Psychologische und neuropsychologische Diagnostik
- Sprachlich-kommunikative Diagnostik
- Verhaltens-/Interaktionsbeobachtung
- Vertiefende Erhebung der psychosozialen Rahmenbedingungen
- Berücksichtigung von Kriterien der Gesundheits- und Kontextfaktoren nach ICF-CY
- Medizinisch-technische Diagnostik (Neurophysiologie, bildgebende Verfahren, Labor u. a.)
- Heranziehen auswärtiger Befunde/Einschätzungen.

Für die genannten Techniken steht eine große Anzahl oft standardisierter Tests und Untersuchungsverfahren zur Verfügung (➤ Kap. 17). Dabei ist nach dem jeweiligen aktuellen fachlichen Standard der verschiedenen Disziplinen/Berufsgruppen zu verfahren. Sozialpädiater sollen die Indikation und die Prinzipien der wesentlichen sozialpädiatrischen Techniken kennen, die berufsgruppenbezogenen, fachlich anerkannten Techniken beherrschen und besondere Kenntnisse und Erfahrungen in den relevanten Techniken der Gesprächsführung besitzen.

Es ist wesentlich, im Rahmen der sozialpädiatrischen Diagnostik das **Ressourcenprofil** von Kindern/Jugendlichen, Familie, Umwelt und Behandler zu erheben. Dazu gehören neurologisch-somatische, kognitive, emotionale, interaktive und soziale Faktoren. Ziel ist es, unter besonderer Berücksichtigung der Stärken eine Gesamtschau der Möglichkeiten des Patienten und seiner Familie zu entwickeln.

Grundvoraussetzung für eine gelingende und von allen Beteiligten wert geschätzte Diagnostik sind sozialpädiatrische Grundhaltungen und Kompetenzen: anteilnehmendes Interesse am Kind/Jugendlichen in seiner Lebenssituation, Sensibilität für familiäre Belastungen und psychosoziale Risikofaktoren, entwicklungspädiatrische und transkulturelle Kompetenz sowie Bereitschaft und reale Kooperation mit anderen Berufsgruppen und Institutionen auf Augenhöhe unter Beachtung der jeweiligen Kompetenzen und Zuständigkeiten.

Aus sozialpädiatrischer Diagnose und Ressourcenprofil wird unter Heranziehung von Fachwissen und Erfahrung der beteiligten Berufsgruppen über Entwicklungsverlauf und prognostische Faktoren und in Abstimmung mit den Kindern und Jugendlichen (soweit möglich) und ihren Bezugspersonen ein **Behandlungsplan** erstellt. Dieser wird in Abständen – durch Verlaufsdiagnostik gestützt – fortgeschrieben bzw. mit Erreichung der festgelegten Ziele ggf. abgeschlossen.

2.4.2 Sozialpädiatrische Therapie

Die **Therapieindikation** ergibt sich aus Störungsprofil, Ressourcen und prognostischer Einschätzung, bezogen auf das Kind/den Jugendlichen und seine Familie. Diese werden in den Therapieprozess einbezogen, auch um eigene Aktivitäten und Problemlösestrategien zu fördern.
Bei der Festlegung der **Therapieziele** steht der Patient in seinem Lebensumfeld im Mittelpunkt: Es werden somatisch-funktionelle Aspekte (Alter, Art, Kompensation und Auswirkung der Störung), psychoemotionale Faktoren (Motivation, Emotion, Intelligenz) und der psychosoziale Bereich (Beziehung Patient-Eltern-Therapeut, soziales und familiäres Umfeld, soziokulturelle Faktoren, Kontakte im Fördernetzwerk, gesellschaftliche Teilhabe) berücksichtigt.
Die sozialpädiatrische Therapie beinhaltet die allgemein üblichen Therapieziele Heilung, Linderung und Vorbeugung von Krankheit, Entwicklungsstörung, Behinderung und deren Komplikation, darüber hinaus im Besonderen als Ziele die Verbesserung der Lebensqualität und die Stärkung des Selbstwertgefühls, der Selbstbestimmung und der Adaptation und die gesellschaftliche Integration und Teilhabe (Partizipation) der behandelten Kinder/Jugendlichen und ihrer Familien. Zur Beschreibung dieser letztgenannten Ziele bietet die von der WHO 2001 publizierte **ICF** (Internationale Klassifikation der Funktionsfähigkeit, Behinderung und Gesundheit) und deren Version für Kinder **(ICF-CY)** eine sehr differenzierte Orientierung (➤ Kap.16).
Neben den genannten übergeordneten Therapiezielen existieren **spezifische sozialpädiatrische Therapieziele,**

- Bewegungskompetenz
- Handlungs- und Wahrnehmungskompetenz
- Kommunikationskompetenz
- Soziale und emotionale Kompetenz
- Elterliche Kompetenz
- Soziale Integration
- Bewältigung der Krankheit/Situation.

Dabei sind für jedes Kind/jeden Jugendlichen in seiner jeweiligen Entwicklungs- und Lebenssituation individuelle Therapieziele in Abstimmung mit dem betroffenen Kind/Jugendlichen und seinen Eltern festzulegen und in angemessenem zeitlichen Abstand fortzuschreiben.
Therapiebereiche sind je nach Problemlage und örtlichen Gegebenheiten in alphabetischer Reihenfolge: ärztliche Therapie (inkl. medikamentöser Therapie), Diätberatung, Ernährungstherapie, Ergotherapie, Heilpädagogik, Hilfsmittelversorgung, Logopädie, Physiotherapie, Psychotherapie, Sozialpädagogik/-arbeit (➤ Kap. 37).
Therapieindikation, -ziele und -modalitäten (Therapieintensität, angewandte Verfahren/Techniken, Dauer, Pausen, Ende) und die **Therapieevaluation** werden i. d. R. interdisziplinär festgelegt und müssen Patient, Eltern und andere Bezugs- und kooperierende Fachpersonen einbeziehen. Dazu sind eine sorgfältige Dokumentation und die Kommunikation der diagnostischen Ergebnisse, des Behandlungsplans und Therapieverlaufs an die Beteiligten erforderlich.
Es versteht sich, dass je nach Art, Schwere und Dauer der Störungen einfachere oder komplexere Vorgehensweisen nötig sind, die unter Berücksichtigung der regionalen Gegebenheiten unterschiedliche Ebenen des medizinischen, pädagogischen und sozialen Systems allein, abgestuft oder in verschiedenen Kombinationen einbeziehen müssen.

2.5 Sozialpädiatrische Perspektiven

Erfreulicherweise scheint die Einsicht in die Bedeutung individualmedizinischen und bevölkerungsbezogenen sozialpädiatrischen Denkens und Handelns zuzunehmen. Verschiedene Studien und die praktische Erfahrung zeigen, dass bei ganz unterschiedlichen Erkrankungen rein kindzentrierte Therapien im Vergleich mit einer umfassenden Behandlung unter Einschluss von Eltern und sozialem Umfeld deutlich schlechter abschneiden. Zahlreiche Befragungen haben ergeben, dass Eltern und Patienten sozialpädiatrische Kompetenz und sozialpädiatrisches Handeln erwarten. Die überall in Deutschland bestehende große Nachfrage nach Behandlungen in den Sozialpädiatrischen Zentren und die langen Wartezeiten sind ein klarer Beleg für den Bedarf. Er muss und wird künftig auch verstärkt in anderen Versorgungsformen erfüllt werden.

Sozialpädiatrische Aspekte nehmen inzwischen einen hohen Anteil unter den Fortbildungsangeboten ein. Kinder- und Jugendärzte, Psychologen, Therapeuten, Pädagogen und Erzieher erkennen deshalb – insbesondere auch bei langjähriger Kenntnis des familiären und sozialen Lebensumfelds – bei Kindern, Jugendlichen und ihren Familien mittlerweile neue Gesundheitsgefährdungen und Morbiditäten früher. Sie sind dafür prädestiniert, für das einzelne Kind/den Jugendlichen und seine Familie angemessene sozialpädiatrische Konzepte zu entwickeln und dabei neben Ressourcen auch Risiken für die kindliche Gesundheit und Entwicklung frühzeitig zu erkennen (z. B. Armut, Marginalisierung, elterliche Überforderung, übermäßiger Medienkonsum, Vernachlässigung, drohende oder manifeste Misshandlung etc.).

Eine sozialpädiatrische Diagnostik und Beratung in der kinder- und jugendärztlichen Praxis kann – oft unter Einbeziehung anderer niederschwelliger Angebote (z. B. Frühförderung, psychologische und Erziehungsberatungsstellen, Kindergarten, Schule, medizinische Therapeuten) – früh präventiv oder kurativ wirksam seinn. Bei Bedarf können weitere Einrichtungen (sozialpädiatrische Zentren, andere Fachärzte, Kinderkliniken, Kinder- und Jugendpsychiatrie, andere spezifische ambulante und stationäre Behandlungseinrichtungen, Sozial- und Jugendamt, Gericht etc.) eingeschaltet werden.

Sozialpädiater werden künftig verstärkt im Rahmen ihres gesellschaftlichen Auftrags Verantwortung übernehmen und empirisch und wissenschaftlich gesichert über die neuen Morbiditäten informieren. Darüber hinaus werden gemeinwesenbezogene, gesundheitswissenschaftlich orientierte Aktivitäten mehr Bedeutung gewinnen. Sie müssen politisch und finanziell stärker unterstützt werden.

Es muss empirisch-wissenschaftlich begründet unter Einbeziehung sozialpädiatrischer Kenntnisse und Erfahrungen über die Vor- und Nachteile sowie die spezifischen Anforderungen an Kinderkrippen, Ganztagsschulen, Integration/Inklusion in Kindertagesstätten und Schulen und andere Formen institutioneller Betreuung verhandelt werden, damit bestmögliche kind- und familiengerechte Ergebnisse zustande kommen.

Auf der Basis einer hypothesengeleiteten und zielorientierten Gesundheitsberichterstattung für Kinder/Jugendliche auf nationaler und regionaler Ebene zu wichtigen und spezifischen Erkrankungen und Gesundheitsmaßnahmen können Programme der Gesundheitsförderung und Erziehung entwickelt, begleitet und wissenschaftlich bewertet und die strukturelle und inhaltliche Betreuung bestimmter Zielgruppen verbessert werden.

Wichtige **künftige Aufgaben der Sozialpädiatrie** sind:

- Krankheitsprävention und Gesundheitsförderung bei Kindern/Jugendlichen
- Unterstützung/Stärkung sozial benachteiligter Kinder/Jugendlicher und ihrer Familien
- Prävention/Behandlung der neuen Morbiditäten im Kindes-/Jugendalter im familiär/gesellschaftlichen Kontext

- Förderung der Integration/Inklusion von Kindern/Jugendlichen mit besonderen Bedürfnissen
- Ausbau sozialpädiatrischer Qualifikationen in der Kinder- und Jugendmedizin in Praxis, Klinik, Lehre und Forschung
- Kapazitätserweiterung und finanzielle Absicherung sozialpädiatrischer Institutionen
- Qualitätssicherung in der diagnostischen und medizinisch-therapeutischen (Heilmittel) Arbeit und der Rehabilitation
- Vernetzung von gesellschaftlichen Aktivitäten zugunsten des körperlichen, geistigen, seelischen und sozialen Wohls von Kindern/Jugendlichen
- Interessenvertretung von Kindern/Jugendlichen und ihren Familien in der Öffentlichkeit und Politik („**child advocacy**").

Literatur

American Academy of Pediatrics, Committee on Psychosocial Aspects of Child and Family (1993) The pediatrician and the „new morbidity". Pediatrics 92: 731–733.

Bode H (2004) Migrantenkinder. Pädiatrische Praxis 65: 289–297.

Bode H (2003) Sozialpädiatrie in der Kinder- und Jugendmedizin. Inhalte, Aufgaben und die Notwendigkeit einer spezifischen Qualifikation. Pädiatrische Praxis 63: 687–696.

Bundesarbeitsgemeinschaft Rehabilitation (2008) Gemeinsames Rahmenkonzept für die Durchführung stationärer medizinischer Maßnahmen der Vorsorge und Rehabilitation für Kinder und Jugendliche. Frankfurt. http://www.bar-frankfurt.de/upload/Rahmenkonzept.72dpi_435.pdf

Deutsche Akademie für Kinder- und Jugendmedizin: Die DAKJ-Zukuftsstudie. http://dakj.de/pages/posts/analyse-und-zukunftsszenario-der-flaechendeckenden-versorgung-der-kinder-und-jugendlichen-in-deutschland---begruendungsdokumente-157.php

Deutsche Gesellschaft für Sozialpädiatrie und Jugendmedizin (2002) Grundlagen und Zielvorgaben für die Arbeit im Sozialpädiatrischen Zentrum (SPZ) – Strukturqualität, Diagnostik und Therapie – „Altöttinger Papier". Kinderärztliche Praxis 73: 498–515.

Deutsches Institut für Medizinische Dokumentation und Information, DIMDI (2005) ICF-Internationale Klassifikation der Funktionsfähigkeit, Behinderung und Gesundheit. http://www.dimdi.de/dynamic/de/klassi/downloadcenter/icf/endfassung/icf_endfassung-2005-10-01.pdf

Ellsäßer G, Böhm A, Kuhn J, Lüdecke K, Rojas G (2002) Soziale Ungleichheit und Gesundheit bei Kindern: Ergebnis und Konsequenzen aus den Brandenburger Einschulungsuntersuchungen. Kinderärztliche Praxis 73: 248–257.

Hollenweger J, Kraus der Camargo O (2011) ICF-CY: Internationale Klassifikation der Funktionsfähigkeit, Behinderung und Gesundheit bei Kindern und Jugendlichen. Bern: Huber

Schlack HG, von Kries R, Thyen U (2009) Sozialpädiatrie – Gesundheitswissenschaft und pädiatrischer Alltag. Heidelberg: Springer

Schlack HG (2013) Pädiatrische Primärprävention durch vorausschauende Beratung. Kinderärztliche Praxis 84: 86–88.

Spencer N et al (2005) Social Pediatrics. J Epidemiol Community Health 59: 106–108.

Straßburg HM, Dacheneder W, Kress W (2012) Entwicklungsstörungen bei Kindern – Grundlagen für eine interdisziplinäre Betreuung. 5. Aufl. München: Elsevier Urban und Fischer.

3 Geschichte der Sozialpädiatrie

Hans-Michael Straßburg

3.1 Einleitung

Die Anfänge einer **„sozialen Pädiatrie"** sind äußerst vielschichtig und können nachfolgend nur stichwortartig behandelt werden – insbesondere ist vor 1900 die Geschichte der allgemeinen Kinderheilkunde und die der Sozialpädiatrie nicht zu trennen. Dennoch können für das heutige Verständnis von Sozialpädiatrie (bzw. sozialer Pädiatrie) einige wesentliche Quellen identifiziert werden.

Seit Urzeiten haben die Menschen versucht, die Entwicklung ihrer Kinder in bestimmter Weise zu beeinflussen, sie zu erziehen. Wichtige Grundlagen hierzu finden sich bei vielen Schriftstellern der Antike (Plato, Aristoteles), des Mittelalters (Thomas von Aquin) und der beginnenden Neuzeit, z. B. bei A. Comenius, 1592–1670. Dieser entwickelte nicht nur Grundlagen für den heutigen Schulunterricht, sondern begründete auch „Mütterschulen" und setzte sich für die Unterweisung von „Schwachsinnigen" ein. Weitere wichtige Vertreter einer Pädagogik, die auch medizinische Gesichtspunkte berücksichtigte, waren u. a. J. J. Rousseau (1712–1778), besonders mit seinem Erziehungsroman „Emile" 1752, der Schweizer Pädagoge J. H. Pestalozzi (1746–1827), F. Fröbel (1782–1852), der 1840 den ersten Kindergarten in Bad Blankenburg/Thüringen eröffnete, und der Begründer der Diakonie J. H. Wichern (1808–1881). Als besonders wichtige europäische Vertreter seien hier noch der polnische Kinderarzt und Erzieher Janusz Korczak (1878–1942) und die italienische Ärztin und Pädagogin Maria Montessori (1879–1952), die 1907 das erste Kinderhaus in Rom errichtete, genannt.

Ein wesentlicher Ursprung der Kinderkrankenhäuser waren die vor allem in Italien und Frankreich verbreiteten Findelhäuser, in denen Säuglinge und Kleinkinder in großer Zahl aufgezogen wurden. Hier hat sich besonders Vinzenz von Paul (1581–1660) in Paris verdient gemacht, während in Deutschland wegen der hohen Sterblichkeit in Masseneinrichtungen bis ins 19. Jahrhundert uneheliche Kinder vor allem in Pflegefamilien aufwuchsen. Schon früh war bekannt, dass das Schicksal dieser Kinder einerseits von der Ernährung, am besten durch eine Amme in den ersten Lebensmonaten, und andererseits von den günstigeren Umfeldbedingungen auf dem Land im Vergleich zur Stadt abhängig war.

J. D. Frank (1745–1821) führte in seinem „System der medizinischen Polizey" die staatliche Fürsorge auch für Kinder in Deutschland ein. Hierzu gehörten u. a. Anweisungen für die Prävention von Krankheiten, die richtige Ernährung und die Unfallverhütung.

Sozialpädiatrie in der Praxis. http://dx.doi.org/10.1016/B978-3-437-31630-3.00003-1
Copyright © 2014 Elsevier GmbH. All rights reserved.

Auch der Begründer vieler verschiedener medizinischer Wissenschaftsbereiche im 19. Jahrhundert, **Rudolf Virchow** (1821–1902), kann als einer der Gründungsväter der Sozialmedizin bzw. der sozialen Pädiatrie in Deutschland angesehen werden. Er hat sich u. a. mit vielen epidemiologischen und sozialhygienischen Fragen beschäftigt, die Wasser- und Abwasserversorgung in der Reichshauptstadt Berlin organisiert und eine Begutachtung des preußischen Schulsystems vorgenommen. Von ihm stammt der Ausspruch: „Die Medizin ist eine soziale Wissenschaft und die Politik ist weiter nichts als Medizin im Großen."

Noch 1894 riet E. Henoch seinem Nachfolger O. Heubner bei der Übernahme der Abteilung für Kinderheilkunde an der Charité in Berlin, die Säuglingsabteilung wegen der dort herrschenden untragbaren hygienischen Verhältnisse und einer Mortalität von über 70 % (!) ganz zu schließen.

Ab der Mitte des 19. Jahrhunderts haben sich engagierte Bürger und in zunehmendem Maße auch Ärzte mit der medizinischen Betreuung von Risikokindern befasst. Arthur Schlossmann richtete 1897 das erste Säuglingsheim in Dresden ein, bald darauf erfolgte die Errichtung des Kaiserin-Auguste-Victoria-Hauses in Berlin-Charlottenburg. Yussuf Ibrahim förderte ab 1917 in Jena die Ausbildung zur Kinderkrankenschwester. 1892 wurde die erste Mütterberatungsstelle in Paris vor allem für Risiko- und frühgeborene Säuglinge eingerichtet.

3.2 Sozialpädiatrie um 1900

Ab 1900 und besonders in der Zeit des Ersten Weltkriegs erlebte die Kinderheilkunde in Deutschland einen enormen Aufschwung – bis 1920 gab es an praktisch jeder Universität eine Klinik für Kinderheilkunde. Die Gesunderhaltung der Kinder galt nun nicht mehr als Privatangelegenheit, sondern war ein gesamtgesellschaftliches politisches Anliegen.

Ab Ende des 19. und zu Beginn des 20. Jahrhunderts haben sich vor allem auch jüdische Kinderärzte besonders um den Auf- und Ausbau sozialpädiatrischer Einrichtungen verdient gemacht. Neben dem bereits erwähnten Arthur Schlossmann waren dies u. a. Hugo Neumann und Heinrich Finkelstein in Berlin und Max Taube in Leipzig. Gustav Tugendreich aus Berlin hat in einem ausführlichen Handbuchartikel zusammen mit M. Mosse am Beispiel der kindlichen Tuberkulose die Zusammenhänge von Krankheiten und sozialer Lage publiziert. Die altersstandardisierte Tuberkulosesterberate zwischen 1840 und 1970 kann als Beispiel für das Zusammenwirken äußerer Lebensumstände und medizinischer Erkenntnisse eindrucksvoll angeführt werden (➤ Abb. 3.1).

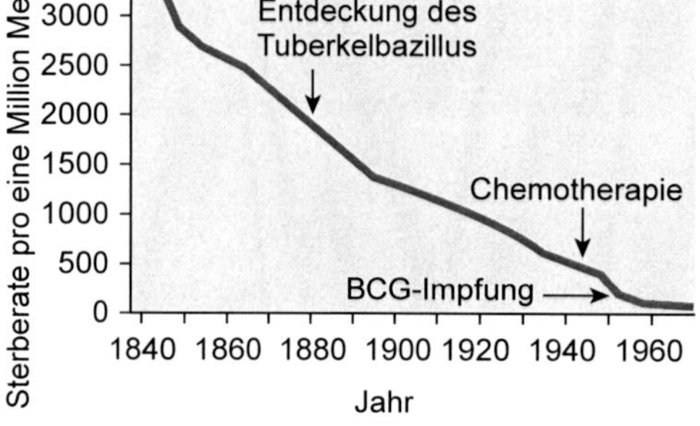

Abb. 3.1 Altersstandardisierte Tuberkulosesterberate in England und Wales zwischen 1840 und 1970 als Beispiel für das Zusammenwirken äußerer Lebensumstände und medizinischer Erkenntnisse (modifiziert nach R. Beaglehole et al. 1997)

Ähnliche Daten lassen sich für praktisch alle anderen Infektionserkrankungen im Kindesalter bzw. für die Säuglings- und Kindersterblichkeit insgesamt erstellen.
Weitere Ursprünge der heutigen Sozialpädiatrie sind umfangreiche Arbeiten zur Entwicklungspsychologie, wie sie u. a. von Alfred Binet (1857–1911), Arno Gesell (1880–1961) und Jean Piaget (1896–1980) begründet wurden, auf die hier nicht näher eingegangen werden kann.

3.3 Die Deutsche Vereinigung für Säuglingsschutz

Am 18. Februar 1909 wurde in Berlin die **„Deutsche Vereinigung für Säuglingsschutz"** gegründet. Sie sah ihre Hauptaufgabe in einer Bekämpfung der Säuglingssterblichkeit und dem Aufbau von Säuglingseinrichtungen in Deutschland. Gründungsvorsitzender war der Direktor des Kaiserin-Auguste-Victoria-Hauses, der Reichsanstalt zur Bekämpfung der Säuglingssterblichkeit (KAVH) in Berlin, Arthur Keller, der 1903 bereits die „Monatsschrift für Kinderheilkunde" begründet hatte.
Besonders während und nach dem Ersten Weltkrieg wurden die sozialen Probleme als Ursache für Krankheiten und Entwicklungsstörungen zunehmend offenkundig. Deshalb erfolgte 1920 die Umbenennung der Gesellschaft in „Deutsche Vereinigung für Säuglings- und Kleinkinderschutz" (➤ Abb. 3.2). Ihre Mitglieder, praktisch alle führenden Direktoren von Kinderkliniken und öffentlichen Gesundheitseinrichtungen, engagierten sich vor allem bei der Veröffentlichung von Ernährungs- und Gesundheitsberatungen, der Prävention von Krankheiten und Forderungen zur allgemeinen Verbesserung der Lebensumstände von Kindern. Beispielhaft seien „Die 10 Gebote für die junge Mutter" genannt, vielfältige Aufforderungen zum Stillen und zur Herstellung hygienisch einwandfreier Nahrung, Hinweise wie z. B. „Licht, Luft und Sonne dienen deinem Kind", „Vermeidet zu warme Kleidung und pralle Hitze" oder Ratschläge wie „Dein Kind ist kein Schaustück" und „Suche in regelmäßigen Abständen den Kinderarzt und die Beratungsstelle auf". Ganz wesentlich prägend für die bürgerliche Gesellschaft wirkte das weit verbreitete Büchlein des Berliner Pädiaters Adalbert Czerny (1863–1941), „Der Arzt als Erzieher des Kindes", das zwischen 1911 und dem Ende des Zweiten Weltkrieges in mehr als 10 Auflagen erschien.
1930 gründete der Direktor der Dortmunder Kinderklinik, **Stefan Engel** (1878–1968), zusammen mit E. Nassau (1888–1974) die **„KINDERÄRZTLICHE PRAXIS"** als Zeitschrift für alle praktisch tätigen Kinderärzte. Von ihm stammt der Ausspruch „Jeder Kinderarzt, der seine Aufgabe voll erfasst, muss gleichzeitig Sozialarzt sein".
Ein weiteres Beispiel herausragender sozialpädiatrischer Aktivitäten sind die Aufklärungsschriften des Mannheimer Kinderarztes Eugen Neter (1876–1966), die z. T. auch in der weit verbreiteten Zeitschrift „Gartenlaube" erschienen.
In den 1920er Jahren beschäftigte sich u. a. Meinhard von Pfaundler (1872–1947) aus München mit der kindlichen Entwicklung. Yussuf Ibrahim (1877–1953) begründete mit seinen Arbeiten über die Ursachen von Entwicklungsstörungen die deutsche Neuropädiatrie und leistete einen wichtigen Beitrag zum Aufbau des Berufsbilds der Kinderkrankenschwester.

Abb. 3.2 Frühes Merkblatt der Deutschen Vereinigung für Säuglings- und Kleinkinderschutz [1]

3.4 „Sozialpädiatrie" in der NS-Zeit

Mit Beginn der Herrschaft der Nationalsozialisten änderte sich die kinderärztliche Versorgung in Deutschland grundlegend. Über 50 % der damals tätigen Kinderärzte waren Juden, viele von ihnen hatten sich mit der Behandlung armer Bevölkerungsschichten, der Prävention und der Betreuung von Kindern mit Entwicklungsauffälligkeiten aller Art beschäftigt.
1934 ging die Deutsche Vereinigung für Säuglings- und Kleinkinderschutz in die „Reichsarbeitsgemeinschaft Mutter und Kind" auf. Hier war vor allem Fritz Rott (1878–1959) als Leiter des

Organisationsamtes Säuglings- und Kinderschutz in Berlin von Bedeutung, der sich weiter mit Aufklärungs- und Präventionsthemen befasste.

Im Vordergrund des Interesses stand nicht mehr das individuelle kranke Kind, sondern „der gesunde Volkskörper als Ganzes", Opfer war das chronisch kranke, das behinderte und deformierte Kind. Der Schritt zur sogenannten „**Euthanasie**", der brutalen Tötung behinderter Menschen, die ab 1939 stattfand und an der mehrere große Kinderkliniken und ihre Leiter beteiligt waren, war dann nicht mehr groß. Über 5.000 Kinder wurden bis 1941 systematisch in der **Aktion T4,** die u. a. von der Universitäts-Kinderklinik Leipzig gesteuert wurde, ermordet. Durch völlig unzureichende Ernährung und medizinische Versorgung sind bis 1945 noch viele weitere tausend Kinder und Jugendliche gestorben. Dem gegenüber stand die zunehmende Zusammenarbeit zwischen Kinderärzten und der Hitlerjugend sowie anderen Parteiorganisationen mit dem Ziel der „Abhärtung und Straffung" der deutschen Jugend. Eindrucksvoll wird von G. Aly in seinem Buch „Die Belasteten" die komplexe Problematik der Tötung behinderter Menschen dargestellt. Eine differenzierte medizinhistorische Aufarbeitung der sozialen Pädiatrie in Deutschland zwischen 1933 und 1945, aber auch der Rolle ehemaliger Anhänger des Nationalsozialismus in der Pädiatrie nach dem Zweiten Weltkrieg ist ein aktuelles Thema der **Historischen Kommission der DGKJ.**

3.5 Neubeginn nach 1945

Auf der Jahrestagung der Gesellschaft für Kinderheilkunde in Göttingen unter H. Kleinschmidt (1885–1977) wurde 1948 die Deutsche Vereinigung für Säuglings- und Kleinkinderschutz wieder gegründet. Auf den nun folgenden Jahrestagungen waren u. a. Säuglingsfürsorge, Schulgesundheit, Erziehungsprobleme, Ausbildung von Kinderkrankenschwestern sowie die Kinder- und Jugendfürsorge Hauptthemen. 1953 wurde die Gesellschaft in „Deutsche Vereinigung für Gesundheitsfürsorge des Kindesalters" umbenannt, und es wurde eine enge Zusammenarbeit mit dem Ausschuss für Fragen der Schulgesundheitspflege und dem Paritätischen Wohlfahrtsverbandes gepflegt.

Anfang der 1960er Jahre entstand in der Poliklinik der Universitäts-Kinderklinik in München eine selbstständige Abteilung für „Prophylaktische Pädiatrie" unter Leitung von **Theodor Hellbrügge** (geb. 1919). Von ihm wurde 1967 ein Aufsatz über „Schwerpunkte der Sozialen Pädiatrie im Kleinkindes- und Schulalter" im Deutschen Ärzteblatt publiziert. Hierin wurden die speziellen Gefährdungen, aber auch die großen Chancen der Kindheit dargestellt, die spezifische Kenntnisse des behandelnden Arztes verlangten. Wesentliche Grundlage der Erziehung des Kindes ist die Familie, wohingegen durch Massenpflege und mangelnde Zuwendung, wie dies in den Säuglings- und Kinderheimen üblich war, schwerwiegende und irreparable Gesundheitsschäden entstehen. Als besondere „Katastrophe für Kinder" wurde von ihm die Mütterarbeit bezeichnet. Weiterhin wurden Überforderungen in der Schule und fehlende Spiel- und Sportzeiten angeprangert. Zur Verbesserung der Kindergesundheit wurden regelmäßige Vorsorgeuntersuchungen in definierten Zeiten, z. B. mit einem Tag, einer Woche, 6 Wochen, 3 Monaten, 9 Monaten, 15 Monaten, 2, 3, 4 und 5 Jahren gefordert. Seine Vorstellungen wurden u. a. auch in einem umfangreichen Kapitel „Soziale Pädiatrie" im legendären Handbuch der Kinderheilkunde von H. Opitz und F. Schmid zusammen mit J. Pechstein dargestellt.

1966 wurde die Vereinigung für Gesundheitsfürsorge des Kindesalters unter dem damaligen 1. Vorsitzenden **Kurt Nitsch** in **Deutsche Gesellschaft für Sozialpädiatrie** umbenannt.

1968 wurde das **erste sozialpädiatrische Zentrum** in München als Einrichtung zur interdisziplinären Behandlung von Kindern jeden Alters mit Entwicklungsauffälligkeiten und Behinderungen gegründet.

In der Satzung der Deutschen Gesellschaft für Sozialpädiatrie in ihrer Fassung vom 20.9.1973 steht: „Die Gesellschaft will Lehre und Forschung in der Sozialpädiatrie fördern, Fragen der Anwendung pädiatrischer Erkenntnisse zur Hebung der Gesundheit allgemein, der Gesundheit von Mutter und Kind im Besonderen, klären, die Ergebnisse dieser Arbeit sammeln und den zuständigen Bundes- und Landesbehörden, den Körperschaften und den auf diesem Gebiet tätigen Einzelpersonen (Ärzten, Lehrern, Sozialarbeitern usw.) in geeigneter Form zugängig machen. Teil dieser Aufgabe ist die Aufklärung der Öffentlichkeit über die Gesundheitspflege für Mutter und Kind."
Der Direktor der Universitäts-Kinderklinik Mainz, **Ulrich Köttgen** (1906–1980), stellte 1977 in einem Aufsatz zur Lage der Sozialpädiatrie fest, dass „das soziale Denken und Handeln in Deutschland einen geringen Stellenwert hat. Säuglingssterblichkeit und Kinderunfälle sind immer noch unverhältnismäßig hoch und es fehlt die praktische Ausbildung in Gesundheits- und Krankheitsbelangen in den Sozialberufen." Er warb für eine notwendige Verbindung zwischen medizinischen und sozialen Diensten und prangerte insbesondere das fehlende Bewusstsein für soziale Themen an den deutschen Universitäten an. Dabei verwies er auf die Notwendigkeit von Kosten-Nutzen-Studien, von Auslandskontakten und intensiver Öffentlichkeitsarbeit.
Ende der 1970er und Anfang der 1980er Jahre wurde, ausgehend von einer Anfrage des Vorstands der Gesellschaft für Sozialpädiatrie, in den Gremien der Kinderheilkunde heftig über die Abgrenzung der Sozialpädiatrie von der allgemeinen Pädiatrie diskutiert. So schrieb der damalige Düsseldorfer Ordinarius, G. H. von Harnack (geb. 1917): „Sozialpädiatrie als Beschäftigung mit dem einzelnen Kind ist integrierender Bestandteil der pädiatrischen Tätigkeit schlechthin. Sozialpädiatrie befasst sich darüber hinaus mit den Auswirkungen der Umweltverhältnisse auf die Gesamtheit aller Kinder oder von Kindergruppen. Sozialpädiatrie versucht also, die Umwelt möglichst so zu gestalten, dass Kinder in optimaler Weise gefördert werden."

3.6 1970–2012

Zwischen 1970 und 1992 wurde die Deutsche Gesellschaft für Sozialpädiatrie vor allem von **Kurt Hartung** und Theodor Hellbrügge geprägt. Schwerpunkte ihrer Arbeit waren u. a. die Säuglings- und Kleinkinderfürsorge, der schulärztliche Dienst, das Sammeln epidemiologischer Daten, die Bekämpfung von Infektionskrankheiten, die Förderung von Schutzimpfungen, die Gesundheitserziehung, die Entwicklungsrehabilitation sowie der Jugendarbeitsschutz. Eine wesentliche Errungenschaft war die Einführung der **Früherkennungsuntersuchungen** mit dem gelben Untersuchungsheft, die seit 1968 von den gesetzlichen Krankenkassen in Deutschland finanziert werden. Neben einer Vielzahl von Merkblättern zu Gesundheitsfragen im Kindesalter wurden regelmäßig erscheinende Zeitschriften wie „Sozialpädiatrie in Praxis und Klinik", „Kindergesundheit" und „Kinderkrankenschwester" gegründet. Verschiedene Ausschüsse in einem wissenschaftlichen Beirat der Gesellschaft beschäftigten sich u. a. mit Impffragen, Kinderkrankenpflege und dem bundesweit flächendeckenden Aufbau der Sozialpädiatrischen Zentren, der mit der Aufnahme 1988 in das Sozialgesetzbuch V dann auch seine gesetzliche Grundlage erhalten hat. 1979 erschien erstmals das von M. Manecke herausgegebene Buch „Sozialpädiatrie – Lehrbuch für Studierende und Ärzte".
1981 gab Theodor Hellbrügge zusammen mit seinen Mitarbeitern das Buch „Klinische Sozialpädiatrie – ein Lehrbuch der Entwicklungs-Rehabilitation im Kindesalter" heraus und publizierte in der Reihe „Forschritte der Sozialpädiatrie" viele Aspekte des Fachs.

Unter den Präsidenten **Hans Georg. Schlack,** Bonn (1992–1996), und **Hubertus von Voß,** München (1996–2000), wurden vor allem die sozialpädiatrischen Zentren in Deutschland flächendeckend weiter ausgebaut und wesentliche Akzente in der Prävention und der wissenschaftlichen Darstellung der Sozialpädiatrie gesetzt, u. a. mit der Neugründung der „Kinderärztlichen Praxis" als Zeitschrift für soziale Pädiatrie und Jugendmedizin und offiziellem Organ der DGSPJ sowie der Stiftung des Stefan-Engel-Preises, der alle zwei Jahre auf der Jahrestagung für die beste wissenschaftliche Arbeit zu einem sozialpädiatrischen Thema vergeben wird. Weitere Aufgaben waren die Zusammenführung der Sozialpädiatrie in den alten und neuen Bundesländern und die enge Kooperation mit Elternselbsthilfegruppen, vor allem dem Deutschen Kindernetzwerk in Aschaffenburg. H. G. Schlack und seine Mitautoren haben in dem 1995 erschienenen Buch „Sozialpädiatrie – Gesundheit, Krankheit, Lebenswelten" den Aufgabenbereich des Fachs, wie es heute verstanden wird, erstmals zusammenfassend dargestellt.
Unter den Präsidenten Harald Bode (2001–2008) und Hans-Michael Straßburg (2009–2012) wurden u. a. die Geschäftsstelle dauerhaft nach Berlin verlegt und intensiv, wenn auch bisher ohne Erfolg versucht, eine Zusatzweiterbildung in Sozialpädiatrie zu etablieren. 1983 wurde als Symbol für die Gesellschaft eine französische Medaille mit der beschützenden Hygieia und zwei kindlichen Genien ausgewählt (➤ Abb. 3.3). Nachdem 1996 der Name der Gesellschaft in **Deutsche Gesellschaft für Sozialpädiatrie und Jugendmedizin** geändert wurde, wählte man 2002 als **Logo** ein modernes Symbol mit vielen Assoziationsmöglichkeiten aus (➤ Abb. 3.3). Es soll die drei Säulen der Sozialpädiatrie, die primäre, die sekundäre und die tertiäre Prävention, darstellen – hierunter werden auch Öffentlicher Kinder- und Jugend-Gesundheitsdienst, Sozialpädiatrische Zentren und stationäre Rehabilitation verstanden, es kann aber auch als Symbol für die Familie auf einer gemeinsamen Grundlage interpretiert werden oder für normale, abweichende und gestörte Entwicklung.

a

Deutsche Gesellschaft für Sozialpädiatrie
und Jugendmedizin e. V.

b

Abb. 3.3 a und b Die Symbole (Logos) der DGSPJ

Literatur

Aly G (2013) Die Belasteten – "Euthanasie" 1939–1945 Eine Gesellschaftsgeschichte. Frankfurt: S.Fischer

Czerny A (1911–1942) Der Arzt als Erzieher des Kindes. Wien: Franz Deuticke.

Hartung K (1983) Ein Jahrhundert Sozialpädiatrie. In: Sonderausgabe von Der Kinderarzt.

Hellbrügge T, Pechstein J (1966) Säuglingsfürsorge und Säuglingsschutz: die allgemeine Säuglingsfürsorge. In: Opitz H, Schmid F (Hrsg.) Handbuch der Kinderheilkunde, Bd. 3, Immunologie, Soziale Pädiatrie. Springer, Berlin 351–384.

Hellbrügge T (1966) Soziale und prophylaktische Pädiatrie. In: Keller B, Wiskott A. Lehrbuch der Kinderheilkunde. Stuttgart: Thieme

Hellbrügge, T (1967) Schwerpunkte der Sozialen Pädiatrie im Kleinkindes- und Schulalter. Dtsch Arztebl 64: 811–815, 872–876, 938–941.

Hellbrügge T (2006) Persönliche Mitteilungen.

Hellbrügge T, Pechstein J (1968) Die Soziale Pädiatrie als Träger der ärztlichen Vorsorge für das Kindesalter. Arbeitsmedizin, Sozialmedizin, Arbeitshygiene III: 101–106.

Hellbrügge T (1981) Klinische Sozialpädiatrie – ein Lehrbuch der Entwicklungs-Rehabilitation im Kindesalter. Berlin: Springer.

Köttgen U (1977) Zur Lage der Sozialpädiatrie. Der Kinderarzt 8: 1497–1499.

Lennert T (1994) Die Entwicklung der Berliner Pädiatrie. In: Fischer W, Hierholzer K et al. Akademie der Wissenschaften zu Berlin. Exodus von Wissenschaften aus Berlin. Berlin: W. de Gruyter: 529–551.

Manecke M (1979) Sozialpädiatrie – Lehrbuch für Studierende und Ärzte. München: Urban & Schwarzenberg.

Nissen G (2005) Kulturgeschichte seelischer Störungen bei Kindern und Jugendlichen. Stuttgart: Klett-Cotta.

Pechstein J (2006) Persönliche Mitteilungen.

Peiper A (1951) Chronik der Kinderheilkunde. Leipzig: Thieme.

Schlack, HG (1995) Sozialpädiatrie – Gesundheit – Krankheit – Lebenswelten. Stuttgart: G. Fischer.

Schlack HG (2006) Persönliche Mitteilungen.

Sick F (1988) E. Neter (1876–1966): Ein Beitrag zur pädagogischen Aufgabe des Kinderarztes. Inaugural-Dissertation an der Medizinischen Fakultät Freiburg.

Windorfer A (1983) 100 Jahre Deutsche Gesellschaft für Kinderheilkunde. In: Sonderausgabe von Der Kinderarzt.

II Grundlagen der Sozialpädiatrie

4 Soziale Lage und Gesundheit von Kindern und Jugendlichen

Harald Bode

Alle Angaben dieses Kapitels beziehen sich auf Deutschland.

4.1 Demografische Entwicklung

Im Jahr 2011 betrug die **Bevölkerungszahl** 81,8 Mio. Menschen, darunter 10,8 Mio. Kinder und Jugendliche im Alter von 0–15 Jahren. Dies entspricht 13,2 % der Gesamtbevölkerung. Der Anteil der 0–15-Jährigen an der Gesamtbevölkerung betrug 1980 noch 18,5 %.
Im Jahr 2011 wurden in Deutschland 662.685 Kinder geboren. Zwischen 1960 und 2011 ging die **Geburtenrate** von 17,2 auf 8,1 Geburten je Tausend Einwohner zurück. Damit gehört Deutschland zu den Ländern mit der niedrigsten Geburtenrate in der Welt.
Das durchschnittliche Alter einer Mutter bei der Geburt ihres Kindes lag 2007 bei 30,7 Jahren, die durchschnittliche Kinderzahl je Frau bei 1,36. Die Nettoreproduktionsrate, d. h. die durchschnittliche Zahl der lebend geborenen Mädchen einer Frau, lag 2010 bei 0,67.

Sozialpädiatrie in der Praxis. http://dx.doi.org/10.1016/B978-3-437-31630-3.00004-3
Copyright © 2014 Elsevier GmbH. All rights reserved.

Im Jahr 2012 wurden 106.815 **Schwangerschaftsabbrüche** gemeldet, davon 373 bei Mädchen unter 15 Jahren und 3.462 bei Frauen zwischen 15 und 18 Jahren.
Die Zahl der Todesfälle im Jahr 2011 betrug 852.328. Dieser Sterbeüberschuss wird zurzeit durch **Zuwanderungen** ausgeglichen. Die Nettozuwanderung lag im Jahr 2011 bei 279.000. Die Bevölkerungszahl ist 2011 erstmals seit 2003 wieder gestiegen, im Vergleich zu 2010 um 92.000.
Prognosen zur **Bevölkerungsentwicklung** müssen die prognostizierten Werte für Kinderzahl je Frau, Lebenserwartung und jährliches Wanderungssaldo berücksichtigen. Nach Vorausberechnungen wird die Zahl der Menschen unter 20 Jahren von 16,5 Mio. (20 % der Gesamtbevölkerung) im Jahr 2005 auf 10,1–11,0 Mio. (16 %) im Jahr 2050 abnehmen bei einer Gesamtbevölkerung von dann 64–70 Mio.

4.2 Familienstrukturen

Im Jahr 2011 waren 32,9 Mio. Menschen ledig, 37,8 Mio. verheiratet und 11,0 Mio. verwitwet oder geschieden.
367.816 Ehen wurden im Jahr 2011 geschlossen, davon waren bei 243.290 Ehen beide Ehepartner vor der **Eheschließung** ledig. Bei 43.511 Eheschließungen war ein Partner Ausländer. Das durchschnittliche Heiratsalter betrug bei Männern 33,3 Jahre, bei Frauen 30,5 Jahre.
Im Jahr 2011 erfolgten 187.640 **Ehescheidungen,** darunter 92.892 Ehescheidungen mit minderjährigen Kindern. Insgesamt waren 148.239 minderjährige Kinder von Ehescheidungen betroffen. Im selben Jahr waren 34 % aller Geburten nichtehelich.
2011 lebten in 8,1 Mio. Haushalten (20 %) insgesamt 13 Mio. minderjährige Kinder. Der Anteil von Personen mit minderjährigen Kindern ist in Ostdeutschland geringer als in Westdeutschland.
Die **familiären Lebensformen** sind vielfältig. Die häufigste Familienform mit 71 % waren Ehepaare mit minderjährigen Kindern. Bei 20 % waren Vater oder Mutter alleinerziehend, in 9 % lebten Lebensgemeinschaften mit minderjährigen Kindern zusammen. Etwa 6.000 Kinder unter 18 Jahren wohnten bei gleichgeschlechtlichen Eltern. In 53 % der Familien lebte ein Kind, in 36 % zwei Kinder, in 9 % drei Kinder und in 2 % mehr als 3 Kinder.
Der Anteil von Kindern mit unverheirateten Eltern lag im Jahr 2011 bei 25 %. Er ist in den neuen Bundesländern und Berlin höher als im früheren Bundesgebiet. Im gleichen Jahr wuchsen 2,2 Mio. (15 %) Kinder unter 18 Jahren bei einem allein erziehenden Elternteil auf. Im Jahr 2009 waren 42 % der rund 1,5 Mio. allein erziehenden Elternteile mit minderjährigen Kindern geschieden, 35 % waren ledig, 17 % verheiratet und getrennt lebend, 6 % verwitwet. 58 % der allein Erziehenden lebt mit einem Kind zusammen, 32 % mit zwei und 10 % mit drei oder mehr Kindern. **Allein Erziehende** sind überdurchschnittlich häufig Leistungsempfänger von Hartz IV.
Haushaltseinkommen, Lebensstandard, Lebenszufriedenheit und Wohnungssituation werden von allein erziehenden Müttern ungünstiger eingeschätzt als von verheirateten Müttern. Es gibt Hinweise auf eine schlechtere Gesundheit und kritischere Gesundheitsbewertung allein erziehender Mütter im Vergleich zu verheirateten Müttern. Allein Erziehende leiden stärker unter allgemeinen Beschwerden, Beeinträchtigungen der psychischen Befindlichkeit und ihrer gesundheitsbezogenen Lebensqualität. Allerdings ist die Lebenssituation allein erziehender Mütter heterogen. Risikofaktoren sind schwierige finanzielle Situationen und gesundheitsgefährdende soziale Belastungen.
Die Perspektiven der Kinder allein erziehender Eltern sind nicht unbedingt ungünstig. Soziale Unterstützung durch Erziehungsgeld, institutionelle Betreuung, elternfreundliche Arbeitszeiten und positive gesellschaftliche und individuelle Einstellungen sind hilfreiche Ressourcen.

Viele Kinder leben heute in **Patchwork-Familien.** Genaue Zahlen existieren nicht.
Im Jahr 2011 wurden in Deutschland 4.060 Kinder und Jugendliche adoptiert. Seit 1993 hat sich die Zahl der **Adoptionen** um über 50 % verringert. 58 % der 2011 adoptierten Minderjährigen wurden von einem Stiefelternteil oder von Verwandten als Kind angenommen. Der Anteil der Adoptionen, bei denen Adoptiveltern und Kind bis dahin nicht verwandt waren, lag bei 42 %. 48 % aller 2011 adoptierten Kinder waren unter 6 Jahre alt, 29 % zwischen 6 und 12 Jahre, 24 % 12 bis 18 Jahre. 23 % besaßen nicht die deutsche Staatsangehörigkeit.
Die abnehmende Stabilität von Familien ist ein Element des sozialen Wandels in unserer Gesellschaft.

4.3 Institutionelle Förderung und Lebenswelt

4.3.1 Schule

In Deutschland besteht, im Gegensatz zu anderen Ländern, eine Schulpflicht für alle Kinder, die vor einem Stichtag das 6. Lebensjahr vollendet haben. Die Voll-Schulpflicht erstreckt sich auf 9–10 Schulbesuchsjahre. Im Jahr 2011 besaßen 65,4 Mio. Bundesbürger einen **Schulabschluss.** 25,9 Mio. gaben einen Volks-/Hauptschulabschluss an, 15,5 Mio. einen Realschulabschluss, 5,6 Mio. den einer polytechnischen Oberschule und 19,0 Mio. eine Fachhochschul- oder Hochschulreife. Das relative Sterberisiko von Menschen mit weniger als 9 Ausbildungsjahren beträgt das 1,46-fache dessen von Menschen mit mehr als 15 Ausbildungsjahren.
Daten zum **Schulbesuch** bilden die föderale Struktur des Bildungswesens ab und verändern sich im Laufe der Jahre aus vielfältigen, u. a. demographischen und politischen Gründen. Im Schuljahr 2011/12 besuchten 8,7 Mio. Schüler allgemeinbildende Schulen. Im Bundesdurchschnitt lag im Schuljahr 2011/12 der Anteil von Schülern, die ein Gymnasium besuchten, bei 34,1 % (43,0 % in Hamburg, 25,2 % in Brandenburg). Der Anteil von Schülern ohne Hauptschulabschluss betrug 2010 bundesweit 6,5 % (5,2 % Baden-Württemberg, 12,6 % Sachsen-Anhalt). Weitere Daten zeigt Tabelle 4.1 (➤ Tab. 4.1).
Die Zahl der Schüler mit **sonderpädagogischem Förderbedarf** nimmt seit Jahren zu. Im Schuljahr 2010/2011 wurde er bei 480.000 Schülern festgestellt, d. h. bei 5 % aller Schüler. 378.000 dieser Schüler besuchten eine Förderschule, 102.000 wurden an allgemeinbildenden Schulen unterrichtet. In Schleswig-Holstein besuchten 50,7 % der Kinder mit sonderpädagogischem Förderbedarf eine Förderschule, in Saarland und in Niedersachsen 100 %.

4.3.2 Kindertagesbetreuung

Ab Kindergartenjahr 2013/14 besteht ein Rechtsanspruch auf einen Betreuungsplatz für Kinder ab dem vollendeten ersten Lebensjahr. Die Zahl der Kinder unter 3 Jahren, die in **Kindertageseinrichtungen** oder öffentlich geförderter **Kindertagespflege** betreut werden, hat in den letzten Jahren deutlich zugenommen. Im März 2012 waren es rund 558.000 Kinder (Betreuungsquote 27,6 %). Die Betreuungsquote für Kinder unter einem Jahr betrug 2,8 %, bei Kindern von 1–2 Jahren 28,4 %, bei 2–3 Jährigen 51,1 %. Der Anteil der Kinder unter 3 Jahren, die in diesen Einrichtungen betreut wurden, lag zwischen 18,4 % (Nordrhein-Westfalen) und 57,5 % (Sachsen-Anhalt).

Die Gesamtzahl der Plätze soll von 285.000 im Jahr 2007 auf 780.000 im Jahr 2013 erhöht werden (Betreuungsquote 39 %).
Seit 1996 gibt es in Deutschland einen Rechtsanspruch auf einen Platz im **Kindergarten** für Kinder vom vollendeten dritten Lebensjahr bis zur Einschulung. Die Betreuungsquoten für Kinder von 3 bis unter 6 Jahren in Kindergärten lag 2012 in Deutschland bei 93,4 % mit Werten zwischen 87,4 % (Hamburg) und 96,9 % (Thüringen). In den Bundesländern Berlin, Rheinland-Pfalz, Saarland, Hessen, Niedersachsen und in einigen Kommunen besteht Kostenfreiheit für das letzte Kindergartenjahr.

4.3.3 Wohnung

Die durchschnittliche Wohnfläche je Einwohner betrug im Jahr 2011 in Deutschland 43,0 m^2. Die Werte lagen zwischen 49,7 m^2 (Saarland) und 36,2 m^2 (Hamburg).
Für Asylbewerber beträgt die pro Person zugewiesene Wohn- und Schlaffläche 5–$10^{2.}$

4.3.4 Armut

Von den rund 82 Mio. Einwohnern Deutschlands waren im Januar 2013 41,4 Mio. erwerbstätig und 2,5 Mio. arbeitslos gemeldet. Etwa 20 Mio. Menschen bezogen eine Rente, etwa 24 Mio. hatten als Familienangehörige kein eigenes Einkommen.
Die abnehmende Stabilität von Beschäftigungsverhältnissen und zunehmende Mobilitätsanforderungen sind ein Element des sozialen Wandels in unserer Gesellschaft.
In den letzten Jahren hat das Thema Armut in Deutschland an Bedeutung zugenommen. Beim **Armutsbegriff** sind verschiedene Konzepte zu unterscheiden: **Absolute Armut** bezeichnet das Fehlen der unumgänglichen lebensnotwendigen Grundlagen (Essen, Kleidung, Wohnung etc.) und die daraus resultierende existenzielle Bedrohung. Armut in Deutschland meint dagegen die **relative Armut** im Sinne einer sozialen Benachteiligung im Vergleich zu mittlerem gesellschaftlichen Lebensstandard. Gemäß EU-Vereinbarung besteht ein Armutsrisiko, wenn weniger als 60 % des Medians des Netto-Äquivalenzeinkommens aller Personen des entsprechenden EU-Staates vorliegt. Nach dieser Definition galten 2011 in Deutschland 15,8 % der Bevölkerung als armutsgefährdet. Im Jahr 2011 lag die S80/S20-Rate, d. h. das Verhältnis des Gesamteinkommens des oberen zu dem des unteren Fünftels der Einkommensverteilung bei 4,5. Der **Gini-Quotient** lag bei 29 %. Er ist eine Maßzahl für die Ungleichheit der Einkommensverteilung. Die Werte bewegen sich zwischen 0 % und 100 %. Je ungleicher die Verteilung ist, desto mehr nähert sich der Wert 100 %.
Allein erziehende Frauen erhalten fünfmal, Ausländer doppelt und Kinder und Jugendliche unter 18 Jahren ebenfalls etwa doppelt so häufig **Sozialhilfe** wie die Gesamtbevölkerung. Der Anteil der in Armut lebenden Kinder in Deutschland hat zugenommen. Im Jahr 2010 lebten 15,5 % der Kinder unter 15 Jahren in Deutschland in Familien, die Hartz-IV-Leistungen erhalten (7,3 % Bayern, 34,5 % Berlin). Das Risiko steigt bei Arbeitslosigkeit des Haushaltsvorstandes, allein erziehenden Eltern, drei oder mehr Kindern in der Familie und bei Migrationshintergrund.

4.4 Migration

Internationale Migration bedeutet nach WHO-Definition eine länger als 1 Jahr die Landesgrenzen überschreitende Wohnsitzverlagerung von Personen. Migranten stellen eine **inhomogene Gruppe** dar: Arbeitsmigranten, Studenten, Asylbewerber, Flüchtlinge, unbegleitete Minderjährige, Menschen ohne gesicherten Aufenthaltsstatus („Illegale"), Aussiedler, Eingebürgerte mit ausländischer Herkunft. Migranten unterscheiden sich in Nationalität, sozialer Lage/Schicht, Bildungsstand, Sprachkompetenz, kulturellen Besonderheiten, familiärem Kontext, Länge des Aufenthalts u. a. m. In Deutschland ist über die letzten Jahrzehnte ein Trend von der Arbeitsmigration zur Flüchtlingsmigration und zum Familiennachzug zu beobachten. Als Menschen mit Migrationshintergrund werden Personen bezeichnet, die seit 1950 nach Deutschland zugewandert sind sowie deren Nachkommen.
Die Daten zur Migration in Deutschland im Jahr 2011 geben 16,0 Mio. Menschen mit Migrationshintergrund an. Davon sind 10,7 Mio. Menschen im Ausland geboren. 7,2 Mio. Menschen haben eine ausländische, 8,8 Mio. Menschen die deutsche Staatsangehörigkeit. Die häufigsten Zuwanderungsländer sind die ehemaligen GUS-Staaten (2,4 Mio.), die Türkei (1,5 Mio.) und Polen (1,1 Mio.). Jährlich lassen sich etwa 100.000 Ausländer einbürgern.
Zwischen 1991 und 2011 sind 19,0 Mio. Menschen nach Deutschland zugewandert und 14,4 Mio. abgewandert. Im Jahr 2011 betrug die Zahl der Zuzüge 958.299, die der Fortzüge 678.969. In den letzten 10 Jahren hat die jährliche Zahl der Zuwanderer nach Deutschland zunächst abgenommen, seit 2007 wieder – zuletzt deutlich – zugenommen. Der Anteil an Ausländern in Deutschland lag Ende 2011 je nach Bundesland zwischen 2,0 % (Sachsen-Anhalt) und 13,1 % (Hamburg).
Seit 1.1.2000 erwirbt ein Kind ausländischer Eltern die deutsche **Staatsangehörigkeit** mit seiner Geburt in Deutschland, sofern ein Elternteil seit mindestens 8 Jahren rechtmäßig seinen gewöhnlichen Aufenthalt in Deutschland und eine Niederlassungserlaubnis oder eine unbefristete Aufenthaltserlaubnis hat. Kinder und Jugendliche mit Migrationshintergrund machen in Deutschland etwa 4,5 Mio. der 0–20-Jährigen aus, d. h. 31 % der Altersgruppe. Der Anteil von Kindern unter 10 Jahren mit Migrationshintergrund betrug im Jahr 2012 in Stuttgart 56 %. Als Migrantenkinder werden hier – in Anlehnung an die KiGGS-Studie – Kinder und Jugendliche definiert, die selbst aus einem anderen Land zugewandert sind und von denen mindestens ein Elternteil nicht in Deutschland geboren ist oder von denen beide Eltern zugewandert oder nicht deutscher Staatsangehörigkeit sind.
Die heutige Lebenslage vieler Migranten in Deutschland ist mit der der deutschen Bevölkerung im Jahr 1970 vergleichbar. Migration wirkt als generationenübergreifender Prozess nicht nur auf die Lebenssituation der Eltern, sondern auch ihrer Kinder. Tabelle 4.2 (➤ Tab. 4.2) zeigt die Verteilung der in der KiGGS-Studie untersuchten Migrantenkinder nach Zuwanderergruppen.
Migrantenkinder haben in Deutschland häufiger einen niedrigen **sozialen Status,** sie erfahren eine Bildungsbenachteiligung. Sie besitzen schlechtere Startchancen für ihr berufliches und soziales Leben (➤ Tab. 4.3).
Migrantenkinder sind seltener in Tageseinrichtungen (Kinder unter 3 Jahren 16 % vs. 32 % der Kinder ohne Migrationshintergrund; 3–5 Jährige 87 % vs. 96 %). Bei Schuleingangsuntersuchungen zeigen Kinder mit Migrationshintergrund vermehrt Sprachauffälligkeiten. Ausländerkinder sind an Gymnasien unterrepräsentiert sind, an Haupt- und Sonderschulen hingegen überrepräsentiert.
In der **PISA Studie** 2006 mit Schwerpunkt auf naturwissenschaftlichen Kompetenzen hatten Jugendliche mit Migrationshintergrund deutlich geringere Kompetenzen als solche ohne Migrationshintergrund, besonders, wenn beide Eltern im Ausland geboren waren. Jugendliche der 2. Generation zeigten z. T. schlechtere Leistungen als die der 1. Generation. Die Kompetenzunterschiede waren zu 40–50 % auf Unterschiede im Sprachgebrauch und in der sozialen Herkunft zurückzuführen.

Erfreulicherweise zeigen die Ergebnisse von PISA 2009, dass sich die Leseleistungen von Kindern mit Migrationshintergrund in den letzten Jahren deutlich verbessert haben. Der Ländervergleich der Leseleistungen 2010 zeigte bei der 2. Generation von Migrantenkindern allerdings zwar leicht verbesserte Leseleistungen im Vergleich zur 1. Generation. Die Leistungen waren aber weiterhin deutlich schlechter als bei Kindern ohne Migrationshintergrund. Besonders schwache Leseleistungen zeigten Kinder mit türkischem Migrationshintergrund.
Der **Integrationsindikatorenbericht** 2009 zeigte, dass der Schulerfolg der Kinder mit Migrationshintergrund sank bei geringem Bildungsniveau der Eltern, geringer Bildungsorientierung und wenn in der Familie kein Deutsch gesprochen wurde.

Tab. 4.1 Schülerinnen und Schüler an allgemeinbildenden Schulen (Sekundarstufe I) in Deutschland 2010/2011 (Quelle: Statistisches Bundesamt)

Gymnasien	34,4 %
Realschulen	26,4 %
Hauptschulen	15,9 %
Integrierte Gesamtschulen	11,0 %
Schulen mit mehreren Bildungsgängen	8,4 %
Sonstige	3,9 %

Tab. 4.2 Migrantenkinder nach Zuwanderergruppen (Quelle: KiGGS-Studie)

Zuwanderergruppen	Mutter Anteil (%)	Vater Anteil (%)
Deutschstämmige (Spät-)Aussiedler	31,7	32,4
Familiennachzug	27,8	18,1
Arbeitsmigranten	11,6	16,8
Bürger eines EU-Staates	6,9	7,1
Asylbewerber	5,7	7,3
Sonstige Gruppe	5,4	4,7
Asylberechtigte	4,1	6,1
Kriegsflüchtlinge	3,3	3,8
Kontingentflüchtlinge	2,5	2,4
Studenten	1,0	1,3

Tab. 4.3 Soziodemografische Merkmale von Migrantenkindern und Nichtmigrantenkindern (Quelle: KiGGS-Studie)

		Migrantenkinder %	Nichtmigrantenkinder %
Sozialer Status	Niedrig	53,7	22,6
	Mittel	36,3	47,1
	Hoch	10,0	30,3
Schultyp/-abschluss	Hauptschule	32,7	17,6
	Realschule	22,7	30,5
	Gymnasium	23,0	38,0
Arbeitslosigkeit	Mutter	18,1	6,5
	Vater	19,2	5,5

Wesentlich zur Verbesserung der sozialen Lage von Kindern mit Migrationshintergrund sind frühe Förderung, insbesondere der deutschen Sprache, Ganztagesschulen, Integrationskurse für Eltern; aber auch Offenheit von Erziehern, Pädagogen und Institutionen sowie der betroffenen Familien, die auch Eigenverantwortung für Gesundheit, Bildung und gesellschaftliche Teilhabe ihrer Kinder übernehmen müssen.

4.5 Gesundheitliche Lage

Die Ausgaben für Gesundheit lagen in Deutschland im Jahr 2011 bei insgesamt 293,8 Milliarden Euro. Die **Gesundheitsausgaben** betrugen damit 11,3 % des Bruttoinlandsprodukts. Auf jeden Einwohner entfielen 3.590 Euro. Umgerechnet pro Kopf lag der Anteil für Gesundheitsausgaben bei Kindern und Jugendlichen unter 20 Jahren bei 4,3 % (Jungen) bzw. 4,0 % (Mädchen) des Bruttoinlandsprodukts. Die relativen Ausgaben für die Gesundheit für Kinder und Jugendliche liegen etwa bei 80 % der Menschen im erwerbsfähigen Alter und bei etwa 25 % der über 65-jährigen. Im Jahr 2006 verursachten Kinder und Jugendliche bis 15 Jahre, die 14 % der Gesamtbevölkerung darstellen, lediglich 6,1 % der **Krankheitskosten.** Pro Kopf der unter 15-Jährigen fielen im Jahr 2006 1.260 € Krankheitskosten an. An diesen hatten Erkrankungen der Atemwege (16 %) und psychische und Verhaltensstörungen (11 %) den größten Anteil. Indikatoren der gesundheitlichen Lage sind Lebenserwartung, alters- und krankheitsbezogene Sterblichkeiten, die Häufigkeiten von Erkrankungen oder Entwicklungsstörungen und deren Folgen sowie subjektive Beschreibungen des eigenen Gesundheitszustands.

Die durchschnittliche Lebenserwartung Neugeborener ist von 1970 bis 2011 bei Jungen von 67,4 auf 77,7, bei Mädchen von 73,8 auf 82,7 Jahre angestiegen. Sie liegt im europäischen Vergleich im Mittel. Im Jahr 2011 starben 4.026 Kinder und Jugendliche unter 18 Jahren, davon 2.408 im ersten Lebensjahr. Die meisten **Todesursachen** haben ihre Gründe in der Perinatalperiode.

Im Jahr 2011 lag die perinatale Sterblichkeit (vom Beginn der 22. Schwangerschaftswoche bis zum 7. Lebenstag) bzw. bei einem Geburtsgewicht über 500 g bei 5,3 je tausend Geburten. Im Jahr 2010 trug Frühgeburtlichkeit mit 77 % zur perinatalen Sterblichkeit bei.

Die **neonatale Sterblichkeit** (Sterblichkeit Lebendgeborener in den ersten 4 Lebenswochen) ist abhängig vom Gestationsalter (➤ Tab. 4.4). Sie lag 2009 in Baden-Württemberg über alle Gewichtsklassen bei 1,1 %.

Die Säuglingssterblichkeit lag 2011 bei 3,5 Sterbefällen im ersten Lebensjahr je tausend Lebendgeborenen desselben Jahres. Diese Werte gehören weltweit zu den niedrigsten; sie sind in den letzten Jahrzehnten stark zurückgegangen. Die regionalen Daten zeigen deutliche Ungleichheiten, z. B. für die Geburtsjahrgänge 2006–2008 in Nordrhein-Westfalen 4,5/1.000, in Baden-Württemberg 3,1/1.000, in Bremen 5,2/1.000. Die Werte in den verschiedenen Kreisen unterscheiden sich z. T. erheblich. Die häufigsten Todesursachen im Kindes- und Jugendalter im Jahr 2006 zeigt Tabelle 4.5 (➤ Tab. 4.5).

Behinderungen bei Kindern und Jugendlichen sind in der Schwerbehindertenstatistik (Grad der Behinderung > 50 %) erfasst. Im Jahr 2011 gab es 165.254 Schwerbehinderte unter 18 Jahren.

Tab. 4.4 Neonatale Sterblichkeit (Quelle: Neonatalerhebung Baden-Württemberg 2009)

Gestationsalter	< 24 W	24–25 W	26–27 W	28–29 W	30–31 W	32–36 W	37–41 W	Gesamt
Neonatale Sterblichkeit	35,2 %	10,2 %	4,5 %	4,7 %	2,6 %	0,7 %	0,5 %	1,1 %

Dabei litten 40 % an einer zerebralen Störung, einer Querschnittslähmung oder einer geistig-seelischer Behinderung. Die Zahl der Anerkennungen einer Schwerbehinderung steigt nach dem 5. Lebensjahr deutlich an.

Pflegebedürftigkeit war im Jahr 2009 bei insgesamt 84.649 Kindern und Jugendlichen unter 20 Jahren anerkannt. Tabelle 4.6 (➤ Tab. 4.6) zeigt die Verteilung auf die verschiedenen Altersgruppen und Pflegestufen.

Die **Zahngesundheit** hat sich in den letzten Jahren erheblich verbessert. 12-Jährige hatten 2005 nur noch 0,7 kariöse Zähne, 1997 lag der Wert bei 1,7. 2009 hatten 60 % der 12-Jährigen und 55 % der 15-Jährigen ein kariesfreies Gebiß.

Die Häufigkeit bestimmter Gesundheitsstörungen bei der Einschulungsuntersuchung ist hoch. Sie wird durch die verwendeten Messverfahren und Normkriterien, besonders auch vom Sozialstatus beeinflusst (➤ Tab. 4.7).

Tab. 4.5 Häufigste Todesursachen im Kindes- und Jugendalter (Quelle: Statistisches Bundesamt/GbE des Bundes)

Alter	Ursachen	Anteil %
< 1 Jahr	Perinatale Probleme inkl. Früh-/Mangelgeburtlickeit	50
	Angeborene Fehlbildungen/Chromosomenanomalien	27
	Plötzlicher Kindstod	6
1–15 Jahre	Verletzungen/Vergiftungen	27
	Neubildungen	20
	Angeborene Fehlbildungen/Chromosomenanomalien	12

Tab. 4.6 Zahl der Pflegebedürftigen nach Altersgruppe und Pflegestufe (Quelle: GbE Bund)

	Stufe I	Stufe II	Stufe III	Gesamt
Unter 5 Jahre	7111	3509	764	11.384
5–10 Jahre	12.865	8322	3642	24.829
10–15 Jahre	13.487	7951	5161	26.599
15–20 Jahre	10.403	6434	5000	21.837

Tab. 4.7 Gesundheitsstörungen und Sozialstatus (Quelle: Einschulungsuntersuchung Brandenburg 2001); Angaben in Prozent (%)

Gesundheitsstörung	Sozialstatus niedrig	Sozialstatus mittel	Sozialstatus hoch
Sehstörungen	20,6	14,6	13,0
Sprachstörungen	17,0	6,6	4,1
Beeinträchtigung geistige Entwicklung	13,8	2,5	0,8
Psychomotorische Störungen	7,3	2,5	1,4
Adipositas	6,8	4,9	3,3
Hörstörungen	5,8	5,0	5,5
Einnässen, Einkoten, psychiatrische Erkrankungen	4,2	2,2	2,3
Emotionale/soziale Störungen	4,1	1,1	0,6
Atopische Dermatitis	5,8	7,5	8,5

Ihren gegenwärtigen Gesundheitszustand schätzen Jugendliche der 7.–9. Jahrgangsstufe zu etwa drei Viertel als sehr gut oder gut ein. Allerdings gaben jeweils mehr als 15 % folgende gesundheitlichen Beeinträchtigungen in den letzten 12 Monaten an: Menstruationsbeschwerden, Allergien, Knochenbrüche/Prellungen, Akne, Bronchitis.

Tab. 4.8 Risikofaktoren und Gesundheitsverhalten in Bezug auf personale und soziale Merkmale bei Kindern und Jugendlichen der KiGGS-Studie

	Junge	Mädchen	Migrationshintergrund	Niedriger Sozialstatus
Risikofaktoren				
Rauchen in der Schwangerschaft	↔	↔	k. A.	↑
Alkohol in der Schwangerschaft	↔	↔	↓	↓
Ressourcen				
personal	↑	↓	↓	↓
sozial	↓	↑	↓	↓
familiär	↑	↓	↔	↓
Gewalterfahrung	↑	↓	↑	↑
Gesundheitsverhalten				
Stillen	↔	↔	↑	↓
Mundgesundheit	↓	↑	↓	↓
Gesunde Ernährung	↓	↑	↑ ↓	k. A.
Körperliche Aktivität	↑	↓	↓	↓
Tabakkonsum	↔	↔	↓	↑
Alkoholkonsum	↑	↓	↔	↔
Elektronische Medien	↑ ↓	↑ ↓	↑ Jungen	↑

↑: häufiger/besser; ↓: seltener/schlechter; ↔: gleich (jeweils in Bezug auf die Gesamtpopulation); k. A.: keine Angaben

Tab. 4.9 Körperliche Erkrankungen und psychische Auffälligkeiten in Bezug auf personale und soziale Merkmale bei Kindern und Jugendlichen der KiGGS-Studie

	Junge	Mädchen	Migrationshintergrund	Niedriger Sozialstatus
Körperliche Erkrankungen				
Allergien	↑	↓	↓	↓
Übergewicht	↔	↔	↑	↑
Verletzungen	↑	↓	↑↓	↔
Schmerzen	↓	↑	↓	↑
spezieller Versorgungsbedarf	↑	↓	↓	↔
Psychische Auffälligkeiten				
Verhaltensauffälligkeiten	↑	↓	↑	↑
ADHS	↑	↓	↓	↑
Essstörungen	↓	↑	↑	↑

↑: häufiger/besser; ↓: seltener/schlechter; ↔: gleich (jeweils in Bezug auf die Gesamtpopulation)

4.6 Gesundheitsverhalten

Die **Impfraten** bei Kindern liegen für viele Impfungen über der epidemiologisch wünschenswerten Rate von 90 %. Dennoch existieren – mit regionalen Unterschieden – noch Impflücken. Deutschlandweit betrugen 2010 bei Schuleingangsuntersuchungen die an das Robert-Koch-Institut übermittelten Impfquoten für Diphtherie, Tetanus, Poliomyelitis 95 %, für Pertussis 90 %, Haemophilus influenzae B 93 %, für Hepatitis B 87 %, für Masern, Mumps und Röteln 96 % (Erstimpfung) bzw. 91 % (Zweitimpfung). Die Quoten betrugen für Varizellen 58 % (Erstimpfung) bzw. 24 % (Zweitimpfung), für Meningokokken C 79 % und für Pneumokokken 15 %. Besonders die Wiederholungsimpfungen nach dem 1. Lebensjahr werden nicht ausreichend angenommen.
Die Teilnahmerate an den **Früherkennungsuntersuchungen** betrug 2011 in Baden-Württemberg für die U5 97 %, für die U7 92 %, für die U7a 83 %, für die U9 87 % und für die J1 53 %. Bei 87 % der Einschüler in Brandenburg waren 2011 die U1–U8 vollständig durchgeführt, bei 95 % die U 3–U6.
Kinder und Jugendliche zeigen beim Essverhalten und Ernährungszustand, abgesehen von den ersten 2–3 Lebensjahren, dieselben ungünstigen Verzehrmuster wie Erwachsene.
Im **Bewegungsverhalten** wird eine Abnahme von Leistungen in der Gesamtkörperkoordination und eine Zunahme motorischer Defizite insbesondere bei Stadtkindern festgestellt. Kinder, die nicht regelmäßig Sport treiben, kommen häufiger aus Familien mit niedrigem Sozialstatus, Migrationshintergrund oder den neuen Bundesländern.
Zum **Sexualverhalten** Jugendlicher führt die Bundeszentrale für gesundheitliche Aufklärung (BZgA) regelmäßige Befragungen und Kampagnen durch. Die Zahl der Schwangerschaften bei unter 18-jährigen Mädchen ist von 31.500 im Jahr 1977 auf 4.219 im Jahr 2011 zurückgegangen.
Die HBSC-Studie belegt, daß zwischen 2002 und 2010 der Anteil regelmäßiger Raucher bei 15-Jährigen von 33 % auf 15 %; bei 13-Jährigen von 14 % auf 3 % abgenommen hat. An Gymnasien ist das **Rauchen** weniger verbreitet als an anderen Schultypen.
Im Bericht der Drogenbeauftragten von 2012 wird ein rückläufiger **Alkoholkonsum** von Jugendlichen und jungen Erwachsenen in Deutschland festgestellt. Im Jahr 2004 tranken 21 %, im Jahr 2011 14 % der 12–17-Jährigen 1× pro Woche Alkohol. 2011 gaben noch 15,4 % der minderjährigen Jugendlichen an, im letzten Monat bei einer Gelegenheit mindestens 5 alkoholische Getränke getrunken zu haben (Rauschtrinken). In diesem Jahr kam es bei 1 von 1.000 der 10–14-Jährigen zu einer Krankenhauseinweisung wegen akuter Alkoholintoxikation.
4,9 % der 12–17-Jährigen gaben 2011 an, in den zurückliegenden 12 Monaten illegale Drogen genommen zu haben.

4.7 KiGGS-Studie

Bundesweite epidemiologische Daten zur Gesundheit von Kindern und Jugendlichen in Deutschland wurden zwischen 2003 und 2006 im Kinder- und Jugendgesundheitssurvey des Robert-Koch-Instituts erhoben. An einer repräsentativen Stichprobe (Teilnahmequote 66,6 %, Kinder mit Migrationshintergrund 17 %, Kinder mit anerkannter Behinderung 2 %) von 17.641 Kindern und Jugendlichen im Alter von 0–17 Jahren wurden Daten zu körperlichen Beschwerden und Befindlichkeit, akuten und chronischen Krankheiten, Behinderungen, Gesundheitsrisiken und Unfällen, psychischer Gesundheit, Lebensqualität, Schutzfaktoren und Personalressourcen, Ernährung, Essstörungen und Adipositas, Gesundheitsverhalten und Freizeitaktivitäten, Medikamentenkonsum, Impfstatus, Inanspruchnahme medizinischer Leistungen, Körper-

maße, motorische Leistungsfähigkeit, Sehvermögen, Blutdruck, Blut- und Urindiagnostik sowie Schilddrüsengröße erhoben.
Erste Ergebnisse sind in einem Band des Bundesgesundheitsblattes 2007 zusammengefasst. Sie beruhen auf einer Befragung der Eltern, ab 11 Jahren auch der Jugendlichen, medizinischen Untersuchungen und Tests, einem PC-gestützten Elterninterview sowie Blut- und Urinproben der Kinder und Jugendlichen.

4.7.1 Risiko- und Schutzfaktoren der kindlichen Entwicklung

Rauchen und Alkoholkonsum in der Schwangerschaft zeigen über die Geburtsjahre 1986–2005 keinen sicheren zeitlichen Trend. 17–18 % aller Frauen gaben an, in der Schwangerschaft geraucht zu haben, 14 % konsumierten Alkohol gelegentlich, unter 1 % regelmäßig.
Kinder aus Familien mit niedrigem sozioökonomischem Status wiesen häufiger Defizite in personellen, sozialen und familiären **Ressourcen** auf (➤ Tab. 4.8).
Insgesamt 19,6 % der Jungen und 9,9 % der Mädchen waren Täter, 5,2 % bzw. 3,9 % Opfer von **Gewalthandlungen.** 82,5 % der Mädchen und 67,2 % der Jungen waren in den zurückliegenden 12 Monaten nie an einer Gewalthandlung beteiligt.

4.7.2 Gesundheitsverhalten

Die **Stillhäufigkeit** nahm von 1986 bis 2005 zu. Über alle Geburtsjahrgänge betrug die Stillhäufigkeit 76,7 %, die durchschnittliche Stilldauer 6,9 Monate, die Dauer des vollen Stillens 4,6 Monate. 22,4 % aller Kinder der KiGGS-Studie wurden mindestens 6 Monate lang voll gestillt.
29 % der 9–17-Jährigen putzten täglich einmal oder seltener die **Zähne.** 43 % der 0–2-Jährigen und 7 % der 3–6-Jährigen erhielten Arzneimittel zur Kariesprophylaxe.
Beim **Lebensmittelverzehr** wurde bei etwa der Hälfte der 3–17-Jährigen täglicher Konsum von Brot, Milch und frischem Obst angegeben, 16 % aßen täglich Schokolade, fast 20 % andere Süßigkeiten.
Die Jodversorgung der 6–17-Jährigen zeigte sich anhand des Medianwerts der Jodausscheidung im Urin auf einem relativ niedrigen Wert innerhalb der von der WHO empfohlenen Spanne.

4.7.3 Körperliche Erkrankungen

Akute Erkrankungen zeigten folgende Jahresprävalenz: Atemwegserkrankungen 88,5 %, Magen-Darm-Infekte 56,8 %, Herpes-Virus-Infektionen 12,8 %, Bindehautentzündungen 7,8 %, Harnwegsinfektionen 4,8 %.
Die Lebenszeitprävalenzen für **Kinderkrankheiten** betrugen für: Keuchhusten 8,7 %, Masern 7,4 %, Mumps 4,0 %, Röteln 8,5 %, Windpocken 70,6 %, Scharlach 23,5 %.
Die Lebenszeitprävalenz von **chronischen Erkrankungen** wurde wie folgt angegeben: allergische Erkrankungen insgesamt 22,9 %, obstruktive Bronchitis 13,3 %, Neurodermitis 13,2 %, Heuschnupfen 10,7 %, Skoliose 5,2 %, Asthma 4,7 %, Krampfanfälle 3,6 %, Diabetes mellitus 0,14 %.
Übergewicht (Bodymass-Index BMI > 90. Perzentile) bestand bei 15 %, **Adipositas** (BMI > 97. Perzentile) bei 6,3 % der 3–17-Jährigen.
Unfälle waren bei 15,9 % der 1–17-Jährigen innerhalb eines Jahres aufgetreten.

Die 3-Monats-Prävalenz von Schmerzen betrug bei 3–10-Jährigen 64,5 %, bei 11–17-Jährigen 77,6 %. Ein gesundheitsbedingter **spezieller Versorgungsbedarf** wurde für 16 % der Jungen und 11,4 % der Mädchen mitgeteilt.

4.7.4 Psychische Auffälligkeiten

Laut elterlicher Angaben im SDQ (Strength and Difficulties Questionnaire) waren 11,5 % der Mädchen und 17,8 % der Jungen verhaltensauffällig bzw. grenzwertig auffällig.
Eine **Aufmerksamkeits-Defizit-Hyperaktivitätsstörung** (ADHS) war bereits zuvor bei 4,8 % der Kinder und Jugendlichen von einem Arzt oder Psychologen diagnostiziert worden. Bei weiteren 4,9 % der Kinder und Jugendlichen bestand aufgrund der Elternangaben im SDQ der Verdacht auf eine ADHS.
Essstörungen wurden bei 21 % der 11–17-Jährigen von den Eltern beschrieben.

4.7.5 Inanspruchnahme medizinischer Leistungen

Bei 13,7 % der Kinder gaben die Eltern einen speziellen Versorgungsbedarf an (chronische Gesundheitsprobleme).
Die Kinder-Früherkennungsuntersuchungen wurden bis zur U7 bei über 90 %, bei der U8 bei 89 % und bei der U9 bei 86,4 % der Kinder in Anspruch genommen. 81 % der Kinder nahmen an allen Früherkennungsuntersuchungen U1–U9 teil, 16 % teilweise, 3 % waren nie bei einer Vorsorgeuntersuchung gewesen.
Die jährliche Inanspruchnahme von Kinder- und Jugendärzten lag bei 0–2-Jährigen bei 95,1 %, bei 1–17-Jährigen bei 25,4 %, die entsprechenden Daten für die Inanspruchnahme von Allgemeinmedizinern lauten 11,8 % bzw. 53,1 %.
Unter den Fachärzten wurden im Vorschulalter am häufigsten HNO-Ärzte, im Schulalter Augenärzte und bei 14–17-Jährigen Internisten, Hautärzte und Chirurgen aufgesucht.
Operationen hatten bei 37,1 % der 0–17-Jährigen stattgefunden, in absteigender Häufigkeit Adenotomie, Zirkumzision, Tonsillektomie, Herniotomie und Appendektomie.
Arzneimittel waren bei 50,8 % aller Kinder und Jugendlichen in den letzten 7 Tagen vor der Befragung angewendet worden. Davon waren 58 % vom Arzt verordnet, 25 % ohne Rezept selbst gekauft, 14 % stammten aus sonstigen Quellen, 2 % von Heilpraktikern.

4.7.6 Risikofaktoren für Gesundheitsverhalten und Krankheiten

Tabelle 4.8 und 4.9 (➤ Tab. 4.8, ➤ Tab. 4.9) fassen Zusammenhänge zwischen Risikofaktoren, Gesundheitsverhalten, körperlichen Erkrankungen/Befindlichkeitsstörungen und psychischen Auffälligkeiten mit personalen und sozialen Merkmalen zusammen.
Ein niedriger Sozialstatus ist vermehrt mit **Risikofaktoren für die gesundheitliche Entwicklung** und mit ungünstigerem Gesundheitsverhalten assoziiert. Für Kinder mit Migrationshintergrund gelten diese Zusammenhänge nur bei einzelnen Faktoren, für Jungen und Mädchen ist das Profil für diese Faktoren differenziert.
Mit Ausnahme von Allergien treten körperliche Erkrankungen und psychische Auffälligkeiten bei Kindern und Jugendlichen mit niedrigem Sozialstatus häufiger auf. Für Kinder und Jugendliche

mit Migrationshintergrund trifft dieser Zusammenhang nur teilweise zu. Jungen zeigten häufiger körperliche Erkrankungen und psychische Auffälligkeiten als Mädchen mit Ausnahme von Schmerzen und Essstörungen.

Literatur

Ärztekammer Baden-Württemberg. Neonatalerhebung 2009. http://www.aerztekammer-bw.de/10aerzte/05kammern/10laekbw/10service/40qualitaetssicherung/neonatalerhebung/neonatalerhebung2009.pdf

Beauftragte der Bundesregierung für Migration, Flüchtlinge und Integration (2011) Zweiter Integrationsindikatorenbericht. http://www.bundesregierung.de/Content/DE/_Anlagen/2012/01/2012-01-12-integrationsbericht.pdf?__blob=publicationFile

Bode H, Brunner M (2007) Gutachten über das Modellprojekt zur Neukonzeption Einschulungsuntersuchung. http://www.sozialministerium-bw.de/fm/1442/Gutachten_Neukonzeption_EinschU.pdf

Bode H (2001) Sozioökonomische Aspekte der Zerebralparesen. In: Heinen F, Bartens W (Hrsg) Das Kind und die Spastik – Erkenntnisse der Evidenced-based Medicine zur Zerebralparese. Bern: H. Huber.

Bos W, Hornberg S, Arnold KH, Faust G, Fried L, Lankes EM, Schwippert K, Valtin R (Hrsg.)
(2007) IGLU 2006. Lesekompetenzen von Grundschulkindern in Deutschland im internationalen Vergleich – Zusammenfassung. http://isc.bc.edu/pirls2006/brochure.html.

Bundesamt für Migration und Flüchtlinge (2013) Migrationsbericht 2011. http://www.bamf.de/SharedDocs/Anlagen/DE/Publikationen/Migrationsberichte/migrationsbericht-2011.pdf?__blob=publicationFile

Drogenbeauftragte der Bundesregierung (2012) Drogen- und Suchtbericht 2012. http://www.drogenbeauftragte.de/fileadmin/dateien-dba/Presse/Downloads/12-05-22_DrogensuchtBericht_2012.pdf

Klieme E, Artelt C, Hartig J, Jude N, Köller O, Prenzel M, Schneider W, Stanat P (Hrsg.) (2010) Pisa 2009. Bilanz nach einem Jahrzehnt. Zusammenfassung. http://pisa.dipf.de/de/de/pisa-2009/ergebnisberichte/PISA_2009_Zusammenfassung.pdf

Lampert T, Kurth BM (2007) Sozialer Status und Gesundheit von Kindern und Jugendlichen. Ergebnisse des Kinder- und Jugendgesundheitssurveys (KiGGS). Dtsch Aerztebl 104: B2.593–2.598.

Prenzel M, Artelt C, Baumert J, Blum W, Hammann M, Klieme E, Pekrun R-PISA-Konsortium Deutschland (Hrsg.) (2007) PISA 2006 in Deutschland. Zusammenfassung. http://pisa.ipn.uni-kiel.de/Zusfsg_PISA2006_national.pdf

Razum O, Breckenkamp J (2007) Kindersterblichkeit und soziale Situation: Ein internationales Vergleich. Dtsch Aerztebl 104: B2.599–2.605.

Robert-Koch-Institut (Hrsg) (2005) Gesundheitsberichterstattung des Bundes Heft 4 – Armut bei Kindern und Jugendlichen.

Robert-Koch-Institut (2004) Schwerpunktbericht der Gesundheitsberichterstattung des Bundes. Gesundheit von Kindern und Jugendlichen.

Robert-Koch-Institut (2003) Gesundheitsberichterstattung des Bundes Heft 14 – Gesundheit allein erziehender Mütter und Väter.

Röhrling G, Czypionka T, Riedel M (2010) Gesundheitsausgaben und -indikatoren im Kindesalter. Monatsschrift Kinderheilkunde 158: 338–347

Internetlinks

www.statistik-bw.de
www.destatis.de
www.gbe-bund.de
www.kiggs.de
http://hbsc-germany.de
www.bmfsfj.de

5 Epidemiologie

Robert Schlack

5.1 Einleitung

Als Querschnittsdisziplin in der Kinder- und Jugendmedizin richtet die Sozialpädiatrie ihren Fokus über die Versorgung einzelner Patienten und ihrer Familien hinaus auf die äußeren Bedingungen von Krankheit und Gesundheit und versteht sich damit zunehmend als Gesundheitswissenschaft des Kindes- und Jugendalters (HG Schlack, 2008). Dabei wird der Tatsache Rechnung getragen, dass die Bedeutung chronisch somatischer Krankheiten und psychischer Auffälligkeiten

Sozialpädiatrie in der Praxis. http://dx.doi.org/10.1016/B978-3-437-31630-3.00005-5
Copyright © 2014 Elsevier GmbH. All rights reserved.

im Kindes- und Jugendalter seit den 1950er Jahren zugenommen hat, ein Phänomen, das unter dem Begriff „neue Morbidität" (Palfrey et al. 2005) Bekanntheit erlangt hat. Der offensichtliche Zusammenhang des Morbiditätswandels mit den veränderten und sich verändernden Lebensbedingungen erweitert den Blick über individual-kurative Behandlungsansätze hinaus auf die bevölkerungsmedizinische Ebene, hin zu Fragen der Prävention und Intervention. Welche Rolle kommt dabei der Epidemiologie zu? Ohne eine solide epidemiologische Datengrundlage zur Häufigkeit unterschiedlicher Erkrankungen, zu Fragen der Zu- oder Abnahme von Erkrankungshäufigkeiten, zur Frage, welche Gruppen oder Individuen bevorzugt erkranken oder welche Risiko- oder Protektivfaktoren mit einer Gesundheitsstörung assoziiert sind, ist es kaum möglich, Ressourcen im Gesundheitssystem zielgerichtet und effektiv zu einzusetzen. Epidemiologische Erkenntnisse spielen jedoch auch im klinischen Alltag eine wichtige Rolle, da der Prozess der Diagnostik oder das Stellen von Prognosen zu einem nicht unwesentlichen Teil auf Populationsdaten beruht. Die Epidemiologie kann damit als eine Wissenschaft am Schnittpunkt von öffentlicher Gesundheit (Public Health) und klinischer Praxis bezeichnet werden.

Ziele und Aufgaben der Epidemiologie (nach Last, 2001)

1. Ursachenerkennung von Krankheiten (ätiologische Forschung) einschließlich der Bestimmung von Determinanten für das Erkrankungsrisiko einer Person
2. Bestimmung der Krankheitslast einer Bevölkerung oder von Bevölkerungsteilen
3. Untersuchung des Verlaufs von Gesundheitsproblemen und Bestimmung von Prognosen (z. B. Überlebenszeiten)
4. Evaluation präventiver oder therapeutischer Maßnahmen
5. Bereitstellung wissenschaftlich geprüfter Grundlagen für gesundheitspolitische Entscheidungen.

Für die Beurteilung von Daten und Zahlen zum Krankheitsgeschehen in der gesundheitswissenschaftlichen wie in der öffentlichen Diskussion sind Kenntnisse elementarer epidemiologischer Begriffe und Verfahrensweisen und Datengrundlagen unabdingbar. Ziel dieses Kapitels ist es, einen anwendungsorientierten Überblick über Datenquellen zur Kinder- und Jugendgesundheit (➤ Kap. 5.2), Maßzahlen und Risikomaße (➤ Kap. 5.3), epidemiologische Studientypen (➤ Kap. 5.4), potenzielle Fehlerquellen (➤ Kap. 5.5) und zur Frage der Kausalität epidemiologischer Studienergebnisse (➤ Kap. 5.6) zu geben. Dabei soll anhand von Praxisbeispielen die Bedeutung epidemiologischer Erkenntnisse für den klinischen Alltag, den medizinisch-gesundheitswissenschaftlichen sowie öffentlichen Diskurs aufgezeigt werden. Das Kapitel schließt mit einem Exkurs zu ethischen Fragen in der epidemiologischen Forschung mit Kindern und Jugendlichen (➤ Kap. 5.7).

5.2 Epidemiologische Datenquellen im Bereich der Kinder- und Jugendgesundheit

Wichtige Datenquellen in der Epidemiologie sind Mortalitäts- und Morbiditätsstatistiken. Mortaliätsstatistiken sind öffentlich gut zugänglich und werden jährlich vom Statistischen Bundesamt und (www.destatis.de) und den jeweiligen statistischen Landesämtern herausgegeben. Morbiditätsstatistiken wie z. B. Diagnosestatistiken der Kassenärztlichen Vereinigung, Abrechnungsdaten der Krankenhäuser oder anonymisierte Datensätze von Krankenversicherungen sind hingegen weniger gut zugänglich und vor allem wegen ihrer eingeschränkten Datenqualität für wissenschaftliche Auswertungen häufig nur bedingt geeignet. Eine herausragende epidemiologische Datenquelle in Deutschland für Informationen zur Morbidität sind daher die bevölkerungsbezogenen Gesundheitssurveys des Robert Koch-Instituts (RKI) im Rahmen des vom Bundesministerium für Gesundheit am RKI etablierten Gesundheitsmonitorings (Kurth et al., 2009). Ziel des Gesundheitsmontorings ist es, durch eine regelmäßige Wiederholung von bevölkerungsbezogenen Befragungs- und Untersuchungssurveys mit quer- und längsschnittlichen Komponenten sowohl Aussagen zu Gesundheitstrends in der Gesamtbevölkerung bzw. in Bevölkerungsgruppen als auch Kohortenvergleiche (z. B. mit Studienteilnehmern verschiedener Geburtsjahrgänge) und Analysen individueller Verläufe mit der Möglichkeit der Aufdeckung von Kausalzusammenhängen (➤ Kap. 5.5) zu ermöglichen. Eine wichtige Komponente des Gesundheitsmonitorings ist der Kinder- und Jugendgesundheitssurvey KiGGS (www.kiggs-studie.de). In der Basiserhebung von 2003 bis 2006 wurden insgesamt 17.641 Kinder und Jugendliche im Alter von 0 bis 17 Jahren körperlich untersucht, ihre Eltern und die Kinder und Jugendlichen ab 11 Jahren umfangreich befragt (➤ Kap. 4.7). Der Survey umfasste Blut- und Urinanalysen, die Erhebung von Körpermesswerten sowie die Erfassung von Eckwerten eines breites Spektrums der körperlichen, psychischen und sozialen Gesundheit (Kurth et al., 2008). Der KiGGS wird als Langzeitstudie fortgesetzt (RKI, 2011).

Abbildung 5.1 (➤ Abb. 5.1) veranschaulicht die unterschiedlichen Datenquellen in Deutschland zur Morbidität und Mortalität des Kindes- und Jugendalter. Grundsätzlich lassen sich Systeme zur vollständigen Erfassung (Vollerhebungen) von stichprobenbasierten Erhebungen und Sentinel (= Wächter)-Erhebungen, Daten der amtlichen Statistik und sogenannte Prozessdaten unterscheiden.

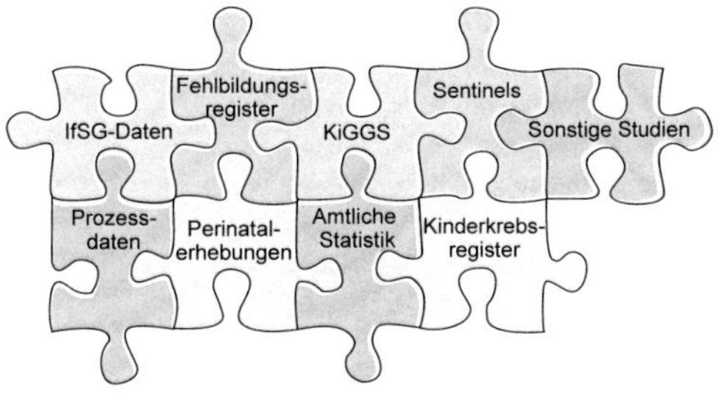

Abb. 5.1 Epidemiologische Datenquellen der Kinder- und Jugendgesundheit (modifiziert nach Kurth und Ziese 2006)

5.2.1 Kompletterfassungen

Die vollständige Erfassung einer Erkrankung wird in der Praxis nie zu 100 % erreicht. Für die vollständige Erfassung von Erkrankungen existieren in Deutschland eine Reihe verschiedener aktiver und passiver Surveillance-Systeme. Ein passives Surveillance-System empfängt Meldungen meldepflichtiger Krankheiten. Die Vollständigkeit der Meldungen wird nicht überprüft, es erfolgt auch keine regelmäßige Abfrage von Institutionen. Auch anhand der veröffentlichten Daten ist die Vollständigkeit nicht überprüfbar; ein Verstoß gegen die Meldepflicht ist in Deutschland nicht strafbewehrt. Ein Beispiel aus dem pädiatrischen Bereich ist das **Kinderkrebsregister** in Mainz, das systematisch bundesweit Daten zu Krebserkrankungen im Kindes- und Jugendalter erfasst und daraus Trend- und Überlebensanalysen sowie Analysen zu regionalen Verteilungen und Zweitmalignomen erstellt (DKKR Jahresbericht 2011). **Fehlbildungsregister** existieren aus Gründen der Finanzierbarkeit in Deutschland nur auf regionaler Ebene. Das Geburtenregister

„Mainzer Modell" (Queiser-Luft et al. 2006) ist auch für andere deutsche Regionen beispielgebend geworden. Hierbei werden sämtliche Neugeborenen einer Region in Monitorstationen nach einem standardisierten Vorgehen von speziell ausgebildeten Pädiatern klinisch und sonographisch untersucht.
Für den Bereich der Infektionskrankheiten regelt seit 2001 das **Infektionsschutzgesetz (IfSG)** die Meldung einer Vielzahl von labordiagnostisch nachgewiesenen Erregern an das Robert-Koch-Institut. Im Infektionsepidemiologischen Jahrbuch meldepflichtiger Krankheiten wird die jeweilige Jahresstatistik auf Bundeslandebene publiziert. Das aktuelle infektionsepidemiologische Geschehen wird wöchentlich bzw. monatlich im Epidemiologischen Bulletin des Robert-Koch-Instituts veröffentlicht. Im Unterschied zu den Meldungen an die Krebs- und Fehlbildungsregister sind Verstöße gegen die Meldepflicht im IfSG sanktionsbewehrt.
Bei aktiven Surveillance-Systemen erfolgt dagegen eine monatliche Abfrage bei potenziell meldenden Institutionen. Bei der **Erhebungseinheit für seltene pädiatrische Krankheiten (ESPED;** www.esped.uni-duesseldorf.de), die von der Forschungsstelle für pädiatrische Epidemiologie bei der DGKJ getragen wird, erhalten alle Chefärzte bzw. Leiter von selbstständigen Abteilungen in deutschen Kinderkliniken sowie ausgewählte Arztpraxen monatlich eine Postkarte, auf der sie gebeten werden mitzuteilen, wie oft eine von derzeit 12 Erkrankungen in ihrer Einrichtung im vorangegangenen Monat beobachtet worden ist. Die Karten sollen auch dann zurückgesandt werden, wenn keine der Erkrankungen in der Klinik beobachtet wurde. Die Rücklaufquoten für das ESPED im Jahr 2011 betrugen über 91 % (ESPED-Jahresbericht 2011).

5.2.2 Sentinel-Erhebungen

Bei Sentinel-Erhebungen werden Erkrankungshäufigkeiten von häufig auftretenden Krankheiten, vor allem Infektionskrankheiten, in einem Netzwerk ausgewählter Beobachtungspraxen erfasst. Derzeit werden im pädiatrischen Bereich z. B. Varizellen-Erkrankungen über eine Sentinel-Erhebung erfasst (www.agmv.de). Seit April 2012 trägt das Robert Koch-Institut die alleinige Verantwortung für die AGV (Arbeitsgemeinschaft Varizellen), deren Finanzierung ausschließlich aus Mitteln des Bundesministeriums für Gesundheit erfolgt. Bei der AGV beteiligen sich derzeit bundesweit 981 Meldepraxen.

5.2.3 Stichprobenbasierte Untersuchungen

Gute Stichproben gleichen der zugrundeliegenden Population (Grundgesamtheit) in möglichst vielen Merkmalen, d. h. sie sind „repräsentativ". Dabei gewährleisten Zufallsverfahren bei der Stichprobenziehung die Repräsentativität am besten. In großen, bevölkerungsrepräsentativen Studien erfolgt die Stichprobenziehung aus pragmatischen Gründen meist in mehreren Stufen. Die Stichprobenziehung für die Basiserhebung des Kinder- und Jugendgesundheitssurveys (KiGGS) erfolgte in zwei Schritten. Zunächst wurden mit einem Zufallsverfahren 167 Städte und Gemeinden unter Berücksichtigung ihrer Größe und ihrer geografischen Verteilung ausgewählt. In einem zweiten Schritt erfolgte eine Zufallsziehung von Personen für jede Geburtskohorte (0 bis 17 Jahre) aus dem Melderegister proportional wurde die Stichprobe (Einzelheiten bei Kamtsiuris et al., 2007). Die Stichprobe ist groß genug, um auch in Teilgruppen Unterschiede von 25 % für ein Gesundheitsproblem mit etwa 1 % Prävalenz statistisch valide ($p < 0{,}05$) zu entdecken. Zu Fehlermöglichkeiten beim Schluss von der Stichprobe auf die Grundgesamtheit siehe Abschnitt (➤ Kap. 5.5.1).

5.2.4 Amtliche Statistik und Prozessdaten

Amtliche Daten zu Geburts- und Sterbefällen werden in Deutschland auf Grundlage der Personenstandsgesetzgebung beim Statistischen Bundesamt registriert und jährlich im Statistischen Jahrbuch veröffentlicht. Auch die Krankenhausdiagnosestatistik gehört zu den amtlichen Daten. Als Prozessdaten bezeichnet man hingegen Daten, die bei Aktivitäten von Akteuren im Gesundheitswesen anfallen, deren Erhebungszweck jedoch nicht primär wissenschaftlich ist. Hierunter fallen z. B. Krankenkassendaten. Häufigkeitsangaben aus amtlichen Statistiken oder Prozessdaten werden auch als **administrative Prävalenzen** bezeichnet. Im Jahr 2013 sorgte der Arztreport der Barmer GEK (Barmer GEK, 2013) mit dem Schwerpunkt Aufmerksamkeitsdefizit-/Hyperaktivitätsstörungen (ADHS) für Diskussion. So wurde von den Herausgebern im Zusammenhang mit einer Zunahme ambulanter F90-Diagnosen bei 0–19-Jöhrigen um etwa 50 % im Zeitraum von 2006–2011 von ADHS als „Modediagnose" gesprochen, was umgehend auf Widerspruch von Kinder- und Jugendpsychiatern und des Berufsverbands der Kinder- und Jugendärzte traf (http://www.zentrales-adhs-netz.de/regional/aktuelles/article/arztreport-2013-der-barmer-gek.html). Eine nicht unerhebliche Einschränkung der Daten besteht darin, dass die Stichprobe ausschließlich aus Versicherten der Barmer-GEK besteht. Es ist zumindest fraglich, inwieweit die Zahlen trotz Adjustierung für Alter und Geschlecht tatsächlich bevölkerungsrepräsentativ sind und nicht die Besonderheiten der kassenspezifischen Versichertenstruktur widerspiegeln. Verallgemeinernde Aussagen aus derartigen Berichten sind daher mit Vorsicht zu betrachten. Für Deutschland repräsentative Ergebnisse zur Ab- oder Zunahme von diagnostizierter ADHS sind im Juli 2014 aus Wiederholungbefragung der KiGGS-Studie (Erhebungszeitraum 2009–2012) zu erwarten (RKI, 2011), die sich dann mit den Ergebnissen verschiedener Versichertenreports vergleichen lassen.

Diskussion: Nehmen Entwicklungsstörungen bei Kindern und Jugendlichen zu?

Nach einer Studie der Betriebskrankenkassen in Bayern in Zusammenarbeit mit dem Berufsverband der Kinder- und Jugendärzte zeigten etwa 20 % einer Stichprobe von insgesamt 7.500 Kindergartenkindern im Alter von 4 bis 5 Jahren Verhaltensauffälligkeiten (PaedNetz Bayern, 2013). In dieser Studie wurden von Kindergärtnerinnen und Kindergärtnern Fragebogen ausgefüllt, die im Rahmen der Vorsorgeuntersuchungen U8 und U 9 von niedergelassenen Pädiatern beurteilt wurden. Tatsächlich entspricht es der Wahrnehmung von praktisch tätigen Kinderärzten, dass Entwicklungsstörungen, Hyperaktivitätsprobleme und seelische Beeinträchtigungen in der täglichen Praxis eine immer größere Rolle spielen. Ob dafür jedoch eine tatsächliche Zunahme psychischer Störungen in der Kinder- und Jugendlichenbevölkerung verantwortlich ist, ist ungewiss. Der US-amerikanische „National Health Interview Survey" der Centers for Disease Control (CDC) berichtet eine Zunahme im untersuchten 10-Jahres-Zeitraum (2000–2010) von 16 % für „childhood disabilities" bei 0 bis 17-Jährigen, worunter körperliche Probleme wie Seh- und Hörstörungen, Asthma, muskuloskelettale Erkrankungen, angeborene Krankheiten und Verletzungen sowie Intelligenzminderung, emotionale und Verhaltensprobleme, ADHS, Sprach- und Lernschwierigkeiten und sonstige Entwicklungsprobleme verstanden wurden. Interessanterweise nahmen die körperlichen Gesundheitsstörungen ab,

während neurologische Entwicklungsstörungen und psychische Gesundheitsstörungen zunahmen – dabei war der größte Anstieg bei Mittelschichtsfamilien zu verzeichnen (Kemp, 2013). Ein Teil des Anstiegs bei den neurologischen Entwicklungsstörungen wird von den Forschern auf die Zunahme von Autismus-Spektrum-Störungen zurückgeführt. Noch drastischer fällt der Befund einer weiteren amerikanischen Repräsentativstudie über den Anstieg elternberichteter ADHS-Diagnosen aus: hier wurde für einen 4-Jahres-Zeitraum (2003–2007) ein Prävalenzanstieg von 21,8 % berichtet (Visser et al., 2010). Bevölkerungsrepräsentative Zahlen für Deutschland sind frühestens 2014 aus der ersten Wiederholungsbefragung der KiGGS-Studie zu erwarten.
Was sind mögliche Gründe für diese Anstiege? Unwahrscheinlich ist, dass sich die realen Prävalenzen von psychischen und Entwicklungsstörungen innerhalb kurzer Zeit derart stark verändert haben. Als eine der möglichen Ursachen kommen z. B. veränderte Lebensbedingungen in Betracht. So konnte in einer Analyse mit den KiGGS-Daten gezeigt werden, dass die Prävalenz von emotionalen und Verhaltensauffälligkeiten und psychosomatischen Problemen in Trennungsfamilien (Eineltern- und Stieffamilien) bei Jungen und Mädchen bis zu dreimal höher war als bei ihren Geschlechts- und Altersgenossen in Kernfamilien mit beiden leiblichen Eltern (R Schlack, 2013). Allerdings sind erhöhte Symptomprävalenzen nicht mit einer tatsächlich höheren Inzidenz und Prävalenz manifester psychischer und Entwicklungsstörungen gleichzusetzen. Die Diagnostik solcher Störungen ist aufwändig und kann im Rahmen großer, bevölkerungsbezogener Studien oft nicht durchgeführt werden. Wahrscheinlicher ist, dass sich neben gestiegenen Erwartungen und Anforderungen an die Kinder und eingeschränkten Bewegungsmöglichkeiten (z. B. im Zusammenhang mit hypermotorischen Verhaltensweisen bei Jungen) eine veränderte prädiagnostische und diagnostische Aufmerksamkeit für Symptome bemerkbar macht (sogenannte „Awareness-Effekte"). Hierbei spielen nicht zuletzt auch Diagnosekriterien eine wichtige Rolle. Der amerikanische Psychiater Allen Frances, der die vierte Ausgabe des Diagnostischen und Statistischen Manuals der American Psychiatric Association (APA) koordiniert hat, führt den starken Anstieg der Prävalenzraten von ADHS, Autismus-Spektrum-Störungen und bipolaren Störungen bei Kindern auf eine Veränderung der Diagnosekriterien beim Übergang vom DSM-III-R zum DSM-IV zurück (Frances, 2013). Frances spricht in diesem Zusammenhang selbstkritisch von drei falschen Epidemien, die dadurch ausgelöst wurden.

5.2.5 Perinatalerhebungen

Perinatal- und Neonatalerhebungen werden von den Ärztekammern auf Bundesland- bzw. Kammerbezirksebene organisiert und sind primär ein Instrument der Qualitätssicherung in der geburtshilflichen Versorgung. Fehlende regionale Vernetzungen erschweren hier jedoch die Auswertungen. Ein sehr umfangreiches internetbasiertes Angebot von Gesundheitsdaten, sowohl von amtlichen als auch von Prozessdaten, bietet das Informationssystem Gesundheitsberichterstattung (IS-GBE) des Bundes (erhältlich unter www.gbe-bund.de).

5.3 Epidemiologische Maßzahlen und Risikomaße

5.3.1 Maße der Erkrankungshäufigkeit (Morbidität)

Prävalenz

Die Prävalenz ist definiert als die Zahl der erkrankten Personen innerhalb einer Population geteilt durch die Gesamtzahl der Population (siehe Gleichung 1 im Kasten). Je nach dem betrachteten

Zeitraum lässt sich zwischen Punkt-, Perioden- und Lebenszeitprävalenz unterscheiden. Die Punktprävalenz stellt dabei die Prävalenz der Erkrankung zu einem bestimmten Zeitpunkt dar, die Periodenprävalenz die Häufigkeit eines Gesundheitsproblems innerhalb eines definierten Zeitraums (z. B. letzte Woche, letzte 4 Wochen, letzte 12 Monate), die Lebenszeitprävalenz die Häufigkeit, mit der eine Erkrankung oder ein Gesundheitsproblem jemals im Leben der Angehörigen einer bestimmten Population aufgetreten ist.

Inzidenz

Die Inzidenzrate ist definiert als die Rate der Neuerkrankungen in einem bestimmten Zeitraum in einer Population unter Risiko. Konkret bedeutet dies: Im Nenner steht die Anzahl der Personen, die die Krankheit (theoretisch) bekommen können, im Zähler befinden sich diejenigen, die in einem bestimmten Zeitraum tatsächlich erkranken (siehe Gleichung 2 im Kasten). Wichtig ist, dass jedes Individuum aus dem Nenner prinzipiell die gleiche „Chance" hat, in den Zähler zu wandern. Dies erscheint trivial; in der Praxis ist jedoch zu beachten, dass z. B. bei Erkrankungen mit Geschlechtsspezifik wie Hodenkrebs keine Frauen in den Nenner aufgenommen werden. Die Erhebung von Inzidenzraten ist in der Forschungspraxis jedoch nur mit großem Aufwand im Rahmen von Längsschnittstudien möglich.

Beziehung zwischen Prävalenz und Inzidenz

Prävalenz- und Inzidenzdaten haben unterschiedliche praktische Bedeutungen, z. B. für gesundheitspolitische Maßnahmen. Prävalenzraten als Maß der Auftretenshäufigkeit können Anhaltspunkte für die Versorgungsplanung geben, während eine niedrigere Inzidenzrate Ausdruck für den Erfolg eines Präventionsprogramms sein kann. Es besteht jedoch auch eine Beziehung zwischen Prävalenz und Inzidenz: Im Zustand eines angenommenen Fließgleichgewichts (d. h. Neuerkrankungen und Remissionen halten sich die Waage) gilt die Gleichung 3 im Kasten.

Von Bedeutung ist dies bei der Beurteilung von Prävalenzraten chronischer Erkrankungen. So können Fortschritte in der Therapie die Mortalitätsraten bestimmter Erkrankungen senken und so zu einem höheren Vorkommen dieser Krankheiten führen. So senkt z. B. ein gutes Diabetesmanagement die spezifische Mortalität unter Diabetikern und erhöht somit deren proportionalen Anteil in der Gesamtbevölkerung.

Maße der Krankheitshäufigkeit

1. Prävalenz (in %): Zahl der Krankheitsfälle in einer Population zu einem definierten Zeitpunkt × 100 ÷ Gesamtzahl der Population
2. Inzidenz (pro 1000): Anzahl der Neuerkrankungen in einer Population innerhalb eines definierten Zeitraums × 1.000 ÷ Anzahl der Personen unter Risiko
3. Beziehung zwischen Inzidenz und Prävalenz: Prävalenz = Inzidenz × Krankheitsdauer

5.3.2 Maßzahlen der Sterblichkeit

Mortalität

Die Sterblichkeitsziffer oder Gesamtmortalität bezeichnet die Mortalitätsrate aller Todesursachen pro 1.000 Einwohner der Gesamtbevölkerung. Die spezifische Mortalitätsrate bezeichnet dagegen die Mortalitätsrate einer Bevölkerungsgruppe. Die altersspezifische Mortalitätsrate z. B. von Kindern unter 5 Jahren errechnet sich aus der Gesamtzahl von Todesfällen von Kindern unter 5 Jahren innerhalb eines Jahres im Zähler sowie der Gesamtzahl aller Kinder unter 5 Jahren in der Bevölkerung (zur Jahresmitte), multipliziert mit 1.000. Bei der ursachenspezifischen Mortalitätsrate ist die Todes- oder Mortalitätsrate auf eine bestimmte Krankheit eingegrenzt. Ein Beispiel für eine sowohl alterspezifische als auch ursachenspezifische Mortalitätsrate wäre die Jahresmortalitätsrate für Leukämie bei Kindern unter 5 Jahren. Der Zeitraum, für den eine Mortalitätsrate angegeben werden kann, ist arbiträr. Üblich sind Ein-, Fünf- oder Zehnjahreszeiträume.

Die Säuglingssterblichkeit beschreibt die Rate an Todesfällen im ersten Lebensjahr bezogen auf die Anzahl der Lebendgeburten einen Jahrgangs und gilt als Maß für die Qualität der medizinischen Versorgung eines Landes. Bei einer plötzlichen Zu- oder Abnahme der Geburtenrate innerhalb eines Jahrgangs, wie z. B. nach der gesellschaftspolitischen Wende in der damaligen DDR, ist diese Kennzahl fehleranfällig: Bei einer plötzlichen Abnahme wird die Säuglingssterblichkeit über-, bei einer plötzlichen Zunahme unterschätzt. Während sich die Säuglingssterblichkeit in der alten Bundesrepublik zu Beginn 1970er Jahre auf einem relativ hohen Niveau befand (etwa vergleichbar mit der Bulgariens oder Griechenlands), ist sie seither kontinuierlich gesunken. Aktuelle Daten der amtlichen Statistik zeigen, dass es jedoch große regionale Unterschiede bei der Säuglingssterblichkeit in Deutschland gibt (➤ Abb. 5.2). So war die Säuglingssterblichkeit im Jahr 2011 in Bremen mit 6,5 Todesfällen auf 1.000 Geburten fast dreimal so hoch wie in Sachsen oder Thüringen mit 2,3/1.000 Geburten. Bemerkenswert ist, dass sich nach wie vor in den neuen Bundesländern eine deutlich geringere Säuglingssterblichkeit findet als in den alten Bundesländern. Schon zu Vorwendezeiten war die Säuglingssterblichkeit in der damaligen DDR erheblich niedriger als in der alten Bundesrepublik.

Bei der perinatalen Sterblichkeit stehen alle Totgeburten (sofern sie ein Geburtsgewicht von mindestens 500 Gramm erreicht haben) sowie innerhalb der ersten 7 Lebenstage verstorbenen

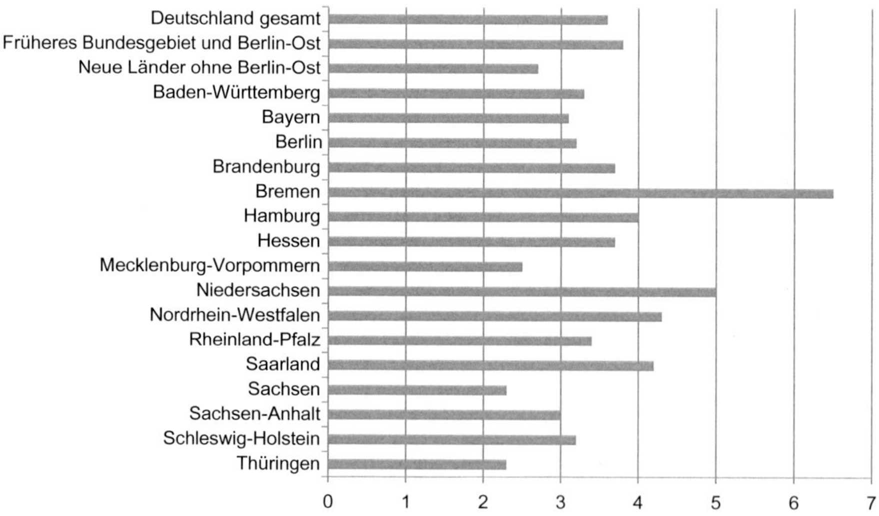

Abb. 5.2 Säuglingssterblichkeit in der Bundesrepublik Deutschland im Jahr 2011 (Todesfälle/1.000 Geburten). Quelle: Statistisches Bundesamt (2013). Statistik der natürlichen Bevölkerungsbewegung 2012

Lebendgeborenen im Zähler und die Gesamtzahl aller Lebend- und Totgeborenen im Nenner. Internationale Vergleiche dieser Kennziffer sind erschwert, da die Definitionen von Totgeburten von Land zu Land unterschiedlich sind. Während in Deutschland das Geburtsgewicht ausschlaggebend ist, erfolgt z. B. in Ungarn, England, Norwegen und Dänemark die Registrierung abhängig vom Gestationsalter des Fötus.

Die häufigsten Todesursachen im ersten Lebensjahr sind Erkrankungen infolge perinataler Probleme sowie plötzlicher Kindstod (SIDS), im Alter von 5–15 Jahren stehen Unfälle, Fehlbildungen und Krebserkrankungen im Vordergrund. Unfälle mit Todesfolge sind in der Todesursachenstatistik des Statistischen Bundesamts nach ICD-Klassifikation dokumentiert. Mit der von der EU initiierten Injury Database (IDB) sollen darüber hinaus Häufigkeit, Hergang und Art von Verletzungen (inkl. Gewaltereignissen und Selbstverletzungen) dokumentiert werden. Die IDB wird in Deutschland derzeit modellhaft in einigen Regionen Brandenburgs, Sachsens, Nordrhein-Westfalens und Niedersachsens aufgebaut (Elsässer, 2011).

5.3.3 Absolutes und relatives Risiko

Für die Berechnung von Risiken werden im Zähler die Anzahl der Ereignisse und im Nenner die Gesamtzahl des interessierenden Kollektivs geführt. Das absolute Risiko drückt damit eine Wahrscheinlichkeit aus. Die Maßzahl zur Abschätzung des absoluten Risikos, unter Exposition an einer bestimmten Krankheit zu erkranken, ist die Inzidenz (➤ Kap. 5.3.1). Das relative Risiko (auch Risk Ratio oder RR) drückt dagegen das Verhältnis zweier Wahrscheinlichkeiten aus: unter Exposition zu erkranken sowie bei Nicht-Exposition zu erkranken. Das relative Risiko kann Werte größer und kleiner als 1 annehmen. Ist der Wert gleich 1, besteht kein Anhaltspunkt für die Annahme, dass ein Zusammenhang zwischen der Erkrankung und dem interessierenden Risiko besteht. Ein RR > 1 deutet auf einen positiven (möglicherweise kausalen) Zusammenhang zwischen Exposition und Erkrankung hin, ein RR < 1 deutet auf eine negative Assoziation, möglicherweise sogar einen protektiven Effekt.

Anders als in Kohorten- oder Interventionsstudien kann das relative Risiko in Fall-Kontroll-Studien nicht direkt berechnet werden, da weder die Inzidenz der exponierten noch die der nichtexponierten Population bekannt sind. Jedoch steht mit der **Odds Ratio** ein Parameter für die Beschreibung einer Risikoerhöhung bzw. Risikoreduktion zur Verfügung, der insbesondere bei seltenen Erkrankungen einen guten Schätzwert für das relative Risiko darstellt. Der Begriff der Odds (=engl.: Chancen) stammt aus dem Wettgeschäft. Bei der Bestimmung der Odds wird die Wahrscheinlichkeit des Eintretens eines Ereignisses zur Wahrscheinlichkeit seines Nichteintretens ins Verhältnis gesetzt, z. B. die Wahrscheinlichkeit, dass Pferd A das Rennen gewinnt (= 60 %) zur Wahrscheinlichkeit, dass Pferd A das Rennen nicht gewinnt (40 %). Die Odds berechnen sich dann auf 60 %/40 % = 1,5. Bei der Odds Ratio werden zwei Odds zueinander ins Verhältnis gesetzt. Bezogen auf ein Erkrankungsrisiko lässt sich die Odds Ratio formulieren als die Chance, dass eine exponierte Person an einer bestimmten Krankheit erkrankt, ins Verhältnis gesetzt zur Chance, dass eine nichtexponierte Person erkrankt. Für seltene Erkrankungen, d. h. bei geringen Erkrankungswahrscheinlichkeiten sind OR und RR nahezu identisch. Bei hohen Erkankungswahrscheinlichkeiten liegen jedoch Odds-Ratios deutlich höher als die zugehörigen relativen Risiken. Beträgt z. B. die Wahrscheinlichkeit einer Gesundheitsstörung ohne Exposition 5 % und mit der Exposition 20 %, ist das relative Risiko 0,2/0,05=4 und das zugehörige Odds Ratio 0,25/0,053=4,7. Beträgt jedoch die Erkrankungswahrscheinlichkeit mit Exposition 90 % und ohne Exposition 60 %, resul-

Tab. 5.1 Vergleich von Risiken und Odds (nach Davies et al., 1998)

Risiko	Odds
0,05 oder 5 %	0,053
0,1 oder 10 %	0,11
0,2 oder 20 %	0,25
0,3 oder 30 %	0,43
0,4 oder 40 %	0,67
0,5 oder 50 %	1,0
0,6 oder 60 %	1,5
0,7 oder 70 %	2,3
0,8 oder 80 %	4,0
0,9 oder 90 %	9,0
0,95 oder 95 %	19,0

tieren ein RR von 1,5 und ein OR von 6,0 (vgl. Tabelle 1). Eine Interpretation des Odds Ratio als relatives Risiko, wie sie häufig vorgenommen wird, wird eher keine Auswirkungen auf die qualitative Bewertung der Studienergebnisse haben. Vorsicht ist jedoch bei der quantitativen Bewertung geboten, da bei hohen Ausgangsrisiken und großen Effektstärken eine deutliche Überschätzung des relativen Risikos durch das Odds Ratio auftritt (Davies et al., 1998).

5.3.4 Attributables Risiko

Das attributable Risiko oder Überschussrisiko ist der Anteil eines Erkrankungsrisikos, der einer bestimmten Exposition zuzuschreiben ist. Damit steht das attributable Risiko für die im Idealfall maximal erreichbare Senkung eines Risikos, wenn es gelänge, die Expositionsquelle vollständig auszuschalten und spielt eine besondere Rolle in der klinischen Praxis sowie im Public-Health-Bereich. Für den klinischen Bereich interessiert besonders, welcher Anteil der Gesamtexposition bei exponierten Personen der entsprechenden Exposition angelastet werden kann. Dazu wird die Differenz der Inzidenzen der exponierten Gruppe und der nichtexponierten Gruppe in den Zähler und die Inzidenz der exponierten Gruppe in den Nenner genommen. Interessiert jedoch das attributable Risiko einer Gesamtpopulation – und dies ist häufig der Fall, wenn nach den Potenzialen präventiver Maßnahmen gefragt wird – so ist für die Berechnung eines Präventionspotenzials als Anteilswert die Inzidenz der nichtexponierten von der Inzidenz der Gesamtbevölkerung zu subtrahieren und durch die Inzidenz der Gesamtbevölkerung zu teilen.

5.4 Epidemiologische Studientypen

Abhängig vom Studien- bzw. Erkenntnisinteresse, aber auch von forschungspraktischen Überlegungen wie Durchführbarkeit oder Finanzierbarkeit, finden bei epidemiologischen Untersuchungen verschiedene Studientypen Anwendung. Grundlegende Studiendesigns der **deskriptiven Epidemiologie** sind Korrelations- und Querschnittsstudien, Fall-Kontroll-Studien, Kohortenstudien und Interventionsstudien lassen sich der **analytischen Epidemiologie** zuordnen.

5.4.1 Korrelationsstudien

Bei einer Korrelationsstudie (auch: ökologische Studie) werden Merkmale der Gesamtbevölkerung wie Alter, Geschlecht, Ernährungsgewohnheiten, gesundheitsbezogenes Risikoverhalten wie Rauchen, Alkohol- oder Substanzkonsum u. a. m. in Beziehung zu interessierenden Inzidenz- oder Prävalenzdaten gesetzt. Friedmann (1967) untersuchte den Zusammenhang zwischen Mortalität durch koronare Herzerkrankungen und Zigarettenrauchen in den USA. Hierzu wurden die Sterberaten aus 44 US-Staaten mit dem Pro-Kopf-Verkauf von Zigaretten korreliert. Dabei zeigte sich, dass die Mortalitätsrate für Herz-Kreislauf-Erkrankungen in den Bundesstaaten mit den höchsten Verkaufszahlen am höchsten und in den Staaten mit den niedrigsten Verkaufszahlen am niedrigsten war. Damit trug diese Untersuchung zur Generierung der Hypothese bei, dass das Rauchen ein wesentlicher Risikofaktor für einen fatalen Verlauf von koronaren Herzkrankheiten ist, was später durch eine Vielzahl analytisch-epidemiologischer Studien bestätigt wurde.
Mit dem Design einer Korrelationsstudie lassen sich auch die Ergebnisse epidemiologischer Studien hinsichtlich ihrer Plausibilität zu überprüfen. So fanden Golding et al. (1992) ein annähernd zweifach erhöhtes Krebsrisiko für Kinder bei intramuskulärer versus oraler oder nicht durchgeführter neonataler Vitamin-K-Prophylaxe. Anschließend wurden Korrelationsstudien in Großbritannien und den USA durchgeführt, die die Kinderkrebsinzidenz in diesen Ländern in Beziehung zur Verbreitung von intramuskulärer Vitamin-K-Prophylaxe setzten. Diese ergaben jedoch keinen Anstieg des Krebsrisikos für Kinder im Zusammenhang mit der Einführung bzw. Verbreitung der intramuskulären Vitamin-K-Prophylaxe. Auch weitere analytisch-epidemiologische Studien erbrachten keine Hinweise, die einen Zusammenhang unterstützten, ein solcher konnte sogar nahezu ausgeschlossen werden (v. Kries et al. 1996). Allerdings verblieb ein marginales Leukämierisiko (v. Kries 1998).
Bei Korrelationsstudien werden Aggregatdaten miteinander in Relation gesetzt. Das bedeutet, dass bekannt ist, wie häufig in einer Population das Merkmal A, z. B. Alkoholkonsum, und wie häufig das Merkmal B, etwa Prävalenz von Leberzirrhose, vorkommen. Was also einzig vorliegt, sind Durchschnittswerte des Alkoholkonsums und ebenso der Prävalenz von Leberzirrhose. Wenn bei höherem Alkoholkonsum auch höhere Lebererkrankungshäufigkeiten auftreten, so bedeutet dies keineswegs zwangsläufig, dass hier ein kausaler Zusammenhang bestehen muss. Vielmehr könnte es sich auch um einen Scheinzusammenhang handeln, da es ebenso denkbar ist, dass diejenigen, die keinen Alkohol tranken, an Leberzirrhose erkranken. Man hätte dann den Mitgliedern einer Gruppe („Leberkranke") Merkmale zugeschrieben, die sie als Einzelpersonen gar nicht besitzen. Dies wird als ökologischer Fehlschluss („ecological fallacy") bezeichnet. Vorteile von Korrelationsstudien liegen darin, dass sie vergleichsweise preiswert und schnell durchzuführen sind, da häufig auf bereits vorliegende Daten, z. B. aus der amtlichen Statistik oder Prozessdaten, zurückgegriffen werden kann.

5.4.2 Querschnittsstudien

Bei einer Querschnittsstudie (Survey) werden Exposition und Zielgröße (z. B. Erkrankungsstatus) simultan erhoben. Das bedeutet, dass die gefundenen Krankheitsfälle die prävalenten Fälle sind, die zum Zeitpunkt der Erhebung existieren. In der Regel ist jedoch nicht bekannt, wie lange sie schon existieren. Daher wird dieses Studiendesign auch Prävalenzstudie genannt. Die zurzeit größte und aktuellste Querschnittsstudie zur Kinder- und Jugendgesundheit in Deutschland ist der Kinder- und Jugendgesundheitssurvey (KiGGS) des Robert-Koch-Instituts. In diesem Survey wurden von 2003 bis 2006 erstmalig in Deutschland umfangreiche Daten zur körperlichen, psychischen und sozialen Gesundheit von insgesamt knapp 18.000 bundesweit repräsentativ ausgewählten Kindern und Jugendlichen im Alter von 0–17 Jahren in einem Untersuchungs- und Befragungssetting konjunkt, d. h. auf Individualebene verknüpfbar, erhoben. Außerdem wurden Blut- und Urinanalysen bei ca. 90 % der teilnehmenden Kinder und Jugendlichen durchgeführt. Die KiGGS-Studie erlaubt aktuelle Aussagen zur Prävalenz einer Vielzahl von Parametern der Kindergesundheit wie der Häufigkeit von Übergewicht und Adipositas, ADHS, psychischer Probleme, Unfälle und Verletzungen, körperlicher Aktivität und motorischer Leistungsfähigkeit, Jodversorgung, Impfstatus, sozialer Lage, Inanspruchnahme medizinischer Leistungen und anderem mehr (Bundesgesundheitsblatt 5/6 2007, Kurth et al. 2008). Außer der Beschreibung von Prävalenzen sind mit diesen Daten aber auch Zusammenhangsanalysen und analytische Auswertungen mit komplexeren statistischen Verfahren möglich. Eine Besonderheit ist auch die umfangreiche Erhebung soziodemografischer Daten, die sehr differenzierte Analysen der Verteilung von Gesundheitsproblemen hinsichtlich der sozialen Lage ermöglicht.

Die Stärken einer Querschnittsuntersuchung bestehen darin, dass sie, bei entsprechender Berücksichtigung im Studiendesign, wertvolle Hinweise auf mögliche Risiko-, aber auch Protektivfaktoren liefern kann. Als Nachteil fällt hingegen ins Gewicht, dass sie aufgrund der begrenzten Möglichkeiten, eine zeitliche Abfolge zwischen Exposition und Ausgang herzustellen, für die Untersuchung ätiologischer Beziehungen eher wenig geeignet erscheint. Dies lässt sich am Beispiel der vieldiskutierten Frage nach dem Einfluss von Bildschirmmedien auf die Entstehung von Übergewicht und Adipositas bei Kindern und Jugendlichen veranschaulichen. So wiesen in der KiGGS-Studie 11–17-jährige Jungen, die eine Gesamtnutzungsdauer von Fernsehen/Video, Computer/Internet und Spielkonsole von 6 Stunden und mehr pro Tag angaben, eine Odds Ratio von 2,09, gleichaltrige Mädchen sogar von ca. 3,5 für Adipositas auf (Lampert et al. 2007). Ob Bildschirmmedienkonsum allerdings tatsächlich zur Genese von Adipositas beiträgt oder ob sich adipöse Kinder und Jugendliche z. B. als Folge denkbarer sozialer Ausgrenzung durch Gleichaltrige vor Bildschirmmedien zurückziehen, kann mit diesem Studiendesign nicht geklärt werden.

5.4.3 Fall-Kontroll-Studien

Um eine mögliche Beziehung zwischen einer Exposition und einem Outcome zu untersuchen, werden beim Design einer Fall-Kontroll-Studie eine Gruppe von Personen mit dieser Erkrankung („Fälle") und zu Vergleichszwecken ein Gruppe von gesunden Personen („Kontrollen") ausgewählt. Anschließend wird bestimmt, welcher Anteil der Fälle bzw. welcher Anteil der Kontrollpersonen exponiert war. Dabei wird angenommen, dass im Fall einer tatsächlichen Beziehung zwischen Exposition und Erkrankung die Prävalenz der Exposition in der Anamnese der Fallgruppe höher ist als in der Kontrollgruppe.

Eine der größten Schwierigkeiten bei der Durchführung einer Fall-Kontroll-Studie besteht darin, dass sich Fälle und Kontrollen in personengebundenen Eigenschaften (z. B. Alter, Geschlecht, sozioökonomischer Status, Freizeit- oder gesundheitliches Risikoverhalten) unterscheiden oder zusätzlichen Expositionen ausgesetzt sind, die in Bezug auf die Studie nicht primär interessieren, die jedoch einen potenziell verfälschenden Einfluss auf die Studienergebnisse haben können. Eine Möglichkeit, derartige Verfälschungen zu kontrollieren, besteht darin, den Fällen Kontrollpersonen zuzuordnen, die in diesen bestimmten Merkmalen mit ihnen übereinstimmen. Dieses Verfahren wird als Matching bezeichnet. Beim Gruppenmatching werden die Kontrollpersonen so

ausgewählt, dass der Anteil der Personen mit einem bestimmten Merkmal identisch ist mit dem Anteil der Fälle mit demselben Merkmal. Beim individuellen Matching (auch: Matched-Pairs-Verfahren) wird für jeden Fall eine Kontrollperson ausgewählt, die in einem oder mehreren bestimmten Merkmalen mit dem Fall übereinstimmt. Es können jedoch jedem Fall auch mehrere Kontrollpersonen zugeordnet werden. Inwieweit dies notwendig und möglich ist, muss vom jeweiligen Forscher mit Blick auf sein Studienziel und auf seine Möglichkeiten entschieden werden.
Fall-Kontroll-Studien sind im Vergleich zu prospektiven Längsschnittstudien vergleichsweise preiswert und schnell durchführbar und erbringen schneller Ergebnisse. Sie werden daher oft durchgeführt, wenn die Einschätzung neuer potenzieller Risiko- oder Protektivfaktoren erforderlich ist und nicht abgewartet werden kann, bis ein prospektiver Studienansatz Ergebnisse erbringt.
Fall-Kontroll-Studien sind auch bei der Untersuchung seltener Erkrankungen die Methode der Wahl: Bei einer geschätzten Inzidenz von 1 Promille einer Erkrankung müssten in einer prospektiven Studie jeweils 20.000 Personen mit und ohne Exposition untersucht werden, um Aussagen auf einem 0,05 %-Signifikanzniveau treffen zu können.
Ein Beispiel für eine **Fall-Kontroll-Studie** ist die **Kikk-Studie** des Deutschen Kinderkrebsregisters in Mainz. Bei zwei vorangegangenen Studien zu Krebserkrankungen von Kindern unter 15 Jahren in der Umgebung von Kernkraftwerken, die das Kinderkrebsregister in den Jahren 1992 und 1997 veröffentlichte, gab es den Nebenbefund, dass Leukämien bei Kindern unter 5 Jahren gehäuft im Umkreis von 5 Kilometern von Kernkraftwerken auftraten. Dies führte zu Beunruhigung und dazu, dass die Untersuchung noch einmal mit dem methodischen Ansatz einer Fall-Kontroll-Studie wiederholt wurde (Spix et al. 2007, Kaatsch et al. 2008). Während sich die vorangegangenen Untersuchungen auf den Zeitraum von 1980 bis 1995 erstreckten, wurden nun zusätzlich Daten aus den Jahren 1996 bis 2003 miteinbezogen. Für insgesamt 1.592 Fälle von Kindern unter 5 Jahren, die an Krebs erkrankt waren, wurden insgesamt 4.735 Kontrollen mit demselben Alter und Geschlecht ausgewählt. Es kamen also etwa 3 Kontrollen auf einen Fall. Für Fälle und Kontrollen wurde der Abstand zum nächstgelegenen Kernkraftwerk auf 25 Meter genau bestimmt. Außerdem wurden bei einem Teil der Fälle und Kontrollen per Fragebogen zusätzliche Risikofaktoren als potenzielle Confounder ermittelt. Die Autoren der Studie kamen zu dem Ergebnis, dass sich tatsächlich ein Zusammenhang zwischen dem Wohnen in der Nähe eines Kernkraftwerks (5-km-Zone) und dem Risiko eines Kindes, vor seinem 5. Geburtstag an Krebs zu erkranken, feststellen lässt. Bezogen auf alle Krebserkrankungen bei Kindern in diesem Alter wurde ein attributables Risiko von 0,2 % errechnet, was bei Gültigkeit der zugrunde gelegten Modellannahmen zusätzliche 1,2 Fälle pro Jahr bedeutete. Bezogen auf die Untergruppe Leukämien wurde ein attributables Risiko von 0,3 % errechnet, was zusätzlichen 0,8 Fällen pro Jahr entspräche, jedoch mit Blick auf die kleineren Fallzahlen mit einer erheblichen statistischen Unsicherheit behaftet ist. Ist daher zu schlussfolgern, dass aus Kernkraftwerken Strahlung entweicht, die Krebserkrankungen bzw. Leukämien bei Kindern auslöst? Die Autoren der Studie verneinen dies, da die geschätzte tatsächliche Strahlenbelastung um mehrere Zehnerpotenzen unterhalb des in Deutschland gültigen Grenzwerts für die Belastung in der Umgebung von kerntechnischen Anlagen liegt. Zur Diskussion um die Studienergebnisse ➤ Kap. 5.6.
Fall-Kontroll-Studien werden auch als retrospektive Studien bezeichnet, da sie von Erkrankten bzw. gesunden Personen ausgehen und dann retrospektiv, z. B. durch Befragung, ermitteln, ob eine Exposition bei den Fällen bzw. Kontrollen vorgelegen hat. Damit ist auch zugleich eine Fehlermöglichkeit von Fall-Kontroll-Studien bezeichnet, nämlich das Problem des Erinnerungsvermögens (engl.: Recall). Probleme mit dem Recall können einerseits in Einschränkungen des Erinnerungsvermögens bestehen; dies ist umso häufiger zu erwarten, je weiter die erfragten Ereignisse zurückliegen. Es besteht dann die Gefahr, dass es zu Fehlklassifikationen des Expositionsstatus kommt.

Dies resultiert meist in einer Unterschätzung des tatsächlichen Risikos. Ein anderes Problem ist eine systematische Verzerrung der Erinnerung (Recall Bias), z. B. durch unterschiedliche Erinnerungsvermögen bei Fällen und Kontrollen. Eine mögliche Ursache kann persönliche Betroffenheit sein. Befragte man z. B. Mütter von Kindern, die mit Fehlbildungen geboren wurden (Fälle), und Mütter von gesunden Kindern (Kontrollen) nach Infektionen während der Schwangerschaft, so ist es vorstellbar, dass die Mütter betroffener Kinder in dem Bemühen, eine Erklärung für die Krankheit ihres Kindes zu finden, eine höhere Selbstaufmerksamkeit an den Tag legen und daher dazu neigen könnten, auch leichtere Infekte zu berichten, die Mütter von gesunden Kindern bereits wieder vergessen haben. Das Beispiel macht auch klar, dass die Bekanntheit der Untersuchungshypothese ebenfalls einen Einfluss auf die Erinnerung bzw. die retrospektive Aufmerksamkeit hat. Wenn im Studienkollektiv Hypothesen über die Beziehung von präpartalen Infektionen und Fehlbildungen bekannt sind, wird dies wahrscheinlich bei den betroffenen Müttern zusätzliche Aufmerksamkeit und intensiveres Nachdenken darüber zur Folge haben. Eine Möglichkeit, diesen Fehler zu umgehen, kann darin bestehen, z. B. Patienten- oder Krankenakten zur Erhebung der Exposition zu nutzen. Dies setzt allerdings voraus, dass solche auch verfügbar sind und dass die interessierende Exposition dort zuverlässig dokumentiert ist. Eine weitere Möglichkeit, systematische Verzerrungen zu vermeiden, ist die Verblindung des Untersuchers. Das bedeutet, dass der Datenerheber nicht weiß, ob er Fälle oder Kontrollen bearbeitet. Natürlich müssen auch in Fall-Kontroll-Studien Störgrößen (Confounder) berücksichtigt werden. Dies setzt voraus, dass diese potenziellen Störgrößen bekannt sind und mit erhoben wurden. Durch statistische Adjustierung können dann bei der Datenauswertung unterschiedliche Verteilungen der Störgrößen bei Fällen und Kontrollen mit speziellen mathematischen Verfahren rechnerisch ausgeglichen werden.

5.4.4 Kohortenstudien

Für Kohortenstudien (auch Längsschnitt- oder prospektive Studien) wird eine Gruppe exponierter sowie eine Gruppe nichtexponierter Personen ausgewählt, über einen Zeitraum nachbeobachtet (Follow-up) und anschließend hinsichtlich ihrer inzidenten Erkrankungen (oder Sterberaten) verglichen. Dabei wird von der Annahme ausgegangen, dass die Exposition dann verantwortlich ist für das Auftreten der Krankheit, wenn diese häufiger in der exponierten Gruppe auftritt als in der nichtexponierten. Es gibt grundsätzlich zwei Wege, zu Studienpopulationen für Kohortenstudien zu gelangen. Der erste besteht darin, eine Gruppe exponierter Personen sowie eine Gruppe nichtexponierter Personen auszuwählen und nachzubeobachten. Die zweite Möglichkeit besteht darin, eine definierte Population anhand eines Merkmals auszuwählen und anschließend über einen bestimmten Zeitraum nachzuverfolgen. Diese Merkmale können unterschiedlicher Natur sein; in der Regel handelt es sich jedoch um ein bestimmtes biografisches Ereignis, das bei den beteiligten Individuen mehr oder weniger zum selben Zeitpunkt aufgetreten ist, z. B. Geburtskohorten, Einschulungskohorten, Scheidungskohorten etc.
Kohortenstudien werden außer in der medizinischen Epidemiologie häufig auch zur Beantwortung entwicklungspsychologischer Fragestellungen oder zur Untersuchung spezieller Risikogruppen eingesetzt. Bekannte Längsschnittstudien zu Fragen der Entwicklung im Kindes- und Jugendalter sind z. B. die Züricher Längsschnittstudien zu Wachstum und Entwicklung oder die Mannheimer Risikokinderstudie. Erste Länggsschnittergebnisse aus der KiGGS-Studie sind ab 2014 zu erwarten (Hölling et al., 2012).

An einem hypothetischen Beispiel seien einige Probleme bei der Durchführung von Kohortenstudien aufgezeigt. In einer Studie interessiert der Zusammenhang zwischen Alkoholkonsum und alkoholbedingter Leberschädigung. In diese Studie wird eine Kohorte von Schulanfängern aufgenommen. Bei einem angenommenen Beginn regelmäßigen Alkoholkonsums im Teenageralter würden nach etwa 10 Jahren Exponierte und Nichtexponierte unterscheidbar sein. Bei einer weiterhin angenommenen Latenz von wiederum 10 Jahren für die Entwicklung von alkoholbedingten Leberschädigungen wären erste Ergebnisse der Studie nach etwa 20 Jahren zu erwarten. Mit diesem Beispiel sind bereits wesentliche Probleme von Kohortenstudien angesprochen. So ist die lange Beobachtungsdauer nicht nur häufig ein finanzielles Problem (Personal, Logistik und Strukturen müssen u. U. über Jahrzehnte vorgehalten werden), es wird mit zunehmender Studiendauer auch immer schwieriger, Probanden oder Patienten in der Studie zu behalten, da sich die Studienpopulation durch Wegzug, Desinteresse, natürlichen Tod oder andere Gründe verkleinert (sog. Panelmortalität). Ein Verfahren, die lange Beobachtungszeit zu verkürzen, besteht darin, den Expositionsstatus rückwirkend zu erheben. Man könnte also im hypothetischen Beispiel die Studienteilnehmer befragen, wann und in welchem Umfang sie mit dem Alkoholkonsum begonnen haben. Dieser Studientyp wird retrospektive oder historische Kohortenstudie genannt. Da die Terminologie leider uneinheitlich ist, sei an dieser Stelle noch einmal darauf verwiesen, dass das grundsätzliche Design, der Ausgang vom Expositionsstatus und die anschließende Unterscheidung in von einer Erkrankung Betroffene und Nichtbetroffene das entscheidende Kennzeichen der Kohortenstudie ist. Dies ist sowohl der prospektiven als auch der retrospektiven Kohorte gemeinsam. Im Unterschied dazu wird bei der Fall-Kontroll-Studie stets von den „Fällen" (Betroffene) und den „Kontrollen" (Nichtbetroffene) ausgegangen und im Nachgang der Expositionsstatus bestimmt.
Probleme von Kohortenstudien liegen außer in der Panelmortalität in der Konstanz der Messinstrumente. So führt der soziokulturelle Wandel dazu, dass sich Begrifflichkeiten verändern, was Auswirkungen auf die Güte verwendeter psychometrischer oder soziometrischer Instrumente haben kann. Ähnliches gilt prinzipiell z. B. auch für Weiterentwicklungen von labortechnischen Analysegeräten. Zu nennen sind weiterhin Paneleffekte, d. h. Personen bilden durch wiederholte Befragung oder Untersuchung bestimmte Meinungen oder Eigenschaften erst aus, sowie die Panelerstarrung, die besagt, dass eine Kohorte durch Lebensereignisse wie Heirat, Geburt eines Kindes oder sozioökonomische Veränderungen bei den Probanden ihre Repräsentativität verliert. Dennoch haben Kohortenstudien einen hohen Wert, da sie sich aufgrund der Tatsache, dass in ihnen Veränderungen bzw. Verläufe auf Individualdatenniveau an denselben Studienobjekten feststellen lassen, in besonderer Weise zur Untersuchung kausaler Fragestellungen eignen.

5.4.5 Interventionsstudien

Interventionsstudien sind grundsätzlich prospektive, meist klinische oder präklinische Studien. Hierbei soll der Effekt einer Exposition, z. B. einer Behandlung oder eines neuen Medikaments, gezielt überprüft werden. Im klinischen Setting handelt es sich daher in der Regel entweder um plazebokontrollierte Studien oder um Studien, in denen die Überlegenheit eines neuen Medikaments/einer Therapie gegenüber einem etablierten Medikament/einer Therapie untersucht werden soll. Interventionsstudien erfordern eine besonders sorgfältige und ethisch fundierte Studienplanung, da die Probanden einer Therapie bzw. einem Wirkstoff ausgesetzt werden, der noch nicht hinreichend erforscht ist. Den Patienten dürfen durch die Studienteilnahme keine gesundheitlichen Nachteile entstehen, auch nicht durch das Vorenthalten einer potenziell nützlichen

Behandlung (bei plazebokontrollierten Studien). Dies wird in der Regel dann der Fall sein, wenn genügend Hinweise auf die prinzipielle Unschädlichkeit eines Medikaments vorliegen und andererseits noch Aufklärung über die Wirksamkeit erforderlich ist. Eine Studie weist eine hohe interne Validität (Gültigkeit) auf, wenn die Ergebnisse eindeutig interpretierbar sind und es wenige Alternativerklärungen gibt. Je besser die Untersuchungsbedingungen kontrolliert sind, desto höher ist die interne Validität. Eine hohe interne Validität kann jedoch zu Lasten der externen Validität (= Verallgemeinerbarkeit) gehen, denn je mehr Variablen kontrolliert werden, desto unnatürlicher wird die Studiensituation.

5.5 Potenzielle Fehlerquellen epidemiologischer Studien

5.5.1 Statistische Unschärfe der Messung

Das Problem der statistischen Unschärfe der Messung (oder statistischer Fehler) tritt bei stichprobenbasierten Untersuchungen auf. Dies betrifft die meisten populationsbezogenen Studien. Kennwerte, z. B. Prozent- oder Mittelwerte, aus einer Zufallsstichprobesind jedoch stets mit einer gewissen Unsicherheit in Bezug auf den „wahren", d. h. tatsächlichen Wert in der interessierenden Population behaftet, da ihre konkrete Manifestation ja von eben der gezogenen Stichprobe abhängt. In anderen Worten: Zöge man theoretisch eine andere Stichprobe aus dem gleichen Kollektiv, würden die Prozent- oder Anteilswerte, die man erhielte, möglicherweise von denen der ersten Stichprobe abweichen. Eine Fragestellung, die sich daran anschließt, könnte lauten: Wie können wir sicher sein, dass, wenn bei 10 % der teilnehmenden Kinder in einer stichprobenbasierten Prävalenzstudie die Diagnose einer obstruktiven Bronchitis festgestellt wurde, tatsächlich auch 10 % der Kinder in der zugrunde liegenden Ausgangsbevölkerung (Grundgesamtheit) betroffen sind? Tatsächlich lässt sich dies mit den aus einer Stichprobenuntersuchung gewonnen Ergebnissen niemals genau sagen. Es lässt sich aber ein gewisser Vertrauensbereich (Konfidenzintervall) angeben, in dem der gefundene Wert mit einer gewissen Wahrscheinlichkeit (üblich sind 95 % oder 99 %) liegt[1]. Diese Wahrscheinlichkeiten bezeichnet man als Konfidenzkoeffizienten. Das Konfidenzintervall (KI) umschließt jeweils symmetrisch den entsprechenden Stichprobenkennwert. In unserem hypothetischen Beispiel könnte die Ergebnisdarstellung so lauten: Anteil der Kinder mit obstruktiver Bronchitis ist 10 % (95 % KI: 9,65–10,35). Konfidenzintervalle lassen sich nicht nur für Anteile oder Mittelwerte angeben, sondern auch für Risikomaße wie relative Risiken oder Odds Ratios. Formeln zur Berechnung von Konfidenzintervallen finden sich in den einschlägigen Statistiklehrbüchern. Das Konzept der statistischen Unschärfe (oder des statistischen Fehlers) ist nicht ganz einfach verständlich und daher einer breiteren Öffentlichkeit nur schwierig zu vermitteln. So ist die Schlagzeile: „10 % der Kinder leiden unter obstruktiver Bronchitis" leichter zu transportieren als die Formulierung: „Mit 95-prozentiger Wahrscheinlichkeit haben 10 % der Kinder in Deutschland obstruktive Bronchitis."

[1] Bortz (1999) weist darauf hin, dass es genauer heißen müsste: Die Wahrscheinlichkeit, dass die untersuchte Stichprobe aus einer Population stammt, in der der festgestellte Anteilswert – in unserem Beispiel von 10 % von Kindern mit der Diagnose obstruktive Bronchitis – innerhalb der angegebenen Konfidenzgrenzen liegt, ist 95 (resp. 99) Prozent.

5.5.2 Bias

Als Bias (engl.: Verzerrung) werden systematische Fehler im Design, bei der Durchführung, der Analyse von Studien oder der Veröffentlichungspraxis von Studienergebnissen bezeichnet. Diese Fehler entstehen häufig durch nicht hinreichend sorgfältige Studienplanung, aber auch unzureichende Datenlage, persönliche Betroffenheit oder Probleme während der Feldphase von Studien. Es existieren eine Reihe unterschiedlicher Formen von Bias. Selektionsbias kann bei Fall-Kontroll-Studien auftreten, wenn vorwiegend solche Fälle ausgewählt werden, bei denen die interessierende Exposition vorlag. Eine Sonderform des Selektionsbias bei klinischen Inanspruchnahmestichproben ist der Überweisungsbias (Referral Bias). So wird aufgrund neuerer populationsbezogener Untersuchungen aus den USA vermutet, dass ADHS bei Jungen und Mädchen annähernd gleich prävalent ist (Biederman et al. 2005). Während bei Mädchen der vorwiegend unaufmerksame Subtyp häufiger ist, ist bei Jungen der vergleichsweise besser beobachtbare und auffälligere hyperaktive Subtyp verbreiteter, der häufiger zu Klinikeinweisungen führt. Folglich weisen die Ergebnisse klinischer Studien eine deutlich höhere ADHS-Prävalenz für Jungen aus.

Weiterhin kann es zu Selektionsfehlern kommen, wenn sich Teilnehmer (Responder) von Nichtteilnehmern (Nonrespondern) einer Studie systematisch unterscheiden. Diese Fehlermöglichkeit besteht auch in repräsentativen Querschnittstudien. So ist z. B. bekannt, dass Personen mit niedrigem sozioökonomischen Status oder geringer Bildung generell weniger bereit sind, an Studien teilzunehmen. In der Konsequenz sind Teilnehmer aus der mittleren und oberen Sozialschicht überrepräsentiert, was die Studienergebnisse verzerren kann, da die Gesundheitsbedingungen mit dem Sozialstatus stark variieren. Kurze Nichtteilnehmerbefragungen, bei denen soziodemografische Basischarakteristika erfragt werden, können hier einen groben Anhaltspunkt für mögliche Verzerrungen geben bzw. ggf. ihr Ausmaß abschätzbar machen.

Ein Informationsbias ist ein klassischer Fehler in Fall-Kontroll-Studien. Da die Informationen zum Expositionsstatus retrospektiv erhoben werden, können lückenhafte Dokumente (z. B. Krankenakten, Todesursachenbescheinigung) zu Fehlklassifizierungen hinsichtlich des Expositionsstatus führen.

Werden die Daten durch Befragung erhoben, kann es, wie bereits unter ➤ Kap. 5.4.3 ausgeführt, durch unzureichendes Erinnerungsvermögen oder persönliche Betroffenheit der Probanden zum Recallbias kommen. Als Berichtsbias bezeichnet man hingegen, dass eine Person aus Weltanschauungs-, Einstellungs- oder religiösen Gründen eine ihr bewusste Exposition nicht berichtet. Zum Beispiel ist es vorstellbar, dass eine religiös eingestellte, HIV-infizierte Person eventuelles sexuelles Risikoverhalten in einer Befragungssituation nicht berichtet.

Ein Interviewerbias kann entstehen, wenn Interviewer Kenntnis über die interessierende Hypothese und/oder den Erkrankungsstatus der untersuchten Person haben. Zu verhindern ist dies nur durch Verblindung; sofern dies nicht möglich ist, muss für eine größtmögliche Standardisierung der Interviewsituation und eine intensive Interviewerschulung Sorge getragen werden.

Als Publikationsbias bezeichnet man ein Phänomen, das auftritt, wenn Forscher Ergebnisse, die nicht die Untersuchungshypothese stützen oder ein erhöhtes Risiko aufzeigen, nicht publizieren. Publikationsbias kann auch durch eine (unbewusst) selektive Auswahl von Beiträgen durch die Redaktionen wissenschaftlicher Fachzeitschriften, z. B. aufgrund von Orientierungen am Publikumsinteresse, entstehen.

5.5.3 Confounding

Confounding bezeichnet die Ergebnisverfälschung durch den Einfluss einer oder mehrerer Störgrößen (Confounder). Ein Confounder ist selbst kein Risikofaktor für eine Erkrankung, variiert aber gleichsinnig mit diesem, sodass ein scheinbarer Zusammenhang mit dem Ergebnis (Outcome) entsteht. So könnte z. B. in einer Studie mit Grundschülern festgestellt werden, dass Abstraktionsfähigkeit und sensomotorische Koordination miteinander korrelieren. Tatsächlich verfügen aber ältere Kinder über eine bessere Abstraktionsfähigkeit und auch eine bessere sensomotorische Koordination als jüngere. Der Zusammenhang wäre also durch den Confounder Lebensalter vermittelt. Grundsätzlich gibt es mehrere Möglichkeiten, Confounder zu kontrollieren. Das Matching wurde bereits in ➤ Kap. 5.4.3 besprochen. Durch stratifizierte Analyse oder mathematische Adjustierung im Rahmen multivariater Verfahren wie der logistischen Regressionsanalyse lassen sich Confounder rechnerisch ausgleichen.

5.5.4 Interaktionen

Von einer Interaktion (Wechselwirkung) spricht man, wenn sich zwei Einflüsse (Risikofaktoren) in unerwarteter Weise gegenseitig verstärken **(Synergismus)** oder abschwächen **(Antagonismus).**

5.6 Kausalität in epidemiologischen Studien

Von in epidemiologischen Studien gefundenen statistischen Zusammenhängen allein ist noch nicht auf eine vorhandene Kausalität zwischen Exposition und Ereignis zu schließen. Vielmehr müssen eine Reihe von Kriterien erfüllt sein, die eine kausale Ursache-Folge-Beziehung wahrscheinlich und akzeptierbar erscheinen lassen. Der Epidemiologe und Statistiker Sir Austin Bradford Hill hat 1965 Kriterien für Kausalitätsbeziehungen in epidemiologischen Studien formuliert, die in geringfügiger Modifikation bis heute gültig sind. Voraussetzung für die Annahme einer Kausalbeziehung ist in jedem Fall eine valide statistisch signifikante Assoziation von Risikofaktor und Erkrankung. Bias und Confounding schließen eine valide statistische Assoziation aus. Zu fordern ist weiterhin, dass eine zeitliche Abfolge zwischen der Exposition gegenüber dem Risikofaktor und der Erkrankung besteht, wobei die Exposition der Erkrankung natürlich vorangehen muss. Zeitliche Abfolgen lassen sich am besten in einer prospektiven Kohortenstudie beobachten, da die Expositionserhebung in Fall-Kontroll-Studien und retrospektiven Kohortenstudien mitunter ungenau ist. Ein weiteres Kriterium ist die Stärke des Zusammenhangs, wobei es hier keine verbindliche Konvention gibt. Üblicherweise geht man aber ab einer Verdopplung bzw. Halbierung des Risikos (OR/RR $\geq$ 2 bzw. OR/RR $\leq$ 0,5) von einem starken Zusammenhang aus. Natürlich sind auch kausale Beziehungen unterhalb dieser Schwellen denkbar, eine starke Assoziation bietet aber eine relative Sicherheit gegenüber möglichen unerkannten Störfaktoren. Ein weiterer Hinweis auf Kausalität ist das Vorliegen einer Dosis-Wirkungs-Beziehung: Je höher die Expositionsdosis, desto stärker steigt auch das Erkrankungsrisiko. Bei Fehlen einer solchen Beziehung ist jedoch Kausalität keineswegs ausgeschlossen. Darüber hinaus ist die Replizierbarkeit der Befunde ein Kriterium. Ist die Assoziation kausal, ist zu erwarten, dass sie auch in anderen Studienkollektiven auftritt. Geht die Krankheitshäufigkeit hingegen zurück, wenn das Risiko entfällt, ist dies ebenfalls ein Hinweis. Das biologische Plausibilitätskriterium schließt schließlich eine Kausalbeziehung aus, wenn sich epidemiologische Befunde auf der Grundlage des gegenwärtigen biologischen Verständnisses nicht plausibel erklären lassen. Allerdings muss in Rechnung gestellt werden, dass auch das biologische Verständnis einem zeitlichen und gesellschaftlichen Wandel unterworfen ist.

In der bereits erwähnten Kikk-Studie des Deutschen Kinderkrebsregisters wurde von einer geschätzten Belastung einer Person von 0,0000019 bis 0,0003200 Millisievert (mSV) durch Strahlung in der Luft in einem 5-km-Radius um ein Kernkraftwerk ausgegangen. Der gesetzlich zulässige Höchstwert in der Umgebung von kerntechnischen Anlagen in Deutschland beträgt 0,3 mSV, die natürliche jährliche Strahlenexposition 1,4 mSV. Damit kommt nach Einschätzung der Autoren eine von Kernkraftwerken ausgehende radioaktive Strahlung nicht für die Erhöhung des Risikos für Kinderkrebs- und Leukämieerkrankungen in Frage. Obwohl die Studie weder in der Durchführung noch in der statistischen Auswertung erkennbare methodische Mängel aufwies, geriet sie hinsichtlich ihrer Schlussfolgerung in heftige Kritik (vgl. Richter-Kuhlmann 2008). So wurde kritisiert, dass die Studie nur auf die kleine Zahl von Fällen in der 5-km-Zone um die Kernkraftwerke abhebe und nicht das gesamte Studiengebiet (50-km-Radius) einbeziehe. Unbestritten ist auch die signifikante Abstandsabhängigkeit des Risikos, d. h. je näher die Kinder an einem Kernkraftwerk wohnen, desto höher ist das Risiko für eine Krebserkrankung, auch nach Ausschluss potenzieller Confounder wie z. B. erhöhter Pestizidverbrauch in den Untersuchungsgegenden. Ebenfalls wurde kritisiert, dass Werte zur Abschätzung der Strahlendosis nicht einfach von Erwachsenen auf Kinder übertragen werden dürften bzw. dass auch präkonzeptionelle Belastungen der Eltern zu einer erhöhten Krebsinzidenz der Kinder beitragen könnten. Außerdem wurde darauf hingewiesen, dass die Angaben zu den Emissionen nicht amtlich gemessen, sondern von den Betreibern der Kernkraftwerke zur Verfügung gestellt werden. So sei es keineswegs auszuschließen, dass mehr Radioaktivität freigesetzt wird, als offiziell angegeben. Auch könne die Bevölkerungsdosis nicht direkt gemessen werden, sondern müsse anhand von Modellrechnungen aus den gemessenen Emissionen über den Weg von Atemluft, Nahrung und Trinkwasser simuliert werden. Inzwischen haben die Autoren der Studie die Möglichkeit einer unterschiedlichen Strahlensensibiliät von Kindern und Erwachsenen eingeräumt (Kaatsch, 2009). Die Diskussion um diese Studie macht nicht nur deutlich, dass die epidemiologische Studien, insbesondere bei hochsensiblen Themen, nicht in gesellschaftlich neutralem Raum stattfinden, sondern dass sich gesellschaftliche Auseinandersetzungen bis hinein in die Diskussion um methodische Details und die biologische Plausibilität der Studienergebnisse erstrecken.

5.7 Exkurs: Ethik in epidemiologischen Studien mit Kindern und Jugendlichen

Für Studien mit Beteiligung von Kindern und Jugendlichen gelten aufgrund deren Zughörigkeit zu einer sog. vulnerablen Bevölkerungsgruppe, ihrer altersbedingt begrenzten Informiertheit und Informierbarkeit sowie ihrer eingeschränkten Urteilsfähigkeit besondere ethische und rechtliche Anforderungen. Da es sich bei Minderjährigen um nichteinwilligungsfähige Personen handelt, sind neben generellen individualethischen Grundsätzen wie dem Gebot „nicht schaden“, der Achtung des Selbstbestimmungsrechts des Probanden, der Einhaltung der sogenannten goldenen Regel „Was du nicht willst, das man dir tu, das füg“ auch keinem andern zu“ (Deutsch und Lippert

1998) sowie den sozialethischen Normen der Nichttäuschung und der Einholung des Einverständnisses besondere Schutzbestimmungen zu beachten. Hierzu gehören:

- Individuelles Recht auf Selbstbestimmung, das durch Einwilligung nach umfassender Aufklärung (Informed Consent) gesichert wird
- Objektive Kriterien zum Schutz der Studienteilnehmer (z. B. umfassende Risiko-Nutzen-Abwägung)
- Verfahrensförmige Sicherung durch Einbeziehung einer Ethikkommission (Fröhlich 1999).

Die Erfordernisse klinischer und pharmakologischer Studien mit nichteinwilligungsfähigen Kindern und Jugendlichen sind durch den Gesetzgeber weit besser abgedeckt als die populationsbezogener epidemiologischer Studien. Bei klinischen oder pharmakologischen Studien kann rechtlich von einem ärztlichen Heilversuch ausgegangen werden, wenn eine noch nicht gesicherte Diagnose- oder Therapiemethode zum Einsatz kommt. Entscheidend ist dabei, dass es sich nicht um eine „versuchsweise" Heilbehandlung des Patienten mit ungewissem Ausgang handelt, sondern um die Beschreitung medizinischen Neulands, also einer Behandlung mit neuartigen, noch nicht standardisierten Mitteln. Dabei ist es unerheblich, wenn beim therapeutischen Versuch quasi als „Nebenzweck" wissenschaftliche Interessen verfolgt, die Ergebnisse ausgewertet und zu Forschungszwecken genutzt werden.

In der bevölkerungsbezogenen KiGGS-Studie des Robert-Koch-Instituts sollten Laboruntersuchungen von Blut- und Serumparametern prima facie gesunder Kinder und Jugendlicher Aufschluss über gesundheitliche Risiken wie unentdeckter Nährstoffmangel („hidden hunger"), allergische Sensibilisierung, Herz-Kreislauf-Risikofaktoren, Schilddrüsenhormone, bestehender Impfschutz u. a. m. einzelner Personen sowie deren Verteilung in der entsprechenden Population geben, die durch Befragung und körperliche Untersuchung allein nicht erkannt werden können. Es bestand Unklarheit darüber, ob Eltern stellvertretend für ihre Kinder in eine medizinisch nicht indizierte Punktion einwilligen können. Diese Untersuchungen nicht durchzuführen hätte bedeutet, auf diese Erkenntnisse für die Bevölkerungsgruppe der Kinder und Jugendlichen zu verzichten und diese dadurch vom medizinischen Fortschritt auszuschließen, was unter ethischen Gesichtspunkten ebenfalls problematisch wäre (HG Schlack 1998). Dieses Dilemma ließ sich in der KiGGS-Studie dahingehend auflösen, dass neben dem gruppenspezifischen Nutzen auch ein potenzieller individueller Nutzen für jedes teilnehmende Kinder bzw. Jugendlichen dadurch bestand, dass bei einer Identifikation bislang unentdeckter Gesundheitsprobleme diese einer weiteren medizinischen Behandlung zugeführt werden konnten. Den besonderen Schutzbestimmungen wurde schließlich u. a. durch umfangreiche Aufklärung und Einwilligung, ausführliche Risiko-Nutzen-Abwägung, sehr strenge Anforderungen zum Schutz der Studienteilnehmer wie eigenständiges „Vetorecht" des Kindes, Anästhesierung der Punktionsstelle durch EMLA-Pflaster, nur ein Punktionsversuch, altersabhängiges Stufenmodell für die Entnahme der Blutmenge, schnelle Ergebnisrückmeldung, strikte Datenschutzvorkehrungen sowie der Begutachtung der Studie durch eine Ethikkommission Rechnung getragen (Bergmann et al. 2004).

Literatur

Barmer GEK (2013). Barmer GEK Arztreport 2013. Schwerpunkt: ADHS. Schriften zur Gesundheitsanalyse, Band 18. Siegburg: Asgard.

Bergmann KE, Schlack R, v. Dewitz C, Dippelhofer A, Kurth BM, Eichstädt H (2004) Ethische und rechtliche Aspekte der epidemiologischen Forschung mit Kindern und Jugendlichen in Deutschland am Beispiel des Kinder- und Jugendgesundheitssurveys. Ethik in der Medizin 16: 22–36.

Biederman J, Kwon A, Aleardi M, Chouninard V–A, Marino T, Cole H, Mick E, Faraone SV (2005) Absence of Gender Effects on Attention Deficit Hyperactivity Disorder: Findings in nonreferred Subjects. Am J Psychiatry 162: 1083–89.

Bortz J (1999) Statistik für Sozialwissenschaftler. 5. vollständig überarbeitete Aufl. Berlin, Heidelberg, New York: Springer

Deutsch E, Lippert HD (1998) Ethikkommission und klinische Prüfung. Berlin, Heidelberg, New York: Springer,

Deutsches Kinderkrebsregister (2012.) Jahresbericht/Annual Report 2011. http://www.kinderkrebsregister.de/dkkr/extern/veroeffentlichungen/jahresberichte/index.html, Zugriff am 14.5.2013

Ellsäßer, G (2011) Verletzungsmonitoring – die Injury Data Base (IDB) in Deutschland. Telematik für die Öffentliche Gesundheit. Vortrag auf der 6. Landeskonferenz Telematik im Gesundheitswesen, Potsdam 16.2.2011. www.telemed-initiative.de/konferenzen/copy2/vortrag_ellsaeser. Zugriff am 14.5.2013

Erhebungseinheit für seltene pädiatrische Erkrankungen (ESPED) (o. J.). ESPED-Jahresbericht 2007. http://www.esped.uni-duesseldorf.de/Jabe2011.pdf, Zugriff am 14.5.2013.

Frances, A (2013). Normal. Gegen die Inflation psychiatrischer Diagnosen. Köln: DuMont.

Friedmann GD (1967) Cigarette smoking and geographic variation in coronary heart disease mortality in the United States. J Chronic Dis 20: 769.

Fröhlich U (1999) Forschung wider Willen? Rechtsprobleme biomedizinischer Forschung mit nichteinwilligungsfähigen Personen. Berlin, Heidelberg, New York: Springer.

Golding J, Birmingham K, Greenwood R, Mott M (1992) Childhood cancer, intramuscular vitamin K, and phetidine given during labour. BMJ 305: 341–6.

Hill AB (1965) The environment and disease: association or causation? Proc R Soc Med 58: 295–300.

Hölling H, Schlack R, Kamtsiuris P, Butschalowsky H, Schlaud M, Kurth BM (2012). Die KiGGS-Studie. Bundesweit repräsentative Längs- und Querschnittstudie zur Gesundheit

von Kindern und Jugendlichen im Rahmen des Gesundheitsmonitorings am Robert Koch-Institut. Bundesgesundheitsbl – Gesundheitsforsch – Gesundheitschutz 55: 836–842.

Kaatsch P, Spix C, Schulze-Rath R, Schmiedel S, Blettner M (2008) Leukaemia in young children living in the vicinity of German nuclear power plants. Int J Cancer 122: 721–6.

Kaatsch, P. (2009). Schlusswort. Dtsch Arztebl 106: 394.

Kemp, C (2013). Childhood disability rate jumps 16 % over past decade. AAP News. http://aapnews.aappublications.org. Zugriff 14.5.2013

Kries v R (1998) Neonatal vitamin K prophylaxis: the gordian knot still awaits untying. BMJ 316: 161–162.

Kurth, BM, Lange C, Kamtsiuris P, Hölling H (2009). Gesundheitsmonitoring am Robert Koch-Institut. Sachstand und Perspektiven. Bundesgesundheitsbl – Gesundheitsforsch – Gesundheitschutz 52: 557–570.

Kries v R, Göbel U, Hachmeister A, Kaletsch U, Michelis J (1996) Vitamin K and childhood cancer: a population based case-control study in Lower Saxony, Germany. BMJ 313: 199–203.

Kurth BM, Kamtsiuris P, Hölling H, Schlaud M, Dölle R, Ellert U, Kahl H, Knopf H, LangeM, Mensink GBM, Neuhauser H, Schaffrath Rosario A, Scheidt-Nave C, Schenk L, Schlack R, Stolzenberg H, Thamm M, Thierfelder W, Wolf U (2008) The challenge of comprehensively mapping children's health in a nationwide health survey: Design of the German KiGGS-Study. BMC Public Health 8: 196.

Kurth BM, Ziese T (2006) Die Epidemiologie von Gesundheit und Krankheit. In: Hurrelmann K, Laaser U, Razum O (Hrsg) Handbuch Gesundheitswissenschaften. 4. vollständig überarbeitete Aufl. Weinheim, München: Juventa: 483–510.

Lampert T, Sygusch R, Schlack R (2007) Nutzung elektronischer Medien im Jugendalter. Ergebnisse des Kinder- und Jugendgesundheitssurvey KiGGS. Bundesgesundheitsbl – Gesundheitsforsch – Gesundheitschutz 50: 643–652.

Last JM (2001) A Dictionary of Epidemiology, ed 4. New York: Oxford University Press.

PaedNetz Bayern (2013). Jedes fünfte Kind bereits im Kindergarten auffällig – neue Auswertung von Versichertendaten bestätigt Trend. http://www.paednetz.de/news/. Zugriff 14.5.2013.

Palfrey JS, Tonniges TF, Green M, Richmond J (2005) Introduction: Adressing the millennial morbidity – the context of community pediatrics. Pediatrics 115 :1121–3.

Queiser-Luft A, Schlaefer K, Schicketanz K-H, Spranger J (2006) Fehlbildungen bei Neugeborenen. Dtsch Arztebl 103: A2.646–24.710.

Richter-Kuhlmann E (2008) Kinderkrebs und Atomkraft: Streit um Kausalzusammenhang. Dtsch Arztebl 105: A-1.052.

Schlack HG (1998) Kind/Kindheit. In: Korff W, Beck L, Mikat P (Hrsg) Lexikon der Bioethik, Bd. 2. Gütersloh: Gütersloher Verlagshaus: 362–366.

Schlack HG (2008) Wir brauchen eine pädiatrische Gesundheitswissenschaft! Editorial. Kinderärztliche Praxis 79: 179.

Robert Koch-Institut (Hrsg) (2011). KiGGS – Kinder- und Jugendgesundheitsstudie Welle 1. Projektbeschreibung. Beiträge zur Gesundheitsberichterstattung des Bundes. Berlin: Robert Koch-Institut.

Schlack R (2013) Die Gesundheit von Kindern und Jugendlichen in Eineltern- und Stieffamilien unter besonderer Berücksichtigung von Jungen. Ergebnisse aus dem Kinder- und Jugendgesundheitssurvey (KiGGS). In: Franz M, Karger A (Hrsg) Scheiden tut weh. Elterliche Trennung aus Sicht der Väter und Jungen. Göttingen: Vandenhoek & Rupprecht: 123–145.

Spix C, Schmiedel S, Kaatsch P, Schulze-Rath R, Blettner M (2007) Case-Control Study on childhood Cancer in the Vicinity of Nuclear Power Plants in Germany 1980–2003. Eur J Cancer 44: 275–84.

Statistisches Bundesamt (2013) Bevölkerung und Erwerbstätigkeit, Fachserie 1, Reihe 1.1 (Primärquelle: Statistisches Bundesamt, Statistik der natürlichen Bevölkerungsbewegung). www.gbe-bund.de (Stichwortsuche: Säuglingssterblichkeit→ Dokumentart Tabelle). Abrufdatum: 26.5.2013

Visser, SN, Bitsko, RH, Danlieson, ML, Perou, R, Blumberg, SJ (2010). Increasing prevalence of Parent-reported Attention-deficit/Hyperactivity Disorder Among Children – United States, 2003 and 2007. Morbidity & Mortality Weekly Report 59: 1.439–1.443

6 Versorgungssysteme in Deutschland

Harald Bode

6.1 Medizinisches Versorgungssystem für Kinder und Jugendliche in Deutschland

6.1.1 Öffentlicher Gesundheitsdienst (ÖGD)

Der **Kinder- und Jugendgesundheitsdienst** im ÖGD erfüllt hoheitliche bevölkerungsbezogene Aufgaben in Gesundheitsschutz, Gesundheitsaufsicht und Umwelthygiene. Hierzu zählen Suchtprävention, Prävention von Infektionskrankheiten und betriebsmedizinische Aufgaben in Gemeinschaftseinrichtungen für Kinder und Jugendliche. Zusätzlich wirkt er auf der individualmedizinischen Ebene mit Präventionsangeboten wie Vorsorgeuntersuchungen und Impfungen sowie entwicklungsbegleitenden Untersuchungen von Kindern im Kindergarten- und Schulalter sowie Beratung von Eltern. Diese Angebote richten sich besonders an Familien mit Kindern und Jugendlichen aus sozial benachteiligten Sozialräumen, da diese die „Kommstruktur" des deutschen

Sozialpädiatrie in der Praxis. http://dx.doi.org/10.1016/B978-3-437-31630-3.00006-7
Copyright © 2014 Elsevier GmbH. All rights reserved.

Gesundheitssystems nur eingeschränkt nutzen. Eine weitere Aufgabe sind Begutachtungen bei Fragestellungen aus der Sozial- oder Jugendhilfe und dem Bildungsbereich.
Eine besondere Rolle spielen die **Schuleingangsuntersuchungen** des ÖGD, die leider nicht mehr in allen Bundesländern flächendeckend durchgeführt werden. Sie stellen eine wichtige Datenquelle für die Gesundheitsberichterstattung dar und bieten die Möglichkeit, gezielte Fragestellungen zur Prävalenz bestimmter Erkrankungen und gesundheitsbezogener Verhaltensweisen zu beantworten. Neuerdings werden sie auch zur frühzeitigen Feststellung von Förderbedarf vor der Einschulung genutzt. Bedeutsam ist auch die Rolle des ÖGD in der Beratung von Politik und Verwaltung bei regionalen Vorhaben, die Einfluss auf die Gesundheits- und Entwicklungsbedingungen von Kindern und Jugendlichen haben.
Ausführliche Informationen zum ÖGD finden sich in Kapitel 7 und Kapitel 9 (➤ Kap.7, ➤ Kap.9).

6.1.2 Ambulante medizinische Versorgung

Von den Ende 2011 berufstätigen 12.841 **Fachärzten für Kinder- und Jugendmedizin** arbeiteten 6.647 als Vertragsärzte im ambulanten Bereich. 721 Kinder- und Jugendärzte arbeiteten 2011 als angestellte Ärzte in Praxen oder medizinischen Versorgungszentren. Jährlich erwerben etwa 600 Personen die Facharztanerkennung für Kinder- und Jugendmedizin, davon sind etwa 70 % Frauen. Im Bundesdurchschnitt kommt ein vertragsärztlich tätiger Kinder- und Jugendarzt auf etwa 1.630 Kinder bis 15 Jahre.
Aufgrund eines veränderten Krankheitsspektrums (chronische Erkrankungen, neue Morbiditäten, sozial Benachteiligte) ist von einem erhöhten zeitlichen Betreuungsaufwand je Kind/Jugendlichen auszugehen. Aufgrund des **demographischen Wandels** werden künftig weniger Kindern und Jugendlichen je Kinder- und Jugendarzt versorgt werden. Bei abnehmender Bruttolebensarbeitszeit (Zunahme von Frauen in der Pädiatrie, Teilzeitarbeit, Arbeitszeitverkürzung) wird jedoch vermutet, dass die real zur Verfügung stehende Versorgungskapazität stärker abnehmen wird als die Zahl der und der Zeitaufwand für zu versorgenden Kinder und Jugendlichen.
Die Kinder- und Jugendarztdichte ist regional sehr verschieden. Die flächendeckende ambulante pädiatrische Versorgung ist in manchen Regionen Deutschlands bereits jetzt gefährdet. Man rechnet demzufolge mit einer Verstärkung dieses Trends in den kommenden Jahren.
Niedergelassene Kinder- und Jugendärzte sind in der Regel die ersten professionellen Ansprechpartner der Eltern bei gesundheitlichen Problemen ihrer Kinder. Sie führen außerdem Vorsorgeuntersuchungen, Impfungen und Gesundheitsberatungen durch. Sozialpädiatrische Fragen haben in der kinder- und jugendärztlichen Praxis heute einen hohen Stellenwert. Einige niedergelassene Kinder- und Jugendärzte arbeiten in Schwerpunkten (z. B. Kinderkardiologie, Neuropädiatrie). Viele arbeiten in intra- und interdisziplinären Qualitätszirkeln und Netzwerken mit.
Ende 2011 führten die **Schwerpunktbezeichnungen** Neonatologie 1.071, Kinderkardiologie 357, Neuropädiatrie 430 und Hämatoonkologie 170 Ärzte. Bundesweit sind als Zusatzweiterbildungen im Gebiet Kinder- und Jugendmedizin außerdem anerkannt: Kinderendokrinologie, Kindergastroenterologie, Kindernephrologie, Kinderpneumologie. Offizielle Planungszahlen zum Versorgungsbedarf mit Schwerpunkten und Zusatzweiterbildungen in der Kinder- und Jugendmedizin existieren nicht.
Für die medizinische Versorgung von Kindern und Jugendlichen werden relativ häufig auch die Kompetenzen von Orthopäden, HNO-Ärzten, Augenärzten, Kinderchirurgen und Kinder- und Jugendpsychiatern benötigt.

Kinder- und Jugendärzten obliegt die Koordination der medizinischen Behandlungen und die langfristige familienorientierte Betreuung.
Die chirurgische Betreuung von Kindern kann nur z. T. durch die 494 **Kinderchirurgen** Deutschlands (Stand 2011) erfolgen. Diese sind überwiegend im stationären Bereich tätig. In den letzten Jahren besteht allerdings ein deutlicher Trend hin zur ambulanten Kinderchirurgie.
Die psychiatrische Versorgung von Kindern und Jugendlichen erfolgt durch **Fachärzte für Kinder- und Jugendpsychiatrie und Psychotherapie.** Von den insgesamt 1.806 registrierten Ärzten (Stand 2011) arbeitet etwa die Hälfte im ambulanten Bereich. 3.110 **Kinder- und Jugendlichen-Psychotherapeuten** beteiligten sich 2012 an der ambulanten vertragsärztlichen Versorgung. Es existieren große Unterschiede in der regionalen Versorgungsdichte; eine flächendeckende Versorgung besteht nicht. Nach Krankenkassenangaben erhalten etwa 10 % aller Schulkinder zu irgendeinem Zeitpunkt eine Psychotherapie.
In der ambulanten medizinischen Versorgung haben interdisziplinäre und vernetzte Versorgungsformen eine zunehmende Bedeutung.

6.1.3 Stationäre medizinische Versorgung

Diese erfolgt durch **Kliniken und Abteilungen für Kinder- und Jugendmedizin** bzw. **Kinder- und Jugendpsychiatrie,** z. T. auch durch **kinderchirurgische Abteilungen.** Es bestehen große regionale Unterschiede bei Kapazitäten und Flächendeckung. Zwischen 1993 und 2009 ist die Zahl der Betten in kinder- und jugendärztlichen Abteilungen von 27.354 auf 19.294 und die durchschnittliche Verweildauer von 7,5 auf 4,5 Tage zurückgegangen.
Insbesondere für kleinere Kliniken/Abteilungen wird es zunehmend schwierig, eine auseichend breite fachliche Kompetenz und ein ausgeglichenes Kosten-/Erlös-Verhältnis vorzuhalten. In den kommenden Jahren ist daher ein weiterer Rückgang der Zahl der Kliniken für Kinder- und Jugendmedizin zu erwarten. Für eine **flächendeckende Versorgung** wird eine maximale Entfernung von 30 km zur nächsten stationären Einrichtung für Kinder- und Jugendmedizin und deren Erreichbarkeit in 40 Minuten als ausreichend angesehen.
Etwa 24 % aller stationär behandelten Kinder und Jugendlichen bis 15 Jahren werden in Kliniken der Erwachsenenmedizin betreut (Fehlbelegung).
An 20,5 % der Kinderkliniken gab es 2009 fachlich selbstständige neuropädiatrische Abteilungen, an 59 % neuropädiatrischen Spezialambulanzen.
Eine gewisse Sonderstellung nehmen die etwa 10 Fachkliniken mit sozialpädiatrischem Schwerpunkt ein.

6.1.4 Ambulante Behandlung im Krankenhaus

Hochschulambulanzen (nach § 117 SGB V) nehmen de facto einen besonderen Auftrag in der Krankenversorgung wahr, der über den für Forschung und Lehre erforderlichen Umfang hinausgeht. Der Umfang der in Anspruch genommenen Versorgungsleistungen beruht auf der fachlichen Kompetenz dieser Einrichtungen. Sie erhalten überregionale Zuweisungen durch Vertragsärzte bzw. niedergelassene Fachärzte. Die zugewiesenen Patienten haben oft einen Bedarf an interdisziplinärer und komplexer Diagnostik und Therapie. Die Erlöse aus den sog. Poliklinikpauschalen liegen deutlich unter den durchschnittlich anfallenden Kosten.

An einigen Universitätskliniken haben sich in den letzten Jahren „Zentren für seltene Erkrankungen (ZSE)“ etabliert.

Ambulante Behandlung im Krankenhaus kann auch nach § 116b SGB V als hochspezialisierte Leistung für spezielle Interventionen und seltene Erkrankungen oder Erkrankungen mit besonderem Krankheitsverlauf entsprechend des jeweils gültigen Katalogs erbracht und direkt mit den Krankenkassen abgerechnet werden. Das Verhältnis von Kosten und Erträgen ist hier besser. In der Kinder- und Jugendmedizin ist dieses Konstrukt bislang wenig verbreitet.

Institutsambulanzen sind nach § 118 SGB V zur Teilnahme an der vertragsärztlichen Versorgung ermächtigt. In psychiatrischen Kliniken, psychiatrischen Abteilungen an Allgemeinkrankenhäusern, Universitätskliniken oder Kliniken für Kinder- und Jugendpsychiatrie bieten sie ambulante Leistungen an. Versorgungsschwerpunkte sind schwer und chronisch Kranke im Rahmen der Nachsorge, aber auch Notfallpsychiatrie. Ziel ist die Vermeidung oder Verkürzung stationärer Behandlung und die Sicherstellung einer Behandlung für Patienten, die von niedergelassenen Ärzten nicht ausreichend erreicht werden. Die Vergütung erfolgt außerhalb der vertragsärztlichen Gesamtvergütung direkt durch die Kassen.

Eine **persönliche Ermächtigung** (nach § 31 Ärzte-ZV) von an Kliniken tätigen Ärzten bzw. eine Ermächtigung von Einrichtungen zur ambulanten vertragsärztlichen Versorgung nach §116 und §120 SGB V kann für bestimmte Leistungen bzw. Krankheiten durch den jeweiligen Zulassungsausschuss bei bestehender oder drohender Unterversorgung erteilt werden. Die Vergütung erfolgt innerhalb der vertragsärztlichen Gesamtvergütung.

6.1.5 Sozialpädiatrische Zentren

Die 145 Sozialpädiatrischen Zentren (SPZ) in Deutschland sind nach § 119 und § 43 a SGB V ambulant tätige **interdisziplinäre, ärztlich geleitete Einrichtungen,** in denen auf Zuweisung eines Vertragsarztes Kinder und Jugendliche aller Altersstufen behandelt werden, die aufgrund Art, Schwere oder Dauer ihrer Erkrankung nicht durch geeignete Vertragsärzte oder Frühförderstellen betreut werden können. Sozialpädiatrische Zentren werden von den regional zuständigen Zulassungsausschüssen zur vertragsärztlichen Versorgung ermächtigt. Die Vergütung erfolgt außerhalb der vertragsärztlichen Gesamtvergütung.

In Deutschland werden jährlich 250.000–300.000 Kinder und Jugendliche in diesen Einrichtungen behandelt (Stand 2012). Dabei handelt es sich besonders um Kinder und Jugendliche mit Regulations-, Entwicklungs- oder Verhaltensstörungen, körperlicher Behinderung, Intelligenzminderung sowie Kinder oder Jugendliche mit erheblichen bio-psycho-sozialen Risikofaktoren. Bei chronisch kranken Kindern stehen die begleitenden psychosozialen Probleme im Mittelpunkt. Die Deutsche Gesellschaft für Sozialpädiatrie und Jugendmedizin betrachtet etwa 1 SPZ auf 450.000 Einwohner als die dem aktuellen Bedarf angemessene Planungszahl. Das Bundessozialgericht hat in einem Urteil ein SPZ je 400.000 Einwohner für angemessen erachtet.

Sozialpädiatrische Zentren arbeiten diagnostisch und therapeutisch überwiegend in der Habilitation und der Rehabilitation. Sie sind Koordinationsstelle und stellen Case-Manager in einem Netzwerk von niedergelassenen Ärzten und Therapeuten, Kliniken, Förder- und Beratungsstellen, Kindergärten, Schulen, Ämtern u. a.

Ziele sozialpädiatrischer Arbeit sind, die Kompetenzen des Kindes und seiner Eltern in definierten Zielbereichen zu verbessern, bei der Krankheitsbewältigung zu helfen und zu einer bestmöglichen sozialen Integration und Teilhabe am alters- und entwicklungstypischen sozialen Leben beizutragen.

Dabei erarbeitet ein Team aus Kinder- und Jugendärzten, Psychologen, Sozialpädagogen, Krankengymnasten, Logopäden, Ergotherapeuten, Heilpädagogen und ggf. weiteren Berufsgruppen individuelle Therapieziele in Abstimmung mit den Betroffenen und ihren Eltern. Spezifisches Fachwissen und komplexe Vorgehensweisen sind dabei erforderlich. Sozialpädiatrische Zentren sind meist organisatorisch in Kliniken für Kinder- und Jugendmedizin integriert oder selbstständige Abteilungen in einem Klinikum. Das Altöttinger Papier beschreibt die Qualitätsanforderungen an die Struktur und die Behandlungsabläufe. Zahlreiche Qualitätspapiere der Deutschen Gesellschaft für Sozialpädiatrie und Jugendmedizin beschreiben die inhaltliche Arbeit der SPZ.

6.1.6 Sozialmedizinische Nachsorge

Sozialmedizinische Nachsorge gemäß § 43 Absatz 2 SGB V ermöglicht von der Klinik ausgehend eine **aufsuchende häusliche Betreuung und Hilfe,** bis die Familie von einer anderen Einrichtung, z. B. einer Frühförderstelle, einem SPZ oder einer Ambulanz weiter betreut werden kann.

Zielgruppen sind chronisch kranke oder schwerstkranke Kinder bis zum 14. Lebensjahr, deren Erkrankungen in einer Diagnoseliste aufgeführt sind (z. B. Kinder mit bösartigen Neubildungen, zerebralen Lähmungen, angeborenen Fehlbildungen und Chromosomenanomalien, nach Frühgeburtlichkeit, Organversagen, Verletzungen, Verbrennungen, Vergiftungen, Missbrauch) und Kinder im Finalstadium einer Erkrankung. Die Kinder müssen unmittelbar zuvor in stationärer Krankenhausbehandlung oder Rehabilitation gewesen sein.

Ziele der sozialmedizinischen Nachsorge sind die Verkürzung des stationären Aufenthalts und die Sicherung der anschließenden ambulanten Behandlung. Darüber hinaus soll die individuelle Krankheitsbewältigung, die gesundheitsbezogene Lebensqualität und die Integration der Krankheit in den Alltag gefördert werden. Untersuchungen weisen darauf hin, dass die sozialmedizinische Nachsorge durch Verkürzung der stationären Behandlungszeit und Vermeidung von stationären Wiederaufnahmen auch eine ökonomisch sinnvolle Maßnahme ist.

Sozialmedizinische Nachsorgemaßnahmen umfassen in Ergänzung der bisher schon erbrachten gesetzlich festgelegten stationären und ambulanten Leistungen sowie der häuslicher Krankenpflege und Leistungen der Pflegeversicherung die Analyse des Versorgungsbedarfs, die Koordinierung der verordneten Leistungen sowie Anleitung und Motivierung zur Inanspruchnahme der verordneten Leistungen. Dies trägt den vielen organisatorischen, finanziellen, sozialen und psychischen Belastungen eines Kindes und seiner gesamten Familie im Gefolge einer schwersten akuten oder chronischen Erkrankung Rechnung.

Ein Nachsorgeteam besteht aus Kinderkrankenschwester, Sozialarbeiter/-pädagogen und einem Facharzt für Kinder- und Jugendmedizin. In der Regel wird ein Case-Manager für ein Kind und seine Familie festgelegt. Die Nachsorge erstreckt sich in der Regel auf einen Zeitraum von 6–12 Wochen. Die Maßnahme kann durch den behandelnden Arzt der stationären Einrichtung oder den behandelnden Vertragsarzt verordnet werden; die Kosten müssen von der Krankenkasse übernommen werden, sofern die Anspruchsvoraussetzungen erfüllt sind. Wegen unzureichender Stundensätze, die von den Kassen gezahlt werden, ist die Finanzierung der sozialmedizinischen Nachsorge ein Problem, eine organisatorische Anbindung an eine Klinik für Kinder- und Jugendmedizin und/oder ein Sozialpädiatrisches Zentrum ist zu empfehlen.

6.1.7 Stationäre Vorsorge und Rehabilitation

Die Anspruchsgrundlagen sind für die gesetzliche Krankenkasse in SGB V § 23 Abs. 4 (stationäre medizinische Vorsorgemaßnahmen), § 40 Abs. 2 (stationäre medizinische Rehabilitationsmaßnahmen), § 11 Abs. 3 (Mitaufnahme einer Begleitperson) und § 41 (Mutter-Kind-Kur) festgelegt. Für die gesetzliche Rentenversicherung findet man sie in SGB VI § 31 (stationäre Kinderheilbehandlung).

2009 wurden bei der Deutschen Rentenversicherung etwa 80.000 Anträge auf eine stationäre Rehabilitationsmaßnahme für Kinder oder Jugendliche gestellt, von denen etwa die Hälfte bewilligt wurde. Bei den Krankenkassen lag die Ablehnungsquote für beantragte **Mutter-Kind-Kuren** bei bis zu 50 %.

Für die früher als Kur bezeichnete Versorgungsform der stationären Vorsorge und Rehabilitation gab es im Jahr 2011 in Deutschland 55 Fachabteilungen mit 6.324 Betten. Die durchschnittliche Verweildauer betrug 29,7 Tage. Hauptindikationen für eine stationäre Vorsorge- oder Rehabilitationsbehandlung waren 2010: Asthma und andere Krankheiten des Atmungssystems (31 %), psychische und Verhaltensstörungen (21 %), Adipositas (20 %) und Hauterkrankungen (9 %). Der Altersgipfel bei der Inanspruchnahme lag bei 10–14 Jahren.

Eine besondere Stellung im Rahmen der stationären Rehabilitation nehmen die etwa 10 Kliniken für **neurologischen Rehabilitation** bei Kindern- und Jugendlichen in Deutschland ein. Sie führen nach erworbenen ZNS-Schädigungen im Kindes- und Jugendalter ein komplexes medizinisches und pädagogisches Rehabilitationsprogramm (Phasen A-D) durch. Weitere Informationen finden sich in Kapitel 39 (➤ Kap. 39).

6.1.8 Palliativversorgung/Hospize

Etwa 20.000 Kinder und Jugendliche in Deutschland leiden an schwerwiegenden, zum Tode führenden Erkrankungen. **Palliativversorgung** ist eine multidisziplinäre, aktive und aufsuchende Betreuung dieser Kinder und Jugendlichen und ihrer Familien. Zentrale Aufgaben einer palliativmedizinischen Versorgung sind die möglichst effektive Behandlung von Symptomen und körperliches Wohlbefinden, damit Betroffene ihre Bedürfnisse artikulieren, befriedigen und an der Gemeinschaft soweit möglich und gewünscht teilnehmen können. Die Verarbeitung des Krankheitsgeschehens und der Trauer von Betroffenen und Angehörigen sind wichtige Elemente der Palliativversorgung.

Ende 2009 existierten in Deutschland 10 **Hospize** zur stationären Begleitung sterbender, schwerstkranker oder behinderter Kinder, Jugendlicher und ihrer Familien. Viele betroffene Familien nehmen die Angebote eines Hospizes wiederholt wahr, Nur etwa 10–15 % der in Kinderhospizen betreuten Patienten sterben dort.

Seit 2009 ist für Kinder und Jugendliche die **spezialisierte ambulante Palliativversorgung** festgelegte Kassenleistung (§37b SGB V). Sie soll bei nicht heilbaren fortschreitenden und weit fortgeschrittenen Erkrankungen, die eine besonders aufwändige Versorgung benötigen, eine spezialisierte ärztliche und pflegerische Betreuung in der häuslichen Umgebung sicherstellen.

Schwierigkeiten, eine einvernehmliche und auskömmliche Kostenregelung zu finden, haben dazu geführt, dass bislang erst in wenigen Regionen ambulante pädiatrische Palliativteams ihre mobile Arbeit aufnehmen konnten. Deren Erfahrungen zeigen, dass nicht nur medizinische Themen wie Schmerz, Atemnot, Erbrechen und Krampfanfälle von Bedeutung sind, sondern darüber hinaus ein biopsychosoziales und spirituelles Betreuungskonzept angemessen ist.

6.1.9 Kinderkrankenpflege

Gesundheits- und Kinderkrankenpfleger/-innen sind ganz überwiegend auf Säuglings- und Kinderstationen in Kliniken für Kinder- und Jugendmedizin und anderen Kliniken tätig. Inhalte der Berufsausbildung und berufliche Perspektiven unterliegen einem Wandel bei abnehmender Zahl der Beschäftigten auf Stationen und Zunahme der Beschäftigung im ambulanten Bereich sowie sinkender Kinderzahl und veränderten Morbiditäten.
Die **häusliche Kinderkrankenpflege** (§ 37 SGB V) bedarf einer ärztlichen Verordnung. Hierdurch sollen Krankenhausaufenthalte vermieden oder verkürzt und die ärztliche Behandlung zu Hause gesichert werden. Indikationen sind z. B. häusliche Beatmung oder Sauerstoffbehandlung, Kinder mit parenteraler Therapie, Ernährungsproblemen, extremem Pflegeaufwand, z. B. infolge Immobilität oder Frühgeborene mit komplexen medizinischen Problemen in den ersten Wochen nach Klinikentlassung. Pflegedienste sollen die Pflegekompetenz der Eltern stärken, sie entlasten und unterstützen durch Organisation weiterer Hilfen und die Vermittlung weiterer Kontakte. In Deutschland existieren in unterschiedlicher Trägerschaft etwa 100 Pflegedienste, die sich auf die Kinderkrankenpflege spezialisiert haben. Die Verordnungen für häusliche Kinderkrankenpflege haben in den letzten Jahren stark zugenommen. Die ambulante und stationäre Pflege in der Palliativversorgung ist als weiteres Aufgabengebiet.

6.1.10 Heilmittelerbringer

Im Jahr 2011 wurden insgesamt 33 Mio. Rezepte zur **Heilmittelbehandlung** für einen der 69,6 Mio. Versicherten der GKV von einem der 139.000 Vertragsärzte ausgestellt, davon 20 Mio. Physiotherapie-Rezepte, 2,0 Mio. Ergotherapie-Rezepte und 1,5 Mio. Sprachtherapie-Rezepte.
Innerhalb der GKV betrug 2011 der Heilmittelumsatz 4,7 Milliarden Euro. Ergotherapie und Sprachtherapie bildeten dabei 5,6 % bzw. 3,9 % des Verordnungsgeschehens, ihr Umsatzanteil lag bei 14,1 % bzw. 10,8 %.
Allgemeinmediziner und Orthopäden verordneten überwiegend physiotherapeutische Leistungen (88,3 % der Heilmittelverordnungen). Kinder- und Jugendärzten waren für 26 % der Ergotherapie- und 44 % der Sprachtherapie-Verordnungen verantwortlich.
2011 waren in Deutschland 136.000 Physiotherapeuten tätig (Vollzeitäquivalent 99.000). Es waren 35.500 zur Heilmittelerbringung zugelassen. Die zweitgrößte Gruppe sind die 35.000–50.000 Ergotherapeuten. Die Zahl der Stimm-, Sprech- und Sprachtherapeuten liegt bei etwa 16.000.
Ergotherapie wird insbesondere bei Jungen unter 15 Jahren verordnet. Auch die Sprachtherapie wird häufiger bei Jungen als bei Mädchen verordnet. Seit 2003 zeigt sich ein deutlicher Anstieg der Sprachtherapie- und ein leichter Anstieg der Ergotherapie-Verordnungen in dieser Altersgruppe. Physiotherapeutische Leistungen werden etwa gleich bleibend häufig verordnet.
Der Verordnungsgipfel der **Ergotherapie** liegt bei Kindern im Alter zwischen 5 und 9 Jahren: Ein gutes Drittel aller ergotherapeutischen Leistungen wird für Kinder dieses Alters erbracht. Im Jahr 2011 wurden in dieser Altersgruppe 79 von 1.000 AOK-versicherten Kindern mit Ergotherapie behandelt (Jungen: 109 von 1.000, Mädchen: 37 von 1.000). Im Einschulungsalter von 6 Jahren waren 135 von 1.000 Jungen beim Ergotherapeuten in Behandlung (Mädchen: 54 von 1000). Die Zahl der Verordnungen lag etwa bei 424 (Jungen) bzw. 167 (Mädchen) je 1.000 Jungen bzw. Mädchen dieses Alters.

In der **Sprachtherapie** entfallen knapp 70 % aller Leistungen auf Kinder und Jugendliche bis zu einem Alter von 15 Jahren. Davon erhalten die Jungen bis 14 Jahren, die einen Anteil an allen Versicherten von 6,4 % ausmachen, 42 % aller sprachtherapeutischen Leistungen. Der Verordnungsgipfel der Sprachtherapie lag 2011 bei Kindern im Alter von 6–10 Jahren: Die Hälfte aller sprachtherapeutischen Leistungen wurde für Kinder dieser Altersgruppe erbracht. Bei den sechsjährigen Kindern wurde der Verordnungsgipfel erreicht: Je 1.000 versicherte Jungen wurden etwa 250 sprachtherapeutisch behandelt, für sie wurden 871 Verordnungen ausgestellt. Bei den sechsjährigen Mädchen lag der Anteil der Behandelten bei 16,8 %, die Zahl der Verordnungen bei 548 je 1.000 Mädchen.
Die Maßnahmen der **Physiotherapie** konzentrieren sich – anders als bei Ergo- und Sprachtherapie – auf die höheren Lebensalter. Die meisten Verordnungen erfolgen bei den weiblichen Versicherten in der Altersgruppe 70–79 Jahre. Bei den männlichen Versicherten liegt der Verordnungsgipfel bei 75–85 Jahren. Mit 62,1 % aller Behandlungen erhalten Frauen rund ein Drittel mehr physiotherapeutische Behandlungen als Männer. Die Krankengymnastik bei Kindern hatte 2011 einen Verordnungsanteil von 4,5 %.
Die Ausbildung der Ergo- und Physiotherapeuten sowie Logopäden erfolgt in Deutschland traditionell in Fachschulen. Neuerdings wurden einige akademische Studiengänge für Physio-, Ergo- und Sprachtherapie etabliert, die einen Bachelor-Abschluss ermöglichen.

6.2 Vorschulische, schulische und berufliche Förderung

6.2.1 Krippen

Die **institutionelle Betreuung** von Kindern unter 3 Jahren in Tageseinrichtungen **(Krippen)** wird vermehrt nachgefragt und politisch gefördert. Die **Gründe** dafür sind vielfältig: Auf Seiten der Eltern spielen wirtschaftlich-ökonomische Gründe, der Wunsch nach beruflicher Verwirklichung und Weiterentwicklung der eigenen beruflich-fachlichen Qualifikationen, die Pflege sozialer Beziehungen, soziale Anerkennung, ein gewandeltes Rollenverständnis in der Partnerschaft und gesellschaftliche Erwartungen eine Rolle. Für Politik und Wirtschaft stehen der Mangel an qualifizierten Fachkräften, die Stabilisierung des Sozialsystems (Steuern, Beiträge zu Kranken-, Rentenversicherung u. a.) und die Nutzung menschlicher Ressourcen bei einer infolge der demographischen Entwicklung sinkenden Zahl von Menschen im arbeitsfähigen Alter im Fokus.
Grundsätze der Förderung von Kindern in Tageseinrichtungen sind in § 22 Abs 2. SGB VIII festgelegt. Die Einrichtungen sollen die Entwicklung des Kindes zu einer eigenverantwortlichen und gemeinschaftsfähigen Persönlichkeit fördern, die Erziehung und Bildung in der Familie unterstützen und ergänzen und den Eltern helfen, Erwerbstätigkeit und Kindererziehung besser miteinander zu vereinbaren. Im Kinderförderungsgesetz sind der stufenweise Ausbau von Tageseinrichtungen und der Rechtsanspruch auf einen Krippenplatz für Kinder ab vollendetem 1. Lebensjahr ab August 2013 geregelt.
Die Zahl der **Betreuungsplätze** betrug im März 2012 in Deutschland 558.000. Es bestehen deutliche West-Ost-Unterschiede: In Sachsen-Anhalt besuchten 57,5 % der Kinder unter 3 Jahren eine Krippe, in Nordrhein-Westfalen 18,1 %. Die Zahl der Krippenplätze hat in den letzten Jahren rasch zugenommen.
Intensiver Kontakt zu anderen Kindern ähnlichen Alters in Krippen kann das soziale Lernen und die Entwicklung von Kindern fördern. Eine geringe **Betreuungsqualität** in Krippen erhöht das

Risiko für eine unsichere Mutter-Kind-Bindung. Eine hohe Betreuungsqualität begünstigt die kognitive Entwicklung der Kinder. Mit zunehmender Dauer einer Gruppenbetreuung in Krippen steigt das Risiko für expansives Problemverhalten der Kinder leicht an. Es sind noch erhebliche Anstrengungen erforderlich, um die notwendige Betreuungsqualität sicher zu stellen.
Dabei sind auch medizinische Gesichtspunkte zu berücksichtigen.
Pflege, Ernährung, Hygiene, Infektionsschutz und Sicherheit für Säuglinge und Kleinkinder müssen besonders hohe Anforderungen erfüllen. Der Aufbau einer sicheren Primärbindung vor Eintritt in eine Krippe ist sehr bedeutsam, daher ist eine außerfamiliäre Betreuung im ersten Lebensjahr nicht generell empfehlenswert. Besondere Herausforderungen an die Krippenbetreuung stellen Kinder mit besonderen Bedürfnissen dar, z. B. chronischer Krankheit oder Behinderung, körperlicher oder emotionaler Vernachlässigung, Kinder aus Familien mit sozialer Benachteiligung oder Isolation.
Chronischer Stress durch Trennungserfahrungen kann zu neurobiologischen Veränderungen und sozial-emotionalen Störungen führen, z. B. Regulationsstörungen, Interaktionsstörungen, Depressivität, Übererregbarkeit. Krippenkinder erleiden im Vergleich zu Hauskindern in den ersten 3 Lebensjahren häufiger Infekte. In Krippen betreute Kinder sollten gemäß den gültigen Impfempfehlungen vollständig geimpft sein. Als Einzelkinder aufwachsende Krippenkinder haben ein deutlich niedrigeres Allergierisiko.
Krippen sollten **Qualitätsmerkmale** der Struktur erfüllen. Der Betreuungsschlüssel sollte bei jeweils 1 Betreuerin für 2 Säuglinge von 9–12 Monaten, 3 Kindern zwischen 12 und 24 Monaten und 4 Kindern zwischen 24 und 36 Monaten liegen. Die Qualifikation des Fachpersonals sollte besonders auf die spezifischen Erfordernisse der Altersgruppe und Institution abgestimmt sein, eine Fachhochschulausbildung für einen Teil der Erzieherinnen ist anzustreben. Die Eltern sollten in die Betreuungskonzepte eingebunden sein. Vorteilhaft ist eine gemeinwesenorientierte Aktivität der Krippe im Stadtteil und in der Kommune. Eine betriebsmedizinische Begleitung ist erforderlich.
Es sind noch erhebliche Anstrengungen erforderlich, um zu einer ausreichenden Zahl qualifizierter Erzieherinnen und Tagespflegepersonen zu kommen. Dies trifft auch für **Tagesmütter** zu, die inzwischen eine Qualifizierung in Kindertagespflege in einem 160-stündigen Curriculum nachweisen müssen.

6.2.2 Kindergärten

Die politischen Zuständigkeiten im Bereich **Kindergarten** sind länderspezifisch geregelt. Beispielsweise fällt in Baden-Württemberg die institutionelle Betreuung der unter 3-Jährigen in Kinderkrippen in den Bereich des Sozialministeriums, die der über 3-Jährigen in Kindergärten/-tagesstätten in den des Kultusministeriums.
Die pädagogische Qualität in deutschen Kindergärten ist großenteils mittelmäßig. Ostdeutsche Einrichtungen schneiden im Durchschnitt schlechter ab als westdeutsche **(NUBBEK-Studie).** Als Folge des PISA-Schocks, der Ergebnisse der Schuleingangsuntersuchungen, der Befunde in den Grundschulen u. a. m. wird in zahlreichen Modellprojekten, neuerdings in einigen Bundesländern auch flächendeckend versucht, sprachliche und andere Fähigkeiten bei Kindergartenkindern stärker und besser als bisher zu fördern. Aus methodischen und praktischen Gründen ist es schwierig, einen wissenschaftlichen Beleg für die Wirksamkeit der verschiedenen **Sprachförderprogramme** zu liefern.

Neben Regelkindergärten existieren je nach Bundesland in unterschiedlichem Ausmaß **Sonderkindergärten** (Kindergärten für besonders förderungsbedürftige Kinder). Der zunehmend erkennbare Elternwunsch zur Integration/Inklusion von entwicklungsverzögerten, verhaltensauffälligen oder behinderten Kindern in Regeleinrichtungen hat vielerorts zur Bildung von **integrativen Kindergärten** oder **integrativen Gruppen** geführt. Diese zeichnen sich durch kleinere Gruppengröße von meist etwa 15 Kinder mit einer kleineren Zahl auffälliger/behinderter und einer größeren Zahl unauffälliger/nichtbehinderter Kinder sowie durch zusätzliches, z. T. heilpädagogisches und/oder therapeutisches Personal aus.
Wo dies nicht möglich oder gewünscht ist, stellt für Kinder mit besonderem Förderbedarf die **Einzelintegration** in Gruppen des Regelkindergartens eine Alternative dar. Diese Maßnahme ist von den Eltern beim zuständigen Jugend- und Sozialamt zu beantragen und bedarf der Befürwortung durch ein amtsärztliches Gutachten, in dem eine drohende Behinderung im Sinne des Gesetzes (§ 53 SGB XII, § 35 a SGB VIII) festgestellt wird. Die Bewilligung erfolgt durch das zuständige Jugendamt. Sie kann pädagogische und begleitende Hilfen umfassen.
Die Aufgaben des ÖGD im Kindergartenbereich werden in Kapitel 9 (➤ Kap. 9) behandelt.

6.2.3 Schulen

Die schulischen Strukturen im Primar- und Sekundarbereich in den einzelnen Bundesländern sind verschieden und unterliegen andauernden Veränderungen. Ein steigender Anteil von Schulen bietet eine **Ganztagsbetreuung** an. Diese ist für Kinder und Jugendliche aus sozial benachteiligten Familien eine wichtige komplementäre Fördermöglichkeit und kann für berufstätige Eltern eine Entlastung darstellen. Die fachliche und personelle Zuständigkeit für die **Schulen** liegt bei den Kultusbehörden der Bundesländer und den ihnen nachgeordneten Behörden (z. B. Regierungspräsidien, [Ober-]Schulamt). Dies gilt auch für die Sonderschulen. Die materielle Ausstattung der Schulen ist Aufgabe der unteren Verwaltungsebene (Städte, Landkreise).
Für die Schulbetreuung ist primär die **Grundschule** am Wohnort des Kindes zuständig. Eine Rückstellung vom Schulbesuch kann in der Regel nur dann erfolgen, wenn im darauf folgenden Jahr voraussichtlich die Aufnahme in der Regelschule möglich sein wird.
Für Kinder mit Lernschwächen sind in den Bundesländern unterschiedliche institutionelle Fördermöglichkeiten geschaffen worden (schulvorbereitende Einrichtungen, Diagnoseförderklassen, Präventivklassen, integrative Klassen, Außenklassen, Berufsvorbereitungsjahr u. a. m.). Die Entwicklungen sind auch in diesem Bereich vielfältig und stark länderspezifisch geprägt. Ein **sonderpädagogischer Förderbedarf** wurde 2012 in Deutschland bei 6,2 % aller Kinder festgestellt. Von diesen besuchten 70 % Förderschulen, 30 % allgemeine Schulen. Die **Integration/Inklusion** von Kindern und Jugendlichen mit manifesten oder drohenden Behinderungen in Regelschulen ist je nach Art und Schwere der Behinderung möglich und wird zunehmend gewünscht (Einzelintegration, Integrationsklassen). Neben der Frage der Finanzierung muss geklärt werden, welche Vorleistung die Regeleinrichtungen zu erbringen haben, z. B. Klassengröße und -zusammensetzung, Lehrerqualifikation, Mitarbeit der Eltern, Raumangebote usw., und welche Kinder von einer Integration/Inklusion profitieren. Eine enge Absprache von aufnehmender Schule, Eltern, Kind/Jugendlichem, Therapeuten, Ärzten und sonderpädagogischem Berater ist sehr empfehlenswert. Für finanzielle Aspekte kommen § 53 SGB XII bei körperlicher und geistiger Behinderung, § 35 a SGB VIII bei manifester oder drohender seelischer Behinderung (z. B. Lese-Rechtschreibstörung) zur Anwendung.

In mehreren Bundesländern gibt es **schulvorbereitende Einrichtungen** (SVE) meist ab dem 4. Lebensjahr. Diese sind als schulische Einrichtungen kostenlos – einschließlich des Schulwegtransports – und meist spezialisierten Förderschulen angeschlossen.
In Bayern hat man für die ersten drei Schuljahre anstelle der Schule für Lernbehinderte eine Eingangsstufe als „Diagnose- und Förderklasse" eingeführt. Hier wird Kindern mit Entwicklungsproblemen der Stoff der ersten zwei Grundschuljahre innerhalb von drei Jahren vermittelt, um dann die endgültige Schullaufbahn besser festlegen zu können.
Sonderschulen, sonderpädagogische Förderzentren und schulvorbereitende Einrichtungen gibt es in Deutschland für eine Vielzahl verschiedener Störungen, insbesondere für:
- Störungen der Sprachentwicklung
- Hörbehinderungen bzw. Taubheit
- Sehbehinderungen bzw. Blindheit
- Körperbehinderungen
- Geistige Behinderungen
- Verhaltensstörungen bzw. Erziehungsschwierigkeiten.

Sonderschulen sind meist in der Trägerschaft eines gemeinnützigen Vereins, z. B. dem Bundesverband der Lebenshilfe u. Ä. Sie können die Aufnahme eines Kindes ablehnen, wenn seine Störung nicht ihrem Konzept entspricht. Während die Sonderschulen einerseits einen zunehmend hohen Spezialisierungsgrad aufweisen, wird andererseits versucht, die Spezialisierung der Einrichtung zu überwinden und Kinder mit unterschiedlichen Entwicklungsstörungen gemeinsam zu betreuen. Neuerdings sollen **mobile sonderpädagogische Dienste** eine bessere Verbindung zwischen Förderschule und Regelschule ermöglichen.
In der Regel erfolgt in der Sonderschule eine Betreuung über 10 Jahre. Viele Sonderschulen bieten am Nachmittag eine **Hortbetreuung** an, die über Eingliederungshilfe nach SGB XII finanziert wird. Während dieser Zeit findet häufig auch eine medizinisch begründete Therapie statt. Über die Aufnahme in diese Einrichtung erfolgt ein Feststellungsbescheid, der in der Regel von der Beurteilung eines für die jeweilige Behinderung zuständigen Landesarztes abhängig ist.
Integrative/Inklusive Schulen mit gesunden Kindern und solchen mit unterschiedlichen Entwicklungsstörungen werden erst vereinzelt, meist nach speziellen pädagogischen Ausrichtungen, z. B. nach der Pädagogik von Maria Montessori, betrieben. Oft ist dies nur in den ersten Schuljahren konsequent möglich. In einzelnen Bundesländern werden durch zusätzliche Betreuungsangebote möglichst viele Kinder mit Behinderungen und Entwicklungsstörungen in Regelschulen integriert. Andererseits wird aber auch versucht, zunehmend nichtbehinderte Kinder in Sonderschulen aufzunehmen, was jedoch nur bei einer Lernzielgleichheit, z. B. in Sehbehinderten- und Sprachheilschulen, möglich ist.

6.2.4 Frühförderstellen

Die etwa 750 **Frühförderstellen** in Deutschland sind meist pädagogisch geleitete Einrichtungen, die in der Regel ohne ärztliche Mitwirkung tätig sind. Frühförderstellen bieten bei drohender oder manifester Entwicklungsstörung ambulante oder aufsuchende Hilfe für Familien mit Kindern im Vorschulalter an. Die Beratung und Förderung erfolgt wohnortnah ohne ärztliche Überweisung und ohne Kosten für die Eltern. Versorgungsdichte und organisatorische Abläufe und Zuständigkeiten sind länderspezifisch. Frühförderstellen haben in den letzten Jahren eine zunehmende Bedeutung in der Betreuung von Kindern aus sozial benachteiligten Familien, bei der Bereitstellung früher Hilfen und im Kinderschutz bekommen. Frühförderstellen arbeiten in Kooperation und Vernetzung mit Ärzten, Therapeuten, Kindertagesstätten, Schulen und anderen Einrichtungen.

Allgemeine Frühförderstellen schließen keine Behinderungsarten aus. **Spezielle** Frühförderstellen richten sich an Kinder mit Sprach- oder Sinnesbehinderungen. Sie sind in der Regel an entsprechende Schulen assoziiert.

Etwa ein Drittel der Frühförderstellen sind **interdisziplinäre** Frühförderstellen (§ 30 und § 56 SGB IX). Sie arbeiten unter Beteiligung von Psychologen, Physiotherapeuten, Logopäden und Ergotherapeuten, in manchen Regionen auch von Ärzten. Arbeitsweise, inhaltliche Vorgaben und finanzielle Rahmenbedingungen (Mischfinanzierung durch Sozialhilfeträger und Krankenkassen) sind durch bundesweite Rechtsverordnungen und länderspezifische Verordnungen zwar überwiegend geklärt, die fachlich-inhaltliche Umsetzung des an sich begrüßenswerten Konzepts der „Komplexleistung" vor Ort wird durch restriktive finanzielle Rahmenbedingungen allerdings erschwert.

Nach Hochrechnungen versorgten die allgemeinen Frühförderstellen in Deutschland (ohne Berlin und Rheinland-Pfalz mit reiner Versorgungsstruktur über SPZ) im Jahresverlauf 2006 97.700 Kinder im Alter von 0–6 Jahren, davon erhielten 31 % der Kinder heilpädagogische und medizinisch-therapeutische Leistungen, 57 % ausschließlich heilpädagogische Frühförderung, 13 % ausschließlich medizinisch-therapeutische Leistungen.

Träger der Frühförderstellen sind Kommunen, Landkreise, Behindertenverbände u. a. gemeinnützige Vereinigungen, z. T. sind sie als sonderpädagogische Frühförderstellen den örtlichen Sonderschulen angegliedert. Dies gilt auch für die überregional tätigen speziellen Frühförderstellen für Kinder mit hochgradigen Seh- oder Hörstörungen.

Das in der Frühförderung eingesetzte Methodenspektrum ist außerordentlich breit (z. B. bindungsorientierte Förderung, heilpädagogische Übungsbehandlung, Psychomotorik, Hippotherapie, Elternberatung und -training, spezielle Methoden bei Sinnesbehinderungen, unterstützte Kommunikation). Die Frühförderstellen sind auf Bundesebene in der Vereinigung der interdisziplinären Frühförderung (VIFF) organisiert.

6.2.5 Beratungsstellen

Institutionelle **Erziehungsberatung** ist nach dem Kinder- und Jugendhilfegesetz eine kommunale Pflichtaufgabe. Erziehungsberatungsstellen, schulpsychologische und psychologische Beratungsstellen bieten Kindern, Jugendlichen und Familien Beratung und Therapie bei Fragen der Erziehung, Problemen im Verhalten, emotionalen Störungen, Schulschwierigkeiten, psychosomatischen Störungen und familiären Belastungs- und Krisensituationen an. Regional bestehen zusätzliche Schwerpunkte (z. B. Schreikinder, Kinder psychisch kranker Eltern). Manche Beratungsstellen arbeiten schwerpunktmäßig im **Kinderschutz.** Sie helfen Familien und Kindern auch bei der Bewältigung von Trennungs- und Scheidungsproblemen.

Für das gesamte Kindes- und Jugendalter und z. T. darüber hinaus zuständig sind spezielle Beratungsstellen, z. B. für Menschen mit **Autismus** oder für die **unterstützende Kommunikation.**

Eine Vernetzung der Beratungsstellen mit anderen Bereuungs- und Versorgungseinrichtungen ist hilfreich. Bei Kinderschutz ist dies zwingend erforderlich, z. B. mit den die Kinder betreuenden Vertragsärzten, Kinderschutzgruppen in Kinderkliniken und den zuständigen Jugendämtern.

6.2.6 Berufliche Ausbildung und Förderung

Das komplexe föderal organisierte Bildungswesen in Deutschland sieht nach Abschluss der 10. Klasse (Sekundarstufe 1) verschiedene weitere Bildungsgänge vor: gymnasiale Oberstufe/berufliches Gymnasium; Berufsfachschule, Berufsaufbauschule, Fachoberschule; FOS 13, Fachschule, duale Ausbildung, und verschiedene berufsvorbereitende Maßnahmen.

Das **Berufsvorbereitungsjahr** (BVJ), in manchen Ländern Vorqualifizierungsjahr Arbeit/Beruf (VAB) genannt, ist ein Schultyp für leistungsschwächere Schüler ohne Hauptschulabschluss, die nach der Beendigung oder dem Abbruch der Schule keinen Ausbildungsplatz finden bzw. aufgrund persönlicher Defizite keine Berufsausbildung beginnen können. Eine differenzierte Kompetenzanalyse dient der Teilhabe- und Berufswegeplanung. Die Schüler können durch das BVJ ihre Berufsschulpflicht erfüllen und gleichzeitig unter Umständen den Hauptschulabschluss erwerben. Im BVJ wird berufliches Grundwissen in einem oder mehreren Berufsfeldern vermittelt. Es dient der beruflichen Orientierung, der Erweiterung der Allgemeinbildung und dem Erwerb von Schlüsselqualifikationen, die zu verbesserten Chancen auf einen Ausbildungsplatz führen sollen. In Hauptschulkursen werden Jugendliche, die die allgemeine Schulpflicht ohne Abschluss erfüllt haben, im Zusammenhang mit einer Berufsvorbereitung auf den Erwerb des Hauptschulabschlusses vorbereitet. Die niedrigschwelligen Kurse richten sich vor allem an schulmüde oder schulängstliche Jugendliche.

Weitere, z. T. nur regional vorhandene Fördermöglichkeiten sind berufsvorbereitende Bildungsmaßnahmen (BvB), berufsvorbereitende Einrichtungen (BvE) und die kooperative berufliche Bildung und Vorbereitung auf den allgemeinen Arbeitsmarkt (KoBV).

Die etwa 50 **Berufsbildungswerke (BBW)** wenden sich an junge Menschen mit unterschiedlichen Beeinträchtigungen/Behinderungen, die nach Erfüllung der 10-jährigen Schulpflicht bzw. in einem Alter von mindestens 16 Jahren Unterstützung beim Übergang von der Schule in den Beruf benötigen. BBW bieten Berufsausbildung, Berufsvorbereitung, Berufsfindung und Arbeitserprobung an. Träger der BBW sind gemeinnützige Organisationen, die Finanzierung der Leistungen erfolgt hauptsächlich durch die Bundesanstalt für Arbeit.

Die **geförderte Berufsausbildung** in den BBW richtet sich an junge Menschen nach der Schulpflicht, die ihren Berufswahlprozess abgeschlossen haben und die zum Ausgleich sozialer Benachteiligung oder zur Überwindung individueller Beeinträchtigungen in erhöhtem Maß sozialpädagogische Unterstützung und Begleitung sowie intensive Anleitung bei der schulischen und beruflichen Ausbildung benötigen. Ziel der geförderten Berufsausbildung ist die Erlangung des jeweiligen Berufsabschlusses. Darüber hinaus sollen die jungen Menschen zu einer selbstständigen Lebensführung und eigenverantwortlichen Existenzsicherung befähigt werden. Sie absolvieren ihre Ausbildung in anerkannten Fachberufen im handwerklichen und kaufmännischen Bereich, im Hotel- und Gaststättengewerbe oder im Dienstleistungsbereich.

In der Berufsvorbereitung erhalten junge Menschen, deren Berufswahlprozess weitestgehend abgeschlossen ist, im BBW am Übergang zwischen Schule und Beruf eine vorbereitende Qualifizierung zur Beschäftigung oder Ausbildung. Zur ausbildungs- bzw. beschäftigungsvorbereitenden Qualifizierung erhalten sie Unterstützung durch sozialpädagogische Hilfen. Die jungen Menschen haben die Möglichkeit, sich in einzelnen Berufsfeldern praktisch zu erproben.

Im Rahmen der **berufliche Orientierung** erhalten junge Menschen zwischen Schule und Beruf zum Ausgleich sozialer Benachteiligung oder zur Überwindung individueller Beeinträchtigungen in erhöhtem Maß Unterstützung, Begleitung und intensive Anleitung bei der Entwicklung einer beruflichen Perspektive sowie einer realistischen Berufswege- und Lebensplanung. Der Umfang der Problemlagen bei Beginn der Maßnahme lässt eine Berufsvorbereitung oder eine Ausbildung nicht zu.

6.3 Pflegefamilien, Heime

Die Erziehung von Kindern in Pflegefamilien ist ein Angebot der Kinder- und Jugendhilfe (SGB VIII §§ 27, 33, 36, 37, 44).

Man unterscheidet Kurzzeit-, Wochen- und Vollzeitpflege. Die Dauerpflege hat Vorrang vor der Heimerziehung, ist aber nachrangig gegenüber der Adoption.

Vollzeitpflege ist eine Hilfe zur Erziehung, auf die Personensorgeberechtigte einen Anspruch haben, wenn eine dem Wohl des Kindes und Jugendlichen entsprechende Erziehung nicht gewährleistet und die Hilfe für seine Entwicklung geeignet und notwendig ist. Diese Maßnahme kann z. B. für in ihren Herkunftsfamilien vernachlässigte, misshandelte oder sexuell missbrauchte Kinder und Jugendliche geeignet sein. Sie kommt auch in Frage, wenn Eltern mit der Erziehung ihrer Kinder überfordert sind und deren Verhalten nicht mehr kontrollieren können.

Vollzeitpflege kann eine zeitlich befristete Erziehungshilfe oder eine auf Dauer angelegte Lebensform sein. Im erstgenannten Fall ist prinzipiell die Rückführung des betroffenen Kindes in die Herkunftsfamilie vorgesehen. Eine enge Kooperation zwischen Pflege- und Ursprungsfamilie wird angestrebt. Im zweiten Fall ist die Pflegefamilie eher eine „Ersatzfamilie", in der das Pflegekind möglichst ungestört von den leiblichen Eltern eine kompensatorische, positiv zu bewertende Erziehung erfahren soll.

An der Entscheidung sind die Herkunftsfamilie und das Kind angemessen zu beteiligen. Das Jugendamt muss nach Inpflegegabe eines Kindes ermitteln, ob die Pflegefamilie das Wohl des Kindes wahrt.

Pflegeeltern ist ein Pflegegeld für den Lebensunterhalt und die Kosten der Erziehung zu zahlen. In der Regel bleiben die leiblichen Eltern des Pflegekindes die Inhaber der elterlichen Sorge. Die Pflegeeltern sind berechtigt, die Personensorgeberechtigten in der Ausübung der elterlichen Sorge zu vertreten, sofern diese oder das Vormundschaftsgericht dieses Recht nicht eingeschränkt haben. Pflegeeltern und leibliche Eltern sollen zum Wohl des Kindes zusammenarbeiten. Außerdem soll die Beziehung des Kindes zur Herkunftsfamilie gefördert, sollen durch Beratung und Unterstützung die Erziehungsbedingungen in ihr verbessert werden. Ist eine Verbesserung innerhalb eines vertretbaren Zeitraums nicht erreichbar, soll für das Kind eine auf Dauer angelegte Lebensperspektive erarbeitet werden.

Heime erfüllen wichtige Aufgaben in der Jugend- und Behindertenhilfe. Heimerziehung ist eine Hilfe zur Erziehung in einer Einrichtung über Tag und Nacht oder in einer sonstigen betreuten Wohnform. Sie soll Kinder und Jugendliche durch eine Verbindung von Alltagserleben mit pädagogischen und therapeutischen Angeboten in ihrer Entwicklung fördern. Sie soll entsprechend dem Alter und Entwicklungsstand des Kindes oder des Jugendlichen sowie den Möglichkeiten der Verbesserung der Erziehungsbedingungen in der Herkunftsfamilie eine Rückkehr in die Familie zu erreichen versuchen oder die Erziehung in einer anderen Familie vorbereiten oder eine auf längere Zeit angelegte Lebensform bieten und auf ein selbstständiges Leben vorbereiten. Jugendliche sollen in Fragen der Ausbildung und Beschäftigung sowie der allgemeinen Lebensführung beraten und unterstützt werden (§ 34 SGB VIII).

Heimerziehung will Kindern und Jugendlichen, deren Eltern infolge individueller, sozialer und gesellschaftlicher Probleme mit der Erziehung überfordert sind, für eine bestimmte Zeit einen neuen pädagogisch gestalteten und professionell strukturierten Lebensort bieten. Heime sichern auch den Besuch von Förderschulen (Eingliederungshilfen für behinderte Menschen). Sie bieten darüber hinaus Unterkunft und Betreuung bei auswärtiger Berufsausbildung und Lehrgangsteilnahme (Berufsförderung).

Für schwerbehinderte Kinder und Jugendliche kann eine Heimunterbringung während der Woche oder in Vollzeit zur Betreuung, Pflege und Förderung erforderlich sein und eine notwendige Entlastung der Eltern darstellen.
Zuständig für die Bewilligung eines Heimaufenthaltes sind die Sozial- und Jugendämter, die anfallenden erheblichen Kosten müssen die Träger der Sozial- und Jugendhilfe, d. h. die Regierungen, Landkreise oder kreisfreien Städte übernehmen.
Tabelle 6.1 (➤ Tab. 6.1) fasst einige familienunterstützende und -ergänzende Maßnahmen und die zuständigen Behörden/Kostenträger zusammen.

6.4 Selbsthilfegruppen

Eltern sind primär für die Gesundheit ihrer Kinder zuständig. Sie sind Spezialisten für ihre Kinder und Partner des Arztes bei deren Behandlung. Über chronische und seltene Erkrankungen wissen Eltern oft mehr als viele Ärzte und andere Fachleute. **Elternselbsthilfegruppen** sind in Fachwelt, Öffentlichkeit, Politik akzeptierte und geschätzte Partner. Sie informieren über die jeweilige Erkrankung, spezialisierte Ärzte und Therapeuten, aktuelle Therapien, Hilfsmittel, sozialrechtliche Fragen, Entlastungs- und Erholungsmöglichkeiten und vermitteln Kontakte zu anderen Betroffenen. Daneben unterstützen sie Familien bei der Bewältigung der mit der Erkrankung verbundenen Probleme und sind in der Öffentlichkeit aktiv mit dem Ziel der bestmöglichen Behandlung ihrer Kinder. Sie realisieren und unterstützen Projekte, die der wissenschaftlichen Erforschung der Erkrankungen oder der konkreten Verbesserung der Betreuung der Patienten zugute kommen. Einen Überblick über die zahlreichen Elternselbsthilfegruppen bietet das **Kindernetzwerk**. Die **ACHSE** (Allianz chronischer seltener Erkrankungen) stellt einen Zusammenschluss von Selbsthilfegruppen dar, der zur Verbesserung von Diagnose und fachgerechter Behandlung seltener chronischer Erkrankungen beitragen will.
Eltern von Kindern mit speziellen, insbesondere seltenen Erkrankungen sollten vom Arzt auf die regional und überregional vorhandenen Selbsthilfegruppen zum jeweiligen Krankheitsbild hingewiesen werden.

6.5 Aufgaben der Sozialleistungsträger

Leistungen der gesetzlichen Krankenkassen sind im Sozialgesetzbuch V festgelegt. Sie umfassen: Maßnahmen der Prävention, medizinische Vorsorge und Krankheitsfrüherkennung, ambulante und stationäre Behandlung, Medikamente, Heil- und Hilfsmittel, zahnärztliche und kieferorthopädische Versorgung. Weitere gesetzliche Krankenkassenleistungen sind u. a. nichtärztliche sozialpädiatrische Leistungen, sozialmedizinische Nachsorge, spezialisierte ambulante Palliativversorgung, Maßnahmen der medizinischen Rehabilitation, Haushaltshilfe, häusliche Krankenpflege, Soziotherapie, Hospizleistungen, Mutter-Kind-Kuren, Krankengeld und medizinisch zwingend erforderliche Fahrtkosten.
Leistungen der sozialen Pflegeversicherung sind in SGB XI festgelegt. Sie umfassen u. a. Pflegesachleistungen bzw. Pflegegeld, Urlaubs- und Verhinderungspflege, teilstationäre Pflege, Kurzzeitpflege, Pflegehilfsmittel, technische Hilfen im Haushalt und Pflegehilfsmittel, pflegebedingte Umbaumaßnahmen, soziale Absicherung der Pflegepersonen, stationäre Pflege.

Die **gesetzliche Rentenversicherung** (SGB VI) spielt für Kinder und Jugendliche eine untergeordnete Rolle. Neben Rentenleistungen werden von ihr auch Leistungen zur medizinischen Rehabilitation (➤ Kap. 39) und zur Teilhabe am Arbeitsleben erbracht.

Die Aufgaben der **gesetzlichen Unfallversicherung** (SGB VII) sind Prävention, Rehabilitation und Entschädigung. Versichert sind bei Unfall neben Beschäftigten u. a. Kinder während des Besuchs einer Tageseinrichtung, Schüler während des Schulbesuchs, Lernende während der beruflichen Aus- und Fortbildung und behinderte Menschen, die in anerkannten Werkstätten arbeiten. Auch der Weg zum und vom Ort der Tätigkeit ist mitversichert.

Leistungen der **Jugend- und Sozialhilfe** umfassen nach §§ 27–35 SGB VIII Hilfen zur Erziehung, Erziehungsberatung, Erziehungsbeistandschaft, sozialpädagogische Familienhilfe, heilpädagogische Gruppen, Tagesstätten für außerschulische Tagesbetreuung von Kindern mit besonderem pädagogischen Bedarf, Tagespflege, Vollzeitpflege in Pflegefamilien und die Heimunterbringung sowie im Rahmen der Jugendhilfe die Eingliederungshilfe für seelisch behinderte Kinder und Jugendliche (§35a SGB VIII). Die Eingliederungshilfe für körperlich oder geistig behinderte Kinder und Jugendliche ist Aufgabe der Sozialhilfe (§ 53 und §54 SGB XII).

Das **Versorgungsamt** (Amt für Soziales, Jugend und Familie) bearbeitet die Anträge für einen Behindertenausweis und legt die Schwere der Einschränkung (Grad der Behinderung, GdB) fest. Gesundheitliche Merkmale zur Inanspruchnahme von Nachteilen sind die Merkzeichen G, aG, H, B, RF, Bl, GI (➤ Tab. 6.2). Die Steuervergünstigungen bei Behinderungen müssen beim zuständigen Finanzamt beantragt werden. Die Höhe richtet sich nach dem Grad der Behinderung. Für den Parkausweis ist das Ordnungsamt zuständig. Für Geldleistungen für Blinde (Blindengeld, -hilfe, Landesblindengeld u. Ä.) existieren länderspezifische Regelungen.

Tab. 6.1 Familienunterstützende und -ergänzende Maßnahmen

Maßnahme	Kostenträger/Zuständigkeit
Haushaltshilfe, Mutter-Kind-Kur	Krankenkasse
Pflegegeld/Pflegesachleistung	Pflegekasse
Verhinderungspflege, Kurzzeitpflege	Pflegekasse
Einzelintegration, Eingliederungshilfe	Sozialamt/ Jugendamt
Sozialpädagogische Familienhilfe, heilpädagogische Gruppen, Pflegefamilie, Heimunterbringung	Jugendamt
Familienunterstützende Dienste	Pflegekasse/ Sozialamt
Schwerbehindertenausweis	Versorgungsamt

Tab. 6.2 Behindertenausweis – Merkzeichen

G	Erhebliche Gehbehinderung → Freifahrt mit öffentlichen Nahverkehrsmitteln im gesamten Bundesgebiet
aG	Außergewöhnliche Gehbehinderung → Parkausweis für Behindertenparkplatz
H	Hilflosigkeit → höchstmögliche Steuererleichterung
B	Ständige Begleitung → Freifahrt für Begleitperson in öffentlichen Verkehrsmittel
RF	Befreiung von Rundfunkgebühren, Ermäßigung bei Telefonanschluss
Bl	Blindheit → Blindenhilfe, Freifahrt mit öffentlichen Verkehrsmitteln
Gl	Gehörlosigkeit → Leistungen nach § 19 Abs. 1 SGB IX, z. B. Gebärdendolmetscher

Die **Bundesagentur für Arbeit** unterstützt aufgrund der Vorgaben in SGB III u. a. Jugendliche bei Berufswahl und Arbeitslosigkeit. Sie bietet individuelle Beratung ggf. unter Einbeziehung des ärztlichen oder psychologischen Dienstes, berät über Möglichkeiten der Berufsvorbereitung, Aus- und Weiterbildungsmöglichkeiten und gewährt finanzielle Hilfen z. B. in Form von Ausbildungsgeld und Übergangsgeld. Sie führt durch oder vermittelt Maßnahmen zur Eignungsfeststellung sowie Trainingsmaßnahmen. Sie unterstützt Maßnahmen zur Förderung der Aufnahme einer Beschäftigung und die berufliche Rehabilitation. Sie bietet Menschen mit Behinderung vielfältige Unterstützungen zur Integration in Ausbildung und Arbeit.
Die Koordination und Abstimmung der vielfältigen Aktivitäten der in diesem Abschnitt genannten Personen und Institutionen stellt eine wesentliche sozialpädiatrische Aufgabe dar.

Literatur

Andler W, Lübbers W, Paetzmann-Sietas B, Meyer M, von Seiche J, Schmidt R, Tewes G (2007) Aktuelle Situation der stationären Kinder- und Jugendmedizin in Deutschland. Ergebnisse der 4. Befragung von stationären Einrichtungen der Kinder- und Jugendmedizin. Monatsschr Kinderheilkd 155: 157–166.

Waltersbacher A (2012) Heilmittelbericht 2011. http://www.wido.de/fileadmin/wido/downloads/pdf_heil_hilfsmittel/wido_hei_hmbericht2011_1211.pdf

Bundesministerium für Familie, Senioren, Frauen und Jugend (2009) 13. Kinder- und Jugendbericht. http://www.bmfsfj.de/RedaktionBMFSFJ/Broschuerenstelle/Pdf-Anlagen/13-kinder-jugendbericht,property=pdf,bereich=bmfsfj,sprache=de,rwb=true.pdf

Bundesministerium für Familie, Senioren, Frauen und Jugend (2012) Familienreport 2012. http://www.bmfsfj.de/RedaktionBMFSFJ/Broschuerenstelle/Pdf-Anlagen/Familienreport-2012,property=pdf,bereich=bmfsfj,sprache=de,rwb=true.pdf

Bundesregierung (2012) Lebenslagen in Deutschland. Entwurf des 4. Armuts- und Reichtumsberichts. http://www.sozialpolitik-aktuell.de/tl_files/sozialpolitik-aktuell/_Politikfelder/Einkommen-Armut/Dokumente/Entwurf%204.%20Armutsbericht%20der%20Bundesregierung%2017.9.2012.pdf

Deutsche Akademie für Kinder- und Jugendmedizin (2012) Thesen zur Analyse und Zukunftsszenario der flächendeckenden medizinischen Versorgung der Kinder und Jugendlichen in Deutschland – Konsequenzen der demographischen Entwicklung. http://dakj.de/pages/posts/analyse-und-zukunftsszenario-der-flaechendeckenden-versorgung-der-kinder-und-jugendlichen-in-deutschland-115.php

Deutsche Gesellschaft für Sozialpädiatrie und Jugendmedizin (2002) Grundlagen und Zielvorgaben für die Arbeit im Sozialpädiatrischen Zentrum (SPZ) – Strukturqualität, Diagnostik und Therapie – „Altöttinger Papier". Kinderärztliche Praxis 73: 498–515.

Führer M (2011) Kinderpalliativmedizin. Monatsschr Kinderheilkd 159: 583–596

Heinrich J, Koletzko B (2008) Kindergesundheit und Kinderbetreuung bei unter 3-Jährigen. Monatsschr Kinderheilkd 156: 562–8.

Hollmann H, Bode H (2007) Wie viele Sozialpädiatrische Zentren braucht das Land? Kinderärztliche Praxis 78: 276–279.

Horacek U, Böhm R, Klein R, Thyen U, Wagner F (2008) Qualitätskriterien zur institutionellen Betreuung von Kindern unter 3 Jahren (Krippen). Kinderärztliche Praxis 79: 312–316.

Pfeuffer F, Engel H, Engels D (2008) Datenerhebung zu den Leistungs- und Vergütungsstrukturen in der Frühförderung behinderter und von Behinderung bedrohter Kinder. Abschlussbericht. Köln.

Podeswik A, Porz F, Groeger K, Thyen U (2009) Sozialmedizinische Nachsorge für schwer und chronisch kranke Kinder. Eine neue Leistung der Krankenkassen. Monatsschr Kinderheilkd 157: 129–135.

Robert-Koch-Institut (Hrsg) (2008) Gesundheitsberichterstattung des Bundes. Heft 41 – Psychotherapeutische Versorgung.

Straßburg HM, Zeitler P (2007) Stellungnahme der DGSPJ zur ganztägigen Fremdbetreuung von Säuglingen und Kleinkindern. Kinderärztliche Praxis 78: 176–178.

Internetadressen

www.arbeitsagentur.de
www.bundesaerztekammer.de
www.bundesverband-kinderhospiz.de
www.dgspj.de
www.familienhandbuch.de
www.gsnp.de
www.kindernetzwerk.de
www.kita-gesundheit.de
www.nubbek.de
www.sozialgesetzbuch.de

7 Gesundheitsförderung und Prävention

Renate Klein und Ulrike Horacek

Sozialpädiatrie in der Praxis. http://dx.doi.org/10.1016/B978-3-437-31630-3.00007-9
Copyright © 2014 Elsevier GmbH. All rights reserved.

7.1 Gesundheit von Kindern und Jugendlichen

Psychisches Wohlbefinden und körperliche Gesundheit sind das Ergebnis von und wesentliche Voraussetzung für eine erfolgreiche Alltagsbewältigung. Gesundheit ist sowohl existenzielles Gut jedes Einzelnen als auch wichtiger Wertschöpfungsfaktor einer Gesellschaft.
Die Weltgesundheitsorganisation hat 1946 in ihrer Verfassung **Gesundheit als Zustand des vollständigen körperlichen, geistigen und sozialen Wohlbefindens** und nicht als die Abwesenheit von Krankheit und Gebrechen definiert (Constitution of the World Health Organisation). Deter-

minanten für Gesundheit und Krankheit liegen nach dieser Definition somit auf individueller wie auch lebensumfeldbezogener Ebene.

Diese Sichtweise entspricht umfassend dem Ansatz der Sozialpädiatrie. Die Gesundheit und gesunde Entwicklung von Kindern wird in besonderem Maße beeinflusst von ihrem **Lebensumfeld,** weil Kinder essenziell auf die elterliche und/oder institutionelle Fürsorge angewiesen sind. Damit kommt den ökonomischen und psychosozialen Bedingungen in der Familie eine große Bedeutung zu, interagieren sie doch mit den beim Kind vorhandenen ererbten und erworbenen Faktoren.

Diese Sichtweise macht aber auch die Schwierigkeit offensichtlich, in Ermangelung komplexer Daten und zeitlicher Trends die Gesundheit von Kindern und Jugendlichen umfassend zu beschreiben. In westlichen Industrienationen hat sich die Lebenssituation von Kindern und Jugendlichen in den letzten hundert Jahren durch veränderte wirtschaftliche Rahmenbedingungen, erfolgreiche Bekämpfung von Infektionskrankheiten durch Hygienemaßnahmen und Impfungen sowie eine gute Gesundheitsversorgung nachhaltig verbessert. Säuglingssterblichkeit sowie Todesfälle durch vermeidbare Krankheiten haben abgenommen.

Das **Krankheitsspektrum** von Kindern und Jugendlichen hat sich verändert – von akuten hin zu chronischen Erkrankungen, von körperlichen hin zu mehr psychosomatischen und psychosozialen Krankheitsbildern. Dazu gehören die Zunahme von Krankheiten des allergischen Formenkreises, Störungen des Ernährungsverhaltens, der Sinneskoordination und des Bewältigungsverhaltens (Schlack 2004, 2009). Viele dieser „neuen" Morbiditäten müssen im Kontext des kindlichen Lebensumfelds gesehen werden.

Dies belegen auch Untersuchungen, wie die Unicef-Studien zu Kindergesundheit (Unicef, 2013,), die Ergebnisse des Kinder- und Jugendgesundheitssurveys des Robert-Koch-Instituts (Robert-Koch-Institut 2007 und 2008, Lange et al. 2007), die Untersuchungen der Öffentlichen Kinder- und Jugendgesundheitsdienste, verschiedene Längsschnittstudien (Grimmer et al. 2008) sowie eine Studie des DAKJ zum ambulanten Versorgungsspektrum (Fegeler, U et al 2011), Die Gesundheit der Kinder ist eng verknüpft mit der **sozialen Lage** der Familien, aber auch von weiteren Faktoren abhängig (➤ Kap. 4; Kap. 13). Eine Vervielfachung der Gesundheitsprobleme zeigt sich in besonders belasteten Gruppen. Kinder mit Migrationshintergrund und aus finanziell schwachen und bildungsarmen Schichten sind in besonderem Maße betroffen (Borde 2005, RKI 2007). Kinder mit chronischen Krankheiten und Behinderungen können ihrerseits zur Belastung der Familie beitragen und negative Entwicklungen fördern (Engelbert 1999). Die Mehrzahl der Untersuchungen zur Gesundheit von Kindern basiert sowohl auf Befunden als auch auf individueller Einschätzung. Die Abhängigkeit der Gesundheit von der sozialen Lage belegen folgende Beispiele: Je niedriger der Sozialstatus ist, umso

- häufiger rauchen Frauen in der Schwangerschaft,
- schlechter schätzen Eltern den subjektiven Gesundheitszustand ihrer Kinder ein,
- weniger sind Kinder in Sportvereinen aktiv,
- stärker sind Kariesprävalenz sowie Übergewicht und Adipositas,
- häufiger sind Sprachentwicklungsstörungen und Koordinationsstörungen.

Die Folgen von Entwicklungsrückständen und Anpassungsstörungen sind weitreichend: Sie mindern Bildungschancen und berufliche Perspektiven und befördern das generationsübergreifende Beharren in belasteten Lebenslagen oder sogar die weitere Verschlechterung: Insbesondere die Zahl jüngerer Kinder, die aus belasteten Familien in Obhut genommen werden, steigt stetig an. Chronische Krankheiten mit dem Risiko einer Progredienz und Entwicklung von Folgeerkrankungen schränken individuelle Lebensplanungen ein und führen zu verstärkter Inanspruchnahme der Gesundheitsversorgung und frühzeitigem Ausscheiden aus dem Erwerbsleben. Damit wirken sich die neuen Gesundheitsstörungen in erheblichem Maß negativ auf die gesellschaftliche

Entwicklung aus: Sie verursachen hohe Sekundärkosten in Jugendhilfe, Bildungssystem und in den sozialen Sicherungssystemen und beeinträchtigen damit die Perspektiven jedes im globalen Wettbewerb stehenden Landes.
Die neuen, in erheblichem Ausmaß auftretenden Gesundheitsstörungen bei Kindern und Jugendlichen erfordern neue Ansätze der **Prävention,** welche die unterschiedlichen Lebensentwürfe mitberücksichtigen müssen. Vorrangig gilt es, die Kompetenz der Eltern in der Förderung der Gesundheit und Entwicklung des Kindes frühzeitig zu stärken und Kindergarten/Schule sowie das weitere Umfeld in diese Aufgabe einzubinden.
Es ist offensichtlich, dass dies nicht in der individuellen Gesundheitsversorgung leistbar ist, sondern eine gesellschaftliche Herausforderung und ein wichtiges Anliegen öffentlicher Gesundheit und Gesundheitswissenschaft (Public Health) darstellt. Die Kinder- und Jugendpolitik hat mittlerweile erkannt, dass nur gesamtgesellschaftliches Handeln den Kontext von sozialer Lage und Entwicklungschancen verändern kann: Die Kinder- und Jugendhilfe wird verstärkt präventiv tätig, u. a. werden mit der Etablierung der Frühen Hilfen und dem Ausbau der U3-Betreuung familienunterstützende und ergänzende Angebote etabliert, z. B. in Form der Bildungsprogramme für Kindertageseinrichtungen und vermehrter Ganztagsbetreuung auch in Schulen. Integration und Umsetzung der UN-Konvention zur Inklusion sind vorrangige politische Ziele (14. Kinder- und Jugendbericht 2012).

7.2 Schutzfaktoren und Risikofaktoren für die kindliche Entwicklung

Die kindliche Entwicklung wird sowohl von den beim Kind vorhandenen (biologischen oder psychologischen) Merkmalen als auch seinem Lebensumfeld bestimmt. Dazu gehören neben der Familie und ihrer Lebenssituation das soziale Umfeld ebenso wie die verfügbaren institutionellen Strukturen und die gesellschaftlichen und kulturellen Rahmenbedingungen.
Die komplexen Wechselwirkungen zwischen Individuum und Lebensumfeld können sich positiv oder negativ, d. h. als Schutz- oder Risikofaktoren, auf die Entwicklung auswirken. Es gibt keine eindeutigen linearen Zusammenhänge zwischen Risiko- und Schutzfaktoren; sie stehen in kumulativer Wechselwirkung. Geschlecht, Alter und Abfolge von Risiken beeinflussen individuelle Entwicklungsprozesse. Risiko- und Schutzfaktoren zu erkennen, ist unabdingbar für die Entwicklung präventiver Ansätze.
Die Art der Intervention wird sich darauf auszurichten müssen, wo sie am meisten Erfolg verspricht. Dies kann/muss in einem Fall beim Kind sein, ein anderes Mal im Umfeld.

7.2.1 Modelle der Entwicklung und Sozialisation

Das Zusammenwirken verschiedener Faktoren auf die kindliche Entwicklung wurde von der Entwicklungsforschung in verschiedenen Modellen beschrieben. Das **ökologische Modell** von Bronfenbrenner sieht den Menschen als Gestalter seiner Entwicklung und sich selbst reflektierendes Wesen, das sich ein Bild von seiner Umwelt aufbaut und flexibel modifiziert. Mensch und Umwelt (Lebensraum) beeinflussen sich gegenseitig (transaktional).
Ökologische Umwelten (Settings) wie Elternhaus oder Kindergarten sind physikalische/soziale Milieus, in denen personenunabhängig ein bestimmtes Verhaltensmuster auftritt. Entwicklung

wird als Prozess der Anpassung zwischen aktivem, sich entwickelndem Menschen und Eigenschaften der wahrgenommenen Lebensbereiche definiert. Positive Aktionen führen zu positivem Feedback und begünstigen weitere positive Entwicklungen im Sinne einer gegenseitigen Interaktion (lachende Mutter – Kind lacht – Mutter ist glücklich).
Ausgehend vom Mikrosystem Familie bekommt der Mensch Zugang zu neuen Lebensbereichen (ökologische Übergänge). Die Beteiligung an mehreren Lebensbereichen ist umso entwicklungsförderlicher, je mehr Kommunikation zwischen den Bereichen besteht (Bronfenbrenner 1976, 1994).
Das eher starre, an Lebensräumen orientierte ökologische Modell wurde durch **systemtheoretische Ansätze** weiterentwickelt, indem die durch die individuellen Merkmale des Kindes und soziale Interaktionen induzierten Feedbackmechanismen differenzierter beschrieben wurden und der **gesellschaftliche** Hintergrund stärkere Beachtung fanden. Wesentliches Merkmal moderner Gesellschaften ist, dass die Ganzheitlichkeit, die ein Kind zu Beginn des Lebens erfährt, im Laufe des Heranwachsens durch diese Interaktionen in unterschiedliche Funktionen aufgesplittet wird. Präventive Ansätze müssen diese Aspekte stärker berücksichtigen.

7.2.2 Ressourcen, Resilienz, Schutzfaktoren

Positive Entwicklungen von Kindern auch unter schlechten sozialen Bedingungen werden seit langem beobachtet und in verschiedenen Längsschnittstudien beschrieben. So zeigten Emmy Werners Ergebnisse einer Langzeitstudie auf der Hawaii-Insel Kauai, begonnen in den 1950er Jahren, dass sich ein Drittel der Kinder mit psychosozialen Risikofaktoren (chronische Armut, Geburtskomplikationen, geringes Bildungsniveau der Eltern, chronische familiäre Disharmonie) positiv entwickelten. Auch im Alter von 40 Jahren schauten sie hoffnungsfroh in die Zukunft und lebten in stabilen Ehen. Ihre beruflichen und schulischen Leistungen waren z. T. besser als die der Mehrheit der Kinder, die ohne nennenswerte Risiken aufgewachsen sind (Werner 1992).
In Deutschland befassten sich die Rostocker Längsschnittstudie (Meyer-Probst et al. 1984), die Mannheimer Längsschnittstudie (Laucht et al. 1998) und die Mannheimer Risikokinderstudie (Ihle et al. 2004) mit den Auswirkungen protektiver und schädlicher Aspekte auf unterschiedliche Entwicklungsbereiche. Der Kinder- und Jugendsurvey des Robert-Koch-Instituts (KIGGS) 2003–2006 hat dies schwerpunktmäßig für die psychische Gesundheit erforscht (Ravens-Sieberer et al. 2007).
Ressourcen. Die individuelle Entwicklung von Kindern ist abhängig von Ressourcen, mit denen das Kind innere oder äußere Anforderungen bewältigen kann. Zu unterscheiden sind individuelle Ressourcen (genetische Disposition, biologische Faktoren) und umfeldbezogene Ressourcen (Familie, Bildungseinrichtung, soziodemografische Variablen). Zusammen bilden sie interaktional die beim Kind aktuell verfügbaren Faktoren, welche die Entwicklung unterstützen. Gleichzeitig entstehen durch eigene Aktivitäten Einstellungen, die zu einer selbstgestalteten Umwelt führen.
Resilienzen. Hierunter versteht man bereichsspezifische Ressourcen, die durch Interaktion mit der Umwelt erworben werden. Resilienz ist die Fähigkeit, erlernte Mechanismen zur Bewältigung alterstypischer Aufgaben trotz schwieriger Umstände zu aktivieren (Petermann und Schmidt 2006); sie wird auch als Widerstandskraft trotz widriger Bedingungen aufgrund bestimmter Schutzfaktoren definiert.
Schutzfaktoren. Dies sind vor dem Auftreten von Risikofaktoren vorhandene individuelle oder Umfeldmerkmale (Resilienzen), die interaktiv als Puffer gegen Risikofaktoren wirken und die Entstehung psychischer Störungen verhindern oder abmildern. Fehlen die Schutzfaktoren, kommt der Risikofaktor voll zum Tragen. Ihre Wirkungsweise selbst bedarf noch weiterer Forschung.

Eine besondere Bedeutung als Schutzfaktor haben personenbezogene Ressourcen, die die Entwicklung unterstützen und die aktuell zur Verfügung stehen. Sie äußern sich u. a. in positiven Temperamenteigenschaften, die bei anderen Menschen positive Reaktionen auslösen. Des Weiteren zählen dazu positive Selbsteinschätzung, positives Sozialverhalten (emotionale Kompetenz) und internale Kontrollüberzeugungen (Laucht et al. 1998). Der Medizinsoziologe Antônôvsqî hat letztere Eigenschaften in seinem Modell der **Salutogenese** beschrieben: Er benennt als Schutzfaktoren das Gefühl von Verstehbarkeit (sense of comprehensibility), von Bewältigbarkeit (sense of manageability) und von Bedeutsamkeit (sense of meaningfulness) (Antônôvsqî und Franke 1997). Das Geschlecht spielt bei den Individualressourcen eine große Rolle; Mädchen sind allgemein körperlich robuster, sozial differenzierter und zeigen teilweise bessere kognitive Fertigkeiten.
Aber auch die Eigenschaften des sozialen Umfelds oder interaktionale Dimensionen können die Auftretenswahrscheinlichkeit einer positiven/gesunden Entwicklung fördern. Emmy Werner benennt als nachhaltige Schutzfaktoren im Umfeld eine enge Bindung zu wenigstens einer Bezugsperson sowie gute Schulbildung der Mutter und ihr kompetenter Umgang mit dem Kind. Die Mannheimer Längsschnittstudie weist als Schutzfaktoren u. a. das Vorhandensein eines guten Freundes/einer Freundin bei Achtjährigen aus sowie gute Beziehungen von Jungen zu einem Lehrer mit 8, 13 und 18 Jahren.
Ergebnisse aus den o. g. und weiteren Studien zeigen Schutzfaktoren in Abhängigkeit von Alter und Geschlecht des Kindes (➤ Tab. 7.1):

Merke

Die Resilienzforschung belegt: Je besser Menschen die eigenen Ressourcen kennen und sich eigenaktiv handelnd verhalten können, desto leichter ertragen sie Belastungen. Damit ist es ein wichtiges Ziel von Entwicklungsförderung und Bildung, Menschen zu befähigen, diese Ressourcen zu entwickeln und sie zu kennen.
Für Kinder ist darüber hinaus das elterliche Bewältigungshandeln bedeutsam. Sie werden entlastet durch Eltern, die sich selbstwirksam fühlen und damit Vorbild für Kinder sind und die die Fähigkeit vermitteln, Konflikte gewaltlos zu bewältigen sowie Verantwortung zu übernehmen.
Geschwister und andere Familienmitglieder können zur Resilienzentwicklung beitragen.
Auch Bildungsprogramme in Kindergarten und Schule sind im Sinne von Kompetenzorientierung auf Resilienzförderung angelegt, indem sie Kinder in ihrer Entwicklung unterstützen und ihnen Verantwortung übertragen.

Tab. 7.1 Ressourcen in der Person des Kindes (aus: Petermann et al. 1998)

Alter	Mädchen	Jungen
	Vornehmlich persönliche Eigenschaften bedeutend	Vornehmlich emotionale Unterstützung durch die Familie
Kleinkind 1.–2. Lebensjahr	Umgängliches Temperament	Höheres Bildungsniveau, positive mütterliche Interaktion, familiäre Stabilität
Mittlere Kindheit 2.–10. Lebensjahr	(Nonverbale) Problemlösefertigkeiten und das Rollenvorbild einer Mutter mit einem Schulabschluss und ständiger Berufstätigkeit	Emotionale Unterstützung durch die Familie, Anzahl der Kinder in der Familie, Anzahl Erwachsener außerhalb des Haushalts, mit denen das Kind gerne verkehrt
Späte Jugend	Hohe Selbstachtung, internale Kontrollüberzeugung und realistisches Bildungsziel	Vorhandensein eines Lehrers als Mentor oder Rollenvorbild und regelmäßige Aufgaben und Verantwortung im Familienalltag

7.2.3 Risikofaktoren, Vulnerabilität

Risikofaktoren. Risikofaktoren sind Faktoren, welche die Entwicklung des Kindes negativ beeinträchtigen können und die Wahrscheinlichkeit des Auftretens psychischer und psychosomatischer Störungen erhöhen. Sie können unterschieden werden in biologische und psychologische Risiken sowie eingreifende biografische und traumatische Erlebnisse. Auch hier sind kindliche und umfeldbezogene Faktoren zu unterscheiden (➤ Tab. 7.2).

Generelle Aussagen zur Auswirkung von Risikofaktoren lassen sich aus den Studien bisher nicht ableiten. Es besteht eine große Variabilität der Entwicklungsergebnisse und des Entwicklungstempos. Die größte Gefährdung geht von der psychosozialen Risikobelastung aus.

Vulnerabilität (Verletzlichkeit). Vulnerable Personen sind besonders leicht emotional verwundbar und entwickeln eher psychische Störungen. Jeder Mensch durchläuft in seinem Leben mehrere vulnerable Phasen, wie z. B. die Pubertät, in denen eine erhöhte Gefahr besteht, eine psychische Störung zu entwickeln. Die Belastung kann aufgrund verfügbarer Bewältigungs-/Coping-Strategien gemindert oder unbedeutsam gemacht werden.

Vulnerable Kinder und Jugendliche in psychosozial belastetem Kontext sind besonders gefährdet in ihrer Entwicklung, dies gilt z. B. für Migrantenkinder mit schlechter Integration oder Kinder suchtkranker oder psychisch kranker Eltern. Vulnerable Säuglinge und Kleinkinder entwickeln bei lang anhaltenden Stresssituationen tiefgreifende Störungen der sozial-emotionalen Entwicklung. Gleichzeitig führt ein damit einhergehender erhöhter Kortisol- und Adrenalinspiegel zu einer Adaptation zentraler hormoneller Steuerungssysteme, die nicht mehr reversibel ist. Die ständig erhöhte Stressbereitschaft ist Ursache von Anpassungsstörungen, psychosomatischen Erkrankungen und langfristig u. a. von Herz-Kreislauf-Störungen.

Merke

Auch wenn die Konstellation von Risiko- und Schutzfaktoren keine Vorhersagen für die individuelle Entwicklung zulässt, so zeigen sich doch wichtige Zusammenhänge zwischen Risikofaktoren und Entwicklung sowie gesundheitlichem Outcome: Kinder aus sozial benachteiligten Schichten haben mehr gesundheitliche Probleme und nehmen präventive Maßnahmen seltener wahr. Kinder mit Migrationshintergrund haben weniger Zugang zu Bildung und zu Gesundheitsversorgung (KIGGS, 2007, RKI 2009, 14. Kinder- und Jugendbericht, 2012).

Die Ergebnisse der Resilienzforschung belegen die Notwendigkeit, über die Betrachtung des Individuums und der individuellen Förderung dem Lebensumfeld des Kindes sehr viel mehr Aufmerksamkeit zu widmen, angefangen von der Familie über Kindergarten, Schule, Peergroup bis hin zur Arbeitswelt. Auch Unterstützungsangebote, wie nachbarschaftliche Gemeinschaften und andere Nischen, mit bestimmten Unterstützungskonstellationen sollten im öffentlichen Bewusstsein Beachtung finden.

Tab. 7.2 Kindliche und umfeldbezogene Risikofaktoren

Kind	**pränatal:** Genetische Ursachen, Stoffwechsel- u. Organkrankheiten, Infektionen, Schädigung durch Alkohol, Nikotin, Drogen, Strahlenschäden, Unfälle, mechanische Traumen **perinatal:** Frühgeburt, Hypoxämie unter Geburt, Hyperbilirubinämie, Hirnblutungen, Traumen **postnatal:** Gedeihstörungen, Anfallsleiden, folgenschwere Unfälle, Organerkrankungen, Entwicklungsverzögerungen, Traumata wie sexueller Missbrauch, Naturkatastrophen, Migration

Tab. 7.2 Kindliche und umfeldbezogene Risikofaktoren (Forts.)

Eltern/Familie	Frühe Familiengründung, postpartale Depression der Mutter, Geburt als schwere/traumatisierende Erfahrung, chronische Überforderung Mutter (Eltern), Alleinerziehende, Altersabstand zwischen Kindern < 18 Monate, behindertes Geschwisterkind, akute, belastende Lebensereignisse, materielle Belastungen (Arbeitslosigkeit, enges Wohnumfeld), fehlende familiäre Unterstützung, Dissozialität, sonstige abnorme Erziehungsbedingungen
Soziales Umfeld	Fehlende familiäre Unterstützung, dissoziales Umfeld
Gesellschaftliche Faktoren	Unzureichende Rahmenbedingungen, fremder soziokultureller Kontext, fehlende/abweichende Normen und Werte

7.3 Gesundheitsförderung, Gesundheitserziehung, Prävention

7.3.1 Gesundheitsförderung

Gesundheitsförderung umfasst alle Maßnahmen, die auf die Veränderung und Förderung des Gesundheitsverhaltens des Individuums und der Gesellschaft insgesamt sowie auf die Lebensverhältnisse abzielen. Sie setzen bei der Analyse und Stärkung der Gesundheitsressourcen und -potenziale des Menschen und auf allen gesellschaftlichen Ebenen an. Gesundheitsförderung ist nicht (allein) auf das Vermeiden spezifischer Krankheitsrisiken ausgerichtet (Bröskamp-Stone et al. 1998). Als wichtigstes Dokument zur Gesundheitsförderung gilt die **Ottawa-Charta,** die die Weltgesundheitsorganisation (WHO) auf ihrer ersten Internationalen Konferenz zur Gesundheitsförderung 1986 verabschiedet hat und die 1989 als Teil der Resolution der Weltgesundheitsversammlung angenommen wurde (www.searo.who.int). In ihr werden die Ziele und Strategien zur Förderung der Gesundheit und Schaffung bzw. Erhaltung fördernder Lebensumwelten benannt, basierend auf dem erweiterten Gesundheitsbegriff: „Gesundheitsförderung zielt auf einen Prozess, allen Menschen ein höheres Maß an Selbstbestimmung über ihre Gesundheit zu ermöglichen und sie damit zur Stärkung ihrer Gesundheit zu befähigen. Um ein umfassendes körperliches, seelisches und soziales Wohlbefinden zu erlangen, ist es notwendig, dass sowohl einzelne als auch Gruppen ihre Bedürfnisse befriedigen, ihre Wünsche und Hoffnungen wahrnehmen und verwirklichen sowie ihre Umwelt meistern bzw. verändern können. In diesem Sinne ist die Gesundheit als ein wesentlicher Bestandteil des alltäglichen Lebens zu verstehen und nicht als vorrangiges Lebensziel. Gesundheit steht für ein positives Konzept, das in gleicher Weise die Bedeutung sozialer und individueller Ressourcen für die Gesundheit betont wie die körperlichen Fähigkeiten. Die Verantwortung für Gesundheitsförderung liegt deshalb nicht nur bei dem Gesundheitssektor, sondern bei allen Politikbereichen und zielt über die Entwicklung gesünderer Lebensweisen hinaus auf die Förderung von umfassendem Wohlbefinden hin." (Ottawa-Charta zur Gesundheitsförderung, übersetzt nach WHO Regional Office for Europe 1998; www.euro.who.int).
Die WHO weist darin auf die grundlegenden Bedingungen für Gesundheit hin:

- Frieden
- Angemessene Wohnbedingungen
- Bildung
- Ernährung
- Einkommen
- Stabiles Ökosystem

- Sorgfältige Verwendung vorhandener Naturressourcen
- Soziale Gerechtigkeit
- Chancengleichheit.

In weiteren Erklärungen zur Gesundheitsförderung wurden die Grundvoraussetzungen für Gesundheit um soziale Sicherheit, soziale Beziehungen, Handlungskompetenzen (empowerment) von Frauen und die Achtung der Menschenrechte ergänzt. Als größte Bedrohung für Gesundheit wird Armut durch zunehmende **soziale Ungleichheiten** innerhalb und zwischen Ländern genannt. „Andere Faktoren, die die Gesundheit beeinflussen, sind rasche und häufig ungünstige soziale, ökonomische und demografische Veränderungen, welche die Arbeitsbedingungen, das Lernumfeld, Familienstrukturen und den kulturellen und sozialen Aufbau von Gemeinschaften beeinträchtigen. Frauen und Männer sind von diesen Entwicklungen unterschiedlich betroffen. Diese Wandlungsprozesse formen weltweit die Wertvorstellungen, Lebensweisen und Lebensbedingungen der Menschen über die gesamte Lebensspanne. Die Gefährdung von Kindern, die Ausgrenzung von Minderheiten, von Menschen mit Behinderungen sowie von Indigenen hat zugenommen."

Die Feststellungen in den Erklärungen der WHO zur Gesundheitsförderung treffen auch auf Deutschland zu. Demografische Entwicklung, auseinanderbrechende Familienstrukturen, Vorrang von Ökonomisierung und Verlust gemeinsamer Normen und Werte führen zu gesellschaftlichen Verwerfungen. Die zunehmende soziale Ungleichheit belegt u.a. die OECD-Studie 2008 „Mehr Ungleichheit trotz Wachstum?": Der Anteil der Kinder mit Armutsrisiko stieg in Deutschland von 1995 bis 2007 von 7 % auf 16 % an und hat sich bis 2010 je nach Datenquelle auf 17–20 % erhöht mit Verschärfung der **regionalen Unterschiede:** während in Berlin nahezu ein Drittel der Kinder von Armut betroffen sind, sind es in Bayern ca. 10 %. (4. Armutsbericht 2013). Deutschland liegt hinsichtlich relativer Kinderarmut nach der Unicef-Studie 2013 auf einem Mittelplatz. Besonders betroffen sind Alleinerziehende und Familien mit mehreren Kindern. Armut ist eng verbunden mit schlechterer Gesundheit und schlechten Bildungschancen für Kinder und Jugendliche. Aber auch durch Migration und unzureichende Integration kommt es zu höherer gesundheitlicher Belastung (RKI 2008, Oberwöhrmann et al 2013)

Vorrangige Handlungsfelder sind nach der Ottawa-Charta:

- Entwicklung einer gesundheitsfördernden Gesamtpolitik (staatliches Handeln)
- Förderung der Entwicklung persönlicher Kompetenzen (empowerment)
- Schaffung gesundheitsfördernder Lebenswelten (Setting-Ansatz)
- Neuorientierung des Gesundheitssektors auf diesen Bereich
- Förderung des aktiven Einbringens in Gemeinschaftsaktionen.

Gesundheitsförderung wurde in Deutschland in den oben beschriebenen Bereichen in unterschiedlicher Tiefe seit Mitte der 1980er Jahre umgesetzt und hat in den letzten Jahren an Dynamik gewonnen. Da sie in vielen Bereichen mit präventiven Maßnahmen verbunden ist, wird die erfolgte Umsetzung im Folgenden gemeinsam aufgelistet. In der Praxis wird der Begriff der Gesundheitsförderung zunehmend mit der primären Prävention zusammen genannt und teilweise auch unter Prävention subsumiert.

7.3.2 Gesundheitserziehung

Gesundheitserziehung oder Gesundheitsbildung sind Begriffe aus der Gesundheitspädagogik, die im Alltag oft neben der Gesundheitsförderung gebraucht und aus unterschiedlicher Sicht (Gesundheit, Bildung) verschieden definiert werden. Historisch begründen sie sich aus der Volksbelehrung.

Nach der Definition der WHO umfasst Gesundheitserziehung wissenschaftlich begründete Maßnahmen der Erziehung und Bildung zur Beeinflussung gesundheitsrelevanten Verhaltens, d. h. bewusst gestaltete Lernmöglichkeiten, die gewisse Formen der Kommunikation einschließen und zur Verbesserung der Gesundheitsalphabetisierung (health literacy) entwickelt wurden; letztere schließt die Erweiterung von Wissen und die Entwicklung von Alltagskompetenzen (life skills) ein, die individueller und kollektiver Gesundheit förderlich sind. Gesundheitsbildung bzw. -erziehung befasst sich nicht nur mit der Verbreitung von Informationen, sondern auch mit der zur Gesundheitsverbesserung notwendigen Stärkung von Motivationen, Kompetenzen und Vertrauen (Selbstwirksamkeit) (WHO 1998).

7.3.3 Gesundheitsschutz

Gesundheitsschutz umfasst alle Maßnahmen, die eine Gefährdung der Gesundheit verhindern und berührt im Prinzip alle Lebensbereiche. Maßnahmen der Gesundheitserziehung, der Gesundheitsförderung und der Prävention tragen zum Gesundheitsschutz bei.
Gesundheitsschutz kann individuell oder allgemein ausgerichtet sein. Im individuellen Bereich kommt er insbesondere im Arbeitschutz sowie beim persönlichen Risikoverhalten (Drogenkonsum, Autofahren, sportliche Aktivitäten) zum Tragen. Im öffentlichen Bereich umfasst er Maßnahmen der Gesundheitsüberwachung und des Umweltmonitorings, des Infektionsschutzes, der allgemeinen und speziellen Hygieneüberwachung, des Verbraucherschutzes, des Arbeitsschutzes und Umweltschutzes.

7.3.4 Prävention

Prävention soll Krankheiten durch gezielte Maßnahmen verhindern oder verzögern oder in ihrem Verlauf abmildern. Voraussetzung ist die Kenntnis der pathogenetischen Zusammenhänge.
Die älteste, meistgenutzte Einteilung präventiver Interventionen ist das dreistufige Modell nach Gerald Caplan (1964). Es wurde um differenzierende Aspekte ergänzt:
Primäre Prävention will Erkrankungen durch Vermeiden von Exposition oder personengebundene Maßnahmen verhindern. Ziel ist, die Inzidenz von Erkrankungen zu senken.
Primäre Prävention wird in zwei Teilbereiche untergliedert: Primordiale Maßnahmen sind unspezifische Maßnahmen, die sich an eine gesunde Gesamtpopulation wenden (Beispiele: bauliche Standards zur Lärmreduzierung; Impfungen). Primäre Maßnahmen richten sich an gesunde Personen, die bereits erkennbare Risikofaktoren besitzen, also an sogenannte Risikogruppen.
Sekundäre Prävention will Krankheiten frühzeitig erkennen und einer Behandlung zuführen (Kinderfrüherkennungsuntersuchungen, Krebsvorsorgeuntersuchungen, Gesundheitsuntersuchungen). Ziel ist, die Inzidenz manifester oder fortgeschrittener Erkrankungen zu senken. Dies setzt voraus, dass der Krankheitsprozess umkehrbar ist, dass verlässliche Screeningtests sowie wirksame Behandlungsmöglichkeiten existieren (WHO 2002).
Tertiäre Prävention soll bei eingetretener Krankheit eine Verbesserung des Gesundheitszustands herbeiführen oder eine Verschlechterung des Krankheitsverlaufs bzw. Folgeschäden und Behinderungen verhindern (Rehabilitation, Habilitation). Tertiäre Maßnahmen umfassen kurative und therapeutische Verfahren bei chronisch fortgeschrittenen Erkrankungen. Der tertiäre Bereich

nach Caplan wird im erweiterten Modell um die **quartäre Prävention** ergänzt und hat die Sicherung der verbleibenden Lebensqualität zum Ziel.
Prävention setzt – ebenso wie Gesundheitsförderung – beim Individuum und beim Lebensumfeld an.
Verhaltensprävention ist darauf gerichtet, gesundheitsfördernde Verhaltensweisen zu stärken oder schädliche Verhaltensweisen des Einzelnen zu verhindern oder abzulegen (z. B. Rauchen, Bewegungsmangel).
Die **Verhältnisprävention** setzt auf übergeordneten Ebenen an, Voraussetzungen für gesundes Verhalten zu schaffen. Verhältnisprävention bedeutet somit die gesundheitsfördernde Ausgestaltung des Lebensumfeldes der Menschen (z. B. Luftreinhaltung, gesunde Wohnformen, Verfügbarkeit von Kindertageseinrichtungen und Sportangeboten) und der gesellschaftlichen Rahmenbedingungen.

7.4 Evaluation von Gesundheitsmaßnahmen – Health-Outcome-Modelle

Health-Outcome-Modelle dienen zur Bewertung der verschiedenen Ansätze und Ziele präventiver Maßnahmen.
Das wichtigste bevölkerungsmedizinische Ziel von Prävention ist die Verminderung von Krankheit, Behinderung und Tod in einer Gesellschaft (Inzidenzabsenkung). Maßnahmen der Prävention sollen an der Krankheitslast, am Nutzen präventiver Ansätze, ihrem Aufwand (Kosten) und möglichen Folgewirkungen orientiert sein.
Damit sind weitere Ziele verbunden: Die Senkung der Krankheitslast (weniger Krankheiten, längere Lebenserwartung) soll zu einer Verbesserung der Lebensqualität der Bevölkerung beitragen. Der medizinische Nutzen sekundär präventiver Maßnahmen soll sich in geringeren Behandlungskosten für Krankheiten zeigen. Durch Krankheitsvermeidung wird die Arbeitsproduktivität eines Volkes erhöht und damit Gesundheit zu einem wichtigen Standortfaktor im weltweiten Wettbewerb. Demgegenüber können sich die Gesamtkosten für Gesundheit durch Maßnahmen zur Verbesserung der Lebensqualität und längere Lebenserwartung erhöhen.
Die gesundheitlich erwarteten Erträge werden als Expected Health Outcome bezeichnet (Schwartz et al. 1998). Unter bevölkerungsmedizinischen Gesichtspunkten sind in der Prävention insbesondere die Krankheiten von Bedeutung, die eine hohe Krankheitslast in der Zielpopulation verursachen und durch Prävention reduzierbar sind. Ferner soll der Aufwand für die Prävention in einem positiven Verhältnis zum Nutzen stehen und die präventiven Maßnahmen zu keinen unbeabsichtigten Folgewirkungen führen.
Wichtige Aspekte für die Umsetzung von Präventionsprogrammen sind Größe und Erreichbarkeit der Zielgruppe, die Art der Interventionsstrategie und deren Implementation in der Routine. Üblicherweise werden zur Planung Erkenntnisse aus der Evaluation von Pilotprojekten und aus der Literatur herangezogen.
Präventionsziele können kurz-, mittel- oder langfristig ausgerichtet sein; letzteres gilt insbesondere in der Prävention chronischer Erkrankungen. Je weiter Präventionsziele entfernt liegen, umso geringer ist der individuelle Anreiz, für eine entfernt liegende Lebensphase gesundheitlich vorzubeugen.

Beispiele für Präventionsziele

- Krankheitslast impfpräventabler Erkrankungen verringern: Impfungen bei Kleinkindern sind eine individuelle, bevölkerungsmedizinisch kurz- und mittelfristig wirksame Präventionsmaßnahme. So reduzierte sich z. B. durch Pertussisimpfung der Anteil von Pertussiserkrankungen, Komplikationen und Todesfällen in dieser Altersgruppe deutlich. Die Nebenwirkungen der neuen antigenhaltigen Impfstoffe sind im Gegensatz zu den früheren Ganzkeimimpfstoffen gering.
- Frühe Vermeidung von Adipositas und ihren Folgekrankheiten: Die Maßnahmen müssen im Lebensumfeld des Kindes greifen und gemeinsam gelebt werden: in der Familie, Gemeinschaftseinrichtung, Kommune. Wichtige Aspekte sind u. a. Ernährung, Bewegung, Freizeitgestaltung, Medienkonsum. Health-Outcome und Kosten-Nutzen-Effekt (s. u.) hängen von der Nachhaltigkeit angestrebter Lebensstiländerungen und Inzidenzsenkung der Folgekrankheiten ab.
- Vermeidung von Tabakkonsum: Der gesellschaftliche Konsens, Rauchen als Gesundheitsgefährdung anzuerkennen, Einschränkungen im öffentlichen Bereich staatlicherseits vorzuschreiben sowie begleitende Maßnahmen der Rauchentwöhnung haben bereits zum Rückgang des Tabakkonsums bei Jugendlichen geführt. Ein Monitoring des längerfristigen Health-Outcomes ist erforderlich.

7.4.1 Ergebnisbewertung

Für die Ergebnisbewertung ist die Bestimmung des Nutzens (utility) entscheidend: Dazu werden Effektivität und Effizienz bestimmt.

Die Effektivität (Wirksamkeit) einer Maßnahme wird am **Outcome** gemessen, bezogen auf die vorher definierten Ziele:

Die Health-Outcome-Maße beziehen sich vor allem auf die Effektgröße der Intervention, der Dauer bis zum Einritt des Effekts und seiner Dauer sowie der Implementation (Penetration) in der Zielgruppe (Schwartz et al. 1998).

Zur Beschreibung der Wirksamkeit auf die Dauer von Effekten und auf die Implementation in der Zielgruppe unterscheidet man folgende Outcome-Maße:

- direkt auf die Gesundheit bezogene (Gewicht, Ernährungsstatus)
- auf intermediär wirksame Risikofaktoren bezogene (Verhalten, Exposition)
- auf die Verteilung von Gesundheitseffekten bezogene (z. B. sozial Benachteiligte)
- zur Erfassung von Sekundärleistungen (Vorsorgen, Operationen)
- zur Dauer von Effekten (Einritt und Dauer, abhängig von Akzeptanz, Risikofaktoren).

Die **Effizienz** stellt die Kosten dem Nutzen der Maßnahmen gegenüber. Entsprechende Rechenmodelle der Kosten-Nutzen-Relation berücksichtigen direkte, indirekte und intangible Kosten (Kosten-Nutzen-Analysen, Lebensqualitätsstudien).

7.4.2 Krankheitskostenstudien

Gemessen werden die ökonomischen Auswirkungen einer Erkrankung unter Berücksichtigung aller Kosten und Konsequenzen, d. h. die durch sie verursachten sozialen Lasten, um Ressourcen für medizinische Therapie und soziale Versorgung besser planen zu können, aber auch daraus abgeleitet präventive Ansätze zu begründen und zu entwickeln (Scucs 1997).

7.4.3 Lebensqualitätsstudien

Messungen der Lebensqualität befassen sich mit den Auswirkungen einer Erkrankung oder medizinischen Dienstleistung auf das soziale Wohlbefinden, den physischen und emotionalen Status, die intellektuelle Leistungsfähigkeit und das allgemeine Wohlbefinden einer Person. Messinstrumente sind u. a. Indikatoren, Skalen, Batterien. Idealerweise umfassen sie all diese Kategorien. Eine Übersicht über wichtige Messinstrumente liefert die OLGA-Datenbank (On-Line Guide to Quality-of-Life Assessment, www.olga-qol.com). Für Kinder und Jugendliche steht seit einigen Jahren als international standardisiertes Fragebogeninstrument der KINDL-R-Fragebogen zur gesundheitsbewussten Lebensqualität zur Verfügung. Zur Bewertung der gesundheitsbewussten Lebensqualität von Kindern und Jugendlichen in Deutschland wurde dieser Fragebogen an der Normstichprobe des Kinder- und Jugendgesundheitssurveys untersucht und für die Population der Kinder und Jugendlichen in Deutschland Referenzdaten ermittelt (Ravens-Sieberer et al. 2007).

Messungen der Lebensqualität werden vorwiegend bei schweren chronischen oder lebensbegrenzenden Erkrankungen, bei palliativer Behandlung oder beim Therapievergleich hinsichtlich Lebensqualität durchgeführt (Scucs 1997).

7.4.4 Health Impact Assessment (HIA)

Einen wichtigen präventiven Ansatz verfolgt das **Health Impact Assessment.** HIA ist eine systematische Bewertung präventiver Politiken jedweder Art auf die Gesundheit der Bevölkerung oder Bevölkerungsgruppen.

Im Sinne einer Gesundheitsverträglichkeitsprüfung, besser Gesundheitsfolgenabschätzung, sollen vorweg negative Einflüsse abgeschätzt und verhindert und damit gesundheitsbewusstere Entscheidungen ermöglicht werden. Sie basiert auf der Theorie der Gesundheitsdeterminanten, d. h. den Faktoren, die die Gesundheit des Individuums beeinflussen (Weinbrenner et al. 2007). HIA sollte idealerweise allen Entscheidungen vorangestellt werden.

7.5 Ausgewählte Handlungsfelder und Maßnahmen der Gesundheitsförderung und Prävention für Kinder und Jugendliche

7.5.1 Gesundheitsfördernde Gesamtpolitik

In einer Reihe gesetzlicher Regelungen wurden gesundheitsfördernde und präventive Ansätze aufgegriffen. Deren Umsetzung erfolgt durch die entsprechenden Ressorts und weitere spezialisierte Institutionen.

Gesetzliche Regelungen

Für die gesetzliche Krankenversicherung wurden ab 1993 in § 20 des **Sozialgesetzbuches V (SGB V)** Gesundheitsförderung und primäre Prävention individuell und im Setting Kindergarten und

Schule mit allerdings geringer finanzieller Ausstattung (2,74 Euro pro Versichertem/Jahr 2006 mit Anpassung enstpr. Bezugsgröße nach § 18 SGB IV, für 2013 auf 3,01€ pro Versichertem) als Soll-Aufgaben definiert und in Handlungsempfehlungen der Krankenkassen näher ausgeführt (Leitfaden Prävention 2010). Die Förderung der gesundheitlichen Selbsthilfe ist in § 20c SGB V festgeschrieben, die Übernahme der Impfstoffkosten für die von der Ständigen Impfkommission am Robert-Koch-Institut (STIKO) empfohlenen Impfungen ist nach § 20d SGB V seit 2006 Pflichtleistung. § 21 SBG V ermöglicht seit 1993 gruppenprophylaktische Maßnahmen zur Verhütung von Zahnerkrankungen durch die niedergelassene Zahnärzteschaft in Kooperation mit dem Öffentlichen Gesundheitsdienst in Kindergärten und Schulen. § 22 SGB V sieht die zahnärztliche Individualprophylaxe ab dem 6. Lebensjahr vor. Als Maßnahmen der sekundären Prävention sollen die Kinderfrüherkennungsuntersuchungen (Vorsorgeuntersuchungen) nach § 26 SGB V eine frühzeitige Intervention durch festgelegte ärztliche und zahnärztliche Untersuchungen ermöglichen. Die in der Mehrzahl der Länder erlassenen gesetzlichen Regelungen zur Verbesserung der Teilnahme an den Früherkennungsuntersuchungen sollen erreichen, dass diese Prävention allen Kindern zugutekommt und damit Vernachlässigung und ggf. Misshandlung vorgebeugt werden kann. Sozialpädiatrische Zentren nach §§ 43a und 119 SGB V erfüllen zudem Aufgaben der tertiären Prävention für behinderte Kinder und Jugendliche; sozialmedizinische Nachsorge nach § 43 Satz 2 SGB V ermöglichst das case management bei schweren Erkrankungen. § 30 SGB IX definiert, dass durch medizinische und heilpädagogische Maßnahmen die Teilhabe an der Gesellschaft gefördert werden soll. Interdisziplinär arbeitende Frühförderstellen und sozialpädiatrische Zentren erbringen medizinische und heilpädagogische Leistungen der ambulanten Frühförderung als Komplexleistung („alles aus einer Hand“) für Kinder bis zum Schuleintritt. Auch die Regelungen zur Versorgung mit Heil- und Hilfsmitteln nach §§ 31–34 SGB V beinhalten präventive Maßnahmen.

Mit dem 2012 in Kraft getretenen Bundeskinderschutzgesetz (BKiSchG) sollen wesentliche Elemente einer frühen Prävention bundesweit umgesetzt und eine verbindliche Kooperation zwischen Gesundheitswesenund Jugendhilfe etabliert werden.

Weitere Gesetze mit Regelungen zu Gesundheitsförderung, Prävention und Gesundheitsschutz für Kinder und Jugendliche enthalten das SGB VIII (**Kinder- und Jugendhilfegesetz KJHG**) und die Ausführungsgesetze der Länder, z. B. die Regelungen zu Kinderbetreuungseinrichtungen, ferner die Schulgesetzgebung, das Familienrecht, die Gesetze über den Öffentlichen Gesundheitsdienst, das **Infektionsschutzgesetz,** die Regelungen zum Verbraucherschutz, Umweltrecht, Arbeitsschutz- und Unfallgesetzgebung. Entschließungen der Gesundheitsministerkonferenzen (2000, 2006, 2008, 2012) und der Kultusministerkonferenzen (1992 2006, 2012) betonten die Notwendigkeit von Gesundheitsförderung und Prävention bei Kindern und Jugendlichen angesichts gesundheitlicher Problemlagen, demografischer Entwicklung und begrenzter Ressourcen und fordern alle gesellschaftlichen Kräfte zu mehr Engagement auf. Lehrpläne und Satzungen der Träger von Kindertageseinrichtungen wurden entsprechend ergänzt, z. B. zu Bewegung, gesunder Ernährung und Unfallverhütung.

Die Forderung aus der Fachwelt an die Politik, die Aktivitäten zur Prävention innerhalb der sozialen Sicherungssysteme (der gesetzlichen Kranken-, Unfall-, Pflege- sowie Rentenversicherung) mit denen der öffentlichen Hand in einem **Präventionsgesetz** verbindlich zusammenzuführen und auch die finanziellen Ressourcen festzuschreiben, wurde bislang nicht erfüllt. Nach Scheitern eines Entwurfs aus 2002 wird 2013 der Entwurf eines Gesetzes zur Förderung der Prävention beraten, der mehr Prävention allein innerhalb der gesetzlichen Krankenversicherung vorsieht und vorrangige Gesundheitsziele qualitätsgesichert innerhalb des Gesundheitsversorgungssystems umsetzen will, aber u. a. wegen der Begrenzung auf den Medizinbereich vielfach kritisiert wird.

Strukturen und Maßnahmen

Die öffentliche Gesundheitsfachverwaltung auf den Ebenen des Bundes, der Länder und der Kommunen hat Gesundheitsförderung und Prävention in ihren Aufgabenkatalog aufgenommen. Sie initiiert Maßnahmen der Gesundheitsförderung und Prävention und kooperiert themenbezogen mit unterschiedlichen Akteuren. Informationen bieten z. B. die Internetportale oder Pressestellen der Bundes- und Länderministerien, die für Gesundheit, Jugend, Bildung, Ernährung, Verkehr oder Umwelt zuständig sind, sowie die örtlichen Gesundheitsämter.

Zur Umsetzung gesundheitsfördernder Maßnahmen wurden in Bund und Ländern mit der Bundeszentrale für gesundheitliche Aufklärung (BZgA) und den Landesvereinigungen für Gesundheitsförderung Institutionen geschaffen, um Gesundheitsförderung voranzubringen. Sie sind vorrangig für die Durchführung von Kampagnen und Netzwerkbildung zuständig. Daneben sind weitere Institutionen und Kooperationsverbünde in der Erstellung von Konzepten und der Umsetzung von Projekten engagiert, z. B. die Bundesvereinigung für Gesundheitsförderung und Prävention e. V. (BVPG), die Deutsche Arbeitsgemeinschaft für Jugendzahnpflege (DAJ) und entsprechende Landesarbeitsgemeinschaften, Deutsche Gesellschaft für Ernährung (DGE), die Plattform Gesundheit und Ernährung e. V., IN FORM – Deutschlands Initiative für gesunde Ernährung und mehr Bewegung, die Deutsche Krebshilfe und deutsche Krebsgesellschaft sowie freie Vereine zur Gesundheitsförderung. Die gesetzlichen Krankenkassen und privaten Krankenversicherer führen eigene zielgruppenspezifische Aktionen zur Gesundheitsförderung durch.

Mit der Einrichtung von Lehrstühlen für Public Health wurde auch die wissenschaftliche Befassung mit den Bedingungen der Gesundheit der Bevölkerung initiiert und verschiedene Forschungsprojekte, u. a. des BMBF, der BZgA, der Bertelsmann- und der Robert-Bosch-Stiftung, initiiert. So hat im Rahmen der Forschungsförderung des BMBF und des BMG das Koordinationsprojekt „Kooperation für nachhaltige Prävention (KNP)“ über 60 Projekte zur Gesundheitsförderung von 2004 bis 2012 evaluiert und zur konzeptionellen Weiterentwicklung von Gesundheitsförderung und Prävention beigetragen (www.KNP.de).

Gesundheitsziele

Aufgrund der Initiative der WHO von 1978, die „Gesundheit für alle“ als erstes weltweites Zielprogramm verabschiedete und das in Europa in das Programm „Gesundheit für alle“ ab 1980 übernommen wurde, beschloss die Gesundheitsministerkonferenz 1999 auch für Deutschland, nationale Gesundheitsziele zu definieren. Auf der Basis der Gesundheitsberichterstattung erarbeiteten Bund, Länder und die Gesellschaft für Versicherungswissenschaft und Gestaltung e. V. (GVG) in einem konsentierten Dialog vorrangige Gesundheitsziele. Sie sind als „verbindliche Vereinbarungen der verantwortlichen Akteure im Gesundheitssystem ein Instrument der Gesundheitspolitik im Rahmen von Public-Health-Ansätzen, um Verbesserungen der Gesundheitssituation der Bevölkerung zu erreichen“. Diese Ziele sollen in einem kontinuierlichen Prozess durch koordinierendes Vorgehen und unter größtmöglicher Partizipation aller Beteiligten in Bund und Ländern in Settings vorangebracht werden. Querschnittsanforderungen sind die Stärkung der Selbsthilfe, Bürger- und Patientenorientierung, Prävention auch bei Krankheitszielen, Evidenzbasierung, sektorale Verzahnung und Integration, gesundheitliche Chancengleichheit und Gender Mainstreaming. Im Fokus eines dieser Ziele - „Gesund aufwachsen: Ernährung, Bewegung, Lebenskompetenz“ – steht die Gesundheit von Kindern und Jugendlichen. Aktivitäten aus diesem Zielbereich werden in unterschiedlicher Tiefe und mit unterschiedlichen Kooperationspartnern teils regional, teils landesweit umgesetzt.

Finanzierung von Gesundheitsförderung und Prävention

An der Finanzierung der Maßnahmen für Gesundheitsförderung und Prävention sind verschiedene Träger beteiligt, u. a. gesetzliche Krankenversicherung, Pflegeversicherung, Unfallversicherung, Rentenversicherung, private Krankenversicherung, öffentliche Hand und Arbeitgeber. Private Ausgaben gehen ebenfalls in die Kostenrechnung ein. Die Ausgaben für Gesundheitsförderung betrugen 2011 6,009 Milliarden Euro, davon 2,942 Milliarden finanziert durch die GKV, 849 Millionen durch die öffentliche Hand, 1,319 Milliarden über private Ausgaben und 821 Millionen durch Arbeitgeber. Neben Gesundheitsförderung zählen allgemeiner Gesundheitsschutz, Maßnahmen der Krankheitsfrüherkennung sowie Gutachten und Koordination zu den Leistungen der Prävention. – Mit 11,082 Milliarden Euro insgesamt beträgt er 3,8 % aller Gesundheitsausgaben (Statistisches Bundesamt 2012), der Anteil für die Gesundheitsförderung allein beträgt 2,05 %. Die Anteile für beide Bereiche sind in den vergangenen Jahren gesunken.

7.5.2 Gesundheitsförderung und Prävention im Gesundheitssystem

In das Gesundheitssystem werden besonders hohe Erwartungen gesetzt, initiierend, koordinierend und in Kooperation mit anderen präventive gesundheitliche Aspekte im Lebensumfeld der Menschen voranzubringen. Dies gilt sowohl für die individuelle Gesundheitsversorgung als auch für den Öffentlichen Gesundheitsdienst.

Die angesichts der Herausforderungen unzulängliche Verankerung von Prävention in der Sozialgesetzgebung trägt entscheidend dazu bei, dass im Alltag der Gesundheitsversorgung (in Praxis und Klinik) die Maßnahmen zur Gesundheitsförderung und primären Prävention völlig unzureichend sind. Oftmals werden sie als Handlungsnotwendigkeiten auch nicht wahrgenommen. Dies gilt auch für weite Kreise der ärztlichen Standespolitik. Forderungen der Pädiatrie nach institutionalisierter Gesundheitsberatung, z. B. im Rahmen der Kinderfrüherkennungsuntersuchungen, sind Beratungsgegenstand des Gemeinsamen Bundesausschusses. Weder die Ausbildung von speziellen Fachkräften noch die Umsetzung von Maßnahmen sind finanziert. Aktivitäten in den Settings Kindergarten, Schule oder Kommune beruhen mehrheitlich auf persönlichem Engagement. Lediglich Interdisziplinäre Frühförderstellen, Sozialpädiatrische Zentren und Rehabilitationseinrichtungen für Kinder und Jugendliche arbeiten unter stringenten Rahmenbedingungen kooperativ und präventiv an gesundheitsfördernden Rahmenbedingungen im Lebensumfeld des Kindes.

Die Handlungsmöglichkeiten des Öffentlichen Gesundheitsdienstes sind überwiegend auf die Gesundheit der ganzen Bevölkerung oder von Bevölkerungsgruppen, d. h. die öffentliche Gesundheit ausgerichtet. Präventions- und Schutzmaßnahmen zielen darauf ab, Gesundheit als öffentliches Gut zu schützen, gefährdenden Faktoren entgegenzuwirken und gesundheitsförderliches Verhalten in allen Bezügen zu stärken. Nach den ÖGD-Gesetzen der Länder hat der ÖGD an der Schaffung gesundheitsfördernder Rahmenbedingungen mitzuwirken und die Gesundheit zu schützen (➤ Kap. 9). Die Kinder- und Jugendgesundheitsdienste können diesem Anliegen aufgrund unterschiedlicher Strukturen und Kompetenzen nicht überall in wünschenswertem Umfang nachkommen. In Kooperation mit einer Vielzahl von Akteuren vor Ort und den Einrichtungen für gesundheitliche Aufklärung tragen sie allerdings dazu bei, den Stellenwert gesundheitlicher Prävention im öffentlichen Bewusstsein zu steigern. Beispiele sind die Kampagnen zur Prävention von HIV und AIDS, zu Impfungen oder die Netzwerke für mehr Kinderschutz, aber auch ihre Beratungen im Rahmen der jahrgangsbezogenen sozialpädiatrischen Untersuchungen.

7.5.3 Gesundheitsförderung und Prävention bei Kindern und Jugendlichen

Die Gesundheit von Kindern und Jugendlichen ist ein Schwerpunkt aller Institutionen zur Gesundheitsförderung und Prävention. Ihre Aktivitäten sind dabei mehrheitlich auf die Eltern als wichtigste Bezugspersonen und das Setting Familie ausgerichtet, indem sie verschiedene gesundheitliche Aspekte aufgreifen (Unfallprävention, gesunde Ernährung, Impfen). Ältere Kinder und Jugendliche werden direkt angesprochen (HIV und AIDS, Drogen, Gewalt). Eine weitere Zielgruppe stellen die Settings Kindertageseinrichtung und Schule dar, für die Konzepte und Materialien zur Bearbeitung bestimmter Themen bereitgestellt werden (u. a. zu gesundem Frühstück, Zahngesundheit, Bewegung).
Eine konsequente Umsetzung präventiver Maßnahmen in den Settings Kindergarten und Schule findet nur unzulänglich statt. **Settingansätze** gibt es aufgrund zwar vielerorts; sie sind jedoch aus finanziellen Gründen meist als Projekte angelegt und in der Regel zeitlich befristet. Von der Gesundheitsversorgung wurden Gesundheitsförderung und Prävention sowohl bei der individuellen Beratung als auch in Settings nur zögerlich aufgegriffen. Eine erhebliche Zahl von Projekten, die im Laufe der Jahre von unterschiedlichen Akteuren auf den Weg gebracht wurden, wurde ohne hinreichende Qualitätssicherung und Evaluation durchgeführt. Verstetigung und Nachhaltigkeit fehlen.

Altersbezogene Inhalte von Gesundheitsförderung und Prävention

Je nach Alter der Kinder und Jugendlichen gibt es sowohl in der individuellen Gesundheitsberatung von Eltern sowie Kindern und Jugendlichen als auch im Setting Gesundheitsthemen, die generell vermittelt werden und im Alltag Beachtung finden sollten (➤ Tab. 7.3).

Informationsmaterialien

Zu den verschiedenen Themen, für Kinder und Jugendliche unterschiedlicher Altersgruppen, für Eltern sowie zur pädagogischen Vermittlung der Themen sind **Informationsmaterialien** (Flyer, Broschüren, Filme und Präsentationen bis hin zu Ausstellungen) wertvolle Hilfen. Dabei spielen Internet, Apps und soziale Netzwerke bei der Informationsübermittlung eine zunehmend wichtige Rolle. Informationsmaterialien zu vielen Aspekten der Gesundheitsförderung und Prävention erstellen u. a. die Bundeszentrale für gesundheitliche Aufklärung (BZgA), das Robert-Koch-Institut (RKI), die Deutsche Gesellschaft für Kinderheilkunde und Jugendmedizin (DGKJ), die Bundesarbeitsgemeinschaft (BAG) Mehr Sicherheit für Kinder e. V., die Bundesvereinigung für Gesundheitsförderung und Prävention e. V. (BVPG), das Deutsche Grüne Kreuz, die Krankenkassen und Unfallkassen, Gesundheitsministerien, Gesundheitsämter sowie Wohlfahrtsverbände und Vereine.

Impfungen

Impfungen gehören weltweit zu den wirkungsvollsten Maßnahmen der medizinischen Prävention. Durch Impfungen konnte eine Reihe von schweren Krankheiten, die zu bleibenden Schädigungen führen oder gar tödlich verlaufen, zurückgedrängt (Diphtherie, Poliomyelitis) oder (wie die Pocken) sogar ausgerottet werden. Überwiegend nützen Impfungen damit sowohl der Gesundheit des Einzelnen als auch der Bevölkerung insgesamt. Eine Durchimpfung von 95 % gegen die einzelnen Infektionskrankheiten (Herdenimmunität) ist erforderlich, um von Mensch zu Mensch übertragbare Infektionskrankheiten nachhaltig zurückzudrängen. Lediglich die Impfun-

Tab. 7.3 Altersbezogene Inhalte von Gesundheitsförderung und Prävention

Säuglings- und Kleinkindalter	• Altersentsprechende Bedürfnisse und Versorgung: Ernährung, Schlaf-Wach-Rhythmus, Schlafumgebung, SIDS-Prävention, Hygiene, Unfallprävention, Allergieprävention, Zahngesundheit und Kariesprävention, Verhalten bei häufigen Krankheiten, Impfungen, Früherkennungsuntersuchungen, geschlechtsspezifische Aspekte und kindliche Entwicklung • Seelisch-soziale Entwicklung: Eltern-Kind-Interaktion, Zuwendung und Bindungsverhalten, Förderung von Selbstvertrauen und Eigenverantwortung, Stressbewältigung, Verhalten gegenüber Fremden • Entwicklungsförderung: Förderung der Motorik und Koordination, der Sprache, Einsatz von Spielen und Spielverhalten, Medieneinsatz, ggf. Einleitung notwendiger Fördermaßnahmen • Nachfrage nach sozialem Umfeld (Wohnen, Familie, soziale Interaktion, Krippenbesuch oder Tagespflege) • Soforthilfe-Maßnahmen am Kind, Straßenverkehr (Sicherungssysteme/Rückhaltesysteme im Auto) • Kinder mit chronischen Erkrankungen und/oder Entwicklungsverzögerungen: in Kooperation mit weitergehender Beratung und Betreuung in SPZ/Kinderklinik • Erkennen von Hinweisen auf Vernachlässigung und Misshandlung bzw. unmittelbare Einschaltung der Jugendhilfe bei Gefahr im Verzug • Benennen von Ansprechpartnern bei gesundheitlichen Notfallsituationen und familiären Krisensituationen • Hinweis auf Programme zur Prävention: Vorausschauende Gesundheitsberatung, Elternschulen, kommunale Projekte
Grundschulalter	• Altersentsprechende Bedürfnisse und Versorgung: Ernährung, Tagesrhythmus, Medienkonsum, Bewegungsförderung, Unfallprävention, Allergieprävention, Kariesprävention, geschlechtsspezifische Aspekte, Entwicklung und Entwicklungsförderung • Nachfrage nach sozialem Umfeld: Wohnen, soziale Interaktion in Schule und Freizeit, Freunde, Hobbys, Verhalten im Straßenverkehr (Fahrradhelm), Verkehrserziehung • Psychosoziale Belastungen von Kind und/oder Eltern
Späte Kindheit und Jugend	• Beratung für Eltern und Jugendliche • Gesunde Ernährung, Bewegungsförderung, Medienkonsum und Schutz vor Lärmbelastung, Drogenkonsum, Körperhygiene, Impfungen und Schutz vor Infektionskrankheiten, Sexualität und Sexualverhalten, Umgang mit Aggression und Prävention von Gewalt • Entwicklungsphasen – Pubertät mit körperlichen, psychosozialen und geschlechtsspezifische Aspekten, Loslösung • Beratung hinsichtlich weiterer Aspekte im Rahmen der Jugendgesundheitsuntersuchung, Nachfrage nach sozialem Umfeld (Wohnen, soziale Interaktion in Schule und Freizeit, Hobbys), • Berufswunsch und mögliche bestehende Risiken – (Entlassschüleruntersuchung, Arbeitsschutzuntersuchung)
Jedes Alter	• Erfragen psychosozialer Belastungen von Kind und/oder Eltern, Hinweis auf Beratungs- und Unterstützungsmaßnahmen sowie präventive Maßnahmen für Eltern/Frauen: Kurmaßnahmen für (Mütter/Vater und) Kind, Erziehungsberatungsstellen, Eheberatungsstellen, Frauenhaus • Bei Vorhandensein von Hinweisen auf Gewalt jedweder Form (Vernachlässigung, Misshandlung, Missbrauch, durch elterliche Partnerschaftsgewalt): Hinweis auf Beratungs- und Hilfsangebote bzw. unmittelbare Einschaltung des Jugendamtes bei Gefahr im Verzug

Tab. 7.4 Zeiträume zur Durchführung der Früherkennungsuntersuchungen und jeweilige Toleranzgrenzen

Untersuchungszeitraum		**Toleranzgrenze**	
U2	3.–10. Lebenstag	U2	3.–14. Lebenstag
U3	4.–5. Lebenswoche	U3	3.–8. Lebenswoche
U4	3.–4. Lebensmonat	U4	2.–4 ½. Lebensmonat
U5	6.–7. Lebensmonat	U5	5.–8. Lebensmonat
U6	10.–12. Lebensmonat	U6	9.–14. Lebensmonat
U7	21.–24. Lebensmonat	U7	20.–27. Lebensmonat
U7a	34.–36. Lebensmonat	U7a	33.–38. Lebensmonat
U8	46.–48. Lebensmonat	U8	43.–50. Lebensmonat
U9	60.–64. Lebensmonat	U9	58.–66. Lebensmonat

gen gegen Tetanus und FSME dienen ausschließlich dem individuellen Schutz. Verschiedene Impfungen, z. B. gegen Masern, Mumps und Röteln, können bei Ungeimpften bei rechtzeitiger Anwendung auch nach Kontakt mit einem Erkrankten noch vor dem Ausbruch der Infektionskrankheit schützen (Riegelungsimpfung).
Die bevölkerungsmedizinische Bedeutung von Impfungen belegen die Regelungen im Infektionsschutzgesetz (§20 IfSG): Alle Ebenen der Gesundheitsfachverwaltung, d. h. Bundesgesundheitsministerium, Robert-Koch-Institut, Gesundheitsministerien der Länder und Gesundheitsämter, sind zur Aufklärung über die Möglichkeiten der Verhütung von Infektionskrankheiten durch Impfungen verpflichtet. Kindertageseinrichtungen und Schulen sollen gemeinsam mit dem jeweiligen Gesundheitsamt die Eltern über einen altersentsprechenden Impfschutz aufklären.

Impfempfehlungen der STIKO

Die Ständige Impfkommission am Robert-Koch-Institut (STIKO) spricht als fachliches Gremium die Impfempfehlungen für Deutschland aus; sie werden entsprechend den epidemiologischen Erfordernissen und der Verfügbarkeit von Impfstoffen gegen einzelne Infektionskrankheiten aktualisiert und im Epidemiologischen Bulletin" des Robert-Koch-Instituts veröffentlicht. Sie beinhalten auch Maßnahmen der spezifischen Prophylaxe für Index- und Kontaktpersonen und Hinweise zur Durchführung der Impfung. Die Empfehlungen der STIKO haben Leitliniencharakter und sind medizinischer Standard für die Impfaufklärung (Bundesgerichtshof 2000), d. h. der Impfarzt muss bei der Impfaufklärung darauf hinweisen.
Im Impfkalender der STIKO werden die Standardimpfungen für Säuglinge, Kinder, Jugendliche und Erwachsene mit dem empfohlenen Impfalter und den Mindestabständen zwischen den Impfungen dargestellt (www.impfen-info.de; www.rki.de).
Zu den aktuellen Standardimpfungen gehören im frühen Säuglings- und Kleinkindalter die Impfungen gegen Diphtherie, Tetanus, Pertussis, Poliomyelitis, Haemophilus influenzae b, Hepatitis B und Pneumokokken. In Planung ist auch die Impfung gegen Rotavirus. Gegen Ende des ersten Lebensjahres stehen Impfungen gegen Masern, Mumps, Röteln, Varizellen und Meningokokken an. Die Impfungen sind in empfohlenen Abständen aufzufrischen bzw. zu wiederholen. Für Mädchen zwischen 12 und 17 Jahren wird zusätzlich die Impfung gegen humane Papillomaviren (HPV) empfohlen. Ein Nachholen versäumter Impfungen ist möglich.
Daneben werden weitere Impfungen nach Indikation angeraten, z. B. gegen FSME, Hepatitis A, Influenza. Eine Indikation kann durch besondere Exposition oder spezielle Risiken am Ausbildungs- oder Arbeitsplatz vorliegen.

Öffentliche Impfempfehlung

Die Bundesländer sprechen öffentliche Empfehlungen zu Schutzimpfungen (und Maßnahmen der spezifischen Prophylaxe) aus. Der Staat rät den Bürgern, sich aus öffentlichem Interesse impfen zu lassen und damit nicht nur sich selbst zu schützen, sondern auch zum Schutz der Allgemeinheit beizutragen. Dieses „Sonderopfer" des Impflings im Interesse der Allgemeinheit führt zu dem Aufopferungsanspruch des Staates: er haftet bereits bei der Möglichkeit eines durch eine empfohlene Impfung eingetretenen Schadens. Die Empfehlungen der meisten Länder beziehen sich auf die von der STIKO empfohlenen Standard- und Indikationsimpfungen, beinhalten aber darüber hinaus teilweise weitere Impfungen oder weitere Indikationen für von der STIKO empfohlene Impfungen, z. B. wird die Impfung gegen Influenza für die gesamte Bevölkerung öffentlich empfohlen.

Kostentragung und Durchführung von Impfungen

Kostenträger der Standard- und vieler Indikationsimpfungen sind nach § 20d Abs. 1 SGB V die Krankenkassen. Einzelheiten zu Voraussetzungen, Art und Umfang der Leistungen bestimmt der Gemeinsame Bundesausschuss in der Schutzimpfungsrichtlinie nach § 92 SGB V Abs. 1 Nr. 3 auf der Grundlage der Empfehlungen der Ständigen Impfkommission beim Robert-Koch-Institut gemäß § 20 Abs. 2 des Infektionsschutzgesetzes unter besonderer Berücksichtigung der Bedeutung der Schutzimpfungen für die öffentliche Gesundheit. Zu Änderungen der Empfehlungen der STIKO hat der Gemeinsame Bundesausschuss innerhalb von drei Monaten nach ihrer Veröffentlichung eine Entscheidung zu treffen; kommt diese nicht termin- oder fristgemäß zustande, dürfen insoweit die von der STIKO empfohlenen Schutzimpfungen erbracht werden. Bei Reisen ins Ausland sind die Kosten für Impfungen vom Impfling selber zu tragen, ausgenommen dann, wenn durch Einschleppung einer Infektionskrankheit auch eine Gefährdung für die deutsche Bevölkerung besteht. Für Impfungen als Maßnahme des Arbeitsschutzes zahlt der Arbeitgeber. Die Krankenkassen können über Satzungsleistung die Kosten für weitere Impfungen übernehmen.
Im Kindes- und Jugendalter erfolgen Impfungen heute **nahezu** vollständig durch das ambulante Versorgungssystem. Der ÖGD informiert über Impfungen in Kindergemeinschaftseinrichtungen; er kann nach § 20 Abs. 5 IfSG unentgeltlich Impfungen anbieten. Dabei kann der ÖGD dies in Kooperation mit den Krankenkassen umsetzen: die Krankenkassen haben nach § 20d Abs. 3 SGB V im Zusammenwirken mit den Behörden der Länder, die für die Durchführung von Schutzimpfungen nach dem Infektionsschutzgesetz zuständig sind, gemeinsam und einheitlich Schutzimpfungen ihrer Versicherten zu fördern und sich durch Erstattung der Sachkosten an den Kosten der Durchführung zu beteiligen. Dazu sind Rahmenvereinbarungen zu schließen. In den einzelnen Bundesländern gibt es dazu unterschiedliche Kooperationsvereinbarungen.
In Vereinbarungen nach § 132e SGB V können die Krankenkassen zudem auch Verträge mit anderen ärztlichen Leistungserbringern, so auch dem ÖGD, zur Durchführung der Impfungen abschließen.

Impfstoffe

Die meisten Impfungen werden in Form von Kombinationsimpfungen verabreicht und sind gut verträglich. Die Impfstoffe sind in der Regel europa- oder weltweit zugelassen. Als deutsche Zulassungsbehörde überprüft das Paul-Ehrlich-Institut in Langen (PEI) die für den deutschen Markt bestimmten Impfstoffchargen. Alle Impfungen sind im Impfbuch mit Name des Impfstoffs, Chargennummer und Unterschrift des Impfarztes zu dokumentieren.

Um Nebenwirkungen und Impfkomplikationen bei der breiten Anwendung eines Impfstoffs in der Bevölkerung zu erfassen und ggf. alle Ärzte rechtzeitig im Sinne der Gefahrenabwehr zu warnen, ist der impfende Arzt verpflichtet, dem PEI nach § 77 Arzneimittelgesetz (AMG) unerwünschte Nebenwirkungen nach einer Impfung zu melden und das Gesundheitsamt nach § 6 (3) IfSG zu informieren, um eine differenzialdiagnostische Abklärung einzuleiten. Die Meldungen über unerwünschte Nebenwirkungen sind auf der Homepage des PEI einsehbar (www.pei.de).

Früher Impfschutz für Kinder und Jugendliche

Grundsätzlich sollten Impfungen so früh wie möglich durchgeführt werden, um den Impfling zu schützen. Die wiederholten Ausbrüche von impfpräventablen Krankheiten in Kindergärten und Schulen, besonders der Masern, führen zunehmend zur Forderung, den Besuch der Kindertageseinrichtungen von einem altersabhängigen Impfschutz bzw. dem Impfschutz gegen einzelne Erkrankungen wie Masern abhängig zu machen, wie dies in einigen Nachbarländern praktiziert wird. Durch den zunehmenden Ausbau der Krippenbetreuung wird diese Forderung verstärkt. Verschiedene Kindergartengesetze der Länder schreiben im Rahmen der ärztlichen Aufnahmeuntersuchung zumindest die Überprüfung des Impfschutzes vor, um damit auch auf das Schließen vorhandener Impflücken hinzuwirken.
Umfassende und aktuelle Daten zum Impfschutz der Bevölkerung, auch dem Impfschutz von Kindern und Jugendlichen, gibt es in Deutschland nicht. Daten der Kassenärztlichen Vereinigungen sind auf die gesetzlich Versicherten beschränkt. Lediglich über den Impfschutz der Schulanfänger liegen seit 2005 bundesweit Angaben vor, die im Rahmen der Einschulungsuntersuchung bzw. Schulaufnahme durch das Gesundheitsamt erhoben, beim Robert-Koch-Institut zusammengeführt und jährlich veröffentlicht werden. Einzelne Bundesländer bzw. einzelne Gesundheitsämter erfassen den Impfschutz der Kinder und Jugendlichen darüber hinaus in unterschiedlichen Altersklassen. So melden z. B. die niedergelassenen Ärzte in Schleswig-Holstein die Impfungen im Rahmen der Kindergartenaufnahmeuntersuchung an das dortige GesundheitsministeriumIn den jährlichen Berichten über die Schuluntersuchungen finden sich auch aktuelle Daten zum Impfschutz vom KiTa-Alter bis zu den Kindern der 8. Klassen (http://schleswig-holstein.de/MSGFG/DE/Service/Broschueren/PDF/schuleinguntber2010__blob=publicationFile.pdf S. 105) Alle vorhandenen Daten zeigen, dass der Impfschutz im Kindes- und Jugendalter noch unzureichend ist, z. B. für die zweite MMR-Impfung und die Auffrischimpfungen gegen Diphtherie, Tetanus, Polio und Pertussis (Reiter S, Poethko-Müller C 2009). Die Defizite sind in der Regel nicht in der generellen Ablehnung des Impfens zu sehen, wie sie von der kleinen, aber über Internet und Presse präsenten Gruppe der Impfgegner bzw. Impfkritiker praktiziert wird. Vielmehr fehlt es an der kontinuierlichen, auf Nachhaltigkeit angelegten Kooperation aller für das Impfen Verantwortlichen. Ansprache und Aufklärung der Eltern über das Impfen im Rahmen der Früherkennungsuntersuchungen oder sonstiger Arztkontakte sowie durch den öffentlichen Gesundheitsdienst gemeinsam mit Kindergärten und Schulen im Auftrag des Infektionsschutzgesetzes können konkret zum besseren Impfschutz beitragen. Die Impfaufklärung ist auch eine Aufgabe der Bundeszentrale für gesundheitliche Aufklärung (www.bzga.impf-info).

Nationaler Impfplan und Nationale Impfkonferenzen

Zur Verbesserung des Impfwesens in Deutschland werden auf Beschluss der Gesundheitsministerkonferenz seit 2009 in zweijährigem Abstand nationale Impfkonferenzen durchgeführt. Dieses Forum soll den Austausch der Akteure im Bereich des Impfens fördern. In dem 2012 veröffentlichten nationalen Impfplan werden alle Aspekte des Impfwesens beschrieben und Vorschläge zur strukturellen und konzeptionellen Verbesserungen daraus abgeleitet. Vorrangiges Impfziel für Deutschland ist die Verbesserung des Impfschutzes gegen Masern und Röteln, um das Ziel der WHO, Masern und Röteln bis 2015 zu eliminieren, zu erreichen. (www.nationaler Impfplan.de)

7.5.4 Kinderfrüherkennungsuntersuchungen (Vorsorgeuntersuchungen)

Rechtliche Grundlagen

Seit 1971 haben nach § 26 Abs. 1 Satz 1 SGB V versicherte Kinder bis zur Vollendung des sechsten Lebensjahrs Anspruch auf Untersuchungen sowie nach Vollendung des zehnten Lebensjahres auf eine Untersuchung zur Früherkennung von Krankheiten, die ihre körperliche oder geistige Entwicklung in nicht geringfügigem Maße gefährden. Die übrigen Krankenversicherer (u. a. Verband der privaten Krankenversicherer und Beihilfe) lassen diese Regelungen gegen sich gelten. Inhalt, Umfang und Dokumentationspflichten der einzelnen Untersuchungen legt der Gemeinsame Bundesausschuss nach § 92 SGB V in den Richtlinien zur Früherkennung von Krankheiten bei Kindern bis zur Vollendung des sechsten Lebensjahres (Kinderrichtlinien) fest. Eine seit 2003 laufende Überarbeitung der Kinderrichtlinien umfasst die Bereiche allgemeine Organisation des Programms (z. B. Untersuchungsintervalle), Standardisierung der klinischen Untersuchung und Nutzenbewertungen für spezifische Screeningmaßnahmen.

Screening

Ein wichtiger Bestandteil der Vorsorgeuntersuchungen sind Screeningmaßnahmen, die der gesamten Zielgruppe – hier Kindern zu verschiedenen Alterszeitpunkten – angeboten werden, um mögliche Anzeichen für das Vorliegen einer Krankheit bzw. von Risikofaktoren vor Ausbruch von Krankheitssymptomen zu erkennen und möglichst frühzeitig eine Behandlung einleiten zu können.

Ein bevölkerungsbezogenes Screeningprogramm ist nur dann sinnvoll und zielführend, wenn die Krankheit für die Gesundheit der Bevölkerung von Bedeutung ist und gut bzw. bei früherem Erkennen deutlich besser behandelbar ist. Das Testverfahren muss eine hohe Sensitivität und Spezifität aufweisen, d. h. der Test soll die gesuchte Erkrankung bzw. die bestehenden Risikofaktoren mit möglichst großer Sicherheit nachweisen oder ausschließen können. Die Untersuchung soll darüber hinaus zeit- und kostengünstig sein und den zu Untersuchenden wenig belasten. Eine möglichst hohe Teilnahme ist anzustreben.

Im Rahmen der Vorsorgeuntersuchungen werden Neugeborene am dritten bzw. vierten Lebenstag im Rahmen der U2 auf angeborene Stoffwechseldefekte und endokrine Störungen gescreent. Seit 1.1.2009 wird auch das Neugeborenenhörscreening durchgeführt, um schwere Hörstörungen zu erkennen. Es findet spätestens bis zum dritten Lebenstag statt. Bei Frühgeborenen soll die Untersuchung spätestens zum Zeitpunkt des errechneten Geburtstermins, bei kranken oder mehrfach behinderten Kindern unter Beachtung der Zusatzstörungen und notwendigen klinischen Maßnahmen spätestens vor Ende des 3. Lebensmonats erfolgen. Das Screening auf eine angeborene Hüftdysplasie ist in der vierten und fünften Lebenswoche im Rahmen der U3 vorgesehen.

Untersuchungsprogramm

Seit Juli 2008 sind insgesamt 10 Vorsorgeuntersuchungen von der Geburt bis zum Alter von fünfeinhalb Jahren sowie eine weitere zwischen 12 und 14 Jahren in definierten Zeiträumen leistungsrechtlich verankert (➤ Tab. 7.4). Darüber hinaus werden drei weitere Untersuchungen (U10, U11, J2) von einzelnen Krankenkassen als Satzungsleistung angeboten Jede Untersuchung hat ihre besonderen Schwerpunkte, orientiert an den Zeitpunkten, zu denen bestimmte Krankheiten erkannt werden können bzw. an den Meilensteinen der Entwicklung. Die Untersuchungen auf Zahn-, Mund- und Kieferkrankheiten gehören bis zur Vollendung des 6. Lebensjahres zum Leistungsumfang und können durch Ärzte oder Zahnärzte erbracht werden. Eine orientierende Entwicklungsdiagnostik ist seit 1.1.2009 Bestandteil aller Vorsorgeuntersuchungen. Die Einführung einer U10 im zehnten Lebensjahr ist in Planung.

Bislang ist eine Gesundheitsberatung nicht vorgesehen, wird aber zunehmend von der Praxis eingefordert und ist Beratungsgegenstand im Rahmen der Überarbeitung der Kinderrichtlinien. Die Bundeszentrale für gesundheitliche Aufklärung hat ein Begleitheft zu den einzelnen Untersuchungen erstellt, das die Eltern auf die nächsten Entwicklungsschritte und Maßnahmen zur Förderung der Gesundheit und Entwicklung hinweist.

Die Kinderrichtlinien verpflichten die Ärztinnen und Ärzte seit 2009 zudem, bei Hinweis auf eine Gefährdung des Kindeswohls die entsprechenden Maßnahmen zu veranlassen. Ein in die einzelnen Untersuchungen integriertes Screening auf psychosoziale Belastung, das entsprechende Risikofaktoren im Lebensumfeld des Kindes oder auch vom Kind ausgehende Belastungsaspekte erfasst, wird zurzeit ebenfalls im Gemeinsamen Bundesausschuss beraten.

Die Teilnahme an den Untersuchungen ist freiwillig. Über die Jahre hat sich gezeigt, dass Eltern das Vorsorgeangebot besonders in den ersten beiden Jahren in hohem Maße in Anspruch nehmen, dass aber danach die **Teilnahmeraten** besonders in sozialen Brennpunkten und bei Kindern mit Migrationshintergrund deutlich zurückgehen.

Im Rahmen der – angesichts gravierender Fälle von Kindesvernachlässigung mit Todesfolge – seit 2006 in Deutschland geführten Diskussion um mehr Kinderschutz wurde den Vorsorgeuntersuchungen bis zum Schuleintritt ein großer Stellenwert beigemessen: Die regelmäßigen ärztlichen Untersuchungen sollen dazu beitragen, eine erhöhte Gefährdung des Kindeswohls zu erkennen und frühzeitige Hilfen einzuleiten. Daher hat mittlerweile die Mehrheit der Bundesländer gesetzliche Regelungen erlassen, mit denen die Teilnahme der Kinder an den Vorsorgeuntersuchungen verbessert werden soll. Durch individuelle Einladungsschreiben werden Eltern in 13 Bundesländern zur Teilnahme motiviert. Ärztinnen und Ärzte werden in einigen Ländern verpflichtet, die Teilnahme des Kindes an den einzelnen Untersuchungen an eine zentrale Stelle zu melden. Dort erfolgt ein Abgleich mit den kommunalen Meldedaten. Bei fehlender Arztmeldung werden Eltern durch Erinnerungsschreiben und nachgehende Intervention des Gesundheitsamtes oder des Jugendamtes an die Teilnahme erinnert. Restriktionen gibt es nicht. Die gesetzlichen Krankenkassen haben sich nach § 26 Abs. 3 SGB V an der Verbesserung der Teilnahme an den Untersuchungen zu beteiligen. Dazu schließen sie Rahmenvereinbarungen mit den entsprechenden öffentlichen Stellen ab. Die Maßnahmen zur Verbesserung der Teilnahme an den Vorsorgeuntersuchungen sind idealerweise eingebunden in ein Konzept „Früher Hilfen“ zur Beratung und Unterstützung der Familien. Die Länderergebnisse zeigen, dass die Teilnahme an den Untersuchungen deutlich gesteigert werden konnte (Hock et al., 2012). Ob und inwieweit die Kinderuntersuchungen dazu beitragen, Gefährdungen des Kindeswohls frühzeitig zu erkennen, bedarf einer sorgfältigen Analyse und kann noch nicht abschließend beantwortet werden. Dabei sollten aus sozialpädiatrischer Sicht nicht nur die gravierenden Fälle von Misshandlung oder Vernachlässigung Beachtung finden, sondern auch die bessere Forderung der Gesundheit und Entwicklung in die Bewertung eingehen.

7.6 Gesundheitsförderung für Kinder und Jugendliche in Settings

In Kindheit und Jugend erfolgt die Verhaltensprägung. In Anbetracht der weltweit zunehmenden Zahl zivilisationsbedingter Erkrankungen steht diese Altersgruppe international im Fokus präventiver Maßnahmen. In Deutschland werden im Setting Kindergarten, Schule und Kommune viele Ansätze verwirklicht, Kinder unabhängig von sozialen Determinanten zu erreichen. Auch die Familien als wichtigster Ort der Gesundheitsförderung werden in unterschiedlichen Modellen angesprochen. Eine systematische Institutionalisierung von Gesundheitsförderung und Prävention ist noch nicht erreicht. Der Nachweis nachhaltiger Wirksamkeit der Maßnahmen ist nur punktuell und projektbezogen erbracht. (➤ Abb. 7.1, ➤ Abb. 7.2)

7.6.1 Gesundheitsförderung in der Familie

Die Familie ist der primäre Ort der Sozialisation. In ihrem Kontext macht das Kind seine ersten prägenden Erfahrungen und verbleibt in der Regel über Jahre des Heranwachsens in ihrem Einflussbereich. Die Möglichkeit und Qualität eines die Gesundheit fördernden Aufwachsens hängt von einer Reihe von Faktoren ab, die in unterschiedlichem Maße belastend oder stützend sein können. Neben der Familienkonstellation, der Qualität der elterlichen Partnerschaft und weiteren Beziehungen innerhalb der Familie, dem Vorhandensein oder Fehlen stützender Familienmitglieder oder Verwandter beeinflussen insbesondere wirtschaftliche Rahmenbedingungen, Bildungsstand und Zugang zur Gesundheitsförderung die Chancen der Umsetzung im Lebensalltag. Eine schwer zu erreichende Gruppe stellen Familien mit Migrationshintergrund dar. Mittlerweile haben ein oder beide Elternteile in jeder vierten Familie mit minderjährigen Kindern ihre Wurzeln in einem anderen Kulturkreis (Bundesregierung 2008). Auch wenn viele Kinder bereits in Deutschland geboren sind und Familien teilweise schon in der dritten Generation hier leben, ist die Integration der Eltern, insbesondere der Mütter, häufig nicht erreicht. Da diese aber in der Regel die Kinder betreuen, tragen allein schon die bestehenden Sprachbarrieren dazu bei, dass Informationen zur Förderung der Gesundheit nicht an sie herangetragen werden können und tradiertes Verhalten zum Tragen kommt. Oftmals behindern auch patriarchale Strukturen die Kommunikation über gesundheitliche Themen. Leben diese Familien noch zusätzlich in relativer Isolation oder in restriktiven Communities, bedarf es großer Anstrengungen, westliche Konzepte der Gesundheitsförderung an sie heranzutragen.

Gesellschaftlicher Wandel und Zugang zu Inhalten der Gesundheitsförderung

In unserer Gesellschaft haben Selbstbestimmung und Selbstverwirklichung in den letzten Jahrzehnten einen hohen Stellenwert erlangt: Kinder sind nicht mehr primäres Lebensziel und in unserer Gesellschaft häufig nicht mehr präsent: Die Bruttoreproduktionsrate mit 1,36 in 2012 ist eine der weltweit niedrigsten. Viele junge Familien wollen in eigenen Haushalten leben, nicht mehr in

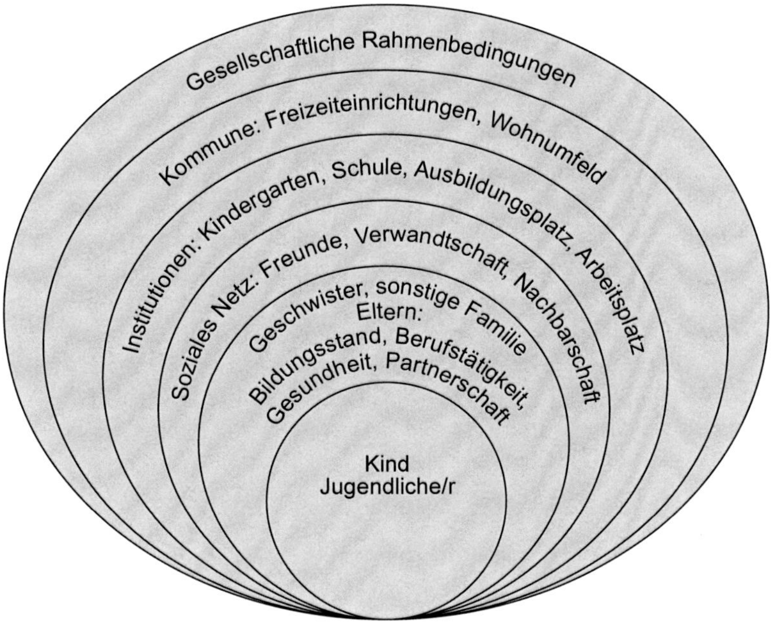

Abb. 7.1 Einflussbereiche auf die Entwicklung von Kindern und Jugendlichen (modifiziert nach Schlack 2000)

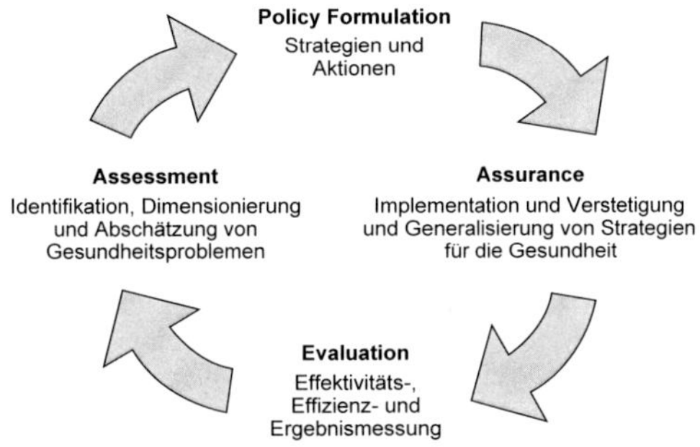

Abb. 7.2 Gesundheitsberichterstattung als unabdingbarer Bestandteil des Aktionskreises Public Health (mod. nach Rosenbrock 1995)

generationenübergreifenden Systemen; andere werden ungewollt durch die Arbeitsplatzentwicklung auseinandergerissen. Dies alles trägt dazu bei, dass grundlegendes Wissen um gesundheitliche Bedürfnisse von kleinen Kindern nicht mehr im Alltag an die Generation der jungen Eltern vermittelt und danach gelebt wird. Im Bildungssystem hat das systematische Erlernen und Erfahren wesentlicher lebenspraktischer Inhalte (gesunde Ernährung und Kochen, allgemeines Wissen um Gesundheit und Hygiene) keinen Platz (mehr). Über Fernsehen, Internet, soziale Netzwerke und Zeitschriften erworbenes Teilwissen und Verunsicherung durch unterschiedliche Expertenmeinungen kennzeichnen heute überwiegend den Wissensstand der jungen Erwachsenen über Gesundheit im Allgemeinen und über die gesundheitlichen Bedürfnisse von Kindern im Besonderen (14. Kinder- und Jugendbericht). Ein Lehrfach „lebenspraktisches Lernen" über alle Klassenstufen hinweg könnte u. a. dazu beitragen, mehr Wissen über Gesundheit und präventives Verhalten in den verschiedenen Lebensphasen in der Gesellschaft zu verankern.

Selbst in der Gesundheitsversorgung in Schwangerschaft und bei der Geburt gibt es keinen systematischen und institutionalisierten Ansatz, Eltern auf den Umgang mit dem Kind und seine Bedürfnisse vorzubereiten. Geburtsvorbereitungskurse sind freiwillig und mittelschichtorientiert, in ihren – nicht vorgegebenen – Inhalten sind Themen zum Kind und seinen Bedürfnissen eher die Ausnahme. Eine Vielzahl von Broschüren und Elternratgebern für die jungen Familien zeigt glücklich lachende, gepflegte und ausgeruhte Eltern, geht aber kaum auf die Alltagsprobleme ein. Auch in der pädiatrischen Betreuung fehlt bislang eine systematische Gesundheitsberatung als Bestandteil der Kinderuntersuchungen. Dies setzt sich in der gesamten Gesundheitsversorgung fort. Elternkurse im Säuglings- und Kleinkindalter durch freie Träger werden eher selten und mittelschichtorientiert wahrgenommen. Der Öffentliche Gesundheitsdienst, hier vor allem der Kinder- und Jugendgesundheitsdienst mit seinen unterschiedlichen Strukturen, bietet regional individuelle oder themenspezifische Elternberatungen an und kooperiert oftmals darüber hinaus mit Kindertageseinrichtungen und Familienzentren. Eine systematische Gesundheitsförderung in Kindertageseinrichtungen und Schulen, über die Gesundheitswissen in die Familien hereingetragen wird, fehlt bislang.
Aus sozialpädiatrischer Sicht sollte allen werdenden Eltern und jungen Familien systematisch Wissen um die gesundheitlichen Bedürfnisse von Säuglingen und Kleinkindern vermittelt werden. Eltern müssen Kompetenzen erlangen, ihre Kinder in den verschiedenen Situationen und Entwicklungsphasen zu begleiten und ggf. Hilfen anzufordern. Mit dem Auf- und Ausbau von Frühen Hilfen und Netzwerken für Eltern, die zunächst modellhaft und seit 2012 auf der Basis des Bundeskinderschutzgesetzes bundesweit etabliert werden, besteht die große Chance, durch systematische Hinweise auf Angebote von Beratung und Unterstützung und durch individuelle Hilfen insbesondere bei Familien mit unzureichenden Ressourcen wesentliche Verbesserungen erreichen zu können (s. Frühe Hilfen).

7.6.2 Ansprechpartner für Gesundheitsförderung in Familien

Gesundheitsversorgung

In der Schwangerschaft sind Frauenärzte und Hebammen wichtige Ansprechpartner der Frauen. Im Rahmen der medizinischen Betreuung können psychosoziale Belastungen systematisch erfragt (z. B. fehlende finanzielle Mittel für ein Kinderzimmer, Suchtproblematik, Partnerschaftskonflikte) und präventiv die Rahmenbedingungen für das Kind verbessert werden. Mit der Einführung verbindlicher Module zu Bedürfnissen des Kindes und zur Gesundheitsförderung in die Geburtsvorbereitungskurse könnte die Wissensvermittlung systematisiert werden. Anreize wie Bonussysteme könnten mehr Frauen bzw. werdende Eltern zur Teilnahme motivieren. Geburtshilfliche Kliniken und Kinderkliniken sollten systematisch entsprechende Inhalte an die Eltern Neugeborener vermitteln. Ein frühes Beispiel von Gesundheitsförderung in Familien aus einer Geburtsklinik heraus ist das Konzept der vorausschauenden Gesundheitsberatung von Bergmann et al. (2004), das Eltern von der Geburtsklinik bis ins zweite Lebensjahr Beratungen in definierten Abständen zu Themen wie Ernährung, Karies- und Allergieprävention anbietet.
Die individuelle Gesundheitsförderung ist in besonderem Maße Aufgabe des betreuenden Kinder- und Jugendarztes. Bereits beim ersten Kontakt mit der Familie muss die Förderung der Gesundheit und Entwicklung des Kindes angesprochen werden, denn gerade in der ersten Zeit sind Eltern besonders motiviert, das Beste für ihr Kind zu tun. Die Inhalte, die an die Eltern vermittelt werden sollen, müssen klar, einfach und nachvollziehbar sein.

Das von der BZgA entwickelte Begleitheft zu den Kindervorsorgeuntersuchungen bietet einen guten Ansatz: Zu jeder Untersuchung gehört eine spezifische Broschüre, die die wichtigsten altersbezogenen Präventionsmaßnahmen im Kontext der Entwicklung aufgreift. In der Beratung sollten diese Aspekte zumindest vom Arzt oder der Helferin angesprochen und auf den Text hingewiesen werden. Dass die Broschüre bislang nur in Deutsch vorliegt, ist bedauerlich.

Gesundheitsförderung darf Eltern nicht überfordern: Wichtig ist, Eltern auch zu entlasten und ihnen Hilfe anzubieten, wenn es Schwierigkeiten gibt, ihnen aber auch Mut zu machen, nicht aufzugeben. Auch hier sind soziale und ökonomische Faktoren oft limitierend; letztlich ist z. B. eine gesunde Ernährung mit bewusstem Einkauf und Selbstzubereitung der Speisen zeitaufwändiger und oft etwas teurer. Sich als Familie mit knappen Mitteln zu outen, erfordert Problembewusstsein, Selbstwirksamkeitsgefühl und ausreichendes Selbstbewusstsein.

„Frühe Hilfen" – aufsuchende Arbeit durch Familienhebammen, sozialmedizinische Assistentinnen und (Familien-) Kinderkrankenschwestern

Um ein gesundes Aufwachsen von Kindern auch in belasteten Familien zu ermöglichen, wurden – befördert durch die 2005 einsetzende politische Diskussion um mehr Kinderschutz – seit 2007 verstärkt regionale Netzwerke gebildet, in denen Gesundheits- und weitere Hilfesysteme zusammenarbeiten. Belastete Familien werden frühzeitig – meist im Rahmen der gesundheitlichen Betreuung – identifiziert und Beratung und Unterstützung angeboten. Als niedrigschwelliges Angebot suchen Familienhebammen, sozialmedizinische Assistentinnen oder Kinderkrankenschwestern die Familien nach Bedarf im häuslichen Umfeld auf und regen dort verhältnis- und verhaltenspräventive Maßnahmen an. In über 10 Modellprojekten wurden länderübergreifend in den Jahren 2008 bis 2010 unterschiedliche Strategien zu Angeboten im Bereich der Frühen Hilfen und deren Akzeptanz und Wirksamkeit erprobt und – finanziert durch Bundesmittel über das Nationale Zentrum Frühe Hilfen (NZFH) wissenschaftlich begleitet. Die Ergebnisse fanden teilweise ihren Niederschlag im **Bundeskinderschutzgesetz,** das zum 1.1.2012 in Kraft trat. Pflicht zur Beratung werdender Eltern über Angebote der Beratung und Hilfen vor Ort und Netzwerkbildung sind die wesentlichen Elemente, die die Jugendhilfe vor Ort umzusetzen hat. Die Umsetzung wird durch eine Cofinanzierung des Bundes flankiert. Damit besteht die große Chance, bundesweit die Frühen Hilfen auszubauen. belastete Familien frühzeitig zu erreichen und ihnen Beratung und passende Unterstützung zu vermitteln. Das Nationale Zentrum Frühe Hilfen"(NZFH), eine Kooperationseinrichtung von BZgA und Deutschem Jugendinstitut (DJI), ist die nationale Plattform für Informationssammlung, Beratung und Erfahrungsaustausch, Koordination und Evaluation im Bereich der Frühen Hilfen (www.NZFH.de).

Elternschulen, Familienbildung, Volkshochschulen

Die vielerorts von unterschiedlichen Akteuren angebotenen Elternkurse in den ersten Lebensmonaten sind inhaltlich umfassend, z. T. überfordern sie allerdings selbst engagierte Eltern. Belastete Familien und Familien mit Migrationshintergrund werden über konventionelle Bildungsträger kaum erreicht. Die wenigsten Elternkurse zielen darüber hinaus primär auf das Thema „Gesundheitsförderung" ab, sondern mehr auf Stärkung der Erziehungskompetenz, auf Sprachförderung, auf das Schaffen guter Bedingungen für Lernerfolge vor und während der Schulzeit. Hier gilt es, für das ambulante Versorgungssystem ebenso wie für den Öffentlichen Gesundheitsdienst, mit den Anbietern vor Ort das Einbringen gesundheitlicher Aspekte in diese Kurse abzustimmen. Gleichzeitig sollten diese Kurse zum besseren Erreichen belasteter Familien an zielgruppenspezifischen Treffpunkten wie Familienzentren, Communities und anderen Örtlichkeiten angeboten werden.

Kommunale Einrichtungen

Wesentliche Themen der Gesundheitsförderung für Kinder lassen sich besonders gut und nachhaltig im Gemeinwesen, im regionalen Bezug angehen. Familienzentren, Kindertageseinrichtungen, Sportvereine und andere kommunale Einrichtungen für Eltern und Kinder sind häufig der Ausgangspunkt elterlichen Engagements für die Verbesserung gesundheitlicher Rahmenbedingungen (z. B. Gestaltung von Kinderspielplätzen, Erhalt von Schwimmbädern) und motivieren auch andere Eltern. Im sozialen Miteinander machen die Kinder authentisch vermittelte Erfahrungen; die Identifikation mit dem Thema wird befördert. Die Einrichtungen greifen ihrerseits Gesundheitsförderung auch systematisch auf, indem sie Eltern z. B. für ausreichende Bewegungserfahrungen und ausgewogene Ernährung sensibilisieren.

Zugang zu Familien mit Migrationshintergrund

Ein besonderer Zugang zu Familien mit Migrationshintergrund ist über die Kontaktaufnahme mit den Repräsentanten oder Autoritätspersonen ihrer Gemeinschaft, so z. B. mit dem Imam in türkischen Communities möglich. Diese Personen können allgemeine Gesundheitsthemen und kommunale (Bildungs-)Angebote, wie z. B. Sprachkurse für Mütter, in die Familien ihres Kulturkreises hineintragen. Wichtige Kontaktpersonen der einzelnen ethnischen Gruppen sind in Ausländerbeiräten der Kommunen engagiert oder über diese zu ermitteln.

7.6.3 Gesundheitsförderung in Kindertageseinrichtungen

Das Kinder- und Jugendhilfegesetz (KJHG) verpflichtet die Länder als Träger der öffentlichen Jugendhilfe, Kindern ab dem vollendeten dritten Lebensjahr Förderung und Betreuung in einer Tageseinrichtung anzubieten. Mit dem Gesetz über Finanzhilfen des Bundes zum Ausbau der Tagesbetreuung für Kinder unterstützt der Bund die Länder beim Ausbau von Betreuungsplätzen für ein Drittel aller unter Dreijährigen bis 2013. Damit sollen Betreuung und Bildung der Kinder vom ersten Lebensjahr bis zur Einschulung zunehmend auch durch Gemeinschaftseinrichtungen geleistet werden. Je nach sozialem Milieu, Herkunft und Veranlagung der Kinder ergeben sich daraus Chancen und Risiken im Hinblick auf Gesundheit, Entwicklung und Förderung, die eine begleitende Forschung notwendig erscheinen lassen. Beim Ausbau dieser weitergehenden Betreuungsstrukturen sollte gesundheitlichen und entwicklungsbezogenen Bedürfnissen junger Kinder Rechnung getragen werden. Allerdings werden im Vorschulalter nach wie vor nicht alle Kinder erreicht; bis zu 10 % der Kinder eines Jahrgangs besuchen keine Kindertageseinrichtung, ein weiterer Teil nur ein oder zwei Jahre lang.

(Gesetzliche) Regelungen zur Gesundheit und Gesundheitsförderung

Die Gesetze der Länder über Kindertageseinrichtungen nach § 24 SGB VIII (KJHG) und die entsprechenden Ausführungsbestimmungen beinhalten in unterschiedlicher Tiefe Regelungen zu gesundheitsfördernden Maßnahmen. Die Träger der Einrichtungen (Kirchen, Betriebe, Kommunen, Privatpersonen) werden zur Schaffung gesundheitsfördernder Rahmenbedingungen verpflichtet. Dazu gehören gesunde Ernährung, Bewegungsförderung, Entwicklungsförderung und

Unfallschutz. Einige Länder haben mit den Trägern Bildungsprogramme vereinbart, in denen konkrete Handlungsfelder festgeschrieben sind: Zusammenarbeit mit dem Elternhaus und Angebote der Elternbildung, Maßnahmen zur Förderung der verschiedenen Entwicklungsbereiche, kontinuierliche Beobachtung der Entwicklung, Gesundheitsförderung, Zusammenarbeit mit Gesundheitsamt und Jugendhilfe, Fortbildungsinhalte für Erzieherinnen.
Eine **ärztliche Aufnahmeuntersuchung** im Hinblick auf gesundheitliche Eignung und Überprüfung des Impfstatus sowie Ergänzung fehlender Impfungen, in neueren gesetzlichen Regelungen der Länder auch des Nachweises der Teilnahme an den altersentsprechenden Vorsorgeuntersuchungen, dient der individuellen Gesundheitsvorsorge. Gesundheitliche Risiken sollen der Kindertageseinrichtung mitgeteilt werden (Vorliegen von Allergien, Impfhindernisse, chronische Krankheiten, laufende Therapien). In Elterninformationen im Rahmen der Aufnahme weisen die Einrichtungen auch auf gesundheitliche Vorgaben hin: Kinder sollten die Einrichtung bei schweren Infektionskrankheiten nicht besuchen. Nach dem Infektionsschutzgesetz sind Eltern darüber hinaus verpflichtet, der Einrichtung umgehend definierte Infektionskrankheiten mitzuteilen.

Wege zur Förderung der Gesundheit in Kindertageseinrichtungen

Gesundheitsförderung im Setting Kindergarten muss vorrangig im Kindergartenalltag stattfinden und in die Tagesstruktur eingebunden werden. Gesunde Ernährung umfasst z. B. das Kennenlernen gesunder Nahrungsmittel, die gemeinsame Zubereitung und Einnahme der Mahlzeiten, das Erfahren und Reflektieren von Ess- und Tischkultur – auch im kulturellen Kontext – sowie das Abwaschen. Weitere Aspekte sind ausreichende Bewegungs- und Alltagserfahrungen, Konditionieren von präventivem Verhalten wie z. B. Zähneputzen, Stärkung der Persönlichkeitsentwicklung im Hinblick auf geringe Suchtgefährdung, Erziehung zu positivem Sozialverhalten im Hinblick auf Ressourcen und Unterstützung durch die Gemeinschaft.
Im Kindergartenalter ist die **Mitwirkung der Eltern** in der Gesundheitsförderung unerlässlich. Deshalb ist es eine wesentliche pädagogische Aufgabe der Einrichtungen, Eltern für die Gesundheits- und Entwicklungsbedarfe ihrer Kinder zu sensibilisieren und sie in der Ausübung der elterlichen Verantwortung zu unterstützen. Individuelle Hilfestellung kann durch Hinweis auf externe Stellen (u. a. auf kinder- und jugendärztliche Kompetenz aus Praxis, Klinik, Sozialpädiatrische Zentren [SPZ] und Öffentlichem Kinder- und Jugendgesundheitsdienst [KJGD]) gegeben werden. Themenbezogene Elternabende, die Fragen und Diskussion ermöglichen, können Problembewusstsein und Handlungswünsche wecken. Sie können dazu beitragen, in der Elternschaft „Verbündete" für entsprechende Anliegen der Erzieher zu finden und Gesundheitsförderung in der Einrichtung konkret anstoßen. Moderation über externe Fachleute, die auch im Projektmanagement erfahren sind, erleichtert die Implementation von Gesundheitsförderprojekten, ihre Evaluation und (Nach-)Steuerung. Konzepte für Maßnahmen der Gesundheitsförderung halten unterschiedliche Institutionen bereit (u. a. BZgA, Krankenkassen). Wegen zusätzlicher Aufwendungen und damit verbundenen Kosten sind solche Projekte leider meist nur befristet umsetzbar.

Gesundheitsförderung und Prävention in Kinderkrippen

Besondere Anforderungen hinsichtlich Gesundheitsförderung und Prävention werden an die **Betreuung von Krippenkindern** gestellt: Dazu gehören spezifische Kenntnisse der Erzieherinnen über altersentsprechende Bedürfnisse der Krippenkinder, eine dem Alter der Kinder angemessene Ausstattung der Einrichtung und angemessener Betreuungsschlüssel sowie die enge Zusammenarbeit der jeweiligen Betreuungspersonen mit den Eltern. Auf das Papier über die Anforderungen an die Qualität der institutionellen Krippenbetreuung der DGSPJ wird verwiesen (DGSPJ 2008).

Im Auftrag des BMFSFJ wurde 2010 bis 2012 ein Curriculum „Gesundheitsförderung für Kinder unter drei Jahren in der Kindertagesbetreuung“ zur Etablierung einheitlicher Standards erarbeitet (www.kinder-gesund-betreut.de).
Die Beratung zur Schaffung gesundheitsfördernder Rahmenbedingungen – allgemeine Hygiene, Spiel-, Essens- und Ruhebereiche, Essenszubereitung – und die Überwachung der Einhaltung sind Aufgaben des Gesundheitsamtes.

7.6.4 Gesundheitsförderung im Setting Schule

Schule eignet sich für Gesundheitsförderung und Prävention in besonderer Weise, weil alle Kinder und Jugendlichen, unabhängig von ihrer Herkunft und sozialen Lage, über Jahre in dieser Struktur leben und daher die Gesamtheit dieser Altersstufe erreicht werden kann.
Mit der allgemeinen Schulpflicht, geregelt in den Verfassungen der Länder, greift der Staat wesentlich in das Elternrecht auf Erziehung ihres Kindes ein. Damit übernimmt er Verantwortung sowohl für die Bildungsinhalte und die (damit zusammenhängende) volkswirtschaftliche Entwicklung als auch für die Rahmenbedingungen des mindestens neunjährigen Schulalltags.

Rahmenbedingungen für Schulen

Schulen befinden sich in Deutschland überwiegend in öffentlicher Trägerschaft, in der Regel der Kommunen, Sekundarschulen oftmals auch der Kreise als Kommunalverbände.
Die Träger sind für Errichtung, Gestaltung und Instandhaltung von Schulgebäuden und Außenanlagen zuständig. Sie regeln Essensangebote (z. B. Auswahl eines Caterers für Pausenverpflegung und Mittagessen), Angebot und Ausstattung von Kantinen, Aufenthaltsräumen und Pausenhallen, Spiel- und Bewegungsangebote auf Schulhöfen und die Reinigung von Schulräumen und Toilettenanlagen.
Bildungsinhalte und Personalpolitik stehen dagegen unter staatlicher Regie. Je nach schulpolitischer Ausrichtung in den einzelnen Bundesländern sind die Schulen in ihrer Organisation und konzeptionellen Ausrichtung mehr oder weniger frei: Für die meisten öffentlichen Schulen gibt es erst wenig Gestaltungsraum, andere entscheiden bereits über Personaleinstellung, Tagesstruktur (Unterrichtsbeginn, Pausen) und die Ausrichtung auf bestimmte Themenschwerpunkte. Innerhalb der Schule bestimmen Schulleitung, Elternvertretung, Schülervertretung und die gemeinsame Schulkonferenz über wesentliche Themen.

Schulgesundheit als Wirtschaftsfaktor

Gesundheit als wesentliche Voraussetzung für Leistungsvermögen und Leistungsbereitschaft in der Schule zu fördern und zu erhalten, muss ein tragendes Element schulischer Bildung und der Gestaltung des Schulalltags sein.
Die Weltgesundheitsorganisation hat der Schulgesundheit als einer die gesamte Gesellschaft erreichenden und den gesundheitlichen und ökonomischen Outcome prägenden Faktor eine besondere Bedeutung zuerkannt. Mit der „Global School Health Initiative“ von 1992, basierend auf der Ottawa-Charta von 1986, wurden die Forschung, die Erarbeitung von Programmen zur Gesundheitsförderung, die Zusammenarbeit zwischen Gesundheits- und Bildungseinrichtungen und die Bildung von Netzwerken zum Erfahrungsaustausch zwischen den Schulen als Strategien benannt.

Die europäische Union hat diese Aspekte aufgegriffen und mit dem „Network of Health Promoting Schools" der WHO, der Europäischen Kommission und des Europarates fast alle europäischen Staaten erreicht.
Die Kultusministerkonferenz hat 1992 eine länderübergreifende Übersicht zur Situation der Gesundheitserziehung in der Bundesrepublik erstellt und die Förderung der Gesundheit in Schulen als Ziel festgelegt. Mit dem Beschluss vom November 2012 konkretisiert sie die Maßnahmen, die erforderlich sind, die Schule zur gesundheitsfördernden Schule zu entwickeln und über Partizipation eine positive Wirkung auf intermediäre Faktoren zu erreichen. Mittlerweile gibt es in allen Bundesländern Kooperationen zwischen Schulen, Kommunen, Krankenkassen und anderen Unterstützern. Viele Schulen haben eigene Leitbilder zur Schulgesundheit erstellt.

Schule: Arbeitsplatz und Ort für Gesundheitsbildung

Unter strukturellen Aspekten ist Schule der Arbeitsplatz für Lehrer und Schüler. Die Arbeitsschutzvorschriften gelten allerdings nur für Lehrer. Die Schulgesundheitspflege ist in den Ländern als Aufgabe des ÖGD ausgewiesen (➤ Kap. 9). Unter den Gesichtspunkten des Arbeitsplatzes Schule engagiert sich der ÖGD als betriebsmedizinischer Dienst im Sinne der arbeitsplatzbezogenen Hygiene und hat Zugang zu allen relevanten Bereichen. Zu seinen Aufgaben gehören die regelmäßige Begutachtung der altersgerechten Ausgestaltung von Anlagen und Räumlichkeiten, Schulmöbeln und Toilettenanlagen, der Licht- und Temperaturverhältnisse in Klassenräumen, Kantinenausstattung, der Essensangebote sowie der Pausenregelung. Notwendige Verbesserungen werden dem Schulträger und der Schulleitung mitgeteilt.
Beim Arbeitsplatz Schule sind Maßnahmen des Arbeits- und Gesundheitsschutzes zu beachten, die abträgliche Entwicklungen verhindern und zur Förderung der Gesundheit und Leistungsbereitschaft in der Schule beitragen können.
Gesundheitsfördernde Schulen haben darüber hinaus zum Ziel, ein Setting zu schaffen, das zur Stärkung der gesundheitsbezogenen Lebenstüchtigkeit beiträgt.
Die **Gestaltung des gesunden Arbeitsplatzes** im Sinne einer gesundheitsfördernden Schule erfordert es, gemeinsam mit Lehrern, Schulleitung, Eltern und Schulträgern Abläufe und Strukturen in den Schulen aktiv zu erfragen, zu analysieren und Verbesserungen abzustimmen. Diese können Pausenregelungen, Essens- und Sportangebote oder die Einrichtung von Arbeitsgemeinschaften betreffen, aber auch die Kommunikation untereinander oder die Einbindung von Schüler- und Elternvertretern in Entscheidungsprozesse und die Anwesenheit der Eltern bei Elternabenden. Die Gesamtheit der Eltern wird je nach Schulform und sozialem Einzugsgebiet kaum erreicht. Wichtig sind die frühzeitige Einbindung aller Lehrer sowie Angebote zur Fortbildung und Supervision. Ziel ist es, in einem kooperativen Vorgehen ein gutes Arbeitsklima und soziales Miteinander für Lehrer und Schüler zu schaffen, das auch durch die Mitwirkung Externer wie Sportvereinen oder Kulturschaffenden weiter verbessert werden kann.
In einer solchen Atmosphäre wird es auch leichter als in starren Schulstrukturen gelingen, wichtige **Gesundheitsthemen** (Sexualität, Kontrazeption, Schutz vor Infektionskrankheiten, Impfungen, Stressbewältigung, gesunde Ernährung) altersentsprechend an die Kinder und Jugendlichen heranzutragen, z. B. durch Schüler an Projekttagen und/oder externe Expertise. Das Prinzip der **Peereducation** sollte im Sekundarschulbereich aktiv genutzt werden: Heranwachsende und Jugendliche sind besonders aufgeschlossen gegenüber Gleichaltrigen; sie haben ähnliche Bedürfnisse, Erfahrungen und Kommunikationsmuster und genießen einen Vertrauensvorschuss, der aus dieser Nähe und Ähnlichkeit resultiert. Vor diesem Hintergrund gibt es viele Bemühungen, in der Peergroup anerkannte, besonders geschätzte Personen als Multiplikatoren für eine gute Sache zu

nutzen. In Grundschulen dagegen ist die enge Zusammenarbeit mit den Eltern und deren Engagement für die Umsetzung gesundheitsfördernder Maßnahmen unerlässlich.
Programme und Konzepte der BZgA, der Krankenkassen oder verschiedener Stiftungen unterstützen vielfältig zu spezifischen Gesundheitsthemen.
Der **Stärkung der Eigenverantwortlichkeit** der Schüler und damit auch ihrer Selbstbefähigung als über die Schule hinausreichendes Ziel von Gesundheitsförderung und Gesundheitsbildung ist im normalen Schulalltag schwieriger umzusetzen und in keiner Schulform und Altersstufe bislang ausreichend. Wichtige Impulse für das einzelne Kind bzw. den Jugendlichen kann z. B. soziales Engagement in einer Schul- oder Arbeitsgemeinschaft liefern, das die Schüler verantwortlich einbindet und auch das soziale Miteinander stärkt (Beispiel Unicef-Schulen, die sich mit selbst organisiertenKulturprojekten/Aktionen für Kinder in Entwicklungsländern engagieren) oder das o. g. Engagement von Jugendlichen in der Peereducation.
Der Öffentliche Gesundheitsdienst unterstützt Gesundheitsförderung und Prävention durch Beratung der Schulen und Eltern (Elternabende), durch Lehrerfortbildung und Beförderung der Themen z. B. in der Schulkonferenz sowie durch Mitwirkung an der Gestaltung von Lehrplänen und Ausbildungsinhalten der Lehrer. Viele Gesundheitsthemen, die in Tabelle 7.3 (➤ Tab. 7.3) für das Grundschul-, spätere Schulalter und jedes Alter genannt sind, sollten als Public-Health-Anliegen begriffen und im schulischen Kontext angesprochen und vertieft behandelt werden.

7.7 Gesundheitsplanung, Gesundheitsberichterstattung

7.7.1 Organisation der Gesundheitsplanung

Gesetzgebung des Bundes und der Länder

Die Rahmenbedingungen für unser Gesundheitssystem, d. h. für die individuelle Gesundheitsversorgung und die für die Gesundheit der Bevölkerung wichtigen Rahmenbedingungen werden nach Artikel 74 Nr. 19 des Grundgesetzes (GG) vom Bund gesetzlich geregelt. Dies betrifft die Sozialgesetzbücher – hier maßgeblich SGB V, VI, IX und XII, die Maßnahmen gegen gemeingefährliche oder übertragbare Krankheiten bei Menschen und Tieren, die Zulassung zu ärztlichen und anderen Heilberufen und zum Heilgewerbe sowie das Recht des Apothekenwesens, der Arzneimittel, der Medizinprodukte, der Heilmittel, der Betäubungsmittel und der Gifte regeln; in Artikel 19a GG ist auch die wirtschaftliche Sicherung der Krankenhäuser und die Regelung der Krankenhauspflegesätze dem Bund übertragen. Die Länder sind für die Umsetzung der Bundesgesetze im Bereich Gesundheit zuständig und erlassen eigene Gesetze, z. B., zu Infektionsschutz und Hygiene, zum Rettungswesen, zu weiteren Gesundheitsberufen, zur Weiterbildung der Gesundheitsfachberufe, zu den Heilberufekammern und für den Krankenhausbereich.

Gesetzliche Krankenkassen – sonstige Formen der Krankenversicherung

Vorrangig wird die **individuelle Gesundheitsversorgung** durch das Gesetz über die gesetzliche Krankenversicherung (SGB V) geregelt: Weitere Regelungen finden sich im SGB VI (gesetzliche Rentenversicherung), SGB XI (soziale Pflegeversicherung), SGB VII (gesetzliche Unfallversiche-

rung) und im SGB IX, dem Gesetz zur Rehabilitation und Teilhabe behinderter Menschen (➤ Kap. 42).
Zum gesetzlich versicherten Personenkreis nach SGB V gehören ca. 90 % der Bevölkerung in Deutschland. Die gesetzliche Krankenversicherung ist eine solidarisch finanzierte Pflichtversicherung in Abhängigkeit vom Einkommen und allen Menschen in Deutschland zugänglich. Im SGB V werden die Grundzüge der Behandlung, die Zusammenarbeit der Sektoren des Gesundheitswesens, die Vergütung der Leistungserbringer und die Organisation der Krankenkassen geregelt. Die verschiedenen gesetzlichen Krankenkassen sind einheitlich im Spitzenverband der Krankenkassen organisiert, der auf Bundesebene u. a. die Vereinbarungen mit den Leistungserbringern abschließt. Über den vorgegebenen Leistungskatalog hinaus können die einzelnen Krankenkassen freiwillige Leistungen in ihren Satzungen festlegen. Unterschiedliche Zugangsregeln gelten für die private Krankenversicherung (PKV), Heilfürsorge, Beihilfe etc. Ihr Leistungsspektrum orientiert sich an dem der GKV, liegt aber oftmals darüber bzw. kann auch Leistungseinschränkungen beinhalten.

Stationäre Versorgung – Land, Krankenkassen, Träger

Die Länder führen in ihrem Bereich in mehrjährigen Intervallen die Krankenhausplanung mit den öffentlich-rechtlichen, frei-gemeinnützigen und privaten Trägern und den gesetzlichen Krankenkassen durch. Dabei vereinbaren sie Versorgungsbereiche mit den notwendigen Versorgungsstufen, die Entwicklung der medizinischen Disziplinen und Bettenkapazitäten jedes Krankenhauses über den Planungszeitraum sowie die Zahl der Ausbildungsplätze für die Gesundheitsfachberufe. Sie beteiligen sich an den Krankenhausinvestitionen. Die Krankenkassen schließen üblicherweise mit den im Krankenhausplan aufgenommenen Krankenhäusern Versorgungsverträge zur Finanzierung von Behandlungsleistungen ab.

Ambulante Versorgung – Kassenärztliche Vereinigung, Krankenkassen

Die Planung und Sicherstellung der Versorgung im ambulanten Bereich ist Aufgabe der 17 Kassenärztlichen Vereinigungen (KVen). Sie sind Körperschaften des öffentlichen Rechts und unterstehen der Rechtsaufsicht des jeweiligen Bundeslandes. Sie legen fest, wie viele Kinder- und Jugendärzte, Allgemeinmediziner sonstige Fachärzte einer Disziplin sowie Psychotherapeuten und Kinder- und Jugendlichenpsychotherapeuten in welchen Bereichen eine Kassenzulassung erhalten. Analog regeln dies die Kassenzahnärztlichen Vereinigungen für ihren Bereich. Sozialpädiatrische Zentren, die fachlich-medizinisch unter ständiger ärztlicher Leitung stehen und die Gewähr für eine leistungsfähige und wirtschaftliche sozialpädiatrische Behandlung bieten, können vom Zulassungsausschuss zur ambulanten sozialpädiatrischen Behandlung von Kindern ermächtigt werden. Die KVen verwalten mit den Krankenkassen die für den ambulanten Bereich zur Verfügung stehenden Finanzen, indem sie den einzelnen Facharztgruppen bestimmte Budgets zuweisen. Seit einigen Jahren können die Krankenkassen im Rahmen der hausarztzentrierten Versorgung („Hausarztmodell") nach § 73b SGB V Direktverträge mit verschiedenen Gemeinschaften (u. a. Berufsverbänden bei definierter Repräsentanz im Versorgungsbereich oder medizinischen Versorgungszentren) abschließen.

Leistungskatalog – Kassenärztliche Bundesvereinigung, Gemeinsamer Bundesausschuss

Die Planung der Mittel und Methoden, die in der medizinischen Diagnostik, Therapie und Rehabilitation eingesetzt werden, ist Aufgabe der Kassenärztlichen Bundesvereinigung (KBV). Im

Gemeinsamen Bundesausschuss (GBA) beraten Ärzte und Spitzenverband der Krankenkassen, welche Verfahren in den Leistungskatalog der gesetzlichen Krankenkassen aufgenommen werden. Die übrigen Krankenversicherer (Beihilfe, Heilfürsorge, private Krankenversicherer) lassen die Regelungen des GBA üblicherweise gegen sich gelten. Die vertiefte Prüfung neuer Verfahren vor Zustimmung des GBA führt das Institut für Qualität und Wirtschaftlichkeit im Gesundheitswesen (IQWiG) durch.
Die approbierten ärztlichen, zahnärztlichen und psychotherapeutischen Berufe steuern den Einsatz der Mittel und Verfahren, sind allerdings durch Höchstmengenverordnungen, die vom GBA festgeschrieben werden, begrenzt. Weitere Beschränkungen ergeben sich im ambulanten Bereich durch von KVen festgelegte Budgets sowie im stationären Bereich durch die in den Versorgungsverträgen vereinbarten Regelungen.

Versorgungsplanung auf Landesebene – Integrierte Versorgung

§ 90a SGBV enthält die Möglichkeit, ein gemeinsames Gremium aus Vertretern des Landes, der Kassenärztlichen Vereinigung, der Landesverbände der Krankenkassen sowie der Ersatzkassen und der Landeskrankenhausgesellschaft sowie weiteren Beteiligten zu bilden, um Empfehlungen zu sektorenübergreifenden Versorgungsfragen abzugeben. Die Abstimmung ist sinnvoll und notwendig, um angesichts der Herausforderungen durch die zunehmend ältere Bevölkerung, Fachkräftemangel und begrenzte ökonomische Ressourcen die flächendeckende medizinische Versorgung der Bevölkerung weiterhin sicherzustellen.

Heilberufekammern für approbierte Berufe

Die approbierten Berufe sind Pflichtmitglieder der jeweiligen Ärzte-, Zahnärzte- bzw. Psychotherapeutenkammern. Die Kammern sind Körperschaften des öffentlichen Rechts und unterstehen der Rechtsaufsicht des jeweiligen Bundeslandes. Ihre Tätigkeit wird in den Heilberufekammergesetzen der Länder geregelt: Sie sind u. a. für die Weiter- und Fortbildung, das Berufsrecht und die Versorgung ihrer Mitglieder zuständig. Die Bundesärztekammer ist ein freiwilliger Zusammenschluss der Landesärztekammern. Sie vertritt nach außen die berufsständischen Interessen gegenüber Politik und Gesellschaft und nimmt in ärztlichen Richtlinien u. a. zu grundsätzlichen medizinischen und ärztlichen Fragen Stellung. Gleiches gilt für die Bundesvereinigungen der übrigen approbierten Berufe.

Gesundheitsfachberufe

Für die durch Bundesgesetze geregelten Gesundheitsfachberufe gibt es keine analogen Strukturen; sie sind in Fachgesellschaften, Berufsverbänden und Dachverbänden organisiert. Sie unterliegen keiner eigenständigen Gesundheitsplanung. Für pflegende Berufe wird Schleswig-Holstein gem. Landtagsbeschluss vom 14.12.2012 als erstes Bundesland eine Landespflegekammer einrichten. Weitere werden folgen.

Patientenorganisationen/Gesundheitliche Selbsthilfe

Die Beteiligung der Patientenorganisationen und Vertretern der gesundheitlichen Selbsthilfe an der Gesundheitsplanung ist in § 140f SGB V festgeschrieben.

7.7.2 Gesundheitsberichterstattung

Grundlegende Voraussetzung für eine zielgerichtete und ressourcenschonende Gesundheitsplanung, für gesetzliche Vorgaben, Vorsorge- und Kontrollsysteme, um eine Verbesserung der Gesundheit und der Lebensqualität zu erreichen, ist das Wissen um den Gesundheitszustand, die Gesundheitsprobleme und die Gesundheitsversorgung in einer Bevölkerung.
Gesundheitsberichterstattung ist die Sammlung, Beschreibung und Bewertung relevanter Daten über die Gesundheit einer Bevölkerung.
Gesundheitsberichterstattung beinhaltet periodische Berichte über den Gesundheitszustand der Bevölkerung, über bedeutsame Gesundheitsrisiken und erreichte oder angestrebte Strukturmerkmale. Weitere Inhalte sind Präventions- oder Versorgungsziele, Ressourceneinsatz sowie Leistungen und Ergebnisse des Gesundheitswesens.
Im Gegensatz zu epidemiologischen Studien ist damit Gesundheitsberichterstattung als epidemiologische Surveillance verantwortlich für Trendmonitoring, die Aufdeckung von Problemen im Gesundheitswesen und die Erstellung von Hypothesen, die dann in epidemiologischen Studien überprüft werden.
Gesundheitsberichterstattung sammelt epidemiologische Daten, analysiert und verbreitet diese und bewertet sie im Zusammenhang mit vorher definierten Gesundheitsindikatoren. Als effektives Surveillancesystem integriert sie Daten aus Epidemiologie, Klinik, Verhaltensforschung, Demografie, Soziologie und anderen Fachrichtungen, um adäquate Information für gesundheitspolitische Entscheidungsträger bereitzustellen (Fiedler 1993).
In Deutschland wurde – u. a. auf der Basis von Empfehlungen der WHO – erst in den 1980er Jahren mit dem systematischen Aufbau der heutigen Gesundheitsberichterstattung auf Beschluss der Gesundheitsminister von Bund und Ländern begonnen. Auslösend war das Fehlen aussagekräftiger Daten zur Gesundheit und zu gesundheitlichen Problemen der Bevölkerung vor dem Hintergrund zunehmender Erkrankungszahlen (insbesondere zivilisationsbedingter chronischer Krankheiten) und steigender Gesundheitskosten. Die bislang verfügbaren amtlichen Statistiken, wie Statistik zur Säuglingssterblichkeit, Seuchenstatistik, Todesursachenstatistik sowie einzelne Landeskrebsregister, waren nicht geeignet, die Zusammenhänge zwischen Lebensweisen und Krankheiten zu beschreiben. Auch als Basis für die Weiterentwicklung der Gesundheitsversorgung und zur Etablierung sinnvoller Präventions- und Evaluationsprogramme waren sie völlig unzulänglich.
Um aussagekräftige Daten aus unterschiedlichen Bereichen zu erhalten, wurden alle Datenhalter im Gesundheitsbereich zur Lieferung der erforderlichen Daten gesetzlich verpflichtet (z. B. Infektionsschutzgesetz, Krankenhausrecht, Sozialgesetzbücher, Statistikrecht, Krebsregistergesetze). Dabei wurden in verschiedenen Bereichen neue Datenerhebungen verpflichtend (ICD-10, Krankenhausdiagnosestatistik, DRG-Dokumentation) und die Gesundheitsberichterstattung eine Aufgabe des Öffentlichen Gesundheitsdienstes auf den Ebenen von Bund, Ländern und Kommunen.

Methoden der Datensammlung

Für die Gesundheitsberichterstattung werden verschiedene Instrumente eingesetzt:

Indikatorenkatalog der Länder und Gesundheitsberichterstattung des Bundes

Die Basis stellt eine kontinuierliche Datensammlung in den Ländern dar, die im Indikatorenkatalog der Länder beschrieben ist. Nach Themenfeldern werden die für die Gesundheitsplanung wichtigsten Daten aus unterschiedlichen Datenquellen zusammengeführt und bedarfsgerecht angepasst.

Themenfelder des Indikatorenkatalogs:
Gesundheitspolitische Rahmenbedingungen: Gesetze, demografische Daten wie Bevölkerung, Schwangerschaften und Geburten, Lebensgemeinschaften, soziale Lage, Aus- und Weiterbildung des Gesundheitspersonals, Arbeitsmarkt, soziale Sicherung (gesetzliche Kranken-, Pflege-, Unfall-, und Rentenversicherung, PKV, Sozialhilfe)
Gesundheitliche Lage: Sterblichkeit, Gesundheitszustand, Beschwerden, Behinderungen, Krankheitsfolgen, Behandlungsfolgen
Gesundheitsverhalten und Gefährdungen: Lifestyle, Impfungen, Gewalt, Umwelt, Arbeitswelt, Unfälle, sonstiges
Krankheiten und Gesundheitsprobleme: u. a. Herz- und Kreislauferkrankungen sowie Erkrankungen weiterer Organsysteme, bösartige Neubildungen, Infektionen, Berufskrankheiten, Verletzungen mit Auflistung der jeweils häufigsten Krankheitsbilder pro Gruppe
Gesundheitsversorgung: Gesundheitsschutz mit ÖGD und Arbeitsschutz, Vorsorge und Früherkennung (Teilnahmeraten), Beschäftigte und Einrichtungen der Gesundheitsversorgung, Arzneimittelversorgung, Heil- und Hilfsmittel, medizinische Verfahren, medizinische Untersuchungen und Behandlungen
Ausgaben, Kosten und Finanzierung: u. a. Ausgaben für verschiedene Sektoren, Kosten der verschiedenen Sektoren, Einkommen der verschiedenen Gesundheitsberufe, Preise für Behandlungen, Zuständigkeiten und Anteile der Finanzierung.
Diese Daten werden in den Ländern z. T. in Gesamtberichten oder Teilberichten veröffentlicht, z. T. über das Internet zur Verfügung gestellt.
Ausgewählte Daten werden auf der Bundesebene zusammengetragen und vom Robert-Koch-Institut in Zusammenarbeit mit dem Statistischen Bundesamt für die Gesundheitsberichterstattung des Bundes aufbereitet.
Viele Informationen sind im Internet als Online-Datenbank in tabellarischer und grafischer Form verfügbar (www.gbe-bund.de), dort werden auch Daten der OECD und der WHO dargestellt.

Surveys

Mittels einer repräsentativen Untersuchung werden Informationen zum Gesundheitszustand der Bevölkerung/einer Bevölkerungsgruppe gesammelt. Die Datenerhebung kann durch Befragung (schriftlich, telefonisch oder direkt) erfolgen und Untersuchungen einschließen. Surveys eignen sich für ein Monitoring von Gesundheitsproblemen, wenn die zu erhebende Problematik in einer Mindestzahl vorkommt. Sie sind auch für zeitliche Trends gut geeignet. Das Robert-Koch-Institut hat 1998 einen ersten Bundesgesundheitssurvey (BGS) für den Zeitraum von Oktober 1997 bis März 1999 zum Gesundheitszustand der deutschen Bevölkerung im Auftrag des Bundesgesundheitsministeriums durchgeführt. 7.124 Personen im Alter von 18 bis 79 Jahren wurden befragt, Größe, Gewicht und Blutdruck wurden ermittelt sowie Blut und Urin untersucht. Der BGS 98 wurde durch vier telefonische Surveys in den Jahren 2003 bis 2006 ergänzt und mit der Studie zur Gesundheit Erwachsener DEGS 1 (2008–2011) fortgesetzt,
Von 2003 bis 2006 wurde der **Kinder- und Jugendgesundheitssurvey (KiGGS)** durchgeführt. Fast 18.000 Probanden im Alter von 4 bis 17 Jahren wurden befragt und untersucht. Auch in den Ländern und Kommunen werden Surveys zur Identifizierung von Gesundheitsproblemen eingesetzt (z. B. Nordrhein-Westfalen, Trier). Seit 2009 wird über einen Telefonsurvey in regelmäßigen Abständen ein Monitoring zur Entwicklung der Gesundheitsparameter durchgeführt. Das RKI veröffentlicht diese Daten zur „Gesundheit in Deutschland aktuell" (GEDA) mit unterschiedlichen Schwerpunkten (zuletzt GEDA März 2012- bis März 2013 mit 26.000 Interviewten).

Weitere Datenerhebungen zur Gesundheit von Kindern und Jugendlichen

Die Kinder- und Jugendgesundheitsdienste der Gesundheitsämter führen auf der Basis der Ländergesetze sozialpädiatrische Untersuchungen an bestimmten Altersklassen durch. Dabei werden epidemiologisch relevante Daten ganzer Alterskohorten erhoben und z. T. auf der Ebene des Landes zusammengeführt. In allen Bundesländern werden Daten von Schulanfängern standardisiert erfasst, entweder im Halbjahr vor Schulbeginn oder im Rahmen einer vorgezogenen Einschulungsuntersuchung mit 4½ bis 5 Jahren. Auch wenn diese Untersuchungen in Ausnahmefällen von niedergelassenen Ärztinnen und Ärzten (Bayern) durchgeführt werden, erfolgen eine entsprechende Dokumentation und Auswertung der Daten. Darüber hinaus stehen durch die Tätigkeit der Kinder- und Jugendgesundheitsdienste in unterschiedlichem Umfang Daten aus dem Kindergartenbereich, dem Grundschul- und Sekundarschulbereich, aus Sonderschulen sowie aus Entlassuntersuchungen zur Verfügung. Aufgrund verschiedener Definitionen und Schwerpunkte der Datenerhebung sind allerdings nur wenige Daten bundesweit vergleichbar. Letzteres gilt für die Daten zu Impfungen, die nach §34 Abs. 10 IfSG beim RKI jährlich zusammengeführt werden, für die Teilnahme an den Vorsorgeuntersuchungen sowie partiell für Gewicht und Größe (Moss et al 2012). Derzeit gibt es Bemühungen der Länder, weitere Datensätze bundeseinheitlich zu erfassen und sie auch mit den Daten des Bundesgesundheitssurveys langfristig vergleichbar zu machen. Dies ist regional und lokal wichtig, um die Dimension eines Gesundheitsproblems im Bundesvergleich einzuordnen und Entscheidungsträger fundierter auf negative Abweichungen hinweisen zu können. So haben die Länder 2013 einen Vorschlag zur Erfassung des Migrationshintergrundes bei der Einschulungsuntersuchung erarbeitet, der in der Integrationsberichterstattung zur Darstellung der Gesundheit und gesundheitlichen Versorgung von Kindern mit Migrationshintergrund benötigt wird.

Gesundheitsberichterstattung – Grundlage für eine bessere Gesundheitsversorgung

Die bewertende Darstellung der definierten Gesundheitsdaten mit Ursachen, Risikogruppen (Alter, soziale Schicht, regional zeitlichen Trends, Migrationshintergrund), Präventions- und Behandlungsmöglichkeiten, Kosten und Finanzierung sowie Darstellung der Defizite in Prävention und Versorgung sowie Handlungsoptionen zur Verbesserung ist Inhalt der Gesundheitsberichterstattung.

Der Bund hat einen ersten umfassenden Gesundheitsbericht 1998 veröffentlicht, in dem die Krankheiten mit der größten Krankheitslast beschrieben sind. Dieser Bericht enthält aber auch eine Gesamtschau der wichtigsten Länderindikatoren aus den übrigen Themenfeldern. (Gesundheitsbericht des Bundes 1998). Eine Aktualisierung erfolgte 2006. In spezifischen Themenheften erfolgt eine umfassende Aufbereitung. Als „GBE kompakt" werden aktuelle Fragestellungen und Fakten in vierteljährlich erscheinenden Publikationen allgemeinverständlich dargestellt. Darüber hinaus gibt es Beitrage zur Gesundheitsberichterstattung des Bundes. Das Informationssystem der Gesundheitsberichterstattung (IS-GBE) liefert als Online-Datenbank gesundheitsrelevante Informationen zu allen Themenfeldern der Gesundheitsberichterstattung

Datennutzung: Die im Gesundheitsbericht des Bundes zusammengetragenen Fakten haben vieles bewegt: Sie haben u. a. dazu beigetragen, vorrangige Gesundheitsziele mit allen in der Gesundheitsversorgung Verantwortlichen zu definieren, die bundesweit umgesetzt werden sollen. Das Wissen um vorrangige Gesundheitsprobleme und Strategien zur Verbesserung fand seinen Niederschlag in den verschiedenen Bestrebungen für ein Präventionsgesetz. Die Daten zu Krankheiten und Krankheitsfolgen und ihre gesundheitsökonomische Dimension haben in der Gesundheitsversorgung die Entwicklung von Disease-Management-Programmen (DMP) wesentlich befördert. In diesem Kontext haben Daten zur Entwicklung der Versorgungsstruktur dazu beigetragen, die Versorgungsplanung nach § 90 a SGB V als gemeinsame Aufgabe von Kassenärztlicher Vereinigung, Krankenhausgesellschaft, gesetzlichen Krankenkassen, Land und Selbsthilfe anzugehen.

Am Beispiel Impfen wird die Bedeutung von GBE für Gesundheitsplanung und Gesundheitsversorgung deutlich. Die Untersuchungen der Kinder- und Jugendgesundheitsdienste in den Schulen wie auch des Kinder- und Jugendgesundheitssurveys zeigten und zeigen regional unterschiedlich große Impflücken bei Kindern und Jugendlichen bei den Auffrischimpfungen gegen Diphtherie, Tetanus und Pertussis. Besonders große Impflücken bestehen z. T. weiter bei der (Wieder)-impfung gegen Masern. Masernausbrüche in schlecht durchimpften Regionen waren die Folge. Die Gesundheitsberichterstattung auf Länder- und Bundesebene hat mit Aufzeigen der Impflücken, der Auflistung vermeidbarer Krankheitskosten und Aufzeigen von Interventionsmöglichkeiten dazu beigetragen, dass der Gesetzgeber mit § 20d SGB V seit 2004 eine Kooperationsmöglichkeit zwischen GKV und ÖGD in den Leistungskatalog der Krankenkassen aufgenommen hat: Zur Förderung der Schutzimpfungen tragen die gesetzlichen Krankenkassen die Kosten für Impfstoffe, der ÖGD führt die Impfungen durch. So kann er im Rahmen seiner gemeinwesenorientierten Arbeitsweise z. B. in Schulen Impfungen anbieten und erinnert mit diesem Vorgehen gleichzeitig alle Eltern daran, den Impfschutz ihrer Kinder beim Kinder- und Jugendarzt überprüfen zu lassen. Die bereits seit Jahren in einigen Ländern laufenden Kooperationen zwischen GKV und ÖGD belegen die Effektivität der Strategie zur Schließung von Impflücken.

Sowohl kommunal als auf Länderebene werden Berichte oder Teilberichte über den Gesundheitszustand veröffentlicht. Auch diese Berichte sollen Entscheidungsträger in Kommunen und auf Landesebene auf Defizite aufmerksam machen und Handlungsbedarfe aufzeigen.

Gesundheitsberichterstattung für Kinder und Jugendliche

Zu Kindern und Jugendlichen wurde 2004 vom RKI ein Schwerpunktbericht: „Gesundheit von Kindern und Jugendlichen" herausgegeben (RKI 2004). Auch die im Kinder- und Jugendgesundheitssurvey von 2003 bis 2006 erhobenen Daten werden in dieser Form seit 2007 aufbereitet und in Schwerpunktberichten veröffentlicht.

Die von den Kinder- und Jugendgesundheitsdiensten der Gesundheitsämter erhobenen Daten werden ebenfalls seit Jahren in den Kommunen und teilweise auch auf Länderebene in Berichten oder Teilberichten über die Gesundheit der untersuchten Kohorte bekannt gemacht. Neben den politischen Entscheidungsträgern im Gesundheits- und Bildungsbereich und den Kommunen als Schulträger ist die Zielgruppe des Berichts in der Regel weiter gefasst: Eltern und Elternvertretungen, Ärztinnen und Ärzte in der Versorgung vor Ort und in der Versorgungsplanung, Krankenkassen, Kindertageseinrichtungen als Einrichtungen für Bildung, Betreuung und Erziehung – auch in Gesundheitsfragen – sowie interdisziplinäre Frühfördereinrichtungen und SPZ in der Versorgung behinderter Kinder. Der Gesundheitszustand einer Alterskohorte sollte in jeder einzelnen Kommune bekannt und mit anderen vergleichbar sein. Dies kann dazu beitragen, Gesundheitsförderung im kommunalen Setting, in Zusammenarbeit von Schule, Sportvereinen, weiteren Einrichtungen der Jugendförderung und Gesundheitsversorgung zusammenzubringen.

Das Ziel von Gesundheitsberichterstattung ist nur dann erreicht, wenn es gelingt, aus den gewonnen Daten Erkenntnisse zum Gesundheitszustand im regionalen Bezug und zeitlichen Trend zu gewinnen, vorhandene Defizite aufzuzeigen, Entscheidungsträger zur Behebung dieser Defizite zu veranlassen und die Effizienz der eingeleiteten Maßnahmen wiederum zu bewerten.

Literatur

Antônôvsqî A, Franke A (1997) Salutogenese: zur Entmystifizierung der Gesundheit. Tübingen: Dgvt-Verlag.

BGH-Urteil vom 15. Februar 2000 – Az. VI ZR 48/99; NJW 2000: 1.784–1.788; www.bundesgerichtshof.de/Entscheidungen

Bergmann L, Bergmann K, Richter R, Hinkel J, Dudenhausen J (2005) Über die Beeinflussbarkeit der Entwicklung von Säuglingen und Kleinkindern: Ergebnisse einer Studie zur vorausschauenden Beratung junger Eltern mit ihrem ersten Kind. Kinderärztliche Praxis Sonderheft: 18–22.

Borde T, David M (Hrsg) (2005) Kinder und Jugendliche mit Migrationshintergrund. Frankfurt: Mabuse-Verlag.

Bronfenbrenner U, Ceci SJ (1994) Nature-Nurture Reconceptualized in Developmental Perspective: A Bioecological Model. Psychological Review 101(4) 568–586.

Bronfenbrenner U (1976) Ökologische Sozialisationsforschung. Stuttgart: Klett-Cotta

Bröskamp-Stone U, Kickbusch I, Walter U (1998) Gesundheitsförderung und Prävention. In: Schwarz FW, Badura B, Leidl R. Das Public-Health-Buch. München: Urban und Fischer: 141–143.

Bengel J, Strittmacher R, Willmann H (2001) Was erhält Menschen gesund? Antonowskys Modell der Salutogenese. In: Bundeszentrale für gesundheitliche Aufklärung (BzGA) (Hrsg) Forschung und Praxis der Gesundheitsförderung, Band 6. Köln: BzGA.

Bundesministerium für Familie, Senioren, Frauen und Jugend: 14. Kinder – und Jugendbericht 2012;

Caplan G (1964) Principles of preventive Psychiatry. New York, Basic Books.

Engelbert A (1999) Familien im Hilfenetz: Bedingungen und Folgen der Nutzung von Hilfen für behinderte Kinder. Weinheim: Juventa: 12–15.

Fiedler AH (1993) Aufbau, Organisation und Evaluation eines Gesundheitsberichterstattungssystems. Ges. Wesen 55: 358.

GKV-Spitzenverband: Leitfaden Prävention. GKV-Spitzenverband 2010

Grimmer Y, Vitt J, Jennen-Steinmetz C, Becker K,·Schmidt MH, Laucht M (2008) Hohe Persistenz von Übergewicht bei Kindern der Mannheimer Risikokinderstudie, Monatsschr Kinderheilkd 156: 365–370.

Hock S et al. Zum aktuellen Stand der Kindervorsorgeprojekte, Gesundheitswesen, © Georg Thieme Verlag KG Stuttgart · New York: Online-Publikation: 2013: DOI HYPERLINK „http://dx.doi.org/" http://dx.doi.org/10.1055/s-0032-1331731

Ihle W, Esser G, Laucht M, Schmidt MH (2004) Depressive Störungen und aggressiv-dissoziale Störungen im Kindes- und Jugendalter: Prävalenz, Verlauf und Risikofaktoren. Bundesgesundheitsbl – Gesundheitsforsch – Gesundheitsschutz 47: 728–735.

Lange M, Kamtsiuris P, Lange C, Schaffrath Rosario A, Stolzenberg H, Lampert T (2007) Messung soziodemographischer Merkmale im Kinder- und Jugendgesundheitssurvey (KiGGS) und ihre Bedeutung am Beispiel der Einschätzung des allgemeinen Gesundheitszustands. Bundesgesundheitsbl – Gesundheitsforsch – Gesundheitsschutz 50: 578–589.

Laucht M, Esser G, Schmidt MH (1998) Risiko- und Schutzfaktoren der kindlichen Entwicklung – Empirische Befunde. Zeitschrift für Kinder- und Jugendpsychiatrie und Psychotherapie 26: 6–20.

Meyer-Probst B, Teichmann H, Heider B(1984) Risiken für die Persönlichkeitsentwicklung im Kindesalter: Rostocker Längsschnittstudie, Stuttgart: Thieme.

Moss A, Klenk J, Simon K, Thaiss H, Reinehr T, Wabitsch M (2012):Declining prevalence rates for overweight and obesity in German children starting school Eur J Pediatr 171: 289–299

Oberwöhrmann S, Bettge, Herrmann S, Meinlschmidt G: Migrationshintergrund als Einflussfaktor auf die kindliche Entwicklung im Einschulungsalter – ein multivariates Modell. Gesundheitswesen 2013; 75(04): 203–209

Paulus, P. (2011). Schule, Gesundheit, Bildung. Neue Perspektiven für die schulische Gesundheitsförderung. Prävention, Zeitschrift für Gesundheitsförderung, 4, 99–104

Paulus P, Brückner G (Hrsg) (2000) Konzept der gesundheitsfördernden Schule des WHO-Regionalbüros Europa 1997 – Wege zu einer gesünderen Schule: Handlungsebenen – Handlungsfelder – Bewertungen. Tübingen: 6.

Petermann F, Schmidt MH (2006) Ressourcen – ein Grundbegriff der Entwicklungspsychologie und Entwicklungspathologie? Kindheit und Entwicklung 15: 118–127.

Petermann F, Kusch M, Niebanck K (1998) Entwicklungspsychopathologie. Weinheim: Beltz: 221.

Ravens-Sieberer U, Eller U, Erhard M (2007) Gesundheitsbezogene Lebensqualität von Kindern und Jugendlichen in Deutschland. Bundesgesundheitsbl – Gesundheitsforsch – Gesundheitsschutz 50: 810–818.

Ravens-Sieberer U, Wille N, Bettge S, Erhart M (2007) Psychische Gesundheit von Kindern und Jugendlichen in Deutschland. Ergebnisse der bella-Studie im Kinder- und Jugendgesundheitssurvey, Bundesgesundheitsbl – Gesundheitsforsch – Gesundheitsschutz 5/6: 871–878.

Reiter S, Poethko-Müller C: Aktuelle Entwicklung von Impfquoten und Impflücken bei Kindern und Jugendlichen in Deutschland. Bundesgesundheitsbl- Gesundheitsforsch – Gesundheitsschutz 2009 · 52: 1.037–1.044,

Robert-Koch-Institut; Ergebnisse des Kinder- und Jugendgesundheitssurveys; Bundesgesundheitsbl – Gesundheitsforsch – Gesundheitsschutz 2007; Heft 5/6 50.

Robert-Koch-Institut Lebensphasenspezifische Gesundheit von Kindern und Jugendlichen in Deutschland, (2008) Berlin: 88

Robert-Koch-Institut (Hrsg), Bundeszentrale für gesundheitliche Aufklärung (Hrsg) (2008) Erkennen-Bewerten-Handeln: Zur Gesundheit von Kindern und Jugendlichen in Deutschland: Berlin 2008: 21

Robert-Koch-Institut: Migration und Gesundheit, Berlin 2008

Rosenbrock R(1995) Public Health als soziale Innovation. Ges. Wes. 57: 140–144.

Schlack HG (2004) Neue Morbidität im Kindesalter – Aufgaben für die Sozialpädiatrie. Kinderärztliche Praxis 5: 292–298.

Schlack H. Thyen U, von Kries R 2009 Sozialpädiatrie. Gesundheitwissenschaft und pädiatrischer Alltag, Heidelberg:182

Schwartz FW, Walter U (1998) Gesundheitsförderung und Prävention. In: Schwartz FW, Badura B, Leidl R. Das Public-Health-Buch. München: Urban & Fischer: 151–164.Scucs TD (1997) Medizinische Ökonomie. München: Urban und Vogel: 60–68.

Ständige Konferenz der Kultusminister der Länder, Empfehlungen zur Gesundheitsförderung und Prävention in der Schule vom 15.11.2012: www.kmk.de

Statistisches Bundesamt, Ausgaben Gesundheit 2006, Statistisches Bundesamt, Wiesbaden 2008: Gesundheitsausgaben 2006 nach Leistungsarten und Ausgabenträgern. www.destatis.de/jetspeed/portal/search/results.psml (letzter Zugriff 3.3.2009)

Weinbrenner S, Wörtz M, Busse R (2007) Gesundheitsförderung im europäischen Vergleich. Berlin: Kompart: 24.

Werner EE, Smith RS (1992) Overcoming the Odds: High Risk Children from Birth to Adulthood. Cornell University Press.

WHO (1998) Glossar Gesundheitsförderung. Genf.

Internetlinks

www.anschub.dewww.BAEK.de
www.bvpraevention.de
www.bzga.de
www.dgkj.de
www.dgspj.de
www.euro.who.int
www.fruehehilfen.de

www.G-BA.de
www.gbe-bund.de
www.gesundheitsziele.de
www.gkv-spitzenverband.de
www.iqwig.de
www.KBV.de
www.kmk.de
www.KIGGS.de
www.kindersicherheit.de
www.oecd.org
www.olga-qol.com
www.opus-nrw.de
www.rki.de
www.searo.who.int
www.who.int/school_youth_health

8 Spezielle Präventionsmaßnahmen

Bernd Wilken

8.1 Einleitung

Der Begriff **Prävention** leitet sich ab vom lateinischen praeventum = zuvorkommen und bedeutet heute im Wesentlichen Krankheitsvermeidung. Die Ziele der ersten bekannten Präventionsprogramme waren andere als heute, z. B. eine effektive Seuchenbekämpfung im 19. Jahrhundert im Gegensatz zur Bekämpfung von Übergewicht im 21. Jahrhundert. Erkenntnisse wurden – und werden – jedoch lange Zeit fehlinterpretiert oder gar ignoriert, da sie nicht dem aktuellen Stand der Forschung entsprachen oder politisch nicht erwünscht waren. Als Beispiel mag die Geschichte von Ignaz Semmelweis (1818–1865) dienen, der die Ursache des sog. Kindbettfiebers entdeckte und als erster Arzt entsprechende Hygienevorschriften erließ. Die Desinfektion der Instrumente und das Waschen der Hände vor jeder Untersuchung senkte 1848 die Mortalität der schwangeren Frauen von 12 auf 1,5 %. Trotz dieser Ergebnisse wurde seine Arbeit über viele Jahre nicht anerkannt und das Waschen der Hände als schlicht überflüssig angesehen. Erst 1882 wurde die Bedeutung der Hygienevorschriften erkannt, Semmelweis und seine Arbeit erhielten öffentliche Anerkennung und Desinfektion wurde als Präventionsansatz eingeführt.

Die Bedeutung von Präventionsprogrammen hat sich stetig weiterentwickelt und niemand zweifelt mehr an ihrem Nutzen. Niedergelassene Kinderärzte und Sozialpädiater widmen etwa 30–40 % ihrer Tätigkeit Präventionsaufgaben.

Wir unterscheiden heute drei Formen der Prävention. Durch **primäre Prävention** sollen schädliche Faktoren und Einflüsse bereits vor Wirksamkeit bekämpft und damit die Entstehung von Krankheiten unterbunden werden. Damit sind somatische und psychosozial bedingte Krankheiten und Störungen gleichermaßen gemeint. Die Maßnahmen setzen vor dem Eintreten einer

Sozialpädiatrie in der Praxis. http://dx.doi.org/10.1016/B978-3-437-31630-3.00008-0
Copyright © 2014 Elsevier GmbH. All rights reserved.

Krankheit ein und sollen ein Neuauftreten einer Erkrankung verhindern. Programme der Primärprävention richten sich auf Impfungen, schulische Programme und Suchtprävention.
Der Begriff **sekundäre Prävention** beschreibt die Notwendigkeit, Krankheiten und Störungen in der kindlichen Entwicklung möglichst früh zu entdecken, um Auswirkungen zu verhindern oder zu minimieren. Dazu ist es notwendig, entsprechende Risikogruppen zu identifizieren. Darüber hinaus dient die Sekundärprävention auch der Verhinderung von Chronifizierung von Krankheiten oder Störungen. Zielgruppe sind Kinder die gefährdet sind und die sich als Gesunde identifizieren. Etwas unsauber davon abzugrenzen ist die **tertiäre Prävention;** gemeint ist eine Verhinderung oder Verringerung von Folgeschäden einer bereits bestehenden Erkrankung (➤ Kap. 7). Schritte der Prävention sind in Tab 8.1 zusammengefasst (➤ Tab. 8.1).
Präventionsprogramme sind stark vom historischen und gesellschaftlichen Kontext abhängig und damit von der Bereitschaft der potenziell Betroffenen, an dem Programm zu partizipieren. Darüber hinaus sind Programme von der Häufigkeit des Auftretens von Erkrankungen oder Störungen, den pathophysiologischen Grundlagen, dem natürlichen Verlauf und der Behandelbarkeit einer Erkrankung abhängig.
Präventionsprogramme werden in **spezifische und unspezifische Maßnahmen** eingeteilt. Spezifische Programme werden nach Zielgruppen definiert, z. B. Impfprogramme bei Kleinkindern, Folsäureprophylaxe bei Schwangeren oder Verhinderung von Nikotinkonsum bei Jugendlichen. Unspezifische Intervention ist nicht auf eine bestimmte Zielgruppe abgestimmt und dient allgemein der Gesundheitsförderung.
Unterschieden wird zwischen einer Veränderung des Verhaltens und der Verhältnisse. Das Erkennen, dass das eigene Verhalten in der Zukunft zu Gesundheitsstörungen führt ist für Kinder- und Jugendliche schwer zu erfassen. Ebenso ist es nicht ohne weiteres möglich die Lebensbedingungen z. B. Familiensituation, Stressabbau, Umweltbedingungen zu verändern und den eigentlichen Bedürfnissen anzupassen. Im Gegensatz zu klar messbaren Gesundheitsrisiken, wie z. B. die Gefährdung von Kindern im Straßenverkehr, ist die Messbarkeit bei psychosozialen Risiken schon sehr viel schwerer möglich und es ist zu erwarten, dass sich diese durch ein Präventionsprogramm nur sehr schwer korrigieren -lassen. Schwierig ist dabei insbesondere, dass objektiv gesundheitsgefährdende Verhaltensweisen zum einen zur lebensgeschichtlichen Entwicklung (z. B. Risiken bei bestimmten Sportarten) von Jugendlichen gehören, zum anderen die schädigende Wirkung sich erst nach vielen Jahren zeigt, z. B. beim Ernährungs- und Bewegungsverhalten.
Präventionsansätze sind immer eng verknüpft mit Annahmen über Ursachen von Gesundheitsstörungen und Risikofaktoren im Verlauf des Lebens eines Kindes. Diese Annahmen müssen permanent evaluiert und der aktuellen Datenlage angepasst werden.

8.2 Grundlagen für Präventionsprogramme

Präventionsprogramme arbeiten immer über verschiedene Mechanismen. Sie sollen die eigene Kompetenz der Kinder und Jugendlichen stärken, damit gesundheitsschädliches Verhalten minimiert wird. Gesetzliche Grundlagen sind notwendig (z. B. Altersbeschränkungen beim Erwerb von Alkohol). Ein positiver Anreiz für das einzelne Kind bzw. Jugendlichen ist notwendig (z. B. es ist cool nicht zu rauchen) um zu motivieren.
Allerdings haben sich die Lebensgrundlage und die Lebenswirklichkeit von Kindern und Jugendlichen haben sich in den letzten Jahren erheblich verändert. Präventionsprogramme sind aber immer träge und können nicht flexibel einzelnen gesellschaftlichen Strömungen angepasst werden.

Bewegungsmangel, die Bedeutung von Medien im Alltag, Migration und die unsichere soziale Situation sind nur einige Beispiele für die rasanten Veränderungen. Es resultiert eine hohe Anzahl von Störungen der Entwicklung und des Verhaltens. Diese Tatsache prägt den Begriff der „Neuen Morbidität“. Präventionsprogramme müssen also auf die aktuellen **Lebensverhältnisse** von Kindern und Jugendlichen mit dem Ziel der Veränderung abgestimmt sein. Von besonderer Bedeutung sind dabei die interdisziplinäre Vorgehensweise und die **Vernetzungsstrukturen** zwischen allen Beteiligten. Erste und wichtige Anlaufstelle ist der niedergelassene Kinderarzt, der die Risikofaktoren erkennt, die entsprechenden Netzwerkstrukturen kennt und idealerweise die Koordination der einzelnen Unterstützung- und Beratungsangebote übernimmt. Neben dem ersten Ziel, der Stärkung der Familien, ist ein weiterer Schwerpunkt der Maßnahmen aufgrund der Erreichbarkeit von Kindern und Jugendlichen dabei in Kindergärten und Schulen zu sehen. **Gesundheitserziehung** in diesen Institutionen hat in erster Linie das Ziel, schädigungsvermeidendes Verhalten zu stärken und Bewusstsein für Risiken zu schaffen. Darüber hinaus müssen materielle, psychosoziale und soziokulturelle Benachteiligungen abgebaut werden. Kinder und Jugendliche leben klar im „Hier und Jetzt“, d. h. Motivation und Bewältigungsstrategien müssen überzeugend und verinnerlichbar sein. Das Durchhalten eines Verhaltens, z. B. einer Diät, muss für den Jugendlichen klar zu einem Vorteil führen, das Nichtdurchhalten muss verknüpft sein mit negativen Konsequenzen, die höher bewertet werden. Grundsätzlich müssen Präventionsprogramme also Lebensfreude vermitteln und den veränderten Lebensumständen von Kindern und Jugendlichen angepasst sein. Es ist wirklichkeitsfremd, z. B. Jugendliche davon überzeugen zu wollen, dass der Umgang mit Computern schädlich ist. Sinnvoller ist es, das Bewusstsein für einen vernünftigen Umgang mit Medien zu schaffen und das Verständnis für die Gefahren der Nutzung zu stärken.
Das Konzept der **Salutogenese** bekommt im Bereich der Prävention von Krankheit eine zunehmende Bedeutung (➤ Kap. 7). Salutogenese ist die Wissenschaft von der Entstehung von Gesundheit. Damit wird ein komplett anderes Konzept verfolgt, als die einfache Vermeidung und Erklärung von Krankheit. Gesundheit ist dabei ein attraktives Ziel, die eigenen Ressourcen werden betont, der Blick nicht auf Defizite gerichtet (z. B. in der motorischen Entwicklung eines Kindes). Eigenaktivität wird betont, Lösungen werden selbst erarbeitet, Möglichkeiten evaluiert.

8.3 Spezielle Präventionsmaßnahmen

8.3.1 Schwangerschaft, Säuglings- und Kleinkindperiode

Besondere Bedeutung kommt der Prävention in den frühen Phasen des Lebens zu. Die **Folsäurezufuhr** während der Schwangerschaft mit 0,4 mg täglich vermindert die Inzidenz von Neuralrohrdefekten signifikant (➤ Kap. 32). Die Folsäurebehandlung präkonzeptionell ist aber auch ein Beispiel für eine unbefriedigend verlaufende Aufklärungskampagne. Die Totalprävalenz von Neuralrohrdefekten hat sich nicht verändert. In Dänemark wurde 1999 und 2001 je eine Kampagne zur Folsäureprophylaxe durchgeführt (Olsen et al. 2008) mit nur einer diskreten Verbesserung der Häufigkeit von Folsäurezufuhr bei Frauen, die eine Schwangerschaft planen. Unter den Frauen, die ungeplant schwanger geworden waren, nahm nur eine kleine Gruppe Folsäure. Besonders ungenügend war der Anteil der Frauen, die Folsäure genommen hatten, bei sehr jungen Frauen und Frauen aus einem bildungsfernen Umfeld. Die Erfahrungen in Deutschland sind sehr ähnlich, nur etwa 7 % der beteiligten schwangeren Frauen in einer europaweiten Studie erreichten die empfohlene Folsäuredosis (Frank et al. 2008). Bei Frauen im gebärfähigen Alter mit einer Epilepsie, die mit Valproinsäure

behandelt werden, sollte grundsätzliche eine Folsäureprophylaxe durchgeführt werden, da durch die antikonvulsive Behandlung das Risiko für Neuralrohrdefekte noch einmal erhöht ist. Abhilfe könnte durch Anreicherung von Nahrungsmitteln mit Folsäure erreicht werden. Diskutiert wird in einigen europäischen Ländern die Supplementation von Folsäure durch die Anreicherung in Mehl.
Ein weiterer wesentlicher Baustein der Prävention in der frühen Phase des Lebens ist die regelmäßige **Schwangerenvorsorge.** Die regelmäßige Beratung, z. B. über das Risiko von Alkohol und Nikotin in der Schwangerschaft, ist in einer vertrauensvollen Umgebung notwendig. Nach wie vor werden die Risiken von Alkohol in der Schwangerschaft bagatellisiert, etwa 58 % der schwangeren Frauen geben an, gelegentlich Alkohol zu trinken. Jährlich werden in Deutschland etwa 10.000 Kinder, die unter den Folgen des fetalen Alkoholsyndroms (➤ Kap. 29) leiden, geboren. Es besteht also weiterhin hoher Beratungsbedarf. Vor oder zu Beginn der Schwangerschaft sollten Laboruntersuchungen erfolgen, wie die Bestimmung der Blutgruppe und des Rötelntiters, weiterhin sollte der Infektionsstatus für Hepatitis B und C sowie HIV geklärt werden. Verlaufsuntersuchungen zur Feststellung des Gedeihens des Feten und der Gesundheit der Mutter sind sinnvoll und werden im Mutterpass dokumentiert. Eine vertrauensvolle und kompetente Begleitung der Schwangerschaft hat sich bewährt, daher ist es schwer verständlich, dass immer mehr Leistungen der Vorsorge nicht von den Krankenkassen übernommen werden. Schlecht oder gar nicht begleitete Schwangerschaften können rasch in Notsituationen münden, die mit hohem Morbiditätsrisiko für die betroffenen Kinder einhergehen.
Vitamin K ist kaum plazentagängig und auch in der Muttermilch nur in geringen Konzentrationen vorhanden. Daher ist eine zusätzliche **Vitamin-K-Prophylaxe** nach der Geburt sinnvoll. Vitamin K ist ein fettlösliches Vitamin und ein wesentlicher Kofaktor für die Blutgerinnungsfaktoren II, VII, IX und X. Das Risiko einer Vitamin-K-Mangelblutung ist insbesondere bei frühgeborenen Kindern deutlich erhöht. Die Blutungen können z. T. wie bei der Hirnblutung lebensgefährlich sein oder zu Folgeschäden führen. In Deutschland erfolgt bei reifgeborenen Kindern eine Vitamin-K-Zufuhr von je 1 × 2 mg nach der Geburt, in der 1. Lebenswoche und nach 4 Wochen. Anfang der 1990er Jahre wurde ein Zusammenhang von hoher Vitamin-K-Dosis und erhöhter Leukämierate diskutiert, dieser Zusammenhang ließ sich nicht bestätigen. Die Diskussion hat aber in dieser Zeit zu einer deutlich erhöhten Inzidenz für Hirnblutungen wegen unterlassener Vitamin-K-Prophylaxe bei früh- und neugeborenen Kindern geführt.
Eine gesunde und optimale **Ernährung des Säuglings** dient der Vermeidung von Mangelzuständen und sorgt für ein gutes Gedeihen des Kindes. Es gibt keinen Zweifel, dass **Stillen** diese Grundvoraussetzungen ideal erfüllt. Muttermilch ist hinsichtlich Zusammensetzung und Absorption jedem kommerziell erhältlichen Produkt überlegen; Vitamine und Spurenelemente sind in der Muttermilch ausreichend vorhanden. Bedacht werden muss aber immer auch die Ernährungssituation der Mutter, so kann es z. B. bei veganischer Ernährung zu schwerwiegenden **Vitamin-B12-Mangelerscheinungen** kommen, die Symptome wie bei einer neurodegenerativen Erkrankung verursachen können. Vitamin D_3 ist das physiologisch beim Menschen vorkommende Vitamin D. Nach Aufnahme über die Darmschleimhaut erfolgt mit Hilfe von 7-Dehydrocholesterin und ultraviolettem Licht die Umwandlung in die aktive Komponente. Damit handelt es sich eher um ein Hormon und nicht um ein Vitamin. Der **Bedarf an aktivem Vitamin D3** ist während des Wachstums erhöht, daher sollte Vitamin D_3 ab dem 14. Lebenstag mit 500 IE tgl. im ersten Lebensjahr substituiert werden. Frühgeborene Kinder haben einen erhöhten Bedarf und bekommen 1.000 IE als Substitution. Der Bedarf von Klein- und Schulkindern wird über die Eigensynthese gedeckt. Ein Mangel an Vitamin D_3 führt zu Rachitis, möglicherweise ist ein Mangel aber auch ein Risikofaktor für die Entstehung weiterer Erkrankungen, da Vitamin D_3 Einfluss auf die Hemmung der Zellproliferation und Apoptose haben kann.

Ergänzt werden sollte die Gabe von 500 IE Vitamin D_3 laut Empfehlung der Deutschen Gesellschaft für Kinder- und Jugendmedizin um 0,25 mg **Fluorid** pro Tag für das erste Lebensjahr. Bis zum etwa 36. Lebensmonat sollte die Fluoridgabe mit 0,25 mg fortgesetzt und das Kind an das regelmäßige Zähneputzen gewöhnt werden. Zusätzlich gilt für die Kariesprophylaxe, dass das Nuckeln oder Schnullern vermieden werden sollte, insbesondere mit zuckerhaltigen Beigaben. Die Kinder sollten eine gesunde Ernährung erhalten und eine gute Zahnpflege einhalten. Die Fluoridbeimischung zur Zahnpasta wird verschiedentlich empfohlen, die Wirksamkeit dieser Maßnahme ist aber nicht belegt und zurzeit kann keine generelle Empfehlung ausgesprochen werden (Marinho et al. 2003).

Die Grundlagen für eine gesunde und bewusste Ernährung werden sehr früh gelegt. **Ernährung** ist nicht nur die Versorgung mit Eiweiß, Kohlenhydraten, Fetten, Vitaminen, Mineralien und Spurenelementen in gesunder und ausgewogener Zusammensetzung, sondern auch ein pädagogischer Auftrag für die Eltern. An dieser Stelle sind Verhältnisse änderbar, Ernährung leistet einen wichtigen Beitrag für die Gesundheit und die Belastbarkeit eines Menschen, gerade Kinder benötigen eine gesunde Ernährung für z. B. die Ausbildung des Immunsystems und den Knochenaufbau. Wichtig sind daher gemeinsame Mahlzeiten mit kalorienarmen Getränken, viel Obst und Gemüse, mäßiger Einsatz von tierischen Produkten wie z. B. Fleisch, Wurst und Milchprodukten, sehr zurückhaltender Einsatz von zuckerhaltigen Lebensmitteln. Kinder und Jugendliche müssen erleben, dass Lebensmittel mit hoher Nährstoffdichte ohne vernünftigen Sättigungswert (Fastfood-Produkte) das Wohlbefinden beeinträchtigen und die Gesundheit schädigen.

Eine unterstützende Funktion kommt im Rahmen der aufsuchenden Hilfe im Kontext der Aktion „ frühe Hilfen" dem Familienhebammen Konzept zu. Windorfer A. (2009) evaluierte in 3 Kommunen die Daten von 590 betreuten Frauen mit ihren Kindern. Die Zielgruppe waren minderjährige Mütter, alleinerziehende Mütter und Mütter in schwierigen sozialen oder psychologischen Lebenssituationen. Bei etwa 80 % dieser betreuten Frauen ergaben sich Verbesserungen in vielen Bereichen des Alltags. Neben dem Abbau von Ängsten z. B. vor Überforderung, gelang die Vermittlung einer angemessen pflegerischen Versorgung des Säuglings und Verbesserung des Wissens um eine gute gesundheitliche Situation des Kindes. Hier kam das Konzept der Salutogenese zur Anwendung, indem vorhandene Ressourcen genutzt und ausgebaut wurden (➤ Kap. 7).

Eine weitere wichtige Forderung ist die **Reduzierung von Umgebungsgiften.** Kinder sind keine kleinen Erwachsenen; was zunächst wie eine Plattitüde klingt, wird oft nicht ausreichend umgesetzt. Kinder haben eine erhöhte Vulnerabilität gegenüber Umgebungsgiften, z. B. in der Luft, Substanzen im Material von Spielzeug und selbstverständlich auch gegenüber belastetem Wasser. Über die Auswirkungen von toxischen Substanzen, z. B. Pestiziden in der Nahrung, auf den kindlichen Organismus ist wenig bekannt (Landrigan et al. 2003). Im Verhältnis zur Körperoberfläche benötigen Kinder mehr Flüssigkeit und Nahrung als Erwachsene, darüber hinaus sind die Stoffwechselwege noch unreif, was zusammen zu einer erhöhten Belastung führt. Der Schutz von Kindern gegen toxische Substanzen ist eine wesentliche Herausforderung unserer Gesellschaft. Jedes Jahr werden hunderte neue chemische Substanzen zum Einsatz gebracht, deren Risiko für Kinder nicht adäquat evaluiert ist. Die gängige Testung der Toxizität von Substanzen muss den Besonderheiten von Kindern angepasst werden. Einzelheiten finden sich z. B. im **Kinder-Umwelt-Survey** und im Informationsangebot Allergie, Umwelt und Gesundheit (www.allum.de, ➤ Kap. 13).

8.3.2 Plötzlicher Säuglingstod (Sudden Infant Death, SID)

Das unerwartete und nicht erklärbare plötzliche Versterben von Säuglingen ist hochdramatisch und für die Eltern eine Tragödie. Definiert ist das Ereignis als Ausschlussdiagnose nach Abklärung zahlreicher Differenzialdiagnosen wie Unfälle, Misshandlung, Infektionen und Stoffwechselstörungen. Am häufigsten ist der **plötzliche Säuglingstod** zwischen dem zweiten und vierten Lebensmonat. Aber auch jenseits des ersten Lebensjahres kann SID eintreten. Es gibt eine Jungenwendigkeit mit 60 % zu 40 %. Etwa 75 % der Todesfälle ereignen sich in den Wintermonaten. Das Gesamtrisiko liegt bei 0,04 % und ist damit die häufigste Todesursache jenseits der Neugeborenenperiode. Für die Prävention ist es wichtig, über die Risikofaktoren und mögliche protektive Faktoren gut orientiert zu sein. Die Präventionskampagnen haben zwischen 1991 und 2005 zu einer Reduktion von SID um 76,8 % geführt (➤ Abb. 8.1). Dieses Ergebnis ist jedoch nicht optimal, in den Niederlanden ist die Aufklärung konsequenter umgesetzt und mit aufsuchenden Hilfen verbunden, die SID-Reduktion liegt dort zwischen 1987–2005 bei 87 %. Dies gelang insbesondere durch die aufsuchenden Hilfen von Hebammen, die in die Familien gehen und konsequent über die Vorbeugung aufklären und bei der Umsetzung helfen.

Die Informationen für die Eltern müssen in der Geburtsklinik und von den betreuenden Hebammen in der Nachsorge kommuniziert und die entsprechenden, vorhandenen Flyer weitergegeben werden. Wichtig erscheint es, dabei keinen Druck auszuüben, z. B. auf Eltern, die rauchen und damit das Risiko für einen SID erhöhen, sondern Hilfen anzubieten. Wesentliche **Risikofaktoren** für Säuglinge, einen SID zu erleiden, sind das Schlafen in Bauchlage oder auf der Seite, Überhitzung im Schlaf, z. B. durch Decken oder Kopfkissen und zu warme Kleidung, frühes Abstillen, Schlafen im Bett der Eltern und Rauchen. Rauchen erhöht das Risiko um das 8,9-fache, daher sind Informationen und Hilfsangebote für die Eltern bei diesem Punkt ausgesprochen wichtig. Protektiv können sich Impfungen (besonders gegen Pertussis) – Senkung des SID-Risiko um den Faktor 1,85 – und die Benutzung eines Schnullers auswirken. **Präventionsmaßnahmen** sind daher:

- Konsequent rauchfreie Umgebung
- Verzicht von Nikotin auch in der Schwangerschaft
- Schlafen im eigenen Bett in Rückenlage ohne Decke oder Kissen, sondern im Schlafsack
- Feste, luftdurchlässige Matratze
- Zimmertemperatur 16–18 °C
- Den Müttern ist unbedingt zum Stillen zu raten.

Diese primäre Prävention ist schon sehr erfolgreich, aber noch nicht erfolgreich genug. Aufklärung sollte z. B. auch über das Fernsehen erfolgen.

Wirkungsvolle Prävention ist insbesondere das Zeigen und Vormachen, wie Risikofaktoren vermieden werden können. Es fällt den Eltern leichter, Strategien, die sie gesehen und erlebt haben, umzusetzen. Umso überraschender ist es, dass in vielen Kinderkliniken mit Neonatologie Kinder nicht auf dem Rücken zum Schlafen gelagert werden und keine Schlafsäcke vorhanden sind. Dieses Problem sollte im Sinne der primären Prävention rasch behoben werden.

8.3.3 Impfungen

Impfungen gehören zu den primären Präventionen und richten sich an die Gesamtbevölkerung mit dem Ziel, vermeidbare Erkrankungen zu eliminieren (➤ Kap. 7). Impfungen stellen eine besonders erfolgreiche Präventionsmaßnahme gegen eine Vielzahl von Infektionserkrankungen dar.

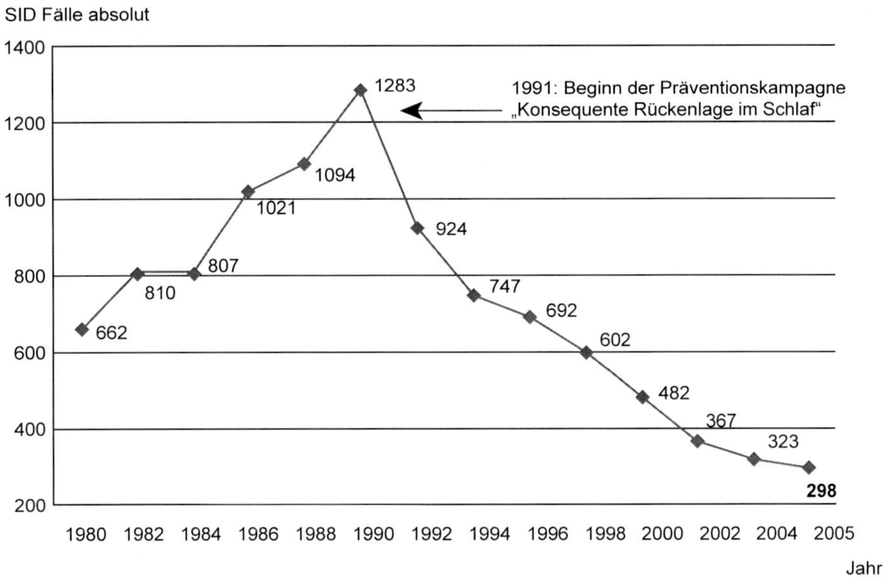

Abb. 8.1 Absolute Häufigkeit des plötzlichen Kindstodes in Deutschland vor und nach der Präventionskampagne zur Vermeidung der Bauchlage im Schlaf

Tab. 8.1 Schritte der Prävention (modifiziert nach Hurrelmann)

	1. Schritt	2. Schritt	3. Schritt	4. Schritt	5. Schritt
Präventionszeitpunkt	Gesundheit	Erkennbare Risikofaktoren	Frühzeichen einer Erkrankung	Manifeste Erkrankung	Langfristige Krankheitsfolgen
Zielgruppe	Gesamtbevölkerung	Risikogruppe	Akut Kranke	Chronisch Kranke	Chronisch Kranke
Präventionsziel	Verhütung von Krankheiten oder Störungen	Gezielte Vorbeugung durch Behandlung der Risikofaktoren	Frühe Behandlung der Krankheitsursache	Behandlung und Heilung der fortgeschrittenen Krankheit	Besserung der Lebensqualität, Vermeidung von Folgen
	Unspezifische Prävention	Spezifische Prävention	Kurative Therapie	Kurative Therapie	Rehabilitation Partizipation
Beispiel	Impfprogramme	Früherkennungstests	Gezielte medizinische oder psychologische Behandlung	Gezielte medizinische oder psychologische Behandlung	Wiedereingliederung, Hilfsmittel

Die Impfempfehlungen werden permanent von der **Ständigen Impfkommission (STIKO)** aktualisiert und auf dem wissenschaftlich neuesten Stand gehalten. Der hohe Kosten-Nutzen-Effekt ist belegt, Impfungen tragen direkt zur Kostensenkung im Gesundheitssystem bei. Als äußerst erfolgreiche Prävention erwies sich die erst seit einigen Jahren eingeführte Impfung gegen Haemophilus influenzae b (Hib). Haemophilus meningitis und H. epiglottitis werden nur noch sehr selten beob-

achtet. Das Ziel der vollständigen Eradikation von z. B. Masern wurde aber bislang nicht erreicht. Im Vergleich zu den Vereinigten Staaten und anderen europäischen Ländern sind die **Durchimpfungsraten** schlecht, es existieren Impflücken. Es wird immer wieder von lokalen Masernendemien in Deutschland berichtet. Die Durchimpfungsraten für Tetanus, Diphtherie und Poliomyelitis liegen deutlich über 90 %. Bei Pertussis liegt die Rate bei 89 % in den neuen Bundesländern und bei 83 % in den alten Bundesländern. Damit sind die Durchimpfungsraten keinesfalls erfreulich hoch. Bei der Masern-Mumps-Röteln-Impfung ist die Durchimpfungsrate deutlich schlechter, lediglich 56–59 % der Kinder erhalten die zweite MMR-Impfung (Dippelhofer et al. 2002). Dies reicht für die angestrebte **Eradikation der Masern** mit notwendigen Raten von 95 % nicht aus. Laut der „Erhebung seltener pädiatrischer Erkrankungen" (ESPED-Erhebung) wurden 2003 25 Kinder mit Masern stationär behandelt, 2004 11 Kinder, 2005 25 Kinder und 2006 116 Patienten. Die Rate für die tödliche Masernkomplikation subakute sklerosierende Panenzephalitis (SSPE) lag bei 2–6 Kinder pro Jahr. Die Daten der ESPED-Erhebung zeigen sehr deutlich, dass die Maserninfektion eine versorgungsrelevante und potenziell tödliche Infektion darstellt. Während die Impfungen gegen Tetanus, Diphtherie und Poliomyelitis gut akzeptiert sind, gibt es Vorbehalte insbesondere gegen die Masernimpfung. Breite Aufklärung über die z. T. gravierenden Folgen der Maserninfektion ist unabdingbar. Insbesondere muss über die Tatsache aufgeklärt werden, dass gerade Säuglinge, die sich mit Masern anstecken, besonders gefährdet für eine schwerwiegende Komplikation sind. Die Schließung der entsprechenden Impflücke sollte vom Öffentlichen Gesundheitsdienst spätestens bei der Einschulungsuntersuchung positiv beeinflusst werden. Bei weiter fehlender Akzeptanz sollte die Impfung mit Eingang in die Schule zur Pflicht gemacht werden, wie dies in anderen europäischen Ländern und in den USA bereits der Fall ist.
Die Ergebnisse und Akzeptanz der „neueren" Impfungen – gegen Haemophilus influenzae b, Pneumokokken und Hepatitis B – sind gut, die Kosten-Nutzen-Relation ebenfalls. Für die Varizellenimpfung müssen die Daten über die Akzeptanz noch abgewartet werden. Die Impfquote liegt bei 85 % und ist damit noch nicht ausreichend. Hochgerechnet werden so etwa 611.000 Fälle von Varizelleninfektion verhindert und damit etwa 4.700 gravierende Komplikationen vermieden (Banz et al. 2003).

8.3.4 Adipositas

Adipositas bei Kindern stellt ein zunehmendes medizinisches und volkwirtschaftliches Problem dar. Immer mehr Kinder und Jugendliche sind zu dick und haben bei ihrer Lebensführung keine Chance abzunehmen. Das Problem wird von den Eltern oft ignoriert oder bei ebenfalls bestehender Adipositas nicht als Krankheit wahrgenommen (➤ Kap. 1, ➤ Kap. 35). Der Teufelskreis aus Bewegungsmangel und Hyperalimentation bzw. falscher Ernährung ist nur schwer zu durchbrechen. Die Langzeitfolgen für die Gesundheit sind enorm, die Folgekosten für die Gesellschaft sehr hoch. Ist Adipositas überhaupt eine Erkrankung oder eine gesellschaftliche Erscheinung? Wenn Adipositas als Krankheit interpretiert würde, müssten die Krankenkassen auch für Schulungsprogramme aufkommen, was sie aber nur für einzelne Programme tun. Das Problem nimmt zu und ist ein Beispiel für die Notwendigkeit einer primären Prävention, aktuell wird „nur" sekundäre Prävention mit Identifizierung von Risikogruppen und Schulungsprogrammen für übergewichtige Kinder und Jugendliche betrieben. Die Prävalenzdaten sind erschreckend hoch und zeigen im Trend, wie wichtig eine primäre Prävention ist. Ende der 1990er Jahre waren in den USA 14 % der Kinder und Jugendlichen adipös, im Jahr 2004 bereits über 17 %! Die Daten aus Deutschland deuten einen ähnlichen Trend an. Die bundesweite Kinder- und Jugendgesundheitsstudie des Robert Koch-Instituts (KiGGS) ergab, dass ungefähr 9 % der Kinder im Alter zwischen 3 und 6 Jahren übergewichtig und 2,9 % adipös sind. Bei den 7–10-Jährigen bestand Übergewicht bei 15 % und Adipositas bei 6,4 % der Kinder. Von den 14–17-jährigen Jugendlichen waren 17 % übergewichtig und 8,5 % adipös. Die gesellschaftspolitische Relevanz ist mit der Tatsache belegt, dass schon heute 7 % der Krankheitskosten, die die Krankenkassen zu tragen haben, direkt auf Adipositas zurückzuführen sind. Da die Behandlungserfolge sehr begrenzt sind, und längerfristig nur wenige Kinder ihr zunächst vermindertes Körpergewicht halten ist Adipositas als chronisches Problem zu definieren. Eine sehr frühzeitige **Adipositasprophylaxe** ist daher dringend notwendig, in der Realität haben wir aber lediglich sekundäre oder z. T. nur tertiäre Prävention zur Verfügung. Die sekundäre Prävention wird sehr schnell zu einer tertiären Prävention, da die **Schulungsprogramme** mit Sport, medizinischer und psychologischer Betreuung sowie Ernährungsberatung nur die Kinder erreichen, bei denen Leidensdruck besteht. Darüber hinaus werden bestimmte Gruppen der Gesellschaft, wie z. B. Migranten, nur sehr unzureichend erreicht. Die Erfolgsquote ist dementsprechend niedrig und die Rate der Kinder und Jugendlichen, die die Programme abbrechen, hoch. Stationären Programmen fehlt der Nachweis der Nachhaltigkeit der Maßnahme. Die Kinder und Jugendlichen verlieren zwar gut an Gewicht, zeigen aber ohne die enge Betreuung sehr schnell einen Reboundeffekt mit rascher Gewichtszunahme.

Die Ernährungsgewohnheiten von Kindern werden maßgeblich von den Eltern bestimmt, erlernte Verhaltensmuster lassen sich nur sehr schwer wieder ändern. So hat eindeutig ein hohes Gewicht der Eltern einen ungünstigen Einfluss auf die Prognose der Adipositasbehandlung bei einem Kind oder Jugendlichen. Die Programme müssen daher immer die Eltern einbeziehen und möglichst früh einsetzen.

Wichtige **Parameter,** die in einem Präventionsprogramm beeinflussbar sind, sind z. B. das Stillen der Kinder und das Vorbildverhalten der Eltern. Der Prozentsatz der Kinder und Jugendlichen, die Lebensmittel mit hoher Energiedichte (Snacks) während des Fernsehens zu sich nehmen, steigt mit der Zeit, die sie vor dem Fernseher verbringen. Die Dauer des Fernsehkonsums, die auch mit der Häufigkeit der Benutzung von elektronischem Spielzeug korreliert, ist schon bei Vorschulkindern ein erheblicher Risikofaktor für die Entwicklung von Übergewicht (Kalies et al. 2001).

Die **Schulungsprogramme** für Kinder und Jugendliche sollten folgende 5 Module beinhalten:

1. Ernährungsberatung
2. Bewegung – Sportangebot
3. Verhaltenstherapeutische Modifikation des Körpergefühls
4. Medizinische Beratung
5. Betreuung und Einbindung der Eltern.

Einige wenige Programme werden auch in überregionalen Netzwerken angeboten. Als Beispiele dafür seien Programme wie Moby Dick, PowerKids oder FITOC genannt.

Wichtig wäre eine primäre Prävention, deren einzelne Bausteine aber gesellschaftspolitisch schwer durchzusetzen sind. Eine einfache Möglichkeit ist die Deklarationspflicht mit Warnhinweisen auf bestimmten Lebensmitteln und Getränken sowie lebensmittelrechtliche Vorschriften im Hinblick auf die Zusammensetzung von Nahrungsmitteln oder das Verbot von Werbeblocks mit falschen Werbeversprechen für Lebensmittel in Kindersendungen. Höhere Besteuerung für bestimmte Lebensmittel und Getränke wären möglich, bei sog. Alcopops ja bereits erfolgreich durchgeführt. Eine weitergehende Maßnahme wäre die Abschaffung des nachmittäglichen Fernsehprogramms. Die Aufklärung über diese Probleme sollte sehr früh beginnen und schon Inhalt der Gespräche bei den Vorsorgeuntersuchungen sein.

Ziel der Prävention muss es aber sein, zu mehr Bewegung im Alltag zu motivieren. Es ist wichtig, Kinder und Jugendliche dazu anzuregen, Wege aus eigener Kraft zu erledigen, Treppen zu steigen

statt Fahrstuhl zu fahren und zu laufen statt mit Bahn/Bus zu fahren, zu spielen statt vor dem Fernseher oder dem Computer zu sitzen. Wenn man das Essverhalten von Kindern langfristig ändern will, reichen keine theoretischen Regeln. Kinder müssen Ernährungsumstellung „erleben“, am eigenen Körper erfahren, verstehen und langfristig einüben. Erst dann kann man hoffen, dass die Kinder ihre Ernährung eigenverantwortlich und gesundheitsorientiert gestalten.
Die Plattform Ernährung und Bewegung e. V. versucht, alle relevanten gesellschaftlichen Kräfte, die sich für einen gesundheitsförderlichen Lebensstil einsetzen, zu vereinigen.

8.3.5 Unfälle

Alle 2 Minuten verletzt sich in Deutschland ein Kind, etwa 380 Unfälle jährlich bei Kindern bis einschließlich 14 Jahre enden tödlich, die häufigsten Ursachen sind Ertrinken, Stürze, Vergiftungen, Feuer und Verkehrsunfälle. Die Verteilung auf die verschiedenen Ursachen ist altersabhängig, so stehen bei sehr kleinen Kindern Ertrinkungsunfälle und Ersticken im Vordergrund, bei den 5–15-Jährigen sind Verkehrsunfälle am häufigsten (Borse et al. 2008). Absolut gab es 2007 160.944 Unfälle im Verkehr und 982.056 Unfälle in der Schule oder im Kindergarten; es ereigneten sich 527.000 häusliche Unfälle. Insgesamt kam es zu 167.0000 Unfällen bei Kindern (Statistisches Bundesamt). Etwa 195.000 Kinder mussten in Deutschland im Jahr 2007 wegen einer Unfallverletzung stationär behandelt werden. Der Anteil von Kindern mit Langzeitschäden ist nicht bekannt. Primäre Prävention muss sich daher an die Gesamtbevölkerung richten und kann nicht nur allein mit Aufklärungskampagnen und Informationen zur Sicherheit von Kindern arbeiten, sondern es bedarf auch gesetzlicher Regelungen, um das Risiko für einen Unfall für ein Kind nachhaltig zu senken. In den USA konnte die Rate der Ertrinkungsunfälle bei Kleinkindern erst deutlich gesenkt werden, als in den sogenannten „Sonnenscheinstaaten“ eine **Einzäunungspflicht für private Schwimmbassins und Teiche** per Gesetz eingefordert wurde. Die rechtlichen Auflagen z. B. zur Verkehrssicherheit sind in Schweden umfangreicher und bilden die Unfallgefahr im häuslichen und Freizeitbereich besser ab, in Schweden liegt die Rate der tödlichen Unfälle bei Kindern und Jugendlichen um zwei Drittel unter der deutschen Quote. Verbesserungsbedarf besteht z. B. bei der Einführung einer **Helmpflicht für Radfahrer.** Eine weitere gesetzliche Präventionsmaßnahme wäre die Einzäunungspflicht für private Swimmingpools und Teiche, um das Risiko für Ertrinkungsunfälle wirksam zu senken. Als weitere wichtige Vorgabe ist das **Verbot** der Herstellung und des Vertriebs von **Lauflernhilfen** – des sogenannten „Gehfrei“ – zu fordern. Diese Maßnahme würde zu einer Senkung der gefährlichen Stürze im Haushalt beitragen. Giftige Substanzen, leicht entflammbare Stoffe und Arzneimittel sollten nur noch mit einer entsprechenden **Kindersicherung** auf dem Markt sein. Die bestehenden Kindersicherungen bei giftigen Substanzen und Arzneimitteln, falls überhaupt vorhanden, lassen sich von 4-jährigen Kindern problemlos öffnen. Hier ist eine gesetzliche Regelung im Interesse der primären Prävention dringend erforderlich.

8.3.6 Verhaltensstörungen

Störungen des Sozialverhaltens sind vielfältig und oft unscharf definiert. Daher sind die Prävalenzzahlen auch je nach Studie sehr unterschiedlich. In den Studien des Kinder- und Jugendgesundheitssurveys waren 11,5 % der Mädchen und 17,8 % der Jungen grenzwertig oder manifest verhaltensgestört (Hölling et al. 2007). In dieser Studie wurde als Instrument der SDQ zugrunde gelegt. Zum Beispiel Costello et al. (2003) fanden bei Jungen eine Prävalenz für Verhaltensstörungen von 4.2 % und bei Mädchen von 1.2 %, dabei wurde der DSM IV zugrunde gelegt. Die Störungen des Sozialverhaltens sind unterteilt in externalisierendes und internalisierendes Verhalten, externalisierend: aggressiv-dissoziales Verhalten, oppositionelles Verhalten, Aufmerksamkeitsdefizit/Hyperaktivitätsstörung (ADHS); internalisierend: Angststörungen und Depression (➤ Kap. 18, ➤ Kap. 22). Risikofaktoren für Verhaltensstörungen sind wissenschaftlich definiert und damit die Grundlage für Präventionsprogramme.

Risikofaktoren:

1. Geringe kognitive Möglichkeiten
2. Störungen der Eltern-Kind Interaktion
3. Nicht konstruktives Erziehungsverhalten mit negativen Erfahrungen für das Kind, fehlende eindeutige Regeln
4. Psychische Erkrankungen, Suchtmittel bei den Eltern
5. Familiäre Konflikte (Überforderung)
6. Schwieriger Sozialstatus der Familien, mangelnde Schulbildung, finanzielle Schwierigkeiten
7. Fehlende Einbindung in eine sog. „peergroup"

Die Verhaltensschwierigkeiten ziehen oft lebenslange Probleme nach sich, mit fehlender psychischer Stabilität und fehlendem prosozialen Verhalten. Für die Entwicklung einer primären Präventionsstrategie ist es wichtig, Risikofaktoren und **protektive Faktoren** zu identifizieren, um wirkungsvolle Programme zu beginnen. Die Präventionsstrategien müssen auf die identifizierten Risikofaktoren abzielen, dabei geht es für die Kinder und Jugendlichen um die Bewertung einer entstehenden sozialen Situation, im nächsten Schritt um die Suche nach einer Problemlösung, der eigenen Wutkontrolle, Entwicklung von Empathie und das Einschätzen des Verhaltens eines Gegenüber. Zurzeit befinden wir uns im Bereich der tertiären Prävention und sind damit viel zu spät; erst wenn Kinder manifest auffällig sind, greifen Hilfen und auch dann meist auch nur zögerlich. Psychische Gesundheit misst sich daran, wie ein Mensch sich selber als Individuum wahrnimmt und sich von seiner Umwelt reflektiert bekommt. Die förderlichen Einflüsse auf soziales Verhalten können dabei sehr unterschiedlich sein, was den Erfolg der Interventionsprogramme erschwert. Masten (2001) teilt die protektiven Faktoren in **personale, familiäre und soziale Komponenten** ein, dabei sind personale Komponenten definiert als Merkmale der Persönlichkeitsentwicklung, wie z. B. gutes Selbstwertgefühl, gute Begabungsausstattung, Kontaktfähigkeit und positives Lebensgefühl. Die familiären Komponenten lassen sich in erster Linie unter dem Begriff des stabilen und positiv emotionalen familiären Umfeldes zusammenfassen. Unter sozialen Komponenten sind Aspekte wie Freundschaft und Integration in der sog. Peergroup subsumiert. **Risikofaktoren** sind demgemäß eine geringere Ausstattung mit den beschriebenen Schutzfaktoren, aber auch Armut und instabile Familien. Gerade Kinder an der Schwelle zur Pubertät und Jugendliche geraten bei Defiziten ihrer protektiven Komponenten schnell in einen Teufelskreis aus sozialen Misserfolgen und inadäquaten Problemlösungen, die die sozialen Bindungen immer weiter stören und eine Entwicklung bis hin zur Dissozialität fördern.

Präventionsprogramme zur Stärkung der psychischen Stabilität bedürfen nicht nur der Kenntnis der protektiven und Risikofaktoren, sondern müssen auch sehr eng an das Lebensalter und die intellektuellen Möglichkeiten der Kinder und Jugendlichen angepasst werden. Einzelne Programme zur primären Prävention existieren und werden zunehmend eingesetzt. So wird z. B. „Faustlos" bereits in Kindergärten und 1. Grundschulklassen erfolgreich eingesetzt. Ziel ist dabei Impulskontrolle und adäquater Umgang mit Wut und Ärger für die Kinder. Ein wichtiger Bestandteil der Prävention ist dabei die Entwicklung von sozialen Netzwerken, um nicht nur an einer Stelle anzusetzen, und die Möglichkeit der Mitgestaltung durch die „Betroffenen". Kinder und Jugendli-

che müssen die Bedrohlichkeit ihres Tuns für ihre psychische und physische Gesundheit „erfahren“, nur dann sind Verhaltensmodifikationen möglich.
Noch früher setzen **sekundäre Präventionsprogramme** wie „Starke Kinder – Starke Eltern“ oder „Stepping Stones und Triple P“ ein, sie sind ein frühes Elterntraining, wenn ein problematisches Erziehungsverhalten identifiziert wird. Stepping Stones Triple P richtet sich an Eltern mit entwicklungsgestörten Kindern, das macht die Validierung dieser Interventionsmethode schwierig, bedingt durch sehr unterschiedliche Störungen sind die Ergebnisse schlecht vergleichbar. Es gibt aber einige Studien die eine hohe Zufriedenheit der Eltern mit diesem Programm belegen und einen positiven Einfluss auf die soziale Adaptation zeigen.

8.3.7 Funktionelle motorische Leistungsfähigkeit

Man muss akzeptieren, dass sich die Lebenswirklichkeit von Kindern geändert hat. Kinder gehen nicht einfach nach draußen, spielen und probieren sich dabei aus. Mittelschichtkinder werden nach Kindergarten und Schule zum Sport oder zu musikalischer Erziehung gebracht, Kinder aus schwächeren Familien beschäftigen sich früh und lange mit Medien. Der Anteil der Kinder mit Entwicklungsauffälligkeiten hat erheblich zugenommen (Scheidt-Nave et al. 2007), die Zahl der Verordnung therapeutischer Maßnahmen wie Logopädie und Ergotherapie steigt. In einer Kindergartenuntersuchung zeigten bereits 17 % der 3–6-Jährigen therapeutisch bedeutsame Auffälligkeiten, 18 % wurden als grenzwertig auffällig beschrieben (Heinrichs 2006). Zur Zeit der Einschulungsuntersuchung erhielten 137 von 1.000 Jungen und 52 von 1.000 Mädchen regelmäßig Ergotherapie (Bode 2008). Dies ist Ausdruck von Störungen der komplex-motorischen Bewegungsabläufe und der Handlungsplanung von Grob- und Feinmotorik. Nur sehr frühe Sport- und Bewegungsprogramme, die alle Kinder erreichen, können helfen. Auch bei diesem Problem ist eine primäre Prävention notwendig, die nicht als Therapie, sondern als verbindliches Angebot, z. B. im Kindergarten, die motorischen Fähigkeiten der Kinder schult. Es gibt keine gute Risikogruppenidentifizierung und damit keine Ansatzpunkte für eine sekundäre Prävention. Zurzeit agieren wir nur mit tertiärer Prävention; wenn die Kinder auffällig werden, bekommen sie eine spezifische Therapie, die aber oft viel zu spät einsetzt, wodurch Auswirkungen der motorischen Probleme in den ersten Schulklassen zu beobachten sind.

8.3.8 Drogen und Suchtmittel

Drogen- und Suchtmittelkonsum ist ein entweder unterschätztes oder ignoriertes Problem bei Kindern und Jugendlichen. Die Erstkonsumenten werden immer jünger, es gibt aber keine Früherkennung- und Interventionsprogramme, Versorgungsstrukturen fehlen. Kinder- und Jugendärzten kommt eine bedeutende Rolle bei der Früherkennung der Risikokollektive zu.
Einstiegssubstanzen sind in erster Linie Alkohol und Nikotin. Verharmlosung und leichter Zugang führen zu einem fehlenden Risikobewusstsein der Jugendlichen. Etwa 20 % der Jugendlichen zwischen 12 und 17 Jahren rauchen regelmäßig, die Zahl der stationären Aufnahmen von Jugendlichen mit **Alkoholintoxikation** nimmt drastisch zu. Es entwickelt sich eine Komorbidität, insbesondere mit psychischen Phänomenen und dem Risiko des Umstiegs auf sogenannte harte Drogen.
Das Konzept der Prävention setzt auf 3 Bausteine: 1. Die Drogenprävention mit abschreckender Information und gezielter Aufklärung. 2. Die Suchtprävention ist ein weiterer Schritt, dazu gehört

das Verstehen von Motiven der Jugendlichen und Risikoverhalten das Sucht verursacht. 3. Förderung der eigenen Kompetenz und Etablierung stabiler sozialer Netzwerke soll Rückzug in Sucht vermeiden. Auch dabei wird verstärkt auf die Ressourcen der Jugendlichen gesetzt, um Gesundheitsförderung zu erreichen.

Der Gesetzgeber ist gefordert, den **Nichtraucherschutz** konsequent umzusetzen und das Werbeverbot für Tabakwaren und Alkohol flächendeckend zu etablieren. Die bestehenden Zugangsbeschränkungen zum Alkohol müssen auch wirklich umgesetzt und überprüft werden.

Programme zur Nikotinprävention wie das des hessischen Kultusministeriums sollten flächendeckend eingesetzt werden. Das Programm macht Nikotinprävention zum Schulkonzept und damit zur elementaren Aufgabe der Schule. Es bietet Hilfen zum Ausstieg für Schülerinnen und Schüler, aber auch für Lehrerinnen und Lehrer. Seit 2002 existiert in Hessen ein landesweiter Nichtraucherwettbewerb der Schulen der jedes Jahr unter dem Motto „Klasse ist rauchfrei – rauchfrei ist Klasse" angeboten wird.

Alkoholmissbrauch ist ebenfalls ein gravierendes Problem unter Jugendlichen. 2009 wurden 26.428 Jugendliche im Alter von 10–20 Jahren wegen einer Alkoholintoxikation stationär behandelt. In verschiedenen Präventionsstudien werden die Risikofaktoren identifiziert, dazu gehört unter anderem auch, dass eine gute sozioökonomische Situation der Eltern den Alkoholkonsum von Jugendlichen eher begünstigt. Weitere Faktoren sind das Trinkverhalten der Eltern, Erziehungsstrategien, Alter bei Einstieg in den Alkoholkonsum, Ärger in der Schule, schlechte Begabungsausstattung. Für die Prävention ist die Definition des riskanten Alkoholkonsums und die Identifizierung der Risikogruppen wichtig. Stärkere Kontrolle der gesetzlichen Vorgaben, z. B. auch eine höhere Besteuerung von Alkohol und Programme in der Schule sind hilfreich und zeigen in den Studien eine gute Wirksamkeit.

8.3.9 Fazit

Prävention will einen unerwünschten Zustand oder ein unerwünschtes Verhalten verhindern. Daher sind gerade sekundäre Präventionsprogramme mit der Aussicht auf eine gesundheitliche Konsequenz oft sehr schwer vermittelbar, besonders an Jugendliche die eine noch sehr unfertige Vorstellung von Zukunft haben. Daher ist das Konzept der **Salutogenese** (➤ Kap. 7, ➤ Kap. 14) möglicherweise in Kombination mit dem Präventionsgedanken eine sinnvolle Ergänzung um eigene Kompetenzen und Ressourcen der Kinder und Jugendlichen zur Gesundheitsförderung des Einzelnen einzusetzen.

Es gibt eine Vielzahl von Präventionsprogrammen, die überregional eingesetzt werden und gut publik sind, wie z. B. die Kampagne zur HIV-Prävention. Alle zur Verfügung stehenden Medien werden eingesetzt und national bekannte Persönlichkeiten wie Schauspieler und Fußballspieler beteiligen sich an dieser Kampagne. Für viele weitere Aspekte der Gesundheitserziehung von Kindern und Jugendlichen bedarf es jedoch noch regionaler Programme und des Willens, etwas zu ändern. Gesetzliche Vorgaben sind erforderlich, wie z. B. die Forderung nach einer Impfpflicht. Eine weitere zentrale Forderung betrifft eine Kindergartenpflicht, zumindest ein Jahr vor Beginn der Schule. Dies alleine reicht aber nicht aus, insbesondere in die Ausbildung der Erzieher und Erzieherinnen im Kindergarten muss investiert werden, pädagogische Programme und Förderkonzepte sind eine Voraussetzung dafür. Es muss ein gesellschaftliches Anliegen sein, den Fehlentwicklungen von Kindern bei schwierigen familiären sozialen Möglichkeiten und Defiziten entgegenzuwirken.

Ein Beispiel für eine Fehlentwicklung mag das Rauchen dienen, ein Hinweis auf die gesundheitlichen Folgen in der Zukunft wird kaum helfen, Jugendliche davon abzuhalten, wenn Rauchen gleichgesetzt wird mit Begriffen wie Erwachsensein, Stärke, Unabhängigkeit und Anerkennung durch Freunde bedeutet. Ein Präventionsprogramm muss hier ansetzen und es muss gelingen, in dieser Gruppe das Rauchen mit negativen Attributen zu belegen, und das gegen den Widerstand der Tabakindustrie. Die Programme dürfen sich aber nicht nur an eine Generation richten, denn auch die Eltern sind Zielgruppe.
Gesundheit darf nicht definiert werden als die Abwesenheit von Krankheit, sondern muss ein erstrebenswerter Zustand sein, der aktiv erworben wird und mit positiven Attributen belegt ist.

Literatur

Banz K, Wagenpfeil S, Neiss A, Goertz A, Staginnus U, Vollmar J, Wutzler P (2003) The cost-effectiveness of routine childhood varicella vaccination in Germany. Vaccine 7: 1.256–1.267.

Barkmann C, Schulte-Markwort M (2006) Psychosoziale Lebenssituation und Gesundheitsprobleme bei Kindern und Jugendlichen in der Bundesrepublik Deutschland. Prax Kinderpsychol Kinderpsychiatr 55: 444–458.

Bode H (2008) Heilmittel – Zahlen, Entwicklungen, Hintergründe. Kinderärztliche Praxis 79: 336–346.

Borse NN, Gilchrist J, Dellinger AM, Rudd RA, Ballesteros MF, Sleet DA (2008) CDC Childhood Injury Report. Patterns of unintentional injuries among 0–19 years olds in the United States, 2000–2006. Atlanta GA: Centers for Disease Control and Prevention, National Center for Injury Prevention and Control.

Dippelhofer A, Meyer C, Kamtsiuris P, Rasch G, Reiter S, Bergmann KE (2002) Erste Ergebnisse zum Impfstatus aus der Pilotphase des Kinder- und Jugendgesundheitssurveys. Bundesgesundheitsbl – Gesundheitsforsch-Gesundheitsschutz 45: 332–337.

ESPED-Erhebung für seltene pädiatrische Erkrankungen. www.esped.uni-duesseldorf.de/Jahresberichte

Franke C, Verwied-Jorky S, Campoy C, Trak-Fellermeier M, Decsi T, Dolz V, Koletzko B (2008) Dietary Intake of natural Sources of Docosahexaenoic Acid and Folat in Pregnant Women of Three European Cohorts. Ann Nutr Metab 53: 167–174.

Heinrichs N (2006) The effects of two different incentives on recruitment rates of families into a prevention program. J Prim Prev 27: 345–365.

Hölling H, Erhart M, Ravens-Sieberer U, Schlack R (2007) Verhaltensauffälligkeiten bei Kindern und Jugendlichen. Erste Ergebnisse aus dem Kinder- und Jugendgesundheitssurveys (KiGGS). Bundesgesundheitsbl – Gesundheitsforsch-Gesundheitsschutz 50: 784–793.

Kalies H, Koletzko B, Von Kries R (2001) Übergewicht bei Vorschulkindern. Der Einfluß von Fernseh- und Computerspiel-Gewohnheiten. Kinderärztliche Praxis 4: 227–234.

Landrigan PJ, Kimmel CA, Correa A, Eskenazi B (2003) Children's Health and the Environment. Public Health Issues and Challenges for Risk Assessment. Environ Health Perspect 112: 257–265.

Marinho VCC, Higgins JPT, Logan S, Sheiham A (2003) Fluoride toothpastes for preventing dental caries in children and adolescents. Cochrane Database of Systematic Reviews.

Masten AS (2001) Resilienz in der Entwicklung: Wunder des Alltags. In: Rösler et al. (Hrsg) Entwicklung und Risiko. Perspektiven einer klinischen Entwicklungspsychologie. Stuttgart: Kohlhammer: 192–219.

Olsen SF, Knudsen VK (2008) Folic acid for the prevention of neural tube defects: the Danish experience. Food Nutr Bull 29 (Suppl) S205-S209.

Probst P, Spreitz J (2008) Empirische Studie zur Sozialen Validität des präventiven Trainingsprogramms Stepping Stones Triple P für Eltern von Kindern mit Entwicklungsbehinderungen. Heilpädagogische Forschung 34: 114–131.

Ravens-Sieberer U, Witte N, Bettge S, Erhart M (2007) Psychische Gesundheit von Kindern und Jugendlichen in Deutschland. Ergebnisse aus der Bella-Studie im Kinder- und Jugendgesundheitssurveys (KiGGS). Bundesgesundheitsbl – Gesundheitsforsch-Gesundheitsschutz 50: 871–878.

Scheidt-Nave C, Ellert U, Thyen U, Schlaud M (2007) Prävalenz und Charakteristika von Kindern und Jugendlichen mit speziellem Versorgungsbedarf im Kinder- und Jugendgesundheitssurveys (KiGGS) in Deutschland. Bundesgesundheitsbl – Gesundheitsforsch-Gesundheitsschutz 50: 750–756.
Schlack HG (2008) Eine kranke Generation? Psychosoziale Gründe der neuen Morbidität. Kinderärztliche Praxis 79: 209–214.
Windorfer A (2009) Prävention von Kindesvernachlässigung durch aufsuchende Arbeit der Familienhebammen. Prax Kinderpsychol Kinderpsychiat 58: 368–383.

Internetlinks

www.umweltbundesamt.de

9 Öffentlicher Gesundheitsdienst

Ulrike Horacek, Renate Klein, Heidrun Thaiss

Sozialpädiatrie in der Praxis. http://dx.doi.org/10.1016/B978-3-437-31630-3.00009-2
Copyright © 2014 Elsevier GmbH. All rights reserved.

9.1 Aufgaben des öffentlichen Gesundheitsdienstes und der Gesundheitsfachverwaltung auf Bundes-, Landes- und kommunaler Ebene

Der Öffentliche Gesundheitsdienst (ÖGD) umfasst die Gesundheitsfachverwaltung auf der Ebene des Bundes, der Länder und der Kommunen. Seine Aufgaben sind die Beobachtung, Begutachtung und Wahrung der **gesundheitlichen Belange der Bevölkerung** und die Beratung der Träger öffentlicher Aufgaben in gesundheitlichen Fragen einschließlich Planungs- und Gestaltungsaufgaben, Gesundheitsförderung und der gesundheitlichen Versorgung, der öffentlichen Hygiene, der Gesundheitsaufsicht sowie der Verhütung und Bekämpfung von Krankheiten.
Zu den Planungs- und Gestaltungsaufgaben zählen u. a. die Gesetzgebung hinsichtlich Strukturen und Finanzierung der Gesundheitsversorgung, des Infektionsschutzes, der Umweltmedizin, der Wasser- und Abwasserhygiene, des Bestattungswesens und der Festschreibung eigener Aufgaben, aber auch die weitere Sozialgesetzgebung. In Kooperation mit anderen Trägern ist der ÖGD für die Schaffung erforderlicher Strukturen der gesundheitlichen Versorgung zuständig. Darüber hinaus hat er die gesundheitliche Entwicklung in den verschiedenen Bevölkerungsgruppen zu beobachten, zu bewerten und die Träger öffentlicher Aufgaben in der Beachtung und Weiterentwicklung gesundheitlicher Belange zu beraten. Er hat die Umsetzung der gesetzlichen Regelungen zu beaufsichtigen und die für die Durchführung seiner Aufgaben notwendigen Strukturen zu schaffen.

9.1.1 Gesundheitsfachverwaltung auf Bundesebene

Zu den Einrichtungen des Öffentlichen Gesundheitsdienstes auf Bundesebene gehören das Bundesministerium für Gesundheit und eine Reihe nachgeordneter Behörden. Seiner Fach-und Dienstaufsicht unterstehen das Robert-Koch-Institut in Berlin (RKI), das Paul-Ehrlich-Institut, das Bundesamt für Sera und Impfstoffe (PEI), das Deutsche Institut für Medizinische Dokumentation und Information (DIMDI), die Bundeszentrale für gesundheitliche Aufklärung in Köln (BZgA) und das Bundesinstitut für Arzneimittel und Medizinprodukte in Bonn (BfArM). Über das Bundesversicherungsamt als eigene rechtsfähige Anstalt (BVA) führt es die Fachaufsicht.
Das **Bundesgesundheitsministerium** ist in Deutschland im Rahmen der konkurrierenden Gesetzgebung zuständig für die Sozialgesetzgebung, u. a. hinsichtlich der gesetzlichen Krankenversicherung Sozialgesetzbuch (SGB V), der sozialen Pflegeversicherung (SGB XI), sowie für die Regelung der Rahmenbedingungen im Krankenhauswesen. Des Weiteren obliegt ihm die Gesetzgebung im Apotheken-, Arzneimittel- und Medizinprodukterecht. Es ist überwiegend zuständig für die Berufsgesetze im Gesundheitswesen, so für alle approbierten Berufe und die Gesundheitsfachberufe. Zu seinem Regelungsbereich gehören des Weiteren u. a. das Infektionsschutzgesetz, das Gewebe- und Transplantationsgesetz, das Gendiagnostikgesetz, das Embryonenschutzgesetz und das Stammzellgesetz
Das **Robert-Koch-Institut** (RKI) ist die zentrale Einrichtung des Bundes im Bereich der öffentlichen Gesundheit zur Erkennung, Verhütung und Bekämpfung von Krankheiten. Es bewertet, analysiert und erforscht dabei Krankheiten von hoher Gefährlichkeit, weitem Verbreitungsgrad oder großer öffentlicher oder gesundheitspolitischer Bedeutung, wie HIV/AIDS, Krebs, Allergien und Infektionskrankheiten.

Als fachliche Leitinstanz für die Umsetzung des Infektionsschutzgesetzes gibt es Merkblätter und Empfehlungen heraus und führt Surveys und wissenschaftliche Forschung durch. Als spezifische Einrichtungen sind die Ständige Impfkommission (STIKO), die Krankenhaushygienekommission (KRINKO) sowie die Kommission für Antiinfektiva, Resistenz und Therapie (ART) dort angesiedelt. Die STIKO spricht als fachliches Gremium die Impfempfehlungen für Deutschland aus; diese werden entsprechend den epidemiologischen Erfordernissen und der Verfügbarkeit von Impfstoffen gegen einzelne Infektionskrankheiten aktualisiert. Die KRINKO gibt umfassende fachliche Empfehlungen im Bereich der Hygiene, die ART zu allgemeinen Grundsätzen für Diagnostik und antimikrobielle Therapie.

Ein wichtiges Arbeitsfeld des Robert-Koch-Instituts ist die Gesundheitsberichterstattung auf Bundesebene: Es koordiniert Daten der Länder, z. B. der Krebsregister in der Dachdokumentation Krebs, und führt eigene Untersuchungen durch. Mit dem Kinder- und Jugendgesundheitssurvey (KiGGS) wurden 2003 bis 2006 umfangreiche Daten zur Gesundheit von Kindern und Jugendlichen bundesweit erhoben. Die Ergebnisse sind Grundlage aktueller Gesundheitsplanungen.

Das **Paul-Ehrlich-Institut, Bundesamt für Sera und Impfstoffe (PEI)** ist verantwortlich für die Arzneimittelsicherheit (immun-)biologischer Präparate im Human- und Veterinärbereich. Dazu gehören die Zulassung und regelmäßige Überprüfung (Chargenprüfung) von Impfstoffen, Sera, bestimmten In-vitro-Diagnostika mit hohem Risikopotenzial, Immundiagnostika und Blutprodukten und die damit verbundene prüfungsbegleitende Forschung. Das PEI ist die zentrale Meldebehörde für Zwischenfälle nach dem Arzneimittel- (AMG) und dem Transfusionsgesetz (TFG). Es führt vertiefte Forschung z. B. im Bereich der angewandten Virologie, der Gentherapie, der DNA-Vakzine oder der Entwicklung von Alternativen zum Tierversuch durch. Als nationale Behörde ist es Ansprechpartner der europäischen Arzneibehörde (European Medicines Agency, EMEA).

Das Deutsche Institut für Medizinische Dokumentation und Information (DIMDI) stellt Informationssysteme für alle Bereiche des Gesundheitswesens bereit: Es verwaltet 70 medizinische Datenbanken, u. a. zu Fachliteratur, toxikologischen Fakten, Arzneimitteln (nach § 67 AMG), Medizinprodukten, Health Technology Assessment (HTA), aktuellen Meldungen. Es ist Herausgeber der deutschen Versionen von medizinischen Klassifikationen wie ICD-10, ICF, Operationsschlüssel (OPS) und veröffentlicht ATC (Acute Toxic Class, Methode zur Prüfung von Substanzen auf ihre Giftigkeit), MeSH (Medical Subjects Headings, Thesaurus, den die National Library of Medicine in den USA erstellt und kontinuierlich pflegt) und UMDNS (Universal Medical Device Nomenclature System, codiertes Klassifikationssystem für Medizinprodukte). Darüber hinaus stellt das DIMDI Informationen zur Telematik im Gesundheitswesen zur Verfügung. Als nationale Behörde ist es Ansprechpartner der entsprechenden europäischen Behörden. Dazu gehört u. a. Eudamed als European Database on Medical Devices, eine zentrale europäische Datenbank für Medizinprodukte.

Die **Bundeszentrale für gesundheitliche Aufklärung** (BZgA) ist die Fachoberbehörde für Gesundheitsförderung. Zu ihren Aufgaben gehört die Entwicklung von Strategien zur gesundheitlichen Aufklärung und Prävention für die der Prävention zugänglichen und am weitesten verbreiteten Krankheiten. Schwerpunktthemen sind Rauchen, übermäßiger Alkoholkonsum, Über- und Fehlernährung, Infektionskrankheiten (HIV und AIDS – „Gib AIDS keine Chance"), Suchtvorbeugung („Kinder stark machen") und gesundheitliche Vorsorge („Ich geh zur U und Du?"). Sie führt in Kooperation mit anderen Akteuren (Kassen, Bund, Ländern) Kampagnen und Projekte durch mit dem Ziel der Information und Motivation zu gesundem Verhalten. Auch die Aufklärung über und Motivation zur Organspende sowie Blut- und Plasmaspende gehören zu den Aufgaben der BZgA. Sie kooperiert auf internationaler Ebene mit vielen Institutionen: EU, WHO, IUHPE (International Union for Health Promotion and Health Education) und berät die Gesellschaft für technische Zusammenarbeit (GTZ). Der vom Bundeskabinett am 20.3.2013 beschlossene Entwurf eines Gesetzes zur Förderung der Prävention sieht für die BZgA eine hervorgehobene

Rolle vor: sie soll kassenübergreifende Leistungen zur Primärprävention, schwerpunktmäßig in Settings, gestalten und umsetzen.
Das **Bundesinstitut für Arzneimittel und Medizinprodukte** (BfArM) ist zuständig für die Zulassung von Arzneimitteln und Medizinprodukten einschließlich klinischer Prüfungen sowie die Registrierung von homöopathischen Arzneimitteln. Es hat die Risikobewertung von Arzneimitteln (Pharmakovigilanz) und Medizinprodukten (Herzschrittmacher, Computertomografie, Implantate) durchzuführen. Als Bundesopiumstelle überwacht es den legalen Umgang mit Betäubungsmitteln und Grundstoffen (Drogenausgangsstoffe) und führt das Substitutionsregister. Als nationale Institution arbeitet es mit den europäischen Arzneibehörden zusammen.
Das **Bundesversicherungsamt** (BVA) führt die Aufsicht über bundesunmittelbare Träger und Einrichtungen der Sozialversicherung.
Zu seinen Aufgaben gehören u. a. die Prüfung, die Geschäfts-, Rechnungs- und Betriebsführung der bundesunmittelbaren Krankenversicherungsträger und deren Pflegekassen, die Durchführung des Risikostrukturausgleichs, Verwaltung des Ausgleichsfonds, die Durchführung des Finanzausgleichs in der sozialen Pflegeversicherung, die Zulassung von Behandlungsprogrammen für chronisch Kranke (DMP-Programme) sowie die Verwaltung des ab 2009 greifenden Gesundheitsfonds. Es wirkt u. a. mit bei Finanzschätzungen in der Rentenversicherung, bei der Regelung der Verhältnisse zur Deutschen Post AG (Renten-Service), bei der Abwicklung der Bundeszuschüsse, den Erstattungen des Bundes, Fragen der Datenverarbeitung und ist für die Zahlung des Mutterschaftsgeldes nach § 13 Abs. 2 Mutterschutzgesetz zuständig.
Das höchste Selbstverwaltungsgremium im deutschen Gesundheitswesen wird durch den **Gemeinsamen Bundesausschuss** (G-BA) repräsentiert. Er steht als eigenständige juristische Person unter der Rechtsaufsicht des Bundesministeriums für Gesundheit und trifft Entscheidungen zu Fragen der gesundheitlichen Versorgung im Bereich der gesetzlichen Krankenversicherung. Im Kontext der Umsetzung des GKV-Modernisierungsgesetzes wurde zum 1.6.2004 das **Institut für Qualität und Wirtschaftlichkeit im Gesundheitswesen (IQWiG)** mit Sitz in Köln gegründet. Diese Einrichtung unterstützt die Entscheidungsfindung im G-BA durch Recherchen, Gutachten, Bewertung evidenzbasierter Leitlinien, Empfehlungen zu Disease-Management-Programmen, Nutzen-Bewertung von Arzneimitteln etc.; sie arbeitet beratend, empfehlend und nicht Beschluss fassend.

9.1.2 Gesundheitsfachverwaltung auf Länderebene

Auf der Ebene der Länder sind die **Gesundheitsministerien** oberste Landesgesundheitsbehörden. Ihnen zugeordnet sind Fachbehörden unterschiedlicher Art, z. B. **Landesgesundheitsämter, Medizinaluntersuchungsstellen** und **Landesprüfungsämter.**
Die Länder sind nach dem Grundgesetz zuständig für gesetzliche Regelungen im Bereich der Gesundheit. Dazu zählen die Gesetze über den Öffentlichen Gesundheitsdienst, die Einrichtung und Aufgaben der Krebsregister, das Friedhofs-, Bestattungs- und Leichenwesen, über das Krankenhauswesen einschließlich der Krankenhausplanung sowie den Rettungsdienst und Katastrophenschutz.
Die Landesgesundheitsbehörden sind als Gesundheitsministerien der Länder darüber hinaus zuständig für die Umsetzung vieler ressortspezifischer Bundesgesetze; dort sind den Ländern entweder konkrete Aufgaben zugeschrieben, z. B. im Infektionsschutzgesetz, oder sie werden verpflichtet, nähere Regelungen in eigenen Ausführungsgesetzen zu treffen, z. B. Ausführungsgesetze zum Transplantationsgesetz.
Derzeit und in Zukunft kommen auf die Landesministerien große Herausforderungen zu in der Umsetzung der Forderungen aus der UN-Behindertenrechtskommission. Soweit gesundheitliche Belange

angesprochen sind, müssen sich die Landesgesundheitsministerien in der Pflicht sehen, eine landesrechtliche Grundlage für Maßnahmen zur Inklusion behinderter und kranker Menschen zu schaffen und ihre Umsetzung voranzutreiben. Hier geht es auch um die Schaffung von Rahmenbedingungen, die zur schulischen Förderung entwicklungsbeeinträchtigter und behinderter Kinder in allgemeinen Schulen aufgebaut werden sollen; dafür haben die jeweiligen Landeskultusministerien rechtliche Grundlagen zu erstellen und Kommunen in der Umsetzung zu unterstützen (➤ Kap 13.6).
Landesgesundheitsbehörden setzen auch weitere Regelungen der europäischen Union in Landesgesetze um. Im Bereich der staatlich anerkannten Gesundheitsberufe sind sie für das Ausbildungs- und Prüfungswesen zuständig. Für die Gesundheitsfachberufe erlassen sie Regelungen zur Weiterbildung und Berufsordnungen. Sie führen die Rechtsaufsicht über die Ärzte-, Zahnärzte-, Psychotherapeuten- und Apothekerkammern und legen in Heilberufekammergesetzen und Verordnungen in Abstimmung mit den Kammern die Prinzipien ärztlicher Berufsausübung, die Aufgaben der Kammern und die Weiterbildungsbestimmungen fest. Darüber hinaus haben sie die Rechtsaufsicht über die landesunmittelbaren gesetzlichen Krankenkassen.
Im Rahmen der Umsetzung der Gesetze über den Öffentlichen Gesundheitsdienst sind sie als zentrale Stelle u. a. zuständig für Maßnahmen der Gesundheitshilfe und Prävention, der Kinder- und Jugendgesundheitspflege einschließlich der Zahngesundheitspflege, Maßnahmen des Infektions- und Umweltschutzes, für die Gesundheitsberichterstattung sowie die Kooperation mit den verschiedenen Akteuren der gesundheitlichen Versorgung. Zur Verbesserung der Gesundheit der Bevölkerung – bezogen auf bestimmte Bevölkerungsgruppen oder ein bestimmtes Gesundheitsziel – initiieren sie Maßnahmen in Form von Kampagnen, sie können aber auch dafür erforderliche Strukturen aufbauen, stärken und weiterentwickeln. Die Gesundheitsministerien sind zur Durchführung präventiver Maßnahmen darüber hinaus in den Landesarbeitsgemeinschaften oder -vereinigungen für Gesundheitsförderung und für Jugendzahnpflege engagiert. Sie unterstützen die Einrichtungen der gesundheitlichen Selbsthilfe in vielerlei Form.
Sie beraten politische Gremien und andere Ressorts in Fragen der Gesundheit und führen die Fachaufsicht über die nachgeordneten Behörden und die unteren Gesundheitsbehörden.
In den **nachgeordneten Landesbehörden,** die in größeren Bundesländern auch bei den Bezirksregierungen angesiedelt sind, sind länderunterschiedlich verschiedene Behörden integriert: Landesprüfungsämter sind für die Durchführung der staatlichen Prüfung in den approbierten und Gesundheitsfachberufen und die staatliche Anerkennung der Berufsabschlüsse zuständig; sie führen auch die Gleichwertigkeitsprüfungen bei den Gesundheitsberufen aus anderen Staaten durch. Auf Landesebene ist auch die Aufsicht über die Umsetzung des Apothekenrechts, des Arzneimittelrechts, der Regelungen im Bereich der Medizinprodukte und des Arbeitsschutzes (staatlicher Gewerbearzt) angesiedelt. Landesmedizinaluntersuchungsstellen sind für die Durchführung von mikrobiologischen Untersuchungen, die Umsetzung des Meldewesens und fachliche Beratung der Gesundheitsämter zuständig. Weitere Einrichtungen, denen oftmals die epidemiologischen Krebsregister zugeordnet sind, befassen sich mit Gesundheitsberichterstattung. Teilweise wird auch der Abgleich der Teilnahme an den Früherkennungsuntersuchungen für Kinder und am Mammographiescreening in Landeseinrichtungen durchgeführt.

9.1.3 Gesundheitsfachverwaltung auf kommunaler Ebene

Auf der Ebene der Kommunen sind die unteren Gesundheitsbehörden angesiedelt. Sie sind traditionell als **Gesundheitsamt** den Landkreisen zugeordnet, in größeren Städten auch der Stadt

(München, Augsburg, Stuttgart) und bis auf Bayern in der Trägerschaft der Kommunen. Ihre Aufgaben sind im SGB V, in den Gesetzen über den Öffentlichen Gesundheitsdienst sowie einer Reihe von Spezialgesetzen und -verordnungen wie Infektionsschutzgesetz, Bestattungsgesetzen, Schulgesetzen, Kinderbetreuungs- und Kinderschutzgesetzen beschrieben. Die unterschiedlichen Arbeitsgebiete im Bereich Hygiene mit Infektions- und Umweltschutz, Kinder- und Jugendgesundheitshilfe, Gesundheitsförderung und Prävention sowie sozialpsychiatrische Dienste erfordern ein breites Spektrum an Qualifikationen: Medizin und Pflege mit Fachrichtung Hygiene, Öffentlicher Gesundheitsdienst, Pädiatrie, Psychiatrie, Sozialarbeit und Rechts- und Verwaltungswissen sind im Gesundheitsamt vertreten. Die sich oftmals überschneidenden Arbeitsgebiete erfordern eine enge interdisziplinäre Zusammenarbeit. Gleichzeitig wird damit die Schnittstellenfunktion des Gesundheitsamtes zu anderen Institutionen und Disziplinen offensichtlich.
Im Rahmen des Infektionsschutzes führen die Gesundheitsämter die infektionshygienische Aufsicht über alle Einrichtungen des Gesundheitswesens und alle Gemeinschaftseinrichtungen (Kindergärten, Schulen, Altenheime, Asylbewerbereinrichtungen, Justizvollzugsanstalten). Sie überprüfen die Qualität der Badegewässer und die Einhaltung der Trinkwasserqualität. Sie sind Meldestelle für Ärzte und Labore nach dem Infektionsschutzgesetz, ermitteln Infektionsquellen und veranlassen die Orts- oder Kreispolizeibehörden, Maßnahmen zur Verhütung und Weiterverbreitung von Infektionskrankheiten (z. B. Salmonellen, Influenza) zu treffen. Sie sind für die Prävention sowie die individuelle Beratung und ggf. Behandlung von sexuell übertragbaren Erkrankungen und Tuberkulose zuständig. Durch Information und eigene Impfaktionen tragen sie zur Verbesserung des Impfschutzes bei. Sie beraten die Träger öffentlicher Belange bei Bauvorhaben in allen Fragen der Gesundheit, z. B. des gesundheitlichen Umweltschutzes, der Infektionshygiene oder altersentsprechender Belange. Darüber hinaus sind sie themenbezogen primäre Ansprechpartner für die Bevölkerung und sonstige Institutionen (u. a. Verhalten während einer Hitzewelle, Lärmschutz).
Sozialpsychiatrische Dienste an den Gesundheitsämtern sind sowohl in der Krisenintervention als auch in der Eingliederung psychisch Kranker tätig. In enger Kooperation mit Akteuren der Gesundheitsversorgung und (psycho-)sozialen Diensten in der Kommune arbeiten sie an der Verbesserung der sozialpsychiatrischen Versorgung in der Gemeinde.

9.2 Kinder- und Jugendgesundheitspflege

9.2.1 Allgemeine Aufgaben

Die Kinder- und Jugendgesundheitspflege umfasst in den meisten Ländergesetzen die gesundheitliche Prävention und den Schutz der Kinder vor gesundheitlicher Gefährdung in Kooperation mit anderen Akteuren der gesundheitlichen Versorgung; dabei ist der Kinder- und Jugendgesundheitsdienst (KJGD) in der Regel subsidiär tätig. Neben den Gesetzen für den Öffentlichen Gesundheitsdienst legen die Gesetze über Kindertagesbetreuungseinrichtungen (siehe Jugendhilfeportal), die Schulgesetze der Länder (siehe deutscher Bildungsserver) sowie spezifische weitere Regelungen (z. B. Landesverordnungen für schulärztliche Aufgaben) nähere Einzelheiten zu den Tätigkeiten des KJGD fest.
Der Kinder- und Jugendgesundheitsdienst ist für die Beratung von Kindergemeinschaftseinrichtungen in Fragen der Gesundheitsförderung und gesundheitsförderlichen Rahmenbedingungen im Setting zuständig. Er führt systematische sozialpädiatrische Untersuchungen bei Kindern verschiedener Altersklassen durch. Überwiegend werden in den Bundesländern alle Schulanfänger

untersucht bzw. eine vorgezogene Untersuchung aller Vier- bis Fünfjährigen durchgeführt. Teilweise erfolgen weitere Untersuchungen in Kindergärten, in Grundschulen oder zum Schulabschluss. Die Kinder- und Jugendgesundheitsdienste arbeiten mit den sozialpädiatrischen Zentren, Frühfördereinrichtungen und schulpsychologischen Diensten bei der Inklusion chronisch kranker und behinderter Kinder in Kindertageseinrichtungen und Schulen zusammen.

Die Begleitung dieser Kinder und Jugendlichen in Fördereinrichtungen und die Verbesserung des Impfschutzes bei Kindern und Jugendlichen durch Information der Eltern in Kindergemeinschaftseinrichtungen sowie Impfangebote vor Ort sind weitere Aufgabenschwerpunkte. Die KJGDs sind höchst unterschiedlich in die Umsetzung „Früher Hilfen" bei erkennbarer gesundheitlicher Gefährdung von Neugeborenen eingebunden, ebenso in die nachgehenden Interventionen bei versäumter Teilnahme an den Früherkennungsuntersuchungen einschließlich des Neugeborenenscreenings.

Zur Verbesserung der Mundgesundheit führen die Jugendzahnärztlichen Dienste gemeinsam mit den niedergelassen Zahnärzten in Kindergärten und Schulen für Kinder von 3 bis 12 Jahren gruppenprophylaktische Maßnahmen (richtiges Zähneputzen, zahngesunde Ernährung, Untersuchung der Zähne) durch; in Einrichtungen für Behinderte und sozialen Brennpunkten ist die Altersgrenze auf 16 Jahre angehoben. Diese Aufgabe ist in § 21 SGB V als gemeinsame Aufgabe des Öffentlichen Gesundheitsdienstes, der niedergelassenen Zahnärzteschaft und der Krankenkassen festgeschrieben; damit geht auch eine gemeinsame Finanzierung der Maßnahmen durch die öffentliche Hand und die gesetzlichen Krankenkassen einher. Die Umsetzung der zahnärztlichen Gruppenprophylaxe ist in den Ländern und Kommunen unterschiedlich geregelt und oftmals nur teilweise realisiert.

Die Daten, die bei der Untersuchung der Kinder und Jugendlichen erhoben werden, sollten idealerweise für die Beratung politischer Gremien (Kreisjugendhilfeausschuss, Jugend- und Sozialausschüsse der Kommunen) und als Grundlage weiterer Gesundheitsplanung genutzt, aber auch betreuenden Ärzten, Schulen, Eltern bzw. Elternvertretungen bekannt gegeben werden. Durch Zusammenführung auf Landesebene ist ein überregionaler Vergleich möglich. Bundesweit sind nur wenige einheitlich erhobene Daten verfügbar, so zum Impfschutz oder zur Teilnahme an Früherkennungsuntersuchungen. Es sind jedoch Bestrebungen in Gang, einheitliche Untersuchungsinstrumente und Dokumentationssysteme in den Ländern zu etablieren.

9.2.2 Schulgesundheitspflege in Deutschland und im benachbarten Ausland

In Deutschland begründet sich die Schuleingangsuntersuchung als wesentliche Schwerpunktaufgabe der Schulgesundheitspflege nach wie vor zunächst mit der Haftungsverpflichtung des Staates für mögliche gesundheitliche Schädigungen durch den gesetzlich vorgeschriebenen Schulbesuch (Gardemann 2001). Die Aufgaben der Schulgesundheitspflege sind später ausführlicher dargestellt. Ähnliche Strukturen der Schulgesundheitspflege wie hierzulande finden sich auch in unseren Nachbarländern.

In der öffentlichen Schulgesundheitspflege in den Nachbarländern sind überwiegend multiprofessionelle Teams tätig, in denen auch Pflegekräfte wesentliche Aufgaben übernehmen und teilweise mit niedergelassenen Ärztinnen und Ärzten zusammenarbeiten.

Die Schulgesundheitspflege in **Österreich** ist im österreichischen Schulgesetz und den Gemeindesanitätsdienstgesetzen geregelt; danach sind in allen Schulen Schulärzte anzustellen, die für re-

gelmäßige Untersuchungs- und Beratungstätigkeiten zur Verfügung stehen sollen. Die bisherige geteilte Zuständigkeit der Kommunen für die sogenannten Eignungsuntersuchungen und der Länder für Maßnahmen der Prävention und Gesundheitsförderung wird derzeit in einzelnen Bundesländern durch ein neues Schularztsystem ersetzt: Ein beim Land angestellter Arzt übernimmt sämtliche Aufgaben in der Schule. Dazu gehören jährliche Untersuchungen aller Schüler einschließlich der Dokumentation der Befunde, die fortlaufende Überwachung des Gesundheitszustands sowie das Angebot von Sprechstunden für Lehrer, Eltern und Schüler, weiterhin die Mitwirkung an gesundheitsbezogenen Projekten und Konferenzen, an vorbeugenden Maßnahmen zur Förderung und Erhaltung der Gesundheit der Schüler und die Mitarbeit bei der Bekämpfung von Infektionskrankheiten.
In der **Schweiz** befinden sich Schulwesen und Schulgesundheitsfürsorge in der Zuständigkeit der Kantone. In Kooperation mit niedergelassenen Kinder- und Jugendärzten sind haupt- und nebenamtliche Schulärzte mit unterschiedlichen Aufgabenschwerpunkten für die gesundheitlichen Belange der Kinder und Jugendlichen zuständig.
In den **Niederlanden** sind Jugendärzte und Jugendpflegekräfte mit Aufgabenschwerpunkten im Bereich individueller Prävention und gruppenorientierter Präventivfürsorge tätig; wesentliche Aufgaben werden dabei von Pflegekräften übernommen.
In **Frankreich** sind nicht nur Pflichtimpfungen vor Schulbesuch vorgeschrieben, sondern auch regelmäßige schulärztliche Betreuung. Zwischen fünf und sechs Jahren sowie zum Ende der Grundschulzeit, vor der Entscheidung über die weitere schulische Laufbahn, erfolgen umfassende Untersuchungen durch staatlich bestellte Schulärzte. Diese betreiben zudem gemeinsam mit Kinderkrankenschwestern und Sozialarbeitern Gesundheitsförderung, Prävention und Gesundheitsschutz in der Schule. Der Schularzt arbeitet eng mit den niedergelassenen Kollegen zusammen.
In **Dänemark** werden Funktionsuntersuchungen durch besonders weitergebildete Gesundheitsschwestern (auch in Schulen) und medizinische Vorsorgeuntersuchungen durch ärztliche Mitarbeiter angeboten.

9.3 ÖGD-spezifische Arbeitsweisen

9.3.1 Grundsätzliches zur Public-Health-Orientierung

Der Öffentliche Gesundheitsdienst hat sich auf allen Ebenen sowohl als Dienstleister für Bürger und Kommune, aber auch als ein Organ des staatlichen Wächteramts zu verstehen. Auf der einen Seite sind Public-Health-Ziele und -Visionen zu verfolgen, ohne die es nicht geht, auf der anderen Seite ist die zunehmende Verknappung finanzieller Mittel in den kommunalen Haushalten zu verkraften. Welche Rolle dem ÖGD im Gemeinwesen vor allem im Hinblick auf Prävention und Wertschöpfung für die Kommune zukommen kann, wird später in diesem Kapitel ausgeführt.
Gleichzeitig ist der kommunale ÖGD aber auch eigenständiger Teil der Verwaltung einer Gebietskörperschaft. Als solcher unterliegt er den Grundsätzen des Verwaltungshandelns, die den Erfordernissen Rechtmäßigkeit, Notwendigkeit und Verhältnismäßigkeit der Mittel zu folgen haben. Öffentliche Gesundheitsverwaltung muss aber auch modernes Verwaltungsmanagement leben und sich als „lernende Organisation" begreifen. Vokabeln wie Kunden- und Mitarbeiterzufriedenheit, Dienstleistungsorientierung, Zeitmanagement, Prozessoptimierung, Personalentwicklung, modernes Finanzmanagement, Zielvereinbarungen und Zielerreichungsgrad gehören schon lange nicht mehr zum Fremdwörterschatz, kommen aber in unterschiedlicher Tiefe und Breite zum Einsatz.

Qualitätssicherung bekommt hier eine zunehmende Bedeutung (➤ Tab. 9.1). Es interessiert den kommunalen Arbeitgeber nicht nur als Finanzier des eigenen Personals, welche Leistungen mit welchen Ressourcen erbracht werden. Interkommunale Vergleiche, Kosten-Nutzen- und -Wirkungsanalysen sind heute leichter möglich und zugänglich. Standards der Leistungserbringung müssen gemeinsam entwickelt, konsensual festgelegt und ihre Einhaltung überprüft werden. Entscheidenden Einfluss haben dabei die Kriterien, die zur Definition von Qualität der erbrachten Leistung herangezogen werden, und diese können sich aus den verschiedenen Perspektiven heraus sehr unterschiedlich darstellen.
Daraus ergibt sich die Notwendigkeit, Qualitätsmanagement auf den Ebenen von Konzeption und Standardisierung, Organisation, Durchführung, Erhebung und Dokumentation von Befunden, Auswertung, Nutzbarmachen von Ergebnissen, Selbstevaluation und Zielüberprüfung im Sinne eines Regelkreises umzusetzen.

9.3.2 Arbeitsweisen in der Kinder- und Jugendgesundheitspflege

Kinder- und Jugendgesundheitsdienste als Organe des Öffentlichen Gesundheitswesens befinden sich auf dem Weg von einem medizinischen Fachdienst zu einem multiprofessionellen Dienst mit hohem Abstimmungsbedarf, von einem staatlichen Gesundheitsdienst zu einem kommunalen, gemeinwohlorientierten Gesundheitsdienstleister. Weg von Kontrolle und Überwachung richten sich sowohl gesetzlicheVorgaben, als auch gesellschaftspolitische Bedingungen stärker aus auf die Mitgestaltung gesundheitlicher Rahmenbedingungen und auf die Optimierung der Versorgung.
Diese Paradigmenwechsel bringen unweigerlich Risiken, aber auch Chancen mit sich, die es im Sinne einer optimierten Gesundheitsversorgung zu nutzen gilt.
Aufgrund seiner epidemiologischen und sozialkompensatorischen Funktion bildet der Bereich Kinder- und Jugendgesundheit des Öffentlichen Gesundheitsdienstes einen Kernbereich gemeinwesenorientierter sozialpädiatrischer Arbeit in den Kommunen und ist unverzichtbarer integraler Bestandteil der Gesundheitsvorsorge und -fürsorge für Kinder und Jugendliche
Handlungsmöglichkeiten sind durch den Kontakt mit Einrichtungen und Schlüsselpersonen gegeben. Zu Ersteren zählen einerseits **Kindertageseinrichtungen,** in denen im Gefolge des Rechtsanspruchs auf einen Kindergartenplatz die weitaus größte Zahl aller drei- bis sechsjährigen Kinder und ihre Familien zumindest indirekt erreichbar sind. In den meisten Kinderbetreuungsgesetzen wird den Erzieherinnen eine hohe Mitverantwortung für die Gesundheitsversorgung der Kinder zugeschrieben, ein Bereich, in dem Unterstützung durch KJGDs gewünscht wird. Sie wird vornehmlich durch Multiplikatorenschulung, durch fachliche Beiträge in der Elternarbeit oder durch eigene subsidiäre Tätigkeit geleistet.
Auf bevölkerungsbezogener, kommunaler Ebene ergeben sich weiterhin durch ein kontinuierliches Monitoring (standardisierte Erfassung, Dokumentation und Auswertung schulrelevanter Befunde) Impulse für eine gesundheits-, entwicklungs- und damit chancenfördernde Gestaltung der vorschulischen Bedingungen.
Zusammengefasst können so in interdisziplinärer Zusammenarbeit die Rahmenbedingungen für eine gesunde Entwicklung, für Integration und Inklusion beeinträchtigter Kinder und für einen frühen Abbau gesundheitlicher und sozialer Entwicklungsrisiken optimiert werden.
In den Bundesländern besteht eine große Heterogenität in Bezug auf die Struktur und Ausrichtung der Kinder- und Jugendgesundheitsdienste (KJGD) in den unteren Gesundheitsbehörden.

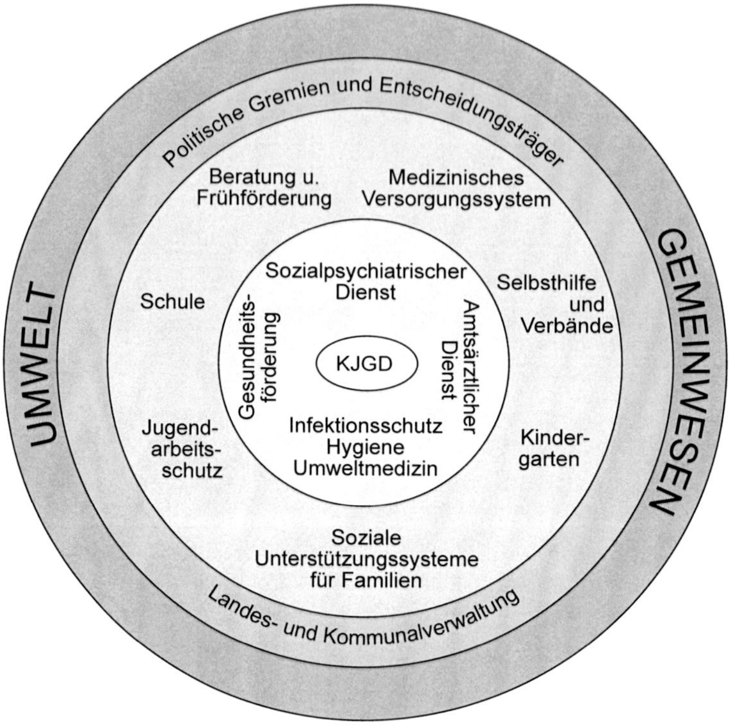

Abb. 9.1 Handlungsfelder und Kooperationsbezüge des Kinder- und Jugendgesundheitsdienstes im ÖGD

Tab. 9.1 Qualitätsdefinition der Schuleingangsuntersuchung aus unterschiedlichen Perspektiven

Institution	Qualitätsinteresse
Schulleiter	Plausibilität und Verwertbarkeit des Gutachtens als Entscheidungshilfe; komplette Vermittlung schulrelevanter Befunde und Förderbedarfe; umsetzbare Hinweise auf den Umgang mit evtl. gesundheitlichen Beeinträchtigungen des Kindes
Eltern	Terminvergabe, Wartezimmergestaltung, Beratungstiefe
Kind	Untersuchungsatmosphäre
Niedergelassener Arzt	Klare Vermittlung von Abklärungsbedarfen; Rollenklarheit
Untersuchungsteam	Rückkopplung der Ergebnisse, Arbeitszufriedenheit; Vermeiden von Monotonie
Gesundheitsberichterstattung	Vollständige und korrekte Dokumentation; Relevanz der Daten für Eltern, Schule, Kommune und Gesundheitsversorgung
Kooperationspartner	Erfülltes Interesse an versorgungsbezogenen oder epidemiologischen Fragestellungen
Kommune	Ökonomie/Ressourcenschonung; Sicherstellung der nachgehenden Fürsorge; Verwertbarkeit der Ergebnisse für gesundheitspolitische Entscheidungen

Insofern sind verbindliche Aussagen darüber, welche Aufgaben in welchem Ausmaß übernommen werden, nicht generalisiert möglich. Letztlich sind sie abhängig vom Bedarf, der finanziellen Situation und der gesundheitspolitischen Prioritätensetzung in der jeweiligen Kommune. ➤ Abb. 1 zeigt Handlungsfelder und Kooperationsbezüge des Kinder- und Jugendgesundheitsdienstes im ÖGD (➤ Abb.1).

Sozialpädiatrische Kompetenzen und Erfahrungen sind jedoch – gleich an welcher Stelle – wesentliche Voraussetzungen und spezifische Merkmale, die Mitarbeiter der KJGD in ihre Arbeit einbringen. Relevant sind darüber hinaus Systemkenntnis und Kenntnis der Hintergründe und Zusammenhänge des sozialen Leistungsrechts.

Zur **Strukturqualität** der KJGD gehört in der Regel eine multidisziplinäre Ausstattung, die sich abbildet in einer Vielzahl von Professionen: Kinderärzte, Kinder- und Jugendpsychiater, Kinderkrankenschwestern/sozialmedizinische Assistentinnen, Sprachberater, Psychologen, Sprachvermittler und andere mehr. Das Kompetenznetz erweitert sich um die Fachbereiche und Professionen, die innerhalb einer unteren Gesundheitsbehörde – allerdings in unterschiedlichem Umfang – vorhanden sind: Sozialpsychiatrische Dienste, Fachleute des Infektions- und Gesundheitsschutzes, Ernährungsberater, Gesundheitsförderer, psychologische und Entwicklungsberatungsstellen etc.

So können z. B. Sozialpsychiatrische Dienste niedrigschwellig Hilfe leisten, wenn sich herausstellt, dass Entwicklungsprobleme eines Kindergartenkindes ursächlich in einer psychischen Krankheit oder Suchtmittelabhängigkeit eines Elternteils begründet sind. Sozialmedizinische Assistentinnen (in der Regel Kinderkrankenschwestern mit einer sozialmedizinischen Zusatzausbildung an der Akademie für öffentliches Gesundheitswesen) haben die Möglichkeit der aufsuchenden Versorgung. Sie können im Rahmen von Hausbesuchen nachsorgend betreuen und so weitgehend sicherstellen, dass Hilfen für Eltern und Kinder dort ansetzen, wo sie gebraucht werden; im Einzelfall kann das lediglich die Organisation eines Kinderarzttermins sein, in anderen Konstellationen eine langfristige Begleitung. Viele in dieser Form und in diesem Umfang Unterstützung benötigende Familien sind dem örtlichen Jugendamt als (Hoch-)Risikofamilien bereits bekannt oder werden von dort vermittelt. Für Eltern kranker und behinderter Kinder oder für aus anderen Gründen stark belastete, überforderte Eltern kann zusammen mit dem örtlichen Sozialamt eine sozialpädagogische Familienhilfe oder ein familienunterstützender Dienst bereitgestellt werden, wenn es die Situation erfordert. Grundsatz bleibt stets, möglichst viel Elternkompetenz zu mobilisieren, aufzubauen und zu erhalten, die Elternverantwortung zu stärken, zu unterstützen oder zu ergänzen. Die Familie steht unter dem Schutz des Grundgesetzes und ist nach der deutschen Verfassung nur dann durch institutionelle Fremdleistung zu ersetzen, wenn dies unvermeidbar ist. Eine solche Situation kann darin bestehen, dass das Kindeswohl gefährdet und diese Gefährdung nicht anders abwendbar ist, hier also der Kindesschutz nach sorgfältiger Abwägung das höhere Rechtsgut darstellt.

Die Kooperation mit den niedergelassenen (Kinder-)Ärzten erfuhr im vergangenen Jahrzehnt eine stetige Verbesserung. Beruhend auf gegenseitiger Wertschätzung und Abgrenzung der Zuständigkeiten entstanden Leitlinien für die Zusammenarbeit, die insbesondere „zur Sicherung einer effektiven und effizienten Zusammenarbeit im Rahmen der ärztlichen präventiven Angebote in der Schule“ dienen sollen. Unberechtigtes Konkurrenzdenken wurde weitestgehend abgebaut, zumal viele Aktivitäten der KJGD gerade darauf abzielen, Kinder und Jugendliche dem Versorgungssystem zuzuführen. Hier sind vor allem Impfmotivationskampagnen und Maßnahmen zur verbesserten Inanspruchnahme der Kinder- und Jugendvorsorgeuntersuchungen zu nennen.

Eine **enge Abstimmung** mit den niedergelassenen Kinder- und Hausärzten der Familien stellt eine weitere wesentliche Ressource der KJGD dar. Auf der Grundlage eines Konsenspapiers, das die

Zuständigkeiten der Vertragskinder- und -jugendärzte und der Ärzte im Öffentlichen Gesundheitsdienst voneinander abgrenzt und sinnvoll aufeinander abstimmt, gelingt die systemübergreifende Zusammenarbeit vor Ort in der Regel gut. Sie kommt den Kindern und Familien im Einzelfall genauso zugute wie der Gesamtbevölkerung, da auch in Qualitätszirkeln und anderen Gremien Public-Health-orientiert zusammengearbeitet wird.

Konsenspapier über die Kooperation und Zuständigkeiten der Kinderärztinnen und Kinderärzte in der Praxis und im Öffentlichen Gesundheitsdienst (DGSPJ 1996, H. Schirm)

In der Präambel beschreibt das Papier die sich ergänzenden Arbeitsfelder und Spezifika von Kinder- und Jugendärzten in Praxis und ÖGD. Es zielt darauf ab, auf dem Boden der Respektierung der jeweiligen Zuständigkeiten und Möglichkeiten eine vertrauensvolle, sich ergänzende Zusammenarbeit in der Gesundheitsversorgung für Kinder und Jugendliche zu befördern und Brüche zwischen den Versorgungssektoren zu minimieren.

Kernsätze:

- „Der Öffentliche Kinder- und Jugendgesundheitsdienst ist integraler und unverzichtbarer Bestandteil der pädiatrischen Versorgung.
- Der Öffentliche Kinder- und Jugendgesundheitsdienst bietet vorrangig in Gemeinschaftseinrichtungen systematische Gesundheitsvorsorge an.
- An Schulen sollen schulärztliche Sprechstunden angeboten werden, aus denen heraus ggf. Schüler in vertragsärztliche Behandlung weitergeleitet werden.
- Um die ärztliche Betreuung von Kindern in Einzelfällen zu sichern, sollte der Kinder- und Jugendgesundheitsdienst notwendige vertragsärztliche Behandlung veranlassen und sich über die Umsetzung empfohlener Maßnahmen vergewissern.
- Der Kinder- und Jugendgesundheitsdienst sollte in die Lage versetzt werden, verstärkt aufsuchende Gesundheitshilfe zu leisten.
- Der Kinder- und Jugendgesundheitsdienst nimmt an der vertragsärztlichen Versorgung grundsätzlich nicht teil. Sofern im Rahmen kompensatorischer Gesundheitssicherungsbemühungen des Öffentlichen Gesundheitsdienstes in Ausnahmefällen subsidiär ärztliche Leistungen aus dem vertragsärztlichen Leistungskatalog erbracht werden, sollen diese von den Krankenkassen finanziert werden. Der Kinder- und Jugendgesundheitsdienst muss verstärkt zur gesundheitlichen Lage von Kindern und Jugendlichen berichten. Die niedergelassenen Kinderärztinnen und -ärzte sollen ihn dabei unterstützen.

Der Berufsverband der Ärzte für Kinderheilkunde und Jugendmedizin Deutschland e. V. und .... die Deutsche Gesellschaft für Sozialpädiatrie und Jugendmedizin e. V. ... werden innerhalb ihrer beruflichen Organisationsstrukturen dafür Sorge tragen, dass auch auf regionaler Ebene eine gegenseitige Kommunikation gesichert ist."

Eine wesentliche Ressource und gleichzeitig Tätigkeitsschwerpunkt ist die **Gesundheitsberichterstattung** als unverzichtbarer, integraler Teil des Aufgabenspektrums, das der öffentliche Gesundheitsdienst (ÖGD) nutzbringend einbringen kann. Einschulungsuntersuchungen werden in fast allen Kommunen des Landes einem gleichen Modell folgend durchgeführt; die Befunde werden weitgehend standardisiert erfasst, dokumentiert und bewertet. Durch überregional zusammengefasste Auswertungen werden valide aussagekräftige interkommunale Vergleiche des Gesundheits- und Entwicklungszustands der Einschulungskinder ermöglicht. Nicht zuletzt auf dieser Basis können prioritäre Handlungsbedarfe in der Kommune – auch kleinräumig – identifiziert werden. So wird es ermöglicht, Interventionen und Projekte in der Kommune bedarfsorientiert zu implementieren; die Allokation von Ressourcen vor Ort kann verantwortlich und zielorientiert erfolgen. Darüber hinaus kann der ÖGD, eine ausreichende Personalausstattung vorausgesetzt, örtliche Interventionen und Maßnahmen evaluieren und durch Nachsteuerung in ihrer Effektivität und Nachhaltigkeit steigern.

9.4 Spezielle Ansätze und Aufgaben

9.4.1 1. Allgemeine Aufgaben in der institutionellen Kinderbetreuung

(Krippen, Horte, Kindergärten, Kindertageseinrichtungen, Kindertagespflege)

Erklärtes Ziel der Bundesregierung ist es, vorschulische Betreuung und Bildung für Kinder von 0 bis 6 Jahren zunehmend in **Gemeinschaftseinrichtungen** zu ermöglichen. Daraus ergeben sich neue Risiken und Chancen im Hinblick auf Gesundheit, ihren Schutz und ihre Förderung.
Das Staatswesen übernimmt indirekt, Jugendhilfe und Träger direkt Verantwortung für Rahmenbedingungen, die den gesundheitlichen und entwicklungsbezogenen Bedürfnissen junger Kinder entsprechen. Einige Vorgaben dazu sind in den Kinderbetreuungsgesetzen der Länder formuliert, in der Regel aber wenig konkretisiert.
Wie kann hier **Gesundheitsförderung** umgesetzt werden? Grundsätzlich lassen sich zwei Möglichkeiten unterscheiden:

- Zum einen lässt sich das Setting Kindertagesstätte als Ort der Umsetzung von Gesundheitsförderung betrachten und leben. Raum und Rahmen zu geben für gesunde Ernährung, für ausreichende Bewegungs- und Alltagserfahrungen, Konditionieren von präventivem Verhalten wie z. B. Zähneputzen, Stärkung der Persönlichkeitsentwicklung im Hinblick auf geringe Suchtgefährdung, Erziehung zu positivem Sozialverhalten im Hinblick auf Ressourcen und Unterstützung durch die Gemeinschaft seien hier exemplarisch genannt.
- Zum anderen ergeben sich Zugänge und Einflussmöglichkeiten auf die Eltern, deren Kinder die Einrichtung besuchen. Eine wesentliche institutionelle pädagogische Aufgabe besteht darin, Eltern für Gesundheits- und Entwicklungsbedarfe ihrer Kinder zu sensibilisieren und sie in der Ausübung der elterlichen Verantwortung zu unterstützen. Wenn es darum geht, Bewusstsein für gesundheitsförderliche Bedingungen und Verhaltensweisen zu entwickeln, muss man aber bereit und in der Lage sein, über das individuelle Interesse und den individuellen Ansatz hinaus zu abstrahieren. An dieser Stelle kann Unterstützung durch externe Fachleute sinnvoll eingebracht werden. Themenbezogene Elternabende, die ausreichend Raum für Fragen und Auseinandersetzung geben, können hilfreich sein: Sie können Problembewusstsein und Handlungswünsche wecken, sie können dazu beitragen, in der Elternschaft „Verbündete" für entsprechende Anliegen der Erzieher zu finden und Gesundheitsförderung in der Einrichtung konkret anzustoßen. Moderation durch externe Fachleute, die auch im Projektmanagement erfahren sind, erleichtert die Implementation von Gesundheitsförderprojekten, ihre Evaluation und (Nach-) Steuerung.

Besondere Bedeutung kommt dem Bahnen und Konditionieren **präventiver Verhaltensweisen** zu. Dazu gehört wesentlich die Inanspruchnahme der Vorsorgeleistungen, die das Versorgungssystem bereithält, insbesondere die von der Ständigen Impfkommission empfohlenen Schutzimp-

fungen sowie die Krankheitsfrüherkennungsuntersuchungen für Kinder; diese werden in den letzten Jahren verstärkt im Kontext der Bemühungen gesehen, Kindswohlgefährdungen rechtzeitig zu erkennen (Thaiss et al, BGBl 2010 Früherkennungs-untersuchungen als Instrument im Kinderschutz. Erste Erfahrungen der Länder bei der Implementation appellativer Verfahren)
So haben in den vergangenen Jahren 13 der 16 Bundesländer Verfahren etabliert, mit denen Eltern zu den anstehenden Früherkennungsuntersuchungen eingeladen, bei Nichtteilnahme erinnert und ggfls. dem Gesundheits- oder Jugendamt der jeweiligen Kommune gemeldet werden. Im direkten Kontakt wird dann über Sinn und Ablauf der Untersuchung informiert, zur Inanspruchnahme motiviert und weitere Hilfsangebote vor Ort bekannt gemacht. Die bundesweiten Ergebnisse zeigen, dass wesentliche Erfolge in der Steigerung der Inanspruchnahmeraten liegen (Hock et al 2013, Gesundheitswesen). Insbesondere werden mit diesen Verfahren bisher schwer zugängliche Bevölkerungsgruppen erreicht und ein positiver Effekt auf die Kindergesundheit erzielt.

Programm der Bundeszentrale für gesundheitliche Aufklärung (BZgA) im Jahre 2005: „Ich geh' zur U – und du?"
Eltern werden in den Tageseinrichtungen gezielt auf anstehende Vorsorgeuntersuchungen U7 und U8 angesprochen und zur Teilnahme motiviert; die Inanspruchnahmeraten werden erfasst und dokumentiert und die teilnehmenden Kinder mit einem T-Shirt belohnt, das den Titel der Kampagne als Aufschrift trägt. Zwischen den offiziell beteiligten Kindertageseinrichtungen gibt es eine Art Wettbewerb im Hinblick darauf, wie viel Prozent der Untersuchungsberechtigten von U7 und U8 Gebrauch machten, eine zusätzlich aktivierende und Anreiz schaffende Komponente. In der Tat waren in den bislang beteiligten Einrichtungen deutliche Teilnahmesteigerungen zu verzeichnen. Als Sekundäreffekt gab es eine positive Konnotation der Vorsorgemaßnahmen: Man sprach darüber, Eltern reagierten durchweg positiv auf die Hinweise der Erzieher, die Einrichtungen konnten sich über eine Profilbildung und ein positives Image in der Öffentlichkeit freuen.

9.4.2 Betriebsmedizinische Aufgaben in Gemeinschaftseinrichtungen

Aufgrund veränderter Entwicklungs- und Lebensbedingungen kommt „betriebsmedizinischen Aufgaben" für Kinder und Jugendliche eine zunehmende Bedeutung zu. Als „Betriebe" in diesem Kontext haben Gemeinschaftseinrichtungen zu gelten, in denen **Kinder und Jugendliche** institutionell betreut und gefördert werden, die also einen Spiel-, Lern- und schulischen Arbeitsplatz darstellen; auch Internate und Heime als außerfamiliärer ergänzender oder ersetzender Lebensmittelpunkt gehören dazu. Adressaten für diesen Aufgabenbereich sind einerseits Kinder und Jugendliche selbst, in der heutigen Zeit mit zunehmender schulischer Ganztags- und Kindergartenbetreuung von sehr jungen Kindern sogar die Altersspanne von 6 Monaten bis 18 Jahre umfassend. Zum anderen werden diejenigen Personen erreicht, die als **professionelle Betreuer und Lehrer** in diesen Institutionen tätig sind. Orte der Betreuung, Förderung und (schulischen) Lehrens sind auch und gerade im Kontext der betriebsmedizinischen Aufgaben als Setting zu verstehen, in dem individuelle und allgemeine gesundheitsschützende und -fördernde Maßnahmen umzusetzen sind.
Die Verantwortung des Gemeinwesens bzw. Staates für diese Aufgaben ist in vielen Landesgesetzen auf kommunale Gesundheitsdienste übertragen, die zum einen vor Ort tätig sind, zum anderen zumindest über grundsätzliche fachliche Qualifikationen und Kompetenzen verfügen. Ganz

unproblematisch gestaltet sich die Umsetzung nicht, da es systembedingte Überschneidungen mit anderen Zuständigkeiten gibt, die sich durch unterschiedliche Anstellungsträger und Personalverantwortungen für die Beschäftigten ergibt. So erfolgt z. B. die betriebsmedizinische Versorgung von Lehrern in NRW seit Februar 2000 auf gruppenvertraglicher Grundlage durch den Berufsgenossenschaftlichen Arbeitsmedizinischen Dienst (BAD).

In welcher Weise können nun kommunale Kinder- und Jugendgesundheitsdienste ihren Beitrag zum Gesundheitsschutz und zur Gesundheitsförderung in den verschiedenen Settings leisten? Welche settingbezogenen Parameter können, welche sollten unbedingt beeinflusst werden?

Wesentliche Grundvoraussetzung ist die Präsenz vor Ort. Betriebsmedizinische Aufgaben sind keine „Schreibtischtätigkeiten" in einer Amtsstube, sondern erfordern unabdingbar Kenntnis der Verhältnisse in den Einrichtungen. Insofern ist das Herzstück betriebsmedizinischer Tätigkeit ein dort **regelmäßig und anlassbezogen stattfindendes Beratungsangebot** für Kinder, Jugendliche, Erzieher, Lehrer, Fachberater, Träger der Einrichtung und Erziehungsverantwortliche.

Im weitaus umfangreichsten Bereich Schule wird dieses Angebot u. a. in Form von **Schulsprechstunden** realisiert. Ähnliche kinder- und jugendärztliche Sprechstunden haben sich in einigen Kindertageseinrichtungen etabliert, wo sich Bedarf und personelle Möglichkeiten des ÖGD decken. Die Nachfrage ist vor dem Hintergrund positiver Erfahrungen steigend, da es gelingt, Kinder und Familien zu erreichen, die – aus unterschiedlichsten Gründen – einen erschwerten Zugang zum Gesundheitsversorgungssystem haben. Beispielhafte Inhalte einer solchen Sprechstunde in Kindertageseinrichtungen sind Information, Motivation und Empfehlung zur Teilnahme an den Krankheitsfrüherkennungsuntersuchungen, insbesondere zur U7a, U8 und U9. Durch das bereits erwähnte Programm der Bundeszentrale für gesundheitliche Aufklärung (BZgA) „Ich geh' zur U – und du?" konnten im Aktionszeitraum von 2004 bis 2010 über 100 Kindertageseinrichtungen erreicht werden. Bei der Umsetzung des Programms in Nordrhein-Westfalen wurden die Eltern derjenigen Kindergartenkinder, die nicht an den altersentsprechenden Vorsorgeuntersuchungen teilgenommen hatten, in den Einrichtungen gezielt angesprochen und zur Teilnahme motiviert; wenn nötig erfolgte eine Terminvereinbarung beim niedergelassenen Kinder- und Jugendarzt. Für einige Familien wurde die Teilnahme an notwendigen Untersuchungen dadurch erleichtert, dass Geschwisterkinder an diesem Termin in der Einrichtung mitbetreut oder gar die Eltern begleitet wurden. Diese Hilfestellungen führten zu messbaren Steigerungen der Inanspruchnahme von U8 und U9, und das vor allem für ein außerhalb der Einrichtung schwerer zugängliches Kollektiv.

Aus den Erfolgsfaktoren dieses Projekts lassen sich weitere sinnvolle betriebsmedizinische Aufgaben ableiten. Im genannten Beispiel werden diejenigen Erzieherinnen und Erzieher besonders engagiert vorgehen, die über Zweck und Inhalt der Kindervorsorgeuntersuchungen und anderer Präventionsleistungen wie Impfungen ausreichend informiert sind. Das Infektionsschutzgesetz hat die Aufklärung über die Prävention von Infektionskrankheiten und einen altersentsprechenden Impfschutz als gemeinsame Aufgabe von Gesundheitsamt und Gemeinschaftseinrichtung festgeschrieben; hier gilt es, die Unterstützung der Erzieherinnen konsequent zu gewinnen und gemeinsam Eltern für einen besseren Impfschutz ihrer Kinder zu motivieren. Kinder- und Jugendärzte aus dem Öffentlichen Gesundheitsdienst übernehmen in der Gesundheitsförderung und Gesundheitsbildung wesentliche Aufgaben: Sie informieren die Berufsgruppen über entsprechende Inhalte, stehen für Fragen und Austausch zur Verfügung und stärken sie damit im Umgang mit gesundheitsbezogenen Fragestellungen. Auf diesem Weg kann eine nachhaltige Kompetenzressource aufgebaut werden, die allerdings einer stetigen Aktualisierung bedarf. Voraussetzung ist auch, dass Träger der Einrichtungen bereit sind, mit zu investieren – im Sinne der Profilbildung und des zunehmenden Wettbewerbs wird dies zunehmend häufig als Chance gesehen.

Das im Folgenden beschriebene Projekt zielt nicht primär auf die quantitative, sondern die qualitative Verbesserung von **Kindervorsorgeuntersuchungen** ab. Damit ist hier der Informationsgewinn des Kinderarztes über psychische und Verhaltensbesonderheiten des Kindes gemeint, der durch Mitberücksichtigung von standardisierten Erzieherbeobachtungen gewährleistet wird.

Erzieherbeobachtungsbogen „Seelisch gesundes Kind"

Bei den zunehmend in den Blickpunkt rückenden kinderärztlichen Früherkennungsuntersuchungen werden die „neuen Morbiditäten" vermutlich untererfasst. Vor diesem Hintergrund wurde im Kreis Recklinghausen ein Kooperationsprojekt entwickelt, das der besseren Früherkennung von psychischen Verhaltens- und Entwicklungsbeeinträchtigungen dient. Kernstück ist ein auf einer Vorlage von Esser basierender Beobachtungsbogen, den die Erzieher mit Einverständnis der Eltern in den Tageseinrichtungen ausfüllen, den Eltern in einem erläuternden Gespräch übergeben und sie motivieren, diesen dem Kinderarzt bei der anstehenden U8 oder U9 zur Verfügung zu stellen. Auf diese Weise kann der Kinderarzt Beobachtungen und Erfahrungen in seine Beurteilung des Kindes einfließen lassen, die ihm sonst nicht oder nicht vollständig zugänglich wären. Die Auswertung zeigt, dass auf diese Weise auch internalisierende, die psychische Entwicklung eines Kindes gefährdende Verhaltensweisen hinreichend früh thematisiert werden, die ansonsten im Gegensatz zu externalisierenden Formen wegen ihres geringeren sozialen Störpotenzials keine Beachtung erführen. Die Schlüsselposition der Erzieher bei früher Verhaltensbeobachtung, die Verknüpfung mit einer verankerten SGBV-Leistung lassen sich so wirksam nutzen und fördern Austausch und partnerschaftliche Zusammenarbeit zwischen den Akteuren in der medizinischen und pädagogischen Betreuung, in der privaten und öffentlichen Gesundheitsvorsorge.

Die Erfahrungen zeigen aber auch immer wieder, dass es Erziehungsberechtigte gibt, die sich ihrer Verantwortung für das Wohl des Kindes durchaus bewusst sind, jedoch nicht über entsprechende Planungs- und Handlungskonzepte verfügen. In der täglichen Begegnung in den Tageseinrichtungen entwickeln Erzieher ein Gespür dafür, welche Eltern in der konkreten Umsetzung von Gesundheitsfürsorge und -vorsorge überfordert sind und entsprechender Hilfen bedürfen; diese sind aber nicht immer von Seiten der Einrichtung leistbar und auch nicht ihr originäres Geschäft. Die Zusammenarbeit mit den örtlichen Gesundheitsämtern ermöglicht den gezielten, bedarfsgerechten Einsatz von **sozialmedizinischen Assistentinnen** – soweit (noch) vorhanden. Im Sinne von Public Health Nurses können diese die Eltern unterstützen und in der Wahrnehmung der Verantwortung für ein gesundes Aufwachsen und Entwickeln ihrer Kinder begleiten. Ganz konkret kann eine solche Unterstützung darin bestehen, dass man einen Vorsorgetermin beim Kinderarzt für die betreffende Familie vereinbart und/oder die Organisation der Betreuung von kleineren Geschwisterkindern organisiert, in Ausnahmefällen selbst übernimmt. Sie kann auch in der Vermittlung und Organisation eines Ersttermins bei einer Frühförderstelle bestehen.

Diejenigen Familien, in denen basale Grundfertigkeiten fehlen, den Alltag zu gestalten bzw. Tagesstrukturen aufzubauen und zu erhalten, können **sozialpädagogische Familienhilfen** eingesetzt werden. Diese sind in der Regel beim örtlichen Sozial- oder Jugendamt angesiedelt und konkret in den Haushalten tätig. Mittel- und langfristiges Ziel ihrer Arbeit ist Hilfe zur Selbsthilfe, das Mobilisieren von Ressourcen, das Empowerment der betroffenen Familien. Denn letztlich geht es darum, die Eltern zu befähigen, ohne externe Unterstützung ihren Aufgaben nachzukommen.

Gute Chancen, das Gesundheitsmanagement für die Kinder aus diesen Familien zu verbessern, bietet eine enge Kooperation zwischen sozialpädagogischer Familienhilfe (SPFH) und sozialmedizinischer Assistentin (SMA). Die Arbeitsschwerpunkte der SPFH liegen nicht primär und nicht direkt auf Gesundheitsbedürfnissen von Kindern, die wiederum im Zentrum der Bemühungen der SMA stehen. Insofern können sich beide Professionen sinnvoll ergänzen. Besonders zu beachten in der (Zusammen-)Arbeit ist die Gefahr, Familien mit vielen verschiedenen Ansprechpartnern eher zu irritieren oder zu überfordern; besondere Sensibilität und klare Rollendarstellung und -einhaltung sind unabdingbar.

Es ist jedoch nicht zu verhehlen, dass das Ziel des Empowerments nicht immer realistisch ist und erreicht werden kann, eine Erkenntnis, die sich oft erst nach längerem Einsatz einer sozialpädagogischen Familienhilfe ergibt. Das Instrumentarium der Jugendämter umfasst einen breiten Fächer von Maßnahmen der „Hilfen zur Erziehung". Wenn die beschriebenen Möglichkeiten ausgeschöpft sind, kommen letztlich nicht mehr ergänzende und unterstützende, sondern **familienersetzende Maßnahmen** zum Tragen.

Kinder- und Jugendgesundheitsdienste nehmen ihren betriebsmedizinischen Auftrag für Gemeinschaftseinrichtungen aktiv wahr; sie erfüllen ihn auch darin, dass sie aktiv mit Erziehern, Kindern und Eltern eine gesundheitsförderliche Umgebung gestalten, die dem hohen gesellschaftlichen Stellenwert von Gesundheit, vorschulischer Entwicklungsförderung, Erziehung und Bildung gerecht wird. Sie unterstützen und ergänzen die Elternarbeit zu gesundheitsbezogenen Themen (Elternnachmittage, Aktionstage, Projekte etc.).

Je jünger die Kinder sind, desto vorrangiger ist unbeeinträchtigte Gesundheit als Voraussetzung für gute Entwicklung und gelingendes Lernen. Insofern erweist sich der institutionelle Zugang zu den Tageseinrichtungen für Kinder als besonders vorteilhaft, da zum einen eine hohe Zahl Kinder erreicht und echte Synergieeffekte erzielt werden können. Insbesondere wurden Präventionsstrategien mit spezifischen und niederschwelligen Zugangswegen in das Blickfeld gerückt und als eine Gemeinschafts- und Querschnittsaufgabe des selbst verwalteten Sozial- und Gesundheitswesens sowie der Jugendhilfe angesehen.

Die in diesem Kapitel beispielhaft vorgestellten Projekte verfolgen nicht nur Ziele der allgemeinen Gesundheitsförderung und der Verhältnis- und Verhaltensprävention. Mit unterschiedlichen Schwerpunkten zielen sie darauf ab, soziale Benachteiligung als Risikofaktor für ungünstige Entwicklung und Gesundheit soweit wie möglich auszugleichen, Entwicklungsgefährdungen frühzeitig zu erkennen, um ihnen entgegenwirken zu können, und dabei auch spezifische Fördermaßnahmen im Setting zum Einsatz zu bringen.

9.4.3 Gesundheitsbezogene Aus- und Weiterbildung für Multiplikatoren

Nachhaltige Wirkungen können dadurch erzielt werden, dass die **Gesundheitskompetenzen** von Schlüsselpersonen in der gesundheitlichen, pädagogischen und sozialen Versorgung ergänzt, gestärkt und aktualisiert werden. Hier können Kinder- und Jugendgesundheitsdienste wertvolle Arbeit leisten. So sind z. B. Curricula besonders gefragt, durch die Erzieherinnen und Erzieher in ihren Fähigkeiten geschult werden, Gesundheits- und Entwicklungsbeeinträchtigungen frühzeitig zu erkennen und damit umzugehen. Für Jugendämter kann der ÖGD Beiträge zur Ausbildung von Tagesmüttern leisten, um diese auf ihre Aufgaben optimal vorzubereiten, weiter zu qualifizieren und einen Beitrag zu ihrer Zertifizierung leisten, so wie es in einigen Bundesländern angestrebt wird.

Erzieher-Curriculum des Gesundheitsamts Kreis Recklinghausen

Seit 2002 wird mit zweijähriger Unterbrechung ein Fortbildungscurriculum für Erzieherinnen und Erzieher in Kindertageseinrichtungen angeboten. Die Veranstaltungsreihe ist modular aufgebaut und trägt den Titel: „Früherkennung von Verhaltensauffälligkeiten und Entwicklungsstörungen bei Kindern im Kindergartenalter" und soll die Adressaten im Erkennen, Einschätzen und dem Umgang stärken. Die Inhalte sind an einem Runden Tisch mit Fachberatern, Trägervertretern und Multiplikatoren zusammengestellt worden und werden veränderten Bedarfen angepasst. Die Organisation und Evaluation übernimmt das Gesundheitsamt, das auch die Honorarkosten für externe Referenten trägt, die sich aus Kollegen der örtlichen Kinderklinik, aus niedergelassenen Logopäden und Motopäden sowie Psychologen aus dem Beratungsdienst zusammensetzen. Teile des Curriculums beziehen sich auf die Gesprächsführung mit Eltern und haben es vielfach erleichtert, auch kritische Punkte im Verhalten der Kinder anzusprechen und gemeinsam nach Lösungen zu suchen. So gelingt es, praxisnahe Inhalte zu vermitteln und verstärkt umsetzungsorientierte Hilfen und Impulse zu geben. Die Resonanz ist unverändert gut. Als „Nebenwirkung" verbessert sich der Austausch zwischen den Akteuren, die Kindergartenkindern pädagogisch fördernd bzw. medizinisch-therapeutisch begegnen.

Für Pädagogen im schulischen Bereich ist es ähnlich hilfreich, wenn im Rahmen ihrer Fortbildungsangebote vor Ort (wie z. B. in der schulinternen Lehrerfortbildung) Kinder- und Jugendärzte aus dem Öffentlichen Gesundheitsdienst zur Verfügung stehen, um aktuelle Themen aufzugreifen. So ist die Entwicklung eines Grundverständnisses des ADHS, seiner multimodalen Behandlung und des pädagogischen Beitrags seit langem und immer wieder Inhalt. Der Umgang mit chronisch kranken Kindern im Kontext Krippe, Kindertagesstätte und Schule erfordert zunehmend Kenntnisse über den Umgang mit Medikamenten und Hilfsmitteln oder dem Verhalten in Notfallsituationen bei Diabetes mellitus, Asthma bronchiale, Anfallserkrankungen oder Allergien. Für spezielle Schulungen des Personals stehen den KJGD wertvolle Materialien der BZgA und der Fachgesellschaften zur Verfügung.

9.4.4 Gesundheitskonferenz und Gesundheitsberichterstattung

Kinder- und Jugendgesundheitsdienste initiieren und moderieren vielfach „Runde Tische" zur Kindergesundheit. Dort arbeiten lokale Akteure in der gesundheitlichen, pädagogischen und sozialen Versorgung der Vorschulkinder zusammen. Probleme werden identifiziert, Lösungsstrategien entwickelt und gemeinsam verbindlich umgesetzt. Kommunale **Gesundheitskonferenzen** als etabliertes Netzwerk haben häufig Initiativen zu den unterschiedlichsten Themen der Kindergesundheitsförderung – nicht zuletzt auch im Setting Kindertageseinrichtung – initiiert und/oder koordiniert. Wo erforderlich und möglich, werden die angestoßenen Maßnahmen aktiv ärztlich begleitet.

In einigen Kommunen existieren verbindliche gesundheitsförderliche Gesamtkonzepte, die von den Jugendämtern in Kooperation mit den Gesundheitsämtern – teilweise auch in umfassenderen Netzwerken – schon seit Jahren im Alltag von Tageseinrichtungen für Kinder etabliert sind. Das „Empowerment" engagierter Erzieherinnen und auch von Eltern soll hier betont werden. Dabei zielen sämtliche Maßnahmen gleichermaßen auf die Schaffung gesundheitsförderlicher Faktoren, die Stärkung gesund erhaltender Faktoren, eine umfassende Vernetzung sowie das Ziel, Kinder und Eltern zu eigenverantwortlichem Handeln zu befähigen. Dort wo Eltern die Verantwortung für ihre Kinder nicht oder nicht in ausreichendem Maße wahrnehmen können, erhalten sie die notwendigen Hilfen.

Gesundheitsberichterstattung ist ein unverzichtbarer, integraler Teil des Tätigkeitsspektrums des Öffentlichen Gesundheitsdienstes (ÖGD). So werden z. B. Einschulungsuntersuchungen in vielen Bundesländern einem einheitlichen Modell folgend durchgeführt; die Befunde werden weitgehend standardisiert erfasst, dokumentiert und z. T. wissenschaftlich bewertet. Durch überregional zusammengefasste Auswertungen werden aussagekräftige Vergleiche des Gesundheits- und Entwicklungszustands der Kinder ermöglicht. Nicht zuletzt auf dieser Basis können prioritäre Handlungsbedarfe identifiziert werden. So wird die Implementation bedarfsorientierter Interventionen und Projekte in der Kommune gefördert; die Allokation von Ressourcen vor Ort kann verantwortlich und zielorientiert erfolgen. Darüber hinaus kann der ÖGD örtliche Interventionen und Maßnahmen evaluieren und durch Nachsteuerung in ihrer Effektivität steigern. Daten der kommunalen Gesundheitsdienste dienen in der Gesamtschau als Grundlage gesundheitspolitischen Handelns. So werden z. B. in Schleswig-Holstein seit 1989 jahrgangsübergreifend etwa 23.000 Kinder untersucht und die Befunde standardisiert erfasst. In einem jährlichen landesweiten Bericht werden die Daten seit 1999 wissenschaftlich begleitet und für die Landesregierung kommentiert (Berichte der Schuleingangsuntersuchungen Schleswig-Holstein http://schleswig-holstein.de/MSGFG/DE/Gesundheit/SchulUntersuchung/Schuleingangsuntersuchung_node.html)

Zusammenfassend lässt sich sagen, dass der öffentliche Gesundheitsdienst über eine breite Palette von Kompetenzen, Handlungsmöglichkeiten, Zugangswegen, epidemiologischen Daten, Planungsgrundlagen, Qualitätsmerkmalen und Ideen verfügt, derer sich eine Kommune nutzbringend bedienen kann.

9.4.5 Mandatsfunktion für Belange der Kindergesundheit

In fast allen Kommunen wird die Frage, welche Rahmenbedingungen erforderlich sind, damit Kinder sich als Individuum und als Teil einer Gemeinschaft optimal entwickeln können, intensiv diskutiert. Ohne Zweifel können nur gut vernetzte multiprofessionelle Ansätze die gesundheitlich-somatischen, sozial-interaktiven, psycho-emotionalen und pädagogisch-kognitiven Optionen des jeweiligen Kindes sichern. Diese kommunalen Hilfen müssen zielgerichtet bedürftige Kinder erreichen – und das sind Kinder, in deren Ursprungsfamilien z. B. Bildungsferne, Langzeitarbeitslosigkeit, chronische Krankheit oder Behinderung zu besonderen Entwicklungsrisiken geführt haben. Ob und welche Versorgungsdefizite in den Fokus genommen werden, ist abhängig von der politischen Willensbildung vor Ort. Sozial- und Jugendhilfeausschüsse, kommunale Arbeitsgruppen zur psychosozialen Versorgung nutzen häufig Anregungen und Vorschläge aus dem Öffentlichen Gesundheitsdienst.

Die Kenntnis von und die Zusammenarbeit mit den lokalen Akteuren in der gesundheitlichen, pädagogischen und sozialen Versorgung von Kindern und Familien stellt eine wesentliche Ressource der KJGD dar. Multiplikatoren und Angebote vor Ort sind in der Regel bekannt, Kommunikations- und Kooperationsstrukturen sind aufgebaut und werden genutzt. Der Gemeinwesenbezug wird in der Arbeit der KJGD zum Nutzen der Kinder und Familien in der Kommune gelebt und gestaltet. Hier seien insbesondere die vielen Projekte zur **stadtteilbezogenen Gesundheitsförderung** genannt, bei denen der Öffentliche Gesundheitsdienst häufig Initiator, oft Koordinator, zumeist aktiver Partner war und ist.

In welcher Weise sich die Mandatsfunktion leben und für die Kindergesundheit in der Kommune wertschöpfend nutzen lässt, wird später weiter konkretisiert. Eine sachverständige und auf Erfahrungen vor Ort fußende Fachberatung politikberatender und -gestaltender Gremien stellt eine wesentliche Public-Health-Aufgabe dar.
In den kommunalen Spitzenverbänden auf Landes- und Bundesebene, Städtetag, Landkreistag, Städte- und Gemeindebund werden Anliegen gebündelt und an die Regierungsorgane weitergegeben. Die Verbände ihrerseits informieren die Kommunen über Zielsetzungen und Gesetzesvorhaben der Landes- bzw. Bundesregierung und geben ihnen Gelegenheit zur Positionierung.

9.5 Gemeinwesenorientierte Aufgaben des Kinder- und Jugendgesundheitsdienstes

9.5.1 Grundsätzliches zur Public-Health-Orientierung

Ein Spezifikum des Öffentlichen Gesundheitsdienstes sind seine Handlungsmöglichkeiten außerhalb der Individualmedizin, also Ansätze, die auf öffentliche Gesundheit ausgerichtet sind. Gesundheit im Public-Health-Sinn gilt als öffentliches, wichtiges und nicht zuletzt teures Gut, sodass Maßnahmen zur Förderung letztlich eine Wertschöpfung für das Gemeinwesen darstellen.
Der ÖGD bringt sich auf örtlicher Ebene ein in Fragen zur Gesundheits- und Umweltverträglichkeit und wird bei Projekten mit Öffentlichkeitsbezug regelhaft an der **Bauleitplanung** beteiligt. Dadurch erhält er die Möglichkeit, regionale Vorhaben und Maßnahmen auch im Sinne der Kinder- und Jugendgesundheit günstig zu beeinflussen. Dabei wird es nur in geringem Maße um grundsätzliche Ablehnung oder Zustimmung gehen, sondern vielmehr um eine von fachlicher Kompetenz und kommunaler Mitverantwortung getragene Beratung und Unterstützung der primär geforderten Entscheidungsträger; diese sind nicht nur in der eigentlichen Gebietskörperschaft angesiedelt, sondern auch auf der überörtlichen Ebene von Genehmigungs- und Aufsichtsbehörden.
Im Kontext der Gesundheitsförderung stehen also **Präventions- und Schutzmaßnahmen** im Mittelpunkt, die darauf abzielen, Gesundheit als öffentliches Gut zu schützen, gefährdenden Faktoren entgegenzuwirken und gesundheitsförderliches Verhalten in allen Bezügen zu stärken. Dem ÖGD kommt dabei zugute, dass er per se durch seine Einbettung in die Kommune Kontakte zu Einrichtungen, Akteuren und Schlüsselpersonen des gesundheitlichen Versorgungssystems besitzt. Seine Ansätze und sein Tätigwerden sind nicht vorrangig oder ausschließlich auf eigene Leistungserbringung, sondern darüber hinausgehend auf **Koordination, arbeitsteilige Kooperation und Vernetzung** angelegt. Der im März 2013 vom Bundeskabinett beschlossene Entwurf für ein Gesundheitsförderungs- und Präventionsgesetz zielt weniger auf diese konkreten kommunalen Handlungsmöglichkeiten ab, als auf die Rolle von den Krankenkassen und der Bundeszentrale für gesundheitliche Aufklärung.
Erstes und elementares Bezugsfeld für das Kind ist die Familie. Eine stabile primäre Bindung, ein Grundverständnis für kindliche Entwicklung und die Bereitschaft und Fähigkeit zur Übernahme von Verantwortung sind wegbereitend für ein gesundes Aufwachsen.
Präventiv orientiertes Verhalten ist jedoch nach wie vor stark abhängig vom sozioökonomischen Status. Besonders eindrucksvoll – und vordergründig – wirkt er sich auf die Inanspruchnahme von Krankheitsfrüherkennungsuntersuchungen bei Kindern aus. Würden sich die Teilnahmequo-

ten für U8 und U9 noch weiter angleichen, so wären zusätzliche individuelle und Public-Health-Effekte im Hinblick auf Chancengleichheit für Kindergesundheit zu erwarten.
Das Schaffen und Pflegen gesundheitsförderlicher Rahmenbedingungen in der Familie erfordert Information, Motivation und Handlungsfähigkeit. Auch hier wirken soziale und ökonomische Faktoren oft limitierend; letztlich ist z. B. eine gesunde Ernährung auf Dauer zeitaufwändiger und meist etwas kostenintensiver. Kochkurse für Eltern, die preis- und gesundheitsbewusstes Kochen vermitteln, haben zunehmend Konjunktur, erreichen allerdings auch wiederum nur diejenigen, die über Problembewusstsein, Handlungsfähigkeit und Selbstwirksamkeitsgefühl verfügen und hinreichend selbstbewusst sind, sich als Familie mit knappen Mitteln erkennen zu geben.

9.5.2 Spezielle Aufgaben der Kinder- und Jugendgesundheitsdienste

Curriculare Gesundheitserziehung in Schulen und örtlichen Bildungseinrichtungen

Jede Mutter, jeder Vater hat eigene Erfahrungen mit dem System Schule und ist für einen langen Zeitraum darin erreichbar gewesen. Schulische Curricula zur Gesundheitsförderung sollten nicht nur auf die eigene Gesundheit des Schülers abzielen, sondern ihn auch auf seine spätere Aufgabe als Vater/Mutter/Sorgeberechtigte(r) bzw. -verantwortliche vorbereiten. Es sollte sich die gute Praxis weiter durchsetzen, dass elementare Kenntnisse über die kindliche Entwicklung ebenso schulisch vermittelt werden wie Inhalte der Gesundheitsfürsorge, der Säuglingspflege und -ernährung und fördernden Erziehung. Hier kann der KJGD sowohl durch konkrete Angebote unterstützen als auch langfristig im Rahmen seiner Möglichkeiten und Zugänge auf eine (festere) Verankerung dieser Themen hinwirken. Leider finden diese Inhalte nicht in allen Schultypen gleichermaßen Berücksichtigung, nicht zuletzt, weil ihre Beherrschung fälschlicherweise einfach vorausgesetzt wird und Jugendliche ungern zugeben, in etwas „Alltäglichem“ unsicher zu sein. Hingegen gibt es, auch befördert durch das Vorherrschen von generationensepariertem Wohnen und von Einkindfamilien, einen besonderen Bedarf an Transfer früher stärker tradierter Kenntnisse und Erfahrungen.
Auch in außerschulischen Bildungseinrichtungen wie Familienbildungsstätten, Volkshochschulen, Elternschulen in unterschiedlicher Trägerschaft kann der ÖGD vor Ort gesundheitsbezogenes Wissen vermitteln und gesundheitsförderliche Verhaltensweisen stärken. Viele Programme der Elternbildung und -schulung sind in den letzten Jahren mit großem Nachdruck entwickelt worden. Allerdings zielen die wenigsten auf das Thema genuiner Gesundheitsförderung ab, sondern mehr auf Stärkung der Erziehungskompetenz, auf Sprachförderung, auf das Schaffen guter Bedingungen für Lernerfolge vor und während der Schulzeit etc. Es ist jedoch nicht leicht, Inhalte und Maßnahmen der Gesundheitsförderung allgemein zu vermitteln bzw. so zu operationalisieren, dass sie in einem institutionellen, abstrakten Rahmen „gelehrt“ werden können. Das Interesse der Eltern richtet sich zudem primär darauf aus, wovon ihr Kind möglichst unmittelbar profitiert. Zum Nutzerkreis freiwilliger Familienbildungsangebote dürften auch eher Menschen mit hoher Eigenkompetenz und -motivation gehören. Gleichwohl mag es ungenutzte Potenziale geben, wo sich Angebot und mögliche Vermittler in räumlicher Nähe befinden und wo durch gute Kommunikation das Themenspektrum auf Inhalte der Gesundheitsförderung abgestellt werden kann.

Beratung werdender Eltern

Einige Ländergesundheitsgesetze verpflichten den Öffentlichen Gesundheitsdienst zur Beratung Schwangerer. In der Praxis gestaltet es sich nicht immer einfach, Zugang zu diesem Personenkreis zu bekommen. Kooperationsverbünde mit niedergelassenen Gynäkologen, Hebammen und Geburtskliniken erleichtern den Zugang und die effektive, sich ergänzende Beratung sehr, stellen aber immer noch eher die Ausnahme dar. Ärzte und Kliniken begreifen die Zusammenarbeit mit einem „Amt" nicht immer als Chance, und auch Eltern begegnen Mitarbeitern oft skeptisch und vorurteilsbeladen. Einen Einstieg kann die curriculare Beteiligung an Geburtsvorbereitungskursen der Kliniken oder der Familienbildungsstätten, Volkshochschulen etc. darstellen, wenngleich hier vornehmlich gut informierte und a priori präventionsorientierte Bevölkerungskreise teilnehmen.

In der Einzelsituation umfassen die Beratungsinhalte in der Schwangerschaft zum einen „klassische" Themen wie gesundheitsförderliches Verhalten (Vorsorgetermine, Ernährung, Substitution von Spurenelementen, Infektionen und Impfungen, Gymnastik etc.). Die Fragen werdender Mütter beziehen sich aber auch auf aktuell bestehende gesetzliche Regelungen zu Mutterschutz, Familienfördermaßnahmen, zu Profilen von Entbindungskliniken, zur Hebammenbetreuung etc. In Einzelfällen wird vorgeschlagen, Beratung in einer Schwangerschaftskonfliktberatungsstelle zu suchen oder Hilfen in anderen Bereichen bis hin zur Schuldnerberatung in Anspruch zu nehmen. Je nach Selbsthilfekompetenz der betroffenen Eltern können Termine und Maßnahmen konkret vermittelt oder die Inanspruchnahme durch organisatorische Hilfen ermöglicht werden. Falls erforderlich, kann die Familie im Sinne eines Case-Managements weiter unterstützt und begleitet werden. Durch den zunehmenden Ausbau Früher Hilfen in den vergangenen Jahren wurden deutlich mehr unterstützungsbedürftige Familien identifiziert, sodass diese Leistungen – zumindest dort, wo diese Strukturen bereits mit Leben gefüllt waren-, verstärkt eingefordert und in Anspruch genommen wurden.

Frühe Hilfen

Frühe Hilfen sind lokale und regionale Unterstützungssysteme für Eltern und Kinder von der Schwangerschaft bis zum Alter von drei Jahren, in manchen Fällen auch bis zum Schulalter. Sie wollen niedrigschwellig und nicht-stigmatisierend alle Eltern erreichen (primäre universelle Prävention), insbesondere aber jene mit speziellem Hilfebedarf, z. B. in sozialen oder gesundheitlichen Problemlagen (allein stehende, psychisch kranke, abhängige, materiell belastete) oder auch bildungsferne Eltern (sekundäre selektive Prävention). Ihre Angebote sollen in partizipativem Ansatz alltagspraktische Unterstützung leisten, die Ressourcen der Familien stärken und ihre Beziehungs- und Erziehungskompetenzen fördern. So können letztlich elterliche Bindungsfähigkeit gefördert, Risiken für das Kindeswohl minimiert und das gesunde Aufwachsen in der Familie und dem Umfeld des Kindes gefördert werden (s. Definition wissenschaftlicher Beirat NZFH 2009). Wichtig ist die Unterscheidung von Maßnahmen des intervenierenden Kinderschutzes (sog. „§ VIII a-Fälle").

In Netzwerken Früher Hilfen arbeiten alle Akteure verschiedener Hilfesysteme fachlich koordiniert zusammen. Neben Familienhebammen und vergleichbaren Berufen des Gesundheitswesens (z. B. Familiengesundheits- und Kinderkrankenpflegerinnen und -pfleger, Sozialmedizinischen Assistentinnen), Psychologen, (Sozial-)Pädagogen, Sozialarbeitern, niedergelassenen Kinder-, Allgemein- und Frauenärzten sind auch ehrenamtliche Mitarbeiter eingebunden.

Da in den Frühen Hilfen Akteure mehrerer (Sozialleistungs-)Systeme wie Jugendhilfe und Gesundheitswesen, Sozial- und Behindertenhilfe mit unterschiedlichen Systemlogiken und verschiedenen Problemlösungsansätzen eng und abgestimmt agieren (müssen), erfordert dies verbindliche Kooperationsvereinbarungen und Regeln. Erschwerend kommt hinzu, dass die Partner unterschiedlichsten Institutionen, Einrichtungen oder Trägern angehören oder freiberuflich tätig und ihre Leistungen in verschiedenen Sozialgesetzbüchern abgebildet sind. Kinder- und Jugendgesundheitsdiensten kommt hier eine bedeutende Funktion zu, da sie für die Jugendhilfe (unter anderem aus Fallkonferenzen in der der Begutachtung) bereits bekannte kommunale Partner im Rahmen der Frühförderung, der Eingliederungshilfe und des Kinderschutzes sind.

Gesetzlich verankert sind die Frühen Hilfen im Rahmen des Gesetzes zur Stärkung eines aktiven Schutzes von Kindern (BKiSchG), das seit dem 1.1.2012 in Kraft ist. Im Gesetz zur Kooperation und Information im Kinderschutz (KKG) regelt die Bundesinitiative Frühe Hilfen und Familienhebammen über eine Verwaltungsvereinbarung den Einsatz der Bundesmittel, die ab 2016 in einem Fonds den Projekten in den Kommunen dauerhaft zur Verfügung stehen.

Das Nationale Zentrum Frühe Hilfen (NZFH), 2007 als nachgeordnete Behörde in Trägerschaft der BZgA und des Deutschen Jugendinstituts (DJI) gegründet, koordiniert, evaluiert und begleitet die Bundesinitiative wissenschaftlich (www.fruehehilfen.de/wir-ueber-uns/aufgaben/). Es erstellt Kompetenzprofile für Familienhebammen, vergleichbare Gesundheitsberufe und Netzwerkkoordinatoren und unterstützt die Qualifikation der Akteure.

Im kommunalen Kontext sind bundesweit bereits vielfältige Modellprojekte Früher Hilfen etabliert und wirksam (STEEP, WIEGE, Frühstart, Guter Start ins Kinderleben, PFIFF, Keiner fällt durchs Netz, Schutzengel). Bedeutende Wirksamkeitskriterien sind dabei Interdisziplinarität und hohe Professionalität.

Projekte in Familienzentren und im Stadtteil

Wesentliche Themen der Gesundheitsförderung für Kinder lassen sich besonders gut und nachhaltig im Gemeinwesen, im regionalen Bezug angehen. Insbesondere das Hinwirken auf ausreichende Bewegungserfahrungen und ausgewogene Ernährung geschieht oft ausgehend von Kristallisationspunkten wie z. B. Familienzentren als besonders exponierte Kindertageseinrichtungen oder im Kooperationsverbund mit anderen Angeboten für Eltern und Kind. Hier lassen sich Eltern besonders gut ansprechen, lassen sich Sensibilität, Interesse und Handlungskompetenz für die Gestaltung möglichst gesunder Lebensverhältnisse wecken. Auf kommunaler Ebene wird beispielsweise Mitverantwortung übernommen für die Planung von Kinderspielplätzen, Wohnanlagen und Freizeitstätten etc. In Bezug auf Ernährung lassen sich Angebote im Stadtteil nutzen: So kann z. B. ein Bäcker Kinder in seine Backstube einladen, um ihnen zu zeigen, wie Brot gebacken wird, kann sie dabei mithelfen lassen, ihnen verdeutlichen, welche Zutaten benötigt werden – so lässt sich besser „begreifen“ und wertschätzen, welchen Wert gesunde Lebensmittel haben.

Besonders vorteilhaft wirkt es sich aus, wenn die Eltern der in Familienzentren betreuten Kinder selbst entsprechende Angebote machen oder gestalten. Im sozialen Miteinander machen die Kinder authentisch vermittelte Erfahrungen; das Gesundheitsbewusstsein und die Identifikation mit dem Thema werden früh und effektiv befördert.

Aufsuchende Arbeit von Familienhebammen und sozialmedizinischen Assistentinnen

Die gesetzliche Verankerung, aufsuchend tätig werden zu können, erlaubt vielfach den dringend notwendigen direkten Kontakt zu Risikofamilien. Hier geht es also vorrangig um sekundäre Prävention unter Nutzung besonderer Zugangsmöglichkeiten. Im Rahmen des § 8a Kinder- und Jugendhilfegesetz (SGB VIII) ist es nicht selten erforderlich, eine gemeinsame Risikoeinschätzung bezüglich der Kindswohlgefährdung vorzunehmen, die nicht zuletzt wesentliche familiäre und häusliche Kontextfaktoren berücksichtigen muss. Eine gute Kooperation zwischen Jugendhilfe- und Gesundheitssystem kann sich hier besonders bezahlt machen und zu einem sich ergänzenden, gut abgestimmten Vorgehen im

Sinne der Kinder beitragen. Fachliche sozialmedizinische Qualifikation und vor allem ein ausgeprägtes Fingerspitzengefühl sind unabdingbar, um im konkreten und übertragenen Sinne die Tür geöffnet zu bekommen. Aufgaben, die später in diesem Kapitel beschrieben sind, gestalten sich vielfach durch den aufsuchenden Zugang leichter und effektiver, wenn es gelingt, eine Vertrauensbasis herzustellen.
Ein ganz anders gelagertes Beispiel für den Nutzen aufsuchender Arbeit stellen Maßnahmen dar, die im Kontext der Vermeidung des plötzlichen Kindstods zum Tragen kommen: Hebammen oder sozialmedizinische Assistentinnen haben Zugang zu den Wohnungen der Familien und gehen mit den Sorgeberechtigten die Punkte durch, die für eine gesunde Schlafumgebung zu beachten sind. Vor Ort – unter Berücksichtigung der dortigen Gegebenheiten – können verhältnis- und verhaltenspräventive Maßnahmen am besten angesprochen und angestoßen werden (Schlafstelle des Kindes, Ausstattung des Kinderbettchens, Benutzen von Schlafsäcken, Stabilisierung der Rückenlage etc.). Zu verschiedenen Themen der Kindersicherheit wird – in Abhängigkeit von den entwicklungsbedingten Risiken – gleichfalls Beratung im häuslichen Umfeld angeboten. Hier gelingt es dann auch leichter, Väter mit einzubeziehen, ganz konkret z. B. bei der Planung und Anbringung von Herd- oder Treppengittern.

Impfen und Infektionsschutz

Für einen ausreichenden Impfschutz im Public-Health-Sinne trägt der Öffentliche Gesundheitsdienst eine wesentliche Mitverantwortung und verfügt über spezifische Handlungsmöglichkeiten. Inhaltliche Grundlage sind die Empfehlungen der Ständigen Impfkommission (STIKO) in der jeweils aktuellen Version. Da sich die Frequenz der Überarbeitungen und damit Modifikationen und Ergänzungen in den letzten Jahren stark erhöht hat, gilt es zunächst, diese Empfehlungen über das Vertragsarztsystem hinaus bekannt zu machen. Adressaten sind dabei Erzieher, Lehrer, Hebammen, Multiplikatoren, Eltern und Jugendliche selbst, die in einer jeweils auf ihre Belange und Bedürfnisse abgestimmten Weise informiert werden müssen. Jede Änderung der STIKO-Empfehlungen zieht eine Überarbeitung der Impfinformationsbroschüren für Eltern und andere Zielgruppen nach sich. In der heutigen Zeit gelingt es zunehmend, Informationssuchende über entsprechend eingerichtete Webseiten des Gesundheitsamts zu erreichen, mit Verlinkungen auf eine „Kindergesundheitsseite" oder eine Seite „Aktuelles zum Impfen". Nutzerbefragungen zeigen, dass Impfinformationen ein vielfach abgerufenes Angebot darstellen, das einen insbesondere an offizieller, interessenneutraler Information interessierten Kreis erreicht. Nicht selten möchte man sich den vom niedergelassenen Kinderarzt verfolgten Impfplan bestätigen lassen oder letztlich eine Zweitmeinung einholen.
In Zeiten, in denen die Inanspruchnahme von Vorsorgeleistungen als Indikator für elterliche Fürsorge herangezogen wird, bekommt die Information der Betreuungsfachkräfte in Gemeinschaftseinrichtungen und Institutionen zunehmende Bedeutung. Insofern ist die Nachfrage nach Informationsangeboten über Schutzimpfungen deutlich wahrnehmbar gestiegen. Nicht selten werden von den Kinder- und Jugendgesundheitsdiensten z. B. Erziehercurricula zu den Themen Schutzimpfungen, neue Vorsorgeuntersuchungen für Kinder etc. angeboten. Dies geschieht auch vor dem Hintergrund des Infektionsschutzgesetzes (IfSG) § 34 (10), um Erzieher zu befähigen, informiert und motivierend auf die Eltern zuzugehen und so für die gemeinsame Aufgabe nach § 34 (10) IfSG hinreichend gerüstet zu sein.
Manche Berufsgruppen sind gleichwohl schwierig zu erreichen. Nicht immer ist es leicht, Hebammen, die in einer sensiblen Phase das Gesundheitsverhalten in Familien entscheidend prägen können, anzusprechen, um Impfempfehlungen zu vermitteln und plausibel zu machen. Wie niedergelassenen Kollegen ab und an beklagen, stehen Ratschläge der Hebammen zum Impfverhalten nicht immer in Übereinstimmung mit den fachlichen Empfehlungen.
In einigen Bundesländern wird die Fortbildungspflicht für Hebammen speziell formuliert und die Kontrolle darüber in den Aufgabenbereich der Medizinalaufsicht für nichtärztliche Heilberufe bei

den unteren Gesundheitsbehörden gelegt. Die einzelne Hebamme erbringt den Nachweis über definierte Fortbildungsstunden, die im Nachhinein auf ihre Eignung hin zu bewerten sind. Vielerorts hat man sich von Seiten der unteren Gesundheitsbehörden entschlossen, eigene Curricula zur Hebammenfortbildung zu entwickeln und anzubieten. Dies bietet natürlich eine optimale Gelegenheit, neutrale Fachinformationen in diese wichtige Berufsgruppe zu transportieren.

Ähnlich wie bei Vorsorgeuntersuchungen ist das Wissen um die Sinnhaftigkeit das eine, die konkrete Umsetzung wiederum etwas ganz anderes. Insbesondere für impfpräventable Krankheiten, deren individualpräventiver Aspekt im Vergleich zum Gesundheitsschutz einer sensiblen Bevölkerungsgruppe oder zur Erzielung einer ausreichenden Durchimpfungsrate in der Kommune zurücksteht, macht es Sinn, vor Ort subsidiäre Impfangebote bereitzuhalten. Vorteilhaft ist zum einen die Erreichbarkeit derjenigen Familien, für die der Zugang zum Versorgungssystem aus verschiedenen Gründen erschwert ist. Zum anderen bringen sich Erzieher und Lehrer unterstützend ein und ermöglichen einen gebündelten, effektiven Ressourceneinsatz. Berichte und eigene Erfahrungen belegen immer wieder, wie erfolgreich solche Impfaktionen sein können. Der Einsatz vor Ort wirkt vielfach besonders überzeugend und ist mit der traditionellen Rolle des Öffentlichen Gesundheitsdienstes kompatibel. Andrerseits gelingt es effektiv, Impflücken dort zu schließen, wo es mit einfachen Hinweisen und informellen Mitteln nicht gelingt. Die Präsenz des „Impfarztes" vor Ort in diesem Kontext kann auch als betriebsmedizinische Vorsorgemaßnahme betrachtet und genutzt werden.

Bei der Einführung der Hepatitis-B-Impfempfehlung für Jugendliche im Oktober 1995 wurden in der gymnasialen Oberstufe Projekte entwickelt, die als Peer education projects gestaltet wurden und ausgesprochen erfolgreich waren. Der Grundgedanke lautet, dass sich Jugendliche am besten durch ihresgleichen überzeugen und motivieren lassen, und das auch und gerade im Hinblick auf Gesundheitsverhalten. Von der Impfnotwendigkeit primär überzeugte „meinungsbildende und -führende" Schülerinnen und Schüler übernahmen die Aufgabe, stufengleiche oder jüngere Schüler zu informieren und sie zum Zweck der Impfung an den Hausarzt zu verweisen. Als Nebeneffekt wurden gesundheitspädagogische Grundfertigkeiten vermittelt und Präventionsinhalte curricular, zum Teil fächerübergreifend, aufgearbeitet.

Schüler informieren Schüler zur Hepatitis-B-Impfung: Gesundheitsförderprojekt im Kreis Minden-Lübbecke

Bei der Erhebung von Impfdaten in den 9. Klassen zeigte sich eine deutliche Impflücke für Hepatitis B. Eine schriftliche Impfempfehlung stellte keine ausreichende Motivation dar; Elternwille und Lehrerratschläge wurden nicht unbedingt befolgt, sodass der Zugang über Multiplikatoren aus den „eigenen Reihen" gesucht wurde. Diese Peers

galt es zunächst zu gewinnen, zu informieren und auszubilden. Es wurde beispielhaft erprobt, mit welchen Methoden die Jugendlichen ihre Botschaft an die Gleichaltrigen weitergeben konnten. Das entwickelte Verfahren erwies sich als erfolgreich im Hinblick auf einen deutlichen Anstieg der Durchimpfungsrate gegen Hepatitis B.

Ähnliche Aktivitäten erweisen sich erfolgreich beim Einsatz junger pädagogischer Fachkräfte des Vereins „Mehr Zeit für Kinder e. V.", die altersadäquat aufbereitete Informationen in die 6. und 7. Klassenstufen tragen, um die aktuell noch unbefriedigende Inanspruchnahmerate der Jugendgesundheitsuntersuchung J1 zu steigern.

Das **Infektionsschutzgesetz** schreibt in § 34 (11) vor, den Impfschutz der Schulanfänger zu erfassen und die Informationen bundesweit beim Robert-Koch-Institut zusammenzuführen. Dies erfolgt in der Regel im Rahmen der Einschulungsuntersuchungen, bei denen der Impfstatus des

Kindes überprüft wird. Durch diese Maßnahme wird, über den individuellen Hinweis zur Schließung evtl. vorhandener Impflücken hinaus, ein epidemiologischer Nutzen erzielt: Bei einem verpflichtenden Regelanlass werden Daten eines ganzen Einschulungsjahrgangs erfasst. So bildet sich regional als auch bundesweit ein kompletter Querschnitt ab, der wiederum im zeitlichen Verlauf als Längsschnitt untersucht wird, wodurch ein „Impfmonitoring" entsteht. Es können interkommunale und regionale Vergleiche gezogen werden, die entsprechendes Nachsteuern durch Impfpromotion und Impfangebote gezielt ermöglichen.
Als Public Health-Organ hat ein Gesundheitsamt bei Infektionskrankheiten und im **Ausbruchsmanagement** besondere Bedeutung und Pflichten nach dem Infektionsschutzgesetz. So stellen z. B. in Kinderbetreuungseinrichtungen, Schulen und Heimen Krankheiten mit großem Ansteckungspotenzial und Ausbrüche besondere Herausforderungen dar, zu deren Bewältigung die Institutionen Unterstützung benötigen. Der Staat steht hier auch in einer Verpflichtung, denn auf dem Boden der Schulpflicht hat er dafür Sorge zu tragen, dass bei Wahrnehmung einer staatsbürgerlichen Pflicht keine vermeidbaren Gesundheitsschädigungen auftreten.
Für das Management der weitaus größten Gruppe banaler Infektionskrankheiten, vornehmlich solcher der oberen Luftwege, sind die Erzieherinnen in der Regel gut gerüstet. Mit den Eltern bestehen in der Regel Absprachen, unter welchen Bedingungen die Betreuung eines leicht erkrankten oder in der Rekonvaleszenz befindlichen Kindes sinnvoll und möglich ist; in die Kindergartenaufnahmeverträge werden oft verbindliche Regelungen dazu aufgenommen. Dass ein akut erkranktes Kind mit Fieber und reduziertem Allgemeinzustand keinen Kindergarten besuchen sollte und davon auch nicht profitieren würde, sollte eine Selbstverständlichkeit sein; für diese Situation, die bei einem Kind im Kindergartenalter etwa drei- bis fünfmal jährlich auftreten dürfte, gilt es, alternative Betreuungslösungen vorzubereiten und abzusprechen. In den Schulen gelingt es aus unterschiedlichen Gründen nicht immer, Eltern zu überzeugen, erkrankte Kinder zum eigenen Wohl wie auch zum Schutz vor Ansteckung der Mitschüler zu Hause zu lassen.
Für eine Reihe von Infektionskrankheiten schreibt der § 34 (1–5) des IfSG ein **Besuchsverbot** sowie eine spezielle **Mitteilungspflicht der Eltern** an die Gemeinschaftseinrichtung und eine Meldepflicht der Einrichtung an das Gesundheitsamt vor; diese Meldepflicht geht über das Spektrum der Meldepflicht des Arztes nach § 6 (1–2) IfSG hinaus und bezieht sich überwiegend auf Krankheiten bzw. Parasitosen mit hohem Übertragungsrisiko beim engen Zusammensein der Betreuten und dem Personal in den Einrichtungen. Für Ausscheider bestimmter Infektionserreger gibt es analoge Regelungen. Zu den Gemeinschaftseinrichtungen zählen solche, in denen überwiegend Säuglinge, Kinder und Jugendliche betreut werden, insbesondere Kinderkrippen, Kindergärten, Kindertagesstätten, Kinderhorte, Schulen oder sonstige Ausbildungseinrichtungen, Heime und Ferienlager.
Die Tatsache, dass eine Meldepflicht teilweise sogar schon bei Verdacht besteht, begründet sich in der Chance, individuelle und kollektive Schutzmaßnahmen nutzen zu können. Diese können generell präventiver Natur sein und auf die Vermeidung von Ansteckung, von schweren Krankheitsverläufen und auf die Begrenzung der Ausbreitung abzielen.

Beispiele

a. Bei einem dreijährigen Kindergartenkind, dessen älterer Bruder die Grundschule besucht, besteht Masernverdacht. Bis zur Bestätigung oder dem Ausschluss der Diagnose sind zunächst ansteckungsgefährdete Kontaktpersonen zu identifizieren und ihr Impfstatus zu überprüfen. Ungeimpfte können vorübergehend vom Besuch der Gemeinschaftseinrichtung ausgeschlossen werden. Wird der Verdacht ausreichend früh bekannt, kann ggf. eine Riegelungsimpfung in der Einrichtung durchgeführt und die Infektionskette damit wirksam unterbrochen werden.

b. Nach wie vor stellt Läusebefall die häufigste Parasitose im Kindergarten dar. Bei Häufung oder wiederholtem Befall ist eine Meldung erforderlich. Für die Erzieher ist es nicht immer leicht, alle befallenen Kinder zu identifizieren und die Eltern zu einer konsequent und lege artis durchgeführten Behandlung zu veranlassen. Bei Häufung oder wiederholtem Befall kann der ÖGD effektiv unterstützen. Betroffene Kinder müssen so lange der Einrichtung fernbleiben, bis nach ärztlichem Urteil keine Weiterverbreitung mehr zu erwarten ist.

Die in Gemeinschaftseinrichtungen nach den in § 34 (1–3) genannten Krankheiten/Parasitosen kann erfolgen, wenn „nach ärztlichem Urteil eine Weiterverbreitung der Krankheit nicht mehr zu befürchten" ist. Dazu hat das Robert-Koch-Institut Empfehlungen herausgegeben. Sie beschreiben den Rahmen, innerhalb dessen im Einzelfall Abwägungen zu treffen sind. Bei Scharlach beispielsweise kann ein nach Antibiotikagabe rasch fieberfreies, unbeeinträchtigtes Kind den Kindergarten sehr viel schneller wieder besuchen als ein Kind, das nicht in dieser Weise behandelt wurde und bei dem mit einer längeren Ansteckungsfähigkeit gerechnet werden muss.

Hinweise für Ärzte, Leitungen von Gemeinschaftseinrichtungen und Gesundheitsämtern zur Wiederzulassung in Schulen und sonstigen Gemeinschaftseinrichtungen des Robert-Koch-Instituts (RKI)

Die jeweils aktuelle Fassung, zuletzt vom Juli 2006, findet man auf der Homepage des RKI unter „Infektionsschutz – Ratgeber für Ärzte". Auszug aus der aktualisierten Fassung vom Juli 2006, Erstveröffentlichung im Bundesgesundheitsblatt 44 (2001).

In der Präambel wird darauf hingewiesen, dass beim Ausschluss vom Besuch einer Gemeinschaftseinrichtungen oder der Wiederzulassung eine Güterabwägung vorzunehmen ist.

(Als Gemeinschaftseinrichtungen gelten hier „Einrichtungen, in denen überwiegend Säuglinge, Kinder oder Jugendliche betreut werden, insbesondere Kinderkrippen, Kindergärten, Kindertagesstätten, Kinderhorte, Schulen oder sonstige Ausbildungseinrichtungen, Heime, Ferienlager und ähnliche Einrichtungen".)

„Dem Anspruch der Allgemeinheit, vor Ansteckung geschützt zu werden, stehen das Recht des Einzelnen auf Bildung und die Grundsätze der Notwendigkeit und der Verhältnismäßigkeit der Mittel gegenüber. Als Kriterien der Abwägung können gelten

- Schwere, Behandelbarkeit und Prognose der zu verhütenden Krankheit,
- tatsächlich beobachtete Übertragungen unter den Bedingungen der jeweiligen Einrichtung und
- alternative Möglichkeiten des Infektionsschutzes wie hygieneorientiertes Verhalten, Chemoprophylaxe oder Impfungen.

Bevor ein Ausschluss von Personen aus einer Gemeinschaftseinrichtung aus Gründen des Infektionsschutzes veranlasst wird, sollte stets geprüft werden, ob die Belastungen, die beispielsweise in einer Familie durch Ausschluss eines Kindes aus dem Kindergarten entstehen, vermieden werden können und ob das Ziel einer Verhütung von Infektionen nicht auch durch Aufklärung über Infektionswege, hygienische Beratung und gegebenenfalls durch detaillierte Anweisungen des zuständigen Gesundheitsamtes erreicht werden kann. Diesen Ausführungen liegt der Rechtsgedanke des § 34 Abs. 7 Infektionsschutzgesetz zugrunde. Das Bundesverwaltungsgericht hat zum Betretungsverbot (im Rahmen eines Masernausbruchs) 2011 ein Grundsatzurteil gesprochen (BVerwG 3 C 16.11. vom 22.3.2011: Schulbetretungsverbot bei Auftreten von Masern an benachbarter Schule).

Für folgende Infektionskrankheiten werden im Folgenden detaillierte Empfehlungen gegeben:

- Cholera
- Diphtherie

- Enteritis durch enterohämorrhagische E. coli (EHEC)
- Virusbedingte hämorrhagische Fieber (VHF)
- Haemophilusinfluenzae-Typ-b-Meningitis
- Impetigo contagiosa (ansteckende Borkenflechte)
- Keuchhusten
- Ansteckungsfähige Lungentuberkulose
- Masern
- Meningokokken-Infektionen
- Mumps
- Paratyphus
- Pest
- Poliomyelitis
- Scabies (Krätze)
- Scharlach oder sonstige Streptococcus-pyogenes-Infektionen (Streptokokken-Angina)
- Shigellose
- Typhus abdominalis
- Virushepatitis A oder E
- Windpocken
- Kopflausbefall
- Infektiöse Gastroenteritis (Besonderheiten für Kinder im Vorschulalter)
- Bakterielle Enteritiden z. B. durch Salmonellen, Campylobacter, Yersinia enterocolitica
- Virale Gastroenteritiden.

Die **lebensmittelhygienische Überwachung** von Gemeinschaftseinrichtungen erfolgt durch die Lebensmittelüberwachungsbehörden bzw. Veterinärämter. Sie überprüfen das Einhalten grundlegender Regeln zur Lagerung und Verarbeitung von Lebensmitteln, die für die Gemeinschaftsverpflegung herangezogen werden. Beschäftigte müssen nachweisen, dass sie über entsprechende Verhaltensweisen eine „Belehrung" nach § 43 des Infektionsschutzgesetzes erfahren haben. Eine der wichtigsten Grundregeln besagt, dass Beschäftigte im Umgang mit sensiblen Lebensmitteln oder Bedarfsgegenständen nach § 42 IfSG, die in Einrichtungen zur Gemeinschaftsverpflegung arbeiten, bei akuter Durchfallsymptomatik oder entzündlichen Hautveränderungen umgehend ihren Arbeitgeber zu informieren; sie dürfen erst dann wieder an ihrem Arbeitsplatz tätig werden, wenn eine Weiterverbreitung des Erregers nicht mehr zu befürchten ist.

Das Thema **Alltagshygiene** sollte auch in der frühkindlichen Gesundheitserziehung nicht vernachlässigt werden. Eine der wichtigsten und effektivsten Maßnahme zur Verringerung von Infektionsrisiken stellt das Händewaschen dar, zu dem Kinder konsequent angeleitet und aufgefordert werden sollten.

Mitmachaktion: „Händewaschen – na klar!"

Das von örtlichen Sponsoren unterstützte Kooperationsprojekt wurde erfolgreich in den Tageseinrichtungen des Landkreises Ahrweiler durchgeführt. Ziel war es, das Händewaschen als elementare und wichtige Maßnahme zur Infektionsprävention zu vermitteln und als Alltagsroutine zu fördern. Mit Schwarzlichtboxen und „Zauberlösung" wurde den Kindern deutlich gemacht, wie sich gründliches Händewaschen auswirkt.

Kinder- und Jugendzahnärztliche Public-Health-Aufgaben in der Kommune

In der Regel sind Parameter der Zahngesundheit Indikatoren für die allgemeine gesundheitliche und die soziale Situation. Statistisch und epidemiologisch wird die Zahngesundheit vorzugsweise nach einem von der WHO vorgegebenen Index erfasst. Dieser **DMFT-Index** gibt an, wie viele Zähne (T = teeth) einer Person oder Population akut kariös (D = decayed), wegen Karies gefüllt (F = filled) oder extrahiert worden bzw. ausgefallen sind (M = missing). Nach einer bundesweiten Studie der Deutschen Arbeitsgemeinschaft für Jugendzahnpflege (DAJ) wiesen im Jahre 2004 6–7-jährige Kinder im Durchschnitt einen DMFT-Wert von 2,16 auf. Dies beschreibt einen Rückgang der Kariesprävalenz von 25 % im vorangegangenen Jahrzehnt; aber im August 2012 von der DAJ veröffentlichte Ergebnisse beziffern die Inzidenz frühkindlicher Karies immer noch mit 10 bis 15 %. Diese Zahlen stellen Bevölkerungsmittelwerte dar; wie differenzierte Analysen nachweisen, profitieren von den Erfolgen sozial Benachteiligte wesentlich weniger. Kaum ein anderer messbarer Gesundheitsparameter steht in vergleichbar starkem Zusammenhang mit der sozialen Situation; schlechte DMFT-Befunde korrelieren mit sozialer Benachteiligung, die Parameter können nahezu als „Sozialindikator“ herangezogen werden. Zahngesundheit von Kindern ist in besonders hohem Maße von der Präventionsorientierung der Eltern abhängig. Gutes Präventionsverhalten schlägt sich in gesunder Ernährung, Durchführung und später konsequenter Anleitung zur Mundhygiene und in regelmäßigen Zahnarztbesuchen nieder. Diese drei Voraussetzungen auf Dauer zu schaffen, ist für benachteiligte oder überforderte Eltern sehr viel schwieriger als für Familien mit guten Ressourcen. Zahnerhaltung und Zahnersatz verursachen darüber hinaus hohe Gesundheitskosten. Da die Kostenübernahme für Zahnersatz durch die GKV bereits bei Kindern mit der regelmäßigen Inanspruchname von individueller Vorsorge gekoppelt ist, ist die Förderung der Zahn- und Mundgesundheit bei Kindern und Jugendlichen und die Vorbeugung ernährungs- und pflegebedingter Erkrankungen der Zähne und des Zahnhalteapparats von individuellem und gesundheitsökonomischem Interesse. Dies findet sowohl in der Aufgabenwahrnehmung des ffentlichen Gesundheitsdienstes als auch in der Sozialgesetzgebung seinen Niederschlag.

Nach § 21 SGB sollen niedergelassene Zahnärzte, öffentlicher Gesundheitsdienst und gesetzliche Krankenkassen in Kindergärten und Schulen gemeinsam **Maßnahmen zur Förderung der Mund- und Zahngesundheit** bei Kindern von 3 bis 12 Jahren, in sozialen Brennpunkten und Einrichtungen für Behinderte bis 16 Jahren durchführen. Sie umfassen die Untersuchung der Mundhöhle, Erhebung des Zahnstatus, Zahnschmelzhärtung, Ernährungsberatung und Mundhygiene. Für Kinder mit besonders hohem Kariesrisiko sind spezifische Programme zu entwickeln. Die Kostentragung erfolgt gemeinsam durch die Länder und die gesetzlichen Krankenkassen. In Rahmenvereinbarungen der Landesarbeitsgemeinschaften für Jugendzahnpflege (LAJ) sollen die Formen der Zusammenarbeit, Finanzierung, Dokumentation und Kontrolle der Effektivität der Maßnahmen festgelegt werden. Die Umsetzung der gruppenprophylaktischen Maßnahmen erfolgt in den Ländern unterschiedlich.

Alle Kinder, die motorisch schon dazu in der Lage sind, sollten schon in Kindertageseinrichtungen zum richtiges Zähneputzen angeleitet werden und regelmäßiges Zähneputzen nach den Mahlzeiten in den Kindergartenalltag integriert werden, sodass es für die Kinder zu einer guten Gewohnheit und Selbstverständlichkeit wird.

Das Hinführen zu gesundem Ernährungsverhalten ist ein wesentliches Ziel von Gesundheitsförderung im Kindergarten und steht in engem Zusammenhang mit Prävention von Milchzahnkaries. Zahnärzte, Zahnarzthelferinnen und Prophylaxeassistentinnen aus dem Öffentlichen Gesundheitsdienst und/oder niedergelassenen Bereich suchen die Einrichtungen auf, führen modellhaft

Zahnputzaktionen durch, erklären kindgerecht Grundbegriffe und vermitteln den Kindern altersentsprechendes Hintergrundwissen. Nachhaltige positive Effekte werden erzielt, wenn sich die Erzieherinnen der Sache engagiert und kontinuierlich annehmen.
Um den Kindern die eventuell vorhandene Angst vor dem Zahnarztbesuch zu nehmen oder gar nicht erst entstehen zu lassen, wird vielerorts angeboten, sich in den Räumen des „Schulzahnarztes" oder des die Einrichtung betreuenden niedergelassenen Zahnarztes mit dem Behandlungsstuhl, den Instrumenten und Gegebenheiten vertraut zu machen. Gerade Kindergartenkinder nehmen dieses Angebot gern wahr. Der Besuch kann von den Erzieherinnen pädagogisch vor- und nachbereitet werden, er kann als Gruppenerlebnis gestaltet und erfahren werden. Ein Lieblingsstofftier oder eine Puppe kann als „Behandlungsobjekt" des Kindes dienen, das dann die Rolle des Zahnarztes übernimmt.
Das Thema Zahngesundheit lässt sich gut in **Elternabende** einbringen, bei denen der „Kindergartenzahnarzt" mit seiner Kompetenz und Erfahrung zur Verfügung steht und die Eltern berät. Fragen gelten häufig auch der Supplementierung mit Fluoriden, Vorbeugung von Zahnfehlstellungen und durch Beläge verursachte Zahnfleischentzündungen.
Erzieherinnen sind erfahrungsgemäß interessiert an Maßnahmen zur Erhaltung und Förderung der Zahngesundheit, nicht zuletzt, weil sie diese wesentlich beeinflussen können. Ihre Umsetzung erfordert gesundheitspädagogische Grundfertigkeiten und kann effektiv unterstützt werden durch den Zahnarzt aus dem Öffentlichen Gesundheitsdienst; für Multiplikatorenschulungen und fachliche Beratungen der Erzieher sollte er zur Verfügung stehen.
Je nach landesspezifischer Ausgestaltung der **Gruppenprophylaxe** sind auch die regelmäßigen zahnärztlichen Untersuchungen Bestandteil der Gruppenprophylaxe und werden von den Akteuren gemeinsam getragen. Als hilfreich erweisen sich örtliche Arbeitskreise oder Landesarbeitsgemeinschaften zur Jugendzahnpflege, in denen ÖGD, Vertreter der Vertragszahnärzte und Kassen zusammenarbeiten. Nach den Prinzipien der Bedarfsorientierung und der sich ergänzenden arbeitsteiligen Aufgabenerfüllung lassen sich Maßnahmen zur Prävention und Gesundheitsförderung gut organisieren, sowohl als kontinuierliche als auch als modellhafte Projekte. Gruppenprophylaktische Maßnahmen werden dabei teilweise durch Zahnärzte des ÖGD, teils durch den Einrichtungen zugewiesene Obleute aus dem niedergelassenen Bereich durchgeführt. In einzelnen Ländern führt der ÖGD allein die zahnärztlichen Untersuchungen durch.
Es handelt sich um standardisierte und kalibrierte Untersuchungen und Befunderhebungen bei Kindern und Jugendlichen, um den individuellen Behandlungs- und Prophylaxebedarf festzustellen. Sie erfolgen vornehmlich in den Kindertageseinrichtungen, in den Eingangsklassen im Primarstufenbereich und bedarfsabhängig in weiterführenden Schulen. Förderschulen stehen hier besonders im Fokus.
Ziel ist es, die Kinder dem Versorgungssystem zuzuführen, wo es notwendig erscheint. Eltern werden über das Untersuchungsergebnis schriftlich informiert und auf eine eventuelle Behandlungsnotwendigkeit hingewiesen.
Die zahnärztlichen Untersuchungsbefunde liefern eine Datenbasis für Epidemiologie und Gesundheitsberichtserstattung und ermöglichen ein Monitoring der Zahngesundheit auf kommunaler Ebene. Durch die Zusammenführung der Daten auf Landesebene sind regionale Vergleiche möglich.
Gleichzeitig dienen die auf diese Weise gewonnenen Aussagen als Steuerungsinstrument für die Gruppenprophylaxe, denn sie schaffen die Voraussetzungen, um Ressourcen gezielt einzusetzen und Bedarfe regional oder zielgruppenspezifisch zu bündeln.
In mehrjährigen Abständen werden zur Überprüfung der Effektivität der gruppenprophylaktischen Maßnahmen bundesweit epidemiologische Begleituntersuchungen (5 %-Stichproben zur

Untersuchung der Zahngesundheit in verschiedenen Altersstufen) durchgeführt, koordiniert durch die Bundesarbeitsgemeinschaft für Jugendzahnpflege (DAJ). Auch wenn sich durchweg gute Fortschritte in der Verbesserung der Mund- und Zahngesundheit von Kindern und Jugendlichen in Deutschland aufzeigen lassen, gibt es dennoch deutliche Unterschiede zwischen den einzelnen Bundesländern auf, und Deutschland wird im internationalen Vergleich lediglich auf einen Platz im mittleren Drittel verwiesen.

In einigen Kommunen wird die Einschulungsuntersuchung mit einem zahnärztlichen Check und einer individuellen Beratung verbunden. So lässt sich gut vermitteln, dass die Gesamtgesundheit der einzuschulenden Kinder als wichtige Ressource für gesundes und erfolgreiches Lernen angesehen und wertgeschätzt wird. Ein ganzheitliches Gesundheitsverständnis und -bewusstsein vermittelt sich im Zusammenhang mit einem wichtigen Schritt, an einer wichtigen Übergangsstelle im Leben der Kinder.

Darüber hinaus ist es Ziel der Bemühungen, Jugendliche zu motivieren, selbst Verantwortung für ihre Zahngesundheit zu übernehmen. Leider erweist sich der Zugang zu ihnen schwieriger, wenn er nicht über Regelanlässe oder am Thema interessierte Lehrer ermöglicht wird. Denkbar sind zahnärztliche Check-Untersuchungen auch im Hinblick auf relevante Kieferfehlstellungen; Auffälligkeiten im Screening sollten zur Vorstellung beim Kieferorthopäden Anlass geben. Jugendliche haben zu diesen Themen häufig Fragen und Beratungsbedarf.

Kariespräventionsprojekt in einer Förderschule

Die bekannte kariesprotektive Wirkung von Fluor wird in einer Modellschule mit hoher Kariesfrequenz im Rahmen einer eigenverantwortlichen und betreuten Gruppenprophylaxe genutzt, um den Kariesbefall unabhängig von der häuslichen Compliance der Schüler zu senken. Die Elternschaft befürwortet die Maßnahmen, Lehrer unterstützen sie aktiv. Bei den teilnehmenden Schülern ist langfristig eine Reduktion des Kariesrisikos zu erwarten, darüber hinaus ein verstärktes Bewusstsein für präventives Verhalten und eine Erhöhung des Selbstwirksamkeitgefühls.

Denkbar, aber zu selten umgesetzt ist eine curriculare Beteiligung zu Themen der Mundgesundheit in der Sekundarstufe, im Fach- oder fachübergreifenden Unterricht.

Gutachterliche Tätigkeiten stellen einen weiteren Aspekt der täglichen Arbeit der Zahnärzte im Gesundheitsamt dar. Gutachtenauftraggeber sind die eigenen Kommunen als Sozialleistungsträger nach dem Asylbewerberleistungsgesetz und als Dienstherr nach dem Beamtenrecht sowie weitere Behörden der öffentlichen Verwaltung. Kenntnisse des Beihilferechts, der Anspruchsvoraussetzungen im Sozialhilferecht und der gesetzlichen Krankenversicherung sind unabdingbar.

9.6 Individualmedizinische Aufgaben des Kinder- und Jugendgesundheitsdienstes

9.6.1 Untersuchung und Beratung von Kindern und Jugendlichen

Schuleingangsuntersuchungen

Diese traditionelle Kernaufgabe der Kinder- und Jugendgesundheitsdienste setzt an der wichtigen Schnittstelle des Übergangs von vorschulischer zu schulischer Betreuung und Bildung an. Die

Begründungszusammenhänge für diese in vielen Bundesländern als Pflichtaufgabe formulierte Untersuchung liegen zum einen auf der individualmedizinischen Ebene: Sie dient der Erfassung schulrelevanter Beeinträchtigungen, dem Einleiten notwendiger medizinischer, fördernder Maßnahmen, dem Anstoßen evtl. notwendiger sozialer Unterstützungsmaßnahmen, dem Vermitteln von Hilfen und somit der Minderung der Auswirkung sozialer Benachteiligung auf Gesundheits- und Bildungschancen. Entscheidend ist hierbei die Beratung des pädagogischen Personals im Hinblick auf den Umgang mit chronisch kranken und (entwicklungs-)beeinträchtigten Kindern. Darüber hinaus lässt sich die Einschulungsuntersuchung verstehen als arbeitsmedizinische Untersuchung des Kindes an seinem künftigen Arbeitsplatz Schule; nicht zuletzt übernimmt der Staat mit dem Aussprechen der Schulpflicht die Verantwortung dafür, dass dem Schüler in Erfüllung der Schulpflicht kein vermeidbarer gesundheitlicher Schaden entsteht. Auf der Grundlage der Behindertenrechtskonvention der Vereinten Nationen ergibt sich eine besondere Herausforderung: es ist der Anspruch zu verwirklichen, möglichst allen entwicklungsgestörten, behinderten und kranken Kindern die Förderung in einer allgemeinen Schule zu ermöglichen. Dazu ist eine umfassende sozialpädiatrische Expertise vor Schulbeginn erforderlich und, darauf aufbauend, ein sorgfältiges gemeinsames Überlegen, welche individuellen Hilfsmittel, Unterstützungsmaßnahmen und sonstigen Rahmenbedingungen nötig sind und wie sie in der Schule der Wahl angemessen umgesetzt werden können. Für das System Schule stellt die Schuleingangsuntersuchung, deren Ergebnis in einem schulärztlichen Gutachten zur Einschulung niedergelegt wird, eine Entscheidungs- und Planungsgrundlage dar, die sich u. a. in der Klassenaufteilung und Ressourcenverteilung innerhalb der Schule und in den Förderplänen niederschlägt. Vorteilhaft ist der Nebeneffekt, über einen gesamten Altersquerschnitt epidemiologisch aussagekräftige Daten zur Kindergesundheit zu erhalten, die sich für die kommunale Gesundheitsplanung nutzen lassen. Häufig werden mit dem Regelanlass der Einschulungsuntersuchung Befragungen verknüpft; beispielweise lassen sich durch Elternbefragungen bei der Einschulung Trends über aktuelle vorschulische und schulbezogene Betreuungswünsche als Planungsgrundlage für Jugendhilfe ermitteln.

Im Rahmen der Schuleingangsuntersuchung werden Feststellungen zur „Schulfähigkeit" getroffen. Während vor einigen Jahrzehnten das Hauptziel dieser Untersuchung noch darin lag, vor dem Besuch der Gemeinschaftseinrichtung Schule ansteckende Krankheiten wie z. B. Tuberkulose oder Besiedelung mit Kopfläusen auszuschließen, hat sich ein stufenweiser Bedeutungswandel abgezeichnet: Es ging dann darum, festzustellen, ob das Kind die für den Schulbesuch ausreichende „Schulreife" aufwies, d. h. ob und inwieweit das Kind die für den Besuch der ersten Klasse einer Regelgrundschule erforderlichen gesundheitlichen und entwicklungsbezogenen Voraussetzungen mitbrachte. Im weitesten Sinne wurde überprüft, ob ein Kind (schon) in die Schule „passt". Die Rahmenbedingungen und Anforderungen in der Schuleingangsphase waren damit implizit als relativ starr und unflexibel zu verstehen, zielten also wenig auf individuelle Besonderheiten und Bedürfnisse des einzelnen Schülers ab. In unterschiedlichem Ausmaß hat es schulpolitische Paradigmenwechsel in den Bundesländern gegeben, die eine neue Sichtweise der „Schulfähigkeit" und damit auch der Zielsetzung der Schuleingangsuntersuchung zur Folge hatten. Größere Gestaltungsspielräume in der Organisation des Primarunterrichts, schulpolitische Lockerung der Rahmenbedingungen und der pädagogischen Konzepte und Schwerpunkte sowie umfangreichere Möglichkeiten binnendifferenzierenden Vorgehens führen dazu, dass ein Erstklässler nicht mehr mit einem starren, sondern weitgehend mit einem an seinen individuellen Bedürfnissen und Möglichkeiten ansetzenden System konfrontiert ist. Hinzu kommt in vielen Bundesländern eine Flexibilisierung des Einschulungsalters, die eine über die Entwicklungsvarianzen sechsjähriger Kinder hinausgehende zusätzliche Inhomogenität der Schülerschaft einer Eingangsklasse nach sich zieht. Diesen Herausforderungen stellt sich Schule unter der pädagogischen Vorstellung, dass frühes Lernen besonders effektiv in altersgemischten Gruppen gelänge, eine Erfahrung, die durchaus nicht unter allen pädagogischen Rahmenbedingungen gleichermaßen ausgeprägt zu beobachten ist.

Diese auf pädagogischen Konzepten begründeten veränderten schulischen Rahmenbedingungen haben große Auswirkungen auf die Definition der **Schulfähigkeit**. Die Fragestellungen, die sich bei der Einschulung ergeben, dienen dazu, zu erfassen, wo Schule das Kind „abzuholen" hat, d. h. wie sein persönliches Entwicklungsprofil aussieht und wo seine individuellen Schwächen und Stärken liegen. Klassenzusammenstellung und Allokation besonderer Ressourcen wie z. B. spezieller Förderangebote richten sich danach aus. Vom Grundsatz her wird jedoch jedes Kind – dem gesetzlich verbrieften Recht auf schulische Bildung und der Schulpflicht folgend – als „schulfähig" betrachtet, spätestens wenn es zum Stichtag das sechste Lebensjahr vollendet hat. Insofern beschreibt der Begriff Schulfähigkeit keine zu bejahende oder verneinende Option mehr, sondern meint über den allgemeinen Sprachgebrauch hinaus Regelschulfähigkeit als Aufnahmefähigkeit der zuständigen Grundschule, die zudem vielerorts über die bestehenden Schulbezirksgrenzen hinaus frei von den Eltern ausgewählt werden kann. Nur in wenigen Bundesländern stehen noch vereinzelt sogenannte Schulkindergärten zur Verfügung, die generell Kindern mit Entwicklungsverzögerungen für ein weiteres Jahr vor und außerhalb der ersten Klasse einen institutionellen Erfahrungs- und Lernrahmen geben. Sie wurden abgelöst durch eine Flexibilisierung der Schuleingangsphase, in der Kindern die individuell erforderliche Zeit gewährt wird, sich mit Grundlagen und Grundinhalten der ersten schulischen Lernens vertraut zu machen. Vor diesem Hintergrund sind Schuleingangsuntersuchungen keineswegs weniger wichtig geworden, auch wenn sie im Regelfall nicht mehr zu einer Ja-/Nein-Entscheidung des aufnehmenden Schulleiters führen.

Da Vorsorgeuntersuchungen im Rahmen der gesetzlichen Krankenversicherung nicht primär darauf ausgerichtet sind, Ausgangsbedingungen für schulisches Lernen zu erfassen und zu kommunizieren, ersetzen sie die Schuleingangsuntersuchung nicht. Eine fundierte Entwicklungsbeurteilung im Hinblick auf förderrelevante schulische Belange abzugeben, die darüber hinaus prognostische Aussagen zur Regelschulfähigkeit zunehmend junger Kinder trifft, ist eine wesentliche sozialpädiatrische Herausforderung für Kinder- und Jugendärzte im Öffentlichen Gesundheitsdienst. Eine gute Kenntnis der rechtlichen Rahmenbedingungen, der Schullandschaft vor Ort, der pädagogischen Konzepte und der Möglichkeiten zur Binnendifferenzierung sind für diese Aufgabe unerlässlich. Dass ein enger Austausch zwischen dem System pädagogischer Förderung und dem der gesundheitlichen Versorgung qualitätsfördernd ist, steht außer Frage, setzt aber auch eine hohe Abstimmungsbereitschaft und ausreichende Zeitressourcen aller Beteiligten voraus. Dies gilt nicht zuletzt für die Belange chronisch kranker Kinder, die es mit ihren gesundheitlichen Handicaps zu integrieren gilt und für die möglichst förderliche Lernbedingungen zu schaffen sind. Auch hier erfüllt der Schularzt eine Vermittlerfunktion zwischen Eltern, Schule und niedergelassenem Kinder- und Jugendarzt, z. B. bei Themen des Notfallmanagements von asthmakranken Kindern, bei der Vermeidung von Hypoglykämien bei diabeteskranken Schülern, bei der Verhaltensbeobachtung und beim Einsatz unterstützender schulischer Hilfen für AHDS-betroffene Kinder.

Zur Beurteilung der Schulreife haben sich körperliche Reifeparameter als wenig diskriminativ erwiesen, um schulreife und nicht schulreife Kinder voneinander zu unterscheiden. Als wesentlich hingegen haben sich Motivations- und Lernbereitschaft des Kindes erwiesen, die dem Kind dazu verhelfen, seine Fähigkeiten unter Beweis stellen zu wollen, sich selbst bestätigend einzusetzen und dabei eine gewisse Anstrengungsbereitschaft und Ausdauer an den Tag zu legen. Nicht nur auf dem Boden individueller Faktoren wie Temperament, kognitive Ausstattung, Frustrationstoleranz, sondern ganz stark auch im Kontext der familiären Situation, der sozialen Bedingungen

und vorschulischer Lernerfahrungen ist der Begriff der **„Schulbereitschaft"** entstanden. Ihre Beurteilung kann wesentlich ergänzt und gestützt werden durch eine Einschätzung der Erzieher, die das Kind längere Zeit im Kindergarten betreut haben und in unterschiedlichsten Situation beobachten und erleben konnten.
Grundvoraussetzungen und Grundfertigkeiten für einen erfolgreichen Schulstart umfassen körperliche, psychosoziale, motivationale und soziale Bereiche. Es wurden und werden unzählige Operationalisierungsvorschläge und Kriterienkataloge entwickelt, die wiederum an epidemiologischen Beobachtungen adaptiert werden und neue Bewertungen und Prioritäten erhalten. So wird z. B. heute niemand mehr ernsthaft erwarten und fordern, dass ein sechsjähriges Kind für die Dauer einer gesamten Unterrichtsstunde nicht von seinem Stuhl aufstehen darf. Selbst grundlegende Defizite im Sprechen und Verstehen der Unterrichtssprache werden heute nur bei besonderer Ausprägung oder unzureichenden Ressourcen zur Rückstellung vom Besuch der ersten Klasse führen, insbesondere dann, wenn es sich mutmaßlich um ein normal begabtes, „schulbereites" Kind handelt.
Am Übergang von der vorschulischen zur schulischen Betreuung und Förderung ist der Kinder- und Jugendgesundheitsdienst oft Wegbereiter und Moderator von individuellen oder systembezogenen Entscheidungsprozessen. Dazu prädestinieren ihn seine kinder-(fach)ärztliche Expertise, sein Erfahrungshorizont, bewährte, vertrauensvolle Kooperationsstrukturen und die Kenntnis von aktuellen (schul-)rechtlichen Rahmenbedingungen und pädagogischen Leitlinien.
Die sozialen und ökonomischen Folgen für die Opfer einer inadäquaten Schullaufbahnentscheidung werden in der Mannheimer Risikokinderstudie folgendermaßen beschrieben: Aus falscher Weichenstellung in der Schullaufbahn resultieren ca. dreimal so häufig fehlende Ausbildungsabschlüsse (bei jedem 6.), ca. dreimal so oft langdauernde Arbeitslosigkeit (etwa bei jedem 7.), ca. dreimal so häufig kinder- und jugendpsychiatrische Behandlung (bei jedem 2.!) und ca. viermal so häufig Delinquenz (bei jedem 5.).

Entlassschüleruntersuchung

In vielen Bundesländern bietet der ÖGD – auf gesetzlicher Grundlage oder einem definierten Bedarf entsprechend – Untersuchungen der Schüler im vorletzten Jahr vor Schulentlassung an. Individualmedizinisch betrachtet können so Jugendliche in einer weichenstellenden Lebensphase erreicht werden, die sich sonst Präventions-, Beratungs- und Versorgungsangeboten nur schwer bzw. anlassbezogen zuwenden, und dies in einer Entwicklungs- und Lebensphase, in der sich häufig Gesundheitsstörungen mit hoher individueller und gesellschaftlicher Relevanz manifestieren.
Die Hauptintention dieser Untersuchung liegt darin, Jugendliche noch vor Ausbildungsbeginn zu erreichen und sie einem Screening in Bezug auf berufswahlrelevante gesundheitliche Risiken zuzuführen. Die erste jugendarbeitsschutzgesetzliche Untersuchung kommt in der Regel, falls sie wegen des Alters bei Ausbildungsbeginn überhaupt noch zum Tragen kommt, zu spät, um absehbaren Fehlentscheidungen im Hinblick auf die Berufswahl vorzubeugen. Besondere Schwerpunkte liegen auf der Erfassung von orthopädischen Belastungseinschränkungen und allergischen Diathesen. In der Regel wird bei anamnestischem und/oder klinischem Verdacht eine weitere Diagnostik beim Kinder- und Jugendarzt oder Hausarzt empfohlen. Wenn geeignete Behandlungs- oder Arbeitsschutzmaßnahmen zur Verfügung stehen, deren Beachtung einen gesundheitlich bedingten Ausbildungsabbruch oder gar eine Berufskrankheit unwahrscheinlich macht, wird man sicherlich nicht vom „Traumberuf" abraten müssen. Andererseits mag es in vielen Fällen früh genug möglich sein, z. B. einer jungen Hauptschülerin mit atopischem Handekzem eine andere berufliche Alternative als die der Friseurin anzuraten.
Ein effektiver Beitrag zur Reduktion von gesundheitlich bedingten Ausbildungsabbrüchen und Berufskrankheiten ist also durch den Zugang bereits in der neunten Klasse, möglichst im schulcurri-

cularen Kontext der Berufsvorbereitung, gezielter möglich, damit auch die Reduktion der „Fehlbelegung“ wertvoller Ausbildungsplätze durch längerfristig gesundheitlich ungeeignete Berufseinsteiger. Die Entlassschüleruntersuchung bietet zudem eine hervorragende Gelegenheit, den Impfschutz einer schwer zugänglichen Altersgruppe zu verbessern. Nicht zuletzt sind Daten zur Gesundheitssituation, zum gesundheitsbezogenen Lebensstil, zum Präventions- und Risikoverhalten und zur gesundheitlichen Versorgung dieser Altersgruppe schwer zu gewinnen und als Begleitprodukt der Entlassschüleruntersuchung recht willkommen.

9.6.2 Zielgruppen- und bedarfsorientierte Untersuchungs- und Beratungsangebote

Ohne Anspruch auf Vollständigkeit sind in der nachfolgenden Übersicht vornehmlich individualmedizinisch definierte Aufgaben der Kinder- und Jugendgesundheitsdienste dargestellt. Die meisten Gesundheitsdienstgesetze beauftragen den ÖGD, „Regeluntersuchungen“ in eigener Zuständigkeit durchzuführen, sofern er einen entsprechenden Bedarf sieht. Diese Formulierungen sind nicht ganz unproblematisch, vor allem, wenn man berücksichtigt, dass die Institution, die den Bedarf definieren soll, gleichzeitig Leistungserbringer ist und sich in Abhängigkeit von den lokalen gesundheitspolitischen Beschlussfassungen und Prioritätensetzungen befindet.
Schwerpunktmäßig geht es hier um institutionen- oder settingabhängige Jahrgangs- oder andere Querschnittsuntersuchungen in Schulen, Kindertagesstätten und Einrichtungen der Jugendhilfe. In diese Kategorie fallen zum Beispiel Untersuchungen von Viertklässlern, die man vor dem Wechsel zur weiterführenden Schule mit diesem Angebot noch gut erreicht. Schul- und förderrelevante Befunde fallen nicht selten zum ersten oder sogar zweiten Mal nach der Einschulungsuntersuchung auf, wenn keine adäquaten Maßnahmen eingeleitet oder konsequent verfolgt wurden.
Des Weiteren sind vielfach engmaschige Untersuchungsangebote in Förderschulen geboten, zumal hier überzufällig häufig Schüler mit ungünstigen sozialen und gesundheitlichen Ressourcen gefördert werden, deren Zugang zu Regelversorgungsangeboten aus mehreren Gründen erschwert ist.

Zielgruppen- und bedarfsorientierte Untersuchungs- und Beratungsangebote des Kinder- und Jugendgesundheitsdienstes im ÖGD:

- Schwangeren- und Elternberatung mit dem Schwerpunkt aufsuchender Tätigkeit (z. B. durch sozialmedizinische Assistentinnen)
- Untersuchungen und Beratungen im Vorschulalter:
  - Standardisiertes Screening subsidiär zur bestehenden kinderärztlichen Vorsorge mit dem Ziel der Primärprävention und der zeitgerechten Förderung
  - Besonders intensive Betreuung der Kinder in heilpädagogischen bzw. integrativen Kindertageseinrichtungen
  - Settingbezogene Fragestellungen im Kontext von Inklusion
  - Dabei Vernetzung mit den Angeboten der kassenärztlichen Versorgung
- Untersuchung und Beratung der 15–16-jährigen Jugendlichen zum Ende der Schullaufbahn:
  - Aufsuchendes, niedrigschwelliges Untersuchungs- und Beratungsangebot unter besonderer Berücksichtigung der berufswunschrelevanten Gesundheitsrisiken (Konzepte zur Weiterentwicklung der Entlassschüleruntersuchung und zur Zusammenführung mit der Erstuntersuchung nach dem Jugendarbeitsschutzgesetz werden in einigen Landesministerien diskutiert)

- Untersuchung und Beratung behinderter und von Behinderung bedrohter Kinder und Jugendlicher:
  - Sozialpädiatrische Beratung zur Inklusion und optimierten Teilhabe und Rehabilitation im Sinne des SGB IX
  - Dabei Wahrnehmung einer Mittler- und Wegweiserfunktion durch das Netz von Beratungs- und Betreuungseinrichtungen.

9.6.3 Sozialmedizinische Begutachtungen

Hier handelt es sich um Tätigkeiten im Auftrag anderer Institutionen, zusammenfassend um ärztliche Begutachtungen, Atteste und Stellungnahmen, die für den Auftraggeber eine plausible, nachvollziehbare Grundlage für sozialrechtliche Leistungen und Kostenträgerschaften darstellen sollen. Auf kommunaler Ebene werden die Auftraggeber zumeist durch die örtlichen Sozialämter, Jugendämter, Ausländerbehörden, Gerichte, Arbeitsagenturen und Landschaftsverbände repräsentiert. Letztlich geht es um die Legitimation individueller Leistungsansprüche gegenüber dem Sozial-oder Jugendhilfeträger, der kommunalen Familie, dem Staats- und Solidarsystem per se. Anlässe, Auftraggeber und Fragestellungen sind vielfältig; der sozialpädiatrische Gutachter ist in einem klar vermittelten und gelebten Rollenverständnis gegenüber dem Begutachteten gefordert; er ist zu Neutralität gegenüber den individuellen Interessen des Begutachteten verpflichtet, in Abwägung gegenüber denen der Solidargemeinschaft und der zu schützenden Solidargüter. Darüber hinaus muss er in der Lage sein, entscheidungsrelevante sozialmedizinische Sachverhalte herauszukristallisieren, transparent, plausibel, unter Beachtung der datenschutzrechtlichen Erfordernisse gegenüber dem Auftraggeber nachvollziehbar zu formulieren, den aktuellen Rechtsrahmen zu kennen und zu berücksichtigen. Letztlich sollte bei der Allokation von Solidargütern der öffentlichen Verwaltung eine interessenneutrale Schlüsselfunktion zukommen.

Gutachterliche Stellungnahme für die Jugendhilfe

Im Rahmen des § 35 a SGB VIII (Kinder- und Jugendhilfegesetz) bittet das örtliche Jugendamt um ein Gutachten. Es braucht eine Entscheidungsgrundlage, um über die Kostenübernahme für spezifische LRS-Förderung durch ein privates Institut zu befinden. Antragsteller sind die Eltern eines Viertklässlers mit einer als Teilleistungsstörung diagnostizierten Legasthenie. Nachdem sich diese den schulischen Fördermaßnahmen nicht hinreichend erschloss, wurden die Eltern auf das Förderinstitut hingewiesen und zur Kontaktaufnahme mit dem Jugendamt veranlasst. Im Anamnesegespräch und bei der orientierenden Exploration stellt sich heraus, dass die lange bestehende Problematik zu einer sekundären Neurotisierung geführt hat. Vor diesem Hintergrund spricht der Gutachter die Empfehlung einer befristeten Fördermaßnahme aus, da eine seelische Behinderung droht.

9.6.4 Schulsprechstunden

Schulsprechstunden sind Bestandteil der betriebsmedizinischen Aufgaben der Kinder- und Jugendgesundheitsdienste der Gesundheitsämter, verdienen aber auch Beachtung unter individualmedizinischen Gesichtspunkten.

Der Zugang zu Jugendlichen mit ihren gesundheitlichen Bedürfnissen ist bekanntermaßen schwer, trotz aller Bemühungen um Professionalisierung und Profilierung der Jugendmedizin in Deutschland. Andrerseits sind Jugendliche und Heranwachsende heutzutage nicht weniger krank, sondern lediglich anders: Folgen von risikobereitem Verhalten, neue Konsummuster, insbesondere psychische und psychosomatische Symptomkomplexe erfordern Beachtung. Nun werden ältere Kinder und Jugendliche am Übergang zur Eigenverantwortung dort am besten durch Angebote erreicht, wo sie sich naturgemäß aufhalten; die Etablierung eines sozialmedizinischen Beratungsangebots für Jugendliche im „originären Lebensraum Schule" ist immens wichtig, da Jugendliche medizinische Hilfen im Bereich der ärztlichen Primärversorgung selten in Anspruch nehmen, insbesondere keine, die präventiv orientiert sind. Gerade an der Schwelle zur Eigenverantwortung auch im Hinblick auf die Gesundheitsfürsorge und -vorsorge widerspricht es ihrer Selbstdefinition häufig, externe Hilfe in Anspruch zu nehmen, es sei denn, sie wird von der Peergroup akzeptiert. Für viele Jugendliche darf das gesundheitliche Problem dort und nur dort angesprochen und diskutiert werden in dem Versuch, sich von der Erwachsenenwelt und vor allem der Welt der „Professionellen" abzugrenzen. Unter dem Schutzgebot der Schweigepflicht vertrauen sich Jugendliche Kinder- und Jugendärzten in Schulsprechstunden an und konfrontieren sie mit Fragen und Problemen, die nicht immer sensu strictu medizinischen Hintergrund haben. Oft geht es um eine zweite Meinung, eine Einschätzung, ob denn diese oder jene Beschwerden einer weiteren (fachärztlichen) Abklärung bedürfen. Nicht selten werden pubertätsspezifische Störungsbilder wie Essstörungen oder Krankheitsbilder wie Morbus Scheuermann erstmals in einer Schulsprechstunde thematisiert. Dabei werden Erstkontakte oft durch befreundete Schülerinnen und Schüler gebahnt und begleitet. Bevorzugt in Stadtteilen mit besonderem Erneuerungsbedarf, hohem Migrantenanteil und in Förderschulen erweisen sich Schulsprechstunden als probates Angebot, um Jugendliche mit ihren gesundheitlichen Problemen zu erreichen und individuell anzusprechen. Individuelle Beratung zu Präventionsmaßnahmen, Hilfen bei der Bewältigung altersspezifischer Entwicklungsaufgaben, fachliche Einschätzung eventuell vorhandener gesundheitlicher Beschwerden und die Vermittlung von Unterstützungsangeboten und Hilfen können Inhalte von Schulsprechstunden sein. Voraussetzung für das Gelingen ist die Fähigkeit des Schularztes, sich auf die besonderen Bedürfnisse des Schülers einzulassen und durch professionelle Ansprechbarkeit in der Kontinuität ein persönliches Vertrauensverhältnis aufzubauen.

9.6.5 Case-Management

In der ambulanten und stationären Gesundheitsversorgung ist die sozialmedizinische Nachsorge, d. h. die Abstimmung der notwendigen medizinischen und sozialrechtlichen Maßnahmen und die Unterstützung bei der Krankheits(folgen-)bewältigung seit 2006 für schwer erkrankte Kinder und Jugendliche eine zeitlich begrenzte Leistung der Krankenkassen nach § 34 SBG V. In Richtlinien sind die Vorgaben definiert. Für chronische Erkrankungen und Behinderungen sind fallbetreuende Professionelle nicht dauerhaft institutionalisiert. Für medizinische Belange steht der niedergelassene Kinder-und Jugendarzt in seiner Hausarztfunktion zur Verfügung. Dennoch gibt es darüber hinaus Bedarf an Bündelung von entwicklungs- und gesundheitsbezogenen Maßnahmen, z. B. im Rahmen der Umsetzung eines Förderkonzepts oder eines Hilfeplans. Selbst oder gerade wenn Aufgaben und Rollen klar verteilt und verbindliche Absprachen getroffen worden sind, ist eine Koordination geboten. Eltern brauchen einen Ansprechpartner, der mit allen Beteiligten kommuniziert und wenn nötig Impulse gibt, Ziele oder Vorgehensweisen zu modifizieren oder zu korrigieren. Dies setzt klare Kommunikationsstrukturen voraus und das Vorhandensein einer (finanzierten) personellen Ressource.

Mitarbeiterinnen und Mitarbeiter von Sozialpädiatrischen Zentren und anderen Nachsorgezentren können diese Funktion des Case-Managements erfüllen, desgleichen sozialmedizinische Assistentinnen des Öffentlichen Gesundheitsdienstes.

Betreuung einer Mutter im Teenageralter durch eine sozialmedizinische Assistentin
Die sechzehnjährige Heike ist der Kinder- und Jugendärztin des Gesundheitsamts aus der Schulsprechstunde heraus bekannt. Sie hat mehrfach eine Freundin begleitet, die sich wegen einer kombinierten Ess- und Körperschemastörung an die Schulärztin gewandt hat und jetzt fachärztlich betreut wird. Erstmals wendet sich Heike selbst und direkt an die Schulärztin, da sie annimmt, schwanger zu sein. Die Schulärztin bespricht mit Heike, wie zunächst der Sachverhalt geklärt werden kann und vermittelt den Erstbesuch beim Gynäkologen und bei der Beratungsstelle Pro Familia. Heike entscheidet sich bewusst für das Kind, zumal sie auf Unterstützung ihrer Familie hoffen kann. Nach komplikationsloser Schwangerschaft kommt ein gesundes Mädchen durch Spontangeburt zur Welt. In der Betreuung des Kindes erhält Heike die Unterstützung der sozialmedizinischen Assistentin, vermittelt durch die Schulärztin, und wendet sich vertrauensvoll mit Fragen zur Entwicklung und zum Umgang mit dem Kind an sie. Dadurch wird die gesundheitliche Betreuung von Mutter und Kind nach der Geburt ergänzt. Ein halbes Jahr nach der Geburt gelingt Heike der angestrebte Schulabschluss.

Weitere Anlässe für ein Case-Management können sein:

- Begleitung einer Familie mit unzureichenden Ressourcen nach Mehrlingsgeburt oder nach Geburt eines Frühgeborenen mit hohem medizinischen Betreuungsaufwand
- Hilfen für eine Schülerin, die im Rahmen einer Schulsprechstunde über Gewalterfahrungen im häuslichen Umfeld berichtet
- Mitversorgung eines Kleinkinds als Tochter einer allein stehenden, akut psychisch erkrankten Mutter
- Reintegration eines Schulkindes nach Erstmanifestation eines Diabetes mellitus, instabiler Stoffwechsellage und hohem Abstimmungsbedarf mit der betreuenden Fachambulanz.

Grundsätzliches Ziel aller Bemühungen ist es, sich als Case-Manager möglichst bald (weitgehend) entbehrlich zu machen, d. h. die jeweils betreuten Menschen an adäquate Einrichtungen des Versorgungssystems zu vermitteln, sie in ihr Umfeld zu (re-)integrieren, Compliance und adäquate Krankheitsbewältigung zu fördern und vor allem möglichst langfristig das größtmögliche Potenzial an Eigenkompetenz und Selbstwirksamkeitserfahrung zu mobilisieren. An dieser Stelle wird besonders deutlich, wie wichtig für den Erfolg eines Case-Managers die Einbettung in ein interdisziplinäres Team und eine handlungsfähige Struktur sind. Nur so kann seine Arbeit verschiedenste Betreuungsanliegen erfüllen und in der Summe zur gesamtgesellschaftlichen Wertschöpfung beitragen.

Literatur

Altgeld T (2003) Gesundheitsziel Kindergesundheit. Bundesgesundheitsblatt 46 (2): 120–127.

Arbeitsgemeinschaft der Spitzenverbände der Krankenkassen, Bundesverband der Unfallkassen, beratende Kommission der Spitzenverbände der Krankenkassen für Primärprävention und betriebliche Gesundheitsförderung (2004) Empfehlungen zur Gesundheitsförderung in Schulen. Bergisch Gladbach und München.

Autorengruppe Bildungsberichterstattung (2012) Bildung in Deutschland 2012. Ein indikatorengestützter Bericht mit einer Analyse zur kulturellen Bildung im Lebenslauf. Kultusministerkonferenz, Bundesministerium für Bildung und Forschung

Bergmann L, Bergmann K et al. (2005) Über die Beeinflussbarkeit der Entwicklung von Säuglingen und Kleinkindern. Ergebnisse einer Studie zur vorausschauenden Beratung junger Eltern mit ihrem ersten Kind. Kinderärztliche Praxis Sonderheft: 18–22.

Büsching U (2005) Positionspapier Handbuch „Arzt und Schule". Kinder- und Jugendarzt 1: 8–10.

Büsching U, Paulus P, Schirm H (2005) Handbuch: Arzt & Schule, Lübeck: Hansisches Verlagskontor.

Bundesregierung/Sachverständigenkommission (30.1.2013) Bericht über die Lebenssituation junger Menschen und die Leistungen der Kinder- und Jugendhilfe in Deutschland – 14. Kinder- und Jugendbericht – Deutscher Bundestag, Drucksache 17/12.200

BZgA Bundeszentrale für gesundheitliche Aufklärung (2001) Chronische Erkrankungen als Problem und Thema in Schule und Unterricht – Handreichung für Lehrerinnen und Lehrer der Klasse 1–10. order@bzga.de

Ceglarek P et al. (2006) Gesundheitsförderung in Brandenburger Kindertagesstätten, Potsdam: Universitätsverlag Potsdam. http://dnb.ddb.de

Deutsche Arbeitsgemeinschaft Jugendzahnpflege (Juni 2012) Frühkindliche Karies: Zentrale Inhalte der Gruppenprophylaxe für unter 3-Jährige

Eaton K et al. (2006) Youth Risk Behavior Surveillance – United States 2005. Journal of School Health. 76 (7) 353–372.

Esser G Ballaschk K et al. (2005) Determinanten sozialer Fertigkeiten von Kindern. Ergebnisse einer prospektiven Längsschnittstudie. Kinderärztliche Praxis Sonderheft: 42–47.

Gerlinger T, Babitsch B et al (2012) Situation und Perspektiven von Public Health in Deutschland-Forschung und Lehre. Gesundheitswesen 2012; 74: 762–766

Daseking M et al. Vorhersage von schulischen Lernstörungen durch SOPESS. Gesundheitswesen 2011;73: 650–659

Gruno D, Trojan A (2002) Deutliche Unterschiede zwischen Status Quo und Wunschbild. Dtsch Arztebl 25 C: 1.370–1.374.

Hock S, Berchner Y, Blankenstein O, Buschbaum T, Ellsäßer G, Heuermann MW, Klein R, Kolbow D, Metzner F, Röhlich-Pause K, Thaiss H, Untze P, Zimmermann E, Kieslich M (2013) Zum aktuellen Stand der Kindervorsorgeprogramme. Ergebnisse des ersten bundesweiten Arbeitstreffens 2011 in Frankfurt am Main. Gesundheitswesen 75 (03): 143–148

Hölling H, Schlack R (2008) Psychosoziale Risiko- und Schutzfaktoren für die psychische Gesundheit im Kindes- und Jugendalter – Ergebnisse aus dem Kinder- und Jugendgesundheitssurvey (KiGGS). Gesundheitswesen 70: 154–163.

Hölling H, Schlack R et al (Juni 2012) Die KiGGS-Studie. Bundesgesundheitsbl 2012. 55: 836–842

Landesinstitut für Gesundheit und Arbeit NRW (2009) Sozialpädiatrisches Entwicklungsscreening für Schuleingangsuntersuchungen – SOPESS. Theoretische und statistische Grundlagen zur Testkonstruktion, Normierung und Validierung

Langness A, Richter M, Hurrelmann K (2005) Gesundheitsverhalten im Jugendalter: Ergebnisse der internationalen „Health Behavior in School-aged Children"-Studie. Gesundheitswesen 67: 422–431.

Laucht M, Esser G et al. (2000) Längsschnittforschung zur Entwicklungsepidemiologie psychischer Störungen. Zielsetzung, Konzeption und zentrale Ergebnisse der Mannheimer Risikokinderstudie. Zeitschrift für Klinische Psychologie und Psychotherapie 29: 246–262.

Liebelt J (1998) Angewandtes Qualitätsmanagement. Gesundheitseinrichtungen als lernende Organisationen. Heidelberg: Springer.

Meireis H (2000) Auftrag des öffentlichen Kinder- und Jugendgesundheitsdienstes. Kinderärztliche Praxis 5: 287–293.

Müller W, Bunte A (2000) Qualitätsmanagement im kommunalen Gesundheitsdienst, Berichte und Materialien der Akademie für Öffentliches Gesundheitswesen Düsseldorf, Bd. 16.

Rattay P (2002) Psychosoziale und medizinische Beratung in der Schule. Bundesgesundheitsblatt 45: 430–437.

Rosenbrock R (2001) Was ist New Public Health? Bundesgesundheitsblatt 44 (8) 753–762.

Schirm H (1996) Konsenspapier über die Kooperation und Zuständigkeiten der Kinderärztinnen und Kinderärzte in der Praxis und im Öffentlichen Gesundheitsdienst. Sozialpädiatrie und Kinderärztliche Praxis 1: 52.

Thaiss H, Klein R, Schumann EC, Ellsäßer G, Breitkopf H, Reinecke H, Zimmermann E (2010) Früherkennungsuntersuchungen als Instrument im Kinderschutz. Erste Erfahrungen der Länder bei der Implementation appellativer Verfahren Bundesgesundheitsblatt 53: 1.029–1.047

Internetlinks

www.BfArM.de
www.bva.de
www.bzga.de
www.dimdi.de
www.pei.de
www.rki.de
www.bvpg.de
www.daj.de
www.kinder-jugendgesundheit.de
www.aerzte-oegd.de
www.kita-gesundheit.de

10 Jugendhilfe

Peter Borusiak

Die Jugendhilfe arbeitet überwiegend auf der Grundlage des **SGB VIII,** auch als **Kinder- und Jugendhilfegesetz (KJHG)** bezeichnet. Die entsprechenden rechtlichen Grundlagen werden in Kapitel 42(➤ Kap. 42) aufgeführt. Die letzte wesentliche Überarbeitung wurde 2006 verabschiedet. Im Vergleich zu der vorher gültigen Version ergab sich ein Paradigmenwechsel mit erheblicher Bedeutung: Während vorher im Jugendwohlfahrtsgesetz von 1961 die Wächterfunktion der staatlichen Gemeinschaft im Vordergrund stand, wurde bei dem neuen Gesetz das Leistungsanrecht in den Mittelpunkt gestellt. Ferner trat zum 1. Januar 2012 das neue **Bundeskinderschutzgesetz (BKiSchG)** in Kraft. Es sollte eine Stärkung der Handlungs- und Rechtssicherheit, eine Verbesserung der „Frühen Hilfen" und Schaffung verlässlicher Netzwerke erreicht werden (➤ Kap. 9, Kap. 42).

10.1 Aufgaben der Jugendhilfe

Aktuell wird für jedes Kind bzw. Jugendlichen im SGB VIII das Recht auf Förderung seiner Entwicklung und auf Erziehung genannt. Für Eltern sind Pflege und Erziehung der Kinder gleichzeitig natürliches Recht und ihnen obliegende Pflicht. Diesbezüglich hat die staatliche Gemeinschaft eine Wächterfunktion. Jugendhilfe soll die individuelle und soziale Entwicklung junger Menschen fördern und Benachteiligungen vermeiden und abbauen, Kinder und Jugendliche vor Gefahren für ihr Wohl schützen und positive Lebensbedingungen für junge Menschen und ihre Familien erhalten oder schaffen. Auch die Beratung und Unterstützung von Eltern oder anderen Erziehungsberechtigten gehört zu den gesetzlich definierten Aufgaben.

Merke

Aufgaben der Jugendhilfe:
- Tageseinrichtungen für Kinder
- Kinder- und Jugend(sozial)arbeit, Förderung von Jugendverbänden
- Erziehungsberatung und -unterstützung
- Betreuung von Pflege- und Adoptivkindern

Sozialpädiatrie in der Praxis. http://dx.doi.org/10.1016/B978-3-437-31630-3.00010-9
Copyright © 2014 Elsevier GmbH. All rights reserved.

- Partnerschafts-, Trennungs- und Scheidungsberatung
- Mitwirkung in Verfahren vor dem Familiengericht
- Inobhutnahme.

Leistungsberechtigt sind primär die Kinder bzw. Jugendlichen, ihre Eltern und Erziehungsberechtigten. Die kulturellen Bedürfnisse und Besonderheiten junger Menschen sollen berücksichtigt und die Gleichberechtigung von Jungen und Mädchen gefördert werden. Für die Aufgaben der Jugendhilfe wurden im Jahr 2011 insgesamt 30,5 Milliarden Euro ausgegeben – 2004 waren es noch 20,7 Milliarden Euro.
Im Rahmen des Personensorge- und Umgangsrechts (z. B. bei Trennung oder Scheidung der Eltern) werden umfangreiche Hilfsangebote von Beratung (Eltern, Kinder, Großeltern), über sozialrechtliche Leistungen, fachspezifische Beratungsangebote und Erarbeitung von Umgangsregelungen bis hin zu Vermittlung an Therapeuten und Rechtsanwälte vorgehalten. Dies bedingt ein hohes Maß an Fachwissen auf verschiedenen Ebenen, was nur durch den Einsatz multiprofessioneller Teams gewährleistet werden kann.

10.2 Hilfen zur Erziehung

In der täglichen Arbeit im Bereich der Sozialpädiatrie ergeben sich häufig Kontakte über den Aufgabenkreis der Erziehungsberatung und -unterstützung. Hierbei sind einige vorgegebene Arbeitsgrundlagen der Jugendhilfe zu beachten. So gilt grundsätzlich das sog. **Subsidiaritätsprinzip:** Es muss prinzipiell die weniger eingreifende Maßnahme zuerst erfolgen (ambulant vor stationär etc.). Dies ist auch bei der Intensität der in Tabelle 10.1 (➤ Tab. 10.1) aufgezählten Möglichkeiten der Unterstützung zu beachten. Die meisten Angebote der Kinder- und Jugendhilfe sind auf die Mitwirkung und Beteiligung der Eltern bzw. Erziehungsberechtigten angewiesen.
Bei längerfristigen Maßnahmen ist immer ein Hilfeplangespräch erforderlich, das zusammen mit den Sorgeberechtigten und dem Kind oder Jugendlichen geführt werden muss. Wenn sich Kinder oder Jugendliche selber direkt an das Jugendamt wenden, kann eine Beratung auch ohne Wissen der Eltern erfolgen. Dabei kann das Kind auch von einer vertrauten Person seiner Wahl zur Unterstützung zu den Gesprächen begleitet werden.

Tab. 10.1 Formen der Jugendhilfe

Arbeitsformen	Angebote	Hauptzielgruppen
Familien-unterstützende Hilfen	Erziehungsberatung	Alle Altersgruppen
	Soz.-päd. Familienhilfe	Familien mit jüngeren Kindern
	Soz. Gruppenarbeit	Ältere Kinder/Jugendliche
	Erziehungsbeistände	Ältere Kinder/Jugendliche
Familien-ergänzende Hilfen	Gemeinsame Wohnformen für Väter/ Mütter & Kinder	Alleinerziehende Eltern mit Kindern < 6 Jahre
	Tagesgruppen	Kinder bis 14 Jahre
	Soz.-päd. Tagespflege	Kinder im Vor-/Grundschulalter
Familien-ersetzende Hilfen	Vollpflege	V. a. jüngere Kinder
	Heimerziehung/sonstige Wohnformen	Kinder/Jugendliche/junge Volljährige
	Intensive soz.-päd. Einzelbetreuung	Jugendliche/Heranwachsende

Die Angebote unterscheiden sich stark hinsichtlich ihres Charakters. Die **Beratungsstellen** sind ein eher niederschwelliges Angebot, da zumeist der „Amtscharakter" fehlt und Beratungen auch anonym ohne Erfassung offizieller Daten und entsprechende Meldung erfolgen können und nicht kostenpflichtig sind. Bei den Beratungsstellen handelt es sich um ein Hilfsangebot auch mit potenziell therapeutischem Auftrag, das sich in einer multidisziplinären Besetzung der meisten Beratungsstellen (Psychologin, Sozialarbeiter/-pädagogin, Psychotherapeuten) widerspiegelt.

Die **Erziehungsbeistandschaft** ist ein kindzentriertes Angebot, oft zur Vermeidung von Fremdunterbringung und (Wieder-)Herstellung tragfähiger Familienbeziehungen. Die Einzelfallarbeit mit dem Kind oder Jugendlichen (aber auch die Moderation des Verhältnisses zur Herkunftsfamilie i. S. einer familienberatenden Ausrichtung) ist ebenfalls vom Konzept her auf die Mitarbeit der Beteiligten ausgerichtet. Sie kann also ggf. einseitig von den Erziehungsberechtigten beendet werden, wenn nicht das Kindswohl unmittelbar gefährdet ist.

Die **sozialpädagogische Familienhilfe** arbeitet familienzentriert unter der Idee der Betreuung innerhalb der Familie mit der Möglichkeit der Änderung des Systems. Sie ist langfristig angelegt und erfordert ebenfalls die Bereitschaft der Familie zur Mitwirkung. Die Arbeit an der Stärkung der Problemlösungskompetenz im Alltag findet mit dem Schwerpunkt der Anleitung zum konkreten Handeln im Gegensatz zu den typischerweise „sprachfixierten" Beratungsangeboten i. S. einer Basisarbeit statt.

Erst wenn diese Maßnahmen der familienunterstützenden Hilfen ausgeschöpft sind, können familienergänzende oder familienersetzende Hilfen zum Einsatz kommen, solange nicht das Kindswohl gefährdet ist.

Weitere Aufgaben der Jugendhilfen mit engem Bezug zur sozialpädiatrischen Arbeit – insbesondere Ausführungen zur Eingliederungshilfe bei (drohender) seelischer Behinderung nach § 35a und zur Inobhutnahme – werden in Kapitel 42 erläutert (➤ Kap. 42).

10.3 Strukturen der Jugendhilfe

Die Jugendhilfe ist überwiegend eine kommunale Aufgabe. Seitens des Bundes erfolgt die grundlegende Gesetzgebung. Ferner werden länderübergreifende Aktivitäten in der Jugendhilfe, Initiativen und Modellprojekte gefördert und alle vier Jahre ein Kinder- und Jugendbericht der Bundesregierung erstellt und veröffentlicht. Auf Landesebene wurden die gesetzlichen Rahmenbedingungen durch eigene Landesgesetze ergänzt und erweitert (z. B. für die Kindertagesbetreuung). Die Bundesländer unterhalten Landesjugendämter, die die kommunalen Träger der Jugendhilfe durch Beratung und Fortbildungen unterstützen. Auf kommunaler Ebene werden die Leistungen durch die Jugendämter der Städte oder Landkreise erbracht. Die Aufgaben des Jugendamtes werden durch den Jugendhilfeausschuss koordiniert. Dieser setzt sich zu zwei Fünfteln aus freien Trägern (Jugendverbände, Wohlfahrtsverbände, Religionsgemeinschaften, Vereine) und zu drei Fünfteln aus Vertretern des Kommunalparlaments zusammen. Der Jugendhilfeausschuss befasst sich mit allen Angelegenheiten der Jugendhilfe, insbesondere mit der Beratung von Problemlagen junger Menschen und Familien, Vorschlägen für die Weiterentwicklung der Jugendhilfe, der Jugendhilfeplanung und der Förderung der freien Jugendhilfe.

Auch Träger der freien Jugendhilfe können Leistungen erbringen, was erwünscht ist und zu passenden Angeboten für möglichst breite Bevölkerungsschichten durch die verschiedenen Werteorientierungen, Arbeitsformen, Angebote und Möglichkeiten führen soll. Die öffentliche und freie Jugendhilfe sollen partnerschaftlich zusammenarbeiten, wobei den freien Angeboten Vorrang eingeräumt wird: Wenn geeignete Angebote von anerkannten Trägern der freien Jugendhilfe bestehen oder geschaffen werden können, soll die öffentliche Jugendhilfe von eigenen Maßnahmen absehen. Die Leistungsberechtigten haben grundsätzlich ein Wahlrecht hinsichtlich Einrichtungen und Dienst der verschiedenen Träger, aber auch in Bezug auf die konkrete Gestaltung der Hilfe, sofern dies nicht mit „unverhältnismäßigen Mehrkosten“ verbunden ist.

10.4 Datenschutz

Der Datenschutz ist im Bereich der Kinder- und Jugendhilfe geregelt, wobei Daten nur zu dem Zweck erhoben werden dürfen, der zur Erfüllung der jeweiligen Aufgabe erforderlich ist. Dem Vertrauensschutz zwischen den betroffenen Kinder und Jugendlichen und den Mitarbeitern des Jugendamtes wird ein hoher Stellenwert eingeräumt. Nach dieser Regelung dürfen Sozialdaten, die einem Mitarbeiter eines Trägers der öffentlichen Jugendhilfe zum Zweck persönlicher und erzieherischer Hilfe anvertraut worden sind, nur unter restriktiven Voraussetzungen weitergegeben werden. Auch eine Zusammenführung von Daten, die zur Erfüllung unterschiedlicher Aufgaben erhoben wurden, ist nur unter bestimmten Voraussetzungen möglich. Dieser besondere Vertrauensschutz macht dennoch im Einzelfall, z. B. in den Fällen, in denen die Gefahr der Vernachlässigung oder Gewaltanwendung gegen Kinder bzw. Jugendliche besteht, eine Datenübermittlung möglich. Im gleichen Rahmen, in dem etwa Ärzte trotz Schweigepflicht ein erfahrenes Geheimnis preisgeben dürfen, dürfen auch die Mitarbeiter des Jugendamtes anvertraute Daten weitergeben. Das neue Bundeskinderschutzgesetz hat die Möglichkeiten erweitert. So wird bei einem Umzug der Familie sichergestellt, dass das neue Jugendamt die notwendigen Informationen vom bisher zuständigen Jugendamt bekommt, die es braucht, um das Kind wirksam zu schützen.
Unter Beachtung des Grundsatzes der Verhältnismäßigkeit ist dabei auch die Einschaltung der Polizei zur Abwendung einer drohenden Gefahr zulässig. Die Abwägung kann im Einzelfall schwierig sein. Dabei ist jedoch zu berücksichtigen, dass es sich nicht um die Abwägung zwischen der Einhaltung des Datenschutzes einerseits und der Hilfe für ein Kind andererseits handelt, sondern dass das Recht auf Vertrauen eines Kindes bzw. Jugendlichen gegen die Erforderlichkeit einer konkreten Hilfe abzuwägen sind.
Der früher z. T. schwierige Informationsaustausch zwischen Ärzten und anderen Mitarbeitern im sozialpädiatrischen Bereich einerseits und Jugendhilfe andererseits wurde durch das Bundeskinderschutzgesetz neu geregelt. Hilfreich ist in jedem Fall eine formale Entbindung von der Schweigepflicht. Ist eine solche nicht zu erhalten, muss die oben erwähnte Abwägung der Verhältnismäßigkeit erfolgen. Ärzte haben aber inzwischen zur Einschätzung der Kindeswohlgefährdung gegenüber dem Träger der öffentlichen Jugendhilfe Anspruch auf Beratung durch eine insoweit erfahrene Fachkraft. Sie sind zu diesem Zweck befugt, dieser Person die dafür erforderlichen Daten zu übermitteln; vor einer Übermittlung der Daten sind diese zu pseudonymisieren.

Literatur

Jordan E, Maykus S, Stuckstätte EC (2012) Kinder- und Jugendhilfe: Einführung in Geschichte und Handlungsfelder, Organisationsformen und gesellschaftliche Problemlagen. 3. Aufl. Weinheim: Beltz Juventa.
Münder J, Meysen T, Trenczek T (2012) Frankfurter Kommentar SGB VIII: Kinder- und Jugendhilfe. 7. Aufl. Baden-Baden: Nomos

Rätz R, Schröer W, Wolff M (2009) Lehrbuch Kinder- und Jugendhilfe. Grundlagen, Handlungsfelder, Strukturen und Perspektiven. Weinheim: Beltz Juventa.
Wiesner R (2011) SGB VIII – Kinder- und Jugendhilfe. 4. Aufl. München: Beck Juristischer Verlag.

Internetlinks

http://www.bmfsfj.de/
http://www.fruehehilfen.de/

11 Selbsthilfeorganisationen

Peter Borusiak

11.1 Entstehung

Nach dem allgemeinen Umdenkprozess der 1960er Jahre entwickelte sich die Selbsthilfe im heutigen Verständnis. Es entstand ein neuer Gesundheitsbegriff, der eine aktive, eigenverantwortliche Rolle des **mündigen Patienten** fördert. In Deutschland spielte der Arzt und Psychoanalytiker M. L. Moeller (1937–2002) eine wichtige Rolle bei der Etablierung von Selbsthilfegruppen. 1982 wurde schließlich die Deutsche Arbeitsgemeinschaft Selbsthilfegruppen e. V. gegründet. Anfangs bestand die Zielgruppe überwiegend aus Erwachsenen, wobei etwa in den letzten 20 Jahren auch die Eltern- bzw. Familienselbsthilfe zunehmende Beachtung findet. Im Sinne eines partnerschaftlichen Verhältnisses zu Fachärzten für Kinder- und Jugendmedizin sind Selbsthilfegruppen zu Säulen der medizinischen Versorgung geworden.
Das Verhältnis zwischen Selbsthilfegruppen und Ärzten war von Anfang an allerdings nicht spannungsfrei. War doch hiermit die Situation gegeben, dass Ärzte – teilweise nach eigenem Verständnis – Kompetenz abgeben mussten und den Patienten eine neue Rolle zukam. In der Anfangszeit akzeptierten viele Ärzte nur schwer, dass Selbsthilfegruppen und Elterninitiativen auch aus Unzufriedenheit mit der Behandlung durch Ärzte entstanden. Hier kam es in den Anfängen auch zu Auseinandersetzungen aufgrund gegenseitiger Provokationen. So wurde seitens der Selbsthilfeinitiativen das übliche Vorgehen der Ärzte kritisiert und Außenseitermethoden propagiert, andererseits aber den Betroffenen auch Kompetenzen abgesprochen. Auch führten Streitigkeiten zwischen einzelnen Selbsthilfegruppen für die gleiche Erkrankung zu erheblichen Irritationen. Insgesamt hat sich aber durch einen gegenseitigen Lernprozess ein fruchtbarer Fortschritt ergeben. Die meisten Ärzte haben erkannt, dass nicht sie allein Spezialisten auf allen Gebieten der Behandlung und Begleitung von Kindern mit Behinderung und chronischen Erkrankungen sein können.

Sozialpädiatrie in der Praxis. http://dx.doi.org/10.1016/B978-3-437-31630-3.00011-0
Copyright © 2014 Elsevier GmbH. All rights reserved.

11.2 Stellenwert

Familien werden durch die Erkrankungen und Behinderungen in ihrem Alltag, im Entwicklungsprozess und in der Lebensplanung erheblich beeinträchtigt. Die Einschränkungen mit der Gefahr der Isolation betreffen die Familie als Ganzes. Vielfach ist die Fähigkeit zur **Selbsthilfe** im Sinne des **Selbstwirksamkeitskonzepts** der beste, nicht selten der einzige Weg, die Integration zu ermöglichen. Ein Engagement in oder der Besuch einer Selbsthilfegruppe zeigt die Motivation und den Willen, aktiv mit der Erkrankung oder Behinderung umzugehen und sich nicht resignativ „dem Schicksal zu ergeben". A. Antonovsky (1923–1994) entwickelte das Konzept der **Salutogenese,** in dem er der Frage, warum Menschen gesund bleiben, den Vorrang vor der Frage nach den Ursachen von Krankheit gab. Im Fokus stehen die Bedingungen und Faktoren, die Gesundheit erhalten und bewahren. Von hohem Gewicht ist die Überzeugung, dass das eigene Leben bedeutungsvoll ist und bewältigt werden kann. Die Arbeit von **Selbsthilfegruppen** stützt sich oft auf eine solche Sichtweise. Die Gruppen bilden einen Rahmen, in dem individuelle Fähigkeiten entwickelt und unterstützt werden, um mit einer chronischen Krankheit oder einer Behinderung umzugehen.

Chronische Krankheiten im Kindes- und Jugendalter nehmen zu. Auch ist heutzutage durch die verbesserten therapeutischen Möglichkeiten die Lebenserwartung vieler Kinder mit chronischen Erkrankungen und Behinderungen höher als noch vor wenigen Jahrzehnten. Für eine ganze Reihe von Symptomenkomplexen stehen diagnostische Möglichkeiten zur Verfügung, die eine spezifische Zuordnung ermöglichen – sei es z. B. auf Ebene der Stoffwechselerkrankungen, sei es auf molekulargenetischer Basis. Auch Spezialisten ist es kaum noch möglich, in allen Bereichen den Überblick über die neuesten Entwicklungen auf diagnostischem und therapeutischem Gebiet zu behalten.

Dass aber auch heutzutage noch Vorbehalte auf ärztlicher Seite bestehen, mögen folgende Daten zeigen. In praktisch kaum einem der gängigen Lehrbücher der Kinderheilkunde findet sich der Begriff „Selbsthilfe(-gruppe)" oder „Elternselbsthilfe" in den Inhaltsverzeichnissen. Eine Umfrage, die im Juli und August 2008 vom Kindernetzwerk e. V. unter 53 Mitgliedsorganisationen durchgeführt wurde, hat auch für Kinderärzte wenig schmeichelhafte Ergebnisse erbracht: Fast zwei Drittel aller Befragten gaben an, nicht ausreichend über die Krankheit, Entwicklungsstörung oder Behinderung ihrer Kinder informiert worden zu sein. Neben anderen auch positiv vermerkten Ergebnissen fiel die Antwort auf die Frage, ob Eltern nach einer Diagnosestellung von den behandelnden Kinder- und Jugendärzten über Elternselbsthilfegruppen informiert worden sind, ernüchternd aus. Drei Viertel aller Eltern haben solche unterstützenden Hinweise auf weiterführende Adressen vermisst.

11.3 Wissenschaftliche Untersuchungsergebnisse

Aufgrund der zunehmenden Bedeutung der Selbsthilfe wird auch von wissenschaftlicher Seite versucht, die Auswirkungen von Selbsthilfegruppen, Selbsthilfeorganisationen und einzelnen Selbsthilfeaktivitäten im Rahmen der Forschung zu evaluieren. Nur wenige Aussagen sind im Sinne methodisch guter Studien klar belegt. Ein Überblick – auch über die Probleme dieser Forschungsrichtung – wurde 2009 veröffentlicht (Slesina 2009). Ziele mit eher direkter Wirkung, wie die Unterstützung anderer Mitglieder, Informationsweitergabe, selbstständiger Umgang mit der Krankheit bzw. Behinderung und gemeinsame Freizeitgestaltung, aber auch Einstellungsände-

rungen bei Betroffenen und im sozialen Umfeld sowie Interessenvertretung für Betroffene, können in recht hohem Maße umgesetzt werden. Die Veränderung von Institutionen und Einstellungsänderungen bei Professionellen werden dagegen in deutlich geringerem Maß erreicht. Weitere Forschungsergebnisse zeigen, dass die Beratung von Betroffenen, die durch Betroffene erfolgt, nach ganz spezifischen Mustern erfolgt, die eine eigene Mischung von Information und sozialer Unterstützung beinhalten, wie sie von professionellen Beratern nicht reproduziert werden kann (Borgetto 2002). Andere Untersuchungen zeigen positive Wirkungen von einzelnen Seminar- und Fortbildungsangeboten, Sportgruppen und therapeutischen Gruppen. Vergleichbare Untersuchungen zu weiteren Tätigkeitsfeldern der Selbsthilfe stehen allerdings noch aus. Nach den neueren Erkenntnissen zeichnet sich eine langsame aber stetige
Verbesserung im gegenseitigen Verständnis und der Kommunikation zwischen Ärzten und Selbsthilfegruppen ab.

11.4 Bedeutung

Im Zeitalter des Internets sind umfangreiche Recherchemöglichkeiten für Patienten und ihre Angehörigen gegeben. Der aufgeklärte und mündige Patient, der sich nicht selten schon vor dem Arztbesuch genau über (mögliche) Erkrankungen informiert hat, ist im Arbeitsalltag nichts Ungewöhnliches mehr. Viele der geschätzten bis zu 100.000 **Selbsthilfegruppen** haben einen eigenen Internetauftritt, über den ausführliche Informationen zu beziehen sind. Das Internet kann und wird allerdings nicht in direkte Konkurrenz zu einer auf Vertrauen basierenden Arzt-Patienten- bzw. Arzt-Eltern-Beziehung treten, wobei die Wertung der verschiedenen Informationen in Bezug auf den jeweiligen Patienten wichtig ist. Das Internet eröffnet jedoch zusätzliche Informationsmöglichkeiten. Auch (Eltern-)Selbsthilfegruppen verstehen sich nicht als Konkurrenz zum behandelnden Arzt, sondern als **Spezialisten für bestimmte Erkrankungen** oder Behinderungen, wobei Informationsaustausch, gegenseitige emotionale Unterstützung und Alltagsstrategien im Mittelpunkt stehen. Dies reicht von Spezialwissen bei seltenen Erkrankungen über Tipps und Tricks bei der sinnvollen Auswahl und dem Einsatz von Hilfsmitteln bis hin zu Hinweisen im Umgang mit Behörden und Anträgen. Ferner vertreten (Eltern-)Selbsthilfegruppen in unterschiedlichem Maß Interessen der Betroffenen nach außen, was von Öffentlichkeits- und Aufklärungsarbeit bis zur Vertretung politischer Interessen und auch der Anregung und Unterstützung von Forschungsprojekten reichen kann.
Die aktuelle Bedeutung der Elternselbsthilfegruppen für verschiedene Zielgruppen kann folgendermaßen dargestellt werden: Betroffene Kinder und Jugendliche, Geschwister, Eltern und weitere Angehörige finden altersgerecht und allgemeinverständlich aufbereitete Informationen über die Erkrankung und z. B. bei seltenen Erkrankungen auch Adressen von ausgewiesenen Experten auf dem Gebiet. Es werden Kontakt- und Austauschmöglichkeiten mit anderen Betroffenen und ein Erfahrungsaustausch bzgl. Alltagsproblemen angeboten. Ärzte können ebenfalls v. a. bei seltenen Erkrankungen wertvolle Informationen erhalten und weitere Ansprechpartner erfahren. Positive Beispiele der Kooperation im Alltag können dazu beitragen, gegenseitige Vorurteile abzubauen. Die Öffentlichkeit – allgemein Interessierte, Verbände, Politiker oder öffentliche Einrichtungen – kann sachliche neue Einblicke und Informationen über die Belange und Probleme der Betroffenen bekommen. Hierdurch können Ideen für Kooperationen auf verschiedenen Ebenen entstehen.

11.5 Strukturen

Die meisten Elternselbsthilfegruppen existieren als eingetragener Verein, wobei die Ämter ehrenamtlich ausgefüllt werden. Auch „lose" Zusammenschlüsse ohne Vereinsstrukturen kommen vor. Die Benennungen, Strukturen und Arbeitsweisen unterscheiden sich teilweise, wobei keine ein-

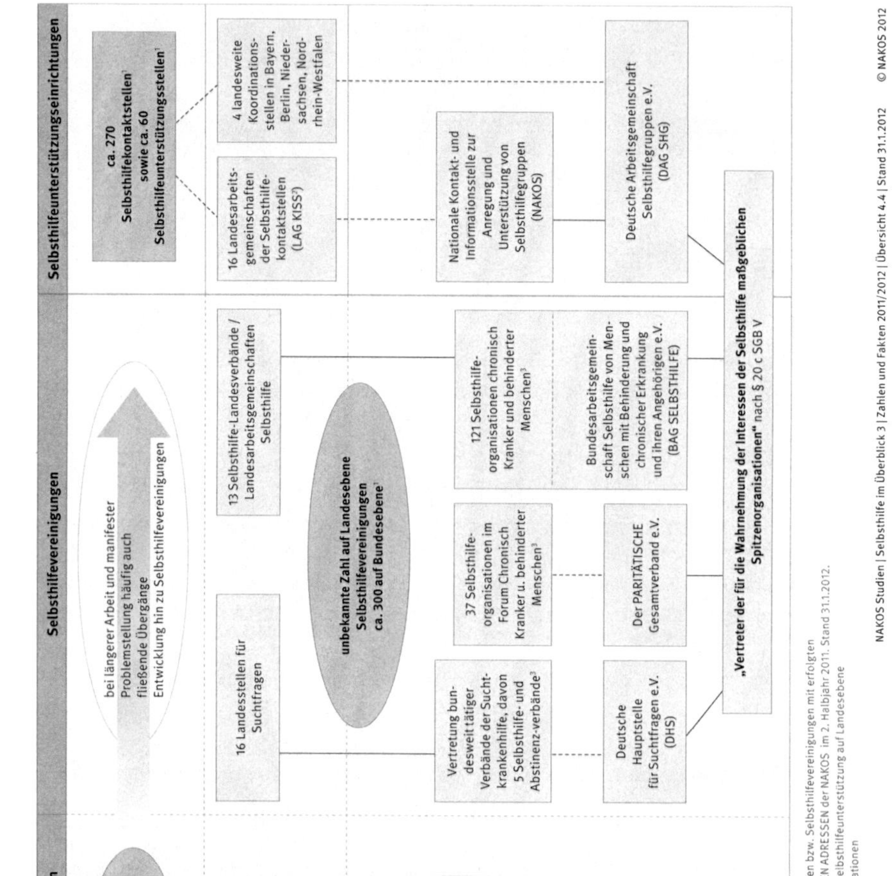

Abb. 11.1 Selbsthilfelandschaft in Deutschland (Quelle: © NAKOS [Nationale Kontakt- und Informationsstelle zur Anregung und Unterstützung von Selbsthilfegruppen])

heitliche Nomenklatur besteht. Ein großer Teil der Gruppen ist auch in anderen Netzwerkstrukturen vertreten. Regionale Arbeitsgemeinschaften oder Kontaktstellen fassen thematisch unterschiedliche Selbsthilfegruppen auf regionaler Ebene zusammen, um gesundheitliche und soziale Probleme aus verschiedenen medizinischen und sozialen Bereichen vor Ort aufzugreifen. Selbsthilfeorganisationen vertreten z. T. auf Bundesebene spezifische Krankheitsbilder oder Behinderungen. Oftmals arbeiten ausgewiesene Experten in einem wissenschaftlichen Beirat mit und unterstützen die Arbeit. Darüber hinaus gibt es mehrere Dachorganisationen, die ebenfalls auf Bundesebene aktiv sind. Etwa die Hälfte der Selbsthilfegruppen sind freie, nicht organisierte Selbsthilfegruppen, die keiner überregionalen Selbsthilfeorganisation angehören (➤ Abb. 11.1).
Im Jahr 2000 wurde die **Selbsthilfeförderung** im SGB V durch die gesetzlichen Krankenkassen verpflichtend geregelt und diese wurden zur Erstellung gemeinsamer Fördergrundsätze mit den Organisationen der Selbsthilfe verpflichtet. Mit drei maßgeblichen Spitzenorganisationen der Selbsthilfe, die die Interessen auch auf Bundesebene bündeln, wurden an der Ausgestaltung der Fördergrundsätze gearbeitet: die Bundesarbeitsgemeinschaft Hilfe für Behinderte, der Deutsche Paritätische Wohlfahrtsverband und die Deutsche Arbeitsgemeinschaft Selbsthilfegruppen. Diese stellen seitdem die Vertreter der Selbsthilfe auf Bundesebene dar.
Die Position der Selbsthilfeorganisationen wurde durch das Gesundheitsmodernisierungsgesetz gestärkt: Über ihre Dachorganisationen haben sie Mitspracherechte in wichtigen Fragen der Gesundheitsversorgung und wirken als Patientenvertreter im Gemeinsamen Bundesausschuss und seinen einzelnen Ausschüssen mit. Im SGB V wurde auch über § 20c eine obligate Mitfinanzierung durch die Krankenkassen einschließlich der Höhe der Gesamtfinanzierung pro Versichertem festgelegt. Weitere finanzielle Unterstützung erhalten die Selbsthilfeorganisationen auch aus Mitgliedsbeiträgen, Spenden und von Rentenversicherungen (Landesversicherungsanstalten, Bundesversicherungsanstalt).
Mehrere überregionale Organisationen versuchen Hilfestellungen zur Orientierung bei den verschiedenen Angeboten zu geben. Für Selbsthilfegruppen allgemein ist hier die **NAKOS (Nationale Kontakt- und Informationsstelle zur Anregung und Unterstützung von Selbsthilfegruppen)** zu nennen. Hilfreich sind zahlreiche Informationsangebote, wie z. B. auch ein Leitfaden zum kritischen Umgang mit Informationen aus dem Internet. Als zentraler Ansprechpartner für Elternselbsthilfegruppen hat sich das **Kindernetzwerk e. V.** etabliert. Aufgrund dieser zentralen Rolle soll hierauf im folgenden Abschnitt ausführlicher eingegangen werden.

11.6 Kindernetzwerk

Das „Kindernetzwerk e. V. – für Kinder, Jugendliche und (junge) Erwachsene mit chronischen Krankheiten und Behinderungen" wurde 1992 gegründet und hat sich in den vergangenen mehr als 20 Jahren zu einer bundesweit vernetzenden Institution für betroffene Eltern von Kindern und Jugendlichen mit chronischen Erkrankungen, Entwicklungsverzögerungen und Behinderungen entwickelt.
Ende 2011 waren über 22.000 Einzeladressen und rund 170.000 vernetzte Adressen, sowie Informationen zu mehr als 2.200 Erkrankungen bei Kindern und jungen Menschen in der Datenbank des Kindernetzwerks erfasst. Ein Teil dieser Adressen kann direkt über das Internet abgefragt werden. Sollten Informationen nicht über das Internet zu finden sein, steht als weitere Möglichkeit das Servicetelefon des Kindernetzwerks zur Verfügung. Die Datenbank bietet einen Überblick über Elternselbsthilfegruppen, Elternadressen, Bundesverbände und Selbsthilfeorganisationen

mit regionalen Anlaufstellen, Schwerpunktkrankenhäuser und Zentren, spezielle Reha-Einrichtungen für kranke Kinder und Jugendliche, weiterführende Literaturangebote zu spezifischen Krankheitsbildern und eine Fülle weiterer Informationen, besonders bei seltenen Erkrankungen. Vereinzelt werden auch Forschungs- und Modellprojekte sowie Stiftungen und Fördervereine für spezielle Krankheitsbilder vorgestellt. Als weiterer Service werden 6- bis 10-seitige patientenorientierte Krankheitsbeschreibungen (bislang nur für einen kleineren Teil der Schlagworte) und auch „Erst-Info-Pakete" angeboten. Therapeutische Auskünfte werden nicht erteilt.
Das Kindernetzwerk fungiert auch als Informationsstelle für Fachleute. Gerade bei seltenen Erkrankungen können Kinderärzte, andere Mediziner, nichtärztliche Therapeuten oder auch Krankenkassen Erstinformationen und weiterführende Adressen beziehen.
Auch auf politischer Ebene vertritt das Kindernetzwerk die Belange von Kindern und Jugendlichen mit seltenen oder schwerwiegenden Erkrankungen/Behinderungen. So wurde z. B. 2002 im **„Berliner Appell"** eine Reihe von Vorschlägen und Forderungen an die zuständigen Politiker und Vertreter der Kostenträger formuliert. Bei diesen Forderungen ging es um Verbesserungen für Kinder und Jugendliche, die genetisch bedingte Fehlbildungen mit z. T. schweren und schwersten Behinderungen haben, die über längere Zeiträume chronisch krank oder durch Entwicklungsstörungen beeinträchtigt sind. Im Jahr 2012 erfolgte nach 10 Jahren eine konsequente Fortschreibung mit dem 2. Berliner Appell. Es wurden die unten aufgeführten zehn Punkte benannt, die jeweils mit konkreten inhaltlichen Herausforderungen und Handlungsschritten hinterlegt sind.

2. Berliner Appell

- Alle Kinder, Jugendliche und junge Erwachsenen mit besonderen Bedürfnissen haben das Recht auf eine individuelle Versorgung und Förderung
- Alle Kinder, Jugendliche und junge Erwachsenen mit besonderen Bedürfnissen haben das Recht auf eine individuelle Beratung
- Alle Kinder, Jugendliche und junge Erwachsenen mit besonderen Bedürfnissen haben das Recht auf eine individuelle Bildung
- Alle Familien mit Kindern, Jugendlichen und jungen Erwachsenen mit besonderen Bedürfnissen werden unterstützt bei der Verwirklichung dieser Rechte
- Alle Kinder, Jugendlichen und jungen Erwachsenen mit besonderen Bedürfnissen werden zum Leben dieser Rechte selbst ermutigt
- Wir alle sind im Umgang mit den Menschen mit besonderen Bedürfnissen angehalten zur Verwirklichung dieser Rechte
- Kindernetzwerk e. V. und seine Mitglieder legen den Schwerpunkt ihrer Arbeit auf die Verwirklichung dieser Rechte
- Die erlebte Kompetenz der Selbsthilfe muss generell besonders eingebunden werden in die Entwicklung, Umsetzung und Vernetzung von qualifizierten und familienorientierten Konzepten und Modellen
- Die Rechte von Kindern, Jugendlichen, jungen Erwachsenen mit besonderen Bedürfnissen und ihrer Familien müssen verwirklicht werden
- Die Verantwortlichen in Politik, im Sozial- und Gesundheitswesen sind zum Wohl aller Kinder, Jugendlicher und junger Erwachsener sowie ihrer Familien verpflichtet zur Verwirklichung dieser Rechte

Literatur

Borgetto B (2002) Selbsthilfe im Gesundheitswesen. Bundesgesundheitsbl – Gesundheitsforsch – Gesundheitsschutz 45: 26–32.

Schmid R (2009) Eltern-Selbsthilfegruppen – Wer hilft weiter? Lübeck: Schmidt-Römhild.

Slesina W, Fink A (2009) Kooperation von Ärzten und Selbsthilfegruppen. Bundesgesundheitsbl – Gesundheitsforsch – Gesundheitsschutz 52: 30–39

Internetlinks

www.bag-selbsthilfe.de
www.dag-selbsthilfegruppen.de
www.kindernetzwerk.de
www.nakos.de

III Normale Entwicklung und ihre Variabilität

12 Normale Entwicklung und ihre Variabilität

Hans-Michael Straßburg

12.1 Einführung

Als **Entwicklung** bezeichnet man alle Veränderungen, die innerhalb eines bestimmten Zeitraums zu struktureller und funktioneller Differenzierung führen. Formen und Ordnungen gehen mit innerer Notwendigkeit auseinander hervor. Die Abfolge der kindlichen Entwicklung führt zur Vervollkommnung, d. h. zur stetigen Annäherung an ein der Entwicklung innewohnendes Ziel. Entwicklung beinhaltet somit Reifung, deren Richtung bestimmt ist, aber auch Entfaltung von Fähigkeiten, die durch Anlagen vorgegeben sind und von Umwelteinflüssen modifiziert werden.
Die Entwicklung beginnt beim Menschen mit der Vereinigung von Ei- und Samenzelle, erstreckt sich über das gesamte Leben und endet für das Individuum mit dem Tod. Somit ist Entwicklung die Reifung eines in den Grundzügen vorbestimmten, aber offenen Systems.
Kenntnisse über die Eigenheiten der Entwicklung eines Kindes und den Einfluss der psycho-sozialen Umwelt darauf sind von großer Bedeutung für die allgemeine Gesundheitsförderung und Maßnahmen zur Verhütung von Erkrankungen **(primäre Prävention),** die Früherfassung von Krankheiten, Verhaltens- und Entwicklungsstörungen **(sekundäre Prävention)** sowie für die Vermeidung von Verschlimmerungen bzw. die Rehabilitation **(tertiäre Prävention)** (➤ Kap. 7).
Bei der Beurteilung von Entwicklung kann man sich einerseits an Normen orientieren, die im Rahmen empirischer Messungen an einem definierten Kollektiv ermittelt werden können. Andererseits ist gerade die Entwicklung des Menschen durch eine große Variabilität praktisch aller Bereiche gekennzeichnet, sodass alle Aussagen über eine „normale" oder „abnorme" Entwicklung differenziert bewertet werden müssen. Häufig erfolgt die Bestimmung eines Entwicklungspara-

Sozialpädiatrie in der Praxis. http://dx.doi.org/10.1016/B978-3-437-31630-3.00012-2
Copyright © 2014 Elsevier GmbH. All rights reserved.

meters bei einem Kind in der Absicht, Aussagen zur zukünftigen Entwicklung, d. h. zur Prognose zu treffen. Hierbei ist umgekehrt proportional zum Alter große Vorsicht geboten. In der Regel sollten Entwicklungsparameter nicht isoliert, sondern im Rahmen eines **Spektrumkonzepts** gemessen und bewertet werden. Zur richtigen Beurteilung der Entwicklung ist vor allem in den ersten Lebensjahren eine sorgfältige kinderärztliche Untersuchung unverzichtbar. Ende 2008 wurde in der Schweiz die Einführung der Zusatzweiterbildung „Entwicklungspädiatrie" als Pendant zur Sozialpädiatrie beschlossen. Auch in der deutschen Sozialpädiatrie ist die Beschäftigung mit allen Aspekten der Entwicklung des Kindes bis zum Erwachsenenalter ein zentrales Anliegen, allerdings soll mit dem Begriff „sozial" auch die Einbindung des sich entwickelnden Individuums in den sozialen und ökologischen Kontext zum Ausdruck gebracht werden.

12.2 Der Einfluss sozialer Faktoren als Erklärung für Entwicklungsstörungen

Grundsätzlich sollte jedes Krankheitssymptom nur im Rahmen eines **bio-psycho-sozialen Konzepts** erklärt werden. Verschiedene ungünstige Faktoren aus der Vorgeschichte eines Kindes, z. B. besondere Krankheiten bei anderen Familienangehörigen, Vorerkrankungen der Mutter, negative Einflüsse während der Schwangerschaft, der Geburt und der Neugeborenenperiode sowie bestimmte Untersuchungsbefunde in den ersten Lebensmonaten werden als **Risikofaktoren** bezeichnet. Hierunter werden Einflussgrößen verstanden, die mit großer Wahrscheinlichkeit durch ihre Folgen die Entwicklung eines Kindes beeinträchtigen können. Andererseits muss man sich aber auch immer fragen, was die Ursache eines solchen Risikofaktors ist, inwieweit eine Erholung möglich ist und ob die Summierung von Risiken die Sicherheit der Vorhersage erhöht.

Nach epidemiologischen Erhebungen gelten bis zu 20 % aller Kinder in Deutschland als „Risikokinder", was z. T. als Stigmatisierung angesehen werden kann. Deshalb sollte bei einer differenzierten Beurteilung der Entwicklung eines Kindes dieser Begriff nur mit kritischer Zurückhaltung verwendet werden. H. F. Prechtl hat neben den Risikofaktoren den Begriff der **perinatalen Optimalität** bzw. der reduzierten Optimalität eingeführt, um die Feststellung von „Normalität" zu umgehen.

Ungünstige psychosoziale Faktoren für die Ausbildung von Entwicklungsstörungen nach dem 1. Lebensjahr sind:

- Beziehungsstörungen der Eltern nach der Geburt
- Depressivität und Resignation der Mutter
- „Vernachlässigende Expansivität" des Vaters
- Relativ kurze Zeit des Zusammenlebens der Eltern vor der Geburt des Kindes
- Höheres Alter der Eltern bei der Geburt (mehr als 35 Jahre)
- „Teenager-Eltern"
- Niedriger Bildungsstand der Mutter
- Armut der Eltern.

Günstig wird die kindliche Entwicklung durch folgende Faktoren beeinflusst:

- Stabile Partnerschaft
- Ausreichende soziale und ökonomische Ressourcen
- Orientierung der Eltern an festgelegten Normen
- Konsequente Erziehungshaltung.

Die Förderung der körperlichen Beziehungen (z. B. Hautkontakt, Streicheln, Schmusen) ab dem Säuglingsalter gilt als wichtige Grundlage für die Entwicklung eines Geborgenheitsgefühls. Für die

Erklärung von Entwicklungsstörungen als Folge einer unzureichenden Betreuung des Kindes sind aber viele andere biologische und psychosoziale Einflüsse, die **primäre Verletzlichkeit,** die in der Familie vorhandenen Fördermöglichkeiten **(Ressourcen)** und die Fähigkeiten des Kindes, seine eigenen Stärken einzusetzen **(Resilienz),** von Bedeutung (➤ Kap. 7).

12.3 Interaktionsmodelle

Verschiedene theoretische Ansätze versuchen, die kindliche Entwicklung in einem Interaktionsmodell der Wechselwirkung zwischen Kind und Umwelt zu erfassen. Danach wirkt die Umwelt unablässig auf das Kind ein, aber auch das Kind auf seine Umwelt, und die kindliche Entwicklung wird als ein Kontinuum von Veränderungen betrachtet.
Für die Prävention ist der **ökologische Ansatz** von Bedeutung, wonach die Vernetzung innerhalb der Familie sowie der Familie mit ihrer Umwelt ins Zentrum der Überlegungen gestellt wird.
In den ersten Lebensjahren ist die Entwicklung eines Kindes in hohem Maße von den Eltern und anderen Hauptbezugspersonen abhängig, die emotionale Stabilität und kontinuierliche Anregung gewährleisten ➤ (Kap. 13.1). Der Einfluss der Eltern setzt sich aus der genetischen Anlage, die sie dem Kind vererbt haben, den das Wachstum und die Entwicklung des Kindes beeinflussenden Faktoren und der Gestaltung des Umfeldes, in dem das Kind aufwächst, zusammen. In fast allen Entwicklungsstudien der letzten Jahre konnte nachgewiesen werden, dass die Eltern sowohl über die Anlage als auch die Umwelt auf das Kind einwirken, wobei vor allem der sozioökonomische Status, der von der schulischen Ausbildung der Mutter und dem Berufsstand des Vaters bestimmt wird, von besonderer Bedeutung ist (➤ Abb. 12.1). Daneben spielen aber auch somatisch prägende Einflüsse in der Schwangerschaft sowie im ersten Lebensjahr eine wichtige Rolle (Schwangerschaftsdiabetes, Ernährungsgewohnheiten der Mutter, Ernährung des Kindes im ersten Lebensjahr usw.). Die Ökologie der kindlichen Entwicklung hat Largo in einem Schema zusammengefasst (➤ Abb. 12.2).

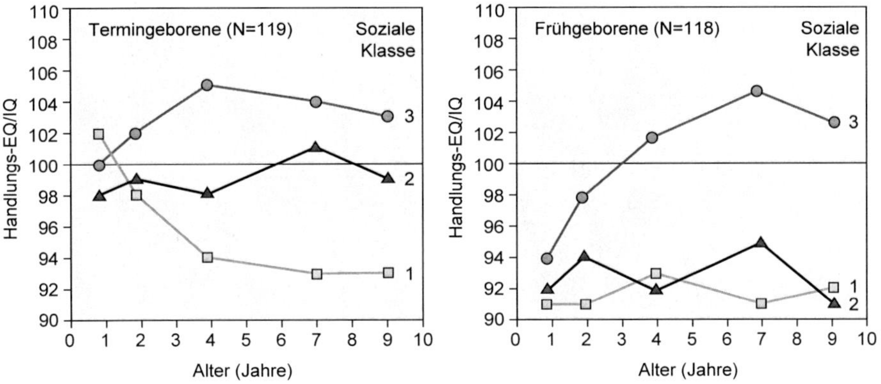

Abb. 12.1 Die Bedeutung der sozialen Klasse bei reifen und frühgeborenen Kindern in der Gesamtentwicklung während der ersten 10 Lebensjahre. 1 = niedrige Sozialschicht, die Mutter hat keinen qualifizierten Schulabschluss; 2 = mittlere Sozialschicht, u. a. qualifizierter Hauptschulabschluss der Mutter; 3 = obere Sozialschicht, mittlere Reife oder Abitur bei der Mutter. Demnach ist vor allem bei Frühgeborenen < 1.500 g die soziale Herkunft von besonderer Bedeutung (nach Largo et al. 1998)

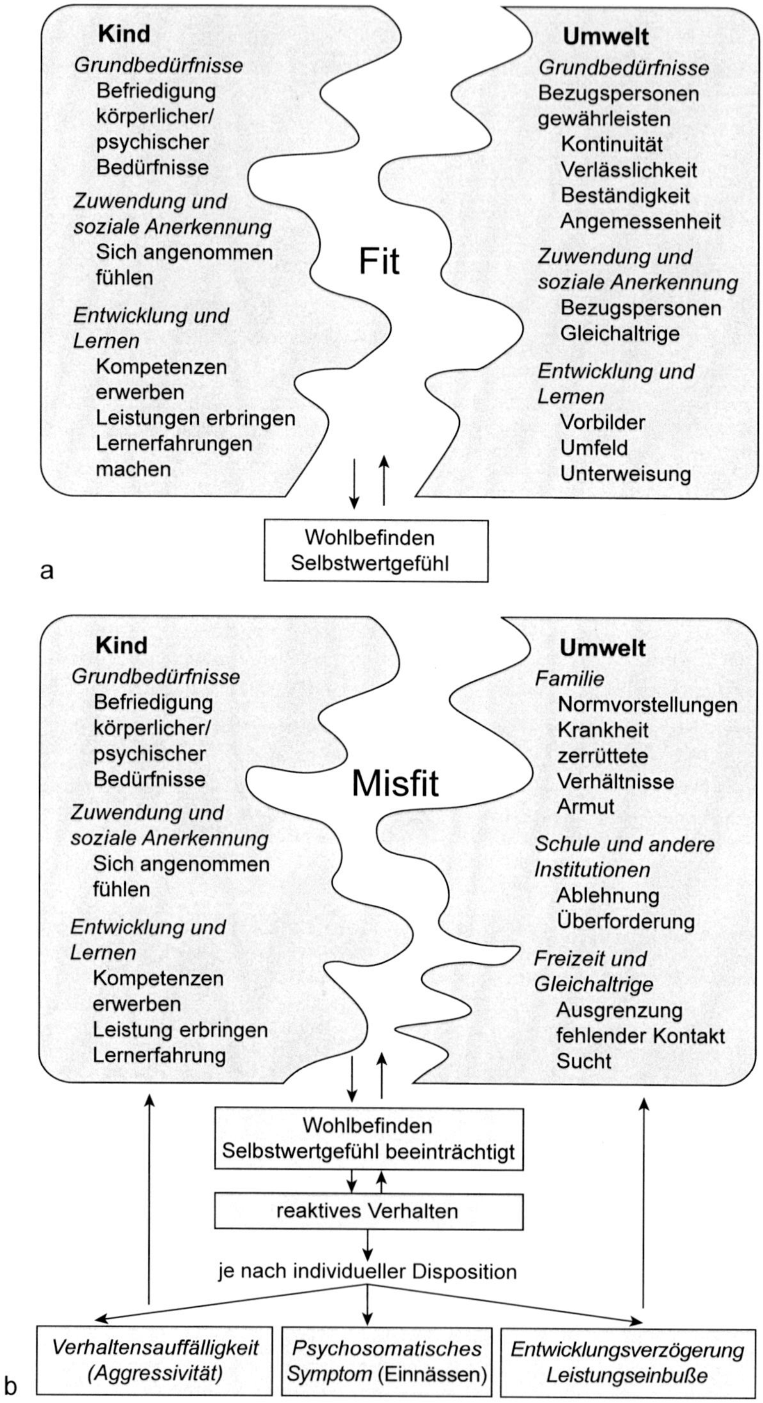

Abb. 12.2 Zusammenfassende Darstellung der Beziehung zwischen Kind und Umwelt bei guter **a** und schlechter **b** Konstellation (nach Largo 2004)

12.3.1 Misfit-Konzept

Die meisten Verhaltens- und viele Entwicklungsstörungen können gerade in den ersten Lebensjahren auf eine unzureichende Übereinstimmung in den Fähigkeiten und Bedürfnissen des Kindes mit Anforderungen und Erwartungen seiner Umwelt zurückgeführt werden, wobei das Kind zunehmend als aktiver Gestalter seiner Entwicklung und seiner Umwelt verstanden wird („Misfit"). Dies ist u. a. auch für therapeutische Ansätze bei Entwicklungsauffälligkeiten von großer Bedeutung.

Jedes Kind hat seine ihm eigenen Schwachstellen, oftmals aber auch individuelle Fähigkeiten, die bei den üblichen Entwicklungsbeurteilungen nur unzureichend berücksichtigt werden. So ist es für das schulische Fortkommen ein großer Unterschied, ob das Kind unmusikalisch und ein schlechter Zeichner ist oder ob es Probleme beim Lesen und Rechnen hat.

Nicht nur Defizite des Kindes, auch überdurchschnittliche Begabungen können Ursache von Verhaltensstörungen sein. Ebenso kann eine langsame Entwicklung mehr Sicherheit vermitteln und eine hastige, zu schnelle Entwicklung mit vielen anderen Problemen verbunden sein.

Besonders eindrucksvoll können die Zusammenhänge von primären Anlagen, sekundären Prägungen und Umwelteinflüssen z. B. auf Temperament und Verhalten in sorgfältig analysierten **Zwillingsstudien** festgestellt werden.

Gerade in den letzten Jahren haben sich viele gesellschaftspolitische Diskussionen über den Einfluss der primären Bindung und der weiteren Entwicklung des Menschen z. B. im Rahmen der außerfamiliären Fremdbetreuung vor dem 3. Lebensjahr (Krippendiskussion) ergeben. Näheres hierzu wird an verschiedenen Stellen, u. a. in Kapitel 13 (➤ Kap. 13), ausführlich dargestellt. Kein Zweifel besteht heute daran, dass zwischen dem 3. und 6. Lebensjahr in den Kindergärten bei Erfüllung von Mindeststandards eine wesentliche Förderung der Gesamtentwicklung stattfinden kann. Dabei sind für das Kind sowohl die Erfahrungen mit Gleichaltrigen im freien Spiel als auch die pädagogischen Anregungen durch die Erzieherinnen von wesentlicher Bedeutung (Stuhlkreis). Durch gut dokumentierte Beobachtungen können dabei die Erzieherinnen im Kindergarten in der Regel wertvolle Grundlagen für die Entwicklungsbeurteilung und die Früherkennung von Verhaltensauffälligkeiten, Entwicklungsstörungen und chronischen Krankheiten legen.

12.4 Grundbegriffe der körperlichen Entwicklung

Die Strukturen und Fähigkeiten des Menschen entwickeln sich aufgrund eines genetisch festgelegten Bauplans und in Abhängigkeit von exogenen Faktoren. Bei Kindern und Jugendlichen werden nach der Geburt folgende **Entwicklungs-** und **Altersstufen** unterschieden:

- **Neugeborenenperiode:** 1. bis 4. Lebenswoche
- **Säuglingsalter:** 5. Lebenswoche bis Ende des 12. Lebensmonats
- **Perinatalzeit:** offiziell die Zeit zwischen der 28. Schwangerschaftswoche (bzw. der extrauterinen Lebensfähigkeit eines Kindes) bis eine Woche nach Geburt
- **Frühgeborenes:** Geburt vor der vollendeten 37. Schwangerschaftswoche
- **Sehr Frühgeborenes:** Geburtsgewicht zwischen 1.000 und 1.499 g bzw. Gestationsalter 28–32 Schwangerschaftswochen
- **Extrem Frühgeborenes:** Geburtsgewicht unter 1.000 g bzw. Geburt vor der 28. Schwangerschaftswoche
- **Kleinkindalter:** 2. bis 6. Lebensjahr

- **Schulalter:** 7. Lebensjahr bis zum Eintritt der Pubertät
- **Pubertät:** Zeitspanne vom Auftreten der ersten sekundären Geschlechtsmerkmale bis zum Eintritt der körperlichen Geschlechtsreife
- **Jugendlichenalter (Adoleszenz)**: 14. bis 18. Lebensjahr bzw. Abschluss der Geschlechtsreife bis zum Abschluss des Körperwachstums.

Das durchschnittliche **Geburtsgewicht** eines gesunden, reifen Neugeborenen beträgt derzeit etwas weniger als 3.400 g (3. bis 97. Perzentile = 2.500 bis 4.600 g), wobei Jungen etwas schwerer als Mädchen sind.

Das Geburtsgewicht wird

- bis zum 5. Lebensmonat etwa verdoppelt,
- bis zum Ende des 1. Lebensjahres verdreifacht (ca. 10 kg),
- bis zum 6. Lebensjahr versechsfacht (ca. 20 kg),
- bis zum 12. Lebensjahr verzwölffacht (ca. 40 kg).

Das Schädelwachstum wird vor allem durch das Gehirnwachstum bestimmt. Der maximale fronto-okzipitale (d. h. um Stirn und Hinterhaupt gemessene) **Kopfumfang** stellt für die Beurteilung von Störungen der Gehirnentwicklung besonders in den ersten Lebensjahren eine äußerst wichtige Messgröße dar. Bei Mädchen liegen die Kopfumfangswerte etwa 1–2 cm unter denen von Jungen.

Weitere Messparameter der körperlichen Entwicklung sind Oberarmlänge, Taillen- und Hüftumfang, Ellbogenbreite, Hautfaltendicke und anderes.

Für die Beurteilung der körperlichen Entwicklung des Kindes sind standardisierte Diagramme (sog. **Somatogramme**) für Körpergewicht, Körperlänge und Kopfumfang in Abhängigkeit vom Lebensalter unerlässlich. Dabei werden in der Regel die 50. Perzentile als Scheitelpunkt der Gauß'schen Verteilungskurve, die 3. und 97. Perzentile als die Begrenzung von zwei Standardabweichungen nach oben und unten sowie die 10., 25., 75. und 90. Perzentile angegeben. In der Regel findet das Wachstum innerhalb eines Perzentilen-Korridors statt. Werden Perzentilen gekreuzt oder Messwerte außerhalb der zwei Standardabweichungen festgestellt, ist dies Anlass zu genaueren Untersuchungen und sich daraus ableitenden Konsequenzen.

Zur Definition von Über- und Untergewicht ist die Berechnung des sogenannten **Body-Mass-Index** (BMI = Körpergewicht in kg geteilt durch das Quadrat der Körpergröße in kg/m^2) unerlässlich. Die Lehre von Wachstum und Entwicklung der Kinder wird als **Auxologie** bezeichnet.

Das biologische Alter eines Kindes kann z. B. durch die Bestimmung der Knochenreifung im Röntgenbild der linken Hand nach Greulich und Pyle, das Wachstum der Zähne oder die Beurteilung der Pubertätsstadien nach J. M. Tanner erfolgen.

12.5 Motorische Entwicklung

Neben den körperlichen Entwicklungsparametern werden je nach Alter sehr unterschiedliche Aspekte zur Einschätzung der Entwicklung eines Kindes berücksichtigt. Vor allem in den ersten Lebensmonaten spielt die Beurteilung der Motorik eine wesentliche Rolle (➤ Abb. 12.3). Dabei unterscheidet man grob(besser: groß-)motorische Fähigkeiten wie Drehen, Sitzen, Krabbeln, Stehen und Gehen von feinmotorischen, wie z. B. der Handgeschicklichkeit, der Mimik und Gestik. Neben spontanen Bewegungen werden in unterschiedlicher Form provozierte Bewegungsabläufe, z. B. die Lagereaktionen in Folge von Positionsänderungen des Kindes im Raum, als Entwicklungskriterium herangezogen. Da jede motorische Äußerung in komplexer Weise mit verschiede-

nen sinnesphysiologischen Erfahrungen wie der Oberflächen- und Tiefensensibilität, dem Sehen, Hören und vestibulären Einflüssen verbunden ist, spricht man besser von **Sensomotorik.**
Nach Prechtl werden bei der Beurteilung der Motorik im frühen Säuglingsalter verschiedene Aktivitätsgrade unterschieden, z. B. Schlaf, ruhiges Verhalten im Wachzustand, aktives Verhalten und Schreien. Mit dem von ihm entwickelten Konzept der Beurteilung der spontanen Körperbewegungen („**General Movements**") können im frühen Säuglingsalter wesentlich bessere prognostische Aussagen über die zukünftigen motorischen Fähigkeiten des Kindes getroffen werden als mit den früher häufig eingesetzten Lagereaktionen.
Aufgrund der Erkenntnis, dass „Normwerte" zur motorischen Entwicklung gerade in den ersten zwei Lebensjahren aufgrund der großen Variabilität problematisch sind, hat sich das Konzept der Entwicklungsbeurteilung nach **„Grenzsteinen"** bewährt. In den Bereichen motorische Entwicklung, Sprachentwicklung, Sozialisation und emotionale Entwicklung sowie Handlungsplanung werden dabei verschiedene Fähigkeiten des Kindes zusammengefasst, die von 90 % aller Kinder bis zu einem bestimmten Zeitpunkt erreicht werden. Beispiele hierfür sind in Tabelle 8.1 zusammengefasst.
Die Entwicklung eines Kindes lässt sich in vielen, z. T. sehr ausführlich standardisierten Verfahren dokumentieren. Beispiele hierfür finden sich in ➤ Kap. 14 und ➤ Kap. 17.

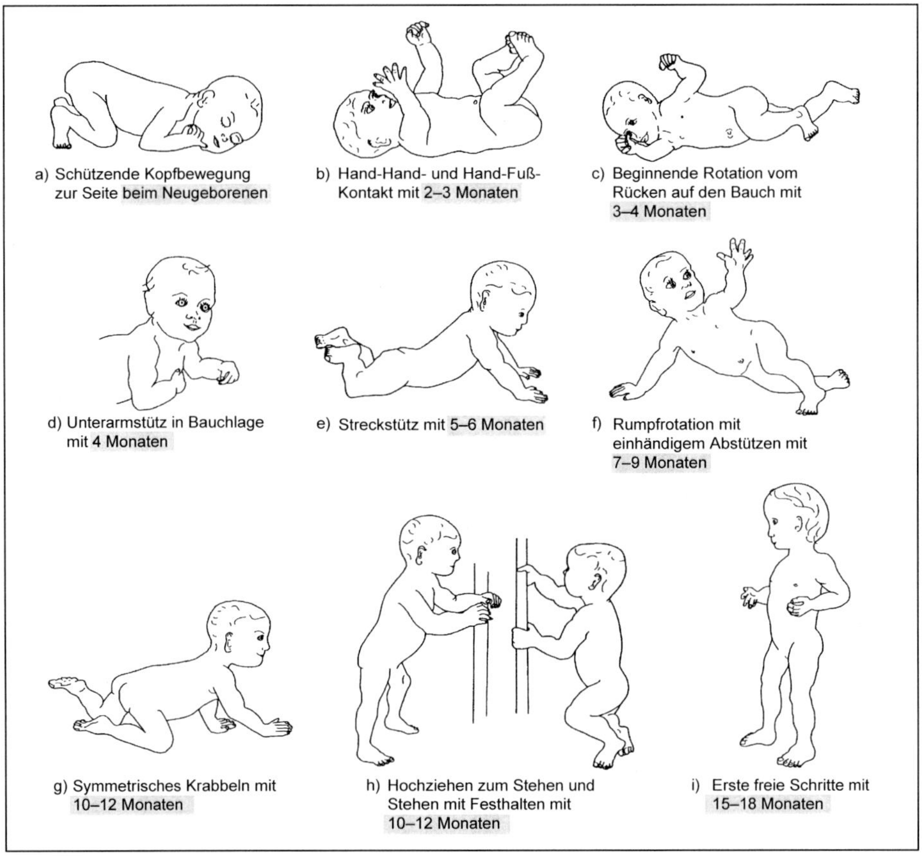

Abb. 12.3 Motorische Entwicklungsschritte [3]

Tab. 12.1 „Grenzsteine" der motorischen, sprachlichen und sozialen Entwicklung von Säuglingen und Kleinkindern bis zum 18. Monat (entsprechend der 90. Perzentile)

Motorische Entwicklung	
1. Lebenswoche (U2)	Trinken ohne Probleme, kräftiges Schreien
4. Lebenswoche (U3)	In Bauchlage wird der Kopf mindestens 3 Sek. angehoben
12. Lebenswoche (U4)	Sicheres Kopfheben in Bauchlage, Rumpf gerade
Ende 6. Lebensmonat (U5)	Sichere Kopfkontrolle bei jedem Lage- und Haltungswechsel
9. Monat	Dreht vom Rücken auf den Bauch, dreht um die eigene Achse, robbt vorwärts oder rückwärts
12. Monat (U6)	Kriecht bzw. krabbelt koordiniert auf Händen und Füßen, steht mit Festhalten
15. Monat	Kommt vom Stehen mit Festhalten alleine wieder zum Sitzen
18. Monat	Geht frei und sicher
Sprachentwicklung	
Ende 4. Lebenswoche (U3)	Kurze gutturale Laute
Ende 12. Lebenswoche (U4)	Spontanes Lallen
6. Lebensmonat (U5)	Vokalisiert auf Ansprache („baby talk")
9. Monat	Silbenkette mit „a" (wawa-rara)
12. Monat (U6)	Gezielter Einsatz von Doppelsilben mit „a" – Mama, Papa
18. Monat	Sinngemäße Verwendung einzelner Worte (Opa, Auto, Ball)
Handlungsplanung	
3. Monat (U4)	Spielt mit den eigenen Fingern
6. Monat (U5)	Greift gezielt palmar
9. Monat	Scherengriff (d. h. Daumen und Zeigefinger werden flach gegeneinander geführt)
Sozialisation und emotionale Entwicklung	
Ende 4. Lebenswoche (U3)	Lässt sich durch Aufnehmen und Ansprechen beruhigen
Ende 12. Lebenswoche (U4)	Lächelt Gesicht an
6. Monat (U5)	Freut sich über Zuwendung, Ansprechen, Anlachen
9. Monat	Unterscheidet zwischen bekannten und fremden Personen
12. Monat (U6)	Enge emotionale Bindung an Bezugspersonen
15. Monat	Ahmt Gestik (Winken, Backe Kuchen) und Laute nach
18. Monat	Versteht direkte Gebote und Verbote, macht Wünsche deutlich

12.6 Kognitive Entwicklung

Neben den motorischen Entwicklungsparametern ist die Feststellung der mental-kognitiven Entwicklung von überragendem praktischen Interesse, aber auch aufgrund der bestehenden Variabilität und vielen Störfaktoren bei der Messung besonders schwierig. Auch wenn heute eine Vielzahl verschiedener Ursachen für signifikante Störungen der mental-kognitiven Entwicklung festgestellt werden können, ist der Stellenwert vieler prä-, peri- und postnataler Parameter nicht immer eindeutig festzulegen. Dies gilt insbesondere auch für psycho-sozio-emotionale Belastungen.
Zur Beurteilung der mental-kognitiven Entwicklung wird im Rahmen der psychologischen Testdiagnostik das **Konstrukt der Intelligenz** bestimmt. Nach ICD-10 werden Menschen mit einem Intelligenzquotienten mehr als zwei Standardabweichungen unter dem Mittelwert, d. h. unter 70, als intelligenzgemindert eingestuft, d. h. bei ihnen liegt eine **geistige Behinderung (= Intelligenzminderung,**

intellectual disability, mental deficiency, mental retardation) vor. Hiervon müssen **umschriebene Entwicklungsstörungen** mit signifikanten Einschränkungen in bestimmten Bereichen, z. B. der Sprache, der Motorik, des Lesens, der Rechtschreibung oder des Rechnens, abgegrenzt werden (➤ Kap. 19, ➤ Kap. 20). Voraussetzung für diese Diagnose ist ein IQ über 70 und eine Abweichung in dem gestörten Bereich von > 1,5 Standardabweichungen. Oft wird auch der Begriff einer **Teilleistungsstörung** verwendet, unter dem man eine Abweichung der Fähigkeiten unabhängig von der Gesamtintelligenz versteht. Bei Menschen mit einem IQ zwischen 70 und 85 wird oftmals von einer **Entwicklungsstörung mit unterdurchschnittlicher kognitiver Leistungsfähigkeit** bzw. von **Lernstörung** oder **Lernbehinderung** gesprochen. Die Begriffe Teilleistungsstörung und Lernstörung kommen in der ICD-10 nicht vor. Obwohl es sich um eine in der ärztlichen, psychologischen und pädagogischen Praxis äußerst wichtige Gruppe handelt, gibt es hierfür keine scharfen Definitionen.
In vielen wissenschaftlichen Studien, vor allem sorgfältig durchgeführten **Längsschnittstudien** wie der **Zürcher Longitudinalstudie,** hat man differenzierte Zusammenhänge zwischen Entwicklungsparametern in den ersten Lebensmonaten und dem Intelligenzquotient nach dem 6. Lebensjahr festgestellt. Ohne auf weitere Einzelheiten eingehen zu können, lässt sich sagen, dass bei der überwiegenden Zahl der Kinder bereits am Ende des 1. Lebensjahres sowohl eine normale Intelligenz als auch eine bleibende Intelligenzminderung festgestellt werden können. Dabei hat sich die Beurteilung der Motorik als prognostisch weniger aussagekräftig erwiesen als die Beurteilung der Sprache und vor allem der Handlungsplanung. Andererseits muss aber auch festgestellt werden, dass Risikokinder, z. B. solche aus psychosozial belasteten Familien oder ehemalig extrem Frühgeborene, am Ende des ersten Lebensjahres z. T. in ihrer prospektiven Entwicklung noch nicht abschließend beurteilt werden können.

12.7 Sozio-emotionale Entwicklung

Bereits beim Feten lassen sich in vielfältiger Weise emotionale Reaktionen, z. B. Änderungen der Bewegungen oder des Herzschlags, nachweisen, meist in Zusammenhang mit der Gefühlssituation der Mutter. Für die Beurteilung der Gesamtentwicklung eines Kindes ist die Beobachtung der **emotionalen Reaktivität** und des **Sozialverhaltens** von besonderer Bedeutung, ohne dass dabei aber immer feste Zeiten angegeben werden können. Hierzu gehören u. a.:

- Sich-Trösten-lassen bereits beim Neugeborenen
- Erstes reaktives Lächeln mit ca. 2 Monaten
- Mitlachen
- Entspannen bei der Pflege
- Stabiles Bindungsverhalten
- Fremdelreaktion
- Anregen eines Spiels
- Konstruktives Spielverhalten
- Differenziertes Mitfühlen können.

Die **Theory of Mind** bezeichnet die Fähigkeit eines Menschen, Gefühle, Bedürfnisse, Ideen, Absichten, Erwartungen und Meinungen von anderen Menschen zu erkennen. Diese offensichtlich angeborene Fähigkeit wird nicht nur als die Grundlage für alle sozialen Interaktionen der Umwelt sondern auch für das Wissen eines Menschen über die Merkmale der Welt angesehen. Hierbei spielt das System der sog. **Spiegelneurone** eine wichtige Rolle. Störungen der sozio-emotionalen Entwicklung können z. B. bei Kindern mit Erkrankungen aus dem Autismus-Spektrum unabhängig von deren Intelligenz auftreten.

12.8 Entwicklung vegetativer Funktionen

In den ersten Lebensmonaten ist das **Schlaf-Wach-Verhalten** des Säuglings noch unregelmäßig, Schlaf- und Wachphasen wechseln innerhalb weniger Stunden ab, meist benötigt er 1–2 Mahlzeiten in der Nacht. Ab dem 3. Monat wird der Schlaf-Wach-Rhythmus zunehmend zirkadian, d. h. er gleicht sich dem Tag-Nacht-Rhythmus an. Das Schlafbedürfnis des jungen Säuglings liegt meist bei über 16 h pro Tag, einen Mittagsschlaf benötigt das Kind in der Regel bis zum 4. Lebensjahr. Es besteht aber eine große Variabilität sowohl für die Schlafdauer als auch den Schlafrhythmus.
Auch das **Schreiverhalten** des gesunden Säuglings ist sehr variabel. Viele Kinder sind in den ersten Lebenswochen relativ ruhig und haben besonders zwischen dem 2. und 5. Lebensmonat längere

Tab. 12.2 Einige Fähigkeiten nach dem Grenzsteinkonzept zum Zeitpunkt der Früherkennungsuntersuchungen U7 bis U9

U7 (21–24 Lebensmonate)
• Geht sicher • Bückt sich und richtet sich wieder auf • Greift gezielt nach Rosine • Hilft beim An- und Ausziehen • Spricht > 50 Worte (mit 24 Monaten)
U7a (34–36 Lebensmonate)
• Hüpft beidbeinig • Steht kurz auf einem Bein • Steigt Treppen im Schrittwechsel • Putzt Zähne • Wäscht und trocknet Hände • Malt Kreuz und Kreis nach • Baut Turm aus > 4 Würfel • Spricht in > 3-Wort-Sätzen • Nennt > 4 Farben • Spielt Rollenspiele
U8 (43–48 Lebensmonate)
• Steht > 5 Sek. auf einem Bein • Kann sich alleine an- und ausziehen • Spricht in ganzen Sätzen grammatikalisch korrekt • Malt beginnend einen Menschen • Benutzt Ich-Form • Kann sich konzentrieren • Spielt einfache Gesellschaftsspiele • Ist sauber und trocken
U9 (60–64 Lebensmonate)
• Spricht fließend in ganzen Sätzen • Hüpft auf einem Bein • Kann Fahrrad ohne Stützräder fahren • Macht Hampelmann-Sprung nach • Malt Drei- und Viereck nach • Kann Frage nach abstraktem Begriff beantworten („Was ist Glück?") • Erfasst Mengen von > 10 • Kann unbekannte Worte nachsprechen

Phasen mit vermehrter Unruhe und Schreien. Im Tagesablauf ist der Säugling vor allem am Nachmittag und Abend unruhig. Schreien kann vielfältige Ursachen haben. Es ist physiologisch bei Hunger, Kälte, unzureichender Kleidung etc., primär pathologisch bei definierten Erkrankungen, z. B. des Gehirns und der Gesamtentwicklung (Cri-du-Chat-Syndrom), und sekundär bei einer Vielzahl von Erkrankungen, z. B. einer Darminvagination, einer Hodentorsion, einer Fraktur, einer Otitis u. v. m. Ob vermehrtes oder gar **exzessives Schreien** vorliegt, ist einerseits sehr vom subjektiven Empfinden der Eltern abhängig, andererseits von der jeweiligen Definition. Nach Wessel wird eine Dauer des Schreiens von > 3 h am Tag für > 3 Tage in der Woche in > als 3 Wochen gefordert, doch werden diese Kriterien selten erreicht. Die Diagnostik und Behandlung von exzessiv schreienden Säuglingen und ihren Eltern erfordert ein großes Maß an Einfühlungsvermögen, Differenzierung und Wachsamkeit. Das größte Risiko für einen vermehrt schreienden Säugling besteht im Erleiden eines Schütteltraumas bzw. anderer Formen der Kindesmisshandlung (➤ Kap. 18, ➤ Kap. 36). Ob es sich um eine, meist vorübergehende, Normvariante, einen Hinweis für eine Entwicklungsauffälligkeit oder eine spezielle Erkrankung, z. B. eine u. U. vorübergehende Nahrungsmittelunverträglichkeit, einen gastro-ösophagealen Reflux oder eine Interaktionsstörung zwischen der Mutter und dem Kind, handelt, muss in Abhängigkeit vom klinischen Bild und vom Verlauf beurteilt werden. Hier muss vor voreiligen Festlegungen und einseitigen Behandlungsvorschlägen gewarnt werden.
Auch die Kontrolle der **Blasen- und Mastdarm-Funktion** ist sehr variabel. In der Regel lernt ein Kind bis zum 3. Geburtstag, tagsüber auf die Toilette zu gehen, in einigen Kulturen werden die Kinder schon am Ende des 1. Lebensjahres sauber. Bei Störungen der Sauberkeitsentwicklung sollten nicht nur psychische Einflussfaktoren, sondern auch verschiedene organische Ursachen berücksichtigt werden (➤ Kap. 24).

12.9 Passagere neurologische Auffälligkeiten der ersten Lebensmonate

Bis zu 30 % aller Säuglinge haben in den ersten Lebensmonaten eine **Asymmetrie,** die im Schädel-, Hals- und Rumpfbereich oder den Extremitätenbewegungen betont sein kann. Die Ursachen sind vielfältig, am häufigsten sind sie durch intrauterine Fehlhaltungen und die überwiegende Rückenlage bedingt. Eltern sind besonders durch eine einseitige Abflachung des Hinterhauptes beunruhigt (Lage-Plagiocephalus) Meist bildet sich die Symptomatik spontan nach dem 6. Lebensmonat zurück. Sie kann auch Ausdruck vielfältiger Krankheiten sein, z. B. von hirnorganischen Anomalien, aber auch durch Skelett-Fehlbildungen sowie muskuläre, gastro-intestinale, okuläre und andere Störungen verursacht werden. Entscheidend ist neben der frühzeitigen Feststellung und dem Ausschluss ernsthafter Grunderkrankungen die richtige Lagerung und die Förderung möglichst vielfältiger aktiver Bewegungen, insbesondere der Rumpfrotation nach beiden Seiten und der Stabilisierung der Bauchlage im wachen Zustand.
Ebenso häufig und in der Regel harmlos ist eine passagere muskuläre Übererregbarkeit (**Hyperexzitabilität**), die sich in Kinnkloni oder fahrigen Bewegungen der Extremitäten, z. B. im Sinne von spontanen Auslösungen der Moro-Reaktion äußern kann und besonders bei Frühgeborenen in den ersten Lebensmonaten vorkommt.
Schwierig kann die richtige Zuordnung von Säuglingen mit einem deutlich erniedrigten Grundtonus der Muskulatur sein (**muskuläre Hypotonie**). Neben physiologischen und passageren Formen (z. B. im Schlaf) müssen vor allem neuromuskuläre Störungen sowie Anomalien des ZNS und der Genetik abgegrenzt werden (z. B. Myopathien, spinale Muskelatrophie, Prader-Willi-Syndrom

und vieles mehr). Auch ein deutlich erhöhter Muskeltonus (muskuläre Hypertonie) oder ein Wechsel von vermindertem und erhöhtem Tonus in Verbindung mit abnormen Bewegungen (muskuläre Dystonie mit Dyskinesen) kann passager sein, aber auch durch sehr unterschiedliche, z. T. ernsthafte Erkrankungen verursacht werden. Da sich viele dieser Symptome spontan zurückbilden und meist mit keinen bleibenden Gesundheitsbeeinträchtigungen verbunden sind, muss der Einsatz therapeutischer Maßnahmen immer wieder kritisch hinterfragt werden. Während die Verordnung von z. B. Physiotherapie zur Anleitung der Eltern im Umgang mit ihrem Kind und zur kurzfristigen Verlaufskontrolle meist sinnvoll ist, gibt es für die Wirksamkeit vieler anderer Empfehlungen, z. B. von Vojta-Therapie, Manualmedizin, Osteopathie, kraniosakraler Therapie usw. keine befriedigenden Beweise (➤ Kap. 37, ➤ Kap. 38).
Der Begriff **„drohende Behinderung"**, der früher vor allem aus sozialrechtlichen Gründen bei entwicklungsauffälligen Säuglingen und Kleinkindern immer wieder verwendet wurde, um z.B. spezielle Förder- und Behandlungsmaßnahmen zu rechtfertigen, sollte vermieden werden, da er völlig unspezifisch ist und meist nur zu einer vermehrten Verunsicherung beiträgt. Die auffälligen Symptome sollten nach den Kriterien der MBS beschrieben werden und ggf. den Eltern auch klar gesagt werden, dass eine definitive Aussage zur Prognose noch nicht möglich ist.

12.10 Kleinkind- und Vorschulalter

Im Säuglingsalter werden die Grundlagen für die weitere Entwicklung gelegt, die sich in der Regel nach dem 1. Geburtstag zeigt, z. B. mit der raschen Zunahme des Wortschatzes und dem zunehmend prosozialen Verhalten.
Ab dem 2. Lebensjahr können neben Beobachtungsparametern auch sogenannte **Entwicklungsaufgaben** eingesetzt werden, worunter Fähigkeiten des Kindes in Abhängigkeit von seiner soziokulturellen Umgebung verstanden werden können, bei denen es eigenständig Aufgaben übernehmen soll. Hierzu können z. B. gehören:

- Akzeptieren von Ruhe- und Schlafzeiten
- Mithelfen beim An- und Ausziehen
- Selbstständiges Ess- und Trinkverhalten
- Stillsitzen können
- Ordnen von Spielsachen und Kleidern
- Zähneputzen
- Toilettengang
- Übernehmen einfacher Aufträge
- Nicht dazwischen reden
- Akzeptieren von Verboten (z. B. nicht fernsehen)
- Hilfe im Haushalt, z. B. beim Tisch decken, Spülmaschine ausräumen, Betten machen, Schuhe putzen usw.
- Akzeptieren von Regeln, z. B. im Straßenverkehr
- Zunehmende Selbstständigkeit
- Prosoziale Verhaltensweisen, z. B. gegenüber Schwachen oder Behinderten
- Berücksichtigung unterschiedlicher Lebensgeschichten
- Akzeptanz von gemeinsamen Aktivitäten in Kindergarten und Schule u. v. m.

Das Alter, in dem diese Aufgaben von dem Kind erfüllt werden, ist naturgemäß variabel und viele Eltern beklagen erhebliche Probleme, wenn sie diese einfordern. Die Trennschärfe zwischen einem noch normalen Verhalten und definitiven Verhaltensauffälligkeiten oder sogar -störungen ist nicht scharf. Die häufigsten Begründungen der Eltern zur Vorstellung ihrer Kinder sind Auffälligkeiten in der Sprachentwicklung, aber auch Verhaltensprobleme, motorische Ungeschicklichkeiten und eine verzögerte Sauberkeitsentwicklung. In jedem Fall sollten diese Sorgen nicht vorschnell bagatellisiert werden. Durch die Einführung der Früherkennungsuntersuchung U7a bei 3jährigen als Regelleistung der gesetzlichen Krankenversicherungen ist die Beurteilung von Entwicklungsauffälligkeiten und eine frühzeitige Intervention wesentlich besser möglich (➤ Kap. 14; ➤ Kap. 18; ➤ Kap. 37; ➤ Kap. 38).

12.11 Schulalter

Alle Kinder haben das Recht, aber auch die Pflicht, die **Schule** zu besuchen. Eine Zurückstellung vom Schulbesuch kann nur durch gesonderten Antrag erfolgen.

Die Entwicklung eines Kindes bleibt auch im Schulalter in einem hohen Maße davon abhängig, inwieweit sich das Kind emotional und sozial geborgen fühlt, weiterhin sind die Eltern seine Hauptbezugspersonen. Das Kind ist aber nicht mehr auf ihre ständige Präsenz angewiesen, braucht jedoch für sein psychisches Wohlbefinden die Gewissheit, jederzeit Zugang zu seinen Eltern oder einer anderen kontinuierlich ansprechbaren Bezugsperson zu haben. Viele Studien haben übereinstimmend gezeigt, dass außerfamiliäre Bezugspersonen die sozio-emotionale Entwicklung von Kindern, die in zerrütteten Familien leben, positiv beeinflussen können (z. B. Kaoai-Studie von Werner und Smith 1989, 1992). Im Verlauf der Schulzeit werden die Beziehungen zu gleichaltrigen Kindern für das psychische Wohlbefinden eines Kindes immer wichtiger und es orientiert sich in seinem Lernverhalten zunehmend an Lehrern und Mitschülern. Dabei können sowohl äußere Einflüsse, wie Trennung der Eltern oder Krankheiten in der Familie, Armut, Änderung der Umgebung durch Umzug und vieles mehr die Entwicklung beeinflussen, aber auch verschiedene krankheitsbedingte bzw. körperliche Belastungsfaktoren des Kindes (z. B. eine Zöliakie, Migräne, atopische Dermatitis, Asthma bronchiale, Minderwuchs, Intelligenzminderungen u. v. m.). Im Rahmen ausführlicher Untersuchungen an Pflege-, Adoptions- und Zwillingskindern konnte immer wieder festgestellt werden, dass das Kind kein passives Wesen ist, das durch die Umwelt geformt wird. Es ist aktiv, indem es selektive Erfahrungen macht und auf seine Umwelt Einfluss nimmt.

So können durch die Ausbildung von speziellen Fähigkeiten (Hobbys, sportliche Aktivitäten, Musik, Lesen) einerseits und durch die Einwirkung belastender Faktoren (Fernsehkonsum, Videospiele usw.) andererseits sehr eindrucksvolle Beeinflussungen der kindlichen Entwicklung erfolgen.

Bis zu Beginn der Schulzeit wird dies zweifelsohne überwiegend von den Eltern bestimmt, mit Beginn der Pubertät bestimmen die selektiven Eigenaktivitäten des Kindes zunehmend sein Beziehungs- und Lernverhalten.

Durch die zunehmende **Inklusion** von Kindern mit speziellen Bedürfnissen bzw. erhöhtem Förderbedarf in den Regelschulen ergeben sich für diese eine Vielzahl von neuen Chancen aber auch von Problemen, die einen besonderen Beratungs- und Betreuungsbedarf haben (➤ Kap. 4; ➤ Kap. 9; ➤ Kap. 13).

12.12 Adoleszenz

Zwischen dem 12. und 16. Lebensjahr verändert sich das Beziehungsverhalten vieler Jugendlicher grundlegend, sie lösen sich oftmals abrupt von ihren Eltern und wenden sich anderen Bezugspersonen, meist Gleichaltrigen, zu. Ihr emotionales und soziales Wohlbefinden hängt nun in einem hohen Maße von der Akzeptanz ihrer **Peergroup** ab. Die sozialen Wertvorstellungen in der Schule, der Umgang der Schüler miteinander sowie der Lehrer untereinander und mit den Schülern wirken sich nachhaltig auf das Verhalten und die Leistungsfähigkeit der Jugendlichen aus. Daneben spielen die aktive Gestaltung der Freizeit, die dabei bestehenden Rahmenbedingungen, die finanziellen Möglichkeiten und die Grenzen eine zunehmende Rolle. Dennoch haben gerade auch in dieser Zeit die **Eltern** eine wichtige Funktion. Dialog und Auseinandersetzung mit den Erwachsenen bleiben für den Jugendlichen von besonderer Bedeutung. Fehlt der elterliche Rückhalt, können andere Bezugspersonen wie Lehrer oder Jugendorganisationen diese stabilisierenden Funktionen übernehmen.

Die Adoleszenz ist eine Altersperiode des Umbruchs und birgt zahlreiche psychosoziale Risiken, die sich in z.T. transitorischen, z.T. aber auch permanenten Verhaltensauffälligkeiten und Entwicklungsstörungen, Delinquenz, Drogenabhängigkeit und inadäquaten schulischen Leistungen sowie beruflichen Qualifikationen äußern können. Wichtige Schutzfaktoren gerade in dieser Umbruchszeit sind vor allem emotionale und soziale Geborgenheit mit mindestens einer verlässlichen Bezugsperson, die nicht unbedingt ein Elternteil sein muss, und der Erziehungsstil.

Die Entwicklung eines Kindes kann nachhaltig gestört werden, wenn die Eltern in ihrer Erziehungshaltung unzuverlässig und unberechenbar sind, z.B. wegen psychischer Schwierigkeiten oder Alkoholismus. Andererseits können Erwachsene, die mit den Anforderungen in Familie und Beruf zurechtkommen, für die Jugendlichen ein positives Vorbild werden. Wird den Kindern in der Familie, der Schule oder im Verein Verantwortung übertragen, erweisen sie sich in der Regel in der Adoleszenz als psychisch stabiler und belastbarer als Kinder, die verwöhnt und ohne altersentsprechende soziale Verpflichtungen aufwachsen.

Die Entfaltung eigener Fähigkeiten und das Aneignen von Wissen wirken sich positiv auf das Selbstwertgefühl aus. Gute Schulleistungen haben einen stabilisierenden Einfluss auf das psychische Wohlbefinden.

Jugendliche, die von ihren Altersgenossen (Peers) nicht oder nur in Randgruppen angenommen werden, sind besonders empfänglich für Verhaltensauffälligkeiten, psychosomatische Störungen oder Delinquenz. Eine offenbar zunehmend häufige Variante von Verhaltensproblemen bei Jugendlichen ist das Mobbing.

Zunehmend mehr Jugendliche haben chronische Krankheiten, Behinderungen und psychische Störungen. Diese benötigen gerade auch in der Übergangsphase zum Erwachsenenalter, der sog. **Transition,** eine intensivere Betreuung und eine gut organisierte Übergabe von der Kinder- und Jugendmedizin zur Erwachsenenmedizin (➤ Kap. 4; ➤ Kap. 9; ➤ Kap. 13).

Literatur

Baumann T (2013) Atlas der Entwicklungsdiagnostik – Vorsorge-Untersuchungen von U1–U10/J1. Stuttgart: Thieme.

Bergmann KE, Bergmann RL, Ellert U, Dudenhausen JW (2007) Perinatale Einflussfaktoren auf die spätere Gesundheit – Ergebnisse des Kinder- und Jugendgesundheitssurveys (KiGGS) Heidelberg: Springer: 670–676.

Largo RH (2000) Kindliche Entwicklung und psychosoziale Umwelt. In: Schlack HG. Sozialpädiatrie – Gesundheit, Krankheit, Lebenswelten. München: Urban & Fischer: 7–25.

Largo RH (1998): Babyjahre – die frühkindliche Entwicklung aus biologischer Sicht – das andere Erziehungsbuch. München: Piper.

Largo RH (2004) Kinderjahre – die Individualität des Kindes als erzieherische Herausforderung. München: Piper.

Largo RH, Beglinger M (2009) Schülerjahre – wie Kinder besser lernen. München: Piper.

Pikler E (1988) Lass mir Zeit . Die selbständige Bewegungsentwicklung des Kindes bis zum freien Gehen. München: Pflaum

Stolzenberg H, Karl H, Bergmann KE (2007) Körpermaße bei Kindern und Jugendlichen in Deutschland – Ergebnisse des Kinder- und Jugendgesundheitssurveys (KiGGS) Heidelberg: Springer: 659–669.

Straßburg HM, Dacheneder W, Kreß W (2012) Entwicklungsstörungen bei Kindern – Praxisleitfaden für die interdisziplinäre Betreuung. München: Elsevier.

Straßburg HM, Mäder B, Sporrer T (2013) Die Dokumentation der Entwicklung von 2- und 3-jährigen Kindern in Tageseinrichtungen. Frühe Bildung 2 (3), 1–9

Werner EE, Smith RS (1992) Overcoming the odds. High risk children from birth to childhood. Ithaca: Cornell University Press.

Winterhoff M (2009) Tyrannen müssen nicht sein – warum Erziehung allein nicht ausreicht – Auswege. Gütersloh: Gütersloher Verlagshaus.

IV Einfluss des Lebensumfeldes

IV

13 Einflüsse des Lebensumfeldes auf Entwicklung, Verhalten, Gesundheit und Teilhabe

Harald Bode

13.1 Eltern/Familien

Eltern sind in der Regel für Kinder der zentrale Bezugspunkt. Die Fürsorge von Eltern für ihre Kinder beinhaltet vielfältige **Aufgaben.** Eltern bzw. Familien sollen die Erfüllung der physischen Bedürfnisse der Kinder garantieren. Dazu gehören Schutz vor Risiken und Gefahren unterschiedlichster Art, angemessene Ernährung und Wohnung sowie Gesundheitsfürsorge. Familien sollen den Entwicklungs-, Verhaltens- und emotionalen Bedürfnissen der Kinder Rechnung tragen. Dazu gehören der Aufbau einer sicheren Bindung, die Anregung und Förderung kindlicher Interessen, erzieherische Maßnahmen und auch emotionale Zuwendung. Familien helfen durch Einüben in der Familie sowie Bereitstellung und Vermittlung sozialer Beziehungen bei der Sozialisierung ihrer Kinder, damit sie zu konstruktiven und verantwortungsvollen Mitgliedern der Gesellschaft heranwachsen.
Instabile **Familienstrukturen** gefährden diese familiären Funktionen. Die traditionelle Familienstruktur ist heute nur noch eines von verschiedenen Lebensmodellen. Familienzusammensetzungen und Rollenzuweisungen verändern sich in der Gesellschaft und auch innerhalb einer Familie (Wohnortwechsel, Wochenendbeziehungen, Partnerwechsel, gleichgeschlechtliche Eltern). Der Anteil von Ein-Eltern-Familien hat stark zugenommen. Hier übernehmen meist die Mütter den

Sozialpädiatrie in der Praxis. http://dx.doi.org/10.1016/B978-3-437-31630-3.00013-4
Copyright © 2014 Elsevier GmbH. All rights reserved.

überwiegenden Anteil elterlicher Aufgaben. Sie sind durch gleichzeitige Berufstätigkeit doppelt belastet. Teilweise übernehmen auch Großeltern oder andere Familienmitglieder elterliche Rollen und Aufgaben. In einer großen Zahl von Familien sind beide Eltern berufstätig. Die genannten Familienstrukturen können zu vermehrtem innerfamiliären Stress, reduzierten Erholungsphasen und einem Mangel an positiven kindbezogenen Aktivitäten führen, sodass die Eltern nur noch eingeschränkt über Energien verfügen, sich ihren vielfältigen Fürsorgeaufgaben zu stellen. Qualitativ hochwertige Kindertageseinrichtungen mit angemessenen Öffnungszeiten sind deshalb heute ein dringendes Erfordernis.

Die Erfüllung der elterlichen Aufgaben ist unter den heutigen gesellschaftlichen Umständen schon für Eltern gesunder Kinder keineswegs selbstverständlich. Akute, insbesondere jedoch chronische Erkrankungen, Entwicklungsstörungen oder Behinderungen eines Kindes verschärfen den familiären Stress. Die Erfüllung der elterlichen Aufgaben erfordert bei diesen Kindern oft Hilfestellungen seitens des Umfeldes der Familie, verschiedener Institutionen und sozialer Sicherungssysteme.

Sozialpädiatrische Elternbegleitung arbeitet mit einem systemischen und ressourcenorientierten Ansatz. Dabei müssen die drei Ebenen der familiären Lebenssituation, d. h. Kind, Eltern und Umfeld, betrachtet werden, um angemessene Interventionsansätze zu finden. **Ressourcen der Eltern** sind ihr intuitives Wissen über angemessene Verhaltensförderung und Erziehung des Kindes, die elterliche Persönlichkeit, positive Lebenserfahrungen sowie die Unterstützung durch andere. Beeinträchtigungen der Eltern beinhalten Erkrankungen, eingeschränkte intellektuelle, soziale oder emotionale Kompetenzen, mangelndes Wissen und soziale Isolation.

Kinder **chronisch oder psychisch kranker Eltern** müssen besondere Herausforderungen erfüllen. Sie übernehmen z. T. pflegerische Aufgaben. Von ihnen wird häufig überdurchschnittliche Selbstständigkeit und Autonomie **(Parentifizierung)** erwartet. Besonders jüngere Kinder sind durch krankheitsbedingte Instabilitäten in familiären Abläufen und – Beziehungen leicht überfordert. Kinder in alkoholbelasteten Familien werden durch ein stark wechselndes, z. T. unberechenbares Verhalten des betroffenen Elternteils stark verunsichert, erleben negative Rollenvorbilder und stehen in Loyalitätskonflikten. Das **Kindeswohl** kann durch mangelnde oder vernachlässigte elterliche Fürsorge und Förderung gefährdet sein (➤ Kap 36). Viele Kinder psychische kranker Eltern zeigen im Verlauf ihrer Entwicklung psychische Auffälligkeiten, insbesondere Störungen der Emotionen oder des Sozialkontaktes.

Viele Eltern zeigen eine **intuitive Kompetenz** für die Entwicklungsbedürfnisse ihrer Kinder. Bei mangelnden praktischen Vorbildern oder infolge der verwirrenden Vielfalt der Vorschläge in unserer informationsübersättigten Gesellschaft ist Elternkompetenz jedoch keineswegs mehr selbstverständlich. Eltern von Kindern mit chronischen Erkrankungen, Entwicklungsstörungen oder Behinderungen stehen vor zusätzlichen Schwierigkeiten. Sie müssen Kenntnisse erlangen über die spezifischen Entwicklungs-, Förder- und Therapiebedürfnisse ihrer Kinder sowie adäquate Techniken und Materialien unter Berücksichtigung des kindlichen Entwicklungsstandes, seiner Stärken, Schwächen und individuellen Bedürfnisse. Die Förderung der Elternkompetenz ist ein zentrales Anliegen jeder Förderung und Behandlung. **Elternschulungsprogramme** können dazu beitragen (Empowerment).

Zu den **Ressourcen eines Kindes oder Jugendlichen** gehören z. B. gute kognitive Fähigkeit, positive emotionale Eigenschaften und interaktive Stärken. Beeinträchtigungen der Kinder beinhalten die verschiedenen Formen von Entwicklungsstörungen, Behinderungen, Begleiterkrankungen und reaktive Störungen, ggf. auch akute Erkrankungen.

Die elterliche Einschätzung der Fähigkeiten, Entwicklungsstand, Krankheiten oder Beeinträchtigungen ihres Kindes und ihre Reaktion auf diese Probleme können sich von denen Professioneller stark unterscheiden. Sind elterliche Erwartungen hoch und das Kind entwickelt sich langsam, werden die Eltern sehr besorgt sein. Sind die elterlichen Erwartungen dagegen niedrig, können

Eltern auch bei deutlichen Entwicklungsauffälligkeiten durchaus zufrieden sein. Gleiches gilt für den Therapieerfolg, der von Eltern oft höher als von Therapeuten eingeschätzt wird. Ein Feedback über Therapieziele und Therapieerfolg zwischen Ärzten und Fachleuten erhöht den Therapieerfolg in den Augen der Eltern und in objektiven Testverfahren.
Die von Eltern behinderter Kinder für sich selbst und ihre Kinder wahrgenommene **Lebensqualität** wird besonders durch die Faktoren emotionale Stabilität, Mobilität und Autonomie des behinderten Kindes bestimmt. Dies zeigt wesentliche Zielbereiche einer an den Bedürfnissen von Familien orientierten Förderung und Behandlung.

13.2 Trennung, Scheidung, Patchwork-Familie, Alleinerziehende

Die gesetzlichen Bestimmungen in Deutschland sehen bei **Trennung oder Scheidung** der Eltern die gemeinsame elterliche Sorge für das Kind als Regelfall vor. Im Vordergrund steht das Kindeswohl, nicht das Elternwohl. Bei getrennten Wohnungen müssen „erhebliche" Fragen der Person und Vermögenssorge einvernehmlich gelöst werden, Angelegenheiten des täglichen Lebens durch das Elternteil, bei dem sich das Kind aufhält (➤ Kap. 42).
Dies bedeutet für Fragen der **Gesundheitsfürsorge,** dass diagnostische und therapeutische Maßnahmen im Rahmen banaler/nicht schwerwiegender Erkrankungen und auch Früherkennungsuntersuchungen bei Zustimmung des Elternteils, bei dem sich das Kind aufhält, durchgeführt werden können. Gleiches gilt für Notfallmaßnahmen. Die Zustimmung beider Eltern sollte zu schwerwiegenden diagnostischen oder therapeutischen Eingriffen und zu Langzeit-Therapien eingeholt werden.
In aller Regel hat das Kind das Recht auf einen **Umgang** mit beiden Eltern, beide Eltern die Pflicht zum Umgang mit dem Kind, ggf. unter Mitwirkung Dritter. Dabei besteht Anspruch auf Beratung und Unterstützung, z. B. seitens der Jugendhilfe. Wesentliche gesetzliche Rahmenbedingungen finden sich im Bürgerlichen Gesetzbuch (BGB) in § 1564 ff. und § 1626 ff. sowie im Sozialgesetzbuch (SGB) VIII §§ 17–20.
Kinder in Scheidungsfamilien sind mit vielfältigen Problemen konfrontiert. Häufig kommt es zu fortgesetzten Streitigkeiten der Eltern oder Auseinandersetzungen über Sorgerecht und Besuchsregelungen. Nicht wenige Eltern zeigen in dieser Lebensphase emotionale oder psychische Störungen und Beeinträchtigungen ihrer elterlichen Funktionsfähigkeit. Sie erleben wenig Unterstützung von außerhalb der Kernfamilie und erfahren erhebliche ökonomische Belastungen. Die Kinder leiden oft unter einem Verlassenwerden durch einen Elternteil. Sie müssen nicht selten eine Beziehung zu einem neuen Partner eines Elternteils aufbauen, was Chancen, aber auch Probleme mit sich bringen kann.
Scheidungsfolgen für die Eltern beinhalten oft eine einschneidende Verminderung der finanziellen Möglichkeiten, die sich aus eingeschränkten oder fehlenden Arbeitsmöglichkeiten, sozialer Isolation, Schwierigkeiten in der Kinderbetreuung und ggf. neuen Wohnverhältnissen ergibt. Die Eltern sind psychisch belastet durch konfliktreiche Interaktionen, depressive Störungen, Probleme und Zeitbedarf für die Beziehung zu einem neuen Partner. Dies führt nicht selten zu einer Abnahme der elterlichen Funktionsfähigkeit.
Die Scheidungsfolgen für die Kinder können ebenfalls gravierend sein. Die elterliche Zuwendung ist nicht selten zeitlich vermindert, die Kinder fühlen sich häufig allein gelassen. Großeltern oder andere Bezugspersonen übernehmen neue Aufgaben.

Allerdings kann die Beendigung bzw. Klärung einer langfristig konflikthaften Elternbeziehung für Kinder auch entlastend sein und neue Chancen eröffnen. Bei Scheidung der Eltern sind bei den Kindern kurz- und langfristige Effekte, die alters- und z. T. geschlechtsabhängig sind, zu unterscheiden.

Kinder im Alter von etwa 2–5 Jahren reagieren häufig mit Regression, Irritabilität oder Schlafstörungen. Im Alter von 6–8 Jahre zeigen Kinder evtl. offene Trauer, Zurückweisung und erleben Loyalitätskonflikte. 9–12-Jährige zeigen oder fühlen Ärger über einen oder beide Elternteile, Adoleszente übertragen die Erfahrungen sorgenvoll auf die eigene Zukunft, entwickeln Depressionen oder ausagierendes Verhalten.

Typische **klinische Symptome** von Kindern/Jugendlichen, bei denen an elterliche Konflikte, Trennung oder Scheidung gedacht werden muss, sind: Regulationsstörungen, psychosomatische Beschwerden, Verhaltens- oder emotionale Störung, Störung des Sozialkontakts, Aufmerksamkeitsstörungen, Rückgang schulischer Leistung, posttraumatisches Stresssyndrom, dissoziative Störungen. Langfristig ist die psychiatrische Morbidität von Kindern/Jugendlichen aus Trennungs-/Scheidungsfamilien erhöht.

Professionelle werden mit unterschiedlichen Szenarien konfrontiert. Chronische Erkrankungen des Kindes, insbesondere bestimmte Behinderungsformen und Verhaltensstörungen, belasten Ehen und führen gehäuft zu Trennung oder Scheidung. In anderen Fällen führen eheliche Konflikte zu einer erhöhten Inanspruchnahme medizinischer Leistungen, einmal aus Sorge um das Kind, aber auch aufgrund erhöhter Morbidität. Eine neue Betreuungs- und insbesondere Wohnsituation erfordert nicht selten auch ein neues medizinisches und gesundheitsförderndes Setting. Unstrittig zeigt sich jedoch auf der anderen Seite, dass eine „gelungene Scheidung" die Lebenssituation und die zuvor bestehenden klinischen Symptome von Kindern im Kontext elterlicher Konflikte deutlich verbessern kann.

Partner und Kinder bewältigen Trennungs- und Scheidungsprozesse unterschiedlich. Ein langjähriger Verarbeitungsprozess ist die Regel. Die emotionale Bedeutung beider Eltern für das Kind hält an, der biologische Vater bleibt auch bei Wiederverheiratung der Mutter relevant. Das Selbstbild von Kindern entspricht auch heute nicht der Ein-Eltern-Familie.

Insgesamt müssen elterliche Konflikte, Scheidung und Scheidungsfragen als Risiko für die körperliche und psychische Gesundheit eines Kindes betrachtet werden.

Kinder mit geistiger oder schwerer Mehrfachbehinderung haben besondere Probleme in der Verarbeitung familiärer Belastungssituationen, auch wegen ihrer oft eingeschränkten sprachlichen Möglichkeiten und eines begrenzten Verhaltensrepertoires.

Die Inhalte professioneller Beratung für Sorgeberechtigte in Trennungs- und Scheidungssituationen sollten sich am Entwicklungsstand der Kinder orientieren. Im **Vorschulalter** ist den Eltern zu empfehlen, häusliche Routinen beizubehalten und das Leben der Kinder (Bezüge und Tagesabläufe) möglichst stabil zu gestalten. Die elterliche Liebe sollte sicher vermittelt und die kindliche Unschuld an der Scheidung klargestellt werden. Klare Absprachen bei Abwesenheit des versorgenden Elternteils sind zwingend erforderlich, regelmäßige Besuche beim anderen Elternteil in aller Regel wünschenswert. Hierbei sind verbindliche Absprachen wichtig und müssen eingehalten werden. Vorschulkindern sollte in derartigen Übergangssituationen vermehrt Zeit und Zuwendung vor dem Einschlafen gewidmet werden, bei Trennungsängsten ist Begleitung, ggf. professionelle, anzuraten.

Im **Grundschulalter** ist Kindern zu vermitteln, dass sie positive Gefühle beiden Eltern gegenüber haben dürfen. Der Betreuungswechsel zwischen den Eltern sollte selbstverständlich, verbindlich und konstruktiv stattfinden. Sorgen des Kindes über sich oder die Eltern sind nicht selten, sollten erkannt und in geeigneter Form reduziert werden.

Kinder im Alter von etwa **9–12 Jahren** negieren nicht selten äußerlich ihre Probleme mit der elterlichen Trennungssituation. Sie nehmen die elterlichen Konflikte mit Sorge und dem Gefühl einer gewissen Machtlosigkeit wahr, versuchen z. T. auch mit sensibler Zuwendung oder Hilfsangeboten einen positiven Beitrag zu leisten. Kinder sollten nicht in elterliche Dispute einbezogen werden. Schulische und außerschulische Aktivitäten sind zur Stabilität wichtig und sollten gefördert werden.

Auch **Jugendliche** in Trennungs- und Scheidungssituationen brauchen Schutz und Bestätigung für ihre Ziele und Aktivitäten. Eltern sollten Jugendliche nicht für eigene Zwecke oder die eigene psychische Stabilisierung instrumentalisieren. Professionelle sollten prüfen, ob Jugendliche ein offenes Gespräch wünschen.

Die langfristige Prognose, insbesondere der psychischen Entwicklung von Kindern in Trennungs- und Scheidungssituationen hängt wesentlich von der Lösung bzw. Persistenz elterlicher Konflikte ab. Wesentlich sind weiterhin die psychologische Integrität, die emotionale und physische Verfügbarkeit des betreuenden Elternteils und die Qualität der elterlichen Betreuung. Eine konstante Beziehung und eine gefühlte Akzeptanz seitens des nicht betreuenden Elternteils sind höchst bedeutsam, aber leider oft nicht stabil gewährleistet. Die Erfüllung der kindlichen Entwicklungsbedürfnisse, sozial unterstützende Netzwerke und positive Resilienzfaktoren seitens des Kindes wirken sich günstig auf dessen langfristige Entwicklung aus.

Nicht selten wird ein **elterliches Entfremdungssyndrom** beobachtet. Dies beinhaltet den Beziehungsabbruch seitens eines Elternteils. Es kommt zur bewussten oder unbewussten Beeinflussung von Kindern durch signalisierte Verstimmung oder verbale Verunglimpfung oder Herabsetzungen bis hin zur massiven Ablehnung eines Elternteils. Gelegentlich werden auch Professionelle in einem solchen Kontext instrumentalisiert. Es ist in solchen Situationen wichtig, sich auf das Kindeswohl und nicht die elterlichen Wünsche zu konzentrieren und soweit möglich sich auf objektive Tatbestände zu beziehen. So müssen z. B. Angaben über Kopf-/Bauchschmerzen oder Schlafstörungen des Kindes, Einnässen etc. nach Besuch beim anderen Elternteil ernst genommen werden, dürfen jedoch nicht vorschnell zu einer einseitigen Parteinahme seitens des Professionellen führen.

Folgende **Hilfen** können Professionelle bei Trennung/Scheidung vermitteln oder anbieten:

- Präventive Beratung
- Ärztliche Diagnostik und Behandlung
- Individualtherapie (Kind/Eltern)
- Mediation mit Kindern
- Gruppen (Kind/Eltern)
- Familientherapien
- Psychiatrische Klinik (Tagesklinik, stationär)
- Sozialpädagogische Familienhilfe
- Heilpädagogische Tagesstätte
- Schule für Erziehungshilfe
- Heime.

In Streitfällen sind Verfahrenspfleger und **familiengerichtliche Maßnahmen** notwendig, die gelegentlich die Übertragung der elterlichen Sorge auf ein Elternteil erforderlich machen. Der Wille des Kindes sollte nach einer persönlichen Anhörung in die Entscheidung einbezogen werden. Entscheidungen zur elterlichen Sorge seitens der Gerichte müssen die Eignung des Elternteils zur Erziehung des Kindes, Faktoren wie Kontinuität und Umfeld sowie die Wohn- und Lebensverhältnisse berücksichtigen, jedoch nicht das Elternwohl, die Berufstätigkeit eines Elternteils oder die Gründe einer Scheidung. Die Trennung von Geschwistern und ihre Aufteilung auf verschiedene

Elternteile sollte allenfalls bei Heranwachsenden und nur auf deren Wunsch hin erfolgen. Verbindliche Besuchsvereinbarungen leiten sich aus dem Recht und der Pflicht des elterlichen Umgangs (außer bei bestimmten Fällen der Kindeswohlgefährdung) ab.
Neue Partnerschaften oder Wiederverheiratung eines Elternteils führen zu den heute verbreiteten **Patchwork-Familien.** In diesen haben Eltern häufiger Probleme mit ihren neuen Rollen als die Kinder. Jüngere Kinder adaptieren sich in solchen Situationen in der Regel besser, insbesondere, wenn durch bessere elterliche Fürsorge oder glückliche Eltern/Stiefeltern die familiären Funktionen wieder stabilisiert sind. Das Leben vieler Kinder wird in Patchwork-Familien durchaus bereichert, allerdings sind Konflikte von Stiefkindern und Stiefeltern gerade bei Adoleszenten nicht ungewöhnlich. Stiefeltern können jedoch auch positive Einflüsse auf die Lebenskonzepte von Jugendlichen haben. Probleme in Patchwork-Familien entstehen dann, wenn Konflikte zwischen alten und neuen familiären Zeremonien oder Werten auftreten, unrealistische Erwartungen bestehen und innerhalb der neuen Familie Spannungen oder Furcht vor einer erneuten Destabilisierung auftreten. Auch erhebliche geografische Veränderungen mit Verlust bisheriger sozialer Netzwerke und Beziehungen können eine Belastung darstellen.
Die meisten **allein erziehenden Eltern** sind Mütter, die oft in ökonomisch belasteten Situationen leben. Die Perspektive für Kinder allein erziehender Eltern ist jedoch nicht unbedingt ungünstig. Zu einer positiven Entwicklung der Kinder tragen neben mütterlichen Faktoren solche der sozialen Unterstützung wesentlich bei. Als Beispiele seien Erziehungsgeld, institutionelle Betreuung, familienfreundliche Arbeitszeiten und Teilzeitstellen sowie positive gesellschaftliche Einstellungen genannt.

13.3 Adoption und Pflege

Adoption ist die Annahme an Kindes statt. Adoptivkinder sind formal und gesetzlich alleinige Kinder ihrer sozialen Eltern. Sie haben dieselben Rechte wie leibliche Kinder. Adoptiveltern sind sorgeberechtigt und tragen den Unterhalt für das Kind allein. Die psychische Realität ist allerdings eine andere: Adoptierte Kinder haben Herkunftseltern und soziale Eltern.
Adoption ist ein lebenslanger Prozess. In der Mehrzahl der Fälle gelingt er Eltern und Kindern in befriedigender Weise. Auf die komplexe Befindlichkeit von leiblichen und Adoptiveltern, deren Motive für die Adoption, Vorerfahrung, Partnerschaftssituation, Lebenssituation und Lebensplanung kann hier nicht eingegangen werden. Adoptierte Kinder haben z. T. kaum Schwierigkeiten, z. T. deutliche Probleme, mit der Adoptionstatsache zurechtzukommen. Manche Kinder grübeln über Fragen nach den Motiven der leiblichen Eltern, deren Verbleib, Persönlichkeit und anderes. Adoptierte Kinder entwickeln häufiger Probleme in sozialen Beziehungen, bei der Akzeptanz von Disziplin und Grenzsetzungen, haben Lern- und Identitätsprobleme. Ihr Risiko für eine psychiatrische Erkrankung ist gegenüber der Normalbevölkerung etwa verdoppelt. Mitverantwortlich sind hierfür auch die Lebensbedingungen der Kinder vor der Adoption, z. B. medizinische Probleme, Deprivation, fehlende Bindungsentwicklung, externe Bezugspersonen, Gewalterfahrung.
Es wird empfohlen, Kinder graduell über die Tatsache ihrer Adoption in realistischer und verständlicher Weise zu informieren. Kinder haben das Recht, die leiblichen Eltern kennen zu lernen. Dies sollte jedoch bei Inkognito-Adoptionen möglichst erst im Erwachsenenalter realisiert werden. Professionelle sollten den Adoptionsstatus der von ihnen betreuten Kinder und Jugendlichen kennen, damit zusammenhängende Probleme erkennen, beraten und ggf. geeignete lokale Hilfsangebote (Beratungsstellen, Jugendamt) nennen.

Den Adoptiveltern sollten die medizinische Vorgeschichte des adoptierten Kindes so umfassend wie möglich bekannt sein, ebenfalls, soweit mit der Inkognito-Situation vereinbar, auch die sozialen Bedingungen. Beides ist bei der Adoption von Kindern aus anderen Herkunftsländern häufig nur lückenhaft oder gar nicht bekannt. In diesen Fällen empfiehlt sich eine gründliche initiale pädiatrische körperliche Untersuchung mit Eintreffen im neuen Heimatland, die häufig die Suche nach speziellen Infektionserkrankungen (z. B. Tbc, Hepatitis B, HIV, Parasiten) einschließt. In den ersten Jahren sind bei diesen Kindern relativ engmaschige medizinische Verlaufsbeurteilungen empfehlenswert, ggf. ergänzt durch medizinisch-therapeutisch-pädagogische Fördermaßnahmen und eine regelmäßige Elternberatung.

Pflegeeltern sind Vertragspartner des Jugendamtes und erbringen als Privatleute eine Dienstleistung für die Herkunftseltern des Kindes, die finanziell entlohnt wird. Pflegekinder leben z. T. vorübergehend, z. T. dauerhaft in der Pflegefamilie, die zur Zusammenarbeit mit der Herkunftsfamilie verpflichtet ist. Pflegeeltern haben Anspruch auf fachliche Hilfe und Begleitung, die vom Jugendamt gewährt wird.

Pflegekinder profitieren von einer aktiven Zusammenarbeit zwischen Pflegefamilie und Herkunftseltern. Interessenwidersprüche, Identitäts- und Loyalitätskonflikte können zu erheblichen emotionalen Belastungen der Kinder führen. Pflegeeltern müssen mit unterschiedlichen Bewältigungsversuchen der Pflegekinder zurechtkommen. Diese glorifizieren ihre Herkunftseltern, wenden sich von ihnen ab oder haben unrealistische Zukunftswünsche bezogen auf die Herkunftseltern. Grundsätzlich ist ein regelmäßiger Umgang mit leiblichen Geschwistern und Eltern vorteilhaft.

Pflegeeltern haben eine besondere Rolle im Dreiecksverhältnis Kind, leibliche und Pflegeeltern. Sie können Pflegekindern für ihre körperliche, emotionale und soziale Entwicklung entscheidende Hilfe leisten. Dabei müssen sie nicht selten mit einer mangelnden Wertschätzung der Herkunftseltern leben. Pflegeeltern von Dauerpflegekindern haben als „soziale Eltern" im Vergleich zu normalen Eltern zusätzliche Aufgaben. Sie sollen dem Kind bei der Lösung von Konflikten helfen, Reifungsprozesse begleiten und den Kindern durch Einbeziehung auch der leiblichen Eltern die Möglichkeit vermitteln, beide Welten miteinander zu verknüpfen.

Behinderte Kinder sind häufiger in Pflege als nichtbehinderte Kinder. In Großbritannien sind 28 % aller Pflegekinder behindert. Gründe hierfür sind elterliche Schuldgefühle oder Überlastung bei der Pflege des behinderten Kindes, aber auch emotionale Probleme, nicht selten auch Folgen von elterlicher Trennung oder Scheidung.

Die Mehrzahl der Pflegekinder kommt wegen körperlicher Misshandlung, sexuellem Missbrauch oder Vernachlässigung in Pflege. Bestimmte Entwicklungsstörungen, Regulationsstörungen im Säuglingsalter, Verhaltensstörungen und Intelligenzminderung erhöhen das Misshandlungsrisiko deutlich. Andererseits befindet sich auch eine nicht geringe Zahl primär regelrecht entwickelter Kinder nach stattgehabter Misshandlung oder Vernachlässigung zu ihrem Schutz in institutioneller Pflege oder bei Pflegefamilien.

Die Untersuchungen an **rumänischen Heimkindern** belegen eindrucksvoll, dass Kinder unter ungünstigen Bedingungen institutioneller Pflege ein hohes Risiko für die Entwicklung körperlicher, sprachlicher, intellektueller und psychischer Störungen tragen. Bei unter ungünstigen institutionell betreuten Waisenkindern ließen sich im Vergleich zu Waisenkindern in Pflegefamilien anomale Hirnaktivitätsmuster im Kernspintomogramm und verkürzte Telomere an den Chromosomen nachweisen. Neben biologischen Faktoren sind andere Faktoren für diese Störungen verantwortlich wie Mangelernährung oder mangelhafte Pflege mit der Folge häufiger Infektionserkrankungen auch die emotionale Vernachlässigung bzw. fehlende Bindungsentwicklung bei wechselnden, oft wenig sensitiven Bezugspersonen. Studien in England haben gezeigt, dass die

Schäden bei einer relativ kurzen Zeit in ungünstiger institutioneller Pflege (< 2 Jahre) durch anschließende stabile und angemessene Betreuung von Pflegeeltern zumindest teilweise wieder ausgeglichen werden können. Die Sicherung eines stabilen, emotional sensitiven und anregenden Umfeldes ist hierfür ein wesentlicher Faktor. Die Untersuchungen zeigten auch, dass die Langzeiteffekte **psychosozialer Deprivation** auf die psychische Entwicklung offensichtlich größer als die der Mangelernährung waren. Die negativen Effekte betrafen auch ein geringeres Kopfwachstum und eine niedrigere Intelligenz.

13.4 Unterschiedliche Kulturen und Gemeinschaften, Migranten

Migration findet weltweit statt und hat es schon immer gegeben. Geschichte und Gegenwart belegen, dass Migration als **Risiko,** aber auch als **Chance** für die Migranten und die sie aufnehmenden Gesellschaften betrachtet werden muss. Verschiedene Faktoren beeinflussen die Lebenslagen von Menschen mit Migrationshintergrund und stehen in vielfältigen **Wechselbeziehungen:** Gesundheit (körperlich/psychisch), Bildung/Sprachkompetenz, Rechtsstatus, Arbeit/Beruf, materielle Ressourcen, Wohnsituation, soziale Netze und gesellschaftliche Teilhabe. Es ist daher nicht ohne weiteres zulässig, bestimmt Befunde bei individuellen oder Gruppen von Migranten ausschließlich auf die Tatsache des Migrationshintergrundes zurückzuführen.

Deutschland ist ein Einwanderungsland. Im Jahr 2011 hatten 19,5 % der in Deutschland lebenden Menschen einen Migrationshintergrund. Kinder- und Jugendärzte müssen sich jetzt und künftig verstärkt mit Kompetenz auf die Bedürfnisse und Probleme von Kindern und Jugendlichen aus Zuwanderungsfamilien einstellen.

Migranten sind in den Aufnahmegesellschaften häufig sozial benachteiligt. Zum Verlust der gewohnten Umgebung, Kultur, Sprache und des Familien- uns Freundeskreises kommen oft Einkommensarmut, ungünstigere Wohnverhältnisse, niedrigere Bildungsabschlüsse, Gleichgültigkeit, Ablehnung, Diskriminierung, ja sogar Ausbeutung zusammen. Die Ursachen dafür sind vielfältig. Migranten und auch Migrantenkinder sind in ihren sozialen Beziehungen nicht selten eingeschränkt, oft fehlen bedarfsgerechte kulturelle Angebote. Menschen mit Migrationshintergrund haben in Deutschland tendenziell ein niedrigeres Bildungsniveau und ein damit einhergehendes höheres Armutsrisiko. Eine besondere Risikogruppe stelle unbegleitete jugendliche Flüchtlinge und „illegale“ Migranten dar.

Gelungene Migration ist jedoch auch als Chance zu betrachten. Migranten können besondere Erfahrungen und Qualitäten gewinnen in der Bewältigung schwieriger Situationen, die als unsicher, benachteiligend oder ausgrenzend erlebt werden. Im günstigen Fall können Migranten problemlos mit mehreren Sprachen umgehen. Sie sind mit verschiedenen kulturellen Traditionen vertraut und besonders befähigt, neue Zukunftsperspektiven zu entwickeln.

Ein Migrationshintergrund allein ist keine spezifische Krankheitsursache und kein Krankheitsrisiko.

Die gesundheitliche Situation von Migrantenkindern wird durch komplexe Wechselwirkungsprozesse verschiedener Faktoren des Lebens in der Fremde beeinflusst.

Sprachbarrieren stellen ein oft entscheidendes Problem dar. Die Ursachen sind vielschichtig. In Deutschland werden sowohl die Muttersprache als auch die Zweitsprache Deutsch bei Migrantenkindern oft nur in ungenügender Weise gefördert. Konkret besteht in Deutschland ein Mangel an muttersprachlichem Unterricht. Die mangelnde Sprachförderung führt dazu, dass sowohl die

Deutsch- wie auch die Muttersprachenkenntnisse oft nur für die Alltagskommunikationen ausreichen. Es besteht eine **doppelte Halbsprachigkeit,** die für die Erreichung von Bildungszielen im Regelschulbereich oder gar für weitergehende Bildungsabschnitte nicht ausreicht. Die Ergebnisse der Iglu- und Pisa-Studien belegen dies eindrücklich.

Unzureichende Informationen über Funktion und Angebote des Gesundheitswesens sowie über die Systeme der sozialen Sicherung führen nicht selten dazu, dass Migrantenfamilien und ihre Kinder bei der Inanspruchnahme medizinischer oder sozialer Leistungen benachteiligt werden. Migranten haben häufig auch Probleme mit den hierzulande starren institutionellen und therapeutischen Rahmenbedingungen (Formblätter, Antragsverfahren, definierte Sprech-/Therapiezeiten). Zuspätkommen oder Nichteinhalten von Terminen sind die Folge und werden von Professionellen als Noncompliance wahrgenommen.

Religion, Weltbild und Wertvorstellungen können in der Praxis ein weites Spektrum an Verständigungsproblemen und Interessenkonflikten schaffen. Besonders häufige Reibungspunkte sind Kommunikationsprobleme, Speisevorschriften, Schamgefühl und unterschiedliche Krankheitskonzepte. Die rituelle Beschneidung von Neugeborenen jüdischer Eltern und von Säuglingen oder Kleinkindern muslimischer Eltern wurde zuletzt in der Öffentlichkeit und auch unter Kinder- und Jugendärzte kontrovers diskutiert.

Das **Menschenbild** vieler Migranten ist oft geprägt durch eine besondere Achtung von Autoritäten und die Befolgung ihrer Anweisungen. Sie erwarten eine klassisch paternalistische Arzt-Patienten-Beziehung. Partnerschaftlichkeit, Selbstständigkeit und Patientenautonomie werden dadurch erschwert.

Die **Familie** ist in Migrantenkulturen häufig eine wichtige Ressource. In vielen Familien existieren klare Rollenverteilungen: Der Vater ist die sich nach außen darstellende Autorität, die Mutter hat die Verantwortung für Erziehung, Ernährung und Hygiene. Diese Rollenzuweisungen können dazu führen, dass die Mütter nur ihre Muttersprache beherrschen und wichtige medizinische und Förderempfehlungen nicht verstehen bzw. umsetzen können. Dies kann ernste medizinische Konsequenzen haben oder zumindest eine zusätzliche Benachteiligung der Migrantenkinder darstellen. In traditionell orientierten muslimischen Familien sind die Rollenzuweisungen klar strukturiert, hierarchisch und autoritär. Die Selbstständigkeit und Unabhängigkeit der Kinder wird wenig gefördert, ältere Geschwister sollen den jüngeren als Vorbild dienen. Verwandtschafts- und Nachbarschaftsbeziehungen sind bedeutsam. Die Werte Ehre und Würde spielen eine große Rolle. Traditionelle Formen der Ehevermittlung und Eheschließung mit nicht primär durch die künftigen Ehepartner betriebenen Partnerfindung und Verwandtenehen sind noch gegenwärtig. Nicht selten führt die Diskrepanz zwischen den klassischen Rollenverteilungen und den hierzulande üblichen bei Migrantenfamilien in den Paarbeziehungen und bei Eltern-Kind-Interaktionen, insbesondere mit Jugendlichen zu Konflikten und Loyalitätsproblemen. Diskordante Wertvorstellungen, Erziehungsstile und -ziele können Einfluss auf das psychische Befinden und das Lernverhalten von Kindern und Jugendlichen haben.

Psychosoziale Risikofaktoren werden bei Migrantenkindern in Deutschland gehäuft beobachtet. Sozialkompensatorische institutionelle Angebote wie Tagesstätten, Beratungsstellen oder die Erziehungshilfe werden seltener genutzt. Die Freizeitsituation, besonders von Mädchen, ist nicht bedarfsgerecht. Nach relativ oft gelingender Integration im Vorschul- und Grundschulbereich setzen bei Migrantenkindern mit Beginn der Sekundarschule häufig verstärkte Segregationstendenzen ein. Bei Jugendlichen mit niedrigem Sozialstatus wird vermehrt gesundheitsriskantes Verhalten beobachtet. Migrantenkinder besuchen überproportional häufig Sonderschulen (15 % versus 3,8 % eines Geburtsjahrgangs bei Kindern ohne Migrationshintergrund). In Sonderschulen ist die muttersprachliche Förderung besonders gering.

Migrantenfamilien konfrontieren Professionelle mit spezifischen Werten und Lebensweisen, auf die geachtet und einfühlsam eingegangen werden sollte.
Krankheiten werden in anderen Kulturen und Ethnien häufig ganzheitlich wahrgenommen oder magisch/religiös interpretiert. Naturwissenschaftliche, organ- oder molekularbezogene **Krankheitskonzepte** sind manchen Migranten völlig fremd. Psychische Probleme werden in der islamischen Tradition und Kultur als Krankheitsform weitgehend tabuisiert. Häufig wird versucht, diese mit magischen oder religiösen Konzepten zu eliminieren. Bei psychosozialen Störungen wird über lange Zeit ausschließlich die Ressource Familie mit ihren spezifischen Strukturen, Stärken, aber auch Grenzen in Anspruch genommen. Erst bei einer krisenhaften Zuspitzung der Probleme werden Ärzte, Psychologen, Psychotherapeuten oder andere Fachleute konsultiert.
Im Zusammenhang mit unterschiedlichen Krankheitskonzepten sind die **therapeutischen Konzepte** bei Menschen verschiedener Kulturen unterschiedlich. Oft erweisen sich türkische Patienten bei der Medikamenteneinnahme als vorbildlich, bei der Ergotherapie leisten sie Widerstand bzw. setzen diese nicht in geeigneter Weise um, da das dahinterstehende therapeutische Konzept ihnen nicht hinreichend vermittelt werden konnte.
Vielen Menschen mit Migrationshintergrund ist der **Präventionsgedanke** eher fremd. Man geht zum Arzt, wenn man krank ist.
Migrantenfamilien übertragen häufig die Möglichkeiten des **heimatlichen Medizinsystems** auf das hiesige und nehmen dann die ihnen zustehenden Hilfen nicht in Anspruch. Fehlende Inanspruchnahme rührt z. T. auch aus der Angst vor der Gefährdung des Status im Zuwanderungsland her. Sie kann auch Folge von Kommunikations- und Verständigungsproblemen sein.
Auch die **Einstellung zu Behinderungen** ist kulturabhängig. Während in einigen Gesellschaften z. B. Unfruchtbarkeit als magisch interpretierte, schwerwiegende körperliche Behinderung angesehen und mit starken Stigmen besetzt wird, betrachtet man sie in Deutschland als eine medizinisch behandelbare Erkrankung. Demgegenüber wird bereits eine leichte geistige Behinderung eines Kindes für Familien in unserer Gesellschaft oft als Kränkung und hohe Belastung empfunden. In Gesellschaften mit weniger komplex ausgerichtetem Bildungs- und Beschäftigungssystem gelten solche Menschen als „etwas schwer von Begriff". Sie werden angelernt und erfüllen oft Aufgaben im Familienverband oder der örtlichen Gemeinschaft, wo sie akzeptiert sind und nicht als behindert gelten.
Ein besonderes, auch juristisches und ethisches Dilemma ist die oft nur mangelhaft mögliche **Aufklärung** über eingreifende medizinische oder operative Maßnahmen, die Eltern von Migrantenkinder immer wieder nicht in angemessener Weise zukommt.
Verschiedene Studien weisen auch auf eine **gesundheitliche Benachteiligung** von Migrantenkindern hin. Hierfür sind wahrscheinlich soziodemografische und Migrationseffekte verantwortlich, die sich z. T. schwer trennen lassen. Migrantenkinder tragen höhere Gesundheitsrisiken (häufiger Übergewicht, mehr Fernsehkonsum, seltenere Inanspruchnahme von Früherkennungsuntersuchungen), zeigen häufiger Verhaltensprobleme und erhalten zumindest teilweise eine schlechtere medizinische Versorgung (schlechtere Stoffwechseleinstellungen, seltener Diagnose und Behandlung eines ADHS). Andererseits bestehen teilweise eine höhere Impfbereitschaft und eine geringere Atopieneigung. Detaillierte Ergebnisse hierzu lieferte die **KiGGS-Studie** (➤ Kap. ➤ 4.7).
Genetische Erkrankungen treten bei Kindern mit bestimmtem Migrationshintergrund deutlich häufiger als in der übrigen Bevölkerung auf (Konsanguinität, höhere genetische Prävalenz in bestimmten Populationen). Die genetische Beratung von Eltern aus anderen Kulturkreisen und anderen familiären Traditionen und Rollenzuweisungen von Mann und Frau wirft komplexe Fragen auf. Soziale Faktoren spielen als Ursache für die erhöhte Prävalenz neuropädiatrischer Erkrankungen bei Familien mit Migrationshintergrund eine Rolle.

Komplexere „westliche" Behandlungen werden z. T. nicht oder ungenügend umgesetzt. Dies wird z. B. bei speziellen Diäten zur Behandlung neurometabolischer Erkrankungen, aber auch bei der medikamentösen Behandlung von Epilepsien und anderen chronischen Erkrankungen beobachtet. Eine transkulturelle Kinder- und Jugendmedizin sollte

- die gleichberechtigte Teilhabe von Migrantenkindern an einer bestmöglichen medizinischen Versorgung sicherstellen,
- gesundheitlichen Schaden durch unterlassendes oder krankheitsförderndes Verhalten von Migrantenfamilien verhindern,
- die Migrantenfamilien befähigen, sich selbst umfassend um die gesundheitlichen Belange ihrer Kinder zu kümmern.

Dazu muss **transkulturelle Kompetenz** bereits in der Ausbildung der verschiedenen Berufsgruppen gelernt werden. In der Praxis müssen kulturspezifische Gesichtspunkte respektvoll erfragt, bedacht und berücksichtigt werden. Dazu gehören ein angemessenes Begrüßungsritual, die Klärung des Vorstellungsanlasses und des Therapieauftrages unter Berücksichtigung der Individualität der Familie, ihres Sprachniveaus, Bildungs- und Sozialstatus. Die im deutschen Medizinsystem übliche Ausrichtung auf Effizienz, gutes Zeitmanagement mit Pünktlichkeit und Sachorientiertheit muss mit den Bedürfnissen der Menschen aus anderen Kulturkreisen abgestimmt werden, für die traditionell das Gespräch und die Beziehungsebene besonders wichtig sind. Bei Familien mit Migrationshintergrund, die hiesige Versorgungsstrukturen nicht nutzen, sollte eine aufsuchende Gesundheitsbetreuung medizinisch indizierte Maßnahmen in der Lebenswelt der Familien umsetzen. Zur Kenntnis des Bedarfs ist eine sozialraumbezogene Gesundheitsberichterstattung mit Berücksichtigung des Migrationshintergrundes der Familien erforderlich.
Andererseits sollten Eltern mit Migrationshintergrund ihre Krankheitskonzepte, ihre Bedürfnisse und Erwartungen an das hiesige Medizinsystem artikulieren, sich aktiv um das Verständnis der für sie relevanten medizinischen Zusammenhänge bemühen, sich offen und konstruktiv in die Therapie einbringen und eigene Sprachbarrieren abbauen. Die Grenzen kulturellen Verständnisses sind dort erreicht, wo die körperliche oder seelische Unversehrtheit eines Kindes oder Jugendlichen bedroht ist.

13.5 Benachteiligte Kinder und Jugendliche – Armut

Armut und **soziale Benachteiligung** (z. B. infolge Zugehörigkeit zur Unterschicht, Bildungsferne, sozialer Isolation oder Migration) führen zu einer erhöhten gesundheitlichen Gefährdung von Kindern und Jugendlichen und ihren Eltern. Hierfür sind u. a. ungünstige Wohn- und Lebensbedingungen, nachteiliges Gesundheitsverhalten und schlechtere gesundheitliche Versorgung verantwortlich. Kinder und Jugendliche aus sozial benachteiligten Gruppen nehmen seltener an den Früherkennungsuntersuchungen U1 bis U9 teil, sie haben häufiger Unfälle, Karies und Übergewicht. Jugendliche aus Armutsgruppen rauchen häufiger, putzen sich seltener die Zähne, treiben seltener Sport, sehen häufiger fern, ernähren sich ungesünder und haben früher den ersten sexuellen Kontakt. Ihr subjektives Wohlbefinden ist im Mittel schlechter, sie zeigen häufiger emotionale, soziale Verhaltens- und psychosomatische Störungen. Ungünstige soziale und ökonomische Lebensbedingungen sind ein Risikofaktor für r Sprach- und Lernstörungen, Störungen des Sozialverhaltens und für Misshandlungen. Sie bedingen einen höheren Förderbedarf. Eine **Frühintervention** („early intervention") bei Kindern aus sozial benachteiligten Familien gehört zu den

Tab. 13.1 Sozialer Gradient der Morbidität und Mortalität im Kindesalter. Vervielfachung bei Kindern der untersten gegenüber der obersten Sozialschicht (Ellsäßer 2002)

Parameter	Faktor
Geburtsgewicht < 2.500 g	1,4
Säuglingssterblichkeit	2,7
Kindersterblichkeit (1–14 Jahre)	2,2
Tod durch Unfall	3,8
Hyperaktivität	3,2
Dissoziales Verhalten	4,0
Adipositas	3,3
Sprech- und Sprachstörungen	4,2
Lernschwächen	14,6

effektivsten und wirtschaftlich lohnendsten Maßnahmen des Sozial- und Gesundheitssystems (➤ Kap. 6, Kap. 38). Bei Frühgeborenen unter 1.500 g Geburtsgewicht sind im Vorschulalter Intelligenz und sprachliche Fertigkeiten enger mit dem soziökonomischen Hintergrund als mit der neonatalen neurologischen Morbidität assoziiert.
Soziale Benachteiligung ist eine wichtige Ursache für Dissozialität und Delinquenz bei Jugendlichen. Tabelle 13.1 zeigt den sozialen Gradienten der Morbidität und Mortalität im Kindesalter.
Bei internationalen Vergleichen haben Nationen mit der größten Ungleichheit in Einkommen und sozialen Chancen unabhängig vom durchschnittlichen nationalen Wohlstand oder Bruttosozialprodukt die schlechteste Kindergesundheit. Die generelle Verfügbarkeit von Impfungen, einer medizinischen Grundversorgung, ausreichender Ernährung und nicht verseuchtem Trinkwasser senkt Morbidität und Mortalität zahlreicher pädiatrischer Erkrankungen.
Die Häufigkeit bestimmter Arten von Zerebralparesen (das „CP-Panorama") in einer Bevölkerungsgruppe stellt möglicherweise einen Indikator der sozioökonomischen Entwicklung bzw. der medizinischen Ressourcen/Versorgungsqualität der entsprechenden Gesellschaft dar.

13.6 Inklusion

Die gemeinsame Förderung und Erziehung von Kindern ohne und mit Behinderung in Kindergärten und Schulen hat durch das in Kraft treten der UN – Behindertenrechtskonvention (➤ Kap 46.3) in Deutschland einen zunehmenden Stellenwert bekommen. Während im englischen Text der Behindertenrechtskonvention von „inkl. education system" gesprochen wird, wird dieser Anspruch in der deutschsprachigen Fassung mit der Bezeichnung „integratives Bildungssystem" abgeschwächt. Bei der **Integration** werden Kinder ohne und mit sonderpädagogischem Förderbedarf unterschieden, **Inklusion** bedeutet dagegen Heterogenität als Normalfall, eine positive Deutung von Andersheit, damit einen „ Kindergarten bzw. eine Schule für alle" mit Aufhebung des Zwei-Gruppen-Ansatzes. Inklusion im weiteren Sinne beschränkt sich nicht auf den medizinisch bzw. pädagogisch gängigen Behindertenbegriff, sondern schließt Menschen mit jeglichen – z. B. auch sozialen oder migrationsbedingten – Benachteiligungen ein.
Vollständige Inklusion ist als Idee und Handlungsmaxime wertvoll, in der Realität schwerlich erreichbar. Alle Menschen befinden sich in ihren konkreten Lebenssituationen an unterschiedlichen Stellen innerhalb eines Kontinuums zwischen Inklusion und Exklusion.

Gemeinsames Leben und Lernen behinderter und nichtbehinderter Kinder und Jugendlicher kann für alle wichtige soziale Erfahrungen ermöglichen. Inwieweit der akademische Erfolg und die langfristigen beruflichen Möglichkeiten für alle Beteiligten dadurch verbessert werden, bedarf einer differenzierten Prüfung. Die empirische Datenlage hierzu ist noch nicht sehr breit.

Integrative Förderung in Kindergärten hat in den letzten Jahren rasch an Verbreitung zugenommen. Mit Unterstützung durch pädagogische und begleitende Hilfen und großem Engagement der Erzieherinnen und Eltern gelingt es häufig, Kinder mit Entwicklungs- oder Verhaltensstörungen bzw. Behinderungen mit deutlichen Vorteilen für ihre Entwicklung gemeinsam mit Kindern ohne Beeinträchtigungen zu fördern. Dies kann im Rahmen einer Einzelintegration oder in integrativen Einrichtungen mit zusätzlichem, z. B. heilpädagogischem Personal und ggf. sonderpädagogischer Beratung erfolgen.

In Baden-Württemberg hat zwischen 2007 und 2010 bei abnehmender Zahl der Vorschulkinder die Zahl der Kinder mit geistiger, körperlicher oder mehrfacher Behinderung, die Eingliederungshilfe für den Besuch eines Kindergartens/einer Kindertagesbetreuung beanspruchen, um fast 50 % zugenommen bei gleich bleibender Zahl von Kindern in Sonderkindergärten. Die Gesamtzahl der Kinder, die sonderpädagogischen Förderbedarf bzw. Eingliederungshilfe beanspruchten, hat mithin bei sinkender Kinderzahl deutlich zugenommen.

Integrativer/Inklusiver Unterricht in Schulen für Kinder mit chronischen Erkrankungen oder Behinderungen ist zunehmender Elternwunsch und gesellschaftliche Aufgabe.

Der Behindertenbegriff der Behindertenrechtskonvention ist ein offener, an der **Teilhabe** orientierter Begriff. Er umfasst für den schulischen Bereich behinderte Schülerinnen und Schüler ohne und mit **sonderpädagogischem Förderbedarf.** Mithin müssen nicht nur Kinder und Jugendliche mit motorischer, geistiger oder Lernbehinderung, Seh-, Hör- oder Verhaltens-/Entwicklungsstörungen, sondern auch solche mit chronischen Erkrankungen Unterstützung für ihre vollständige Teilhabe in der Schule erhalten.

Laut Nationalem Bildungsbericht „Bildung in Deutschland 2012" wurden 70 % der Schüler mit sonderpädagogischem Förderbedarf (6,2 % aller Kinder) in Förderschulen, nur knapp 30 % in allgemeinen Schulen gefördert. Die Inklusionsrate ist altersabhängig: In Kindertagesstätten liegt sie bei gut 60 %, in Grundschulen etwa bei einem Drittel, in weiterführenden Schulen bei 15 %. Die Quote der Förderschüler lag in Deutschland im Schuljahr 2010/2011 mit 4,2 % in einer vergleichbaren Größenordnung wie in Finnland (3,8 %), Dänemark (4,4 %) und in der Schweiz (5,4 %). Die Quote der in Regelschulen geförderten Kinder war doppelt so hoch wie 2000/2001. In den Förderschwerpunkten Sprache, geistige, emotionale und soziale Entwicklung ist die Förderschul-Besuchsquote in den letzten Jahren aber stetig angewachsen. Die sogenannte Exklusionsquote ist daher von 4,6 % (2000/2001) auf 4,8 % (2011/2012) angestiegen.

Es sind deutliche Unterschiede zwischen verschiedenen Regionen und Bundesländern festzustellen. Während in Bremen 55,5 % und in Schleswig-Holstein 54,1 % der Kinder mit sonderpädagogischem Förderbedarf inklusiv beschult wurden, waren es in Bayern 22,4 % und in Niedersachsen 11,1 %. In den Schwerpunktregionen für die Modellprojekte schulischer Inklusion in Baden-Württemberg wurden für das Schuljahr 2011/12 bei 71 % der Anträge für die Kinder ein sonderpädagogischer Förderbedarf festgestellt. Von diesen Kindern besuchten anschließend 74 % eine Sonderschule und 26 % den gemeinsamen Unterricht an allgemeinen Schulen.

Die Bundesländer haben aufgrund ihrer Kulturhoheit unterschiedliche Formen der schulischen Integration/Inklusion entwickelt und für diese verschiedene, immer wieder sich verändernde Bezeichnungen gefunden, die nur noch für Spezialisten durchschaubar sind. Auch das regionale schulische Angebot ändert sich dauernd. Eine bundesweite Übersicht hierzu würde den Rahmen dieser Arbeit sprengen. Der (sozial)pädiatrisch tätige Arzt bleibt gefordert, mit dem jeweiligen Stand der schulischen Vielfalt in seiner Region vertraut zu sein.

Inklusive Bildung mit gemeinsamem Lernen aller Schüler hat neben einem anerkannten Schulabschluss als Ziel den individuellen Bildungserfolg, die Persönlichkeitsentwicklung, die Fähigkeit zu selbst bestimmtem Leben mit aktiver Teilhabe an der Gesellschaft. Inklusive Pädagogik arbeitet zielgleich und zieldifferent, mit individuellen Förderplänen und angepassten Lehr- und Lernmitteln. Ein **Nachteilsausgleich** kann individuelle Leistungen vergleichbar machen. Personen unterschiedlicher Professionen und Qualifikationen müssen im Unterricht und außerhalb kooperieren. Die rechtlichen Rahmenbedingungen finden sich in den Sozialgesetzbüchern V, VII, IX, XII und verschiedenen Verordnungen (➤ Kap. 42). Kostenträger sind insbesondere die Krankenkassen, die Sozialhilfeträger und die Schul- bzw. Kultusbehörden.

Lehrer stehen heute vor der komplexen Aufgabe, der zunehmenden Heterogenität ihrer Schüler gerecht zu werden. Der zusätzlichen neuen pädagogischen Herausforderung eines integrativen/inklusiven Unterrichts behinderter oder chronisch kranker Kinder fühlen sich manche Lehrer nicht gewachsen. Sonderpädagogische Inhalte sollten verstärkt in die Aus- und Weiterbildung für Lehrer an allgemeinen Schulen integriert werden.

Der Einfluss chronischer Erkrankungen wie Diabetes, Epilepsie oder ADHS und von Behinderungen der verschiedenen Ausprägungen auf den Alltag von **Kindern und Jugendlichen** in der Schule ist erheblich. Kinder und Jugendliche mit solchen Beeinträchtigungen leiden bei krankheits- oder behinderungsbedingtem Ausschluss von schulischen Aktivitäten, aber auch an Ausgrenzung durch Mitschüler. Chronische Krankheiten oder Behinderungen erfordern Kenntnisse der Lehrer und anderer Bezugspersonen zu den Besonderheiten und Bedürfnissen des individuellen Kindes und zum Umgang mit bestimmten medizinischen Problemen bis hin zu Notfallsituationen (➤ Kap. 35). Die Gabe von Medikamenten, spezielle Diäten, förderliche und riskante Aktivitäten, insbesondere bei Schulausflügen und Klassenfahrten, ärztliche, pflegerische und therapeutische Kontaktpersonen, Aspekte der Schweigepflicht, Aufklärung der Lehrer, Mitschüler, Eltern der anderen Kinder u. a. müssen im Einzelnen geklärt werden. Verhaltensbesonderheiten erfordern bei anderen Kindern psychologisch basierte Erziehungs- und Lernkonzepte. Die Gesundheit und die Teilhabe am schulischen Leben müssen soweit möglich für jedes Kind in individuellen Absprachen optimal entwickelt werden. Dazu gehören auch bauliche Maßnahmen und Fragen der Gestaltung der Schulräume (➤ Kap. 9). Vertrauensvolle Absprachen können Ängste von Eltern und Lehrern abbauen helfen.

Kinder und Jugendliche mit **chronischen Erkrankungen** können – mit wenigen Ausnahmen – eine allgemeine Schule besuchen. Dazu sind eine gute medizinische Betreuung, Schulung der Betroffenen und konkrete Informationen für die Schule wichtige Bedingungen.

Die Erfahrungen zeigen, dass integrativer/inklusiver Unterricht bei Kindern mit **Behinderungen** unter den heutigen Bedingungen am besten in der Grundschule/Primarschule gelingt. Kinder mit guten sozial- emotionalen und kognitiven Ressourcen werden sich dabei leichter tun als andere. Dies gilt auch für Kinder mit Körperbehinderungen. Mit zunehmendem Alter wird ein integrativer/inklusiver Unterricht bei Kindern mit geistiger Behinderung schwieriger. Sprachstörungen und Verhaltensauffälligkeiten verschwinden zwar nicht komplett mit zunehmendem Alter, haben jedoch oft einen abnehmende Einfluss auf Kommunikation und soziale Aspekte im Unterricht, sodass integrativer/inklusiver Unterricht – evtl. mit individuellem Nachteilsausgleich – mit zunehmendem Alter eher besser gelingt. Die meisten Kinder mit hochgradigen Sehstörungen und ein Teil der Kinder mit schwerwiegenden Hörstörungen profitiert von den besonderen Rahmenbedingungen der eigens für sie existierenden Sonderschulen. Für Schwermehrfachbehinderte mit hohem medizinisch-therapeutischem und pflegerischem Betreuungsaufwand dürfte ein integrativer/inklusiver Unterricht nur in Einzelfällen sinnvoll sein.

Vor dem Hintergrund sinkender Schülerzahlen öffnen sich Schulen vermehrt für integrative/inklusive pädagogische Konzepte, auch um den Schulstandort zu sichern oder attraktiv zu machen. Dies sollte genutzt werden im Interesse der betroffenen Schüler mit chronischen Erkrankungen oder Behinderungen. Eine Überforderung der Schüler, Lehrer und Schulen durch zu wenig bedachte Schulentscheidungen, unzureichende flankierende Maßnahmen oder gar ein unbedachter Abbau sonderpädagogischer Schulen und – Förderangebote darf nicht stattfinden. Auch in Zukunft werden Eltern für ihr Kind z. T. einen integrativen/inklusiven Unterricht, z. T. einen Unterricht in einer Sonderschule wünschen. Eine kompetente pädagogische und medizinisch-sozialpädiatrische Beratung bei der Schulwahl ist unabdingbar, vor allem bei gestärktem Elternwahlrecht. Integrativer/Inklusiver Unterricht sollte mit Augenmaß stattfinden und schrittweise ausgeweitet werden. In der Begleitung von Inklusionsmaßnahmen und -Prozessen ist Achtsamkeit vonnöten. Inklusion kostet Geld und ist keine Mogelpackung, um Einsparungen zu verdecken.

13.7 Umweltbelastungen

ALLUM (www.allum.de) ist ein kinderärztliches Informationsportal für Fragen zu Allergie, Umwelt und Gesundheit. Zielgruppe sind Eltern, Patienten und interessierte Fachleute.
Die Aussagen des folgenden Abschnitts wurden weitgehend diesem Portal entnommen.
Verschiedene Faktoren sprechen für eine besondere Empfindlichkeit von Kindern und Jugendlichen gegenüber Umwelteinflüssen und deren herausragende langfristige Bedeutung für die Kindergesundheit: **Embryonen, Feten und Kinder entwickeln sich in körperlicher und funktioneller Hinsicht.** Diese Entwicklung ist störanfällig. Besondere anatomische und physiologische Verhältnisse in der Pränatalzeit und im Säuglings- und Kleinkindesalter führen zu erhöhter Toxizität bestimmter Stoffe. **Der Mensch steht am Ende einer Nahrungskette,** während derer sich Schadstoffe und Rückstände, insbesondere langlebige fettlösliche Substanzen, im Fettgewebe und in der Muttermilch anreichern können. Kinder sind körperlich aktiver, sie trinken und atmen im Verhältnis zum Körpergewicht mehr als Erwachsene, sie halten sich öfter und länger draußen auf. **Kinder haben zumeist eine lange Lebenszeit vor sich.**
Der **Kinder-Umwelt-Survey** (KUS) ist ein Modul des Kinder- und Jugendgesundheitssurveys (KiGGS). Die Teilnehmer stellen eine zufällige Teilstichprobe aus dem KiGGS dar. Ziele des KUS waren die Erfassung, Bereitstellung, Aktualisierung und Bewertung repräsentativer Daten über die Exposition deutscher Kinder mit Umweltschadstoffen. Da Kinder im Alter zwischen 6 und 14 Jahren auch im Umwelt-Survey 1990/92 untersucht wurden, lassen sich für diese Altersgruppe zeitliche Trends und damit der mögliche Erfolg umweltpolitischer Maßnahmen erfassen.
Bei den zwischen 2003 und 2006 untersuchten 3–14-jährigen Kindern fand man deutlich geringere Belastungen durch **Blei, Quecksilber, PAH und PCB** (polychlorierte Biphenyle) als 1990/92. Vergleichbar niedrige Bleiwerte im Blut wurden nur in Schweden gemessen. Die Konzentrationen von PCB im Blut sind bei den 13–14-Jährigen immer noch deutlich höher bei gestillten im Vergleich zu nicht gestillten Kindern. Die PCB-Belastung sinkt jedoch mit dem Alter. Überschreitungen der Grenzwerte für Quecksilber gab es nur bei zwei Kindern, für Cadmium bei einem Kind. Rund 10 % der getesteten Kinder sind gegenüber mindestens einem der untersuchten Schimmelpilze sensibilisiert. Die Belastung der Kinder durch das **Passivrauchen** ist seit 1990/92 nicht gesunken, sondern eher gestiegen. Etwa 50 % der Heranwachsenden leben demnach zurzeit in Haushalten mit mindestens einer rauchenden Person. Die Benzolbelastung war in der Studie in der Hälfte der untersuchten Raucherhaushalte höher als der EU-Grenzwert für Außenluft. Die Menge

an Cotinin – einem Abbauprodukt des Nikotins – im Urin deutet darauf hin, dass die Kinder heute sogar stärker durch Passivrauchen belastet sind als früher. Bereits 13 % der 8–14-Jährigen (d. h. Kinder und Jugendliche in einem Alter, in dem sie noch keiner Diskomusik ausgesetzt sind) zeigen einen **Hörverlust** von mehr als 20 Dezibel (mindestens in einer Frequenz auf mindestens einem Ohr), 2,4 % von mehr als 30 Dezibel. Im Trinkwasser von Haushalten mit 3–14-jährigen Kindern wurde bei 0,4 % der Haushalte die Grenzwerte der Zufallsstichproben für Blei überschritten, bei 1 % der Haushalte gab es Überschreitungen bei Kupfer und bei 1,8 % bei Nickel.

Die Daten des aktuellen Kinder-Umwelt-Surveys (KUS) belegen, dass die Umwelt in den letzten Jahrzehnten in vielerlei Hinsicht sauberer geworden ist. Die Belastung mit umweltmedizinisch relevanten Stoffen wie Blei, Quecksilber, PAH und PCB hat weiter abgenommen. Deutlich ist auch der Rückgang von persistierenden Organochlorsubstanzen in der Muttermilch. Trinkwasser und Nahrungsmittel sind von hoher Qualität, auch die Außenluft ist in vielen Parametern besser geworden. Einige Umweltstoffe gelten aufgrund ihrer Persistenz in den Umweltmedien als bedenklich: Beispiele sind polyzyklische Moschusverbindungen, Phthalate, Flammschutzmittel und perfluorierte Tenside. Insgesamt verursachen heute besonders Lifestyle-Faktoren bedrohliche Umweltbelastungen für Kinder und Jugendliche: Konsum legaler (Alkohol) und illegaler Drogen, Aktiv- und Passivrauch, übermäßige UV-Exposition, Hörverlust durch Diskomusik.

Human-Biomonitoring. Die Kinderärztliche Beratungsstelle für Umweltmedizin (gemeinnützige Kinderumwelt GmbH) als Betreiber von www.allum.de hat in 2011 eine interaktive Animation zu Grundlagen und Grundbegriffen des *Human-Biomonitoring* als eLearning-Modul erstellt. Das Modul zeigt auf, wie zuverlässige Daten zur Belastung des Menschen mit Umweltstoffen erhoben werden können und welche Personen besonders gefährdet sind (Link: www.allum.de/hbm).

Im Folgenden wird auf ausgewählte Umweltbelastungen eingegangen.

Amalgam. Nach Auffassung des bei der Europäischen Kommission angesiedelten wissenschaftlichen Ausschusses (European Commission Scientific Committee on Emerging and Newly Identified Health Risks, SCENIHR) ist sowohl die Versorgung mit Dentalamalgam als auch die Versorgung mit alternativen Materialien zur Wiederherstellung der Zahngesundheit geeignet. Beide Versorgungsarten können in seltenen Fällen lokale Reaktionen hervorrufen. Hinweise auf eine systemische Wirkung liegen nicht vor.

Alkohol. Alkohol, von der werdenden Mutter in der Schwangerschaft konsumiert, führt zum embryofetalen Alkoholsyndrom, einer der häufigsten Ursachen von Intelligenzminderung. Das lebenslang prägende Krankheitsbild, das auch mit vielen körperlichen Erscheinungen (z. B. geringes Geburtsgewicht, kleiner Kopfumfang, angeborene Herzfehler) einhergeht, wird in Kapitel 14 besprochen (➤ Kap.29). Alkoholabstinenz in der Schwangerschaft ist die einzige Möglichkeit einer sicheren Prävention dieser häufigen, wenn auch nicht immer diagnostizierten umweltbedingten Schädigung von Kindern.

Aus einmaligem oder gelegentlichem exzessiven Alkoholkonsum (Rauschtrinken, Komasaufen) erwächst in der Regel keine dauerhafte Abhängigkeit. Bedeutung, Ausmaß und die gesundheitlichen, psychischen und sozialen Folgen dauerhaften Alkoholkonsums von Kindern und Jugendlichen werden jedoch stark unterschätzt. Handlungsbedarf besteht besonders bei etwa 20 % der stationär akut wegen Alkoholintoxikation aufgenommenen Jugendlichen, bei denen Alkohol als „Problemlöser" dient. Notwendig ist in diesen Fällen die Zusammenarbeit verschiedener Fachleute (Psychologen, Heilpädagogen, Sozialarbeitern) mit den betroffenen Patienten und ihren Familien in Kooperation mit zusätzlichen Helfersystemen wie dem Jugendamt, Diensten der Suchtprävention oder Erziehungsberatungsstellen. Bewährt haben sich Präventionsprogramme wie z. B. HaLT (Hart am LimiT).

Elektrosmog. Der Begriff „Elektrosmog“ ist irreführend. Manchmal sind Felder der elektrischen Hausinstallation, ein anderes Mal Hochspannungsleitungen oder Bahnstrom, Rundfunk- und Fernsehsender und neuerdings auch Mobilfunk, DECT-Telefone, WLAN und Bluetooth gemeint. In Fachwelt und Öffentlichkeit werden insbesondere Befindlichkeitsstörungen („Elektrosensibilität“) und Krebsentstehung durch elektrische und magnetische Felder kontrovers diskutiert. Eine epidemiologische Studie erbrachte Hinweise für ein erhöhtes Risiko von Kindern, an Leukämie zu erkranken, wenn sie dauerhaft niederfrequenten magnetischen Feldern im häuslichen Umfeld ausgesetzt waren. Nach Einschätzung der Strahlenschutzkommission könnten dadurch weniger als 1 % der Leukämiefälle bei Kindern erklärt werden. Bisher ist kein Wirkungsmechanismus bekannt, der einen Zusammenhang zwischen Krebserkrankungen und elektromagnetischen Feldern erklärt. Das von Hochspannungsleitungen ausgehende elektrische Feld wird durch Mauerwerk effektiv abgeschirmt. Das magnetische Feld lässt sich dagegen nur mit hohem Aufwand abschirmen. Aus Gründen der vorsorgenden Medizin erscheint es ratsam, Flächen unter Hochspannungsleitungen nicht für den Bau von Gebäuden, in denen sich Nutzer lang andauernd aufhalten, zu nutzen. Auch für niederfrequente Felder gilt der allgemeine strahlenhygienische Grundsatz, dass unnötige Expositionen vermieden und unvermeidbare Expositionen so gering wie möglich gehalten werden sollten.

Feinstaub und Smog. Die Schädlichkeit von Smog (Gemisch im Wesentlichen von Ozon, nitrosen Gasen, Schwefeldioxid und Feinstäuben) ist seit langem bekannt und belegt. Es gibt keine unschädliche Feinstaubkonzentration. Schwebestäube, Feinstäube und ultrafeine Partikel werden als Verursacher von Lungenkrebs, Atemwegs- und Herz-Kreislauf-Erkrankungen sowie -todesfällen angesehen. Kinder halten sich länger in der Außenluft auf als Erwachsene und atmen bezogen auf ihr Körpervolumen mehr Außenluft ein. Luftverunreinigung (Smog) ist bei Kindern assoziiert mit Frühgeburtlichkeit, vermindertem Geburtsgewicht und frühkindlichen Todesfällen aufgrund von Atemwegserkrankungen.

Die Feinstaubbelastung ist ein Gesundheitsproblem ersten Ranges, das quantitativ viel bedeutender ist als die meisten anderen umweltmedizinischen Probleme, auch deshalb, weil sehr große Teile der Bevölkerung davon betroffen sind. Hier besteht dringender Handlungsbedarf.

Lärm. Neben dem physikalischen Schallpegel spielt beim Lärmempfinden auch die subjektive Bewertung eine wesentliche Rolle. Relevante Lärmquellen sind Verkehrs- und Freizeitlärm. Lärm kann zu Störungen bei der Kommunikation, Beeinträchtigung von Erholung und Entspannung, Herabsetzung des psychischen und körperlichen Wohlbefindens, der Konzentration und Leistungsfähigkeit, Schlafstörungen und zu erhöhten Konzentrationen an Adrenalin, Noradrenalin und Kortisol im Blut führen.

Kinder sind in jedem Alter Lärmquellen ausgesetzt. Im Kleinkindalter spielen Rasseln, Spielzeuge und Musikinstrumente für Kinder eine Rolle, später Spielzeugpistolen, Computer- und Konsolenspiele, schließlich tragbare Musikabspielgeräte und Diskotheken. Neben der Schallpegelhöhe spielt auch die Expositionsdauer eine Rolle. Lärmmessungen in Diskotheken und bei Life-Musikveranstaltungen zeigen, dass Musikschallpegel zwischen 90 und 110 dB erreicht werden. Ein Trend zu niedrigeren Werten ist bisher nicht erkennbar. Unter diesen Bedingungen kann es zu irreversiblen Schädigungen der für die Sinneswahrnehmung zuständigen Haarsinneszellen und zu Tinnitus und Hörverlust kommen. Männer und Jungen leiden häufiger an Hörverlust als Frauen und Mädchen.

Maßnahmen zum Schutz gegen Lärm umfassen Ohrstöpsel, Verwendung von Kopfhörern, Hören von Musik nicht mit voller Lautstärke, Vermeidung von lautem Spielzeug wie Rasseln oder Spielzeugpistolen, Benutzung lärmarmer Geräte und Maschinen, Einhaltung von Nacht- und Ruhezeiten u. a. m.

Mobilfunk und WLAN. Angesichts der verbreiteten Nutzung von Mobiltelefonen und Smartphones durch Kinder und Jugendliche und der heutigen Telefoniergewohnheiten („home zones", „Flatrates") wird die Frage diskutiert, ob Kinder und Jugendliche auf Mobilfunkfelder empfindlicher als Erwachsene reagieren. Dabei steht die Handynutzung im Vordergrund, bei der es zu einer meist kurzzeitigen, unter ungünstigen Bedingungen aber vergleichsweise hohen lokalen Exposition am Kopf kommen kann. Die beim Mobilfunk verwendeten Felder dringen nur geringfügig in den Kopf ein. Sie werden durch Ohr, Kopfhaut und Schädelknochen stark absorbiert. Bislang lässt sich keine besondere Empfindlichkeit von Kindern gegenüber mobilfunkbedingten Feldern belegen. Allerdings besteht noch Forschungsbedarf, u.a. zum Zusammenhang zwischen Handynutzung und Hirntumoren. Die Kinderärztliche Umweltmedizinische Beratungsstelle (Kinderumwelt) und auch das Bundesamt für Strahlenschutz raten zu einer bedachtsamen Nutzung von Mobiltelefonen durch Kinder und Jugendliche. Dies geschieht aus präventivmedizinischen Gründen, eine wissenschaftliche Grundlage für diese Empfehlung gibt es nicht.
Nach heutigen Kenntnissen gibt es keine Gründe, den Aufenthalt von Kindern und Jugendlichen in der Umgebung von Mobilfunkbasisstationen einzuschränken. Bluetooth-Geräte und WLAN-Anwendungen stellen bei üblicher Nutzung kein erkennbares Risiko für die Gesundheit dar. Das Bundesamt für Strahlenschutz rät dennoch zu einem vorsorglichen Umgang mit dieser Technologie. Im Mai 2011 wurden hochfrequente Felder, wie sie beim Mobilfunk, aber z.B. auch bei Rundfunk und Fernsehen benutzt werden, von der IARC (International Agency for Research on Cancer, Lyon) als „möglicherweise krebserzeugend" (Gruppe 2 B) eingestuft.
Nahrungsmittel. Mangelhafte Wasserqualität, Kalorienmangel, Eiweiß-, Eisen- und Vitaminmangel gehören weltweit zu den häufigsten Ursachen für Erkrankungen bei Kindern. Gerade in den ersten Lebensjahren ist eine ausgewogene Mischkost für die Entwicklung von großer Bedeutung (➤ Kap.8.3).
Nahrungsmittelallergene. Im Kindesalter werden 95 % der Nahrungsmittelallergien von nur wenigen Allergenen ausgelöst: Milch, Hühnerei, Nüsse, Soja, Weizen, Fisch. Bei der Mehrzahl der Kinder, die sich dagegen sensibilisieren, verlieren sich die spezifischen Symptome im weiteren Verlauf wieder. Wer auf ein Nahrungsmittel allergisch reagiert, sollte es meiden – was im Alltag nicht einfach ist. In Fertignahrungsmitteln sind Nahrungsmittelallergene sehr häufig versteckt vorhanden.
Ozon. Ozon ist natürlicher Bestandteil der Atmosphäre. Es wird neben anderen Substanzen unter dem Einfluss von Sonnenlicht gebildet und dem photochemischen Smog zugerechnet. Die Ozonkonzentration an heißen Sommertagen steigt vormittags bis zum frühen Nachmittag an und bleibt bis in den Abend hinein konstant. Danach überwiegt der Ozonabbau. Der Ozonabbau wird durch Luftschadstoffe gefördert. Darum ist der nächtliche Abfall der Ozonkonzentration in Ballungsgebieten größer ist als in Reinluftgebieten. Ozonmaximalwerte werden wesentlich von der Wetterlage mitbestimmt. Im Innenraum ist die Ozonkonzentration bei Schönwetterlagen wesentlich geringer als in der Außenluft.
Höhere Ozonkonzentrationen reizen die Schleimhäute und führen zu Augenbrennen, trockenem Husten und vor allem bei körperlicher Belastung zu Beklemmungsgefühlen. Kinder reagieren nicht empfindlicher auf Ozonexposition als Erwachsene. Bei Ozonkonzentrationen von mehr als 180 Mikrogramm pro Kubikmeter sollten länger dauernde intensive körperliche Beanspruchungen vermieden werden. Schulsport und Wettkämpfe sollten an ozonreichen Sommertagen in den Morgenstunden stattfinden. Kinder können in der Regel auch bei Konzentrationen bis 360 Mikrogramm pro Kubikmeter im Freien spielen, da sie sich bei den zumeist hohen Sommertemperaturen ohnehin körperlich schonen.

Durch **Passivrauchen** werden Menschen mit einer Vielzahl schädlicher Stoffe exponiert. **Tabakrauch** ist der wichtigste und verbreiteste Innenraumschadstoff. Bis zu 50 % der Kinder sind zuhause Passivrauch ausgesetzt. Rauchen ist die häufigste Ursache für Wohnungsbrände.

Mütterlicher Tabakkonsum in der Schwangerschaft führt häufiger zu verlangsamtem fetalen Wachstum, vermindertem Lungenvolumen, vorzeitigem Blasensprung und Frühgeburtlichkeit. Viele Stoffe aus dem Zigarettenrauch gegen in die Muttermilch über und liegen dort z. T. in gleicher Konzentration wie im mütterlichen Blut vor (Nikotin, Dioxin, Benzpyrene, Nitrosamine, Schwermetalle). Mütterliches Rauchen beeinträchtigt das Saugvermögen des gestillten Kindes. Das Kind nimmt weniger an Gewicht zu, zeigt häufiger Unruhe, Koliken und Erbrechen. Während der Stillperiode sollte daher nicht geraucht werden, zumindest nicht in der Nähe des gestillten Kindes. Passivrauchen im Säuglings- und Kleinkindalter verdoppelt das Risiko für einen plötzlichen Kindstod (➤ Kap. 8.4) erhöht die Rate von Infekten der unteren Atemwege und führt zu häufigeren Klinikaufenthalten. Es treten häufiger Mittelohrentzündungen und Tonsillitiden auf, die Asthmahäufigkeit bei über 3-Jährigen ist um 50–70 % höher.

Schadstoffe in der Raumluft. Innenräume sind die überwiegenden Aufenthaltsorte von Kindern und Jugendlichen. Die Innenraumluftqualität wird durch die Güte der einströmenden Außenluft, durch Schadstoffquellen, schadstoffabsorbierende Materialien im Innenraum und die Häufigkeit des Luftwechsels bestimmt. Stoßlüftung ist besonders effektiv.

In den letzten Jahren ist in der Innenraumluft eine Zunahme schwerflüchtiger organischer Verbindungen zu verzeichnen. Diese Produkte werden als „lösemittelarm" oder „lösemittelfrei" deklariert. Der Ersatz „chemischer Produkte" durch „Naturprodukte" führt nicht notwendig zur Verminderung der Schadstoffe, sondern oft nur zu einer Veränderung des Schadstoffspektrums.

Viele Innenraumluftschadstoffe sind Folge unkritischer Anwendung von Holzschutz- und Schädlingsbekämpfungsmitteln, Elektroverdampfern, Teppichausrüstung, übermäßigem Einsatz von Reinigungs- und Pflegemitteln sowie von Hobbytätigkeiten. Pflanzen und Zimmerspringbrunnen können sich zu Schimmel-/Allergenquellen entwickeln. Duftlampen, Räucherstäbchen, Raumluftsprays, Duftgels und Toilettensteine haben Einfluss auf die Qualität der Innenraumluft. Baubedingte Schadstoffquellen umfassen Dämmstoffe, Fassadenbehandlungsmittel, holzschutzmittelbehandeltes Holz, mineralische Werkstoffe, Fußbodenbeläge, Heizstätten mit offenem Feuer und Klimaanlagen.

Maßnahmen zur Verbesserung der Innenraumluftqualität sind in erster Linie präventiv: Es sollten keine Materialien mit hohem Emissions- bzw. Schadstoffpotenzial in den Innenraum gebracht werden. Die Raumbeladung mit Möbeln und Textilien sollte ein vernünftiges Maß nicht übersteigen. Gute Lüftungsmöglichkeiten tragen zu einem schadstoffarmen, ausgeglichenen Raumklima bei. Pflege- und Reinigungsmittel sowie Anstrichstoffe sollten sorgsam ausgewählt und überlegt eingesetzt werden. Tiere tragen durch Hautschüppchen und andere Absonderungen dazu bei, dass allergiegefährdete Personen tatsächlich erkranken. In Innenräumen sollte nicht geraucht werden, insbesondere nicht in Gegenwart von Säuglingen und Kindern.

Ultraviolette Strahlung. UV-Strahlung kann je nach Anwendungsart und Dosis eine gesundheitsfördernde oder eine gesundheitsschädigende Wirkung besitzen. UV-B-Strahlung fördert die Bildung des Provitamins D_3. Hierfür genügt es, für etwa 15 Minuten am Tag Gesicht und Hände dem Tageslicht auszusetzen. UV-Strahlung wird auch bei einer Reihe von Krankheiten therapeutisch eingesetzt. Zu den akuten Wirkungen der UV-Strahlung der Sonne und auch der Solarien gehören Sonnenbrand, Hautbräunung, Wirkungen auf das Immunsystem und phototoxische/photoallergische Reaktionen auf UV-A-Strahlung. Die wichtigsten chronischen Wirkungen sind vorzeitige Hautalterung, eine Trübung der Augenlinse und Hautkrebs.

UV-bedingte Gesundheitsschäden lassen sich am einfachsten dadurch vermeiden, dass der Aufenthalt in der Sonne begrenzt wird. Insbesondere während der Mittagsstunden sollten Schattenplätze aufgesucht und eine geeignete UV-dichte Sommerkleidung getragen werden. Geeignete Sonnenbrillen können die Augen vor Schäden schützen. Babys sollten in den ersten 12 Monaten gar nicht der Sonne ausgesetzt werden, später sollte neben einer Kopfbedeckung auch ein Sonnenschutzmittel mit Lichtschutzfaktor 30 oder höher angewendet werden. Kinder sollten durch einen Sonnenschirm o. Ä. vor direkter Sonnenstrahlung geschützt werden. In Übereinstimmung mit internationalen Empfehlungen empfiehlt die Strahlenschutzkommission, künstliche UV-Strahlung zu kosmetischen Zwecken, zur Steigerung des allgemeinen Wohlbefindens oder zur nichtmedizinischen Gesundheitsprophylaxe (Solarien) wegen der damit verbundenen Risiken nicht zu nutzen.

Gen-Umwelt-Interaktionen. Seit längerer Zeit ist bekannt, dass genetisch bedingt eine unterschiedliche Sensibilität gegenüber bestimmten Umweltnoxen besteht. In den letzten Jahren mehren sich weitere Hinweise, dass bestimmte Störungen häufiger auftreten, wenn Individuen mit bestimmten genetischen Eigenschaften diesen Noxen ausgesetzt sind. So wurde festgestellt, dass Kinder ein höheres Risiko für die Entwicklung eines Aufmerksamkeitsdefizitsyndroms (ADHS) bzw. eines dissozialen Verhaltens zeigten, wenn ihre Mütter Träger bestimmter zu ADHS disponierender genetischen Polymorphismen waren und in der Schwangerschaft Alkohol bzw. Nikotin konsumierten.

13.8 Elektronische Medien

Der Alltag von Kindern und Jugendlichen wird heute wesentlich durch elektronische Medien geprägt. Fast alle Kinder und Jugendlichen in Deutschland haben zu Hause Zugang zu einem Fernseher und zu Computern mit Internetanschluss. Die Mehrzahl der Kinder und Jugendlichen besitzt ein eigenes Handy oder Smartphone. Bereits Babies und Kleinkinder wachsen mit Kinderliedern und Geschichten von Kassetten und CD's auf.

Kleinkinder erlangen im Umgang mit elektronischen Medien auf spielerische Weise erstaunliche Fähigkeiten. Etwa 40 % der 6–23 Monate alten Kinder können einen Fernseher selbst einschalten und Programme mit der Fernbedienung umschalten, etwa 40 % der 2–3-Jährigen ein Video oder eine DVD selbstständig einlegen. Viele Kinder- und Jugendliche übertreffen frühere Generationen im Umgang mit Medien bei Weitem. Der kompetente Umgang mit Medien gehört heute wie Lesen, Schreiben und Rechnen zu den erforderlichen Kulturtechniken. Vor- und Nachteile elektronischer Medien werden in der Öffentlichkeit und von Fachleuten kontrovers diskutiert. Übermäßiger Medienkonsum ist unstrittig schädlich.

Die **Nutzung** der Medien in Zeitumfang und Qualität ist problematisch. Laut KiGGS-Studie (➤ Kap.4.7) verbringen 11–17-jährige Jungen im Durchschnitt 3,8 Stunden täglich mit Fernsehen, Video, Computer, Internet und Spielkonsole, Mädchen 2,7 Stunden. Zwei Drittel der Jungen beschäftigen sich damit 3 Stunden und mehr, ein Drittel 5 Stunden und mehr. Bei Mädchen liegen die entsprechenden Anteile bei 43 % und 18 %. Bei Jungen steigern ein niedriger Sozialstatus, der Hauptschulbesuch, Migrationshintergrund und eine Wohnung in Ostdeutschland den Medienkonsum z. T. deutlich. Bei Mädchen wirken diese Variablen mit Ausnahme des Migrationshintergrunds in gleicher Richtung. Die an Computern verbrachte Zeit wird überwiegend mit Computerspielen und in „Chatrooms" verbracht. Zahlreiche Fernsehprogramme ermöglichen Kindern den freien Zugang zu zweifelhaften oder gefährdenden Filmen. Videos aus Internet und auf Handys mit gewalttätigem oder pornografischem Inhalt sind unter Kindern und Jugendlichen weit verbreitet. Allerdings sind heute elektronische Medien zur Informationsgewinnung unverzichtbar und für die schulische Bildung eine reguläres Hilfsmittel geworden.

Die virtuelle Welt der Medien kann biologische, neurobiologische, psychologische und soziale **Folgen** haben und die körperliche, geistige und seelische Gesundheit von Kindern und Jugendlichen beeinflussen. Folgen erhöhten oder unangemessenen Medienkonsums können Schlafstörungen, Essstörungen, Bewegungsmangel, Hang zur Selbstüberschätzung, eingeschränkter Realitätsbezug, soziale Isolierung, Störungen der Aufmerksamkeit, Sprache oder Motorik, Hyperaktivität, Lern- und Leistungsstörungen, Schulversagen, internalisierende oder externalisierende Störungen, Spielsucht, Überschuldung u. a. m. sein. Das Risiko für Übergewicht bei Kindern, die mehr als vier Stunden am Tag fernsehen, ist fast dreimal so hoch wie das Risiko der Kinder, die weniger als zwei Stunden am Tag vor dem Fernseher sitzen. Kinder, die ein eigenes TV-Gerät im Zimmer haben, sind besonders gefährdet, später dick zu werden. Kinder aus Migranten- und Unterschichtfamilien stellen eine besondere Risikogruppe dar, da in diesen Familien überdurchschnittlich viele Medien konsumiert werden. Die Folgen von überzogenem Medienkonsum im frühen Kindes- und Schulalter sind nur schwer zu korrigieren.

Etwa 14 % der 14–24-Jährigen sollen einen problematischen und 2,5 % einen abhängigen Internetgebrauch zeigen, der sich insbesondere auf Computerspiele oder schriftliche Kommunikationsformen (email, Chatten, Internetforen) bezieht.

Problematisches Nutzungsverhalten zeigt sich bei:

- Computerspiel zum Abbau von Streß oder negativen Gefühlen
- Unwiderstehlichem Verlangen nach und gedanklicher Vereinnahmung durch Computer-/Internetnutzung
- Fehlender Kontrolle über die Dauer der Nutzung
- Einschränkung von sozialen Kontakten (Freundschaften, Familie, Hobbies)
- Abfall schulischer/akademischer Leistungen
- Vernachlässigung von Ernährung, Körperhygiene, Schlafmangel
- Finanziellen Verlusten
- Psychischen Entzugserscheinungen wie Unruhe oder Traurigkeit

Abhängiger Internetgebrauch wird von Fachleuten auf eine Störung der Impulskontrolle oder als Folge anderer psychischer Störungen angesehen. Andere sehen Computer-/Internetsucht als eigenständige klinisch relevante Diagnose an. Für diese Störung existieren erst wenige Therapieangebote.

Wegen der unerwünschten Folgen von TV-/Medienkonsum für die körperliche und geistige Entwicklung von Kindern und Jugendlichen ist Prävention erforderlich. Medienerziehung ist heute eine wichtige Erziehungsaufgabe. Eltern können und sollen dabei eine Vorbildfunktion haben. Die Entwicklung von **Medienkompetenz** ist dabei entscheidend: Mediale Produkte sollten in ihren ästhetischen Formen erkannt und gedeutet, Handlungsabläufe und Medienformen verstanden und interpretiert sowie Realität und Fiktionalität unterschieden werden können (Wahrnehmungskompetenz). Kinder sollten lernen, sich vor Medieneindrücken zu schützen, deren Einfluss auf ihre Stimmungen erkennen, sich bedürfnis- und interessengemäß zwischen Medien und Medienprodukten entscheiden und über ihre Medienerfahrungen kommunizieren können (Nutzungskompetenz). Schließlich sollten sie Wissen über die Entstehung von Medien und ihrer Produkte besitzen, Medien kritisieren und selbst gestalten und sie als eigene Ausdrucksform nutzen können (Handlungskompetenz). Eltern, Schulen, Betreuungseinrichtungen, Medien und Medienwissenschaften können einen Beitrag zum Erwerb von Medienkompetenz leisten.

Professionelle sollten bei Konsultationen regelmäßig eine Medienanamnese erheben und Eltern, Kinder und Jugendliche frühzeitig (z. B. bei Vorsorgeuntersuchungen, Beratungen, Sprachtherapien) auf die **Gefahren übermäßigen Medienkonsums** hinweisen. Dabei muss der Umgang insbesondere mit dem Medium Fernsehen von den Eltern häufig erst selbst gelernt werden.

Regeln, die den Fernseh- und Medienkonsum von Kindern begrenzen und regulieren:

- Kinder unter drei Jahren sollten überhaupt nicht fernsehen.
- Bei Kindern von drei bis sechs Jahren sollte die Fernsehzeit höchstens zweimal 30 Minuten am Tag betragen, im Grundschulalter 60 bis 90 Minuten.
- In den ersten zehn Lebensjahren gehört kein eigener Fernseher ins Kinderzimmer.
- Geschenke, die den Medienkonsum steigern, sollten gründlich überdacht werden.
- Für Grundschulkinder kann ein Handy sinnvoll sein, ein Smartphone ist nicht notwendig
- Eltern sollten so lange wie möglich Kenntnis über und Interesse an den Medienaktivitäten ihrer Kinder haben und mit diesen über Auswahl, Inhalte, Nutzen und Risiken kommunizieren.
- Finanzielle Belastungen von Kindern und Jugendlichen durch Mediennutzung ist durch geeignete Maßnahmen zu begrenzen.
- Eltern sollten als Vorbilder ihren eigenen Medienkonsum einschränken und Medienkompetenz erwerben.
- Mit zunehmendem Alter sollte das Kind in die Festlegung der Regeln für die Mediennutzung und die Auswahl der Medienprodukte einbezogen werden.
- Eltern sollten vielseitige Aktivitäten der Kinder als Alternativen zum Medienkonsum fördern und gemeinsam mit diesen praktizieren.
- Detaillierte Hinweise für Eltern zur Mediennutzung und -erziehung liefert die Bundeszentrale für gesundheitliche Aufklärung (BzGA).

Weitere Präventionsmaßnahmen sind die politisch zu setzenden Rahmenbedingungen des Jugendschutzes. Die kommerziellen Interessen der Medienhersteller und -anbieter müssen dem Ziel einer gesunden Entwicklung von Kindern und Jugendlichen untergeordnet bzw. verpflichtet werden. Die öffentliche Aufklärung über Nutzen und Risiken der Mediennutzung, muss in geeigneter Weise und spezifisch für verschiedene Zielgruppen erfolgen. Flächendeckende Angebote guter Ganztagseinrichtungen und sinnvoller alternativer Freizeitaktivitäten (z. B. Sport, Musik, kreative Gruppen) sind Möglichkeiten, übermäßigen und schädlichen Medienkonsum einzudämmen.

13.9 Familienunterstützende Maßnahmen

Familienunterstützende Maßnahmen sind von wesentlicher Bedeutung bei der Stärkung der personalen und familiären Ressourcen. Sie können einen wichtigen Beitrag zu einer positiven Entwicklung, zur Gesundheit und zur gesellschaftlichen Teilhabe von Kindern und Jugendlichen leisten. Derartige Maßnahmen können durch Familienangehörige, Freunde oder Bekannte, durch Arbeitgeber, Vereine, ehrenamtlich Tätige, Selbsthilfegruppen und Spendenaktionen erfolgen. Sie umfassen materielle und/oder persönliche Zuwendungen (Betreuung, Beratung, Vermittlung von Hilfen u. a.).

Familienunterstützende institutionelle Angebote bestehen seitens der verschiedenen Kindertageseinrichtungen und Schulen.

Es besteht allgemein oder unter bestimmten Voraussetzungen Rechtsanspruch auf familienunterstützende Maßnahmen seitens der Sozialhilfeträger, d. h. Krankenkassen, Renten-, Pflege und Unfallversicherung (➤ Kap. 42). Die Leistungen der **Kinder- und Jugendhilfe** nach Sozialgesetzbuch (SGB) VIII umfassen auch familienunterstützende, familienergänzenden und familienersetzende Hilfen (➤ Kap.10).

Zu den **familienunterstützenden Hilfen** gehören die Regelungen zur Jugendarbeit, Jugendsozialarbeit und zum Jugendschutz (§§ 11–14). Maßnahmen zur allgemeinen Förderung der Erziehung

in der Familie umfassen Familienbildung (§ 16), Erziehungsberatung (§ 16), Familienerholung (§ 16) und andere allgemeine Fördermaßnahmen (§ 16). Hilfen in besonderen Lebenslagen umfassen Partnerschafts- und Trennungsberatung (§17), Beratung und Unterstützung Alleinerziehender (§ 18), gemeinsame Wohnformen für Alleinerziehende (§ 19), Beratung und Versorgung der Kinder in Notsituationen (§ 20) und die Unterstützung bei der Unterbringung zur Erfüllung der Schulpflicht (§21).
Familienergänzende Hilfen beziehen sich auf Tageseinrichtungen (§ 22: Krippe, Kindergarten, Hort), Tagespflege (§ 23) sowie Hilfe durch Erziehung (§ 27). Diese kann durch Erziehungsberatung (§ 28), soziale Gruppenarbeit (§ 29), Erziehungsbeistand/Betreuungshilfe (§ 30), sozialpädagogische Familienhilfe (§ 31), Tagesgruppen (§ 32) erfolgen. **Familienersetzende Hilfen** umfassen Vollzeitpflege (§ 33), Heimerziehung (§ 34) und Intensivbetreuung (§ 35).
Für die praktische Arbeit in der Sozialpädiatrie bedeutsam ist die Eingliederungshilfe für seelisch behinderte Kinder und Jugendliche (§ 35a).
Personensorgeberechtigte und Kind oder Jugendlicher sind vor der Entscheidung des Jugendamtes über die Inanspruchnahme einer Hilfe und vor einer notwendigen Änderung der Hilfe zu beraten und auf mögliche Folgen für die Entwicklung des Kindes oder Jugendlichen hinzuweisen. Werden bei der Durchführung der Hilfe andere Personen, Dienste oder Einrichtungen tätig, sind sie am Hilfeplan zu beteiligen (§ 36).

13.10 Krankheitsbewältigung von Kindern und Jugendlichen und ihren Eltern/ Bezugspersonen

13.10.1 Allgemeines

Schwere akute Krankheiten, Unfälle, aber auch chronische Erkrankungen oder Behinderungen stellen im Leben eines Kindes/Jugendlichen und seiner Eltern/Bezugspersonen ein einschneidendes Ereignis dar, das vorübergehend, langfristig oder gar definitiv die Koordinaten der Lebensentwürfe grundlegend verändert. Die Bewältigung dieser neuen Situation stellt eine besondere Herausforderung an die Betroffenen dar und erfordert achtsame Begleitung und ggf. Unterstützung.
Die personale **Krankheitsbewältigung** (Coping) verläuft in **Phasen:** Ungewissheit, Gewissheit/ Schock, Verleugnung, Aggression/Depression, Verhandeln, Annahme, Aktivität. Diese Phasen sind individuell unterschiedlich gestaltet. Beschwerden oder Funktionsstörungen stehen mit dem individuellen Coping auf der kognitiven, emotionalen und Handlungsebene in wechselseitiger Beziehung.
Die Handlungsebene betrifft das Krankheitsverhalten. Bereits ein Frühgeborenes kann Schmerzempfindungen durch Mimik, Bewegung und vegetative Zeichen verdeutlichen. Ein Kleinkind verhält sich bei Fieber oder Kopfschmerzen still und zieht sich zurück.
Auf der emotionalen Ebene stellt sich das individuell sehr unterschiedliche Krankheitserleben dar.
Auf der kognitiven Ebene der Krankheitsbewältigung entwickelt das Individuum abhängig von seinem kognitiven Reifegrad und seines Wissens subjektive Krankheitstheorien.
➤ Tabelle 13.2 zeigt typische Formen der personalen Krankheitsbewältigung. Dabei sind die Reaktionen der von der Erkrankung/Behinderung selbst betroffenen Kinder/Jugendlichen und der Eltern/Bezugspersonen nicht immer identisch.

Tab. 13.2 Formen der personalen Krankheitsbewältigung

Handlungsbezogen	Kognitionsbezogen	Emotionsbezogen
Ablenkendes Anpacken Altruismus Aktives Vermeiden Körperliche Reaktionen Kompensation Konstruktive Aktivität Konzentration, Entspannung Sozialer Rückzug Solidarisieren Krankheitsbezogenes Zupacken Zuwendung	Ablenken Aggravieren Akzeptieren Dissimulieren Haltung bewahren Humor, Ironie Problemanalyse Relativieren Religiosität Grübeln Sinngebung	Hadern Selbstbedauern Emotionale Entlastung Isolieren, Unterdrücken Optimismus Passive Kooperation Resignation, Fatalismus Selbstbeschuldigung Schuld zuweisen Wut ausleben

Das objektive Ausmaß der Beeinträchtigung durch eine chronische Erkrankung oder Behinderung korreliert nicht eindeutig mit der subjektiv durch Kind/Jugendlichen oder Eltern/Bezugspersonen wahrgenommenen.

Die **Anpassungsfähigkeit** von Kindern/Jugendlichen an chronische Erkrankungen/Behinderungen ist sehr unterschiedlich ausgeprägt. Je früher im Leben eine chronische Krankheit/Behinderung auftritt desto eher gelingt eine konstruktive Adaptation. Sehr viele chronisch kranke oder behinderte Kinder führen ein erfülltes und fröhliches Leben, selbst bei schweren und schwersten Beeinträchtigungen. Lange Phasen diagnostischer Unsicherheit, Art und Weise der Diagnose- und Prognosevermittlung durch Professionelle, häufige oder traumatisierende Untersuchungen/Behandlungen stellen **Belastungsfaktoren für die Krankheitsbewältigung** dar. Die zuletzt genannten Faktoren können z. T. durch geeignete kind- und elternorientierte Maßnahmen (Aufklärung, Rollenspiel, Vermittlung von Sicherheit und Verlässlichkeit durch Eltern/Bezugspersonen) vermieden werden. Besonderes Einfühlungsvermögen erfordert die Begleitung von Kindern mit progredienten und lebensverkürzenden Erkrankungen. Persönlichkeitseigenschaften des Kindes (z. B. Optimismus, geringe Ängstlichkeit, aktives und prosoziales Verhalten) haben Einfluss auf die kindlichen Copingprozesse. Dysfunktionelle Krankheitskonzepte von Kindern sollten erkannt und ggf. richtig gestellt werden (falsche, magische oder irrationale Annahmen). Dabei hilft ein dem Entwicklungsstand angemessener Umgang in Offenheit, Ehrlichkeit und Respekt. Die Krankheitsbewältigung seitens der Eltern und die Einstellung von Bezugspersonen und Mitgliedern der Peergruppe zu Krankheit/Behinderung sind für eine erfolgreiche oder ungünstige Krankheitsbewältigung des Kindes oder Jugendlichen mit entscheidend. ist Auf Schuldgefühle seitens der Eltern und deren Krankheitskonzepte ist zu achten. Die Integration und Teilhabe am gemeinsamen Leben mit gesunden Kindern ist oft hilfreich, selbst wenn dies nur partiell möglich ist. Im Einzelfall und bei ungünstigen Umständen können dadurch aber auch Belastungen und Nachteile entstehen.

Professionelle müssen sich bei jedem Kontakt mit chronisch kranken/behinderten Jugendlichen und ihren Bezugspersonen dieser Probleme bewusst sein, sich adäquat verhalten und zur Entwicklung konstruktiver Lösungsmöglichkeiten beitragen. Hierfür sind ausreichend Zeit, Vertrauen und Empathie erforderlich. Eine auf dieser Ebene gelingende Interaktion führt zu einem besseren Gesundheits- und Funktionszustand sowie zu Behandlungszufriedenheit der Betroffenen und einer höheren Zufriedenheit der Professionellen.

Tab. 13.3 Krankheitsbewältigung bei Kindern und Jugendlichen – hilfreiche Maßnahmen

Altersabhängige Informationsvermittlung
Beschäftigung, Ablenkung, Normalität
Förderung emotionaler Ausdrucksmöglichkeiten
Aktivierung von Ressourcen
Steigerung des Selbstwertgefühls
Entwicklung von Copingstrategien
Integration hilfreicher Erfahrungen
Krisenintervention
Motivationsförderung
Training sozialer Fähigkeiten
Förderung alltagsrelevanter Fähigkeiten (statt defizitorientiertem Training)

➢ Tabelle 13.3 zeigt, welche Maßnahmen bei der persönlichen Bewältigung von Krankheiten/ Behinderungen im Kontext der verschiedenen Entwicklungsaufgaben von Kinder und Jugendlichen hilfreich sind.

13.10.2 Konzepte von Gesundheit, Krankheit und Tod bei Kindern/Jugendlichen

Die alters- und entwicklungstypischen **Krankheitskonzepte** spielen bei der Krankheitsbewältigung eine Rolle. Professionelle sollten im Umgang und bei einer altersgemäßen Erklärung über Krankheitsursachen und therapeutische Maßnahmen hierauf Rücksicht nehmen.

Konzepte von Kindern und Jugendlichen zu Gesundheit und Krankheit sind von ihrem kognitiven Entwicklungsstand (4 Stadien nach der Theorie von Piaget), aber auch von ihrem erworbenen allgemeinen und spezifischen Krankheitswissen abhängig (Ansatz von Carey). Sie sollten Professionellen bekannt sein und werden im Folgenden kurz dargestellt. Sie sollten bei der Begleitung der Kinder/ Jugendlichen angemessen berücksichtigt und ggf. beim einzelnen Kind/Jugendlichen erfragt werden.

Gesundheit ist für Kinder im Vorschulalter (präoperationales Stadium nach Piaget, 2–6 Jahre) schwierig oder gar nicht zu definieren. Im Grundschulalter (konkret-operationales Stadium, 7–11 Jahre) wird Gesundheit als positiver Zustand wahrgenommen, der zu körperlicher Aktivität befähigt. Die Reversibilität von Gesundheit wird erkannt. Im Jugendalter (formal-operationales Stadium, ab 12 Jahre) wird Gesundheit wahrgenommen als langfristige positive Qualität, die sich in physischem, psychischem und sozialem Wohlbefinden sowie in Leistungsfähigkeit ausdrückt.

Das **Körperwissen** im Vorschulalter ist rudimentär und beschränkt sich auf sichtbare bzw. wahrnehmbare Körperbestandteile. Im Grundschulalter können mehr Körperbestandteile in ihrer Position richtig bezeichnet werden. Erste Funktionszuordnungen werden klar. Jugendliche kennen weitere Körperbestandteile und zunehmend auch Organsysteme. Funktionen werden auch unter zellulären und biochemischen Gesichtspunkten weiter differenziert. Nach dem 14. Lebensjahr bleibt der Kenntnisstand über den eigenen Körper bis ins Erwachsenenalter oft unverändert.

Krankheitskonzepte beziehen sich auf Entstehungsbedingungen, Verlauf, Behandlungsmöglichkeiten und Sinn der Erkrankung. Bei Vorschulkindern variiert je nach Kind und Krankheit die Kenntnis von Krankheitsursachen (keine Vorstellung, magische oder globale Vorstellung, Bestrafung für Fehlverhalten, korrekte Vorstellung). Vorschulkinder stellen allgemeine Krankheitssymptome in den Mittelpunkt. Nur bei häufig auftretenden spezifischen Erkrankungen können sie Symptome nennen.

Tab. 13.4 Vorstellungen zum Tod bei Kinder (Quelle: Internationale Gesellschaft für Sterbebegleitung und Lebensbeistand e. V.)

Entwicklungsalter (Jahre)	Vorstellung
2–3	Keine Vorstellung
3–4	Ereignis, das Trennungsängste hervorruft. Tod als Schlafzustand. Abwesenheit
4–5	Unklare Vorstellung. Dunkelheit, Bewegungslosigkeit
5–6	Bestrafung für böse Taten, Vorstellung des Begraben werdens
6–7	Körper zerfällt nach dem Tod
8–9	Sterben und Tod sind noch unannehmbar, Glauben an Unsterblichkeit
9–10	Sachlich nüchtern. Unumkehrbarkeit. Wissen, dass es sie auch betrifft

Im frühen Vorschulalter wird das Verhalten, später die Ansteckung als Ursache erkannt. Faktoren wie Schuld und Strafe können eine Rolle spielen. Zum Krankheitsverlauf und spezifischen Behandlungskonzepten bestehen kaum Vorstellungen, das Einhalten gewisser Regeln wird als positiv gekannt. Im Grundschulalter können Kinder neben äußeren objektiv sichtbaren auch nicht sichtbare Krankheitszeichen benennen. Als Ursachen für Erkrankungen wird eine Kontamination durch direkten Kontakt, Einatmen oder Schlucken angesehen. Viren und Bakterien erscheinen als wichtigste Krankheitserreger. Weitere Ursachen sind Lebensstil (Unachtsamkeit, Ernährung, Kleidung), Fehlfunktion innerer Organe/System, Verhalten, Einstellungen und Gefühle. Differenziertere Verursachungskonzepte zu bestimmten spezifischen Erkrankungen entstehen abhängig vom erworbenen Wissen. Erste Erkenntnisse zu Krankheitsprozessen, deren Beeinflussbarkeit und Prävention von Krankheiten entstehen. Bei Jugendlichen differenzieren sich die Krankheitskonzepte weiter und beinhalten physiologische und psychologische Aspekte. Die Konzepte werden abstrakter und beziehen neben Ansteckung, externen Ursachen, Aspekten des Lebensstils, der Fehlfunktion innerer Organe/Systeme auch Verhalten, Einstellungen, Gefühle und Sonstiges wie Stress, Prädisposition, Konstitution, Bestimmung mit ein. Zu spezifischen Erkrankungen und ihrer Behandlung bestehen meist weiterhin kaum Kenntnisse des wissenschaftlichen Hintergrunds. Medizinische Behandlungen werden als Zusammenwirken von Körper, Medikation, weiteren Behandlungsmaßnahmen und auch Selbstbehandlung mit dem Ziel der Genesung/Heilung verstanden.
Die Vorstellungen zum **Tod** unterliegen bei Kindern dem Entwicklungsstand entsprechend einem Wandel (➤ Tab. 13.4). Erfahrungen im Umgang mit Kindern mit lebensbegrenzenden Erkrankungen legen nahe, mit Kindern und Jugendlichen das Gespräch über ihren bevorstehenden Tod zu suchen oder zumindest offen zu sein für entsprechende Gesprächssignale und konkrete Anliegen der Betroffenen.

13.10.3 Spezifische Probleme der Krankheitsbewältigung Jugendlicher

Chronische Erkrankungen/Behinderungen nehmen wesentlichen Einfluss auf die Auseinandersetzung Jugendlicher mit ihren Entwicklungsaufgaben und Problemen: Identitätsfindung, Autonomie, Berufs- und Lebensplanung (➤ Kap. 35).
Defizite von Körperfunktionen, motorischen, kognitiven oder anderen Leistungen sowie eine selbst wahrgenommene oder von anderen mitgeteilte geringere Attraktivität beeinflussen Selbstbild, Identitätsfindung und Rollenverhalten nachhaltig. Dies gilt z. B. für Jugendliche mit motori-

schen Problemen wie Spina bifida oder Hemiparese ebenso wie für Jugendliche mit Erkrankungen wie Kleinwuchs, Mukoviszidose, angeborenen Herzfehlern oder chromosomalen Anomalien.
Die **Autonomieentwicklung** von Jugendlichen, die Lösung von den Eltern ist vielfach erschwert. Sorgen der Eltern können sachlich berechtigt oder Ergebnis eines langjährigen Verarbeitungsprozesses sein und zu einer „Infantilisierung" des Jugendlichen mit übermäßiger elterlicher Kontrolle führen. Für einen Jugendlichen mit Epilepsie stellt der abendliche Kino-, Gaststätten- oder Discobesuch mit Gleichaltrigen ein Problem oder Risiko dar. Bei lern- und insbesondere geistig behinderten Jugendlichen fällt es den Eltern oft besonders schwer, die notwendige Autonomieentwicklung ihrer Kinder in einem realistischen Rahmen zu akzeptieren und zu fördern. Möglichkeiten der Ablösung vom Elternhaus sollten auch hier als Perspektive rechtzeitig, also vor dem 18. Lebensjahr, angesprochen werden.
Nicht wenige Jugendliche neigen zu Verhaltensbesonderheiten. Sie zeigen sich Erwachsenen gegenüber oppositionell, sind im Alltag oft sprunghaft, labil, leicht verletzbar und neigen zu riskantem Verhalten. Dies erschwert eine normale medizinische Behandlung gerade chronischer Erkrankungen deutlich. Die Compliance/Adhärenz zu regelmäßiger und zeitgerechter Medikamenteneinnahme, zu adäquaten Ernährungsregimes, aber auch zur Wahrnehmung ärztlich-therapeutischer Konsultationen und Maßnahmen kann erheblich gestört sein. Dies kann z. B. zu erneutem Auftreten epileptischer Anfälle, mangelhafter Blutzuckereinstellung bei Diabetes mellitus oder der Verschlechterung einer Skoliose bzw. Bewegungsstörung führen. Wenn bei Jugendlichen die eigenen oder die externen (z. B. elterlichen) Erwartungen mit der realen Funktionsfähigkeit nicht übereinstimmen, können **dysfunktionelle Verarbeitungsstrategien** die Folge sein. Es kommt zu psychosomatischen und Störungen des Sozialverhaltens, emotionalen Störungen, Lern-/Leistungsstörungen, depressiver Störung, Substanzmissbrauch, risikoreichen Verhaltensweisen.
Jugendliche mit chronischen Erkrankungen/Behinderungen zeigen häufiger als gesunde Gleichaltrige internalisierende Lösungsversuche. Behinderte finden in unserer Gesellschaft meist Kontakt zu anderen Behinderten. Es fällt ihnen deutlich schwerer, sexuelle Erfahrungen zu sammeln oder eine dauerhafte Partnerschaft einzugehen. Vielfältige Integrationsbemühungen sind erforderlich, stoßen jedoch immer wieder an Grenzen.
Chronische Erkrankungen/Behinderungen beeinflussen in vielen Fällen die Berufswahl und Lebensplanung. Individuelle Einschränkungen müssen berücksichtigt werden. Trotz vielfältiger sozialgesetzlicher Erleichterungen fällt es oft schwer, einen angemessenen Arbeitsplatz zu finden.

13.10.4 Krankheitsbewältigung bei Eltern/Bezugspersonen

Verschiedene Modelle beschreiben **typische Abläufe** der Krisen- und Trauerverarbeitung. Bekannt geworden ist die sogenannte „ Trauerspirale". Eine schwerwiegende medizinische Erkrankung oder eine bleibende Behinderung können bei Betroffenen und Bezugspersonen typische Bewältigungsprozesse auslösen, die oft aufeinander folgen: Ungewissheit (Unwissenheit, Unsicherheit, Unannehmbarkeit), Gewissheit, Aggression, Verhandlung, Depression, Annahme, Aktivität, Solidarität. Die Kenntnis dieser Abläufe erleichtert das Verständnis für Äußerungen und Verhaltensweisen der Kinder- und Jugendlichen bzw. deren Bezugspersonen. Der Kinder- und Jugendarzt sollte auf die Abläufe feinfühlig eingehen.
Eltern oder andere Bezugspersonen können für Kinder und Jugendliche mit chronischen Erkrankungen/Behinderungen eine wesentliche Unterstützung (Ressource) darstellen durch ihr intuitives Wissen, ihre Persönlichkeit (Motivation, Optimismus, emotionale Kompetenz, Frustrationstoleranz und anderes), ihre positiven Lebenserfahrungen und durch die Einbeziehung in soziale Netzwerke.

Die Art der **Diagnosevermittlung** durch Professionelle an Eltern ist sehr bedeutsam für deren Bewältigung einer schwerwiegenden Erkrankung/Behinderung ihrer Kinder. Unsicherheiten bei der Mitteilung können sich auf Eltern übertragen und deren funktionelle Krankheitsbewältigung nachhaltig beeinträchtigen.

Eine **schwerwiegende medizinische Diagnose** sollte Eltern in einem ruhigen Raum mit ausreichend Zeit und ohne Störungen und Unterbrechungen in sachlicher und empathischer Weise vermittelt werden. In einer ersten Phase („Schock") zeigen Eltern häufig eine reduzierte Aufnahme- und Verarbeitungskapazität. Sie nehmen nur einen Bruchteil der vermittelten Informationen auf oder richten ihren Fokus auf einzelne verbale oder nonverbale Äußerungen. Eltern sollten in ihrer Trauer, Wut und Verzweiflung nicht allein gelassen werden. Gegebenenfalls ist spezielle psychologische oder seelsorgerische Hilfe empfehlenswert.

Auch bei **nicht lebensbedrohlichen, aber die Lebensperspektive nachhaltig beeinflussenden eindeutigen medizinischen Diagnosen** (z. B. Chromosomenanomalien, Fehlbildungen, Epilepsien) ist ein persönliches Gespräch soweit möglich mit beiden Eltern unbedingt empfehlenswert. Insbesondere betroffene Neugeborene und Säuglinge sollten dabei zur Bekräftigung von Akzeptanz und Bindung soweit möglich anwesend sein. Es ist nützlich, die fachlichen Informationen für die Eltern kompetent und in einer den Eltern angepassten, verständlichen Sprache darzubieten und durch Bilder (CT, MRT, Abbildungen in Büchern, Videos), Schemazeichnungen oder schriftliche Informationen zur Mitnahme zu ergänzen. Es ist wichtig, genügend Zeit für Fragen der Eltern zu haben und auf diese in verständlicher Form einzugehen. Das Angebot eines weiteren Gesprächs in angemessenem zeitlichen Abstand erscheint sinnvoll.

Bei **eindeutiger prognostischer Diagnose** (z. B. geistige Behinderung, Zerebralparese) sollte der Zeitpunkt der Diagnosevermittlung dem Alter des Kindes, der Sicherheit der Prognose, den Erwartungen, Ängsten und Bewältigungskompetenzen der Eltern angepasst sein. Die Information sollte taktvoll und realistisch erfolgen, Hoffnungen geben und lassen mit Hinweis auf die Variabilität der Entwicklung im vorgegebenen Spektrum und auf neue realistische Ziele. Den Eltern sollten Wege zu hilfreicher eigener Aktivität für das Kind und sie selbst aufgezeigt werden. Andere Ressourcen zur Bewältigung der neuen Situation sollten benannt und einbezogen werden.

Bei **unsicherer Diagnose und Prognose** (z. B. Kindern mit Entwicklungsstörungen im frühen Alter) sollte die ätiologische Diagnostik abgestuft, zeitgerecht und nicht überstürzt erfolgen. Sie sollte die Wünsche und Befürchtungen der Eltern berücksichtigen. Eine therapierelevante Diagnostik muss hingegen zügig angegangen werden.

Der persönliche **Bewältigungsprozess** (Coping) von Eltern chronisch kranker oder behinderter Kinder dauert oft Jahre oder Jahrzehnte und wird von potenziellen Krisen begleitet. Dieser Prozess beeinflusst Stimmung und Verhalten der Eltern, die Compliance/Adhärenz zu Therapieterminen und -inhalten, die Stabilität der Beziehung zwischen Eltern und Professionellen. Er hat aber auch Auswirkungen auf die elterliche Sensitivität dem Kind gegenüber, die Eltern-Kind-Interaktion sowie auf die Selbstsicherheit der Eltern, die wahrgenommene Selbstwirksamkeit, Stress, mögliche psychische und soziale Probleme (z. B. Depressivität, Alkoholismus, Trennung der Partner). Am elterlichen Bewältigungsprozess können Partner, weitere Familienangehörige, Freunde, Elterngruppen und Professionelle beteiligt sein. Sie können in positiver oder in ungünstiger Form auf ihn Einfluss nehmen. Dabei gibt es kein absolut richtiges Coping. Vielmehr können individuell unterschiedliche Wege ein gelingendes Coping darstellen. ➤ Tabelle 13.5 zeigt günstige, ➤ Tabelle 13.6 ungünstige Copingstrategien von Eltern.

Der elterliche Copingprozess ist ein wesentlicher Aspekt einer längerfristigen professionellen Begleitung. Professionelle können einen Beitrag zum gelingenden Coping von Eltern/Bezugspersonen leisten, indem sie günstige Verarbeitungsstrategien bestärken bzw. helfen, diese zu erkennen und umzusetzen. Generell ist es vorteilhaft, zu einer Annahme der Probleme und einer aktiven

Tab. 13.5 Günstige Copingstrategien von Eltern

Suche nach gesicherten Informationen
Aktive Mitgliedschaft in Selbsthilfegruppen
Erprobung neuer Versorgungskonzepte
Teilnahme an wissenschaftlichen Studien
Förderung der Integration von Kind und Familie
Soziale Einbindung und Unterstützung
Akzeptanz des Problematik, realistische Ziele
Bestmögliche Normalität
Freiraum für eigene Bedürfnisse
Aufbau neuer Lebensperspektiven (Entwicklungschancen, Wertorientierung)

Tab. 13.6 Ungünstige Copingstrategien von Eltern

Maximale Therapieintensität und -frequenz
Anfälligkeit für einfache Konzepte mit Heilscharakter
Anfälligkeit für „Propheten" einer guten Prognose
Vorwürfe an Ärzte/Therapeuten
Doktor-/Therapeuten-Hopping
Sozialer Rückzug/Selbstvorwürfe
Extreme Anspruchshaltung → Widersprüche, Sozialgerichtsverfahren
Vernachlässigung von Geschwistern
Blockierung der Beziehung zum Kind
Verlust von Zuversicht und Vertrauen
Belastungen der Partnerschaft

Haltung („fighting spirit“) zu gelangen. Dazu sollten die vorhandenen Probleme des Kindes und der Eltern/Bezugspersonen analysiert und gemeinsam sinnvolle Zielsetzungen erarbeitet werden. Hierbei müssen innere und äußere Ressourcen erkannt und genutzt werden und Eltern/Bezugspersonen die Erfahrung der Selbstwirksamkeit machen. Die Steigerung der Selbstwirksamkeit (Empowerment), die Teilhabe am Leben in der Gemeinschaft und Lebensqualität der betroffenen Kinder/Jugendlichen und ihrer Familien sind oberste Ziele der professionellen Begleitung.

Ärztliche Gespräche mit **verwaisten Eltern** sind wichtig und verlangen Zeit, Ruhe und Einfühlungsvermögen. Im übertragenen Sinn sollte den Eltern die Hand gereicht werden („nicht davor oder dahinter, sondern neben den Eltern stehen“). Ungünstig sind Äußerungen, die etwas Positives am Tod des Kindes finden, die auf gesunde Kinder/Geschwister hinweisen (das verstorbene Kind ist nicht ersetzbar!) oder dass man wisse, was die Eltern empfinden. Hinweis auf einen problematischen Bewältigungsprozess sind elterliche oder auch ärztliche Äußerungen, dass es besser sei, wenn das Kind tot ist als wenn es behindert gewesen wäre.

Literatur

Ball J (2004) Untersuchung und Erfassung von kindlichen Krankheitskonzepten im Grundschulalter. Marburg: Dissertation.

Beauftragte der Bundesregierung für Migration, Flüchtlinge und Integration (2010) 8. Bericht der Beauftragten der Bundesregierung für Migration, Flüchtlinge und Integration über die Lage der Ausländerinnen und Ausländer in Deutschland.

http://www.bundesregierung.de/Content/DE/_Anlagen/2010/2010-07-07-langfassung-lagebericht-ib.pdf?__blob=publicationFile

Beauftragte der Bundesregierung für Migration, Flüchtlinge und Integration (2011) Zweiter Integrationsindikatorenbericht.
http://www.bundesregierung.de/Content/DE/_Anlagen/2012/01/2012-01-12-integrationsbericht.pdf?__blob=publicationFile

Bode H (1998) Diagnostik bei Entwicklungsstörungen. Pädiatrische Praxis 54: 41–47.

Bode H (2002) Die Bedeutung der Eltern für eine erfolgreiche Frühförderung. Frühförderung interdisziplinär 21: 88–95.

Bode H (2004) Migrantenkinder. Pädiatr Prax 65: 289–297. Kinder mit Lernstörungen und Behinderungen in integrativen Schulen oder in Sonderschulen? Die Sichtweise von Eltern und Fachleuten. Klin Padiatr 225: 57–63

Bundesamt für Migration und Flüchtlinge (2013) Migrationsbericht 2011.
http://www.bamf.de/SharedDocs/Anlagen/DE/Publikationen/Migrationsberichte/migrationsbericht-2011.pdf?__blob=publicationFile

Bundesministerium für Familie, Senioren, Frauen und Jugend (2009) 13. Kinder- und Jugendbericht.
http://www.bmfsfj.de/RedaktionBMFSFJ/Broschuerenstelle/Pdf-Anlagen/13-kinder-jugendbericht,property=pdf,bereich=bmfsfj,sprache=de,rwb=true.pdf

Deutsche Gesellschaft für Sozialpädiatrie und Jugendmedizin. Pressemappe
http://www.presseportal.de/pm/55202

Ellsäßer G, Böhm A, Kuhn J, Lüdecke K, Rojas G (2002) Soziale Ungleichheit und Gesundheit bei Kindern. Ergebnisse und Konsequenzen aus den Brandenburger Einschulungsuntersuchungen. Kinderärztliche Praxis 73: 248–257.

Fröhlich J, Lehmkuhl G (2012) Computer und Internet erobern die Kindheit. Stuttgart:Schattauer

Gehrmann J, Sumargo S (2009) Kinder psychisch kranker Eltern. Monatsschr Kinderheilkd 157: 383–394.

Hart H (2008) Disabled children living away from home in foster care and residential settings. Dev Med Child Neurol 50: 885.

Hollmann H, Kretzschmar R, Schmid RG (2009) Qualität in der Sozialpädiatrie. Bd. 1. 2. Aufl. Altötting: RS-Verlag.

Ipsiroglu OS, Bode H (2005) Transkulturelle Pädiatrie. Eine Einführung. Monatsschr Kinderheikld 153: 8–15.

Lampert T, Sygusch R, Schlack R (2007) Nutzung elektronischer Medien im Jugendalter. Ergebnisse der KiGGs-Studie. Bundesgesundheitsbl – Gesundheitsforsch – Gesundheitsschutz 5/6: 643–652.

Piaget, J. (1978) Das Weltbild des Kindes. dtv/Klett-Cotta München.

Sherry SN, Nickman SL (1999) Adoption and foster family care. In: Levene M, Carey W, Crocker AC (ed) Developmental-Behavioral Pediatrics. Philadelphia: WB Saunders.

Sonuga-Barke EJS, Beckett J, Castle J et al. (2008) Is subnutrition necessary for a poor outcome following early institutional deprivation? Dev Med Child Neurol 50: 664–671.

Internetlinks

www.adoption.de
www.allum.de
www.bzga.de
www.buergerliches-gesetzbuch.info
www.pisa.ipn.uni-kiel.de
www.sozialgesetzbuch.de

V Diagnostik

14 Entwicklungsdiagnostik im Rahmen der Früherkennungsuntersuchungen in der kinder- und jugendärztlichen Praxis

Folkert Fehr und Norbert Knieß

14.1 Aufgaben der Früherkennungsuntersuchungen

Kinder und Jugendliche haben einen gesetzlichen Anspruch auf Früherkennungsuntersuchungen (*§26, SGB V;* U1-U9, J1), die einen wesentlichen Beitrag zur Krankheitsprävention in dieser Altersgruppe leisten. Dies beruht zunächst auf einem **primärpräventiven Ansatz** durch Beratung der Eltern, Impfungen und medikamentöse Prophylaxe (Vitamin K, D, Fluorid). Beratungsinhalte sind dabei u. a. Ernährung, Zahngesundheit, Entwicklungsförderung, Unfallvermeidung, Mediennutzung, Sexual- und Suchtverhalten. Im Sinne der **Sekundärprävention** sollen organische Erkrankungen, Entwicklungs- und Verhaltensstörungen frühzeitig aufgedeckt und einer Behandlung zugeführt werden, welche eine Progredienz oder Chronifizierung verhindert (z. B. mittels Hüft-, Hör- und Sehscreening oder Elternfragebogen). Eine zunehmende Anzahl chronisch kranker Kinder bedarf zudem einer regelmäßigen vorausschauenden Beratung und Betreuung, die Rückfällen und Folgeschäden vorbeugt (**Tertiärprävention**).
In Hinblick auf eine frühe Erfassung von elterlicher m Unterstützungs- und Hilfebedarf sowie zur Vermeidung von Vernachlässigung, Misshandlung oder Missbrauch des Kindes wurde ein verbindliches Einlade- und Meldewesen für die Früherkennungsuntersuchungen in der Mehrzahl der Bundesländer gesetzlich (siehe Landesgesetze zum Kinderschutz) verankert. Eine Rechtspflicht zur Teilnahme an der Früherkennung besteht jedoch nach wie vor nicht.
Hauptsächliche Faktoren, die einer Inanspruchnahme der Früherkennungsuntersuchungen entgegen stehen, stellen ein niedriger sozioökonomischer Status, mangelnde Sprachkenntnisse der Eltern, ausländische Staatsangehörigkeit und ein höheres Alter des Kindes/Jugendlichen dar.

Sozialpädiatrie in der Praxis. http://dx.doi.org/10.1016/B978-3-437-31630-3.00014-6
Copyright © 2014 Elsevier GmbH. All rights reserved.

Tab. 14.1 Bezeichnung und Altersbereich der Früherkennungsuntersuchung in Deuschland

Vorsorge	Altersbereich	Gültigkeitsbereich
U1	Geburt	alle Versicherten
U2	1. Woche (3.–10. Lebenstag)	
U3	1 Monat (4.–5. Lebenswoche)	
U4	3 Monate (3.–4. Monat)	
U5	6 Monate (6.–7. Monat)	
U6	1 Jahr (10.–12. Monat)	
U7	2 Jahre (21.–24. Monat)	
U7a	3 Jahre (34.–36. Monat)	
U8	4 Jahre (46.–48. Monat)	
U9	5¼ Jahre (60.–64. Monat)	
U10	7–8 Jahre	Selektivverträge mit Krankenkassen für U10/U11.
U11	9–10 Jahre	Geplant ist *eine* Vorsorge im Grundschulalter für alle Versicherten
J1	13 Jahre (12–14 Jahre)	alle Versicherten
J2	16–17 Jahre	Selektivverträge mit Krankenkassen

14.2 Untersuchungstermine

Früherkennungsuntersuchungen finden zu einem festgelegten Alter statt (➤ Tab. 14.1). Die Altersbereiche der U1–11 und J1–2 können aus der Tabelle 14.1 entnommen werden. Die U10–11, sowie J2 werden bisher ausschließlich in Selektivverträgen mit Krankenkassen angeboten. Aufgrund der bestehenden Morbiditäten gibt es einen Bedarf, Früherkennungsuntersuchungen auch für diese Altersgruppen allgemein zu etablieren. Für das Grundschulalter wurde eine entsprechende Gesetzesänderung durch die Bundesregierung 2013 auf den Weg gebracht (Neues Präventionsgesetz). Früherkennungsuntersuchungen werden im Vorschul- und Grundschulalter zum ganz überwiegenden Teil von niedergelassenen Kinder- und Jugendärzten (KJÄ) durch-geführt. Durch den Berufsverband der Kinder- und Jugendärzte (BVKJ) wurde 2012 ein erweitertes Vorsorgeheft vorgestellt, welches die Inhalte und Dokumentation aller Früherkennungsuntersuchungen neu strukturiert. Wesentlicher Bestandteile sind dabei entwicklungsneurologische (Grenzsteine der Entwicklung), sowie psychosoziale und emotionale Befunde, die zusätzlich, ausgehend von einer ausführlichen Anamnese, erhoben werden. Als Instrumente dienen eine Checkliste zur Erfassung von familiären Belastungsfaktoren (U1/2), Elternfragebogen zur Verhaltensregulation (U3–U6), Mannheimer Elternfragebogen (U7–U11), sowie Mannheimer Jugendlichen-Fragebogen (J1/J2) und die direkte Beobachtung der Mutter/Bezugsperson-Kind-Interaktion. Auch wenn diese Erweiterung der Früherkennung zunächst nur in einem Selektivvertrag des BVKJ mit einer einzelnen Krankenkasse bundesweit realisiert wird, wird damit ein zukünftig anzustrebender allgemeiner Standard definiert (➤ Tab. 14.2).

14.3 Ablauf

Zur Beurteilung der Entwicklung werden gemäß dem Grenzsteinprinzip für die Bereiche Grob- und Feinmotorik, Perzeption/Kognition, Spiel, Soziale/emotionale Kompetenz, Sprache, sowie Interaktion/Kommunikation Entwicklungsitems vorgegeben, die von 90–95 Prozent der jeweili-

gen Altersgruppe erfüllt werden. Bei Nichterreichen dieser Grenzsteine ergibt sich daher in Verbindung mit der Anamnese und klinischen Untersuchung im Rahmen der Früherkennung der Verdacht auf eine Entwicklungsverzögerung oder -störung.
Werden im Gespräch mit der Mutter oder einer anderen Bezugsperson Sorgen um die Entwicklung des Kindes oder auffällige Beobachtungen des Umfelds (z. B. Kindergarten, Schule) mit Krankheitswert dem Arzt mitgeteilt, müssen diese ernst genommen werden.
Der Kinder- und Jugendarzt sollte daraufhin eine weiterführende Entwicklungsdiagnostik in der Praxis unter Berücksichtigung des sozialen Umfelds und der Ressourcen der Familie einleiten. Aus dieser ergibt sich dann als therapeutische Konsequenz eine Beratung der Bezugspersonen, ggf. Behandlung oder Weiterüberweisung des Kindes oder Jugendlichen. Die Behandlung kann in Form von häuslicher Förderung, Förderung im Sozialraum (KiTa, Schule), Physiotherapie, Ergotherapie, Logopädie, Heilpädagogik, Frühförderung, Psychotherapie, Früher Hilfen und familienstützender Maßnahmen des Jugendamtes, sowie Vorstellung in einem Sozialpädiatrischen Zentrum, bei einem Neuropädiater, einem Kinder- und Jugendpsychiater oder in einer Kinderklinik bestehen und als „empfohlene Maßnahmen" im neuen Vorsorgeheft dokumentiert werden (➤ Kap.6, Kap.7, Kap.10, Kap.37, Kap.38).
Insbesondere die Komplexität oder Schwere einer Störung, sowie ein schwieriges soziales Umfeld sollten den Kinder- und Jugendarzt zu einer frühzeitigen Anbindung des Kindes an eine Stelle der nächsthöheren Versorgungsstufe, z. B. ein SPZ, veranlassen.
Aufgrund der Kontinuität der Betreuung von Kindern zu einem sehr frühen Zeitpunkt ihrer Entwicklung bieten die Früherkennungsuntersuchungen den Kinder- und Jugendärzten eine große Chance, Fehlentwicklungen rechtzeitig zu erkennen und durch geeignete Förderung oder andere Maßnahmen mögliche Folgekomplikationen wie psychische Belastungen, Verhaltensstörungen und Schulprobleme abzumildern oder gar zu vermeiden.

14.4 Inhalte der Früherkennungsuntersuchungen

In den Tabellen (➤ Tab. 14.3 und ➤ Tab. 14.4) werden einige ausgewählte Testverfahren exemplarisch genannt, die aufgrund ihrer guten Validität bei ökonomischer Handhabung für niedergelassene Kinder- und Jugendärzte besonders geeignet erscheinen.
Ein Teil dieser Verfahren wurde dabei bereits in Selektivverträgen mit Krankenkassen in Kinder- und Jugendarzt-Praxen etabliert und erprobt (SBE-2-KT, SSV, BUEVA, BUEGA, MEF, SDQ).

Tab. 14.2 Inhalte der Früherkennungsuntersuchungen U1-J2 gemäß des erweiterten Vorsorgehefts des BVKJ und einige therapeutische Konsequenzen bei Auffälligkeiten

U1 bis J2	Jede Früherkennungsuntersuchung bietet Anlass, von, mit und über Kinder, Jugendliche und deren Familien und Sozialräume zu lernen. Sie besteht aus einer ausführlichen Anamnese (Belange der Versorgung, Entwicklung, Hinweise auf mögliche Erkrankungen), einer klinischen Untersuchung (inkl. Körpermaße, neurologischer Befund, Entwicklungsstand, Seh- und Hörvermögen, Hinweise auf Vernachlässigung oder Misshandlung), Überprüfung des Impfstatus und präventiver alters-abhängiger Beratung der Eltern/des Kindes oder Jugendlichen. Abhängig vom Bedarf und erhobenen Befund erfolgt eine Beratung der Eltern, die Einleitung einer weiterführenden Diagnostik (in der KJA-Praxis selbst oder durch Überweisung/Einweisung) und/oder Therapie mit Verordnung von Medikamenten, Heil- oder Hilfsmitteln, einer integrierten Frühförderung, Psychotherapie oder weiterer Maßnahmen.

Tab. 14.2 Inhalte der Früherkennungsuntersuchungen U1-J2 gemäß des erweiterten Vorsorgehefts des BVKJ und einige therapeutische Konsequenzen bei Auffälligkeiten (Forts.)

	Prophylaxe/Beratung	***Verfahren***	***Konsequenz bei Auffälligkeiten (z. B.)***
U1	Postnatale Adaptation, Vitalität und Reife des Neugeborenen, Ausschluss von angeborenen Fehlbildungen oder Erkrankungen, familiäre Belastung *Prophylaxe/Beratung:* Vitamin K, ggf. Hepatitis-B-Simultanimpfung	APGAR, Nabelschnur-pH, Reifezeichen nach Lubchenko/Ballard, Checkliste zur Erfassung familiärer Belastungsfaktoren	Überwachung in der Geburtsklinik, Stationäre Aufnahme in eine Kinderklinik, Hebamme, frühe Hilfen
U2	Neugeborenen-Hörscreening, Erweitertes NG-Stoffwechselscreening, Pupillenreflexe, Fundusreflex, evtl. frühes Hüftscreening, evtl. POx-Screening auf kongenitalen Herzfehler, familiäre Belastung *Prophylaxe/Beratung:* Vitamin K, Vitamin D-/Fluoridprophylaxe, Stillberatung, Unfallprävention, postpartale Depression bei der Mutter, u. a. m.	TEOAE u. o. AABR, konventionelles Labor, Tandemmassenspektrometrie, Lichtreaktion, Brücknertest, Hüftsonographie, Pulsoxymetrie, Checkliste zur Erfassung familiärer Belastungsfaktoren	Pädaudiologie, spezif. Therapie, Augenarzt, Kinderorthopädie, UKG, Kinderkardiologie, Hebamme, frühe Hilfen
U3	Hüftscreening, Überprüfung des Hörscreening, Pupillenreflexe, Fundusreflex, Interaktion Mutter/Bezugsperson-Kind, Regulationsstörung und familiäre Belastung *Prophylaxe/Beratung:* Vitamin K, Vitamin D-/Fluoridprophylaxe, ggf. 2. aktive Hepatitis-B-Impf., Impfempfehlungen n. STIKO, Stillempfehlung, SIDS-Prävention, Notfallversorgung, Postpartale Depression, u. a. m.	Hüftsonographie, Befund-Dokumentation, Lichtreaktion, Brücknertest; Beobachtung, Elternfragebogen zur Verhaltensregulation,	Kinderorthopädie, Pädaudiologie, Augenarzt, Familienhebamme, frühe Hilfen, Ambulanz für frühe Regulationsstörungen
U4	Grenzsteine der Entwicklung, Hörscreening, Sehscreening (u. a. Fixieren, Motilität), Interaktion Mutter/Bezugs-person-Kind, Regulationsstörung und familiäre Belastung *Prophylaxe/Beratung:* Vitamin D-/Fluoridprophylaxe, Impfungen n. STIKO, Stillen und Ernährung (Beikost) Karies, Unfallprävention, u. a. m.	gemäß Vorsorgeheft, Reaktionsaudiometrie, z. B. Hochtonrassel, Lichtquelle, Objekt im Gesichtsfeld, Hornhautreflex; Beobachtung, Elternfragebogen zur Verhaltensregulation	Entwicklungsdiagnostik, ggf. Physiotherapie, HNO, Augenarzt, Familienhebamme, frühe Hilfen, Ambulanz für frühe Regulationsstörungen
U5/U6	Grenzsteine der Entwicklung, Hörscreening, Sehscreening (u. a. Strabismus), Interaktion Mutter/Bezugsperson-Kind, Regulationsstörung und familiäre Belastung *Prophylaxe/Beratung:* Vitamin D-/Fluoridprophylaxe, Impfstatus n. STIKO bzw. Impfungen n. STIKO, Ernährung, Karies, Schlafen, Unfallprävention, Sprachförderung u. a. m.	gemäß Vorsorgeheft, Verhaltensaudiometrie, z. B. Hochtonrassel, Lichtquelle verfolgen, Hornhautreflex, Ab-/Aufdecktest, (evtl. Brücknertest); Beobachtung, Elternfragebogen zur Verhaltensregulation	Entwicklungsdiagnostik, ggf. Physiotherapie, integrierte Frühförderung, SPZ, HNO, Augenarzt, Familienhebamme (U5), frühe Hilfen, Ambulanz für frühe Regulationsstörungen

Tab. 14.2 Inhalte der Früherkennungsuntersuchungen U1-J2 gemäß des erweiterten Vorsorgehefts des BVKJ und einige therapeutische Konsequenzen bei Auffälligkeiten (Forts.)

	Prophylaxe/Beratung	***Verfahren***	***Konsequenz bei Auffälligkeiten (z. B.)***
U7	Grenzsteine der Entwicklung, z. B. Sprachentwicklung, psychoemotionale Entwicklung, Hörscreening, Sehscreening (u. a. Strabismus, Stereosehen), Interaktion Mutter/Bezugsperson-Kind *Prophylaxe/Beratung:* Fluoridprophylaxe, Kontrolle Impfstatus n. STIKO, ggf. Nachimpfung, Unfallprävention, Erziehungsstil, Medien u. a. m.	gemäß Vorsorgeheft u. ggf. FRAKIS-K, Mannheimer Elternfragebogen (MEF für die U7/U7a), Verhaltensaudiometrie, z. B. Hochtonrassel, Hornhautreflex, Ab-/Aufdecktest, Lang-Stereotest II, (evtl. Brücknertest); Beobachtung	Entwicklungsdiagnostik, evtl. Heilmittel, integrierte Frühförderung, Heilpädagogik, SPZ, HNO, Augenarzt, frühe Hilfen, Ambulanz f. frühe Regulationsstörungen, Erziehungsberatung, SPZ
U7a	Grenzsteine der Entwicklung, z. B. Sprachentwicklung, psychoemotionale Entwicklung, Hörscreening, Sehscreening (u. a. Sehprüfung monokular), Interaktion Mutter/Bezugsperson-Kind *Prophylaxe/Beratung:* Indikationsimpfung (FSME) und Impfstatus nach STIKO, Erziehungsstil, Sprachförderung, Medien, Sauberkeitsentwicklung, u. a. m.	gemäß Vorsorgeheft, ggf. SBE-3-KT[3], ggf. Artikulation, z. B. nach Kottmann, Mannheimer Elternfragebogen (MEF für U7/U7a), Verhaltensaudiometrie, z. B. Hochtonrassel, Hornhautreflexbild, Lang-Stereotest II o. TNO-Test o. ä., Lea-Hyvärinen Test, H-Test o. ä.; Beobachtung	Entwicklungsdiagnostik, evtl. Heilmittel (z. B. Logopädie), integrierte Frühförderung, Heilpädagogik, SPZ, HNO, Augenarzt, Frühe Hilfen, Ambulanz f. frühe Regulationsstörungen, Erziehungsberatung,
U8/U9	Grenzsteine der Entwicklung, z. B. Sprachentwicklung, z. B. Motorik (Fein-, Graphomotorik), z. B. kognitive Entwicklung, z. B. psycho-emotionale Entwicklung, z. B. sozio-emotionale Kompetenz, psychosomatische Beschwerden, Hörscreening, Sehscreening (z. B. Strabismus, Stereosehen, Visus, Farbensehen), Blutdruckmessung, Urinstatus *Prophylaxe/Beratung:* Impfungen n. STIKO, Ernährung und Bewegung, Medien, Erziehungsstil, Unfallprävention, u. a. m.	gemäß Vorsorgeheft, ggf. SSV [4], Artikulation, z. B. nach Kottmann, Nachzeichnen v. geometrischen Formen, ggf. Haus-Baum-Mensch-Zeichnung, Mannheimer Elternfragebogen (MEF U8/U9), Ton- oder Sprachaudiometrie, z. B. Pilotenhörtest, Hornhautreflex, Ab-/Aufdecktest, TNO-Stereotest, Lea-Hyvärinen-Test, HOTV-Test o. ä., (ggf. Brücknertest und Farbtafeln n. Ishihara) manuell o. maschinell, Teststreifen für Eiweiß, Glukose, Erys.	Entwicklungsdiagnostik, evtl. Heilmittel (z. B. Logopädie, Ergotherapie), Heilpädagogik, integrierte Frühförderung, Erziehungsberatung, integrativer Kindergarten, SVE, Schulwahl, SPZ, KJP, HNO, Augenarzt, Kinderkardiologie, Kontrolle, ggf. Kindernephrologie
U10/U11	Entwicklungsauffälligkeiten, Lernstörungen, Aufmerksamkeitsstörung, Verhaltensauffälligkeiten, sozio-emotionale Kompetenz, psychosozial-emotionale Entwicklung, psychosomat. Beschwerden, Hörscreening, Sehscreening (z. B. Strabismus, Stereosehen, Visus usw.), Blutdruckmessung *Prophylaxe/Beratung:* Impfungen n. STIKO (U11), Ernährung und Bewegung, Medien, Hörschutz, Sucht- und Unfallprävention u. a. m.	gemäß Vorsorgeheft, Mannheimer Elternfragebogen, (MEF U10/U11), Ton- oder Sprachaudiometrie, Hornhautreflexbild, Ab-/Aufdecktest, TNO-Stereotest, Refraktometer, Farbtafeln n. Ishihara manuell o. maschinell	weiterführende Diagnostik, SPZ/KJP, Erziehungsberatung, Psychotherapie, HNO-Arzt, Augenarzt, Kinderkardiologie

Tab. 14.2 Inhalte der Früherkennungsuntersuchungen U1-J2 gemäß des erweiterten Vorsorgehefts des BVKJ und einige therapeutische Konsequenzen bei Auffälligkeiten (Forts.)

	Prophylaxe/Beratung	*Verfahren*	*Konsequenz bei Auffälligkeiten (z. B.)*
J1/J2	Entwicklungsauffälligkeiten, Schulleistungen, Verhaltensauffälligkeiten, sozio-emotionale Entwicklung, psychosozial-emotionale Kompetenz, psychosomatische Beschwerden, Hörscreening, Sehscreening (z. B. Visusprüfung, Stereosehen, Farbensehen), Blutdruckmessung; nur J1: Laborscreening *Prophylaxe/Beratung:* Impfstatus/Impfung nach STIKO, Ernährung und Bewegung, Medien, Hörschutz, Suchtprävention, Sexualität, Antikonzeption; J2: Beruf, u. a.	gemäß Vorsorgeheft, Mannheimer Jugendlichen-Fragebogen (MEFJ1/J2), Tonaudiometrie, Hornhautreflexbild, Ab-/Aufdecktest, TNO-Stereotest, Refraktometer, Farbtafeln n. Ishihara manuell o. maschinell, kleines Blutbild	weiterführende Diagnostik, SPZ/KJP, Psychotherapie, HNO-Arzt, Augenarzt, Kinderkardiologie

Tab. 14.3 Screeningtestverfahren/Fragebogen in der Kinder- und Jugendarzt-Praxis

Testverfahren	Entwicklungsbereich	Einsatz in Selektivverträgen
SBE-2-KT[1] Frakis-K[2] SBE-3-KT[3]	Expressive Sprache	BKK-Starke Kids, Bayern
SSV[4]	Expressive und rezeptive Sprache	BKK-Starke Kids, Bayern
Mannheimer Elternfragebogen (MEF), Strength and Difficulties Questionnaire (SDQ)	Psychosozial-emotionale Entwicklung, psychosomatische Beschwerden, Aufmerksamkeit, Psychosozial-emotionale Entwicklung, Aufmerksamkeit/Hyperaktivität	AOK-Patientenzentrierter Versorgungsvertrag, Bayern. Techniker-Krankenkasse, Barmer-GEK und LKK bundesweit. BKK-Starke Kids, Bayern

1 Sprachbeurteilung durch Eltern – Kurztest – für die U7
2 Fragebogen zur frühkindlichen Sprachentwicklung-Kurzform
3 Sprachbeurteilung durch Eltern – Kurztest – für die U7a
4 Sprachscreening für das Vorschulalter, Kurzform des SETK 3–5.

Tab. 14.4 Basisdiagnostikverfahren in der Kinder- und Jugendarzt-Praxis

Testverfahren	Entwicklungsbereich
MFED 1., 2.–3. Lj. Münchener funktionelle Entwicklungsdiagnostik für das 1., 2.–3. Lebensjahr	Laufalter, Krabbel-, Sitzalter, Handgeschicklichkeit, Perzeption, expressive und rezeptive Sprache, Sozialalter und Selbstständigkeit
ET 6–6 Entwicklungstest 6 Monate bis 6 Jahre	Hand- und Körpermotorik, Kognition, expressive und rezeptive Sprache, soziales Verhalten, soziale und emotionale Entwicklung
SETK 2 Sprachentwicklungstest für Kinder mit 2 Jahren	z. B. Untertests für rezeptive Sprache

Tab. 14.4 Basisdiagnostikverfahren in der Kinder- und Jugendarzt-Praxis (Forts.)

Testverfahren	**Entwicklungsbereich**
Hase Heidelberger auditives Screening in der Einschulungsdiagnostik	Phonolog. Arbeitsgedächtnis, allgemeine Sprachleistungsfähigkeit, Früherkennung Legasthenie
BUEVA II Basisdiagnostik umschriebener Entwicklungsstörungen im Vorschulalter II	Nonverbale und verbale Intelligenz, expressive Sprache, Arbeitsgedächtnis, Aufmerksamkeit, (Visuomotorik, Artikulation)
FEW 2 Frostig Entwicklungstest der visuellen Wahrnehmung 2	Visuelle Wahrnehmung und Visuomotorik
BUEGA Basisdiagnostik umschriebener Entwicklungsstörungen im Grundschulalter	Nonverbale und Verbale Intelligenz, Aufmerksamkeit, Schulische Fertigkeiten wie Lesen, Rechtschreiben, Rechnen

14.5 Bedeutung

Durch die regelmäßige Untersuchung und Beobachtung einer verhältnismäßig großen und wenig vorselektierten Anzahl von Kindern und Jugendlichen bestimmten Alters erwirbt der niedergelassene Kinder- und Jugendarzt im Laufe seines Berufslebens eine entwicklungsdiagnostische Kompetenz, die ihm durch die Kenntnis vieler Entwicklungsverläufe häufig auch eine prognostische Einschätzung der Befunde ermöglicht und sein tägliches Handeln leitet.

Allerdings können gerade die ärztlichen Kompetenzen, die umfassende Früherkennungsuntersuchungen erfordern, in der Weiterbildung zum Kinder- und Jugendarzt in einer überwiegend von Neonatologie und Akuterkrankungen geprägten Klinik für Kinder- und Jugendmedizin häufig nicht ausreichend vermittelt werden.

Damit bleibt ein wesentlicher Teil der konkreten Ausgestaltung der Früherkennung, der Diagnostik und Behandlung von Entwicklungs-, Verhaltens- und psychischen Störungen beim niedergelassenen Kinder- und Jugendarzt dem eigenen Engagement nach der Weiterbildung überlassen. Es existieren zu wenig verbindliche Standards des diagnostischen Prozedere, vorzuhaltender objektiver Verfahren der Entwicklungsdiagnostik und z. T. auch Therapieempfehlungen für die niedergelassene Kinder- und Jugendarzt-Praxis.

Ferner sind die niedergelassenen Kinder- und Jugendärzte (5.621, Mitgliederstatistik des BVKJ, Stand 31.3.2013) bezüglich ihres Praxisalltags eine durchaus heterogene Gruppe. Faktoren wie Praxisstandort (städtisch, ländlich, sozialer Brennpunkt), Patienten-Fallzahl, Weiterbildungsweg (SPZ-, Neuropädiatrie-Erfahrung), Fortbildungen (psychosomatische Grundversorgung, Entwicklungsdiagnostik), Schwerpunktpädiater versus Allgemeinpädiater oder Einzel- versus Gemeinschaftspraxis/Praxisgemeinschaft (51 % bzw. 49 %) nehmen nicht unerheblichen Einfluss auf die Ressourcen und Qualität einer Patientenversorgung bei den KJÄ auf dem Gebiet der Entwicklungsdiagnostik und Sozialpädiatrie.

In letzter Zeit gibt es Anstrengungen, Entwicklungsdiagnostik in der Praxis der niedergelassenen Kinder- und Jugendärzte weitergehend zu beschreiben und ein Stufenkonzept der Entwicklungsdiagnostik zu etablieren (➤ Abb. 14.1).

Im Rahmen einer interdisziplinären, Verbände übergreifenden Arbeitsgruppe zur Entwicklungsdiagnostik in der Kinder- und Jugendarztpraxis (IVAN) wurde durch die Deutsche Gesellschaft

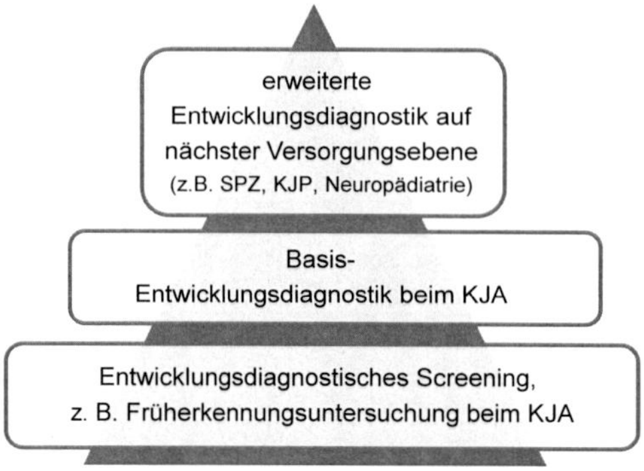

Abb. 14.1 Stufenkonzept der Entwicklungsdiagnostik in der Praxis der KJÄ

für Sozialpädiatrie und Jugendmedizin (DGSPJ) eine Diskussion um Standards auf diesem Gebiet für die niedergelassenen Kinder- und Jugendärzte initiiert.
Unbeschadet der überragenden Bedeutung einer ausführlichen Anamnese und der Erhebung auffälliger Befunde aus den Elternfragebogen haben sich in vielen Kinder- und Jugendarzt-Praxen auch Testverfahren im Rahmen der Entwicklungsdiagnostik etabliert, die aufgrund des standardisierten Vorgehens und der prognostischen Aussagekraft einen sinnvollen Beitrag zur Bewertung des Entwicklungsstandes eines Kindes bereits auf der Stufe der Grundversorgung leisten können. Erst- und Verlaufsbefunde können objektiviert werden. Testergebnisse helfen, den Eltern Einschätzungen zu begründen und darzulegen und mit anderen Professionen ins Gespräch zu kommen. Ihre Anwendung setzt einschlägige Kenntnisse über Durchführung und Evaluation von Verfahren, Rahmenbedingungen von Verhaltensbeobachtungen und beurteilungen, möglicher Ergebnisfehler/-verzerrungen sowie Gütekriterien inkl. Aussagekraft (Validität) der Verfahren voraus (➤ Kap.17). Zusätzlich sind spezifische Fertigkeiten erforderlich, um die jeweiligen Testverfahren angemessen zu indizieren, durchzuführen oder bei Delegation zu supervidieren, zu interpretieren und in der Gesamtschau der Befunde zu gewichten. Schließlich ist eine Haltung des Testleiters nötig, die die Bedürfnisse der Kinder respektiert und eine „Balance zwischen freundlicher Aufgeschlossenheit und aufgabenbezogener Sachlichkeit" ermöglicht (M Mickley). Auch wenn die Durchführung von Testdiagnostik in der KJA kinder- und jugendärztlichen Praxis allein noch nicht die Kompetenz auf diesem Gebiet widerspiegelt, haben Pädiater der Grundversorgung die Herausforderung angenommen, sich und ihre Teams dazu ausbilden lassen, diese Möglichkeiten zum Nutzen der Kinder bereitzuhalten.
Entwicklungs- und Verhaltensstörungen (bzw. neue Morbiditäten) nehmen bereits heute einen großen Raum in der alltäglichen Grundversorgung von Kindern und Jugendlichen in einer kinder- und jugendärztlichen Praxis ein. Diese zeitaufwendigen ärztlichen Leistungen sind aufgrund der derzeitigen Vergütungsstruktur in der kassenärztlichen Versorgung (mit Orientierung an der Fallzahl, – nicht an Einzelleistungen orientiert) allerdings wirtschaftlich für die Praxen nur in Selektivverträgen mit einzelnen Krankenkassen annähernd darzustellen.
So besteht eine Aufgabe der wissenschaftlichen und berufspolitischen kinder- und jugendärztlichen Verbände für die allgemeinpädiatrische Grundversorgung in näherer Zukunft sicher darin, die große Variabilität der Durchführung von Entwicklungsdiagnostik in der Kinder- und Jugendarzt-Praxis durch eine klare Struktur von Screening, Basisdiagnostik und definierten Bindungsstellen zur nächst höheren Versorgungsebene zu ersetzen und dafür in der kassenärztlichen Versorgung die wirtschaftlichen Voraussetzungen zu schaffen.

Literatur

Baumann, Thomas (2007): Atlas der Entwicklungsdiagnostik. Vorsorgeuntersuchungen von U1 bis U10/J1, 2.Auflage, Georg-Thieme-Verlag, Stuttgart

Bergmann, Rolf (2013): Förderung der Teilnahme an den Kinder-Früherkennungs-untersuchungen in Thüringen, Durchführung und vorläufige Ergebnisse des Einladungs-, Erinnerungs- und Meldeverfahrens für die Früherkennungs-untersuchungen U3–U9, http://www.laek-thueringen.de/ (21.4.2013)

Diagnostische Verfahren, Testkatalog 2011, Pearson Clinical & Talent Assesment, Frankfurt

Dresden International University (2012): Epidemiologische Daten in der Pädiatrie. Analyse des Versorgungsspektrums niedergelassener Kinderärzte

Endauswertung: 2.Quartal 2010 bis 1.Quartal 2011. Abschlussbericht.

Fegeler U, Jäger-Roman E, Martin R, Nentwich HJ (2011) Warum kommen Kinder und Jugendliche in die Praxis der Kinder- und Jugendärzte. Erste Zwischenergebnisse. (Vortrag)

Hagen B, Strauch S (2011) Jugendgesundheitsuntersuchung J1: Auswertung von Daten aus der KiGGS-Umfrage, Dtsch Arztebl Int 2011; 108(11): 180–6;

Jeuk, J., Schäfer, J. (Hrsg.), (2012): Deutsch als Zweitsprache in Kindertageseinrichtungen und Schulen. Aneignung, Förderung, Unterricht. Klett-Verlag, Stuttgart

Klinik und Poliklinik für Kinder- und Jugendpsychiatrie, Psychosomatik und Psychotherapie der LMU München, (11.4.2013)

http://www.kjp.med.uni-muenchen.de/sprachstoerungen

Pawils S, Metzner F, Schwinn A, Lietz J, Ravens-Sieberer U (2012) Ergebniszusammenfassung der Evaluation des Erinnerungs- und Meldewesens für die Früherkennungsuntersuchungen U6 und U7 in Hamburg (Stand Mai 2012)

Richtlinien des Bundesausschusses der Ärzte und Krankenkassen über die Früherkennung von Krankheiten bei Kindern bis zur Vollendung des 6. Lebensjahres

(„Kinder-Richtlinien“), Fassung vom 26. April 1976, zuletzt geändert am

16.Dezember 2010, veröffentlicht im Bundesanzeiger 2011; Nr. 40: S. 1.013

in Kraft getreten am 12.März 2011und

Bekanntmachung eines Beschlusses des Gemeinsamen Bundesausschusses

über eine Änderung der Kinder-Richtlinien: Verdacht auf Kindesmisshandlung, vom 21.2.2008, BAnz. Nr. 57 (S. 1.344) vom 15.4.2008

Testkatalog 2012/2013, Testzentrale Göttingen, Hogrefe Verlag, Göttingen

15 MBS: Mehrdimensionale Bereichsdiagnostik und Behandlung in der Sozialpädiatrie

Helmut Hollmann

Ärztliche Aufgabe in der Entwicklungsdiagnostik und Sozialpädiatrie ist die korrekte Diagnosestellung als Grundlage für die Behandlungsplanung. Die so genannte **„Ganzheitliche" Sichtweise** gründet dabei auf dem Verständnis von wechselseitiger Bedingung der Merkmalsausprägung zwischen genetischer Disposition und Anlage einerseits sowie Einflüssen aus der Umgebung andererseits, die fördernd oder hemmend, positiv oder negativ sein können. Durch den aktuellen Stand der Neurowissenschaften und der Entwicklungspsychologie ist diese Auffassung von Entwicklung, Störung und Krankheit als bio-psycho-soziales Modell gut begründet. Ebenso verweisen Humangenetik und epigenetische Prägung auf die permanente wechselseitige Beeinflussung von Individuum und Umwelt. Epidemiologisch und soziologisch hat der bundesweite repräsentative Kinder- und Jugend-Gesundheitssurvey „KiGGS" 2007 nochmals die seit Jahrzehnten bekannte enorme Auswirkung der sozialen Kontextfaktoren auf Entwicklung, Verhalten und Gesundheit belegt. Das bio-psycho-soziale Modell ist elementare Grundlage im fachlich-wissenschaftlichen Selbstverständnis der Sozialpädiatrie. Es beeinflusst sowohl das Vorgehen in der Diagnostik als auch die Planung von entwicklungsunterstützenden Interventionen in Absprache mit den primären Bezugspersonen.

15.1 Die Mehrdimensionale Bereichsdiagnostik

Die Mehrdimensionale Bereichsdiagnostik in der Sozialpädiatrie sieht die systematische Erfassung aller für das Kind relevanter Ebenen vor. Ausgehend von der biografischen Anamnese mit vertiefender Exploration von krankheits- und störungsbildspezifischen anamnestischen Daten weist die MBS im EKPSA-Schema fünf Bereiche diagnostisch aus:
E ntwicklung und Intelligenz
K örperlich-neurologischer Befund
P sychischer Befund und Verhalten
S oziale Kontextfaktoren
A bklärung der Ätiologie

Sozialpädiatrie in der Praxis. http://dx.doi.org/10.1016/B978-3-437-31630-3.00015-8
Copyright © 2014 Elsevier GmbH. All rights reserved.

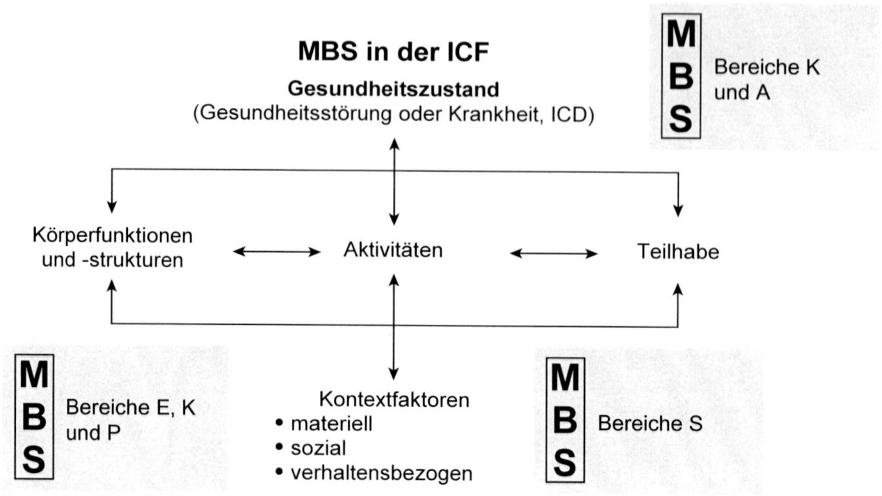

Abb. 15.1 MBS in der ICF

Die differenzierte diagnostische Zuweisung nach der **Internationalen Klassifikation der Krankheiten ICD der WHO** ermöglicht eine Annäherung an den wichtigsten Behandlungsauftrag für gravierend entwicklungsgestörte oder behinderte Kinder unter Einbezug der Familie: Erreichung der Realisation von Lebensqualität, größtmöglicher Selbstständigkeit und sozialer Teilhabe.
Insbesondere der Hinweis auf die notwendige Klärung der Ursache einer Krankheit, Störung oder Behinderung führt über die Beschreibung in der ICD hinaus und fordert die Berücksichtigung des Bedingungsgefüges der Faktoren, die zur klinischen Ausprägung der Symptomatik beitragen.
Die Rahmenbedingungen und Kontextfaktoren werden in der **Internationalen Klassifikation der Funktionsfähigkeit, Behinderung und Gesundheit bei Kindern und Jugendlichen ICF-CY der WHO** zusammen mit individualspezifischen Aspekten erfasst (➤ Kap. 16). Deshalb sieht die WHO neben der diagnostischen Ausweisung nach ICD komplementär die Angaben zu den 4 Komponenten der ICF-CY vor. Demzufolge muss die sozialpädiatrische Behandlungsplanung nicht einzelne Symptome oder Diagnosen berücksichtigen, sondern die Lebenssituation des Kindes und seiner Familie in der Gesamtheit erfassen.
Da aktuell noch keine im klinischen Alltag einsetzbare Version der ICF-CY zur Verfügung steht, werden im **Sozialpädiatrischen Bericht (Arztbrief)** (➤ Kap. 40.2) übergangsweise Kernbegriffe und Formulierungen ausgewiesen. Zu den oben bezeichneten fünf Dimensionen der Mehrdimensionalen Bereichsdiagnostik kommt in der Kurz-Angabe zur Behandlungsplanung somit noch der Bereich hinzu:
T eilhabe-Faktoren in Anlehnung an Kriterien und Terminologie der ICF.

15.2 Behandlungsplanung

Bei der **Behandlungsplanung** sind **Ressourcenprofile** (➤ Kap. 37.1.6) von Patient, Bezugspersonen und weiterer Umgebung wie Institutionen oder Therapeuten zu berücksichtigen.

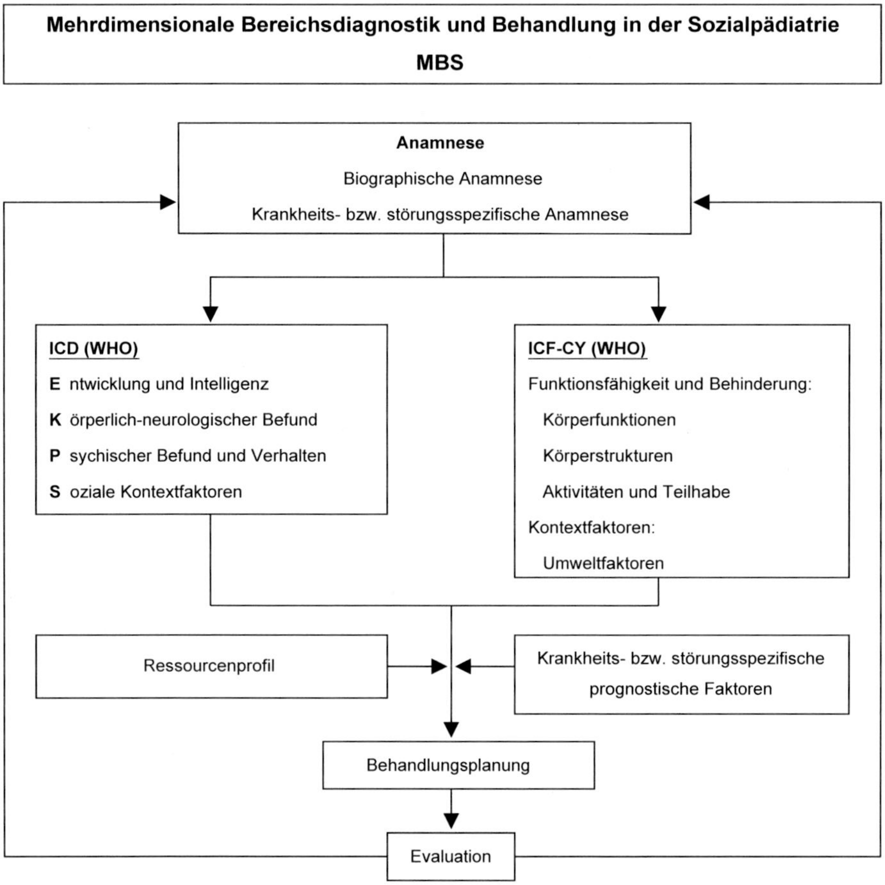

Abb. 15.2 Mehrdimensionale Bereichsdiagnostik und Behandlung in der Sozialpädiatrie MBS

Ebenso müssen **krankheits- und störungsspezifische prognostische Faktoren** einbezogen werden. Insgesamt stehen somit neben kurativen Maßnahmen zur Behandlung von einzugrenzenden Krankheits- und Störungsbilder die Aspekte der Teilhabe im Zentrum. Dies gilt besonders für Kinder und Jugendliche mit Behinderungen, chronischen Krankheiten oder psychischen Störungen.

In regelmäßigen Abständen ist eine **Evaluation** (➤ Kap. 37.2.1) der eingeleiteten Maßnahmen erforderlich, um die Adaptation des Behandlungsplans zu erreichen. Hierbei muss je nach Verlauf und Situation auch die mehrdimensionale Bereichsdiagnostik mit dem EKPSA-Schema wiederholt werden, um die diagnostische Klassifikation zu ergänzen oder zu erweitern.

Das System der Mehrdimensionalen Bereichsdiagnostik und Behandlung in der Sozialpädiatrie ist langfristig angelegt. Es berücksichtigt damit die Erfordernisse von entwicklungsgestörten, chronisch kranken oder behinderten Kindern und Jugendlichen. Für die Bezugspersonen bietet es die Möglichkeit, Beratung und Unterstützung jeweils im erforderlichen Umfang zu gestalten.

Literatur

Hollmann H, Kretzschmar C, Schmid RG (2003) Mehrdimensionale Bereichsdiagnostik Sozialpädiatrie MBS. Qualität in der Sozialpädiatrie Band 1. www.dgspj.de

Mendes U, Mundt D, Spörkmann KH, Hollmann H (2013) Mehrdimensionale Bereichsdiagnostik und Behandlung in der Sozialpädiatrie MBS – Einführung und Glossar. Qualität in der Sozialpädiatrie Band 4 (in Vorb.). www.dgspj.de

16 Internationale Klassifikation der Funktionsfähigkeit, Behinderung und Gesundheit: Version für Kinder und Jugendliche (ICF–CY)

Olaf Kraus de Camargo

16.1 ICF und ICF–CY

Störungs- und Krankheitsbilder in der Sozialpädiatrie werden mehrdimensional beurteilt (➤ Kap. 15). Die Behandlungsplanung erfordert genaue Kenntnis der vorliegenden Störungen, Einschränkungen und Barrieren sowie der Ressourcen und (sozialen) Kontextfaktoren. Das sozialpädiatrische Beurteilungssystem spiegelt damit ein umfassendes bio-psycho-soziales Verständnis von Entwicklung, Gesundheit und Krankheit (vgl. ➤ Kap. 37). Damit greift es den Paradigmenwechsel von einem ausschließlich biomedizinischen Krankheitsmodell hin zu einem ganzheitlichen Bedingungsgefüge von Gesundheit, Individuum und Umfeld auf.

- Im Gesundheitssystem hat die Klassifikation von Krankheitszuständen eine zentrale Bedeutung. Die Kodierung nach der Internationalen Klassifikation der Krankheiten (ICD) ist Grundlage für einheitliche Definition, Statistik und Abrechnung. Gesundheitsinformationen sollen in einer standardisierten Form vergleichbar gemacht werden.

Diese Klassifikation kann aber nur einen Teil des Gesamtbildes darstellen. Deshalb hat die Weltgesundheitsorganisation (WHO) 2001 mit der Klassifikation der Funktionsfähigkeit, Behinderung und Gesundheit (ICF) eine Möglichkeit geschaffen, den Gesundheitszustand eines Menschen mit all seinen Auswirkungen und Wechselwirkungen im Kontext der Lebenswelt in der dieser Mensch lebt, zu beschreiben. Das Grundverständnis der Teilhabe ist zeitgleich 2001 wesentlich auch in die Formulierung des Sozialgesetzbuches IX in Deutschland eingeflossen.

Seit 2011 steht die Kinder- und Jugendversion der ICF (Children and Youth, ICF–CY) in der deutschen Übersetzung zur Verfügung. Somit können jetzt im sozialpädiatrischen Bereich die ätiologisch orientierten Zuschreibungen von Krankheit, Störung oder Entwicklungsabweichung kombiniert werden mit der Beschreibung von Funktionsfähigkeit oder Behinderung, jeweils bezo-

Sozialpädiatrie in der Praxis. http://dx.doi.org/10.1016/B978-3-437-31630-3.00016-X
Copyright © 2014 Elsevier GmbH. All rights reserved.

gen auf die Lebenswelt (Kontext). Die Klassifikationssysteme der ICD und ICF-CY sind komplementär.

Der Grundaufbau der ICF umfasst zwei Teile, jeder mit zwei Komponenten:

- Teil 1: Funktionsfähigkeit und Behinderung
  - Körperfunktionen und -strukturen
  - Aktivitäten und Teilhabe
- Teil 2: Kontextfaktoren
  - Umweltfaktoren
  - Persönliche Faktoren.

Jede Komponente kann mit positiven oder negativen Begriffen beschrieben werden. Innerhalb der 4 Komponenten umfasst die weitere Unterteilung verschiedene so genannte Domänen, und jede Domäne besteht schließlich aus Kategorien. Gesundheitsbezogene Zustände eines Individuums können durch die adäquate Auswahl von „Codes" einzelner Kategorien beschrieben werden. Dabei erfolgt eine graduelle Bewertung der Ausprägung dieser „Codes" mittels eines sogenannten „qualifiers".

Der zentrale Begriff in der ICF ist die „funktionale Gesundheit".

Funktionale Gesundheit

Eine Person ist funktional gesund, wenn

- vor ihrem gesamten Lebenshintergrund (Konzept der Kontextfaktoren)
- ihre körperlichen Funktionen (einschließlich des mentalen Bereichs) und Körperstrukturen allgemein anerkannten Normen entsprechen (Konzepte der Körperfunktionen und –strukturen),
- sie nach Art und Umfang das tut oder tun kann, wie es von einem Menschen ohne Gesundheitsproblem erwartet wird (Konzept der Aktivitäten) und
- sie ihr Dasein in allen Lebensbereichen, die ihr wichtig sind, in der Art und dem Umfang entfalten kann, wie es von einem Menschen ohne Schädigungen der Körperfunktionen/–strukturen und Aktivitätseinschränkungen erwartet wird (Konzept der Teilhabe).

Bei Kindern und Jugendlichen sind Wachstum und Entwicklung die zentralen Parameter zur Unterscheidung von vergleichbaren Gesundheitscharakteristika bei Erwachsenen. Es resultieren Besonderheiten in Motorik, Sprache und Kognition sowie beim Verhalten.

Der dynamische Prozess der kindlichen Entwicklung verläuft zeitlich individuell verschieden und weist somit eine erhebliche Variationsbreite auf. Abweichungen können eine Verzögerung darstellen, oder sie sind Ausdruck eines grundlegenden Problems und führen zu einer Störung. Die Problemursache kann anlagebedingt im Kind selbst liegen oder ergibt sich aus Einflüssen der Umwelt. Insbesondere das junge Kind ist in hohem Maße von Umgebungsfaktoren abhängig. Über die Zeit hinweg besteht ein dynamischer Prozess nicht nur hinsichtlich Entwicklungsfunktionen, sondern auch in Bezug auf Autonomie und soziale Partizipation. Diese Reifung und Differenzierung erfolgt in ständigem Wechselspiel mit der familiären, sozialen und sächlichen Umgebung.

Dabei ist die Lebenswelt von Kindern in besonderem Maße bestimmt von Faktoren, die sie in ihrer nächsten Umgebung erfahren. Mit zunehmendem Alter treten Einflussgrößen aus weiterer Distanz hinzu (bioökologisches Konzept nach Bronfenbrenner, 1977; (➤ Kap. 37).

Die Informationen, auf die sich die Beurteilung stützt, müssen unmittelbar erhoben sein. Dazu kommen Beobachtung, Interview und/oder persönliche fachliche Evaluation in Betracht. Bei Kindern sind die Bezugspersonen in die Erhebung integriert; Interviews von Kindern und Jugendlichen müssen durchgeführt werden, wo immer möglich.

Die Evaluationsmerkmale der ICF verwenden neutrale Begriffe in der Terminologie zur qualitativen Beschreibung.

Evaluation in der ICF mit „qualifiern": Beurteilungs- und Quantifizierungsmerkmale

0:	keine (nicht vorhanden / ohne./ unerhebliche)	Schädigung, Schwierigkeit, Barriere
1:	leichte (schwach / gering)	Schädigung, Schwierigkeit, Barriere
2:	mäßige (mittel / ziemlich)	Schädigung, Schwierigkeit, Barriere
3:	erhebliche (hoch / äußerst)	Schädigung, Schwierigkeit, Barriere
4:	komplette (total / voll)	Schädigung, Schwierigkeit, Barriere
8:	nicht spezifiziert	
9:	nicht anwendbar	

In der ICF wird Behinderung allgemein definiert als eine negative Wechselwirkung zwischen einer Person und ihren Kontextfaktoren. Der Gesundheitszustand einer Person mag dabei hinsichtlich der Definition nach ICD einer Krankheit, Schädigung oder Störung entsprechen mit Auswirkungen auf die Funktionsfähigkeit der Person. Speziell liegt eine Behinderung dann vor, wenn sich diese negative Wechselwirkung auf die Teilhabe in einem Lebensbereich auswirkt (z.B. in der Kindertagesstätte, der Schule oder beim Spiel bzw. Freizeitaktivitäten). Die Untergliederung der Kontextfaktoren in Umweltfaktoren und persönliche Faktoren bringt zum Ausdruck, dass es auch von persönlichen Neigungen und Interessen abhängig ist, welche Aktivitäten für den Einzelnen Bedeutung haben. Es geht also nicht um die Erfüllung absoluter Normen; die ICF ist kein Test- oder Prüfungsverfahren. In der praktischen Anwendung werden die persönlichen Faktoren jedoch nicht klassifiziert, da sie stark von kulturellen Werten abhängig sein können.
Behinderungen sind nicht absolut, sondern auf Lebensbereiche bezogen. Die neutrale deskriptive Terminologie der ICF in der Festlegung von Abweichungen verändert dabei die Sichtweise auf den betroffenen Menschen. Nicht mehr die Orientierung auf Defizite und entsprechende Negativ-Ausweisung steht im Vordergrund, sondern die Bestandsaufnahme der Fähigkeiten und Fertigkeiten in Hinblick auf Aktivitäten und Partizipation.
In der Kinder- und Jugendversion ICF-CY wurden zahlreiche zusätzliche Kategorien in den Bereichen Körperfunktionen und -strukturen, Aktivitäten und Teilhabe sowie bei den Kontextfaktoren eingeführt. Damit wird eine Beschreibung der funktionalen Gesundheit auch von Säuglingen und Kleinkindern ermöglicht.

16.2 Die ICF-CY in der Sozialpädiatrie

Die theoretische Grundlage bei der Behandlung von Krankheiten mit Auswirkungen auf die Entwicklung ist im Selbstverständnis der Sozialpädiatrie die Berücksichtigung des Zusammenspiels von Umwelt (insbesondere sozialem Kontext) und Krankheits- bzw. Störungszuständen. Dies entspricht dem bio-psycho-sozialen Modell von Gesundheit (► Abb. 16.1). Auch in der praktischen sozialpädiatrischen Arbeit ist die Erfassung des Lebenshintergrundes der Patienten, der Ressourcen und Barrieren sowie prognostischer krankheitsspezifischer Faktoren ein fester Bestandteil (➤ Kap. 15). Durch die systematische Nutzung der ICF-CY wird es nun möglich, diese Informationen in einer einheitlichen Nomenklatur festzuhalten. Dies ermöglicht eine umfassende Darstellung des Gesundheitszustandes des einzelnen Kindes und Jugendlichen. Gerade unter dem Aspekt einer familien- bzw. patientenzentrierten Betreuung ist eine solche detaillierte Darstellung für die Planung von Hilfen und Interventionen sinnvoll und hilfreich.

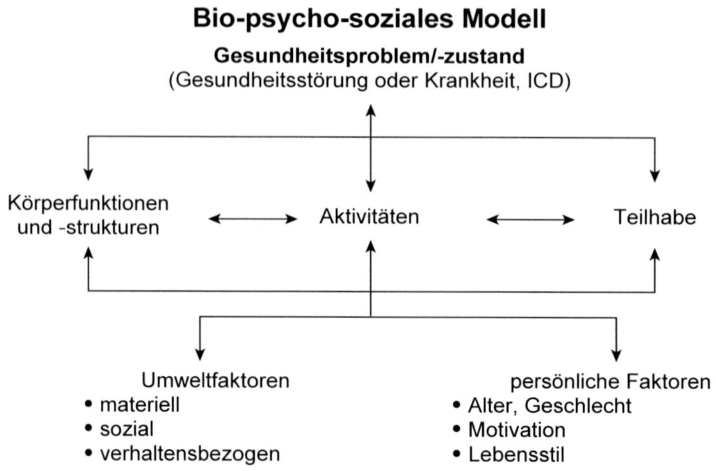

Abb. 16.1 Das bio–psycho–soziale Modell von Gesundheit ist die Grundlage für die Struktur der ICF – die Grafik verdeutlich die wechselseitige Abhängigkeit von Gesundheitszustand und Kontext mit Einfluss auf Körperfunktionen, –strukturen, Aktivitäten und Teilhabe (WHO, 2005)

Mit der ICF–CY ist es möglich, das Kind im Kontext der Familie, mit seinen Entwicklungsbesonderheiten und bezüglich des Ausmaßes der Teilhabe zu beschreiben. Eine Besonderheit im Vergleich zur Erwachsenenversion ist dabei das Konzept der Entwicklungsverzögerung in Abgrenzung zur definierten Störung. Entwicklungsverzögerungen können in einzelnen Domänen der ICF dargestellt werden. Die ICF–CY bietet die Möglichkeit, dies in den „qualifiern" mit „Verzögerung" darzustellen, anstatt in den Abstufungen von „leicht" bis „vollkommen" geschädigt (0–4). Die Funktionsfähigkeit von Kindern hängt dabei sehr stark von der Interaktion mit der Familie und den Bezugspersonen ab. Verständnis und Definition von Störungen der Entwicklung und des Lernens werden hierdurch grundlegend verändert (Waber, 2011).
Für die Planung von therapeutischen Interventionen hat dies zur Folge, dass auch die Definition von Therapiezielen nicht mit einer Normalisierung von Körperfunktionen gleichgesetzt werden muss, sondern die Ziele einzelner Kinder und Familien sehr individuell sein können und in Abhängigkeit von den Kontextfaktoren formuliert werden müssen. Für die Operationalisierung solcher Art von Therapiezielen eignen sich Verfahren wie das COPM (Canadian Occupational Performance Measure) und das PEGS (Perceived Efficacy and Goal Setting). Diese Verfahren orientieren sich an den Erwartungen und Ansprüchen der Kinder an ihre Alltagsfertigkeiten und dienen daher dazu, die Therapieinhalte auf bedeutsame Ziele für die Kinder abzustimmen.

Fallbeispiel

(Kraus de Camargo 2009, modifiziert nach Zimmerman–Targownik und Targownik 2007)
Pascal ist 8 Jahre alt. Als er 3 Jahre alt war, erlitt er einen Ertrinkungsunfall. Die Folge war eine hypoxische Hirnschädigung mit mehrwöchigem Koma. Im Verlauf entwickelte Pascal eine zerebrale Bewegungsstörung, charakterisiert durch eine erhöhte Spastik der Extremitäten mit Rumpfhypotonie sowie einer erheblichen Dysarthrie. Pascal kann nur mit Unterstützung im Rumpf gehen und dies auch nur in seiner vertrauten Umgebung über kurze Distanzen. Er benötigt Hilfe bei der Nahrungsaufnahme, das Essen muss ihm gereicht werden, er trinkt aus einem Becher mit Trinkhilfeaufsatz. Es gelingt ihm, sich über einen Sprachcomputer zu verständigen. Pascal besucht ein Förderzentrum und

hat dort viele Freunde. Im häuslichen Umfeld wird er durch seine Eltern und seine Geschwister unterstützt, die ihn bei vielen Freizeitaktivitäten mitnehmen und integrieren. Pascals Eltern sind froh und dankbar um jeden kleinen Fortschritt, den er erreicht. Sie stellen sich auf seine körperlichen Einschränkungen ein und versuchen, ihn zu unterstützen, wo es möglich ist. Pascals größter Wunsch ist es, Laufen zu können. Als Hilfsmittel stehen Pascal sein Sprachcomputer, ein Therapiestuhl, ein Rollstuhl, eine Trinkhilfe und ein Pflegebett zur Verfügung.
Stefan ist 12 Jahre alt. Auch er erlitt einen Ertrinkungsunfall im Alter von 3 Jahren. Die Folge war eine hypoxische Hirnschädigung mit mehrwöchigem Koma. Im Verlauf entwickelte er eine zerebrale Bewegungsstörung, charakterisiert durch eine erhöhte Spastik der Extremitäten mit Rumpfhypotonie sowie einer erheblichen Dysarthrie. Stefan kann an einer Hand gehalten gehen. Er schläft in einem niedrigen Bett ohne Füße, aus dem er sich selbstständig herausrollen kann. In seinem Zimmer bewegt er sich durch Krabbeln fort. Er benötigt Aufsicht und etwas Führung bei der Nahrungsaufnahme. Er trinkt selbstständig aus einem Becher mit Trinkhilfeaufsatz. Er kann sich nur wenig über einen Sprachcomputer verständigen und hat wenig Gelegenheit, dies zu Hause zu üben. Im häuslichen Umfeld wird Stefan durch seine Eltern, seinen kleinen Bruder und die weitere Familie unterstützt, die ihn bei vielen Freizeitaktivitäten mitnehmen und integrieren. Stefans Eltern machen sich große Vorwürfe, dass sie nicht besser aufgepasst haben und er deshalb den Ertrinkungsunfall erlitten hat. Sie ermöglichen ihm alle denkbaren Therapien im In– und Ausland und nehmen dafür erhebliche finanzielle Einbußen in Kauf. Sie sehen die bisherige Entwicklung als Beweis dafür, dass eine intensive Förderung und Forderung des Jungen sich lohnt. Aus diesem Grund lehnen sie auch den Gebrauch des Sprachcomputers ab, da sie möchten, dass Stefan wieder sprechen lernt. Stefan hat aufgrund seiner zahlreichen Therapien keine Zeit, freundschaftliche Kontakte zu anderen Kindern aufzubauen oder zu pflegen. Über seine Wünsche ist nichts bekannt, da er sich nicht ausreichend artikulieren kann. Als Hilfsmittel verfügt er über einen Sprachcomputer und eine Trinkhilfe.
Welche Diagnosen können gestellt werden?
Pascal: Erworbene zerebrale Bewegungsstörung mit Spastik (G82.4), Sprachstörung (R48.2), globale Entwicklungsstörung bei Intelligenzminderung (F78.0), Ataxie (R27.0), Z.n. Ertrinkungsunfall (W69), Z.n. Wachkoma (U99.0!), Z.n. Herzstillstand (I46.0), hypoxischer Hirnschaden (G93.1).
Stefan: Erworbene zerebrale Bewegungsstörung mit Spastik (G82.4), Sprachstörung (R48.2), globale Entwicklungsstörung bei Intelligenzminderung (F78.0), Ataxie (R27.0), Z.n. Ertrinkungsunfall (W69), Z.n. Wachkoma (U99.0!), Z.n. Herzstillstand (I46.0), hypoxischer Hirnschaden (G93.1).
Erläuterung: Hinsichtlich ihrer organischen Diagnosen unterscheiden sich die beiden Kinder nicht. Erst durch die Beschreibung der funktionalen Gesundheit, -selbst ohne genaue Differenzierung des Ausmaßes der Einschränkungen/Schädigungen, in „qualifiern"- können bedeutsame Unterschiede erkannt werden, insbesondere im Bereich der Teilhabe und der Kontextfaktoren.
Pascal:
Eingeschränkte Körperfunktionen:
Gelenkkontrakturen (b729), Bewegungsfunktionen (b730), Sprachfunktionen (b320), mentale Funktionen wie Gedächtnis, Aufmerksamkeit (b140–fff).
Geschädigte Körperstrukturen:
Gehirn (s110), Muskulatur (s7).
Eingeschränkte Aktivitäten und Teilhabe:
Mobilität (d450), Kommunikation (d330–d332), Lernen (d199), ATL (Nahrungsaufnahme/Essen, Hygiene, Anziehen, Aufräumen) (d230, d510–d571).
Wesentliche Kontextfaktoren (Barrieren und Förderfaktoren):
Barrieren: Stufen zu Hause (e1551), Pflegebett (e1150) , Rollstuhl (e1201).
Erläuterung: Die Stufen erschweren die Fortbewegung mit Unterstützung und mit einem Rollstuhl. Das Pflegebett behindert Pascal, indem er nicht selbstständig aus dem Bett herauskommen oder hineinkommen kann. Pascal kann den Aktiv–Rolli nicht selbst in Bewegung setzen, daher muss er geschoben werden und ist abhängig von anderen Personen.

Förderfaktoren: Sprachcomputer (e1251), Einstellung der Eltern und Geschwister (e410).
Erläuterung: Der Sprachcomputer ermöglicht es Pascal seine Bedürfnisse und Wünsche zu äußern. Die Eltern und Geschwister haben eine positive Einstellung dazu und unterstützen Pascal in seiner Kommunikation. Des Weiteren unterstützt die Familie Kontakte zu anderen Kindern, Pascal hat viele Freunde.
Welche Interventionen könnten dazu beitragen, die Aktivitäten und Teilhabe zu verbessern?
Motorik fördern (Physiotherapie) mit dem Ziel „Laufen", Verordnung eines Rollators und eines E-Rollis, ggf. Rampen zu Hause.
Erläuterung: Laufen ist ein Ziel, das Pascal selbst äußert. Mit Veränderungen im Umfeld und gezielter Therapie besteht die Möglichkeit einer Fortbewegung durch Laufen im häuslichen Umfeld. Gleichzeitig hat Pascal die Fähigkeit, einen Joystick zu bedienen, und könnte so die Kontrolle über einen E-Rolli gewinnen, was ihm mehr Unabhängigkeit sowohl im Haus wie auch außer Haus bringen würde.
Stefan:
Eingeschränkte Körperfunktionen:
Bewegungsfunktionen (b730), Sprachfunktionen (b320), mentale Funktionen wie Gedächtnis, Aufmerksamkeit (b140–fff).
Geschädigte Körperstrukturen:
Gehirn (s110), Muskulatur (s7).
Eingeschränkte Aktivitäten und Teilhabe:
Kommunikation (d330–d332, d360), Freundschaften schließen (d7200).
Wesentliche Kontextfaktoren (Barrieren und Förderfaktoren):
Barrieren: Altbau (e1550), Therapieintensität (e450, e410), Einstellungen der Eltern (e410).
Erläuterung: Der Altbau mit Treppen erschwert es Stefan selbstständig ins Freie zu gelangen, was anderen Kindern in seinem Alter möglich ist. Durch die Einstellung der Gesundheitsberufe (e450) und der Eltern erhält Stefan eine sehr hohe Therapieintensität mit täglichen Terminen, wodurch andere soziale Aktivitäten und die Möglichkeit, Freundschaften zu schließen eingeschränkt ist. Die Möglichkeiten des Sprachcomputers sind durch die mangelnde Akzeptanz der Eltern eingeschränkt.
Förderfaktoren: Therapien (e5800), Kinderzimmer (e1551).
Erläuterung: Das vorhandene System an Interventionen und Therapien unterstützt vor allem Stefans motorische Funktionen. Sein Kinderzimmer ist so eingerichtet, dass er sich darin mit größtmöglicher Unabhängigkeit fortbewegen kann.
Welche Interventionen könnten dazu beitragen, die Aktivitäten und Teilhabe zu verbessern?
Stefans Ziele herausfinden, Logopädie, Sprachcomputer, Familientherapie mit dem Ziel der Verarbeitung des Unfalls.
Erläuterung: Auch wenn intensiv Therapie betrieben wird, ist nicht klar, inwieweit die Ziele mit Stefans eigenen Wünschen übereinstimmen. In den laufenden Therapien muss daher versucht werden, diese Wünsche zu ergründen. Mit einer zusätzlichen logopädischen Behandlung und intensiverem Einsatz des Sprachcomputers könnten Stefans kommunikative Fähigkeiten und damit seine Teilhabe sowohl am familiären Leben wie auch im Kontakt mit anderen Kindern verbessert werden. Um ihm ausreichend Zeit für soziale Kontakte zu verschaffen, müssten wahrscheinlich einige der therapeutischen Aktivitäten beendet werden, was wiederum nur mit einer Veränderung der Einstellung der Eltern möglich ist. Dies wäre das Ziel einer familientherapeutischen Intervention.
FAZIT:
Über die Beschreibung der funktionalen Gesundheit mittels der ICF ergeben sich unterschiedliche Bilder zweier Kinder mit der gleichen Diagnose. Daraus leiten sich auch unterschiedliche Ansätze für die Intervention ab. In der Praxis wird durch eine direkte Beteiligung der Eltern bei der Beschreibung ihres Kindes mit der ICF die Komplexität der Therapieplanung für alle Beteiligten deutlich; realistische Ziele können einvernehmlich vereinbart werden.

16.3 Praktische Anwendung der ICF-CY

Die ICF-CY ist kein Definitions- oder Prüf-Instrument. Deshalb gibt es keine typischen ICF–Ziffern für bestimmte Krankheitsbilder wie Epilepsien, Zerebralparesen, Sprachentwicklungs- oder emotionale Störungen. Vielmehr wird durch die Auswahl von Kodierungen der individuelle Zustand von Funktion und Lebenssituation beschrieben. Hierzu steht eine große Anzahl von Verschlüsselungsziffern in den einzelnen Bereichen zur Verfügung.

Es ist jedoch möglich, sich aus der Vielzahl der möglichen Kategorien jene herauszusuchen, denen aufgrund der bestehenden Erkrankung besondere Aufmerksamkeit gelten soll, so wie aufgrund der klinischen Erfahrung bei der Anamneseerhebung bestimmte zielgerichtete Fragen je nach den berichteten Beschwerden gestellt werden. Bei systematischer Umsetzung erhält man Gruppen oder „sets" aus Kategorien der unterschiedlichen Domänen der ICF wie den Körperfunktionen und –strukturen, den Aktivitäten und der Teilhabe sowie den Umweltfaktoren. Diese „sets" werden als „core–sets" ((XE:"ICF:core-sets")) bezeichnet und sollen besonders häufig betroffene Bereiche im Rahmen einer bestimmten Erkrankung darstellen. Daneben können Zusammenstellungen in „check-lists" ((XE:"ICF:check-lists")) den Zugang zum komplexen System der ICF-CY erleichtern. Eine interdisziplinäre und verbändeübergreifende Arbeitsgruppe mit Beteiligung der DGSPJ hat altersbezogene ICF-CY Checklisten zusammengestellt, die im Internet abrufbar sind:

http://www.dgspj.de/index.php?option=com_content&view=category&id=:icf-cy&Itemid =&layout=blog).

Bei allen Überlegungen, die Anwendbarkeit der ICF–CY mittels „core–sets" oder „Checklisten" zu vereinfachen, muss stets bedacht werden, dass es darum gehen muss, mit der ICF–CY das Kind zu beschreiben, mit allen seinen Eigenarten und Besonderheiten, und nicht ein Krankheitsbild. Somit müssen auch Kategorien zur Verfügung stehen, die vielleicht nicht typisch oder häufig bei Patienten mit einem bestimmten Krankheitsbild sind, jedoch für den einzelnen Patienten relevant sind. Dies gilt insbesondere für die Umweltfaktoren, sowohl in Bezug auf die Barrieren wie auch die Förderfaktoren. Insofern können Untergruppen der ICF immer nur eine Orientierung sein. In Zukunft werden daher Beurteilungen mittels der ICF mit großer Wahrscheinlichkeit rechnergestützt erfolgen, um einen raschen Zugriff auf die notwendigen Kategorien zu gewährleisten.

16.4 Messinstrumente zur differenzialdiagnostischen Beurteilung von Interventionseffekten

Erhalten Kinder mit Entwicklungsstörungen oder Behinderungen therapeutische Interventionen nach dem Konzept der ICF–CY, müssen auch die Therapieverläufe entsprechend beurteilt werden. Da das Konzept an sich noch relativ neu ist, sind geeignete deutschsprachige Verfahren zur Evaluation des therapeutischen Vorgehens so gut wie nicht auf dem Markt.

Ein wesentlicher Aspekt bei der Beurteilung der Funktionsfähigkeit ist vor allem das Ausmaß der Teilhabe, die wiederum stark von den Umweltfaktoren abhängig ist. Inzwischen gibt es auch erste Verfahren, die es ermöglichen, diese Aspekte der sozialpädiatrischen Befundung standardisiert zu erheben, wie z.B. mittels des PEM-CY (Participation and Environment Measure for Children and Youth) (http://participation-environment.canchild.ca/en/index.asp). Eine gute Übersicht zu Test-

und Untersuchungsverfahren und ihrer Zuordnung zu den Bereichen der ICF findet sich im Handbuch „Measures for Children with Developmental Disabilities framed by the ICF-CY" (Hrsg. Annette Majnemer)

Eine sehr direkte und einfach anzuwendende, jedoch wenig standardisierte Methode ist die Beurteilung des Zielerreichungsgrades (goal–attainment–scaling, GAS) ((XE:"GAS:goal-attainment-scaling")). Dabei wird vor Beginn der Therapie ein individuell mit dem Kind/den Eltern vereinbartes Ziel festgelegt sowie der Zeitrahmen zur Erreichung dieses Ziels bestimmt. Nach Ablauf dieses Zeitrahmens wird beurteilt, inwieweit das Ziel erreicht ist. In der Literatur wird dazu kritisch bemerkt, dass gerade bei Kindern nicht immer klar ist, wie diese zu den Zielformulierungen gelangen und welche Inhalte sie tatsächlich mit den formulierten Zielen verbinden. Generell stellt sich auch die Frage, wie man mit unterschiedlichen Zielvorstellungen von Eltern und Kind umgehen soll. Treten hier große Diskrepanzen auf, ist es aber sicherlich hilfreich, diese schon vor oder zu Beginn einer therapeutischen Intervention zu kennen. So können Schwierigkeiten in der Zusammenarbeit und beim Aufbau einer therapeutischen Beziehung sowie Frustrationen vermieden und ggf. andere, besser geeignete Maßnahmen und Vorgehensweisen überlegt werden.

Eine weitere Möglichkeit besteht in der Nutzung des CAPE (Children's Assessment of Participation and Enjoyment) ((XE:"CAPE:Children's Assessment of Participation and Enjoyment")) ((XE:"Children's Assessment of Participation and Enjoyment, CAPE")). In diesem Verfahren werden Kinder und Jugendliche von 6–21 Jahren anhand von 55 Punkten zu ihren Aktivitäten im Alltag und der Präferenz für bestimmte Aktivitäten befragt. So kann festgestellt werden, ob die Kinder z.B. durch therapeutische Unterstützung in die Lage versetzt wurden, an mehr oder an bestimmten Aktivitäten teilzuhaben, die ihnen wichtig sind. Es wird dabei nicht beurteilt, wie gut sie diese Aktivitäten durchführen. Das wäre das Ziel von klassischen Methoden der Therapieevaluation.

Unter der Vorstellung, dass die Teilhabe ganz wesentlich zur Lebensqualität, insbesondere der krankheitsbezogenen Lebensqualität beiträgt, kann es auch sinnvoll sein, diese mit Hilfe von speziellen Fragebögen zu beurteilen. Im europäischen Raum steht hierzu der Fragebogen „Kidscreen/Disabkids" auch in deutscher Sprache zur Verfügung (The KIDSCREEN Group Europe 2006; http://www.kidscreen.org). Zunehmend wird man in Zukunft diese Aspekte bei der Beurteilung von Interventionen und Hilfemaßnahmen bei Kindern mit Behinderungen und ihren Familien in Betracht ziehen müssen, um den Interessen der Kinder im Sinne der funktionalen Gesundheit gerecht zu werden.

Dies soll nicht bedeuten, dass auf klassische Messinstrumente der Körperfunktionen wie z.B. der Aufmerksamkeitsleistung, der Visuomotorik oder Gangfunktionen verzichtet werden kann. Diese Messwerte können wertvolle Hinweise für die Art der zu wählenden therapeutischen Ansätze liefern und helfen bei der Beurteilung, ob diese Körperfunktionen dadurch beeinflussbar sind. Sie können aber nicht das Ziel der Intervention an sich sein.

16.5 Bedeutung der ICF für die interdisziplinäre Arbeit

Kinder mit Entwicklungsstörungen und Behinderungen sind häufig in vielfältigen Bereichen betroffen und werden somit von unterschiedlichen Fachpersonen betreut. Dies ist besonders in den Sozialpädiatrischen Zentren und Interdisziplinären Frühförderstellen der Fall, deren Personalstruktur das bereits widerspiegelt. Die erhobenen Befunde werden in Teambesprechungen oder

Fallkonferenzen ausgetauscht. Es hat sich gezeigt, dass die Nutzung der Struktur und der Nomenklatur der ICF besonders hilfreich ist, um das Verständnis der einzelnen Berufsgruppen untereinander zu verbessern. Denkansatz und Nomenklatur stellen eine disziplinübergreifende gemeinsame Arbeitsgrundlage dar. So ist es möglich, eine besser auf die individuellen Bedürfnisse des Patienten und der Familie abgestimmte Therapie- bzw. Förderempfehlung auszusprechen. Die Behandlungsplanung wird neben funktionellen Gesichtspunkten schwerpunktmäßig auf die Realisation von Eigenaktivität und Teilhabe unter Berücksichtigung von Kontextfaktoren ausgerichtet. Dazu eignen sich die Prinzipien der ICF–CY hervorragend, und es ist wichtig, diese gemeinsame Sprache im Team-Austausch mit Ärzten, Psychologen, Therapeuten und Pädagogen zu beherrschen.

Für die interne Strukturierung von Abläufen und Zuständigkeiten kann die ICF ebenfalls hilfreich sein, da mit ihr abgebildet werden kann, in welchen Bereichen Redundanzen bestehen und wo es weiterer fachlicher Qualifikationen bedarf, um alle relevanten Aspekte eines Patienten und seiner Familie erfassen zu können. Ebenso können auf diese Art und Weise transdisziplinäre Teams speziell für bestimmte Funktionsbereiche zusammengestellt werden.

Literatur

Bronfenbrenner, U. and P. A. Morris (1998). The ecology of developmental processes. Handbook of child psychology: Volume 1: Theoretical models of human development (5th ed.).
W. D. R. M. Lerner. Hoboken, NJ, US, John Wiley & Sons Inc: 993-1028.

Coster, W., et al. (2011). "Psychometric evaluation of the Participation and Environment Measure for Children and Youth." Dev Med Child Neurol 53(11): 1030-1037.

Hollenweger J, Kraus de Camargo O (Hrsg., 2011) ICF-CY: Internationale Klassifikation der Funktionsfähigkeit, Behinderung und Gesundheit bei Kindern und Jugendlichen. Verlag Hans Huber.

King GA (2007) Measuring children's participation in recreation and leisure activities: construct validation of the CAPE and PAC. Child Care, Health and Development 33(1): 28.

Kraus de Camargo O (2007) Die ICF–CY als Checkliste und Dokumentationsraster in der Praxis der Frühförderung. Frühförderung Interdisziplinär 26: 158–166.

Kraus de Camargo, O. and N. Fayed (2013). 'Health status' and the usefulness of the ICF framework: Clinical and Program perspectives. Health, Participation and Quality of Life in Young People with Neurodevelopmental Conditions: Theory, Concepts, Evidence and Practice.
G. M. Ronen and P. L. Rosenbaum. London, Mac Keith Press.

Law M, Baptiste S et al. (2005) Canadian Occupational Performance Measure COPM. Ottawa: CAOT Publications ACE.

Majnemer, A., Ed. (2012). Measures for Children with Developmental Disabilities framed by the ICF-CY, Wiley - McKeith Press.

Schuntermann, M. F. (2009). Einführung in die ICF - Grundkurs, Übungen, offene Fragen.
Heidelberg, ecomed MEDIZIN.

Steenbeek D (2007) Goal attainment scaling in paediatric rehabilitation: a critical review of the literature. Dev Med Child Neurol 49(7): 550.

Waber, D. P. (2011). Rethinking Learning Disabilities. New York, London, The Guilford Press.

Zimmerman–Targownik P, Targownik, D (2007) Sekunden, die mein Leben verändern. Teil 3/3: Das Unglück. Film; WDR Köln.

WHO ICF browser - Englisch:
http://apps.who.int/classifications/icfbrowser/

ICF-CY Checklisten - Deutsch:
http://www.bvkm.de/Aktuelle_Meldungen/ICF-Checklisten

http://www.dgspj.de/index.php?option=com_content&view=category&id=80:icf-cy&Itemid=78&layout=blog
Fragebogen zur Teilhabe (PEM-CY) sowie weitere links und infos zum Thema Teilhabe - Englisch:
http://participation-environment.canchild.ca/en/index.asp
WHO ICF e-learning tool - Englisch:
http://p.ideaday.de/104.2/icf/

17 Fragebögen und Tests

Harald Bode

Sozialpädiatrie in der Praxis. http://dx.doi.org/10.1016/B978-3-437-31630-3.00017-1
Copyright © 2014 Elsevier GmbH. All rights reserved.

17.1 Allgemeines

Eine fundierte interdisziplinäre Diagnostik auf wissenschaftlicher Grundlage ist wesentliches Merkmal einer qualitativ hochwertigen sozialpädiatrischen Behandlung von Kindern mit drohenden oder manifesten Behinderungen, Entwicklungs-, Verhaltens- und emotionalen Störungen. Diagnostische Methoden in der Sozialpädiatrie beziehen sich auf die verschiedenen Ebenen der Mehrbereichsdiagnostik in der Sozialpädiatrie (MBS): **E**ntwicklung/Intelligenz, **K**örperlich-Neurologischer Befund, **P**sychischer Befund, **S**oziale Begleitumstände, **A**etiologie, **T**eilhabe (Merke: **EKPSA+T**).
In diesem Kapitel wird auf die verschiedenen Methoden, die der Abklärung der Ätiologie einer Störung dienen, nicht im Einzelnen eingegangen.
Neben einer ausführlichen **biografischen Anamnese** gehört die **körperliche und neurologische Untersuchung** im klassischen Sinn zu jeder diagnostischen Erstabklärung. Die verschiedenen Methoden zur Überprüfung des Seh- und Hörvermögens, die je nach Alter und Fragestellung in einem gestuften Ablauf zum Einsatz kommen müssen, können hier nicht besprochen werden.
Die Methoden der Elektrophysiologie (Elektroenzephalogramm, visuell, akustisch und somatosensibel evozierte Potenziale, transkranielle Magnetstimulation, Nervenleitgeschwindigkeitsmessung, Elektromyogramm) werden in Lehrbüchern der Neuropädiatrie dargestellt. Häufig kommen bei der Abklärung der Ätiologie **bildgebende Verfahren** (Sonografie von Gehirn, Rückenmark oder Muskulatur, Magnetresonanztomografie, gelegentlich Computertomografie oder spezielle Röntgenuntersuchungen) zum Einsatz.
Die Vielzahl der aus Blut, Urin und Liquor möglichen **Laboruntersuchungen** und **genetischen Tests** verlangt genaue Kenntnisse der verschiedenen Krankheitsbilder und der Aussagekraft der Untersuchungen. Eine abgestufte Diagnostik sollte sich am klinischen Verdacht, der Häufigkeit der Störungen, dem Untersuchungsaufwand und der Belastung des Kindes durch die Untersuchung orientieren. Hierzu wird auf pädiatrische und neuropädiatrische Lehrbücher verwiesen.
Die in diesem Kapitel dargestellten diagnostischen Methoden beziehen sich auf die Ebenen Entwicklung/Intelligenz, psychischer Befund, soziale Begleitumstände und Teilhabe.
Fragebögen richten sich an Eltern, Erzieher, Pädagogen, Therapeuten und andere Bezugspersonen, u. U. an die Kinder/Jugendlichen selbst. Sie müssen standardisiert und möglichst normiert sein und konkrete Fragen beinhalten. Zahlreiche Fragebögen mit sehr unterschiedlichen Güteeigenschaften (➤ Kap. 17.2.1) sind im Umlauf. Die Validität von Angaben in Fragebögen ist kritisch zu beurteilen. Die Wertungen in Fragebögen sind in der Regel subjektiv. Nicht selten werden bei verschiedenen Personen diskrepante Einschätzungen gefunden. Es empfiehlt sich daher, soweit möglich verschiedene Informationsquellen heranzuziehen und die Einschätzungen aus Fragebögen mit den Ergebnissen objektiver Tests zu vergleichen.
Fragebögen sind ein hilfreiches diagnostisches **Screening**instrument in der **Entwicklungsdiagnostik (➤** Kap.14). Sie beziehen sich auf die verschiedenen Aspekte der kindlichen Entwicklung. Es ist nicht zulässig, eine Diagnose allein aufgrund von Daten aus Fragebögen zu stellen.
Die Verwendung von Fragebögen ist unverzichtbar zur Beurteilung des **Verhaltens,** wenn die Eigenschaften oder Verhaltensweisen in standardisierten Testsituationen nicht mess- oder beobachtbar sind, z. B. das Verhalten zu Hause oder in der Schule. Auch für die Beurteilung des psychischen Befindens, der sozialen Begleitumstände und der Teilhabe am Leben in der Gemeinschaft sind Fragebögen erforderlich.
Die Durchführung von **Entwicklungs- und Intelligenztests** erfordert professionelle Expertise, die Beachtung ethischer Standards bei der Testuntersuchung sowie die Berücksichtigung der Rechte der Probanden und der anderen am Testprozess Beteiligten. Die Tests müssen in einer

kindgerechten Umgebung mit entsprechendem Material, in angemessener Sprache und in freundlicher Interaktion erfolgen. Die Testdauer muss den Möglichkeiten des Kindes angepasst sein, Ablenkung ist zu vermeiden. Besonders bei jungen Kindern muss das diagnostische Vorgehen flexibel an die kindlichen Möglichkeiten und Bedürfnisse angepasst werden. Besonders bei behinderten Kindern und Jugendlichen ist große Sorgfalt und eine faire Untersuchung und Bewertung ggf. mit Adaptation des Verfahrens an die speziellen Beeinträchtigungen erforderlich. Sprachliche und kulturelle Besonderheiten sind bei manchen Kindern mit Migrationshintergrund zu berücksichtigen. Faktoren, die die Testergebnisse beeinflussen können, sind z. B. Tagesform, Motivation, sozial-emotionale Faktoren, Übung, Vorerfahrungen, Geschick des Untersuchers, An- oder Abwesenheit der Eltern/Bezugspersonen, Kontextbedingungen. Die Auswertung und insbesondere die Bewertung der Ergebnisse muss unter Berücksichtigung der Testsituation und im Gesamtzusammenhang erfolgen. Neben Defiziten sollten auch individuelle Stärken des Kindes und protektive Faktoren des Umfeldes identifiziert werden. Durch diese Vorgehensweise wird das Risiko falsch positiver Testergebnisse, die zu Verunsicherung, Stigmatisierung und unnötigen oder fehlerhaften therapeutischen oder pädagogischen Konsequenzen führen, verringert.
Testverfahren werden in der Praxis oft mit recht unterschiedlichem Qualitätsanspruch eingesetzt, z. T. durch wenig geschultes Personal. Dies ist kritisch zu bewerten. Regelmäßiger Erfahrungsaustausch und kollegiale Supervision der Untersucher sind zu fordern. Sozialpädiatrische Zentren mit ihrer interdisziplinären Expertise stellen eine hohe fachliche Qualifikation bei Indikation, Auswahl, Durchführung und Bewertung von Testverfahren sicher. Hierbei spielen Psychologen eine besonders wichtige Rolle.
Insbesondere im Säuglings- und Kleinkindalter ist die strikte Trennung verschiedener **Entwicklungsdimensionen** artifiziell (z. B. motorische und mentale Funktionen). Auch später ist eine strikte Trennung z. B. der Dimensionen Sprache, Aufmerksamkeit und Intelligenz in Tests nicht immer vollständig gegeben oder überhaupt möglich.
In der Diagnostik von Entwicklungs-, Verhaltens- und anderen psychischen Störungen können kategoriale und dimensionale Klassifikationsverfahren unterschieden werden. Der **kategoriale Ansatz** geht davon aus, dass sich bestimmte Störungen als klare Entitäten wie somatische Erkrankungen voneinander unterscheiden lassen. Die Klassifikationssysteme der ICD-10 und der DSM-IV beruhen auf diesem Prinzip. Der Realität von Entwicklungs-, Verhaltens- und anderen psychischen Störungen kommt der **dimensionale Ansatz** näher. Hierbei wird eine Störung angenommen, wenn die Ausprägung des Merkmals in einem bestimmten Ausmaß von der Norm abweicht. Beispiel für ein dimensional definiertes Merkmal ist die Aufmerksamkeitsstörung. Sie erfordert als Kriterium einen bestimmten Ausprägungsgrad des Merkmals Unaufmerksamkeit.
Wesentliche **Aufgaben der diagnostischen Methoden** in der Sozialpädiatrie sind:

- Objektivierung und Beschreibung der Entwicklung, Intelligenz inkl. neuropsychologischer Funktionen, des emotionalen Befindens und des Verhaltens inkl. Kommunikation und Interaktion, der sozialen Begleitumstände und der Teilhabe eines Kindes/Jugendlichen
- Objektivierung und Beschreibung von Ressourcen, Rückständen, individuellen Profilen, Störungen und des Verlaufs
- Einschätzung der Prognose
- Erkennung von Interventions- oder Unterstützungsbedarf bei Kind/Jugendlichem, Familie und Umfeld
- Erkennung von Beratungs-, oder Schulungsbedarf der Bezugspersonen
- Entscheidungshilfe für die Indikation zu professioneller Förderung/Therapie
- Entscheidungshilfe für die Wahl von Kindergarten, Kindertagesstätte, Schule, Ausbildung
- Erstellung eines Behandlungsplans und dessen Anpassung im zeitlichen Verlauf.

Die in Kapitel 17.3–17.15 (➤ Kap. 17.3) beschriebenen diagnostischen Verfahren stellen eine Auswahl der in der Sozialpädiatrie gebräuchlichen Fragebögen und Tests dar. Es besteht kein Anspruch auf Vollständigkeit. Die Ausführungen beruhen im Wesentlichen auf den Angaben der Anbieter der jeweiligen Verfahren. Die Einschätzung des Einsatzbereichs entspricht der praktischen Erfahrung des Autors. Detaillierte Bewertungen hinsichtlich Gütekriterien, Normen und Aussagekraft der Verfahren sprengen den Rahmen dieses Kapitels. Dazu sei auf die Manuale der Verfahren verwiesen. Eine umfassendere Einschätzung findet man in einem Qualitätspapier der DGSPJ (http://www.dgspj.de/media/QZTestverfahren.pdf). Standards der psychologischen Diagnostik in SPZ sind in einem weiteren Qualitätspapier festgehalten (http://www.dgspj.de/media/QZ Psychologische Diagnostik Endversion.pdf).
Die Durchführung einer Diagnostik setzt eine vorangehende Aufklärung und Zustimmung der Sorgeberechtigten und soweit möglich des Kindes/Jugendlichen voraus. Eine einfühlsame Mitteilung der Ergebnisse trägt dazu bei, mit Kind und Eltern eine gute Kooperation zu etablieren und Zukunftsperspektiven zu eröffnen.

17.2 Methodische Grundlagen

17.2.1 Gütekriterien

Fragebögen und psychologische Tests (im Folgenden als „Verfahren" zusammengefasst) sollten bestimmten Gütekriterien genügen, die ihre Qualität anzeigen. Die Hauptgütekriterien sind Objektivität, Reliabilität und Validität.

Objektivität. Unter der Objektivität eines Verfahrens versteht man die Unabhängigkeit der Ergebnisse von den Rahmenbedingungen, z. B. von den räumlichen Bedingungen oder dem Versuchsleiter. Die **Durchführungsobjektivität** bezeichnet die Unabhängigkeit der Testergebnisse von der Person des Versuchsleiters. Sie erfordert eine Standardisierung (s. u.) der Testsituation und eine möglichst geringe soziale Interaktion zwischen Versuchsleiter und Testperson. **Auswertungsobjektivität** bedeutet, dass gleiches Verhalten einer Testperson stets auf gleiche Weise ausgewertet wird. Die Objektivität bei projektiven Tests ist eher gering, bei Intelligenztests durch die standardisierte Auswertung eher hoch. **Interpretationsobjektivität** bedeutet, dass gleiche Testwerte auf die gleiche Weise interpretiert werden, z. B. durch die Festlegung von Grenzwerten zwischen unauffällig und auffällig oder durch die Zuordnung von Testwerten zu den Kategorien durchschnittlich, unter-/überdurchschnittlich etc. Die Interpretationsobjektivität ist bei projektiven Tests gering und bei Intelligenztests hoch.

Reliabilität. Reliabilität (Zuverlässigkeit) bezeichnet die formale Genauigkeit der Messung eines Merkmals durch ein Verfahren. Reliable Ergebnisse sind frei von Zufallsfehlern, d. h. bei Wiederholung eines Verfahrens unter gleichen Rahmenbedingungen würde das gleiche Ergebnis erzielt. Die Schwierigkeit und Trennschärfe in Testfragen/-aufgaben und die daraus abgeleitete Homogenität der Skalen eines Verfahrens bestimmen die Reliabilität eines Tests wesentlich. Die Reliabilität kann mit verschiedenen Methoden gemessen werden. Die **Paralleltest-Reliabilität** gibt an, ob ein vergleichbares Messverfahren identische Ergebnisse liefert. Bei der **Split-Half-Reliabilität** werden die Ergebnisse in zwei Mengen unterteilt. Bei hinreichend großer Ergebnismenge sollten die Durchschnittswerte und weitere statistische Kenngrößen gleich sein. Die **Retest-Reliabilität** zeigt, ob eine Wiederholung der Messung bei Konstanz der zu messenden Eigenschaft die gleichen Messwerte liefert. Für viele Tests ist eine Wiederholung nach dem Test-Retest-Verfahren nur theoretisch möglich, da die mit dem Test einhergehenden Erinnerungseffekte das Ergebnis beeinflussen. Die in Bezug auf dieselben Testpersonen ermittelte Übereinstimmung zwischen Beurteilern bezeichnet man als **Interrater-Reliabilität**. Reliabilitätswerte unter 80 erlauben nur bedingt diagnostisch relevante Aussagen. Verfahren, die in den ersten Lebensjahren angewandt werden, weisen solche Werte selten auf.

Validität. Die Validität eines Verfahrens gibt an, ob es tatsächlich das gewünschte Merkmal misst (z. B. Intelligenz oder Impulsivität). **Inhaltsvalidität** wird angenommen, wenn ein Verfahren zur Messung eines bestimmten Konstrukts bzw. Merkmals alle Aspekte dieses Konstrukts ausreichend erfasst, z. B. Lese- oder Rechenfähigkeit. **Konstruktvalidität** bezieht sich auf die Zulässigkeit von Aussagen aufgrund der Operationalisierung des dahinter liegenden Konstrukts (z. B. Aufmerksamkeit). Als Indikatoren der Konstruktvalidität gelten die konvergente und die divergente Validität. Konvergente Validität besteht, wenn die Messdaten von Verfahren, die dasselbe Konstrukt abbilden, hoch miteinander korrelieren. Diskriminante Validität besteht, wenn die Messdaten von Verfahren, die verschiedene Konstrukte abbilden, nur gering miteinander korrelieren. **Kriteriumsvalidität** bezieht sich auf den Zusammenhang zwischen den Ergebnissen des Verfahrens und einem externen empirischen Kriterium. Von innerer Kriteriumsvalidität wird gesprochen, wenn als externes Kriterium ein anderer, als valide anerkannter Test herangezogen wird. Sofern als externes Kriterium die objektive (z. B. physikalische) Messung eines Merkmals oder ein Expertenrating herangezogen wird, spricht man von äußerer Kriteriumsvalidität. Bei Kindern unter drei Jahren ist die **prognostische Validität** auch standardisierter Tests eingeschränkt (z. B. hinsichtlich kognitiver Fähigkeiten im Schul- und Erwachsenenalter). Das macht bei entsprechender Fragestellung Verlaufsuntersuchungen erforderlich.
Weitere Gütekriterien von Verfahren sind ihre Standardisierung und Normierung.

Standardisierung. Die Standardisierung eines Verfahrens beinhaltet eine präzise schriftliche Anleitung für die Durchführung, einheitliches Testmaterial und eine einheitliche Protokollierung und Auswertung nach festgelegten Kriterien.

Normierung. Sie erfolgt durch die Erhebung von Messwerten mit dem Verfahren an einer hinreichend großen, repräsentativen Stichprobe (Normstichprobe). Da sich manche Merkmale von Normstichproben im Verlauf von Jahren verändern (z. B. Intelligenz), ist es wichtig, Verfahren mit möglichst aktueller Normierung zu verwenden. Nach 10–15 Jahren sollte eine Überprüfung der Normen und ggf. eine Neunormierung des Verfahrens erfolgen.

17.2.2 Weitere Eigenschaften von Fragebögen und Tests

Utilität. Sie bezeichnet die Nützlichkeit/Tauglichkeit für die Beantwortung einer speziellen Fragestellung.

Testfairness. Sie zeigt an, ob Personengruppen mit unterschiedlichem Alter, Geschlecht, sprachlichen Hintergrund, sprachlichen Fähigkeiten oder bestimmten Behinderungen gleich behandelt werden bzw. die gleichen Chancen auf ein entsprechendes Testergebnis haben. Dies ist u. a. bei Kindern mit Sprach-, Lese-, Rechtschreib- oder Rechenstörungen zu beachten.

Testökonomie. Sie beschreibt den Aufwand des Verfahrens (Zeit für Einarbeitung, Durchführung und Auswertung, Materialaufwand) im Verhältnis zu seinem Nutzen.

Transparenz. Sie ist gegeben, wenn das Verfahren verständliche Instruktionen beinhaltet und die Testpersonen sich vorher ausreichend mit dem Verfahren vertraut machen können.

Unverfälschbarkeit. Sie ist gegeben, wenn die befragte Person/Testperson ihre Ergebnisse nicht gezielt steuern oder verfälschen kann.

Zumutbarkeit. Sie zeigt an, ob das Verfahren die befragte Person/Testperson in zeitlicher, psychischer und körperlicher Hinsicht nicht über Gebühr beansprucht.

17.2.3 Bewertung der Ergebnisse von Verfahren

Bei der Beurteilung der kindlichen Entwicklung werden häufig die Begriffe Meilen- oder Grenzstein benutzt.

Ein **Meilenstein** ist eine entwicklungsabhängige Fähigkeit, die in einer Population von 50 % der Kinder bis zu einem bestimmten Lebensalter erreicht wird. Ein **Grenzstein** ist eine Fähigkeit, die von 90 % (95 %) der Kinder bis zu einem bestimmten Lebensalter erreicht wird.

Entwicklungstests basieren häufig auf diesen Kennzahlen. Das Grenzsteinprinzip hat sich bewährt, um Kinder als entwicklungsverzögert zu identifizieren. Diese Kinder benötigen eine weitergehende Diagnostik hinsichtlich der Ursache ihrer Entwicklungsverzögerung. Eine Therapieindikation ist nicht automatisch gegeben, wenn ein Kind in einem oder mehreren Entwicklungsbereichen einen Grenzstein nicht erreicht.

Viele Entwicklungstests geben als Ergebnis ein **Entwicklungsalter** an. Es bezeichnet das Alter, in dem bestimmte Fähigkeiten von 50 % der gesunden Kinder einer Population beherrscht werden. Daraus ergibt sich der Entwicklungsquotient (Entwicklungsalter × 100 ÷ Lebensalter).

Viele Messwerte/Merkmale in Biologie und Medizin zeigen bei hinreichend großer Zahl von Messungen/Individuen eine **Normalverteilung.** Beispiele sind die Körpergröße und die Intelligenz. Die Werte verteilen sich symmetrisch in charakteristischer Weise um einen Mittelwert (MW). Die Standardabweichung (SD) stellt eine Eigenschaft der jeweiligen Normalverteilung dar. Bei Intelligenzmessungen und anderen psychologischen und Entwicklungstests wird der Mittelwert des Normalen häufig als ein Wert von 100 definiert mit einer Standardabweichung von 15. Abbildung 17.1 (➤ Abb. 17.1) zeigt, dass die Intelligenz von 68,26 % der normalen Individuen (Normstichprobe) im Bereich des MW ± 1 SD und von 95,54 % im Bereich des MW ± 2 SD liegt.

Andere Skalierungen von Normalverteilungen benutzen die Z-Skala. Dabei wird der MW als 0 definiert, eine SD beträgt ± 1. Bei der T-Skala wird der MW als 50 definiert, eine SD führt dann zu den Werten 40 bzw. 50.

Tabelle 17.1 zeigt die korrespondierenden Werte für die verschiedenen Skalierungen von normalverteilten Test-/Messwerten.

Der Prozentrang (PR) ist ein Kennwert bei Merkmalen/Messwerten, die nicht in einer Normalverteilung angeordnet sind. Er zeigt an, wie viele Individuen für dieses Merkmal/diesen Messwert das

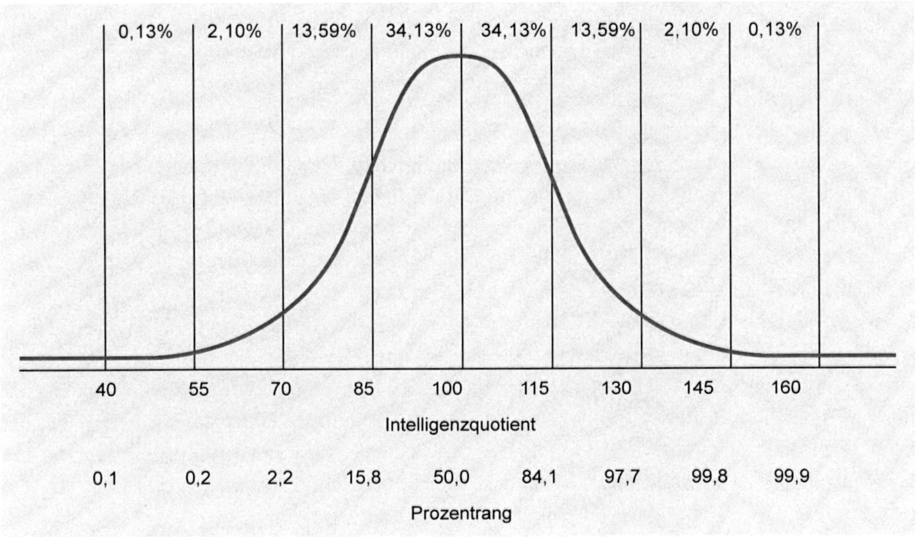

Abb. 17.1 Normalverteilung des Intelligenzquotienten

Tab. 17.1 Korrespondierende Werte verschiedener Skalierungen einer Normalverteilung

Mittelwert/ Standardabweichung	- 3SD	- 2 SD	- 1 SD	MW	+ 1 SD	+ 2 SD	+ 3 SD
Prozentrang	0,1 %	2,2 %	15,8 %	50 %	84,2 %	97,7 %	99,9 %
IQ-Skala	55	70	85	100	115	130	145
Z-Skala	- 3	- 2	- 1	0	+ 1	+ 2	+ 3
T-Skala	20	30	40	50	60	70	80
C-Skala	0	1	3	5	7	9	10

gleiche oder ein kleineres Resultat zeigen. Ein PR = 50 entspricht dem Median für dieses Merkmal. Bei manchen Messwerten, z. B. den Körpermaßen, wird statt des Prozentrangs mit gleicher Bedeutung der Begriff „Perzentile" verwendet.
Die **Klassifikation** von Test- oder Messwerten als normal, „auffällig" oder pathologisch ist je nach Verfahren unterschiedlich, vom Klassifikationsziel und z. T. auch vom Untersucher abhängig. Sie muss bei der Interpretation von Prävalenzangaben, z. B. von Entwicklungs-, Verhaltens- und psychischen Störungen, und bei der Entscheidung für Förder- und Therapiemaßnahmen unbedingt berücksichtigt werden. Dazu die folgenden Beispiele:

Körpermaße. Körperlänge und Kopfumfang werden als normal eingestuft, wenn sie zwischen der 3. und 97. Perzentilen liegen, jenseits dieser Perzentilen spricht man von Kleinwuchs/Hochwuchs bzw. Mikro-/Makrozephalie. Übergewicht besteht bei einem Bodymass-Index (BMI) > 90. Perzentilen, Adipositas bei einem BMI > 97. Perzentilen. Intelligenzwerte. Sie werden nach der international gebräuchlichen ICD-10 im Bereich von 70–130 (MW ± 2 SD) als normal angesehen. Erst bei Werten von 50–69 (< MW – 2 SD) spricht man von leichter Intelligenzminderung, bei IQ-Werten 35–49 von mittelschwerer und bei IQ-Werten von 20–34 von schwerer Intelligenzminderung. Die Klassifikation der Intelligenzminderung orientiert sich mithin eher an klinisch-lebenspraktischen Kriterien und nicht an der Zahl der Standardabweichungen.

Entwicklungstests und psychologischen Verfahren. Bei diesen wird der Normalbereich (Durchschnittsbereich) häufig als MW ± 1 SD angegeben. Werte von mehr als 1 SD unter dem MW sind dann unterdurchschnittlich, Werte von mehr als 2 SD unter dem MW weit unterdurchschnittlich. Diese Klassifikation führt dazu, dass die Ergebnisse von 15,8 % der Probanden als unterdurchschnittlich klassifiziert werden. Vergleichbar ist die 15. Perzentile eines Merkmals als Grenze zwischen normal und auffällig. Diese Grenze könnte z. B. ein Kriterium für speziellen pädagogischen Förderbedarf in Kindergarten/Schule darstellen.
Bei manchen Fragestellungen, z. B. nach einer **Therapieindikation** im Sinne einer Heilmittelanwendung bei Kindern mit Entwicklungsstörungen, schließt die Grenzsetzung der 15. Perzentilen zu viele Individuen ein. Viele Kinder würden dann ohne Notwendigkeit therapiert. Ihre Fähigkeiten sind zwar unterdurchschnittlich, haben jedoch nicht den Charakter einer Störung. Bei der Frage einer Therapieindikation bei kindlichen Entwicklungsstörungen kann die Grenze von < – 1,5 SD bzw. < 10. Perzentile eine Orientierungshilfe darstellen. Sie muss allerdings auf ihre Berechtigung ja nach Kind und Fragestellung kritisch geprüft werden.

17.2.4 Statistische Kenngrößen bei Screeningverfahren

Screeningverfahren dienen der Erfassung bestimmter Erkrankungen, Entwicklungs- oder Verhaltensstörungen in großen Populationen (z. B. Neugeborenenscreening, Schuleingangsuntersuchungen, Teile der KiGGS-Studie, Eltern-, Erzieherfragebögen). Hierfür bedarf es zeitökonomischer Verfahren. Im Screening auffällige Befunde bedürfen beim Individuum einer Überprüfung durch ein differenziertes (konfirmatorisches) Verfahren. Die Güte des Screeningverfahrens ergibt sich aus dem Vergleich des Ergebnisses mit dem eines konfirmatorischen Verfahrens (Goldstandard). Dies wird durch die folgenden statistischen Kenngrößen (➤ Tab. 10.2) beschrieben.
Die **Sensitivität** eines Verfahrens bezeichnet die Wahrscheinlichkeit, einen tatsächlich positiven Sachverhalt auch durch ein positives Testergebnis zu erkennen. Die Sensitivität gibt also den Anteil der richtig positiv erkannten Sachverhalte an der Gesamtheit der tatsächlich positiven Sachverhalte an. Mit der Falschnegativ-Rate addiert sich die Richtigpositiv-Rate zu 100 % der tatsächlich positiven Sachverhalte.
Die **Spezifität** eines Verfahrens gibt die Wahrscheinlichkeit an, mit der ein tatsächlich negativer Sachverhalt auch durch ein negatives Testergebnis erkannt wird. Die Spezifität gibt damit den Anteil der richtig negativ erkannten Sachverhalte an der Gesamtheit der tatsächlich negativen Sachverhalte an. Mit der Falschpositiv-Rate addiert sich die Richtignegativ-Rate zu 100 % der tatsächlich negativen Sachverhalte.
Die Sensitivität von Screeningverfahren ist abhängig vom untersuchten Merkmal. Sie liegt bei Stoffwechselstörungen (biochemischer Parameter) über 98 %, bei Hörstörungen (apparative physikalische Messung) über 90 %, bei Entwicklungsstörungen (Fragebögen, orientierende Tests) bestenfalls bei 70 %.
Auch die Spezifität eines Screeningverfahrens für Entwicklungsstörungen liegt bestenfalls zwischen 70 und 80 %. Eine Steigerung der Sensitivität eines solchen Screeningverfahrens geht zu Lasten einer sinkenden Spezifität. Durch die Verschiebung der Grenze zwischen normal und nicht

Tab. 17.2 Statistische Kenngrößen

	Test positiv	Test negativ	gesamt Zeile
Störung/Krankheit	a=richtig positiv	b=falsch negativ	a+b
Normal/Gesund	c=falsch positiv	d=richtig negativ	c+d
Gesamt Spalte	a+c	b+d	

Sensitivität=a/(a+b); Spezifität=d/(c+d); Positiver Vorhersagewert=a/(a+c); Negativer Vorhersagewert=d/(b+d)

normal werden mehr Personen vom Screening als auffällig eingestuft, d. h. es erfasst zwar mehr richtig positive Fälle, es kommt aber auch zu mehr falsch positiven Bewertungen.

Der **positive Vorhersagewert** eines Verfahrens bezeichnet die Wahrscheinlichkeit, dass ein positives Ergebnis einem tatsächlich positiven Sachverhalt entspricht. Sie gibt also den Anteil der richtig positiv erkannten Ergebnisse an der Gesamtheit aller positiven Ergebnisse des Verfahrens an. Beispielsweise gibt der positive Vorhersagewert eines Screeningverfahrens (z. B. Hörscreening, Fragebögen zur Entwicklungsbeurteilung) an, wie viele Personen, bei denen eine Hörstörung/Entwicklungsstörung mittels des betreffenden Verfahrens festgestellt wurde, auch tatsächlich eine Hörstörung haben/in ihrer Entwicklung gestört sind.

Der **negative Vorhersagewert** eines Verfahrens bezeichnet die Wahrscheinlichkeit, dass ein negatives Ergebnis einem tatsächlich negativen Sachverhalt entspricht. Sie gibt also den Anteil der richtig negativ erkannten Ergebnisse an der Gesamtheit aller negativen Ergebnisse des Verfahrens an. Beispielsweise gibt der negative Vorhersagewert eines Screeningverfahrens (z. B. Hörscreening, Fragebögen zur Entwicklungsbeurteilung) an, wie viele Personen, bei denen keine Hörstörung/Entwicklungsstörung mittels des betreffenden Verfahrens festgestellt wurde, auch tatsächlich keine Hörstörung haben/in ihrer Entwicklung unauffällig sind.

Während Sensitivität und Spezifität feste testabhängige Kenngrößen sind, hängen positiver und negativer Vorhersagewert eines Testverfahrens von der Prävalenz der Störung in der untersuchten Population ab. Tabelle 17.2 fasst die besprochenen statistischen Kenngrößen zusammen.

17.3 Entwicklung

17.3.1 Fragebögen

EFkE (Elternfragebogen zur kindlichen Entwicklung)

- Alter: 1.–6. Lebensjahr
- Einsatz: Kindertagesstätten, Kindergärten, Frühförderstellen, kinderärztliche Praxen,
- Ziel: Screeningmaßnahme zur Diagnostik von Entwicklungs- und Teilleistungsstörungen im Kleinkindalter mit Hilfe von Elternangaben.

Der Fragebogen umfasst 270 Fragen zu 8 Teilgebieten der kindlichen Entwicklung, die von den Eltern durch ankreuzen (ja/nein) beantwortet werden: soziale Entwicklung, Selbstständigkeit, Grobmotorik, Feinmotorik, Sprachausdruck, Sprachverständnis, Buchstaben, Zahlen und 30 Fragen zu speziellen Gesundheits- und Verhaltensproblemen. Für die Altersstufen 11–12 Monate (U6), 23–24 Monate (U7), 36–59 Monate (U8) und 60–67 Monate (U 9) erfolgt die Auswertung anhand von Perzentilenwerten der Normstichprobe.

EDB (Entwicklungsdokumentationsbogen für Eltern)

- Alter: 1. Lebensjahr
- Einsatz: kinderärztliche Praxen
- Ziel: Screeningmaßnahme zur Diagnostik von Entwicklungsstörungen im Säuglingsalter mit Hilfe von Elternangaben.

Der Fragebogen umfasst 48 Fragen zu den Bereichen Fortbewegung, Verhalten bei Nahrungsaufnahme, Verhalten bei Pflege, Entwicklung der Handlungsfähigkeit, Sprachentwicklung. Die 3., 50., und 97. Perzentile eines Normkollektivs ist für jedes Item angegeben.

17.3.2 Screening

Denver-Entwicklungsskalen

- Alter: 0–6 Jahre
- Einsatz: Kindertagesstätten, Kindergärten, Frühförderstellen, kinderärztliche Praxen
- Ziel: Screening auf Entwicklungsstörungen im Säuglings- und Kleinkindalter.

Der Denver-Test basiert auf dem Prinzip der Stufenleiter der Entwicklung. Es untersucht die Entwicklungsbereiche Grobmotorik, Sprache, Feinmotorik-Adaptation, Sozialkontakt. Die 105 Aufgaben werden mit steigendem Schwierigkeitsgrad adaptiv an die kindlichen Fähigkeiten gestellt. Etwa die Hälfte der Aufgaben wird über die Eltern erfragt. Die Materialien sind selbst zu besorgen. Zu jeder Aufgabe wird die 25., 50., 75. und 90. Perzentile der Normstichprobe angegeben. Eine nicht erfüllte Aufgabe zum Zeitpunkt der 90. Perzentile wird als Verzögerung angesehen.

MZT (Mann-Zeichen-Test)

- Alter: 3–14 Jahre
- Einsatz: Psychologen, Kindertagesstätten, Kindergärten, Frühförderstellen, kinderärztliche Praxen
- Ziel: diagnostisches Hilfsmittel zum Entwicklungsscreening und zur Frage der Schulreife.

In der Neufassung wurde das Testverfahren den heutigen gesellschaftlichen Lebensumständen angepasst, da sich diese auch auf das Malverhalten der Kinder ausgewirkt haben. Der Testauftrag lautet: „Male einen Menschen, so gut du kannst." Nach festgelegtem Verfahren wird die Zahl der Körperteile addiert und daraus der Mann-Zeichen-Quotient bzw. das Mann-Zeichen-Alter bestimmt. Das überarbeitete Verfahren ist geeignet, im Rahmen einer Testbatterie erste Aussagen zu bestimmten Aspekten der Wahrnehmungsentwicklung eines Kindes zu treffen.

EVU – Erweiterte Vorsorgeuntersuchung

- Alter: 3.–64. Lebensmonat
- Einsatz: kinderärztliche Praxen
- Ziel: Entwicklungsbeurteilung jeweils gekoppelt an die Früherkennungsuntersuchungen U4–U9.

Die EVU stellt ein entwicklungsneurologisch und entwicklungspsychologisch basiertes Screeningverfahren in Ergänzung zu den kinderärztlichen Vorsorgeuntersuchungen U4 bis U9 dar. Die Dimensionen Motorik, Sprache, Kognition und sozial-emotionale Entwicklung werden erfasst. Es werden Grenzwerte für die Summenwerte der einzelnen Dimensionen und ein Gesamtgrenzwert angegeben. Ergänzende Informationen werden mit altersentsprechenden Elternfragebögen erhoben. Als Referenzwert für jede Altersgruppe wird die 10. Perzentile der Normstichprobe angegeben.

17.3.3 Tests

GES (Griffiths-Entwicklungsskalen)

- Alter: 1–24 Monate sowie ältere Kinder mit Retardierungen oder Behinderungen
- Einsatz: Frühförderstellen, sozialpädiatrische Zentren, neuropädiatrische Ambulanzen, Nachsorgeeinrichtungen, Forschung
- Ziel: Erstellung eines Entwicklungsprofils, Feststellung von Entwicklungsverzögerungen, -störungen.

Der GES basiert auf dem Prinzip der Stufenleiter der Entwicklung. Mit ihm lässt sich der Entwicklungsstand insgesamt und nach fünf Funktionsbereichen differenziert feststellen: Motorik, Persönlich-Sozial, Hören und Sprechen, Auge und Hand sowie Leistungen. Die 208 Testaufgaben mit steigendem Schwierigkeitsgrad werden adaptiv an die kindlichen Fähigkeiten gestellt. Es wird ein Gesamt-Entwicklungsquotient (EQ) sowie ein Entwicklungsquotient für jede Unterskala berechnet, woraus sich ein Entwicklungsprofil ergibt. Die GES geben bei Vorliegen von Retardierungen in Unterskalen oder der Gesamt-Entwicklung konkrete Hinweise zur weiterführenden Diagnostik, zur Beratung der Eltern sowie für die Planung und Durchführung gezielter Frühfördermaßnahmen. Durch wiederholte Untersuchungen kann der Entwicklungsverlauf spontan oder unter Therapie beurteilt werden. Eine Auflistung der prozentualen Lösungshäufigkeiten für jede einzelne Testaufgabe, vergleichbar mit Perzentilen zur Wachstumsbeurteilung, kann für die quantitative Entwicklungsbeurteilung benutzt werden.

Bayley II (Bayley Scales of Infant and Toddler Development II)

- Alter: 1–42 Monate
- Einsatz: Frühförderstellen, sozialpädiatrische Zentren, neuropädiatrische Ambulanzen, Nachsorgeeinrichtungen, Forschung
- Ziel: Feststellung der motorischen und mentalen Entwicklung, von Entwicklungsverzögerungen, -störungen.

Die Bayley-II-Skalen basieren auf dem Prinzip der Stufenleiter der Entwicklung. Die deutsche Bearbeitung erfasst zwei grundlegende Fähigkeitsbereiche mit der kognitiven (Mental Development Index – MDI) und der motorischen Skala (Psychomotor Development Index – PDI). Die Aufgaben der kognitiven Skala überprüfen frühe Gedächtnisleistungen, Habituation, Problemlösefähigkeiten, frühe Zahlkonzepte, Klassifikation und Kategorisierungsfähigkeit, Vokalisation und sprachliche Kompetenzen sowie frühe sozial-kommunikative Fähigkeiten. Die Aufgaben der motorischen Skala erfassen die zunehmende Haltungskontrolle sowie die fein- und grobmotorische Koordination. Diese umfasst die Bewegungskontrolle beim Krabbeln, Kriechen, Sitzen, Stehen, Gehen und Rennen. Die motorische Skala überprüft außerdem die feinmotorische Manipulation beim Greifen, den altersgemäßen Gebrauch von Stiften und das Imitieren von Handbewegungen. Die Aufgaben sind nach Altersgruppen geordnet. Das Verfahren wird verbreitet bei der Nachsorge ehemaliger Frühgeborener eingesetzt und hat sich zur Planung von früher Förderung bewährt. Allerdings existieren bislang keine deutschen Normen.

Bayley III (Bayley Scales of Infant and Toddler Development III)

- Alter: 1–42 Monate
- Einsatz: Frühförderstellen, SPZ, neuropädiatrische Ambulanzen, Nachsorgeeinrichtungen, Forschung
- Ziel: Feststellung der motorischen und mentalen Entwicklung, von Entwicklungsverzögerungen, -störungen

Der Bayley III Entwicklungstest ist eine Weiterentwicklung des Bayley II Test und weist wie dieser kein einheitliches Theoriegebäude auf. Die kognitive Skala umfasst 91 Aufgaben, die Sprachskala (rezeptiv und expressiv) 97 Aufgaben, die Motorikskala (Feinmotorik, Grobmotorik) 138 Aufgaben. In begleitenden Fragebögen können die sozial-emotionale Entwicklung und Alltagsfertigkeiten erfasst werden. Die Testaufgaben werden in Stufenleiterform angeordnet, d. h. die Einstiegsaufgaben beginnen mit dem Alter des Kindes. Entsprechend der Ergebnisse in den ersten Testaufgaben wird auf der Stufenleiter vor- oder zurückgegangen. Eine Schulung des Untersuchers ist unbedingt erforderlich. Für die Gesamtskalen werden aus den Rohwerten standardisierte Werte (Mittelwerte und Standardabweichungen), Prozentränge und Konfidenzintervalle ermittelt, für die Subskalen können standardisierte Werte und das Entwicklungsalter berechnet werden. Bislang liegen nur eine englischsprachige Version und US-amerikanische Normen vor. Kinder erreichen im Bayley III – Test höhere Werte auf der kognitiven Skala als im Bayley II-Test, sodass mit dem Bayley III möglicherweise Entwicklungsretardierungen nicht erkannt werden.

MFED 1, MFED 2–3, MFED 4–6 (Münchener Funktionelle Entwicklungsdiagnostik)

- Alter: 1., 2–3. Jahr, 4.–6. Jahr
- Einsatz: kinderärztliche Praxen, Frühförderstellen, sozialpädiatrische Zentren
- Ziel: Erstellung eines Entwicklungsprofils, Feststellung von Entwicklungsverzögerungen, -störungen.

Die MFED basieren auf dem Prinzip der Stufenleiter der Entwicklung. Sie ermöglichen eine differenzierte Erfassung des kindlichen Entwicklungsstandes in verschiedenen Funktionsbereichen (1. Lebensjahr: Krabbeln, Sitzen, Laufen, Greifen, Perzeption, Sprechen, Sprachverständnis, Sozialverhalten; 2. und 3. Lebensjahr: Statomotorik, Handmotorik, Wahrnehmungsverarbeitung, Sprechen, Sprachverständnis, Selbstständigkeit, Sozialverhalten). Für die verschiedenen Bereiche kann für das 1.–3. Lebensjahr das jeweilige Entwicklungsalter bestimmt werden. Es wird die 90. (bzw. 95.) Perzentile der Altersnorm als Basis für die Einschätzung der Entwicklungsstandes empfohlen. Auf die Bestimmung eines globalen Entwicklungsalters oder eines Entwicklungsquotienten wird dabei verzichtet.
In der MFED 4–6 werden Feinmotorik, visuelle Intelligenz, Sprache, Logik, Zahlenverständnis und Selbstständigkeit geprüft. Zusätzlich wird ein Elternfragebogen eingesetzt. Das Entwicklungsprofil des einzelnen Kindes kann anhand der Perzentilen der Normstichprobe dargestellt werden. Aus den Testprofilen der MFED ergeben sich Hinweise auf die Ursachen der Retardierung, Ansatzpunkte zur Beratung der Eltern und zur Planung des therapeutischen Vorgehens.

TÜKI (Tübinger Luria-Christensen Neuropsychologische Untersuchungsreihe für Kinder)

- Alter: 5–16 Jahre
- Einsatz: sozialpädiatrische Zentren, neuropsychologische Diagnostik und Therapie, Forschung
- Ziel: umfassende Diagnostik und Differenzialdiagnose neuropsychologischer Störungen und ihrer spezifischen Lokalisation, Planung von Therapiemaßnahmen, zur Therapieverlaufskontrolle.

Die TÜKI gliedert sich in folgende 15 Untersuchungsbereiche: Gesamtkörperkoordination; motorische Funktionen der Hände; orale Praxie; sprachliche Regulation motorischer Vollzüge; akustisch-motorische Koordination; höhere hautkinästhetische Funktionen; Stereognosie; höhere visuelle Funktionen; räumliche Orientierung; räumliches Denken (Mosaiktest); rezeptive Sprache; expressive Sprache; Lernprozess (Wortreihe); mnestische Prozesse und Denkprozesse.

ET 6–6 (Entwicklungstest für Kinder von 6 Monaten–6 Jahren)

- Alter: 0,6–5,11 Jahre
- Einsatz: kinderärztliche Praxen, Frühförderstellen, sozialpädiatrische Zentren
- Ziel: differenzierte Erfassung von normaler Entwicklung, Entwicklungsdefiziten und individuellen Stärken.

Der ET 6–6 berücksichtigt konzeptuell die Variabilität normaler kindlicher Entwicklung und orientiert sich am Grenzsteinprinzip. Im Rahmen des ET 6–6 werden sieben Entwicklungsbereiche geprüft: Körpermotorik, Handmotorik, kognitive Entwicklung (Gedächtnis, Handlungsstrategien, Kategorisieren, Körperbewusstsein), Sprachentwicklung (rezeptiv, expressiv), Sozialentwicklung, emotionale Entwicklung; zusätzlich im Vorschulalter: Subtest Nachzeichnen. In der Auswertung wird bereichsspezifisch der prozentuale Anteil der gelösten Aufgaben auf altersspezifische Mittelwerte und Standardabweichungen bezogen. Als Ergebnis erhält man ein Entwicklungsprofil, das eine Einschätzung individueller Stärken oder Defizite und eine Visualisierung typischer Muster ermöglicht. Darüber hinaus wird zur spezifischeren Beurteilung eine Betrachtung des Testergebnisses auf Itemebene empfohlen.

BUEVA 4–6 (Basisdiagnostik für umschriebene Entwicklungsstörungen im Vorschulalter)

- Alter: 4–6 Jahre
- Einsatz: kinderärztliche Praxen, Frühförderstellen, sozialpädiatrische Zentren
- Ziel: Erkennung von Teilleistungsstörungen vor Schulbeginn.

In der BUEVA wurden besonders aussagekräftige Untertests aus bewährten Testverfahren zusammengestellt und so modifiziert, dass den Besonderheiten der Kinder im Vorschulalter Rechnung getragen wird. Bei Vierjährigen wird die allgemeine Intelligenz, die Artikulation, die expressive und die rezeptive Sprache sowie die Visuomotorik erfasst, bei Fünfjährigen wird zusätzlich die Aufmerksamkeitsleistung mit zwei Untertests erhoben. Ihr Differenzierungsoptimum weist die Testbatterie im subnormalen Leistungsbereich auf.

BUEGA 6–11 (Basisdiagnostik Umschriebener Entwicklungsstörungen im Grundschulalter)

- Alter: Klassen 1–5
- Einsatz: kinderärztliche Praxen, schulpsychologischer Dienst, sozialpädiatrische Zentren
- Ziel: Erkennung von Teilleistungsstörungen und Aufmerksamkeitsstörungen im Grundschulalter.

Folgende Leistungsbereiche werden geprüft: verbale Intelligenz, nonverbale Intelligenz, expressive Sprache, Lesen, Rechtschreibung, Rechnen und Aufmerksamkeit. Aus den Testergebnissen wird ein Gesamtwert gebildet, der aufgrund seiner hohen Korrelation mit Schulleistungen für die Schullaufbahnberatung herangezogen werden kann.

➤ Tabelle 17.3 fasst weitere Eigenschaften der besprochenen Verfahren zur Entwicklungsdiagnostik zusammen.

Tab. 17.3 Entwicklung – Fragebögen und Tests

Verfahren	Gütekriterien	Normen	Bearbeitungszeit (min)	aussagekräftig
EfkE	ja	ja	40	ja
EDB	ja	ja	15	ja
Denver	nein	ja	15–25	mäßig
Mann-Z-Test	nein	ja	5–10	mäßig
EVU	ja	ja	10–20	ja
Griffith	ja	ja	25–45	ja
Bayley II	ja	nein*	25–60	ja
Bayley III	ja	nein*	50–90	fraglich
MFED	nein	ja	50	ja
TÜKI	ja	ja	120	ja
ET 6–6	nein	ja	12–50	ja
BUEVA 4–6	ja	ja	20–25	ja
BUEGA 6–11	ja	ja	40–60	ja

* keine deutschen Normen

17.4 Intelligenz

17.4.1 Fragebogen

KOPKI 4–6 (Fragebogen zur Erfassung kognitiver Prozesse bei 4- bis 6-jährigen Kindern)

- Alter: 4,0–6,0 Jahre
- Einsatz: kinderärztliche Praxen, Psychologen, Erzieherinnen, Frühförderstellen, Beratungsstellen, SPZ
- Ziel: Erfassung kognitiver Prozesse auf Basis der Elterneinschätzung

Der KOPKI 4–6 ist ein Screeningverfahren. Er hilft bei der Indikationsstellung für globale und spezifische Förderung kognitiver Fähigkeiten sowie zur Verlaufsbeurteilung. Der Elternfragebogen hat 98 Items, die 6 Skalen zugeordnet werden können. Ergänzend besteht eine Symptomliste, eine Frage zur globalen Einschätzung des Kindes und allgemeine anamnestische Fragen. Zu beachten ist das häufig divergierende Urteil von Eltern und Erziehern.

17.4.2 Allgemeine Tests

K-ABC (Kaufman Assessment Battery for Children)

- Alter: 2,6–12,5 Jahre
- Einsatz: Psychologen, Frühförderstellen, Beratungsstellen, sozialpädiatrische Zentren, Forschung
- Ziel: Individualtest zur Messung von Intelligenz und erworbenen Fertigkeiten.

Die Grundlage der K-ABC ist die Definition der Intelligenz als Fähigkeit, Probleme durch geistiges Verarbeiten zu lösen. Bei der Diagnose steht der Prozess der Lösungsfindung und nicht der Inhalt der Aufgabe im Vordergrund. Die K-ABC besteht aus 16 Untertests, von denen je nach Alter

maximal 13 durchgeführt werden. Die K-ABC ist in vier Skalen gegliedert: Skala einzelheitlichen Denkens – SED (Untertests: Handbewegungen, Zahlen nachsprechen, Wortreihe), Skala ganzheitlichen Denkens – SGD (Untertests Zauberfenster, Wiedererkennen von Gesichtern, Gestaltschließen, Dreiecke, bildhaftes Ergänzen, räumliches Gedächtnis, Fotoserie). Diese beiden Skalen werden als Skala intellektueller Fähigkeiten – SIF zusammengefasst. Die Fertigkeitenskala – FS gliedert sich in die Untertests Wortschatz, Gesichter und Orte, Rechnen, Rätsel, Lesen/Buchstabieren, Lesen/Verstehen. Die sprachfreie Skala besteht aus einer Kombination der genannten Untertests. Damit ist eine Abschätzung der intellektuellen Fähigkeiten von fremdsprachlichen, hör- oder sprachentwicklungsgestörten und autistischen Kindern möglich.
Inzwischen wird wegen nicht mehr ganz aktueller Normierung des K-ABC bevorzugt der HAWIK – IV eingesetzt.

HAWIK-IV (Hamburg-Wechsler-Intelligenztest für Kinder – IV)

- Alter: 6–16,11 Jahre
- Einsatz: Psychologen, Frühförderstellen, Beratungsstellen, sozialpädiatrische Zentren, Forschung
- Ziel: Intelligenztest zur Erfassung allgemeiner und spezifischer intellektueller Fähigkeiten.

Der HAWIK-IV basiert auf der Definition von Intelligenz als Fähigkeit, zweckvoll zu handeln, vernünftig zu denken und sich mit der Umwelt wirkungsvoll auseinanderzusetzen. Er enthält insgesamt 15 Untertests. Die Erfassung der fünf Intelligenzwerte Sprachverständnis, wahrnehmungsgebundenes logisches Denken, Arbeitsgedächtnis, Verarbeitungsgeschwindigkeit sowie des Gesamt-IQ-Wertes ermöglicht ein differenziertes Bild des kognitiven Entwicklungsstandes. Zusätzlich können auf der Untertestebene weitere Profilanalysen und damit sehr differenzierte Aussagen über Stärken und Schwächen eines Kindes erfolgen. Prozessanalysen liefern Hinweise für eine gezielte Therapieplanung.

HAVIVA III (Hannover-Wechsler-Intelligenztest III für das Vorschulalter)

- Alter: 2,6–6,11 Jahre
- Einsatz: Psychologen, Frühförderstellen, Beratungsstellen, sozialpädiatrische Zentren
- Ziel: Untersuchung allgemeiner und spezifischer intellektueller Fähigkeiten.

Der HAVIVA basiert auf der gleichen Definition von Intelligenz wie der HAWIK-IV. Es können der Verbal-IQ, der Handlungs-IQ, der Gesamt-IQ sowie ein Quotient für die Verarbeitungsgeschwindigkeit und die allgemeine Sprache gebildet werden. Für die beiden Altersgruppen 2,6–3,11 und 4,0–6,11 steht jeweils eine altersgerechte Version zur Verfügung.

WPPSI – R III (Wechsler Preschool and Primary Scale of Intelligence – Revision III)

- Alter: 3–7,2 Jahre
- Einsatz: Psychologen, Frühförderstellen, Beratungsstellen, SPZ
- Ziel: Messung des kognitiven Entwicklungsstandes im Vorschulalter

Mit dem WPPSI-III werden allgemeine und spezifische intellektuelle Fähigkeiten erfasst. Es lassen sich der Gesamt-IQ, der Verbal- und der Handlungsteil, die Verarbeitungsgeschwindigkeit und eine allgemeine Sprachskala berechnen. Das Verfahren besteht aus 14 Untertests (Kerntests, optionale und zusätzliche Tests): Mosaik-Test, allgemeines Wissen, Matrizen-Test, Wortschatz-Test,

Bildkonzepte, Symbol-Suche, Begriffe erkennen, Symbole kodieren, Allgemeines Verständnis, Bilder ergänzen, Gemeinsamkeiten finden, passiver Wortschatz, Figuren legen, aktiver Wortschatz. Neben IQ-Werten können Prozentränge berechnet werden. Die Wechsler Skalen sind sehr verbreitet.

IDS (Intelligence and Development Scales – Intelligenz- und Entwicklungsskalen für Kinder von 5–10 Jahren)

- Alter: 5,0–10,11 Jahre
- Einsatz: Psychologen, Frühförderstellen, Beratungsstellen, SPZ
- Ziel: Feststellung eines Intelligenzwertes bzw. eines Entwicklungsprofils

Es werden spezifische Stärken und Schwächen in 6 Funktionsbereichen untersucht: Kognition, Psychomotorik, Sozial-Emotionale Kompetenz, Mathematik, Sprache, Leistungsmotivation. Der IDS eignet sich für folgende Fragestellungen: Minder-, Hochbegabung, Entwicklungsstörungen der Sprache, der Motorik, der schulischen Fertigkeiten, Fragen der geeigneten Schulform, hyperkinetische Störungen, Asperger-Syndrom, Störungen des Sozialverhaltens, emotionale Störungen.

17.4.3 Nonverbale Tests

CFT 1 – R (Grundintelligenztest Skala 1 – Revision)

- Alter: 5,3–11,11 Jahre
- Einsatz: Kindergärten, Beratungslehrer, Psychologen, Ärzte, Frühförderstellen, Beratungsstellen, SPZ
- Ziel: kultur- und sprachfreie Intelligenzmessung

Der CFT 1-R basiert auf dem Konzept zweier Intelligenzformen (flüssige und kristalline Intelligenz). Flüssige Intelligenz als Fähigkeit komplexe Beziehungen in neuartigen Situationen wahrzunehmen wird als weitgehend kultur- und sprachfreie Fertigkeit angesehen, kristalline Intelligenz umfasst Faktenwissen. Der CFT 1-R ist eine Weiterentwicklung des CFT 1. Er ermöglicht formallogische Denkprobleme nonverbal zu lösen. Der 1. Teil besteht aus den Untertests Substitutionen, Labyrinthe und Ähnlichkeiten, der 2. teil aus den Untertests Reihen fortsetzen, Klassifikationen und Matrizen. Im 1. teil wird die wahrnehmungsgebundene Leistung unter Zeitdruck gemessen, im 2. Teil das figurale denken. Die Gesamtleistung bildet die Grundintelligenz (gf-Faktor) ab.

CFT 20 (Grundintelligenztest Skala 2)

- Alter: 8,7–18 Jahre; 18–70 Jahren (Teil 1); Wortschatz- und Zahlenfolgentest von 8,7–15,6 Jahre
- Einsatz: Beratungslehrer, Psychologen, Beratungsstellen, sozialpädiatrische Zentren
- Ziel: Verwendung in der Schullaufbahn- und Berufsberatung, Hochbegabtendiagnostik, bei förderdiagnostischen Anliegen sowie bei klinisch definierten hirnorganischen Störungen.

Der CFT 20 erfasst das allgemeine intellektuelle Niveau (Grundintelligenz) im Sinne der flüssigen Intelligenz als weitgehend kultur- und sprachfreie Fertigkeit. Der CFT 20 besteht aus zwei gleichartig aufgebauten Testteilen mit je vier Untertests: Reihenfortsetzen, Klassifikationen, Matrizen und topologische Schlussfolgerungen. Das Testkonzept wurde durch Aufnahme zweier schulnaher Ergänzungstests erweitert: Durch Wortschatz- und Zahlenfolgenaufgaben können Elemente der Verarbeitungskapazität erfasst werden. Sie stellen eigenständige Teiltests dar, die unabhängig vom CFT 20 durchgeführt werden. Eine Revision (CFT 20-R) ist seit 2008 verfügbar.

CPM (Coloured Progressive Matrices)

- Alter: 3,9–11,8 Jahre
- Einsatz: Psychologen, Frühförderstellen, Beratungsstellen, sozialpädiatrische Zentren
- Ziel: kultur- und sprachfreie Intelligenztestung.

Der CPM erfasst die allgemeine und flüssige Intelligenz, d.h. Fähigkeiten zum Analogieschlussdenken, Regelerkennen, Prinzipienanwenden und induktiven und räumlichen Denken. Er besteht aus drei Sätzen zu je 12 Aufgaben mit steigendem Schwierigkeitsgrad. Es müssen komplexe Muster richtig vervollständigt werden, andere Aufgaben sind durch visuelle Vergleichsprozesse und logisch-induktive Strategien zu lösen.

SON-R 2½–7(Snijders-Oomen Non-verbaler Intelligenztest 2½–7)

- Alter: 2,6–7,11 Jahre
- Einsatz: Psychologen, Frühförderstellen, Beratungsstellen, sozialpädiatrische Zentren
- Ziel: sprachunabhängige Intelligenzdiagnostik, vor allem für Kinder mit Sprachentwicklungs-, Hörstörungen oder Intelligenzminderungen sowie Kinder mit nichtdeutscher Muttersprache.

Der SON-R 2½–7 ist nicht ausdrücklich an einem gängigen Intelligenzkonzept orientiert. Er untersucht folgende Bereiche: visuomotorische und perzeptive Fähigkeiten, räumliches Verständnis, Erkennen von Ordnungsprinzipien sowie die Fähigkeit zum abstrakten und konkreten Denken. Er beinhaltet die sechs Untertests: Mosaike, Kategorien, Puzzles, Analogien, Situationen und Zeichenmuster. Ergebnisse sind der Gesamt-Intelligenzquotient (SON-IQ) sowie Intelligenzwerte für die Denkskala und die Handlungsskala. Außerdem kann im Sinne eines Entwicklungsalters das Referenzalter eines Kindes bestimmt werden. Der SON-R 2½–7 weist einige leistungs- und konzentrationsfördernde Besonderheiten auf: genaue verbale und nonverbale Instruktionen, Feedback nach jeder Aufgabe, Demonstration richtiger Lösungen und adaptives Testen.

SON-R 5½–17 (Snijders-Oomen Non-verbaler Intelligenztest 5½–17)

- Alter: 5,6–17 Jahre
- Einsatz: Psychologen, Frühförderstellen, Beratungsstellen, sozialpädiatrische Zentren
- Ziel: sprachunabhängiger Intelligenztest für Kinder, die in der verbalen Kommunikation behindert sind oder Kinder, in deren Herkunftsfamilie nicht Deutsch gesprochen wird.

Der SON-R 5½–17 greift kein spezifisches Intelligenzkonzept auf. Er besteht aus sieben Subtests, die sich inhaltlich in vier Gruppen einteilen lassen: abstraktes Denken (Kategorien, Analogien),

Tab. 17.4 Intelligenz-Tests

Verfahren	Gütekriterien	Normen	Bearbeitungszeit (min)	aussagekräftig
KOPKI 4–6	ja	ja	15–25	ja
K-ABC	ja	ja	30–90	mäßig
HAWIK-IV	ja	ja	60–90	ja
HAVIVA III	ja	ja	30–90	ja
WPPSI – RIII	ja	ja	20–50	ja
IDS	ja	ja	90–120	ja
CFT 1-R	ja	ja	30–60	mäßig
CFT 20	ja	ja	25–55	mäßig
CPM	ja	ja	20–30	mäßig
Snijders-Oomen	ja	ja	45–90	ja

konkretes Denken (Situationen, Bildgeschichten), räumliches Vorstellungsvermögen (Mosaike, Zeichenmuster) und Perzeption (Suchbilder). Das Verfahren ermöglicht Angaben zum Zusammenhang der Testwerte bei Kindern mit Migrationshintergrund, sozialökonomischen Variablen und Gehörlosigkeit. Tabelle 17.4 (➤ Tab. 17.4) fasst weitere Eigenschaften der besprochenen Verfahren zur Intelligenzdiagnostik zusammen.

17.5 Motorik

Neurologische Untersuchung nach Touwen

- Alter: 3–12 Jahre
- Einsatz: Ärzte
- Ziel: Identifikation geringer neurologischer Funktionsstörungen.

Als Bestandteil der entwicklungsneurologischen Untersuchung werden standardisiert quantitative und qualitative Befunde in Ruhe und mit Funktionsprüfungen erhoben:
Gehen (Haltung, Gangbild, Abrollen der Füße, Seiltänzergang, Fersengang, Zehengang),
Stehen (Haltung, Positionsversuch in Pro-/Supination, assoziierte Bewegungen, Finger-Nase-Versuch, Finger-Finger-Versuch, Finger-Daumen-Versuch, Finger-Folge-Versuch, Kreisversuch, Schulterschlag, Einbeinstand, Hüpfen), Rumpf (Wirbelsäule, Bauchhautreflexe, Galant-Reflex, Kremasterreflex), Kopf (Gesichtsmuskulatur, Augen, Ohren, Mund), Sitzen (Haltung, Schulterschlag, Tonus, Bewegungsradien, Kraft, Kicken), Reflexe (Muskeleigenreflexe, Fußklonus, Plantarreaktion, Masseter-Reflex), Hirnnervenfunktionen, Liegen (Bein-/Fußhaltung in Rücken- und Bauchlage, Knie-Hacken-Versuch, Aufsetzen ohne Hilfe).

General Movements nach Prechtl

- Alter: Früh- und Termingeborene, Säuglinge bis ca. 20. Woche
- Einsatz: Physiotherapeuten, Ärzte, Früh- und Neugeborenenstationen, sozialpädiatrische Zentren, Nachsorgeeinrichtungen, Forschung
- Ziel: Diagnostik autonomer Bewegungen, Prognostik normaler/gestörter motorischer Entwicklung.

General Movements (GM) sind von der 9. Schwangerschaftswoche bis zur 21. Woche nach der Geburt nachweisbar. Mit dem Auftreten von Willkürbewegungen verschwinden sie.
GM sind große Bewegungen, die den ganzen Körper einbeziehen: Kopf, Rumpf, Arme und Beine. Diese Muster sind prinzipiell in der ganzen Nachweisperiode identisch, zeigen aber gewisse Charakteristika in verschiedenen Zeitphasen: „preterm" (bis zur 36.–38. SSW); „writhing" oder Drehbewegungen (38. SSW bis 6–8 Wochen nach Geburt); „fidgety" (6.–20. Woche). Die Qualität der GM dient als Parameter zur Beurteilung des aktuellen Hirnfunktionszustands und hat prognostische Bedeutung. Normale GM sind flüssige, elegante, glatt gerundete Bewegungen und kontinuierliche „Fidgety Movements" (FM) von geringer Amplitude, moderater Geschwindigkeit und variabler Ausprägung. Fehlende oder abnorme FM (verkrampft – synchronisiert) sind hoch indikativ für eine schlechte Prognose, normale FM sind ein ausgezeichneter Hinweis für einen neurologisch günstigen Verlauf. Die Beurteilung erfolgt aufgrund klinischer Beobachtung bzw. Videodokumentation. Eine gründliche Schulung ist erforderlich.

KTK (Körper-Koordinationstest für Kinder)

- Alter: 5–14 Jahre
- Einsatz: Pädagogen, Therapeuten
- Ziel: Messung des Entwicklungsstandes der Gesamtkörperbeherrschung und -kontrolle von Kindern ohne und mit Behinderungen.

Der KTK besteht aus den vier Untertests Balancieren rückwärts (BR), Monopedales Überhüpfen (MÜ), Seitliches Hin- und Herspringen (SH) und Seitliches Umsetzen (SU), die sämtlich das Merkmal Gesamtkörperbeherrschung erfassen. Es existieren Standardwerte (MW = 100, SD = 15) und Prozentrangwerte für 5- bis 14-Jährige.

MOT 4–6 (Motoriktest für vier- bis sechsjährige Kinder)

- Alter: 4–6 Jahre, bei behinderten Kindern bis 7–8 Jahre
- Einsatz: Pädagogen, Therapeuten
- Ziel: Beurteilung der Grob- und Feinmotorik von Kindern ohne und mit Behinderungen.

Der MOT 4–6 besteht aus 18 Testaufgaben, die folgende Dimensionen der Motorik überprüfen: gesamtkörperliche Gewandtheit und Beweglichkeit; feinmotorische Geschicklichkeit; Gleichgewichtsvermögen; Reaktionsfähigkeit; Sprungkraft und Schnelligkeit; Bewegungsgenauigkeit; Koordinationsfähigkeit. Es liegen Normen aus der Zeit vor 1987 in Halbjahresschritten für den Altersbereich von 4,0–6,11 Jahren vor, die möglicherweise heute nicht mehr zutreffen, da die motorischen Fähigkeiten in diesem Alter nachgelassen haben.

Movement ABC II (Movement Assessment Battery for Children – 2. Auflage)

- Alter: 3–16,11 Jahre
- Einsatz: Therapeuten, Frühförderstellen, sozialpädiatrische Zentren
- Ziel: Untersuchung der motorischen Leistungsfähigkeit für Kinder und Jugendliche.

In jeder Altersgruppe werden acht Aufgaben gestellt, mit denen drei Komponenten motorischer Funktionen untersucht werden können: Handgeschicklichkeit, Ballfertigkeit, Fähigkeit zur statischen und dynamischen Balance. Für alle drei Komponenten sowie den gesamten Testwert werden dem Alter angepasste Standardwerte und Prozentränge angegeben.

Zürcher Neuromotorik-Test

- Alter: 5–18 Jahre
- Einsatz: entwicklungs-, sozialpädiatrische Zentren, Forschung
- Ziel: Beurteilung von motorischen Ungeschicklichkeiten und neurologischen Störungen.

Bestandteile des Zürcher Neuromotorik-Tests sind im Stehen: repetitives Fußklopfen, alternierendes Zehen-Hacken-Klopfen, repetitives Handklopfen, alternierende Pro- und Supination des Unterarms, repetitive Fingerbewegungen, sequentielle Fingerbewegungen; am Tisch: Steckbrett; im Stehen: statische Balance (Einbeinstand), dynamische Balance (Hüpfen), Diadochokinese; Stressgehen: gehen auf Zehen, Fersen, Außen- und Innenseite des Fußes. Das standardisierte Testverfahren bestimmt Leistungsfähigkeit (Zeit) und Qualität (Bewegungsqualität, Mitbewegungen). Es gibt getrennte Normwerte für Jungen und Mädchen. Die Auswertung einiger Aufgaben erfolgt getrennt nach dominanter und nichtdominanter Hand.

GMFM (Gross Motor Function Measure)

- Alter: 1–18 Jahre
- Einsatz: Physiotherapeuten, sozialpädiatrische Zentren, neuropädiatrische Ambulanzen, Forschung
- Ziel: Bestimmung grobmotorischer Fähigkeiten bei Kindern mit Zerebralparese und Schädel-Hirn-Traumen sowie deren Veränderungen.

Der GMFM untersucht die Kinder standardisiert nach 5 Subskalen: A. Liegen und Drehen; B. Sitzen; C. Krabbeln und Knien; D. Stehen; E. Gehen, Rennen und Hüpfen. Die ursprüngliche Version beinhaltet 88 Items, eine spätere Modifikation 66 Items. Die Bewertung jeder Aufgabe erfolgt in 4 Stufen. Es wird ein Prozentwert der maximal erreichbaren Punktzahl errechnet. Ein motorisch normal entwickeltes Kind erfüllt mit etwa 5 Jahren alle Items des GMFM vollständig. Eine gründliche Schulung ist erforderlich.

GMFCS (Gross Motor Function Classification System)

- Alter: 1–18 Jahre
- Einsatz: Physiotherapeuten, sozialpädiatrische Zentren, neuropädiatrische Ambulanzen, Nachsorgeeinrichtungen, Forschung
- Ziel: Klassifikation von Schweregraden der Bewegungsstörungen bei Kindern mit Zerebralparese.

Die GMFCS-Klassifizierung unterscheidet fünf Stufen (I–V). Die Grundlage der Klassifizierung sind funktionale Fertigkeiten des Kindes, Bedarf an Hilfsmitteln einschließlich der Fortbewegungsmittel (Rollator, Krücken, Stöcke, Rollstuhl), weniger die Bewegungsqualität. Der GMFCS-Klassifizierung nach schließt die Stufe-I-Kinder ein, die es vermögen, ohne Unterstützung zu laufen, die jedoch keine komplexeren motorischen Aufgaben meistern. Der Stufe V entspricht ein Kind mit sehr beschränkten Möglichkeiten der selbstständigen Fortbewegung, auch bei Anwendung von Hilfsmitteln, mit geringer Kopf- und Rumpfkontrolle. Da die motorische Entwicklung altersabhängig ist, wird jede Stufe der Klassifizierung für die jeweilige Altersgruppe (bis 2 Jahre, 2–4 Jahre, 4–6 Jahre, 6–12 Jahre) gesondert beschrieben.

AHA (Assisting Hand Assessment)

- Alter: 18 Monate–12 Jahre
- Einsatz: Ergotherapeuten, Physiotherapeuten
- Ziel: Messung der Handfunktion bei Kindern mit unilateraler Bewegungsstörung.

Das AHA ist ein standardisiertes Testverfahren, in dem Spielmaterial eingesetzt wird, das beidhändiges Manipulieren erfordert. Dabei wird die Effektivität der betroffenen Hand (assisting hand) bewertet und mit der der gesunden Hand verglichen. Die beobachtete Ausführung der Spielsituation wird mittels Videoanalyse ausgewertet. Das AHA besteht aus 22 Items mit einer 4-Punkte-Skala. Das AHA wird bei Kindern mit Plexusparese oder Zerebralparese eingesetzt.

MACS (Manual Ability Classification System)

- Alter: 4–18 Jahre
- Einsatz: Ergotherapeuten, sozialpädiatrische Zentren, neuropädiatrische Ambulanzen, Forschung
- Ziel: Beurteilung der Handfunktion bei Kindern mit Zerebralparese.

Die MACS-Klassifizierung unterscheidet fünf Stufen (I–V). Die Grundlage der Klassifizierung sind funktionale Fertigkeiten des Kindes. Die MACS-Klassifizierung schließt in Stufe-I-Kinder ein, die Objekte leicht und erfolgreich „händeln", jedoch geringe Beeinträchtigungen zeigen (z. B. bei sehr kleinen, schweren oder zerbrechlichen Objekten). Der Stufe V entspricht ein Kind ohne aktive Handfunktion mit vollständigem Hilfsbedarf.

HDT – Hand-Dominanz-Test

- Alter: 6–10 Jahre
- Einsatz: Pädagogen, Ergotherapeuten, Ärzte
- Ziel: Feststellung bzw. Ausprägung der Händigkeit.

Der HDT ist ein Papier-und-Bleistift-Verfahren, das die Leistungsüberlegenheit einer Hand prüft. Der HDT besteht aus drei Untertests: Spuren nachzeichnen, Kreise punktieren und Quadrate punktieren. Die Leistungsüberlegenheit einer Hand wird rechnerisch bestimmt, wobei die Handdominanz theoretisch von -100 (extreme Linkshändigkeit) bis +100 (extreme Rechtshändigkeit) variieren kann.

17.6 Visuelle Funktionen

FEW-II (Frostigs Entwicklungstest der visuellen Wahrnehmung II)

- Alter: 4–9 Jahre
- Einsatz: Ergotherapeuten, Psychologen, Frühförderstellen, sozialpädiatrische Zentren
- Ziel: Beurteilung der kindlichen Entwicklung hinsichtlich der visuellen Wahrnehmung.

Der FEW-II hat konzeptionelle Mängel und methodische Schwierigkeiten des FEW überwunden, indem er motorikfreie (motorikreduzierte) und motorikabhängige Anteile unterscheidet. Er besteht aus 8 Subtests: 1. Auge-Hand-Koordination, 2. Lage im Raum, 3. Abzeichnen, 4. Figur-Grund, 5. Räumliche Beziehungen, 6. Gestaltschließen, 7. Visuo-motorische Geschwindigkeit, 8. Formkonstanz. Jeder der acht Subtests misst einen Typ visueller Wahrnehmungsfähigkeit, klassifizierbar als Lage im Raum, Formkonstanz, räumliche Beziehungen oder Figur-Grund. Der

Tab. 17.5 Motorik und visuelle Wahrnehmung – Tests

Verfahren	Gütekriterien	Normen	Bearbeitungszeit (min)	aussagekräftig
Touwen	nein	nein	10–15	ja
Prechtl	ja	qualitativ	25–50	ja
KTK	ja	ja	20	mäßig
MOT 4–6	ja	veraltet	15–20	ja
Movement ABC II	ja	ja	20–30	ja
LOS KF 18	ja	veraltet	30	ja
Zürcher NMT	ja	ja		ja
GMFM	ja	-	45–60	ja
GMFCS	ja	-	< 5	ja
AHA	ja	ja	10–15	ja
MACS	ja	-	< 5	ja
HDT	nein	veraltet	10	mäßig
FEWII	ja	ja	30–50	ja

FEW-II ist die leicht modifizierte deutsche Bearbeitung des Developmental Test of Visual Perception 2 (DTVP-2). Tabelle 17.5 (➤ Tab. 17.5) fasst weitere Eigenschaften der besprochenen Verfahren zur Diagnostik von Motorik und visuellen Funktionen zusammen.

17.7 Sprache

17.7.1 Fragebögen

ELFRA 2 (Elternfragebogen für die Früherkennung von Risikokindern)

- Alter: 24 Monate (U7)
- Einsatz: kinderärztlichen Praxen, Frühförderung, Forschung
- Ziel: Früherkennung von Risikokindern, insbesondere bzgl. Der Sprachentwicklung.

Der ELFRA 2 für 24 Monate alte Kinder misst den produktiven Wortschatz, Syntax und Morphologie, er dient der Identifikation der „late talker". Bei Kindern mit Entwicklungsstörungen kann der Elternfragebogen auch noch später eingesetzt werden. Es existiert eine Kurzversionen für die kinderärztliche Praxis. Dem Verfahren liegen Ratgeber für Eltern von Kindern mit einer Sprachentwicklungsverzögerung bei.

ELAN (Eltern Antworten – Elternfragebogen zur Wortschatzentwicklung im frühen Kindesalter)

- Alter: 16–26 Monate
- Einsatz: Frühförderstellen, Sprachtherapeuten, kinderärztliche Praxen, Forschung
- Ziel: Erfassung des expressiven Wortschatzes zur frühen Identifikation von Risikokindern für eine Sprachentwicklungsstörung.

Der von den Eltern auszufüllende Fragebogen besteht aus einem ausführlichen anamnestischen Teil sowie aus einer Checkliste mit 250 Wörtern, die unterschiedlichen Wortarten und -feldern zuzuordnen sind. Ergänzungen zum individuellen Sprachgebrauch des Kindes können vorgenommen werden.

SBE-2-KT (Elternfragebogen zur Früherkennung von Late-Talkers)

- Alter: 21–24 Monate
- Einsatz: Frühförderstellen, Sprachtherapeuten, kinderärztliche Praxen, Forschung
- Ziel: Früherkennung von Kindern mit Sprachentwicklungsverzögerungen.

Der Fragebogen besteht aus einer Liste von 57 Wörtern und einer Frage zu Mehrwortsätzen. Fragebogen und Handbuch sind kostenlos aus dem Internet zu beziehen (http://www.kjp.med.uni-muenchen.de/sprachstoerungen/SBE-2-KT.php)

17.7.2 Screening

SSV (Sprachscreening für das Vorschulalter)

- Alter: 3–5,11 Jahre
- Einsatz: Sprachtherapeuten, Psychologen, Kindergärten und andere vorschulische Einrichtungen, Frühförderstellen, kinderärztliche Praxen
- Ziel: Screening eines Risikos für eine Störung der Sprachentwicklung.

Es werden für den gestörten Spracherwerb als grundlegend betrachtete Aspekte erfasst: bei Dreijährigen das phonologisches Arbeitsgedächtnis für Nichtwörter (PGN) und die morphologische Regelbildung, bei Vier- und Fünfjährigen PGN und Satzgedächtnis.
Das SSV stellt eine Kurzform des SETK 3–5 dar und teilt mit diesem Gesamttest inhaltliche und methodische Eigenschaften.

HASE (Heidelberger auditives Screening in der Einschulungsdiagnostik)

- Alter: ab 4 Jahren
- Einsatz: Sprachtherapeuten, Psychologen, Kindergärten, andere vorschulische Einrichtungen, Frühförderstellen, kinderärztliche Praxen, SPZ, Einschulungsuntersuchungen
- Ziel: Screening auf Sprach- bzw. Schriftspracherwerbsstörungen.

HASE enthält vier Aufgabengruppen: Nachsprechen von Sätzen, Nachsprechen von Kunstwörtern, Wiederholen von Zahlenfolgen und Erkennen von Wortfamilien. Diese Aufgaben haben sich als Indikatoren für die allgemeine Sprachleistungsfähigkeit sowie für die phonologische Schleife des Arbeitsgedächtnisses erwiesen, die für die Erkennung von Sprach- und Schriftspracherwerbsstörungen relevant sind. Die Items wurden von einer professionellen Sprecherin auf Tonträger gesprochen.

HVS (Heidelberger Vorschulscreening zur auditiv-kinästhetischen Wahrnehmung und Sprachverarbeitung)

- Alter: 5–7 Jahre
- Einsatz: Sprachtherapeuten, Psychologen, Phoniater, Frühförderstellen, sozialpädiatrische Zentren
- Ziel: Screening auf Spracherwerbsstörungen.

Es werden sprachanalytische und auditiv-kinästhetische Wahrnehmungsfähigkeiten untersucht, die im Zusammenhang mit Sprachentwicklungsverzögerungen/-störungen, Artikulationsstörungen, phonologischen Störungen, Leserechtschreibstörungen und Hörstörungen stehen. Das Verfahren untersucht: auditive Merkspanne, expressive Anlautanalyse, Silben segmentieren, phonematische Differenzierung, Artikulomotorik, Wortfamilien erkennen, Reimwörter erkennen. Die Items wurden von einer professionellen Sprecherin auf Tonträger gesprochen.

SELDAK (Sprachentwicklung + Literacy bei deutschsprachig aufwachsenden Kindern)

- Alter: 4–7 Jahre
- Einsatz: Kindergärten und andere vorschulische Einrichtungen
- Ziel: Beobachtungsbogen zur Sprachstandserhebung bei deutschsprachig aufwachsenden Kindern.

Im ersten Teil des Beobachtungsbogens werden sprachrelevante Situationen, im zweiten Teil sprachliche Kompetenzen im engeren Sinn wie Wortschatz und Grammatik erfasst. Der Bogen ist für Kinder mit Deutsch als Erstsprache einsetzbar. Im Begleitheft werden im ersten Teil theoretische Grundlagen und Konzeption des Bogens dargestellt, der zweite Teil enthält Anregungen für die Förderung der Lese- und Schreibfähigkeit.

SISMIK (Sprachverhalten und Interesse an Sprache bei Migrantenkindern in Kindertageseinrichtungen)

- Alter: 3½ Jahre bis Schulalter
- Einsatz: Beobachtungsbogen für Kindergärten und Kindertagesstätten
- Ziel: systematische Begleitung der Sprachentwicklung von Migrantenkindern.

Der Bogen enthält Fragen zur kindlichen Sprache und Literacy (kindliche Erfahrungen rund um Buch-, Erzähl-, Reim- und Schriftkultur). Ein Begleitheft gibt praktische Hinweise für das Vorgehen bei der Beobachtung; verschiedene Ebenen der Nutzung (z. B. Reflexion des pädagogischen Angebots, Zusammenarbeit im Team); Anleitung für die qualitative und quantitative Auswertung des Bogens (mit Fallbeispielen); Hinweise für die Förderung.

SCREEMIK 2 Screening der Erstsprachfähigkeit bei Migrantenkindern (Russisch-Deutsch, Türkisch-Deutsch)

- Alter: 4–5,11 Jahre
- Einsatz: Lehrer, Sprachheilpädagogen, Sprachtherapeuten, Ärzte, sozialpädiatrische Zentren
- Ziel: Einschätzung der Sprachfähigkeit bei Migrantenkindern mit den Erstsprachen Russisch oder Türkisch.

Das Screening überprüft die phonetisch-phonologische, morphologisch-syntaktische und semantisch-lexikalische Ebene in der Erstsprache, ist kindgerecht als Spiel und für die Praxis konzipiert und einfach in der Durchführung und Bedienung. Es ermöglicht durch die computergestützte Umsetzung deutschsprachigen Fachpersonen ohne Kenntnisse der Erstsprache des Kindes und ohne externen Übersetzer eine differenzialdiagnostische Abgrenzung zwischen erwerbsbedingten Auffälligkeiten in der Zweitsprache Deutsch und einer Sprachentwicklungsstörung, die sich sprachübergreifend in allen Sprachen eines mehrsprachigen Kindes äußert. Das PC-gestützte Testverfahrens wird für die Praxis als ökonomisch und praktikabel beschrieben.

17.7.3 Tests

SETK 2 (Sprachentwicklungstest für zweijährige Kinder)

- Alter: 2,0–2,11 Jahre, ältere Kinder mit Entwicklungsstörungen (u. a. Hörschädigungen, geistige Behinderung, Autismus, sensorische Behinderungen)
- Einsatz: Sprachheilpädagogen, Sprachtherapeuten, Psychologen, Frühförderstellen, sozialpädiatrische Zentren
- Ziel: Untersuchung des sprachlichen Entwicklungsstandes und Abklärung von Entwicklungsstörungen im sprachlichen Bereich.

Der SETK 2 erfasst Wort- und Satzverständnis sowie -produktion anhand von Bildvorlagen. In vier Untertests (Verstehen I, Verstehen II, Produktion I und Produktion II) werden die rezeptive und produktive Sprachverarbeitungsfähigkeit bestimmt.

SETK 3–5 (Sprachentwicklungstest für drei- bis fünfjährige Kinder)

- Alter: 3–5,11 Jahre, ältere Kinder mit Entwicklungsstörungen
- Einsatz: Sprachheilpädagogen, Sprachtherapeuten, Psychologen, Frühförderstellen, sozialpädiatrische Zentren

- Ziel: Untersuchung des sprachlichen Entwicklungsstandes und Abklärung von Entwicklungsstörungen im sprachlichen Bereich.

Der SETK 3–5 erfasst mit vier Untertests bei den dreijährigen Kindern und mit fünf Untertests bei den vier- bis fünfjährigen Kindern rezeptive und produktive Sprachverarbeitungsfähigkeiten sowie auditive Gedächtnisleistungen. Im Einzelnen werden untersucht: Satzverständnis, Enkodierung semantischer Relationen, morphologische Regelbildung, phonologisches Arbeitsgedächtnis, Gedächtnisspanne für Wortfolgen, Satzgedächtnis. Mit dem SETK 3–5 ist es möglich, das erreichte Sprachentwicklungsniveau festzustellen und in einen kausalen Erklärungszusammenhang mit auditiven Gedächtnisleistungen zu bringen.

AWST-R (Aktiver Wortschatztest für 3- bis 5-jährige Kinder – Revision)

- Alter: 3–5,5 Jahre
- Einsatz: Sprachheilpädagogen, Sprachtherapeuten, Psychologen, Frühförderstellen, sozialpädiatrische Zentren
- Ziel: Individualtest zur Beurteilung des expressiven Wortschatzumfangs im Kindergartenalter.

Als Bildvorlagen dienen fotografische Darstellungen von insgesamt 51 Substantiven und 24 Verben. Die quantitative Auswertung zur Beurteilung des expressiven Wortschatzes wurde um eine qualitative Auswertung erweitert (z. B. Wortbedeutungswissen). Die qualitative Auswertung erweist sich als nützlich zur Therapieplanung und -gestaltung. Ein unterdurchschnittliches Testergebnis kann bedeuten: (1) Indikationsstellung zu einer logopädischen Therapie, (2) Objektivierung von Wortschatzdefiziten in der deutschen Sprache z. B. bei Migrantenkindern, damit (3) Ausgangspunkt für lexikalische Sprachfördermaßnahmen in Frühfördereinrichtungen, Kindergärten, vorschulischen Einrichtungen oder Beratungsstellen.

LUT (Lautunterscheidungstest für Vorschulkinder)

- Alter: 4–7 Jahre
- Einsatz: Vorschulerzieher, Sprachtherapeuten, Psychologen, Frühförderstellen, sozialpädiatrische Zentren
- Ziel: Gruppenverfahren zur Überprüfung der Lautunterscheidungsfähigkeit.

Die Kurzform (LUT) besteht aus 17 Aufgaben, in denen jeweils ein repräsentativ ausgewählter Laut überprüft wird. Mit seiner Hilfe wird geprüft, ob die Lautunterscheidungsleistung eines Vorschulkindes seiner Altersnorm entspricht. Gleichzeitig kann beurteilt werden, ob bei mangelhafter Lautbildungsleistung eine Teil- oder Hauptursache eine Lautunterscheidungsschwäche ist. Mit der diagnostischen Testform (DLUT), die aus 32 Aufgaben besteht, kann bei Kindern, deren Lautunterscheidungsleistung mangelhaft entwickelt ist, die Lautunterscheidungsleistung exakt und systematisch überprüft und somit eine Grundlage zur gezielten Förderung gefunden werden.

TROG-D (Test zur Überprüfung des Grammatikverständnisses)

- Alter: 3–11 Jahre
- Einsatz: Sprachtherapeuten, sozialpädiatrische Zentren
- Ziel: Überprüfung des Grammatikverständnisses.

Der TROG-D ist ein rezeptiver Sprachtest, der das Verständnis für grammatikalische Strukturen des Deutschen, die durch Flexion, Funktionswörter und Satzstellung markiert werden, überprüft. Er beinhaltet eine quantitative und eine qualitative Analyse der Probleme. Das Verfahren besteht aus 84 Testitems, die mit je 3 Ablenkern versehen sind. Zum Testsatz muss aus 4 Bildern das passende gefunden werden.

HSET (Heidelberger Sprachentwicklungstest)

- Alter: 3–9 Jahre, bei Kindern mit Entwicklungs- und Lernstörungen auch noch in höheren Altersstufen
- Einsatz: Sprachheilpädagogen, Sprachtherapeuten, Psychologen, Frühförderstellen und sozialpädiatrische Zentren
- Ziel: differenzierenden Erfassung der sprachlichen Fähigkeiten.

Der HSET ist ausführlich und umfassend. Er besteht aus 13 Untertests, die über die Diagnose des beobachtbaren Sprachverhaltens hinaus ermöglichen, Aussagen über die dem Verhalten zugrunde liegenden Wissensvoraussetzungen und Verarbeitungsmechanismen zu treffen. Die Untertests sind: 1. Verstehen grammatischer Strukturen, 2. Singular-Plural-Bildung, 3. Imitation grammatikalischer Strukturformen, 4. Korrektur semantisch inkonsistenter Sätze (nur für Kinder über 5 Jahre), 5. Bildung von Ableitungsmorphemen, 6. Benennungsflexibilität (nur für Kinder über 5 Jahre), 7. Begriffsklassifikation, 8. Adjektivableitung, 9. In-Beziehung-Setzung von verbaler und nonverbaler Information, 10. Enkodierung und Rekodierung gesetzter Intention (nur für Kinder über 5 Jahre), 11. Satzbildung, 12. Wortfindung, 13. Textgedächtnis.

PET (Psycholinguistischer Entwicklungstest)

- Alter: 3–10 Jahre
- Einsatz: Sprachheilpädagogen, Sprachtherapeuten, Psychologen, Frühförderstellen und sozialpädiatrische Zentren
- Ziel: Ermittlung spezifischer sprachlicher Fertigkeiten und Störungen von gesunden und lernbehinderten Kindern.

Der PET besteht aus folgenden Untertests: Wortverständnis; Bilder deuten; Sätze ergänzen; Bilder zuordnen; Gegenstände beschreiben; Gegenstände handhaben; Grammatik-Test; Wörter ergänzen; Laute verbinden; Objekte finden; Zahlenfolgen-Gedächtnis und Symbolfolgen-Gedächtnis. Die Testergebnisse sind Ausgangspunkt für die Gestaltung eines individuellen Trainingsprogramms.

P-ITPA (Potsdam-Illinois Test für Psycholinguistische Fähigkeiten)

Alter: 4 Jahre bis Ende der 5. Schulklasse
Einsatz: Sprachheilpädagogen, Sprachtherapeuten, Psychologen, Frühförderstellen, SPZ
Ziel: Feststellung von Auffälligkeiten im sprachlichen und schriftsprachlichen Bereich
Der P-ITPA ist aus dem Psycholinguistischen Entwicklungstest (PET) hervorgegangen. Geprüft werden: verbale Intelligenz, Wortschatz, Expressive Sprache, Phonologische Bewusstheit, Verbales Kurzzeitgedächtnis, Lesen (sinnhaft und sinnfrei), Rechtschreiben (Sinnhaft und sinnfrei). Er ermöglicht die Identifikation von Entwicklungsstörungen als Voraussetzung für die Einleitung geeigneter Interventionen.

H-LAD (Heidelberger Lautdifferenzierungstest)

- Alter: Kinder der Klassenstufen 1–4
- Einsatz: Sprachheilpädagogen, Sprachtherapeuten, Legasthenietherapeuten, sozialpädiatrische Zentren

Tab. 17.6 Sprache – Fragebögen und Tests

Verfahren	Gütekriterien	Normen	Bearbeitungszeit (min)	aussagekräftig
ELFRA 2	ja	ja	10–20	ja
ELAN	ja	ja	25	ja
SBE-2-KT	ja	ja	10	ja
SSV	ja	ja	10	ja
HASE	ja	ja	10	ja
SCREEMIK 2	ja	ja	15–20	ja
SETK 2	ja	ja	15–25	ja
SETK 3–5	ja	ja	20–30	ja
AWST-R	ja	ja	15	ja
LUT	ja	ja	12	ja
TROG-D	ja	ja	10–20	ja
HSET	ja	ja	40–80	ja
PET	nein	veraltet	15–20	ja
P-ITPA	ja	ja	20–60	ja
H-LAD	ja	ja	15–25	ja

- Ziel: Identifikation von Lautdiskriminationsfehlern innerhalb der Sprach- und Legastheniediagnostik.

Im H-LAD wird die Lautdifferenzierungsfähigkeit untersucht. Minimalpaare sollen auditiv als gleich oder ungleich erkannt und anschließend nachgesprochen werden. Die zu unterscheidenden Laute bestehen in Konsonanten mit gleicher Artikulationsstelle (z. B.: d/t, g/k) und Konsonanten mit unterschiedlicher Artikulationsstelle (z. B. t/k, m/n). Die Minimalpaare enthalten sowohl reale Wörter als auch Kunstwörter. In der Lautanalyse müssen die ersten beiden Konsonanten von Wörtern mit Konsonantenhäufungen genannt werden. Die Tests wurden professionell auf einem Tonträger gesprochen. Durch die Systematik des Testaufbaus lassen sich spezifische therapeutische Interventionen ableiten. Tabelle 17.6 (➤ Tab. 17.6) zeigt weitere Eigenschaften der besprochenen Verfahren zur Sprachdiagnostik.

17.8 Aufmerksamkeit

Das ADHS als Konstrukt mit unterschiedlichen neuropsychologischen Modellen von Aufmerksamkeit und unterschiedlichen Kriterien in ICD-10 und DSM-IV sowie der Notwendigkeit einer Beurteilung in unterschiedlichen Lebensbereichen lässt sich nur durch eine Kombination verschiedener diagnostischer Verfahren wahrscheinlich machen. Daher besitzt jedes einzelne Verfahren nur eine mäßige Aussagekraft (➤ Tab. 17.7).

17.8.1 Fragebögen

CRS (Conners-Rating Skalen)

- Alter: 3–17 Jahre
- Einsatz: Psychologen, Ärzte
- Ziel: Erfassung problematischen Verhaltens, insbesondere von Symptomen des ADHS.

Es existieren Eltern-, Lehrer-, Erzieherversionen sowie Selbstbeurteilungsbögen. Die verbreitete Kurzform erfasst in einem 4-stufigen Verfahren anhand von 10 Fragen wesentliche Aspekte des kindlichen Problemverhaltens. Ab 15 Punkten wird die Diagnose ADHS unterstützt.

Du Paul-ADHD Rating Skala IV

- Alter: 5–17 Jahre
- Einsatz: Psychologen, Ärzte
- Ziel: Erfassung problematischen Verhaltens, insbesondere von Symptomen des ADHS.

Auf Basis der DSM-IV-Klassifikation werden anhand von 18 Fragen mit 4-stufigen Antwortmöglichkeiten Eltern bzw. Lehrern nach den Kernsymptomen des ADHS (Aufmerksamkeitsstörung, Hyperaktivität und Impulsivität) befragt. Die Diagnose ADHS wird unterstützt, wenn jeweils in den Bereichen Unaufmerksamkeit bzw. Hyperaktivität/Impulsivität bei 6 von 9 Fragen das Problemverhalten häufig oder sehr häufig beschrieben wird.
Weitere Fragebögen zur ADHS-Diagnostik (FBB-HKS, SBB-HKS, TRF) sind in Kapitel 17.12 aufgeführt (➤ Kap. 17 .12).

17.8.2 Tests

ZVT (Zahlen-Verbindungstest)

- Alter: 8–95 Jahre
- Einsatz: Psychologen
- Ziel: Diagnostik von Hirnleistungsstörungen, schulische Differenzialdiagnostik, Entwicklungsdiagnostik.

Der ZVT dient der Erfassung der basalen, allen Intelligenzleistungen zugrunde liegenden, weitgehend milieuunabhängigen und genetisch bedingten kognitiven Leistungsgeschwindigkeit. Diese korrespondiert mit Fähigkeiten wie flüssige Intelligenz und Bearbeitungsgeschwindigkeit. Der Test ist sehr ökonomisch durchzuführen.

D2 (Test d2 – Aufmerksamkeits-Belastungs-Test)

- Alter: ab 9 Jahre
- Einsatz: Psychologen
- Ziel: Untersuchung der Aufmerksamkeit und Konzentrationsfähigkeit.

Der Test d2 misst Tempo und Sorgfalt des Arbeitsverhaltens bei der Unterscheidung ähnlicher visueller Reize (Detail-Diskrimination von d-p mit unterschiedlicher Zahl und Anordnung von vertikalen Strichen darüber/darunter) und ermöglicht damit die Beurteilung individueller Aufmerksamkeits- und Konzentrationsleistungen. Absicherung der Testgütekriterien, einfache Anwendung und geringer Aufwand an Zeit und Material sind Gründe dafür, dass der Test d2 zu den am häufigsten verwendeten psychodiagnostischen Verfahren in Deutschland gehört.

TPK (Testreihe zur Prüfung der Konzentrationsfähigkeit)

- Alter: 2.–6. Klasse
- Einsatz: Sonderpädagogen, Psychologen, Forschung
- Ziel: Prüfung der Konzentrationsfähigkeit.

Die Testreihe zur Prüfung der Konzentrationsfähigkeit ermöglicht die Erfassung von Leistungsmenge, Leistungsgüte und Stabilität bei wechselnden konzentrativen Anforderungen im zeitlichen Umfang einer knappen Schulstunde. Die Entwicklung der TPK orientiert sich an der im Schulunterricht geforderten Konzentrationsleistung. Die TPK besteht aus drei unterschiedlich strukturierten Untertests. Die Kinder müssen einen Text abschreiben, aus einer vorgelesenen Geschichte Tiernamen wiedergeben und einfache Additions- und Subtraktionsaufgaben lösen. Es können insgesamt sieben Messwerte erhoben werden, die eine qualitative Profildarstellung ermöglichen. Aufgrund der Iteminhalte ist das Verfahren jedoch für Kinder mit Lese-Rechtschreib-Schwäche und Dyskalkulie nicht geeignet.

CPT (Continuous Performance Test)

Alter: 4,0–20,11 Jahre
Einsatz: Psychologen, Beratungsstellen, SPZ
Ziel: Erfassung exekutiver Funktionen
Der CPT ist ein computergestütztes Verfahren zur Erfassung der selektiven Aufmerksamkeit, der Daueraufmerksamkeit und des impulsiven Verhaltens. Der Proband hat auf eine Taste zu reagieren, wenn auf dem Bildschirm auf ein „O" ein „X" folgt. Verschiedene Fehlerarten (ausbleibende bzw. falsche Reaktionen) werden bestimmt. Aufmerksamkeitsgestörte Kinder zeigen längere Reaktionszeiten und eine höhere Zahl falscher Reaktionen.

TAP (Testbatterie zur Aufmerksamkeitsprüfung)

- Alter: 6–89 Jahre
- Einsatz: Psychologen
- Ziel: differenzierte Diagnostik von Aufmerksamkeitsstörungen.

Die TAP ist eine PC-gestützte neuropsychologische Testbatterie. Auf einfache Reaktionsparadigmen ist selektiv auf gut diskrimierbare, sprachfreie Reize durch einen einfachen Tastendruck zu reagieren. Kriterien für die Leistungsfähigkeit sind die Reaktionszeit und eventuell begangene Fehler. Das Testverfahren umfasst folgende 13 Subtests: Alertness, Arbeitsgedächtnis (in 3 Schwierigkeitsabstufungen), Augenbewegung, crossmodale Integration, Daueraufmerksamkeit (in 2 Varianten), Flexibilität (in 6 Varianten), Gesichtsfeld- bzw. Neglectprüfung (in 5 Varianten), geteilte Aufmerksamkeit (in 6 Varianten), Go-/Nogo-Test (in 2 Varianten), Inkompatibilitäten, verdeckte visuelle Aufmerksamkeitsverschiebung, Vigilanztest (in 3 Varianten), visuelles Scanning.

KITAP (Kinderversion der Testbatterie zur Aufmerksamkeitsprüfung)

- Alter: 6–10 Jahre
- Einsatz: Psychologen
- Ziel: differenzierte Diagnostik von Aufmerksamkeitsstörungen.

Die KITAP wurde nach den gleichen Prinzipien entwickelt, die der Erwachsenenform (TAP) zugrunde liegen. Die Testauswahl orientierte sich an den zentralen Verfahren der TAP, die in eine kindgerechte Form umgesetzt wurden. Die KITAP („Das Schloss der Geister") umfasst folgende Items: Ablenkbarkeit („Das fröhliche und das traurige Gespenst"), Alertness („Die Hexe"), Daueraufmerksamkeit („Der Tanz der Geister"), Flexibilität („Das Haus der Drachen"), Geteilte Aufmerksamkeit („Die Eulen"), Go/Nogo („Die Fledermaus"), Vigilanz („Der Spiegel"), Scanning („Der Ausflug der Hexen"). Tabelle 17.7 (➤ Tab. 17.7) zeigt weitere Eigenschaften der besprochenen Verfahren zur Aufmerksamkeitsdiagnostik.

Tab. 17.7 Aufmerksamkeit – Fragebögen und Tests

Verfahren	Gütekriterien	Normen	Bearbeitungszeit (min)	aussagekräftig*
CRS	ja	ja	5	mäßig
duPaul	ja	ja	5	mäßig
FBB-/SBB-HKS	ja	ja	10	mäßig
TRF	ja	ja	10	mäßig
ZVT	ja	ab 8 J.	5–10	mäßig
D2	ja	ab 9 J.	5	mäßig
TPK	ja	ja	45	mäßig
CPT	ja	ja	15	mäßig
TAP/KITAP	ja	ab 6 J.	60–90	mäßig

* in Bezug auf ADHS

17.9 Lesen

17.9.1 Vorläuferfähigkeiten

BISC (Bielefelder Screening zur Früherkennung von Lese-Rechtschreibschwierigkeiten)

- Alter: Vorschulkinder zu Beginn oder Mitte des letzten Vorschuljahres
- Einsatz: Sprachheilpädagogen, Sprachtherapeuten, Kindergärten, Frühförderstellen
- Ziel: Identifizierung von Vorschulkindern mit einem Risiko zur Ausbildung von Lese- Rechtschreibschwierigkeiten.

Das Verfahren basiert auf der Annahme, dass eine nicht ausreichend ausgebildete phonologische Bewusstheit, Probleme im Abruf aus dem Langzeitgedächtnis, beim phonetischen Rekodieren im Kurzzeitgedächtnis und in der Steuerung der visuellen Aufmerksamkeit für die Ausbildung von Lese-Rechtschreib-Schwierigkeiten verantwortlich sind. Diese Vorläuferfertigkeiten werden in 9 Untertests geprüft: Reimen, Silben segmentieren, Laut-zu-Wort (Phonologische Bewusstheit); schnelles Benennen Farben, schnelles Benennen Wissen (Abrufgeschwindigkeit aus Langzeitgedächtnis), Nachsprechen von Pseudowörtern (Gedächtnisspanne und Artikulationsgenauigkeit), Wort-Vergleich-Suchaufgabe (visuelle Aufmerksamkeit). Eine zweimalige Messung wird empfohlen. Aus den Ergebnissen lassen sich Risikopunkte und daraus die Möglichkeit von Lesen-Rechtschreib-Schwierigkeiten ermitteln sowie unmittelbar Schlüsse für Förderungen ziehen.

BAKO 1–4 (Basiskompetenzen für Lese-Rechtschreib-Leistungen)

- Alter: jeweils Ende der 1.–4. Klasse
- Einsatz: Beratungslehrer, Sprachheilpädagogen, Sprachtherapeuten, Schulpsychologen
- Ziel: Überprüfung der phonologischen Bewusstheit von Kindern im Grundschulalter für eine frühzeitige Diagnostik von Lese-Rechtschreib-Schwächen.

Das Verfahren besteht aus insgesamt 74 Aufgaben, die sich sieben Subtests zuordnen lassen: Pseudowort-Segmentierung, Vokalersetzung, Restwortbestimmung, Phonemvertauschung, Lautkategorisierung, Vokallängenbestimmung und Wortumkehr. Durch einfaches Auszählen der richtigen Antworten wird der Gesamttestwert bestimmt. Zusätzlich kann für die erreichten Punktewerte auf Subtestebene ein Leistungsprofil erstellt werden, das Hinweise auf mögliche Problembereiche liefert.

17.9.2 Lesefähigkeit

ELFE 1–6 (Leseverständnistest für Erst- bis Sechstklässler)

- Alter: 1.–6. Schulklasse
- Einsatz: Lehrer, Psychologen
- Ziel: Erkennen von Defiziten im Leseverständnis.

ELFE 1–6 prüft sowohl basale Lesestrategien als auch die Fähigkeit zum Verstehen von Sätzen und Texten. Das Leseverständnis wird auf den folgenden Ebenen erfasst: Wortverständnis (Dekodieren, Synthese), Lesegeschwindigkeit (Schwelle der visuellen Worterkennung, nur in der Computerversion verfügbar), Satzverständnis (sinnentnehmendes Lesen, syntaktische Fähigkeiten), Textverständnis (Auffinden von Informationen, satzübergreifendes Lesen, schlussfolgerndes Denken). Im Vordergrund steht die Erfassung des Leseverständnisses, nicht des orthografischen Wissens oder der Artikulationsfähigkeiten. Für die Klassenstufen 5 und 6 hat der Test den Charakter eines Screeningverfahrens. Das ELFE-Trainingsprogramm knüpft direkt an ELFE 1–6 an.

ZLT II (Zürcher Lesetest II)

- Alter: 1.–8. Klassenstufe
- Einsatz: Beratungslehrer, (Schul)Psychologen, SPZ
- Ziel: Diagnostik und Verlaufskontrolle von Lesestörungen

Der ZLT II überprüft in 7 Untertests mit Arbeitsblättern und Lesekarten die Lesefertigkeit mit den Aspekten Lesegenauigkeit, Automatisierungsgrad, auditive Merkfähigkeit, Benenngeschwindigkeit und phonologische Bewusstheit im weiteren Sinn. Individuelle Fehlerprofile können erstellt und für Förderempfehlungen genutzt werden. Es werden T-Werte und Prozentränge angegeben. Mit Ausnahme des Untertests „schnelles Benennen", der auch Anforderungen an den Wortschatz stellt, gelten die Normen auch bei Kindern mit einer anderen Muttersprache als deutsch.

SLRT II (Salzburger Lese- und Rechtschreibtest II)

- Alter: 1.–6. Klasse
- Einsatz: Beratungslehrer, (Schul)Psychologen, SPZ
- Ziel: Erfassung von Lese- und Rechtschreibstörungen

Der Ein-Minuten-Leseflüssigkeitstest ermöglicht eine separate Diagnose zweier wesentlicher Teilkomponenten des Wortlesens: Defizite in der automatischen, direkten Worterkennung und Defizite des synthetischen, lautierenden Lesens. Der Rechtschreibtest beurteilt Schwächen in der lautgetreuen Schreibung, der orthographisch korrekten Schreibung und der Groß- und Kleinschreibung. Damit ergibt sich eine Basis für detaillierte Förderpläne.

17.10 Rechtschreibung

DRT 1-DRT 5 (Diagnostischer Rechtschreibtest 1.–5. Klasse)

- Alter: 1.–5. Klasse
- Einsatz: Beratungslehrer, Sonderschullehrer, Psychologen, sozialpädiatrische Zentren
- Ziel: quantitative Einstufung der Rechtschreibleistung und qualitative Analyse von Fehlerschwerpunkten.

Der DRT besteht aus Lückentexten, in die der Schüler Wörter zunehmender Schwierigkeit einfügen muss. Für jede Klassenstufe existiert ein eigenes Aufgabenheft. Im DRT 4 und DRT 5 werden z. B. die Lautunterscheidung, Lautnachbarschaften (st/sp, pf, qu), Dehnung/Doppelung, Ableitung (durch Verlängern, von der Grundform, vom Stamm ä und a), die Vorsilbe ver-, Groß- und Kleinschreibung geprüft. Die Testhefte sind nach den Regeln der neuen Rechtschreibung aufgestellt. Die Hauptaufgabe des DRT 1-DRT 5 ist es, gezielte schulischer Förderungsmaßnahmen auf der Basis der Fehleranalyse zu indizieren und deren Erfolg zu kontrollieren.

HSP 1–9 (Hamburger Schreibprobe 1–9)

- Alter: Mitte der 1. Bis Ende der 9. Klasse
- Einsatz: Beratungslehrer, Sonderschullehrer, Psychologen, SPZ
- Ziel: Erfassung der Rechtschreibfähigkeiten

Die als Gruppen- oder Einzeltest durchführbare Probe dient der Einschätzung individueller und klassenbezogener Leistungen. Bestimmt werden die Zahl der richtig geschriebenen Wörter und Grapheme sowie richtige und falsch angewandte Rechtschreibstrategien. Damit werden Grundlagen für gezielte Unterrichtsmaßnahmen und eine Einzelförderung geliefert. Es bestehen Vergleichswerte für den sonderpädagogischen Bereich sowie für großstädtische Ballungsgebiete und Stadtstaaten.

17.11 Rechnen

ZAREKI – R (Testverfahren zur Dyskalkulie bei Kindern – revidierte Fassung)

- Alter: 2.–4. Klasse
- Einsatz: Beratungslehrer, (Schul)Psychologen, SPZ
- Ziel: qualitative und quantitative Prüfung wesentlicher Aspekte der Zahlenverarbeitung und des Rechnens

Der ZAREKI – R ermöglicht qualitative und quantitative Einblicke in wesentliche Aspekte der Zahlenverarbeitung und des Rechnens und ermöglicht darauf basierende differenzielle Hilfen. Enthält die Subtests: Abzählen, Zählen rückwärts mündlich, Zahlen schreiben, Kopfrechnen, Zahlenlesen, Anordnen von Zahlen auf einem Zahlenstrahl, Zahlen nachsprechen vorwärts und rückwärts, Zahlenvergleich (Worte), perzeptive Mengenbeurteilung, kognitive Mengenbeurteilung, Textaufgaben, Zahlenvergleich (Ziffern). Vorbereitende Übungsaufgaben sind gegeben und leicht zu lösende Einstiegsaufgaben erleichtern den Zugang für ängstliche/leistungsschwache Kinder.

DEMAT 1+/DEMAT 4 (Deutscher Mathematiktest für die ersten Schulklassen)

- Alter: Ende 1./Anfang 2. Klasse bzw. 4. Halbjahr der 4. Schulklasse
- Einsatz: Beratungslehrer, (Schul)Psychologen, SPZ
- Ziel: Überprüfung der mathematischen Kompetenz in Bezug auf die bundesweit gültigen Lehrpläne

Erfassung der Rechenleistung, auch einer ganzen Schulklasse. Im DEMAT 1+ erfassen die Subtests Mengen-Zahlen, Zahlenraum, Addition, Subtraktion, Zahlenzerlegung, Zahlenergänzung, Teil-Ganzes-Schema, Kettenaufgaben, Ungleichungen und Sachaufgaben. Im DEMAT 4 werden

Tab. 17.8 Lesen, Rechtschreiben und Vorläuferfähigkeiten – Tests

Verfahren	Gütekriterien	Normen	Bearbeitungszeit (min)	aussagekräftig
BISC	ja	ja	20–25	mäßig
BAKO 1–4	ja	ja	30	mäßig
ELFE	ja	ja	10–15	ja
ZLT II	ja	ja	15–35	ja
SLT II	ja	ja	20–30	mäßig
DRT 1–5	ja	ja	30–60	ja
HSP 1–9	ja	ja	20–30	ja
ZAREKI-R	ja	ja	15–30	ja
DEMAT 1+/4	ja	ja	20–35	ja
RZD 2–6	ja	ja	30–45	ja

Arithmetik (Zahlenstrahlen, Additionen, Subtraktionen, Multiplikationen, Divisionen), Sachrechnen (Größenvergleiche, Sachrechnungen), und Geometrie (Lagebeziehungen, Spiegelzeichnungen) getestet. Individuelle Stärken, Schwächen, Empfehlungen für die Förderung und die weitere Beschulung (DEMAT 4) sind ableitbar.

RZD 2–6 (Rechenfertigkeiten- und Zahlenverarbeitungs-Diagnostikum für die 2. Bis 6. Klasse)

- Alter: Ende der 2. Bis Mitte der 6. Klasse
- Einsatz: Beratungslehrer, Sonderschullehrer, Psychologen
- Ziel: Rechentest, Hinweise auf basale Störungen der Informationsverarbeitung.

Es werden basale Zahlenverarbeitung und Rechenfertigkeiten erfasst. Im Einzelnen werden Zählfertigkeiten, Zahlenwissen, visuell-räumliche Mengenaspekte, Kopfrechnen, schriftliches Rechnen, Textaufgaben sowie das Wissen und flexible Anwenden von Rechenregeln überprüft. Das Diagnostikum beurteilt Arbeitsgüte und Arbeitsgeschwindigkeit. Tabelle 17.8. (➤ Tab. 17.8) zeigt weitere Eigenschaften der besprochenen Verfahren zur Diagnostik von Lesen, Rechtschreiben und Vorläuferfähigkeiten.

17.12 Psychische Entwicklung/Verhalten

SDQ (Strengths & Difficulties Questionnaires)

- Alter: 3–16 Jahre
- Einsatz: Therapeuten, Psychologen, Ärzte, sozialpädiatrische Zentren, Reihenuntersuchungen in Praxis und Forschung
- Ziel: Screening auf Verhaltensauffälligkeiten.

Der SDQ ist ein kurzer Fragebogen, der 25 positive und negative Eigenschaften auf einer dreistufigen Skala erfragt. Durch Addition aller Antworten lässt sich ein Gesamtproblemwert ermitteln, durch Addition bestimmter Einzelfragen ergeben sich die Dimensionen emotionale Probleme, Verhaltensauffälligkeiten, Hyperaktivität, Probleme mit Gleichaltrigen, prosoziales Verhalten. Es existieren ein Elternfragebogen, ein Lehrerfragebogen und ein Selbstfragebogen. Der SDQ ist kostenlos im Internet in vielen Sprachen verfügbar (http://www.sdqinfo.com/b1.html).

VBV 3–6 (Verhaltensbeurteilungsbogen für Vorschulkinder)

- Alter: 3–6 Jahre
- Einsatz: Psychologen, Ärzte, sozialpädiatrische Zentren
- Ziel: Screeninginstrument zur Erfassung von Verhaltensauffälligkeiten auf der Grundlage des Urteils von Eltern und Kindergartenerzieherinnen.

Die Items des VBV sind folgenden vier Dimensionen zugeordnet: sozial-emotionale Kompetenzen, oppositionell-aggressives Verhalten, Aufmerksamkeitsschwäche/Hyperaktivität vs. Spieldauer und emotionale Auffälligkeiten. Eltern- und Erzieherfragebogen liegen in getrennter Form vor, beide werden durch eine Symptomliste ergänzt. Das Verfahren kann zur Diagnosestellung, Therapieplanung und Therapiekontrolle eingesetzt werden.

CBCL 4–18 (Child Behavior Checklist – Elternfragebogen über das Verhalten von Kindern und Jugendlichen 4–18)

- Alter: 4–18 Jahre
- Einsatz: Psychologen, sozialpädiatrische Zentren, Kinder- und Jugendpsychiatrie u. a.
- Ziel: Elterneinschätzung hinsichtlich Kompetenzen und Problemen im Verhalten des Kindes.

Die CBCL 4–18 umfasst 118 Items, die anhand einer dreistufigen Skala beurteilt werden müssen. Die Auswertung umfasst 3 Kompetenzskalen (Aktivität, soziale Kompetenz und Schule) und 8 beurteilungsübergreifende Syndrome (sozialer Rückzug, körperliche Beschwerden, Angst/Depressivität, soziale Probleme, schizoid/zwanghaft, Aufmerksamkeitsstörung, delinquentes Verhalten, aggressives Verhalten), bei denen ein Vergleich über Eltern-, Lehrer- und Selbstbeurteilungsfragebogen möglich ist. Aus den Syndromskalen werden Skalen zu internalisierenden und externalisierenden Störungen sowie ein Gesamtwert für Problemverhalten gebildet.

Der **CBCL 1½–5** umfasst 99 Problemitems, von denen 50 dem CBCL für ältere Kinder entsprechen. Daraus werden sieben Problemskalen (emotionale Reaktivität; ängstlich/depressiv; körperliche Beschwerden; sozialer Rückzug; Schlafprobleme; Aufmerksamkeitsprobleme und aggressives Verhalten) sowie drei übergeordnete Skalen gebildet, die externalisierende Auffälligkeiten, internalisierende Auffälligkeiten und Gesamtauffälligkeit abbilden.

Der **TRF (Lehrerfragebogen über das Verhalten von Kindern und Jugendlichen)** ist von 5–18 Jahren anwendbar. Er beruht auf den gleichen Prinzipien wie der CBCL und beinhaltet 93 Probleme, die im CBCL 4–18 erfragt werden. Er enthält zusätzlich 27 Fragen, die auf die Schulsituation und das Lehrer-Schüler-Verhältnis gerichtet sind.

Der **YSR (Fragebogen für Jugendliche – Youth Self-Report)** dient bei Jugendlichen von 11–18 Jahren zur Selbstbeurteilung und nutzt den Fragenpool des CBCL. 16 Fragen aus dem CBCL wurden durch Fragen ersetzt, die sich auf sozial erwünschtes Verhalten beziehen.

DISYPS-KJ (Diagnostiksystem für psychische Störungen im Kindes- und Jugendalter nach ICD-10/DSM-IV)

- Alter: 11–18 Jahre
- Einsatz: Psychologen, sozialpädiatrische Zentren, Kinder- und Jugendpsychiatrie
- Ziel: Diagnostiksystem zur Erfassung psychischer Störungen bei Kindern und Jugendlichen entsprechend den Diagnosekriterien von ICD-10 und DSM-IV.

Das DISYPS-KJ umfasst die Störungsbereiche hyperkinetische Störungen (HKS), Störungen des Sozialverhaltens (SSV), Angststörungen (ANG), depressive Störungen (DES), tiefgreifende Entwicklungsstörungen (TES), Tic-Störungen (TIC), Störungen sozialer Funktionen (SSF), d. h.

Bindungsstörungen und Mutismus. Dabei werden drei Beurteilungsebenen miteinander kombiniert. Die klinische Beurteilung erfolgt anhand der Diagnose-Checklisten (DCL). Die Einschätzung durch Eltern, Lehrer und Erzieher kann anhand von Fremdbeurteilungsbögen (FBB) vorgenommen werden. Kinder und Jugendliche im Alter von 11–18 Jahren können sich selbst anhand von Selbstbeurteilungsbögen (SBB) einschätzen. Die Instrumente des Diagnostiksystems erleichtern die kategoriale Diagnostik nach ICD-10 und DSM-IV und ermöglichen eine differenzierte dimensionale Beschreibung psychischer Auffälligkeiten.

DISYPS II (Diagnostiksystem für psychische Störungen im Kindes- und Jugendalter nach ICD-10/DSM-IV)

DISYPS-II ist der Nachfolger des DISYPS-KJ. Es umfasst die Störungsbereiche: Aufmerksamkeitsdefizit-/Hyperaktivitätsstörungen, Störungen des Sozialverhaltens, Angststörungen, depressive Störungen, Zwangsstörungen, tiefgreifende Entwicklungsstörungen, Tic-Störungen, Störungen sozialer Funktionen, d.h. Bindungsstörungen und Mutismus. Im DISYPS II sind zusätzlich die neuentwickelte Diagnose-Checkliste für Zwangsstörungen und der neue Fremdbeurteilungsbogen für Vorschulkinder mit Aufmerksamkeitsdefizit-/Hyperaktivitätsstörung enthalten. Darüber hinaus wurden die DISYPS-Fragebögen überarbeitet, vereinfacht und durch Kompetenzitems ergänzt sowie psychometrisch überprüft und teilweise neu normiert. Tabelle 10.9 zeigt weitere Eigenschaften der besprochenen Fragebögen zur Diagnostik von Störungen des psychischen Entwicklung und des Verhaltens.

VFE (Verhaltensfragebogen für Entwicklungsstörungen)

- Alter: ab 4 Jahre bis zum Erwachsenenalter
- Einsatz: Psychologen, SPZ, Einrichtungen für Behinderte
- Ziel: Erfassung von Verhaltensauffälligkeiten bei Menschen mit geistiger Behinderung

Der VFE ist die deutsche Bearbeitung der australischen Developmental Behaviour Checklist. Der Fragebogen liegt in Versionen für Eltern/primäre Bezugspersonen (VFE-E; 96 Items), einer Lehrerversion (VFE-L; 94 Items) und einer Erwachsenen-Version (VFE-ER; 107 Items) vor. Es können ein Gesamtproblemwert sowie 5 Subskalen-Werte bestimmt werden: disruptiv/antisoziales Verhalten, Selbstabsorbierung, Kommunikationsstörung, Angst, soziale Beziehung. Für die Elternversion existiert eine Normierung an einer deutschen Stichprobe.

Tab. 17.9 Psychische Entwicklung/Verhalten – Fragebögen

Verfahren	Gütekriterien	Normen	Bearbeitungsdauer (min)	aussagekräftig
SDQ	ja	ab 6 J.	5	ja
VBV 3–6	ja	ja	20–40	ja
CBCL 4–8	ja	ja	15	ja
DISYPS	ja	ja	je nach Fragebogen	ja
DISYPS II	ja	ja	je nach Fragebogen	ja
VFE	ja	ja	15	ja

17.13 Autismus

17.13.1 Screening

FSK (Fragebogen zur Sozialen Kommunikation – Autismus Screening)

- Alter: ab 4 Jahren (Entwicklungsalter 2 Jahre)
- Einsatz: Psychologen, SPZ, Kinder-und Jugendpsychiatrie
- Ziel: Erfassung von abnormen sozialen Interaktions- und Kommunikationsmustern sowie stereotypen Verhaltensweisen im Vorfeld einer eingehenderen klinischen Diagnostik.

Der FSK ist ein Screening-Instrument bei Kindern und Jugendlichen mit Verdacht auf Autismus oder eine andere Störung des autistischen Spektrums. Er umfasst 40 binäre Items. Die Auswertung erfolgt über den Summenwert der als auffällig kodierten Items. Es liegen Werte für die Abgrenzung von Autismus/autistisches Spektrum und anderen psychischen Störungen sowie Normalität vor. Durchführung und Auswertung nehmen insgesamt ca. 20 Minuten in Anspruch.

17.13.2 Interview/Test

ADI-R (Autismus – Diagnostische Interview-Revision)

- Alter: ab Entwicklungsalter 2 Jahre
- Einsatz: Psychologen, SPZ, Kinder-und Jugendpsychiatrie
- Ziel: Erfassung und Differenzialdiagnostik von Störungen des autistischen Spektrums.

Der ADI-R beinhaltet in einem Interview 93 Items zur frühkindlichen Entwicklung, zu Spracherwerb und möglichem Verlust von sprachlichen Fertigkeiten, verbalen und non-verbalen kommunikativen Fähigkeiten, Spiel- und sozialem Interaktionsverhalten, stereotypen Interessen und Aktivitäten sowie komorbiden Symptomen (Aggression, Selbstverletzung, Epilepsie). Informanten sind in der Regel die Eltern oder seltener andere Bezugspersonen, die mit der Entwicklung und aktuellen Symptomatik des Probanden sehr gut vertraut sind. Die Auswertung und Interpretation der Ergebnisse erfolgt über die Verrechnung einer Auswahl von Items in einem empirisch generierten diagnostischen Algorithmus, der sich den Klassifikation nach ICD-10 und DSM-IV-TR orientiert. Das ADI-R verlangt eine gute Vertrautheit mit dem Erscheinungsbild des Autismus und dem Interviewprotokoll. Je nach Ausprägung der Symptomatik liegt die Bearbeitungsdauer zwischen 1½ und 3 Stunden.

ADOS (Diagnostische Beobachtungsskala für Autistische Störungen)

- Alter: etwa ab Entwicklungsalter 1 Jahr
- Einsatz: Psychologen, SPZ, Kinder-und Jugendpsychiatrie
- Ziel: Erfassung von Kommunikation, sozialer Interaktion und Spielverhalten oder Fantasiespiel bei Kindern und Erwachsenen, bei denen das Verdacht auf eine autistische oder eine andere tiefgreifende Entwicklungsstörung besteht.

In der wird ADOS wird in einem strukturierten Verfahren unter Verwendung verschiedener Gegenstände in Abhängigkeit vom Alter und Sprachniveau des jeweiligen Patienten eine von vier Untersuchungsstrategien (Modulen) gewählt, um anhand von gezielt inszenierten spielerischen Elementen, Aktivitäten und Gesprächen für die Diagnose des Autismus relevante Sachverhalte und Symptome prüfen zu können. Auffälligkeiten der sozialen Interaktion und rezipro-

ken Kommunikation im Sinne des Autismus werden in einer strukturierten Ratingskala festgehalten. Das Verfahren stellt hohe Ansprüche an die Kompetenz des Untersuchers, sodass eine Einweisung in das Instrument empfehlenswert ist. Die Bearbeitungsdauer eines Moduls umfasst 30–45 Minuten.
Die Diagnosekonvergenz der deutschsprachigen Versionen des ADI-R und der Diagnostischen Beobachtungsskala für Autistische Störungen (ADOS) beträgt 76 %.

17.14 Funktionsfähigkeiten/Lebensqualität

WeeFIM® (Functional Independence Measure)

- Alter: 6 Monate–7 Jahre, bis 21 Jahre bei behinderten Kindern/Jugendlichen
- Einsatz: bislang überwiegend in der neurologischen Rehabilitation
- Ziel: Bestimmung der Funktionsfähigkeit in relevanten Alltagsbereichen.

WeeFIM beinhaltet eine Skala von 18 Items essenzieller Alltagsfunktionen. Die Hauptbereiche sind Selbstversorgung (8 Items: Essen, Waschen, Zähneputzen, Haarpflege, Anziehen, Baden, Toilettenbenutzung, Kontinenz), Mobilität (5 Items: Stuhl – Aufstehen/Hinsetzen, Toilette – Hinsetzen/Aufstehen, Dusche/Badewanne – Ein- bzw. Aussteigen, selbstständige Mobilität in Innenräumen/im Freien, Treppensteigen) und Kognition (5 Items: Sprachverständnis, expressive Sprache, soziale Interaktion mit Gleichaltrigen, Problemlösen, Gedächtnis). Jedes Item wird auf einer 7 – stufigen Skala (1= komplette Unterstützung; 7= komplette Unabhängigkeit) bewertet. Die Einschätzung erfolgt durch Interview oder Beobachtung.

PEDI (Pediatric Evaluation of Disability Inventory)

- Alter: 6 Monate–7 Jahre, bis 21 Jahre bei behinderten Kindern/Jugendlichen
- Einsatz: bislang überwiegend in der neurologischen Rehabilitation
- Ziel: Bestimmung der Funktionsfähigkeit in relevanten Alltagsbereichen.

PEDI bestimmt die Funktionsfähigkeiten eines Kindes/Jugendlichen in den Domänen Selbstversorgung (15 Bereiche), Mobilität (13 Bereiche) und soziale Funktionen (13 Bereiche). Die Fähigkeit/Unfähigkeit der Ausführung wird in jedem Bereich für mehrere Items abgefragt. Die Einschätzung erfolgt durch Interview oder Beobachtung.

COPM (Canadian Occupational Performance Measure)

- Alter: alle Altersstufen
- Einsatz: Ergotherapeuten, sozialpädiatrische Zentren
- Ziel: Beurteilung der Patienten-/Elternwahrnehmung der Funktionsfähigkeit des Patienten.

COPM ist ein standardisiertes, semistrukturiertes Interview, das Veränderung im Therapieverlauf abbilden soll. Es ermöglicht einen patienten- bzw. elternzentrierten Therapiezugang orientiert an deren Bedürfnissen/Prioritäten.

KINDL-R (Revidierter KINDer Lebensqualitätsfragebogen)

- Alter: 4–16 Jahre
- Einsatz: Psychologen, Ärzte, Forschung
- Ziel: Erfassung der gesundheitsbezogenen Lebensqualität bei Kindern und Jugendlichen.

Tab. 17.10 Funktionsfähigkeit/Lebensqualität – Fragebögen

Verfahren	Gütekriterien	Normen	Bearbeitungszeit (min)	aussagekräftig
WeeFIM®	ja	ja	20	ja
PEDI	ja	ja	45–60	ja
COPM	ja	nein	20–40	ja
KINDL-R	ja	ja	10	ja

Dies ist ein Fragebogen mit 24 Fragen, der sowohl von Kindern und Jugendlichen als auch von deren Eltern ausgefüllt werden kann. Der KINDL-R-Fragebogen liegt für unterschiedliche Alters- und Entwicklungsstufen vor: 4–7 Jahre, 8–11 Jahre, 12–16 Jahre sowie in einer Elternversion. Er ist für gesunde und erkrankte Kinder verwendbar (krankheitsübergreifender Ansatz) und kann durch krankheitsspezifische Module zu den Indikationen erweitert werden: Adipositas, Asthma bronchiale, Neurodermitis, Krebserkrankung, Diabetes, Onkologie, Spina bifida, Epilepsie. Aktuelle Normwerte sind veröffentlicht.
Tabelle 17.10 (➤ Tab. 17.10) zeigt weitere Eigenschaften der besprochenen Verfahren zur Diagnostik von Funktionsfähigkeit und Lebensqualität.

17.15 Elterliche Belastung/Bewältigungsverhalten

FaBel (Familien-Belastungs-Fragebogen)

Der FaBel ist die deutsche Version der Impact on Family Scale. Er dient zur Selbstbeurteilung der Auswirkung chronischer Erkrankungen und Behinderungen im Kindesalter auf die Familie. Der Fragebogen enthält 33 Likert-skalierte Items zur Erfassung der täglichen sozialen Belastung der Eltern, der persönlichen Belastung, der Belastung der Geschwisterkinder, der finanziellen Belastung, der Bewältigungsprobleme mit der Belastung sowie eines Summenwertes (total score) der Belastungen insgesamt. Der FaBel-Fragebogen zeigt eine gute Akzeptanz durch die Befragten, eine akzeptable Konstruktvalidität, gute interne Konsistenz und Hinweise auf diskriminante Validität. Die psychometrische Struktur des Verfahrens konnte an einem anderen Kollektiv von Kindern mit chronischen Erkrankungen bestätigt werden. Die Gütekriterien sind zufriedenstellend. Die Bearbeitungsdauer liegt bei 10–15 Minuten. Der FaBel-Fragebogen ist zur Erfassung der familiären Belastung bei chronischen Erkrankungen und Behinderungen von Kindern und Jugendlichen geeignet.

SOEBEK (Soziale Orientierungen von Eltern behinderter Kinder)

Der SOEBEK ist das erste normierte deutschsprachige Verfahren zur Erfassung des Bewältigungsverhaltens von Eltern behinderter Kinder. Ziel ist die Beurteilung von Bewältigungsreaktionen, wie sie sich in der subjektiven Wahrnehmung der in ihren sozialen Kontext eingebundenen Elternperson abbilden. Der Fragebogen besteht aus 58 Items, die sich in die Skalen «Intensivierung der Partnerschaft», «Nutzung soziale Unterstützung», «Fokussierung auf das behinderte Kind», «Selbstbeachtung/Selbstverwirklichung», «Stressbelastungsskala» sowie in Einzelfragen zu thematischen Schwerpunkten und Fragen zur Zufriedenheit mit sozialer Unterstützung gliedern. Der SOEBEK kann bei Eltern geistig und/oder körperlich behinderter Kinder zwischen dem 1. und 14. Lebensjahr eingesetzt werden. Anwendungsbereiche des Verfahrens finden sich in der Diagnostik, Therapieplanung und Therapieverlaufskontrolle. Darüber hinaus dient es der Evaluation rehabilitativer Maßnahmen und kann als Screening-Instrument in der Erziehungsberatung und Forschung eingesetzt werden. Die Gütekriterien sind zufriedenstellend. Es werden Prozentrangnormen, nach Geschlecht der Eltern und teilweise nach Altersgruppen bzw. Behinderungstyp des Kindes getrennt, angegeben. Die Bearbeitungsdauer liegt bei etwa 20 Minuten.

Literatur

Aksu (Hrsg) (2011) Neuropädiatrie. 4. Aufl. Bremen: Uni-Med.

Brandt I, Sticker J (2001) GES Griffiths Entwicklungsskalen, Deutsche Bearbeitung, 2. Aufl. Göttingen: Beltz Test.

Daseking M, Janke N, Petermann F (2006) Intelligenzdiagnostik. Monatsschr Kinderheilkd 154: 314–319.

Esser G, Wyschkon A (2000) Umschriebene Entwicklungsstörungen. In: Petermann F. Lehrbuch der klinischen Kinderpsychologie und -psychotherapie. Göttingen: Hogrefe: 409–429.

International Test Commission (2000) Internationale Richtlinien für die Testanwendung, Deutsche Fassung ZPID.

Ireton H (1992) Child Development Inventory Manual. Minnesota: Behavior Science Systems Inc: 1–39.

Largo RH, Caflisch J, Hug F, Muggli K, Molnar A, Molinari L, Sheehy A, Gasser T (2001) Neuromotor development from 5 to 18 years: Part 1: Timed performance. Dev Med Child Neurol 43: 436–443.

Largo RH, Caflisch JA, Hug F, Muggli K, Molnar A, Molinari L (2001) Neuromotor development from 5 to 18 years: Part 2: Associated movements. Dev Med Child Neurol 43: 444–453.

Lipsius M, Petermann F, Daseking M (2008) Wie beeinflussen Testleiter die HAWIK-IV-Befunde? Kindheit und Entwicklung 17: 107–117.

Macha T, Petermann F (2006) Psychologische Tests in der Pädiatrie. Monatsschr Kinderheilkd 154: 298–304.

Palisano R, Rosenbaum P, Walter S, Russell D, Wood E, Galuppi B (1997) Development and reliability of a system to classify gross motor function in children with cerebral palsy. Dev Med Child Neurol 39: 214–223.

Petermann F, Macha T (2005) Psychologische Tests für Kinderärzte. Göttingen: Hogrefe.

Prechtl HFR (1990) Qualitative changes of spontaneous movements in fetus and preterm infant – a marker of neurological dysfunction. Early Human Development 23: 151–158.

Reuner G, Pietz J (2006) Entwicklungsdiagnostik im Säuglings- und Kleinkindalter. Monatsschr Kinderheilkd 154: 305–313.

Rosenbaum P (2006) The Manual Ability Classification System (MACS) for children with cerebral palsy: scale development and evidence of validity and reliability. Dev Med Child Neurol. 48: 549–54.

Russell DJ, Rosenbaum PL, Cadman D, Gowlan C, Hardy S, Jarvis S (1989) The Gross Motor Function Measure: a means to evaluate the effects of physical activity. Dev Med Child Neurol 31: 341–352.

Russell DJ, Rosenbaum PL, Avery LM, Lane M (2006) GMFM und GMFCS – Messung und Klassifikation motorischer Funktionen: Übersicht – Handbuch – CD-ROM. Bern: Hans Huber.

Rydz D et al. (2005) Developmental Screening. J Child Neurol 20: 4–18.

Straßburg, HM, Dacheneder W, Kreß W (2012) Entwicklungsstörungen bei Kindern – Praxisleitfaden für die interdisziplinäre Betreuung. 5. Aufl. München: Urban & Fischer.

Touwen BCL (1982) Die Untersuchung von Kindern mit geringen neurologischen Funktionsstörungen. Stuttgart: Thieme.

Touwen BCL, Prechtl HFR (1970) The neurological examination of the child with minor nervous dysfunction. Clinical Developmental Medicine 38.

Wright FV, Boschen KA (1993) The Pediatric Evaluation of Disability (PEDI): Validation of a new functional assessment outcome instrument. Canadian Journal of Rehabilitation 7: 41–42.

Internetlinks

www.caot.ca/copm
www.efke-online.de

www.entwicklungsdiagnostik.de
www.general-movements-trust.info
www.kindl.org
www.pearsonassessment.de
www.psytest.net/
www.sdqinfo.com/b1.html
www.testzentrale.de/
www.testinformationen.de/

VI Störungen der Entwicklung, des Verhaltens und der Gesundheit

18 Verhaltensstörungen in der frühen Kindheit

Hans-Michael Straßburg

18.1 Interaktions- und Regulationsstörungen

Hierunter versteht man **Verhaltensstörungen des Säuglings- und Kleinkindes,** insbesondere in Form von vermehrtem **Fütter/Ess-, exzessiven Schrei- und Schlafstörungen.** Aber auch andere Formen von Verhaltensauffälligkeiten, z. B. ausgeprägte Stimmungsschwankungen, überwiegend ruhig-apathisch-abweisendes Verhalten, das oft stundenlange Ausführen gleichförmiger Bewegungen (Jaktationen), vermehrtes Trotzverhalten und erhöhte Aggressivität werden hierunter subsumiert. Um den Begriff er Regulationsstörung nicht zu häufig zu verwenden, hat man sich in einer Leitlinien-Kommission der AWMF geeinigt, nur beim Vorliegen von mindestens 2 der genannten Symptome davon zu sprechen.

Fallbeispiel

4 Monate alter Säugling, erstes Kind, die Mutter raucht, auch während der Schwangerschaft, vorzeitige Wehen. Bei Geburt sekundäre Wehenschwäche, Entwicklung mittels Vakuum-Extraktion, Apgar 6/8/9. Geburtsgewicht 2.900 g. In den ersten 3 Lebenswochen ruhig, dann zunehmende Unruhe und Schreien, besonders ab dem Nachmittag und in der Nacht. Besserung beim Hochnehmen und nach Absetzen von Stuhl. Wegen V. a. mangelnde Sättigung Zufüttern von Formula-Milch, gelegentlich auch Tee. Gelegentliches Spucken, auch mehrere Stunden nach den Mahlzeiten.
Die Mutter wirkt unsicher im Umgang mit dem Säugling, leidet unter Schlafmangel. Sie war bereits bei verschiedenen Vertretern von Heilberufen, äußert Angst, das Kind würde sich nicht normal entwickeln. Der Vater sei während der Woche auf Montage.
Befunde: eutroph, erhöhte Erregbarkeit, Fäusteln, Rumpfasymmetrie, Überstreckung, aber guter Blickkontakt, lächelt reaktiv, zeigt keine abnormen General Movements.
Diagnose: persistierend exzessives Schreien (R68.1G, F98.2V)
Depression der Mutter zur Abklärung (F 32.0).
Maßnahmen:
- Bei der klinischen Untersuchung des Säuglings auf Zeichen der normalen Entwicklung hinweisen
- Evt. Sonografie von Kopf und Abdomen
- Ernährungsberatung, Entscheidung für Stillen oder Flaschenernährung
- Möglichst Stillen ad libitum
- ggf. Eindicken der Nahrung bzw. Umstellung auf eine Antireflux-Milch

Sozialpädiatrie in der Praxis. http://dx.doi.org/10.1016/B978-3-437-31630-3.00018-3
Copyright © 2014 Elsevier GmbH. All rights reserved.

- Besprechen der Vor- und Nachteile des Einwicklens mit einem Tuch („Pucken")
- Evt. Einschaltung einer Familienhebamme, einer ambualnten Kinderkrankenschwester oder einer anderen Mitarbeiterin der Frühen Hilfen
- Angebot zur jederzeitigen stationären Aufnahme
- Kurzfristige Terminvergabe bei der Physiotherapeutin des SPZ zur Anleitung im Handling und zur Entwicklungsbegleitung
- Vereinbarung regelmäßiger Gesprächstermine zusammen mit dem Vater
- Gegebenenfalls Mitbeurteilung der Mutter durch einen Erwachsenen-Psychiater
- Untersuchung der Schilddrüsenfunktion der Mutter
- Kurzfristiges Angebot für eine Wiedervorstellung
- Kein Rauchen in der Wohnung
- Evt. Vermittlung an eine Spezialambulanz für Schreibabies.
- Bei einer Wiedervorstellung nach 6 Wochen hat sich die Symptomatik deutlich zurückgebildet, die Mutter fühlt sich der Betreuung des Kindes gewachsen

In Bezug auf das **Schlafbedürfnis** und das **Schreien** besteht eine hohe **Variabilität.** Primär ist immer das normale körperliche und psychosoziale Gedeihen der Kinder das entscheidende Kriterium dafür, ob eine Gesundheitsstörung im engeren Sinne vorliegt. Nicht selten wird auf Grund exakter Dokumentationen, z. B. anhand eines Schlaf- oder Schreiprotokolls, deutlich, dass sich das kindliche Verhalten noch im Rahmen der normalen Variabilität bewegt. Dann ist es umso wichtiger, die psychische Situation der Eltern, insbesondere der Mutter, genauer zu analysieren. Häufig liegt vor allem eine erhebliche Verunsicherung vor, die oft schon vor der Geburt eingesetzt hat. Es ist eine sehr verantwortungsvolle Aufgabe, ggf. in Zusammenarbeit mit anderen Berufsgruppen, z. B. auch der Erwachsenenpsychiatrie, zu klären, ob eine psychische Erkrankung bei einem oder gar beiden Eltern vorliegt (z. B. eine Depression oder andere Psychosen, eine Abhängigkeit von Alkohol oder Drogen oder eine andere Persönlichkeitsstörung) und eine adäquate Betreuung einzuleiten.
Aber auch die **Umfeldbedingungen** des Kindes – vor allem die Aufenthaltsorte zu Hause – müssen in Bezug auf Lärm, Reizstoffe und Sauberkeit überprüft werden. Nicht selten kann durch die Beobachtung des Kindes im Rahmen einer stationären Klinikaufnahme rasch geklärt werden, ob ein diesbezüglicher Zusammenhang besteht. Auch Videoaufzeichnungen der Eltern von zu Hause können bei der Beurteilung hilfreich sein.
Nach Ausschluss organischer Ursachen, was aber z. B. bei Ess-, Schrei- und Schlafstörungen nicht einfach ist, müssen Interaktionsstörungen im engeren Sinne in Betracht gezogen werden. Hier können zum einen psychische Erkrankungen eines Elternteils, z. B. eine Depression der Mutter, oder anlagebedingte Verhaltensstörungen des Säuglings, aber auch Gewalt, Alkohol oder andere Drogen in der Familie eine Rolle spielen. Es muss immer bewusst sein, dass diese Kinder in erhöhtem Maße gefährdet sind, eine Kindesmisshandlung, vor allem in Form des **Schütteltraumas** (➤ Kap. 36) zu erleiden.
Während die meisten Fütter-, Schrei- und Schlafstörungen des frühen Säuglingsalters vorübergehend sind und vor allem einer psychosozialen Stützung der Familie bedürfen, muss bei Persistenz bzw. späterem Auftreten der Symptome auch an gravierende Erkrankungen, z. B. aus dem Autismus-Spektrum, an Magen-Darm-Erkrankungen wie eine Zöliakie oder eine definierte Nahrungsunverträglichkeit, aber auch an maligne Erkrankungen, z. B. Tumoren des Hirnstamms, gedacht werden.
Vor allem bei extrem Frühgeborenen können langwierige erhebliche Fütterstörungen bestehen. Dann muss sorgfältig abgewogen werden, ob eine Sondenernährung mit intensivem Füttertraining, psychotherapeutische Maßnahmen bei Mutter und Kind oder die Anlage einer perkutanen

endoskopischen Gastro- oder Jejunostomie (PEG-Sonde) indiziert ist, die nach Besserung der Symptome wieder zurückverlagert werden kann. Es existieren zunehmend Berichte darüber, dass nach dem 4. Lebensjahr durch die Behandlung mit Wachstumshormon bei diesen Kindern eine deutliche Besserung der Essstörung erreicht werden kann.
In Einzelfällen kann durch eine differenzierte Analyse der Eltern- bzw. Mutter-Kind-Interaktion mittels spezieller Videotechniken (face to face), z. B. während der Füttersituation, ein Ansatz für gezielte verhaltenstherapeutische Interventionen im Sinne eines Feinfühligkeits-Trainings gefunden werden. Dies sollte möglichst in speziellen Ambulanzen bzw. im Rahmen eines stationären Aufenthalts in einer dafür eingerichteten Mutter-Kind-Station geschehen (➤ Kap. 37; ➤ Kap. 38).

18.2 Weitere Verhaltensstörungen

Zunehmend werden in Deutschland im Rahmen der „**Frühen Hilfen**" flächendeckend Projekte zur möglichst frühen Identifikation von Risiko-Kindern und ihren Familien etabliert (➤ Kap. 9).
Weitere Verhaltensstörungen, die vor dem 6. Lebensjahr diagnostiziert werden können, sind:

- Depressive Störung
- Angststörung
- Anpassungsstörung
- Posttraumatische Belastungsstörung
- Bindungsstörung
- Hyperkinetische Störung/Aufmerksamkeits-Defizit-Hyperaktivitätsstörung
- Störung des Sozialverhaltens mit oppositionellem Verhalten.

Zusätzlich muss immer an eine globale Entwicklungsstörung und an eine beginnende Erkrankung aus dem Bereich der Autismus-Spektrum-Störungen gedacht werden (➤ Kap. 19, ➤ Kap. 21)
2013 wurden nach umfangreichen Diskussionen AWMF-Leitlinien für die Erkennung und Behandlung von psychischen Störungen bei Kindern zwischen 0 und 5 Jahren konsentiert. Ob diese Bemühungen zur Verbesserung von Früherkennung und -therapie von Verhaltens- und Interaktionsstörungen im Säuglings- und Kleinkindalter wirklich mit nachhaltigen Verbesserungen der Gesamtentwicklung verbunden sind, muss in Langzeit-Versorgungsstudien multizentrisch nachgewiesen werden.

Literatur

Cierpka M (2012) Frühe Kindheit 0–3 Jahre. Beratung und Psychotherapie für Eltern mit Säuglingen und Kleinkindern, Heidelberg: Springer

Papousek M, Schieche M, Wurmser H (2004)Regulationsstörungen der frühen Kindheit. Frühe Risiken und Hilfen im Entwicklungskontext der Eltern-Kind-Beziehungen. Bern: H. Huber

Straßburg HM, Dacheneder W, Kreß W (2012) Entwicklungsstörungen bei Kindern. Praxisleitfaden für die interdisziplinäre Betreuung. München: Elsevier

Von Gontard A (2010) Säuglings- und Kleinkindpsychiatrie – Ein Lehrbuch. Stuttgart: Kohlhammer

Ziegenhain U (2007) Förderung der Beziehungs- und Erziehungskompetenz bei jugendlichen Müttern. Prax Kinderpsychol Kinderpsychiatr 8 660–675

www.awmf.org/uploads/tx_szleitlinien/028-

19 Intelligenzminderung

Hans-Michael Straßburg

19.1 Definition

Eine häufige Ursache für eine sozialpädiatrische Betreuung im Kindes- und Jugendalter ist eine **Intelligenzminderung.** Unter Berücksichtigung aller Faktoren handelt es sich hierbei um die Gesundheitsstörung, die die höchsten Kosten für die Gesellschaft verursacht, ohne dass offiziell darüber viel gesprochen wird. In den ersten Lebensjahren wird die Diagnose meist nur bei vordergründig erkennbaren, schwerwiegenden Entwicklungsstörungen im Sinne einer Mehrfachbehinderung verwendet. Die klinisch erkennbare Symptomatik wird in der Regel auf Grund der zunehmenden Diskrepanz zu einer normalen Entwicklung immer deutlicher („Entwicklungs-Schere"), dennoch wird mitunter Eltern auch vermittelt, dass sich Auffälligkeiten auch „auswachsen" oder durch spezifische Therapie-Maßnahmen beseitigt werden können. Erst ab dem 5. Lebensjahr, also mit den Überlegungen zur Einschulung wird dann die Persistenz der Entwicklungsstörung zunehmend bewusst. Einzelheiten zu den Möglichkeiten der Behandlung und Intervention sind im Kapitel 37.6 dargestellt. (➤ Kap. 37.6)
Intelligenzminderungen sind nach einer Definition der WHO von 1974 vor allem durch eine weit unterdurchschnittliche Intelligenz gekennzeichnet und von einer Art oder einem Ausmaß, dass eine medizinische Behandlung oder andere spezielle Vorsorgen bzw. Übungen notwendig und sinnvoll sind. Definitiv spricht man von einer Intelligenzminderung, wenn der Intelligenzquotient, der mit einem aktuellen standardisierten Verfahren bestimmt wurde, mehr als zwei Standardabweichungen unter dem statistischen Mittelwert von 100 liegt, d. h. weniger als 70 beträgt (= weit unterdurchschnittliche Intelligenz).

Definition Intelligenzminderung

Geistige Behinderung ist eine Unfähigkeit, charakterisiert durch
- signifikante Begrenztheit intellektueller Funktionen
- deutliche Einschränkung der Verhaltensanpassung
- reduzierte konzeptionelle, soziale und praktisch-adaptive Fertigkeiten
- Auftreten während der Entwicklung (vor dem 18. Lebensjahr)

(American Association of Mental Retardation AAMR, 2002)

Sozialpädiatrie in der Praxis. http://dx.doi.org/10.1016/B978-3-437-31630-3.00019-5
Copyright © 2014 Elsevier GmbH. All rights reserved.

Sozialrechtlich wird dies mit einer **Geistigen Behinderung (GB)** gleichgesetzt. Im angloamerikanischen Schrifttum wird der inkorrekte Begriff einer **„mental retardation“** zunehmend von dem einer **intellectual disability** ersetzt. Durch diese Definitionen kommen jedoch vielfältige weitere Gesichtspunkte, insbesondere die sozialen, sprachlichen und motorischen Kompetenzen, nicht ausreichend zum Ausdruck.
Alle Definitionen sind deskriptiv für einen bestimmten Zustand. Sie erlauben keine Aussage zur Ätiologie und Pathogenese unter medizinischen Aspekten.
In Analogie zum Intelligenzquotienten (IQ) wurde von C. Bühler und H. Hetzer ein sogenannter **Entwicklungsquotient** (EQ) vorgeschlagen. Hierzu wurde die Entwicklung verschiedener Bereiche, z. B. der Sinneswahrnehmung, der Körperbewegung, des Sozialkontakts, des Lernens, der Materialbetätigung und der geistigen Produktivität, erfasst (➤ Kap. 17).

Intelligenzminderung und Altersäquivalent bei Erwachsenen

- **Leichtgradig:** IQ 50–69 9 bis 12 Jahre
- **Mittelgradig:** IQ 35–49 6 bis 9 Jahre
- **Schwergradig:** IQ 20–34 3 bis 6 Jahre
- **Schwerstgradig:** IQ < 20 unter 3 Jahre

Der Intelligenzquotient allein ermöglicht keine sinnvolle Untergruppierung von Menschen mit geistiger Behinderung!

Wichtige zusätzliche Aspekte zu dem Konstrukt, das heute als Intelligenz verstanden wird, wurden von dem Schweizer Psychologen Jean Piaget (1896–1980), von dem russischen Neuropsychologen und Psychiater A. Lurija (1902–1977) und dem Amerikaner D. Wechsler (1886–1981) beigetragen. Heute wird nach dem sogenannten CHC-Modell von R. B. Cattell, J. L. Horn und J. B. Carroll Intelligenz in einer komplexen Weise definiert. Dabei wird unter **fluider Intelligenz** z. B. verstanden:

- Über Neues nachdenken
- Probleme lösen
- Konzepte formulieren
- Mit Abstraktionen, Regeln und logischen Beziehungen umgehen.
- Durch hirnorganische Funktionsstörungen eher beeinträchtigt.

Im Gegensatz dazu wird als **kristalline Intelligenz** verstanden:

- Breites und tiefes Wissen einer Kultur
- Fähigkeit, die durch den Einsatz von Bildung und Erfahrung gebildet wird
- Behalten von Informationen
- Fähigkeit, vor allem verbal zu kommunizieren
- Vernünftig mit den vorausgegangenen Lernerfahrungen umzugehen.

19.2 Epidemiologie der Intelligenzminderung

Etwa 3 % aller Kinder haben eine Intelligenzminderung im Sinne einer geistigen Behinderung mit einem IQ unter 70, knapp 20 % davon haben einen IQ unter 50, etwa 80 % einen IQ zwischen 50 und 69. Das Verhältnis von Jungen und Mädchen beträgt 1,4:1.
Bis zu 10 % der Bevölkerung haben einen IQ zwischen 70 und 84 (unterdurchschnittliche Intelligenz), was auch als unterdurchschnittliche kognitive Leistungsfähigkeit bzw. **„Lernbehinderung“**

bezeichnet wird. Diese große Gruppe von Menschen hat im Alltagsleben, vor allem bei der Ausbildung und Berufsfindung, signifikant häufiger vielfältige Probleme, ohne dass diese in der Regel als Gesundheitsstörung anerkannt werden. Viele dieser Menschen haben auch z. B. auf Grund einer Überforderung sekundär psychische Probleme.
Bei mindestens 4 % aller Menschen besteht eine **umschriebene Entwicklungsstörung,** d.h. nur in einem definierten Bereich der Entwicklung besteht eine Abweichung von 1,5 Standardabweichungen bei einer insgesamt normalen Gesamtintelligenz (➤ Kap. 17).

19.3 Ätiologie der Intelligenzminderung

Etwa 15 % der Intelligenzminderungen sind durch Chromosomenanomalien, mehr als 15 % durch monogene Erkrankungen, ca. 12 % durch primäre ZNS-Fehlbildungen und Dysmorphie-Syndrome und weitere 10 % durch entwicklungshemmende Einflüsse in der Schwangerschaft zu erklären, (➤ Kap. 29; ➤ Kap. 32; ➤ Kap. 33).
10–15 % der Intelligenzminderungen müssen heute besonders bei sehr und extrem Frühgeborenen durch Komplikationen am Ende der Schwangerschaft (Kap. 30) und ca. 10 % durch Komplikationen während der Geburt, die zur Minderperfusion des Gehirns oder Sauerstoffmangel führen, erklärt werden, weniger als 10 % sind durch erworbene Hirnschäden nach der Geburt verursacht.
Auch bei umfangreicher Diagnostik können ca. 15 % der schweren mentalen Entwicklungsstörungen ätiologisch nicht geklärt werden.
Differenzialdiagnostisch muss bei Kindern und Jugendlichen, bei denen der Verdacht auf das Vorliegen einer Intelligenzminderung besteht, an viele verschiedene Ursachen gedacht werden.

Differenzialdiagnose bei Verdacht auf Intelligenzminderung
- Hörstörungen
- Auswirkungen einer Epilepsie
- Deprivation
- Emotionalstörung
- Posttraumatische Belastungsstörung
- Frühkindlicher Autismus

Es ist eine der wichtigsten Aufgaben des Kinder- und Jugendarztes, den Eltern klare Aussagen zum Entwicklungsstand ihres Kindes zu machen bzw. weitere Untersuchungen einzuleiten, um die Ursache von Entwicklungsauffälligkeiten abklären zu können (➤ Kap. 14). In einzelnen Fällen lassen sich durch frühzeitige Behandlungsmaßnahmen signifikante Verbesserungen erreichen (z. B. bei einer Epilepsie, einem Hydrocephalus oder einer Hörstörung).
Eltern wünschen oftmals die Gewissheit einer Diagnose, um sich mit den daraus ergebenden Konsequenzen auseinanderzusetzen, z. B. beim Vorliegen einer Chromosomenanomalie oder spezieller genetischer Veränderungen. Die richtige Diagnose kann eine differenzierte humangenetische Familienberatung (z. B. beim Fragiles-X-Syndrom) ermöglichen, es ergeben sich u. U. juristische Konsequenzen (z. B. bei einem gesicherten intranatalen Sauerstoffmangel) und es können aufgrund einer eindeutigen Diagnose gezielt komplexe Förder- und Behandlungsmaßnahmen eingeleitet werden (➤ Kap. 37; ➤ Kap. 38).

Eine Herausforderung für die Gesellschaft ist, es, für alle Menschen mit unterdurchschnittlicher Intelligenz die Voraussetzungen für ihre **Inklusion** in Schule, Beruf und Alltagsleben zu schaffen (➤ Kap. 13.6). Ein besonderes Problem ist weiterhin die **Transition,** d. h. der Übergang ins Erwachsenenalter. Für die ganzheitliche Betreuung dieser Menschen setzt sich vor allem die Bundesarbeitsgemeinschaft der Ärzte für Menschen mit geistiger oder mehrfacher Behinderung (BAG e. V.) ein.

19.4 Situation in der Gesellschaft

Menschen mit Intelligenzminderung und ihre Familien haben Anspruch auf eine besondere Fürsorge der Gesellschaft, sinnvolle Förder- und Therapiemaßnahmen (➤ Kap. 37, ➤ Kap. 38) und eine ausführliche Rechtsberatung (➤ Kap. 13, ➤ Kap. 42, ➤ Kap. 44).

Neben der adäquaten Betreuung und Versorgung von Menschen mit Intelligenzminderung in unserer Gesellschaft ist die allgemeine und spezielle **Prävention** (➤ Kap. 8) zur Vermeidung von Entwicklungsstörungen und Behinderungen ein großes Anliegen der Sozialpädiatrie, insbesondere durch

- adäquates Ernährungs- und Bewegungsverhalten vor und während einer Schwangerschaft (z. B. Folsäure!)
- regelmäßige Teilnahme an den Schwangerschafts-Vorsorgeuntersuchungen
- Vermeidung von Frühgeburt
- Sicherstellung optimaler Geburtsbedingungen
- bestmögliche postnatale Betreuung des Kindes und seiner Eltern
- und bei einem Index-Fall in der Familie durch
- eine differenzierte Analyse der Ursachen und ggf. eine qualifizierte humangenetische Beratung.

Trotz der zunehmend differenzierteren Möglichkeiten der pränatalen, insbesondere genetischen Diagnostik muss immer wieder darauf hingewiesen werden, dass es keine Sicherheit dafür gibt, dass die geistige Entwicklung eines Kindes normal ablaufen wird.

Literatur

Michaelis R, Niemann G (2010) Entwicklungsneurologie und Neuropädiatrie. Grundlagen und diagnostische Strategien. Stuttgart: Thieme

Neuhäuser G, Steinhausen HC, Häßler F, Sarimski K (2013) Geistige Behinderung. Grundlagen, Erscheinungsformen und klinische Probleme, Behandlung, Rehabilitation und rechtliche Aspekte. Stuttgart: Kohlhammer

Sarimski K, Steinhausen HC (2008) Psychische Störungen bei geistiger Behinderung. Göttingen: Hogrefe.

Straßburg HM, Dacheneder W, Kreß W (2012) Eintwicklungsstörungen bei Kindern. Praxisleitfaden für die interdisziplinäre Betreuung. München: Elsevier

20 Umschriebene Entwicklungsstörungen (Teilleistungsstörungen) einschließlich Sprachentwicklungsstörungen

Hans-Michael Straßburg und Harald Bode

20.1 Einleitung

Der Begriff der **umschriebenen Entwicklungsstörung** geht von isolierten und umschriebenen Funktionsstörungen im Laufe der kindlichen Entwicklung aus, deren Grundlage im Wesentlichen eine Funktionsstörung des Zentralnervensystems ist, die zu einer Störung der Informationsverarbeitung führt. A. R. Lurija und J. Graichen bezeichnen als **Teilleistungsstörung** umschriebene Einschränkungen sehr unterschiedlicher Hirnfunktionen, die aus dem übrigen Leistungsniveau bzw. Entwicklungsstand eines Kindes herausfallen. Dabei handelt es sich nicht um eine einheitliche Krankheitsentität. Voraussetzungen für die Diagnosestellung einer umschriebenen Entwicklungsstörung sind einerseits Teilleistungsstörungen um mehr als 1,5 Standardabweichungen vom Altersnormwert, andererseits ein Gesamt-IQ über 70. Aber auch Menschen mit einer Intelligenzminderung können Teilleistungsschwächen bzw. -störungen haben. In der ICD-10 werden die umschriebenen Entwicklungsstörungen wie folgt eingeteilt:

- F80 umschriebene Entwicklungsstörung des Sprechen und der Sprache
- F80.0 Artikulationsstörung
- F80.1 expressive Sprachstörung
- F80.2 rezeptive Sprachstörung
- F80.3 erworbene Aphasie mit Epilepsie
- F80.8 sonstige Entwicklungsstörungen des Sprechens und der Sprache
- F80.9 nicht näher bezeichnete Entwicklungsstörungen des Sprechens und der Sprache
- F81 umschriebene Entwicklungsstörung schulischer Fertigkeiten
- F81.0 Lese- und Rechtschreibstörung
- F81.1 isolierte Rechtschreibstörung
- F81.2 Rechenstörung
- F81.3 kombinierte Störung schulischer Fertigkeiten
- F81.8 sonstige Entwicklungsstörungen schulischer Fertigkeiten
- F81.9 nicht näher bezeichnete Entwicklungsstörung schulischer Fertigkeiten

Sozialpädiatrie in der Praxis. http://dx.doi.org/10.1016/B978-3-437-31630-3.00020-1
Copyright © 2014 Elsevier GmbH. All rights reserved.

- F82 umschriebene Entwicklungsstörung der motorischen Funktionen
- F83 kombinierte umschriebene Entwicklungsstörungen.

Je jünger ein Kind ist, umso weniger ist es möglich, umschriebene Hirnfunktionen eindeutig zu diagnostizieren und die Entwicklungsauffälligkeiten von einer globalen Intelligenzminderung abzugrenzen. Umschriebene Entwicklungsstörungen müssen von spezifischen Leistungsausfällen nach anfänglich normaler Entwicklung unterschieden werden. Hier spricht man bei Störungen im Handlungsablauf eher von einer Apraxie bzw. Dyspraxie, bei erworbenen Sprachstörungen von einer Aphasie bzw. Dysphasie und bei erworbenen Wahrnehmungsstörungen von einer Agnosie bzw. Dysgnosie. Allerdings wird in der Literatur der Begriff der Entwicklungsdyspraxie auch bei primär ungeschickten Kindern verwendet.

Weiterhin gibt es immer wieder Überschneidungen mit anderen Diagnosebezeichnungen, z. B. dem **MCD-Syndrom** (minor oder minimal cerebral dysfunction), dem POS (psycho-organisches Syndrom), DCD (developmental coordination disorder), verschiedenen Formen von zentralen Wahrnehmungs- und Verarbeitungsstörungen (auditiv, taktil-kinästhetisch, vestibulär, visumotorisch) oder einer „spezifischen Lernstörung". Die letztgenannten Begriffe sind in der ICD-10 nicht aufgeführt. Demgegenüber wird bei sehr vielen Kindern und Jugendlichen das ADHS diagnostiziert, bei dem nicht selten Überlappungen mit den umschriebenen Entwicklungsstörungen bestehen.

Über die Häufigkeit von Teilleistungsstörungen liegen sehr unterschiedliche Angaben vor, was mit der ungenauen Definition und den Problemen der Abgrenzung zur Normalität bzw. zu umschriebenen anderen Krankheiten und Störungen zusammenhängt. In der gründlichsten epidemiologischen Studie, der Mannheimer „Kurpfalz"-Studie, geht man von einer Häufigkeit von 4–5 % der Kinder eines Jahrgangs aus, die eine umschriebene Entwicklungsstörung der schulischen Fertigkeiten haben. Bei den **expressiven Sprachentwicklungsstörungen** geht man nach dem 3. Lebensjahr von ca. 8 % aus, Artikulationsstörungen und auch motorische Entwicklungsstörungen sind sicher häufiger.

Die **Ursachen** von umschriebenen Entwicklungsstörungen sind vielfältig. Besonders die Lese-Rechtschreib-Störung ist häufig genetisch determiniert, sie kann bei mehreren Mitgliedern einer Familie in unterschiedlicher Ausprägung vorliegen. Aktuell werden Veränderungen auf Chromosom 6 als Ursache für Störungen der akustischen Verarbeitung und auf Chromosom 15 als Ursache für Störungen der visuellen Verarbeitung diskutiert. Bei schweren expressiven Sprachstörungen hat man Deletionen im sogenannten **FOXP2-Gen** auf Chromosom 7 nachweisen können. Darüber hinaus können organisch bedingte Hirnstörungen (z. B. ein Hydrozephalus oder eine Balkenanlagestörung), unterschiedliche Sehstörungen (z. B. latentes Schielen) und Störungen der Blickbewegungen sowie Epilepsien (Rolando-Epilepsie, Landau-Kleffner-Syndrom) die Symptomatik verstärken.

Über den Einfluss von **Umweltfaktoren** auf die mental-kognitive Entwicklung gibt es sehr unterschiedliche Ansichten. Neben ideologischen Erklärungsversuchen können im Einzelfall allergische Dispositionen, Reizüberflutung durch Fernsehen und Computer, Schlafmangel, spezifische Ernährungsstörungen (z. B. Zöliakie), Toxine (Blei, Quecksilber) und vieles mehr eine Rolle spielen.

20.2 Störungen der Sprach- und Sprechentwicklung

Harald Bode

Umschriebene Entwicklungsstörungen des Sprechens und der Sprache (UES) treten bei 5–10 % aller Kinder im Vorschulalter auf und sind die häufigste Ursache für die Vorstellung eines Kindes mit Entwicklungsauffälligkeiten in einem Sozialpädiatrischen Zentrum. Jungen sind fast 3-mal häufiger als Mädchen betroffen. Es handelt sich um sozial relevante Störungen, bei denen der normale Spracherwerb von frühen Entwicklungsstadien an beeinträchtigt ist. Die Störungen können nicht direkt neurologischen Störungen oder Veränderungen des Sprachablaufs, sensorischen Beeinträchtigungen, Intelligenzminderung oder Umweltfaktoren zugeordnet werden. Umschriebene Entwicklungsstörungen des Sprechens und der Sprache ziehen oft sekundäre Folgen nach sich, wie Schwierigkeiten beim Lesen und Rechtschreiben, Störungen im Bereich der zwischenmenschlichen Beziehungen, im emotionalen und Verhaltensbereich.

Leitsymptome einer umschriebenen Entwicklungsstörung des Sprechens und der Sprache sind neben der außerhalb der Norm liegenden Sprachfertigkeit ein altersentsprechendes Kommunikationsbedürfnis, eine relativ ungestörte nichtverbale Kommunikation, eine primäre Störung mit stetigem Verlauf und eine Besserung mit zunehmenden Alter.

Die ICD-10 unterscheidet: Artikulationsstörungen (F 80.0), expressive Sprachstörungen (F 80.1) und rezeptive Sprachstörungen (F 80.2). Die Sprachwissenschaften unterscheiden: phonetisch-phonologische Störungen (Aussprachestörungen), Störungen von Lexikon und Semantik (Wortschatz, Wortbedeutungen), morphologisch-syntaktische Störungen (Dysgrammatismus), Störungen der Pragmatik (Dialogkompetenz)

Die **Ursachen der UES** sind heterogen und meist multifaktoriell. Intrauterine oder perinatale Hirnschäden sind nach heutigem Kenntnisstand von geringer Bedeutung, **Hörstörungen** müssen immer ausgeschlossen werden. Genetische Faktoren spielen eine entscheidende Rolle. Neuropathologisch wurden in Einzelfällen subtile Störungen beschrieben. Im Elektroenzephalogramm finden sich relativ häufig epileptische Foci (Rolando-Fokus), deren Wertigkeit nicht gesichert ist. Neuropsychologisch wird angenommen, dass aufgrund gestörter basaler sprachbezogener Prozesse der sprachliche Input nicht optimal aufgenommen, verarbeitet und repräsentiert wird. Dies führt zu einem langsameren und mühsameren Spracherwerb.

In der **Pathogenese** sind Gen-Umwelt-Interaktionen bedeutsam. Mangelnde sprachliche Anregung, fehlerhafte Sprachvorbilder, ungünstige Eltern-Kind-Kommunikation, Mehrsprachigkeit und Deprivation können bei entsprechender Disposition zur Manifestation einer umschriebenen Entwicklungsstörung des Sprechens und der Sprache beitragen. Bevor die Diagnose einer UES gestellt wird, sind verschiedene andere Ursachen von Sprach- und Sprechstörungen auszuschließen.

Differenzialdiagnosen umschriebener Entwicklungsstörungen des Sprechens und der Sprache sind:

- Hörstörungen
- Erkrankungen peripherer Sprechorgane (z. B. Lippen-Kiefer-Gaumenspalten, Gaumen-, Kiefer-, Zahnanomalien, **Adenoide**)
- Intelligenzminderung. Bei einigen Syndromen mit IQ-Minderung gehört eine Sprachstörung zum Verhaltensphänotyp
- Akute, chronische, degenerative neurologische Erkrankungen
- Epileptische Hirnfunktionsstörung (Landau-Kleffner-Syndrom)
- Psychogene Faktoren: z. B. emotionale Störungen, Mutismus
- Tiefgreifende Entwicklungsstörungen/Autismus-Spektrum
- Mangelnde sprachliche Anregung
- Zweisprachigkeit.

Die **sozialpädiatrische Diagnostik bei UES** ist in Tabelle 20.1 (➤ Tab. 20.1) dargestellt. Sie bezieht sich auf die Störungen des Sprechens und der Sprache sowie die häufigen Komorbiditäten (z. B. umschriebene Entwicklungsstörung der Motorik, Wahrnehmungsstörung, emotionale Störung, Aufmerksamkeitsdefizit/Hyperaktivitätssyndrom, Lese-Rechtschreib-Schwäche).

Tab. 20.1 Sozialpädiatrische Diagnostik bei umschriebenen Entwicklungsstörungen des Sprechens und der Sprache

Anamnese	**Kommunikationsfähigkeit des Kindes**
	Verlauf der Sprachentwicklung
	Motorische, kognitive, soziale und emotionale Entwicklung
	Bei Zweisprachigkeit: Erwerb und Kompetenz in der Erstsprache
	Aktuelles und früheres Sprachmilieu des Kindes
	Familiäre Belastung mit Sprach-, Lese-/Rechtschreibstörungen
	Elternfragebögen zur Sprachentwicklung
Klinischer Befund	Interaktion und Kommunikation (nonverbal und verbal)
	Spontansprache (expressiv, rezeptiv)
	Bei Zweisprachigkeit: sprachlicher Befund in der Erstsprache (soweit möglich)
	Klinisch-pädiatrischer und neurologischer Befund
	Motorische, kognitive, emotionale Entwicklung und Verhalten (Komorbiditäten)
	Hörprüfung
Testverfahren	Sprachscreeningverfahren
	Sprachtests
	Intelligenztests
	Ggf. Lese- und Rechtschreibtests
Sonstige Untersuchungen	Schlaf-EEG

Die Indikation für eine **Sprachtherapie** besteht, wenn – unabhängig vom Alter – das Kind in seinen alters- und entwicklungstypischen Fähigkeiten der sprachlichen Kommunikation und Verständlichkeit signifikant beeinträchtigt ist. Dies gilt auch für Kinder mit Intelligenzminderung, Störungen aus dem Autismus-Spektrum oder des Gehörs. Eine Sprachtherapie ist umso früher indiziert, je ausgeprägter die Sprachentwicklungsstörung ist und je mehr Sprachbereiche betroffen sind. Konkrete Entscheidungshilfen finden sich in den Leitlinien der Fachgesellschaften.

Bei harmonisch retardierten Kindern vom Säuglingsalter bis zum Schuleintritt besteht die Indikation für eine pädagogische **Frühförderung** zur Anregung der Kommunikation und Sprechfreude. Bei diesen Kindern ist eine Sprachtherapie medizinisch nicht notwendig.

Kinder mit leichteren – meist soziogenen – Auffälligkeiten in der Sprachentwicklung und Kinder mit spätem Zweitspracherwerb benötigen eine adäquate **Sprachförderung** in Kindergarten/Kindertagesstätte (weitere Ausführungen zur Förderung und Therapie siehe ➤ Kap. 37 und ➤ Kap. 38).

Bei **Zweisprachigkeit** bestehen Indikationen bei nicht altersgerechter Muttersprache, altersgerechter Muttersprache und – trotz ausreichender natürlicher Förderung – Störungen in der deutschen Sprache, die sich nicht durch Interferenzen mit der Erstsprache erklären lassen, und evtl. bei andauernden und ausgeprägten Interferenzerscheinungen. Bei Zweisprachigkeit ist eine Sprachtherapie in der Regel nicht indiziert bei weitgehend altersgerechter Muttersprache. In diesen Fällen ist eine natürliche Förderung von Deutsch als Zweitsprache in Frühförderung, Kindergarten/Kindertagesstätte und Schule erforderlich.

Die familiären Bedingungen und das Lebensumfeld des Kindes müssen in der Sprachtherapie berücksichtigt werden. Bei jeder Sprachtherapie ist eine begleitende Sprachförderung in häuslicher Umgebung durch die wichtigsten Bezugspersonen bzw. in Kindertagesstätte/Kindergarten nötig. Dabei ist eine sensible Anpassung an den jeweiligen Stand der kindlichen Sprachentwicklung und die Möglichkeiten von Eltern und anderen Bezugspersonen erforderlich.

Eine sinnvolle Gesamtförderung von Kindern mit umschriebenen Sprachentwicklungsstörungen von erheblichem Ausmaß ist im Vorschulalter in speziellen schulvorbereitenden Einrichtungen (SVE) und Sprachheilkindergärten bzw. im Schulalter in speziellen Sonderschulen durch Sprachheilpädagogen möglich. Meist ist eine zusätzliche Sprachtherapie dann nicht mehr indiziert.

Fallbeispiel

Ein 3-jähriger Junge wird in einem SPZ wegen verzögerter Sprachentwicklung vorgestellt.
Im Alter von 2 Jahren spricht er erst 10 Wörter, die zudem schlecht verständlich sind. Eine Hörstörung wird ausgeschlossen. In der Familie väterlicherseits haben mehrere Mitglieder einen Sprachheilkindergarten besucht, die Mutter hat als Kind eine logopädische Behandlung erhalten und später eine Lese-Rechtschreib-Störung entwickelt.
Bei der Untersuchung zeigt der Junge eine altersgerechte Motorik, ein interessiertes Spiel- und Interaktionsverhalten und kommuniziert mit differenzierten Gesten und Lautbildungen, die nur den Eltern verständlich sind. Das Sprachverständnis erscheint altersgerecht. Auch die kognitiven Fähigkeiten in der Entwicklungstestung und das EEG sind altersgemäß. Es wird eine schwere expressive Sprachstörung diagnostiziert und eine Sprachtherapie mit Anleitung der Eltern und des Kindergartens verordnet. 6 Monate später spricht der Junge Zweiwortsätze, die Lautbildung zeigt Fortschritte bei noch stark beeinträchtigter Verständlichkeit. Unter fortlaufender Sprachtherapie mit jeweils dem Entwicklungsstand angepassten Therapiezielen auf den Ebenen Aussprache, Wortschatz/Wortbedeutungen, Grammatik und dialogischer Kompetenz macht der Junge langsame und kontinuierliche Fortschritte bis zum Eintritt in den Sprachheilkindergarten mit 4 ½ Jahren. Die sprachfreie Intelligenztestung ergibt zu diesem Zeitpunkt einen IQ-Wert von 102. Während der 2 Jahre im Sprachheilkindergarten zeigt die Sprachentwicklung bei regelmäßigen Vorstellungen im SPZ kontinuierliche Fortschritte. Bis zum 8. Lebensjahr werden zeitweise zusätzliche logopädische Therapieblöcke für spezifische Ziele (Lautbildung, Wortstellung im Satz) verordnet. Zu keinem Zeitpunkt zeigt der Junge emotionale oder Verhaltensstörungen. Er lässt sich durch das Sprachvorbild seiner 2 Jahre jüngeren Schwester positiv motivieren und hat mehrere Freunde. Nach 2 Jahren Sprachheilschule wechselt der Junge erfolgreich in die Regelschule. Er erhält dort eine zusätzliche Förderung wegen einer inzwischen festgestellten Lese-Rechtschreib-Störung. Mit Noten von 2 in Mathematik und 3–4 in Deutsch wechselt der Junge nach der 4. Klasse auf eine Realschule.

20.3 Lese-Rechtschreibstörung (Legasthenie) und andere Formen der umschriebenen Entwicklungsstörungen

Viele Kinder haben in der Schule Probleme mit dem Lesen und dem Diktat, dies kann aber verschiedene Ursachen haben. Dabei sollte primär zwischen einer unspezifischen Lese-Rechtschreib-Schwäche, z. B. auf Grund mangelnder Deutsch-Kenntnisse oder einer Verminderung der Gesamt-Intelligenz und einer Lese-Rechtschreib-Störung im engeren Sinn unterschieden werden.

Fallbeispiel

Vorstellung eines 8jährigen Knaben wegen Schulproblemen. 4. von 5 Kindern. Der Vater berichtet, dass auch er in der Schule Probleme beim Lesen und bei der Rechtschreibung gehabt habe, Diktate seien wiederholt mit „ungenügend" bewertet worden. Er habe dies aber durch gute Leistungen in

Mathematik und Sachkunde ausgleichen können. Nach dem Hauptschulabschluss habe er eine Lehre als Mechaniker absolviert und später wegen guter Berufsleistungen eine Fachhochschule besucht. Jetzt arbeitet er erfolgreich als Ingenieur in einem Betrieb, habe aber immer noch Schwierigkeiten mit der Rechtschreibung.

Der Sohn hatte eine normale SS und Geburt sowie eine regelrechte motorische Entwicklung. Sprechen erster Worte mit 22 Monaten, Noch zu Beginn des 3. Lebensjahres habe er sich überwiegend mit Mimik und Gestik verständlich gemacht, was als „Sprechfaulheit" angesehen wurde. In der Schule habe er keine wesentlichen Probleme bei der Vermittlung von Sachthemen und in Mathematik, tue sich aber sehr schwer mit dem Lesen und mache im Diktat extrem viele Fehler.

Regelrechter körperlicher Untersuchungsbefund, lediglich Probleme bei komplexen Koordinationsaufgaben, z. B. dem Hampelmannsprung und der schnellen Diadochokinese, kein Hinweis für Seh- oder Hörstörung. Im Gesamt-IQ durchschnittliches Ergebnis mit heterogenem Profil, im Bielefelder Screening (BISC) bereits deutliche Auffälligkeiten, die sich im diagnostischen Rechtschreibetest und im Lesetest deutlich bestätigen.

Diagnose: Lese-Rechtschreibstörung (F 81.0)

Vorgehen: Ausstellung einer Bescheinigung für die Schule. Einleitung einer psychologischen Übungsbehandlung. Der 4 Jahre jüngere Bruder, der ebenfalls eine Sprachentwicklungsstörung habe, soll vor allem ein Trainingsprogramm der phonologischen Bewusstheit erhalten.

Zunehmend häufig wird vor allem bei Mädchen eine **umschriebene Entwicklungsstörung des Rechnens (Dyskalkulie)** diagnostiziert, wobei hier neuropsychologisch differenziert werden muss.

Von einer **umschriebenen Entwicklungsstörung der Motorik** sollte man nur dann sprechen, wenn eine signifikante Abweichung motorischer Fähigkeiten in einem standardisierten Test, z. B. dem M-ABC nachgewiesen wurde und trotz differenzierter Diagnostik keine befriedigende Erklärung für die motorischen Ungeschicklichkeiten festgestellt wurden, z. B. eine Zerebralparese, eine Muskelerkrankung o. a.

Problematisch ist die Diagnosestellung einer **umschriebenen Entwicklungsstörung der Wahrnehmung (= zentrale Verarbeitungs- und Wahrnehmungsstörung),** z. B. im Bereich des visuellen, auditiven oder taktil-kinästhetischen Systems. Während von pädagogischer und therapeutischer Seite immer wieder diese Diagnosen gestellt werden, können sie nach den Kriterien der sozialpädiatrischen Mehrbereichsdiagnostik bzw. der ICD-10 nur nach ausführlicher neuropädiatrischer, pädaudiologischer und neuropsychologischer Untersuchung als Hypothese postuliert werden. Ein Teil dieser Kinder profitiert von speziellen Hilfen, z. B. Seh- und Hörhilfen, vor allem aber von sonderpädagogischen Maßnahmen.

Die Betreuung und Behandlung von Kindern mit umschriebenen Entwicklungsstörungen ist sehr stark vom Erziehungssystem und den kulturellen Besonderheiten einer Gesellschaft abhängig. Allein in Deutschland gibt es in den einzelnen Bundesländern unterschiedliche Konzepte und Vorgehensweisen (➤ Kap. 6.3, ➤ Kap. 37, ➤ Kap. 38).

Einerseits wird versucht, die Persönlichkeit der Kinder zu stabilisieren und ihre Probleme zu „umschiffen", andererseits werden spezifische Lernprogramme angeboten. Auf der einen Seite wird gefordert, die Kinder in Regeleinrichtungen zu integrieren und ihre umschriebenen Schwächen, z. B. im Diktat oder Rechnen, zumindest in den ersten Schulklassen nicht zu benoten, andererseits wird befürwortet, frühzeitig eine sonderpädagogische Betreuung in speziellen Klassen einzuleiten. Dies gilt vor allem auch für Kinder mit Migrationshintergrund, bei denen die Beurteilung der Sprachfähigkeit erschwert ist.

Die Indikation für Heilmitteltherapien, vor allem Ergotherapie, bei umschriebenen Entwicklungsstörungen wird immer wieder diskutiert. In eindeutig diagnostizierten Fällen kann die Behandlung einer umschriebenen Entwicklungsstörung der schulischen Fähigkeiten wegen „drohender seelischer Behinderung" nach § 53 SGB XII durch das Jugendamt in Form von zusätzlichen sonderpädagogischen Maßnahmen mitfinanziert werden. Leider weigern sich die gesetzlichen Krankenkassen immer noch Behandlungen nur für umschriebene Entwicklungsstörungen schulischer Fähigkeiten zu übernehmen, da dies ihrer Ansicht nach eine Aufgabe des Schulsystems sei.
Die **Prognose** von Kindern mit umschriebenen Entwicklungsstörungen ist sehr unterschiedlich: besonders bei Sprech- und Sprachentwicklungsstörungen gibt es einen hohen Prozentsatz von Normalisierungen bzw. deutlichen Verbesserungen („**late bloomers**"). Auch bei den umschriebenen Entwicklungsstörungen schulischer Fähigkeiten gibt Beispiele für eine gute Integration in die Gesellschaft und eine Kompensation vorhandener Probleme, z. T. durch überdurchschnittliche Leistungen in anderen Bereichen (Handwerker, Künstler, Ingenieure, Mathematiker). Bei einem Teil der Betroffenen ist die Entwicklung jedoch weniger günstig: Sie haben lebenslang soziale Schwierigkeiten, z. B. in Beruf und Familie, häufiger psychiatrische Auffälligkeiten und neigen vor allem im jugendlichen Alter vermehrt zu Dissozialität und Delinquenz, was im Erwachsenenalter jedoch wieder zurückgeht.

Literatur

Buschmann A, Joos B, Rupp A, Dockter S, Blaschtikowitz H, Heggen I, Pietz J (2008) Children with developmental language delay at 24 months: result of a diagnostic work-up. Dev Med Child Neurol 50: 223–229.

Deutsche Gesellschaft für Kinder- und Jugendpsychiatrie und -psychotherapie (2007) Leitlinie: umschriebene Entwicklungsstörungen der Sprache: www.leitlinien.net

Deutsche Gesellschaft für Phoniatrie und Pädaudiologie (2008) Leitlinien Sprachentwicklungsstörung, Stimmstörung, Stottern, Poltern, Näseln/Gaumenspaltensprache, Schluckstörungen: www.leitlinien.net

Deutsche Gesellschaft für Sozialpädiatrie und Jugendmedizin (2008) Umschriebene Entwicklungsstörungen des Sprechens und der Sprache – Indikationen zur Verordnung von Sprachtherapie. http://www.dgspj.de/lllogo.php

Fegeler U, Jäger-Roman E (2008) Sprachförderung in Berliner KiTas – ermutigende Ergebnisse einer Breitenförderung in sozialen Brennpunkten. Kinder- und Jugendarzt 38: 15–17.

Hellbrügge T, Schneeweiß B. (Hrsg.) (2013) Sprache, Kommunikation und soziale Entwicklung – frühe Diagnostik und Therapie. Stuttgart, Klett-Cotta

Irblich D, Renner G (Hrsg.) (2009) Diagnostik in der klinischen Kinderpsychologie. Die ersten sieben Jahre. Göttingen, Hogrefe

Law J, Garrett Z, Nye C (2003) Speech and language therapy interventions for children with primary speech and language delay or disorder. The Cochrane Collaboration. http://www.cochrane.org/reviews

Rosenkötter H (2012) Neuropädagogik der Wahrnehmung und Motorik. Entwicklung und Bildung in der frühen Kindheit. Stuttgart: Kohlhammer

Straßburg HM, Dacheneder W, Kreß W (2012) Entwicklungsstörungen bei Kindern. Praxisleitfaden für die interdisziplinäre Betreuung. München: Elsevier

Straßburg HM, Ottensmeier H (2012) „Wahrnehmungsstörungen" bei Kindern – Versuch eines Brückenschlags zwischen medizinischer, psychologischer und pädagogischer Bewertung. Teil 1 und 2.Praxis Ergotherapie 25: 44–50 und 92–94

Suchodoletz W von. (2010) Therapie von Entwicklungsstörungen – was wirkt wirklich? Göttingen, Hogrefe

Szagun G. (2006) Sprachentwicklung beim Kind. Göttingen, Beltz

21 Autismus-Spektrum-Störung

Hans-Michael Straßburg

21.1 Einleitung

Autistische Verhaltensweisen sind ein häufiges Symptom bei Kindern mit Entwicklungsstörungen und Behinderungen. Nach neueren Angaben kommt eine Autismus-Spektrum-Störung bei 1 von 150 Kindern vor.

Als Autismus-Erkrankungen im engeren Sinne werden folgende Formen unterschieden:

Frühkindlicher Autismus (Kanner-Syndrom) mit schwerwiegender Ausprägung, einer Häufigkeit von ca. 2 : 1000, bei einer Bevorzugung des männlichen Geschlechts von 3:1 und einem Beginn meist vor dem 2. Lebensmonat

Asperger-Syndrom mit einer Häufigkeit von mehr als 8 : 10.000, einer Bevorzugung des männlichen Geschlechts von 10:1 und einem klinischen Beginn, der meist erst nach dem 3. Lebensjahr eine definitive Diagnose erlaubt. Das Asperger-Syndrom wird als Störung der sozialen Interaktion mit Pedanterie, eingeengtem Interesse, Sprach- und Sprechproblemen sowie Auffälligkeiten in der nichtverbalen Kommunikation, z. B. Mimik, Körperbewegung und Verhalten, bei normaler bis überdurchschnittlicher Intelligenz definiert

Atypischer Autismus, bei dem die Symptomatik in den verschiedenen Bereichen nicht vollständig ausgebildet ist.

21.2 Symptome

Wesentliche **Symptome** für die Diagnose einer **Autismus-Spektrum-Störung** sind:

- Qualitative Beeinträchtigungen der sozialen Interaktion in mindestens zwei der folgenden Bereiche:
  - Beeinträchtigung von Blickkontakt, Mimik und Gestik
  - Unfähigkeit, entwicklungsgemäße Beziehungen zu Gleichaltrigen aufzubauen
  - Mangel, spontan Freude, Interesse oder Erfolge mit anderen zu teilen
  - Mangel an sozio-emotionaler Gegenseitigkeit

Sozialpädiatrie in der Praxis. http://dx.doi.org/10.1016/B978-3-437-31630-3.00021-3
Copyright © 2014 Elsevier GmbH. All rights reserved.

- Qualitative Beeinträchtigungen der Kommunikation in mindestens einem der folgenden Bereiche:
  - Verzögertes Einsetzen oder völliges Ausbleiben der Entwicklung gesprochener Sprache
  - Deutliche Beeinträchtigung der Fähigkeit, ein Gespräch zu beginnen oder fortzuführen
  - Stereotyper oder repetitiver Gebrauch der Sprache
  - Fehlende Übernahme von entwicklungsgemäßen Rollen
- Beschränkte Verhaltensweisen, Interessen und Aktivitäten in mindestens einem der folgenden Bereiche:
  - Umfassende Beschäftigung mit einer oder mehreren Interessen von abnormem Inhalt und Intensität
  - Auffälliges starres Festhalten an bestimmten nicht funktionalen Gewohnheiten
  - Motorische Manierismen
  - Ständige Beschäftigung mit Teilen von Objekten.

Darüber hinaus können unspezifische Probleme wie Ängste, Schlafstörungen, Essstörungen, Wutausbrüche, Aggressionen und selbstverletzendes Verhalten sowie fehlende Lernbereitschaft bei z. T. guten visuellen und visuomotorischen Fähigkeiten und erstaunlichen Gedächtnisleistungen auftreten.

21.3 Ursachen

Ursachen und Entstehung des Autismus sind unklar, sicher handelt es sich aber nicht um ein einheitliches Krankheitsbild. Bei der frühkindlichen Form besteht eine hohe Übereinstimmung bei monozygoten Zwillingen, das Risiko für Geschwister ist oft mehr als 50-fach erhöht. Mittlerweile sind über 20 verschiedene Genanomalien als mögliche Ursache beschrieben worden, aber auch teratogene Einflüsse und Virusinfektionen während der Schwangerschaft werden als mögliche Ursache vermutet. Früher postulierte Verhaltensstörungen der Eltern oder Deprivationstraumen werden ebenso nicht mehr ernsthaft als Ursache diskutiert wie Folgeveränderungen nach Impfungen. Zusätzliche psychosoziale und emotionale Belastungen in und außerhalb der Familie können die Symptomatik verstärken.

Bei der frühkindlichen Form lassen sich mit den heute zur Verfügung stehenden diagnostischen Methoden in ca. 40 % der Fälle definierbare Erkrankungen, z. B. Hirnanlagestörungen wie kortikale Dysplasien, primärer Mikrozephalus, Balkenagenesie, Kleinhirnanomalien, genetische Veränderungen wie das Fragiles-X-Syndrom, Stoffwechselstörungen usw., nachweisen. In weiteren 50 % der Fälle können gravierende Hirnfunktionsstörungen festgestellt werden, davon in bis zu 90 % der Fälle unterschiedliche Intelligenzminderungen, in 30–40 % eine erhöhte zerebrale Anfallsbereitschaft, in 20 % zentrale Hörverarbeitungsstörungen und in 10 % Sehstörungen. Darüber hinaus werden häufig Störungen der Groß- und Feinmotorik, der Visuomotorik, der Tast- und Gleichgewichtsfunktion sowie weitere Teilleistungsstörungen diagnostiziert. Aktuell werden Funktionsstörungen der Spiegelneurone vor allem für Störungen des Empathie-Empfindens verantwortlich gemacht (➤ Kap. 12.7).

21.4 Diagnostik und Therapie

Bei der **Diagnostik** stehen wie üblich die Anamneseerhebung, die Verhaltensbeobachtung und die klinische Untersuchung des Kindes im Vordergrund. Man sollte bei eindeutiger klinischer Symptomatik zumindest ein EEG mit Einschlafen, eine Hörtestung, einen Sehtest mit Gesichts-

feldbestimmung, eine Chromosomenanalyse einschließlich der Molekulargenetik auf Fragiles-X-Syndrom, eine Kernspintomographie und Untersuchungen auf organische Säuren im Urin sowie eine Laktatbestimmung in Serum und Liquor durchführen.
Bei der psychologischen Diagnostik sollten primär die üblichen Entwicklungs- und Intelligenztests eingesetzt werden, die gerade bei autistischen Verhaltensstörungen jedoch sehr unterschiedliche Ergebnisse erbringen. Weitere psychologische Tests müssen in Abhängigkeit von der Symptomatik von erfahrenen klinischen Psychologen durchgeführt werden, z. B. die diagnostische Beobachtungsskala für autistische Störungen (ADOS-Test) (➤ Kap. 17). Patienten mit Autismus sind in der Regel mehrfachbehindert – seelisch, geistig und auch körperlich –, sodass ihnen Hilfe nach dem SGB XII § 53 zusteht. Eine einheitliche **Therapie** ist bei der Autismus-Spektrum-Störung nicht möglich (➤ Kap. 37, ➤ Kap. 38), spezifische Maßnahmen sollten nur bei nachgewiesener Indikation eingesetzt werden, z. B. eine Diät bei angeborener Stoffwechselstörung, Antiepileptika bei nachgewiesener Epilepsie. Eine ganzheitliche Behandlung sollte durch ein spezialisiertes, interdisziplinäres Team erfolgen, bei der die Individualität des Patienten und seiner Familie in ausreichendem Maße berücksichtigt werden kann. Im Vordergrund stehen dabei:

- Familienorientierte praktische Hilfen mit einer **Strukturierung von Umgebung und Tagesablauf**
- Kontinuität der Bezugspersonen und des Aufenthaltsorts
- Vermeidung abrupter Veränderungen
- Regelmäßige körperliche Aktivierung, z. B. durch sportliche Betätigung im Freien
- Förderung spezifischer Interessen, z. B. von Musik, Malen oder Basteln
- Aufzeigen langfristiger Perspektiven
- Beantragung eines Schulbegleiters
- Spezifische Trainingsmethoden (➤ Kap. 37, ➤ Kap. 38)

Unter **elektivem Mutismus** versteht man Verhaltensstörungen, bei denen der Patient nur mit bestimmten Personen und unter besonderen äußeren Bedingungen spricht. Oft bestehen zusätzliche Sprech- und Sprachprobleme, z. B. auch in Form von Teilleistungsstörungen. In der Familienanamnese finden sich gehäuft psychiatrische Erkrankungen.

21.5 Verhaltenstherapeutisch orientierte Programme bei Autismus

Hierfür sind z. B. TEACCH (Treatment and Education of Autistic and related Communication handicapped Children), EIBI (Early Intensive Behavioral Intervention) oder ABA (Applied Behavioral Analysis) geeignet. Außerdem hat sich bei vielen Patienten mit Autismus ein kontinuierlicher **Schulbegleiter** bewährt, der über das Jugendamt nach SGB VIII § 35a beantragt werden kann.
Darüber hinaus sind vor allem psychologische Maßnahmen wie unterstützende Psychotherapie für den Patienten und seine Eltern, evtl. auch eine Familientherapie sowie verhaltenstherapeutische Maßnahmen mit Aneignung spezieller Kommunikationsmethoden indiziert. Hierbei werden in letzter Zeit zunehmend verschiedene elektronische Kommunikationssysteme mit Erfolg eingesetzt (unterstützte Kommunikation).
Die **Prognose** des Autismus ist vor allem von der Intelligenzentwicklung abhängig: Bei einem IQ unter 50 vor dem 8. Lebensjahr ist die Prognose schlecht, bei einem IQ über 70 vor dem 8. Lebensjahr eher günstig. Diese menschen können unter Umständen einen evtl. auch ungewöhnlichen

Beruf erlernen und dort ihr Spezialwissen einsetzen. Langfristig sind 60 % der Betroffenen vollständig auf fremde Hilfe angewiesen, 25 % machen kontinuierliche Fortschritte und werden teilweise selbstständig, 10–20 % verschlechtern sich in der Adoleszenz, 10 % werden überwiegend selbstständig.

Literatur

Keysers C (2013) Unser empathisches Gehirn. Warum wir verstehen, was andere fühlen. München: C. Bertelsmann

Noterdaeme M, Freisleder FJ (2010) Entwicklungsstörungen im Überblick. München: Kienzle Edition Zuckschwerdt

Remschmidt H (2005) Kinder- und Jugendpsychiatrie – eine praktische Einführung. Stuttgart: Thieme

Sarimski K, Steinhausen HC (2008) Psychische Störungen bei geistiger Behinderung. Göttingen: Hogrefe.

Straßburg HM, Dacheneder W, Kreß W (2012) Entwicklungsstörungen bei Kindern. Praxisleitfaden für die interdisziplinäre Betreuung. München: Elsevier

22 Psychische Störungen bei Schulkindern und Jugendlichen

Hans-Michael Straßburg und Harald Bode

22.1 Einleitung

Psychische Störungen bei Kindern und Jugendlichen sind eine großes Problem, womit sich die gesamte Gesellschaft auseinandersetzen muss und das in erheblichem Maß zu Verunsicherungen und Grundsatzdiskussionen Anlass gibt. Sie kommen sicher gehäuft bei Kindern vor, die auch andere Entwicklungsauffälligkeiten und Behinderungen haben. In diesem Kapitel werden die wichtigsten Krankheitsbilder vorgestellt. Ob die Zahl von Kindern mit definierten Verhaltensstörungen in Deutschland in den vergangenen Jahren wirklich signifikant angestiegen ist oder ob ihre Zahl nur durch erhöhte Empfindsamkeit, Verunsicherungen und erweiterte Betreuungsangebote so deutlich angestiegen ist, lässt sich nicht sicher feststellen.

Sozialpädiatrie in der Praxis. http://dx.doi.org/10.1016/B978-3-437-31630-3.00022-5
Copyright © 2014 Elsevier GmbH. All rights reserved.

22.2 Aufmerksamkeits-Defizit-Hyperaktivitäts-Syndrom (ADHS)

Fallbeispiel

Vorstellung eines 9-jährigen Knaben wegen erheblicher Schulprobleme und Verhaltensauffälligkeiten.

Kind einer bei Geburt 19-jährigen Mutter, die in der Schwangerschaft rauchte und wiederholt seelischen Belastungen ausgesetzt war. Regelrechte Geburt, 3 Wochen gestillt, ab der 5. Woche vermehrte Unruhe und ausgeprägtes Schreien, vor allem am Abend. Selbstständiges Gehen mit 11 Monaten, Sprechen erster Worte mit 19 Monaten, Benutzung eines Gehfrei. Mutter seit dem 2. Lebensjahr berufstätig, Trennung der Eltern im 3. Lebensjahr, Mutter jetzt wieder verheiratet. Alle 14 Tage ist das Kind am Wochenende bei seinem leiblichen Vater. Bereits im Kindergarten immer wieder Verhaltensprobleme, nahm am Stuhlkreis nicht teil, war eigenwillig und aggressiv.

Jetzt Besuch der 3. Klasse, stört immer wieder den Unterricht, redet dazwischen, hatte wiederholt tätliche Auseinandersetzungen mit Schulkameraden. Die Leistungen in Mathematik und Sachkunde sind im Durchschnitt, in Deutsch unterdurchschnittlich, besondere Probleme hat er mit dem Diktat. Im Sport trotz guter motorischer Fähigkeiten wegen mangelnder Kooperation bei den Spielen keine guten Noten.

Kann bei den Mahlzeiten nicht ruhig am Tisch sitzen, hat große Probleme mit den Hausaufgaben, lässt sich immer wieder ablenken, will sich hauptsächlich mit Computerspielen beschäftigen, schaut bis zu 2 Stunden pro Tag fern.

Bisher seit über 1 Jahr Ergotherapie ohne genaue therapeutische Zieldefinition und ohne Elternanleitung und Aufsuchen der Beratungsstelle für Eltern.

Bei der körperlichen Untersuchung bis auf grenzwertiges Übergewicht keine wesentlichen Auffälligkeiten.

In den von der Mutter, den Lehrern und der Ergotherapeutin ausgefüllten DISYPS- und Conners-Fragebögen deutlich erhöhte Werte.

Diagnosen:

- Aufmerksamkeitsstörung mit Hyperaktivität (F 90.0G)
- Störung des Sozialverhaltens bei vorhandenen sozialen Bindungen (F 91.1G)
- V. a. umschriebene Entwicklungsstörung schulischer Fähigkeiten (F81.9V).

Maßnahmen:

- Psychologische Diagnostik, dabei Feststellung einer durchschnittlichen kristallinen Intelligenz mit besseren Werten im Handlungs- als im Verbalteil. Grenzwertige Befunde bei der Überprüfung der Rechtschreibung
- Ausführliche Besprechung der Ergebnisse mit der Mutter
- Vereinbarung eines verhaltenstherapeutischen Konzepts
- Einleitung einer Behandlung mit retardiertem Methylphenidat
- Vermittlung einer Hortbetreuung am Nachmittag
- Vermittlung einer Aufnahme im Sportverein
- Kurzfristige Wiedervorstellungen mit Gesprächsangeboten.

22.2.1 Symptome

Vermehrte motorische Unruhe ist vor allem im Kleinkindesalter ein physiologisches Symptom. Dennoch hat sich in den vergangenen Jahrzehnten gezeigt, dass es neben der kindertypischen, sehr variablen Bewegungsunruhe ein Störungsbild gibt, das zunehmend als eigenständiges Krank-

heitsbild abgegrenzt wird und das vor gut 150 Jahren erstmals eindrucksvoll von dem Frankfurter Psychiater Heinrich Hoffmann im „Zappelphilipp" beschrieben wurde. Dabei werden Störungen der Aufmerksamkeit, vermehrte motorische Aktivität sowie Störungen der Impulskontrolle als wichtigste Symptome angegeben. Vor allem bei Mädchen wird ein Subtyp der Aufmerksamkeitsstörung ohne Hyperaktivität beschrieben.
Eine erste Einschätzung ist mit Fragebögen, die von Eltern, Lehrern oder Therapeuten ausgefüllt werden, möglich: Die Häufigkeit des ADHS bei Kindern zwischen dem 5. und 15. Lebensjahr wird in der Literatur sehr unterschiedlich angegeben, in Deutschland kann man von einer Prävalenz von ca. 4 % ausgehen, wobei Jungen deutlich häufiger betroffen sind.

22.2.2 Ursachen

Die **Ursachen** für ein ADHS sind komplex: Zum einen besteht oft eine deutliche genetische Komponente, die sich unter anderem auch in Familien- und Zwillingsstudien nachweisen lässt. Zurzeit geht man davon aus, dass aufgrund unterschiedlicher genetischer Anomalien verschiedene Bereiche des Dopaminstoffwechsels vor allem im rechten Frontalhirnbereich verändert sind. Bei einigen Kindern konnten aber auch hirnorganische Schädigungen, z. B. ein Zustand nach ZNS-Infektion, periventrikuläre Veränderungen nach Frühgeburt, nach Intoxikation oder Stoffwechselstörungen festgestellt werden. Daneben spielen bei einem großen Teil der Kinder psychosoziale Probleme eine wesentliche Rolle, die oft mit ungünstigen familiären Situationen in Zusammenhang stehen. Unklar ist dabei, ob diese äußeren Faktoren die Symptomatik ursächlich erklären oder ob sie nicht auch Ausdruck dafür sind, dass zumindest ein Elternteil primär entsprechende Veranlagungen hat.
Bei einem Teil der Kinder bestanden bereits in der Schwangerschaft Probleme, häufig liegen auch Hinweise für leichtere Komplikationen vor, unter und nach der Geburt. Bis zu 60 % der Kinder waren als Säuglinge auffällig, überwiegend mit vermehrtem Schreien und Bewegungsunruhe, selten auch durch auffällige Bewegungsarmut als sogenannte „pflegeleichte" Säuglinge. Gehäuft lassen sich bei ADHS-Kindern umschriebene Entwicklungsstörungen überwiegend der Sprache nachweisen. Sie sind vermehrt unfallgefährdet, fallen durch expansive, oft ziellose Verhaltensweisen, eine hohe Ablenkbarkeit, Störungen der Grob- und Feinmotorik sowie mangelnde Impulskontrolle auf. Sie zeigen verminderte Frustrationstoleranz, Ich-Schwäche, Kontaktschwäche und vermehrtes Vermeidungsverhalten. Sie können sich häufig nicht an Regeln halten, wollen nicht schmusen, zeigen eine vermehrte Reizbarkeit, können Wesentliches von Unwesentlichem nicht unterscheiden. Verbal sind sie oft deutlich besser als bei schriftlichen Aufgaben; in der Schule spielen sie häufig den „Klassenkasper".
Bei den Müttern der Kinder bestehen häufig z. T. unbewusste Schuldgefühle, bei den Vätern eine erhöhte Ablehnungsrate, oft zeigt zumindest ein Elternteil auch ein auffälliges Verhalten. Viele Kinder haben chronische Schlafstörungen, ggf. auch verstärkt durch Obstruktion der oberen Luftwege oder eine Allergiedisposition. In über 15 % der Fälle lassen sich vor allem in der Einschlafphase EEG-Veränderungen, z. B. ein zentro-temporaler Sharp-Wave-Fokus, nachweisen.
Bei der psychologischen Beurteilung von Kindern mit ADHS sollte möglichst immer eine IQ-Bestimmung erfolgen. Nicht wenige der Kinder werden chronisch über-, selten aber auch unterfordert. Außerdem sind differenzierte Überprüfungen des Schulleistungsstandes und Untersuchungen auf das Vorliegen von Teilleistungsstörungen notwendig.

22.2.3 Therapie- und Betreuungsmaßnahmen

Es gibt keine monokausale **Behandlung des hyperkinetischen Syndroms,** immer müssen mehrdimensionale Strategien und eine psychotherapeutische Unterstützung der Familie eingesetzt werden. Primär kann durch verhaltenstherapeutische Maßnahmen eine Besserung versucht werden. Erwünschtes Verhalten wird dabei durch positive Verstärkung wie Lob und vermehrte Aufmerksamkeit beantwortet, bei negativen Verhaltensmustern sollten klare Absprachen zu den Konsequenzen getroffen werden, z. B. „time out" beim Ausrasten. Es sollten Ziele definiert werden, die in kleinen Schritten angegangen werden. Dabei können Entspannungsübungen und autogenes Training versucht werden, eventuell Musiktherapie, Psychomotorik, sensorische Integration oder bestimmte Sportarten (z. B. fernöstliche Kampfsportarten, Mannschaftssportarten) (➤ Kap. 37).
Ganz wichtig ist eine **Elternberatung,** die vor allem die Akzeptanz des Kindes mit seinen Symptomen sowie eine Entlastung von Schuldgefühlen zum Ziel haben sollte. Aber auch Lehrer und Erzieher sollten in eine Beratung mit einbezogen werden, damit sie das Kind nicht als Störobjekt, sondern als besonders hilfsbedürftige Person betrachten. Dabei können konkrete Richtlinien zum Umgang mit dem betroffenen Kind vermittelt werden: so kann es häufig positiv angesprochen werden und kleine Aufgaben übertragen bekommen. Darüber hinaus können Krisenbewältigungsstrategien besprochen werden. Hyperkinetische Kinder profitieren von speziellen Angeboten im Unterricht und in der Freizeit, z. B. vermehrter körperlicher Betätigung („bewegte Schule"), von Kletterübungen, Ausdauertraining, Mannschaftssportarten usw. In speziellen Kursen können Eltern verhaltenstherapeutische Maßnahmen im Umgang mit ihrem Kind, z. B. nach dem australischen Triple P (positive parenting program), vermittelt werden. Unter Umständen kann auch im Rahmen einer genauen **Ernährungsanamnese** ein Zusammenhang mit bestimmten Nahrungsmitteln, z. B. Schokolade, Süßigkeiten, Soja, Kuhmilch, Eier und Fast Food, festgestellt werden. Die immer wieder versuchten speziellen Diäten bei ADHS haben sich langfristig nicht bewährt, kurzfristige Erfolge sind eher unspezifischen Einwirkungen, z. B. der vermehrten Zuwendung, zuzuschreiben. Ganz wichtig ist es, für ausreichenden und geregelten Nachtschlaf und für eine ausreichende Trinkmenge am Tag zu sorgen.
Seit 1937 ist bekannt, dass sogenannte Weckamine, chemische Abkömmlinge des Amphetamins wie Methylphenidat, dessen Verordnung nur nach dem Betäubungsmittelgesetz bei Kindern und Jugendlichen zwischen 6 und 18 Jahren erfolgen darf, eine deutliche positive Wirkung auf das hyperkinetische Verhalten haben. Ein Ansprechen ist primär nicht vorhersehbar und hängt auch von der gewählten Dosierung ab. In 30 % der Fälle tritt eine dramatische Verbesserung ein, in 30 % der Fälle ist eine sichere Wirkung nicht nachzuweisen. Sehr wahrscheinlich beruht die Wirkung auf einer Beeinflussung des Dopaminstoffwechsels im Gehirn. Als Alternative zu einer Behandlung mit Methylphenidat wird der Noradrenalin-Wiederaufnahmehemmer Atomoxetin, der nicht betäubungsmittelpflichtig ist, eingesetzt (➤ Kap. 37, ➤ Kap. 38). Sehr selten ist die Gabe von anderen Psychopharmaka indiziert, in jüngster Zeit werden auch positive Ergebnisse von EEG-Biofeedback-Behandlungen berichtet. 70 % der Kinder mit hyperkinetischem Syndrom zeigen auch als Erwachsene noch Auffälligkeiten.

22.3 Störungen des Sozialkontakts

Heinrich Hoffmann (1809–1894) hat in seinem Buch „Der Struwwelpeter" die Geschichte des bösen Friederich als die eines Jungen mit einer Störung des Sozialverhaltens dargestellt. Hierbei han-

delt es sich um Verhaltensweisen, durch die altersgemäße Normen, Regeln oder Rechte anderer Menschen beeinträchtigt werden. Andere Bezeichnungen sind soziale Anpassungsstörung, Dissozialität oder antisoziales Verhalten, es bestehen Übergänge zur Delinquenz und zur Verwahrlosung. In der Kriminalstatistik 2007 wurden 102.012 Kinder unter 10 Jahren und 277.447 Jugendliche in Deutschland als Tatverdächtige registriert, wobei die Zahl der Kinder in den letzten Jahren leicht rückläufig ist, genauere Aufschlüsselungen liegen hierzu nicht vor.
Steinhausen (2006) hat folgendes zweidimensionale Modell für Störungen des Sozialverhaltens vorgeschlagen:

22.3.1 Symptome

In der ICD-10 wird **aggressiv-dissoziales Verhalten** in der Familie (F91.0) von solchem in der Gruppe bei **fehlenden sozialen Bindungen** (F91.1) und bei **vorhandenen sozialen Bindungen** (F91.2) unterschieden. F91.3 ist für oppositionell-aufsässige Kinder unter 10 Jahren vorgesehen. Diese Verhaltensweisen sind bei Kindern mit Entwicklungsauffälligkeiten aller Art insgesamt sicher häufiger, es bestehen Komorbiditäten mit dem ADHS, Teilleistungsstörungen und Intelligenzminderungen, aber auch Gemeinsamkeiten mit emotionalen Störungen und Störungen aus dem Autismus-Spektrum. Knaben sind insgesamt deutlich häufiger betroffen. Nach Ansicht mehrerer Autoren werden besonders in den letzten Jahren und zunehmend auch bei insgesamt normal entwickelten Jungen und Mädchen antisoziale Verhaltensweisen und Eigenwilligkeiten beobachtet. Winterhoff (2008) nennt diese Kinder „Tyrannen" und erklärt das Verhalten u. a. mit fehlender pädagogischer Kompetenz, übermäßiger Identifikation der Eltern mit dem Kind und den negativen Einwirkungen der modernen Welt, insbesondere den elektronischen Spielmaterialien, dem Fernsehen und Videospielen.
Viele dieser Kinder waren im Kleinkindalter bereits umtriebig und unruhig, haben wenig an Spielen mit Gleichaltrigen teilgenommen, konnten bei Gesellschaftsspielen nicht verlieren und zeigten früh sowohl in der Familie als auch in Kindergarten und Grundschule ein eigenwillig-oppositionelles Verhalten.
Bei vielen dieser Kinder und Jugendlichen lassen sich Zeichen psychischer Überforderung feststellen, oft bestehen unbewältigte Belastungen und psychische Traumatisierungen. Andererseits müssen vielfältige andere **Ursachen** berücksichtigt werden, z. B.:

- Genetische Einflüsse, die oft schwer von Umweltbedingungen, z. B. bei Drogen, Gewalt und Armut in der Familie, getrennt werden können
- Hormonelle Einflüsse, vor allem der Einfluss der Sexualhormone
- Hirnorganische Veränderungen, besonders im Bereich des Frontal- und des Temporallappens
- Anomalien im neuronalen Transmittersystem, besonders des Serotonins
- Schlafmangel
- Verschiedene chronische Organerkrankungen, z. B. Z. n. Hirn-OP, Epilepsie, Zöliakie, Migräne usw.

Fallbeispiel
Vorstellung eines 8-jährigen Jungen wegen immer wieder auftretenden Verhaltensstörungen in der Schule. Er würde Anweisungen der Lehrerin ignorieren, plötzlich durch die Klasse laufen, schlage Mitschüler ohne Grund und habe Mädchen am Rock gezerrt. Wenn er zur Rede gestellt werde, würde er

seine Verfehlungen nicht eingestehen, akzeptiere keine pädagogischen Maßnahmen und benutze üble Schimpfworte. Die Schulleistungen sind durchschnittlich.
Beide Eltern sind berufstätig, eine 2 Jahre jüngere Schwester macht keine Probleme.
Bis zum Kindergartenalter normale Entwicklung, dort hatte er bereits wenige Spielpartner, störte im Stuhlkreis, verweigerte Anordnungen und beteiligte sich nicht an gemeinsamen Spielen.
Während der Anamneseerhebung im SPZ mit den Eltern malt er differenziert einen gepanzerten Polizeiwagen mit vergitterten Fenstern, „um die Verbrecher abzutransportieren".
Bei der klinischen Untersuchung lassen sich keine Auffälligkeiten feststellen, ein EEG ist normal. Die psychologischen Leistungstests zeigen ein normales Profil.
Den Eltern und dem Kind wird eine primär ambulante Psychotherapie angeboten, die Langzeitprognose ist noch offen.

22.3.2 Behandlung

Die **Behandlung** von Kindern mit Störungen des Sozialverhaltens im Rahmen der Sozialpädiatrie ist sicher von vielen regionalen und individuellen Einzelaspekten abhängig und sollte immer in einem engen Netzwerk verschiedener Betreuungsstrukturen erfolgen. Wesentlich ist auch, ob es sich um ein primär normal entwickeltes Kind handelt oder ob Entwicklungsstörungen im weitesten Sinne nachweisbar sind.
Am wichtigsten ist ohne Zweifel die Prävention bzw. die möglichst frühzeitige Einleitung von Therapiemaßnahmen, die einerseits innerhalb der Familie, aber auch in den Betreuungseinrichtungen erfolgen. Hier haben sich verhaltenstherapeutische Konzepte, z. B. in Anlehnung an das in Australien etablierte Programm **Triple P (positive parenting program)** sehr bewährt, in denen Eltern pädagogische Ratschläge für den Alltag vermittelt bekommen. Weiterhin sind Verbesserungen des Lebensumfeldes wichtig, vor allem wenn Trennungsprobleme, Armut, soziale Isolation, chronische Erkrankungen bei anderen Familienmitgliedern u. v. m. bestehen. Wenn irgend möglich, sollten diese Kinder an Gemeinschaftsveranstaltungen teilnehmen, vor allem in Form von regelmäßigem Sport, aber auch aktive Musikausübung, Theaterspiel, Tanzen oder andere Hobbys sollten unterstützt werden.
Bei persistierenden Symptomen können klassische Psychotherapie, Psychomotorik, Psychodrama, Musiktherapie u. v. m. eingesetzt werden. Hier ist meist eine Weiterbehandlung durch Kollegen der Kinder- und Jugendpsychiatrie indiziert.
Nur in sehr schweren Fällen können nach sorgfältiger Abwägung Psychopharmaka wie Methylphenidat, Risperidon, Serotonin-Reuptake-Hemmer oder Lithium eingesetzt werden (➤ Kap. 37).

22.4 Emotionale Störungen

Harald Bode

Emotionale Störungen im Kindesalter stellen Verstärkungen normaler Entwicklungstrends hinsichtlich Ausprägung, Dauer und Alter dar. Sie führen zu deutlichen Problemen in sozialen Funktionen. Eine emotionale Störung mit Trennungsangst (F 93.0) wird dann diagnostiziert, wenn die Furcht vor Trennung den Kern der Angst darstellt, wenn eine solche Angst erstmals während der frühen Kindheit auftritt und länger als 4 Wochen anhält. Auch bei einer emotionalen Störung mit sozialer Ängstlichkeit (F 93.2) treten die Ängste bereits in der frühen Kindheit auf und sind ungewöhnlich stark. Auf phobische Störungen (F 93.1) des Kindesalters im engeren Sinne wird in diesem Leitfaden nicht weiter eingegangen.

22.4.1 Symptome

Symptome der emotionalen Störung mit Trennungsangst sind:

- Unrealistische, anhaltende Besorgnis über mögliches Unheil, das der Hauptbezugsperson zustoßen oder zur Trennung des Kindes von ihr führen könnte
- Andauernder Widerwille oder Weigerung, zum Kindergarten/zur Schule zu gehen, um bei der Bezugsperson oder zu Hause bleiben zu können
- Anhaltende, unangemessene Angst, allein oder ohne eine Hauptbezugsperson zu Hause zu sein oder außer Haus schlafen zu gehen
- Die Trennung betreffende, wiederholte Albträume
- Wiederholtes Auftreten körperlicher Beschwerden (Übelkeit, Bauchschmerzen, Erbrechen oder Kopfschmerzen) vor oder während der Trennung von einer Hauptbezugsperson
- Extremes, wiederkehrendes Unglücklichsein in Erwartung, während oder unmittelbar nach der Trennung (Unglücklichsein, Schreien, Wutausbrüche, Anklammern, Apathie oder sozialer Rückzug).

Die Symptome können unterschiedliche Schweregrade zeigen. Nicht selten treten begleitend Störungen des Sozialverhaltens oder ein ADHS auf.

Symptome der emotionalen Störung mit sozialer Ängstlichkeit sind:

- Anhaltende Ängstlichkeit und vermeidendes Verhalten in Situationen, in denen das Kind auf fremde Personen (auch Gleichaltrige) trifft
- Befangenheit, Verlegenheit, Sorge, ob das eigene Verhalten gegenüber Fremden angemessen ist
- Deutliche Beeinträchtigung/Verminderung sozialer Beziehungen
- Unglücklichsein in neuen Situationen (Weinen, Schweigen, Rückzug)
- Befriedigende Beziehungen zu Familienmitgliedern und gut bekannten Gleichaltrigen.

Für die **Entstehung** der emotionalen Störungen mit Trennungsangst bzw. mit sozialer Ängstlichkeit bedeutsam sind neben Persönlichkeits- bzw. Temperamentseigenschaften ängstigende Lebensereignisse in Familie oder Umfeld des Kindes und der elterliche Erziehungsstil (überprotektiv, Angst auslösend). Ähnliche Verhaltensweisen bei Geschwistern oder anderen Kindern können als Modell wirken. Genetische Faktoren und eine ängstlich-unsichere Bindung werden als bedeutsam angesehen. Überforderungssituationen können bei Kindern mit umschriebenen Entwicklungsstörungen oder unterdurchschnittlicher Intelligenz Ängste auslösen.

Die häufigste Form der emotionalen Störung mit Trennungsangst in der Praxis ist die **Schulphobie.** Leitsymptome sind häufige und wechselnde unspezifische körperliche Beschwerden, die dazu führen, dass die Kinder mit Wissen der Eltern nicht in den Kindergarten/die Schule gehen. Die wesentliche Ursache der Schulphobie liegt nicht im Bereich von Kindergarten/Schule, sondern in einer übermäßig engen Bindung des Kindes an die Hauptbezugsperson, in der Regel die Mutter. Von der Schulphobie abzugrenzen sind andere Formen der Schulverweigerung: Bei der Schulangst bestehen Ängste des Kindes im Zusammenhang mit der Schule (Leistungsprobleme bei Prüfungen, Angst vor Lehrern oder Schülern). Das Schulschwänzen ist eine Form dissozialen Verhaltens, das in der Regel ohne Wissen der Eltern stattfindet und nicht angstbegründet ist.

In der **Anamnese** sind für die emotionalen Störungen bzw. Ängste ursächliche Umfeldfaktoren, z. B. in Kindergarten oder Schule, belastende innerfamiliäre Faktoren und kindliche Bedingungen zu erfragen. Die **Elternberatung** umfasst vordringlich die Erkennung und den Abbau dysfunktionaler familiärer Interaktionsmuster (sekundärer Krankheitsgewinn) und der Angst unterstützenden Verhaltensweisen (überprotektives Verhalten).

22.4.2 Behandlung

In der **Behandlung** sollte die Autonomie der Kinder gestärkt werden, indem man ihnen etwas zutraut und zumutet.
Bei der Schulphobie muss zunächst Kind und Eltern klar vermittelt werden, dass keine körperliche Erkrankung vorliegt. Dennoch sollten die Symptome akzeptiert und ernst genommen werden. Eine vorübergehende Reduktion der täglichen Schulstunden kann sinnvoll sein. Die Stundenzahl sollte nach einem transparenten Programm schrittweise und in Absprache mit dem Kind, aber konsequent wieder auf die reguläre Zahl gebracht werden. Eine längere Befreiung vom Schulbesuch ist nicht hilfreich. Enge Absprachen von Eltern, Schule, ggf. Therapeut sind dabei erforderlich. Bei starker oder anhaltender Problematik sind familien- und verhaltenstherapeutische Behandlungsverfahren zu empfehlen. Eine kinder- und jugendpsychiatrische Beurteilung sollte dann erwogen werden. Die Beratung oder Behandlung kann nur erfolgreich sein, wenn bei der Hauptbezugsperson eine ernsthafte Bereitschaft und Fähigkeit besteht, die Autonomieentwicklung des Kindes aktiv zu unterstützen.

Fallbeispiel

Ein 5-jähriger Junge wird wegen immer wieder auftretender Bauchschmerzen, die gelegentlich mit Erbrechen einhergehen, erstmals bei einem Kinderarzt vorgestellt. Er war deshalb bereits zweimal stationär in einer Kinderklinik. Die dortige Abklärung hatte keine pathologischen Befunde erbracht. Der Junge ist das einzige Kind seiner Eltern. Die Mutter ist nicht berufstätig und hat den Jungen mit 4 ½ Jahren erstmals stundenweise in einen Kindergarten gegeben. Der Junge fordert stets die Anwesenheit der Mutter im Kindergarten und schreit heftigst, sobald die Mutter den Gruppenraum verlassen will. Der Junge ist bislang erst wenige Male mehr als eine halbe Stunde ohne die Mutter im Kindergarten geblieben. Auch die Untersuchung beim Kinderarzt erbringt keinen pathologischen körperlichen Befund. Der Junge ist motorisch, sprachlich und kognitiv altersgerecht entwickelt. In einem ausführlichen Gespräch erfährt der Kinderarzt, dass die Mutter- Kind-Bindung von Anfang an ungewöhnlich eng war.
Die Mutter schildert sich selbst als sehr ängstlich, sie habe wenige soziale Kontakte außerhalb der Kernfamilie. Sie sieht in einer eigenen Berufstätigkeit große Schwierigkeiten für sich und ihren Sohn. In einer familientherapeutischen Beratung werden mit beiden Eltern die Problematik überprotektiven Verhaltens und der sekundäre Krankheitsgewinn des Kindes herausgearbeitet. In einer psychotherapeutischen Behandlung der Mutter kann als gemeinsames Behandlungsziel die Autonomieentwicklung des Sohnes erarbeitet werden. Das mit Kindergarten, Mutter und Therapeut abgesprochene verhaltenstherapeutische Programm für den Jungen mit schrittweiser Verlängerung der Besuchszeit im Kindergarten
dazu, dass die emotionale Störung mit Trennungsangst nach einigen Monaten abklingt und der Junge ohne weitere oder erneute Ängste den Eintritt in die Schule meistert.

22.4.3 Sozialpädiatrische Behandlung bei emotionalen Störungen mit Trennungsangst oder sozialer Ängstlichkeit

Diagnostik:
- Allgemeine medizinische und biografische Anamnese
- Spezifische auslösende Situationen, Symptome, Folgen
- Körperliche Untersuchung (Verletzungszeichen)
- Untersuchung von Seh- und Hörvermögen
- Verhaltensbeobachtung
- Beurteilung von Verhalten und sozialer Anpassung (Fragebögen)
- Testung (umschriebene Entwicklungsstörungen, Intelligenz, Aufmerksamkeit)
- Informationsgewinnung aus Kindergarten oder Schule.

Therapie:
- In leichten Fällen: Elternberatung (gemeinsame Störungskonzepte und Behandlungsziele)
- Veränderungen des Umfeldes (Wechsel von Bezugsgruppen, Kindergarten, Schule)
- Maßnahmen zur Verbesserung von Autonomie, Selbstbewusstsein und Selbstwirksamkeit
- In schwereren oder anhaltenden Fällen: zusätzlich ambulante Verhaltens- und/oder Familientherapie
- Behandlung begleitender umschriebener Entwicklungsstörungen.

22.5 Hochbegabung

Harald Bode

22.5.1 Symptomatik

Eine **Hochbegabung** liegt bei einem Intelligenzquotienten von mindestens 130 vor, d. h. bei 2,2 % der Bevölkerung. Hochbegabung entsteht als Folge und in Wechselwirkung von genetischer Veranlagung und sozialem Umfeld. Sie muss nicht unbedingt mit großem schulischem oder beruflichem Erfolg einhergehen. Dieser kann sich nur einstellen, wenn Motivation, fördernde Umgebung und Stressbewältigungskompetenz hinzukommen.

Mögliche **Zeichen** einer Hochbegabung sind:
- Großes Detailwissen und sehr gutes Verständnis von Zusammenhängen
- Ungewöhnlich ausgeprägter Wortschatz und sprachlicher Ausdruck
- Frühes Interesse an Büchern, die weit über dem Altersniveau liegen
- Bevorzugung selbstständiger Arbeit
- Starke Vertiefung in bestimmte Probleme
- Perfektionistische Ansprüche, hohe Ziele
- Sehr weite intellektuelle Entwicklung, alterstypische emotionale Entwicklung
- Ständiges kritisches Hinterfragen von Autoritäten
- Wahl deutlich älterer Freunde
- Sehr individualistisches Verhalten
- Tendenz, Situationen allein bestimmen zu wollen
- Bevorzugung verbaler gegenüber körperlicher Auseinandersetzungen
- Langeweile oder Arbeitsverweigerung bei Routineaufgaben

- Kein Interesse an altersgemäßer Beschäftigung/am Schulstoff der Jahrgangsstufe
- Stören anderer Kinder, um Aufmerksamkeit zu erlangen
- Rolle als Streber oder Besserwisser
- Außenseiterposition, Isolation, Alleinsein.

Nicht selten wird von Eltern ein außergewöhnliches Verhalten ihres Kindes als Zeichen einer Hochbegabung interpretiert. Diese Annahme ist nicht immer zutreffend, sie bedarf einer Objektivierung durch einen Intelligenztest mit aktueller Normierung (➤ Kap. 17).

Hochbegabung und **ADHS** können sich in ähnlichem Verhalten äußern. ADHS tritt bei Hochbegabten ähnlich häufig wie bei Normalbegabten auf. Wenn statt einer vorliegenden Hochbegabung ein ADHS diagnostiziert wird, können sich Motivation, Arbeitshaltung und Konzentration durch anhaltende Unterforderung verschlechtern.

Das **Asperger-Syndrom,** eine Erkrankung aus dem Formenkreis der **Autismus-Spektrum-Störungen,** kann mit einer evtl. umschriebenen Hochbegabung einhergehen (➤ Kap. 21). Oftmals können die Betroffenen dank ihrer hohen Intelligenz ihre Schwächen kompensieren, sodass das Syndrom unerkannt bleibt. Bei fehlender Intelligenztestung werden Kinder/Jugendliche mit Asperger-Syndrom aber auch gelegentlich fälschlicherweise als geistig behindert eingestuft und nicht angemessen gefördert.

Hochbegabte Kinder durchlaufen die gleichen Entwicklungsschritte wie Normalbegabte, die kognitiven allerdings deutlich schneller. Hochbegabung ist per se keine Störung. Hochbegabte Jugendliche sind genauso beliebt wie Normalbegabte. Einige tendieren zu wenigen, exklusiven Freundschaften. Eine Diskrepanz zwischen kognitivem Entwicklungsstand und insbesondere emotionalen und sozialen Kompetenzen kann allerdings zu Missverständnissen und Problemen mit dem Umfeld (Isolation, störendes Vergalten, Aggression, Mobbing) führen. Häufig wird eine Hochbegabung erst festgestellt, wenn soziale oder psychische Probleme mit Krankheitswert auftreten.

Wenn die Hochbegabung nicht erkannt wird oder eine angemessene Förderung ausbleibt, kann es zu einer **Unterforderung** kommen, insbesondere in der Schule. Vor allem Jungen stören den Unterricht, um Aufmerksamkeit zu erhalten. Mädchen versuchen eher, sich ihrer Umgebung anzupassen, da sie die Schuld für ihre Lage bei sich selbst sehen. Sie arbeiten absichtlich langsam oder machen Fehler und werden dann nicht als hochbegabt erkannt. Jugendliche neigen dazu, ihre Langeweile demonstrativ zu zeigen oder sich einer anderen Beschäftigung zu widmen. Eine andauernde Unterforderung führt zu Motivationsmangel, Leistungsverweigerung und evtl. dazu, dass die Leistungen in Schule und Ausbildung hinter dem kognitiven Potenzial zurückbleiben („underachievement").

Die **Förderung** Hochbegabter ist wichtig und muss individuell festgelegt werden. Sie kann durch eine den Interessen des Kindes oder Jugendlichen entsprechende Anreicherung des Lern- und Erfahrungsangebots in Kindergarten, Schule und Freizeit (z. B. Bücher zum Lesen für Vorschulkinder, Musik, Sprachen, Computer, Schach, Kurse und Arbeitsgemeinschaften, Wettbewerbe, Austauschprogramme, Kooperation mit Universitäten) oder durch Beschleunigung der Schullaufbahn (vorzeitige Einschulung, Überspringen von Klassenstufen, flexibler Unterricht in höheren Klassen) erfolgen. Dabei ist neben der kognitiven Förderung auch die emotionale und soziale Entwicklung zu berücksichtigen. Auf mögliche emotionale Belastungen eines hochbegabten Schülers, z. B. durch Überspringen einer Klasse, ist zu achten. Bei reaktiven oder begleitenden Verhaltensstörungen sind pädagogische oder (schul-)psychologische Hilfen sinnvoll. Es gibt einzelne spezialisierte Schulen, Verbände und Einrichtungen, die Hochbegabtenförderung betreiben (➤ Kap. 38).

Fallbeispiel

Ein 7-jähriger Junge wird wegen unruhigem und störendem Verhalten, mangelnder Kooperationsbereitschaft in der 2. Grundschulklasse und partieller Schulverweigerung vom Kinderarzt in ein SPZ überwiesen. Die bisherigen Vorsorgeuntersuchungen waren unauffällig, die körperliche Untersuchung regelrecht. Nach Angaben der Eltern hat der Junge eine unauffällige medizinische Vorgeschichte und eine regelrechte motorische Entwicklung. Er habe sich bereits mit 3 Jahren für die Lesebücher seiner älteren Schwester interessiert und diese mit knapp 5 Jahren komplett lesen können. Der Junge kooperiert in der orientierenden Entwicklungsdiagnostik zunächst wenig, gibt an, dass die Schule langweilig sei. Angesprochen auf seine Interessen berichtet er sehr ausführlich, präzise und mit hohem sprachlichem Geschick über das Spiel mit seinen elektronischen Baukästen. Die psychologische Testung mittels HAWIK IV ergibt einen Gesamt-IQ von 142. Eine besondere Stärke zeigt sich in den Untertests, die das wahrnehmungsgebundene logische Denken erfassen, auch die Ergebnisse in den das Arbeitsgedächtnis messenden Untertests sind überdurchschnittlich. Die Angaben von Eltern und Lehrern in Fragebögen zu Zeichen eines ADHS (duPaul – ADHD Rating Skala IV, FBB-HKS) ergeben geringe, nicht signifikante Hinweise auf Hyperaktivität und Störungen der Aufmerksamkeit. Nach ausführlichen Beratungen und Informationen von Eltern und Schule unter Einbeziehung des schulpsychologischen Dienstes wird mit Zustimmung des Jungen ein Wechsel in die 3. Grundschulklasse mit begleitender Eltern- und Lehrerberatung vorgenommen. Er erhält als spezielle Aufgaben die durch einen Lehrer begleitete Betreuung der Naturkundesammlung seiner Schule. Außerdem werden die Teilnahme an einer Elektronik-AG in der benachbarten Stadt und der Besuch einer Sportgruppe mit Gleichaltrigen organisiert. Eine Behandlung mit Methylphenidat wird nicht als indiziert erachtet.

Einige Monate später berichten die Eltern dem Kinderarzt, dass die eingangs genannten Verhaltensprobleme des Jungen rasch abgeklungen seien.

22.5.2 Sozialpädiatrische Behandlung bei Verdacht auf Hochbegabung

Diagnostik.

- Medizinische und biografische Anamnese
- Körperliche Untersuchung inkl. Seh- und Hörvermögen
- Untersuchung der motorischen Fähigkeiten
- Psychologische Testung (Intelligenz, umschriebene Entwicklungsstörungen, Aufmerksamkeit)
- Beurteilung der emotionalen Entwicklung und sozialen Adaptation (Anamnese, Fragebögen)
- Informationsgewinnung aus Kindergarten oder Schule.

Therapie. Nur bei nachgewiesener Hochbegabung und Störungen der emotionalen Entwicklung oder der sozialen Adaptation erforderlich:

- Anreicherung des Lernangebots, Beschleunigung der Schullaufbahn
- Beratung der Eltern und der Erzieher/Pädagogen
- Vermittlung der Adressen von Selbsthilfegruppen und spezialisierten Institutionen zur Förderung Hochbegabter
- Gegebenenfalls Psychotherapie (Einzel- oder Gruppentherapie)
- Gegebenenfalls Behandlung von begleitenden Störungen (z. B. motorische Entwicklungsstörung, ADHS, Autismus-Spektrum-Störungen, Asperger-Syndrom).

22.6 Psychische Störungen bei geistiger und mehrfacher Behinderung

Harald Bode

22.6.1 Ursachen und Verbreitung

Psychische Störungen bei Kindern/Jugendlichen mit Behinderungen sind häufig. Bei körperlichen Behinderungen, isolierten Sinnesbehinderungen oder äußerlich sichtbaren Fehlbildungen und jeweils relativ normalen kognitiven Fähigkeiten erwachsen Probleme aus der selbst wahrgenommenen, von der „Norm" abweichenden äußeren Erscheinung, den nicht altersentsprechenden Fähigkeiten oder Erfahrungsmöglichkeiten, aus den Reaktionen der Umgebung und eingeschränkter sozialer Teilhabe. Diese Probleme sind aufgrund der normalen kognitiven Fähigkeiten von den Betroffenen und den Bezugspersonen relativ gut zu erkennen und zu verstehen und damit der sprachlichen/psychotherapeutischen Begleitung zugänglich. Trotzdem besteht oft eine erhebliche und anhaltende subjektive Belastung der Kinder/Jugendlichen. Sie zeigen reaktive psychische Auffälligkeiten, die auch die Bezugspersonen belasten.

An dieser Stelle soll etwas ausführlicher auf Kinder/Jugendlichen mit geistiger oder mehrfacher Behinderung eingegangen werden, bei denen psychische und Verhaltensstörungen besonders häufig auftreten, oft schwieriger zu verstehen sind und nicht selten therapeutisch eine relative Ratlosigkeit verursachen. Diese Störungen entstehen überwiegend als Folge einer dysfunktionellen Interaktion des Kindes/Jugendlichen und seines Umfeldes. Interdisziplinäre Konzepte sind zur Prävention, Diagnostik und Behandlung erforderlich. Es bestehen heute noch erhebliche Lücken in der kompetenten Versorgung dieser Kinder/Jugendlichen.

Geistig behinderte Kinder/Jugendliche besitzen aufgrund ihrer genetisch-biologischen Ausstattung oft Eigenschaften, die ihnen die Interaktion mit Bezugspersonen und der Umwelt erschweren. Dabei sind für verschiedene biologisch-genetische Störungsbilder unterschiedliche und z. T. sehr charakteristische Kompetenzprofile (**„Verhaltensphänotypen"**) beschrieben worden, die keineswegs streng mit den Intelligenzquotienten korrelieren. Die soziale Interaktion und Teilhabe eher ermöglichende Eigenschaften besitzen z. B. viele Kinder mit Trisomie 21, während Kinder mit Fragiles-X-Syndrom oder Prader-Willi-Syndrom nicht selten problematische soziale Verhaltensweisen zeigen.

Geistig behinderte Kinder/Jugendliche haben häufig Probleme in folgenden Bereichen:

- Informationsverarbeitung
- Arbeitsspeicher für das Gedächtnis
- Erkennen relevanter Informationen
- Auswahl/Kontrolle von Problemlösungsstrategien
- Ausdauer
- Komplexe Aufgaben
- Handlungsplanung
- Praktische Fähigkeiten und Fertigkeiten
- Bindungsfähigkeit
- Erkennen der Gefühle anderer
- Verständnis, Perspektivenübernahme, Problemlösefähigkeit
- Kommunikationsfähigkeit
- Soziale Interaktion
- Selbstregulation
- Selbstkonzept.

Die Probleme in den genannten Bereichen bedingen eingeschränkte Anpassungs- oder Bewältigungsfähigkeiten der Kinder/Jugendlichen in der Interaktion mit Bezugspersonen und Umfeld. Hieraus können ungünstige soziale Erfahrungen erwachsen (z. B. Über- oder Unterforderung, Stress, gelernte Hilflosigkeit, negative Konditionierungen). Die psychischen Störungen oder Verhaltensauffälligkeiten der Kinder/Jugendlichen mit geistiger oder mehrfacher Behinderung sind damit häufig ganz wesentlich verursacht durch nicht an ihre Bedürfnisse angepasste Verhaltensweisen von Bezugspersonen (Eltern, Erzieher, andere Kinder) oder Strukturen/Prozesse in ihrer Lebenswelt (Familie, pädagogische und therapeutische Einrichtungen).
Psychische und Verhaltensauffälligkeiten/-störungen treten wahrscheinlich bei mindestens einem Drittel der geistig oder mehrfach behinderten Kinder/Jugendlichen auf. Sie werden bei stärkerer Intelligenzminderung und in inadäquater Umgebung (große Gruppen, Heime mit Personal ohne Fachkompetenz und Empathie) häufiger beobachtet. Die Abgrenzung zur „Normalität" ist naturgemäß schwierig, unscharf und durch die Maßstäbe des Umfeldes mitbestimmt.

22.6.2 Symptomatik

Häufige psychische und **Verhaltensstörungen** bei geistig oder mehrfach behinderten Kindern/Jugendlichen sind:
- Essstörungen
- Schlafstörungen
- Autismus-Spektrum-Erkrankungen
- Hyperkinetische/impulsive Störungen
- Störungen des Sozialverhaltens: oppositionelles, aggressives Verhalten (z. B. selbst-, fremdverletzend)
- Depressive Störungen (sozialer Rückzug)
- Emotionale Störungen: Trennungsangst, Phobien, Angst vor Neuem
- Zwangsstörungen
- Distanzlosigkeit (→ Risiko für sexuellen Missbrauch)
- Spezifische Störungen: Stereotypien, PICA (anhaltender Verzehr nicht essbarer Substanzen), Kotschmieren, taktile/optische Selbststimulation, exzessive Masturbation u. a.

22.6.3 Diagnostik

Die **Diagnostik** mit Fragebögen für die Betroffenen oder durch Selbstauskünfte ist kaum standardisierbar und reliabel. Für die standardisierte Befragung von Bezugspersonen eignen sich z. B. der „Verhaltensfragebogen für Behinderte" oder die „Nisonger Child Behavior Rating Form". In der Diagnostik spielt die genaue Beobachtung des Verhaltens und der auslösenden Faktoren/Trigger eine wesentliche Rolle. Eine Videodokumentation kann sehr hilfreich sein. Sinn, inhaltliche Botschaft und mögliche Bedeutung des auffälligen Verhaltens sollten im Kontext des Lebensumfeldes und der Verhaltensweisen der Bezugspersonen interdisziplinär zu entschlüsseln versucht werden. Dazu ist langjährige Erfahrung hilfreich. Eine Zuordnung der Symptomatiken zu den F-Diagnosen der ICD-10 ist von begrenztem Wert.

22.6.4 Prävention

Zur **Prävention** psychischer und Verhaltensstörungen bei geistig oder mehrfach behinderten Kindern/Jugendlichen sollte deren Selbstwertgefühl und erkannte Selbstwirksamkeit gestärkt, eine positive Beziehung zwischen Eltern bzw. Erziehern und Kind/Jugendlichem unterstützt, die soziale Teilhabe in Kindertagesstätte, Schule und Freizeit gefördert und die emotionalen, kognitiven und pragmatischen Bewältigungskräfte der Familien gestärkt werden. Angemessene Gruppengrößen bzw. Personalschlüssel, qualifizierte Erzieher/Pädagogen und den spezifischen Bedürfnissen des Kindes angepasste Lebenswelten (klare, einfache und bekannte Routinen, konstante Örtlichkeiten und Personen, angepasstes Reiz-, Spiel- und Förderangebot, Empathie u. a.) sind erforderlich. Die Integration/Inklusion in integrative oder Regelkindertagesstätten oder -schulen kann bei geeigneten Bedingungen für viele Kinder empfohlen werden.

22.6.5 Behandlung

Die **Behandlung** psychischer und Verhaltensstörungen bei geistig oder mehrfach behinderten Kindern/Jugendlichen setzt eine genaue Diagnostik nach den ausgeführten Prinzipien voraus. Organische Erkrankungen, Schmerzen oder Sinnesbehinderungen (Seh-, Hörstörungen) müssen ausgeschlossen bzw. ursächlich oder symptomatisch behandelt werden. Die angeführten Prinzipien der Prävention sollten konsequent umgesetzt werden. Die Behandlung erfordert ein sehr individualisiertes Vorgehen, welches das Kind und seine Umfeld gleichermaßen einschließt. Oft ist eine Kombination pädagogischer, psychologischer und medizinischer Vorgehensweisen sinnvoll, die in interdisziplinärer Kooperation entwickelt und im Verlauf angepasst werden sollten.
Häufig lassen sich Störungen durch Änderungen des Umfeldes erklären bzw. durch Anpassung an die Bedürfnisse der Kinder/Jugendlichen günstig beeinflussen. Kommunikationsförderung (z. B. unterstützte Kommunikation) und adäquate kognitive, emotionale und soziale Angebote erleichtern die Teilhabe, reduzieren Stress und steigern das Wohlbefinden der Kinder/Jugendlichen. Reichen diese Maßnahmen nicht aus, können verhaltenstherapeutische Techniken positive Verhaltensweisen verstärken bzw. die häufiger zu beobachtende positive Verstärkung unerwünschter Verhaltensweisen durch Bezugspersonen z. B. durch das „Time-out-Prinzip" abbauen helfen.
Eine Pharmakotherapie ist als unterstützende Maßnahme zu verhaltenstherapeutischen Maßnahmen manchmal nützlich und als Ultima Ratio indiziert bei Selbst- oder Fremdgefährdung, bei massiver Belastung des Kindes/Jugendlichen oder der Umgebung durch das gestörte Verhalten (z. B. Schlafstörung, Tag-Nacht-Umkehr, exzessive Stereotypien bei schwerster geistiger Behinderung) oder wenn der Verbleib des Kindes/Jugendlichen in Familie, Kindertagesstätte oder Schule infolge einer schwersten Verhaltensstörung gefährdet ist. Als mögliche Stoffgruppen kommen je nach Problemstellung (insbesondere atypische) Neuroleptika, Antidepressiva, Stimulanzien, Sedativa, Antiepileptika oder Melatonin in Frage. Ihr Einsatz bedarf entsprechender Erfahrung und sollte bei längerfristiger Anwendung regelmäßig auf seine Notwendigkeit hin überprüft werden (➤ Kap. 37).
Eine große Bedeutung hat die Arbeit mit den Eltern (ggf. auch den Erziehern/Pädagogen), die infolge der psychischen oder Verhaltensstörungen der Kinder/Jugendlichen oft massiv belastet, überfordert oder auch von Schuldgefühlen geplagt sind. Die elterlichen Bewältigungskräfte sollten gestärkt, ihr intuitives Verhalten dem Kind gegenüber verbessert und die Interaktion mit dem

Kind durch Beratungen ggf. mit Videounterstützung gelingender werden. Zusammenfassend können folgende **Behandlungsmaßnahmen** sinnvoll und wirksam sein (➤ Kap. 37, ➤ Kap. 38):

- Ausschluss/Behandlung körperlicher Erkrankungen/Schmerzen/Sinnesbehinderungen
- Änderungen des Lebensumfeldes
- Beratung, Wechsel der Einrichtung
- Integration/Inklusion in Regeleinrichtungen
- Basale Stimulation/Kommunikation
- Entspannungstechniken (taktil, vestibulär, optisch, akustisch)
- Kommunikationsförderung
- Adäquates Reiz-, Spiel- und Förderangebot
- Musiktherapie/kreatives Malen/Werken
- Tiergestützte Therapien
- Verhaltenstherapeutisch orientierte Maßnahmen
- Spezifische Programme (z. B. TEACCH)
- Pharmakotherapie
- Elternarbeit.

Fallbeispiel

Ein 8-jähriges Mädchen mit schwerer Intelligenzminderung ohne Sprachfähigkeit infolge einer chromosomalen Störung hört innerhalb weniger Monate auf zu laufen, verliert das Interesse an Spielzeug, wird sehr unruhig und unzufrieden und schlägt sich zunehmend heftig auf die Augen. Es kommt zu äußerlich sichtbaren Verletzungen im Gesichtsbereich. Das Kind kann nur noch mit Armschienen und kontinuierlicher Anwesenheit einer Bezugsperson bzw. im Bett angeschnallt betreut werden. Der Augenarzt stellt eine traumatische Erblindung des rechten Auges und einen grauen Star, eine schwere Glaskörperblutung und eine teilweise Netzhautablösung des linken Auges fest. Nach ausführlichen Verhaltensbeobachtungen unter Einbeziehung der Einschätzungen der Mutter und Verlaufsbeobachtungen wird der Entschluss gefasst, durch eine Augenoperation (Kunstlinse, künstlicher Glaskörper und Verplombung der Netzhaut) links die Restsehfähigkeit dieses Auges zu sichern. Bereits am Tag nach der erfolgreichen Operation ist das Kind erkennbar ruhiger und zufrieden und interessiert sich wieder für Spielzeug und Umfeld. Einige Tage später macht es an der Hand der Mutter wieder erste Schritte.
Durch den infolge der chromosomalen Störung entstandenen grauen Star hatte das Mädchen seine Sehfähigkeit verloren. Die Verhaltensstörungen (Hyperaktivität, Verlust an Interessen, Autostimulation durch Schlagen auf Gesicht/Augen) waren die Folge und ließen sich durch eine operative Wiederherstellung der Sehfähigkeit beseitigen.

Literatur

Ettrich C, Ettrich KU (2006) Verhaltensauffällige Kinder und Jugendliche. Berlin: Springer.
Hartmann T (2004) ADHS als Chance begreifen – Nennen wir es das Edison-Gen. Lübeck: Schmidt-Römhild.
Herpertz-Dahlmann B, Rösch S (2007) Entwicklungspsychiatrie – biopsychologische Grundlagen und die Entwicklung psychischer Störungen. Stuttgart, Schattauer
Remschmidt H (2005) Kinder- und Jugendpsychiatrie – eine praktische Einführung. Stuttgart, Thieme
Steinhausen HC, Rothenberger A, Döpfner M (2010) Handbuch ADHS. Stuttgart, Kohlhammer
Straßburg HM, Dacheneder W, Kreß W (2012) Eintwicklungsstörungen bei Kindern. Praxisleitfaden für die interdisziplinäre Betreuung. München: Elsevier

23 Epilepsien

Hans-Michael Straßburg

23.1 Einleitung

Unter einer Epilepsie versteht man wiederholt auftretende, vom Gehirn ausgehende anfallsartige Zustände, denen abnorm synchrone, hochfrequente elektrische Entladungen von Nervenzellen zugrunde liegen. Sie sollten von Gelegenheitsanfällen (z. B. zerebralen Anfällen bei Fieber) und nur im EEG nachweisbaren Zeichen einer zerebralen Anfallsbereitschaft (hypersynchroner Aktivität) abgegrenzt werden. Viele Menschen mit mentalen Entwicklungsstörungen zeigen EEG-Veränderungen, insbesondere eine Verlangsamung des Grundrhythmus, Herdveränderungen und hypersynchrone Potenziale. Seit Urzeiten sind Anfälle mit vielen mystischen Vorstellungen und Ängsten verbunden.

Sie sind ein häufiges Phänomen, ca. 5 % aller Menschen haben während ihres Lebens mindestens einen zerebralen Anfall, überwiegend im Kindesalter. 15 % der Menschen mit leicht verminderter Intelligenz leiden an einer Epilepsie und über 30 % der Menschen mit einem IQ unter 50. Andererseits gibt es aber auch Menschen mit Epilepsie, die überdurchschnittlich intelligent sind.

Fallbeispiel

Erstvorstellung eines 3-jährigen Knaben wegen Z. n. zerebralem Anfall von mehr als 10 min Dauer. Die bisherige Entwicklung war verzögert, im ersten Lebensjahr fragliche kurze Abwesenheitszustände, selbstständiges Gehen mit 2,9 Jahren, Sprechen erster Worte mit 3,6 Jahren.

Bei der Untersuchung des körperlich normal entwickelten Kindes fallen an der Haut mehrere 1 × 2 cm große pigmentarme Areale auf sowie einzelne rote, nicht wegdrückbare Papeln im Bereich beider Wangen. Es lassen sich keine neurologischen Ausfälle nachweisen, das Kind ist vermehrt umtriebig und schwer zu führen.

Unter dem Wood-Licht, einer UV-Lampe, leuchten die Hautareale deutlich hell auf.

Sonographisch werden eine Vergrößerung beider Nieren und einzelne echogene Areale im Herzen festgestellt, das EEG zeigt einen pathologischen Befund mit langsamen und steilen Wellen über der rechten > linken Hemisphäre.

Im zu einem späteren Zeitpunkt angefertigten zerebralen MRT werden multiple subependymale Knoten an beiden Seitenventrikeln und Strukturanomalien in beiden Großhirnhemisphären (Hamartome) nachgewiesen.

Sozialpädiatrie in der Praxis. http://dx.doi.org/10.1016/B978-3-437-31630-3.00023-7
Copyright © 2014 Elsevier GmbH. All rights reserved.

Diagnosen:
- Tuberöse Sklerose Bourneville Pringle (Q 85.1G)
- Lokalisationsbezogene, symptomatische Epilepsie (G 40.2G).

Therapie:
- Einleitung einer antikonvulsiven Behandlung, primär mit Vigabatrin
- Regelmäßige Kontrollen der Sehfunktionen
- Einleitung einer interdisziplinären Frühförderung
- Beantragung eines Behindertenausweises
- Kontaktvermittlung mit der Eltern-Selbsthilfegruppe
- Regelmäßige Kontrollen der Nieren- und Herzfunktion
- Regelmäßige sonographische Kontrollen der Nieren und des Herzens.

Es bestehen keine Hinweise für andere betroffene Familienmitglieder, sodass von einer Neumutation ausgegangen wird. Bei Zunahme der zerebralen Tumoren muss eine Behandlung mit dem Medikament Everolimus erwogen werden.

Die Prognose in Bezug auf die geistige Entwicklung ist eingeschränkt, allerdings werden nicht selten deutliche Stabilisierungen beobachtet. Ein wesentliches chronisches Gesundheitsproblem können vor allem die Nieren bereiten.

23.2 Ursachen

Die **Ursachen** zerebraler Anfälle sind vielfältig, oftmals liegen mehrere Störungen gleichzeitig vor (multifaktorielle Genese):

- Genetische Veranlagung
- Hirnorganische Veränderungen, z. B. Fehlbildungen der Hirnrinde (= kortikale Dysplasien)
- Sekundäre ZNS-Schädigungen, z. B. nach Sauerstoffmangel, durch Unfälle, Infektionen und Blutungen
- Fieber
- Schlafmangel
- Stoffwechselstörungen, z. B. Unterzuckerung (Hypoglykämie), Elektrolytstörungen, spezielle andere Stoffwechselstörungen (z.B. Speicherkrankheiten und Mitochondriopathien)
- Tumoren
- Intoxikationen
- Äußere Reize, z. B. Flackerlicht, Hyperventilation und anderes.

Die Zusammenhänge zwischen einer Epilepsie und genetischen Anomalien sind vielfältig, zunehmend lassen sich bei Epilepsien ohne hirnorganische Veränderungen Störungen der Ionenkanäle in den Nervenzellmembranen und Neurotransmitteranomalien nachweisen.

23.3 Symptomatik

Epielpsien können mit sehr unterschiedlichen Symptomen auftreten. Unter **Semiologie** versteht man die Beurteilung der klinisch erkennbaren zerebralen Anfälle. Folgende Typen werden unterschieden:

- **Primär generalisierte Anfälle,** entweder in Form von nur wenige Sekunden dauernden Abwesenheitszuständen oder in Form von generalisierten, mehrere Minuten dauernden tonisch-klonischen Anfällen
- **Fokale Anfälle** oder Partialanfälle, bei denen nur umschriebene Körperregionen betroffen sind, die entweder einfach fokal, z. B. im Gesicht oder an den Händen, oder komplex fokal mit Bewusstseinsstörung einhergehen
- Außerdem können fokale Anfälle sekundär generalisiert auftreten.

Die genaue Beschreibung eines Anfallsgeschehens, evtl. auch die Dokumentation mittels Video, ist für die Beurteilung oft unentbehrlich.

Wichtigste diagnostische Methode zur Erkennung einer erhöhten zerebralen Anfallsbereitschaft ist das EEG, das bei allen Kindern mit globalen Entwicklungsstörungen mindestens einmal abgeleitet werden sollte. Hierzu ist vor allem auch bei jüngeren Kindern eine artefaktfreie Ableitung durch erfahrenes Personal und eine fachkundige Auswertung Voraussetzung. Bei jedem Patienten sollten klassische Provokationsmethoden wie Öffnen und Schließen der Augen, wiederholte Lichtreize und bei ausreichender Mitarbeit Hyperventilation eingesetzt werden. Bei vielen Patienten lässt sich erst im Rahmen einer Schlaf-EEG-Ableitung ein auffälliger Befund, z. B. in Form von fokaler hypersynchroner Aktivität, nachweisen.

Unter **symptomatischen Epilepsien** versteht man alle Formen mit nachweisbarer hirnorganischer Ursache, unter **idiopathischen Epilepsien** solche mit nicht erkennbaren hirnorganischen Veränderungen und überwiegend genetischer Disposition. Nach der aktuellen Nomenklatur der Internationalen Liga gegen Epilepsie (ILAE) erfolgt nur noch eine Unterteilung in genetisch, strukturell-metabolisch und unbekannt. Im Neugeborenenalter können zerebrale Anfälle sehr unterschiedliche klinische Symptome, z. B. diskrete tonische Streckzustände, Atempausen, Bradykardien und Zyanose-Zustände, verursachen.

Am häufigsten treten zerebrale Anfälle im Kleinkindalter bei Fieber auf. Wenn sich keine hirnorganische Ursache nachweisen lässt bzw. eine familiäre Disposition vorliegt, spricht man von **Fieberkrämpfen.** Nur sehr selten kommt es zum Übergang von meist langdauernden Fieberkrämpfen in schwer behandelbare Epilepsieformen, evtl. auch mit sekundärer Intelligenzminderung. Bei immer mehr Patienten wird dabei die Diagnose eines **Dravet-Syndroms,** einer genetisch bedingten schweren Epilepsie mit progredientem Verlust zerebraler Funktionen gestellt. Dies wurde vor allem auch bei Menschen mit angeblichem „Impfschaden", z. B. nach einer Keuchhusten- oder Masernimpfung in den ersten Lebensjahren nachgewiesen.

Während Myoklonien ein unspezifisches Symptom darstellen, das sowohl bei gesunden Kindern als auch bei progredienten neurodegenerativen Erkrankungen auftreten kann, treten **Absencen** in Form von kurzen Abwesenheitszuständen mit charakteristischen 3/sec spike-waves im EEG überwiegend bei primär normal entwickelten Kindern zwischen 5 und 12 Jahren auf.

Häufigste Ursache für fokale Anfälle sind die genetisch determinierten zentro-temporalen Sharp-waves-Anfälle oder die **Rolando-Epilepsie,** die mit charakteristischen EEG-Veränderungen einhergehen. Ursache hierfür sind funktionelle Reifungsstörungen der Ionenkanäle, insbesondere der Natriumionenkanäle, die in seltenen Fällen auch mit unterschiedlich ausgeprägten Formen der Intelligenzminderung assoziiert sein können.

Ein besonderes Problem stellt die **BNS-(Blitz-Nick-Salaam-)Epilepsie** oder **West-Syndrom** im Säuglings- und Kleinkindesalter dar. Hierbei kann es aus völlig unterschiedlichen Gründen zu progredienten Hirnfunktionsstörungen mit oft bleibender mentaler Beeinträchtigung kommen. Typisch sind multifokale irreguläre hypersynchrone Potenziale im EEG im Sinne einer **Hypsarrhythmie.** Bei persistierenden BNS-Anfällen muss praktisch immer mit schweren bleibenden

Hirnfunktionsstörungen gerechnet werden, sodass eine konsequent ausführliche Diagnostik und frühe Therapieeinleitung notwendig sind.
Lässt sich das West-Syndrom therapeutisch nicht beherrschen, kommt es zum Übergang in das **Lennox-Gastaut-Syndrom,** bei dem neben ausgeprägten kleinen und großen Anfällen praktisch immer auch eine bleibende schwere mentale Entwicklungsstörung besteht.
Liegt bei einer ausgeprägten EEG-Veränderung im Sinne einer Rolando-Epilepsie auch eine Sprachentwicklungsstörung vor, spricht man von einem **Landau-Kleffner-Syndrom** (epileptische Aphasie), bei im Schlaf nachweisbaren häufigen hypersynchronen Potenzialen ohne klinische Anfälle auch vom **ESES-Syndrom** (= electrical status in slow wave sleep).
Umfangreiche psychologische Untersuchungen bei Patienten mit Epilepsie haben gezeigt, dass es keine typische „epileptische Persönlichkeit" gibt.
Wesensänderungen bei Menschen mit Epilepsien können durch

- genetische Faktoren,
- primäre Hirnschädigung,
- psychoreaktive Störungen,
- medikamentöse Effekte,
- und anfallsbedingte Hirnfolgeschäden, z. B. eine Hippocampussklerose,

erklärt werden.
Bei anfallsartigen Zuständen muss aber immer auch an nichtzerebrale, d. h. nicht primär vom Gehirn ausgehende Ursachen gedacht werden, z. B. an:

- Hirnstamm-bedingte Atempausen
- Stoffwechselstörungen wie Hypoglykämie und Hypokalzämie
- Hypotone Blutdruckregulationsstörungen (=Synkope)
- Respiratorische Affektsynkopen
- Herzrhythmusstörungen
- Intoxikationen
- Zerebrale Durchblutungsstörungen
- Anfallsartiger Schwindel
- Migräne
- Psychogene (= dissoziative) Anfälle mit bizarrer Symptomatik. Sie liegen nicht selten auch zusätzlich bei einer echten Epilepsie vor
- Tic-Syndrome
- Narkolepsie (Schlafsucht) und verschiedenes mehr.

23.4 Therapie

Wichtigstes Ziel bei jeder Epilepsieform ist es, die optimale Gesamtentwicklung des Patienten zu gewährleisten. Deshalb kann bei nachweislich selten auftretenden nächtlichen Anfällen, z. B. der Rolando-Epilepsie, u. U. auf eine medikamentöse Behandlung verzichtet werden. Andererseits dient eine Behandlung aber auch der Vermittlung von Sicherheit und der Vermeidung von zusätzlichen Gefährdungen durch die Anfälle, z. B. Stürze.
Seit über 100 Jahren kann durch verschiedene Medikamente die Anfallsbereitschaft eines Patienten signifikant reduziert werden. Der akute Anfall wird immer noch am besten mit Benzodiazepinen, z. B. Diazepam, Clonazepam oder Lorazepam, unterbrochen. Eine medikamentöse Dauerbehandlung wurde früher vor allem mit Brom, Phenobarbital, Hydantoin und Carbamazepin durch-

geführt, was nicht selten mit Nebenwirkungen, vor allem vermehrter Müdigkeit, eingeschränktem Lernvermögen, Hautausschlägen, Leberveränderungen u. a., verbunden war. Wirksamstes Mittel sowohl bei generalisierten großen Anfällen als auch bei kleinen Anfällen (Absencen) ist auch heute noch Valproat, das aber wegen seiner seltenen, u. U. aber gravierenden Lebertoxizität und anderer Nebenwirkungen nicht immer gerne verordnet wird. Bei fokalen Epilepsien haben sich heute vor allem Oxcarbazepin, Sultiam und Levetiracetam bewährt, bei generalisierten Epilepsien außer Valproat Ethosuximid, Lamotrigin und Topiramat.
Daneben stehen eine Vielzahl weiterer Medikamente, aber auch epilepsiechirurgische Maßnahmen sowie spezielle Diätformen (ketogene Diät) zur Verfügung und durch spezielle elektrische Stimulationsbehandlungen des Vagusnervs kann eine signifikante Reduktion der Anfälle erreicht und somit u. U. auch eine normale Lebensqualität als Hauptziel der Therapie gewährleistet werden.
Zusätzlich sind bei allen Menschen mit Epilepsie die Vermeidung von Toxinexposition (Alkohol, Drogen, Nikotin etc.), ausreichender Schlaf, körperliche Bewegung und die Vermeidung von psychischem Stress zu empfehlen. Bei schweren Epilepsien können verhaltenstherapeutische Lernprogramme (z. B. **FAMOSES**) und Kontakte zu **Selbsthilfegruppen** (➤ Kap. 11) sinnvoll sein. Immer wieder muss darauf hingewiesen werden, dass die alleinige Diagnose einer Epilepsie nicht ausreichend ist (➤ Kap. 37).

Literatur

Aksu F (2011) Neuropädiatrie – Diagnostik und Therapie neurologischer Erkrankungen im Kindes- und Jugendalter. Bremen: Unimed

Heinen F, Böhmer J, Hufschmidt A et al. (2009) Pädiatrische Neurologie in Diagnose und Therapie. Stuttgart: Kohlhammer

Heinen F, Krieg S, Borggräfe I (2012) Neuropharmakotherapie und klinische Systematik. Stuttgart: Kohlhammer

Neubauer BA, Hahn A (2012) Dooses Epilepsien im Kindes- und Jugendalter. 12. Auflage Berlin: Springer

Straßburg HM, Dacheneder W, Kreß W (2012) Eintwicklungsstörungen bei Kindern. Praxisleitfaden für die interdisziplinäre Betreuung. München: Elsevier

24 Ausscheidungsstörungen

Hans-Michael Straßburg

24.1 Enuresis

Unter einer Enuresis versteht man das nicht kontrollierte Einnässen bei Kindern. Es ist ein häufiges Symptom, das bei normal entwickelten 8-jährigen Kindern noch zu 1–2 % vorkommt. Bei schweren Entwicklungsstörungen, insbesondere im mental-kognitiven Bereich, bestehen sehr viel häufiger Störungen der Blasen- und Mastdarmkontrolle. Aber auch bei vielen Kindern mit leichteren Entwicklungs- und Verhaltensauffälligkeiten finden sich häufig verschiedene Formen der Enuresis.

24.1.1 Formen der Enuresis

Die **primär isolierte Enuresis nocturna** ist die häufigste Form des Einnässens und überwiegend genetisch bedingt, da sich die nächtliche Reduktion der Nierenproduktion durch das Hormon Adiuretin noch nicht eingespielt hat. Hier ist in der Regel eine abwartende Haltung, ggf. auch verhaltenstherapeutische Maßnahmen und in Ausnahmen die Gabe von Desmopressin-Tabletten indiziert.
Die **idiopatische Dranginkontinenz** ist Folge einer Funktionsstörung der Blasenöffnung. Therapeutisch ist hierbei ein Blasentraining am Tag, evtl. auch eine Verhaltenstherapie (Klingelhose), in seltenen Fällen die medikamentöse Behandlung mit Oxybutinin indiziert.
Bei der **Sphinkter-Detrusor-Dyssynergie** kommt es zu einem mangelhaften Zusammenspiel von Blasenschließmuskel und „Entleerungsmuskel". Hierbei müssen vor allem neurologische Störungen, insbesondere im Bereich der unteren Rückenmarksabschnitte, ausgeschlossen werden.
Eine **sekundäre Enuresis nocturna** ist öfter mit ausgeprägteren psychischen Problemen, z. B. bei schwereren Familienkonflikten oder organischen Erkrankungen (z. B. des Rückenmarks), verbunden. Häufig zeigen diese Kinder auch Teilleistungsstörungen und nässen nicht nur nachts, sondern auch tags ein. Bei der sekundären Enuresis, bei der zumindest vorübergehend eine Kontrolle der Blasenfunktion bestand, muss immer auch an eine organische Grunderkrankung, z. B. eine Nieren- oder Harnwegserkrankung, einen beginnenden Diabetes mellitus oder insipidus sowie kaudale Rückenmarkserkrankungen (Tethered-Cord-Syndrom), gedacht werden.
Als Therapie werden, wenn organische Ursachen ausgeschlossen sind, vor allem eine psychosoziale Entlastung und eine kindgerechte Psychotherapie empfohlen.

Sozialpädiatrie in der Praxis. http://dx.doi.org/10.1016/B978-3-437-31630-3.00024-9
Copyright © 2014 Elsevier GmbH. All rights reserved.

24.2 Enkopresis

Eine Störung der Stuhlentleerung kann entweder in Form einer **Obstipation** oder des Einkotens **(Enkopresis)** auftreten. Besonders häufig ist die Obstipation bei Mehrfachbehinderten mit eingeschränkter Mobilität und Intelligenzminderung. Auch hier müssen sowohl organische als auch funktionelle Ursachen berücksichtigt werden. So sollten bei jedem Kind mit chronischer Obstipation einerseits eine Schilddrüsenunterfunktion und Nebenwirkungen von Medikamenten (Antiepileptika, Psychopharmaka), andererseits Fehlfunktionen im Enddarm, insbesondere eine Aganglionose (Morbus Hirschsprung) oder eine neuronale Kolondysplasie, berücksichtigt werden. Auch Anlagestörungen oder Erkrankungen des zentralen Nervensystems, insbesondere des kaudalen Rückenmarks, können Ursache einer chronischen Obstipation sein. Daneben können eine Obstipation und ein sekundäres Stuhlschmieren (Enkopresis) auch im Rahmen psychischer Belastungen, emotionaler Vernachlässigung, von Fehlernährung und allgemeiner Überforderung auftreten. Therapeutisch sind neben Verbesserungen der psychosozialen Situation ballast- und flüssigkeitsreiche Kost, ausreichende Bewegung und die Verabreichung von verschiedenen Quell- oder Gleitmitteln (Laktulose, Makrogol) indiziert, u. U. auch von stuhlregulierenden Suppositorien.

Literatur

Equit M, Sambach H, Niemczyk J, von Gontard A (2012) Ausscheidungsstörungen bei Kindern und Jugendlichen. Ein Therapieprogramm zur Blasen- und Mastdarmschulung. Göttingen: Hogrefe

Straßburg HM, Dacheneder W, Kreß W (2012) Eintwicklungsstörungen bei Kindern. Praxisleitfaden für die interdisziplinäre Betreuung. München: Elsevier

25 Kopfschmerzen und andere Somatisierungsstörungen

Hans-Michael Straßburg

25.1 Definitionen

Hierunter werden Gesundheitsstörungen zusammengefasst, bei denen die organischen Ursachen der klinischen Symptome gegenüber den psychischen im Hintergrund stehen. Die ICD-10 unterscheidet zwischen den **dissoziativen Störungen** bzw. Konversionsstörungen einerseits (F 44.0) und den **somatoformen Störungen** andererseits (F 45.0). Erstere kommen vor allem bei Kindern und Jugendlichen ab dem 10. Lebensjahr vor, wobei Mädchen deutlich häufiger betroffen sind. Symptome sind vor allem neurologische Ausfälle, z. B. Gangstörungen, Lähmungen, Krampfanfälle, sensible Ausfälle und Sehstörungen. Unter letzteren werden alle Formen von organbezogenen Beschwerden zusammengefasst, die kein objektiv nachweisbares Korrelat haben und von Kopfschmerzen über Bauchschmerzen und Herz-Kreislauf-Symptome bis zu Atemstörungen reichen können und auch bereits bei Kleinkindern auftreten.

Somatisierungsstörungen können sowohl bei normal entwickelten Kindern auftreten, sind aber insgesamt gehäuft bei Kindern mit Entwicklungs- und Verhaltensproblemen aller Art zu finden und neigen zur Chronifizierung bzw. zu Rezidiven.

Grundvoraussetzung für die Diagnose ist immer eine differenzierte klinische Diagnostik mit besonders ausführlicher Anamneseerhebung und dem bestmöglichen Ausschluss einer behandelbaren organischen Ursache der Beschwerden. Durch die Verbesserungen in der Diagnostik sind viele Störungsbilder, denen früher eine psychogene Ursache zugeschrieben wurde, zunehmend pathophysiologisch erklärbar, z. B. Bauchschmerzen bei Nahrungsmittelunverträglichkeit (Gluten, Laktose, Milcheiweiß), Sensibilitäts- und Visusstörungen beim ersten Schub einer Multiplen Sklerose, rezidivierende Kopfschmerzen bei chronisch erhöhtem intrakraniellem Druck usw.

Andererseits treten **Konversionsstörungen** besonders auch bei Jugendlichen mit organischen Dispositionen, z. B. chronischen Schmerzen, Z. n. Fraktur, Z. n. Schiel-OP oder einer primären Epilepsie auf. Die Diagnose kann dann nicht nur durch den Ausschluss organischer Erklärungen gestellt werden, sondern muss gezielt durch typische psychologische Befunde bestätigt werden. So besteht häufig ein subjektives Überforderungserleben, es gibt Vorbilder für die gewählten Symptome im näheren Umfeld, von den Betroffenen werden primär relevante Probleme geleugnet und

Sozialpädiatrie in der Praxis. http://dx.doi.org/10.1016/B978-3-437-31630-3.00025-0
Copyright © 2014 Elsevier GmbH. All rights reserved.

es besteht ein Mangel an eigener Betroffenheit. Es lässt sich aber auch oft ein primärer oder sekundärer Krankheitsgewinn nachweisen, z. B. eine vermehrte Zuwendung.

25.2 Symptome

Typische **Symptome** sind z. B.:
- Eine evtl. nur umschriebene Gedächtnislücke (Amnesie)
- Stupor oder Trancezustand ohne entsprechende objektiv-klinische Befunde (Pupillenreaktion, EEG-Veränderungen)
- Bewegungsstörungen, z. B. eine Unfähigkeit zu Stehen oder Gehen, ohne sich beim Zusammensacken zu verletzen und mit normal auslösbaren Eigenreflexen
- Sprechstörungen ohne andere neurologische Ausfälle
- Sensibilitätsstörungen, die nicht durch periphere oder zentrale Nervenläsionen zu erklären sind
- Visusstörungen bei normalen Pupillenreaktionen und visuell evozierten Potenzialen
- Persönlichkeitsstörungen
- Krampfanfälle bis hin zu statusähnlichen Zuständen bei normalem EEG u. v. m.

Typischerweise erfolgt wegen der z. T. bizarren Symptome ein häufiger Arztwechsel. Die primäre Abklärung sollte in der Regel im Rahmen eines stationären Aufenthalts in einer Kinderklinik unter Einbeziehung verschiedener Spezialisten erfolgen. Wenn die Symptome nur zu bestimmten Zeiten und an bestimmten Orten auftreten, hilft oft eine Videodokumentation. Bei der psychologischen Diagnostik ist das Heidelberger Dissoziations-Inventar (HDI) u. U. hilfreich.

Auf die im Kindesalter vielfältigen **somatoformen Störungen** kann hier nur kurz eingegangen werden. Sie können neben unterschiedlichen organischen Mitursachen u. a. durch Ängste, z. B. vor einer Trennung der Eltern, Schulphobie, Überforderung, Mobbing durch Mitschüler („Bullying") und vieles mehr bedingt sein. Wie bei den dissoziativen Störungen ist die gründliche Suche nach organischen Ursachen immer vorrangig, diese muss aber auch einmal abgeschlossen werden. Bei chronischen Kopf- und Bauchschmerzen haben sich z. B. das konsequente Führen eines Schmerzkalenders und das Malen eines Selbstbildnisses im Zustand mit und ohne Schmerzen als wichtiges Kriterium zur Differenzierung zwischen einer echten Migräne und Spannungskopfschmerzen erwiesen.

In Langzeituntersuchungen wurde festgestellt, dass bei 15 % der Patienten mit dissoziativen Symptomen im weiteren Verlauf doch eine organische Ursache feststellbar ist, z. B. eine neurodegenerative Erkrankung oder eine Psychose.

25.3 Therapie

In der **Therapie** ist es primär wichtig, dass die Symptome ernst genommen werden und dass dem Patienten die Möglichkeit gegeben wird, z. B. bei einer psychogenen Lähmung ohne Gesichtsverlust durch eine regelmäßige Physiotherapie die Symptomatik zu überwinden. Darüber hinaus ist eine ausführliche Aufklärung der Eltern und des Kindes über die klinischen Zusammenhänge notwendig, u. U. auch eine klassische Psychotherapie im Rahmen eines stationären Aufenthalts. Besonders wichtig ist für viele Kinder eine Entlastung, vor allem in der Schule, und die Beibehaltung eines regelmäßigen Tagesablaufs mit festen Zeiten für Lernen und für Erholung und Entspannung. Reflexionen über die symbolische Bedeutung der Störungsmuster sind in der Regel nachrangig, Medikamente meist entbehrlich.

Fallbeispiel

12-jähriges Mädchen mit seit mehreren Wochen bestehenden Bewegungs- und Empfindungsstörungen des rechten Arms und der rechten Hand. Primär verzögerte Sprachentwicklung, in der Grundschule Probleme mit dem Lesen und der Rechtschreibung, jetzt Besuch der 5. Realschulklasse. Eltern beide berufstätig, Alkoholprobleme beim Vater, ein jüngerer Bruder ist geistig behindert. Z. n. apoplektischem Insult bei der Großmutter.

Klinisch besteht ein Übergewicht, die Stimmung ist verschlossen-gleichgültig, normale Orientierung. Subjektive Angabe von strumpfförmiger Sensibilitätsstörung distal vom rechten Ellbogen, der rechte Arm wird leicht gebeugt gehalten, Adiadochokinese rechts, beim Gehen vermindertes Mitschwingen, der Muskeltonus des rechten Arms ist leicht erhöht, die Eigenreflexe seitengleich auslösbar, keine Pyramidenbahnzeichen, seitengleiches monopedales Hüpfen.

Normalbefunde für EEG, somato-sensibel evozierte Potenziale des Nervus medianus bds. und MRT von Kopf und oberer Wirbelsäule. Liquor o. B.

Diagnose: dissoziative Störung mit Parese des rechten Armes (F 44.4G).

Sozialpädiatrische Therapie:

- Nach Abschluss der organischen Diagnostik kurzfristige Übernahme in eine regelmäßige Physiotherapie 3-mal pro Woche
- Psychologische Diagnostik mit Neuropsychologie und Intelligenzmessung. Es besteht ein heterogenes Leistungsprofil, überwiegend im mittleren und unteren Normbereich
- Parallel dazu regelmäßige Gesprächstermine mit den Eltern und der Patientin durch Arzt und Psychologen
- Einleitung familienunterstützender Maßnahmen durch den Sozialdienst
- Umschulung auf eine Hauptschule
- Vermittlung von Freizeitangeboten
- Nach Rückbildung der Bewegungsstörung Wiedervorstellung in zunehmend größeren Zeitintervallen.

Literatur

Aksu F (2011) Neuropädiatrie – Diagnostik und Therapie neurologischer Erkrankungen im Kindes- und Jugendalter. Bremen: Unimed

Brunner R (2012) Dissoziative und Konversionsstörungen (Manuale psychischer Störungen bei Kindern) (2012) Heidelberg: Springer

Noeker M (2007) Funktionelle und somatoforme Störungen im Kindes- und Jugendalter. Göttingen: Hogrefe

26 Schlafstörungen

Hans-Michael Straßburg

26.1 Einleitung

Grundsätzlich gibt es eine große Variabilität in der für Wohlbefinden und Gesundheit notwendigen Schlafdauer. Andererseits können sehr unterschiedliche Ursachen für Störungen des Schlafverhaltens in jedem Alter auftreten und zusätzliche erhebliche Gesundheitseinschränkungen verursachen. An dieser Stelle sollen einerseits organische, andererseits psychoemotionale Gründe für Schlafstörungen vor allem auch bei chronisch kranken und behinderten Kindern dargestellt werden.

Im ersten Lebensjahr können Schlafstörungen im Rahmen von sogenannten **Regulationsstörungen** der Interaktion zwischen Kind und Eltern auftreten. Hierfür gibt es sehr verschiedene Erklärungsansätze, die einerseits mit dem Temperament des Kindes, dem Verhalten und der Stimmungslage der Eltern, überwiegend der Mutter, und den suboptimalen äußeren Gegebenheiten (schlechte Luft, Unruhe, nicht optimale Temperatur usw.) begründet werden.

26.2 Ursachen

Am häufigsten sind Schlafstörungen bei Kindern ein vorübergehendes Phänomen. Andererseits können vielfältige organische und funktionelle Störungen als Erklärung vermutet werden. Bei den nicht-organischen Ursachen können einerseits psycho-emotionale Belastungen, Angst und Verunsicherungen, andererseits eine nicht kindgerechte Umgebung z. B. durch Rauchen in der Wohnung, laufenden Fernseher, Streitigkeiten der Eltern oder ein nicht adäquates Bett eine Rolle spielen.

Zu den **organischen Ursachen** für Schlafstörungen, insbesondere bei von jungen und behinderten Kindern, gehört zum einen die Obstruktion der oberen Luftwege, insbesondere ab Ende des ersten Lebensjahrs, in Form von vergrößerten **Adenoiden.** Typischerweise schnarchen diese Kinder, verbunden mit relativem Austrocknen des Mundes. Sie werden dann nachts wach und verlangen nach einer Trinkflasche, was jedoch die Grundproblematik nur unzureichend beeinflusst. Im Rahmen von meist viralen Infekten oder bei allergischer Disposition können sich diese Symptome deutlich verstärken. Daneben können auch vielfältige Anlagestörungen des Gesichtsschädels und der Zähne für eine Atemstörung mit konsekutivem Schlafmangel verantwortlich sein.

Sozialpädiatrie in der Praxis. http://dx.doi.org/10.1016/B978-3-437-31630-3.00026-2
Copyright © 2014 Elsevier GmbH. All rights reserved.

Weitere Ursachen können unterschiedliche Formen von Ernährungsstörungen, insbesondere der **gastro-ösophageale Reflux** sein. Durch eine Überladung des Magens während des Tages mit Nahrung oder anatomische Veränderungen, z.B. nach Operation einer Ösophagus-Atresie, kommt es hauptsächlich bei jungen Säuglingen und bei Behinderten jeden Alters infolge einer funktionellen Verschlussstörung des Mageneingangs zu rezidivierenden Reflux-Episoden, die zu einem typischen pressenden Exspirium oder zu langdauernden Schreiphasen Anlass geben können. Auch Nahrungsmittelunverträglichkeiten (Kuhmilchprotein- oder Laktoseintoleranz) können nächtliche Unruhezustände verursachen. Darüber hinaus können chronische Hauterkrankungen (atopische Dermatitis), chronische Entzündungen, Hernien und vieles mehr für Schlafstörungen verantwortlich sein. Bei Jugendlichen sollte auch an eine beginnende Depression oder an seltene Erkrankungen wie das restless leg Syndrom gedacht werden.

Da Schlafstörungen bei Kindern und Jugendlichen häufig mit **psychischen Belastungen** und **emotionalen Störungen** unterschiedlicher Art verbunden sind, ist bei länger als 2 Monaten bestehenden Störungen eine genauere Diagnostik notwendig. Hier können neben einer subtilen Anamneseerhebung und dem weitgehenden Ausschluss organischer Ursachen das Führen eines Schlafkalenders und ggf. eine Beobachtung in einer neutralen Umgebung, z. B. im Rahmen eines stationären Aufenthalts hilfreich sein.

Bei Unruhezuständen aus dem Schlaf heraus sollte der **Pavor nocturnus,** das sind Angstzustände, an die sich ein Kind später nicht mehr erinnert, vom Albtraum getrennt werden. Es sollte aber auch an **schlafassoziierte zerebrale Anfälle,** z. B. an die zentro-temporale Sharp-Wave-Epilepsie (Rolando-Epilepsie), oder an Frontallappen-Epilepsien gedacht werden (➤ Kap. 23). Nicht selten zeigen ältere Säuglinge, Kleinkinder und meistens entwicklungsgestörte größere Kinder in Müdigkeits- und Einschlafphasen rhythmische Körperbewegungen mit vegetativen Symptomen wie vermehrtem Schwitzen im Sinne von Stereotypien, die in der Regel keiner besonderen Therapie bedürfen. Bei Schlafstörungen von Patienten mit hirnorganischen Symptome, insbesondere des Hirnstamms, sollte eine polysomnographische Untersuchung, d. h. eine kontinuierliche Aufzeichnung mehrerer biologischer Parameter während des Schlafens in einem akkreditierten Schlaflabor durchgeführt werden. So können z. B. am besten obstruktive und zentrale Apnoen und ihr wahres Ausmaß festgestellt werden.

26.3 Therapie-Prinzipien

Für die Entwicklung eines geregelten Schlaf-Wach-Rhythmus sind verschiedene neuronale Kerngebiete im Hypothalamus zuständig, die wiederum die zyklische Ausschüttung des in der Epiphyse gebildeten Hormons **Melatonin** auslösen.

Bevor vor allem auch bei Kindern und Jugendlichen mit Intelligenzminderung sedierende Medikamente zum Schlafen verabreicht werden, sollten immer wieder organische Ursachen und Verbesserungen der äußeren Schlafbedingungen berücksichtigt werden.

Wichtigste Voraussetzungen für einen ausreichenden Schlaf des Kindes sind ein regelmäßiger Tagesrhythmus, eine stringente Vorgehensweise am Abend („Einschlaf-Zeremonie"), Ruhe, Verzicht auf Fernsehen u. Ä., vor allem aber Maßnahmen, um dem Kind Sicherheit und Geborgenheit zu vermitteln. Dies kann z. B. durch die Betreuung in einer heilpädagogisch orientierten Beratungsstelle erreicht werden.

Besonders bei seh- und mehrfachbehinderten Patienten kann der von Hell- und Dunkelphasen beeinflusste Tag-Nachtrhythmus gestört sein – hier kann eine abendliche Gabe von **Melatonin** (2–8 mg) sinnvoll sein.

Literatur

Wiater A, Paditz, E, Hoch B Erler T (2010) Spezialreport Gesundheitsökonomie – Zur Kosten- und Erlössituation der pädiatrischen Polysomnografie im Vergleich zum Erwachsenenalter. Dresden: Kleanthes Verlag

Fricke L, Lehmkuhl G (2009) Schlafstörungen im Kindes- und Jugendalter – ein Therapiemanual für die Praxis. Göttingen: Hogrefe

Jenni OG, Benz C (2007) Schlafstörungen. Pädiatrie up2date 4: 309–333

27 Muskelerkrankungen

Hans-Michael Straßburg

27.1 Krankheiten mit Muskelhypotonie

Eine große Zahl völlig unterschiedlicher Zustände und Krankheitsbilder kann für eine Muskelhypotonie verantwortlich sein (➤ Tab. 27.1) – oft ist die Muskelhypotonie nur ein Teilaspekt bei einem Säugling mit allgemeiner Entwicklungsverzögerung. Nachfolgend können nur wesentliche Aspekte der Differenzialdiagnostik und Grundprinzipien der Behandlung angesprochen werden. Im klinischen Sprachgebrauch werden die Begriffe Normotonie, Hypertonie, Hypotonie und Dystonie verwendet, obwohl exakte Messungen des Muskeltonus vor allem im Kindesalter nur schwer möglich sind.

Die muskuläre Hypotonie im Säuglings- und Kleinkindalter ist ein häufiges, sehr komplexes Symptom, dessen Einschätzung sehr von der Erfahrung des Untersuchers abhängig ist. Wesentlich zur richtigen Einschätzung sind Angaben über die zeitliche Entwicklung, ob sie konstant, progredient oder regredient ist. Weiterhin können Aussagen über die intrauterinen Kindsbewegungen, über das Trinkverhalten, die Atmung, vermehrtes Speicheln, Zyanosezustände und das Schlafverhalten weiterhelfen. Bei der Diagnostik ist die Beachtung auch diskreter Dysmorphiezeichen, die Überprüfung der Eigenreflexe, die Beobachtung der spontanen Bewegungsabläufe vor allem an den Extremitäten, der Kontaktaufnahme und die mental-kognitive bzw. sprachliche Entwicklung von wesentlicher Bedeutung.

Die **Diagnostik** muss sehr vom klinischen Erscheinungsbild abhängig gemacht werden, hilfreich sind initial vor allem die Sonografie des Schädels und der Muskulatur sowie die Blutuntersuchung mit Bestimmung der **Kreatinkinase.** Weitergehende Untersuchungen, insbesondere eine Elektroneurographie, eine Elektromyographie, Kernspintomographien von Gehirn und Rückenmark sowie molekulargenetische Untersuchungen oder eine Biopsie von Muskelgewebe oder anderen Geweben, müssen sehr von der klinischen Symptomatik abhängig gemacht werden.

Tab. 27.1 Ursachen für Muskelhypotonie bei Kindern

Physiologisch	**Frühgeborenes, Schlaf, Entspannung**
Genetische Veränderung	Chromosomopathien, z. B. Down-Syndrom. Monogenetische Erkrankungen, z. B. Prader-Willi-Syndrom, Angelman-Syndrom, Rett-Syndrom usw.
Hirnorganische Veränderungen	Hydrozephalus, hypoxisch-ischämische Enzephalopathien, Hirntumoren, neurokutane Syndrome

Sozialpädiatrie in der Praxis. http://dx.doi.org/10.1016/B978-3-437-31630-3.00027-4
Copyright © 2014 Elsevier GmbH. All rights reserved.

Tab. 27.1 Ursachen für Muskelhypotonie bei Kindern (Forts.)

Physiologisch	Frühgeborenes, Schlaf, Entspannung
Stoffwechselstörungen	Amino- und Organoazidurien, peroxisomale Störungen wie Adrenoleukodystrophie, Hormonstörungen, Elektrolytstörungen, CDG-Syndrom usw.
Spinale Ursachen	Spinale Muskelatrophien, Anlage- und Ausreifungsstörungen im Sinne einer Arthrogryposis, Rückenmarktumoren, Myelitiden
Neuronale Ursachen	Genetisch bedingte Ausbildungs- und Ausreifungsstörungen des peripheren Nervensystems, z. B. hereditäre sensomotorische Neuropathien, entzündliche Neuropathien, neuromuskuläre Übertragungsstörungen, z. B. Myasthenien
Muskelanlagestörungen	Primäre Fehlanlage von Muskelgruppen, Strukturanomalien der Muskulatur
Myositiden	Immunerkrankungen, bakterielle Infektionen
Metabolische Myopathien	Verschiedene Enzymdefekte, Mitochondriopathien, kongenitale Myasthenie, kongenitale Muskeldystrophie, X-chromosomale Muskeldystrophie, myotone Dystrophie, Gliedergürteldystrophie etc.
„Benigne" passagere beinbetonte Muskelhypotonie	„Hosenrutscher", „sensitiver Perfektionist" usw.

27.2 Muskeldystrophie Duchenne

Fallbeispiel

Vorstellung eines 4,6-jährigen Knaben wegen einer zufällig im Rahmen einer präoperativen Blutuntersuchung festgestellter Erhöhung der **Kreatinkinase (CK)** im Blut auf > 5.000 U/l. Bei genauerem Nachfragen berichten die Eltern, dass ihr Sohn in letzter Zeit nicht mehr so bewegungsfreudig und von Anfang an „sprachfaul" gewesen sei und nur wenig mit anderen Kindern spiele. In der Familienanamnese findet sich kein Hinweis für andere betroffene Angehörige.

Befunde:

- Normale Größe und Gewicht mit auffallend verdickten, in der Konsistenz verhärteten Waden. Sprache einfach, wenig aktives Sprechen, leichter Dysgrammatismus. Angedeutetes Abkippen des Beckens beim Stehen und Gehen, vermehrte Lendenlordose mit vorgewölbtem Abdomen und vorgekipptem Becken. Gowers-Manöver positiv, d. h. bei Aufstehen vom Boden muss sich der Patient an den Oberschenkeln abstützen
- Im Ultraschall diffuse Erhöhung der Echogenität in den rumpfnahen Muskeln
- Molekulargenetisch lässt sich eine typische Deletion im Dystrophin-Gen nachweisen
- Bei der Bestimmung der Kreatinkinase im Serum der Mutter ist der Wert leicht erhöht.

Diagnosen:

- Muskeldystrophie Duchenne in der beginnenden klinischen Manifestation (G 71.0G)
- Expressive Sprachentwicklungsstörung (F 80.1G).

Vorgehen im SPZ:

- Aufklärung der Eltern über das Krankheitsbild
- Mitteilung, dass die Mutter als Überträgerin angesehen werden muss
- Einleitung einer logopädischen Behandlung
- Beantragung eines Behindertenausweises
- Beantragung einer heilpädagogischen Einzelintegration im Kindergarten
- Förderung jeglicher motorischer Aktivitäten, Vermeidung von Spitzfußhaltung, z. B. beim Sitzen, Empfehlung, immer wieder zu stehen

- Vermittlung einer Kontaktaufnahme mit der lokalen Geschäftsstelle der Deutschen Gesellschaft für Muskelkranke
- Gemeinsame Kontrolluntersuchung mit dem pädiatrischen Orthopäden zur genauen Beurteilung des Ausmaßes der Kontrakturen, besonders im Bereich des Beckens, der Wirbelsäule und der Beine. Ansprechen auf einen Zeitpunkt zu Operation nach Rideau zur Kontrakturlösung
- Besprechung, ab wann eine Therapie mit Methylprednisolon oder Deflazacort eingeleitet werden könnte
- Psychosoziale Stützung der Familie
- Vereinbarung regelmäßiger Kontrolltermine im Abstand von 6 Monaten im SPZ
- Überlegung, ob die Anmeldung in einem experimentellen Therapieprojekt sinnvoll sein kann.

Die **Duchenne-Muskeldystrophie** ist die wichtigste angeborene Muskelerkrankung. Mit 1 : 3000 ist sie die häufigste und zugleich auch bösartigste, d. h. rasch fortschreitende Muskelerkrankung, die praktisch nur bei Knaben auftritt, da sie X-chromosomal vererbt wird. Ursächlich fehlt das für die Stabilität der Zellmembran notwendige Protein Dystrophin.
Die Symptomatik beginnt schleichend, meist zwischen dem 3. und 5. Lebensjahr mit zunehmender Schwäche der Beckengürtelmuskeln, einer zunehmenden Lordose und Spitzfußhaltung bei auffallend hypertrophen Wadenmuskeln. Nicht selten bestehen gleichzeitig auch Sprachentwicklungsauffälligkeiten. Typisch sind in den ersten Lebensjahren exzessiv hohe Werte der Kreatinkinase und deutlich erhöhte Werte der Leberenzyme GOT und GPT. In den meisten Fällen kann heute mittels der Molekulargenetik eine definitive Diagnose erstellt werden, nur in Ausnahmefällen ist noch eine Muskelbiopsie zur Diagnosesicherung notwendig.
Eine kausale Behandlung ist nicht möglich, neben der Physiotherapie kann durch eine Kortikoidbehandlung und spezielle Operationen, z. B. nach Rideau, die Gehfähigkeit noch um einige Jahre verlängert werden. In der Regel sind ab dem 10. Lebensjahr die Patienten auf die Verwendung eines Elektrorollstuhls angewiesen, ab dem 15. Lebensjahr können primär im Schlaf zunehmende Atemprobleme auftreten, die mit Hilfe von Atemhilfsgeräten (CPAP) behandelt werden können. Ob bei ausgewählten Patienten mit einer Gentherapie (**Exon-Skipping**) eine bleibende Verbesserung erreicht werden kann, muss abgewartet werden.
Neben der Adaptation der Behandlung an das jeweilige Krankheitsstadium (➤ Kap. 13.1) ist die Elternselbsthilfe (➤ Kap. 11) und -entlastung von wesentlicher Bedeutung.

27.3 Andere Muskelerkrankungen

Die zweithäufigste Gruppe angeborene Muskelerkrankung ist die Gruppe der **spinalen Muskelatrophien.** Bei einer Manifestation bereits im ersten Lebensjahr (=infantile Form) besteht in der Regel eine rasche Progredienz, sodass die Kinder früh an einer zunehmenden Ateminsuffizienz sterben. Bei der klinischen Manifestation im Kleinkind und Schulalter gibt es unterschiedliche Verläufe. Vor allem aufgrund ihrer normalen bis überdurchschnittlichen Intelligenz können sich die Patienten trotz schwerer körperlicher Einschränkungen oft erstaunlich gut mit Hilfe verschiedener Hilfsmittel, insbesondere elektronischer Geräte zurechtfinden. Besonders durch atemunterstützende Maßnahmen kann eine deutliche Verbesserung der Lebensqualität und eine Lebensverlängerung erreicht werden.
Die dritthäufigste Muskelerkrankung ist die **myotone Muskeldystrophie** Curschmann-Steinert, bei der neben einer allgemeinen Muskelhypotonie Störungen des Endokriniums, Linsentrübun-

gen, IQ-Minderung, Skoliose, Herz-Rhythmus-Störungen, chronische Obstipation und Blasenentleerungsstörungen auftreten.
Zunehmend können bei allgemeiner Muskelhypotonie ab dem Säuglingsalter auch durch molekulargenetische Untersuchungen die Ursachen festgestellt werden, z. B. bei einer **myotubulären Myopathie** oder einer Aktin-Myopathie. Auch selten Stoffwechselerkrankungen, z. B. eine **Mitochondriopathie,** können die Erklärung für Funktionsstörungen der Muskulatur sein. Sehr selten sind Übertragungsstörungen an der motorischen Endplatte im Sinne einer **Myasthenie** bei jungen Kindern, während sie vor allem bei weiblichen Jugendlichen öfter vorkommen. Diese Erkrankung kann durch das Medikament Physostigmin und eine radikale Thymusoperation oft sehr gut kausal behandelt werden.

Literatur

Aksu F (2011) Neuropädiatrie – Diagnostik und Therapie neurologischer Erkrankungen im Kindes- und Jugendalter. Bremen: Unimed

Heinen F, Böhmer J, Hufschmidt A et al. (2009) Pädiatrische Neurologie in Diagnose und Therapie. Stuttgart: Kohlhammer

Heinen F, Krieg S, Borggräfe I (2012) Neuropharmakotherapie und klinische Systematik. Stuttgart: Kohlhammer

Straßburg HM, Dacheneder W, Kreß W (2012) Entwicklungsstörungen bei Kindern. Praxisleitfaden für die interdisziplinäre Betreuung. München: Elsevier

28 Infantile Zerebralparese (ICP)

Hans-Michael Straßburg

28.1 Definitionen

Die infantile Zerebralparese (= ICP, zerebrale Kinderlähmung oder Little'sche Erkrankung) ist ein zentrales Symptom bei vielen Bewegungsstörungen im Kindesalter und findet sich vor allem bei ehemaligen Frühgeborenen oder bei Kindern mit schweren Komplikationen während der Geburt, aber auch bei Anlagestörungen des Gehirns, Stoffwechselerkrankungen, genetischen Anomalien und entzündlichen Erkrankungen. In den ersten 6 Lebensmonaten kann eine Zerebralparese schwer zu erkennen sein. Unter einer infantilen Zerebralparese versteht man heute kein einheitliches Krankheitsbild, sondern ein sehr vielfältiges klinisches Bild mit Störungen der Körperhaltung und der Fortbewegung, die bleibend, aber nicht unveränderlich sind und auf eine abgeschlossene Läsion während der prä-, peri- oder postnatalen Entwicklungsperiode zurückgeführt werden können. In der Regel liegt die Schädigung im Bereich des ersten Motoneurons, d. h. sie betrifft die für die Bewegungssteuerung zuständigen Nervenzellen und Nervenbahnen im zentralen Nervensystem. Häufig ist die Zerebralparese mit anderen Entwicklungsstörungen, vor allem Sinnes-, Wahrnehmungs- und Sprachstörungen, Intelligenzminderung, zerebralen Anfällen und Verhaltensauffälligkeiten kombiniert, sie ist damit ein Beispiel für eine **Mehrfachbehinderung.** Heute wird eine manifeste Zerebralparese bei 1–2 von 1000 Kindern diagnostiziert.

Es werden folgende **Formen** der ICP unterschieden:

1. Bilaterale Zerebralparese (= spastische Tetraparese oder Tetraplegie) mit entweder bein- oder armbetonter Lähmung aller 4 Extremitäten
2. Unilaterale Zerebralparese (= spastische Hemiparese oder Hemiplegie) mit halbseitiger Lähmung
3. Dyston-dyskinetische Zerebralparese (= Athetose oder Chorea)
4. Die sehr heterogenen hypotonen und ataktischen Zerebralparesen
5. Mischformen.

Die Einteilung des Schweregrads einer Zerebralparese erfolgt nach der Gross Motor Function Measure (GMFM) bzw. dem GrossMotor Functional Classification System (GMFCS).

Sozialpädiatrie in der Praxis. http://dx.doi.org/10.1016/B978-3-437-31630-3.00028-6
Copyright © 2014 Elsevier GmbH. All rights reserved.

28.2 Symptome

Typische **Symptome** der Zerebralparese sind:

- Muskelspastik mit erhöhter Muskelanspannung
- Störungen der Zielmotorik
- Verminderung der Kraft
- Steigerung der Eigenreflexe
- Pathologische Reflexe, z. B. positive Pyramidenbandzeichen (Babinski, Rossolimo)
- Persistenz spezieller Bewegungsmuster, z. B. des asymmetrisch-tonischen Nackenreflexes (ATNR).

Man unterscheidet nach der GMFCS verschiedene **Schweregrade:**

1. Nur bei temporeichen Bewegungen erkennbar
2. Freies Gehen möglich, jedoch deutliche funktionelle Beeinträchtigungen
3. Kein freies Gehen, aber Robben oder Krabbeln möglich
4. Keine aktive Fortbewegung möglich
5. Keine aktive Bewegung möglich.

Außerdem müssen bei der ICP auch andere motorische Störungen, z. B. der Mund- und Handfunktionen, berücksichtigt werden.

Fallbeispiel

Vorstellung eines 1,7-jährigen Kindes.

Geburt mit 33 Gestationswochen bei fast fehlendem Fruchtwasser. Geburtsgewicht 1.800 g. APGAR nach 5 min 8, nach 10 min 9, Nabelschnur-Arterien pH 7,20. 3 Std. nach der Geburt zunehmende Atemnot, maschinelle Beatmung für 4 Tage. Deutliche Trinkschwäche, im weiteren Verlauf auffallende Schlaffheit der Beine.

Bakteriologisch und serologisch keine Auffälligkeiten.

Zerebrale Sonografie am 2. LT orientierend regelrecht.

Bei Entlassung aus der Klinik im zerebralen Ultraschall Nachweis von unregelmäßig begrenzten echofreien Zonen periventrikulär.

Trotz intensiver Physiotherapie keine altersentsprechende motorische Entwicklung, zunehmende Strecktendenz der Beine, Neigung zur asymmetrischen Überstreckung, zeitweilige Fütter- und Schlafprobleme. Bei der Untersuchung ist der Kopfumfang 1 cm unter der 3. Perzentile, überwiegender Beugetonus der Arme und hypertone Streckung beider Beine. Kein freies Stehen, kein Gehen, in Bauchlage Streckstütz. Keine harmonischen Rotationsbewegungen der Hände. Gesteigerte Eigenreflexe an den Beinen > Armen, Rossolimo-Zeichen als Hinweis für eine Störung der kortiko-spinalen Bahnen positiv. Deutlich Abspreizhemmung im Hüftgelenk von 40° bds. Ausgeprägter Strabismus, leichter Nystagmus, nicht konstante Blickfolge.

Diagnosen:

- Frühgeborenes 33 SSW (P 07.1Z)
- Spastische Diparese (G 80.1G)
- Periventrikuläre Leukomalazie (P 91.2G)
- Mikrozephalie (Q 0.2G)
- Strabismus intermittens (H 50.3G)
- Amblyopie (H 53.0V).

Vorgehen:

- Vermittlung der Diagnose einer bleibenden motorischen Bewegungsstörung
- Förderung jeglicher motorischer Eigenaktivitäten des Kindes, z. B. durch häufige Lagerung auf einer Krabbeldecke

- Vorstellung beim Augenarzt mit Sehschule zur Visusbestimmung und Therapieeinleitung
- Komplexbehandlung mit Physiotherapie und Heilpädagogik über eine interdisziplinäre Frühförderstelle
- Röntgenaufnahme des Beckens zur Beurteilung der Hüftgelenke
- Beantragung eines Behindertenausweises.
- Kontroll-Untersuchung im SPZ alle 3 Monate

28.3 Ätiologie

Eine typische Ursache für das Bild der spastischen Tetraparese des Frühgeborenen ist die **periventrikuläre Leukomalazie,** die durch eine hypoxisch-ischämische Schädigung des zerebralen Marklagers (= weiße Substanz) vor der 32. Schwangerschaftswoche erklärt werden kann. Daneben lassen sich verschiedene andere Durchblutungsstörungen, z. B. **ischämische Infarkte** im Bereich der Capsula interna, Grenzzoneninfarkte im Bereich der Hirnrinde oder hypoxische Schädigungen der Basalganglien bei akuten Sauerstoffmangelzuständen des reifen Neugeborenen, unterscheiden.
Traumatische Hirnschädigungen, z. B. unter der Geburt, aber auch als Folge von Kindesmisshandlung, Unfällen und Operationen können ebenfalls das Bild einer Zerebralparese verursachen. Daneben muss jedoch auch an zerebrale Fehlbildungen (kortikale Dysplasien, Schizenzephalie), ZNS-Infektionen, Hirntumoren, Vergiftungen, Autoimmunerkrankungen, Stoffwechselerkrankungen, z. B. der Mitochondrien, und neurodegenerative Erkrankungen gedacht werden. Zunehmend häufig werden auch genetisch bedingte Systemerkrankungen des zentralen Nervensystems, z. B. mit Störungen der Myelinisierung, als Ursache einer Zerebralparese erkannt.
Bis in die 1960er Jahre war eine häufige Ursache für eine dyston-dyskinetische Zerebralparese der Neugeborenenikterus mit einer durch Bilirubin bedingten Schädigung der Stammganglien. Heute ist dieses Symptom viel seltener und eher durch akute Sauerstoffmangelzustände unter der Geburt einerseits, aber auch durch postentzündliche oder genetische Erkrankungen (Chorea minor nach Streptokokkeninfekten), Störungen des Neurotransmitters Dopamin (Torsionsdystonie Segawa) oder Ernährungsstörungen (Sutcliff-Sandifer-Syndrom bei gastro-ösophagealem Reflux) zu erklären.

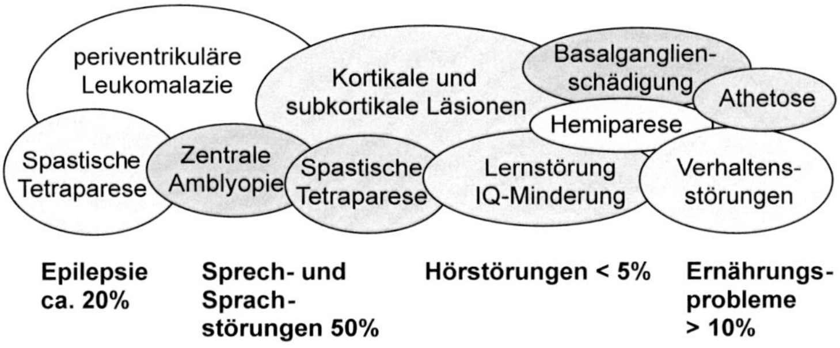

Abb. 28.1 Das Konzept zu Pathogenese und Klinik von zentralen Entwicklungsstörungen bei hirnorganischen Läsionen, modifiziert nach M. Bax [3]

Häufig bestehen bei Patienten mit Zerebralparesen zusätzlich weitere Erkrankungen verschiedener Organsysteme, z. B. des Sehens, des Hörens, der Verdauung, der Atemwege, des Skelettsystems sowie vielfältige pflegerische Probleme.

28.4 Therapieprinzipien

Die **Therapie** der Zerebralparese muss immer multidisziplinär sein und ist eine typische Aufgabe eines sozialpädiatrischen Zentrums, daneben aber auch eine wesentliche Herausforderung des gesamten Gesundheitssystems. Mit Hilfe moderner Behandlungsmethoden, insbesondere der Physiotherapie und gezielten Medikamentengaben, vor allem die intramuskuläre Injektion von Botulinum-Toxin A, kann nicht selten eine Teilhabe am Alltagsleben erreicht werden. In schweren Fällen kann eine intraspinale Baclofen-Pumpenbehandlung eine Reduktion der spastischen Tonuserhöhung herbeiführen. Zusätzlich können verschiedene Hilfsmittel eingesetzt und Stabilisierungen durch z. T. sehr aufwändige orthopädische Operationen erreicht werden (➤ Kap. 37). Keine Methode ist aber in der Lage, auch bei frühestmöglicher Behandlung eine manifeste Zerebralparese zu heilen.

Spezielle sozialpädiatrische Aufgaben sind neben der Indikationsstellung zu den verschiedenen Therapiemaßnahmen und einer umfassenden psychosozialen Diagnostik u. a. die Unterstützung bei der Entscheidung über die Betreuung in Kindergarten und Schule, die Entlastung der Eltern und ggf. von Geschwistern sowie Hilfe bei den vielen Beantragungen und eine frühzeitige Einleitung der **Transition.**

Literatur:

Aksu F (2011) Neuropädiatrie – Diagnostik und Therapie neurologischer Erkrankungen im Kindes- und Jugendalter. Bremen: Unimed

Heinen F, Böhmer J, Hufschmidt A et al. (2009) Pädiatrische Neurologie in Diagnose und Therapie. Stuttgart: Kohlhammer

Heinen F, Krieg S, Borggräfe I (2012) Neuropharmakotherapie und klinische Systematik. Stuttgart: Kohlhammer

Panteliadis C (2011) Cerebral palsy. A multidisciplinary approach. Oberhaching: Dustri

Straßburg HM, Dacheneder W, Kreß W (2012) Eintwicklungsstörungen bei Kindern. Praxisleitfaden für die interdisziplinäre Betreuung. München: Elsevier

Suchodoletz W von (Hrsg) (2010) Therapie von Entwicklungsstörungen – was wirkt wirklich? Göttingen: Hogrefe

29 Angeborene und erworbene ZNS-Erkrankungen

Hans-Michael Straßburg

29.1 Hirnfehlanlagen

Durch die mittlerweise bei jedem Kind mit globaler Entwicklungsstörung zu fordernde Bildgebung der Hirnstrukturen kann in vielen Fällen eine Hirnfehlanlage erkannt werden. Ob diese auch für die klinische Symptomatik verantwortlich ist, muss jedoch im Einzelfall geklärt werden. Sichere Ursache für eine schwere Intelligenzminderung sind u. a. eine **Holoprosenzephalie** und eine **Lissenzephalie.** Ein **Balkenmangel (Corpus-callosum-Agenesie)** kann mit schweren Entwicklungsstörungen verbunden sein, ebenso eine **Schizenzephalie,** beides kann aber auch klinisch nahezu inapparent sein. **Kortikale Dysplasien** spielen vor allem bei therapieresistenten Epilepsien eine wichtige Rolle, da geklärt werden muss, ob hier ein epilepsiechirurgischer Eingriff indiziert ist.

Eine **Mikrozephalie,** d. h. ein altersbezogener Kopfumfang unter der 3. Perzentile, geht sehr häufig mit einer Intelligenzminderung einher. Ursächlich kommen genetische Anomalien, primäre Fehlbildungen, teratogene Einflüsse, insbesondere Alkohol und Drogen, Stoffwechselstörungen (eine PKU, evtl. auch nur bei der Mutter), Infektionen (z. B. Röteln, Zytomegalie und HIV) sowie Fehlernährung in Frage. Eine sekundäre Mikrozephalie mit normalem Kopfumfang bei Geburt kann Hinweis für eine intranatale Hirnschädigung sein, kommt aber z. B. auch beim Rett-Syndrom vor.

Eine **Makrokranie** (d. h. ein übergroßer Kopf) kann Ausdruck einer Erweiterung der inneren Hirnkammern im Sinne eines Hydrozephalus internus sein. Häufigste Ursachen hierfür sind Hirnanlagestörungen, z. B. eine Chiari-II-Malformation bei einer Meningomyelozele oder beim Dandy-Walker-Syndrom sowie ein Zustand nach intrakranieller Blutung, seltener nach Infektionen oder bei Hirntumoren.

Eine **Megalenzephalie** (ein übergroßes Gehirn) kann im Rahmen verschiedener Syndrome (Sotos-Syndrom), die z. T. auch mit allgemeinen Entwicklungsstörungen einhergehen, bei Speichererkrankungen (Mukopolysaccharidose) und bei Gefäßanomalien (Vena-Galeni-Angiom) auftreten.

Sozialpädiatrie in der Praxis. http://dx.doi.org/10.1016/B978-3-437-31630-3.00029-8
Copyright © 2014 Elsevier GmbH. All rights reserved.

Anomalien der Kopfform, meist bedingt durch vorzeitigen Verschluss der Schädelnähte (**Schädelnaht-Synostosen, Kraniostenosen**), kommen isoliert oder bei genetisch definierten Syndromen (Apert, Crouzon, Saethre-Chotzen u. v. m.) vor. Die Entwicklung der Kinder ist von der Diagnose und einer komplexen interdisziplinären Therapie abhängig.

29.2 Teratogene Hirnschäden

Das wichtigste Beispiel einer toxisch bedingten Entwicklungsstörung ist das **embryofetale Alkoholsyndrom** (= Alkohol-Spektrum-Störung, ASS).
Typische klinische Zeichen hierfür sind:
- Minderwuchs
- Untergewicht
- Mikrozephalie bereits vor der Geburt
- Störung der allgemeinen Entwicklung mit muskulärer Hypotonie und Hyperexzitabilität
- Typische Gesichtsdysmorphie mit niedriger Stirn, flachem, verkürztem Nasenrücken, schräger Lidachse, kurzer Lidspalte, flachem Filtrum und schmalem Oberlippenrot
- Verschiedene innere Fehlbildungen, z. B. des Herzens, des Urogenitaltrakts und im Analbereich.

Die Kinder gelten als aggressionsarm und fröhlich, wirken zutraulich bis distanzlos, unruhig bis rastlos und sind sehr stimmungslabil. Ein linearer Zusammenhang zwischen der Trinkmenge und dem Grad der Schädigung lässt sich nicht nachweisen, wohl aber eine Korrelation mit dem Ausmaß der mütterlichen Alkoholkrankheit. Ob der Konsum von Alkohol der Mutter während der Schwangerschaft wirklich die häufigste Ursache von Lernstörungen der Kinder darstellt, ist umstritten, da viele andere Einflussfaktoren (Nikotin, Medikamente, harte Drogen, Ernährungsstörungen, genetische Dispositionen usw.) die Ausprägung der Hirnentwicklungsstörungen mit beeinflussen.
Harte Drogen, vor allem Heroin und Kokain, können zu unterschiedlichen Hirnfehlbildungen in der Frühschwangerschaft führen und verursachen Entzugssymptome des Neugeborenen mit Trinkstörungen, Hyperexzitabilität und einer deutlich erhöhten Gefahr eines plötzlichen Kindstods. Nikotin, viele Medikamente, z. B. Thalidomid, Marcumar, Hydantoin, Valproat, Chemotherapeutika und radioaktive Strahlen können in unterschiedlicher Ausprägung die Entwicklung eines Kindes schädigen, ebenso kann ein schlecht eingestellter Diabetes mellitus der Mutter die Kindsentwicklung beeinflussen.

29.3 Intrauterine Infektionen

Bei einer **Röteln-Virus-Infektion** im 1. Schwangerschaftsdrittel kommt es in bis zu 90 % der Fälle zu einer Schädigung des Kindes mit den typischen Symptomen von Innenohrschwerhörigkeit, Herzfehler, Linsentrübung und Intelligenzminderung. Während dies durch eine Impfung sicher verhindert werden kann, ist auch heute eine Schädigung des Feten durch Zytomegalie-, Varizellen-, Epstein-Barr-, HI- und Parvo-Viren möglich. Außerdem kann eine Infektion des Feten mit Toxoplasmose oder Treponema pallidum (Lues) bleibende Schäden verursachen.

29.4 Perinatale Hirnschäden

Hierunter werden unterschiedliche Schädigungen, insbesondere durch eine Sauerstoffmangelversorgung des Gehirns verstanden. Im Einzelnen unterscheidet man:

- **Pränatale Schäden** (zu jedem Zeitpunkt der Schwangerschaft bis Beginn der Eröffnungswehen bzw. Eröffnung der Fruchtblase)
- Perinatale Schäden zwischen der Lebensfähigkeit des Feten (22. SSW) und 8 Tagen nach der Geburt
- **Intranatale Schäden** zwischen Geburtsbeginn und Abnabelung
- **Neonatale Schäden** zwischen der Abnabelung und dem Ende der 4. Lebenswoche
- Postnatale Hirnschäden.

Pränatale Hirnschäden sind vor allem die Folge einer Sauerstoffmangelversorgung des Kindes bei Plazenta- und Nabelschnuranomalien, z. B. vorzeitiger Plazentalösung, EPH-Gestose, Mehrlingsschwangerschaften mit feto-fetaler oder feto-maternaler Transfusion, Blutgruppenunverträglichkeit und immunologischen Störungen.

Von besonderer Bedeutung ist die **periventrikuläre Leukomalazie,** die typische Hirnschädigung von Frühgeborenen mit einer Schwangerschaftsdauer unter 34 Wochen (➤ Kap. 30). Aus wahrscheinlich unterschiedlicher Ursache, unter anderem in Folge von Entzündungsreaktionen in den Gefäßen und einer verminderten Sauerstoffversorgung des Feten, entstehen typische Schädigungen im periventrikulären Marklager, die sich typischerweise oft erst 2 bis 4 Wochen nach der Geburt mittels zerebraler Ultraschalldiagnostik erkennen lassen (**„white matter disease“**). Wahrscheinlich spielen die Faktoren, die eine Frühgeburtlichkeit verursachen, auch bei der Entstehung der periventrikulären Leukomalazie die entscheidende Rolle. Typische Folgen sind einerseits Schädigungen der kortiko-spinalen Bahnen mit einer beinbetonten spastischen Zerebralparese, andererseits Schädigungen des Tractus opticus mit zentralen Sehstörungen, vor allem aber auch Schädigungen des frontalen Marklagers mit den für viele Frühgeborene typischen, unterschiedlich ausgeprägten Intelligenzeinschränkungen vor allem in Bezug auf logisches Denken und Abstraktionsfähigkeit.

Hirninfarkte beim reifen Neugeborenen entstehen z. B. durch einen embolischen Verschluss der Arteria cerebri media, z. B. bei einem Herzfehler mit Thrombenbildung, bei unterschiedlichen Traumen mit Defekten von Arterienwänden (Dissektion), bei Gefäßanlagestörungen, bei Infektionen, bei gesteigerter Gerinnung und eventuell auch durch primär in der Plazenta gebildete Thromben oder einen abgestorbenen Zwillingsfeten. Klinisch können in manchen Fällen schon in den ersten Lebenstagen zerebrale Anfälle beobachtet werden, während sich die meist halbseitigen armbetonten spastischen Bewegungsstörungen erst innerhalb der ersten Lebensmonate zunehmend ausbilden.

Schwere Sauerstoffmangelzustände (ungenau = Asphyxie) direkt vor und unter der Geburt führen in der Regel zu unterschiedlich ausgeprägten Hirnschädigungen, entweder im Grenzbereich zwischen den großen Hirnarterien mit entsprechenden fokalen Ausfällen oder zu Schädigungen der Stammganglien mit der Ausbildung einer dyston-dyskinetischen Zerebralparese. Die schwerste Ausprägung ist die globale Hirnschädigung mit porenzephaler Defektbildung im gesamten Hirngewebe, sekundären Hirnparenchymblutungen und Ausbildung einer inneren und äußeren Atrophie.

Fallbeispiel

Vorstellung eines 5,3-jährigen Knaben zur weiteren Behandlung. In der Schwangerschaft übermäßige Gewichtszunahme der Mutter, protrahierte Geburt bei Schulterdystokie, Geburtsgewicht 5.100 g. Apgar nach einer Minute 1, nach 5 Minuten 3, nach 10 Minuten 6. Nabelschnurarterien-pH 6,85.

Intubation nach 20 Minuten, Verlegung auf eine pädiatrische Intensivstation. Dort maschinelle Beatmung, Auftreten von zerebralen Anfällen nach 12 Stunden, pathologische Werte für die Leber- und Nierenfunktion, pathologische Gerinnung.
Stationäre Behandlung bis zum 4. Lebensmonat. Massive Ernährungs- und Schlafprobleme, häufiges Schreien. Ausgeprägte Störung der psychomotorischen Entwicklung, bis zum Ende des 1. Lebensjahres kein aktives Drehen. Im MRT mit 8 Monaten Nachweis von Gliosen im Globus-pallidus-Bereich beidseits.
Die Eltern geben an, durch die erhebliche Unruhe und die Ernährungsprobleme des Jungen stark belastet zu sein. Sie haben bis vor wenigen Monaten regelmäßig eine intensive Physiotherapie nach Vojta durchgeführt, jetzt wird versucht, den Patienten in einem Stehständer zum Stehen zu bringen. Er wird vormittags in einer Schulvorbereitenden Einrichtung (SVE) mit Förderschwerpunkt im Bereich der geistigen Entwicklung betreut.
Befunde:
Mit 14,2 kg (2 kg unter der 3. Perzentile) und 108 cm (10. Perzentile) dystropher, unruhiger Knabe. Kopfumfang mit 50,8 cm auf der 10. Perzentile. Auf der Untersuchungsliege ausgeprägte Überstreckung des Rumpfes und der rechten Extremitäten bei Beugung der linken Extremitäten mit häufigen Dyskinesien beider Arme und des rechten Beins. Gute Blickfolge, ausgeprägtes Grimassieren, artikuliert Kehllaute, z. T. einzelne, schwer verständliche Worte. Eigenreflexe allseits gesteigert, positive Pyramidenzeichen an Händen und Füßen. Im Sitzen werden rhythmische Bewegungen mit dem linken Fuß ausgeführt.
Diagnosen:
- Dyston-dyskinetische Zerebralparese (G80.3G)
- Intranatale hypoxisch-ischämische Hirnschädigung (P21.0Z)
- Ausgeprägte Essstörung (F 98.2G)
- Dystrophie (E 45G)
- Verdacht auf gastro-ösophageale Reflux-Krankheit (K21.1V)
- Verdacht auf heterogenes Begabungsprofil bei leichter Intelligenzminderung (F78.0V).

Vorgehen:
- EEG: kein pathologischer Befund
- Breischluck-Untersuchung: Nachweis eines gastro-ösophagealen Refluxes
- Behandlung mit einem Antazidum
- Empfehlung zur Anlage einer perkutanen endoskopischen Jejunotomie (PEJ-Sonde). Kontaktaufnahme mit einer Sonderschule für Körperbehinderte
- Beantragung eines adäquaten Rollstuhls
- Konzentration der Förderung auf die aktiven Fähigkeiten mit dem linken Fuß
- Anmeldung zur stationären Behandlung in einer kinderneurologisch-sozialpädiatrischen Klinik mit dem Ziel einer Verbesserung der Alltagsfunktionen und der Kommunikation mit apparativen Hilfen
- Neue Beantragung des Behindertenausweises unter besonderer Berücksichtigung des Zusatzmerkmals aG
- Beantragung eines Therapie-Tandemanhängers für ein Fahrrad unter Berücksichtigung der Ziffern aus dem ICFCY.

Hirnblutungen treten vor allem bei Frühgeborenen mit weniger als 32 Schwangerschaftswochen auf. Sie sind in der Regel Folge einer bereits intrauterinen Hypoxie mit Ischämie vor allem im sogenannten Keimlager, einem besonders fragilen Gewebe am Boden der Seitenventrikel. Hier lassen sich meist am zweiten oder dritten Lebenstag initial kleine Blutungen, die sich vergrößern können und in das Ventrikelsystem oder das Hirnparenchym einbrechen, nachweisen. Die subependymalen Blutungen des Frühgeborenen verursachen keine typischen klinischen Symptome,

sie sind allerdings ein Hinweis dafür, dass ein erhöhtes Risiko für eine Störung der Gesamtentwicklung besteht. Insgesamt ist die Inzidenz der Hirnblutungen von Frühgeborenen durch die modernen Möglichkeiten der Perinatalmedizin, u. a. auch die Behandlung des Atemnotsyndroms mit Surfactant, deutlich geringer geworden.
Hirnblutungen beim Reifgeborenen sind heutzutage sehr selten, sie können traumatisch durch einen Einriss der Gefäße während der Geburt zustande kommen, aber auch spontan als Ausdruck einer Blutgerinnungsstörung, z. B. einem Vitamin-K-Mangel oder einer Thrombozytopenie. Weiterhin können Hirnblutungen bei Neugeborenen auch Folge einer Gefäßfehlanlage sein.

29.5 Postnatale Hirnschäden

Hier sei vor allem auf das Problem der Hirnschädigung durch Unfälle (Stürze, Verkehrsunfälle), besonders aber auf die unter Umständen gravierenden Hirnschädigungen in Folge eines **Schütteltraumas bei Kindesmisshandlung** hingewiesen (➤ Kap. 36).
Glücklicherweise sind in Deutschland **bakterielle Meningitiden** in Folge der Impfungen nur noch sehr selten die Ursache von bleibenden Hirnschäden, Gleiches gilt für die subakut sklerosierende Panenzephalitis nach Masern-Wildvirusinfektion. Dieser Zustand kann aber auch bald durch eine zunehmende Impfmüdigkeit der Bevölkerung wieder verloren gehen.
Gesicherte **Hirnschäden in Folge einer Impfung** sind nur nach der Pocken-Vakzination und bei früheren Keuchhusten-Impfungen, die in den 70iger Jahre eingestellt wurden, beschrieben. Ursächliche Zusammenhänge von unterschiedlichen Entwicklungsstörungen, Autismus und ADHS mit anderen Impfungen, insbesondere gegen Masern, Keuchhusten, Poliomyelitis, Diphtherie, Hepatitis B und Tetanus sind, auch wenn sie in früheren Jahren anerkannt wurden, sehr fraglich und werden in neueren Studien als extrem unwahrscheinlich angesehen.
Auch nach der Geburt erworbene toxische Hirnschäden, z. B. durch Blei, organische Lösungsmittel oder Pestizide, sind ausgesprochen selten.
Ebenso ist eine Entwicklungsstörung in Folge einer Mangelfunktion der Schilddrüse (**Kretinismus bei Hypothyreose**) heute auch aufgrund des Hypothyreosescreenings im Neugeborenenalter extrem selten.
Hirntumoren können in jedem Lebensalter auftreten, bei Kleinkindern sind es vor allem das Medulloblastom, der primitiv-neuroektodermale Tumor (PNET), Astrozytome, Ependymome und Teratome, die in unterschiedlichen Hirnregionen entstehen können. Die häufigsten Primärsymptome sind Entwicklungsstillstand, chronische Kopfschmerzen und Nüchternerbrechen sowie Sehstörungen. Ihre Prognose ist von der Operabilität, z. T. von den Möglichkeiten einer Nachbehandlung mittels Bestrahlung, Chemotherapie und Stammzelltransplantation abhängig.

Literatur

Aksu F (2011) Neuropädiatrie – Diagnostik und Therapie neurologischer Erkrankungen im Kindes- und Jugendalter. Bremen: Unimed
Jansen O, Stephani U (2007) Fehlbildungen und frühkindliche Schädigungen des ZNS. Stuttgart: Thieme
Michaelis R, Niemann G (2010) Entwicklungsneurologie und Neuropädiatrie. Grundlagen und diagnostische Strategien. Stuttgart: Thieme

Neuhäuser G, Steinhausen HC, Häßler F, Sarimski K (2013) Geistige Behinderung. Grundlagen, Erscheinungsformen und klinische Probleme, Behandlung, Rehabilitation und rechtliche Aspekte. Stuttgart: Kohlhammer

Straßburg HM, Dacheneder W, Kreß W (2012) Eintwicklungsstörungen bei Kindern. Praxisleitfaden für die interdisziplinäre Betreuung. München: Elsevier

Volpe JJ. (2008) Neurology of the newborn. Philadelphia: WB Saunders

30 Verlaufsbegleitung nach Frühgeburt

Hans-Michael Straßburg

30.1 Einleitung

Frühgeborene sind Neugeborene mit einem **Gestationsalter** von weniger als 37 vollendeten Schwangerschaftswochen. Heute erfolgt die Reifebestimmung eines Kindes neben den anamnestischen Angaben der Mutter durch die Messungen der verschiedenen Körpergrößen des Feten. Bei jedem Neugeborenen sollte eine klinische Reifebestimmung, z. B. nach dem Schema von Farr, erfolgen, bei dem Veränderungen der Haut, der Ohren, der Brustdrüsen und des äußeren Genitales unabhängig vom Allgemeinzustand des Kindes festgestellt werden.

Durch Verbesserung der intensivmedizinischen Behandlungen können heute Frühgeborene mit einem Gestationsalter von 23 Wochen bzw. unter 500 g Geburtsgewicht – wenn auch mit vielfältigen Problemen verbunden – überleben. Bei einem Geburtsgewicht zwischen 1.000 und 1.500 g (bzw. einem Gestationsalter zwischen 28 und 32 Schwangerschaftswochen) spricht man von einem **sehr Frühgeborenen,** bei Kindern mit einem Geburtsgewicht unter 1.000 g (bzw. einem Gestationsalter unter 28 Schwangerschaftswochen) von **extrem Frühgeborenen.** Mittlerweile haben mehrfach Kinder mit einem Geburtsgewicht von ca. 300 g überlebt.

30.2 Komplikationen der Frühgeburtlichkeit

In epidemiologischen Erhebungen über Kinder mit Entwicklungsstörungen, insbesondere auch Kinder mit schwerer Mehrfachbehinderung, spielt die Frühgeburtlichkeit eine wesentliche Rolle. Dabei weisen die Befunde der vergangenen 20 Jahre darauf hin, dass Frühgeborene mit einem Gestationsalter über 32 Wochen eine insgesamt günstige Langzeitprognose haben, vorausgesetzt sie haben keine zusätzlichen Erkrankungen. Bis Anfang der 1990er Jahre war bei Frühgeborenen mit einer Schwangerschaftsdauer unter 32 Schwangerschaftswochen in 10 % der Fälle mit der Entwicklung einer Zerebralparese (➤ Kap. 28) und in 20–30 % mit einer mentalen Entwicklungsstörung (➤ Kap. 19) nach dem 5. Lebensjahr zu rechnen, in über 50 % der Fälle bestanden Schulprobleme. Die Schädigungsmöglichkeiten sind vor allem bei extrem Frühgeborenen sehr vielfältig und in Tabelle 30.1 (➤ Tab. 30.1) zusammengestellt.

Sozialpädiatrie in der Praxis. http://dx.doi.org/10.1016/B978-3-437-31630-3.00030-4
Copyright © 2014 Elsevier GmbH. All rights reserved.

Tab. 30.1 Komplikationen und Spätfolgen bei sehr und extrem Frühgeborenen

Frühkomplikationen	Spätfolgen
Zerebrale Hypoxie und Ischämie	Mentale Entwicklungsstörung, Bewegungsstörungen
Periventrikuläre Leukomalazie	Spastische beinbetonte Zerebralparese, zentrale Sehstörung, Lern- und Verhaltensstörungen
Neugeborenenkrämpfe	Symptomatische Epilepsie
Intraventrikuläre Blutungen	Hydrocephalus internus
Schädigung der Augen	Sehverlust, Retinopathia praematurorum (Netzhaut- und Glaskörper-Vernarbungen), Schielen, Kurzsichtigkeit
Atemstörung	Chronische Bronchitis, Stenosen der oberen Luftwege, broncho-pulmonale Dysplasie
Nekrotisierende Kolitis	Kurzdarmsyndrom, Fehlernährung, Ileus, cholestatische Lebererkrankung, Gedeihstörung, Leberzirrhose
Mangelernährung	Osteomalazie, Anämie, Dystrophie, Wachstumsstörung, IQ-Defizit
Sozialer Stress	Kindesmisshandlung und -vernachlässigung, Gedeihstörung
Zerebrale Unreife	Apnoen, plötzlicher Kindstod
Andere Störungen	Gehäufte Infektionen, gastro-ösophagealer Reflux, fehlender Verschluss des Ductus Botalli, Hämangiome, genetisch oder teratogen bedingte Anlagestörungen, Infektionsfolgen

Fallbeispiel

Vorstellung eines 5 Monate alten männlichen Säuglings. Ehemaliges Frühgeborenes von 28 + 6 Schwangerschaftswochen mit einem Geburtsgewicht von 810 g. Die Schwangerschaft war durch vorzeitige Wehen und Fieber der Mutter 2 Wochen vor dem Geburtstermin kompliziert. Apgar 6/8/8 NSApH (Nabelschnur-Arterien-pH) 7,14. Kopfumfang 28 cm. Mechanische Atemhilfe, ausgeprägte Trinkschwäche, allseits schlaffer Muskeltonus, rezidivierende Apnoen. In der zerebralen Sonografie am 4. Lebenstag vermehrte Echogenität periventrikulär, sonst regelrecht. Entlassung aus dem Kinderkrankenhaus nach 4 Monaten mit liegender naso-gastraler Sonde. Ambulante Versorgung zu Hause durch eine Kinderkrankenschwester.

Befunde:

Gewicht 2.840 g, Kopfumfang mit 34 cm korrigiert 2,5 cm unter der 3. Perzentile. Muskeltonus allseits vermindert, Kopf zur Seite gewendet, gleichmäßige Streckbewegung der Beine, keine sichere Blickfolge, vermehrtes Fäusteln. Eigenreflexe an den Beinen mehr als in den Armen gesteigert, positives Trömmner- und Rossolimo-Zeichen. In Bauchlage allenfalls schützende Kopfwendung zur Seite, bei Traktion deutlich verminderte Kopfkontrolle.

Sonografie des Schädels: erweiterte Seitenventrikel besonders frontal und okzipital mit irregulären Wandstrukturen, periventrikulär multiple echofreie Areale und vermehrte Echogenität.

Besprechung der Befunde und vorsichtige, ausführliche und ehrliche Vermittlung der Diagnosen:

- Schwere zerebrale Bewegungsstörung im Sinne einer beginnenden spastischen Zerebralparese (G80.0G)
- Periventrikuläre Leukomalazie (P91.2G)
- Mikrozephalie (Q02G)
- Verdacht auf zentrale Amblyopie (H54.2V)
- Fütterstörung (F98.2G).

Die Ursache der Entwicklungsstörung ist die hirnorganische Schädigung, die mit an Sicherheit grenzender Wahrscheinlichkeit bereits vor der Geburt entstanden ist.

- Baldmögliche Einleitung einer Komplexbehandlung durch die interdisziplinäre Frühförderstelle in Form von Physiotherapie und Heilpädagogik mit Hausbesuch. Zusätzliche Betreuung des Kindes und der Familie durch die mobile Kinderkrankenschwester
- Ansprechen der Anlage einer perkutanen endoskopischen Gastrotomie (PEG-Sonde)
- Genaue Prüfung des Visus durch eine Orthoptistin in einer Sehschule, ggf. Einleitung einer speziellen Frühförderbehandlung bei Sehstörung
- Besprechung, inwieweit die Betreuungsmöglichkeiten in der Familie für den Patienten und seine zwei älteren Geschwister ausreichend sind, ggf. Beantragung einer psychosozialen Familienhilfe
- Beantragung eines Behindertenausweises
- Vereinbarung regelmäßiger Vorstellungen im SPZ in achtwöchigem Abstand, u. a. zur genauen Überprüfung der Hüftgelenksituation.

Neben den genannten organischen Problemen kann vor allem bei extrem Frühgeborenen eine Vielzahl weiterer Komplikationen auftreten, z. B. eine nicht optimale Ernährung, eine Zytomegalieinfektionen durch Muttermilch, psycho-emotionale und sozio-ökonomische Belastungen, sekundäre Essstörungen, Ein- und Durchschlafstörungen und vieles mehr. Dennoch muss aber betont werden, dass vor allem die Prognose der sehr Frühgeborenen sich in den vergangenen 20 Jahren enorm verbessert hat, über 80 % dieser Kinder haben eine langfristig normale Entwicklung. Mittlerweile gibt es auch viele Beispiele von Kindern, die sich mit einem Geburtsgewicht unter 500 g mittelfristig sehr gut entwickelt haben. Bei Mehrlingen besteht allerdings noch ein zusätzliches Risiko.

30.3 Sozialpädiatrische Betreuungskonzepte

In vielen Studien hat sich zeigen lassen, dass die Förderung der psychosozialen Interaktion zwischen den Eltern und dem Kind ab dem frühestmöglichen Zeitpunkt eine besonders wichtige Grundlage für eine Verbesserung der Gesamtprognose darstellt. Darüber hinaus benötigen viele Eltern, insbesondere Mütter von Frühgeborenen spezielle psychotherapeutische Behandlung, nicht nur während der initialen Zeit im Krankenhaus, sondern besonders auch nach der Entlassung des Kindes.

Auch Frühgeborene, die sich primär problemlos entwickeln, zeigen nicht selten mit beginnendem Schulalter auffallende Defizite, besonders bei:

- Expressiver und rezeptiver Sprachdifferenzierung
- Körperwahrnehmung
- Konzentrationsfähigkeit
- Simultaner Informationsverarbeitung
- Sozialer Adaptation
- Entwicklung des Selbstbewusstseins.

Trotz dieser Befunde weisen Erfassungen der Langzeitprognose ehemaliger Frühgeborener aber auch darauf hin, dass diese Menschen eine sehr zufriedenstellende Lebensqualität erreichen und sich langfristig häufig gut im Leben zurechtfinden, einen Schulabschluss erreichen und einen normalen Beruf ausüben. **Frühgeburtlichkeit ist also keine Krankheit, sondern ein „biologischer Unfall"**, der sehr wohl auch mit einer langfristig ganz normalen Entwicklung des Kindes einhergehen kann.

Nach Entlassung aus der stationären Behandlung erfolgt an vielen Stellen eine **sozialmedizinische Nachsorge** durch erfahrene Kinderkrankenschwestern mit Telefonkontakt, Wiedervorstellungsterminen und ggf. Hausbesuchen nach SGB V §43, Abs. 2, z. B. durch Einrichtungen des „Bunten Kreises".

30.4 Die Langzeitprognose von Frühgeborenen

Die **Langzeitprognose** von sehr und extrem Frühgeborenen ist ein wichtiges Qualitätskriterium der primär behandelnden Institutionen. Insgesamt ist die Prognose von vor allem extrem Frühgeborenen besser, wenn sie in einem perinatalen Zentrum geboren und primär medizinisch versorgt werden. Auch Kinder mit einem Gestationsalter zwischen 32 und 36 SSW können langfristig etwas mehr Probleme als reifgeborene Kinder haben, sodass auch diese Eltern u. U. zusätzliche Unterstützung und Betreuung benötigen.

Nach Beschluss des gemeinsamen Bundesausschuss (GBA) ist bei allen Frühgeborenen mit einem Geburtsgewicht unter 1.500 g im Alter von korrigiert 2 Jahren eine standardisierte Nachuntersuchung mit den Bayley Scales verpflichtend vorgesehen. Leider gibt es hierzu noch keine öffentlich zugängige deutschlandweite Auswertung. Bei ca. 30 % der extrem Frühgeborenen lässt sich jedoch auch zu diesem Zeitpunkt keine abschließende Aussage zur Prognose machen, weshalb eine differenzierte und standardisierte Nachuntersuchung mit neuropsychologischen Testuntersuchungen im Alter von frühestens 5 Jahren gefordert wird. Leider werden die erhobenen Daten bisher nur regional gesammelt und ausgewertet.

Die Mitgliedschaft in bundesweit aktiven Elternselbsthilfe-Gruppen, regionale Treffen der Eltern, deren Kinder in einer Kinderklinik betreut wurden, Elternseminare und Kur- bzw. Rehabilitations-Maßnahmen sind für viele Eltern eine Hilfe, mit den Sorgen und Problemen ihrer zu früh geborenen Kinder besser zurecht zu kommen.

In einer umfangreichen Leitlinie der AWMF werden voraussichtlich 2013 die wichtigsten Aspekte für die Betreuung von zu früh geborenen Kindern und ihren Familien zusammengestellt.

Literatur

Aly M (2002) Das Sorgenkind im ersten Lebensjahr – ein Ratgeber für Eltern. Berlin: Springer

Müller-Rieckmann E(2006) Das frühgeborene Kind in seiner Entwicklung. München: E. Reinhardt

Strassburg HM, Leimer S, Platz A (2008) Gesundheitszustand und Lebensqualität bei ehemalig Frühgeborenen mit einem Geburtsgewicht unter 1500 g im Erwachsenenalter in Deutschland. Klin Padiat 220, 61-65

Straßburg HM, Dacheneder W, Kreß W (2012) Entwicklungsstörungen bei Kindern – Praxisleitfaden für die interdisziplinäre Betreuung. München, Elsevier.

Wüsthoff A, Böning V (2005) Frühgeborene – Leben zwischen Hoffnung und Technik. München: Urban & Fischer, 2005

AWMF Registrier-Nr. 071-013: Sozialpädiatrische Nachsorge extrem unreifer Frühgeborener mit einem Geburtsgewicht unter 1000g. S2k Leitlinie in Vorbereitung

31 Seh- und Hörstörungen

Hans-Michael Straßburg

31.1 Sehstörungen

Eine **Sehbehinderung** liegt vor, wenn der Visus (Sehschärfe) des besseren Auges 0,3 oder weniger beträgt. **Blindheit** besteht bei einem Visus von 0,02 oder weniger auf dem besseren Auge.
Bei mehr als 750 Erkrankungen mit Mehrfachbehinderung sind bisher Sehstörungen beschrieben worden. Circa 10 % aller Kinder mit geistiger Behinderung sind sehbehindert oder blind, in 80 % der Fälle besteht bei einer Mehrfachbehinderung mit Sehschädigung eine zerebrale Bewegungsstörung, in 67 % eine Epilepsie. Bei mehrfachbehinderten sehgeschädigten Menschen ist in über 75 % der Fälle die Sehbahn, d. h. das zentrale Nervensystem zwischen Netzhaut und Sehrinde, betroffen.
Eine Visus-Bestimmung kann besonders bei behinderten Kindern erhebliche Schwierigkeiten bereiten. Am besten erfolgt sie mit dem **Preferential-looking-Test,** der von einer erfahrenen **Orthoptistin** durchgeführt werden sollte. Hierbei wird beobachtet, wie ein Kind schwarze Balken auf einer sich bewegenden weißen Tafel erkennen und verfolgen kann.

31.1.1 Ursachen von Sehstörungen

Im Rahmen einer Mehrfachbehinderung sind die **Ursachen** einer zentralen Sehstörung vielfältig, am häufigsten sind Schädigungen der Sehbahnen bei Frühgeborenen, Druckschädigungen der Sehnerven bei erhöhtem Schädelinnendruck und Netzhauterkrankungen. Beispiele für prä- und frühe postnatale Ursachen sind:
- Chromosomenanomalien, z. B. Trisomie 21
- Genetische Erkrankungen, z. B. Leber'sche Amaurose, Albinismus
- Fehlbildungs- und Dysmorphiesyndrome, z. B. CHARGE-Assoziation, Lissenzephalie, angeborene Optikus-Hypoplasien
- Intrauterine Infektionen, z. B. Toxoplasmose oder Zytomegalie

Sozialpädiatrie in der Praxis. http://dx.doi.org/10.1016/B978-3-437-31630-3.00031-6
Copyright © 2014 Elsevier GmbH. All rights reserved.

- Hypoxisch-ischämische Schädigungen der Sehbahnen, vor allem infolge einer periventrikulären Leukomalazie beim Frühgeborenen
- Hydrocephalus internus, z. B. nach Hirnblutung
- Retinopathia praematurorum
- Infektionen, z. B. durch Herpesbakterien oder im Rahmen einer bakteriellen Meningitis
- Schädel-Hirn-Traumen, z. B. ein Schütteltrauma mit Retinablutungen
- Schwere zerebrale Hypoxien, z. B. beim beinahen Kindstod und bei Intoxikationen
- Intrakranielle Tumoren, vor allem ein Kraniopharyngeom
- Intrakranielle Drucksteigerungen, z. B. bei Kraniosynostosen
- Venösen Abflussstörungen und Pseudotumor cerebri
- Stoffwechselstörungen, z. B. Gangliosidose, metochromatische Leukodystrophie und Zeroidlipofuszinose.

Ein **Strabismus** (Schielen) kommt deutlich häufiger bei Kindern mit zentralnervösen Funktionsstörungen, insbesondere den verschiedenen Formen der Zerebralparese, mentalen Entwicklungsstörungen und Hirnstammerkrankungen vor. Außerdem können Refraktionsanomalien von Hornhaut und Linse sowie Anlagestörungen des Bulbus mit Myopie, Hyperopie und Astigmatismus vorliegen.

Ein akut aufgetretenes Lähmungsschielen muss immer rasch abgeklärt werden, z. B. kann ein Hirntumor zugrunde liegen, während seitenwechselnde Schielformen eher eine gute Prognose haben. Wichtigste Behandlungsmethode ist das zeitweise Abdecken des dominanten Auges, um eine bleibende Schwachsichtigkeit des vermehrt schielenden Auges zu vermeiden.

Linsentrübungen (Katarakte) können bei einer Vielzahl von Entwicklungsstörungen auftreten, z. B. bei:

- Chromosomenanomalien
- Kalzium-, Zucker- und Aminosäure-Stoffwechselstörungen
- Pränatalen Infektionen, z. B. Röteln, Zytomegalie und Toxoplasmose
- Myotonen Muskeldystrophien
- Seltenen Syndromen, wie z. B. dem Lowe-Syndrom, zusammen mit Nieren- und Gehirnanomalien.

Häufig kann aber trotz intensiver Diagnostik keine Ursache für angeborene Katarakte gefunden werden. Zur Erhaltung der Sehfähigkeit des Auges ist eine frühzeitige operative Linsenentfernung notwendig.

Eine Vielzahl weiterer Symptome am Auge kann Hinweise auf die Ursache einer Entwicklungsstörung geben. Hierzu gehören u. a.:

- Ptose bei Myopathien, z. B. Mitochondriopathien oder bei Hirnnervenfunktionsstörungen
- Enge Lidspalten (Blepharophimose) bei verschiedenen Syndromen mit einer Entwicklungsstörung
- Zusammengewachsene Augenbrauen (Synophrys) bei Cornelia-de-Lange-Syndrom und bei embryofetalem Alkoholsyndrom
- Hornhautanomalien bei angeborener Lues
- Iris-Anomalien, z. B. Kolobome bei der CHARGE-Assoziation
- Pigmentablagerungen in der Netzhaut (Retinopathia pigmentosa) beim Bardet-Biedl-Syndrom
- Nystagmus bei Amblyopien, Schädigungen im Hirnstammbereich, neurodegenerativen Erkrankungen, z. B. dem Pelizaeus-Merzbacher-Syndrom, Albinismus und anderen Störungen der zentralen Myelinisierung
- Glaukom bei verschiedenen Anlagestörungen der Augen, z. B. dem Axenfeld-Rieger-Syndrom.

31.1.2 Therapie-Prinzipien

Bei allen Kindern wird spätestens im 2. Lebensjahr eine augenärztliche Untersuchung empfohlen, um klinisch nicht erkennbare Sehschwächen rechtzeitig zu erkennen und z. B. mit einfachen Maßnahmen, wie dem zeitweiligen Abdecken des besseren Auges zu behandeln. Unklare Augensymptomen müssen spätestens zu Beginn des 2. Lebensjahrs durch einen erfahrenen Augenarzt, möglichst in Verbindung mit einer Sehschule, abgeklärt sein. Vielen Menschen selbst mit hochgradigen Sehstörungen kann durch spezifische Therapien und Hilfsmittel eine oft erstaunliche Teilhabe am normalen Leben ermöglicht werden. Blinden steht ein Behindertenausweis mit dem Merkzeichen Bl zu.

31.2 Hörstörungen

Hörstörungen können durch Erkrankungen des Mittelohrs, des Innenohrs oder der Hörbahnen bedingt sein. Eine mittelgradige **Schwerhörigkeit** liegt bei einem Hörverlust von 50 Dezibel vor, eine hochgradige bei einem Hörverlust von 60–80 Dezibel. Von **Gehörlosigkeit bzw. Taubheit** spricht man ab einem Hörverlust über 120 Dezibel.

Die Fähigkeit des Hörens ist eine der wesentlichen Grundvoraussetzungen für eine normale Entwicklung, insbesondere zum Erwerb des Sprachverständnisses und der aktiven Sprache. Damit ist aber auch die Ausbildung der mental-kognitiven Fähigkeiten und der sozialen Kompetenz verbunden.

31.2.1 Ursachen von Hörstörungen

Es gibt eine Vielzahl von Ursachen von Hörstörungen im Kindesalter. Während durch Anlagestörungen, intrauterin und in der Neugeborenenphase vor allem die empfindlichen Strukturen des Innenohrs geschädigt werden können, liegt die häufigste Ursache für Hörstörungen nach der Geburt in Funktionsstörungen des Mittelohrs, insbesondere in Form des Paukenergusses.

Seit 2009 sind **Screeninguntersuchungen** zur Erkennung von Hörstörungen des Innenohrs bei Neugeborenen Teil des von den gesetzlichen Krankenkassen finanzierten Untersuchungsprogramms, entweder durch Ableitung der oto-akustischen Emissionen (OAE) oder der akustisch-evozierten Hirnstammpotenziale (➤ Kap. 14) Dennoch müssen auch im weiteren Verlauf immer wieder anamnestische Angaben der Eltern zur Hörreaktion des Kindes und die Sprachentwicklung nach dem 6. Monat mit der Ausbildung von Doppellauten bzw. der sekundären Lallphase berücksichtigt werden. Ab dem 2. Lebensjahr können Hörtests auch mit der Spielaudiometrie durchgeführt werden. Am genauesten ist aber vor allem auch bei mangelnder Mitarbeit der Kinder die Untersuchung mit der **basalen elektrischen Reflexaudiometrie (BERA;** auch Hirnstamm-Reflexaudiometrie). Folgende anamnestische Angaben können ein Hinweis auf Hörstörungen sein:

- Hörstörungen innerhalb der Familie
- Durchgemachte Virusinfektionen in der Schwangerschaft
- Fehlbildungen der Ohren, des Schädels und des Gaumens
- Frühgeborene mit einem Geburtsgewicht unter 1.500 g
- Eher selten sind heute schwerer Sauerstoffmangel in der Perinatalperiode und

- Durchgemachte schwere, insbesondere bakterielle Hirnhaut- und Gehirnentzündungen.

Innenohrstörungen können bei verschiedenen Chromosomopathien sowie vielen genetischen Erkrankungen auftreten. Dabei können sie mit einer Vielzahl unterschiedlicher anderer Organstörungen vergesellschaftet sein, z. B. einer **Hypothyreose,** Nierenfunktionsstörungen, Netzhautschäden, Pigmentierungsstörungen der Haut, Herz-Rhythmus-Störungen, Blutbildveränderungen und vielem mehr. Außerdem können Innenohrschädigungen durch intrauterine Infektionen, z. B. Röteln oder Zytomegalie, durch verschiedene Medikamente, z. B. Aminoglykoside oder andere Toxine, vor und nach der Geburt verursacht sein.
In über 30 % der Fälle kann eine Veränderung im **Connexin-Gen,** die autosomal rezessiv, dominant, X-chromosomal rezessiv vererbt wird, für eine angeborene Hörstörung verantwortlich gemacht werden.

31.2.2 Funktionsstörungen des Mittelohr

Während der ersten Lebensjahre treten bei praktisch allen Kindern zumindest vorübergehend Funktionsstörungen der Mittelohren (Paukenerguss) auf, vor allem in Folge von Flüssigkeitsansammlungen aufgrund von Belüftungsstörungen der Paukenhöhle. Besonders gefährdet sind hierfür alle Kinder mit Lippen-Kiefer-Gaumenspalten und kraniofazialen Dysplasiesyndromen (Apert- und Crouzon-Syndrom) sowie Kinder mit unterschiedlichen Formen von Dysbalancen im Bereich der Gaumenmuskulatur, z. B. auch bei verschiedenen Formen der Zerebralparese und bei neuromuskulären Erkrankungen oder anderen Hypotonieformen. Sehr häufig sind vergrößerte Adenoide (Rachenmandeln) für eine Funktionsstörung der Eustachischen Tube verantwortlich, die neben wiederholten Virusinfekten auch durch Umwelttoxine, vor allem Rauchen in der Wohnung oder im Auto, bedingt sein kann.
Typische Symptome für **vergrößerte Adenoide** sind:

- Offener Mund, besonders während des Schlafens
- Vermehrtes Schnarchen
- Häufiges Erwachen während der Nacht
- Evtl. aber auch Störungen der normalen Schlafrhythmik
- Wiederholte Apnoen, besonders im Schlaf
- Verhaltensprobleme, insbesondere vermehrte Unruhe, Ablenkbarkeit, Quengeligkeit
- Gehäufte Müdigkeit während des Tages
- Gehäufte Infekte der oberen Luftwege und wiederholte Mittelohrentzündungen
- Appetitmangel
- Wachstumsstörungen.

31.2.3 Therapie-Prinzipien

Bei ausgeprägten Hörstörungen von mehr als 80 Dezibel sollte möglichst in den ersten Lebensmonaten die Diagnose erfolgen und bis zum Ende des 1. Lebensjahrs eine Hörgeräteanpassung stattfinden. Bei Belüftungsstörungen des Mittelohrs ist die erste Maßnahme die operative Entfernung der Rachenmandeln. Deutlich vergrößerte Gaumenmandeln (Tonsillen) sollten vor dem 6. Lebensjahr möglichst nicht vollständig entfernt werden, da sie eine wichtige Funktion bei der Infektabwehr ausüben. Gegebenenfalls kann eine operative Verkleinerung mit einer Laser-Tonsillotomie erfolgen.

Bei hochgradiger, vor allem beidseitiger Hörstörung erfolgt heute möglichst beidseits die operative Implantation einer elektronischen **Cochlea (CI).** Anschließend sind intensive Fördermaßnahmen notwendig. Ein spezielles ethisches Problem tritt dann auf, wenn beide Eltern taub sind und die Gebärdensprache verwenden. Sie verweigern u. U. die Cochlea-Implantation bei ihrem Kind und damit seine Inklusion in die Gesellschaft. Gehörlosen steht ein Behindertenausweis mit dem Merkzeichen Gl zu.

Literatur

Aksu F (2011) Neuropädiatrie – Diagnostik und Therapie neurologischer Erkrankungen im Kindes- und Jugendalter. Bremen: Unimed

Straßburg HM, Dacheneder W, Kreß W (2012) Eintwicklungsstörungen bei Kindern. Praxisleitfaden für die interdisziplinäre Betreuung. München: Elsevier

Wirth G (2000) Sprachstörungen, Sprechstörungen, kindliche Hörstörungen. Lehrbuch für Ärzte, Logopäden und Sprachheilpädagogen. Köln: Deutscher Ärzteverlag

32 Hydrozephalus und Meningomyelozele (Spina-bifida-Syndrom)

Hans-Michael Straßburg

32.1 Formen der Myelomeningocele

Bei der Meningomyelocele (MMC, Spina bifida) mit Hydrozephalus handelt es sich um ein **komplexes Fehlbildungssyndrom** infolge einer **Verschlussstörung des Neuralrohrs,** bei der typischerweise im lumbalen oder sakralen Bereich des Rückens eine blasige Vorwölbung oder eine Hautspalte besteht. Diese ist mit unterschiedlich ausgeprägten Veränderungen am unteren Rückenmark verbunden, die u. U. nur sehr diskret mit Störungen der Blasen- und Mastdarmkontrolle sowie der Fußmotorik-Sensibilität ausgeprägt sein können, aber auch verbunden mit hochgradiger Querschnittslähmung bis in den thorakalen Bereich. Neben der spinalen Querschnittssymptomatik muss in über 90 % der Fälle bei der typischen Meningomyelocele mit zerebralen Fehlbildungen im Sinne einer **Chiari-II-Malformation** gerechnet werden, die meist in den ersten Lebenstagen bereits zur Ausbildung eines **Hydrocephalus internus** führt. Daneben bestehen komplexe Fehlbildungen des zentralen Nervensystems, vor allem im Bereich des Hirnstamms und des Kleinhirns, was mittel- und langfristig mit unterschiedlich ausgeprägten weiteren Hirnfunktionsstörungen verbunden sein kann.

Bereits ab der 12. Schwangerschaftswoche kann mittels der Sonografie sowie einer Bestimmung des α-Fetoproteins im Blut der Mutter bzw. genauer im Fruchtwasser in 98 % der Fälle eine pränatale Diagnose erfolgen. Eine definitive Aussage zur Prognose ist leider oft nur schwer möglich.

Man unterscheidet die Spina bifida aperta oder cystica von der Spina bifida occulta sowie seltenere Formen von Rachischisis und Enzephalozelen. Die Gesamtsituation ist einerseits von der Höhe des Neuralrohrdefekts und andererseits von den zusätzlichen zerebralen Veränderungen abhängig.

Die Geburt sollte mittels Sectio erfolgen, die operative Deckung der offenen Zele möglichst in den ersten zwei Lebenstagen. In mehr als 80 % der Fälle muss bei einer MMC mit einem shuntpflichtigen Hydrocephalus internus gerechnet werden, der ganz überwiegend Folge einer Chiari-II-Fehlbildung ist.

Sozialpädiatrie in der Praxis. http://dx.doi.org/10.1016/B978-3-437-31630-3.00032-8
Copyright © 2014 Elsevier GmbH. All rights reserved.

Fallbeispiel

Vorstellung eines 3½-jährigen Knaben im SPZ, da dieser nicht selbstständig gehen kann.

Vorgeschichte:

3. Kind, in der 25. Schwangerschaftswoche im Ultraschall Feststellung eines Hydrocephalus und einer Zele im unteren Rückenbereich. Geburt mittels Sectio. Am 1. Lebenstag operativer Verschluss der Zele, am 10. Lebenstag Anlage eines ventrikulo-peritonealen Shunts. Seither regelmäßige Physiotherapie nach Vojta. Versorgung mit

Unterschenkelorthesen durch einen Orthopäden. Unterstützung der Blasen- und Stuhlentleerung durch Physiotherapie. In letzter Zeit vermehrt nächtliche Unruhe und Überstreckung. Bisher keine Beantragung von Pflegegeld.

Befunde:

Etwas verschlossen wirkender Knabe mit leicht offenem Mund. Strabismus convergens, endständiger Blickrichtungs-Nystagmus, knapp altersentsprechende Sprachentwicklung. Reizlose Narbe lumbal. Steht mit Unterschenkelorthesen, geht am Rollator. Unterschenkelatrophie beidseits, Senkfuß beidseits, Sensibilitätsverlust im Bereich beider Unterschenkel und perianal. ASR beidseits nicht auslösbar, PSR beidseits zweifach positiv, erhöhter Adduktorentonus, beginnende Beugekontraktur im Hüft- und Kniegelenkbereich, beginnende Skoliose. Anus klaffend, Harnträufeln.

Diagnosen:

- Meningomyelocele unterhalb von Segment L4 mit Shunt-versorgtem Hydrozephalus (Q 05.2G)
- Schlafstörung durch Hirnstammirritation (Chiari-Symptomatik) (G 47.2G)
- Verdacht auf Tethered-Cord-Syndrom (Q 06.8V)
- Blasenentleerungsstörung (N 31.2G)
- Obstipation (K 59.0G)
- Beginnende Kontrakturen (M 24.5G)
- Beginnende Skoliose (M 41.4G).

Vorschlag zum weiteren Vorgehen:

- Indikation zur Kernspintomographie kranial und spinal, insbesondere zur Beurteilung von Ventrikelgröße, Stand der Kleinhirntonsillen und des Rückenmarkkonus. In Abhängigkeit davon Indikation zur operativen Revision bei V. a. Tethered Cord
- Bildgebung der ableitenden Harnwege und Einleitung des intermittierenden, Latex-freien Katheterisierens
- Konsequente Behandlung der Obstipation mit diätetischen Maßnahmen und Suppositorien
- Absprache mit Kinderorthopäden, Verordnung von Oberschenkelorthesen
- Konsequente Physiotherapie mit Schwerpunkt bei Steh- und Gehübungen
- Beantragung von Pflegegeld
- Beantragung des Zusatzmerkmals aG im Behindertenausweis
- Beantragung von Einzelintegration im Kindergarten
- Pulsoxymetrische Registrierung des Nachtschlafs, Führen eines Schlafkalenders, ggf. Polysomnographie
- Kontaktaufnahme mit der Selbsthilfegruppe Arbeitsgemeinschaft Spina bifida und Hydrocephalus (ASbH)
- Besprechung der zukünftigen Schulbetreuung
- Planung einer psychologischen Diagnostik spätestens im Alter von 5 Jahren
- Besprechung der Wohnsituation, insbesondere in Bezug auf Treppen, Sanitäranlagen und Schlafmöglichkeit
- Genetische Beratung und Empfehlung der Folsäureprophylaxe vor allem bei der 17-jährigen Schwester
- Regelmäßige Kontrolltermine im SPZ in Abständen von 6 Monaten.

32.2 Komplikationen

Wichtige Zeichen einer **Shunt-Dysfunktion** sind: abnormes Kopfwachstum, gespannte Fontanelle, Sonnenuntergangs-Phänomen, Strabismus, Kopfschmerzen, Trinkschwäche, Visusstörung, Stauungspapillen, Nüchternerbrechen, zerebrale Anfälle.
Zeichen einer **Störung von Hirnstamm und Kleinhirn** im Rahmen der Chiari-II-Malformation können sein: Schlafstörung, Heiserkeit, inspiratorischer Stridor, zentrale und obstruktive Apnoen, Schluckbeschwerden, opisthotone Haltung, zervikale und okzipitale Kopfschmerzen, Schwäche der Hand- und Armmotorik, progrediente Spastik, progrediente Ataxie und Nystagmus.
Nicht selten sind mangelnde Fortschritte bzw. Rückschritte in den motorischen Fähigkeiten bei MMC-Kindern durch das **Tethered-Cord-Syndrom** zu erklären. Hierbei können folgende Symptome auftreten:

- Schmerzen im Bereich des Rückens und der Beine
- Progrediente Bein- und Fußfehlhaltungen
- Progrediente Skoliose
- Zunehmende motorische Ausfälle
- Zeichen einer Spastik, besonders im Bereich der Adduktoren
- Zunehmende sensible Ausfälle
- Blasen- und Mastdarmentleerungsstörungen.

32.3 Therapie-Prinzipien

Blasen-Mastdarm-Störungen sind bei fast allen Formen der MMC zu erwarten und müssen oft z. B. mit intermittierendem Katheterisieren und einer konsequenten Laxanziengabe behandelt werden. In seltenen Fällen sind spezielle urologische Operationen indiziert.
Orthopädische Probleme sind bei MMC-Kindern abhängig vom Lähmungsniveau:

- Bei segmentalen Ausfällen unterhalb von S2 besteht oft nur ein Senkfuß
- Bei Lähmungen in Höhe von S1 bis S2 sollte eine äußere Stabilisierung der Sprunggelenke erfolgen
- Bei Lähmungen unterhalb von L5 sind Unterschenkelorthesen indiziert
- Bei Lähmungen in Höhe von L4 sind Oberschenkelorthesen und nur für längere Strecken ein Rollstuhl indiziert
- Bei Lähmungen oberhalb von L3 sind Oberschenkelorthesen oft mit Beckenteil, Gehstützen oder Rollator indiziert sowie praktisch immer ein Rollstuhl
- Bei thorakaler MMC kann zusätzlich ein Swivel-Walker oder eine reziproke Gehorthese eingesetzt werden.

Die **Intelligenz** ist bei Menschen mit MMC unterschiedlich ausgebildet, bei tiefer Lokalisation ist sie meist ganz normal. Je höher die Rückenmarkfehlbildung und je ausgeprägter die zerebralen Fehlbildungen, umso eher muss mit einer Intelligenzeinschränkung gerechnet werden. Die Ausprägung des Hydrozephalus bei der Geburt hat nur eine geringe prognostische Bedeutung.

32.4 Aspekte der Langzeitbetreuung

Spezielle sozialpädiatrische Aufgaben sind u. a. die langfristige ganzheitliche Begleitung des Kindes und der Eltern, möglichst im Rahmen einer Ambulanz mit mehreren Spezialisten, die Indikation zur Therapie mit Heilmitteln, insbesondere Physiotherapie (➤ Kap. 37), eine ständige Elternanleitung, z. B. beim Katheterisieren, die Beratung beim Besuch von Kindergarten und Schule und die Vermittlung zur Teilnahme an Angeboten der Eltern- und Patientenselbsthilfe (Arbeitskreis Spina bifida Hydrozephalus ASbH).
Mit zunehmendem Alter können neben den orthopädischen Problemen Alterationen im Hirnstammbereich auf Grund der Chiari-Malformation eine immer wichtigere Rolle spielen, z. B. in Form von Atemantriebsstörungen, zentralen Bewegungs- und Sehstörungen.
Die MMC ist ein Beispiel für die Notwendigkeit spezieller Konzepte zur **Transition** vom Jugendlichen- zum Erwachsenenalter einer lebenslangen individuellen Betreuung der betroffenen Menschen in Spezialambulanzen.

Literatur

Aksu F (2011) Neuropädiatrie – Diagnostik und Therapie neurologischer Erkrankungen im Kindes- und Jugendalter. Bremen: Unimed

Heinen F, Böhmer J, Hufschmidt A et al. (2009) Pädiatrische Neurologie in Diagnose und Therapie. Stuttgart: Kohlhammer

Michael T, von Moers A, Strehl AE (2014) Spina bifida. Berlin: De Gruyter

Michaelis R, Niemann G (2010) Entwicklungsneurologie und Neuropädiatrie. Grundlagen und diagnostische Strategien. Stuttgart: Thieme

Straßburg HM, Dacheneder W, Kreß W (2012) Entwicklungsstörungen bei Kindern. Praxisleitfaden für die interdisziplinäre Betreuung. München: Elsevier

33 Chromosomale und andere genetische Anomalien

Hans-Michael Straßburg

33.1 Numerische Chromosomenanomalien

33.1.1 Einteilung und Häufigkeit der Trisomie 21

Die **Trisomie 21** ist die häufigste Form einer Chromosomenanomalie und ist unabhängig von der ethnischen Zugehörigkeit. Das überzählige Chromosom stammt in 90 % der Fälle von der Mutter, bei einem Alter der Mutter über 35 Jahre steigt das Risiko exponentiell an. Die Prävalenz liegt durchschnittlich bei 1 : 660 Lebendgeborenen. Das Risiko bei einer Schwangerschaft im Alter unter 33 Jahren beträgt weniger als 1 : 1000 und steigt bis zum Alter von 40 Jahren auf 1 : 100. 97 % der Patienten haben eine freie Trisomie, bei 3 % besteht eine unbalancierte Translokation, die u. U. von einem Elternteil vererbt werden kann. In seltenen Fällen besteht eine Mosaikkonstellation oder eine Verdreifachung der Region 21q22 mit Translokation auf ein anderes Chromosom. Deshalb sollte bei allen Patienten mit Verdacht auf Down-Syndrom bei fehlendem zusätzlichem Chromosom 21 eine Fluoreszenz-in-vitro-Hybridisierung durchgeführt werden.

Viele Menschen mit Trisomie 21 haben zusätzliche Fehlbildungen oder Erkrankungen, in 30 % der Fälle besteht ein Herzfehler, oft Magen-Darm-Anomalien, Seh- und Hörstörungen. Es besteht ein signifikant höheres Risiko für gravierende Infektionen, besonders im Bereich der Luftwege, für das Entstehen einer Epilepsie und anderer sekundärer Schäden des zentralen Nervensystems sowie einer Leukämie.

Die Entwicklungsmöglichkeiten von Kindern mit Trisomie 21 sind ausgesprochen unterschiedlich, in den ersten Lebensjahren ist vor allem die Sprachentwicklung deutlich verlangsamt, wohingegen die sozio-emotionale Entwicklung und auch die musikalischen Fähigkeiten wesentlich besser ausgebildet sind. Beim Schulwissen bestehen häufig Probleme im konstruktiv-abstrakt-logischen Denken (fluide Intelligenz), andererseits können Menschen mit Down-Syndrom gut einfa-

Sozialpädiatrie in der Praxis. http://dx.doi.org/10.1016/B978-3-437-31630-3.00033-X
Copyright © 2014 Elsevier GmbH. All rights reserved.

che Handlungsaufträge bewältigen, eine Fremdsprache erlernen, Rhythmus-Instrumente spielen oder Schauspielrollen übernehmen.

Fallbeispiel

Erstmalige Vorstellung eines 2 Monate alten Säuglings mit Trisomie 21 zur Planung des Weiteren Betreuungskonzepts.

Vorgeschichte:

1. Kind einer 24-jährigen Frau. Problemlose Geburt, Apgar 6/9/9, Geburtsgewicht 2.730 g. Postnatal 3 Wochen stationäre Behandlung in einer Kinderklinik wegen Trinkschwäche und Obstipation. Kein Hinweis auf Herzfehler.

Befunde:

Gewicht 3.200 g, Länge 47 cm, Kopfumfang 33,4 cm. Typische Zeichen der Trisomie 21 mit größerem Augenabstand, kurzem, kleinem Kopf, nach außen oben ansteigender Lidachse, kleinen Ohren, überstreckbaren Gelenken, 4-Finger-Furche, verbreitertem Abstand zwischen dem ersten und zweiten Zehenstrahl, ausladendem Abdomen. Generalisierte Muskelhypotonie, Mund offen, vorgestreckte Zunge, verminderte Kopfkontrolle bei Traktion, schwach auslösbare Eigenreflexe. Kleiner Penis, Hoden allenfalls 1 ml. Leises Systolikum. Kurze Blickfolge, kein sicherer optokinetischer Nystagmus auslösbar. Keine sichere Reaktion auf Geräusche.

Diagnosen:

- Trisomie 21 (Q 90.9G)
- Nicht altersentsprechende Entwicklung (F 89G)
- Entwicklungsstörung der mundmotorischen Funktion (F 82.0G)
- Mikrozephalie (Q02G)
- V. a. Schwerhörigkeit (H 91.9V).

Weiteres Vorgehen:

- Überprüfung der kinderkardiologischen Befunde
- Kontrolluntersuchungen des Hörens mit BERA und des Sehens
- Abdominelle und zerebrale Sonografie
- Blutbild
- Hüftgelenksonographie
- Empfehlung zur Anmeldung bei der interdisziplinären Frühförderung zur Einleitung einer Komplexbehandlung mit Physiotherapie und Heilpädagogik
- Regelmäßige Gewichtskontrollen
- Diätetische und medikamentöse Förderung der Stuhlentleerung
- Regelmäßige Kontrollen der Schilddrüsenfunktion
- Mitteilung der Adressen des Arbeitskreises Down-Syndrom und der Bundesvereinigung Lebenshilfe für Menschen mit geistiger Behinderung
- Ansprechen der Möglichkeit, einen Behindertenausweis zu beantragen
- Psychische Stützung, insbesondere der Mutter
- Genetische Beratung der Familie.

33.1.2 Komplikationen bei Trisomie 21

Bei Menschen mit Down-Syndrom können eine Vielzahl verschiedener medizinischer Probleme auftreten, u. a.:

- Hörstörungen durch Innen- oder Mittelohrerkrankungen
- Sehstörungen durch Linsentrübung, Retinaveränderungen, Sehnervenveränderungen

- Zerebrale Auffälligkeiten, z. B. unterschiedliche Hirnentwicklungsstörungen in Form von einer praktisch immer vorhandenen Mikrozephalie mit zusätzlichen Veränderungen wie Hydrozephalus, Balkenmangel, kortikale Dysplasien, Epilepsien
- Gehäufte Infekte der oberen und unteren Luftwege
- Unterschiedliche Herzfehler, u. a. ein offener AV-Kanal
- Anomalien des Magen-Darm-Trakts in Form von Stenosen im Bereich des Zwölffingerdarms, Morbus Hirschsprung, neuronaler Kolondysplasie, Pancreas anulare
- Hämatologische Auffälligkeiten, z. B. Anämie, Neigung zur Leukämie
- Nierenfehlbildungen
- Hüftgelenkdysplasien
- Allgemeine Muskelhypotonie
- Periphere Neuropathien
- Instabilität des atlanto-okzipitalen Übergangs etc.

33.1.3 Therapie- und Betreuungs-Prinzipien

Von besonderer Bedeutung ist bei Kindern mit Trisomie 21 die **Förderung der Mundmotorik** und der Sprachentwicklung. Hier kann relativ früh neben der allgemeinen Physiotherapie die orofaziale Regulationsbehandlung nach Castillo Morales eingesetzt werden. Es empfiehlt sich eine frühzeitige Sprachförderung, z. B. auch in Form einer Elternanleitung. Sollte bis zum Ende des 2. Lebensjahres keine aktive Sprachentwicklung einsetzen, empfiehlt sich eine gebärdenunterstützte Kommunikation (GuK), später auch andere EDV-gestützte Verfahren (➤ Kap. 37, ➤ Kap. 38).
Kinder mit Down-Syndrom haben in der Regel einen sehr freundlichen, gutmütigen und sozialen Charakter. Oft profitieren sie vom Zusammensein mit altersentsprechend entwickelten Kindern, allerdings kann es besonders bei schwerwiegenden Entwicklungsstörungen auch zu sekundären Verhaltensproblemen kommen. Menschen mit Down-Syndrom sind vor allem in den ersten Lebensjahren ein gutes Beispiel für die **Inklusion** in Regeleinrichtungen. Sehr sinnvoll ist für Eltern die Mitgliedschaft in der **Bundesvereinigung Lebenshilfe,** der sozialpolitisch wichtigsten Selbsthilfegruppe. Diese organisiert u. a. die vorschulische und schulische Betreuung, Werkstätten und betreutes Wohnen. Leider haben aber nur wenige Menschen mit Trisomie 21 die Möglichkeit, ab dem Jugendlichenalter ein selbstbestimmtes Leben zu führen. Auch nach dem 18. Lebensjahr können sich eine Vielzahl von spezifischen Gesundheitsproblemen verschiedener Organsysteme ergeben, die sinnvoller Weise in spezialisierten Zentren mitbetreut werden sollten. Hierfür setzt sich u. a. die Bundesarbeitsgemeinschaft der Ärzte für Menschen mit geistiger oder mehrfacher Behinderung ein

33.2 Andere numerische Chromosomenaberrationen

Andere Trisomien, z. B. des 13. oder des 18. Chromosoms, sind mit schweren allgemeinen Entwicklungsstörungen und anderen Organmanifestationen verbunden.
Eine Sonderform stellen Aberrationen der Geschlechtschromosomen, insbesondere das Turner-Syndrom (X0) und das Klinefelter-Syndrom (XXY) bzw. das XYY- oder Triple-X(XXX)-Syndrom, dar. Hierbei spielen Intelligenzminderungen in der Regel nicht die vorherrschende Rolle.

33.3 Strukturelle Chromosomenanomalien

33.3.1 Das Fra-X-Syndrom

Die häufigste strukturelle Chromosomenanomalie ist das **Fragile-X-Syndrom** (Martin Bell-Syndrom = FRA-X-Syndrom), das durch eine brüchige Stelle am Ende des langen Armes des X-Chromosoms verursacht wird. Bei Jungen ist es nach der Trisomie 21 die häufigste genetische Ursache einer geistigen Behinderung, die Häufigkeit wird mit 1 : 1200 bei Jungen und 1 : 2400 bei Mädchen angegeben. Typische **Symptome,** die sich oft jedoch erst im Laufe der ersten Lebensjahre ausbilden, sind:

- Relativ großer Kopf
- Großer Gesichtsschädel
- Vermehrte Unruhe ab dem Kleinkindalter
- Sprachentwicklungsstörungen sowohl der expressiven als auch der rezeptiven Sprache
- autistische Verhaltensweisen
- Unterschiedliche Wahrnehmungsstörungen
- Nach der Pubertät auffallend lange Nase, große Ohren und große Hoden.

Bei differenzierter Beobachtung können einige weitere Symptome auffallen, z. B.:

- Vermehrte Echolalie
- Polternde Aussprache
- Vermindertes Sprachverständnis
- Umschriebene Rechenstörung
- Raumorientierungsstörung
- Konzentrationsstörung
- Hyperaktivität
- Impulsivität
- Störungen des Kontaktverhaltens, insbesondere des Augenkontakts
- Koordinationsstörungen von Augen und Hand.

Fallbeispiel

Vorstellung eines 4,2-jährigen Knaben mit verzögerter Sprachentwicklung. Außerdem bestehe bei ihm eine „Wahrnehmungsstörung" und Probleme mit der motorischen Koordination. Bisher durchgeführte Ergotherapie und Logopädie hätten keine befriedigende Therapieerfolge erbracht.

Zweites von drei Kindern, ein älterer Bruder besucht wegen motorischer Koordinations- und Verhaltensprobleme die 1. Klasse einer Schule für Körperbehinderte. Eine 2,3-jährige Schwester sei gesund, sie habe bereits jetzt einen größeren Wortschatz als der Patient.

Geburtsgewicht unseres Patienten 3.210 g. In der Säuglingszeit sei er auffallend ruhig gewesen, selbstständiges Gehen mit 15 Monaten, Sprechen erster Worte mit 26 Monaten. Noch nicht sauber und nicht trocken.

Befunde:

- Größe und Gewicht im mittleren Normbereich, Kopfumfang mit 54,2 cm 0,2 cm oberhalb der 97. Perzentile. Relativ hohe Stirn, relativ große Ohren. Muskulatur kräftig, intern kein auffälliger Organbefund, keine Hautauffälligkeiten. Auffallend ist ein verminderter Blickkontakt mit dem Untersucher. Antwortet in kurzen Sätzen
- Im EEG mit 7–8/sec etwas langsamer Grundrhythmus, sonst regelrecht

- Veranlassung einer molekulargenetischen Untersuchung auf Fragiles-X-Syndrom nach ausführlicher Aufklärung, dass dadurch auch bei anderen Familienmitgliedern genetische Auffälligkeiten festgestellt werden könnten
- Nach Erhalt des positiven Befundes von einer signifikant erhöhten CDG-Repeat-Zahl kurzfristige Vereinbarung eines Termins mit beiden Eltern. Ausführliche Besprechung der genetischen Zusammenhänge. Auch bei dem älteren Bruder wird eine molekulargenetische Untersuchung veranlasst, die ebenfalls einen positiven Befund ergibt.

Diagnosen:
- Fragiles-X-Syndrom (Q 99.2G)
- Leichte Intelligenzminderung ohne wesentliche Verhaltensstörung (F 70.0G).

Vorgehen:
- Ausführliche Aufklärung zum Störungsbild
- Genetische Beratung
- Kontaktaufnahme mit der SVE mit überwiegendem Förderbedarf im Bereich der geistigen Entwicklung
- Fortführung der logopädischen Behandlung
- Beantragung eines Behindertenausweises und von Pflegegeld für beide Söhne
- Kontaktaufnahme mit der Bundesvereinigung Lebenshilfe und der
- Elternselbsthilfegruppe FRA-X
- Differenzierte IQ-Testung im Alter von 5 Jahren.

In der Regel ist die Mutter eines Jungen mit FRA-X-Syndrom klinisch unauffällige weibliche Überträgerin. 30 % ihrer Töchter zeigen leichte Einschränkungen der Intelligenz, vor allem in Teilleistungsbereichen, z. B. in Form einer Dyskalkulie.

Fallbeispiel, Fortsetzung

Nach 3 Jahren stellt die Mutter ihre mittlerweile 6-jährige Tochter mit der Frage vor, ob bei ihr auch eine Veranlagung zum FRA-X-Syndrom vorliegt. Sie ist somatisch altersentsprechend und wirkt etwas schüchtern. Bei einer psychologischen Untersuchung mit Verwendung des HAWIVA-Tests ergibt sich ein etwas inhomogenes Leistungsprofil mit verminderter Leistungsfähigkeit im Bereich des Zahlengedächtnisses und des räumlich-konstruktiven Vorstellungsvermögens. Von einer molekulargenetischen Diagnostik mit Bestimmung der CDG-Repeats wird vorerst abgeraten, da es sich dabei um eine prospektive Untersuchung handeln würde, für die eigene Einsichtsfähigkeit vorhanden sein sollte. Es wird darauf verwiesen, dies z. B. nach dem 12. Lebensjahr mit beiden Eltern und der Tochter nochmals zu besprechen.
Einige Monate später teilt die Mutter mit, dass sie die Untersuchung bei der Tochter durch einen anderen Arzt habe durchführen lassen – bei ihr sei ein Überträger-Status für das Fra-X-Syndrom festgestellt worden. Schulisch gäbe es sonst keine weiteren Probleme.

33.3.2 Andere genetisch bedingte Syndrome mit Intelligenzminderung

Hierzu gehört z. B. das **Rett-Syndrom,** das praktisch nur bei Mädchen vorkommt und mit schwerer Intelligenzminderung und Epilepsie vergesellschaftet ist. Typischerweise entwickelt sich die Symptomatik erst nach dem 6. Lebensmonat mit einem Entwicklungsstillstand, stereotypen Bewegungen vor allem der Hände, einer Einschränkung der Motorik, Schlaf- und Verhaltensstörungen. Die Ursache des Rett-Syndroms ist eine Mutation auf dem X-Chromosom ist (MECP2-Gen).

Das **Prader-Willi-Syndrom** führt zu einer Intelligenzminderung in Kombination mit gesteigertem Essverhalten und muskulärer Hypotonie, besonders im Säuglings- und Kleinkindalter. Zugrunde liegt, ebenso wie beim **Angelman-Syndrom,** eine Deletion auf Chromosom 15. Bei Letzterem bestehen darüber hinaus im Säuglingsalter eine muskuläre Hypotonie mit späterer Ausbildung einer deutlichen Hypertonie, unmotivierte Lachanfälle und verschiedene Epilepsieformen.

Das **Williams-Beuren-Syndrom** ist durch typische Gesichtszüge, ein freundliches Wesen, Minderwuchs, Herz- und Gefäßanomalien, meist in Form einer supravalvulären Aortenstenose sowie Nieren- und Kalziumstoffwechselstörungen gekennzeichnet.

Daneben gibt es eine Vielzahl weiterer mittlerweile klar definierter genetischer Syndrome, z. B. Bardet-Biedl-Syndrom, Coffin-Lowry-Syndrom, Charge-Syndrom, Cornelia-Lange-Syndrom, Noonan-Syndrom, Rubinstein-Taybi-Syndrom, Sjögren-Larsson-Syndrom, Smith-Magenis-Syndrom.

Ein wahrscheinlich immer wieder übersehenes klinisches Syndrom mit Intelligenzminderung, Gesichtsdysmorphien, z. T. mit Lippen-Kiefer-Gaumenspalten, Herzfehler und Immunstörungen ist das 22q11-Deletionssyndrom bzw. das **DiGeorge-Syndrom,** bei dem gehäuft auch unterschiedliche psychiatrische Erkrankungen, z. B. Psychosen, auftreten können.

Für die Therapie und Betreuung dieser Kinder und Jugendlichen gelten einerseits die Prinzipien wie bei der Trisomie 21 (➤ Kap. 37, ➤ Kap. 38), andererseits müssen aber auch viele spezifische Faktoren berücksichtigt werden. Hier wird auf Spezialliteratur verwiesen. Ein konstruktiver Austausch der Erfahrungen findet vor allem in den Elternselbsthilfegruppen statt, die bundesweit von der Organisation Kindernetzwerk e. v. vertreten werden.

Literatur:

Aksu F (2011) Neuropädiatrie – Diagnostik und Therapie neurologischer Erkrankungen im Kindes- und Jugendalter. Bremen: Unimed

Michaelis R, Niemann G (2010) Entwicklungsneurologie und Neuropädiatrie. Grundlagen und diagnostische Strategien. Stuttgart: Thieme

Murken J, Grimm T, Hollinski-Feder E (2006) Taschenlehrbuch Humangenetik. Stuttgart: Thieme.

Neuhäuser G, Steinhausen HC, Häßler F, Sarimski K (2013) Geistige Behinderung. Grundlagen, Erscheinungsformen und klinische Probleme, Behandlung, Rehabilitation und rechtliche Aspekte. Stuttgart: Kohlhammer

Straßburg HM, Dacheneder W, Kreß W (2012) Eintwicklungsstörungen bei Kindern. Praxisleitfaden für die interdisziplinäre Betreuung. München: Elsevier

34 Andere seltene Erkrankungen und Syndrome

Hans-Michael Straßburg

34.1 Neurokutane Syndrome

Die häufigste Erkrankung aus diesem Formenkreis ist die **Neurofibromatose Typ I,** die durch multiple hellbraune Pigmentflecken der Haut (Café-au-Lait-Flecken), Anomalien der Iris und sehr unterschiedliche Tumorbildungen im Bereich des peripheren und zentralen Nervensystems gekennzeichnet ist. In 50 % der Fälle handelt es sich um eine Neumutation. Bei vielen Kindern lassen sich diffuse Gliome des Nervus opticus mit Sehbehinderung, aber auch unterschiedliche Astrozytome an anderer Lokalisation, Meningiome und Sarkome, pathologische Knochenfrakturen und anderes feststellen. Die wichtigste Kontrollmaßnahme bei Patienten mit Neurofibromatose I ist die regelmäßige augenärztliche Prüfung. Weitere Kontrolluntersuchungen müssen vom jeweiligen Entwicklungsstand bzw. dem neurologischen Befund abhängig gemacht werden.

Im Gegensatz dazu ist die **Neurofibromatose Typ II** durch das Auftreten ein- oder beidseitiger Tumoren im Bereich des VIII. Hirnnervs charakterisiert.

Die **tuberöse Sklerose** tritt mit einer Häufigkeit von 1 : 6000 auf, in 85 % der Fälle handelt es sich um eine Neumutation. Typisch ist im Säuglings- und Kleinkindalter das Auftreten einer Epilepsie mit allgemeiner Entwicklungsstörung. An der Haut lassen sich typische weiße Flecken, die u. U. nur mittels UV-Licht (Wood-Lampe) nachgewiesen werden können, sowie später seborrhoische Adenome, besonders im Gesicht, Nierenveränderungen, Herzrhythmusstörungen und verschiedenes mehr nachweisen. Die häufig vorkommenden Hirntumoren können z. T. mit dem Medikament Everolimus erfolgreich behandelt werden.

Das **Sturge-Weber-Syndrom** ist durch kapilläre Hämangiome im Versorgungsbereich des ersten und zweiten Astes des Nervus trigeminus gekennzeichnet und führt in der Regel zu schweren Schäden des Großhirns mit therapierefraktärer Epilepsie. Bei einigen Patienten war eine operative Dekonnektion einer Großhirnhemisphäre zu Behandlung der Epilepsie notwendig.

Sozialpädiatrie in der Praxis. http://dx.doi.org/10.1016/B978-3-437-31630-3.00034-1
Copyright © 2014 Elsevier GmbH. All rights reserved.

34.2 Stoffwechselstörungen

34.2.1 Die Phenylketonurie

1939 beschrieb A. Fölling erstmals ein Krankheitsbild mit schwerer geistiger Entwicklungsstörung und zerebralen Anfällen, bei dem die betroffenen Patienten auffallend blass, meist blond und blauäugig waren. In ihrem Urin konnte eine massive Erhöhung des Phenylketonwerts festgestellt werden, weshalb die Erkrankung **Phenylketonurie** (PKU) genannt wurde. Als Ursache wurde eine Abbaustörung der essenziellen Aminosäure Phenylalanin aufgrund eines Mangels des Enzyms Phenylalanin-Hydroxylase nachgewiesen.
Der deutsche Pädiater H. Bickel konnte zeigen, dass durch eine Diät, in der die Aminosäure Phenylalanin deutlich reduziert ist, die Entwicklung der Kinder signifikant verbessert wird. Heute kann durch die Erkennung der Stoffwechselstörung mit Hilfe des Neugeborenenscreenings bereits in den ersten Lebenstagen eine Phenylketonurie diagnostiziert und durch Einleitung der entsprechenden Diät eine normale Entwicklung des Kindes erreicht werden, die allerdings kontinuierlich überwacht werden muss.
Ein spezielles Problem kann darin bestehen, dass bei einer Mutter eine nicht bekannte, subklinische PKU besteht, die eine intrauterine Wachstumsstörung des Gehirns verursacht.

34.2.2 Andere Stoffwechselerkrankungen

Aktuell sind mehr als 250 Stoffwechselerkrankungen bekannt, die sich vor allem durch Funktionsstörungen des zentralen Nervensystems manifestieren. Hierzu gehören vor allem vielfältige Störungen im Ab- und Aufbau der Amino- und organischen Säuren sowie verschiedener Kohlenhydrate (z. B. der Galaktose). Klinische Hinweise auf das Vorliegen einer Stoffwechselstörung können sein:

- Auffälliger Geruch der Haut und des Urins
- Metabolische Azidose, vor allem nach Beginn der Ernährung und bei Belastungen wie Infekten oder Operationen
- Rezidivierendes Erbrechen
- Gedeihstörungen
- Zerebrale Anfälle
- Mikro- und Makrozephalie
- Intelligenzminderung
- Krisenhafte Verschlechterung bei Infekten, z. B. mit Hypoglykämie u. v. m.

Nachfolgend werden einige sozialpädiatrisch wichtige Stoffwechselstörungen ohne Anspruch auf Vollständigkeit aufgeführt. Nicht bei allen lassen sich mit speziellen Diäten oder Medikamenten so eindeutige Erfolge wie bei der PKU erreichen.
Bei den Erkrankungen der Lysosomen in den Zellen zeigen sich unterschiedliche klinische Erscheinungsbilder mit Ansammlungen verschiedener Stoffwechselprodukte in den Zellen. Das wichtigste Beispiel sind die **Mukopolysaccharidosen,** bei denen komplexe Moleküle in Leber, Knochen, Hirn, Linse, Innenohr und anderen Organen gespeichert werden. In den letzten Jahren wurde bei mehreren Formen der Mukopolysaccharidosen eine aufwändige intravenöse Enzymersatztherapie eingeführt, mit der vor allem die Symptome außerhalb des ZNS positiv beeinflusst werden können.

Bei den **Gangliosidosen,** z. B. der Tay-Sachs'schen Erkrankung, kommt es zu schweren mentalen Abbaustörungen, Erblindung, Ertaubung und Epilepsie. Da sie besonders bei Ashkenasy-Juden häufiger vorkommt, wird Familienmitgliedern, bei denen ein entsprechender Krankheitsfall bekannt geworden ist, von den religiösen Führern (Rabbi) untersagt, zu heiraten. Bei der **metachromatischen Leukodystrophie** tritt vor allem eine progrediente spastische Bewegungsstörung zusammen mit Seh- und Hörstörungen auf. Die verschiedenen Formen des **Niemann-Pick Syndroms** gehören zu den Sphingolipidosen und können z. T. auch durch eine Enzymersatztherapie in ihrer Progredienz aufgehalten werden

Die **Adrenoleukodystrophie** und das Zellweger-Syndrom sind Beispiele für angeborene Erkrankungen der Peroxisomen. Die Hoffnung, dass mit Hilfe einer Diät ohne langkettige Fettsäuren und durch Zufuhr spezieller Öle („Lorenzos Öl") das Krankheitsbild teilweise in seinem Verlauf aufgehalten werden kann, hat sich nicht bestätigt, aber, auch aufgrund des gleichnamigen Hollywood-Films, die Bekanntheit dieser seltenen Erkrankung gesteigert. Es gibt erste erfolgversprechende Behandlungsergebnisse nach Stammzelltransplantation.

Bei sehr unterschiedlichen Formen von Entwicklungsstörungen mit Intelligenzminderung, Erkrankungen der Augen, der Ohren und der Muskulatur sowie seltener auch anderer Organsysteme sollte auch an eine **Mitochondriopathie** gedacht werden. Klinisch lassen sich diese Erkrankungen vor allem durch erhöhte Laktatwerte in Serum und Liquor, den Nachweis von Störungen der Atmungskettenenzyme und neuerdings auch durch Feststellung von Deletionen im mitochondrialen Genom diagnostizieren.

Störungen im Stoffwechsel der Nukleinsäuren können u. a. zu dem X-chromosomal rezessiv vererbten **Lesch-Nyhan-Syndrom** führen, das durch schwere Intelligenzminderung, Dyskinesien und auffallend verminderte Schmerzempfindung gekennzeichnet ist. Deshalb kommt es bei den Betroffenen oft zu schweren Verletzungen und Selbstverstümmelungen, z. B. in Folge von Bisswunden.

Bei vielen Funktionsstörungen der Leber und des zentralen Nervensystems muss auch an das Vorliegen einer Kupferstoffwechselstörung, den **Morbus Wilson,** gedacht werden. Die Diagnose wird durch eine Bestimmung des Kupfertransportproteins Zöruloplasmin im Serum, den Nachweis einer erhöhten Kupferausscheidung und ggf. eine Messung des Kupfers in einer Leberbiopsie bestimmt. Bei dem X-chromosomal rezessiv vererbten **Menkes-Syndrom** liegt eine Kupferverwertungsstörung mit schweren Veränderungen an Gehirn, Knochen, der Haut und den Haaren (kinky hair syndrome) vor. Es muss z. B. bei der Diagnose einer Kindesmisshandlung (shaken baby syndrome) berücksichtigt werden.

Störungen im Fettstoffwechsel, z. B. des Cholesterins, können mit Mikrozephalie, Gaumenspalte, Muskelhypotonie, Wachstumsverzögerung und Verschiedenem mehr vergesellschaftet sein (**Smith-Lemli-Opitz-Syndrom**).

Bei therapieresistenten Epilepsien muss im Zusammenhang mit dem Vorliegen von Seh- und Hörstörungen an eine **Zeroidlipofuszinose** gedacht werden.

Eine sehr komplexe Gruppe von Stoffwechselstörungen wird als **CDG-Syndrom** (= congenital defect of glycosylation) bezeichnet. Charakteristisch hierfür sind Kleinhirn- und Hirnstammanomalien, muskuläre Hypotonie, Veränderungen an Nieren, Leber und Herz sowie unterschiedliche neurologische Auffälligkeiten.

Patienten mit Epilepsie und unterschiedlichen Formen der Intelligenzminderung, die an einem **Glukose-Transport-Protein-Defekt** leiden, können hinsichtlich ihrer Prognose von einer **ketogenen Diät** profitieren. Bei globaler Entwicklungsstörung kann durch eine Kernspinspektroskopie z. B. ein **zerebraler Kreatinmangel** festgestellt werden, dessen Folgen sich bei entsprechender Substitution z. T. deutlich zurückbilden können.

34.3 Bedeutung von Stoffwechselanomalien

Es muss immer bedacht werden, dass viele genetisch-metabolische Erkrankungen auch in subklinischer, d. h. deutlich abgeschwächter Form vorkommen können, was ihre Erkennung sehr erschweren kann und erst mit deutlich verbesserten diagnostischen Methoden ermöglicht.
Bei vielen Menschen mit globalen Entwicklungsstörungen lässt sich noch keine eindeutige Ursache feststellen, obwohl klinische Aspekte für eine eine genetische Anlagestörung sprechen. Weltweit sind in großen Dateien, z. B. OMIM (Online Mendelian Inheritance in Man) und der London Dysmorphology Database mehrere Tausend verschiedene klinische Syndrome beschrieben, die durch Entwicklungsstörungen aller Art und ggf. zusätzliche äußere Merkmale gekennzeichnet sind. Durch die rasanten Verbesserungen der molekulargenetischen Diagnostik lassen sich immer mehr Ursachen auch seltener Syndrome aufklären
Erkennung und Behandlung jeder dieser seltenen Erkrankungen benötigt spezielle Kenntnisse und Erfahrungen. Hierbei spielen die Selbsthilfegruppen, die sich im KINDERNETZWERK zusammengeschlossen haben, eine wichtige Rolle. Die Gruppe ACHSE (Allianz chronischer seltener Erkrankungen) setzt sich speziell für Verbesserungen bei der Behandlung und Erforschung seltener Erkrankungen ein. Hierunter versteht man solche Gesundheitsstörungen, die bei weniger als 5 von 10.000 Menschen auftreten.

Literatur

Aksu F (2011) Neuropädiatrie – Diagnostik und Therapie neurologischer Erkrankungen im Kindes- und Jugendalter. Bremen: Unimed

Heinen F, Böhmer J, Hufschmidt A et al. (2009) Pädiatrische Neurologie in Diagnose und Therapie. Stuttgart: Kohlhammer

Scriver CR, Beaudet AL (2001) The metabolic and molecular basis of inherited diseases. Berlin: Springer

Straßburg HM, Dacheneder W, Kreß W (2012) Eintwicklungsstörungen bei Kindern. Praxisleitfaden für die interdisziplinäre Betreuung. München: Elsevier

Wiedemann HR, Kunze J (2001) Atlas der Klinischen Syndrome. Stuttgart: Schattauer

35 Chronische Erkrankungen

Dirk Schnabel

35.1 Einleitung

Nach Lee (2007) setzt sich die Diagnose „chronische Erkrankung mit speziellem Versorgungsbedarf" aus den folgenden Hauptkriterien zusammen: Dauer der Erkrankung länger als 3 Monate, Einschränkung von altersangemessenen Aktivitäten (fehlende Inklusion) sowie wiederkehrender Bedarf an fachärztlicher Betreuung, die über das Ausmaß der Versorgung gesunder Kinder mit akuten Erkrankungen hinausgeht. Kinder und Jugendliche mit einem derart speziellen medizinischen Versorgungsbedarf weisen eine chronische körperliche, nicht selten zusätzlich eine psychische, entwicklungs- oder verhaltensbedingte Störung auf oder haben ein erhöhtes Risiko für eine solche. Das Resultat ist eine eingeschränkte soziale Teilhabe.

Der Kinder- und Jugendgesundheitssurvey (KIGGS) des Robert-Koch-Institutes ermittelte eine Prävalenz von Kindern und Jugendlichen mit einem speziellen Versorgungsbedarf in Deutschland von 11,4 % bei Mädchen und 16,0 % bei Jungen (Scheidt-Nave 2008).

Das Sozialpädiatrische Zentrum bietet ideale konzeptionelle, personelle und organisatorische Voraussetzungen für Kinder und Jugendliche mit mehrdimensionalen Störungen bei chronischen Erkrankungen.

In dem folgenden Kapitel finden chronische Erkrankungen des Kindes- und Jugendalters Eingang, bei denen die Patienten grundsätzlich oder auch nur für bestimmte Lebensphasen eine interdisziplinäre und multiprofessionelle sozialpädiatrische Betreuung benötigen. Es umfasst epidemiologisch häufige chronische Erkrankungen wie Adipositas, Diabetes mellitus Typ 1, die Gruppe der häufigen chronisch-entzündlichen Erkrankungen, Idiopathische juvenile Arthritis und chronisch-entzündliche Darmerkrankungen, aber auch Erkrankungen mit einem hohen Versorgungsbedarf, wie z. B. onkologische Erkrankungen und Osteochondrodysplasien. Zu Beginn stehen die Wachstumsstörungen. Wachstum darf ohne Einschränkung als ein wichtiger Indikator für das allgemeine Wohlbefinden, für Gesundheit im Kindes- und Jugendalter angesehen werden. Ein

Sozialpädiatrie in der Praxis. http://dx.doi.org/10.1016/B978-3-437-31630-3.00035-3
Copyright © 2014 Elsevier GmbH. All rights reserved.

gestörtes Wachstum geht nicht selten den eigentlich pathognomonischen Symptomen einer chronischen Erkrankung voran.

35.2 Wachstumsstörungen

Das menschliche Wachstum ist ein komplexer, dynamischer, biologischer Prozess, der im Wesentlichen von genetischen, aber auch von hormonellen, ernährungsbedingten und psychischen Faktoren beeinflusst wird. Aber auch chronische Erkrankungen und/oder deren medikamentöse Behandlung können das Wachstum und eventuell sogar die Endgröße beeinflussen. Wichtigste Faktoren zur Beurteilung des Körperwachstums beim Menschen sind Gewicht, Körperlänge und Kopfumfang. Diese Messdaten werden bei jedem Kind in Normkurven, sogenannte Somatogramme, meist im Rahmen der Vorsorgeuntersuchungen eingetragen, und stellen wichtigste Parameter für die Beurteilung der Entwicklung und des Gesundheitszustands dar. Das altersgerechte Wachstum auf dem genetisch vorgegebenen Perzentilenkanal (mittlere Elternzielgröße) stellt einen hervorragenden Indikator für das allgemeine Wohlbefinden des Organismus dar.

Abweichungen von der Norm müssen möglichst frühzeitig erfasst und differenzialdiagnostisch abgeklärt werden. Dies gilt einerseits für Werte außerhalb der Standardabweichungen (mehr als 97 und weniger als 3 %), aber auch für ein sogenanntes Schneiden der Perzentilen innerhalb eines definierten Zeitabschnitts.

Fallbeispiel

Ein 7-jähriges Mädchen wird wegen leichter Entwicklungsauffälligkeiten und Kleinwuchs bei bekanntem Herzfehler vorgestellt. Körperlängenentwicklung verläuft seit dem 1. Lebensjahr 2 cm < 3 Perz. (genet. Zielgröße 25. Perz.), Es finden sich ein hoher Gaumen, Cubiti valgi, ein weiter Mamillenabstand. Feinmotorik scheint gestört, knapp altersentsprechende Sprache.

Im Alter von 2 Lebenswochen Herzgeräusch, Diagnose Aortenisthmusstenose. Es erfolgte eine Ballondilatation und Stenteinlage.

Familiensituation: Alleinerziehende Mutter (vor 8 Jahren aus Kenia nach Deutschland gekommen) mit noch 3 weiteren jüngeren Kindern, in der Schule Lernschwierigkeiten (bes. im Umgang mit Zahlen), zu Hause keine schulische Unterstützung durch die Kindsmutter wegen der 3 Geschwister. Kindsvater entzieht sich den finanziellen Verpflichtungen.

Diagnosen:

- Kleinwuchs (E 34.3G)
- Ullrich-Turner-Syndrom (Q96.9V)
- Aortenisthmusstenose (Q25.1G)
- Anpassungsprobleme an die Übergangsphasen im Lebenszyklus (Z60.0 V)
- Nicht altersentsprechende Entwicklung (F 89.0V)

Vorgehen:

- Besprechen möglicher Ursachen des Kleinwuchses, hochgradiger V. a. Chromosomopathie
- Durchführung einer Chromosomenuntersuchung
- Vereinbarung eines Termins zur ausführlichen psychologischen Entwicklungsdiagnostik
- Nach zytogenetischer Bestätigung Besprechung der Möglichkeit einer Behandlung mit Wachstumshormonen
- Vermittlung an Selbsthilfegruppe
- Angebot einer psychotherapeutischen Begleitung

- Beantragung des Einsatzes einer Sozialpädagogischen Familienhilfe nach §31 SGB VIII
- Beantragung einer Einzelfallhilfe nach §35 SGB XIII
- Fortsetzung der kinderkardiologischen Mitbehandlung

Ein **Kleinwuchs** liegt bei einer Körperlänge unterhalb der 3. Altersperzentile vor, ein **Hochwuchs** bei einer Körperlänge oberhalb der 97. Perzentile. Bei einem Gewicht zwischen der 90. und 97. Perzentile spricht man von **Übergewicht,** oberhalb der 97. Perzentile von **Adipositas,** ein Gewicht unter der 3. Perzentile entspricht einer Dystrophie.

Zur Beurteilung des Körperlängenwachstums müssen sehr unterschiedliche Daten berücksichtigt werden: Länge und Gewicht des Kindes bei Geburt (SGA?), Größe der Eltern, Pubertätsentwicklung der Eltern („Früh- oder Spätentwickler").

Als mittlere genetische Zielgröße wird die Größe des Vaters plus die Größe der Mutter geteilt durch 2 minus 6,5 bei Mädchen bzw. plus 6,5 bei Jungen mit einem Streubereich von ± 8 cm (Mädchen) und ± 10 cm bei Jungen angegeben. Eine Endgrößenbestimmung ist durch eine Röntgendarstellung des Handskeletts mit einer Bestimmung des Knochen- bzw. biologischen Alters etwa ab dem Alter von 7 Jahren möglich.

Kleinwuchs ist zunächst nur ein Symptom. Die Zuordnung zu einem pathologischen Geschehen hängt in erster Linie von den elterlichen Genen ab. Häufigste Ursachen des Kleinwuchses sind der familiäre Kleinwuchs (permanent) und die konstitutionelle Entwicklungsverzögerung von Pubertät und Wachstum (KEV). Bei dieser transitorischen Wachstumsstörung verläuft die Körperlängenentwicklung bereits in der Präpubertät verlangsamt und unterhalb der elterlichen Zielgrößenperzentile. Der Pubertätswachstumsschub tritt verspätet ein, aber die betroffenen Kinder erreichen in der Regel ohne eine hormonelle Intervention eine Endlänge im Elternzielgrößenbereich. Häufig sind ein oder beide Elternteile ebenfalls »Spätentwickler« gewesen.

Liegt ein Kleinwuchs vor, so muss zunächst zwischen einer dysproportionierten und einer proportionierten Form unterschieden werden. Diese Unterscheidung erfolgt durch die Messung der Körperproportionen (Armspannweite, Sitzhöhe, Quotient Ober-/Unterlänge).

Ursache des **dysproportionierten Kleinwuchses** ist zumeist eine primäre Störung von Skelettanomalien. Zurzeit sind über 200 verschiedene genetisch bedingte Anlagestörungen des Skeletts bekannt, Beispiele dafür sind die Achondroplasie und das Klippel-Feil-Syndrom. Auch Wachstumsstörungen infolge von Anlagestörungen der Wirbelsäule und des Rückenmarks (z. B. Meningomyelozele) können hierzu gerechnet werden.

Den proportionierten Kleinwuchs kann man in primäre und sekundäre Formen unterteilen:

Beispiele für **proportionierten primären Kleinwuchs** sind:

- **Ullrich-Turner-Syndrom** = Geschlechtschromosomenstörung bei der zumeist eine XO-Monosomie vorliegt. Typische klinische Zeichen sind u. a. eine Verbreiterung des Halsansatzes (Pterygium), ein tiefer Haaransatz, Nieren- und Gefäßanomalien, Herzfehler. Die Intelligenz ist in der Regel nicht vermindert.
- Daneben gibt es viele andere chromosomale oder monogenetische Anomalien, die mit einem primären Kleinwuchs verbunden sind, z. B. Noonan-Syndrom, Cockayne-Syndrom, Williams-Beuren-Syndrom, Russell-Silver-Syndrom.
- Durch intrauterine Wachstumsverzögerung kann es aus verschiedenen Gründen, z. B. im Rahmen des embryofetalen Alkoholsyndroms oder bei verschiedenen Formen der Plazentainsuffizienz (Rauchen in der Schwangerschaft), zu Kleinwuchsformen, meist verbunden mit einer allgemeinen Dystrophie, kommen. Relativ häufig ist bei Frühgeborenen die intrauterine Wachstumsverzögerung (small for gestational age = SGA).

- Fehlanlage der Hypophyse mit der Folge von das Wachstum beinflussenden Hormonausfällen, u. a. von Wachstumshormonen, TSH und/oder Gonadotropinen, Mutationen/Deletionen in Genen/Transkriptionsfaktoren, die für die Bildung von wachstumsregulierenden Hormonen verantwortlich sind (z. B. GH, β-TSH, KAL1, PROP1, LHX3, HESX1,SOX3).

Unter **sekundärem Kleinwuchs** versteht man eine Wachstumsstörung infolge einer definierten Erkrankung. Beispiele hierfür sind Verdauungsstörungen wie die Zöliakie und die Mukoviszidose, chronische Nieren-, Herz-, Lungen- und Lebererkrankungen sowie erworbene Hormonstörungen, z. B. Schilddrüsenunterfunktion bei Thyreoiditis oder eines Cushing-Syndroms aufgrund eines Kortisol-Exzesses. Der Wachstumshormonmangel kann u. a. Folge eines Schädel-Hirn-Traumas, der Geburt aus Steißlage, durch Hirntumordestruktion direkt oder auch als Therapiefolge von Hirntumoren sein (Bestrahlung, Chemotherapie).

Der **psychosoziale Kleinwuchs** resultiert meist aus einer schweren längerfristig psychosozialen Vernachlässigung, nicht selten assoziiert mit einer Mangel- oder Fehlernährung.

Auch der **Hochwuchs** kann differenzialdiagnostische Probleme verursachen. Am häufigsten liegt bei einer Körperlänge oberhalb der 97. Perzentile eine konstitutionelle Wachstumsbeschleunigung vor. Diese findet sich vor allem bei stark übergewichtigen Kindern (Adiposogigantismus). Die Endgröße dieser Kinder liegt zumeist aber im Normalbereich. Beim familiären Hochwuchs liegt die genetisch determinierte Endgröße oberhalb der Normalverteilung.

Wesentlich seltener sind Hochwuchsformen aufgrund chromosomaler Störungen, wie das Klinefelter-Syndrom (XXY), das XYY-Syndrom, das Triple X-Syndrom.

Es finden sich nur wenige Syndrome, die mit Hochwuchs einhergehen. Diese sind dann meist mit einer Makrozephalus assoziiert und gehen nicht selten mit mentaler Retardierung, Entwicklungs-, Teilleistungssstörungen einher (Sotos, Weaver, Beckwith-Wiedemann). Ein hormoneller Exzess führt selten zu einem Hochwuchs (Pubertas/Pseudopubertas praecox, Gigantismus).

35.3 Adipositas

Adipositas im Kindes- und Jugendalter ist immer eine chronische und komplexe Erkrankung. Der **Body-Maß Index,** errechnet aus der Körperhöhe und dem Körpergewicht, dient dabei als Beurteilungskriterium. In größeren sozialpädiatrischen Einrichtungen sind etwa 60 % der adipösen Kinder und Jugendlichen extrem adipös (BMI oberhalb der 99. Perzentile), bei ca. 30 % besteht bereits ein metabolisches Syndrom als Folge der Adipositas.

Die Adipositas ist immer Folge einer langfristigen positiven Energiebilanz, das heißt, die Energieaufnahme (Mahlzeiten, Getränke, Snacks, Süßigkeiten) stehen einem verminderten Energieverbrauch (Grundumsatz, motorische Grundaktivität, Alltagsaktivität, sportliche Aktivität) gegenüber.

Kinder aus Familien mit niedrigem Sozialstatus und/oder Familien mit Migrationshintergrund haben ein zwei- bis dreifach höheres Risiko für Adipositas im Vergleich zu Kindern mit hohem Sozialstatus und ohne Migrationshintergrund. Häufig sind mangelnde familiäre Ressourcen, Sprachprobleme, andere Vorstellungen zu Gesundheit und Krankheit sowie eine fehlende geregelte Alltagsstruktur dafür mitverantwortlich.

Aber auch bei Kindern und Jugendlichen mit chronischen Erkrankungen besteht nicht selten ein echtes Adipositas-Risiko, das im Zusammenhang mit der Grunderkrankung und/oder der erforderlichen Behandlung zusammenhängt.

Man kann in der sozialpädiatrischen Betreuung vier Grundprinzipien bzw. Krankheitsgruppen unterscheiden, die es chronisch kranken Kindern und Jugendlichen schwer machen, ein normales Körpergewicht zu halten:

Adipositas als Folge einer Medikamenten-(Neben-)Wirkung

Folgende Medikamente können mit einer Gewichtszunahme einhergehen:
Insulin und orale Antidiabetika, Glucocorticoide, Kontrazeptiva, Antipsychotika (z. B. Clozapin, Olanzapin, Risperidon), trizyklische Antidepressiva (Amitriptylin, Imipramin, Nortriptylin), Antikonvulsiva (Valproinsäure, Carbamazepin), Antihypertensiva (Propranolol, Clonidin), Antihistaminika.

35.3.1 Erkrankungen in Kombination mit Adipositas

Das „**Binge Eating Disorder**" **(BED)** ist gekennzeichnet durch regelmäßige Essanfälle mit Kontrollverlust ohne nachfolgendes Erbrechen. Auch Depressionen bei Kindern und Jugendlichen können eine Adipositas nach sich ziehen.
Syndromale und monogene Adipositas
Genetische Formen der Adipositas sind selten und bedürfen einer besonderen therapeutischen Expertise. Bei ca. 5 % der Kinder mit frühmanifester Adipositas (vor dem 3. Lebensjahr) finden sich funktionell relevante Mutationen des MC4-Rezeptors, einer zentralen Stelle im Leptin-Melatonin-Regelkreis der zentralen Appetitregulation.
Häufigste Form der syndromalen Adipositas ist das **Prader-Willi-Syndrom.** Bei dieser komplexen genetischen Erkrankung finden sich eine Muskelhypotonie mit verlangsamter statomotorischer und mentaler Entwicklung sowie eine Sprachentwicklungsstörung. Hinzu kommt eine hypothalamische Störung im Sinne einer gestörten Appetitregulation. Daneben gibt es eine Reihe weiterer Syndrome mit einem erhöhten Adipositas-Risiko. Aber auch nach Operationen von Hirntumoren oder Bestrahlungen des ZNS kann der Hypothalamus, in dem das Zentrum der Appetitregulation liegt, geschädigt werden, sodass es nicht selten zu erheblichen Störungen sowohl in der Energieverwertung als auch in der Energieaufnahme kommt.
Auch bei vielen Erkrankungen, die mit Bewegungsstörungen, einer Muskelschwäche oder insgesamt verminderter Muskulatur einhergehen, ist der Energieverbrauch im Vergleich zu Gleichaltrigen erniedrigt und es kommt vor allem bei Jugendlichen zu einer zunehmenden Gewichtszunahme.
Sozialpädiatrische Zentren mit ihren multidisziplinären Teams können eine kompetente Anlaufstelle für Kinder, Jugendliche und ihre Familien bilden, wenn diese einem allgemeinen Adipositas-Schulungskomplex nicht zugänglich sind (➤ Abb. 35.1). Die erfolgreiche Intervention bei Kindern und Jugendlichen mit morbider Adipositas hängt nicht zuletzt von einem guten Netzwerk aller Beteiligten ab (➤ Abb.35.2).

Fallbeispiel

Vorstellungsanlass:

Vorstellung eines 15-jährigen Jugendlichen mit morbider Adipositas , Ein- und Durchschlafstörungen, (nächtliches) rastloses Umherwandern in der Wohnung, ausgeprägter Interessenverlust, Antriebslosigkeit, exzessiver Medienkonsum, Hänseln und Ausgrenzen in der Schule.

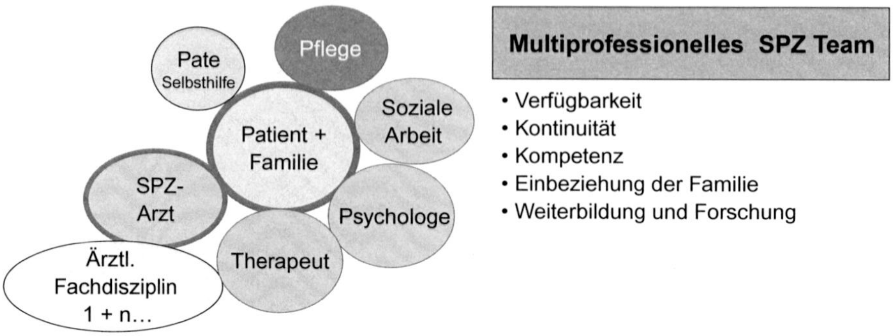

Abb. 35.1 Mögliche Zusammensetzung des multiprofessionellen Teams in dem SPZ für chronisch kranke Kinder und Jugendliche

Abb. 35.2 Externe Verzahnung eines SPZ am Beispiel Adipositas

Unauffälliger Schwangerschafts- und Geburtsverlauf. Altersgerechtes Erreichen der Entwicklungsmeilensteine. Im 5. Lebensjahr bereits BMI > 97. Perz. (frühmanifeste Adipositas). Seither überproportionale Gewichtszunahme bei gutem Längenwachstum. BMI jetzt 58,8 kg/m^2 (>>99. Perz.)
Familiensituation: Adipositas bei den Eltern. Kindsvater seit 10 Jahren wg. Alkoholabhängigkeit von der Familie getrennt, Kindsmutter hat Gelegenheitsjobs. Nachmittags ist der Patient meist alleine zu Hause, hat freien Zugang zur Küche, 2 warme Mahlzeiten täglich (Schule, abends zu Hause), exzessiver Medienkonsum (online-Rollenspiele, Spielkonsolen), hat keine Freunde. Erfolglose Versuche, Gewicht bei extremer Adipositas zu reduzieren (mehrfache mehrwöchige Aufenthalte in Reha-Einrichtungen)

Diagnosen:

- Adipositas permagna (E66.9)
- Metabolisches Syndrom (E88.9)
- Akanthosis nigricans (N83)

- Gynäkomastie (N62)
- Dysthymia (F34.1)
- Generalisierte Angststörung (F41.1)
- Mediensucht (Z72.8)
- Psychosoziale Belastungssituation (F93.0)

Vorgehen:

- Organische Basis-Diagnostik (Hormonstatus, Lipidstatus, oraler Glukose-Toleranztest inkl. Insulinkinetik, EKG, Echokardiografie, Grundumsatzmessung + BIA, Ausschluss monogener und syndromaler Adipositasformen, Langzeit-Blutdruckmessung)
- multiprofessionelle Adipositasdiagnostik (mit der BABELUGA-Baustellen-Methode): Anamnese und Risikogruppeneinschätzung (mit der ABCD-Checkliste); Erhebung des alltäglichen Ernährungs-, Ess- und Bewegungsverhaltens und der psychosozialen Einfluss- bzw. Belastungsfaktoren
- Termin für psychologische Diagnostik u. a. zur Beurteilung der Intelligenz, dem Ausmaß der Depressivität und des Suchtverhaltens
- Einbindung in Sportgruppe für adipöse Jugendliche (Reha-Sportverordnung), Anleitung zur Erhöhung der körperlichen Alltags-Aktivitäten
- Zunächst engmaschige interdisziplinäre Interventionen (2–6 Mal wöchentlich) unter Einbeziehung u. a. von Arzt, Ernährungstherapeut, Sozialpädagogin, Psychologe/Therapeut
- Bei Nichtbesserung Beantragung einer stationären Reha-Maßnahme in Adipositas-Spezialeinrichtung

35.4 Gastroenterologische Erkrankungen

Die chronisch entzündlichen Darmerkrankungen **Morbus Crohn** und **Colitis ulcerosa** gehören sicherlich zu den Erkrankungen im Kindes- und Jugendalter, die eine besondere multiprofessionelle sozialpädiatrische Betreuung bedürfen.

In den letzten 30 Jahren ist es zu einer Zunahme der Inzidenz dieser multifaktoriellen Erkrankungen gekommen. Bis zum 18. Lebensjahr manifestieren sich bis zu 25 % aller Morbus Crohn- und Colitis ulcerosa-Fälle. Der Häufigkeitsgipfel liegt zwischen dem 15. und 25. Lebensjahr. Die Therapie basiert auf einer medikamentösen- und einer Ernährungstherapie.

Etwa 85 % der Morbus Crohn Patienten sind bei Diagnosestellung dystroph, bei den Colitis ulcerosa Patienten immerhin 65 %. Nach einer zumeist längeren stationären Behandlungsphase, in der es gilt, möglichst schnell eine Linderung der Beschwerden zu erreichen und eine Remissionsinduktion zu erzielen, erfolgt dann eine ambulante Weiterbetreuung, mit dem Ziel einer dauerhaften remissionserhaltenden Therapie. Im Vordergrund der Therapieziele des Patienten stehen:

- Schmerz- und Symptomfreiheit
- Eine altersentsprechende körperliche Entwicklung mit einem Aufholwachstum.
- Eine altersentsprechende Pubertätsentwicklung.
- Erreichen einer normalen Lebensqualität, unter möglichst vollständiger Inklusion im Alltag und in der Schule.

Diese Ziele sind zumeist nur durch ein multiprofessionelles Team erreichbar, in dem der pädiatrische Gastroenterologe die Therapie steuert und einer qualifizierten Ernährungsberatung, die die z. T. komplexe Ernährungstherapie (Semi-Elementare-Diät, polymere Spezialnahrungen, Spurenelemente, Vitamine) steuert.

Nicht selten finden sich bei den Patienten, besonders in der Pubertät, z. T. erhebliche psychosoziale Probleme, die aus dem Kleinwuchs, einer verzögerten Pubertätsentwicklung, einer möglichen Stuhlinkontinenz, einem gestörten Körperbild (cushingoider Habitus) oder der Stomata-Anlage resultieren können und zu einer Non-Adherence der komplexen Therapien führen können.
Der Einsatz von Biologika, insbesondere von TNF-alpha-Blockern, hat zu einer steroidsparenden Medikation geführt und somit die Gewichtsentwicklung und das Längenwachstum der jugendlichen Patienten erheblich positiv beeinflusst. Die Einbindung der betroffenen Patienten und ihrer Familien in Selbsthilfegruppen ist aus Krankheits- und Lebensbewältigungsgründen sehr zu unterstützen.
Seit der ersten Lebertransplantation eines Kindes im Jahre 1963 hat sich das operative Vorgehen gerade in den letzten Jahren deutlich verändert. Aus Mangel an Spenderorganen transplantiert man zunehmend erfolgreich sogenannte „Split"-Lebern von lebenden Organspenden, zumeist aus der Familie des betroffenen Kindes oder Jugendlichen. Hauptindikationen zur Lebertransplantation im Kindesalter sind cholestatische Lebererkrankungen, metabolische Erkrankungen, Lebertumore, aber auch das akute
Leberversagen. Nach erfolgreicher Organtransplantation ist in der Langzeittherapie die regelmäßige Einnahme von immunsuppressiven Medikamenten erforderlich. Schwierige familiäre Verhältnisse oder aber auch Compliance-Probleme in der regelmäßigen Medikamenteneinnahme können zu Organfunktionsstörungen oder gar Abstoßungsreaktionen führen.

Fallbeispiel

Erstmalige Vorstellung eines 12 Jahre alten Jungen im SPZ für chronisch kranke Kinder, da dieser zunehmend stärkere Ängste entwickelt wegen einer psychogenen schweren Belastungssituation. Nach Angaben des behandelnden pädiatrischen Gastroenterologen besteht zusätzlich bei den Eltern eine hochgradige emotionale Belastung wegen der rezidivierenden Transplantatabstoßungen ihres Sohnes und der nunmehr möglicherweise anstehenden Nierentransplantation. Des Weiteren ist bei dem Jungen eine sekundäre Enuresis nocturna aufgetreten.
Unauffälliger Schwangerschafts- und Geburtsverlauf, altersgerechte Entwicklung bis plötzlich im 3. Lebensjahr ein Haut- und Sklerenikterus auftrat. Wenige Tage nach ambulanter Vorstellung erfolgte bei zunehmender Vigilanz- und Gerinnungsstörung aufgrund eines akuten Leberversagens eine orthotope Lebersegmenttransplantation. Zunächst war der Verlauf bei guter Organfunktion unauffällig bis dann rezidivierende Cholangitiden auftraten, die zur Anlage einer perkutanen transhepatischen Cholangiodrainage führte.
In letzter Zeit ist es zu mehrfachen stationären Aufenthalten bei Verdacht auf Transplantatabstoßung gekommen. Zudem hat sich in Folge der langjährigen immunsuppressiven Therapie eine chronisch-toxische Nierenschädigung mit nunmehr einem prärenalen Nierenversagen entwickelt. Bei dem Jungen bestehen große Ängste, dass zumindest eine weitere Organtransplantation (Niere) eventuell zusätzlich zu einer erneuten Lebertransplantation bevorsteht. Die Kindseltern haben Sorge, dass für eine erneut erforderliche Organtransplantation das Spenderorgan eventuell nicht rechtzeitig gefunden werden kann.

Diagnosen:

- Zustand nach orthotoper Lebertransplantation (T86.4) bei akutem Leberversagen unklarer Genese (K72.0G)
- Rezidivierende cholestatische Hepatitiden (K75.8G)
- Anastomosenstenose der Gallengänge (K83.0G)
- Anlage einer perkutanen transhepatischen Cholangiodrainage (K83.1G)
- chronische Transplantatrejektion (T86.9G)
- Panzytopenie (D61.9G)

- Aktuell:
- Prärenales Nierenversagen bei chronisch toxischer Nierenschädigung (T86.1G)
- Akute Belastungssituation und multiple Ängste (F.83G)
- Sekundäre Enuresis nocturna (F98.0 V)

Geplantes weiteres Vorgehen:

- Fallkonzeption mit einem multiprofessionellen Team (pädiatrischer Gastroenterologe, Psychologe, Sozialarbeiter) unter Leitung des pädiatrischen Gastroenterologen als Case Manager.
- Diagnostik zur Überprüfung der Leber- und Nierenfunktion sowie die Medikamentenkontrolle der immunsuppressiven Therapie
- Vorstellung des Jungen beim Psychologen wegen seiner Ängste und der aufgetretenen sekundären Enuresis nocturna.
- Gesprächsangebot an die Kindseltern wegen ihrer starken Ängste bei einem Psychologen.

35.5 Skelettdysplasien

Osteochondrodysplasien sind zumeist genetisch bedingte, generalisierte Entwicklungsstörungen des Knorpel-/Knochengewebes. Ihre Inzidenz liegt bei ca. 4 bis 10 Erkrankungen pro 10.000 Neugeborene, wobei in dieser Gruppe mehrere 100 Erkrankungen subsummiert werden. Osteochondrodysplasien umfassen ein Spektrum von Erkrankungen mit unterschiedlichen Schweregraden, von pränatalen letalen Formen bis hin zu nur leicht betroffenen, nahezu symptomfreien Patienten.

Hauptmerkmal der meisten Osteochondrodysplasien ist der **dysproportionierte Kleinwuchs.** Die Diagnostik bei unklaren Krankheitsfällen umfasst radiologische Kriterien als auch eine metabolische Diagnostik. Diagnosesichernd ist die molekulargenetische Analytik, mit der es zunehmend möglich wird, krankheitsauslösende Genveränderungen zu identifizieren. In Kenntnis der ursächlichen molekulargenetischen Mechanismen wird es zukünftig möglich sein, gezielte therapeutische Strategien zu entwickeln und zu etablieren.

Häufigste Osteochondrodysplasie mit einer Frequenz von 1 : 15.000 Geburten ist die Achondroplasie. Die Vererbung ist autosomal dominant, 80–90 % der Erkrankungen sind Resultate von Spontanmutationen. Krankheitsauslösend ist ein Defekt des Knorpelzellrezeptors für den Fibroblastenwachstumsfaktor (FGFR-3), der die Zellteilungsrate in enchondralen Knorpelzellsäulen beeinflusst. Klinisches Haupt-Symptom ist der dysproportionierter Kleinwuchs, bei dem eine Verkürzung der proximalen Extremitäten-Abschnitte imponiert (rhizomeler Kleinwuchs). Die Endgröße bei Frauen liegt bei etwa 125 cm, die der betroffenen Männer bei 130 cm. Im Säuglings- und Kleinkindesalter besteht meist eine ausgeprägte Muskelhypotonie, die zu einem deutlich verspäteten Erreichen der motorischen Meilensteile (Sitzen, freies Laufen) führt. Aus der Wachstumsstörung der Schädelbasis kann sich eine Foramen magnum-Stenose mit lebensbedrohlichen Komplikationen (Rückenmarkkompression, Schlafapnoen) entwickeln. Diese muss dann neurochirurgisch beseitigt werden. Der Entwicklung einer thorakolumbalen Kyphoskoliose sollte möglichst mit frühzeitig einsetzender gezielter Physiotherapie versucht werden, zu begegnen.

Hoffnungsfroh stimmen erste Studienergebnisse bei Mäusen, dass mittels eines Analogon des C-Typs des natriuretischen Peptids, der durch die Mutation des FGFR-3-Gens hervorgerufene Knochen-/Knorpeldefekt kompensiert werden konnte. Klinische Studien an Kindern mit Achondroplasie sind vor kurzem initiiert worden.

Fallbeispiel

Vorstellung eines 6 Monate alten Mädchens mit Achondroplasie.

In letzter Zeit sind der Kindsmutter eine überstreckte Körperhaltung mit Reklination des Kopfes, deutliche Schnarchgeräusche, aber auch kleinere Atempausen, bei ihrer Tochter aufgefallen. Ferner nahm der Kopfumfang überproportional stark zu.

In der 21. Schwangerschaftswoche wurde eine Extremitäten-Verkürzung festgestellt. In der 38. Schwangerschaftswoche erfolgte dann bei einem Geburtsgewicht von 3.200 g und einer Geburtslänge von 48 cm eine Spontangeburt. Postnatal fanden sich Stigmata der Achondroplasie. In der gezielten molekulargenetischen Untersuchung des FGFR-3-Gens fand sich die für die Erkrankung beweisende Mutation im Exon 10 (G380R). Die bisherige Betreuung erfolgte durch den Kinderarzt.

Diagnosen:
- Achondroplasie (Q77.11)
- Makrozephalie (Q75.3)
- Makrozephalie bei erweiterten Liquorräumen (G91.1)
- Obstruktives Schlafapnoe-Syndrom (G47.3)
- Muskelhypotonie (R29.89)
- Hyperlordose LWS-Bereich (M40.46)

Geplantes weiteres Vorgehen:
- Indikation zur kranialen Kernspintomographie mit besonderer Berücksichtigung des kraniozervikalen Übergangs zum Ausschluss eines zu engen Foramen occipitale mit zerviko-medullärer Kompression
- Initiierung einer Polysomnographie zum Ausschluss von obstruktiven Apnoen
- Durchführung von somatisch evozierte Potenziale (SSEP) von Medianus und Tibialis
- bei Vorlage aller Untersuchungsbefunde Fallkonferenz mit pädiatrischen Neurochirurgen zur Frage Dekompressionsoperation

Längerfristige Maßnahmen:
- Durchführung einer objektiven Hörprüfung
- Vorstellung bei der Physiotherapeutin zur Durchführung einer gezielten Physiotherapie
- engmaschige Kontrollen z. B. nach dem Betreuungsprotokoll des BKMF e. V.
- Kontaktaufnahme mit dem Bundesverband kleinwüchsiger Menschen und ihre Familien (BKMF)

35.6 Diabetes mellitus

Der Diabetes mellitus, charakterisiert durch eine **chronische Hyperglykämie** als Folge einer fehlenden Insulinsekretion, einer gestörten Insulinwirkung oder beider, ist die häufigste Stoffwechselerkrankung im Kindes- und Jugendalter. In der Altersgruppe 9 bis 19 Jahre sind 21.000 bis 24.000 Kinder und Jugendliche in Deutschland betroffen. Die Neuerkrankungsrate ist steigend, die Inzidenzrate in Deutschland beträgt 10 bis 20 Kinder pro 100.000 pro Jahr.

Neben genetischen Faktoren scheinen Umweltfaktoren, vor allem Ernährung und Infektionen, eine wesentliche Rolle bei der Entstehung des Autoimmunprozesses, der zu einer Destruktion der insulinproduzierenden Betazellen der Pankreas führt, ursächlich zu sein. Wurde bis Mitte der 70-iger Jahre die Diabetes mellitus Typ 1 Therapie im Kindes- und Jugendalter bei einem einfachen Behandlungsschema durchgeführt, so setzte sich seitdem die intensivierte Insulintherapie mit 4–6 Injektionen durch. Inzwischen werden immer mehr Kinder und Jugendliche mit einer Insulinpumpentherapie behandelt. 2011 waren es bereits 42 % der erkrankten Kinder und Jugendliche. Besonders bei Säuglingen und Kleinkindern mit geringem Insulinbedarf eignet sich die

Insulinpumpe durch seine Feindosierung besonderes. Mittlerweile liegt die Rate der mit einer Insulinpumpe versorgten unter 5-Jährigen bei ca. 75 %. Indikator für eine gute Stoffwechseleinstellung ist nach internationaler Übereinkunft in der Pädiatrie ein HbA1c-Wert unter 7,5 %.
Ein betroffenes Kind oder Jugendlicher in der Familie stellt die Eltern nicht selten vor eine zusätzliche komplexe Lebensaufgabe. Bis weit in das Jugendalter hinein tragen Eltern die Verantwortung für die tägliche Diabetestherapie ihres Kindes. Dabei müssen sie der schwierigen Doppelaufgabe als liebevolle Eltern und Erzieher einerseits und konsequente Therapeuten des Insulinmanagements andererseits gerecht werden. Die chronische Erkrankung des Kindes kann dabei ebenfalls für eine psychische Dauerbelastung für die ganze Familie werden. Evidenzbasierte Leitlinien empfehlen daher eine Langzeitbetreuung durch ein spezialisiertes pädiatrisches Team, das neben Erfahrungen in der somatischen Therapie auch spezifische pädagogische, psychologische und soziale Beratungen anbieten sollte. Diese Möglichkeiten können durch ein sozialpädiatrisches Zentrum mit entsprechender Expertise zur Verfügung gestellt werden. Zu dem komplexen Betreuungskonzept sollten auch unbedingt Schulungsangebote für Vor- und Grundschüler sowie Jugendliche, unter Einbeziehung der Familie, aber auch möglicher Betreuer gehören.
Der Diabetes mellitus Typ 2, bedingt durch die Zunahme der Kinder und Jugendlichen mit morbider Adipositas, wird zunehmend häufiger diagnostiziert. Die Zahl der pädiatrischen Neuerkrankungen hat sich in den letzten 10 Jahren verfünffacht und liegt nach Schätzungen in Deutschland bei etwa 5.000 betroffenen Kindern und Jugendlichen vor. Nicht selten findet man begleitend bereits einen Hypertonus, eine Albumin-Ausscheidung im Urin sowie Retina-Veränderungen.

Fallbeispiel

Vorstellung eines 11 9/12 Jahre alten Mädchens mit Diabetes mellitus Typ1.

Vorgeschichte:

Diabetes mellitus Typ 1 seit dem 3. Lebensjahr, Diagnosestellung bei typischer Symptomatik mit postivem GAD-Nachweis ohne Ketoazidose. Therapie: intensivierte konventionelle Insulintherapie (ICT), auch unter Verwendung von Insulinanaloga. Bisherige Stoffwechseleinstellung gut bis verbesserungsbedürftig (HbA1c-Werte zwischen 7,3 % und 9,3 %). Bisheriger Verlauf ohne Komplikationen und ohne Hinweis für Folgeerkrankungen.
Bisher sonst keine ernsthaften Erkrankungen. Besuch der Regelschule mit guten Schulleistungen. Vater: Adipositas. Kein Diabetes in der Familie. Zwillingsschwester der Patientin: Fallot-Tetralogie. Eltern leben seit dem 2. Lebensjahr der Patientin getrennt.
Aktuell Vorstellung im Alter von 11 9/12 Jahren:
Mutter berichtet, dass seit ein paar Wochen „das Essen problematisch geworden ist" und die Patientin weniger isst. Laut Mutter verschenkt die Patientin ihre Pausenbrote in der Schule und kontrolliert ihr Gewicht auf der Waage. Aktuell sonst keine Beschwerden und keine Probleme mit der Stoffwechselführung und der Insulintherapie. Die Patientin übernimmt, laut Mutter, die Insulintherapieführung schon sehr selbstständig. Mutter (alleinerziehend) aktuell sehr belastet, da die Zwillingsschwester gerade intensivere Betreuung braucht (operativer Eingriff, Psychotherapie).
Befund: Guter Allgemeinzustand, schlank, BMI 15,44 kg/m^2 (12.Perz.), Blutdruck, Puls unauffällig, internistischer Status unauffällig, Spritzstellen ohne pathologischen Befund.
Die Stoffwechsellage ist gut (HbA1c 7,8 %).
Es wurde gemeinsam vereinbart, die Patientin zu kurzfristigen Vorstellungsterminen zu sehen und es wurde die psychologische Mitbetreuung und Diagnostik initiiert.

Verlauf:

Bei den folgenden Vorstellungsterminen zeigte sich ein Gewichtsstillstand (BMI fiel sukzessive auf die 9. Perzentile, 5. und dann 2. Perzentile). Die Stoffwechsellage verschlechterte sich (HbA1c bis 8,8 %).

Befund: Stabiler Allgemeinzustand, reduzierter Ernährungszustand, BMI 14,37 kg/m^2 (2. Perz.), peripher kühle Extremitäten, Blutdruck 95/55, Puls 61/min, internistischer Status unauffällig, Spritzstellen ohne pathologischen Befund.
In der Diagnostik wurden organische Ursachen und Begleiterkrankungen des Untergewichts mittels Laboruntersuchung, Sonografie, EKG, Echokardiographie und EEG ausgeschlossen. Die psychologische und kinder- und jugendpsychiatrische Diagnostik zeigte eine atypische Anorexia nervosa (ohne aktive Maßnahmen zur Gewichtsabnahme).
Diagnosen:
- Diabetes mellitus Typ 1 (E10.91 G)
- Atypische Anorexia nervosa (F50.1 G)
- Anpassungsprobleme an die Übergangsphasen des Lebenszyklus (Z60)
- Getrennte Eltern (Z63)

Geplantes weiteres Vorgehen:
- Zwei- bis vierwöchentliche Vorstellungen zur Gewichtskontrolle
- Initial engmaschige Verlaufskontrollen hinsichtlich der notwendigen Änderungen der Insulintherapie
- Ernährungsberatung mit Schwerpunkt Diabetes mellitus Typ 1 und Anorexie
- Initiierung einer ambulanten kinder- und jugendpsychiatrischen Betreuung (Psychotherapie)
- Kontaktaufnahme mit Selbsthilfegruppen
- Beratung hinsichtlich Unterstützungsmöglichkeiten der Familie durch das Jugendamt
- Bei unzureichender Gewichtszunahme oder Therapieabbruch Initiierung einer stationären kinder- und jugendpsychiatrischen Behandlung
- Im Verlauf kam es zur stetigen Gewichtszunahme (nach sechs Monaten BMI 10. Perz.), unter kontinuierlicher ambulanter Psychotherapie (1x/Woche). Die Stoffwechsellage verbesserte sich (HbA1c zwischen 7,4 und 8,2 %). Die Familie konnte sich nicht entschließen, weitere Unterstützungen in Anspruch zu nehmen.

35.7 Rheumatologische Erkrankungen

In Deutschland leiden ca. 20.000 Kinder und Jugendliche an chronisch-entzündlich-rheumatologischen Erkrankungen. Dazu gehören die juvenile idiopathische Arthritis, juvenile Kollagenosen sowie juvenile Vaskulitiden.
Die **juvenile idiopathische Arthritis (JIA)** ist die häufigste chronisch entzündliche Erkrankung aus dem rheumatischen Formenkreis im Kindes- und Jugendalter. Die JIA ist nicht nur eine Krankheitsidentität, sondern entspricht vermutlich mehreren Erkrankungen, die von der hochinflammatorischen und bedrohlichen systemischen Verlaufsform (Morbus Still) bis zu häufigen, meist günstig verlaufenden oligoartikulären Formen reichen. Ursache des Krankheitsspektrums ist eine entzündliche Erkrankung der Gelenke unklarer Genese. Frühzeitige Entzündungsmediatoren, einschließlich TNF-Alpha, IL6 und IL17, spielen bei der Entstehung und Aufrechterhaltung der Erkrankung eine Rolle. Wichtig zur Erstellung eines gezielten Therapieplanes ist die Bestimmung des Sub-Typs des JIA: juvenile Polyarthritis, Oligoarthritis, systemische Arthritis, Psoriasis-Arthritis, Enthesitis-assoziierte Arthritis. Die Versorgung der JIA Patienten ist komplex und schließt medikamentöse, krankengymnastische, ergotherapeutische und psychosoziale Maßnahmen mit ein. Die multiprofessionelle Betreuung dieser Patienten hat sich in den letzten 10 Jahren verstärkt aus dem stationären Versorgungsbereich in den ambulanten Sektor verlagert. Ein erfolgreiches ambulantes Versorgungskonzept erfordert daher die interdisziplinäre Zusammenarbeit

verschiedenster Professionen in einem Team: pädiatrischer Rheumatologe, spezialisierte Physiotherapeuten und Ergotherapeuten (Verhinderung der Entwicklung von Gelenk-Fehlhaltungen, Fehlstellungen und Kontrakturen), Hilfsmittelerbringer, Sozialdienst. Die Einbindung der Psychologen in diesen Teams ist aufgrund der die Lebensqualität z. T. erheblich beeinträchtigende Erkrankung, aber auch wegen möglicher Compliance-Probleme, besonders in der Pubertät, sinnvoll.
Die medikamentöse Therapie beruht weiterhin auf nichtsteroidale Antiphlogistika, Corticosteroiden, aber auch Methotrexat. Die Einführung der Biologika, insbesondere des TNF-Alpha-Antagonisten, führte zu einer deutlichen Verbesserung der Beschwerdesymptomatik und zu einem Rückgang der medikamentenassoziierten Nebenwirkungen (z. B. cushingoider Habitus).

Fallbeispiel

Vorstellung eines 2.5 Jahre alten Mädchens mit juveniler idiopathische Arthritis.

Vorgeschichte:

Seit 3 Wochen bestehen intermittierendes Fieber, ein flüchtiges Exanthem, zervikale Lymphknoten sowie unklare Schmerzzustände. Bei Erstvorstellung ist eine bisher nicht bemerkte Polyarthritis kleiner und großer Gelenke auffällig. Während eines kurzen stationären Aufenthaltes Ausschluss weiterer Organmanifestationen, infektiöser Ätiologien und weiterer Organmanifestationen. Einleitung einer hochdosierten Steroidtherapie begleitet von einer immunsuppressiven Therapie mit Methotrexat und einer antientzündlichen Therapie mit Indometacin, daraufhin Entlassung in ambulante Betreuung. In der Folge Entwicklung eines Cushing-Syndroms. Bei Dosisreduktion erneuter Schub der Erkrankung, daher Intensivierung der Basismedikation durch Einsatz von Medikamentenkombinationen und Biologika sowie mehrfache intraartikuläre Steroidinstillationen.
Aufgrund der hochaktiven Erkrankung Entwicklung von Gelenkfehlstellung. Im Alter von 11 Jahre beidseitige Volar- und Ulnardeviationen, Schwanenhalsdeformitäten sowie rechtsseitiges Genu recurvatum mit Patellaluxation. Darüber hinaus mäßiggradige Lateralisierung der Manibula und Retrognathie bei Kiefergelenkbeteiligung mit verzögerter Sprachentwicklung.
Aufgrund einer familiären Vernachlässigung mit unregelmäßigen Arztbesuchen und einer relativen Non-Compliance bezüglich der regelmäßigen Medikamenteneinnahme Implementierung einer Familienhilfe sowie einer Einzelfallhilfe. Bei leichter mentalen Entwicklungsstörung Betreuung in einer Förderschule.

Diagnosen:

- Juvenile idiopathische Arthritis mit systemischer Verlaufsform (M08.99G)
- Volar- und Ulnardeviation, Schwanenhalsdeformitäten (M21.99G)
- Genu recurvatum (M21.86G)
- Patelladeformität (M22.8G)
- Mentale Entwicklungsstörung (F79.9G)
- Diuresis nocturna und diurna (F98.0V)

Vorgehen:

- Kontinuierlich medikamentöse orale und parenterale immunsuppressive Therapie. Intermittierende intraartikuläre Steroidinstillationen
- Regelmäßige Mitbetreuung durch Kinderorthopäden und Kinderurologen
- Regelmäßige physiotherapeutische und ergotherapeutische Behandlung
- Zeitweilige logopädische Betreuung
- Ordination von Handlagerungsschienen
- In größeren Abständen Durchführung einer Helferkonferenz zur Koordination der Hilfsangebote
- Beantragung eines Behindertenausweises

35.8 Onkologische Erkrankungen

Fallbeispiel

Vorstellung eines 6-jährigen Mädchens.

Vorgeschichte:

Im Alter von 3.5 Jahren Vorstellung beim Kinderarzt wegen fehlenden Fixierenkönnens von Gegenständen, zudem würde sie beim Laufen um Ecken, diese meist nicht richtig wahrnehmen können. Kein Erbrechen, keine Kopfschmerzen, keine weiteren neurologischen Symptome. Da zudem ein progredienter Kleinwuchs vorliegt, erfolgt die Einweisung in Kinderklinik bei V.a. Hirntumor. Die eiligst initiierte Diagnostik zeigt im MRT-Schädel einen ausgedehnten intra- und suprasellären Tumor mit Kompression der Nn. optici, im CCT stellt sich der Tumor als verkalkend dar. In der augenärztlichen Untersuchung findet sich eine erhebliche Stauungspapille, eine Amblyopie des rechten Auges und einen kaum noch nachweisbaren Visus links. In der Operation findet sich histologisch ein Adamantinomatöses Kraniopharyngeom (WHO Grad I), das aufgrund der Nähe zum Hypothalamus nur subtotal reseziert wird. Von den postoperativ nachzuweisenden Hypophysenhormonausfällen (ADH, TSH, ACTH, Wachstumshormon) wurde zunächst nur der ADH- (Minirin), TSH- (L-Thyroxin) und ACTH-Mangel (Hydrocortison) substituiert. 10 Monate später kommt es wg. eines Rezidivs zur erneuten Operation.

Befund: 5.5 Jahre altes Mädchen, Körperhöhe 90 cm (14 cm < 3.Perz.), BMI 17.3 kg/m^2 (> 97. Perz.), adipös, internistisch unauffällig, präpubertär (B1, P1), ausgeprägter Nystagmus, Fingerperimetrie pathologisch, übriger neurologischer Status unauffällig

Diagnosen:

- Kraniopharyngeom (D44.4G), Zn. subtotaler Tumorresektion
- Amblyopie rechts (H 54.5G), Amaurosis links (H54.4G)
- Nystagmus (H55G)
- Hypophyseninsuffizienz (E23.0G)
- Diabetes insipidus (E23.2G)
- Kleinwuchs (E34.3G)
- Wachstumshormon-Mangel (E23.0V)
- Adipositas (E66.8G)
- Entwicklungsstörung schulischer Fertigkeiten (F81.9V)

Geplantes weiteres Vorgehen:

- Psychologische Diagnostik
- Verordnung von Ergotherapie, u. a. zur Schulung auf Blindenmaterialen
- Vermittlung des Kontakts zum Blinden- und Sehschwachenverband
- Besprechung der anstehenden Beschulung in einer Einrichtung für stark Sehbehinderte
- Endokrinologische Funktionsdiagnostik bei V. a. Wachstumshormon-Mangel
- Ernährungsberatung mit Erhebung des alltäglichen Ernährungs-, Ess- und
- Bewegungsverhaltens
- Beantragung eines Behindertenausweises
- Regelmäßige MRT-Kopfkontrollen
- Regelmäßige augenärztliche Untersuchungen (Visus, Gesichtsfeld, Augenhintergrund)
- Kontaktvermittlung mit Kraniopharyngeom-Selbsthilfegruppe

Jährlich erkranken in Deutschland etwa 15 von 100.000 Kindern unter 15 Jahren an einer Tumorerkrankung. Dies sind 1.700 Neuerkrankungen pro Jahr.

Tumorerkrankungen stellen nach Unfällen die zweithäufigste Todesursache bei Kindern dar. Die häufigsten Krebserkrankungen bei Kindern und Jugendlichen sind Leukämien (34 %), ZNS-Tumore (23 %) und Lymphome (11,5 %).

Tumore des ZNS sind nach den Leukämien die zweithäufigste Neoplasie im Kindes- und Jugendalter. Das mittlere Erkrankungsalter liegt etwa bei 6 Jahren, dabei erkranken Jungen häufiger als Mädchen.
In der Gruppe der Hirntumore finden sich am häufigsten Astrozytome (50 %), das Medulloblastom (20 %), Ependymome (10 %) sowie das Kraniopharyngiom (8 %).
Allgemeinsymptome wie Kopfschmerzen, Erbrechen, Müdigkeit, Somnolenz, Leistungsabfall, Wesensveränderung, Zunahme des Kopfumfangs, Retardierung, Anorexie sowie Kachexie, aber auch fokale Symptome wie eine Stauungspapille, Ataxie, Visusstörungen, Schwindel, Hirnnervenparesen, Paresen, Krampfanfälle, Wachstumsstörungen, Sprachstörungen, Vielfalts- und Sensibilitätsstörungen sollten differentialdiagnostisch an das Vorliegen einer Hirntumorerkrankung denken lassen.
Kraniopharyngeome sind histologisch gutartige Tumore, die im Bereich der Mittellinie aus den Resten des embryonalen Ductus craniopharyngealis (Rathke-Tasche) entstehen. Diese können intra-und/oder supraselIär lokalisiert sein und aus soliden und zystischen Anteilen bestehen. Klinisch finden sich bei Erstdiagnose unspezifische Symptome eines gesteigerten intrakraniellen Drucks wie Kopfschmerzen. Weitere Leitsymptome sind Sehstörungen (62 %–84 %) und endokrine Ausfälle (52 %–87 %). Die progrediente Wachstumsstörung ist häufig ein Frühsymptom des Kraniopharyngeoms.
Therapieentscheidungen müssen sehr individuell getroffen werden. Operative Maßnahmen und Radiotherapie sind dabei die wichtigsten Interventionen. Ist eine vollständige chirurgische Resektion ohne Schädigung, z. B. hypothalamischer Strukturen, nicht möglich, kommt eine nachfolgende Strahlentherapie in Betracht. Als Folge des Tumors bzw. seiner Behandlung treten zumeist eine Hypophyseninsuffizienz, aber auch schwere hypothalamische Funktionsstörungen mit einer ausgeprägten Adipositas auf. Die Hormonausfälle müssen, bis auf die Substitution mit Wachstumshormonen, zügig postoperativ substituiert werden. Bei subtotaler Tumorresektion und nachfolgender Bestrahlung liegt die 10-Jahres-Überlebensrate zwischen 60 % und 80 %.
Die Behandlung von Kindern und Jugendlichen mit Hirntumoren ist nur durch eine interdisziplinäre Zusammenarbeit der Disziplinen Neuropädiatrie, pädiatrische Onkologie, pädiatrische Neurochirurgie, Strahlentherapie, Neuroradiologie, pädiatrische Endokrinologie und Sozialpädiatrie erfolgreich.

Literatur

Gruttadauria S (2009) Pediatric liver transplantation. World J Gastroenterol 15: 648–674

Hiort O, Danne T, Wabitsch M (Hrsg) (2010) Pädiatrische Endokrinologie und Diabetologie. Springer-Verlag, Berlin

Laederich MB, Horton WA (2010) Achondroplasia: pathogenesis and implications for future treatment. Curr Opin Pediatr 22: 516–523

Lee vd JH, Mokkink LB et al. (2007) Definition and measurement of chronic health conditions in childhood – a systemic review. JAMA 297: 2.741–2.751

Minden K, Niewerth M (2012) Rheumakranke Kinder und Jugendliche. Monatsschr Kinderheilkd 160: 237–243

Mohnike K, Klingebiel H, Zabel B (Hrsg) (2001) Achondroplasie und Hypochondroplasie. Palatium Verlag, Mannheim

Müller HL (2013) Childhood craniopharyngeoma. Pituitary 16: 56–67

Radke M (2010) Therapie chronisch-entzündlicher Darmerkrankungen. Monatsschr Kinderheilkd 158: 766–773

Scheidt-Nave C, Ellert U, Thyen U, Schlaud M (2008) Versorgungsbedarf chronisch kranker Kinder und Jugendlicher. Bundesgesundheitsbl-Gesundheitsforsch-Gesundheitsschutz 51: 592–601

Schnabel D (2007). Wachstumsstörungen, in: Lentze MJ, Schaub J, Schulte FJ, Spranger J (Hrsg.). Pädiatrie – Grundlagen und Praxis, 3. Auflage: 603–609. Springer Verlag, Berlin

Tenbrock K (2012) Therapie der juvenilen idiopathischen Arthritis. Monatsschr Kinderheilkd 160: 224–230

Wiegand S, Schnabel D (2013) Betreuung adipöser Kinder und Jugendlicher im SPZ. Kinderärztl Praxis 83: 222–226

Yasoda A, Kitamura H, Fuji T, Kondo E, Murao N, Miura M, Kanamoto N, Komatsu Y, Arai H, Nakao K (2009) Systemic administration of C-type natriuretic peptide as a novel therapeutic strategy for skeletal dysplasias. Endocrinology 150: 3.138–3.144

36 Kindswohlgefährdung und Kindesmisshandlung

Hans-Michael Straßburg

36.1 Geschichte der Kindesmisshandlung

Gewalt gegen Kinder hat es zu allen Zeiten und in allen Kulturen gegeben. Seit Menschengedenken wurden schwächlich und krank erscheinende Neugeborene getötet, schon immer hat es, z. T. auch kulturbedingt, Verstümmelungen von Kindern, besonders von Mädchen, gegeben. Vor allem hatten und haben Kinder an Unterernährung, Krieg und dem Mangel an Sicherung der grundlegenden Lebensbedürfnisse zu leiden. Im Alten Testament finden sich an mehreren Stellen Hinweise, die das Recht auf Züchtigung durch die Eltern, insbesondere den Vater, als Erziehungsmaßnahme befürworten. Bis heute müssen in vielen Ländern Kinder härteste Arbeiten, z. B. in der Landwirtschaft und in kleinen Handwerksbetrieben, z. B. auch beim Teppichknüpfen, verrichten und werden selbstverständlich in der Familie und in der Schule körperlich gezüchtigt. Bis zur Aufklärung war es üblich, dass Kinder schon vom Säuglingsalter an nicht von ihren Eltern betreut wurden. Erst **J. J. Rousseau** hat das Anrecht jedes Kindes auf die Liebe seiner Eltern und auf seiner Entwicklung angemessene ideelle und materielle Lebensumstände eingefordert, allerdings bei seinen eigenen Kindern auch nicht erfüllt.

Sozialpädiatrie in der Praxis. http://dx.doi.org/10.1016/B978-3-437-31630-3.00036-5
Copyright © 2014 Elsevier GmbH. All rights reserved.

Eine 1860 von dem Pariser Rechtsmediziner **A. Tardieu** verfasste Schrift über die Misshandlung von Kindern und die sich daraus ergebenden schweren Gesundheitsschäden blieb unbeachtet. Ein erster Prozess wegen der Misshandlung eines Kindes fand 1890 in New York statt – die Situation des körperlich schwer misshandelten Mädchens wurde juristisch in Analogie zum Tierschutzrecht bewertet.
Sigmund Freud verfolgte in der Frühphase der Entwicklung der Psychoanalyse unter bis heute nicht eindeutig geklärten Umständen seine These, dass viele Formen neurotischer Verhaltensstörungen als Folge von Kindesmisshandlung bzw. sexuellem Missbrauch angesehen werden müssen, nicht mehr weiter.
Erst durch die Publikationen des Radiologen J. Caffey und des Kinderarztes **C. H. Kempe** und seiner Mitarbeiter ab Anfang der 1950er Jahre wurde deutlich, dass die Misshandlung von Kindern, besonders im Säuglings- und Kleinkindesalter, häufig ist, dass sie in allen Schichten und oft unerwartet vorkommt und dass eine große Wiederholungsgefahr besteht.

36.2 Gesellschaftspolitische Aspekte

In den vergangenen 30 Jahren, vor allem aber seit 2006, gibt es aufgrund von z. T. spektakulären Fällen von schwersten Verletzungen und Tötungen von Kindern eine zunehmende Bereitschaft der Gesellschaft, insbesondere auch von Politikern, sich mit der Thematik auseinanderzusetzen. Dennoch herrscht innerhalb der Gesellschaft Uneinigkeit darüber, wer die Hauptverantwortung für die Misshandlung von Kindern trägt und welche praktischen Konsequenzen daraus gezogen werden sollen: Ist es allein die Verantwortung der Eltern oder die Pflicht des Staates, den Schutz der Kinder zu gewährleisten? Inwieweit sind Kinderärzte im Rahmen der Vorsorgeuntersuchungen, Jugendämter, öffentlicher Gesundheitsdienst, andere Arztgruppen, Erzieherinnen in den Kindergärten, Lehrer, Trainingsleiter in Sportgruppen, Leiter von kulturellen Aktivitäten und Freizeitangeboten oder einfach die Nachbarn verantwortlich für das Erkennen von Zeichen einer Kindesmisshandlung und die konsequente Verfolgung der Täter?
Folgende **Institutionen** müssen sich heute mit dem Thema Kindesmisshandlung, -vernachlässigung und sexuellem Missbrauch befassen:
- Ärzte für Kinder- und Jugendmedizin
- Ärzte für Allgemeinmedizin (Familienarzt)
- Ärzte für Kinder- und Jugendpsychiatrie, Psychotherapie und Psychosomatik
- Ärzte für Kinderchirurgie
- alle anderen Arztgruppen
- Institut für Rechtsmedizin
- Jugendamt
- Gesundheitsamt
- Polizei
- Staatsanwaltschaft
- Familiengericht
- Kinderschutzbund
- andere Selbsthilfegruppen
- Beratungsstellen
- Heime für Kinder und Jugendliche
- Versorgungsamt u. v. m.

Ohne Zweifel kann ein frühzeitiges Erkennen und die Einleitung der richtigen Maßnahmen weiteres Leid und evtl. den Tod des Kindes vermeiden helfen. Andererseits können aber durch Fehldeutungen, Unterstellungen, Überbewertungen usw. schwere Schäden im menschlichen Miteinander entstehen. Dies gilt insbesondere auch für das in den vergangenen 20 Jahren zunehmend in den Vordergrund gerückte Thema des sexuellen Missbrauchs.
In diesem Zusammenhang muss auch darauf hingewiesen werden, dass viele der bis Anfang der 1980er Jahre üblichen diagnostischen und therapeutischen Maßnahmen in der Medizin, z. B. die fehlende Schmerzbehandlung bei intensivpflichtigen Neugeborenen, die Durchführung schmerzhafter Prozeduren wie häufige Fersenblutentnahme und arterielle Punktionen, sogar von Operationen ohne Schmerzbekämpfung, vor allem aber auch das Verbot für Eltern, ihre Kinder während des stationären Aufenthalts im Krankenhaus zu besuchen, aus heutiger Sicht als Kindesmisshandlung anzusehen sind.

36.3 Definitionen

Unter **Kindesmisshandlung** versteht man eine nicht zufällige, meist wiederholte, gewaltsame Beeinträchtigung des physischen oder psychischen Wohlergehens von Kindern durch Handlungen oder Unterlassungen. Verursacher sind meist die Eltern, seltener Geschwister, entferntere Verwandte oder der Familie bekannte Betreuer.
Im umfassenden Sinn kann man unter Kindesmisshandlung die Gesamtheit der Lebensbedingungen, der Handlungen und Unterlassungen, die das Recht des Kindes auf Leben, Entwicklung und Förderung beschneiden, verstehen, was sich auf alle Lebenssituationen eines Kindes bezieht und sowohl auf den Einzelfall als auch auf Kinderfeindlichkeit im gesellschaftlichen Sinne anwendbar ist.
Eine **körperliche Misshandlung** liegt vor, wenn durch aktive Handlungen oder deren Unterlassung dem Kind Verletzungen zugefügt werden. Auch spricht man von Kindesmisshandlung, wenn gewalttätiges Verhalten der Eltern und/oder Erzieher ein Grundelement der Kindererziehung ist. Im Bürgerlichen Gesetzbuch steht im § 1631 Abs.2: Kinder haben ein Recht auf gewaltfreie Erziehung: körperliche Bestrafungen, seelische Verletzungen und andere entwürdigende Maßnahmen sind unzulässig.
Strafrechtlich ist die Kindesmisshandlung in § 223 b des StGB folgendermaßen definiert: „Wer Personen unter 18 Jahren…, die seiner Fürsorge oder Obhut unterstehen oder seinem Hausstand angehören, … quält oder roh misshandelt oder wer durch böswillige Vernachlässigung seiner Pflicht, für sie zu sorgen, sie an der Gesundheit schädigt, wird bestraft."
Körperliche Vernachlässigung liegt vor, wenn Eltern oder Erzieher ein Kind nicht seinen Bedürfnissen entsprechend angemessen versorgen. Formen körperlicher Vernachlässigung können z. B. sein, dass das Kind mangelhaft ernährt ist, unzureichend gekleidet und hygienisch schlecht versorgt wird und dass ihm die notwendige medizinische Behandlung vorenthalten wird.
Vernachlässigung ist aber auch der Ersatz elterlicher Zuwendung durch ein übermäßiges Taschengeld, das für Süßigkeiten oder elektronische Unterhaltungsware ausgegeben wird. Vernachlässigung kommt zwar eher in sozialen Armutsschichten vor, wird aber ebenso in reichen Familien gefunden (Wohlstandsverwahrlosung).
Vernachlässigung kann auch vorliegen, wenn eine Mutter die Vorsorgeuntersuchungen in der Schwangerschaft nicht in Anspruch nimmt, während der Schwangerschaft Alkohol, Nikotin und Drogen konsumiert, wenn die Eltern in Anwesenheit ihres Kindes rauchen oder die empfohlenen Impfungen nicht vornehmen lassen.

Emotionale Misshandlung meint eine feindliche oder abweisende Haltung von Eltern oder Erzieher gegenüber einem Kind. Das Spektrum der emotionalen Misshandlung ist außerordentlich breit, wenn man alle Aktionen und Handlungen einbezieht, die die vertrauensvolle Beziehung zwischen Eltern und Kind beeinträchtigen und das Kind in seiner Entwicklung zu einer lebensbejahenden Persönlichkeit behindern.

Sexueller Missbrauch bedeutet, dass ein Kind einer sexuellen Stimulation ausgesetzt wird, die nicht zu seinem Alter, seiner psychosexuellen Entwicklung und seiner Rolle innerhalb der Familie passt. Sexueller Missbrauch kann mit Körperkontakt, mit penetrierendem Körperkontakt unter unterschiedlichen Formen von Zwang, Gewalt und Erpressung, aber auch ohne Körperkontakt stattfinden.

Bei sexuellem Missbrauch werden Kinder in sexuelle Handlungen oder Empfindungen einbezogen, die nicht ihrem Alter entsprechen. Dies kann mit Zwang zu Manipulationen, Zuschauen oder Positionieren verbunden sein. Zusätzlich können in unterschiedlichster Form direkte körperliche Misshandlungen erfolgen. Sexueller Missbrauch ist nach § 176 des Strafgesetzbuches strafwürdig. Durch die zunehmende Vorverlagerung der körperlichen Reife besonders bei Mädchen und die Verbreitung sexualisierter Vorstellungen in den Medien und im Umgang mit Kindern und Jugendlichen insgesamt treten zunehmend Probleme bei der strafrechtlichen Würdigung im Verhalten von Kindern und Jugendlichen über 12 Jahren auf.

Das Risiko für sexuellen Missbrauch innerhalb der Familie ist bei Mädchen in jedem Alter wesentlich höher als bei Knaben, letztere sind aber in Risikogruppen, z. B. Sport- und anderen Freizeitgruppen, selbst in Schuleinrichtungen durch pädophile Erwachsene auch gefährdet.

Zusammengefasst ist das Risiko für eine Kindesmisshandlung immer dann erhöht, wenn Kinder mit schwierigem Temperament und chronischen Gesundheitsproblemen auf überlastete, unkontrollierte und wenig kompetente Erwachsene, am häufigsten die eigenen Eltern, treffen.

Münchhausen-Stellvertreter- (by proxy-) oder **Meadow-Syndrom:** Hierbei werden primär gesunde Kinder meist von ihren Müttern verschiedenen Ärzten vorgestellt und unnötigerweise eingehend untersucht, medikamentös behandelt und sogar operiert. Nicht das Kind, sondern die Mutter ist dabei psychisch krank und benötigt eine entsprechende Behandlung.

Zusammengefasst kann man jede Form von Kindesmisshandlung als ein Scheitern von Beziehungen verstehen. Man fasst dies heute auch unter dem Begriff **Kindeswohlgefährdung** zusammen.

36.4 Epidemiologie

Die Angaben zur Häufigkeit von Kindesmisshandlungen, Kindesvernachlässigungen und sexuellem Missbrauch sind auch 50 Jahre nach der Erstbeschreibung in der medizinischen Literatur noch immer sehr unterschiedlich. Nach Auskunft des Bundeskriminalamtes waren 2006 334 Todesfälle bei Säuglingen und Kindern bis 10 Jahren durch körperliche Gewaltanwendungen zu erklären. 2007 wurden in Deutschland 3.373 Fälle von Kindesmisshandlung und 12.772 Fälle von sexuellem Missbrauch registriert. Aufgrund mehrerer Erhebungen kann man davon ausgehen, dass ca. 2 % der stationären Aufnahmen von Kindern in Kliniken für Kinder- und Jugendmedizin mit Misshandlungen, Vernachlässigungen und sexuellem Missbrauch im Zusammenhang stehen. Hochrechnungen verschiedener Organisationen legen nahe, dass bis zu 150.000 Kinder pro Jahr in Deutschland körperlich misshandelt werden und bis zu 100.000 Kinder und Jugendliche sexuell missbraucht.

Im Gegensatz zu anderen Ländern wie den USA und den Niederlanden gibt es in Deutschland keine offizielle Meldepflicht für Kindesmisshandlung.

In einer Allgemeinpraxis muss bei unter 10-jährigen Patienten in etwa 0,6 % mit einer Kindesmisshandlung gerechnet werden, in der kinderärztlichen Notfallsprechstunde bei Kindern unter 6 Jahren in bis zu 10 % der Fälle. 15 % der Knochenfrakturen bei Kindern bis 18 Monaten sind durch Misshandlungen bzw. mangelnde Aufsicht zu erklären, bei Verbrennungen und Verbrühungen geht man in etwa 10 % der Fälle davon aus.

36.5 Risikofaktoren

Kindesmisshandlung kommt in allen sozialen Schichten sowie allen ethnischen und religiösen Gruppen vor. Ein besonderes Problem kann in unserer multikulturellen Gesellschaft bei unzureichenden Kommunikationsmöglichkeiten bestehen.
Dennoch gibt es eine Reihe von **Risikofaktoren.** Hierzu gehören:

- Kinder aus unerwünschter, komplizierter Schwangerschaft
- Erst- und Letztgeborene
- Kinder mit Fehlbildungen aller Art, z. B. Lippen-Kiefer-Gaumenspalte oder Meningomyelozele
- Kinder mit Hirnschäden, z. B. Zerebralparese, Intelligenzminderung, Entwicklungsstörung, Hörstörung usw.
- Frühgeborene, besonders nach langer Inkubatorpflege
- Kinder mit Bewegungsstörungen, z. B. Muskel- und Nervenerkrankungen
- Kinder, die einnässen und einkoten
- Schreikinder, insbesondere in den ersten 2 Lebensjahren
- Spei- und Spuckkinder
- Kinder mit häufigen und häufig wechselnden Erkrankungen, z. B. Kopfschmerzen, Bauchschmerzen und Ernährungsstörungen
- Kinder mit länger dauernder Unterbrechung der elterlichen Zuwendung, z. B. Stief-, Adoptiv-, Pflege- und Heimkinder
- Kinder mit Schulproblemen, z. B. Schwierigkeiten beim Erlernen des Lesens, Schreibens und Rechnens
- Kinder, die nicht schmusen wollen, z. B. auch bei bestimmten taktilen Empfindungsstörungen.

Auch bei den Eltern, die mit dem misshandelten Kind zum Arzt gehen, lassen sich einige **typische Auffälligkeiten** feststellen:

- Mutter unter 18 Jahren
- Mutter alleinstehend
- Alkohol in der Schwangerschaft
- Drogen in der Schwangerschaft
- Gewalterfahrung bei den Eltern
- Gesetzeskonflikte
- Die Eltern entstammen häufiger gesellschaftlichen Minderheiten mit niedrigerem Ausbildungs- und Sozialniveau
- Die Eltern zeigen vermehrt psychische Auffälligkeiten
- In etwa 30 % der Fälle bestehen Alkoholprobleme
- In etwa 30 % der Fälle Vorstrafen
- Häufig besteht Arbeitslosigkeit
- Häufig sind die Eltern selbst misshandelt worden

- Häufig gibt es eine erhöhte Gewaltbereitschaft innerhalb der Familie, z. B. beim Austragen von Konflikten zwischen den Eltern
- Es besteht eine mangelnde Krisenbewältigung in der Ehe, am Arbeitsplatz und im sozialen Umfeld
- Soziale Isolierung
- Das Kind war nicht erwünscht
- Häufig sind die Eltern sehr jung.

36.6 Diagnostik

36.6.1 Anamnese

Jeder Arzt, insbesondere der Kinder- und Jugendarzt, muss bei jeder Untersuchung eines Kindes an das Vorliegen von Misshandlungsfolgen denken. Dabei muss er sowohl das Kind als auch sein Bezugssystem, d. h. seine Eltern und sein soziales Umfeld, berücksichtigen. Die Entdeckung von Hinweisen für eine Kindesmisshandlung ist vorrangig von der Erfahrung und dem Kenntnisstand des Arztes abhängig. In jedem Zweifelsfall muss deshalb immer eine sehr genaue Dokumentation aller anamnestischen und körperlichen Befunde sowie einer genauen Verhaltensbeobachtung stattfinden.
Gerade die erste Konsultation kann **wichtige Hinweise** erbringen:

- Die Kinder werden nicht direkt nach einem Vorfall, z. B. nach einem Sturz, zum Arzt gebracht
- Die Verletzung kann nicht plausibel erklärt werden
- Es bestehen auffällige Über- und Unterreaktionen der Eltern gegenüber den Krankheitszeichen
- Es besteht eine Diskrepanz zwischen dem Anspruch „perfekte Eltern“ und dem Zustand des Kindes
- Häufig wird der Arzt gewechselt bzw. ein wechselnd besetzter ärztlicher Notdienst in Anspruch genommen
- Häufig besteht ein schlechter Pflegezustand, z. B. der Fingernägel, der Haare oder des Windelbereichs
- Typischerweise werden andere Gründe für den Arztbesuch angegeben, z. B. Bauchschmerzen oder Schlafstörungen
- Es liegen frische und ältere Verletzungen bei dem Kind vor

36.6.2 Erste klinische Befunde

Typische **klinische Befunde** können sein:

- Hämatome, die nicht durch Zufallsunfälle erklärbar sind
- Hämatome als Abdruck von Schlaginstrumenten
- Verletzungen am äußeren Ohr
- Blut hinter dem Trommelfell
- Augenhintergrundsblutung
- Verletzungen um den Mundbereich, an Zungen- oder Lippenbändchen oder am weichen Gaumen
- Abgebrochene Zähne, schlechter Zahnstatus

- Symmetrische Verbrühungen, z. B. an Händen oder Füßen oder am Gesäß durch heißes Wasser
- Umschriebene Verbrennungen, z. B. durch Zigaretten
- Verletzungen im Anogenitalbereich
- Jedes Hämatom und jede Fraktur beim Säugling
- Ältere Verletzungen, die nicht versorgt wurden
- Mangelhafter Ernährungs- und Pflegezustand
- Nicht adäquates Herausputzen des Kindes
- Verhaltensauffälligkeiten des Kindes u. v. m.

Indirekte klinische Symptome bei misshandelten Kindern können sein:

- Vermehrte Unruhe
- Schlafstörungen
- Apathie
- Einnässen
- Einkoten
- Essstörungen
- Übermäßiges Essen
- Rezidivierende Kopfschmerzen
- Bauchschmerzen
- Gespannte Aufmerksamkeit („frozen watchfulness")
- Schulprobleme
- Weglaufen
- Fehlende Freundschaften
- Sexualisiertes Verhalten
- Schwangerschaft.

36.6.3 Vorgehen bei der Untersuchung

Nach einer gründlichen Anamneseerhebung unter Berücksichtigung der sozialen Situation, der Schwangerschaft und des Geburtsverlaufs sowie der bisherigen Entwicklung muss die Darstellung des Vorstellungsgrunds, meist eines angeblichen Unfallhergangs, sehr genau aufgenommen werden. Danach sollte die Untersuchung des völlig entkleideten Kindes erfolgen, wobei besonders bei Verdacht auf sexuellen Missbrauch mit Vorsicht, Diskretion und Geduld vorgegangen werden muss. Größe, Gewicht und Kopfumfang sollten gemessen und in einem Somatogramm erfasst werden. Äußere Verletzungen oder Auffälligkeiten sollten am besten fotographisch so dokumentiert werden, dass auf dem Foto eine zweifelsfreie Identifikation des Kindes, z. B. durch Miterfassen des Kopfes in einer Totalaufnahme und evtl. zusätzliche Detailaufnahmen, möglich ist. Neben der Haut müssen das Skelettsystem, die Abdominalorgane, der Mund, die Ohren und die Augen sowie das Genitale besonders beachtet werden.

Äußere Verletzungen

Während Säuglinge sich von selbst praktisch nie verletzen, kann sich jedes zur Fortbewegung fähige Kind durch nicht mutwillig herbeigeführte Einflüsse meist an exponierten Körperteilen, z. B. dem Kopf, am Rücken und den Knien, verletzen. Hals und Genitalbereich sind in der Regel gut geschützt, das Gesäß wird nur bei erheblichen mechanischen Einflüssen Hautveränderungen zeigen. Oft kann man aufgrund der Abdrücke, meist in Form von Hämatomen und Rissverletzungen

der Haut, Rückschlüsse auf den Unfallhergang bzw. bei der Verletzung verwendete Geräte wie Gürtel, Kleiderhaken, Löffel usw. ziehen. Typisch sind Verletzungen durch Abwehrhaltung des Kindes im Bereich der Unterarme, aber auch am Hinterkopf, im Bereich der Ohren, der Beine, am Rücken und am Gesäß. Prellungen und Zerrungen im Mundbereich führen zu Lippenhämatomen, Einrissen des Frenulums, Lippen- und Zungeneinrissen, Verletzungen des weichen Gaumens, z. B. beim Fütterungskampf. Ein fester Schlag auf den Gehörgang kann zu einer Trommelfellruptur führen.
Verbrühungen und Verbrennungen bei Säuglingen und Kleinkindern sind vor allem in Ländern mit vermehrter Armut und schlechten Wohnverhältnissen häufig, in unserem Kulturkreis ab dem Krabbelalter meist Folge von Sorglosigkeit und Vergesslichkeit der Eltern. Verbrühungen an den Füßen oder am Gesäß, evtl. auch im Gesicht sind verdächtig auf direkte Misshandlung, ebenso multiple runde Hautverbrennungen durch Zigarettenglut oder Hautschäden durch ein heißes Bügeleisen.
Blutungen aus dem Ohr sowie Hämatome im Orbitabereich können Ausdruck einer Schädelbasisfraktur sein. Wichtig ist auch die Beachtung von Blutungen in die Bindehaut der Augen infolge eines erhöhten Venendrucks beim Schütteltrauma, Atemnot bei Rippenfrakturen und abdominale Abwehrspannung bei Einblutungen von Milz-, Leber- oder Gefäßverletzungen.

Sexueller Missbrauch

Klinische Merkmale bei aktuellem **sexuellem Missbrauch:**
- Vaginale Blutungen
- Anorektale Blutungen
- Klaffende Scheide
- Klaffender Anus
- Perianale Rhagaden
- Hämatome im Anogenitalbereich
- Verletztes Hymen
- Defäkationsschmerzen
- Dysurie
- Vulvovaginitis
- Kondylome
- Eitriger Ausfluss aus Urethra
- Bauchschmerzen
- Kopfschmerzen
- Essstörungen/Bulimie.

Schütteltrauma

C. H. Kempe beschrieb als typische Symptome einer **Misshandlung von Säuglingen** das Auftreten von
- Subduralergüssen,
- Knochen- und Gelenkveränderungen im Röntgenbild und
- Retinablutungen.

Bei Letzterem handelt es sich um Folgen des sogenannten **Schütteltraumas** (= shaken baby syndrome), bei dem das Kind meist an beiden Oberarmen gegriffen und so geschüttelt wird, dass der muskulär unzureichend stabile Kopf hin- und hergeschleudert wird. Hierdurch kommt es zu Einrissen der subduralen Hirnhaut und der sie versorgenden Venen, zu Kontusionen der Hirnrinde, evtl. zu venösen Abflussstörungen und zu Einklemmungen im Bereich des Tentoriums. Typischerweise treten gleichzeitig streifenförmige oder flächenhafte Blutungen am Augenhintergrund

auf, die von der Papille in die Peripherie ziehen, evtl. auch Einblutungen in den Glaskörper mit Netzhautablösung und narbigen Schrumpfungen.
Radiologische Veränderungen sind oft mehrzeitig, d. h. mit unterschiedlicher Kallusbildung. Sie können besonders an den Rippen, den Meta- und Epiphysen im Bereich der großen Gelenke und, in der Regel 2–3 Wochen nach dem Trauma, als Verkalkungen des Periosts nachgewiesen werden. Bei jedem Säugling oder Kleinkind, bei dem eine akute oder chronische Kindesmisshandlung vermutet wird, ist die Anfertigung von Röntgenaufnahmen des Thorax und der Extremitäten sowie eine Schädelsonographie und eine Untersuchung des Augenhintergrunds indiziert. In der letzten Zeit hat sich daneben vor allem auch die Kernspintomographie der Hirnstrukturen, aber auch der Extremitäten, des Thorax und des Abdomens zur Darstellung von Gewebsverletzungen bewährt. Weitere Laboruntersuchungen müssen vom klinischen Zustand und den differenzialdiagnostischen Erwägungen abhängig gemacht werden.

Verhaltensauffälligkeiten misshandelter Kinder

Misshandelte Kinder sind zweifellos oft psychisch auffällig. Die Art der Auffälligkeit umfasst das gesamte denkbare Spektrum von kindlichem Verhalten und Fehlverhalten. Es reicht vom aggressiven bis zum völlig in sich zurückgezogenen passiven Kind, vom distanzlosen bis zum extrem ängstlichen Kind, sodass ein bestimmtes Muster nicht beschrieben werden kann. Vielfach ist auch nicht klar zwischen Ursachen und Folgen zu unterscheiden, z. B. wenn ein Kind einnässt. Tut es dies, weil es misshandelt wird, oder wird es deshalb der Misshandlung ausgesetzt? Dennoch gilt als sicher, dass das Misshandlungssyndrom eine Reihe von psychischen Störungen verursacht, die allerdings ebenso Folge von emotionaler Misshandlung oder sexuellem Missbrauch sein können.
Langzeitfolgen von emotionaler Misshandlung und sexuellem Missbrauch sind z. B.:

- Partnerprobleme
- Gestörtes Selbstbewusstsein
- Depression
- Suizid
- Drogen
- Prostitution
- Kriminalität
- Verwahrlosung
- Essstörungen (Anorexie, Adipositas)
- Zwangserkrankungen
- Dissoziative (psychogene) Symptome
- Beibehaltung einer Opferrolle
- Vermehrte Aggressivität.

36.7 Differenzialdiagnostik

Nicht zufällige, d. h. durch Misshandlung zustande gekommene Verletzungen bei einem Kind müssen von einer Vielzahl völlig unterschiedlicher Erkrankungen abgegrenzt werden.
So müssen bei Hautmanifestationen zum einen angeborene Pigmentanomalien („Mongolenflecke"), Hautanlagestörungen wie eine Epidermolysis bullosa und Gefäßveränderungen, vor allem aber auch Gerinnungsstörungen wie eine Thrombozytopenie oder eine Hämophilie A berücksichtigt werden. Hautschwellungen, die oberflächlich von Verletzungsfolgen schwer abgrenzbar sind,

können z. B. auch im Rahmen von Ringelröteln und anderen Infektionen, Ekzemen und mykologischen Erkrankungen auftreten. Besonders differenziert müssen Veränderungen im Anogenitalbereich beurteilt werden, z. B. perianale Kondylome, vaginaler Ausfluss oder Erytheme. Klinische Merkmale bei sexuellem Missbrauch sind in der oben stehenden Auflistung zusammengestellt.
Obwohl selten, müssen bei Subduralergüssen spezielle Ursachen wie Stoffwechselstörungen (Glutarazidurie, Menkes-Syndrom) oder chronische Entzündungen berücksichtigt werden, bei wiederholten Knochenbrüchen muss auch an eine Osteogenesis imperfecta gedacht werden.
Von besonderer Bedeutung ist die Abgrenzung einer Kindesmisshandlung vom plötzlichen Säuglingstod bzw. akuten lebensbedrohlichen Ereignissen (**ALTE-Syndrom** = acute life threating event bzw. near-miss-sudden-infant-death-syndrome). So werden auf der einen Seite durch subtile Untersuchungen, vor allem auch im Rahmen der Obduktion, sehr unterschiedliche Erklärungen gefunden, z. B. anlagebedingte oder erworbene Hirnstammstörungen, schwerwiegende Herzerkrankungen, foudroyante Epilepsien, akute Embolien und schwerwiegende Aspirationen, andererseits existiert für viele Fälle des plötzlichen Säuglingstods auch heute noch keine befriedigende Erklärung. Unverändert muss davon ausgegangen werden, dass ein nicht unbeträchtlicher Teil plötzlich verstorbener Säuglinge und Kleinkinder doch durch äußere Einflüsse, z. B. Ersticken, Strangulation u. Ä., zu Tode gekommen ist. Deshalb ist gerade auch in solchen Fällen eine sehr genaue Untersuchung auf möglicherweise nur diskrete Folgen von Gewaltanwendung unbedingt erforderlich.
Ein besonderer Problembereich sind Kinder mit mentalen Entwicklungs- und Verhaltensstörungen, bei denen immer wieder auch Komponenten einer Kindesmisshandlung mitberücksichtigt werden müssen.

Fallbeispiel

Vorstellung eines 2-jährigen Knaben durch seine Pflegeeltern. Soweit bekannt regelrechte Schwangerschaft und Geburt, Geburtsgewicht 3.040 g, keine besonderen Probleme in der Neugeborenenperiode. Im Alter von 4 Monaten Vorstellung in einer Kinderklinik wegen akuter Atemstörungen. Bei dem somnolenten Säugling wurden im Bereich der Oberarme und des Halses Hämatome gesehen, sonographisch und im Computertomogramm des Schädels wurden subarachnoidale Blutungen nachgewiesen, am Augenhintergrund Retinablutungen. Der Liquor war blutig, der Überstand gelblich mit einem Bilirubin von 3,2 mg %.
Radiologisch konnten an den Rippen mehrere Frakturen unterschiedlichen Alters festgestellt werden. Eine genaue Klärung, wer der leiblichen Eltern für die Misshandlung verantwortlich war, konnte in der Gerichtsverhandlung nicht erreicht werden.
Nach zweimonatiger stationärer Behandlung im Kinderkrankenhaus war die Entwicklung in der Pflegefamilie deutlich verzögert. Erhebliche Probleme mit der Nahrungszufuhr und dem Schlafen. Die leiblichen Eltern haben sich getrennt, aktuell hat die leibliche Mutter die Rückführung des Kindes beantragt. Die Vorstellung im SPZ erfolgt auf Initiative des Jugendamts zur differenzierten Entwicklungsdiagnostik.
Befunde:
Mit 13 kg und 86 cm eutropher Knabe, Kopfumfang mit 46,8 cm 1 cm unter der 3. Perzentile. Guter Pflegezustand, intern keine Auffälligkeiten, Haut regelrecht. Das Kind steht mit Festhalten, vollführt seitwärts erste Schritte, dabei vermehrte Überstreckung der Füße. Vermehrtes Fäusteln, kein Pinzettgriff. Beginnt mit einfachen Doppellauten, nur kurze Blickfolge, keine differenzierte Gestik. Insgesamt ist das Kind sehr umtriebig, fällt viel um. Strabismus divergens. Klinisch keine Anfälle.
Diagnosen:
- Allgemeine Entwicklungsstörung (F79.0G)
- Mikrozephalie (Q02G)

- Störung des Sozialverhaltens (F98.8G)
- Strabismus divergens (H50.1G)
- Abnorme psychosoziale Umstände.

Maßnahmen:

- Genaue Seh- und Hörprüfung
- Einleitung einer Komplexbehandlung durch die zuständige interdisziplinäre Frühförderstelle mit Schwerpunkt im Bereich der Heilpädagogik
- Beantragung von Behindertenausweis und Pflegegeld
- Beantragung einer Ausgleichszahlung beim zuständigen Versorgungsamt nach dem Opferentschädigungsgesetz
- Kontaktaufnahme mit dem zuständigen Jugendamt
- Veranlassung einer sehr sorgfältigen Prüfung der Lebensumstände der leiblichen Mutter
- Videodokumentation beim Zusammenkommen der leiblichen Mutter mit dem Kind
- Vereinbarung regelmäßiger Wiedervorstellung im SPZ zur Beurteilung der weiteren Entwicklung

36.8 Behandlungsprinzipien

Misshandlungen bei Kindern können Ärzten in praktisch allen Berufssparten begegnen, leider wird dies immer wieder bewusst oder unbewusst verdrängt und verleugnet. Jeder Verdacht sollte heute daher von Anamnese und Symptomatik genau erfasst und dokumentiert werden. Ein Schema zur Abschätzung der Gefährdung des Kindeswohls zeigt Abbildung 36.1 (➤ Abb. 36.1). Die ärztliche Schweigepflicht darf nicht als Hindernis für das Verfolgen eines Verdachts vorgenom-

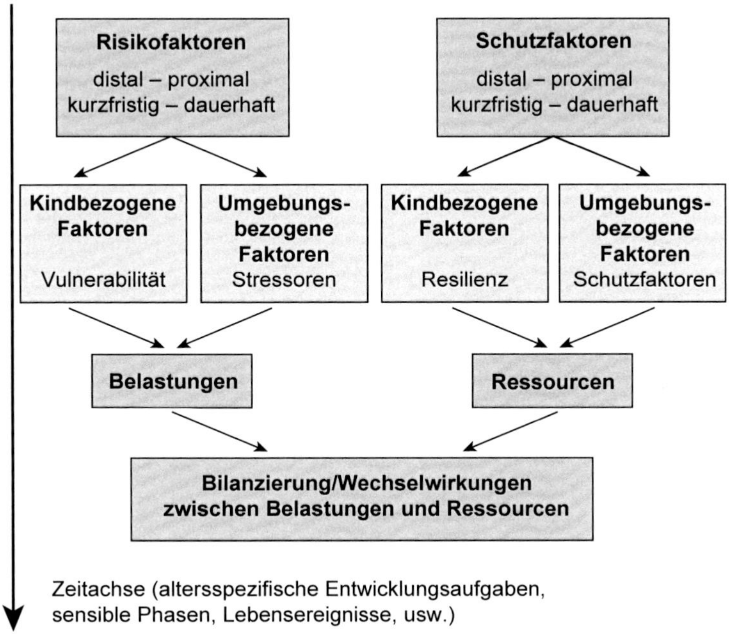

Abb. 36.1 Schema zur Abschätzung der Gefährdung des Kindeswohls

men werden. Der Schutz des Kindes hat auch in der Rechtsprechung den absoluten Vorrang. Dies wird unter anderem im Jugendschutzgesetz sowie in dem seit 2012 gültigen Kinderschutzgesetz (➤ Kap. 42) festgelegt.

In der Regel kann ein einzelner Arzt bzw. eine einzelne Institution beim ernsthaften Verdacht auf das Vorliegen einer Kindesmisshandlung nicht alle notwendigen Maßnahmen ergreifen. Entscheidend

ist, dass in einer interdisziplinären Kooperation von verschiedenen Fachinstitutionen die Problematik des Kindes und seiner Familie aufgegriffen und nach entsprechenden Voruntersuchungen klar und umfassend mit den Sorgeberechtigten angesprochen wird.

Ganz in Abhängigkeit von den vorhandenen Symptomen müssen die verschiedenen Institutionen wie Rechtsmedizin, Jugendamt, Polizei, Staatsanwaltschaft, Kinderschutzbund u. Ä. einbezogen werden. Bei schwerwiegenderen Fällen hat es sich bewährt, dass eine solche **interdisziplinäre Absprache** in Ruhe im Rahmen eines stationären Aufenthalts in einem Krankenhaus stattfindet. Nach der Neufassung der §§ 8 a und 42 kann vom Jugendamt eine bis zu 48 Stunden dauernde Entfernung des Kindes aus der elterlichen Obhut veranlasst werden, bis über ein Familiengericht die weitere Entscheidung über das Aufenthaltsbestimmungsrecht gefällt wird.

Zwar gilt der Grundsatz „Hilfe vor Strafe", aber nicht „Hilfe statt Strafe". Im Vordergrund steht die Sicherstellung des andauernden Schutzes des Kindes, so lange wie möglich auch die Stützung der Familie, z. B. durch psychologische Betreuungsangebote, sozialpädagogische Familienhilfe, Familientherapie und Einzeltherapie. Zunehmend werden dabei auch verhaltenstherapeutische Programme zur Verbesserung der Eltern-Kind-Interaktion, vor allem im Säuglings- und Kleinkindesalter, eingesetzt. Es ist immer problematisch, das Kind aus der Familie herauszunehmen und es in einem Heim oder einer Pflegefamilie unterzubringen. Dennoch sind solche Maßnahmen einschließlich Sorgerechtsentzug und Strafverfolgung des bzw. der Täter u. U. nicht zu vermeiden.

Besonders bei Verdacht auf sexuelle Misshandlung von Kindern hat sich immer wieder gezeigt, dass wohlgemeinte aber suggestive Beeinflussungen des Kindes durch nicht professionelle Helfer eine objektive Erfassung des Tathergangs und eine Verurteilung des Täters unmöglich machen. So mussten wiederholt Tatverdächtige aufgrund von Verfahrensfehlern bzw. Mangel an objektiven Beweisen freigesprochen werden, nicht zuletzt Mitarbeiter in kirchlichen Einrichtungen, Sportvereinen und Heimen. Leider erweist es sich aber auch immer wieder, dass z. B. in Fällen von getrennt lebenden Ehepartnern Anschuldigungen wegen Kindesmisshandlung ohne Grund erhoben und vordergründige Symptome überinterpretiert werden, die eine saubere juristische Aufarbeitung unmöglich machen. Misshandlungen und sexueller Missbrauch von Kindern sollten lebenslang nicht verjähren.

Eine noch nicht befriedigende Situation besteht dann, wenn Kinder nicht direkter Gewalt ausgesetzt waren, wohl aber Gewaltanwendung in ihrem Umfeld erlebt haben, z. B. zwischen den Eltern oder in Kriegsgebieten. Deshalb sollte z. B. in Frauenhäusern und Asylanten-Einrichtungen bei den Kindern besonders auf posttraumatische Belastungsstörungen geachtet werden.

36.9 Prävention

Alle Kinder haben das Recht auf die Erfüllung von Grundbedürfnissen, wie sie u. a. auch in den UN-Kinderrechten zusammengefasst sind. Darüber hinaus können folgende Punkte aufgeführt werden.

Grundbedürfnisse von Kindern:

- Schutz vor Kälte, Nässe und übermäßiger Hitze
- Adäquate Kleidung

- Quantitativ und qualitativ angemessenes Essen
- Schutz vor Unfällen
- Frische Luft
- Ausreichender Schlaf
- Adäquate ärztliche Versorgung bei Erkrankungen
- Vorbeugung gegen Krankheiten (Vitamin D, Impfungen)
- Ausbildung
- Ausreichende aktive Bewegung, Freizeit
- Schutz vor Einflüssen mit nicht altersadäquaten Inhalten, z. B. in Medien.

Wesentlich für eine Verhinderung aller Arten von Kindesmisshandlung sind natürlich **präventive Maßnahmen** ➤ (Kap. 7) Hierzu gehören z. B.:

- Schulung aller potenzieller Eltern über ihre Rechte und Pflichten und die bei jedem Kind möglichen Gesundheitsprobleme, am besten im Rahmen des Schulunterrichts
- Teilnahme an den Schwangerschaftsvorsorgeuntersuchungen
- Regelmäßige Teilnahme an Früherkennungsuntersuchungen durch Kinder- und Jugendärzte, wie dies mittlerweile in mehreren Bundesländern verpflichtend eingefordert wird
- Familiennahe Betreuung von Mutter und Kind in den ersten Lebenswochen und -monaten durch erfahrene Hebammen (Familienhebamme)
- Gegebenenfalls die frühestmögliche Erfassung und regelmäßige Betreuung von Kindern aus Risikofamilien
- Frühzeitige Erziehung der Kinder zu einer selbstbewussten Persönlichkeit mit Anerkennung der Integrität des eigenen Körpers
- Förderung des Selbstbewusstseins, rechtzeitig Nein zu sagen
- Förderung der positiven Vorbildfunktion von Erwachsenen, insbesondere Eltern
- Verbannung von Gewalt aus Familien sowohl zwischen den Erwachsenen als auch zwischen Erwachsenen und Kindern und unter den Kindern
- Sinnvoller Einsatz der Medien, insbesondere der Verzicht auf Gewalt in Fernsehen, Videospielen und Kino vor dem 10. Lebensjahr
- Bereitstellung von ausreichend großen und hygienisch einwandfreien Wohnungen
- Vermeidung/frühzeitige Erkennung aller Arten von Sucht
- Zuteilung von bezahlter Arbeit, zumindest an einen Elternteil
- Ausreichendes Angebot von Kinderkrippen und Kinderhorten, vor allem bei Berufstätigkeit der Mutter
- Angebot eines wohnungsnahen Kindergartens für alle Kinder
- Betreuung in adäquaten Schulen
- Ausreichendes Angebot von Freizeitaktivitäten u. v. m.

Bei den Projekten, die regional in unterschiedlicher Form zur Erfassung von Risikokindern umgesetzt werden, muss allerdings darauf geachtet werden, dass gut geschultes Betreuungspersonal in ausreichender Zahl zur Verfügung steht und dass keine Stigmatisierung Einzelner durch gut gemeinte Einzelkonzepte erfolgt.

Daneben hat sich als wesentlich herausgestellt, dass

- Nachbarn sich untereinander stützen und ansprechen,
- Sorgentelefone benutzt werden können und
- einsichtige Eltern sich in unterschiedlicher Form treffen und Probleme und Erfahrungen austauschen bzw.
- sich vertrauensvoll an ausgebildete Ärzte, Psychologen und Sozialpädagogen wenden können, die ihre Probleme ernst nehmen.

Dies gilt auch für Eltern, die bereits Gewalt gegenüber ihren Kindern eingesetzt haben – z. B. die Selbsthilfegruppe Anonymer Eltern. Dies bedeutet aber auch, dass von der öffentlichen Hand, insbesondere in den Jugendämtern, eine ausreichende Zahl qualifiziert ausgebildeter Fachkräfte tätig ist, die einen möglichst umfassenden Überblick über alle jungen Familien, insbesondere aber die gefährdeten Familien haben.
Eine besondere Problematik stellen Familien dar, in denen ein Elternteil an einer psychischen Erkrankung (Depression, Psychose, Drogenabhängigkeit etc.) leidet, sowie Familien mit Gewalterfahrungen – viele misshandelnde Eltern sind als Kind misshandelt worden. Hier müssen einerseits ambulante sozialpsychiatrische Behandlungsmöglichkeiten in ausreichender Zahl vorhanden sein, andererseits aber u. U. auch stationäre Behandlungsformen, in denen z. B. die psychisch kranke Mutter mit ihrem Kind behandelt werden kann.
Es ist ein Trugschluss, wenn Menschen von vornherein sagen, „ihnen könne so etwas nicht passieren“. Trotz schwerwiegender Misshandlungen haben nicht wenige misshandelnde Eltern auch eine positive, zumindest eine ambivalente Einstellung zu ihrem Kind, vor allem aber besteht aber bei den Kindern trotz schwerwiegender Misshandlungen oft noch eine positive Beziehung zu ihren Eltern. Diese Abhängigkeiten müssen in ihren langfristigen Auswirkungen gerade auch bei Entfernung der Kinder aus der leiblichen Familie sorgfältig bedacht werden.

36.10 Begutachtungsfragen

Immer wieder muss im Rahmen von Gerichtsverfahren durch Begutachtungen festgestellt werden, ob die Anschuldigung, dass eine Misshandlung vorliegt, objektiv zutrifft. Besonders beim sexuellen Missbrauch müssen die Glaubhaftigkeit der kindlichen Angaben und ggf. anderer Zeugen sowie mögliche Folgen aus diesen Aussagen im Rahmen ärztlicher Gutachten geklärt werden. Dabei hat sich gezeigt, dass für körperliche Misshandlungen in der Regel hinreichend sichere Beweismittel vorgelegt werden können, z. B. in Form von Fotodokumentationen, Röntgenbildern, Kernspintomographien und ähnlichem. Besonders beim sexuellen Missbrauch liegen jedoch oft nur unspezifische Hinweise vor, sodass die Aussagefähigkeit eines Kindes nach seiner allgemeinen, insbesondere kognitiven Entwicklung, seiner psychosexuellen Entwicklung und nach ggf. vorliegenden zusätzlichen Verhaltensauffälligkeiten beurteilt werden muss.
Analysen haben ergeben, dass Erwachsene wesentlich häufiger falsche Aussagen vor Gericht machen als Kinder. Dennoch müssen die Aussagemotive, vor allem auch in familienrechtlichen Auseinandersetzungen, z. B. um das Sorgerecht, berücksichtigt werden. Hierbei können u. U. auch indirekte, z. B. projektive Testverfahren (Malen einer verzauberten Familie, Szenotest, Spiel mit anatomischen Puppen und allgemeiner Spielbeobachtungen) herangezogen werden. Man kann davon ausgehen, dass bei einem 3-jährigen Kind eine Erinnerungsfähigkeit für etwa 6 Monate besteht, bei einem 4-jährigen Kind für etwa 14 Monate. Fantasie und Suggestibilität sind gerade im Vorschulalter jedoch individuell sehr unterschiedlich ausgeprägt. Ein wesentliches Problem sind die Befragungstechniken und ihre Dokumentation. So sollen möglichst keine Ja-/Nein-Fragen, keine Suggestivfragen und keine Fragenwiederholungen gegenüber dem Kind eingesetzt werden. In der Regel können solche Untersuchungen nur von sehr erfahrenen Kinder- und Jugendpsychiatern und/oder Kinderpsychologen vorgenommen werden. Andererseits müssen dabei aber auch u. U. neuropädiatrische Aspekte (Vorliegen einer Epilepsie, Ursachen einer Intelligenzminderung etc.) berücksichtigt werden.

Die Befragung der potenziellen Täter bzw. anderer Familienangehöriger sollte möglichst ausführlich und detailliert, u. U. auch mit konfrontierenden Techniken (z. B. entsprechendem Bildmaterial) durchgeführt werden. In Deutschland ist der Einsatz von technischen Hilfsmethoden, z. B. sogenannten „Lügendetektoren", bei denen Hautwiderstand, Puls, Atemfrequenz usw. des potenziellen Täters gemessen werden, nicht zugelassen.

36.11 Betreuungsmöglichkeiten

Primär ist für die Betreuung der Familien, in denen eine Kindswohlgefährdung festgestellt wurde, das Jugendamt zuständig. Daneben gibt es eine Vielzahl von Organisationen, die sich für Kinder einsetzen, die eines Misshandlung erlitten haben bzw. hierfür gefährdet sind, insbesondere den **Deutschen Kinderschutzbund.** Für die **Betreuung von sexuell misshandelten Kindern und Jugendlichen** haben sich verschiedene Organisationen wie „Wildwasser", „Zartbitter" oder „Allerleirauh", Frauenhäuser, Beratungsstellen bei Gesundheitsämtern und vieles mehr bewährt.

Das Thema der Kindesmisshandlung stellt eine ständige gesellschaftliche Herausforderung dar, die in ihrem vollen Umfang erst in den letzten 20 Jahren bekannt wurde. Für jeden Arzt ist es wichtig, Grundkenntnisse zu den verschiedenen Arten der Kindesmisshandlung zu besitzen und im Zweifelsfall Kontakte zu in diesen Fragen spezialisierten Stellen aufzunehmen. Das **Nationale Zentrum Frühe Hilfen** versucht zurzeit, die verschiedenen Initiativen in Deutschland zu koordinieren und wissenschaftlich zu begleiten (www.fruehehilfen.de). Hierzu gehören u. a. die Verpflichtung zur Teilnahme an den Früherkennungsuntersuchungen, bessere Betreuungsmöglichkeiten auf lokaler Ebene und die Begleitung von Familien mit einem erhöhten Risiko für Misshandlung. Hierdurch lassen sich die Probleme zwar nicht beseitigen, erlauben aber sicher einen zunehmend offenen und konstruktiven Umgang.

Literatur

Deegener G, Körner W (2005) Kindesmisshandlung und Vernachlässigung – ein Handbuch. Göttingen: Hogrefe

Deegener G, Körner W (2006) Risikoerfassung bei Kindesmisshandlung und Vernachlässigung – Theorie, Praxis, Materialien. Lengerich: Pabst.

Häßler F, Schepker R, Schläfke D (Hrsg.) (2008) Kindstod und Kindstötung. Berlin: Medizinisch Wissenschaftliche Verlagsgesellschaft

Herrmann B, Dettmeyer R, Banaschak S, Thyen U (2008) Kindesmisshandlung – Medizinische Diagnostik, Intervention, rechtliche Grundlagen. Heidelberg: Springer

Jacobi MG (2008) Kindesmisshandlung und -vernachlässigung – Epidemiologie, Diagnostik und Vorgehen. Bern: Hans Huber

Straßburg HM, Dacheneder W, Kreß W (2012) Entwicklungsstörungen bei Kindern. Praxisleitfaden für die interdisziplinäre Betreuung. München: Elsevier

VII Therapie, Förderung und Rehabilitation

VII

37 Grundlagen der Therapie

Helmut Hollmann

Sozialpädiatrie in der Praxis. http://dx.doi.org/10.1016/B978-3-437-31630-3.00037-7
Copyright © 2014 Elsevier GmbH. All rights reserved.

37.1 Einleitung

Maßnahmen zur **Entwicklungsunterstützung** zählen zum normalen Handlungsrepertoire von Eltern, Erzieherinnen und Lehrerinnen. Sie zielen auf erstmalige oder vertiefende Angebote ab. Alle Kinder profitieren unabhängig vom Entwicklungsstand von der Möglichkeit, vielfältige Erfahrungen machen zu können. So lernen sie spielerisch oder zielgerichtet eigene Fähigkeiten und Grenzen kennen.

Im Gegensatz dazu sollen **therapeutische Maßnahmen** auf solche Situationen beschränkt sein, wo eine Entwicklungsstörung beim Kind vorliegt und durch die zielgerichtete korrigierende Therapie eine Besserung zu erwarten ist. Die Statistiken der Heilmittel-Verordnung verweisen jedoch darauf, dass erheblich mehr Kinder in eine Therapie genommen werden, als aufgrund der Häufigkeit von Entwicklungsstörungen zu erwarten wäre. Offensichtlich besteht eine große Unklarheit über korrekte diagnostische Ausweisung, Notwendigkeit einer therapeutischen Maßnahme und zu erreichende Effekte, was zur Ausweitung der Indikationsstellung führt. Hintergrund ist die

Verunsicherung bis Ängstigung von Eltern und anderen Bezugspersonen durch steigende Anforderungen an Kinder und Jugendliche in Schule und Berufsausbildung. Wider besseres Wissen um die Bandbreite der normalen Entwicklung werden normative und mittenzentrierte Erwartungen zum Maßstab erhoben, denen dann viele Kinder nicht genügen. So führen geringe Auffälligkeiten oder auch Normvarianten zur ungerechtfertigten Vermutung drohender Defizite. Hieraus resultieren Irritation und Suche nach der vermeintlich notwendigen Abhilfe.
Es ist dringend erforderlich, beim Thema der Therapie und Förderung zu einer größeren Sachlichkeit zu finden. Ohne jede Frage steht eine Vielzahl von Interventionsmöglichkeiten zur Verfügung, die bei sinnvollem Einsatz ausgezeichnete Hilfestellungen für Kinder, Jugendliche und deren Bezugspersonen, insbesondere die Eltern, geben können. Davon abzugrenzen ist jedoch ein überschießendes Vorgehen, das für die Kinder durchaus auch Risiken beinhaltet. **„Das Bedürfnis des Kindes, entwicklungsgerechte Erfahrungen zu machen, verträgt sich nicht mit der immer größer werdenden Förderwut.“** (Remo H. Largo, 2000).
Krankheitsbilder, die zu Entwicklungsstörungen führen und in der Sozialpädiatrie von Bedeutung sind, stammen aus den unterschiedlichsten somatischen und psychischen Bereichen. Zu einer möglichst genauen Einschätzung ist deshalb stets eine mehrdimensionale Betrachtung und diagnostische Beurteilung erforderlich. Diese muss durch die Analysen der Ressourcen des Kindes, der Familie und anderer naher Bezugssysteme mit den darin wirkenden Personen ergänzt werden (➤ Kap. 15).
Therapeutische Maßnahmen im medizinischen Sinn bei Entwicklungsstörungen und Behinderungen müssen immer auch psychische und soziale Kontextfaktoren berücksichtigen, um dem übergeordneten Ziel der größtmöglichen individuellen Teilhabe gerecht zu werden. Im Zusammenhang mit Krankheits- und Störungsbildern in der Sozialpädiatrie ist es deshalb günstig, sich von dem Begriff „Therapie“ etwas zu lösen, der im medizinischen Sinn stets mit der Vorstellung der „Heilung“ assoziiert ist, und stattdessen neutraler und damit ergebnisoffener das Spektrum der unterschiedlichen Interventionsmöglichkeiten zu benennen. Gerade in der Sozialpädiatrie, wo in aller Regel unterschiedliche Berufsgruppen interdisziplinär zusammenarbeiten, kommt den nichtmedizinischen Behandlungsansätzen hohe Bedeutung zu. Die ärztliche Gesamtverantwortung bleibt hiervon unberührt.
In diesem und dem folgenden Kapitel wird ein Überblick über sinnvolle **Therapie- und Fördermaßnahmen** in der Sozialpädiatrie als Möglichkeiten der Intervention gegeben.

37.1.1 Entwicklungskorridor als Zielgröße

Entwicklungsstörungen und Behinderungen bei Kindern und Jugendlichen sind Anlass, entwicklungsfördernde Therapiemaßnahmen einzuleiten. Ziel ist dabei die Hinführung zur größtmöglichen Realisierung von **Entwicklungspotenzialen** mit der Hoffnung, eine Annäherung an den Normbereich zu erreichen. Je jünger das Kind ist und je weniger eine primäre zentralnervöse Ursache im Sinne der organischen Störung eruiert werden konnte, umso eher können diese Maßnahmen wirksam sein.
Übergeordnetes Ziel jeder Therapie ist die Erreichung der sozialen Teilhabe und einer bestmöglichen Bewältigung der Situation für den Patienten sowie die Stärkung der elterlichen Kompetenz, um das Kind auf seinem individuellen Entwicklungsweg optimal unterstützen zu können.
Die Vorstellungen über die zu erwartenden Auswirkungen von Therapiemaßnahmen sind eng mit dem Begriff des **„Entwicklungskorridors“** verbunden. Hierunter ist der Bereich jedweder kognitiven, motorischen, sprachlichen, emotionalen oder kreativ-künstlerischen Fähigkeit zu verste-

hen, der in seiner Grundlage und potenziellen Dimension genetisch determiniert ist, in seiner Ausdifferenzierung und dem Nutzungsgrad des Potenzials aber insbesondere von den fördernden oder hindernden Umgebungseinflüssen abhängt. Das Kind mit der überragenden Intelligenz oder dem ungewöhnlichen Talent wird seine Gaben eben nur realisieren können, wenn es in die Situation kommt, diese abrufen zu müssen und gleichzeitig seine Bezugspersonen die besondere Qualität bemerken und in der Ausdifferenzierung entsprechend unterstützen.

Begabung und Umgebungseinfluss

Carl Friedrich GAUSS wurde 1777 als Kind einfacher Eltern geboren. Seine besondere Begabung im Umgang mit Zahlen soll schon im Kleinkindalter aufgefallen sein. Daniel Kehlmann schildert in seinem 2012 auch verfilmten Roman „Die Vermessung der Welt" anschaulich, dass Gauß in der Volksschule eher ein gewisses Außenseiterdasein führte. Die blitzschnelle Lösung der Additionsaufgabe aller Zahlen von 1 bis 100, vom Lehrer Büttner als Dauerbeschäftigung seiner Schülerschar gedacht, führte diesem das Ausnahmetalent vor Augen. In der Folge unterrichtete der Pädagoge den Jungen in besonderer Form, verwandte sich bei Hofe für Gauß und verschaffte ihm so das erste Stipendium. Ohne diese Intervention wäre das mathematische Jahrhunderttalent Gauß aufgrund seiner Herkunft aus einfachsten Verhältnissen zur damaligen Zeit nicht entdeckt und zum Studium gebracht worden.
Eine Ausnahmebegabung hat vermutlich auch **José Antonio ABREU.** Der Komponist aus Venezuela, geb. 1939, studierte parallel Musik, Volkswirtschaft und Jura. Für alle Fächer hatte er eine Professur an der Universität Caracas inne. 1975 initiierte er ein landesweites Musikschul-Projekt, das sich an sämtliche Kinder richtet. **„El Sistema", das nationale Musikprojekt der Kinder- und Jugendorchester Venezuelas,** umfasst heute 250.000 Kindern, 125 Jugendorchester, 57 Kinderorchester und 30 professionelle Sinfonieorchester. „Die Regierung unterstützt mein Projekt genau wegen seiner sozialen Ausrichtung. Der Staat hat sehr gut verstanden, dass das Projekt, wiewohl es mit Mitteln der Musik arbeitet, zuvorderst ein soziales ist: ein Projekt zur Förderung allgemeiner menschlicher Qualitäten. Denn für die Kinder, mit denen wir arbeiten, stellt die Musik fast den einzigen Weg zu einem menschenwürdigen Dasein dar. Armut – das heißt: Einsamkeit, Traurigkeit, Anonymität. Orchester – das heißt: Freude, Motivation, Teamgeist, Streben nach Erfolg. Wir sind eine große Familie auf der Suche nach Harmonie und jenen schönen Dingen, die allein die Musik den Menschen zu bringen vermag." Er zitiert gern Mutter Teresa und sagt: „Es ist auch ein geistlicher Kampf für das Wahre, Schöne, Gute – gegen Not und wirtschaftliche Gier." (aus: www.wikipedia.org).
Ähnlich wie in vielen Ländern der Welt durch ein breitflächiges Angebot die Fußballtalente entdeckt werden, förderte das Projekt in Venezuela musikalische Ausnahmebegabungen zutage. Gustavo Dudamel, Dirigent, Aléxis Cárdenas, Geige, und Edicson Ruiz, Kontrabass sind Beispiele für Karrieren auf Weltniveau. Ruiz wurde mit 17 Jahren als jüngster Musiker seit Bestehen Mitglied der Berliner Philharmoniker. Der Trompeter Francisco Colmenares gewann 2006 den Concours Maurice André in Paris.
Das Musikprojekt in Venezuela verbindet in beeindruckender Weise soziales Engagement, Prävention und Förderung miteinander. Sozusagen als Nebenergebnis ist die Identifikation und systematische Unterstützung von Begabungen entstanden. Ohne dieses von außen induzierte Angebot wäre der Lebensweg der Kinder völlig anders verlaufen.

In derselben Weise wie für normal oder überdurchschnittlich begabte und talentierte Kinder gilt diese Überlegung auch für Kinder mit Entwicklungsstörungen und Behinderungen. Optimale Ausdifferenzierung einer Fähigkeit oder Fertigkeit im Rahmen des individuellen Entwicklungskorridors bedeutet hier, dass sich durch jede fördernde Intervention stets nur eine Optimierung der individuellen Situation im Rahmen der vorgegebenen Möglichkeiten erreichen lässt. Im Zusammenwirken der verschiedenen Entwicklungsbereiche entscheidet sich über die Zeit das Ausmaß der Differenzierungsfähigkeit und damit auch eine mögliche Annäherung an die Normalität.

Letztere ergibt sich stets allenfalls aus dem Verlauf heraus, je nach Zusammentreffen von Bedingungen, kann aber nicht primär als übergeordnetes Therapieziel vorgegeben werden. Therapiemaßnahmen im Kindesalter sind somit kein Training mit garantierter Erfolgsaussicht im Sinne der unbedingten Normalisierung, sondern reflektieren das Bemühen um eine Optimierung der entwicklungsbezogenen Rahmenbedingungen für das Kind, damit es **vorhandene Potenzialitäten** auch tatsächlich abrufen kann.

37.1.2 Humanökologisches Konzept

Kindliche Entwicklung ist Ausdruck eines komplexen Systems mit unterschiedlichen wechselseitigen Einflüssen und Beziehungen. Den zentralen Anteil dieses „Ökosystems" stellen die individuellen Voraussetzungen des Kindes dar, die die Grundlage seiner Entwicklung überhaupt bilden. Diese sind untrennbar mit den familiären und umfeldbezogenen Rahmenbedingungen verknüpft, welche die Rolle des Taktgebers für Tempo und Differenzierungsgrad von Fähigkeiten und Fertigkeiten übernehmen. Weitere Einflussvariablen wie Erkrankungen oder spezifische Entwicklungsbehandlung spielen eine modulierende Rolle.

Sobald bei Kindern eine Auffälligkeit in der Entwicklung auftritt, werden unter der Voraussetzung eines sozial intakten Umfeldes Maßnahmen zur Entwicklungsunterstützung eingeleitet. Diese sind üblicherweise mit den Bezugspersonen und damit der Lebensumwelt des Kindes verknüpft, sodass ein **„System der frühen Intervention"** " resultiert.

Die Notwendigkeit des Zusammenwirkens von medizinisch-funktionellen und heilpädagogischen Interventionen leitet sich aus Studiendaten ab. In einer Metaanalyse zur Effektivität früher Interventionen (Dunst 1989) wurde festgestellt, dass eine intuitive Befürwortungstendenz für Therapiemaßnahmen immer dann besonders deutlich vorhanden ist, wenn die Ausprägung einer Entwicklungsstörung schwer und deren Ursache somatisch begründet ist. Dies steht im Widerspruch zum Wissen darüber, dass die Effizienz der Maßnahmen hingegen dann gut ist, wenn nur ein leichter Ausprägungsgrad der Störung vorliegt und die Ursache im psychosozialen Bereich zu sehen ist. Eine weitere Metaanalyse aus demselben Jahr (Schlack 1989) belegte, dass eine deutliche Abhängigkeit der funktionellen Entwicklung des Kindes von der Mutter-Kind-Interaktion besteht. Diese Erkenntnisse werden durch die Daten im deutschen Kinder- und Jugendgesundheits-Survey (KiGGS) 2007 nachhaltig bestätigt. Auch diese Analysen zeigen eine wesentliche Abhängigkeit der kindlichen Entwicklung von fördernden und hemmenden Milieufaktoren (➤ Kap. 4.7).

Nach übereinstimmender Auffassung in der Entwicklungsforschung hat die Eigenaktivität des Kindes den entscheidenden Anteil an der Realisation seines Entwicklungspotenzials. Dieses **Konzept der genuinen Aktivität** bedeutet, dass durch die Eigenaktivität des Kindes entwicklungsabhängige Fähigkeiten genutzt und ausgestaltet werden. Übergeordnetes therapeutisches Ziel muss deshalb die Schaffung eines **„fördernden Milieus"** sein. Hierüber eröffnen sich Möglichkeiten der Kompensation von Entwicklungsstörungen, während Inaktivität deren Ausprägung verstärkt. Medizinisch-funktionelle Therapien ebenso wie heilpädagogische Förderung können die Eigenaktivität anregen, diese aber nicht ersetzen oder Entwicklungsfortschritte von außen und damit aus Sicht des Kindes passiv induzieren. Kinder mit geringen organischen Belastungsfaktoren und hohen psychosozialen Risiken profitieren deshalb am meisten und auf Dauer von Interventionen. Die Effekte von Therapie und Förderung sind umso deutlicher, je größer zuvor Vernachlässigung und Entwicklungsdeprivation gewesen sind. Bei organisch bedingten Entwicklungsstörungen hingegen sind die Grenzen des erreichbaren Fortschritts wesentlich durch den individuell bestimmten Entwicklungskorridor vorgegeben.

37.1.3 Wissenschaftliche Grundlagen: Therapieforschung

Kindliche Entwicklung ist ein dynamischer Prozess, gekennzeichnet durch individuell variable Verläufe und abhängig von Umgebungseinflüssen. In der Beurteilung spielt die klinische Beobachtung eine herausragende Rolle. Diese wird gestützt durch entwicklungsdiagnostische Maßnahmen nach dem **„Grenzstein-Prinzip“** (➤ Kap. 12.5, ➤ Kap. 17.2) in den ersten Lebensjahren und testdiagnostische Überprüfungen einzelner Funktionsbereiche sowie der Begabung oder schulischer Fertigkeiten im weiteren Verlauf. Es besteht eine große **interindividuelle Variabilität** in Bezug auf das zeitliche Erreichen bestimmter Differenzierungsgrade. Deshalb hat jede diagnostische Einschätzung diese normale Bandbreite der Entwicklung unbedingt zu berücksichtigen.
Neben dieser Variabilität der normalen, ungestörten Entwicklung besteht zusätzlich die Möglichkeit der **intraindividuellen Variation.** Fähigkeiten und Fertigkeiten desselben Individuums sind unterschiedlich ausgebildet im Sinne persönlicher Stärken und Schwächen. Ein zusammenfassender IQ- oder Standardwert bildet das Begabungsprofil nicht ab. Bei ausgeprägten Differenzen spricht man von **dissoziierter Intelligenz** oder **heterogenem Begabungsprofil.**
Es ist dementsprechend schwierig, die Wirksamkeit von funktionsorientierten Therapiemaßnahmen und allgemeinen Förderansätzen unter wissenschaftlichen Gesichtspunkten nachzuweisen. Auch bei der Betrachtung einer einzelnen Funktion, wie dies zu Studienzwecken erforderlich ist, spielen die Wechselwirkungen mit individuellem Entwicklungstempo, emotionaler Befindlichkeit des Kindes und fördernden oder hemmenden Rahmenbedingungen eine entscheidende Rolle. Außerdem muss unterschieden werden zwischen Trainingseffekten, die kurzfristig durchaus messbar sind, und einer nachhaltigen, somit über die Zeit andauernden Veränderung. Bedingt durch die Intervention kann es durchaus zu anderen als den primär intendierten funktionellen Effekten kommen, z. B. in Hinblick auf eine vom Kind erlebte Bereicherung durch das therapeutische Angebot mit dementsprechender Freude und daraus resultierender emotionaler Stabilisierung.
Es gibt aus allen Bereichen der kindlichen Entwicklung vielfältige Berichte über Therapieansätze und -methoden, die auf kurzfristige Effekte hinweisen. Diese Publikationen genügen jedoch ganz überwiegend nicht den unter wissenschaftlichen Gesichtspunkten zu fordernden methodischen und statistischen Anforderungen. Der sog. „Goldstandard“ ist der randomisierte, doppelblind und mit Kontrollgruppe durchgeführte Vergleich. Aufgrund der hier vorliegenden Fragestellungen sind jedoch allenfalls kleine Fallzahlen zu erreichen, und der vorstehend geschilderte Faktor der Entwicklungsvariabilität bleibt wirksam.
In der **Cochrane-Bibliothek zur evidenzbasierten Medizin** finden sich dementsprechend bisher nur Belege für eine Wirksamkeit der **logopädischen Behandlung** in Bezug auf Sprachentwicklungsstörungen bei der Artikulation und in geringerem, weniger gesichertem Ausmaß hinsichtlich der Wortschatzerweiterung. Kein konstant nachweisbarer Einfluss kann gesichert werden bezüglich Störungen in der grammatikalischen Konstruktion und beim Sprachverständnis. Ebenso sind bisher keine übergeordneten und die Nachhaltigkeit belegenden Studien zur **Physiotherapie** und **Ergotherapie** in der Cochrane-Bibliothek vorhanden.
Diese auf den ersten Blick sehr ernüchternde Datenlage beruht auch darauf, dass viele Studien an sehr heterogenen und relativ kleinen Patientengruppen durchgeführt wurden. Die Ziele der Therapeuten und die in den Studien verwendeten Messinstrumente korrespondierten oft wenig. Der Fokus bisheriger Studien lag unter ICF-Gesichtspunkten (➤ Kap. 16) zumeist auf der Ebene der Körperstrukturen und -funktionen und nicht auf der Ebene der Aktivitäten oder der Teilhabe der Patienten. Demnach ist keinesfalls der Rückschluss berechtigt, therapeutische Maßnahmen seien überflüssig und entbehrlich. Der Fokus der Untersuchungen hebt auf den eingangs dargestellten

Wunsch einer **Entwicklungsnormalisierung** ab. Diese Erwartung ist nach heutigem Verständnis jedoch nur bei reifungsbedingten Entwicklungsverzögerungen gerechtfertigt und auch realisierbar, nicht jedoch bei Entwicklungsstörungen im Ausmaß einer manifesten und persistierenden Einschränkung oder Behinderung in einem oder mehreren Funktionsbereichen bzw. bei der kognitiven Verarbeitung. Evaluationsansätze für die oben genannten Therapiemaßnahmen unter dem Gesichtspunkt einer Beeinflussung von Teilhabe und Lebensqualität des Kindes oder auch eines nachhaltig positiven Einflusses auf die familiären Interaktionen fehlen.
Andererseits haben sich in den letzten Jahren basierend auf der Grundlagenforschung insbesondere zur neuromotorischen Entwicklung zunehmend Ansätze in den Vordergrund geschoben, die nicht mehr eine globale und weitreichende Entwicklungsverbesserung als Therapieziel definieren, sondern sich kleinschrittig und auf Einzelfunktionen zentriert der Anbahnung von alltagsrelevanten Fortschritten widmen. Hierzu liegen inzwischen durchaus ermutigende Ergebnisse vor, z. B. im Rahmen von themenzentrierten Maßnahmen zur Funktionsverbesserung bei Hemiparese oder mit dem Vibrationstraining zur Stärkung der Muskel-Knochen-Einheit in Verbindung mit neuromotorischer (Re-)Habilitation.
Somit besteht eine sehr große Spannbreite im Hinblick auf die wissenschaftliche Absicherung von Therapieeffekten, die von der empirischen Einzelfalldarstellung bis hin zu gut belegten Einzelaspekten der entwicklungszentrierten Maßnahmen reicht. Es besteht weiterhin großer Forschungsbedarf, um belastbare Hinweise darauf zu erhalten, welche therapeutischen Maßnahmen und Interventionen mittel- und langfristig für Kinder und Jugendliche mit Entwicklungsstörungen geeignet sind und den hohen Zeitaufwand der Betroffenen und ihrer Familien für Interventionen rechtfertigen.

37.1.4 Medizinische und pädagogische Maßnahmen

Das System der frühen Förderung und Therapie ist in Deutschland sehr stark unter inhaltlich-ideologischen wie finanzierungstechnischen Gesichtspunkten in eine künstliche Dichotomie von Pädagogik und Medizin gedrängt worden. Somit sind medizinisch-funktionelle Behandlungsmaßnahmen und heilpädagogische Vorgehensweisen getrennt zu betrachten, beide jedoch für eine Entwicklungsunterstützung von Kindern in der Sozialpädiatrie unverzichtbar.
Im System des Gesundheitswesens werden die ärztlich zu verordnenden Therapiemaßnahmen unter dem Begriff **„Heilmittel“** zusammengefasst. Dabei dürfen nur solche Maßnahmen verordnet werden, die im offiziellen Heilmittelkatalog der gesetzlichen Krankenversicherungen aufgeführt sind. Einzelheiten des Vorgehens werden für alle Vertragsärzte verbindlich in den Heilmittelrichtlinien geregelt. Diese sehen ausdrücklich eine Nachrangigkeit von Leistungen, die durch die Krankenkassen finanziert werden, gegenüber heil- und sonderpädagogischen Interventionen vor.
In der täglichen Verordnungspraxis spielt jedoch die Erreichbarkeit von Maßnahmen und die Organisation ihrer Finanzierung unter rein pragmatischen Gesichtspunkten eine entscheidende Rolle. Während Heilmittel über eine Vielzahl von Praxen für Logopädie, Ergotherapie und Physiotherapie nahezu flächendeckend erreichbar sind, gilt dies für pädagogische Maßnahmen nicht. Auch die Finanzierungsbeantragung ist für heilpädagogische Maßnahmen ungleich aufwändiger im Vergleich zur formulargestützen ärztlich organisierten Heilmittelverordnung zu Lasten der Krankenversicherungen. **Heilpädagogische Maßnahmen** (➤ Kap. 38) können in nennenswertem Umfang nur institutionell gebunden erreicht werden. Dies ist in den über 800 Frühförderstellen in Deutschland sowie in der Mehrzahl der 145 Sozialpädiatrischen Zentren der Fall.

Das Ansinnen einer eindeutig geregelten Kostentrennung für die Maßnahmen im medizinischen bzw. heilpädagogischen Bereich hat zu dem Bemühen geführt, Aspekte der kindlichen Entwicklung in entsprechende Zuständigkeiten aufzuteilen. Für die Verordnung von Heilmitteln muss demzufolge eine medizinische Indikation durch die entsprechende ärztliche Diagnosestellung vorliegen. Auch in den neueren Vorgaben zu heilpädagogischen Maßnahmen im Rahmen der Interdisziplinären Frühförderung (IFF) wird analog die Ausweisung der Diagnose nach medizinischem Verständnis auf der Basis der ICD-10-Klassifikation verlangt. Es ist aber zu betonen, dass diese Unterteilung der Entwicklungsaspekte ausgesprochen künstlich ist und die untrennbare Einheit der kindlichen Entwicklung missachtet.
Allgemein anerkannte **Therapieprinzipien** lassen sich deshalb unter dem Gesichtspunkt der praktizierten Umsetzung dahingehend zusammenfassen, dass im ärztlichen Bereich ganz überwiegend funktionsorientierte Heilmittel mit Logopädie, Ergotherapie und Physiotherapie verordnet werden, in wesentlich geringerem Umfang in Hinblick auf Verhaltensauffälligkeiten psychotherapeutische Maßnahmen, während im parallel existierenden System der Frühförderung unter (heil-)pädagogischer Leitung vorrangig heilpädagogische Maßnahmen zum Einsatz kommen. Eine Vernetzung beider Bereiche findet systematisch in den Sozialpädiatrischen Zentren statt, zukünftig im Rahmen der seit 2001 gesetzlich definierten „Interdisziplinären Frühförderung" (§ 30 SGB IX) auch vermehrt dort. Die BAG SPZ erarbeitet in Qualitätszirkeln die Rahmenempfehlungen zu Sozialpädiatrischen Vorgehensweisen (Kap.41). In den folgenden Abschnitten wurden wesentliche Ausführungen aus den entsprechenden Qualitätspapieren übernommen.

37.1.5 Systematische sozialpädiatrische Beurteilung

Die Beurteilung von Fragen zur Entwicklung eines Kindes oder Jugendlichen sowie Abweichungen oder Störungen fordert die Integration aller Ebenen des ärztlichen Wissens. Aus der vollständigen Berücksichtigung der individuellen Gegebenheiten ergibt sich die sozialpädiatrische Beurteilung. Als Matrix für den systematischen Ansatz dient die Mehrdimensionale Bereichsdiagnostik und Behandlung in der Sozialpädiatrie MBS (➤ Kap. 15).

Systematische sozialpädiatrische Beurteilung bei Entwicklungsstörungen und Behinderungen

- Biographische Anamnese
- Spezifische Problemanamnese
- Mehrdimensionale Bereichsdiagnostik nach EKPSA-Schema
  Bereich **E**ntwicklung und Intelligenz
  Bereich **K**örperlich-neurologischer Befund
  Bereich **P**sychischer Befund und Verhalten
  Bereich **S**oziale Kontextfaktoren
  Bereich **A**bklärung der Ätiologie
- Ressourcenprofil
- Krankheitsspezifische prognostische Faktoren
- Behandlungsplan
- Evaluation
- Lebensgestaltung und Partizipation mit Kriterien der ICF-CY

Die biografische Anamnese vermittelt Eckpunkte der motorischen, sprachlichen, kognitiven, emotionalen, sozialen und verhaltensbezogenen Entwicklungsgeschichte des Kindes. In der Problemanamnese wird dies zielgerichtet ergänzt um die spezifischen Aspekte der jeweils zur Vorstellung führenden Auffälligkeiten. Die umfassende standardisierte Untersuchung und Verhaltens- und Interaktionsbeschreibung, die so weit wie möglich durch Beobachtung, mindestens aber entsprechende Fragebogen-Verfahren und situationsübergreifend erhoben wird, führt zur Ausweisung der Diagnosen nach dem EKPSA-Schema in der MBS (➤ Kap. 15). Je nach Befund sind dabei auch Maßnahmen zur Abklärung der Ursache einzuleiten, sofern nicht von einer unspezifischen individuellen Disposition auszugehen ist.

Die Erhebung des Ressourcenprofils muss parallel zur Diagnostik nach EKPSA-Schema nicht nur die vorhandenen Begabungen, Stärken und Möglichkeiten des Kindes berücksichtigen, sondern ebenso relevante Faktoren von Eltern, Familie und anderen, für das Kind wesentlichen Umgebungsfaktoren. Ebenfalls parallel muss das ärztliche Wissen um den krankheits- bzw. störungsspezifischen Verlauf die Einschätzung der Prognose bestimmen.

Der Behandlungsplan (➤ Kap. 37.1.7) führt die sich ableitenden Erfordernisse für Interventionen auf der Basis der Diagnostik nach EKPSA-Schema mit den Ergebnissen der Ressourcenanalyse und prognostischen Einschätzung zusammen. Der Entwicklungsverlauf unter dem Einfluss der eingeleiteten Maßnahmen muss regelmäßig überprüft werden; aus dieser Evaluation ergibt sich das Handlungsraster für das weitere Vorgehen (➤ Kap. 37.2.1).

Bei Kindern und Jugendlichen, die gravierende und persistierende Entwicklungsstörungen im Sinne einer Behinderung aufweisen, muss der Fokus aller Interventionen neben einer größtmöglichen Realisation der funktionellen Möglichkeiten auf die Ermöglichung von Teilhabe ausgerichtet sein. Hieraus ergibt sich als ärztliche und sozialpädiatrische Aufgabe die Mitwirkung bei der Erarbeitung einer Perspektive zur Lebensgestaltung insgesamt. Dabei ist die Anwendung von Kriterien der ICF-CY hilfreich, insbesondere wenn bei Behinderungen der Einsatz standardisierter Testverfahren limitiert ist (➤ Kap. 16, ➤ Kap. 17).

37.1.6 Ressourcenprofil

Jede Intervention bei Entwicklungsstörungen richtet sich zunächst auf das zugrunde liegende Defizit oder Problem. Im Anfangsstadium der Behandlung ist oftmals noch nicht absehbar, welches Ausmaß an erwünschter und positiver Veränderung zu erreichen sein wird. Dies hängt ab von der Ausprägung und den Ursachen. Bei organisch bedingter Abweichung der Entwicklung werden Grenzen erreicht, die auch durch ein Maximum an Training und Förderung nicht zu überwinden sind. Deshalb muss sich mittel- und langfristig die Betreuung mehr auf die vorhandenen individuellen Stärken und Fertigkeiten ausrichten.

Die Ressourcen spielen nicht nur in Bezug auf den Patienten selbst eine zentrale Rolle. Ebenso bedeutsam sind sie bei den Eltern und in der Familie, aber auch für die Person der behandelnden Therapeuten wie des Versorgungssystems insgesamt. Sie entscheiden letztlich über das zu erreichende Funktionsniveau und die Lebensqualität.

Die systematische Erhebung und Beachtung der Ressourcen ist deshalb ein eminent wichtiger Baustein des sozialpädiatrischen Beurteilungssystems.

Ressourcenprofile

Abhängigkeit beim Patienten von:
- Krankheitsbild
- Entwicklungsstand
- Intelligenz
- Verhalten
- Selbstständigkeit

Abhängigkeit in der Familie von:
- Familiärer Situation
- Lebenssituation
- Einbindung in Familien- und Bekanntenkreis

Abhängigkeit der Therapeutin von:
- Persönlichkeit
- Qualifikation: Ausbildung und Wissensstand
- Erfahrung
- Einbezug in das Gesamtkonzept der Behandlung

Möglichkeiten der Versorgungssysteme:
- System Gesundheit:
  - Kinder- und Jugend-Früherkennungsuntersuchungen
  - Öffentlicher Kinder- und Jugend-Gesundheitsdienst
  - Funktionell orientierte Therapien (Heilmittel)
  - Hilfsmittel: Rehatechnik, Orthopädietechnik, Seh- und Hörhilfen, Kommunikation
  - SPZ
  - Kinder- und Jugendpsychiatrie
  - Rehabilitation
- System Pädagogik:
  - Frühpädagogik: Frühförderstellen, Interdisziplinäre Frühförderung
  - Heilpädagogik: Montessori, Petö, Pikler
  - Sozialpädagogik
  - Sonderpädagogik
- System Jugendhilfe
  - „Frühe Hilfen"
  - Hilfen zur Erziehung
  - Tagesgruppen
  - Heilpädagogisch-therapeutische Heime
- Sozialrechtliche Ansprüche
  - Schwerbehinderten-Ausweis
  - Pflege-Versicherung
  - Verhinderungspflege

37.1.7 Behandlungsplan

Entwicklungsstörungen unterscheiden sich in Ausprägungsgrad und Dauer der Beeinträchtigung. Das Kind mit Zerebralparese, Trisomie 21 oder Autismus-Spektrum-Störung benötigt andere Maßnahmen als dasjenige mit Artikulationsstörung oder einer Umschriebenen Entwicklungsstörung der Motorik. Aufgabe des niedergelassenen Kinder- und Jugendarztes ist es, im Rahmen der in der Praxis möglichen Basisdiagnostik das Entwicklungsproblem zu erfassen und in der Dimension einzuschätzen. Danach richtet sich das weitere Vorgehen.

Das **Handlungsfeld des niedergelassenen Kinder- und Jugendarztes** zur Intervention bei Entwicklungsstörungen umfasst im Dialog mit den Eltern die 3 Bausteine:

1. Beratung
2. Gestaltung von Förderung und
3. Einleitung von Therapiemaßnahmen.

Es ist ärztliche Verantwortung, die Entwicklung des Kindes mit der Mehrdimensionalen Bereichsdiagnostik in der Sozialpädiatrie (MBS) zu beschreiben, soziale Rahmenbedingungen und Einflussfaktoren zu berücksichtigen und einen vollständigen Behandlungsplan zu erstellen.

Dazu muss der niedergelassene Kinder- und Jugendarzt alle Möglichkeiten kennen, die ihm zur Unterstützung eines Kindes und dessen Familie bei der Bewältigung von Entwicklungsproblemen zur Verfügung stehen. Voraussetzung ist ferner das Wissen um spezifische Verläufe in Abhängigkeit vom jeweiligen Krankheits- und Störungsbild.

Behandlungsplan bei Entwicklungsstörungen und Behinderungen

- Beratung: Information und Anleitung
- Förderung bei Entwicklungsstörungen:
  allgemeine Grundsätze zur Erreichung von Entwicklungsfortschritt
- Therapie: fokussierte und befristete Intervention
- Förderung bei Behinderungen: spezielle Maßnahmen zur Erreichung von
  - Lebensqualität
  - Teilhabe
  - Selbstständigkeit
  - Autonomie

Der Behandlungsplan umfasst deshalb nicht nur die aktuellen Therapiemaßnahmen, sondern beinhaltet auch notwendige Unterstützungsmaßnahmen für die Familie in Sozialrecht und Jugendhilfe. Perspektivplanung hinsichtlich Kindergarten und Schule gehören ebenfalls dazu. Hieraus leitet sich der kontinuierliche Austausch mit den Eltern im ärztlichen Beratungsgespräch ab. Dabei sind Gesichtspunkte der Problembearbeitung und Akzeptanz bei Behinderung ebenso Inhalt wie die antizipierende Beratung und positive Unterstützung zur gelingenden Bewältigung.

Der Behandlungsplan ist somit das Kernstück in der Therapiegestaltung insgesamt.

Ärztliche Entscheidungsrationale bei Entwicklungsstörungen und Behinderungen

- Anamnese: Verlauf und Parameter der Entwicklung
- Untersuchung und Diagnose: präzise Einschätzung des Kindes hinsichtlich Funktionen – Kognition – Verhalten und Persönlichkeit
- Beratung:
  - Antizipation der Entwicklung
  - vorausschauende (Gesundheits-)Beratung
- Behandlung: Überlegung und Abwägung der Interventionsmöglichkeiten;
  - Eigene ärztliche Beratung und Anleitung
  - Delegation

- Effekt und Nutzen beim Kind –
  Belastung der Eltern und Familie (Geschwister!!)
- Einsatz der Maßnahmen: Haus-Frühförderung, Heilpädagogik; Kindergarten, Schule; Funktionstherapie (Heilmittel); Psychotherapie (auch: Eltern-Kind!); Hilfen zur Erziehung (Jugendamt)
- Evaluation

Literatur

Bergmann KE, Thefeld W, Kurth BM (2002) Der Kinder- und Jugendgesundheitssurvey – eine Grundlage für Prävention, Gesundheitsförderung und Gesundheitsziele. Gesundheitswesen 64 Sonderheft 1: S53–S58.

Chess S, Thomas A (1984) Origins and evolution of behavior disorders. New York: Bruner and Mazel.

Dunst CJ, Snyder SW, Mankinen (1989) Efficacy of Early Intervention. In: Wang MC, Reynolds MC, Walberg HJ (eds) Handbook of Special Education: Research and Practice. Vol. 3. Low Incident Conditions. New York: Pergammon.

Largo RH (1999) Kinderjahre. München: Pieper.

Largo RH, Jenni OG (2005) Das Zürcher Fit-Konzept. Familiendynamik 30 (2) 111–127.

Laucht M, Schmidt M, SHG (2004) The development of at-risk children in early life. Educational and Child Psychology 21: 20–31.

Schlack HG (Hrsg) (1998) Welche Behandlung nützt behinderten Kindern? Mainz: Kirchheim.

Schlack HG (Hrsg) (2000) Sozialpädiatrie – Gesundheit, Krankheit, Lebenswelten. 2. Aufl. München: Urban & Fischer.

Schlack HG, Largo RH, Michaelis R, Neuhäuser G, Ohrt B (1994) Praktische Entwicklungsneurologie. München: Hans Marseille.

Spittle AJ, Orton J, Doyle LW, Boyd R (2007) Early developmental intervention programs post hospital discharge to prevent motor and cognitive impairments in preterm infants. Cochrane Database of Systematic Reviews

Suchodoletz Wv (Hrsg.) (2010) Therapie bei Entwicklungsstörungen. Was hilft wirklich? Hogrefe Göttingen

37.2 Heilmittel

37.2.1 Verordnung von Heilmitteln

Heilmittelrichtlinie

Im System des Gesundheitswesens werden die ärztlich zu verordnenden Therapiemaßnahmen unter dem Begriff „Heilmittel" zusammengefasst. Der Anspruch gesetzlich Versicherter ist in SGB V § 32 beschrieben. Dabei dürfen nur solche Maßnahmen verschrieben werden, die den Heilmittelrichtlinien des Gemeinsamen Bundesausschusses (GBA) entsprechen. Regionale Besonderheiten, wie z. B. die Budgetierung, werden für die Vertragsärzte verbindlich durch die jeweiligen Kassenärztlichen Vereinigungen vorgegeben.

Die Heilmittelrichtlinien führen aus: „Heilmittel können zu Lasten der Krankenkassen nur verordnet werden, wenn sie notwendig sind, um

- eine Krankheit zu heilen, ihre Verschlimmerung zu verhüten oder Krankheitsbeschwerden zu lindern,
- eine Schwächung der Gesundheit, die in absehbarer Zeit voraussichtlich zu einer Krankheit führen würde, zu beseitigen,

- einer Gefährdung der gesundheitlichen Entwicklung eines Kindes entgegenzuwirken oder
- eine Pflegebedürftigkeit zu vermeiden oder zu mindern.

Heilmittel dürfen bei Kindern nicht verordnet werden, wenn an sich störungsbildspezifische heilpädagogische/sonderpädagogische Maßnahmen zur Beeinflussung von Schädigungen geboten sind. Sind heilpädagogische/sonderpädagogische Maßnahmen nicht durchführbar, dürfen Heilmittel nicht an deren Stelle verordnet werden. Neben heilpädagogischen/sonderpädagogischen Maßnahmen dürfen Heilmittel nur bei entsprechender medizinischer Indikation außerhalb dieser heilpädagogischen/sonder-pädagogischen Maßnahmen verordnet werden."

Obwohl diese Vorgaben sehr restriktiv klingen, geben die Heilmittelrichtlinien gerade für den pädiatrischen Bereich in den Ausführungen zu Logopädie, Physiotherapie und insbesondere Ergotherapie viele Gestaltungsmöglichkeiten. Auch ist darauf hinzuweisen, dass im indizierten und durch den Befund begründeten Einzelfall jederzeit Verordnungen zulässig sind, die das Regelvolumen überschreiten.

Für die Sozialpädiatrischen Zentren sind die formalen Vorgaben der Heilmittelrichtlinien identisch. Durch die unmittelbare Einbettung in das interdisziplinäre Behandlungskonzept entstehen jedoch zusätzliche Optionen der Ausführung und Therapiegestaltung. Ferner besteht die Möglichkeit zum direkten Austausch mit anderen Professionen im multiprofessionellen Team wie Heilpädagogik, Psychomotorik bzw. Motopädie, Musiktherapie und andere.

Indikation und Zieldefinition in der Therapie

Die Heilmittelrichtlinien geben den formalen Rahmen für die Verordnungen des niedergelassenen Vertragsarztes vor. Für die Sozialpädiatrischen Zentren hat die Bundesarbeitsgemeinschaft der SPZ bereits 2002 das „Rahmenkonzept des therapeutischen Vorgehens" als Teil des „Altöttinger Papiers zur Strukturqualität in den SPZ" definiert, das von der Deutschen Gesellschaft für Sozialpädiatrie und Jugendmedizin publiziert worden ist. Hierbei wurden alle therapeutischen Maßnahmen berücksichtigt, Handlungsanweisungen für die Praxis im SPZ gegeben und ein Standard für die Arbeit der SPZ in Deutschland erreicht.

Als übergeordnete Ziele jeder Therapie sind im „Altöttinger Papier" das Erreichen der sozialen Teilhabe und die bestmögliche Bewältigung der Alltagssituation für den Patienten definiert sowie die Stärkung der elterlichen Kompetenz. Darin eingeschlossen sind die Verbesserung der Bewegungskompetenz, Handlungs- und Wahrnehmungskompetenz, Kommunikationskompetenz sowie der sozialen und emotionalen Kompetenz.

Diese Zielvorgaben gelten uneingeschränkt auch hinsichtlich der Anwendung von Heilmitteln außerhalb der SPZ. Die Indikationsstellung leitet sich aus der ärztlichen Abwägung des individuellen Befundes des Kindes unter Berücksichtigung seiner Lebenssituation ab. Alle Ziele bei der Verordnung von Heilmitteln und der Durchführung von Therapien sind grundsätzlich an den Bedürfnissen des Kindes zu orientieren. Dabei müssen die **Zielsetzungen** folgende Eigenschaften erfüllen:

- abgesprochen mit den Eltern und wenn möglich mit dem Kind
- konkret
- realistisch
- alltagspraktisch
- erreichbar,
- überprüfbar.

Die Unterscheidung zwischen **Nahzielen** und einem übergeordneten **Fernziel** ist sinnvoll, um die therapeutische Ausrichtung für alle Beteiligten (Kind, Eltern, Ärzte, Therapeuten und Pädagogen) jederzeit nachvollziehbar zu machen. Dabei ist streng auf die realistische Formulierung, auch des Fernziels, zu achten.

Die Akzeptanz bestehender und nicht überwindbarer Einschränkungen der Funktionen oder Fertigkeiten des Kindes muss dazu führen, kritisch den potenziellen mittelbaren Nutzen einer langfristig fortgesetzten Therapie abzuwägen, auch mit den Eltern und wenn möglich dem Kind.
Physiotherapie, Logopädie und Ergotherapie sind medizinische Verfahren zur Therapie der kindlichen Entwicklung, wenn Entwicklungsrückstände oder -störungen vorliegen. Eine Quantifizierung ist mit entsprechenden testdiagnostischen Verfahren vorzunehmen. Wenn die Teilhabe alltagsrelevant beeinträchtigt ist, können auch Patienten behandelt werden, deren Testdiagnostikergebnisse noch innerhalb des Grenzbereichs dieser Verfahren liegen. Ein isolierter Störungsbefund in der Testdiagnostik allein begründet umgekehrt nicht automatisch eine Indikation für funktionelle Therapiemaßnahmen. Zukünftig wird es evtl. besser möglich sein, auf der Basis der ICF-CY eine Entscheidung in Grenzbereichen herbeizuführen. Voraussetzung hierfür wäre aber eine im Alltag praktikable und mit vertretbarem Zeitaufwand umzusetzende Einschätzung mit den ICF-Kriterien, was derzeit noch nicht der Fall ist.
Die Notwendigkeit des Einsatzes der Heilmittel ist im Rahmen der Diagnostik durch den niedergelassenen Vertragsarzt hinreichend zu belegen. Neben der Funktionsdiagnostik müssen ergänzend berücksichtigt werden:

- Komorbiditäten beim Kind
- Aspekte der Unterstützung bei umschriebenen Entwicklungsstörungen und ansonsten unbeeinträchtigtem Entwicklungsverlauf
- Kognitive Entwicklung des Kindes
- Einordnung von Funktionseinschränkungen in ein übergeordnetes Krankheitsbild
- Ressourcen des Kindes und der Familie.

Optimal ist dies bei systematischem Vorgehen im Rahmen der Mehrdimensionalen Bereichsdiagnostik in der Sozialpädiatrie (MBS) zu realisieren. Insbesondere bei schwer behinderten Kindern ist der beste Zugangsweg in der Diagnostik die gemeinsame ärztlich-therapeutische Befunderhebung mit Festlegung eines Stufenplans durch den Arzt, auch in Bezug auf den Einsatz der verschiedenen Heilmittel, wie es im SPZ möglich ist.
Bei der Verordnung durch den niedergelassenen Vertragsarzt ist das Heilmittel zu wählen, das die beste Zielführung und größte Effektivität besitzt. Der Arzt gibt die primäre Arbeitsrichtung durch die Auswahl des Heilmittels vor und muss ebenfalls das damit verbundene übergeordnete Behandlungsziel festlegen. Basis hierfür sind die Verordnungsziffern aus dem Heilmittelkatalog. Da diese nicht genügend eindeutig in der Anweisung für die behandelnden Therapeuten sind, muss das angestrebte Behandlungsnahziel vom Arzt ausdrücklich ausformuliert werden. Beispiele hierzu auf Grundlage der Heilmittel-Richtlinien sind unter anderem von der Kassenärztlichen Vereinigung Berlin angegeben unter: http://www.kvberlin.de/20praxis/50verordnung/20heilmittel/verordnungsbeispiele.pdf.
Die Entscheidung über den zeitlichen Umfang wird ebenfalls auf der Grundlage der Heilmittelrichtlinien getroffen.
Der **behandelnde Therapeut** muss in den ersten Behandlungseinheiten die vom Arzt vorgegebene Zielsetzung der Verordnung im Rahmen einer Kurzanamnese und einer gezielten Befunderhebung ergänzend überprüfen. Hierzu zählt auch die Selbstreflexion hinsichtlich der eigenen fachlichen Kompetenz, die zur Durchführung der jeweiligen Behandlung erforderlich ist. Danach ist der Abgleich mit den Erwartungen der Eltern zur Feinjustierung des Behandlungsziels herbeizuführen. Anschließend bestehen für den Therapeuten zwei Möglichkeiten:

- Bestätigung von Befund und Therapieziel, Umsetzung der Heilmittelverordnung wie vom Arzt angesetzt
- Abweichender Befund und/oder verändertes Therapieziel; deshalb direkte Rückkopplung mit dem verordnenden Arzt zwecks Änderung von Zielbereich oder Heilmittel insgesamt.

Muster-Beispiel der KBV für Verordnung Ergotherapie

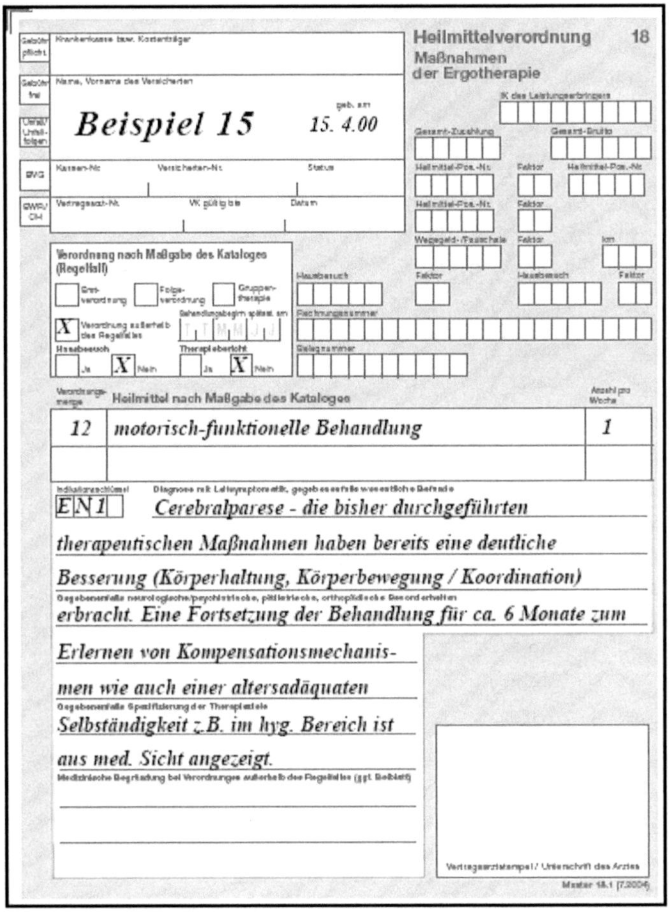

Heilmittelverordnung 18
Maßnahmen der Ergotherapie

Beispiel 15 geb. am 15. 4.00

Verordnung nach Maßgabe des Kataloges (Regelfall)

Heilmittel nach Maßgabe des Kataloges

Verordnungsmenge	Heilmittel	Anzahl pro Woche
12	motorisch-funktionelle Behandlung	1

Indikationsschlüssel: EN1

Diagnose mit Leitsymptomatik, gegebenenfalls wesentliche Befunde: Cerebralparese - die bisher durchgeführten therapeutischen Maßnahmen haben bereits eine deutliche Besserung (Körperhaltung, Körperbewegung / Koordination) erbracht. Eine Fortsetzung der Behandlung für ca. 6 Monate zum Erlernen von Kompensationsmechanismen wie auch einer altersadäquaten Selbständigkeit z.B. im hyg. Bereich ist aus med. Sicht angezeigt.

Medizinische Begründung bei Verordnungen außerhalb des Regelfalles (ggf. Beiblatt)

Vertragsarztstempel / Unterschrift des Arztes

Muster 18.1 (7.2004)

Abb. 37.1 Muster-Beispiel der KBV für Verordnung Ergotherapie

Bei Übernahme der Behandlung konkretisiert der Therapeut auf der Basis der ersten Behandlungsstunden das individuelle Therapieziel mit den Eltern und nach Möglichkeit auch mit dem Kind. Zur Umsetzung im Alltag und zur Anleitung in der häuslichen Entwicklungsunterstützung werden die Eltern und andere Bezugpersonen in die Behandlung einbezogen. Zentrales Anliegen ist die Zielerfüllung, nicht die Korrektur von Symptomen. Dabei hat der Therapeut Gestaltungsfreiheit innerhalb des Verordnungsrahmens und gleichzeitig Verantwortung für die Beachtung der Vorgaben.

Nach Ablauf des verordneten Behandlungszeitraums wird der Abschlussbefund des Therapeuten im Therapiebericht dokumentiert. Dieser kann durchaus auch den Hinweis auf die sofortige oder spätere Fortsetzung der Therapie oder Wahrscheinlichkeit der Notwendigkeit einer erneuten Überprüfung beinhalten.

Das Therapieende wird durch den verordnenden Arzt festgelegt.

Therapiemodalitäten/Therapieintensität

Zu **Beginn der Therapie** steht die Aufklärung der Eltern und ggf. des Kindes durch den Arzt über Ziel, Umfang und Dauer. Ein wesentlicher Aspekt ist außerdem, realistische Möglichkeiten und Grenzen der Therapie zu vermitteln. Weiterhin ist es notwendig, Bezugspersonen zur **Mitarbeit im therapeutischen Prozess** zu gewinnen und gemeinsam Ansätze zur Integrierung der Therapie in den Alltag zu finden. Dies ist wesentliche Aufgabe des behandelnden Therapeuten. Hierbei ist darauf zu achten, dass Eltern nur in begrenztem Maß eine Rolle als Co-Therapeuten einnehmen. Vielmehr sollen sie als Moderatoren angeleitet werden, was die Fokussierung auf Schwächen und Defizite reduziert, optimal vermeidet. Eine geeignete Begleitung der Familie muss deshalb als Teil der therapeutischen Tätigkeiten angeboten werden. Dabei ist es wichtig, ein Verständnis für den Entwicklungsstand und das Verhalten des Kindes/Jugendlichen zu vermitteln. Effekt und Nachhaltigkeit von (funktions-)therapeutischen Maßnahmen sind nur zu erreichen, wenn ein Transfer in den Alltag systematisch erarbeitet wird. Deshalb müssen Eltern einbezogen werden, bei Therapiedurchführung in Institutionen wie Kindergarten oder Schule stellvertretend die dortigen Bezugspersonen. Die einzelne Therapiestunde für sich kann keine Veränderung in der Entwicklung des Kindes bewirken. Dementsprechend ist es ein therapeutischer Kunstfehler, die Mitbeteiligung der Bezugspersonen nicht zu gewährleisten.

Der **therapeutische Prozess** erfordert eine regelmäßige Kontrolle der im Therapieplan festgelegten Zielsetzung sowie der therapeutischen Interaktion. Hierzu ist eine entsprechende Dokumentation unerlässlich. Aus den erhobenen Befunden ergeben sich Ansätze für die Fortführung, Modifikation oder Begrenzung der Therapie. Gegebenenfalls muss auch die Diagnose überprüft bzw. die Diagnostik erweitert werden. Dies gilt auch bzw. in besonderem Maß für Patienten mit Mehrfachbehinderungen. Eine Fortführung von Therapien allein aufgrund des Vorliegens einer Behinderung oder aufgrund von Defiziten ist nicht sinnvoll.

Die **Therapiemodalitäten** müssen dem Störungsbild, Alter und Entwicklungsstand angepasst werden. Zur Steuerung dieses therapeutischen Prozesses sind **interdisziplinäre Fallbesprechungen** unerlässlich. Für den Austausch in der Praxis müssen hierzu grundlegende Möglichkeiten geschaffen werden, am besten über telefonischen Kurzkontakt, mindestens aber durch einen regelmäßigen Therapiebericht mit striktem Bezug auf die Erreichung des vorgegebenen alltagsrelevanten Therapieziels. Im SPZ ist die interdisziplinäre Fallbesprechung bereits vom Grundkonzept her realisiert.

Die **Therapieintensität** ist individuell abhängig u. a. von:

- Stadium und Komplexität der Krankheit, Entwicklungsstörung oder Behinderung
- Sensiblen Phasen der Entwicklung
- Bewältigungsprozessen
- Belastbarkeit (u. a. Erholung, Spiel, räumliche Entfernung)
- Zeitressourcen (Ganztagseinrichtung, berufliche Belastung der Eltern).

Von diesen Überlegungen wird das **therapeutische Vorgehen (Therapie-Setting)** bestimmt, wie z. B.:

- Einzeltherapie
- Gruppentherapie
- Kombinationstherapie
- Phase intensivierter Therapie
- Intervalltherapie
- Therapie- bzw. Therapeutenwechsel
- Therapie mittels Anleitung einer Bezugsperson.

Auf der Basis der Mehrdimensionalen Bereichsdiagnostik in der Sozialpädiatrie (MBS) ist zu entscheiden, ob der Einsatz von
- nur einem Heilmittel,
- mehreren Heilmitteln oder
- eine multimodale Behandlung/Komplextherapie

erforderlich ist. In aller Regel wird einer Abfolge von einzelnen Heilmitteln der Vorzug vor einem simultanen Einsatz mehrerer Heilmittel gegeben, um eine Überlastung des Kindes zu vermeiden. Bei der Komplextherapie kommt v. a. der interdisziplinäre Behandlungsansatz eines multiprofessionellen Teams in Betracht.

Die **Therapiedauer** ist unter Berücksichtigung der Vorgaben der Heilmittelrichtlinien ausschließlich abhängig vom Therapieziel. Dieses ist festzulegen unter Berücksichtigung von:
- Art und Schweregrad der Störung
- Motivation
- Prognose
- Therapieeffekten
- Therapieform.

Therapiepausen sind sinnvoller Bestandteil des Behandlungskonzepts. Unter Verlaufskontrollen dienen sie der:
- Beobachtung der spontanen Weiterentwicklung des Patienten
- Beurteilung des Transfers therapeutischer Inhalte in den Alltag
- Überprüfung der weiteren Therapie-Notwendigkeit
- zeitlichen Entlastung der Familie.

Die Therapie soll dann wieder aufgenommen werden, wenn sie zur Erhaltung bzw. Verbesserung des bisherigen Therapieergebnisses, zur Krankheitsprävention bzw. zur sozialen Integration erforderlich ist und somit zur Verbesserung der Lebensqualität beiträgt.

Aufgabe des Therapeuten bei Übungsbehandlungen ist auch, in Fällen von Behinderung, die nicht überwunden werden kann, die Akzeptanz der Behinderung bei dem Kind, der Familie und dem sozialen Umfeld zu fördern und damit eine Flucht in Therapie zu vermeiden.

Das **Therapieende** ist gegeben bei:
- Erreichen des Therapieziels
- Angemessener Eigenkompetenz
- Mangelnder Therapiemotivation bei Patient und/oder Eltern
- Nicht messbarem Therapieerfolg bzw. bei fehlendem Effekt auf die Teilhabe.

Besonderheiten müssen bei degenerativen Erkrankungen und in der Palliativmedizin berücksichtigt werden.

Anforderungen an funktionell ausgerichtete Therapiemaßnahmen
- Transparenz
- Zusammenarbeit Eltern-Kinderarzt-Therapeut (gemeinsame Sprache, gemeinsame Ziele)
- Ausrichtung der Therapie auf Alltags- und Betätigungsorientierung (Physio- und Ergotherapie) bzw. auf Kommunikation (Logopädie)
- Tragfähige Beziehung zwischen Kind, Therapeuten und Eltern
- Einbeziehung der Eltern in die Therapie mit Stärkung der Elternkompetenz (Empowerment)
- Generalisierung und Transfer in den Alltag
- Sinnvolle Therapieintervalle
- Evaluation.

Wechselwirkungen, unerwünschte Nebenwirkungen und Komplikationen von Therapien

Jede Behandlung birgt die Möglichkeit unerwünschter Wirkungen. Dies gilt auch für die Physiotherapie, Logopädie und Ergotherapie. Es ist prioritäre Aufgabe des Therapeuten, auf entsprechende Anzeichen sorgfältig zu achten und dabei sowohl das Kind selbst als Patient als auch die Eltern im Blick zu haben.

Hinweise seitens des Kindes auf unerwünschte Wirkungen im therapeutischen Prozess sind:

- Verweigerung,
- Rückzugsverhalten
- Interaktionsstörungen.

Daneben müssen weitere Faktoren beachtet werden, insbesondere:

- Zeitliche Überlastung des Kindes
- Fortschreibung von Störungsmustern
- Vernachlässigung differenzialdiagnostischer Schritte
- unbeabsichtigte Förderung einer mangelnden Akzeptanz für die Behinderung des Kindes
- Verschwendung von emotionalen, interaktiven, sozialen, zeitlichen und finanziellen Ressourcen der Familie
- Einschränkung der eigenen Beziehungs- und Handlungskompetenz des Patienten bei Übertherapie
- Psychische Abhängigkeit des Patienten/Angehörigen von den Therapeuten.

Unerwünschte Wirkungen beim Kind müssen die Fortsetzung einer Therapiemaßnahme grundsätzlich infrage stellen. Ansonsten sind Therapiemaßnahmen ggf. unter Risiko-Nutzen-Abwägung zu modifizieren.

Mögliche unerwünschte Effekte von funktionell ausgerichteter Therapie

- Beim Kind:
  - Induktion von Störungsbewusstsein
  - Minderung des Selbstwertgefühls
  - Sekundäre Verhaltensauffälligkeiten
  - Hinderung an freier Beschäftigung
- Bei der Therapeutin:
  - Überforderung
  - Unzufriedenheit
  - Ablehnung von Kind und Eltern
- Bei den Eltern:
  - Verunsicherung
  - Negativer Einfluss auf die Interaktion
  - Belastung, Zeitverlust
  - Verminderte Beachtung und Zuwendung für Geschwister.

Evaluation

Die Überprüfung und Bewertung des Therapieprozesses ist zwingende Aufgabe des behandelnden Therapeuten und des verordnenden Arztes. Letzterer trifft die Entscheidung über Fortsetzung, Therapiepause oder Beendigung und trägt die Gesamtverantwortung (➤ Kap. 37.1.7).

Bei der Evaluation sind die Selbstevaluation des Patienten bzw. seiner Bezugspersonen und die Fremdevaluation des Therapeuten zu berücksichtigen. Die zuvor vereinbarten Inhalte und Therapieziele sind die Grundlage für die Beurteilung. Unter der Voraussetzung einer alltagsrelevanten Planung mit Berücksichtigung der Gegebenheiten bei Kind und Familie müssen Veränderungen in der intendierten Form erkennbar sein. Ein Zeitraum von mindestens 3 Monaten ist zu veranschlagen, um Effekte beurteilen zu können. Dies entspricht etwa 10 Behandlungseinheiten.
Die verbesserte Adaptation an Alltagsanforderungen in Betätigung bzw. Kommunikation ist die Leitlinie insgesamt. Deshalb spielen Aspekte der Zufriedenheit von Kind, Eltern und anderen Bezugspersonen mit der eingetretenen Veränderung eine wesentliche Rolle bei der Evaluation. Hierfür stehen zahlreiche Fragebogeninstrumente zur Verfügung, die die therapeutischen Verlaufsberichte ergänzen. Zukünftig werden dabei Einschätzungen nach der ICF-CY (Internationale Klassifikation der Funktionsfähigkeit, Behinderung und Gesundheit) zunehmend eine Rolle spielen. Die ICF ist aber kein Untersuchungsverfahren, sondern eher Bezugsgröße für Befunde zur umfassenden Einschätzung der sozialen Teilhabe.
Der für die Verordnung der funktionellen Therapie verantwortliche Arzt muss regelmäßig Verlaufsuntersuchungen durchführen. Diese sollen in drei- bis sechsmonatigen Abständen erfolgen und wo immer möglich die Beurteilung des Effektes anhand standardisierter Testverfahren umfassen. Dabei ist nach Möglichkeit auch eine ergänzende Außenevaluation durchzuführen. Hierfür kommen Erzieherinnen im Kindergarten, Lehrerinnen, Betreuer bei Freizeitaktivitäten, Familienhelferin oder Mitarbeiter des Jugendamtes in Betracht. Bei Problemen in der Beurteilung der Therapieeffekte und damit Schwierigkeiten für die weitere Erstellung des Behandlungsplans ist die Überweisung an ein SPZ zur Durchführung einer mehrdimensionalen Bereichsdiagnostik indiziert.

Evaluation von funktionell ausgerichteten Therapiemaßnahmen

- Wie geht es dem Kind?
  (Verhalten, Stimmung, Affekt)
- Zeigt sich ein Effekt der Maßnahme auf die Entwicklung?
  (Erreichungsgrad Therapieziel)
- Ist das Betätigungsproblem/die Verhaltensstörung noch gestaltungs- bzw. behandlungsbedürftig?
- Ist die gewählte Maßnahme ausreichend und geeignet?
- Was ist die Position der Bezugspersonen?
  (Erwartung, Zufriedenheit, Sorge)
- Welche Alternativen bestehen?

Alternativen zum Einsatz von Heilmitteln

Andere, nicht im Heilmittelkatalog erfasste Maßnahmen müssen hinsichtlich Sinn und Effektivität vom verordnenden Arzt mit bedacht und den Eltern zur Anwendung vorgeschlagen werden. Dies ist in den Heilmittelrichtlinien verbindlich festgelegt. Dabei sind die Modalitäten bei der Beantragung, Umsetzung wie ggf. auch einer finanziellen Beteiligung für die Eltern zu beachten. Für Maßnahmen der Heilpädagogik (➤ Kap. 38) bestehen Ansprüche auf der Basis von §§ 55/56 SGB IX, die aber von den Eltern selbst geltend gemacht werden müssen. Auch Maßnahmen der Jugendhilfe müssen je nach Problemlage als Alternative in Betracht gezogen werden (➤ Kap. 12).
Andere Ansätze und Verfahren sind mitunter im Heilmittelkatalog nicht gelistet, auch wenn ihre therapeutische Wertigkeit belegt ist. Hierzu zählt z. B. die Motopädagogik (Psychomotorik) oder

das „Heidelberger Elterntraining“ zur Sprachanbahnung (➤ Kap. 37.2.3). Deshalb erfolgt in der Regel keine Kostenübernahme durch die gesetzlichen Krankenkassen. Die Möglichkeit des praktischen Einsatzes hängt dann im Einzelfall ab von organisatorischen Gegebenheiten sowie insbesondere den finanziellen Ressourcen der Eltern.

Literatur

Hollmann H, Kretzschmar C, Schmid RG (2008) Das Altöttinger Papier und die Mehrdimensionale Bereichsdiagnostik in der Sozialpädiatrie. Qualität in der Sozialpädiatrie. Band 1. www.dgspj.de

Richtlinie des Gemeinsamen Bundesausschusses über die Verordnung von Heilmitteln in der vertragsärztlichen Versorgung (Heilmittel-Richtlinie) in der Fassung vom 20. Januar 2011. www.g-ba.de

37.2.2 Physiotherapie

Grundlagen

Krankengymnastik zählt zu den etabliertesten therapeutischen Maßnahmen überhaupt. Die Manipulation an Knochen und Gelenken zur Mobilisation bei Schwierigkeiten in der Bewegung wird seit alters her praktiziert. Gleiches gilt für die ergänzende Anwendung von physikalischen Maßnahmen mit Wärmebehandlung, Packungen und Massagen.

Die Durchführung von zielgerichteter krankengymnastischer Übungsbehandlung als Behandlungsansatz bei Kindern mit Zerebralparese wurde erstmals von Berta Bobath und Elisabeth Köng propagiert, die Einflussmöglichkeiten bei erwachsenen Patienten beobachtet hatten. Theoretische Überlegungen auf neurophysiologischer Grundlage von Berta und Karel Bobath führten zum nach ihnen benannten **Bobath-Konzept,** entstanden in den 1950er und 1960er Jahren. Es beruht auf den Prinzipien von Hemmung (Inhibition) und Bahnung (Fazilitation). Etwa zehn Jahre später entwickelte Vaclav Vojta aus der Beobachtung provozierbarer motorischer Muster bei Säuglingen, die er als „Reflexkriechen“ und „Reflexumdrehen“ bezeichnete, sein Konzept der **Reflexlokomotion, aus der er die Vojta-Therapie ableitete.** Beiden Ansätzen ist gemeinsam, dass sie ein hierarchisch-deterministisches Verständnis von motorischer Entwicklung repräsentieren und davon ausgehen, über die Übungsbehandlung günstige, die Pathologie ablösende Bewegungsmuster an den motorischen Kortex heranzutragen und dort zu implementieren. Diese frühen theoretischen Vorstellungen von motorischer Steuerung und therapeutischer Intervention wurden gestützt durch empirische Beobachtungen. Diese belegen, dass sich die motorische Gesamtsituation der behandelten Kinder besserte und weniger orthopädische operative Maßnahmen als zuvor erforderlich waren. Das Bobath-Konzept ist seither kontinuierlich erweitert und auch modifiziert worden; es setzt heute stark auf die Unterstützung der Eigenaktivität bei alltagsrelevanten motorischen Betätigungen und hat sich damit deutlich von den ursprünglichen theoretischen Überlegungen gelöst. Die **Vojta-Therapie** wird aufgrund der häufig gerade bei jungen Kindern induzierten aversiven Reaktionen mitunter ergänzt durch lerntheoretisch fundierte Vorgehensweisen der systematischen Verstärkung.

Die Vorgehensweisen, die von Bobath und Vojta als Konzept bzw. Methode erdacht und mit dem damaligen Stand des Wissens begründet worden sind, haben in der Krankengymnastik für Säuglinge, Kinder und Jugendliche große Verbreitung erfahren. Als sog. **„neurophysiologische Techniken“** sind sie explizit in den seit 2004 geltenden Heilmittelrichtlinien des Gemeinsamen Bundesausschusses ausgewiesen.

In deutlicher Diskrepanz hierzu steht die Datenlage hinsichtlich des Nachweises einer spezifischen Wirksamkeit für beide Vorgehensweisen. Beide Ansätze nehmen für sich in Anspruch, eine in Entwicklung befindliche, klinisch also noch nicht fixierte Zerebralparese so beeinflussen zu können, dass im Optimalfall eine Kompensation resultiert. Mindestens sollen aber funktionelle Verbesserungen erreicht werden, die Folge der spezifischen Behandlung sind. Kontrollierte Studien, die diese Annahmen belegen, existieren jedoch bislang nicht. Die genaue Wirkgröße von Physiotherapie bleibt damit unklar. Übereinstimmend belegen allerdings die wissenschaftlichen Publikationen, dass Kinder mit einer Zerebralparese heute einen anderen und günstigeren Entwicklungsverlauf nehmen, als dies bis vor wenigen Jahrzehnten der Fall gewesen ist.

Beurteilung von Motorik

Bewegung wird in Ausmaß und Qualität von jedem Menschen intuitiv beurteilt. Das Erkennen von normalen motorischen Mustern ist ontogenetisch verankert; Abweichungen stellen ein unspezifisches Alarmzeichen im Sinne von potenzieller Andersartigkeit und Gefahr dar.
In der Beurteilung der motorischen Entwicklung des Kindes sowie der Abweichungen von der Normalität spielt die Erfahrung bei der Erkennung entsprechender Muster eine wesentliche Rolle. Um neben zeitlichen und quantitativen Leistungsparametern gerade im Säuglingsalter andere objektivierbare Daten zur Verfügung zu haben, wurde nach systematischen, klinisch zu verwendenden Verfahren gesucht.
Vor dem Hintergrund des damaligen neurowissenschaftlichen Wissens mit Betonung von Reflexen und Entwicklungshierarchien war die Anwendung der Lagereaktionen nach Vojta lange das mit Abstand am stärksten genutzte Instrument. Über die quantifizierbare Beurteilung im Rahmen dieser von extern induzierten motorischen Provokationsuntersuchung und Zuordnung zur „Zentralen Koordinationsstörung“ sollte der Rahmen für beobachtendes Zuwarten, Indikationsstellung für Physiotherapie und -zusammen mit dem klinisch-neurologischen Untersuchungsbefund die frühe Identifikation einer drohenden zerebralparetischen Entwicklung sicher einzuschätzen sein.
In der Folge wurden in den 1980er Jahren sehr viele Säuglinge von ärztlicher Seite als entwicklungsgefährdet und behandlungsbedürftig beurteilt. In der seinerzeitigen Euphorie über vermeintlich heilende physiotherapeutische Behandlungsmöglichkeiten wurde die enorme Diskrepanz zu den gut fundierten Prävalenzzahlen für Zerebralparese zunächst nicht beachtet. Erst infolge der Ernüchterung darüber, dass sich seither keine grundlegende Änderung der Daten ergeben hat, setzte ein Umdenken und die Suche nach besserer Beurteilung ein. Gleichzeitig verwies Michaelis im Rahmen von Verlaufsuntersuchungen auf die „transitorischen neuromotorischen Phänomene“ als zwar von optimalen motorischen Mustern abweichende, gleichwohl physiologische Durchgangsstadien ohne pathologische Bedeutung mit Maximum Mitte des zweiten Lebenshalbjahres.
Von zunehmender Bedeutung wird deshalb die Einschätzung der motorischen Entwicklung im Rahmen beobachtender und videogestützter Verfahren, die ebenfalls seit langem bekannt sind. Die Beschreibung der **„General Movements“** durch Touwen und Prechtl in den 1980er Jahren wurde aber zunächst als zu aufwändig in der alltäglichen Praxis erachtet, zumal mit den Lagereaktionen nach Vojta ein vermeintlich zeitökonomisches und valides Instrument zur Verfügung zu stehen schien. Inzwischen setzt sich aber zunehmend die Erkenntnis durch, dass über die Beschreibung dieser bestimmten, an motorische Entwicklungsperioden gekoppelten motorischen Muster des Säuglings („writhing“ und „fidgety movements“) mit hoher Genauigkeit zumindest eine pathologische motorische Entwicklung vorherzusagen ist (➤ Kap. 17).

Heutiger Untersuchungsstandard ist deshalb die **Beobachtung der motorischen Eigenaktivität** des Säuglings. Mit der so erfolgenden Erhebung des kindlichen Bewegungsprofils (Infant Motor Profil, IMP; Hadders-Algra et al., 2011) ist die Beschreibung der gelingenden Motorik und gleichzeitig Abwesenheit von harten pathologischen Zeichen möglich. Der neurologische Basisbefund mit Reflexstatus, Muskelkraft und -tonus sowie Gelenkbeweglichkeit liefert erst bei Auffälligkeiten wichtige weitere Informationen zur Einschätzung und Zuordnung des Störungsbildes.

Infant Motor Profil IMP

Qualitatives Überprüfungsinstrument zur Beschreibung des motorischen Verhaltens in der Kindheit in 5 Domänen:

- Bewegungs-Variation
- adaptive motorische Variabilität
- Flüssigkeit
- Symmetrie
- Ergebnis-Erreichung

Gestaltung der Physiotherapie

Neben dem Einsatz von Physiotherapie bei früher motorischer Entwicklungsverzögerung im Säuglingsalter oder vermuteter bzw. bereits gesicherter zerebralparetischer Entwicklung gibt es eine Vielzahl von Bereichen, bei denen die Anwendung von Krankengymnastik nach den Vorgaben des Heilmittelkatalogs als sinnvoll angesehen wird. Hierzu zählen:

- Neuromuskuläre Krankheiten
- Orthopädische Krankheitsbilder
- Motorische Ungeschicklichkeit (Dyspraxie)
- Unterstützungsbehandlung bei Mehrfachbehinderung
- Atemtherapie bei pulmonalen Erkrankungen
- Rehabilitative Behandlung nach Erkrankungen oder Unfällen, die mit dem Verlust bereits vorhandener motorischer Fertigkeiten einhergehen.

So breit das Spektrum der Anwendungsmöglichkeiten von Physiotherapie ist, so unterschiedlich sind auch die **Zielsetzungen** im Heilmittelkatalog bei der Behandlung. Hierzu zählen:

- Kontrakturvermeidung und -lösung
- Tonusregulierung
- Funktionsverbesserung bei krankhaften Muskelinsuffizienzen und -dysbalancen
- Funktionsverbesserung von funktionsgestörten Gelenken
- Erhaltung, Förderung und Wiederherstellung der Leistungen der Stütz- und Bewegungsorgane
- Bei Kindern die Erleichterung des Bewegungsablaufs durch Ausnutzung komplexer Bewegungsmuster sowie Bahnung von Innervation und Bewegungsabläufen und Förderung oder Hemmung von Reflexen.

Um eine möglichst effektive Vorgehensweise zu entwickeln, sind Abstimmungen zwischen dem verordnenden Arzt und der behandelnden Therapeutin unabdingbar. In der Praxis erfolgt dies in aller Regel anhand von Verordnung und Therapiebericht, während unter den institutionellen Gegebenheiten des Sozialpädiatrischen Zentrums oder vergleichbarer interdisziplinär arbeitender Spezialsprechstunden bzw. Institutionen die Fallbesprechung und sogar gemeinsame Untersuchung des Patienten unter Einbeziehung der Eltern mit entsprechender Zielfestlegung möglich ist.

Ausgehend von der ärztlichen Befunderhebung und Diagnosestellung lässt sich eine stark fokussierte krankengymnastische Intervention mit eindeutiger Zielsetzung immer dann beschreiben, wenn rehabilitative Aspekte im Vordergrund stehen wie z. B. Mobilisation einer Extremität nach Operation oder Unfall mit Wiedererlangung des vorherigen individuellen Funktionsniveaus. Bei kindlichen Entwicklungsstörungen ebenso wie bei progredienten Erkrankungen (z. B. spinale Muskelatrophie, Mukoviszidose) ist es ungleich schwieriger, operationalisierte und damit auch evaluierbare **Therapieziele** zu beschreiben, die dann auch die Dauer der Maßnahme bestimmen. Deshalb ist es hilfreich und notwendig, eine Konkretisierung über ein kleinschrittiges Vorgehen zu erreichen. Dieses hat den Vorteil, dass gerade bei individuell kaum abzuschätzenden Krankheits- und Störungsverläufen die Möglichkeit besteht, den nahen Bezugspersonen den Zweck der Maßnahme zu verdeutlichen, so eine gute Kooperation aufzubauen und gemeinsam zu einem Weg zu finden, die Dauer von Behandlungsintervallen und ggf. Therapiepausen festzulegen.

Fallbeispiel Physiotherapie

Leon, 8 Monate, erstes Kind der Familie. Hypotonie, Ernährungsproblematik.
Keine organisch relevanten Belastungen aus Schwangerschaft und Geburt, NA-pH 7,26.
Klinisch-neurologischer Status bei der U5 unauffällig bis auf Hypotonie. Eltern stark beunruhigt.
Maßnahmen: Einleitung einer Physiotherapie (Bobath) mit Zielvorgabe: „Bewegungsmotivierende Spiel-Anleitung, therapeutische Hilfestellung bei der Nahrungsaufnahme". Ärztliche Verlaufsuntersuchungen im Abstand von 3 Monaten.
Umsetzung durch eine erfahrene Therapeutin mit zunächst 2 Behandlungseinheiten pro Woche, in der Folgeverordnung nach 5 Wochen Reduktion der Therapie auf einmaligen Wochentermin.
Verlauf: langsame, aber kontinuierliche Fortschritte in der Motorik, neurologischer Befund bis auf abnehmende Hypotonie weiter unauffällig, freies Laufen mit 22 Monaten verzögert erreicht. Physiotherapie in Absprache mit der behandelnden Therapeutin mit knapp 19 Monaten bei bereits sicher einzuschätzendem motorischem Entwicklungsverlauf beendet.
Wesentliches Behandlungselement war die Beruhigung der Eltern, die über die situationsspezifische Anleitung der Mutter zu erzielen war. Insbesondere in der Gestaltung der Nahrungsaufnahme konnte die Therapeutin geeignete Techniken und Vorgehensweisen vermitteln. Durch die Entspannung in einer zentralen Versorgungssituation wurde die Gelassenheit aufgebaut, die auch in der bewegungsmotivierenden Interaktion mit dem Kleinkind weiterhin einen großen Stellenwert hatte.

Die Vorgehensweise bei **chronisch kranken und motorisch oder mehrfach behinderten Kindern und Jugendlichen** ist mit besonderen Schwierigkeiten verbunden. Mit dem Übergang in die schulische Betreuung und damit die sonderpädagogische Förderung, erst recht aber an der Schwelle zum Erwachsenenalter stellt sich die Frage, in welchem Umfang eine Physiotherapie noch erforderlich ist. Gerade bei diesen Patienten besteht in aller Regel eine erhebliche und dauerhafte Diskrepanz zur altersdurchschnittlichen Entwicklung und dem damit verbundenen Funktionsniveau. Hieraus resultiert eine hohe intuitive Bereitschaft, die Notwendigkeit einer Fortsetzung der Behandlung zu bejahen. Dem steht entgegen, dass sich eine spezifische Wirksamkeit auf der Basis wissenschaftlicher Studien nicht belegen lässt. Die Entscheidungsfindung, die im bestehenden System stets durch den zuständigen Arzt in Form seiner Verordnung erfolgen muss, wird häufig zusätzlich belastet durch Diskussionen um die verursachten Kosten und die Zuständigkeit hierfür, was dann in diffuse moralische und ethische Diskussionen mündet. Hintergrund sind meist unvollständige Kenntnisse und falsche Vorstellungen über den individuellen Erkrankungsverlauf und die Einwirkungsmöglichkeiten hierauf durch eine physiotherapeutische Behandlung. In solchen Situationen, die sich für die Beteiligten zu einem Dilemma entwickeln können, ist es

besonders empfehlenswert, durch inhaltliche Aufklärung und gemeinsame Diskussion sowie die Festlegung von sehr gut abgestimmten Erwartungen in Form von operationaliiserten Therapiezielen einen Konsens zu finden. Da eine heilende Wirkung im eigentlichen Sinn nicht attestiert werden kann, kommt den unterstützenden Aspekten die wesentliche Bedeutung zu. Es ist nicht gerechtfertigt, pauschal in einer bestimmten Krankheitssituation oder ab einem bestimmten Alter eine Physiotherapie als ineffektiv zu erklären. Ebenso wenig ist aus der Tatsache der weiterhin vorliegenden Störungsbefunde im Rahmen der Grunderkrankung die dauerhafte Notwendigkeit der Verordnung abzuleiten. Je genauer und alltagsrelevanter die Behandlungsziele definiert sind, umso leichter sind Einschätzungen und Entscheidungen zu erreichen.
Gerade bei **Jugendlichen mit Zerebralparese** sind Mobilität und motorische Aktivierung von hoher Alltagsrelevanz. Oftmals ist bei höhergradigen Funktionseinschränkungen der Übergang zum Rollstuhl vom Patienten selbst erwünscht, da seine bipedale Fortbewegung allenfalls im unmittelbaren und entsprechend barrierefrei ausgestatteten Wohnumfeld von Bedeutung ist. Teilhabe an der Gruppe der Gleichaltrigen ist eher auf angemessene Geschwindigkeit angewiesen, was mit dem Hilfsmittel besser zu erreichen ist. Gleichzeitig belegen Untersuchungen bei Jugendlichen und jungen Erwachsenen mit Zerebralparese, dass ihr allgemeines motorisches Aktivitätsniveau weit hinter allgemeinen Empfehlungen zur körperlichen Betätigung zurück bleibt. Hier kann eine physiotherapeutische Behandlungsmaßnahme mit Anleitung zu einem täglichen Trainingsprogramm sehr hilfreich sein. Ziel ist der Übergang in eine geeignete Sportgruppe.

Manuelle Medizin und Osteopathie

In den letzten Jahren wurden Techniken der **Manuellen Medizin** als Teilbereich der Physikalischen Therapie vermehrt in die Behandlung von zerebralen Bewegungsstörungen einbezogen. Hierbei sind passive oder aktive Mobilisation, Manipulation mit Impulsgebung und Osteopathie zu unterscheiden. Gemeinsam ist ihnen die Behandlung mit der Hand des ausgebildeten Arztes oder Therapeuten. Diagnostik und Therapie beruhen auf biomechanischen und neurophysiologischen Prinzipien, wobei alle Strukturen, die in einer neuroreflektorisch-humoralen Wechselwirkung zum Bewegungsorgan stehen, einbezogen sind. Bei motorischen Entwicklungs- und zerebralen Bewegungsstörungen bei Säuglingen und Kindern spielt die Wirbelsäule eine besondere Rolle mit Beachtung von vertebroviszeralen und viszerokutanen Wechselwirkungen.
In der Manuellen Medizin haben die Techniken der Mobilisation eine lange Tradition; Gleiches gilt für Massagen. Es ist allgemein akzeptiert, dass durch solche therapeutischen Vorgehensweisen grundsätzlich Muskeltonus und Körperhaltung positiv zu beeinflussen sind und zur Linderung von Schmerzsymptomen beigetragen werden kann. Diese Maßnahmen sind somit als Ergänzung zur krankengymnastischen Behandlung im Einzelfall sinnvoll.
Als problematischer muss eine **Mobilisation** mit gleichzeitiger Impulsgebung insbesondere im Bereich der Halswirbelsäule angesehen werden. Genaue Angaben über ernsthafte Zwischenfälle bei Kindern liegen nicht vor, Einzelfallbeschreibungen bei Erwachsenen sehr wohl. Diese Risiken müssen beachtet werden, weshalb üblicherweise vor Manipulation an der Halswirbelsäule eine Röntgen-Zielaufnahme zumindest zum Ausschluss von angeborenen Fehlbildungen durchgeführt wird. Einzelfallbeschreibungen belegen, dass durch die Manipulation die Symptomatik von sogenannten „Schreibabys" schlagartig beendet werden konnte. Es ist aber bis heute nicht möglich, durch entsprechende Untersuchungstechniken die für dieses Vorgehen tatsächlich geeigneten Patienten im Vorfeld zu bestimmen und damit zielgerichtet auszuwählen.
Die postulierten Blockierungen im Bereich des 1. und 2. Halswirbels mit Auswirkungen auf die gesamte Sensomotorik und Entwicklungsbeeinflussung bis in das Schulalter hinein erscheinen

demgegenüber spekulativ und nicht nachvollziehbar. Es ist unverständlich, dass ein peripheres neuro-muskuläres Störungsgebiet einen nachhaltigen Einfluss auf zentral gesteuerte Bewegungsabläufe und erst recht auf kognitive Prozesse haben soll. Die Formulierungen sind so pauschal, dass sie sich auch einer näheren Überprüfung entziehen. Selbst bei einem näherliegenden Zusammenhang von Blockierungen der Halswirbelsäule zu chronischen Kopfschmerzen im Kindes- und Jugendalter konnte bisher kein signifikanter Nachweis der Wirksamkeit für die Manipulation geführt werden.
Nochmals einen anderen Zugangsweg innerhalb der Manuellen Medizin eröffnet die **Osteopathie**. Sie geht davon aus, dass der Körper eine Funktionseinheit darstellt und sich deshalb Störungen in einem Bereich auch auf andere Bereiche auswirken. Durch die Behandlungsmaßnahmen am Knochengerüst und unter Einbeziehung des Bewegungsapparates sollen sich demzufolge auch Störungen des Organismus beheben lassen. Der Therapeut kann mit den Händen die sogenannte Grundspannung von Muskeln, Knochen und Gelenken feststellen und so gestörte Funktionen erkennen. Die Osteopathie geht davon aus, dass sich der Körper bei Störungen grundsätzlich selbst heilt und es nicht möglich sei, ihn von außen zu heilen. Die Techniken der Osteopathie sollen somit die Selbstheilungskräfte aktivieren. Die Ausübung der Osteopathie ist in Deutschland durch das Heilpraktikergesetz geregelt. Von schulmedizinischer und wissenschaftlicher Seite werden die Annahmen der Osteopathie kritisiert. Wie bei den anderen Verfahren fehlen Nachweise für die postulierte Heilungswirkung. Die Publikation einer randomisierten klinischen Studie mit verblindeten Videoaufnahmen zur Wirksamkeit osteopathischer Behandlungen bei Säuglingen mit Haltungsasymmetrien hat deshalb für erhebliche Irritationen gesorgt. In dieser Studie, die hinsichtlich Design und Untersuchungstechnik alle Anforderungen erfüllt, konnte gezeigt werden, dass unter der durchgeführten osteopathischen Behandlung in den ersten Lebensmonaten die Ausprägung einer Haltungsasymmetrie signifikant im Vergleich zu einer Scheinbehandlung vermindert werden konnte (Philippi et al. 2006, 2007). Nicht beantwortet wird dadurch die weitergehende Frage, in wiefern es sich bei solchen Haltungsasymmetrien tatsächlich um Störungen mit Krankheitswert und darauf basierender Behandlungsnotwendigkeit handelt.

Technikunterstützte Physiotherapie

Nachdem die Maßnahmen der traditionellen Krankengymnastik nicht den Zugewinn an Funktionsfähigkeit bewirken können, die bei der Einführung in die Behandlung von Zerebralparesen davon erhofft worden waren, sind in den vergangenen Jahren zunehmend **technische Unterstützungen** für die Physiotherapie in den Brennpunkt des Interesses gerückt. Die Erforschung von „motor patterns“, die sehr spezifisch für einzelne Funktionen und motorische Abläufe sind, hat dazu geführt, dass solchen Bewegungsmustern bei motorischen Trainings zunehmend Aufmerksamkeit gewidmet wird. Dabei wird die Lernfähigkeit des motorischen Kontroll- und Regulationssystems auf spinaler und kortikaler Ebene systematisch durch den Einsatz technischer Verfahren genutzt und unterstützt.
Zunächst wurde das aus der Rehabilitation bekannte **Laufband-Training** eingesetzt. Wegen des notwendigen hohen personellen Einsatzes und ungünstiger ergonomischer Bedingungen für die Therapeuten bei der Durchführung hat sich die Verwendung des Laufbandes als unterstützende Behandlung bei Kindern mit Zerebralparese aber nur bedingt in den Behandlungszentren durchsetzen können. Eine Weiterentwicklung ist der **Lokomat,** wo das Kind mit komplexen technischen Hilfen auf dem Laufband mit PC-Berechnung des individuell erforderlichen Ausmaßes in seiner Bewegung unterstützt wird. Die Kombination des eigenen Gehens mit Vermittlung einer Führung, die im Endergebnis weitestgehend dem normalen Bewegungsmuster entspricht, führt

zu einer signifikanten Abnahme der technischen Unterstützungsnotwendigkeit binnen einer Trainingseinheit von drei Wochen mit täglich einer Stunde Laufbandarbeit. Das Verfahren ist bisher überwiegend bei Kindern mit guten kognitiven Voraussetzungen und beinbetonter bilateraler spastischer Zerebralparese in Anwendung. Die Evaluation dieses „Roboter-Laufbandes" insbesondere hinsichtlich der hiermit zu erreichende Nachhaltigkeit steht noch aus.
Viele Fragen sind auch in Bezug auf den „forced use"-Ansatz nicht geklärt, der bei der **„constraint induced movement therapy" (CIMT)** verfolgt wird. Im Rahmen des physiotherapeutischen Gesamtkonzepts wird bei unilateralen armbetonten spastischen Zerebralparesen die funktionell führende gesunde Hand in ihrem Einsatz in unterschiedlichem Umfang eingeschränkt, sodass das Kind während der Therapiestunden, bei ausreichender Toleranz ggf. auch in seinen Alltagsaktivitäten, systematisch die paretische Hand benutzt. Die hierdurch zu erzielenden positiven Trainingseffekte sind belegt; offen sind die Fragen zum mittel- und langfristigen Gesamterfolg hinsichtlich motorischer manueller Fertigkeiten, insbesondere unter dem Aspekt der Auswirkungen auf die individuell führende Funktionshand.
Einem völlig anderen Forschungsansatz unterliegt der Einsatz des **Vibrationstrainers** „Galileo". Die Knochen-Muskel-Einheit ist bisher in ihrem Zusammenwirken im anatomisch-physiologischen Wechselspiel bei Kindern mit Bewegungseinschränkungen nur unzureichend beachtet worden. Durch den Einsatz des Vibrationstrainers, der aus der Weltraumforschung übernommen wurde und bei Astronauten dem Auftreten von Osteoporose und Muskelatrophie entgegenwirkt, kann ein Zugewinn an Kraft, Ausdauer und Koordination erzielt werden. Die bisher publizierten Ergebnisse belegen die statistische Signifikanz der Effekte. Auch hier steht allerdings der Nachweis der Alltagsrelevanz sowie insbesondere der Nachhaltigkeit aufgrund des erst kurzen Anwendungszeitraums noch aus.

Hilfsmittel in der Physiotherapie

Von größter Bedeutung für Patienten mit Zerebralparese und bleibender motorischer Behinderung ist der sachgerechte Einsatz von Hilfsmitteln. Die optimale Nutzung der technischen Möglichkeiten ist eine Herausforderung für entsprechend qualifizierte Physiotherapeuten in der Zusammenarbeit mit Orthopädie- und Rehatechnikern, Neuropädiatern und Kinderorthopäden. Der Zugewinn an Mobilität, Alltagsbewältigung und Teilhabe kann erheblich sein und entscheidend zur Verbesserung der Lebensqualität beitragen. Eine ausführliche Darstellung findet sich in Kapitel 37.4 (➤ Kap. 37.4).

Physiotherapie und Motopädagogik

Die enge Verbindung von Psyche und Motorik ist aus Beobachtungen im Alltag bekannt. Störungen der emotionalen Befindlichkeit wie Ärger oder Angst, ebenso krankhafte Zustände wie posttraumatische Belastungsstörung oder Depression schränken auch motorische Leistungen ein. Medienwirksam evident wird dies in den unterschiedlichsten Beispielen von Spitzensportlern. In der Medizin umfasst die Erhebung des psychopathologischen Befundes auch die Beurteilung der Psychomotorik.
In den 1950er Jahren war es Ernst Kiphard, der über das Angebot einer motorischen und kreativen, gleichzeitig auf das soziale Miteinander ausgerichteten Betätigung in der Gruppe mit Kindern und Jugendlichen ein neues therapeutisches Angebot schuf. Die von ihm so genannte "Psychomotorik" bewirkte im Kontext mit anderen Maßnahmen positive Veränderungen insbesondere bei Verhaltensauffälligkeiten. Seither ist das Konzept weiter entwickelt und differenziert worden. Die Terminologie mit "Motopädagogik, Motopädie, Mototherapie, Bewegungstherapie und Bewe-

gungspädagogik" verweist auf unterschiedliche Ansätze mit medizinischen, psychotherapeutischen und pädagogischen Bezügen. Diese finden nicht nur im medizinisch-funktionellen Bereich, sondern ebenso in Kindergarten und Schule Verwendung.
Die stets in der Gruppe realisierte Intervention stellt somit eine Brücke dar zwischen den Heilmitteln einerseits und den im nachfolgenden Kapitel dargestellten (heil)pädagogischen Maßnahmen. Die motopädagogische Behandlung ist im Rahmen der Heilmittel-Richtlinien nicht verordnungsfähig, wird aber häufig von Krankenkassen zumindest bezuschusst. In der Versorgungspraxis stellt sie bei regional stark unterschiedlicher Präsenz dennoch eine wichtige Option dar, weil oftmals gerade für emotional beeinträchtigte oder verhaltensauffällige sowie motorisch ungeschickte Kinder im Freizeit-Angebot der Sportvereine die Möglichkeiten fehlen, sich in einem geschützten und begleiteten Rahmen in der Gruppe zu betätigen. Oftmals sind Praxen für Physiotherapie an der Umsetzung beteiligt.

Medikation und Physiotherapie

Schließlich sind die physiotherapeutischen Interventionsmöglichkeiten bei Zerebralparesen mit Spastik entscheidend erweitert worden durch lokale Injektionen von **Botulinustoxin A** und die intrathekale Anwendung von **Baclofen mittels implantierter Pumpe.** Hierdurch werden im Einzelfall optimale Ausgangsbedingungen im Rahmen der jeweils individuell eingeschränkten Möglichkeiten geschaffen, um zielgerichtet einen funktionellen Zugewinn, mindestens aber eine Verbesserung der Pflegbarkeit zu erzielen.
Eine gewisse Bedeutung kommt ferner der Verabreichung von L-DOPA bei bestimmen Formen der Dystonien zu. Zunehmend gewinnt hierbei ebenfalls die lokale Injektionsbehandlung mit Botulinustoxin A an Bedeutung. Schließlich kann eine niedrig dosierte abendliche Verabreichung von Benzodiazepinen, z. B. Tetrazepam, hilfreich sein. Bei allen anderen zentral wirksamen Wirkstoffen (Tolperison, Pridinol, Mephenesin, Methocarbamol) fehlen ausreichende Belege für die therapeutische Wirksamkeit. Neben Baclofen kommen auch Dantamacrin, Tizanidin und Gabapentin in Betracht, haben aber keine Überlegenheit in der Effektivität.
Bei schweren dyston-dyskinetischen Zerebralparesen kann durch Implantation einer Tiefenhirn-Stimulationssonde eine mitunter deutliche Linderung der Symptomatik erzielt werden.

Zukunft der Physiotherapie

Zusammenfassend ist festzuhalten, dass sich die Möglichkeiten der physiotherapeutischen Behandlung in den vergangenen ca. 60 Jahren für Säuglinge, Kinder und Jugendliche mit motorischen Funktionseinschränkungen, v. a. aber im Zusammenhang mit einer Zerebralparese wesentlich erweitert haben. In Zukunft wird es notwendig sein, sich noch bewusster von den „Mythen der Vergangenheit“ zu lösen und auf dem inzwischen erreichten Stand der neurowissenschaftlichen Erkenntnis solche Konzepte zu entwickeln, die insbesondere den Zielen der Alltagsrelevanz und Förderung von Eigenaktivität bei den betroffenen Kindern angemessen Rechnung tragen.
Die Groninger Arbeitsgruppe Entwicklungsneurologie um Mijna Hadders-Algra entwickelt seit 2006 ein neues Interventionssystem. „COPCA – Coping with and Caring for infants with neurological dysfunction – a family centered program" (xe „COPCA“) verknüpft dabei Aspekte der Physiotherapie aus dem Bobath-Konzept mit einer systematischen Elternanleitung. Zielsetzung ist die Unterstützung des Kindes in jeder Form der Eigenaktivität, wobei das Kind als Akteur seiner Entwicklung verstanden wird. Dieses Interventionssystem bei schwerwiegenden Abweichungen in der motorischen Entwicklung stellt somit den ersten konsequent umgesetzten Ansatz dar, das aktuelle Wissen um normale und gestörte Entwicklung in ein dementsprechend gestaltetes prak-

tisches Vorgehen zu transferieren. Die Daten zur Evaluation bis zum Alter von 18 Monaten konnten allerdings bislang nicht die Hypothese stützen, dass in dieser Modifikation des Vorgehens bessere Resultate als in der konventionellen neuromotorisch orientierten Physiotherapie zu erreichen seien. Beide zahlenmäßig kleinen Gruppen – COPCA vs. Traditionelle Physiotherapie schnitten in der motorischen Entwicklung gleich ab.
Bei den etablierten Verfahren ist die Ausrichtung auf Unterstützung der Eigenaktivität des Kindes im Bobath–Konzept bereits überwiegend erfolgt, da es in den letzten Jahren in der skizzierten notwendigen Weise fortgeschrieben worden ist. Im Gegensatz zum Bobath-Konzept, das weltweit Anwendung findet, hat sich die Vojta-Therapie außerhalb von Deutschland nur in geringem Umfang etablieren können. Da eine Überlegenheit dieser in der Durchführung gerade bei Säuglingen eingreifenden Methode wissenschaftlich nicht belegt ist, sollte ihre Anwendung älteren kooperationsfähigen Kindern und Jugendlichen vorbehalten bleiben.
In Verbindung mit neuen technischen und medikamentösen Maßnahmen wird sich das Spektrum der physiotherapeutischen Behandlung in den kommenden Jahren zumindest für eine Teilgruppe von Patienten wesentlich erweitern. Dabei bleibt in der mittel- und langfristigen Evaluation abzuwarten, in welchem Verhältnis Aufwand und Zeit zu den zu erreichenden Gewinnen an Funktion stehen. Es ist ferner absehbar, dass die heute dominierende „low dose“ Methode der wöchentlichen Therapiestunde abgelöst wird durch Intervall-Behandlungen mit intensiven, aber kurzen Therapieblöcken

Einsatz von Physiotherapie bei Störungen der Motorik

- KEINE „Routine" bei Zustand nach Frühgeburt
- Säugling mit Retardierung der Motorik:
  nur bei Verunsicherung der Eltern
- Manifestation einer zerebralen Bewegungsstörung:
  Physiotherapie als mittel- und langfristige Unterstützung ansehen;
  Therapeutin als „case managerin"
- Umschriebene Entwicklungsstörung der Motorik, "Bewegungsungeschicklichkeit":
  Ansatz in der Psychomotorik-Gruppe, sofern keine geeignete, nicht leistungsorientierte Freizeit-Aktivität zur Verfügung steht
- Leben mit Cerebralparese:
  "Integrierter Behandlungsansatz" mit
  - geräteunterstützter Physiotherapie als Intervall-Blockbehandlung,
  - Erhaltungstherapie zuhause mit Supervision,
  - adäquater individueller Hilfsmittel-Versorgung und
  - gleichzeitiger Nutzung medikamentöser Behandlungsansätze

Literatur

AWMF-Leitlinie (2011) Umschriebene Entwicklungsstörungen motorischer Funktionen. www.awmf.org/uploads/tx_szleitlinien/022-017l_S3_Umschriebene_Entwicklungsstörungen_motorischer_Funktionen_2011-08.pdf

Blauw-Hospers CH, de Graaf-Peters VB, Dirks T, Bos AF, Hadders-Algra M (2007) Does early intervention in infants at high risk for a developmental motor disorder improve motor and kognitive development? Neuro SCI Biobehav Ref 31: 1201–12.

Bode H (2002) Indikation zur Behandlung von cerebralen Bewegungsstörungen und Evaluationsergebnisse. In: Aksu F (Hrsg) Aktuelle Neuropädiatrie. Nürnberg: Novartis Verlag.

Borggräfe I, Kumar A, Schäfer JS, Berweck S, Meyer-Heim A, Hufschmidt A, Heinen F. (2007) Robotergestützte Laufbandtherapie für Kinder mit zentralen Gangstörungen. Monatsschr Kinderheilkd 155: (6) 529–34.

Dirks T, Blauw-Hospers CH, Hulshof LJ, Hadders-Algra M (2011) Differences between the family-centered "COPCA" program and traditional infant physical therapy based on neurodevelopmental treatment principles. Phys.Ther. 91: 1,303–22.

Hadders-Algra M (2008) COPCA – Coping with and Caring for infants with neurological dysfunction – a family centered program. In: Cerebralparese und therapeutische Bedrohung – neue Aspekte zur Intervention. Vortrag beim Symposium 30 Jahre Kinderneurologisches Zentrum Bonn

Heineman KR, Hadders-Algra M (2008) Evaluation of neuromotorfunction in infancy – a systematic review of available methods. J Def Behav Pediatr 29: 315–323.

Heineman KR, Middelburg KJ, Bos AF, Eidhof L, La Bastide-Van Gemert S, Van Den Heuvel ER, Hadders-Algra M (2013) Reliability and concurrent validity of the Infant Motor Profile. Dev Med Child Neurol 55(6):539–45.

Hielkema T, Blauw-Hospers CH, Dirks T, Drijver-Messelink M, Bos AF, Hadders-Algra M (2011) Does physiotherapeutic intervention affect motor outcome in high-risk infants? An approach combining a randomized controlled trial and process evaluation. Dev Med Child Neurol. 53(3)

Kanda T, Pidcock FS, Hayakawa K, Yamori Y, Shikata Y (2004) Motor outcome differences between two groups of children with spastic diplegia who received different intensities of early onset physiotherapy followed for 5 years. Brain & Development 26–2: 118–126.

Karch D (2003) Jenseits wissenschaftlicher Studien – Die Notwendigkeit der Behandlung von cerebralen Bewegungsstörungen. In: Korinthenberg R (Hrsg) Aktuelle Neuropädiatrie. Nürnberg: Novartis Verlag: 225–235.

Karch D, Boltshauser E, Göhlich-Ratmann G, Groß-Selbeck G, Pietz J, Schlack HG (2005) Physiotherapie auf neurophysiologischer Grundlage nach Bobath und Vojta bei Kindern mit cerebralen Bewegungsstörungen (unter besonderer Berücksichtigung von infantilen Cerebralparesen). Stellungnahme der Gesellschaft für Neuropädiatrie und der Gesellschaft für Sozialpädiatrie und Jugendmedizin Mai 2005, Kommission der Gesellschaft für Neuropädiatrie zu Behandlungsverfahren bei Entwicklungsstörungen und cerebralen Bewegungsstörungen und Mitwirkung der DGSPJ. http://www.neuropaediatrie.com

Karch D, Boltshauser E, Groß-Selbeck G, Pietz J, Schlack HG (2005) Manualmedizinische Behandlung des KISS-Syndroms und Atlastherapie nach Arlen. Stellungnahme der Gesellschaft für Neuropädiatrie, Kommission zur Behandlungsverfahren bei Entwicklungsstörungen und cerebralen Bewegungsstörungen. http://www.neuropaediatrie.com

Karch D, Groß-Selbeck G, Hanefeld F, Schlack HG, Ritz A (1998) Behandlung motorischer Störungen mit Manueller Therapie. Stellungnahme der Gesellschaft für Neuropädiatrie, Kommission zu Behandlungsverfahren bei Entwicklungsstörungen und cerebralen Bewegungsstörungen. http://www.neuropaediatrie.com

Law MC, Darrah J, Pollock N (2011) Focus on function: a cluster, randomized controlled trial comparing child- versus context-focused intervention for young children with cerebral palsy. Dev.Med.Child Neurol. 53: 621–9.

Möller P, Burkhardt I, Heuer K, Kostede E, Jung M, Westerfeld U, Isebarth-Walther H (2009) Physiotherapie bei zentralen motorischen Störungen in der Sozialpädiatrie. Rahmenempfehlung. Qualitätszirkel der Bundesarbeitsgemeinschaft Sozialpädiatrischer Zentren. www.dgspj.de

Niehaus E (2007) Neurologische Entwicklung von Frühgeborenen mit einem Geburtsgewicht unter 1.500 g von der Geburt bis zum 3. Lebensmonat: Betrachtungen von longitudinaler Stabilität und paralleler Übereinstimmungen verschiedener Testmethoden. Inaug.-Diss., Univ. Lübeck

Philippi H, Faldum A, Schleupen A, Pabst B, Jung T, Bergmann H, Bieber I, Kämmerer C, Dijs P, Reitter B (2006) Infantile postural asymmetry and osteopathic treatment: a randomized therapeutic trial. Dev Med Child Neurol 48: 5–9.

Schönau E (2012) Die Effekte der stationären Intensiv-Rehabilitation mit gerätegestützter integrierter Physiotherapie. Vortrag Kinderneurologisches Zentrum Bonn

Seitz J, Jenni OG, Molinari L, Caflisch J, Largo RH, Latal Hajnal B (2006) Correlations between motor performance and cognitive functions in children born < 1.250 g at school age. Neuropediatrics 37 (1) 6–12.
Stark C, Semler O, Duran I, Stabrey A, Kaul I, Herkenrath P, Hollmann H, Waltz S, Schönau E (2013) Intervall-Rehabilitation mit häuslichem Training bei Kindern mit Zerebralparese. Monatsschr Kinderheilkd; 161:625-632
Stark C, Nikopoulou-Smyrni P, Stabrey A, Semler O, Schoenau E (2010) Effect of a new physiotherapy concept on bone mineral density, muscle force and gross motor function in children with bilateral cerebral palsy. J Musculoskelet Neuronal Interact 2010; 10(2):151–158
Turnball JD (1993) Early intervention for children with or at risk of cerebral palsy. AJDC 147: 54–59.
Ulrich BD (2010) Opportunities for early intervention based on theory, basic neuroscience, and clinical science. Phys Ther 90(12):1868–80.

37.2.3 Logopädie

Grundlagen

Sprachentwicklungsstörungen sind in verschiedener Hinsicht von besonderer Bedeutung. Die **Prävalenz** der sogenannten „spezifischen" Sprachentwicklungsstörungen wird im internationalen Schrifttum mit 5–7 % angegeben und liegt damit hoch. Nimmt man weitere Faktoren hinzu, die die Sprache beeinträchtigen, wie unzureichende Anregung im muttersprachlichen Bereich oder erschwerte Lernprozesse bei Migrationshintergrund und hierdurch oftmals bedingter „doppelter Halbsprachigkeit", liegt der Anteil von Kindern mit Auffälligkeiten in der Sprachentwicklung insgesamt noch wesentlich höher. Im bundesdeutschen Durchschnitt ergeben sich derzeit bei den Einschulungsuntersuchungen Sprachauffälligkeiten bei mindestens jedem viertem Schulanfänger.
Sprachentwicklungsstörungen sind unmittelbar gekoppelt mit Beeinträchtigungen in der Kommunikation, geringeren Lernfortschritten in der Schule und eingeschränkten späteren beruflichen Möglichkeiten. Sie haben eine hohe Assoziation zu Verhaltensauffälligkeiten.
Maßnahmen der Sprachtherapie sind deshalb von großer Bedeutung. Die Heilmittelrichtlinien führen hierzu aus: „Die Sprachtherapie dient der Wiederherstellung, Besserung und dem Erhalt der sprachlichen und kommunikativen Fähigkeiten sowie des Schluckvorgangs. Sie umfasst insbesondere Maßnahmen zum/zur

- Anbahnung sprachlicher Äußerungen
- Aufbau des Sprachverständnisses
- Ausbildung und Erhalt der Lautsprache zur sprachlichen Kommunikation
- Artikulationsverbesserung bzw. Schaffung nonverbaler Kommunikationsmöglichkeiten
- Normalisierung bzw. Verbesserung der Laut- und Lautverbindungsbildung
- Verbesserung bzw. Normalisierung der auditiven Wahrnehmungsfähigkeit
- Aufbau von Kommunikationsstrategien
- Normalisierung des Sprachklangs
- Beseitigung der Dysfunktionen der Kehlkopf- und Zungenmuskulatur
- Besserung und Erhalt des Schluckvorganges."

Sprachstörungen vor Abschluss der Sprachentwicklung stehen in Verbindung mit verschiedenen anderen Diagnosegruppen. So werden z. B. Entwicklungsstörungen der Motorik oder Kognition, frühkindliche Hirnschädigungen, periphere und zentrale Hörstörungen, periphere Anomalien der Sprechorgane, genetisch bedingte Krankheiten, Mehrfachbehinderungen sowie familiäre Sprachschwäche mit Krankheitswert aufgeführt.

Daneben werden Formen der Sprechapraxie und Störungen des Redeflusses (Stottern, Poltern) benannt. Sprechtherapeutische Maßnahmen können aber auch bei krankhaften Störungen des Schluckaktes, z. B. in Verbindung mit infantilen Zerebralparesen und anderen neuropädiatrischen Krankheitsbildern, indiziert sein.

Nach den Empfehlungen des Spitzenverbandes der gesetzlichen Krankenversicherungen 2008 können sehr unterschiedliche Berufsgruppen als **Heilmittelerbringer** zulassungsfähig sein. Hierzu zählen Logopäden, staatlich anerkannte Sprachtherapeuten, medizinische Sprachheilpädagogen sowie Angehörige anderer pädagogischer Berufsgruppen und Ausbildungen, sofern sie bestimmte Voraussetzungen erfüllen. Für alle in der **Sprachtherapie** Tätigen ist es notwendig, über Fachwissen aus der Linguistik, der Sprachheilpädagogik, der Medizin und der Psychologie zu verfügen. Die Ausbildung umfasst neben dem Erlernen von klinisch bewährten Therapiemethoden auch die Behandlung von Patienten unter Supervision.

Umschriebene Entwicklungsstörungen des Sprechens und der Sprache mit Schwerpunkt im expressiven bzw. rezeptiven Bereich repräsentieren die spezifischen Sprachentwicklungsstörungen (specific language impairment, SLI). In der Entwicklungsdiagnostik und Sozialpädiatrie sind neben Artikulationsstörungen ferner Sprachentwicklungsstörungen bei Intelligenzminderung bis hin zur ausbleibenden Entwicklung einer Sprachfähigkeit sowie Sprachauffälligkeiten bei einer anders determinierten Kommunikationsstörung, z. B. im Rahmen von Erkrankungen aus dem Autismus-Spektrum, von Bedeutung. Insbesondere bei Kindern bis zum vollendeten 3. Lebensjahr gibt es daneben die Gruppe der „Späten Sprecher“ („late talkers“), die eine Form der Sprachentwicklungsverzögerung repräsentieren. Der Störungsbegriff in Verbindung mit Sprache soll deshalb erst nach dem vollendeten 3. Lebensjahr verendet werden.

Die Durchführung von Sprachtherapie bei Sprachentwicklungsstörungen ist einem Wandel unterworfen. Wurde früher propagiert, dass ein Beginn kaum vor dem 4. Lebensjahr sinnvoll sei, so haben sich auf der Basis neuer Erkenntnisse über die frühkindliche Sprachentwicklung zunehmend Vorgehensweisen durchgesetzt, die verstärkt die Eltern und andere nahe Bezugspersonen einbeziehen. Darüber ist der Inhalt der **Sprachanbahnung** in den Mittelpunkt des Interesses gerückt. Besonders das „Heidelberger Elterntraining“ (Buschmann 2007) kann auf eine gute wissenschaftliche Datenlage in der Evaluation verweisen; um so paradoxer ist die fehlende Kostenübernahme im Rahmen der Heilmittel. Strukturierte Sprachförderprogramme im Kindergarten gewinnen rasch an Bedeutung.

Grundlage für diese anders ausgerichteten Interventionsansätze der Sprachtherapie sind die Forschungsergebnisse der **Genetik**, die wesentlich auf Befunden der Zwillingsforschung basieren. In der „Twins Early Development Study (TEDS)“ wurde von Plomin und Dale 2001 belegt, dass die Vererbung einen umso höheren Stellenwert hat, je stärker die Ausprägung des Wortschatzdefizits ist (Spinath 2004). Dem gegenüber ist die normale Entwicklung des Wortschatzes in hohem Maße von Umgebungseinflüssen abhängig. Die Kombination von Worten im 2. und 3. Lebensjahr wiederum unterliegt stärkeren genetischen Einflüssen. Auch in der Deutschen Sprachentwicklungsstudie (German Language Development Study, glad-study) ergaben sich entsprechende Hinweise. Seit der Erstbeschreibung des FOXP2-Gens 2001 „FOXP2-Gen“) durch Fisher sind weitere Einzelheiten Gegenstand der laufenden Forschungen. Insbesondere das CNTNAP2-Gen auf Chromosom 7 hat sich als relevant erwiesen. Varianten werden nicht nur mit der spezifischen Sprachentwicklungsstörung in Verbindung gebracht, sondern mit Beeinträchtigungen von Sprache und Kommunikation insgesamt. So sind Assoziationen mit Mutismus sowie den Autismus-Spektrum-Störungen beschreiben, aber auch zum Tourette-Syndrom.

In Bezug auf Umgebungseinflüsse sind Ergebnisse aus den USA zur frühen außerfamiliären Betreuung in Krippen bemerkenswert. Das National Institute for Child Health and Development NICHD verweist darauf, dass es von der Qualität der Betreuung abhängig ist, ob eine **Krippenerziehung**

für die Sprachentwicklung einen Risikofaktor darstellt oder protektiv und fördernd wirksam wird. Als gesichert kann angesehen werden, dass durch die Anleitung von Bezugspersonen Verbesserungen hinsichtlich Wortschatz, der Benutzung von Mehrwort-Äußerungen, in der Verständlichkeit der Äußerungen sowie generell in der Kommunikationsfähigkeit und -bereitschaft zu erzielen sind. Langzeiteffekte hierfür sind aber bisher nicht belegt.
Die spezifische Sprachentwicklungsstörung ist somit nach heutigem Kenntnisstand überwiegend durch genetische Faktoren determiniert. Diese bedingen Schwierigkeiten und Störungen von zentralen Verarbeitungsprozessen und Zugriffsmechanismen.

Ätiologie der spezifischen -primären– Sprachentwicklungsstörung

- genetische Faktoren als Hauptursache
  (bekannte Genorte: auf 7q, 13q, 16q, 19q und 2p)
- mangelhafte auditive Verarbeitung des sprachlichen Inputs
- begrenzte Aufnahmekapazität des phonologischen Arbeitsgedächtnisses
- Probleme mit der Repräsentation des sprachlichen Wissens im Gehirn und mit seinen Verarbeitungs- und Zugriffsmechanismen

Ansätze zu einer **primären Prävention** liegen somit allenfalls in der Vermeidung frühkindlicher Hirnschädigungen und der Frühtherapie von Hörstörungen als zusätzlichen organischen Bedingungsvariablen.
Die Bedeutung von Umwelteinflüssen für spezifische Sprachentwicklungsstörungen ist insgesamt geringer als angenommen, außer in der Auswirkung auf den Umfang des Vokabulars. Es gibt allerdings nachgewiesene Einflussfaktoren, die mit der Person der Mutter in Zusammenhang stehen. Mütterliche Responsivität wirkt sich günstig, Depression der Mutter oder instabile Lebensumstände ungünstig auf die frühe Sprachentwicklung aus.
Wenn also die Ursache für eine Sprachentwicklungsstörung überwiegend genetisch begründet und somit nicht auszuschalten ist, bleibt als therapeutischer Zugangsweg die Modifikation und Optimierung von Umgebungsbedingungen. Die Erhöhung von Menge und Qualität des Sprachangebots sowie insbesondere der Verbesserung des familiären Kommunikationsstils sind hierfür bedeutsam. **Sekundär** sind individuelle Sprachtherapie, Anleitung der Eltern zu sprachförderndem Verhalten einzeln oder in Gruppen mit Anregung speziell zum dialogischen Vorlesen und auch Sprachförderprogramme in Kindereinrichtungen geeignet.

Gestaltung der Sprachtherapie

Einsatz von Sprachtherapie

- Sprachentwicklungs-VERZÖGERUNG:
  Eltern-Anleitung ab Ende 2. bzw. im 3. Lebensjahr
- Sprachentwicklungs-STÖRUNG:
  gezielter Einsatz der Logopädie ab dem 3.–4. Lebensjahr
  unter Einbezug der Eltern bzw. Erzieherinnen;
- Einzel- oder Gruppentherapie
- Umschriebene Lautbildungsstörung:
  Beginn einer individuellen Übungsbehandlung je
  nach Ausprägung nach dem 4.–5. Lebensjahr
  (cave: tägliches Üben notwendig!)

- Redeunflüssigkeiten/Stottern:
- Ansatz je nach Situation;
- systematische Behandlung mit Eigengestaltung ab dem Jugendalter möglich
- Voraussetzung:
  eine relevante Hörbeeinträchtigung muss ausgeschlossen werden, ggf. auch wiederholt im Verlauf

Als **Grundprinzip** sind logopädische Interventionen wie bei den anderen funktionstherapeutischen Maßnahmen auch auf den alltagsrelevanten Zugewinn der Sprech-, Sprach- und Kommunikationsfähigkeit ausgerichtet. Ansatzpunkte können einerseits Basisfertigkeiten wie oral motorische Fähigkeiten oder Verbesserung auditiver Wahrnehmungsfunktionen sein, andererseits sind Ansatzpunkte mit unmittelbarem Sprachbezug zu unterscheiden. Diese zielen auf die Optimierung des sprachlichen Angebots sowohl quantitativ wie qualitativ ab, um den Input für das Kind zu verbessern, oder sollen das Kind zum gesteigerten eigenen aktiven Sprachgebrauch (Output-Ebene) anregen.

Sprachtherapeuten müssen zur Identifikation günstiger individueller Ansatzpunkte auf der Basis der ärztlichen Indikationsstellung eine **Zieldiagnostik** der sprachlichen Funktionen durchführen, um den ärztlich vorgegebenen Behandlungsplan fachlich zu konkretisieren. Hierzu gibt es standardisierte diagnostische Testverfahren sowie Eltern-Fragebögen, die die Altersspanne vom 2. bis 10. Lebensjahr abdecken. Problematisch ist die Erfassung des Sprachentwicklungsstandes mit Testverfahren ab dem 11. Lebensjahr bis zum jungen Erwachsenenalter, da für diesen Altersbereich keine standardisierten Untersuchungsinstrumente zur Verfügung stehen. Danach kommen erneut Methoden der Stotterdiagnostik sowie Sprachleistungstests, die insbesondere bei den neurologischen Krankheitsbildern der Aphasie und Dysarthrie eingesetzt werden, in Betracht.

Die Befunderhebung vor der Aufnahme einer Sprachtherapie soll die expressiven und rezeptiven Fähigkeiten auf den **vier linguistischen Ebenen** möglichst präzise beurteilen.

Linguistische Ebenen

- Lexikalisch-semantische Ebene
- Syntaktisch-morphologische Ebene
- Phonetisch-phonologische Ebene
- Pragmatisch-kommunikative Ebene
  einschließlich Einschätzung der vorsprachlichen Kommunikation und der Eltern-Kind-Interaktion.

Daneben werden stets auch Redefluss (Stottern und Poltern), Stimme und relevante Motorik (Gesichts-, Mund- und Zungenmotorik) mitbeurteilt. Im Einzelfall kommen Überprüfungen auf Störungen der Sensorik sowie der zentralen auditiven Verarbeitung hinzu.

Die Entwicklung der menschlichen Sprache ist ausgehend von einem angeborenen, genetisch determinierten neurokognitiven Inventar an drei **grundlegende Bedingungen** geknüpft. Voraussetzungen sind:

- Ausreichendes Hörvermögen
- Ausreichende Intelligenz
- Ausreichende Anregung im Sinne des Sprachmodells.

Deshalb kann eine Sprachentwicklungsstörung niemals abschließend ausschließlich durch die umschriebene Zieldiagnostik im sprachlichen Bereich beurteilt werden. Vielmehr müssen diese Untersuchungsergebnisse systematisch im Rahmen einer mehrdimensionalen Bereichsdiagnostik (➤ Kap.15) in das allgemeine Entwicklungs- bzw. Intelligenzprofil unter Berücksichtigung von Umgebungsfaktoren eingeordnet werden. Bei jeder nachgewiesenen Sprachentwicklungsstörung im artikulatorischen, expressiven oder rezeptiven Bereich einschließlich der zentral auditiven Verarbeitung muss eine relevante Hörbeeinträchtigung durch die entsprechend differenzierte **pädaudiologische Diagnostik** ausgeschlossen werden. Hierzu zählen als Grundanforderungen:

- Inspektion von Gehörgang und Trommelfell (optimal: mikroskopische Otoskopie)
- Tympanometrie
- Ableitung der otoakustischen Emissionen.

Hinzu kommen die Verfahren der **subjektiven Audiometrie** mit Spiel- und Ablenkaudiometrie bzw. Freifeld- und Kopfhöreraudiometrie. Bei entsprechenden Verdachtsfällen oder unklarem Befund infolge eingeschränkter Kooperation ist die weitere Abklärung in einer Abteilung für Pädaudiologie notwendig, wo als zusätzliches Verfahren ggf. in Sedierung die Hirnstamm-Audiometrie (BERA) durchgeführt werden kann. Gerade bei entsprechend disponierten Kindern mit häufigen Infekten der oberen Luftwege oder Mittelohrentzündungen ist es wichtig, regelmäßig das Hörvermögen zu kontrollieren. Durch rezidivierende Paukenhöhlenergüsse entstehen Beeinträchtigungen der Schallleitung, die zu einem **undulierenden Hörvermögen** führen. Nach derzeitigem Wissensstand ist es nicht erwiesen, dass hieraus allein eine Sprachentwicklungsstörung resultiert; je nach Ausprägung der peripheren Hörbeeinträchtigung bedeutet ein solcher Befund aber für das ohnehin bereits in seiner sprachlichen Entwicklung beeinträchtigte Kind einen zusätzlichen Erschwernisfaktor. Die Indikation zur operativen Sanierung ist im Einzelfall zu stellen.

Schwerhörigkeiten infolge einer Schallempfindungsstörung müssen möglichst früh einer angemessenen apparativen Versorgung mit Hörhilfen (Hörgerät, Cochlea-Implantat) zugeführt werden und bedürfen nachfolgend einer spezifischen sprachtherapeutischen Anleitung.

Die **Indikation** für eine logopädische Behandlung kann prinzipiell dann gegeben sein, wenn ein oder mehrere sprachliche Funktionsbereiche signifikante Entwicklungsstörungen aufweisen. Auf testdiagnostischer Ebene definiert sich dies über entsprechende Unterschiede im Vergleich zu den Intelligenzleistungen des Kindes. Daneben sind aber auch Maßnahmen der Sprachanbahnung bei Kindern im 2. oder 3. Lebensjahr eine mögliche Indikation, genauso wie rehabilitative Maßnahmen nach Unfall oder Erkrankung sowie die Versorgung mit technischen Hörhilfen oder Kommunikationsunterstützung.

Zur Therapie bei **Sprach- und Sprechstörungen** steht eine Vielzahl von Methoden zur Verfügung. Die Auswahl richtet sich nach Art und Ausprägungsgrad der Störung, den Möglichkeiten der Familie, der Leistungsfähigkeit des Kindes und der Einbettung in die familiären und außerfamiliären Betreuungsstrukturen. Das Ziel besteht in jedem Fall in einer Verbesserung auf der pragmatisch-kommunikativen Ebene.

Nach wie vor ist die **Einzeltherapie** in der logopädischen Praxis, die durch eine Heilmittelverordnung des behandelnden Arztes ermöglicht wird, die häufigste Form einer Sprachtherapie. Nach den vorliegenden statistischen Daten liegt der Häufigkeitsgipfel der Behandlung im ausklingenden Kindergarten- sowie Grundschulalter. Die Studiendaten zur Therapie-Evaluation belegen, dass neben der traditionellen Einzeltherapie ebenso Maßnahmen der Gruppenbehandlung sowie insbesondere Elterntrainings zur Sprachförderung in der Effektivität der Unterstützung bei Sprachentwicklungsstörungen gleichrangig wirksam sind.

Wesentlich seltener sind **institutionelle Maßnahmen** in integrativen oder heilpädagogischen Kindergärten, Förderschulen oder medizinischen Einrichtungen wie den SPZ. Spezialisierte medizinische stationäre Behandlungsmöglichkeiten für Kinder mit besonders ausgeprägten Sprachentwicklungsstörungen, häufig in Verbindung mit Komorbiditäten, bestehen in einigen Kliniken. Die klinisch-stationären Maßnahmen stellen eine interdisziplinäre Blockbehandlung dar, die einen multimodalen Therapieansatz verfolgt. Aufgrund des hohen Personal- und Kosteneinsatzes sind diese spezialisierten Komplexleistungen beschränkt auf besonders ausgeprägte Formen der Sprachentwicklungsstörung, bei denen ambulante und sonderpädagogische Maßnahmen keinen ausreichenden Erfolg im Hinblick auf die Teilhabe bewirken konnten.

Stationäre medizinische Behandlung bei Sprachentwicklungsstörungen

- Bonn:
  Kinderneurologisches Zentrum, LVR-Klinik Bonn
  Abteilung für Sprach- und Entwicklungsstörungen
- Frankfurt:
  Clementine Kinderhospital, Abteilung für Neuropädiatrie
- Mainz:
  Klinik für Kommunikationsstörungen der Universitäts-HNO-Klinik

Die **SPZ** nehmen bei Kindern mit Sprachentwicklungsstörungen und assoziierten Entwicklungsabweichungen oder Behinderungen eine besondere Position ein. Ihr ambulanter interdisziplinärer Ansatz ergänzt die vertragsärztliche Versorgung mit spezifischen therapeutischen Angeboten. Dabei kann es notwendig sein, dass die Dauer der Konsultationen, die Therapiefrequenz oder die Gesamt-Behandlungsdauer erheblich von Vorgaben der Heilmittelrichtlinien abweichen. Bei den langfristigen punktuellen Interventionen wird vor allem bei eher älteren Kindern und Jugendlichen die Erstellung eines Gesamtbehandlungsplans unter Berücksichtigung der sonderpädagogischen Fördermöglichkeiten im Vordergrund stehen als eine Fortsetzung der spezifischen Funktionstherapie. In vielen SPZ bestehen darüber hinaus besonders Angebote in Form von Elterntraining, Spezialsprechstunden für die Versorgung mit Kommunikationshilfen im Rahmen der unterstützten Kommunikation oder interdisziplinäre Spezialsprechstunden unter Hinzuziehung externer Spezialisten aus den Bereichen Pädaudiologie, Hörgerätetechnik, Cochlea-Implantation, Dysphagie oder Kieferorthopädie.

Die Datenlage zur **Therapieevaluation** ist hinsichtlich der Wirksamkeit von Sprachtherapie ähnlich gering ausgeprägt wie bei allen anderen Therapiemaßnahmen auch. Allerdings gibt es bedeutsame Unterschiede in Abhängigkeit von den einzelnen Störungsbildern und auch dem Ausprägungsgrad. Die Evaluationen von Sprachtherapie beschränken sich in ihren positiven Aussagen vorwiegend auf die Abhilfe bei Artikulationsstörungen und die Erweiterung des Wortschatzes. Übersichtsarbeiten hierzu sind die Metaanalyse von Law et al. (2004) sowie der Vorbericht des Instituts für Qualitätssicherung und Wirtschaftlichkeit im Gesundheitswesen IQWiG (2008) zur Einschätzung der gesicherten Möglichkeiten von „Früherkennungsuntersuchungen auf Entwicklungsstörung des Sprechens und der Sprache bei Kindern". Insbesondere bei letzterer wird auf der Basis der vorhandenen Datenlage sehr viel Skepsis deutlich, inwiefern durch frühe Interventionen tatsächlich der Entwicklungsverlauf hinsichtlich der Sprache signifikant verändert und verbessert werden kann. Diese Einschätzung ist in Anbetracht der zunehmend deutlich werdenden genetischen Grundlagen bei Störungen der Rezeption und Grammatik nicht überraschend, stellt aber bei Betrachtung der Entwicklung eines Kindes mit einer Spracherwerbsstörung die Nützlichkeit des therapeutischen Handelns dennoch nicht in Abrede. Die Metaanalyse von Law et al. bestätigt, dass eine Sprachtherapie bei Abwesenheit einer ausgeprägten rezeptiven Sprachstörung die

sprachproduktiven Fähigkeiten hinsichtlich Lautbildung, aktivem Wortschatz und Grammatik verbessert. Das Sprachverständnis hingegen scheint kaum therapeutisch beeinflussbar. Es ist ebenfalls anhand der Datenlage nachzuvollziehen, dass längerfristige Angebote effektiver sind als kurzfristige. Bei Kindern mit ausgeprägten Sprachverständnisstörungen sind insgesamt deutlich geringere Therapieeffekte zu erwarten.
Grundsätzlich ist auch die Frage der **Nachhaltigkeit** einer Sprachtherapie noch nicht abschließend zu beantworten. Mittelfristige Nachkontrollen bis zu 5 Jahren nach Therapieende ebenso wie bis zu 2 Jahren nach Elterntraining zeigen länger anhaltende Effekte, während Longitudinalstudien bis ins Erwachsenenalter diese nicht belegen.
In keiner Studie finden sich nachweisbare unterschiedliche Effekte in Abhängigkeit von der eingesetzten **Methode** der Sprachtherapie. Zu beachten sind aber in jedem Fall **mögliche Nebenwirkungen** einer Sprachtherapie. Ritterfeld (2000) hat nachgewiesen, dass die Mitteilung einer Störungsdiagnose das Interaktionsverhalten ungünstig verändert. Erzieherinnen, die eine Gruppe von sprachunauffälligen (!) Kindern unter der vermeintlichen Diagnose „Sprachentwicklungsstörung" zu betreuen hatten, verfielen in ein „Sprach-Lehrverhalten", wurden dadurch in ihrer Interaktion rigider und eingeengter und verhielten sich schlussendlich in der Video-Dokumentation insgesamt weniger sprachförderlich.
Sprache ist die wichtigste Kommunikationsmöglichkeit und differenzierteste Fähigkeit des Menschen. Es ist deshalb sinnvoll, den Spracherwerbsstörungen besondere Aufmerksamkeit zu widmen. Auch wenn die spezifischen Sprachentwicklungsstörungen insbesondere bei stärkerer Ausprägung in hohem Maße genetisch determiniert scheinen und somit einer wie auch immer gestalteten übenden Behandlung nur begrenzt zugänglich sind, ist es dringend erforderlich, in einer von medialen Einflüssen geprägten und stark kommunikationsverarmenden Umwelt mehr Anstrengungen zu unternehmen, bereits im frühen Kindesalter den Spracherwerb zu unterstützen. Zumindest sekundäre Beeinträchtigungen durch Umfeldvariablen können vermieden werden. Einflüsse von Migration mit der Notwendigkeit des Zweitspracherwerbs und generelle Anregungsarmut sind durch Maßnahmen der systematischen Unterstützung und Angebotserweiterung in ihren Auswirkungen zu mindern. Trotz aller Bemühungen der frühen institutionellen Förderung in Krippe und Kindergarten bleibt aber festzuhalten, dass familiäre Einflüsse wesentlich stärker wirksam sind. Somit kommt der zielgerichteten Anleitung des elterlichen Verhaltens höchster Stellenwert zu.

Fallbeispiel Sprachtherapie

Sabrina, Erstvorstellung mit 2 ½ Jahren. Wortschatz ca. 30 Begriffe, stark phonologisch entstellt, häufig nur für die Mutter verständlich.
Drittes Kind der Familie. Beide älteren Brüder verzögerte, letztlich aber nicht behandlungsbedürftig gestörte Spracheentwicklung. In der Familienlinie väterlicherseits verschiedene Angehörige mit Auffälligkeiten in der Sprachentwicklung; Mutter leichte Lese-Rechtschreib-Störung.
Maßnahmen und Verlauf: Ausschluss einer relevanten Hörstörung durch pädaudiologische Diagnostik bei unsicherer Kooperation in der kinderärztlichen Praxis.
Teilnahme der Mutter am „Heidelberger Elterntraining" im regionalen SPZ zur Optimierung des sprachfördernden Verhaltens. Dort auch logopädische Basisdiagnostik mit Bestätigung von altersentsprechenden rezeptiven Sprachleistungen.
Auf Wunsch der Eltern Aufnahme des Mädchens im Alter von gut 3 Jahren im Regel-Kindergarten, den auch der mittlere Bruder noch besuchte. In den Folgemonaten rasche Zunahme des Wortschatzes, beginnende Mehrwort-Verbindungen, aber Verständlichkeit und Kommunikation mit Erwachsenen infolge der ausgeprägten phonologischen Störung weiterhin stark beeinträchtigt. Normale soziale Integration in der Kindergruppe.

Bei der Früherkennungsuntersuchung U8 diagnostische Ausweisungen: „Expressive Sprachentwicklungsstörung, Retardierung im gegenständlichen Malen und bei der Feinmotorik".
Einleitung einer logopädischen Behandlung, Ziel: Unterstützung in der expressiven Sprache, speziell Wortschatz und Lautbildung.
Bei der U9 trotz deutlicher Fortschritte mit Satzbildung weiterhin ausgeprägter Störungsbefund in der expressiven Sprache. Von logopädischer Seite diagnostische Zuordnung als „Zentrale verbal-expressive Dyspraxie". Gleichzeitig Verdeutlichung des Störungsbefundes in der visuellen Verarbeitung mit Unsicherheiten in der Formreproduktion und zeichnerischen Gestaltung.
Therapiepause in der Logopädie, Einleitung einer Behandlungssequenz in der Ergotherapie mit Fokussierung auf visuelle Verarbeitung und Graphomotorik unter Anleitung der Eltern zu spielerischer Förderung im familiären Alltag.
Verlaufsuntersuchung mit 5 ½ Jahren: Fortschritte in allen Störungsbereichen, aber unverändert keine altersdurchschnittlichen Leistungen. Wiedervorstellung im SPZ zur psychologischen Untersuchung und Planung des Vorgehens hinsichtlich der Einschulung.
In der standardisierten Intelligenzdiagnostik zeigt sich ein heterogenes Begabungsprofil. Die sprachlichen Leistungen sind insgesamt durchschnittlich, während Teilbereiche der visuell-schlussfolgernden Anforderungen unterdurchschnittlich ausfallen. Die separat überprüften zentral-auditiven Verarbeitungsleistungen wiederum sind ebenfalls unterdurchschnittlich.
Der mittelfristige Behandlungsplan wird mit den Eltern erörtert. Aufgrund der kombinierten umschriebenen Entwicklungsstörungen bei insgesamt normaler Intelligenz sind erhebliche schulische Lern- und Leistungsprobleme zu erwarten. Dieser Ausgangsposition von Sabrina muss angemessen Rechnung getragen werden. Die Eltern beantragen die Überprüfung des sonderpädagogischen Förderbedarfs. Parallel wird die ergotherapeutische Behandlung nach 6 Monaten beendet und erneut die logopädische Intervention fortgesetzt. Graphomotorik und Grundzüge der speziellen Förderung zur visuellen Verarbeitung sind ausreichend vorhanden, sodass jetzt die Anbahnung der phonologischen Bewusstheit in den Mittelpunkt gerückt wird. Die logopädische Behandlung wird in den folgenden Jahren als Intervalltherapie gestaltet. Hinsichtlich der Beschulung entscheiden sich die Eltern für den „Gemeinsamen Unterricht" in einer Integrationsklasse mit Präsenz einer Sonderpädagogin.
Sabrina lebt weiterhin mit einer ausgeprägten expressiven Sprachentwicklungsstörung. Im Schriftsprach-Bereich manifestiert sich eine Umschriebene Lese-Rechtschreib-Störung. Das Mädchen gerät unter Druck und zeigt Verhaltensauffälligkeiten. Deshalb entschließen sich die Eltern Mitte des zweiten Schulbesuchsjahres zu einem Wechsel mit Übergang auf die Förderschule mit dem Förderschwerpunkt Sprache bei gleichzeitiger Wiederholung des Schuljahres.
Sabrina kommt mit den Anforderungen besser zurecht und stabilisiert sich emotional. Die psychologische Verlaufsuntersuchung im SPZ zeigt bei weiterhin heterogenem Begabungsprofil eine Entwicklungs- und Nachreifungstendenz der problematischen zentralauditiven Verarbeitungsleistungen um die untere Normalbereichsgrenze. Ebenso liegen die Lese- und Rechtschreib-Leistungen in diesem Bereich.
Sabrina wechselt im Alter von knapp 11 ½ Jahren auf die Gesamtschule. Die jeweils für etwa 6 Monate mit gleich langem Pausenintervall durchgeführte logopädische Behandlung wird beendet. In der schulischen Beurteilung wird weiterhin Nachteilsausgleich hinsichtlich der Benotung in den schriftsprachlichen Fächern eingeräumt.
Das Mädchen ist in seiner Sprache weiterhin auffällig mit schwachen Ausdrucksleistungen. Die phonologische Störung besteht in einer undeutlichen Artikulation fort. Für die Alltagskommunikation bedeutet dies jedoch kein bedeutsames Hindernis mehr. Sabrina ist mit ihrer schulischen Situation zufrieden. Sie nimmt an Maßnahmen des Förderunterrichts teil.
Die Eltern haben über inzwischen fast ein Jahrzehnt die Entwicklung ihrer Tochter kompetent unterstützt. Sie wissen um die Notwendigkeit, bei Sabrina andere Erfolgsmaßstäbe anzulegen. Im Alter von 15 Jahren ist die nächste Verlaufsuntersuchung im SPZ vorgesehen. Auf der Basis dieser Befunde sollen dann mit der Jugendlichen selbst sowie den Eltern die Überlegungen einer auf Fähigkeiten und Ressourcen basierenden Planung von Ausbildung und Berufsfindung eingeleitet werden.

Fazit: der Entwicklungsverlauf von Sabrina zeigt exemplarisch die besonderen Herausforderungen bei Umschriebenen Entwicklungsstörungen: Auch wenn das Hauptaugenmerk auf der Sprache lag, wurde durch die differenzierten Beobachtungen und zielgerichteten Untersuchungen der vollständige Umfang des Problems aufgedeckt. Im Zusammenspiel von familiärer Förderung, fokussierter Therapie und angemessener Beschulung konnte das Entwicklungspotenzial des Mädchens realisiert werden. Dennoch bleibt auch die berufliche Eingliederung eine Herausforderung.
Der dosierte und wohlüberlegte Einsatz funktionstherapeutischer Maßnahmen über eine lange Zeit hinweg hat entscheidend zu der im Rahmen der Möglichkeiten positiven Entwicklung des Mädchens beigetragen.

Literatur

AWMF-Leitlinie (2011) Diagnostik von Sprachentwicklungsstörungen (SES), unter Berücksichtigung umschriebener Sprachenentwicklungsstörungen (USES). www.awmf.org/uploads/tx_szleitlinien/049-006l_S2k_Sprachentwicklungsstoerungen_Diagnostik_2012.pdf

Blank C, Petrik A, Rösch B, Voigt F, Lidtke K (2009) Sprachtherapie bei Störungen des Sprechens und der Sprache in der Sozialpädiatrie. Rahmenempfehlung. Qualitätszirkel der Bundesarbeitsgemeinschaft Sozialpädiatrischer Zentren. www.dgspj.de

Bode H (2008) Umschriebene Entwicklungsstörungen des Sprechens und der Sprache – Indikationen zur Verordnung von Sprachtherapie. www.dgspj.de

Boyle J, McCartney E, Forbes J, O'Hare A (2007): A randomised controlled trial and economic evaluation of direct versus indirect and individual versus group modes of speech and language therapy for children with primary language impairment. Health Technol Assess 7: 1–139.

Buschmann A, Joos B, Rupp A, Dockter S, Blaschtikowitz H, Heggen I, Pietz J (2008) Children with developmental language delay at 24 months: result of a diagnostik work-up. Dev Med Child Neurol 50: 223–229.

Institut für Qualität und Wirtschaftlichkeit im Gesundheitswesen (2009) Früherkennungs-untersuchung auf Entwicklungsstörung des Sprechens und der Sprache bei Kindern. Abschlussbericht 17.8.2009. www.iqwig.de/index.541.html

Keilmann A, Kiese-Himmel C (2011) Stationäre Sprachtherapie bei Kindern mit schweren spezifischen Sprachentwicklungsstörungen. Laryngo-Rhino-Otol 2011; 90(11): 677–682

Law J, Garrett Z, Nye C (2004) The efficacy of treatment for children with developmental speech and language delay/disorder: A meta-analysis, Journal of Speech, Language and Hearing Research 47: 924–943.

Ritterfeld U (2000) Zur Prävention bei Verdacht auf eine Spracherwerbsstörung: Argumente für eine gezielte Interaktionsschulung der Eltern. Frühförderung Interdisziplinär 2: 82–87.

Spinath FM, Price TS, Dale PS, Plomin R (2004) The genetic and environmental origins of language disability and ability. Child Development 75: 445–454.

Suchodoletz Wv (2004) Zur Prognose von Kindern mit umschriebenen Sprachentwicklungsstörungen. In: Suchodoletz Wv (Hrsg) Welche Chancen haben Kinder mit Entwicklungsstörungen? Göttingen: Hogrefe:155–199.

Suchodoletz Wv (2013) Sprech- und Sprachstörungen. Leitfaden der Kinder- und Jugend-Psychotherapie, Band 18. Hogrefe Göttingen

Internetlinks

www.giskid.eu

37.2.4 Ergotherapie

Grundlagen

Die **Ursprünge** der „Arbeits- und Beschäftigungstherapie" liegen zu Beginn des 20. Jahrhunderts in der Psychiatrie. Dort erfolgte die Entwicklung des Konzepts der Betätigung als heilendem Mittel unter Berücksichtigung der Erfahrung, dass durch die Aktivierung von Patienten Wohlbefinden indiziert werden kann. Dabei müssen Inhalte und Zielsetzung auf die Bedürfnisse des Individuums in seinem Krankheitserleben abgestimmt werden. In der Folge wurden die Arbeitsfelder auf die Bereich Orthopädie, Neurologie und Pädiatrie ausgeweitet. Der Einbezug von Hilfsmitteln (➤ Kap. 37.4) und insbesondere elektronischer Technik führte zu einer wesentlichen Erweiterung der Möglichkeiten v. a. auch in der Rehabilitation .

Die inhaltlichen **Konzepte** der pädiatrischen Ergotherapie waren eng an die Medizin und die dort vorherrschende Sichtweise des hierarchischen Aufbaus von Entwicklungsstufen angebunden. Erst in den letzten 25 Jahren fand ein Paradigmenwechsel statt mit Übergang zu einer ganzheitlichen, systemischen und dynamischen Auffassung von Krankheitssituationen und den dabei möglichen therapeutischen Wechselwirkungen. Damit einher ging auch die Namensänderung in die heute gebräuchliche Bezeichnung „Ergotherapie".

Der Veränderungsprozess ist noch nicht abgeschlossen. Parallel zur Umorientierung von der medizinischen Diagnose zur Lebenswelt des Patienten und vom Behandeln zur Aktivierung auf der eigenen Handlungsebene des Patienten findet durch Einführung der neuen Studiengänge eine Akademisierung statt. Damit gehen auch Entwicklungsprozesse einher, die zum Aufbau von wissenschaftlich gestützten konzeptionellen Modellen des Vorgehens führen. Diese lösen zunehmend die empirisch geprägten früheren Vorstellungen ab. Ähnlich wie in der Physiotherapie dienen die ergotherapeutischen Praxismodelle der theoretischen Untermauerung des Behandlungsverfahrens und systematisieren gleichzeitig den Behandlungsverlauf und das methodische Vorgehen. Dazu bieten die Konzepte alltagsrelevante und betätigungsorientierte Erhebungsverfahren und Befundinstrumente. Das seit längerem erprobte kanadische **Modell der kognitiven Orientierung** in der Ergotherapie CO-OP findet zunehmend auch in Deutschland Verbreitung.

Der Deutsche Verband der Ergotherapeuten (DVE) hat 2007 folgende **Definition** gegeben: „Ergotherapie unterstützt und begleitet Menschen jeden Alters, die in ihrer Handlungsfähigkeit eingeschränkt oder von Einschränkung bedroht sind, bei für sie bedeutungsvollen Betätigungen mit dem Ziel, sie in der Durchführung dieser Betätigungen in den Bereichen Selbstversorgung, Produktivität und Freizeit in ihrer persönlichen Umwelt zu stärken. Hierbei dienen spezifische Aktivitäten, Umweltanpassung und Beratung dazu, dem Menschen Handlungsfähigkeit im Alltag, gesellschaftliche Teilhabe und eine Verbesserung seiner Lebensqualität zu ermöglichen."

Gestaltung der Ergotherapie

Bei Kindern und Jugendlichen zielt Ergotherapie auf die Herstellung oder Verbesserung von Betätigungen und Handlungen ab, die für die Gesamtentwicklung förderlich sind und somit zur Alltagsbewältigung beitragen. Neben den Behandlungsstunden, in die die Eltern zur Anleitung und spielerischen Übung einbezogen sein sollen, ist je nach Situation im Einzelfall auch der Einsatz von Hilfsmitteln sinnvoll und notwendig.

Wesentliche Elemente bei der Ausgestaltung einer ergotherapeutischen Behandlung sind die ganzheitliche Sicht des Menschen in seiner Umwelt, Klientenzentrierung, Therapiefokus auf bedeutungsvollen Betätigungen mit dem Ziel der persönlichen Weiterentwicklung zu Selbstständigkeit und Handlungsfähigkeit, Aktivierung mit Hilfe zur Selbsthilfe sowie Einsatz von handwerklichen und gestalterischen Medien und Techniken, in der Pädiatrie alters- und entwicklungsbezogen auf spielerischem Niveau. Im Gegensatz zu früheren Vorstellungen eines Aufbaus von Funktionen über die Stärkung von Grundlagen und Basissinnen als „Bottom-up"-Ansatz stehen demzufolge heute mehr Überlegungen im „Top-down"-Ansatz mit konkreter Zielführung im Vordergrund.

Wesentliche Elemente der aktuellen pädiatrischen Ergotherapie
- Betätigung
- Partizipation
- Klientenzentriertheit
- Systemische Sichtweise
- Top-Down-Ansatz

Indikationen zur Ergotherapie bei Kindern und Jugendlichen können aus folgenden Auffälligkeiten resultieren:
- Alltagsrelevante Störung bei der Handlungsplanung und der Objektmanipulation
- räumlich-konstruktive Störungen
- Alltagsrelevante Störung der Fein- und Graphomotorik
- Alltagsrelevante Störung im Bereich der Grobmotorik mit erhöhter Unfallgefahr und sekundärer Ausgrenzung im sozialen Umfeld mit resultierender Einschränkung der Teilhabe
- Verhinderung von Chronifizierung und Schmerzen bei bestehenden gravierenden Entwicklungsstörungen insbesondere im Bereich der Muskeln und Knochen.

In den Heilmittelrichtlinien werden für die Pädiatrie verschiedene Krankheitsbilder unterschieden, bei denen Ergotherapie verordnet werden kann. Neben Erkrankungen des Stütz- und Bewegungssystems sind ZNS-Erkrankungen und/oder Entwicklungsstörungen ausgewiesen, ebenso geistige und psychische Störungen im Kindes- und Jugendalter. Somit bietet sich dem verordnenden Kinder- und Jugendarzt ein ungewöhnlich breites Spektrum der Interventionsmöglichkeiten, das vom ausführenden behandelnden Therapeuten ein hohes Maß an Kompetenz verlangt.
Die ärztliche Verordnung muss den Inhalt **Behandlungsinhalt** konkretisieren. Nicht die Diagnose allein ist Anlass für eine Therapie, sondern das Maß an Beeinträchtigung und Behinderung im Alltag des Kindes. Es gilt zu beachten, dass Ergotherapie kein Ersatz für pädagogische bzw. sonderpädagogische Maßnahmen darstellt, ebenso wenig die Funktion einer psychotherapeutischen Behandlung übernehmen kann und insbesondere kein Heilmittel bei insuffizient wahrgenommenen erzieherischen Aufgaben der Eltern ist.
Der Ergotherapeut vertieft im einleitenden Erstgespräch auf der Basis der ärztlichen Verordnung die Erfassung der Probleme zur Behandlungskonkretisierung mit den Bezugspersonen, möglichst auch dem Kind bzw. Jugendlichen selbst, und klärt die Erwartungen an die Therapie. Dies kann in freier Form erfolgen; besser sind jedoch **standardisierte Messinstrumente** in Form eines halbstrukturierten Interviews, das die Betätigungsprobleme des Kindes, die Art der Betätigungsausführung sowie schließlich seine oder die Zufriedenheit der Eltern mit dem Erfolg der Betätigungsausführung erfasst. Gut geeignet hierfür ist das COPM (Canadian Occupational Performance Measure) . Die adaptierte Version für Grundschulkinder COPMa-kids ist allerdings noch nicht standardisiert. Mithilfe des COPM können alltagsbezogene Betätigungsbedürfnisse formuliert, Therapieziele vereinbart und der Therapieverlauf strukturiert werden. Das Messinstrument kann auch zur Evaluation des Ergebnisses eingesetzt werden, um festzustellen, wie weit sich aus Sicht

des Patienten bzw. seiner Eltern die Betätigungsausführung verändert und verbessert hat. Ergänzend werden wesentliche Umweltbedingungen erhoben sowie im Rahmen einer Analyse der Betätigungsausführung und der Aktivitäten wichtige Faktoren für Motivation und Selbstkonzept des Kindes erfasst. Für diese einleitenden Tätigkeiten und die Entwicklung des primären Behandlungsplans unter ergotherapeutischer Zielformulierung werden in der Regel nicht mehr als zwei Behandlungseinheiten benötigt.

Ergotherapeutische **Ziele** sollen handlungs- und alltagsorientiert sowie überprüfbar sein. Neben den Handlungszielen werden funktionelle Teilziele formuliert. Diese sind durch die Befundaussagen im Rahmen der ärztlichen Veranlassung zu begründen. Handlungsziele und Zieldifferenzierung sollen nach außen transparent gemacht werden. Ziele auf Körperfunktionsebene wie Tonusregulation, Koordination oder Wahrnehmungsverarbeitung sind untergeordnet.

Für die erfolgreiche **Durchführung** einer zielgerichteten ergotherapeutischen Behandlung ist der Aufbau einer tragfähigen Beziehung zwischen dem Kind bzw. Jugendlichen, der Therapeut und den familiären Bezugspersonen die wesentliche Voraussetzung. Das Wesen des Kindes, sein Selbstkonzept und seine Motivation sind die Ausgangspunkte der Therapie. Die Ermöglichung eines positiven Kreislaufs der Selbstwirksamkeit über die Eigenaktivität des Kindes schafft die Voraussetzung für die Einleitung von Lernprozessen. Die Aktivitäten im Rahmen der Therapie sollen so gewählt werden, dass sie am Interesse des Kindes ansetzen und im engen Kontext mit seinen Handlungs- und Alltagsproblemen stehen. Eine unmittelbare Einbindung der Eltern in den Therapieverlauf ist notwendig, um über Verstärkung und positive Konnotation schlussendlich die Generalisierung und damit den Transfer in den Alltag zu Hause, im Kindergarten oder in der Schule zu erreichen.

Gerade in der ergotherapeutischen Behandlung ist es wegen der Vielzahl von Möglichkeiten und Ansatzpunkten wichtig, das **Therapieende** oder zumindest den Beginn einer Therapiepause bei der Gestaltung als Intervall-Behandlung zu definieren. Dies ist umso leichter, je konkreter das Behandlungsziel im Hinblick auf den Alltag des Kindes oder Jugendlichen definiert ist. Da Entwicklung auch im Falle einer Behinderung stets ein dynamischer, dabei allerdings unter Umständen sehr stark zeitlich verzögerter und damit langsamer Prozess ist, bedeutet die Nichterreichung eines zuvor definierten Behandlungsziels in aller Regel, dass dieses nicht optimal gewählt worden ist. Es ist von großer Wichtigkeit für die Eltern, dass der Therapeut vermeidet, durch unscharfe Formulierungen wie „Wahrnehmungsstörung" oder „Körperbewusstsein" dem Eindruck der Unentbehrlichkeit von Ergotherapie Vorschub zu leisten. Bei einer solchen, den aktuellen Auffassungen von Konkretisierung und Zieldefinitionen in der Ergotherapie nicht entsprechenden Vorgehensweise schadet sich der Therapeut letztlich auch selbst, da eine Erwartungshaltung der Heilung entsteht, die gerade bei Behinderungen nicht erreicht wird. Liegen umschriebene Entwicklungsstörungen vor, die wie z. B. eine Störung der Graphomotorik im Vorschul- und frühen Grundschulalter gut für eine fokussierte Ergotherapie geeignet sind, wird das Therapieende über die individuell festzulegende Steigerung der Alltagsbewältigung definiert; auch dies bedeutet keinesfalls zwangsläufig eine Normalisierung. In anderen Bereichen wie Handlungsplanung oder Aufmerksamkeitssteuerung kann die Intervallbehandlung auch über eine längere Zeit hinweg angezeigt sein, wobei wiederum unter dem Blickwinkel der Adaptation an das erforderliche Funktionsniveau im Alltag sich die einzelnen Behandlungsblöcke und deren jeweiliges Ende festlegen lassen.

Veränderungsprozesse beim Kind bzw. Jugendlichen im Rahmen der ergotherapeutischen Behandlung werden ermöglicht durch Entwicklung im Sinne der Reifung und Differenzierung, durch Lernen oder Anpassung sowie schlussendlich bei Störungen infolge von Unfällen oder einschneidenden, die normale Entwicklung unterbrechenden Erkrankungen durch Rehabilitation. Es existiert eine Vielzahl von Konzepten und entwicklungsorientierten Verfahren, bei denen durch externe Stimuli im Rahmen der Behandlung das Kind zu einer selbstmotivierten Handlungsweise angeregt wird. Hierzu zählen als **motorisch-funktionelle und sensomotorische Ansätze:**

- Basale Stimulation nach A. Fröhlich
- Sensorische Integrationsbehandlung nach Jean Ayres
- Neurophysiologischer Entwicklungsaufbau nach T. Pörnbacher
- Konzept nach Marianne Frostig
- Geführte Interaktionstherapie nach Félicie Affolter
- Castillo-Morales-Konzept
- Bobath-Konzept
- Psychomotorik nach J. Kiphard
- Methode nach M. Feldenkrais.

Als **biomechanische Konzepte** zur Verbesserung adaptiver Funktionen sind zu nennen:

- ADL-Training (Aktivitäten des täglichen Lebens)
- Rollstuhl-Training sowie der Umgang mit anderen Hilfsmitteln
- Kompensatorische zielorientierte Ansätze und Übungen
- PC-Training.

Im Rahmen der **psychosozial ausgerichteten Vorgehensweisen** werden sowohl entwicklungsbasierte als auch alltagsorientierte Ansätze über die Beeinflussung von Interaktion und Spiel sowie der Eltern-Kind-Beziehung verfolgt. Hierzu zählen:

- Bobath-Konzept
- Konduktive Förderung nach Petö.

Die **kognitiv-verhaltensbezogenen Konzepte** sind lerntheoretisch basiert und verhaltenstherapeutisch ausgerichtet mit spezifischer Wirkung. Hierzu sind zu zählen:

- Behandlungsansatz nach Polatajko CO-OP (Cognitive Orientation to Daily Occupational Performance; kognitive Orientierung bei der alltäglichen Betätigungsausführung), der im Rahmen der von Kanada ausgehenden Systematisierung durch die CAOT (Canadian Association of Occupational Therapists) zunehmend Verbreitung findet
- Training der Handfunktion bei Hemiparese im Rahmen der CIMT (Constraint Induced Movement Therapy)
- Verschiedene Trainingsprogramme zur Verhaltenssteuerung und Konzentrationsförderung bei ADHS, die auch im Rahmen der ergotherapeutischen Behandlung angewendet werden.

Die **Evidenz** für Maßnahmen der pädiatrischen Ergotherapie ist unter wissenschaftlichen Gesichtspunkten schwach. In der Cochrane-Bibliothek finden sich keine Angaben. Dem steht eine hohe Wertschätzung bei Eltern, Erzieherinnen und Pädagogen gegenüber. Unter Alltagsgesichtspunkten der notwendigen ärztlichen Verlaufsevaluation zur Frage der Fortsetzung oder Beendigung der Behandlung ist generell festzuhalten, dass sich Effekte um so besser nachweisen lassen, je konkreter die Zieloperationalisierung vorher erfolgt ist.

Günstiger ist dementsprechend die Datenlage bei den verhaltenstherapeutischen und auf isolierte motorische Funktionen ausgerichteten Konzepten, die sich gleichzeitig im Übergangsbereich zu Psychotherapie bzw. Physiotherapie befinden.

Nach den Heilmittelrichtlinien werden folgende **Behandlungsmaßnahmen** der Ergotherapie unterschieden:

- Motorisch-funktionelle Behandlung
- Sensomotorisch-perzeptive Behandlung
- Hirnleistungstraining und neuropsychologisch orientierte Behandlung
- Psychisch-funktionelle Behandlung.

Einsatz von Ergotherapie

- Vielfältige Möglichkeiten
- Einsatz ärztlich rational gelenkt je nach individueller Situation des Kindes
- Alltagsrelevante Betätigungsprobleme:
  je konkreter zu benennen, desto wahrscheinlicher Effektivität
- KEINE positiven Effekte für Verhalten belegt, sofern nicht Techniken der Verhaltenstherapie oder Systemischen Therapie eingesetzt werden
  (dann aber: cave „Pseudo-Psychotherapie"!)

Arbeitsfelder in der **motorisch-funktionellen Behandlung** sind Abbau pathologischer Haltungs- und Bewegungsmuster, Aufbau physiologischer Funktionen in Grob- und Feinmotorik, Entwicklung oder Verbesserung der Koordination und der funktionellen Ausdauer, Verbesserung von Gelenkfunktionen einschließlich Gelenkschutz, Desensibilisierung bzw. Sensibilisierung einzelner Sinnesfunktionen, Erlernen von Ersatzfunktionen, Verbesserung der eigenständigen Lebensführung auch unter Einbezug technischer Hilfen.

Die **sensomotorisch-perzeptive Behandlung** umfasst De- bzw. Sensibilisierung einzelner Sinnesfunktionen, Koordination mit Umsetzung und Integration von Sinneswahrnehmungen, Verbesserung der Körperwahrnehmung, Hemmung und Abbau pathologischer Haltungsmuster, Aufbau physiologischer Bewegungen, Stabilisierung sensomotorischer und perzeptiver Funktionen mit Verbesserung der Gleichgewichtsfunktion, Kompensation eingeschränkter praktischer Möglichkeiten durch Erlernen von Ersatzfunktionen, Verbesserung der Mund- und Essmotorik, Verbesserung der eigenständigen Lebensführung auch unter Einbezug technischer Hilfen.

Beim **Hirnleistungstraining** bzw. der **neuropsychologisch orientierten Behandlung** geht es um Verbesserung und Erhalt kognitiver Funktionen wie Konzentration, Merkfähigkeit, Aufmerksamkeit, Orientierung, Gedächtnis, Handlungsplanung oder Problemlösungsstrategien sowie Verbesserung der Grundarbeitsfähigkeiten und der eigenständigen Lebensführung, auch unter Einbezug technischer Hilfen, z. B. (Neuro-)Biofeedback.

Die **psychisch-funktionelle Behandlung** schließlich hebt ab auf Verbesserung und Stabilisierung der psychischen Grundleistungsfunktionen wie Antrieb, Motivation, Belastbarkeit, Ausdauer, Flexibilität, Selbstständigkeit und Selbstvertrauen, Erhöhung der Realitätsbezogenheit, der Selbst- und Fremdwahrnehmung, des situationsgerechten Verhaltens, der sozio-emotionalen Kompetenz und Interaktionsfähigkeit.

Zusammengefasst bietet Ergotherapie eine Fülle von Möglichkeiten. Es obliegt der besonderen Verantwortung von verordnendem Arzt und behandelndem Therapeuten, hier individuell das günstigste Vorgehen unter Einbezug des Kindes und der Eltern zu entwickeln.

Fallbeispiel Ergotherapie

Tom, Erstvorstellung mit knapp 4 Jahren. Allgemeiner Entwicklungsrückstand in Sprache und Motorik, insbesondere aber in der Eigenbeschäftigung. Stark reduzierter Sozialkontakt, perseverierendes und stereotypes Verhalten. Besuch des Regelkindergartens, Eltern nach konkreten Rückmeldungen durch die Erzieherinnen stark beunruhigt.

Erstes Kind der Familie. In der Familie mütterlicherseits sind Angehörige bekannt, die „etws eigenwillige Typen" seien. Die Mutter sieht in der „Zurückhaltung und Schüchternheit" von Tom ähnliche Verhaltensmuster wie bei sich im vergleichbaren Alter. Auch heute benötige sie außerhalb der Familie nicht unbedingt Menschen um sich. In ihrem Denken und Handeln sei sie sehr präzise, als Physikerin penibel.

Bei der Früherkennungsuntersuchung U7a Ausweisung mit „Leichter Entwicklungsverzögerung, Kontrolle nach 6 Monaten." Da die erhoffte Normalisierungstendenz nach Aufnahme in den Kindergarten ausblieb und sich die soziale Kontakt- und Interaktionsproblematik verdeutlichte, Überweisung ins SPZ zur weiteren Abklärung. Dort standardisierte mehrdimensionale Diagnostik, Ausweisung als Autismus-Spektrum-Störung (Hochfunktions-Autismus), knapp durchschnittliche Intelligenz bei rezeptiver Sprachentwicklungsstörung, Störung grob- und feinmotorischer Funktionen mit ausgeprägter Ungeschicklichkeit.
Maßnahmen und Verlauf: umfassende Information der Eltern über das Störungsbild. In der Folge regelmäßige problemzentrierte psychoedukative Anleitung. Parallel Einleitung einer ergotherapeutischen Behandlung, Ziel: Umsetzung des elterlichen Wunsches nach Aufbau von Eigenaktivität bei Tom sowie Unterstützung in der motorischen Entwicklung. Betätigungsanalyse nach COPM mit der Mutter, Identifikation von Spielideen mit Tom. Zeitkorridor 4 Monate, um Eigenaktivität zuhause für 10 Minuten, zeichnerische Gestaltung und Benutzung eines Laufrades zu erreichen.
Einbezug der Ergotherapeutin in die SPZ-Prozessdiagnostik hinsichtlich autistischer Muster von Tom und Erreichung der lenkenden Interaktionsgestaltung durch die Mutter.
Therapie-Evaluation: an einigen Tagen in der Woche ist Tom in der Lage, sich für ca. 10 Minuten zuhause selbst zu beschäftigen. Zeichnerische Gestaltung ist trotz ausreichender graphomotorischer Fertigkeit nicht erreicht worden, dafür aber intensives Schneiden mit der Kinderschere. Das Laufrad wird benutzt, wenn auch mit wenig Begeisterung. Erhebliche Entlastung der Mutter, die in der Kombination der Maßnahmen für sich eine Perspektive und Sicherheit aufgebaut hat.
Weiterer Verlauf: Beibehaltung der ergotherapeutischen Intervention mit kleinschrittiger Adaptation von alltagsrelevanten Betätigungszielen für Tom und gleichzeitigem Einbezug der Mutter unter Rückkopplung mit der behandelnden Psychologin.
Wegen sozialer Überforderung durch die Rahmenbedingungen der Großgruppe Wechsel in einen Integrativen Kindergarten mit gut 4 Jahren. Einbezug der Erzieherinnen in die psychologische Psychoedukation. Austausch zu Fördermöglichkeiten mit der Ergotherapeutin.
Inzwischen sehr gute zeichnerische Fertigkeiten von Tom, bevorzugt mit abstrakten Figuren oder Schienen und Röhren. In der mehrdimensionalen Entwicklungs-Verlaufsdiagnostik mit 6 ½ Jahren noch Schwächen im Sprachverständnis, aber kein Störungsbefund mehr. Heterogenes Begabungsprofil mit insgesamt altersentsprechender Intelligenz und Stärken in einzelnen visuell-schlussfolgernden Bereichen.
Beendigung der ergotherapeutischen Behandlung vor der Einschulung im Alter von 6 ¾ Jahren. Besuch einer kleinen Grundschulklasse im „Gemeinsamen Unterricht". Autismusspezifische Mitbetreuung durch das regionale Autismus-Therapie-Zentrum ATZ. Dennoch zum 2. Schulbesuchsjahr Zuschaltung einer pädagogisch qualifizierten Schulbegleitung als Integrationshilfe notwendig bei weiterhin bestehenden großen Schwierigkeiten in der Regelakzeptanz und hinsichtlich der Gestaltung von sozialen Kontakten. Schwimmen als motorikunterstützende Freizeitaktivität, bald mit Leidenschaft und großer Ausdauer von Tom betrieben. Hierüber auch Anbahnung einzelner freundschaftlicher Kontakte.
Wechsel zum Gymnasium. Klinisches Bild jetzt insgesamt an das Asperger-Syndrom erinnernd mit hervorragenden mathematisch-naturwissenschaftlichen Leistungen, ausgeprägter Diskussionsfreude und Detailverliebtheit, jedoch Schwächen in der schriftsprachlichen Umsetzung. Führende Problematik ist die Tendenz zu ausgeprägtem Eigensinn mit Rücksichtslosigkeit und Impulsdurchbrüchen im familiären Rahmen, was die Eltern bei inzwischen 3 Kindern bis an die Grenzen ihrer Belastbarkeit fordert.
Systematische Fortsetzung der Psychoedukation im SPZ für die Eltern mit verhaltenstherapeutischer Ausrichtung unter Einbezug von Tom als Jugendlichem. Wahrnehmung von spezifischen Trainingsangeboten im ATZ.
Mühsame, letztlich aber gelingende Stabilisierung der familiären Betreuung für Tom bis zum Abitur. Der junge Mann hat wenige belastbare Freundschaften. Seine berufliche Zukunft ist noch unklar. Er weiß um seine Besonderheiten im Leben mit einer autistischen Störung, hat aber eine aktive Befassung damit im Rahmen einer Psychotherapie abgelehnt.

Fazit: Der funktionstherapeutische Ansatz in der Ergotherapie mit konkreten und realisierbaren Zielsetzungen war von zentraler Bedeutung, um einige grundlegende Entwicklungssituationen positiv zu beeinflussen. Vor allem aber gelang es der Mutter, über die vertrauensvolle Beziehung zur Therapeutin auch die Schritte der Psychoedukation konkret einüben und umsetzen zu können. Möglicherweise wurde hier der Grundstein gelegt, um auch in den nachfolgenden Jahren die immer wieder von hohen Belastungen geprägten Lebensphasen mit Tom beharrlich und letztlich erfolgreich zu gestalten.

Literatur

Becker H, Steding-Albrecht U (Hrsg) (2006) Ergotherapie im Arbeitsfeld Pädiatrie. Stuttgart: Thieme.

Blank R, Wellhausen U, Gadegast B, Bisky A, Braunreuther S (2009) Ergotherapie in der Sozialpädiatrie. Rahmenempfehlung. Qualitätszirkel der Bundesarbeitsgemeinschaft Sozialpädiatrischer Zentren. www.dgspj.de

Dernick R (2012) RopE – Ressourcen-orientierte pädiatrische Ergotherapie. 5. erw. Aufl., www.familienergo.de

Gede H, Kriege S. Strebel H, Sulzmann-Dauer I (2007) Kinder zu Wort kommen lassen. Die adaptierte Version des Canadian Occupational Performance Measure für Grundschulkinder (COPM a-kids). Handbuch und Materialien zur praktischen Anwendung. Idstein: Schulz-Kirchner.

Polatajko H (2008): Ergotherapie bei Kindern mit Koordinationsstörungen – der CO-OP-Ansatz. Thieme Stuttgart.

37.3 Psychotherapie

Gestörte Entwicklung und (chronische) Krankheit oder Behinderung bei Kindern und Jugendlichen sind für Familien einschneidende und belastende Erfahrungen. Der stetige medizinische Fortschritt führt auch dazu, dass Familien ihr Leben immer länger auf solche ggf. völlig von der üblichen Situation abweichende Umstände ausrichten und Wege finden müssen, mit den vielfältigen Folgen umzugehen. Psychotherapie kann wichtig sein, um diesen Herausforderungen und Belastungen zu begegnen. Dabei ist ein bio-psycho-soziales Verständnis von Krankheitsentstehung hilfreich, um anhand der umfassenden Kenntnis der körperlichen Grundlagen einerseits den psychisch gestörten Kindern und Jugendlichen die Integration und soziale Teilhabe zu erleichtern, andererseits die Funktionsfähigkeit des Familiensystems für die Bewältigung der emotionalen und praktischen Aufgaben zu unterstützen.

Unter Psychotherapie wird die Heilung und Linderung seelischer Störungen mit Krankheitswert mit überwiegend psychischen Mitteln verstanden. Als seelische Krankheit definieren die Psychotherapierichtlinien des Bundesausschusses der Ärzte und Krankenkassen „krankhafte Störungen der Wahrnehmung des Verhaltens, der Erlebnisverarbeitung, der sozialen Beziehungen und der Körperfunktionen. Es gehört zum Wesen dieser Störungen, dass sie der willentlichen Steuerung durch den Patienten nicht mehr oder nur z. T. zugänglich sind. Als seelische Krankheit gilt auch eine geistige oder seelische Behinderung, bei der Rehabilitationsmaßnahmen notwendig werden."

37.3.1 Versorgungsebenen

Bedingt durch die Aufgabe sozialpädiatrischen Handelns, sich mit den Ursachen und Folgen komplexer Entwicklungsstörungen und Behinderungen zu befassen, ergeben sich auch für das psychotherapeutische Vorgehen teilweise andere Rahmenbedingungen.
Ziel jeder Intervention ist es, die Belange der betroffenen Kinder und Jugendlichen zu erfassen und im Hinblick auf eine optimale Teilhabe im Rahmen der jeweiligen individuellen Möglichkeiten zu verbessern. Die eigene Sicht der Patienten ist wesentlich, aber mitunter nur partiell einzubeziehen; oft stehen diesem Anliegen das Alter des Kindes oder das Ausmaß der Beeinträchtigung entgegen. Deshalb kommt in der Kinder- und Jugendpsychotherapie generell der Zusammenarbeit mit den Bezugspersonen eine besondere Wichtigkeit zu. Motivationale Klärung ebenso wie Festlegung der Zielinhalte und die Entscheidungsbildung zum therapeutischen Setting müssen ersatzweise auf dieser Erwachsenenebene erfolgen. Für die behandelnden Therapeuten bleibt die ethische Gesamtverantwortung in der altersangemessenen Patientenorientiertheit unberührt.
In der Sozialpädiatrie können folgende **Versorgungs- und Interventionsebenen** unterschieden werden:

- Psychotherapeutische Maßnahmen im Kontext der kinder- und jugendärztlichen Regelversorgung
- Psychotherapeutische Maßnahmen im Zusammenhang mit der Behandlung im SPZ
- Zielerreichung durch Fokussierung auf die Eltern, einzeln oder in der Gruppe
- Psychotherapeutische Behandlung des Kindes oder Jugendlichen, einzeln oder in der Gruppe, unter Einbezug der Eltern
- Familientherapeutisches Vorgehen.

Daneben spielen **psychoedukative Maßnahmen,** die sich sowohl an die Eltern als auch an andere Bezugspersonen richten, eine große Rolle.
Der Begriff „Eltern“ wird hier als Kurzform für das System von versorgenden nahen Bezugspersonen verwendet. Viele Familien sind aufgrund ihrer Lebenssituation gerade mit entwicklungsgestörten und behinderten Kindern auf Hilfe und Entlastung angewiesen. Dabei entspricht aufgrund von Trennung und neuen Partnerschaften das Elternpaar häufig nicht mehr den biologischen Eltern. Diese familiären Konstellationen müssen bei psychotherapeutischem Handeln beachtet und in den Behandlungsplan einbezogen werden.

37.3.2 Psychopathologischer Befund

Zur Einschätzung der Gesamtentwicklung ist die **psychometrische Diagnostik** mit Erhebung des Begabungsprofils, ggf. mit verschiedenen Verfahren, notwendig. Je nach Bearbeitungsfähigkeit des Kindes bzw. Jugendlichen sind **Fragebogeninstrumente** zur strukturierten, quantifizierenden und teilweise themenspezifischen Problemerfassung einzusetzen(Kap.17). **Projektive Verfahren** können ergänzend durchgeführt werden. Der **psychopathologische Befund** umfasst die **Beurteilung folgender Bereiche:**

- Orientierung
- Bewusstsein und Wahrnehmung
- Aufmerksamkeit und Konzentration
- Denken und Gedächtnis
- Affekt
- Ich-Erleben.

37.3.3 Abklärung organischer Ursachen

Durch die klinisch-neurologische Diagnostik, ggf. durch weitere technische Maßnahmen, muss stets vor Aufnahme einer Psychotherapie geklärt werden, ob eine fassbare organische Ursache für die geklagten Beschwerden und beobachteten Symptome vorliegt. Falls eine dauerhafte somatische Erkrankung besteht, müssen die Auswirkungen bekannt sein und in das psychotherapeutische Vorgehen einbezogen werden.

37.3.4 Diagnosestellung

Zu Einzelheiten des Vorgehens bei der Abklärung des Verdachts auf das Vorliegen spezifischer Krankheitsbilder wird auf die Leitlinien der AWMF verwiesen, ferner auf die Rahmenempfehlungen zur „Qualität in der Sozialpädiatrie". Entwicklungs- und psychische Störungen des Kindes- und Jugendalters gemäß ICD-10 stellen mögliche Indikationen für eine psychotherapeutische Intervention und Behandlung dar. Die Krankheitsbilder lassen sich dabei verschiedenen, im sozialpädiatrischen Versorgungsalltag wichtigen **Diagnosegruppen** zuordnen:

- Interaktions-, Regulations- und Bindungsstörungen
- Emotionale Störungen
- Expansive Störungen mit Störung des Sozialverhaltens und ADHS
- Zustand nach Trauma, physischer/psychischer Kindesmisshandlung ebenso wie nach Kindesvernachlässigung/-verwöhnung; Kinder psychisch kranker Eltern
- Psychische Störungen bei chronischen (neuro-)pädiatrischen Krankheitsbildern
- Psychische Störungen mit der Notwendigkeit von Hilfe bei der Bearbeitung einer Entwicklungsstörung oder Behinderung
- Phobische Störungen
- Somatoforme Störungen
- Nichtorganische Schlafstörungen
- Essstörungen, insbesondere bei Adipositas
- Enuresis, Enkopresis
- Tiefgreifende Entwicklungsstörungen
- Ticstörungen
- Umschriebene Entwicklungsstörungen (insbesondere schulischer Fertigkeiten, F81) mit sekundären Störungen des Erlebens und Verhaltens.

37.3.5 Durchführung der Psychotherapie

Bei gegebener Indikation zur Psychotherapie dienen die probatorischen Sitzungen der weiteren Abklärung von Krankheitseinsicht und Veränderungsmotivation. Spätestens danach ist verbindlich die Klärung des Anliegens und damit des Auftrags für die Psychotherapie notwendig.
Psychotherapie hat eine klare Zielorientierung auf der Grundlage einer Problemanalyse und Hypothesenbildung. Die zeitliche Limitierung von Psychotherapie ergibt sich aus den Richtlinienverfahren mit festgesetzten wöchentlichen ein- bis mehrmaligen Terminen. Neben inhaltlich fokussierten Kurzinterventionen sind gerade bei der Langzeitbegleitung von Familien mit behinderten Kindern auch niederfrequente psychotherapeutische Maßnahmen von Bedeutung. Das Kontinu-

um psychotherapeutischen Handelns umfasst daneben die psychotherapeutische Grundhaltung im Kontakt mit Patienten allgemein, eine dem gemäße Gestaltung der Sprechstunde und das offene Gesprächsangebot. Voraussetzung für eine solche Grundhaltung ist zumindest eine Basisqualifikation mit tiefenpsychologischen, verhaltenstherapeutischen oder systemischen Kenntnissen. Bei definierten psychischen Krankheiten erfolgt die Einleitung einer Psychotherapie nach den jeweils geltenden Richtlinien. Die Durchführung obliegt entsprechend ausgebildeten Psychologen, Ärzten und Kinder- und Jugendpsychotherapeuten, ggf. auch Ausbildungskandidaten im Rahmen der Supervision.

37.3.6 Psychotherapie in der sozialpädiatrischen Grundversorgung

Beratung und Prävention

Im gegliederten Gesundheitssystem kommt dem niedergelassenen Facharzt für Kinder- und Jugendmedizin die Primärfunktion in der sozialpädiatrischen Grundversorgung auch im Sinne der vorstehend skizzierten psychotherapeutischen Grundhaltung zu, da er für die Beratung im Rahmen der Früherkennungsuntersuchungen und die fachärztliche Begleitung zuständig ist. Dabei sollen Aspekte der emotionalen Entwicklung, des Verhaltens und der seelischen Gesundheit gezielt erörtert werden. Somit kommen dem Vertragsarzt auch die Aufgaben der Prävention in diesen Bereichen zu, die er durch entsprechende Beratungsansätze wahrnehmen muss. Viele niedergelassene Kinder- und Jugendärzte haben deshalb mittlerweile die Zusatzbezeichnung Psychotherapie erworben und sind hierdurch für diese Aufgabenstellungen besonders qualifiziert. Dies ist aber noch nicht durchgängiger Standard.

Psychoedukation

Im Übergangsbereich von spezifischen Verhaltensauffälligkeiten mit Beginn in Kindheit und Jugend sowie Störungen der Regulation, Interaktion und des Sozialverhaltens unterhalb der von der ICD-10 vorgegebenen Schwellenkriterien kann zunächst ein auf Informationsvermittlung und Veränderung zielender Beratungsansatz verfolgt werden. In dieser Psychoedukation wird insbesondere auf die Erläuterung basaler Kenntnisse zu den Ursachen und Wirkzusammenhängen der Störung/Erkrankung abgehoben. Hierfür ist im Rahmen der Basisversorgung der niedergelassene Facharzt für Kinder- und Jugendmedizin zuständig. Dabei können die Termine niederfrequent und bei günstigem Behandlungsverlauf anschließend bedarfsorientiert gestaltet werden. Auch hierbei ist aber die systematische Dokumentation des Behandlungsprozesses hinsichtlich Rückgang der Primärsymptomatik wie auch Beachtung möglicher Komorbiditäten zu fordern. Bei unzureichender Stabilisierung oder Verschlechterung ist die Überleitung in eine psychotherapeutische Behandlung notwendig.
Daneben sind auch andere, nicht primär psychotherapeutische Hilfemöglichkeiten zu nutzen. Hierzu zählen Maßnahmen von Frühförderinstitutionen, Therapeuten, (heilpädagogischen) Kindergärten, Schule und schulpsychologischem Dienst, Familien- und Erziehungsberatungsstellen sowie der Jugendhilfe.

Einbezug der Eltern – Kinder psychisch kranker Eltern

Den Eltern kommt in ihrer Funktion als Mittler für die medizinischen und psychischen Belange für das Kind bzw. den Jugendlichen zentrale Bedeutung zu. Durch die Aufklärung über das vorlie-

gende Störungs- bzw. Krankheitsbild und die vorgeschlagenen psychotherapeutischen Maßnahmen muss das Vertrauen gewonnen werden, um das notwendige therapeutische Bündnis herzustellen. Im Vergleich zu somatischen Erkrankungen ist dies ungleich schwieriger. Das therapeutische Vorgehen muss aber an den Vorstellungen der Eltern orientiert sein bzw. auf eine notwendige Änderung von Einstellungen hinwirken, da anders die Therapie des Kindes oder Jugendlichen nicht in der Alltagsrealität verankert werden kann. In der Regel kann davon ausgegangen werden, dass bei Einverständnis zur Einleitung einer Psychotherapie diese Voraussetzungen erfüllt sind.
Eine Besonderheit in der Praxis des niedergelassenen Kinder- und Jugendarztes stellt der ständige Kontakt zu den familiären Bezugspersonen dar. Psychische Störungen und Erkrankungen bei diesen Erwachsenen können deshalb mit erfasst werden. Aus der Perspektive des Kindes als Index-Patient ist es geboten, hierauf wegen der unmittelbaren negativen Auswirkungen für das Kind bewusst zu achten. Bei der Vermutung einer entsprechenden Diagnose oder dem Wissen darum ist es Aufgabe des Kinder- und Jugendarztes, den Eltern die erforderliche Beratung und Hilfestellung mit Zielsetzung des Aufsuchens geeigneter ärztlicher oder psychologischer Praxen und Kliniken zu geben. Dabei ist im Einzelfall auf eine mögliche Gefährdung des Kindeswohls zu achten.
Dieser Behandlungs- und Versorgungsaspekt für Kinder und Jugendliche in der Obhut psychisch kranker Eltern wird in der Praxis viel zu wenig gewürdigt. Zahlenmäßig kommt ihm eine größere Bedeutung zu als den meisten somatischen Krankheitsbildern. Hochrechnungen (Mattejat 2006) ergaben für Deutschland folgende Daten:

- 740.000 Kinder und Jugendliche mit einem alkohol- oder drogenabhängigen Elternteil
- 270.000 Kinder und Jugendliche mit einem an Schizophrenie erkrankten Elternteil
- 1.230.000 Kinder und Jugendliche mit einem an affektiven Störungen erkrankten Elternteil
- 1.555.000 Kinder und Jugendliche mit einem an Angststörungen erkrankten Elternteil.

Praktisches Vorgehen

- Präventivmaßnahmen („vorausschauende Gesundheitsberatung“) und Psychoedukation in Verbindung mit den pädiatrischen Früherkennungsuntersuchungen
- Beurteilung der psychischen Entwicklung eines Kindes oder Jugendlichen
- Einschätzung der psychischen Verfassung von familiären Bezugs- und Begleitpersonen
- Interaktions- und Gesprächsgestaltung orientiert an oder auf der Basis von psychotherapeutischen Vorgehensweisen
- Bei Verdacht auf das Vorliegen einer psychischen Störung beim Kind oder Jugendlichen: Durchführung der Basisdiagnostik mit Anamnese, Beobachtung des Kindes und seiner Interaktion sowie zumindest orientierender Testdiagnostik zur Erfassung anderer Entwicklungsstörungen
- Einleitung und Durchführung psychotherapeutischer Behandlung bei entsprechender Qualifikation und im Rahmen der zeitlichen Möglichkeiten der pädiatrischen Versorgungspraxis
- Überweisung zur Mit- und Weiterbehandlung bei Fehlen eigener Möglichkeiten.

Zusammengefasst ist es Aufgabe der psychotherapeutischen Grundversorgung in der Sozialpädiatrie, entsprechende Störungs- und Krankheitsbilder zu erkennen, einer angemessenen Behandlung zuzuführen oder bei differenzialdiagnostischen bzw. differenzialtherapeutischen Problemen die Kinder und Jugendlichen an geeignete Praxen oder Kliniken zu überweisen. Gleichzeitig müssen auch psychische Auffälligkeiten und psychiatrische Krankheiten der familiären Bezugspersonen erkannt, nach Möglichkeit die angemessene Beratung bzw. Behandlung induziert und insbesondere Auswirkungen auf das Kind erfasst werden. Die getroffenen Maßnahmen müssen der Entlastung dienen, indem die Familie einbezogen wird.

37.3.7 Psychotherapie im SPZ: Kriterien und Durchführung

Fachliche Voraussetzungen

Wie im „Altöttinger Papier zur Strukturqualität im SPZ“ dargestellt, ist bei den Mitarbeitern eines SPZ eine ausreichende psychotherapeutische Kompetenz erforderlich. Als Minimum bedeutet dies, dass ein Psychologe (Diplom- oder Master-Abschluss) als Psychologischer Psychotherapeut approbiert ist oder ein Arzt über die Zusatzbezeichnung Psychotherapie verfügt. Nach Möglichkeit sollten alle Mitarbeiter eines SPZ zur Erlangung einer psychotherapeutisch geprägten Grundhaltung entsprechend fortgebildet sein.

Anlass für die Überweisung

Psychische Auffälligkeiten und Störungen treten in Verbindung mit Entwicklungsstörungen, Behinderungen und chronischen Krankheiten häufig auf. Gerade in diesem Kontext ist die interdisziplinäre Zusammenarbeit von besonderer Bedeutung. Deshalb ist insbesondere bei Vorliegen eines komplexen Krankheitsbildes mit kombinierten Störungsbereichen im Rahmen der Mehrdimensionalen Bereichsdiagnostik in der Sozialpädiatrie (MBS) die Überweisung in ein SPZ mit entsprechendem multi-professionellem Team angezeigt. Das fallbetreuende Behandlungsteam eines Sozialpädiatrischen Zentrums muss im erforderlichen Umfang informiert sein und sein Vorgehen nicht nur hinsichtlich medizinischer und funktionstherapeutischer Maßnahmen, sondern ebenso in Bezug auf psychotherapeutische Interventionen abstimmen.

Vorgehen

- Sichtung der Vorbefunde im Rahmen der Überweisung durch den niedergelassenen Vertragsarzt
- Interaktions- und Gesprächsgestaltung auf der Basis von psychotherapeutischen Vorgehensweisen
- Klärung von Sichtweise und Anliegen des Kindes bzw. Jugendlichen und seiner Eltern
- Beurteilung der psychischen Entwicklung des Kindes oder Jugendlichen
- Einschätzung der psychischen Verfassung von familiären Bezugs- und Begleitpersonen
- Durchführung der Basisdiagnostik:
  - Biografische Anamnese
  - Beobachtung des Kindes bzw. Jugendlichen und seiner Interaktion
  - Erhebung des psychopathologischen Befundes
  - Testdiagnostik zur Erfassung des Begabungsprofils und Ausschluss anderer Entwicklungsstörungen
  - Einsatz von Fragebogenverfahren, insbesondere wenn die Einschätzung durch extrafamiliäre Bezugspersonen erforderlich ist
- Diagnosestellung auf Basis der Mehrdimensionalen Bereichsdiagnostik in der Sozialpädiatrie (MBS) mit Ausweisung nach ICD-10
- Festlegung des Vorgehens in Absprache mit dem verantwortlichen Arzt des SPZ
- Befundbesprechung mit den Eltern und je nach Anlass und Einsichtsfähigkeit mit dem Kind bzw. Jugendlichen mit dem Ziel der Herstellung des notwendigen therapeutischen Arbeitsbündnisses
- Arztbrief mit Formulierungen, die auch die Weiterleitung an die Eltern zulassen

- Einleitung und Durchführung psychotherapeutischer Behandlung bei entsprechender Qualifikation und im Rahmen der zeitlichen und finanziellen Möglichkeiten des SPZ
- Koordination der Weiterbehandlung bei Fehlen solcher eigener Möglichkeiten
- Evaluation des Behandlungsverlaufs entsprechend den Vorgaben des „Altöttinger Papiers“:
  - Selbstevaluation des Patienten bzw. seiner Bezugspersonen
  - Fremdevaluation durch den Therapeuten
  - Fremdevaluation durch Verlaufsuntersuchungen mit dem verantwortlichen Arzt
  - Außenevaluation durch Einbezug verfügbarer Informationen von anderen Bezugspersonen.

Inhaltliche Abstimmung

Das interne Prozessmanagement muss die gemeinsame Besprechung der Erstfälle und regelmäßige, idealerweise quartalsmäßige Verlaufsbesprechungen umfassen. Optimal zur frühzeitigen fachlichen Erfassung der Mehrdimensionalität von Problemen ist die Gestaltung des Ersttermins im SPZ mit ärztlich-psychologischem Tandem als Gesprächspartner für die Eltern. Umgekehrt eröffnet dies den Eltern einen erleichterten Zugang zur Erörterung psychischer Abweichungen in der Familie und bietet damit Niederschwelligkeit für den Einstieg in einen psychotherapeutisch strukturierten Beratungsansatz.

Einbezug der Bezugspersonen

Die Behandlung in der Sozialpädiatrie muss die Eltern einbeziehen. Auch hier erfolgen im indizierten Einzelfall psychotherapeutische Interventionen. Bei der psychotherapeutischen Behandlung von Bezugspersonen im Rahmen der Behandlung im SPZ muss eine dies begründende Diagnose beim Kind vorliegen, die aus dem unmittelbaren Zusammenspiel mit der interventionsbedürftigen Störung auf der Erwachsenenebene resultiert. Dieses Vorgehen ist ausdrücklich konform mit der Formulierung in § 119 SGB V, wonach die „Art der Krankheit“ einen herausragenden Stellenwert hat. Interaktionale Bedingungen auf dem Hintergrund psychischer Störungen der Eltern sind kaum anders aufgreifbar.

Behandlungsnetzwerk

Die optimale psychotherapeutische Behandlung der Patienten setzt die Kenntnis der regional vorhandenen Kompetenzen und Möglichkeiten voraus. Die Ausbildung sinnvoller Versorgungsnetzwerke kann über interdisziplinäre Qualitätszirkel gelingen.

Übergänge zwischen den beteiligten Professionen und Institutionen müssen definiert und auf Funktionalität überprüft werden. Gerade bei psychisch Kranken kommt der Kontinuität und Zuverlässigkeit von Behandlungsmaßnahmen eine hohe Bedeutung zu. Deshalb ist Kooperation bei der psychotherapeutischen Behandlung von Kindern und Jugendlichen sowie deren Familien von zentraler Bedeutung für das Gelingen des therapeutischen Prozesses.

Je nach Einzelfall sind weitere fachliche Partner wie das Jugendamt, andere Beratungsstellen, aber auch Kindergarten und Schule einzubinden. Es kann erforderlich sein, ein qualifiziertes Case-Management zu installieren; hierbei spielen besonders Sozialpädagogen eine wichtige Rolle.

Sozialpädiatrischer Arztbrief

Der zusammenfassende Arztbrief informiert über die Behandlung im SPZ. Hinsichtlich der Kooperation im psychotherapeutischen Prozess kann der Arztbrief eine wesentliche Rolle übernehmen. Er ist ein wichtiges Instrument zur Dokumentation der Vereinbarungen wie auch zur späteren Kontrolle des Erfüllungsgrades im Verlauf. Zuständigkeiten und Verantwortlichkeiten können so nachvollziehbar festgelegt werden. Durch die Zustellung einer Kopie auch an die Eltern wird deren Position im Behandlungsprozess gewürdigt und ihr Einbezug gewahrt.

Spezielle Behandlungsangebote

Psychotherapeutische Maßnahmen stehen im Sozialpädiatrischen Zentrum stets im Kontext der Entwicklungs- oder Verhaltensauffälligkeit bzw. Behinderung oder chronischen Krankheit des Kindes oder Jugendlichen. Gerade in der Verknüpfung von ärztlichen oder funktionell-therapeutischen Behandlungen einerseits und psychotherapeutischen Interventionen andererseits liegt die Chance der Akzeptanz der Eltern für ein solches stützendes und Ressourcen eröffnendes Vorgehen. Aufgrund der vielfältigen Möglichkeiten des multiprofessionellen Teams in den SPZ werden unterschiedliche Möglichkeiten und spezielle Behandlungsangebote für Kinder, Jugendliche und Eltern vorgehalten. Hierzu zählen die folgenden Bereiche.

Themenzentrierte Therapiegruppen für Kinder oder Jugendliche. Zum Beispiel soziales Kompetenztraining, Konzentrations- und Aufmerksamkeitstraining bei ADHS, Entspannungsverfahren.

Themenzentrierte Therapiegruppen für Eltern. Zum Beispiel Betreuung nach Frühgeburt, bei ADHS.

Interaktionstherapie. Konkrete Elternanleitung unter systemischen oder verhaltenstherapeutischen Vorgehensweisen; videogestützte Intervention.

Eltern-Säuglings-Psychotherapie. Sog. „Schreibaby-Sprechstunde“.

Fokussierte Behandlung bei psychosomatischen Störungen. Zum Beispiel chronische Kopfschmerzen, Adipositas.

Psychotherapie mit geistig behinderten oder autistischen Kindern und Jugendlichen. Besonderes Augenmerk wird je nach Erfahrung und Spezialisierung auf die Durchführung psychotherapeutischer Maßnahmen mit geistig behinderten Kindern und Jugendlichen gelegt. Rückzugsverhalten, fremd- oder autoagressives Verhalten, impulsives Verhalten u. a. stellen Indikationen dar, oft mit Einsatz verhaltenstherapeutischer Maßnahmen. Aufgrund der reduzierten kognitiven Reflexionsmöglichkeiten setzt die Therapie ein besonderes Maß an Einfühlungsvermögen seitens des Psychotherapeuten voraus. In ähnlicher Weise gilt dies für die Psychotherapie mit Kindern und Jugendlichen, die an Erkrankungen aus dem Autismus-Spektrum leiden.

Psychotherapie bei physischer oder psychischer Traumatisierung. Ein spezielles Thema in der kinder- und jugendmedizinischen Praxis wie Akutklinik, das hohe psychotherapeutische Kompetenz erfordert, ist der Umgang mit Misshandlungs- und Missbrauchsfällen. Die subtile Verwahrlosung und Gefährdung von Kindern und Jugendlichen wird häufig erst im Rahmen einer Prozessdiagnostik in der Psychotherapie deutlich, die wegen primär unspezifischer Verhaltensstörungen eingeleitet worden war. Im sozialpädiatrischen Behandlungskontext ist überzufällig häufig mit dem Auftreten solcher Probleme bei den Patienten zu rechnen, die belasteten oder sozial benachteiligten Familien entstammen. Das Kindeswohl hat dabei Priorität vor der Schweigepflicht. In jeder psychotherapeutischen Situation muss eine Prozessbeschreibung für das Krisenmanagement bei der Aufdeckung einer Kindestraumatisierung vorhanden sein, in der das Vorgehen mit Einschaltung geeigneter Fachpersonen geregelt ist. In Einrichtungen wie dem SPZ ist es darüber hinaus empfehlenswert, einen Kinderschutzbeauftragten intern zu qualifizieren. Weitere Erläuterungen finden sich in Kapitel 36 (➤ Kap.36).

Psychotherapeutisch fundierte Begleitung des Akzeptanz- und Bewältigungsprozesses bei der Vermittlung schwerwiegender Diagnosen.

Krisenintervention in familiären Belastungssituationen. Spezielle Gesprächsführung auch mit fokussierter Kurzzeitintervention.

Psychotherapie und Migration. Die stetig steigende Zahl von Menschen mit Migrationshintergrund stellt besondere Anforderungen auch an die Psychotherapie. Psychische Störungen und Krankheiten werden oftmals anders als in der deutschen Kultur bewertet und verarbeitet. Die Daten der KiGGS-Studie (www.kiggs.de) zeigen, dass die Diagnostik von psychischen Probleme sowie deren Behandlung bei Menschen mit Migrationshintergrund unterrepräsentiert ist (➤ Kap. 4.7). Psychotherapeutisches Handeln setzt die Sprache als wesentliches Medium ein, und gerade hier bestehen bei Familien mit Migrationshintergrund besondere Schwierigkeiten. Mangelnde Deutschkenntnisse stellen dann ein zusätzliches Hindernis für die Verständigung und adäquate Therapieangebote dar. Deshalb ist bei Patienten und ihren Familien mit Migrationshintergrund auf die angemessene Versprachlichung von psychischen Problemen mindestens durch Einschaltung professioneller Dolmetscher, im Idealfall aber Übernahme der psychotherapeutischen Behandlung durch muttersprachliche Therapeuten zu achten.

Mittelfristig wird eine Annäherung an die Lösung dieses Problems aber nur gelingen, wenn Maßnahmen der Sprachförderung frühzeitig im Kindergarten und unter Einbezug der Mütter als Nebeneffekt auch die Durchführbarkeit von Psychotherapie erst ermöglichen.

37.3.8 Kriterien für die Auswahl von Therapieverfahren

Es existieren vielfältige psychotherapeutische Methoden und Verfahren. Erstattungsfähig durch die Krankenkassen sind die **Richtlinien-Verfahren** Psychoanalyse, tiefenpsychologisch fundierte Psychotherapie und Verhaltenstherapie. Systemische Therapie und Gesprächspsychotherapie einschließlich ihrer kindspezifischen Modifikationen (personenzentrierte Spieltherapie) stehen ebenfalls, zumindest bezogen auf einzeln indizierte Störungsbilder, vor der Zulassung.

Aus Sicht der im pädiatrischen Bereich Tätigen sind Gestalttherapie, Hypnotherapie, Psychodrama sowie Kunst- und Kreativitätstherapie im Spektrum der Behandlungsmöglichkeiten für Familien wichtig. Die Auswahl von Therapieverfahren richtet sich nach dem diagnostizierten Krankheitsbild. Insbesondere bei externalisierenden Störungen ist der Einsatz von **Verhaltenstherapie** indiziert.

Der niedergelassene Facharzt für Kinder- und Jugendmedizin wendet als Primärarzt Elemente der Verhaltenstherapie in Bezug auf die Kinder und Jugendlichen und der Gesprächstherapie in Bezug auf die Eltern an. Sofern ein aufwändigeres Vorgehen möglich ist, können auch psychodynamisch fundierte Ansätze angewandt werden.

Dies stellt in aller Regel aber bereits den Übergang zu einer Psychotherapie im engeren Sinne dar, die bei entsprechender Zusatzqualifikation unmittelbar in der Praxis möglich ist oder für deren Umsetzung zu niedergelassenen Psychologischen bzw. Kinder- und Jugendlichenpsychotherapeuten überwiesen wird. Bei Komorbidität mit psychiatrischen Erkrankungen oder akuter stationärer Behandlungsnotwendigkeit muss das notwendige Vorgehen eingeleitet werden. Dies gilt insbesondere bei vermuteter oder drohender Suizidalität sowie anderem selbst- und fremdgefährdendem Verhalten.

37.3.9 Empirie und Evidenz

Psychotherapie und psychotherapeutische Interventionen sind zeit- und kostenintensive Behandlungsmaßnahmen. Deshalb ist grundsätzlich eine Evaluation durchzuführen. Die Schwierigkeit

des Wirksamkeitsnachweises ist bekannt, insbesondere was mittel- und langfristige Effekte betrifft. Im sozialpädiatrischen Behandlungskontext sind jedoch ohnehin häufig langfristige Begleitungen des Kindes und seiner Familie aufgrund des Krankheitsbildes notwendig. Deshalb besteht zumindest die Chance einer Annäherung an eine aussagekräftige **Evaluation** durch den Einsatz geeigneter Fragebogeninstrumente (z. B. zur Reduktion der Krankheits- und Störungssymptomatik, Lebensqualität, Alltagsbewältigung etc.). Diese können sowohl auf die Selbst- als auch die Fremdevaluation ausgerichtet sein. Sie ergänzen die eigene Überprüfung des Psychotherapeuten anhand des Erfüllungsgrades der vereinbarten Behandlungsziele im Sinne einer basalen Messung der Interventionen prä/post in Relation zum Anliegen. Dazu muss die Beobachtung der Prozessqualität systematisiert werden, um als gemeinsame Verantwortung von Kind, Eltern und Therapeutin auch ein Abbild der Alltagsrealität zu schaffen.
Mitarbeit („Compliance") und Widerstand, Übertragung und Gegenübertragung sind zentrale **Elemente in der Psychotherapie**. Es obliegt der besonderen Verantwortung des Psychotherapeuten, diese Aspekte im Behandlungsprozess kontinuierlich zu beachten und professionell therapeutisch darauf zu reagieren.
Die Erfolgsmessung psychotherapeutischer Behandlung und Intervention ist auch deshalb schwierig, weil systematische Rückmeldungen durch Dritte fehlen. Die Informationen von Eltern wie überweisenden Kinder- und Jugendärzten sowie anderen extrafamiliären Bezugspersonen müssen deshalb im Rahmen der Fremdevaluation systematisch einbezogen werden.
Aus Sicht der Patienten ist ein Erfolg der Psychotherapie dann gegeben, wenn eine subjektiv ausreichende Problem- und Symptomminderung eingetreten ist. Die Verringerung der Beeinträchtigung durch das Symptom führt zu Teilhabe, Normalität und Eingliederung in die soziale Gemeinschaft
Die bestgestützte Aussage der Psychotherapieforschung bezieht sich auf die zentrale Bedeutung der zwischenmenschlichen Beziehung; die therapeutische Allianz oder das Arbeitsbündnis wird als allgemeiner **unspezifischer Wirkfaktor** bezeichnet.
Daneben finden sich **methodenspezifische Wirkprinzipien.** Für psychodynamische Therapien seien exemplarisch die Formulierung der Übertragungsbeziehung und das Erfassen des zentralen Beziehungskonflikts in den Deutungen des Therapeuten sowie der Zuwachs an Einsicht und Selbstverständnis genannt. Für die Verhaltenstherapie (VT) kann beispielhaft für das Phasenmodell die Verhaltensanalyse sowie die Planung, Auswahl und Durchführung spezieller Methoden angeführt werden.
Als die vier **allgemeingültigen Wirkprinzipien** der Psychotherapie werden heute angesehen:

Allgemeingültige Wirkprinzipien in der Psychotherapie

- Ressourcenaktivierung
- Problemaktualisierung
- aktive Hilfe zur Problembewältigung
- motivationale Klärung

Die Wirksamkeit des psychotherapeutischen Vorgehens ist in unterschiedlichem Ausmaß belegt. Gute Daten existieren hinsichtlich solcher Krankheits- und Störungsbilder, bei denen das Vorgehen der Wahl verhaltenstherapeutisch ausgerichtet ist. Schwächer ist die Datenlage in Bezug auf das tiefenpsychologisch fundierte Vorgehen. Der systemische Ansatz ist inzwischen ebenfalls als Richtlinien-Verfahren zugelassen. In der Sozialpädiatrie ist die Wirksamkeit der systemischen Sichtweise belegt.

37.3.10 Psychotherapie in der Stationären Sozialpädiatrie

An mehreren Zentren in Deutschland besteht die ergänzende Möglichkeit der stationären interdisziplinären sozialpädiatrischen Behandlung. Hier werden Maßnahmen als Block- oder Intensivtherapie im Einzel- oder Gruppensetting durchgeführt.
Psychotherapie spielt hierbei eine wesentliche Rolle, da gerade im stationären Kontext solche Patienten betreut werden, die besonders schwerwiegende, chronifizierte oder komplexe Störungs- und Krankheitsbilder aufweisen. Die Anleitung und Unterstützung der Bezugspersonen ist dabei für den Therapieerfolg von zentraler Bedeutung. Die unmittelbare Möglichkeit, mit den verschiedenen beteiligten Berufsgruppen eng zusammenzuarbeiten und dabei die Familie sehr konkret und zeitlich intensiviert einzubeziehen, optimiert die Rahmenbedingungen für eine erfolgreiche Intervention.
Zu Einzelheiten der stationären sozialpädiatrischen Behandlung wird auf ➤ Kap. 37.5 verwiesen.

37.3.11 Zukünftige Entwicklungen und Notwendigkeiten

Die Verbreitung der „neuen Morbiditäten" mit Schwerpunkt auf chronischen und psychischen Krankheiten erfordert ein verändertes Vorgehen bereits in der Praxis des Kinder- und Jugendarztes. Das Verständnis für familiäre und andere systemische Zusammenhänge ist hierbei wichtig. Damit gewinnt die **qualifizierte Ausbildung** in der Durchführung psychotherapeutischer Maßnahmen zunehmend an Bedeutung. Gleichzeitig bereitet die ausgeprägte Spezialisierung auch in der Pädiatrie erhebliche Schwierigkeiten, diesen Wissenserwerb in einer überschaubareren Ausbildungszeit überhaupt zu leisten.
Es wird notwendig sein, ein **funktionierendes Netzwerk** von Kinder- und Jugendlichenpsychotherapeuten der verschiedenen Provenienzen zu etablieren. Dies ist komplementär zu verstehen zu den Versorgungsnetzwerken der Pädiatrie einerseits und Kinder- und Jugendpsychiatrie andererseits.
In den Sozialpädiatrischen Zentren ist neben Zusatzausbildungen in der Psychotherapie zumindest für den ärztlichen und psychologischen Bereich die Vermittlung eines **psychotherapeutischen Grundverständnisses** für alle dort tätigen Mitarbeiter zu fordern. Die überzufällig häufige Konfrontation mit psychisch kranken Menschen in den SPZ bedingt das Risiko des Scheiterns eines Beziehungsaufbaus bereits in der Diagnostikphase, dem durch entsprechende Weiterbildung professionell vorgebeugt werden muss. Darüber hinaus schützt ein Grundverständnis von Psychopathologie auch die Mitarbeiter selbst vor übermäßigen emotionalen Belastungen als Folge fehlenden Verständnisses für entsprechende Situationen. Erforderlich ist eine stabile innere Haltung, die sowohl Empathie als auch professionelle Distanz sichert. Kollegialer Austausch und regelmäßige Supervisionsangebote dienen diesem Aspekt zusätzlich.
Ähnlich wie in den Sozialpädiatrischen Zentren stellt sich die Situation in den Spezialambulanzen und Polikliniken der pädiatrischen Akutkliniken dar. Auch hier müssen begleitende seelische Entwicklungsstörungen und psychische Auffälligkeiten erkannt und einer angemessenen Behandlung zugeführt werden können. Aufgrund der Zunahme chronischer intern-pädiatrischer Krankheitsbilder ist mit einem Anstieg entsprechender Problemstellungen zu rechnen, auch bei den familiären Bezugspersonen.
Erforderlich ist ferner eine vertiefende **Versorgungsforschung** zum Einsatz von psychotherapeutischen Maßnahmen gerade bei behinderten Kindern und Jugendlichen. Hierfür müssen die notwendigen personellen und finanziellen Ressourcen eröffnet werden.

Literatur

Beelman A, Schneider N (2003) Wirksamkeit von Psychotherapie bei Kindern und Jugendlichen. Eine Übersicht und Metaanalyse zum Bestand und zu Ergebnissen der deutschsprachigen Effektivitätsforschung. Zeitschrift für Klinische Psychologie und Psychotherapie 32 (2) 129–143.

Döpfner M, Lehmkuhl G (2002) Die Wirksamkeit von Kinder- und Jugendlichenpsychotherapie. Psychologische Rundschau 53: 184–193.

Hollmann H, Grigat I, Kamp G, Mock I, Reichelt S, Streeck S, Paulus F (2009) Psychotherapie in der Sozialpädiatrie. Rahmenempfehlung. Qualitätszirkel der Bundesarbeitsgemeinschaft Sozialpädiatrischer Zentren. www.dgspj.de

Petermann F, Reinecker H (2005) Handbuch der klinischen Psychologie und Psychotherapie. Göttingen: Hogrefe.

Schmidtchen S (2001) Allgemeine Psychotherapie für Kinder, Jugendliche und Familien. Ein Lehrbuch. Stuttgart: Kohlhammer.

37.4 Hilfsmittel

37.4.1 Rechtliche Grundlagen

Es ist eine der wesentlichen Errungenschaften der Menschheit, Gegenstände zu Hilfsmitteln umzufunktionieren oder sich bei ihrem Einsatz physikalische Gesetzmäßigkeiten zunutze zu machen, um den Alltag besser bewältigen zu können. Werkzeuge ebenso wie Essbesteck, Flaschenzug und Auto oder Bleistift und elektronische Medien sind aus unserer Welt nicht mehr fortzudenken. Ein Großteil der Menschheit trägt eine Brille.

Unter medizinischen Gesichtspunkten sind Hilfsmittel dazu gedacht, im Einzelfall

- den Erfolg der Krankenbehandlung zu sichern,
- einer Behinderung vorzubeugen oder
- eine Behinderung bei der Befriedigung von Grundbedürfnissen des täglichen Lebens auszugleichen.

Der **Anspruch** gesetzlich Versicherter auf Hilfsmittel wird in Sozialgesetzbuch V § 33 geregelt. In der aktuell seit Februar 2009 gültigen Fassung der Hilfsmittelrichtlinien des Gemeinsamen Bundesausschusses werden Hilfsmittel als sächliche Mittel oder technische Produkte definiert, die individuell gefertigt oder als serienmäßig hergestellte Ware an die Versicherten abgegeben werden. Dabei kann bei Basisprodukten auch eine entsprechende individuelle handwerkliche Zurichtung, Ergänzung oder Abänderung notwendig sein. Gemäß den gesetzlichen Bestimmungen gehören zu den Hilfsmitteln:

- Sehhilfen
- Hörhilfen
- Körperersatzstücke
- Orthopädische und andere Hilfsmittel.

Alle von der Leistungspflicht der gesetzlichen Krankenkassen erfassten Hilfsmittel werden im **Hilfsmittelverzeichnis** aufgeführt. Das Verzeichnis gliedert sich in Produktgruppen. Bei jeder Produktgruppe wird eine systematische Unterteilung in Anwendungsorte, Untergruppen und Produktarten vorgenommen. Für jede Produktart ist ein Indikationsrahmen angegeben. Die Vertragsärzte führen die Verordnung von Hilfsmitteln durch, um dem Patienten eine nach den Regeln der ärztlichen Kunst und dem allgemein anerkannten Stand der medizinischen Erkenntnis ausreichende, zweckmäßige und wirtschaftliche Versorgung mit Hilfsmitteln zukommen zu lassen. Der verordnende Arzt muss den Zustand des Patienten kennen und sich über die persönlichen Lebensumstände informieren. Die Notwendigkeit für die Verordnung von Hilfsmitteln ergibt sich nicht allein aus der Diagnose. Unter **Gesamtbetrachtung** nach Kriterien der ICF mit Berücksichtigung der funktionellen/strukturellen Schädigungen, der Beeinträchtigungen der Aktivitäten (Fähigkeitsstörungen), der noch verbliebenen Aktivitäten und einer störungsbildabhängigen Diagnostik sind

- der Bedarf,
- die Fähigkeit zur Nutzung,
- die Prognose und
- das Ziel

einer Hilfsmittelversorgung auf der Grundlage realistischer, für den Patienten alltagsrelevanter Anforderungen zu ermitteln. Dabei sind die individuellen Kontextfaktoren als Voraussetzung für das angestrebte Behandlungsziel zu berücksichtigen.

In weiteren Ausführungen spezifiziert die Hilfsmittelrichtlinie den Inhalt der Verordnung sowie die Abgabe von Hilfsmitteln. Im Rahmen der ärztlichen Abnahme bei Übergabe an den Versicherten soll der Vertragsarzt prüfen, ob das abgegebene Hilfsmittel seiner Verordnung entspricht und den vorgesehenen Zweck erfüllt, insbesondere dann, wenn es individuell angefertigt oder zugerichtet wurde.

37.4.2 Praktische Umsetzung

Die vorstehenden Ausführungen der Hilfsmittelrichtlinie sind plausibel und scheinen das notwendige Vorgehen bei der Versorgung mit Hilfsmitteln gut zu regeln. In der täglichen Praxis ist jedoch kaum ein Bereich bei der Behandlung von Kindern und Jugendlichen mit Behinderungen so schwierig und mit Streitfällen behaftet wie die Hilfsmittelversorgung. Dies hat verschiedene Gründe: Durch die enorme technische Weiterentwicklung von Hilfsmitteln und Verwendung neuer Materialien bei der Herstellung sind in den letzten Jahren besondere Fortschritte für die Patienten erreicht worden. Gleichzeitig bedingt dies aber auch eine zunehmende Unübersichtlichkeit des Angebots und eine damit einhergehende Steigerung des Preisniveaus. Die Mehrzahl der verordnenden Vertragsärzte, die Therapeuten und die Hilfsmittelversorger können bei dem inzwischen erreichten Angebotsumfang kaum noch einen vollständigen Überblick über sämtliche Versorgungsmöglichkeiten aus eigener Anschauung gewinnen. Folge können suboptimale, die Zielsetzung für den Patienten nicht erfüllende oder auch unwirtschaftliche Hilfsmittelversorgungen sein. Gleichzeitig sind die Rechtsgrundlagen so umfassend geworden, dass den meisten Patienten, aber auch nicht wenigen verordnenden Vertragsärzten diese Materie fremd ist. Auch Vertreter der Krankenkassen als Kostenträger überblicken nicht immer die Rechtslage.

Gerade im Kindes- und Jugendalter, wo zu den häufig vorhandenen individuellen Schwierigkeiten der Adaptation von Hilfsmitteln noch die Dynamik des Wachstums und der Entwicklung hinzukommt, empfiehlt es sich deshalb, die Versorgung mit Hilfsmitteln in Zentren durchzuführen, in denen verschiedene Professionen unmittelbar beim Patienten zusammenkommen können. Hierzu zählen die Behandelnden, die Pflegefachkräfte und die die Versorgung durchführenden Physiotherapeuten, Ergotherapeuten und Logopäden, einschlägig versierte Kinder- und Jugendärzte sowie Kinderorthopäden und die mit der Technik befassten Rehabilitationsberater und Orthopädietechniker. Die Eltern und andere nahe Bezugspersonen geben die wesentlichen Umfeldinformationen zur Alltagspraxis. Eine bedarfsgerechte und wirtschaftliche Hilfsmittelversorgung ist deshalb das Resultat einer aufwändigen Prozesskette, die die folgenden Schritte umfasst:

Prozesskette Hilfsmittelversorgung

- Indikationsstellung
- Definition des Versorgungsziels
- Wahl und Anpassung des Hilfsmittels unter Berücksichtigung der Alternativen, ggf. mit Erprobungsphase
- Zuständigkeits-, Notwendigkeits-, Zweckmäßigkeits- und Wirtschaftlichkeitsprüfung durch den Kostenträger, ggf. mit Prüfung durch den MDK; nachfolgend Zustimmung zur Versorgung mit Kostenregelung
- Anproben mit Anpassung an die individuellen Gegebenheiten des Patienten
- Abnahme durch den verordnenden Vertragsarzt
- Auslieferung an den Patienten.

Auch nach Übergabe des Hilfsmittels an den Patienten und seine Betreuungspersonen sowie die behandelnden Therapeuten sind weitere Funktionsprüfungen unter den Bedingungen des Alltags notwendig, im Einzelfall auch Schulungsmaßnahmen im Rahmen zu verordnender Funktionstherapien.

Im weiteren Verlauf muss das **Erreichen des Versorgungsziels** überprüft und je nach Weiterentwicklung von Funktionen und körperlichen Gegebenheiten beim Patienten fortgeschrieben werden. Komplikationen mit Funktionseinschränkungen können im technischen Bereich auftreten, aber auch trotz sorgfältiger Anpassung in Form von Druckstellen, Hautunverträglichkeiten und anderen körperbezogenen Irritationen. Wenn im Vorfeld der Versorgung nicht wirklich aufmerksam auf die eigene Anwendungsmöglichkeit des Patienten oder auch genügende Assistenz geachtet worden ist, wird dies zu mangelnder Akzeptanz und nachfolgender Nichtbenutzung führen. Solche Fehl- und Überversorgungen sind z. B. bei Stehständern (zu mühsam) und elektronischen Kommunikationshilfen (zu komplex) immer wieder zu beobachten. In solchen Fällen ist nach Möglichkeit eine Weiterverwendung der Hilfsmittel für andere Patienten anzustreben, wobei eine Zweitversorgung in der Kostenbewilligung schwierig zu erwirken ist.

Für eine gelingende Hilfsmittelversorgung bei Kindern und Jugendlichen sind für die Berücksichtigung der individuellen und z. T. behinderungsspezifischen Entwicklungsdynamik umfangreiche Erfahrungen und Kenntnisse erforderlich. Häufig bestimmt sich allein schon daraus die Entscheidung, ob ein Hilfsmittel konfektioniert oder mit individuellen Anpassungen bzw. Zurüstungen zu wählen ist oder nach Maß angefertigt werden muss, was auch eine erhebliche wirtschaftliche Tragweite hat. Mehrfachbehinderte Kinder und Jugendliche erfüllen wegen dieses beträchtlichen Aufwands und der Anforderungen an den Prozess bei der Hilfsmittelversorgung regelhaft die Vorgaben für eine Behandlung im SPZ nach §119 SGB V. Die SPZ sind gleichzeitig aufgrund der vorhandenen Teamstrukturen und interdisziplinären Zusammenarbeit für die Übernahme dieser Aufgabe prädestiniert.

37.4.3 Anforderungen bei der Hilfsmittelversorgung von körper- und mehrfachbehinderten Kindern und Jugendlichen

Die Versorgung mit Hilfsmitteln ist als Bestandteil eines Habilitationskonzepts zu sehen, dessen Ziel die WHO so definiert: „(Re-)Habilitation umfasst den koordinierten Einsatz medizinischer, sozialer, beruflicher, pädagogischer und technischer Maßnahmen sowie Einflussnahmen auf das physische und soziale Umfeld zur Funktionsverbesserung, zum Erreichen einer größtmöglichen Eigenaktivität und zur weitestgehend unabhängigen Partizipation in allen Lebensbereichen, damit der Betroffene in seiner Lebensgestaltung so frei wie möglich wird."

Primär muss sich die Versorgung mit Hilfsmitteln am Alter des Kindes und individuellen und sozialen Bedürfnissen orientieren. Dieser Aspekt verdient besondere Beachtung, da gerade bei kleinen oder schwerbehinderten Kindern und Jugendlichen die Bedürfnisse der versorgenden Bezugspersonen eine wichtige Rolle spielen. Dennoch muss versucht werden, das Kind mit seinem individuellen Bedarf tatsächlich zu erfassen. Wenn eine praktikable deutsche Version der „Internationalen Klassifikation der Funktionsfähigkeit, Behinderung und Gesundheit für Kinder und Jugendliche" (ICF-CY) (➤ Kap. 16) der Weltgesundheitsorganisation WHO vorliegt, wird es möglicherweise einfacher sein, diese Gesichtspunkte mit der Realisierung von Teilhabe besser zu beschreiben und im Verlauf zu dokumentieren.

Neben den Bedürfnissen des Kindes muss sich die Hilfsmittelversorgung an Art und Ausprägungsgrad der Behinderung orientieren. Hieraus leiten sich Informationen über das Ausmaß der Versorgungsnotwendigkeit ab. Die Festlegung der therapeutischen Ziele setzt eine genaue Diagnose sowie klinische Kenntnisse über den individuell zu erwartenden Verlauf der Krankheit voraus.

Schließlich sind auch die Kontextfaktoren, die sowohl als Förderung ebenso wie als Barriere wirken können, bei der Versorgung zu beachten, wobei auch die Hilfsmittel selbst zu den Kontextfaktoren zu rechnen sind. Dies spielt eine große Rolle bei der Betrachtung der Wohnraumsituation oder der geografischen Lage der Wohnung.

Die Aktivitäten des Kindes werden durch Hilfsmittel unterstützt, wenn das Kind in die Lage versetzt wird oder sich seine Chancen dazu verbessern, all die Aktivitäten zu entfalten, die in seinem Alter sowie seiner gesellschaftlichen und sozialen Position üblich und von ihm bzw. seinem Umfeld gewünscht sind. Dazu gehört, dass durch Hilfsmittel die Handlungskompetenz in den **wichtigen Lebensbereichen** gesteigert wird, die zur Integration und Teilhabe zählen. Dies sind:

Prozesskette Hilfsmittelversorgung

- Einbeziehung des Kindes als Subjekt in Gesellschaft und Umwelt
- Positive Beeinflussung von Zufriedenheit und Lebensqualität des Kindes und seiner nahen Bezugspersonen
- Förderung von Wertschätzung und Anerkennung des Kindes
- Ermöglichung bzw. Förderung der Handlungsfähigkeit des Kindes, die die praktische Teilnahme am Alltagsleben und die Übernahme und Erfüllung entsprechender sozialer Funktionen und Rollen unterstützt. Dazu zählen:
  - Lernen und Wissensanwendung
  - Allgemeine Aufgaben und Anforderungen
  - Kommunikation
  - Mobilität
  - Selbstversorgung
  - Häusliches Leben
  - Interpersonelle Interaktionen und Beziehungen
  - Bedeutende Lebensbereiche (u. a. Arbeit, Beschäftigung)
  - Gemeinschaftliches, soziales und staatsbürgerliches Leben.

Bei diesen vielfältigen Aspekten der Teilhabe soll nicht nur die Leistungsfähigkeit (ICF: capacity), sondern v. a. die tatsächliche Ausführung einer Handlung (ICF: performance) gefördert werden.

Die **Körperfunktionen** des Kindes werden von Hilfsmitteln unterstützt, wenn:
- fehlende Funktionen ersetzt oder kompensiert werden
- Restfunktionen unterstützt werden
- die Hilfsmittel vom Kind in die Planung und Kontrolle seiner Handlungen integriert werden.

Die **Körperstrukturen** des Kindes werden von Hilfsmitteln unterstützt, wenn:
- sie präventiv auf Deformitäten wirken
- bestehende Deformitäten zunehmend korrigiert werden
- der Verlauf von primären, sekundären und tertiären Deformitäten verhindert oder verlangsamt wird
- der Verlust von Gliedmaßen ausgeglichen wird.

Hilfsmittel können für die betroffenen Kinder und Jugendlichen keine Heilung bewirken, wohl aber oftmals einen bemerkenswerten Zugewinn an Funktion und Fähigkeit ermöglichen. Hierdurch kommt es bei vielen Patienten zu einer wesentlichen Steigerung der Zufriedenheit und Lebensqualität. In Anbetracht der oftmals hohen Kosten bei einer differenzierten Hilfsmittelversorgung kommt der Expertise des verordnenden und fertigenden Teams eine besondere Verantwortung und damit Bedeutung zu. Gerade auch die Erprobung von Möglichkeiten setzt voraus, dass entsprechende Geräte vorrätig gehalten werden oder kurzfristig probeweise zu beschaffen sind. Für diesen speziellen und technikabhängigen Bereich der Intervention bei behinderten Kindern und Jugendlichen ist deshalb die Bildung von Kompetenzzentren anzustreben, möglichst mit unmittelbarer Einbeziehung der Kostenträger.

Evaluation der Hilfsmittelversorgung

Problemlos nachzuvollziehen ist die Wirksamkeit der Versorgung mit Hilfsmitteln z. B. bei Beeinträchtigungen des Sehens oder Hörens. Hier sind die Effekte auch durchaus objektivierbar. Ganz anders stellt sich die Situation bei Körpergeistiger oder Mehrfachbehinderung dar. Insgesamt ist hier die Wirksamkeit der Hilfsmittelversorgung wenig evaluiert. Dies wäre eine Domäne der Versorgungsforschung, die aber erst am Anfang steht. Solange keine wissenschaftlichen Daten im Sinne einer Evidenz vorliegen, müssen deshalb andere geeignete Vorgehensweisen etabliert werden, um die Versorgung der Patienten mit Hilfsmitteln zu sichern und zu verbessern. Qualitätsstandards in den Behandlungszentren, die empirisch begründet sind, dienen dazu, eine Hilfsmittelversorgung mit von außen nachvollziehbarer Zielsetzung und einem für alle Beteiligten klar dargestellten Ablauf zu gewährleisten. Dies erhöht die Wahrscheinlichkeit für eine sachgerechte Bearbeitung bei den Kostenträgern und somit die zügige Realisation der Verordnung. Derzeit ist aber zu beobachten, dass aufgrund geänderter rechtlicher Grundlagen die Schwierigkeiten noch zunehmen. Einzelverträge mit Versorgern, öffentliche Ausschreibungen und anderes zur Wettbewerbsstärkung mit dem Ziel der Kostenersparnis mögen wirtschaftlich sinnvoll sein, tragen aber häufig aufgrund unzureichender Sachkenntnis der Beteiligten weder zu einer Verbesserung der Versorgungsqualität für den Patienten, noch zur erhofften Kostensenkung bei den Krankenkassen bei.

37.4.4 Übersicht über die zur Verfügung stehenden Hilfsmittel

Die unterschiedlichen Hilfsmittel sind detailliert im Hilfsmittelverzeichnis aufgeführt (www.rehadat.de). Dieser Katalog ist für die Leistungsträger nur eine Auslegungshilfe, kein rechtsverbindlicher Positivkatalog. Deshalb können auch nicht gelistete Hilfsmittel im Zuge einer Einzelfallentscheidung genehmigt werden.

Grundsätzlich werden Hilfsmittel nach der Erhebung des körperlichen Befunds und der damit verbundenen Fähigkeiten und Funktionen sowie individueller Beratung angepasst. Dabei werden folgende **Bereiche und Faktoren** berücksichtigt:
- Kopf-, Rumpf- und Sitzstabilität
- Steh- und Gehfähigkeit
- Greiffunktion
- Sprache, Kognition und Perzeption
- Hören, Sehen, Propriozeption
- Kontrolle der Ausscheidungsfunktionen
- Komorbiditäten (z. B. Epilepsie, pulmonale Erkrankungen, Ernährungsstörungen mit PEG)
- Kontextfaktoren, insbesondere die Unterstützung durch nahe Bezugspersonen und die Motivation des Kindes
- das gegenwärtige Behandlungskonzept insbesondere bei multimodaler Therapie.

Hilfsmittel aus dem Bereich der **Orthopädietechnik** sind:
- Lagerungsorthesen
- Funktionsorthesen
- Rumpforthesen.

Hilfsmittel aus dem Bereich der **Rehabilitationstechnik** sind:
- Hilfen zum Lagern
- Hilfen zum Liegen, Sitzen, Stehen und zur Mobilität
- Hilfen zur Kommunikation, zum Sehen und Hören
- Hilfen zur Dekubitusprophylaxe und bei Inkontinenz.

Außerdem sind Hilfen zur Eingliederung in Form von Personenbegleitung bzw. persönlicher Assistenz möglich.

Hilfsmittel aus dem Bereich der Orthopädietechnik

Lagerungsorthesen.
- Werden während Ruhephasen, v. a. nachts, getragen
- Dienen der Vermeidung von Gelenkfehlstellungen
- Dienen zur Unterstützung der Therapie, werden jedoch meist nicht während Aktivitäten getragen
- Dienen postoperativ zum Erhalt des Operationsergebnisses und zur Kontrakturprophylaxe
- Können zur Korrekturverstärkung mit statischem Quengelgelenk oder dynamischer Feder ausgestattet sein.

Die Zielsetzung der Lagerungsorthesen für Rumpf, obere/untere Extremität muss die Belange des Kindes angemessen berücksichtigen. Das soziale Umfeld sollte den Nutzen der Lagerungsorthesen gewährleisten können. Die Gewährleistung der Korrektur steht im Vordergrund, deshalb sollten innerhalb der Orthesen keine Kompensationsmöglichkeiten gegeben sein. Mögliche Hilfsmittel sind Becken-, Bein- und Fußorthesen mit und ohne Gelenke sowie Lagerungssysteme für Segment- und Ganzkörperlagerung aus Schaumstoff, auch als Pelottensystem.

Funktionsorthesen für die obere und untere Extremität.
- Dienen zur Unterstützung eines möglichst physiologischen und funktionalen Bewegungsablaufs
- Sollen bestehende Funktionen im täglichen Gebrauch erhalten und verbessern
- Sollen wenig wiegen und leicht zu handhaben sein
- Geschwindigkeit und Symmetrie stehen im Vordergrund.

Die Materialauswahl ergibt sich aus der Anforderung an die Orthese hinsichtlich:
- Aktivität, Größe und Gewicht des Kindes
- Funktion und Gebrauchsintensität der Orthese
- Anforderungen an die Hygiene
- Komplikationsrisiko (z. B. bei Sensibilitätsstörungen).

Mögliche Hilfsmittel an den unteren Extremitäten sind Einlagen, Fuß- und Knöchel-Fuß-Orthesen, Ringorthesen, stabilisierende Schuhversorgung mit bettender Einlage, Schafterhöhung, Schaftversteifung, Sohlenverbreiterung oder Orthesen mit gesperrtem oder freiem oberem Sprunggelenk/Kniegelenk/Hüftgelenk (mehrdimensional).

An den oberen Extremitäten kommen in Betracht:
- Silikon-Handorthesen
- Daumenoppositionsorthesen
- Unterarm-Handorthesen
- Oberarm-Unterarm-Handorthesen.

Rumpforthesen. Sie dienen der Unterstützung und Stabilität sowie ggf. Korrektur der Rumpfhaltung (Korsett, Sitzorthesen). Dabei werden bestmögliche Korrektur und Stabilität durch zirkulär wirkende Rumpforthesen erzielt unter Verwendung eines rigiden Materials bei Rotationsabweichungen in der Wirbelsäule. Rumpforthesen können in Form von aktiver Korsettkonstruktion wie z. B. das Cheneau-Korsett eingesetzt werden, was aber einen deutlich erhöhten Aufwand bei der Pflege bedeutet, oder in Form von Miedern.

Hilfsmittel aus dem Bereich der Reha-Technik

Lagerungshilfen. Verschiedene Polster, Kissen, Keile.

Hilfen zum Liegen. Individuell angefertigte Bauch- und Seitlagerungssysteme zur Vermeidung von Kontrakturen oder postoperativ zur Unterstützung des Operationsergebnisses, zur Vermeidung von Druckstellen oder zur Unterstützung des Sekretflusses (Abhusten).

Hilfen zum Sitzen. Sitzschalen nach Maß in unterschiedlicher individueller Fertigungstechnik.

Hilfen zur Unterstützung des Sitzens.
- Gurtsysteme, Beckenbügel, etc. zur Fixierung des Beckens
- Brustgurte, Leibchen, Reklinationspelotten, Therapietisch etc. zur Unterstützung der Aufrichtung
- Kopfstützen zur Führung des Kopfes in einer gewünschten Position.

Hilfen zum Stehen.
- Stehbrett mit oder ohne Aufgurtung und Neigungsmöglichkeit
- Stehständer feststehend, schrägverstellbar in Bauch- oder Rückenlage oder als mobiler Stehständer, der es ermöglicht, durch selbstständiges Betätigen der Antriebsräder Therapie und Mobilität miteinander zu verbinden.

Hilfen zur Mobilität. Zum Beispiel Rampe, Lifter, Therapierad. Krankenfahrzeuge im Sinne von Rollstuhluntergestellen und Gehhilfen, z. B. Adaptiv- und Aktivrollstühle, Elektrorollstühle, elektrische Zusatzantriebe. Rollatoren, Gehwagen, Posteriorwalker, Gehstock u. a. m.

Hilfen zur therapeutischen Bewegung. Zum Beispiel Spreizrad, Rollbrett, Pezziball.

Hilfen zur Lagerung und Pflege. Krankenpflegeartikel, Lagerungshilfen, Pflegehilfsmittel, Badehilfen, Toilettenhilfen, Pflegebetten sind individuell altersentsprechend zu verordnen.

Hilfen bei passagerer oder chronischer Inkontinenz bzw. zur Dekubitusprophylaxe. Je nach Ausmaß der Behinderung sind auch Hilfen bei passagerer oder chronischer Inkontinenz bzw. zur Dekubitusprophylaxe erforderlich.

Hilfen zur Kommunikation. Der Begriff der Kommunikation umfasst im allgemeinen Sprachgebrauch alle Prozesse der Informationsübertragung. Hierunter fallen nicht nur sprachliche und sprachassoziierte Prozesse, sondern auch die Nachrichtenvermittlung über Medien und technische Einrichtungen. Kommunikation ist durch unterschiedliche Arten von Zeichen möglich. Funktionsdefizite in einem oder mehreren dieser Bereiche können durch geeignete Hilfsmittel kompensiert werden. Dabei entsprechen die eingesetzten Hilfsmittel nicht immer unbedingt der Wirkungsweise der natürlichen Körperfunktion.

Kommunikationshilfen im Sinne des Hilfsmittelverzeichnisses sind ausschließlich Gegenstände, die die direkte lautsprachliche und/oder schriftliche Mitteilungsmöglichkeit eines Menschen unterstützen bzw. diese erst ermöglichen.

Ferner zählen dazu Produkte, die aufgrund fehlender Hörfähigkeit benötigt werden, dieses Defizit jedoch nur indirekt ausgleichen, z. B.:

- Tafeln oder Symbolsammlungen mit Symbolen und Worten
- Kommunikationsgeräte mit Schrift- oder Sprachausgabe
- Behindertengerechtes Computerzubehör
- Geräte zur Kommunikationsunterstützung mit taktiler Ausgabe
- Signalanlagen für Gehörlose.

Hilfen zum Sehen. Sehhilfen sind optische bzw. opto-elektronische Vorrichtungen, die zur Korrektur oder dem Ausgleich, der Verbesserung und der Behandlung eines Krankheitszustands des Auges dienen.

Hilfen zum Hören. Hörhilfen sind technische Hilfen, die angeborene oder erworbene Hörfunktionsminderungen, die einer kausalen Therapie nicht zugänglich sind, ausgleichen. Die Schwerhörigkeit allein stellt keine Indikation für eine Hörgeräteausstattung dar, sondern erst der Nachweis, dass durch die ordnungsgemäße Anpassung von Hörgeräten ein Sprachverständnisgewinn oder ggf. eine Verbesserung des Richtungshörens möglich ist. Eine weitere Voraussetzung ist ferner, dass der Versicherte oder eine Bezugsperson in der Lage ist, das Hörgerät sachgerecht zu bedienen.

Hilfen zur Eingliederung in Form von Personenbegleitung; persönliche Assistenz

Im Rahmen der Eingliederungshilfe nach §§ 2, 37 SGB V in Verbindung mit §§ 17, 55 SGB IX kann es notwendig sein, eine Pflegekraft zur Sicherung der ärztlichen Behandlung einzusetzen (Absaugen, Katheterisieren, Sauerstoffversorgung), wenn ein behinderter Mensch eine Regelschule/Regelkindergarten besucht. Auch der Transport von behinderten Menschen von und in die Regeleinrichtung muss gewährleistet sein.

Literatur

Peters H, Happes K, Incoronato P, Schmidt-Ohlemann M, Stein-Bayet B, Steinebach S, Wollstädter C (2009) Hilfsmittelversorgung im Kindes- und Jugendalter in Sozialpädiatrischen Zentren. Rahmenempfehlung. Qualitätszirkel der Bundesarbeitsgemeinschaft der Sozialpädiatrischen Zentren (in Vorbereitung). http://www.dgspj.de

Richtlinie des Gemeinsamen Bundesausschusses über die Verordnung von Hilfsmitteln in der vertragsärztlichen Versorgung (Hilfsmittel-Richtlinie) in der Neufassung vom 15. März 2012. www.g-ba.de

Internetlinks

www.rehakind.com

37.5 Stationäre sozialpädiatrische Behandlung

Die Langfristigkeit von Behandlungsmaßnahmen bei Entwicklungsstörungen und Behinderungen bedingt die Notwendigkeit, diese Interventionen in aller Regel primär ambulant durchzuführen. Daneben gibt es jedoch spezifische Situationen wie auch Therapieinhalte, die sich im ambulanten Vorgehen nicht ausreichend bearbeiten lassen oder sich einem Zugriff entziehen. An mehreren Zentren und Kliniken besteht deshalb neben der ambulanten Betreuung auch die Möglichkeit der stationären interdisziplinären sozialpädiatrischen Block- oder Intensivbehandlung. Dabei können die familiären Bezugspersonen mit aufgenommen werden.

Zentren mit stationärer sozialpädiatrischer Behandlung

- Altötting: Zentrum für Kinder und Jugendliche Inn-Salzach, Kreiskrankenhaus Altötting
- Berlin: Diagnose- und Behandlungszentrum DBZ für Sozial- und Neuropädiatrie, Vivantes-Klinikum Berlin-Neukölln
- Bonn: Kinderneurologisches Zentrum (KiNZ), LVR-Klinik Bonn
- Düren: Kinderklinik, St. Marien-Hospital
- Düsseldorf: Kinderneurologisches Zentrum KNZ, Sana-Krankenhaus Düsseldorf-Gerresheim
- Hamburg: Werner-Otto-Institut
- Iserlohn: Kinderklinik, Evangelisches Krankenhaus Bethanien
- Mainz: Zentrum für Kinderneurologie und Sozialpädiatrie (KiNZ), Rheinhessen-Fachklinik Mainz
- Maulbronn: Klinik für Kinderneurologie und Sozialpädiatrie, Kinderzentrum (kize) Maulbronn
- München: kbo-Kinderzentrum München
- Neunkirchen/Saar: Fachabteilung Kinder- und Jugendmedizin, Marienhausklinik St. Josef Kohlhof
- Neustadt/Holstein: Sozialpädiatrische Fachklinik, Kinderzentrum Pelzerhaken
- Unna: Fachklinik für Kinderneurologie und Sozialpädiatrie, Lebenszentrum Königsborn
- Wesel: Zentrum für Kinder und Jugendliche, Marien-Hospital

Grundlage einer stationären sozialpädiatrischen Behandlung ist die Möglichkeit, auf alle Berufsgruppen zurückzugreifen, die auch im SPZ die Versorgung der Patienten gewährleisten. Hinzu kommen die Mitarbeiterinnen des Pflege- und Erziehungsdienstes, die über eine zusätzliche Qualifikation im Umgang mit entwicklungsgestörten oder behinderten Kindern und Jugendlichen sowie insbesondere auch deren Bezugspersonen verfügen. Die spezifischen Erfordernisse dieser Patientengruppe bedingen anders als oftmals sonst in der klinischen Akutpädiatrie den umfassenden Einbezug der Eltern oder anderer familiärer Begleitpersonen. In aller Regel wird auch die Betreuungs- und Pflegeverantwortung nur partiell an die professionellen Bezugspersonen delegiert; insbesondere auf den sozialpädiatrischen Eltern-Kind-Stationen bleiben die Eltern für die Dauer der Behandlung voll in der Verantwortung und Aufsichtspflicht. Somit kommt gerade den Mitarbeiterinnen des Pflege- und Erziehungsdienstes hier eine besondere begleitende und auch beratende Rolle zu, die nur in enger Abstimmung mit dem Behandlungsteam der Fachtherapeuten gelingen kann. Umgekehrt sind die Beobachtungen in einer alltagsnahen Situation der Stationsgruppe von größter Wichtigkeit für die Einschätzung eines Kindes hinsichtlich Interaktion und sozialer Kompetenz, was in dieser Form beim ambulanten Vorgehen nicht erreicht werden kann. Eine **Indikation** für die stationäre sozialpädiatrische Behandlung bei komplexen Entwicklungsstörungen ist dann gegeben, wenn ambulante Behandlungs- und Beratungsmaßnahmen nicht ausreichend erfolgreich waren oder die diagnostische Zuordnung der Symptomatik im Rahmen der ambulanten Möglichkeiten nicht ausreichend gelungen ist.

Indikationen für die stationäre sozialpädiatrische Behandlung

- Differenzialdiagnostische Klärung bei unzureichender ambulanter Beurteilbarkeit
- Akute Eskalation von spezifischen Verhaltensproblemen bei Behinderung
- Psychotherapie mit familienzentriertem Ansatz
- Umfassende Adjustierung des Behandlungsplans
- Komplexe Versorgung mit Hilfsmitteln
- Spezifische Maßnahmen in Verbindung mit Kindeswohl-Gefährdung (Misshandlung, sexueller Missbrauch)
- Fokussierte diagnosespezifische Behandlungsmaßnahmen

Ziel der stationären Behandlung ist es, Ursachen des bestehenden Störungsbildes sowohl im somatischen als auch psychosozialen Bereich im Rahmen einer mehrdimensionalen Diagnostik zu erkennen, die Symptomatik zu beseitigen, zu kompensieren oder zumindest zu lindern oder ihrer Progredienz entgegenzuwirken. Dabei sind diagnostische wie therapeutische Maßnahmen integrale Bestandteile des Behandlungsprozesses, die sich wechselseitig beeinflussen, mitunter nicht voneinander zu trennen sind und deshalb oft zeitgleich ablaufen müssen.

Die Festlegung der kurzfristigen Behandlungsziele während der stationären Behandlungsphase erfolgt gemeinsam mit den Bezugspersonen, nach Möglichkeit auch mit dem Kind bzw. Jugendlichen selbst. Sehr häufig ist es notwendig, während des stationären Aufenthalts Kontakt herzustellen zu bisher behandelnden Fachkräften bzw. Institutionen wie Kindergarten, Schule oder Jugendamt. Der Austausch trägt dazu bei, ein umfassendes Bild von der bestehenden Problematik zu erhalten und darauf basierend mittel- und langfristige Ziele zu entwickeln. Diese Erarbeitung von Perspektiven im Sinne eines Behandlungsplans trägt dazu bei, dass die Ziele alltagsrelevant und realistisch sind. Neben der Optimierung der individuellen Entwicklungsmöglichkeiten in funktionellen Bereichen der Entwicklung wie Sprache, Motorik und Kognition zählen hierzu im Rahmen der mehrdimensionalen Vorgehensweise ebenso die soziale Eingliederung in Familie, Kindergarten oder Schule, die Herstellung oder Verbesserung der Kommunikationsfähigkeit, die Förderung der größtmöglichen Selbstständigkeit und die Unterstützung der Teilhabe an den altersbezogen üblichen Aktivitäten.

Bei der **Entlassung** liegt aufbauend auf den Erfahrungen in der stationären Behandlung in aller Regel ein Konzept zur Weiterbehandlung vor, das in Absprache mit den Eltern bzw. anderen Bezugspersonen den mittelfristigen Behandlungsplan darstellt. Die stationäre sozialpädiatrische Block- oder Intensivbehandlung ist somit ein Modul innerhalb der langfristigen ambulanten Betreuung, das insbesondere dazu geeignet ist, neben der Anpassung von unmittelbar kindzentrierten Maßnahmen vor allem die Ressourcen im Umfeld zu eröffnen.

Die **Behandlungsmodalitäten** sind in Abhängigkeit vom jeweiligen Auftrag und der teilweise in den verschiedenen Zentren erfolgten Spezialisierung auf bestimmte Bereiche unterschiedlich.

Grundsätzlich werden sie bestimmt von dem Schweregrad und der Komplexität der vorliegenden Störung und dem bisherigen Verlauf. Die stationäre sozialpädiatrische Behandlung ist nach dem Fallpauschalengesetz zur Vergütung von Krankenhausleistungen im DRG-Abrechnungssystem ausgewiesen. Je nach Umfang des Einsatzes der verschiedenen beteiligten Berufsgruppen, der zeitlichen Intensität und dem Vorgehen im Einzel- oder Gruppen-Setting werden hierfür verschiedene Kodierungen (OPS-Ziffer 9–403 mit Untergruppen) eingesetzt. Es erfolgt eine pauschalierte Vergütung bezogen auf den einzelnen Behandlungstag, die jeweils für die behandelnde Klinik individuell mit den Kostenträgern zu verhandeln ist. Insofern unterscheidet sich der stationäre sozialpädiatrische Bereich von der sonst inzwischen im Medizinsystem üblichen landesbezogen einheitlichen Verpreisung von Krankenhausleistungen.
Zum Einsatz kommen funktionell orientierte Therapien der Logopädie, Physio- oder Ergotherapie, heilpädagogisch bzw. kinderpsychotherapeutisch orientierte Behandlungen, kreativtherapeutische Maßnahmen, Medikation und Hilfsmittelversorgung. Die inhaltliche Klammer bildet die ärztliche und psychologische Behandlung, wobei in der Interaktion mit den Bezugspersonen vielfach psychotherapeutische Aspekte wesentlich sind. Ergänzend kommen Informations- und Beratungsleistungen sowie Maßnahmen der Koordination durch Sozialarbeiter bzw. Sozialpädagogen hinzu. Bei längeren stationären Behandlungen ist die regelmäßige Unterrichtung im Rahmen des Besuchs der Schule für Kranke ein sehr wichtiger Faktor. Das Verhalten der Kinder in diesem Umfeld erbringt wertvolle zusätzliche Informationen über das Störungsbild ebenso wie über erreichte therapeutische Effekte.
Bei der Entlassung aus der Behandlung wird mit den Eltern, häufig auch anderen familiennahen Bezugspersonen, ein interdisziplinäres Abschlussgespräch geführt, oft unter Einbezug von externen Fachkräften wie Therapeutin, sozialpädagogischer Familienhelferin oder Lehrerin. Hierin werden neben einer Zusammenschau der erreichten Schritte auch die Vereinbarungen für das weitere Vorgehen im Sinne des Entlassungsmanagements festgelegt.
Die Dokumentation der stationären Behandlung erfolgt im Rahmen eines zusammenfassenden Arztbriefes, der in aller Regel auch den Eltern zur Verfügung gestellt wird.
Die **Evaluation** der zu erreichenden Effekte und damit der Qualität einer sozialpädiatrischen stationären Behandlung ist wie im gesamten Bereich der Entwicklungs- und Psychotherapie schwierig. Geeignet sind systematische Verlaufsuntersuchungen mit Einsatz von standardisierten testdiagnostischen Verfahren zur Erfassung von Veränderungen in den funktionellen Zielbereichen. Notwendig sind ebenfalls den Gütekriterien entsprechende Fragebögen, die von den Bezugspersonen und nach Möglichkeit den betroffenen Kindern bzw. Jugendlichen selbst bearbeitet werden. Zielgrößen sind insbesondere die Teilhabe am altersbezogenen gesellschaftlichen Leben und die Lebensqualität an sich.
Im familiären System kann es eine entscheidende Entlastung und damit Veränderung bedeuten, wenn die Integration der notwendigen Abläufe bei einem schwer behinderten Familienmitglied optimiert worden ist. Dies lässt sich nicht mit den sonst in der Medizin üblichen kurativen Maßstäben abbilden. Damit ist die interdisziplinäre sozialpädiatrische stationäre Krankenhausbehandlung eine spezialisierte Form der Versorgung für Kinder und Jugendliche mit Entwicklungsstörungen und Behinderungen unter Einbezug ihrer Familie. Innerhalb der begrenzten Palette von Möglichkeiten für diese besondere Patientengruppe stellt sie einen essenziellen Bestandteil dar.

Literatur

Diedenhofen C, Schlack HG (1993) Formen und Indikationen stationärer Behandlung in Sozialpädiatrischen Zentren. Sozialpädiatrie in der Pädiatrie für Praxis und Klinik 15 (10) 586–587.

Gleißner U, Johler M, Ney-Wilkens R, Hollmann H, Röhr-Sendlmeier UM (2013) Langzeitstabilität von Pflegeverhältnissen nach stationärer Übergangsbehandlung

Hellmann-Mersch B, Schlack HG (2002) Stationäre Behandlung von Kindern mit psychischer Traumatisierung und Milieuschädigung – Zielsetzung und Katamnese. Kinderärztliche Praxis 6: 404–409.

Karch D, Fricke C, Hameister K, Hasmann R, Hollmann H, Marton MA, Rossa M, Richardt H, Weise S, Voss Hv, Peters H (2009) Stationäre sozialpädiatrische Behandlung. Beitrag zur Qualitätssicherung in sozialpädiatrischen Kliniken. www.dgspj.de/qspz.php

Mattejat F, Remschmidt H (2008) Kinder psychisch kranker Eltern. Dtsch Arztebl 105: 413–418.

37.6 Zwischen Therapie und Förderung: Interventionen bei Geistiger Behinderung

37.6.1 Grundlagen

Für die **Diagnose** einer Intelligenzminderung (vgl. Kapitel 19) (➤ Kap. 19) wird nach der Definition der American Association on Mental Retardation (AAMR, 2002) neben den Ergebnissen der psychometrischen Verfahren die Einschätzung des Niveaus der adaptiven Fähigkeiten einbezogen. Hierzu bietet der ICD 10 keine ausreichenden Klassifikationsmöglichkeiten, wobei durch die ICF-CY (➤ Kap. 16) gute Möglichkeiten der Beschreibung und Kodierung bestehen.
Im pädagogischen Bereich ebenso wie bei den Betroffenen selber wird der Begriff **„Geistige Behinderung"** gegenüber dem medizinischen Terminus „Intelligenzminderung" bevorzugt. Im Behinderungsbegriff ist die Unterstützungsnotwendigkeit in Hinblick auf die Realisation von sozialer Teilhabe impliziert. Beide Begriffe werden häufig auch synonym verwendet.
Eltern intelligenzgeminderter Kinder und Jugendlicher sind oft unsicherer in ihrem Erziehungsverhalten: es fällt ihnen schwerer, kindliche Grundbedürfnisse zu erkennen und angemessen darauf zu reagieren. Entwicklungshemmende Bindungsmuster wie symbiotische Bindungen zu Bezugspersonen, feindselige und / oder vernachlässigende Bindungsmuster; einseitig Erwachsenengesteuerte Interaktionsmuster im Kontext elterlich-familiärer Belastung und resultierende Isolation können im sozialen Bereich Risikofaktoren für die Entwicklung des Kindes insgesamt und hinsichtlich des Auftretens emotionaler Störungen darstellen. Eine wesentliche Rolle spielt dabei auch die Akzeptanz der Diagnose einer Intelligenzminderung durch die Eltern sowie Schuld- und Schamgefühle bei den Eltern. Zudem müssen sich die Eltern im Entwicklungsverlauf ihres Kindes und Jugendlichen mit vielfältigen Sorgen und Belastungen im Umgang mit den kognitiven, körperlichen und psychischen Störungen auseinandersetzen.
Beim niedergelassenen **Kinder- und Jugendarzt** findet die medizinische Grundversorgung für Patienten mit Intelligenzminderung statt. Diese umfasst die Diagnostik und Behandlung akuter Erkrankungen, die Beratung hinsichtlich Impfungen und Ernährung, Vorsorgeuntersuchungen und die Koordination zur Diagnostik und Behandlung bei weiteren Fachärzten. Die Diagnostik und Behandlung von intelligenzgeminderten Patienten verlangt spezifische klinische Kenntnisse und ist zeitaufwändig: Anamnese, gezielte Abklärung organischer Problemen als Auslöser für Verhaltensauffälligkeiten sowie Beratung und Behandlung der Bezugspersonen sind zeitaufwändiger als bei normalintelligenten Patienten und erfordern oft eine hohe Termindichte. Da bei

Patienten mit Intelligenzminderung und komorbiden psychischen Störungen in der Regel eine interdisziplinäre Diagnostik und Behandlung erfolgen muss, empfiehlt sich die frühzeitige Weiterleitung und langfristige Anbindung an spezialisierte Einrichtungen wie Sozialpädiatrische Zentren.

Somatische Störungen und Krankheiten bei Intelligenzminderung; erforderliche Maßnahmen

Versorgung mit Brille und/oder Hörgerät
Pharmakotherapie
Optimierung der antikonvulsiven Dauermedikation bei Epilepsien zur Minimierung von Nebenwirkungen
Schmerzmedikation
Medikation von Reflux / Obstipation
Beachtung von Medikamenteninteraktionen
Optimierte Mundpflege und Zahnsanierung
Adäquate Ernährung, Diäten bei Lebensmittelunverträglichkeiten
Lagerung, Hilfsmittelversorgung und gegebenenfalls Physiotherapie

Bei intelligenzgeminderten Patienten treten in ca. 10 % Hörbehinderungen, in ca. 15 % schwere Sehbehinderungen auf. Diese Störungen sind ebenfalls deutlich schwieriger diagnostizierbar und werden daher oft verspätet erkannt.

37.6.2 Grundzüge der Behandlung: Förderung und Therapie

Bei der **Behandlungsplanung** müssen Eltern und andere Bezugspersonen beteiligt sein. Je jünger und je schwerer beeinträchtigt das Kind ist, desto mehr steht die Beratung der Bezugspersonen im Vordergrund. Kinder und Jugendliche mit ausreichenden kommunikativen Fähigkeiten sollen aktiv an der Therapieplanung beteiligt sein und ihr Einverständnis zu geplanten therapeutischen Schritten geben. Eine vertrauensvolle, emotional Halt gebende Beziehung ist dabei wie bei anderen (psycho-)therapeutischen Maßnahmen zentral.
Eine an die Belastbarkeit und die Fähigkeiten des Kindes oder Jugendlichen angepasste und tolerante **Umgebung** leistet einen entscheidenden Beitrag zum Rückgang psychischer Störungen und problematischen Verhaltens. Wichtig sind eine Vermeidung von Reizüberflutung durch die Schaffung einer strukturierten, überschaubaren Umgebung, die Einhaltung fester, ritualisierter Tagesabläufe, sowie die Ankündigung und Unterstützung bei situativen Übergängen. Wenn ein Förderangebot soweit vereinfacht wird, dass das Kind oder der Jugendliche es auf seinem aktuellen Entwicklungsstand erfasst, wird das Erlernen neuer Fähigkeiten erleichtert.
Die **Selbstbeschäftigung** des intelligenzgeminderten Kindes oder Jugendlichen mit Dingen, die ihm Freude bereiten und die die Eigenmotivation stärken, soll gefördert werden. Dabei muss auf den problemlosen Zugang zu diesen Beschäftigungen geachtet werden. Hilfreich hierbei ist das **Lernen am Modell.** Bei diesem Prinzip ist zu beachten, dass die zu imitierenden Modelle für die Kinder möglichst interessant und anziehend sind. Seinen Fähigkeiten entsprechend sollte dem Kind oder Jugendlichen die Möglichkeit zur Selbst- bzw. Mitbestimmung eingeräumt werden. Häufige Wechsel der **Bezugspersonen** sollten vermieden werden. Gerade bei einer größeren Anzahl von Bezugspersonen sind Absprachen zwischen diesen unabdingbar. Überforderungen und Konflikte der Eltern müssen thematisiert werden, ggf. sollte der Einsatz von Jugendhilfemaßen

(Einzelfallhilfe, Familienhilfe), empfohlen werden. Eine wirksame Unterstützung kann die Einbindung der Eltern in **Selbsthilfegruppen** sein. Diese kann auch die Akzeptanz der Behinderung durch die Eltern fördern.
Eine **institutionelle Unterbringung** kann einerseits aus einer Überforderungssituation und der Notwendigkeit eines anders strukturierten Unterstützungssystems resultieren. Zum anderen kann sie eine Option zur Ablösung eines älteren intelligenzgeminderten Jugendlichen darstellen. Der Prozess einer Unterbringung erfordert immer eine Begleitung der Familie.
Eine Beratung zur Erfüllung der **sozialrechtlichen Ansprüche** (Beantragung eines Schwerbehindertenausweises, Pflegeversicherung) ist in jedem Fall erforderlich. Zusätzlich kann eine Beratung zu Freizeitangeboten und den Möglichkeiten von Kurzzeitpflege notwendig sein.
Der **Einsatz funktioneller Therapien und Übungsbehandlungen** muss alltagsorientiert erfolgen. Als Bezugsgröße ist das Entwicklungsalter des Kindes zu wählen und nicht das chronologische Alter. Eine klare Zieldefinition und eine begrenzte Behandlungszeit mit anschließender Überprüfung der Behandlungserfolge und ihrer Alltagsrelevanz sind notwendig, um unreflektierte Dauerbehandlungen zu vermeiden. Generell sollten Initiative, Reaktionsbereitschaft und Ausdauer des Kindes, seine Toleranz für Umwelteindrücke und. Frustrationen, seine Selbstregulation und die Fähigkeit, soziale Beziehungen einzugehen, gefördert werden. In alle funktionellen Therapien müssen die Eltern bzw. Bezugspersonen eingebunden sein, um einen Transfer von Therapieinhalten in den Alltag sicher zu stellen.
Kinder mit deutlichen Einschränkungen in der Kommunikation profitieren von Visualisierungshilfen und Instrumenten zur **Kommunikationsunterstützung** (z. B. sprachbegleitende Gesten bei der Gebärden unterstützen Kommunikation (GUK), Einsatz von Bildkarten, Talkern etc.).

37.6.3 Psychoedukation und Psychotherapie

Die durch die Behinderung des Kindes hervorgerufene Entwicklungsstörung bedingt Folgebelastungen für die Familie. Eltern benötigen deshalb eine Beratung und bei Bedarf auch eine psychotherapeutische Behandlung. Neben der individuellen Begleitung von Eltern wurde in den letzten Jahrzehnten eine Reihe von **Elterntrainings** entwickelt, die den Eltern den Umgang mit einem schwierigen Kind erleichtern können. Sie beziehen sich aber meist auf allgemeine kindliche Verhaltensprobleme, nur selten auf Kinder und Jugendliche mit Intelligenzminderungen. Typisch für die meisten Elterntrainings ist ein verhaltenstherapeutisch orientiertes Konzept. Metaanalysen belegen die Wirksamkeit von Elterntrainings vor allem im Hinblick auf die soziale und kommunikative Interaktion und auf das konsistente und konsequente Verhalten der Eltern. Weniger systematische Effekte werden für Lernverhalten und soziale Kompetenzen von Kindern beschriebenen. Ein Elterntraining, das speziell für Belange der Eltern behinderter Kinder entwickelt worden ist, ist das **Stepping Stones Triple P Eltern-Gruppentraining** von Sanders. In der Evaluation zeigten sich signifikante Effekte hinsichtlich der Reduktion von Verhaltensauffälligkeiten der Kinder, der Verbesserung der elterlichen Erziehungskompetenz, der Verminderung der psychischen Belastung der Eltern sowie der Steigerung ihres Selbstwirksamkeitserlebens und der Zufriedenheit mit der Elternrolle.
Neben den verhaltenstherapeutischen Vorgehensweisen, die sich in Organisation und Durchführung primär an die Bezugspersonen in Familie und betreuender Institution richten, kann bei ausreichender Einsichts- und Reflexionsfähigkeit auch mit dem Kind oder Jugendlichen selbst psychotherapeutisch gearbeitet werden. In Betracht kommen je nach Alter Spieltherapie und Verhal-

tenstherapie. Bei eingeschränkten kommunikativen Fertigkeiten sind nonverbale Therapieformen, wie Musik- und Kunsttherapie, zu überlegen. Diese können auch im Sinne einer Ressourcenaktivierung wirken. Kinder und Jugendliche mit ausreichenden kommunikativen Fertigkeiten können von einem Training sozialer Kompetenzen im Rahmen einer Gruppentherapie profitieren. Der in der Behandlung autistischer Störungen etablierte **TEACCH-Ansatz** kann auch intelligenzgeminderten Kindern und Jugendlichen ohne autistische Symptomatik darin unterstützen, eine bessere Orientierung im Alltag und eine verbesserte Kommunikation zu erlangen.
Ein besonderes Gewicht kommt der **Verhaltenstherapie** zu mit Techniken zur Verbesserung sozialer Kompetenzen und Problemlösefähigkeiten, des Selbstmanagements und der Selbstinstruktion. Im Rahmen einer verhaltenstherapeutischen Behandlung sollten entsprechend der individuell erstellten Verhaltensanalyse solche Anforderungen, auf die regelmäßig problematisches Verhalten folgt, genau überprüft werden. Einzelne Lernschritte sollten bei der Wiedereinführung möglichst sensibel, klein und gut aufeinander abgestimmt sein. Es muss darauf geachtet werden, dass das Problemverhalten des Kindes nicht ungewollt sozial verstärkt wird. Stattdessen sollen alternative, erwünschte Verhaltensweisen eine positive Konsequenz nach sich ziehen. Der Einsatz einer videogestützten Beratung kann hierbei hilfreich sein. Es müssen individuell wirksame Verstärker eingesetzt werden. Die Verstärkung soll kontinuierlich und kurzfristig auf das Zielverhalten folgen. Ziel ist auch, dass das Kind bzw. der Jugendliche, Überforderungssituationen selbst wahrnimmt, sich entsprechend äußert und ggf. eigene Strategien im Umgang damit entwickelt.
Das therapeutische Vorgehen setzt eine besondere Expertise voraus. Die Grenzen einer psychotherapeutischen Einzelbehandlung eines intelligenzgeminderten Patienten müssen bedacht werden, um eine Überforderung zu vermeiden.
Weitere allgemeine Behandlungshinweise finden sich im Kapitel 22.6. (➤ Kap. 22.6)

37.6.4 Spezielle Behandlungsmaßnahmen bei psychischen und Verhaltensstörungen

Intelligenzgeminderte Kinder und Jugendliche mit psychischen Störungen benötigen eine frühzeitig einsetzende interdisziplinäre sozialpädiatrische Mehrdimensionale Bereichsdiagnostik nach dem EKPSA-Schema (➤ Kap.15). Ätiologisch-diagnostisch und therapeutisch muss der Einsatz medizinischer, psychologischer, sozial- und heilpädagogischer Maßnahmen erwogen werden.

Ärztliche Behandlung

Sofern die ärztliche Diagnostik Hinweise auf somatische Erkrankungen ergibt, sind diese zu behandeln.

Störungsspezifische Behandlung, nicht medikamentös

Eine störungsspezifische Behandlung richtet sich nach der diagnostizierten psychischen Erkrankung und unterscheidet sich prinzipiell nicht von der Behandlung derselben Störung bei nicht intelligenzgeminderten Patienten. Im Folgenden wird auf die Behandlung häufig zur Vorstellung führender Probleme eingegangen.

Expansiv-aggressives und selbstverletzendes Verhalten

(mögliche Kodierung unter ICD 10, F91.1; F91.3; F07.8)
Eine genaue Analyse der Auslöser des (auto-)aggressiven Verhaltens ist wichtig. Häufig wird aggressives Verhalten auf der Basis von Fehlinterpretationen kindlicher Signale durch soziale Zuwendung in den Situationen des Auftretens ungewollt verstärkt. Diese ungewollte Verstärkung soll durch die Verhaltensanalyse aufgedeckt und wenn möglich reduziert werden. Soziale Verstärkung sollen Kinder vermehrt für nicht-aggressive Verhaltensweisen erhalten. Kinder können auch darin unterstützt werden, Wünsche und Aversionen auf anderen Wegen gezielt und eigenständig mitzuteilen. Bei Kindern mit sehr eingeschränkter Kommunikation bedarf es dazu individuell abgestimmter Kommunikationshilfen, z. B. Bildkarten zur Auswahl von Beschäftigungen.
Bei Selbstverletzungen muss auch an interne Verstärkungsprozesse gedacht werden. So reduzieren solche Verhaltensweisen die Schmerzwahrnehmung oder führen zu einer als angenehm erlebten sensorischen Stimulation (Rohjahn et al., 2007). Bei schweren Selbstverletzungen ohne eindeutige soziale Verstärkung sind Schutzmaßnahmen des Kindes indiziert. Der Einsatz eines Helmes oder von Armmanschetten, time-out oder Fixierungen sollten so kurz wie möglich erfolgen, da Kinder durch die Einschränkungen allein keine alternativen Verhaltensweisen erlernen können.

Starke motorische Unruhe und impulsive Verhaltensweisen

(mögliche Kodierung unter ICD 10, F84.4; F90.8)
Bei Kindern mit motorischer Unruhe ist es sinnvoll, ihren erhöhten Bewegungsdrang im Alltag zu berücksichtigen. Dies kann durch das Vorliegen einer gleichzeitig vorhandenen körperlichen Beeinträchtigung erschwert sein. Beispiele für zusätzliche Bewegungsangebote sind: Spaziergänge mit Gehhilfen, der Einsatz behindertengerechter Fahrräder, Bewegen des eigenen Rollstuhls mit Muskelkraft, Bewegung im Wasser und Reiten.
Der Einsatz von Sanduhren oder Symbolkarten kann intelligenzgeminderten Patienten helfen, Impulse nicht sofort auszuagieren bzw. abwarten zu lernen. Wie normal begabte Kinder mit einer ADHS-Symptomatik profitieren auch intelligenzgeminderte Patienten von klar strukturierten Tagesabläufen, klaren Kontingenzen und einer Vermeidung einer zu großen Menge an gleichzeitig zu verarbeitenden Reizen.

Stark Aufmerksamkeit suchendes Verhalten

(mögliche Kodierung unter ICD 10, F93.0; F93.8)
Dieses Verhalten wird häufig ungewollt sozial verstärkt; eine Selbstbeschäftigung des Kindes häufig unzureichend beachtet und wertgeschätzt. Patienten mit diesem Problemverhalten profitieren von einer Förderung selbstständiger Beschäftigung und einer Verbesserung ihrer Fähigkeiten, Bedürfnisse klar zu äußern. Aufmerksamkeit suchendes Verhalten kann für das Kind auch die Funktion haben, Anforderungen zu vermeiden. Genauso kann es Ausdruck von Trennungsängsten oder spezifischen Phobien sein, wie z. B. bei sehr klammerndem Verhalten, Schwierigkeiten, allein in einem Raum zu spielen und Abneigung gegenüber bestimmten Spielsachen oder Alltagsgegenständen. Eine gestufte Exposition in Verbindung mit sicherheitsgebenden Ritualen kann dann hilfreich sein.

Schlafstörungen

(mögliche Kodierung unter ICD 10, F51.0; F51.2)
Wichtig ist es, konstante Schlafens- und Aufstehzeiten festzulegen und ein Zu-Bett-Geh-Ritual einzuhalten. Entspannungshilfen beim Einschlafen können leise Musik, bestimmte Gerüche oder Übergangsobjekte (Kuscheltiere) sein, die individuelle Vorlieben des Kindes berücksichtigen. Bei Ängsten vor Dunkelheit oder Alleinsein können ein Nachtlicht oder eine leicht geöffnete Tür zu einem beleuchteten Raum eine Hilfe sein. Schlafen am Tag (insbesondere am Nachmittag) sollte verhindert werden, damit das Kind am Abend ausreichend müde ist. Einschlafstörungen sollen vor Durchschlafstörungen behandelt werden. Zuwendung in der Einschlafsituation sollte schrittweise reduziert werden. Kinder die das Bett verlassen, sollten ohne verbale Ansprache ins Bett zurückgebracht werden. Bei sehr langer Einschlafzeit ist es z. T. notwendig, die Zu-Bett-Geh-Zeit vorrübergehend so weit nach hinten zu verschieben, dass das Kind aufgrund der dann sehr großen Müdigkeit rasch einschläft und das Bett so zum diskriminativen Stimulus für das Einschlafen wird. Die Zu-Bett-Geh-Zeit kann dann schrittweise wieder nach vorn verlagert werden. Ebenso kann es sinnvoll sein, ein Kind, das sehr früh aufwacht, vorübergehend noch früher zu wecken. Das Wecken wird dann zum konditionierten Stimulus für Aufwachen/Aufstehen. Nachfolgend ist die Zeit dann schrittweise wieder nach hinten zu verschieben.

Fütter- und Essstörungen

(mögliche Kodierung unter ICD 10, F98.2; F98.3; F50.5; F50.8)
Die problemspezifische Analyse der Füttersituation unter Berücksichtigung eventueller orofacialer Probleme, ggfs. unterstützt durch eine Videoaufzeichnung der Mahlzeit erlaubt es, spezifische Strategien für die Behandlung abzuleiten. Voraussetzung für die Behandlung ist das Einhalten regelmäßiger Mahlzeiten (4-5 / Tag). Dazwischen sollte keine Nahrung angeboten werden, das Kind sollte lediglich kalorienarme Getränke erhalten. Die Länge der einzelnen Mahlzeit sollte auf 15-30 min begrenzt sein. Während der Mahlzeit sollte keine Ablenkung des Kindes erfolgen. Auf eine gute und bequeme Sitzposition ist zu achten. Nahrung, die aufgrund des Geschmacks oder der Konsistenz bisher abgelehnt wurde, sollte in kleinen Mengen angeboten werden. Die Akzeptanz wird durch intensives Loben und anschließende Gabe präferierter Nahrungsmittel verstärkt. Beim Auftreten von Rumination, willkürlich herbeigeführtem Erbrechen oder Ausspucken der Nahrung können umschriebene negative Konsequenzen als ultima ratio notwendig werden, z. B. die Gabe einer schlecht schmeckenden Flüssigkeit. Der Einsatz solcher negativer Konsequenzen darf nur bei Versagen anderer Methoden, zeitlich begrenzt und unter erfahrener therapeutischer Anleitung erfolgen.

Störungen der Sauberkeitsentwicklung

(mögliche Kodierung unter ICD 10, F98.0; F98.1; N39.x)
Entwicklungsaufgaben im Bereich der Sauberkeitsentwicklung müssen auf das Entwicklungsalter des Kindes und seine tatsächlichen Fähigkeiten abgestimmt werden. Um intelligenzgeminderte Kinder bei der Sauberkeitsentwicklung zu unterstützen, ist ein kleinschrittiges Vorgehen notwendig. Das Kind muss in engen Zeitabständen (stündlich), auf die Toilette oder einen Topf gesetzt werden. Bei Erfolg werden die Zeitintervalle langsam vergrößert. Der Topf sollte nur vorübergehend, vor allem bei Angst vor der Toilette verwendet werden. Das Kind sollte bequem

und sicher sitzen. Erfolgte Entleerung wird positiv verstärkt. Eine apparative Verhaltenstherapie (Klingelgerät) ist bei nächtlichem Einnässen die Methode der ersten Wahl, bei schwerer intelligenzgeminderten Kindern kann sie auch tagsüber eingesetzt werden. Weitere Details finden sich im Qualitätspapier zu nichtorganischen Ausscheidungsstörungen (zur Publikation eingereicht).

Stereotype Verhaltensweisen und zwanghaftes Verhalten

(mögliche Kodierung unter ICD 10, F84.4; F 98.4; F42.1)
Diese Verhaltensweisen treten bei intelligenzgeminderten Kindern häufig in Verbindung mit freudiger Erregung oder bei Langeweile auf. Wenn stereotypes Verhalten die Alltagsaktivitäten nicht oder nur wenig beeinträchtigt, sollte es als emotionaler Ausdruck von Befindlichkeit oder als Form kommunikativen Ausdrucks toleriert werden. Die gezielte Förderung von Selbstbeschäftigung, die dem Kind Freude bereitet, kann Kindern helfen, Langeweile zu reduzieren. Das Angebot vestibulärer Reize (Schaukeln) kann Jaktationen vermindern. Entspannungstechniken (mit propriozeptiven Reizen) können stereotypes Verhalten vermindern.
Die Behandlung von Zwängen bei Kindern und Jugendlichen mit Intelligenzminderung unterscheidet sich grundsätzlich nicht von dem Vorgehen ohne Vorliegen einer Intelligenzminderung. Zum Einsatz kommen vor allem die gestufte Exposition mit Reaktionsverhinderung sowie Selbstverbalisationen zur Korrektur von Fehlinterpretationen. Wichtig ist eine Psychoedukation für Bezugspersonen, da sie meist in Zwangsrituale eingebunden sind.

Autistische Symptome mit zusätzlich eingeschränkter Kommunikation

(mögliche Kodierung unter ICD 10, F84.x)
Bedingung für eine Beeinflussung problematischer Verhaltensweisen ist bei diesen Patienten, eine positive Interaktion zu schaffen, die zunächst völlig losgelöst von Anforderungen oder Verhaltensbeeinflussung sein sollte. Ziel ist es herauszufinden, worauf das Kind positiv und worauf es aversiv reagiert, was es mag, was es interessiert und welche Art von Kontaktaufnahme es zulässt. Diese Erkenntnisse bilden zusammen mit einer guten Alltagsstrukturierung die Grundlage für eine Anbahnung kommunikativer Prozesse und eine allmähliche Verhaltensänderung. Die Komplexität autistischer Störungen erfordert eine Behandlung über eine längere Zeit. Weitere Ausführungen dazu finden sich im Qualitätspapier Autismus (Fricke et al., 2012).

Soziale Ängste und soziales Rückzugsverhalten, Vermeidung

(mögliche Kodierung unter ICD 10, F93.2)
Soziale Ängste bis hin zu mutistischem Verhalten können Ausdruck erlebter sozialer Ausgrenzung und kognitiver Überforderung sein. Häufig sind sie rückläufig, wenn ein Kind begabungsadäquat gefördert wird und sich in einer seine Besonderheiten akzeptierenden Umgebung befindet. Leicht geistig behinderte Kinder mit ausreichenden kommunikativen Fertigkeiten können von einer Einzelpsychotherapie profitieren, in die Bezugspersonen eingebunden sind. Bei der Indikationsstellung zur Behandlung im Rahmen einer Gruppenpsychotherapie ist zu beachten, dass diese Gruppe möglichst klein und hinsichtlich des Entwicklungsstands der behandelten Kinder und Jugendlichen homogen sein sollte. Patienten laufen sonst Gefahr, erneut in Situationen von Überforderung und Ausgrenzung zu geraten.

Depressive Symptomatik

(mögliche Kodierung unter ICD 10, F32.x; F43.2, F92.0)
Depressive Symptome bei Kindern und Jugendlichen mit Intelligenzminderungen treten meist im Sinne einer Anpassungsstörung auf. Hintergründe dafür können der Verlust von Bezugspersonen, überfordernde Umgebungsbedingungen oder soziale Isolation sein. Intelligenzgeminderte Kinder und Jugendliche sind häufig schlechter in der Lage, ihre Gefühlszustände differenziert zu benennen und bekommen so häufig keine angemessene Unterstützung bei deren Bewältigung. Liegt eine depressive Symptomatik vor, sollten ggfs. ungünstige Umgebungsbedingungen verändert werden. Visualisierungshilfen können die Patienten bei der differenzierteren Wahrnehmung und Mitteilung ihrer Stimmung unterstützen. Bezugspersonen können dann besser darauf eingehen. Lustbetonte Selbstbeschäftigungen und mögliche nützliche Beiträge für die Gemeinschaft sollten gefördert werden.

Störungsspezifische Behandlung, medikamentös

Häufig stellt sich die Frage nach einer ergänzenden psychopharmakologischen Behandlung. Generell gilt, dass anderen, nichtpharmakologischen Maßnahmen der Vorrang gegeben wird und eine Pharmakotherapie in ein Gesamtbehandlungskonzept einzubinden ist.
Bei intelligenzgeminderten Patienten besteht ein höheres Risiko unerwünschter Arzneimittelwirkungen (UAW). Daher sollte mit einer niedrigeren Dosierung als üblich begonnen und die Dosis eines Medikaments langsamer gesteigert werden. Die UAW sind bei diesen Patienten schwieriger zu erkennen. Daher ist eine besonders engmaschige Betreuung unter Einbezug verlässlicher Beobachtungen von Bezugspersonen erforderlich. Zu beachten sind Medikamentenwechselwirkungen, da diese Patienten aufgrund komorbider Erkrankungen oft mehrere Medikamente erhalten. Eine Polypharmazie ist, wenn irgend möglich, zu vermeiden. Bei fehlender Zulassung der Medikamente für den entsprechenden Indikationsbereich, die Altersgruppe der Kinder- und Jugendlichen und bei Dosierungen oberhalb der Zulassungsgrenze (sog. off-label-use) ist es notwendig, die Eltern über die Anwendung außerhalb des Zulassungsbereichs und die Behandlung im Rahmen eines individuellen Heilversuchs zu informieren und ihr schriftliches, widerrufbares Einverständnis einzuholen.
Vor der medikamentösen Behandlung muss ein EKG abgeleitet werden und eine Laboruntersuchung (Routine-Parameter) erfolgen. Im Verlauf der Behandlung müssen zum Ausschluss von UAW je nach Wirksubstanz unter Beachtung ggf. spezifischer Hinweise in der Fachinformation die Laborparameter, Herzfrequenz (ggf. EKG) und Blutdruck sowie die Entwicklung von Länge und Gewicht kontrolliert werden.

Spezielle Medikamente (Auswahl)

Die Auswahl des Medikamentes richtet sich nach der zu behandelnden Symptomatik, den komorbiden Erkrankungen und der Gabe weiterer Medikamente. Die im folgenden dargestellten einzelnen Substanzen stellen eine empirisch gestützte Auswahl dar (weiterführende Literatur Häßler, 2011; Gerlach et al 2009).

Neuroleptika

Hoch- und mittelpotente Neuroleptika sind indiziert bei psychotischen Symptomen, bei impulsiven aggressiven und autoaggressiven Verhaltensweisen, sowie bei ausgeprägten Unruhezuständen.

- **Risperidon** (als Tropfen und Tabletten verfügbar)
Indiziert zur Behandlung von Hyperaktivität, Impulsivität, Irritabilität, (auto-) aggressiven Verhaltensweisen und Stereotypien
Einziges Medikament mit Zulassung zur Therapie aggressiver Verhaltensstörungen bei intelligenzgeminderten Kindern ab dem 5. Lebensjahr für eine Behandlung bis zu 6 Wochen
Seltener Extrapyramidal-motorische Störungen (EPS) als bei Neuroleptika der ersten Generation
- **Zuclopenthixol** (als Tropfen und Tabletten verfügbar)
Indiziert zur Behandlung von Hyperaktivität, Impulsivität und Aggressivität
Keine Zulassung für Kinder
Hinweise auf günstige Nutzen-Risiko-Relation, häufiger anticholinerge und kardiogene UAW

Weitere atypische Neuroleptika wie **Olanzapin, Quetiapin, Aripiprazol** und **Clozapin** sind nur in wenigen Studien untersucht und gelten deshalb nicht als Mittel der ersten Wahl. Aripiprazol hat eine in Studien belegte Wirksamkeit auf Impulsivität, (Auto-)Aggressionen, Unruhe und Stereotypien. Da unter diesem Medikament deutlich weniger UAWs auftreten (insbesondere weniger EPS, Gewichtszunahme, Prolaktinerhöhung), verfügt es über eine günstigere Wirkung-Nebenwirkungsrelation (cave: fehlende Zulassung für Kinder und den Indikationsbereich).

Niedrig potente, stärker sedierende Neuroleptika werden zur Behandlung von Schlafstörungen und Unruhe- und Erregungszuständen eingesetzt. Aufgrund der Vielfalt derverfügbaren Substanzen kann hier nur eine Auswahl dargestellt werden, die sich vor allem an der Zulassung für Kinder und dem Indikationsbereich orientiert.

- **Melperon** (als Lösung und Tabletten verfügbar)
Indiziert zur Behandlung von Unruhe- und Erregungszustände, Schlafstörungen
In diesem Indikationsbereich bei Intelligenzminderung ab 12. LJ zugelassen
Als einziges Neuroleptikum auch antikonvulsiv wirksam
Atypikum, seltener EPS, weniger häufig anticholinerge UAW
- **Chlorprothixen** (als Saft und Dragees verfügbar)
Indiziert zur Behandlung von Unruhe- und Erregungszustände, Schlafstörungen
Zulassung ab 3. LJ
Höhere Wahrscheinlichkeit anticholinerger und kardiogener UAW
- **Pipamperon** (als Saft und Tabletten verfügbar)
Indiziert zur Behandlung von Unruhe- und Erregungszustände sowie Schlafstörungen
Eingeschränkte Zulassung < 18 Jahren
Seltener anticholinerge UAWs, seltener EPS, QT-Zeit-Verlängerungen bei höheren Dosierungen
- **Promethazin** (als Tropfen und Tabletten verfügbar)
Indiziert zur Behandlung von Unruhe- und Erregungszuständen und Schlafstörungen
In diesem Indikationsgebiet ab 2. LJ zugelassen
Häufiger anticholinerge UAW, QT-Zeit-Verlängerung, geringe therapeutische Breite

Stimulanzien

- **Methylphenidat**
Indiziert zur Behandlung von hyperkinetischem Verhalten bei IQ > 50
Response-Rate <20% bei IQ<50, dann auch höhere Rate an UAW sowie paradoxe Wirkungen
Vorteil ist die rasche Einschätzbarkeit von Wirkung und UAW

Auch **Atomoxetin** kann zur Behandlung hyperkinetischen Verhaltens indiziert sein.

Antidepressiva

Aufgrund der hohen Rate an UAW sollten trizyklische Antidepressiva nicht eingesetzt werden. Studien zu selektiven Serotonin-Wiederaufnahmehemmern (SSRI) zeigen keine eindeutige Wirksamkeit auf stereotypes und aggressives Verhalten. **Fluoexetin** besitzt eine Zulassung für den Indikationsbereich Depression für Kinder ab 8. LJ, **Fluvoxamin** ist zugelassen für die Behandlung von Zwangsstörungen ab 8. LJ.

Mood Stabilizer

Valproinsäure, Carbamazepin, Oxcarbazepin und **Lamotrigin**
haben neben der antiepileptischen eine stimmungsstabilisierende Wirkung und sind effektiv in der Behandlung von Impulsivität, (Auto-)Aggressivität und bipolaren Störungen. Insbesondere bei komorbider Epilepsie und einer Verhaltensstörung sollten, wenn möglich, stimmungsstabilisierenden Antiepileptika der Vorrang gegeben werden.

Weitere Psychopharmaka

- **Naltrexon**

Zeigt in Studien mit kleinen Fallzahlen gute Effekte auf Hyperaktivität, Impulsivität, (Auto-) Aggressionen und Stereotypien
Buspiron
Zeigt Effekte in der Behandlung von Angststörungen und oppositionellen Verhaltensweisen

- **Melatonin**

Indiziert zur Behandlung von Schlafstörungen
Keine Sedierung am nächsten Tag
Keine Zulassung für Patienten<18 LJ
Kombination mit SSRI sollte vermieden werden

Spezialisierte (teil-)stationäre Versorgung

Eine (teil-)stationäre Behandlung ist zu erwägen, wenn:

- der adäquate Behandlungsbedarf ambulant nicht eindeutig festgelegt werden kann, weil die diagnostische und / oder ätiologische Einordnung der Symptomatik nicht gelingt, oder
- die Erfolge der ambulanten Behandlungs- und Beratungsmaßnahmen weit hinter den begründeten Erwartungen zurückbleiben, oder
- eine akute Verschlechterung der Symptomatik oder eine Dekompensation der psychosozialen Situation vorliegt und eine Krisenintervention indiziert ist.

Notwendig ist dann die Behandlung in einer sozialpädiatrischen oder kinder- und jugendpsychiatrischen Einrichtung, die mit dieser Thematik Erfahrungen hat und die besondere Bedürfnislage intelligenzgeminderter Patienten und ihrer Familien konzeptionell in einem inklusiven Konzept berücksichtigt. Im Einzelnen bedeutet das:

- Eine enge Verzahnung der neuro- und sozialpädiatrischen mit der kinder- und jugendpsychiatrischen Versorgung
- Zusammenarbeit mit weiteren Fachärzten während einer stationären Behandlung
- Eine Betreuungskontinuität in kleinen geschützten Gruppen, in denen vorhersehbare Abläufe vorherrschen und eine Reizüberflutung vermieden werden kann

- Kein Gruppenwechsel von der tagesklinischen zur vollstationären Behandlung, z. B. feste Tagesgruppen, die bei entsprechendem Bedarf durch eine Nachtbetreuung ergänzt werden können
- Am Alter und Entwicklungsstand der Kinder orientiertes heilpädagogisches und therapeutisches Angebot
- Inklusive Beschulung durch Sonderschulpädagogen
- Die Verfügbarkeit von Eltern oder anderen vertrauten Bezugspersonen in der gesamten Behandlungszeit, um zusätzlichen Stress durch die stationäre Aufnahme zu minimieren und die Bezugspersonen als Experten für die individuellen Besonderheiten der Patienten einzubeziehen
- Ausreichende zeitliche und personelle Ressourcen
- Eine besondere Expertise sowohl für die psychopharmakologische als auch für die psychotherapeutische Behandlung und
- Eine enge Verzahnung der (teil-)stationären Behandlung mit der ambulanten Behandlung, die im Vorfeld der stationären Behandlung, im Verlauf und in der Entlassungsplanung erfolgt

Literatur

Hampel O A, Hasmann SE, Schaadt A-K, Holl R, Petermann F, Hasmann R (2010) Effekte des Stepping Stones Elterngruppentrainings für Familien mit behinderten Kindern. Kindheit und Entwicklung 19, 36-46.

Gerlach M, Warnke A, Wewetzer C (Hrsg) (2004) Neuro–Psychopharmaka im Kindes- und Jugendalter. Grundlagen und Therapie. Wien – New York: Springer.

Häßler F (2011) Intelligenzminderung. Springer-Verlag Berlin Heidelberg

Mendes U, Blank R, Hasmann R, Hollmann H, Mickley M, Peters H, Voigt F (2013)

Psychische Störungen bei Kindern und Jugendlichen mit Intelligenzminderung. Rahmenempfehlung. Qualitätszirkel der Bundesarbeitsgemeinschaft Sozialpädiatrischer Zentren. www.dgspj.de

Sarimski K, Steinhausen HC (2007) Geistige Behinderung und schwere Entwicklungsstörung, Kinder-Diagnostik-System-2, Göttingen: Hogrefe.

Sarimski K, Steinhausen HC (2008) Psychische Störungen bei geistiger Behinderung. In: Döpfner M, Lehmkuhl G & Petermann F (Hrsg.), Leitfaden Kinder- und Jugendpsychotherapie. Göttingen: Hogrefe.

38 Ganzheitliche Fördermaßnahmen

Helmut Hollmann

38.1 Einleitung

Kindliche Entwicklung ist ein dynamischer Prozess. Sie verläuft intrinsisch auf der Grundlage der genetischen Disposition des Gehirns zum Lernen. Dies gilt solange, wie Rahmenbedingungen kindgerecht gestaltet sind und den jeweiligen Alters- und Entwicklungsstand berücksichtigen. Kindliche Entwicklung ist primär unabhängig von speziellen Förder- und Therapiemaßnahmen.
Die **Unterstützung der Entwicklung** wird durch Eltern in sehr unterschiedlicher Form umgesetzt. Grundlegend erfolgt dies durch die Auswahl von Betätigungsmöglichkeiten mit geeignetem Spielzeug oder Handlungsfeldern, Gestaltung von sozialen Situationen und insbesondere die innerfamiliäre Kommunikation. So wird die Ausdifferenzierung kognitiver, motorischer und sprachlicher Funktionen befördert bei gleichzeitiger Modellbildung auch im emotionalen und sozialen Bereich. Eltern verfügen hierzu über ein intuitives Repertoire an kreativen und angemessenen Ideen.
Spezifische Anliegen an eine bestimmte Entwicklung führen zu einem veränderten Verhalten der Bezugspersonen im Sinne der **Förderung.** Unterricht auf einem Musikinstrument, Aufnahme in die Malschule oder den Sportverein sollen Neigungen unterstützen oder besondere Begabungen fördern. Die Betonung und Förderung einer einzelnen Fertigkeit führt dabei zu deren besonders guter Ausprägung. Andere Fähigkeiten und Fertigkeiten stehen zurück. Die breitbasige Entwicklungsunterstützung wird bei der gezielten Förderung zwangsläufig eingeschränkt. Dies ergibt sich automatisch aus der Begrenzung der Flexibilität des Kindes; die zeitlichen Ressourcen auch der Bezugspersonen sind fest verplant. Die Zeitspanne, die das Kind auf dem Sportplatz verbringt, kann nicht zum Lesen verwendet werden, und während des Unterrichts auf dem Instrument und der erforderlichen Übungszeit kann keine soziale Aktivität stattfinden.
Förderung ist immer der Versuch, verschiedene Aspekte und Interessen auf einen gemeinsamen Nenner zu bringen. Die Vorstellungen des Kindes mögen sich im Einzelfall deutlich von den Erwartungen oder Idealen der Bezugspersonen unterscheiden. Wenn hier keine Synthese gefunden werden kann, resultieren emotionale Belastungen und letztlich Konflikte.
Die **Grundsätze für die Förderung** von sich normal entwickelnden oder besonders talentierten Kindern gelten uneingeschränkt ebenso bei der Betreuung entwicklungsgestörter oder behinder-

Sozialpädiatrie in der Praxis. http://dx.doi.org/10.1016/B978-3-437-31630-3.00038-9
Copyright © 2014 Elsevier GmbH. All rights reserved.

ter Kinder und Jugendlicher. Weil gleichzeitig oftmals die Möglichkeit der eigenen Meinungsäußerung und damit altersgemäßen Mitsprache eingeschränkt ist, unterliegt es der besonderen Verantwortung von Fachpersonen in allen Professionen, diese Gesichtspunkte im Rahmen der **Förder- und Behandlungsplanung** zu berücksichtigen. Familiäre Bezugspersonen müssen deshalb in Entscheidungsprozesse systematisch mit einbezogen sein. Mit steigendem Alter ist die Abkehr von der defizitorientierten Aufarbeitung von Rückständen erforderlich. Stattdessen müssen auch bei bleibenden Entwicklungsstörungen im Sinne der Behinderung zunehmend die individuellen Fähigkeiten und Ressourcen in den Mittelpunkt gerückt werden.

Für einen gelingenden Förderprozess ist die Unterstützung im Bewältigungsprozess der Eltern eine zwingende Voraussetzung. Nur wenn Verständnis und Akzeptanz für die tatsächlichen Möglichkeiten und deren Limitierung beim Kind erreicht ist, können die Eltern eine dennoch positive und unterstützende Haltung aufbauen. Diese ist die wichtigste Bedingung für die Entwicklung von Selbstwertgefühl und Zufriedenheit beim Kind oder Jugendlichen. Die erfolgreiche Umsetzung des Entwicklungspotenzials im Rahmen des individuellen Entwicklungskorridors (➤ Kap. 37.1.1) benötigt die Eigenaktivität des Kindes als Motor seiner Entwicklung. Das humanökologische Konzept (➤ Kap. 37.1.2) spielt somit in der Gestaltung von Förderung eine besondere Rolle.

38.2 Heilpädagogik

In der kindlichen Entwicklung liegen Spielen und Lernen eng beieinander. Vor allem in den ersten Lebensmonaten lässt sich der Entwicklungsstand unmittelbar aus dem Spielverhalten ableiten. **Spielen** ist eine lustbetonte zielgerichtete Tätigkeit, die aus eigenem Antrieb erfolgt und nicht auf ein Endprodukt abhebt. Zweck des Spiels sind Erfahrung und Erprobung, damit verbunden sind Kommunikation und Lernen. Der Ausprägungsgrad des Spiels und sein sozialer Bezug sind alters- und umgebungsabhängig. Spielen ist ein transkulturelles Phänomen, das in modifizierter Form in jedem Alter auftritt.

Zu den **Charakteristika** des Spielens (Goetze 2002) zählen:

- Positiver Affekt und Freude
- Freiwilligkeit, Selbstkontrolle und intrinsische Motivation
- Aktivität und Engagement
- Flexibilität und Variation
- Prozess-, nicht Produktorientierung
- „Quasi-Realität".

Nur ein Kind, das sich wohl und geborgen fühlt, spielt. Der Sinn liegt dabei in der Handlung des Spieles selbst. Die Abfolge der verschiedenen Stufen des Spiels ist bei allen Kindern gleich (Largo 2003).

Als **Formen** des Spielens (Mc Mahon 1992) können unterschieden werden:

- Sensorisches und kreatives Spiel
- Motorisches Spiel
- Exploratives Spiel
- Soziales Spiel
- Symbolspiel.

Pädagogik hat den Anspruch, durch das Angebot von Lernprozessen und personalen Bindungen dem Kind zu ermöglichen, individuelle und soziale Fertigkeiten auszubilden, sie verfügbar zu machen und schließlich die Fertigkeiten vor dem Einsatz am erlernten Maßstab des eigenen sozialen

Verhaltens auf ihre mögliche Wirkung hin zu prüfen. Pädagogik ist immer auch Erziehung, d.h. neben der Vermittlung von Wissen und Fertigkeiten sollen auch die Normen des „richtigen Handelns" vermittelt werden.

Heilpädagogik ist der Bereich der Pädagogik, der sich speziell mit entwicklungsgestörten oder behinderten Kindern befasst. Dabei bezieht sich der Begriff des „Heilens" auf die Erziehung im Sinne der Hinführung zu persönlichem Wohlergehen des beeinträchtigten Kindes. Übergeordnetes Ziel ist die Erreichung von Lebensqualität, wozu die Erfüllung menschlicher Grundbedürfnisse, die Teilhabe an Gesellschaft, Bildung und Kultur sowie eine möglichst große Autonomie zählen. Heilpädagogik ist somit ein ganzheitliches Konzept hinsichtlich Betrachtung, Behandlung und Integration des Menschen mit seinen Persönlichkeitsbereichen. Im Zentrum der Heilpädagogik steht die einmalige unverwechselbare Persönlichkeit; erst nachgeordnet werden Auffälligkeiten, Krankheiten und Behinderungen betrachtet. Heilpädagogik bindet neue Erkenntnisse in Bezug auf das geförderte Kind in die bestehende Betreuungssituation ein und stützt sich dabei auf medizinische, psychologische und soziale Aspekte. Heilpädagogik ist eine spezielle Pädagogik, die eine optimale Lernentwicklung des Kindes ermöglichen soll. Die Grundlage für diese positive Veränderung ist eine vertrauensvolle Beziehung. Die Heilpädagogik soll durch eine selbstkritische innere Haltung geprägt sein, von der das Denken und Handeln des Pädagogen getragen wird. Heilpädagogik lenkt den Blick auf die vorhandenen Fähigkeiten und Fertigkeiten des Kindes und will dadurch Stigmatisierungen und Negativzuschreibungen vermeiden. Dabei schenkt sie der Beziehungsdimension besondere Aufmerksamkeit und baut in diesem Rahmen auf gegenseitige Akzeptanz, Begegnung, Verstehen und Kooperation.

Heilpädagogen, die sich um entwicklungsgestörte und behinderte Kinder bemühen, müssen über spezifische Kenntnisse und Methoden verfügen, die der jeweiligen Art der Beeinträchtigung angemessen sind und fachgerecht eingesetzt werden. Im Rahmen einer heilpädagogischen Übungsbehandlung wird über eine längere Zeit kontinuierlich gearbeitet, sodass Kontakte in regelmäßigen Abständen, aber nicht ständig, zustande kommen. Durch die Behandlungsstunden will der Heilpädagoge die Eltern und andere nahe Bezugspersonen des beeinträchtigten Kindes unterstützen und darin befähigen, mit dem Kind zunehmend im vorstehend skizzierten ganzheitlichen Sinne „heilpädagogisch" umzugehen. Der Heilpädagoge will dabei einerseits dem Kind helfen, Defizite auszugleichen und soziales Verhalten aufzubauen. Gleichzeitig will sie aber auch die Familie darin unterstützen, den für sie individuell typischen Bezug zum Kind zu festigen oder überhaupt erst aufzubauen.

Heilpädagogik will das Lernen des Kindes unterstützen. Sie fördert deshalb auf der Grundlage lerntheoretischer Kenntnisse das gerichtete Antriebsverhalten des Kindes durch eine spezifische Motivation. Schwierigkeiten im Handlungsablauf müssen dabei mit dem Kind gemeinsam überwunden und die neuen erlernten Verhaltensweisen gezielt angeboten und verstärkt werden. Dies ist in erheblichem Umfang von einer positiven personalen Bindung abhängig. Bekräftigendes und ermutigendes Verhalten sowie eine freundlich-zugewandte Atmosphäre sind die Voraussetzung dafür. Es ist zu bezweifeln, dass ohne eine positive sozial-emotionale Beziehung überhaupt nennenswerte Entwicklungsfortschritte erzielt werden können. Dies deckt sich mit den Beobachtungen des kindlichen Spiels, das sofort zum Erliegen kommt, sobald die Umgebungssituation in irgendeiner Form nicht mehr entspannt ist.

Die Heilpädagogik hat ihren Ursprung in der **Spielpädagogik** von **Friedrich Fröbel.** Während ursprünglich der Gedanke im Vordergrund stand, das Spiel des Kindes pädagogisch zu nutzen, um es möglichst rasch auf die Notwendigkeiten des Erwachsenenlebens vorzubereiten, formulierte Fröbel erstmals die Idee der Ergänzung und Harmonisierung des Lebens durch Spiel. In der Heilpädagogik ist es sinnvoll, das kindliche Spiel durch besondere Angebote so zu gestalten, dass dadurch dem Kind Möglichkeiten des zusätzlichen Ausdrucks und auch der besonderen Konfliktverarbeitung gegeben werden. Entscheidend ist dabei aber, dass der Charakter des freien Angebots grundsätzlich gewahrt bleibt. In der Umsetzung eines solchen Ansatzes grenzt Heilpädagogik an den tiefenpsychologischen Psychotherapiebereich der nondirektiven Spieltherapie. Diese Kombination wird an verschiedenen Studienorten gezielt gelehrt.

Der Ansatz der Heilpädagogik hat in der frühen Entwicklungsförderung große Verbreitung gefunden, reflektiert er doch alle Erkenntnisse derzeitiger Entwicklungstheorien. Gerade in Frühförderstellen und SPZ spielt deshalb die Heilpädagogik in Verbindung mit funktionstherapeutischen Behandlungsmaßnahmen (Heilmitteln) eine wesentliche Rolle. Das duale Finanzierungssystem in Deutschland, das pädagogische Maßnahmen nicht als Leistungsbestandteil der gesetzlichen Krankenversicherungen anerkennt, behindert die eigentlich notwendigen und sinnvollen komplementären Weiterentwicklungen in Verbindung mit den Heilmitteln. Aspekte der Heilpädagogik gehören längst zum alltäglichen Rüstzeug in Ergotherapie, Physiotherapie und Logopädie.

38.3 Montessori-Pädagogik

Eine ganz andere Richtung der heilpädagogischen Intervention verfolgte Maria Montessori. Sie entdeckte zunächst die fast 100 Jahre zuvor verfassten Schriften der französischen Pädagogen und Ärzte J. Itard und E. Séguin, die ausgehend von einer „physiologischen Erziehung" glaubten, durch die Gestaltung der Umweltbedingungen die Bildung von Intelligenz herbeiführen zu können. Aus der Erkenntnis der Vergeblichkeit dieser Überlegungen formulierten sie erste ganzheitliche Ansätze einer Förderung. Montessori entwickelte aufbauend auf diesen Überlegungen ein breites Programm mit didaktischem Material und verwendete es mit überraschendem Erfolg im Unterricht geistig behinderter Kinder. Das eher statisch ausgelegte Trainingsprogramm wurde dann von ihr ab 1907 auch in der Erziehungsarbeit sich normal entwickelnder Kinder in den Slums von Rom genutzt. Diese Montessori-Heilpädagogik veränderte die Unterrichtsformen und fand eine rasche Verbreitung.

Die Montessori-Methodik ist eng mit dem speziellen Material im Sinne einer psycho-physiologischen Methode verbunden. Das nicht behinderte Kind verwendet das Material in einer ganz anderen Weise als das geistig behinderte. Es erkennt intuitiv die in ihm liegende Aufgabe oder findet diese experimentierend heraus und kann sie dann ohne Hilfestellung der Erwachsenen lösen. So wird das Material für das Kind zu einem Mittel der Spontanbildung. Es weckt seine Selbsttätigkeit, die zu einer außergewöhnlichen und freiwilligen Aufmerksamkeitsleistung führen kann. Durch die Entwicklung eines umfassenden Übungsprogramms mit „Übungen des praktischen Lebens" und dem „didaktischen Material" wurde dieser Aspekt besonders herausgearbeitet.

Kinder mit Entwicklungsstörungen und geistiger Behinderung hingegen benötigen zur Anwendung des Materials die Anleitung durch eine Bezugsperson, um durch die Beschäftigung in der geistigen Entwicklung Fortschritte zu erzielen. Als Bedingung hierfür wird von Montessori die „vorbereitete Umgebung" angesehen, die den erstrebten Übungseffekt weitgehend garantiert und die Eigeninitiative des Kindes trotzdem in diesem Rahmen erhält.

Montessori erdachte ihr pädagogisches Material vornehmlich zum Erwerb von „technischen Fertigkeiten". Die geringere Beachtung der schöpferischen Komponente im Kinderspiel birgt bei ihr die Gefahr in sich, die Eigeninitiative und -aktivität des Kindes zu reduzieren. Bei aller Sinnhaftigkeit des Materials ist somit darauf zu achten, gerade in den Förderprogrammen für behinderte Kinder keine einseitigen Übungseffekte zu induzieren.

Der Satz **„Hilf mir, es selbst zu tun"** ist zu einem Leitmotiv heilpädagogischen Handelns allgemein geworden. Ausgehend vom Kinderzentrum München haben T. Hellbrügge und sein Team den Gedanken der Montessori-Pädagogik auch in Kindergärten und Schulen getragen. Aufgrund der universellen Verwendbarkeit des Materials ist hier eine besondere Möglichkeit der integrativen Förderung für behinderte und nicht behinderte Kinder gegeben.

38.4 Konduktive Förderung nach A. Petö

Im Bereich der heilpädagogisch orientierten Maßnahmen bewegt sich auch die konduktive Förderung, die der ungarische Arzt, Pädagoge und Schriftsteller Andras Petö als ganzheitliches Erziehungssystem ursprünglich für Kinder mit zerebralen Bewegungsstörungen entwickelt hat. Inzwischen wird es auch für Kinder mit anderen Entwicklungsbeeinträchtigungen eingesetzt. Ein zentrales Element ist dabei der „konduktive" Ansatz (aus dem Latein, „zusammenführen"), womit die Integration aller Entwicklungs- und Persönlichkeitsbereiche des Kindes in einem einheitlichen pädagogisch-therapeutischen Ansatz gemeint ist. Die herausragende Person ist die Konduktorin. Ihr muss es gelingen, sehr unterschiedliche Aufgaben oder Kompetenzen zu übernehmen, die sonst von Funktionstherapeuten und Pädagogen, teilweise auch Psychologen ausgeübt werden. Damit wird nicht nur in der theoretischen Auffassung, sondern auch personell ein Zusammenwirken der unterschiedlichen Teilsysteme bei der kindlichen Entwicklung umgesetzt.
Konduktive Förderung betrachtet die Bewegungsstörung nicht isoliert als „behandlungsbedürftige" Behinderung, sondern als komplexe Lernaufgabe, die das Kind mit Hilfe einer kompetenten Bezugsperson, eben der Konduktorin, aktiv handelnd zu bewältigen lernen soll. Ziel ist es, das Kind zur sogenannten **„Orthofunktion"** zu führen. Damit sind die zunehmend selbstständige Bewältigung elementarer Daseinsfunktionen und eine weitgehende Unabhängigkeit von Personen und Hilfsmitteln gemeint. Das Kind soll auf jeden Fall zu Tätigkeiten angeregt werden. Dabei besteht eine zielgerichtete und motivierte Handlung aus sogenannten Operationen als Einzelbausteinen des Bewegungsaktes. Ein weiteres Element in der konduktiven Förderung ist die **Unterstützung der Kommunikation.** Neben diesen sozialen Kontakten, die der Umsetzung der Tätigkeiten und Operationen dienen, ist die Berücksichtigung der Sprache ein wesentliches Element. Die verschiedenen Überlegungen werden unter dem Begriff einer **„Tätigkeitstheorie"** zusammengefasst. Damit ist ein integratives, ganzheitliches System gemeint, das es dem Menschen ermöglicht, seine Persönlichkeit zu entwickeln. Umgekehrt solle eine Störung von Organisation und Programmierung des Bewegungsablaufs automatisch auch zu einer gestörten Persönlichkeitsentwicklung führen. Das wesentliche Merkmal der Tätigkeitstheorie nach Petö wird in dem Kernsatz zusammengefasst: „The brain cannot be influenced from the periphery but only by fullfilling a task" (Cottam und Sutton 1986).
Das System nach Petö stellt eine Zusammenführung von Elementen aus der Physiotherapie mit Ähnlichkeit zur Bobath-Therapie alter Prägung, gleichzeitig heilpädagogischen und lerntheoretischen Aspekten dar. Die Förderung erfolgt in Gruppen von Kindern mit gleichem oder ähnlichem Störungsbild. Dadurch wird ein Lernen am Modell nicht nur über die Konduktorin, sondern auch innerhalb der Gruppe ermöglicht. Kommunikation, Lob und Kritik spielen eine große Rolle, ebenso die Gestaltung der Umgebungsbedingungen. Hilfsmittel werden nur in geringem Maß eingesetzt, da sich die Zielsetzung auf die Eigenaktivität und Entwicklung eigener Bewegungsabläufe ohne Berücksichtigung der Qualität konzentriert. Hierin unterscheidet sich die konduktive Förderung grundlegend von den Vorstellungen der Physiotherapie auf neurophysiologischer Grundlage.

Die Effektivität der konduktiven Förderung nach Petö ist vergleichbar mit den Auswirkungen sonst üblicher Therapiemaßnahmen. In Studien wird die gute sozioemotionale Entwicklung der Kinder hervorgehoben. Risiken liegen möglicherweise in einer stärkeren Belastung und auch Destabilisierung mit Luxationstendenz der Hüftgelenke aufgrund des verminderten Einsatzes von Hilfsmitteln im Vergleich zur konventionellen Physiotherapie. Durch die besondere Ausrichtung auf die Eigenaktivität und Selbstständigkeit der Kinder ist es vorstellbar, dass v. a. kognitiv gut entwickelte zerebralparetische Kinder, die zu einer sinnvollen und aktiven Mitarbeit im Rahmen des Förderprogramms fähig sind, von dieser Vorgehensweise profitieren. Umgekehrt ist der Ansatz weniger geeignet für schwer- und mehrfach behinderte oder sehr junge Kinder.

38.5 Sonderpädagogik

Kinder und Jugendliche mit Entwicklungs- und Verhaltensauffälligkeiten oder mit Behinderungen benötigen für eine positive schulische Entwicklung mehr noch als grundsätzlich alle Kinder solche Rahmenbedingungen, die ihren Erfordernissen ausreichend Rechnung tragen. Dies bedeutet, dass einerseits das Erlernen der Kulturtechniken im jeweils möglichen Umfang gesichert ist, andererseits die Bewältigung von Alltagsanforderungen und Realisation von Teilhabe erreicht wird. Maßnahmen der **Sonderpädagogik** sollen diese sehr differenten Ziele umsetzen.

Das stark gegliederte Schulsystem in Deutschland, zudem noch in jedem Bundesland unterschiedlich dimensioniert, erschwert oft die Suche nach dem geeigneten schulischen Förderort für Kinder und Jugendliche mit besonderen Bedarfen (➤ Kap. 6). Ernsthafte integrative Beschulungsmaßnahmen im Sinne der Inklusion sind immer noch die Ausnahme und enden weiterhin häufig beim Übergang in eine weiterführende Schule. Nur zögerlich wird der längst vorhandenen Kenntnis Rechnung getragen, dass viele Kinder gerade auch im Grenzbereich zur besonderen Fördernotwendigkeit von einem flächendeckenden Unterricht in kleinen Klassen und mit einem Tandem aus Regel- und Sonderpädagoge enorm profitieren. Die hierfür notwendigen Investitionen in langfristig wirksame Personalkosten werden von der Politik gescheut. Daran ändert auch der seit 2009 gesetzlich festgeschriebene Grundsatz der Inklusion in der Praxis noch sehr wenig. Umgekehrt bieten die traditionellen Förderschulen gerade für stark entwicklungsgestörte und mehrfach behinderte Kinder und Jugendliche gut qualifizierte Rahmenbedingungen insbesondere auch hinsichtlich medizinischer Betreuungserfordernisse. Im Zuge einer unzureichend ausgeprägten inklusiven Schulumgebung besteht für diese kleine, aber besonders aufwändige Schülerklientel das Risiko, gänzlich ausgeschult zu werden, wie aktuelle Entwicklungen in einigen Regionen zeigen. Hier verkehrt sich Inklusion in ihr Gegenteil.

So bleibt den sozialpädiatrisch tätigen Kinder- und Jugendärzten auf absehbare Zeit nur die intensive Empfehlung an die Eltern, sich umfassend über die zur Verfügung stehenden und erreichbaren Alternativen zu informieren und dann gemeinsam mit dem verantwortlichen Schulamt eine bestmögliche Kompromisslösung zu finden. In zunehmend häufigeren Fällen muss zusätzlich eine Einzelfallhilfe im Sinne des Schulbegleiters in Anspruch genommen werden. Dies ist nicht nur eine teure, sondern auch eine fachliche unsichere Lösung, weil sehr unterschiedlich ausgebildete und qualifizierte Menschen diese Tätigkeit wahrnehmen. Eine interdisziplinäre Projektgruppe der Stiftung Baden-Württemberg erstellt ab Juli 2013 hierzu eine Rahmenempfehlung.

Es ist die Aufgabe des Kinder- und Jugendarztes, aus ärztlicher und sozialpädiatrischer Sicht möglichst präzise die Bedürfnisse des Kindes oder Jugendlichen zu beschreiben, um so auf der Basis von fundierten Empfehlungen an der Schulentscheidung mitzuwirken (➤ Kap.13.6).

Literatur

Karch D, Hanefeld F, Ritz F, Schlack, HG (1997) Konduktive Förderung nach Petö; Stellungnahme der Gesellschaft für Neuropädiatrie, Kommission zu Behandlungsverfahren bei Entwicklungsstörungen und cerebralen Bewegungsstörungen. In: Boltshauser E, Schmitt B, Steinlin M (Hrsg.) Aktuelle Neuropädiatrie. Nürnberg: Novartis Pharmaverlag: 476–484.

Lichtenstern-Peters E, Diem S, Frieel B, Grödl-Zimmermann P, Hermann S, Müller M, Sürtenich A (2013) Heilpädagogik in der Sozialpädiatrie. Rahmenempfehlung. Qualitätszirkel der Bundesarbeitsgemeinschaft Sozialpädiatrischer Zentren. www.dgspj.de

Voss H von, Blank R (2002) Modellprojekt Petö – Kurzfassung. In: Voss H von (Hrsg) Sozialpädiatrie aktuell. Bd 1. Mainz: Kirchheim-Verlag.

39 Rehabilitation für Kinder und Jugendliche

Johannes Oepen

39.1 Einleitung

Die Entstehung der stationären medizinischen Rehabilitation von Kindern und Jugendlichen als „Kinderkur" bzw. „Kinderlandverschickung" ist begründet in genuin sozialpädiatrischen Überlegungen: Ein wichtiger Aspekt bei den Gründungsaktivitäten bestand darin, das gesundheitliche Elend von Kindern aus Familien der ärmeren Bevölkerungsschichten, wie es im Laufe der Industrialisierung entstanden bzw. öffentlich wahrnehmbar geworden war, zu mindern.
In England hatte – 40 Jahre nach Gründung des ersten Seebades 1750 in Brighton für die wohlhabende Bevölkerung – die industrielle Revolution mit den Folgen der Ballung der Bevölkerung in Städten und dem damit verbundenen sozialen Elend und daraus erwachsenen Krankheiten wie Tuberkulose bzw. „Skrophulose" zur Gründung des ersten Seehospizes 1796 in Margate geführt: „The Royal Sea Bathing Infirmary for the Relief of the Scrophulous Poor of London and all England". Dieses Haus an der Themsemündung mit ganzen 30 Betten wurde Vorbild für die Seehospi-

Sozialpädiatrie in der Praxis. http://dx.doi.org/10.1016/B978-3-437-31630-3.00039-0
Copyright © 2014 Elsevier GmbH. All rights reserved.

ze und Kinderheilstätten in Europa: Inhalation, Solebäder und allgemeine Roborierung waren jedoch mehr als eine Abkehr von „Säftelehre“ und magischen Vorstellungen in der Medizin sowie wirksame Maßnahmen zu Minderung vorherrschender Gesundheitsbelastungen bedingt durch die Pferchungssituationen der Kinder, sondern auch prägend für die bewertende allgemeine Meinung, dass eine „Kur“ in einem solchen Setting kranken Kindern „gut tut“.

Der Ludwigsburger Arzt und Pädagoge A. H. Werner (1808–1882) erwirkte in Deutschland 1862 die Eröffnung des ersten Kindersolbades Bethesda in Bad Friedrichshall-Jagstfeld als „Anstalt für skrophelkranke Kinder über die Dauer der Sommermonate“. – Wohlhabende Familien, deren Eltern im Solbad Kreuznach Linderung ihrer Beschwerden erfahren hatten, waren im Laufe der Jahre dazu übergegangen, auch Kindern armer Leute den Besuch des Bades zu ermöglichen, was, nach kirchlichen Kollekten und Spenden von Großindustriellen, 1876 zur Gründung der Kinderheilanstalt Viktoriastift in Bad Kreuznach führte. In der Folge wurden zahlreiche weitere Kliniken und Heime für Kinder gegründet, vor allem an der See und im Gebirge mit Betonung von Klimafaktoren: Salzhaltige bzw. allergenarme Luft, aber auch Aufnahme in Gruppen, Wanderungen im Gebirge oder am Strand und Bäder in der See und in Solewannen (s. a. Hartung 1988). Diakonisches Werk, Caritas, sowie DRK, DPWV, AWO und später Rentenversicherungen (mit 4 eigenen bis heute bestehenden Einrichtungen) waren neben Privatträgern prägende Organisationen der Leistungserbringer. Die Rentenversicherungen haben sich als Kosten- bzw. Leistungsträger seit den 1920er Jahren engagiert, die Krankenkassen ab den 1930er Jahren – jeweils vom Zeitgeist oder den Nöten der Nachkriegszeit geprägt. Was aus heutiger Sicht gegenüber der Akutmedizin zu sehr in den Hintergrund geriet, war die wissenschaftliche Untersuchung und Weiterentwicklung dieser Hilfen und Therapien.

Obwohl der „Kur“-Ansatz mit Roborierung, Zuwendung und Pflege anfangs wirksam und effizienzorientiert war, geriet er im 20. Jahrhundert zunehmend in die Kritik: Einerseits wurde ihm ein Verharren in wissenschaftlich nicht mit zeitgemäßen Methoden und Vorstellungen zu Wirkungen ortsspezifischer Heilmittel vorgehalten. Andererseits wurden reine Erholungsmaßnahmen in Zeiten des allgemeinen Wohlstands weniger zum Ausgleich von Gesundheitsstörungen erforderlich und mit der modernen Akutmedizin auch die Behandlung chronischer Krankheiten davon unabhängig weiterentwickelt. Ursprünglich psychosomatisch-pädagogische Ansätze gerieten teilweise in den Hintergrund.

Die „Deutsche Gesellschaft für Sozialpädiatrie“ hat seit 1978 die Weiterentwicklung der Kinderkuren mitgeprägt mit dem „Arbeitskreis Kinderkuren“ (heute Fachausschuss „Stationäre Präventions- und Rehabilitationsmaßnahmen für Kinder und Jugendliche“): Unter Leitung von K. Hartung wurden 1982 „Anforderungsprofile“ aus fachlicher Sicht erarbeitet mit Kriterien für die Ausstattung von Kinderkureinrichtungen und einer differenzierten Indikationsliste, die 1998 unter Mitwirkung von H. Goldschmidt und A. Baumann im Konsens mit den Krankenkassen als „Gemeinsames Rahmenkonzept für die Durchführung stationärer medizinischer Maßnahmen der Vorsorge und Rehabilitation für Kinder und Jugendliche“ formuliert und von der BAR (Bundesarbeitsgemeinschaft für Rehabilitation) herausgegeben wurden. Im gleichen Jahr gaben die Rentenversicherungen ihr „Rahmenkonzept und indikationsspezifische Konzepte zur medizinischen Rehabilitation von Kindern und Jugendlichen“ heraus. 2008 konnte unter Mitwirkung der DGSPJ in der Schriftenreihe der BAR erstmals ein „Gemeinsames Rahmenkonzept der Gesetzlichen Krankenkassen und der Gesetzlichen Rentenversicherung für die Durchführung stationärer medizinischer Leistungen der Vorsorge und Rehabilitation für Kinder und Jugendliche“ veröffentlicht werden – redaktionell im Sinne der ICF überarbeitet.

Seit den 1990er Jahren gab es in Deutschland ein Wiederaufleben der Bemühungen um eine wissenschaftliche Fundierung und Weiterentwicklung der Kinderrehabilitation gerade auch mit

Bezug auf das SGB IX, (z. B. „Rehabilitation und Teilhabe – Wegweiser für Ärzte und andere Fachkräfte der Rehabilitation“ 3. Auflage 2005 erstmals mit Beiträgen zur Rehabilitation von Kindern und Jugendlichen).
Seit 2001 hat die Deutsche Gesellschaft für pädiatrische Rehabilitation und Prävention, DGPRP, zur stationären Rehabilitation von Kindern und Jugendlichen Leitlinien veröffentlicht, eine allgemeine Präambel und Leitlinien zu Adipositas, Asthma bronchiale, Diabetes mellitus, pädiatrischer Onkologie, angeborenen und erworbenen Herzerkrankungen, neurologischer Rehabilitation, Psychiatrie, Psychosomatik und Psychotherapie im Kindes- und Jugendalter. Gemeinsam haben DGSPJ und DGPRP (2007) auch allgemeine Rahmenempfehlungen zur ambulanten Rehabilitation von Kindern und Jugendlichen der Deutschen Gesellschaft für Sozialpädiatrie und Jugendmedizin und der Fachgesellschaft für Rehabilitation im Kindes- und Jugendalter unter Moderation der BAR erarbeitet.

39.2 Gemeinsames Rahmenkonzept von Krankenkassen, Rentenversicherungen und Leistungserbringern

Hierin wird betont, dass vor dem Hintergrund der sich verändernden Lebens- und gesundheitlichen Situation von Kindern und Jugendlichen mit Zunahme von chronischen, vor allem psychosomatischen Erkrankungen im Kindes- und Jugendalter nach wie vor erheblicher Bedarf an spezifischen medizinischen Vorsorge- und Rehabilitationsleistungen für diesen Personenkreis besteht. Im gesundheitlichen Versorgungssystem kommt somit der Vorsorge und Rehabilitation bei Kindern und Jugendlichen ein besonderer Stellenwert zu.
Dazu wird verwiesen auf die Ergebnisse der Kinder- und Jugend-Gesundheitssurvey, KiGGS, (Robert-Koch-Institut 2006). Danach wird das Krankheitsgeschehen bei Kindern und Jugendlichen unter Berücksichtigung des Zusammenwirkens von Risiko- und Schutzfaktoren durch eine Verschiebung von akuten zu chronischen Krankheiten sowie von somatischen zu psychischen und psychosomatischen Störungen geprägt (z. B. 17,8 % der Jungen und 11,5 % der Mädchen zeigen Hinweise auf Verhaltensauffälligkeiten oder emotionale Probleme). Diese Aspekte zeigen sich deutlicher, je älter die Kinder werden. Kinder mit Migrationshintergrund und vor allem aus sozial belasteten Familien sind in Gesundheit und Lebensqualität deutlicher gefährdet und betroffen mit Häufung von Risikofaktoren, Unfällen, Umweltbelastungen, psychischen Auffälligkeiten Übergewicht und anderen Krankheiten aber auch schlechterer gesundheitliche Versorgung bzw. Inanspruchnahme der Versorgung.
Das **BAR-Rahmenkonzept von 2008** fordert eine Anpassung der medizinischen Vorsorge- und Rehabilitationsleistungen an einen Behandlungsansatz, der über das Erkennen, Behandeln und Heilen einer Krankheit hinaus krankheitsbezogene physische, psychische und soziale Belastungsfaktoren umfassend berücksichtigt. Dieser Ansatz erfordert die Integration der medizinischen, pädagogischen und sozialen Leistungen mit enger Abstimmung ärztlicher, pflegerischer, physiotherapeutischer und psychotherapeutischer Versorgung auch mit Hilfen zur psychischen Stabilisierung und sozialen Kompetenz. Dabei ist die Verzahnung mit Leistungen nach dem SGB VIII bis heute eine wichtige aber unerfüllte Forderung.
Die besonderen Schwerpunkte und primäre Ziele von Kuration, Vorsorge (primäre und sekundäre Prävention) und Rehabilitation (tertiäre Prävention) sind in der Zuordnung zu Hilfen für chro-

nisch kranke Kinder zu beachten, auch wenn es eine strikte Trennung der verschiedenen Versorgungsbereiche nicht geben kann und soll. Schematisch betrachtet ergeben sich folgende Unterschiede:

- Die **kurative Versorgung** i. S. des SGB V ist im Unterschied zur medizinischen Rehabilitation primär ausgerichtet auf das klinische Bild als Manifestation einer Krankheit bzw. Schädigung und zielt auf Heilung oder Remission (kausale Therapie) oder bei chronischen Krankheiten auf Vermeidung einer Verschlimmerung sowie Linderung der Krankheitsbeschwerden und Vermeidung weiterer Krankheitsfolgen ab. Konzeptionelles Bezugssystem ist vorrangig das biomedizinische Krankheitsmodell der ICD.
- Die **medizinische Rehabilitation** orientiert sich grundsätzlich am bio-psycho-sozialen Modell von funktionaler Gesundheit und deren Beeinträchtigung, das Gesundheit und Krankheit als Ergebnis des Ineinandergreifens physiologischer, psychischer und sozialer Vorgänge beschreibt – gemäß ICF-CY (➤ Kap.16). Auf diese Weise wird sichergestellt, dass neben Erkenntnissen der medizinischen Versorgung auch die gesellschaftlichen Bedingungen, in denen Kinder und Jugendliche leben, Grundlage der medizinischen Rehabilitation sind. Bei Kindern und Jugendlichen geht es neben der Wiederherstellung der Gesundheit immer auch darum, eine absehbare Schwächung des Gesundheits- und Entwicklungs-Potenzials abzuwenden, die individuelle Gesundheit zu fördern und sekundären Erkrankungen vorzubeugen – als körperliche, seelische und geistige Gesundheit.

Im Sinne der ICF ist also Behinderung, vor allem eine Beeinträchtigung der Teilhabe in eine altersangemessene Lebenssituation und altersangemessene Lebensbereiche des Kindes oder Jugendlichen, nicht mehr nur ein personenbezogenes Merkmal, sondern entsteht aus dem ungünstigen Zusammenwirken von gesundheitlichen Problemen einer Person und ihrer Umwelt- unter Berücksichtigung der Sichtweise der einzelnen Person! Das Konzept der Teilhabe bei Kindern und Jugendlichen ist mit Fragen nach Zugang zu Lebensbereichen, Entfaltung von Entwicklungspotenzialen, Ermöglichen eines adäquaten Schulabschlusses sowie einer entsprechenden Berufsausbildung und der Verwirklichung von Autonomie und Gleichberechtigung verknüpft und hat Auswirkungen auf Lebenszufriedenheit, gesundheitsbezogene Lebensqualität, erlebte Anerkennung und Wertschätzung, die für die Kinder bzw. Jugendlichen wichtig sind.

Viele Aktivitäten eines Kindes können aber vor allem anfangs nur mit Hilfe bzw. in Interaktion mit den Bezugspersonen ausgeführt werden! Somit hängt der Erwerb vieler Aktivitäten u. a. mit Auswirkungen auf die Teilhabe von den Kompetenzen, Einstellungen und von den Ressourcen seiner Bezugspersonen ab (Bindungsverhalten, Erziehungsstil, Bildungsnähe, soziale Herkunft, Migrationshintergrund u. ä.): Im Kindes- und Jugendalter sind also umwelt- und personenbezogene Kontextfaktoren von besonders herausragender Bedeutung (➤ Kap. 13).

Das Konzept der medizinischen Rehabilitation beinhaltet somit **5 Handlungsprinzipien:**

- Gleichwertige Berücksichtigung und Einbeziehung somatischer, psychischer, sozialer und pädagogischer Faktoren für den gesamten Rehabilitationsprozess (**integrativer Ansatz**)
- Enge Kooperation mit Familien und ihre routinemäßige Einbeziehung in die Rehabilitation des Kindes/Jugendlichen (**Familienorientierung**)
- Einbezug des unmittelbaren weiteren sozialen Umfeldes (insbesondere Kindergarten, Schule, Ausbildungsstelle, u. U. auch Vereine, Jugendhilfe, Freizeitangebote für Kinder und Jugendliche usw.) in geeigneter Form in die Rehabilitation (**Wohnortbezogenheit**)
- Patientenbezogene Kooperation der Experten aus medizinischen, psychosozialen und anderen Bereichen, die im Rahmen der Rehabilitation tätig werden (**Interdisziplinparität**)
- Die Angebote sollen sich am Bedarf des Kindes/Jugendlichen sowie an der Wirksamkeit und Wirtschaftlichkeit der Maßnahmen orientieren (**evidenzbasierte Rehabilitation**).

39.3 Gesetzliche Grundlagen der ambulanten und stationären Rehabilitation von Kindern und Jugendlichen

Im Rahmen der Rehabilitation und Teilhabe ist den besonderen Bedürfnissen behinderter und von Behinderung bedrohter Kinder bei der Entscheidung über die Leistungen und die Ausführung der Leistungen ausdrücklich Rechnung zu tragen (§§1 Satz2, 4 Abs. 3 SGB IX).

Voraussetzung für die Bewilligung einer Rehabilitationsleistung ist immer eine **Antragstellung** durch den Versicherten (bzw. gesetzliche Vertreter), evtl. auf Empfehlung der betreuenden Kinder- oder Hausärzte. Sind die persönlichen und versicherungsrechtlichen Voraussetzungen bei den gesetzlichen Krankenkassen bzw. Rentenversicherungsträgern erfüllt, können die Rehabilitanden bzw. Angehörigen des Versicherten Rehabilitations-Leistungen in Anspruch nehmen. Antragsformulare sind u. a. bei den Auskunfts- und Beratungsstellen der gesetzlichen Kranken- und Rentenversicherungsträger erhältlich.

Der seit den 1990er Jahren zu beobachtende Shift von der GKV zur DRV entspricht nicht den gesetzlichen Intentionen! Allerdings muss das bei gemeinsamer Planungsgrundlage und Berücksichtigung unterschiedlicher Zielsetzungen auch nicht zu Leistungsverschlechterung für die Rehabilitanden führen. – Andere Leistungsträger (gesetzliche Unfallversicherung) werden von Amts wegen tätig. Die gemeinsamen Servicestellen der Rehabilitationsträger nach §22 Abs. 1 Satz2 Nr. 3 SGBIX (aktuelle Liste der Servicestellen www.bar-frankfurt.de) haben leider oft noch keine ausgewiesene pädiatrische Kompetenz.

Nach Antragstellung prüft der jeweilige Träger zunächst die versicherungsrechtlichen Voraussetzungen, anschließend erfolgen durch die Leistungsträger die Beurteilung von Rehabilitationsbedürftigkeit, -fähigkeit und -prognose sowie die Auswahl einer geeigneten Rehabilitationseinrichtung für Kinder und Jugendliche.

Da Schädel-Hirn-Verletzungen oft durch Unfälle entstehen, kommt hier nach §27 Abs. 1 Nr. 7 SGB VII die gesetzliche Unfallversicherung als Leistungsträger öfter in Betracht: Eine sehr gute Übersicht bietet hierzu u. a. die „Arbeitshilfe für die Rehabilitation und Teilhabe Schädel-Hirnverletzter Kinder und Jugendlicher" (BAR 2007).

Gesetzliche Grundlagen für stationäre medizinische Rehabilitationsleistungen bei Kindern und Jugendlichen (s. a. gemeinsames Rahmenkonzept BAR, 2008) sind:

Sozialgesetzbuch IX – insbesondere:
- Selbstbestimmung und Teilhabe §1 SGB IX
- Leistungen zur Teilhabe §4 SGB IX

Regelungen der gesetzlichen Krankenversicherung
- Stationäre medizinische Vorsorgeleistungen §23 Abs. 4 SGB V
- Stationäre medizinische Rehabilitationsleistungen §40 Abs. 2 SGB V
- Mitaufnahme einer Begleitperson §11 Abs. 3 SGB V
- Reise- und Fahrkosten §§60 SGB V, §53 Abs. 1–3 SGB IX

Regelungen der gesetzlichen Rentenversicherung
- Stationäre Kinderheilbehandlungen §31 Abs. Satz 1 Nr. 4,
- Satz Mitaufnahme einer Begleitperson §31 Abs. 2 Satz 2 SGB VI (KiHB-Richtlinien)1
- Reisekosten §28 SGB VI i. V. m. §53 SGB IX

Schulgesetze der Bundesländer

Je nach Alter des Jugendlichen können auch Leistungen zur Teilhabe am Arbeitsleben in Betracht kommen (Anspruchsgrundlagen §§33 ff. SGB IX). Diese Leistungen werden von der Bundesagentur für Arbeit, der Unfallversicherung, bei selbst versicherten Jugendlichen oder jungen Erwachsenen von der Rentenversicherung, in selteneren Fällen von den Trägern der Kriegsopferfürsorge, der Sozial- und Jugendhilfe erbracht. Träger der Sozialhilfe sind für die Leistungen zur Teilhabe am Leben in der Gemeinschaft als Rehabilitationsträger insoweit zuständig, als nicht vorrangige Zuständigkeit der Unfallversicherung, der Kriegsopfer-Versorgung/-Fürsorge oder der Jugendhilfe gegeben ist. Als Leistungen zur Teilhabe am Leben in der Gemeinschaft kommen auch heilpädagogische Leistungen für Kinder, die noch nicht eingeschult sind (§56 SGB IX), sowie die Versorgung mit anderen als den in §31 SGB IX genannten Hilfsmitteln oder den in §33 SGB IX genannten Hilfen in Betracht.

39.4 Indikationen der ambulanten und stationären Rehabilitation von Kindern und Jugendlichen

Vor jeder Rehabilitationsmaßnahme, muss eine genaue Prüfung der Diagnose, der Voraussetzungen – Reha-Bedürftigkeit, Reha-Fähigkeit – der Reha-Ziele, sowie -Motivation und -Prognose stehen (s. u.). Die medizinische Rehabilitation soll als Teil eines Gesamtrehabilitationsplans gesehen werden, dessen Teile aufeinander aufbauen oder sich ergänzen („Reha-Kette“ oder „Reha-Vernetzung“).

Oft sind Kinder mit chronischen Gesundheitsbelastungen durch die Benennung einer einzigen Diagnose im Sinne der Rehabilitation nicht ausreichend beschrieben. Allerdings wird in der Regel doch ein Schwerpunkt gesetzt, was als Rehabilitationsdiagnose und was als Komorbidität angesehen wird. Die häufigsten **Diagnosen,** die danach Anlass für die Aufnahme eines Kindes oder Jugendlichen zur stationäre Rehabilitation sind, sind auch heute noch Asthma bronchiale/atopische Erkrankungen und Lungenerkrankungen wie Mukovoszidose, aber auch Adipositas, Diabetes mellitus, Entwicklungsverzögerungen und andere chronische Erkrankungen wie Kontinenzstörungen (Enuresis und Enkopresis) sowie nephrologische, neurologische, orthopädische Erkrankungen (z. B. Skoliose) und chronisch entzündliche Darmerkrankungen. Zunehmend werden psychosomatische und psychische Erkrankungen benannt.

39.5 Kontraindikationen

Nicht geeignet ist die stationäre Rehabilitation, wie sie aktuell besteht, für Kinder oder Jugendliche und junge Erwachsene, die nicht gruppenfähig sind, insbesondere aufgrund von Fremd- und Eigengefährdung (auch bzgl. akuter Erkrankungen). Suchterkrankungen stellen für viele Einrichtungen ebenfalls eine Kontraindikation dar.

Hochfrequente Therapien oder Einzelbehandlung können aktuell auch bei gegebener Indikation von den meisten Rehabilitationskliniken nicht geleistet werden,

39.6 Voraussetzungen für die Rehabilitation

Ob rehabilitationsbedürftige Versicherte Leistungen zur Rehabilitation überhaupt beantragen, hängt nicht nur vom „Gesundheitszustand" ab, sondern wird auch von Informationsstand, Motivation und weiteren psychosozialen und beruflichen Faktoren beeinflusst. Eine Verbesserung des Zugangs zur Rehabilitation von rehabilitationsbedürftigen Versicherten erfordert daher die Erfassung und Beseitigung eventuell bestehender Zugangshemmnisse und die besondere Unterstützung von Versicherten in besonderen Lebenslagen (z. B. allein erziehende Mütter und Väter, pflegende Familienangehörige, Langzeitarbeitslose, Menschen mit Migrationshintergrund). Dabei ist der Fortschritt sehr mühsam. Beispielhafte Probleme, die aktuell von Leistungserbringern und Leistungsträgern gemeinsam angegangen werden sollen, sind:

- Vereinfachung des Zugangs durch Überarbeitung des Formularwesens
- Minderung von Kommunikationsprobleme (Sprachbarrieren und Informationslücken u. a.)
- Differente Gesundheits- und Krankheitskonzepte
- Unzureichendes Angebot an entsprechenden Rehabilitationseinrichtungen
- Oft noch unzureichende Erfahrungen z. B. mit Migration und darauf folgende Verzögerung der Inanspruchnahme von Leistungen wegen Angst vor Verlust des Arbeitsplatzes oder des aufenthaltsrechtlichen Status.

Rehabilitationsbedürftigkeit bezieht sich auf eine gesundheitlich bedingte drohende oder bereits manifeste Beeinträchtigung der Teilhabe, die über die kurative Versorgung hinaus den mehrdimensionalen und interdisziplinären Ansatz der medizinischen Rehabilitation erforderlich macht. Rehabilitationsbedürftigkeit kann angenommen werden, wenn

- Folgeschäden einer Erkrankung drohen bzw. bereits eingetreten sind
- die ambulante ärztliche Krankenbehandlung nicht ausreicht, um Schädigungen mit daraus resultierenden Beeinträchtigungen von Aktivitäten und Teilhabe zu beseitigen/zu mindern
- Leistungen zur Krankheitsbewältigung und zur Unterstützung des Krankheits- bzw. Gesundheitsmanagements erforderlich sind.

In den „Leitlinien zur sozialmedizinischen Beurteilung der Rehabilitationsbedürftigkeit" der Deutschen Rentenversicherung Bund sind für diesen Bereich entsprechende Kriterien dargestellt – wenn auch nach eigener Bewertung noch nicht für Kinder und Jugendliche.

Rehabilitationsfähigkeit bezieht sich auf die somatische und psychische Verfassung des Rehabilitanden (z. B. Motivation bzw. Motivierbarkeit und Belastbarkeit, aber auch Fähigkeit aktiv mitzuarbeiten und sich in das Gruppensetting zu integrieren) für die Teilnahme an einer geeigneten Rehabilitation.

Es müssen dafür allgemeine Kriterien erfüllt sein, z. B.:

- ausreichende körperliche und psychosoziale Belastbarkeit
- Voraussetzungen für aktive, entwicklungsgemäße Mitarbeit
- soziale Integrationsfähigkeit (bzgl. Rehabilitationssetting v. a. die Stationsgruppe als alltagsähnliche Modellsituation).

Basis für die Beurteilung sind vorliegende Unterlagen über Schädigungen von Körperfunktion und Körperstruktur sowie über Beeinträchtigungen von Aktivitäten und Teilhabe. Mindestens bestehen diese Unterlagen aus einem ärztlichen Befundbericht (Vordruck bei GKV und DRV), meist vom Kinderarzt, Hausarzt oder Krankenhausarzt, in dem Funktionsbefunde mit Auswirkungen auf den beruflichen und sozialen Bereich aufgeführt sind, sowie in einem Selbstauskunftsbogen mit Eigenangabe der Beeinträchtigungen. Möglichst kommen weitere relevante medizinische/therapeutische Befunde hinzu.

Der Befundbericht für die GKV mit dem Formular 60 kann nur bei Nachweis einer Qualifikation (Rehabilitationsmedizin bzw. entsprechende Kurse) abgerechnet werden. Dieser restriktive Ansatz, die Verordnung von Rehabilitation durch niedergelassene Ärzte an eine Zusatzqualifikation zu binden, ist zwar zur Steuerung einer der Kosten nachvollziehbar, lässt aber eine Lücke zum Anliegen der frühzeitigen Bedarfserkennung. Aus Patientensicht dürfte sich kein Arzt für unzuständig halten, den Kontakt zum Reha-System herzustellen.

Die **Rehabilitationsprognose** ist eine medizinisch begründete Wahrscheinlichkeitsaussage

- für den Erfolg der Rehabilitation auf der Basis der Erkrankung (Schwere, Dauer, Verlaufsform und Risikofaktoren), des Kompensationspotenzials/der Rückbildungsfähigkeit unter Beachtung und Förderung individueller Ressourcen (Rehabilitationspotenzial einschließlich psychosozialer Faktoren)
- über die Erreichbarkeit eines festgelegten Rehabilitationsziels
- durch eine geeignete Rehabilitationsleistung
- in einem notwendigen Zeitraum.

Die Komplexität aller körperlichen, psychischen und sozialen Faktoren und deren wechselseitige Rückwirkung sind dabei entscheidend für die Prognose.

Reha-Motivation: In Leitlinien und Konzepten bzw. Anforderungsprofil wird die Bedeutung der Motivation hervorgehoben. Sie wird in der Regel als Eigenschaft – oder Ressource – der Person beschrieben, z. T. als kontextabhängig (Wechselwirkungen mit Bezugspersonen). Gut herausgearbeitete Aspekte der Reha-Motivation sind: Leidensdruck, Hoffnung auf ein gutes Behandlungsergebnis, Wissen über Rehabilitation, Einstellung zur Rehabilitation, Ausmaß an Eigeninitiative, Bereitschaft zur aktiven Mitarbeit und Änderung des Lebensstils, Gesundheits- und Krankheitskonzepte.

Von Bedeutung für die Umsetzung von Motivation dabei auch körperbedingte Ressourcen/Einschränkungen, Ausmaß der sozialen Unterstützung sowie Änderungsbereitschaft, evtl. „Krankheitsgewinn" aber auch Informationsstand bzw. Skepsis bzgl. der Reha-Maßnahme. Im Rehabilitationsalltag hat sich außerdem bewährt zu unterscheiden zwischen einer mitgeteilten Motivation (z. B. zu Therapieumsetzung, Selbstmanagement, Gewichtsabnahme usw.) und einer zu erwartenden Motivierbarkeit bzw. Fähigkeit, zu einer aktiven Bewältigung zu gelangen (z. B. nach Überwindung der Fremdheitsreaktion): Manche Jugendliche zeigen sich z. B. vor Aufnahme hoch- bis übermotiviert, ohne eine entsprechende Handlungsstabilität schon aufgebaut zu haben; andere sind zunächst abblockend, können aber im Laufe der 4–6 Wochen (bei konsequenter Haltung der Bezugspersonen zuhause) aus dem positiven Erleben der Gruppe eine sehr gute, auf Wahrnehmung erster Teilerfolge basierte Motivation aufbauen.

Die **Dauer** stationärer medizinischer Rehabilitationsleistungen für Kinder und Jugendliche wird nach dem BAR-Rahmenkonzept von 2008 für GKV und DRV aufgrund der medizinischen und entwicklungsspezifischen Besonderheiten von vornherein mit 4 Wochen angesetzt. Eine längere Dauer ist möglich, wenn das Rehabilitationsziel sonst nicht erreicht werden kann – ggf. ist dazu von der Rehabilitationseinrichtung rechtzeitig ein ausführlich medizinisch begründeter Verlängerungsantrag zu stellen. Seit 2012 geben einige Rentenversicherer diagnosebezogener Durchschnittsverweildauern vor. Wird das Kind durch eine Begleitperson betreut, beträgt die Dauer 4 Wochen. Die empfohlene Dauer der Rehabilitationsleistungen ist ein aber nur ein allgemeiner Richtwert, der nach sozialmedizinischer Beurteilung und individuellen Besonderheiten angepasst werden muss.

Die **Reha-Vorbereitung** ist von besonderer Bedeutung: Nicht ausreichende Vorbereitung auf die Vorsorge- oder Rehabilitationsleistungen ist häufig verbunden mit erheblichen allgemeinen Compliance-Problemen, die z. B. eine besonders einfühlsame Hinführung zu den therapeutischen

Abläufen erfordern. Oft ist auch eine Orientierung gebende Unterstützung (Standfestigkeit bzgl. des eingeschlagenen Weges) sowohl durch Bezugspersonen zuhause als auch in der Klinik für die Adaptation bzw. Bewältigung der entwicklungsgerechten Aufgaben im Behandlungsprozess notwendig und hilfreich.

In der Regel sind zur Vor- bzw. Nachbereitung der Rehabilitationsleistung bei Aufnahme und Entlassung eines Kindes oder Jugendlichen Gespräche mit den Bezugspersonen erforderlich – und nach dem Eindruck aus der Praxis leider aus wirtschaftlichen Gründen bisher noch zu selten ausführlich und in zielführender Weise umgesetzt bzw. umsetzbar!

Nach §23 SGB IX ist die **Zusammenarbeit** mit Personen und Stellen von Bedeutung, die eine Rehabilitationsbedürftigkeit rechtzeitig erkennen können, die über die Möglichkeiten der Rehabilitation informieren und dazu anregen, einen Antrag auf Rehabilitationsleistungen zu stellen, sowohl niedergelassene und im Krankenhaus tätige Ärztinnen und Ärzte, als ggf. auch betriebs- und arbeitsmedizinische Dienste, Krankenkassen und ihre Medizinischen Dienste, andere Sozialleistungsträger, gemeinsame Servicestellen der Rehabilitationsträger, die den behinderten oder von Behinderung bedrohten Menschen, ihren Vertrauenspersonen und Personensorgeberechtigten Beratung und Unterstützung anbieten.

Bereits vor Beginn der Rehabilitation sollte nach DRV-Mitteilungen auf den aktiven Charakter der Rehabilitation (z. B. Bereitschaft zur Teilnahme an angesagten Therapien und Aktivitäten) und auf die verhaltensmedizinischen Grundlagen der entsprechenden Therapiekonzepte in geeigneter Weise vorbereitet werden. Dabei sollte unter Einbezug von subjektivem Krankheits- bzw. Gesundheitskonzept der Betroffenen und ihrer Bezugspersonen insbesondere auf den Aufbau von gesundheitsförderndem und krankheitsadäquatem Verhalten als eine zentrale Aufgabe der Rehabilitation hingewiesen werden – z. B. durch Informationsveranstaltungen (Fachambulanzen und Selbsthilfegruppen), Übersendung von Informationsbroschüren (möglichst in der jeweiligen Muttersprache) oder ein Gespräch im Rahmen der sozialmedizinischen Begutachtung. In Einzelfällen und bei bestimmten Indikationen können auch Vorgespräche im Auftrag der Leistungsträger in der Rehabilitationseinrichtung sinnvoll sein, um Zweifel über die Eignung einer Einrichtung auszuräumen oder einen angemessenen therapeutischen Ansatz auszuwählen. Solche Kontakte lassen sich ggf. auch dazu nutzen, diagnostische Erhebungen einzuleiten.

39.7 Ambulante versus stationäre Rehabilitation

In den Grundlagen unterscheiden sich die Rehabilitationsziele bei ambulanter und stationärer Rehabilitation nicht. Vor *jeder* Rehabilitationsmaßnahme, ambulant oder stationär, muss eine genaue Überprüfung der Diagnose, der Voraussetzungen für Rehabilitation (Rehabilitationsbedürftigkeit, -fähigkeit), der Rehabilitationsziele, -motivation und -prognose stehen (s. o.).

Ambulante und stationäre Rehabilitation könnten als Teil eines Gesamtrehabilitationsplans gesehen werden, die aufeinander aufbauen oder sich ergänzen, z. B. im Sinne einer gezielten ambulanten Rehabilitationsvorbereitung oder -einleitung mit konsekutiv stationärer Rehabilitation, oder des Anschlusses ambulanter Rehabilitation an eine stationäre Rehabilitation, um den heimatnahen Transfer zu gewährleisten. Je nach Diagnose, Schweregrad der Erkrankung, Kontextfaktoren und regionaler Angebotsstruktur soll der Schwerpunkt dabei mehr in den Bereich der stationären Rehabilitation bzw. der ambulanten Rehabilitation liegen. Voraussetzung für ambulante Rehabilitation sind in jedem Fall die Sicherstellung der kinder- und jugendmedizinischen hausärztlichen Versorgung außerhalb der Rehabilitationszeiten, ausreichende Mobilität und Erreichbarkeit der Rehabilitations-Einrichtung und Sicherung der Grundversorgung und Pflege durch die Familie.

Die **ambulante Rehabilitation** kann in Betracht kommen:

- Zur Vorbereitung einer stationären, wohnortfernen Rehabilitation
- Anstelle einer stationären Rehabilitation als eigenständiges interdisziplinäres Konzept
- Zur Verkürzung einer stationären Rehabilitation bei ambulanter Fortsetzung eines stationär begonnenen Rehabilitationsprogramms
- Im Anschluss an eine stationäre Rehabilitation unter besonderer Berücksichtigung der Teilhabe am familiären, sozialen und schulischen Leben sowie ggf. Arbeitsleben.

Wegen Besonderheiten, insbesondere der niedrigen Prävalenz und Komplexität vieler Erkrankungen und Behinderungen im Kindesalter sollten in der Regel für den Aufbau einer ambulanten Rehabilitation keine neuen Einrichtungen geschaffen werden, sondern sich unterschiedliche Anbieter zu **regionalen Netzwerken** zusammenschließen. Dazu könnten als Anbieter bzw. Angebote unter anderem zählen: stationäre Reha-Einrichtungen, Sozialpädiatrische Zentren – auch mit stationären Angeboten, mobile Rehabilitations-Angebote, Patientenschulungen durch ermächtigte Ärzte, Schwerpunkt-Ambulanzen und ermächtigte Spezialisten an Krankenhäusern und Universitätskliniken, integrierte Versorgungsmodelle in Verbindung mit Akutkrankenhäusern, niedergelassene Kinder- und Jugendärzte, niedergelassene Facharztpraxen und im Rahmen von DMP ermächtigte Ärzte, niedergelassene Praxen oder Ambulanzen für Kinder- und Jugendpsychiatrie und Psychotherapie mit Ermächtigung zur sozialpsychiatrischen Versorgung, pädagogische Einrichtungen und Behindertenhilfe, Frühförderstellen und niedergelassene Therapeuten.

Eine sozialpädiatrische Behandlung nach § 119 und § 43a SGB V unterscheidet sich von der ambulanten Rehabilitation in ihrer geringeren Behandlungsintensität (Therapieeinheiten pro Woche) und in dem längerfristigen, in der Regel mehr als 6-monatigen Behandlungsverlauf.

Aufgrund ihrer fachlichen Expertise und der vorhandenen interdisziplinären Mitarbeiterstruktur sind SPZ sehr geeignet als potenzielle Leistungserbringer für ambulante Rehabilitation. Sie sollten bei entsprechenden gesetzlichen und ökonomisch vertretbaren Rahmenbedingungen diese sehr sinnvolle und wichtige Leistung anbieten. Dies setzt allerdings voraus, dass an SPZ für die verschiedenen Reha-geeigneten Erkrankungen neue inhaltliche Behandlungskonzepte und -ziele entwickelt und die Behandlungsangebote flexibilisiert werden. Die Aufgaben der SPZ in der ambulanten sozialpädiatrischen Krankenbehandlung nach SGB V und zur Erbringung der Komplexleistung Frühförderung nach SGB IX bleiben von diesen Rahmenempfehlungen unberührt.

Prinzipiell muss nach der gemeinsamen Rahmenempfehlung zur ambulanten Rehabilitation eine Einrichtung nicht alle Leistungen organisatorisch oder räumlich selbst anbieten. Ambulante Rehabilitation in diesem Sinne kann auch in einer Vernetzung vorhandener Angebote und Ressourcen bestehen, muss aber entsprechende Qualifikationsvoraussetzungen für die Durchführung der Rehabilitation beachten, auch räumliche und weitere strukturelle Voraussetzungen. Bei der Nutzung von Netzwerkstrukturen ist eine zentrale Struktur (im Sinne des Anbieters der Leistung) verantwortlich, den Behandlungsplan aufzustellen und die Durchführung der Maßnahme zu kontrollieren.

Insbesondere die Prüfung der Voraussetzungen, der Bedürftigkeit oder der Motivation kann dazu führen, dass eine ambulante Rehabilitation einer stationären Maßnahme primär vorzuziehen ist – nicht selten kommt die stationäre Rehabilitation aber noch immer auch deshalb in Betracht, weil für die so geartete ambulante, wohnortnahe Rehabilitation die Leistungsstrukturen im Umfeld der Familie bisher einfach nicht adäquat, insbesondere entwicklungsangemessen aufgebaut sind!

Die Chancen der **ambulanten Rehabilitation** bestehen insbesondere:
- In der Möglichkeit, über einen größeren Zeitraum den Rehabilitationsprozess zu gestalten und damit auch Lern- und Trainingseffekte zu unterstützen
- Im Bereich Erziehung und Bildung darin, gezielt den Kindergarten, die Schule oder Ausbildungsstelle in der Alltagssituation in den Rehabilitationsprozess mit einzubeziehen, sodass die individuellen Gegebenheiten besser berücksichtigt und die Einrichtungen nachhaltig in den Rehabilitationsprozess eingebunden werden können
- Im Bereich der Familien durch verstärkte Einbindung Familienangehöriger (z. B. Geschwister, Großeltern) in den Rehabilitationsprozess, sofern erforderlich. Insbesondere für Jugendliche ergibt sich die Möglichkeit, die Peergroup als wesentlichen Unterstützungsfaktor für den langfristigen Erfolg der Reha einzubeziehen. und damit die Verbesserung fördernder Kontextfaktoren. Dies betrifft insbesondere die Aspekte Patientenschulung und psychosoziale Maßnahmen.
- Im Bereich der Berufsberatung frühzeitige/langfristige Unterstützung von Jugendlichen bei Berufswahl/Auswahl der Ausbildungsstellen, ggf. gezielte rehabilitative Maßnahmen am Arbeitsplatz (mit kontinuierlicher Begleitung durch die Agenturen für Arbeit).

39.8 Mitaufnahme einer Begleit- bzw. Bezugsperson

Die Mitaufnahme einer Begleitperson wurde bis 2012 in der Regel bei Kindern bis zum Eintritt der Schulfähigkeit von den Leistungsträgern mitgeplant, die Altersgrenze wird seit kurzem auf das vollendete 8. Lebensjahr angesetzt. Bei älteren Kindern und Jugendlichen wird in der Regel die Mitaufnahme einer Begleitperson – auch nach dem gemeinsamen Rahmenkonzept – nicht als erforderlich angesehen, da die Einrichtungen über entsprechend qualifiziertes Pflege- und Betreuungspersonal verfügen, um das Kind im Rehabilitations-Prozess zu betreuen. Eine „Familien-Rehabilitation" im eigentlichen Sinne – z. B. Behandlung (!) der erschöpften Mutter – ist in der Regel nicht vorgesehen, auch wenn sich mit der familienorientierten Rehabilitation (FOR) eine Versorgung zumindest von Kindern mit vor allem onkologischen Diagnosen und nach Transplantation für die Zeit nach der Akuterkrankung etwas Ähnliches etabliert hat (s. a. Verfahrensabsprache zur FOR 2009).

Von den genannten Grundlagen kann im Einzelfall abgewichen werden, wenn bei stationären Rehabilitationsleistungen aus medizinischen Gründen und um das Rehabilitationsziel zu erreichen die Mitaufnahme einer Begleitperson zwingend erforderlich ist, wenn z. B.
- wegen entwicklungsbedingter (!) psychologischer Gründe die Trennung eines Kindes von der Begleitperson die erfolgreiche Durchführung der Maßnahme gefährden würde oder
- das Kind bzw. der Jugendliche wegen einer schweren Behinderung einer ständigen Hilfe bedarf, die von der Vorsorge- oder Rehabilitationseinrichtung nicht erbracht werden kann,
- oder während der stationären Leistung eine Einübung/Anleitung der Begleitperson in therapeutische Verfahren, Verhaltensregeln oder Gebrauch von Hilfsmitteln notwendig ist. Letzteres soll dabei in komprimierter Form in dem zeitlichen Gesamtumfang erfolgen, der für die Einübung/Anleitung notwendig ist (i. d. R. wesentlich kürzer angesetzt als die Leistung für den Patienten selbst). Wenn Einübungen/Anleitungen einer die Behandlung des Patienten begleitenden Person mit vergleichbarer Erfolgsaussicht auch an deren Wohnort möglich sind, dann sollen sie demnach auch dort erfolgen.

39.9 Diagnostik in der Rehabilitationsmedizin

„Die indikationsbezogene, vorsorgespezifische Funktionsdiagnostik und notwendige apparative Ausstattung muss durch die Einrichtung vorgehalten werden oder am Ort in erreichbarer Nähe verfügbar sein. Sie hat die Behandlung nach dem allgemein anerkannten Stand der medizinischen Erkenntnisse zu gewährleisten, wobei die Diagnostik je nach weiterer Spezialisierung der Einrichtung ergänzt werden muss" (Rahmenkonzept, BAR 2008). Aufgabe der medizinischen Rehabilitation damit ist Diagnostik der Erkrankung und der Funktionsstörungen auf der Ebene der Körperstrukturen und -funktionen (falls noch nicht ausreichend erfolgt oder die vorliegenden Befunde nicht mehr aktuell sind), spezielle Leistungsdiagnostik auf der Ebene der Aktivitäten (körperliche, psychische Leistungs-fähigkeit) und psychosoziale Diagnostik als Eingangs-, Verlaufs- und Abschlussdiagnostik, Berücksichtigung individueller Kontextfaktoren (Umwelt- und personbezogene Faktoren) im Hinblick auf ihre unterstützende oder Barrierewirkung. Eine Aufgabe für die nahe Zukunft besteht auch darin, hierfür geeignete Assessments zusammenzustellen bzw. zu entwickeln.

39.10 Ziele der stationären medizinischen Rehabilitationsleistungen

39.10.1 Allgemeine Ziele

Die Charta des Kindes der Weltgesundheitsorganisation (WHO) legt fest, dass jedes Kind Anrecht auf die ihm gemäße ungestörte körperliche, geistige und seelische Entwicklung hat. Entsprechend den vielfältigen Bedingungsfaktoren von Krankheit ist (auch nach dem gemeinsamen Rahmenkonzept) neben den medizinischen und psychosozialen Leistungen vor allem den flankierenden gesundheitsfördernden Leistungen zur Verminderung der Risikofaktoren und zur Verhaltensänderung eine hohe Bedeutung zuzumessen.
Ziel der medizinischen Rehabilitation ist, die krankheitsbedingt drohenden oder bereits manifesten Beeinträchtigungen der Teilhabe am Leben in der Gesellschaft durch frühzeitige Einleitung der gebotenen Rehabilitationsleistungen abzuwenden, zu beseitigen, zu mindern, ihre Verschlimmerung zu verhüten oder ihre Folgen zu mildern. Das Kind oder der Jugendliche soll z. B. durch die Rehabilitation (wieder) befähigt werden, bestimmte Aktivitäten des täglichen Lebens oder spätere oder ggf. bereits vorhandene Erwerbstätigkeit möglichst in der Art und in dem Ausmaß auszuüben, die für sie als „normal" für deren persönlichen Lebenskontext erachtet werden.
Gerade bei Kindern und Jugendlichen müssen neben den körperlichen und psychischen Faktoren bei der angestrebten Verbesserung des Gesundheitszustands auch die sozialen Komponenten berücksichtigt werden, um z. B. Folgeerkrankungen zu vermeiden. Es gilt auch hier, Krankheitssymptome zu mildern, Krankheitsbewältigung zu fördern, eingeschränkte physische und psychische Leistungsfähigkeit zu verbessern durch:

- Vollständige Wiederherstellung des ursprünglichen Niveaus der Aktivitäten/Teilhabe (Restitutio ad integrum)
- Größtmögliche Wiederherstellung des Ausgangsniveaus der Aktivitäten/Teilhabe (Restitutio ad optimum)

- Einsatz von „Ersatzstrategien“ bzw. Nutzung verbliebener Funktionen oder Aktivitäten (Kompensation, Adaptation)
- Anpassung der Umweltbedingungen an die bestehenden Beeinträchtigungen der Aktivitäten oder der Teilhabe.

Das SGB IX kann als gesetzliche Umsetzung der ICF gesehen werden und ist insoweit mit den genannten Zielen in allem kongruent. Nach § 26 Abs. 2 Nr. 7 SGB IX können darüber hinaus ggf. auch die medizinisch-beruflichen Leistungen wie Arbeitstherapie und Belastungserprobung zu den Leistungen zur medizinischen Rehabilitation gerechnet werden: Die Arbeitstherapie dient dem Erlernen des Umgangs mit Werkzeug und Material und der Verbesserung des Arbeitsverhaltens. Die Belastungserprobung prüft die Leistung und Beanspruchung in der später vorgesehenen Tätigkeit, die zeitliche Belastbarkeit und die Verträglichkeit von Arbeitsumgebungseinflüssen.

39.10.2 Ziele der Kosten- bzw. Leistungsträger

Die Rentenversicherungsträger erbringen stationäre medizinische Rehabilitationsleistungen für Kinder und Jugendliche, wenn hierdurch voraussichtlich eine erhebliche Gefährdung der Gesundheit beseitigt oder eine beeinträchtigte Gesundheit wesentlich gebessert oder wiederhergestellt werden kann und dies Einfluss auf eine spätere Erwerbsfähigkeit haben kann. Im Einzelnen gelten die Gemeinsamen Richtlinien der Träger der Rentenversicherung nach § 31 Abs. 1 Satz 1 Nr. 4 SGB VI für Kinderheilbehandlungen (KiHB-Richtlinien) – 2013 werden die Richtlinien überarbeitet, um den aktuellen Anforderungen angepasst zu werden.
Die Krankenkassen erbringen stationäre medizinische Rehabilitationsleistungen für Kinder und Jugendliche nach den Bestimmungen des §40 SGB V, um eine drohende Beeinträchtigung der Teilhabe abzuwenden bzw. eine bereits eingetretene Beeinträchtigung der Teilhabe zu beseitigen, zu vermindern oder deren Verschlimmerung zu verhüten. Medizinische Rehabilitation schließt die Krankenbehandlung mit dem Ziel ein, Krankheit zu erkennen, zu heilen, Verschlimmerung zu verhüten oder Krankheitsbeschwerden zu lindern.

39.11 Therapieplanung

Die Therapie chronischer Erkrankungen muss bestimmte Aspekte berücksichtigen: Eine „Ursache“ ist oft nicht bekannt bzw. die Krankheit ist nicht monokausal zu definieren, meist zeigen sich vielfältige Aspekte, die für Manifestation, Schweregrad und Verlauf relevant sind. Die Krankheit ist in der Regel nicht völlig „umkehrbar“, es gibt keine kausal ansetzende Therapie oder Heilung i. S. einer Restitutio ad integrum. Das Erleben der Unvorhersagbarkeit des Krankheitsverlaufs stellt oft eine eigenständige Belastung dar, die sich ihrerseits wieder auf Verlauf und Krankheitsmanagement auswirkt. Sehr häufig besteht eine dauerhafte Abhängigkeit von medizinischen Spezialisten und anderen Helfern – es kommt leicht zu vielfältigen (negativen) Auswirkungen im Alltag und Verlusten in personeller und sozialer Hinsicht (Freunde, Peerkontakte, schulische und berufliche Integration).
Der **Rehabilitationsplan** soll – unabhängig vom Leistungsträger – soweit wie möglich die Behandlungskonzepte der Vorbehandler berücksichtigen, auf Ergebnissen von Beobachtungen und Diagnostik der Rehabilitationseinrichtung aufbauen und die individuellen Voraussetzungen der Rehabilitanden sowie die besonderen Anforderungen an sie in Schule (bzw. Beruf) und Alltag

einbeziehen. Dabei wird es als notwendig angesehen, den Rehabilitanden (bzw. dessen Vertreter) bei der Erstellung des Rehabilitationsplans entwicklungsangemessen einzubinden. Frühzeitige Planung für die Zeit nach der stationären Rehabilitationsleistung ist ebenfalls Bestandteil des Reha-Plans (Versorgung gewährleistet? Materielle oder betreuende Unterstützung der Familie nötig? Muss das Kind unter Umständen vorübergehend oder auf Dauer aus der Familie genommen werden?).

Bei der Erbringung von **Teilhabeleistungen** ist stets den besonderen Bedürfnissen behinderter und von Behinderung bedrohter Kinder und Jugendlicher mit der Zielsetzung der Förderung von Selbstbestimmung und Teilhabe Rechnung zu tragen. Allgemein gilt eben, „dass Kinder keine kleinen Erwachsenen sind“: Bedürfnisse von Kindern und Jugendlichen sind immer auch alters-, entwicklungsabhängig, geschlechtstypisch und bezogen auf die Ziele der Vorsorge und Rehabilitation von unterschiedlicher Wertigkeit. Sie müssen also in ihrer Bedeutung für die Planung und Durchführung von Hilfen entsprechend berücksichtigt werden. Vor allem sollte Beachtung finden, welche entwicklungsbezogenen Bedürfnisse bzw. Ziele zur Teilhabe des Kindes, Jugendlichen oder jungen Erwachsenen bei der Therapie besonderes Gewicht im Vorsorge- und Rehabilitationsverlauf erhalten sollen (z. B. Entwicklung von Selbstständigkeit, Entdecken und Erforschen der Umwelt, Entwicklung eigener Fähigkeiten und Stärken).

39.12 Therapie

Therapeutische Leistungen werden multiprofessionell von einem Rehabilitationsteam erbracht und enthalten ärztliche, physiotherapeutische, sporttherapeutische, balneologisch-physikalische Leistungen, psychologische Interventionen und Entspannungsverfahren, ggf. Logopädie, Ergotherapie, Kreativtherapie und Maßnahmen des Gesundheitstrainings.

Die Rehabilitationsbehandlung beinhaltet dabei:

- Fortführung (ggf. Anpassung) der medizinischen Therapie und Durchführung von physikalischen, psychologischen und weiteren Therapieleistungen entsprechend den Zielen für eine angemessene Krankheitsbewältigung im Alltag sowie ggf. Training von Restfunktionen und Ausbildung neuer Fertigkeiten zur Kompensation oder Adaptation von beeinträchtigten Funktionen und Aktivitäten (ICF: geistig-seelische und sensorische Funktionen, Stimm- und Sprechfunktionen, Funktionen von kardiovaskulärem, hämatologischem, Immun- und Atmungssystem, Verdauungs-, Stoffwechsel- und endokrinem Systems, Urogenitalsystem und Reproduktion, sowie neuromuskuloskeletale und bewegungsbezogene Funktionen und Funktionen der Haut und Hautanhangsgebilde)
- Erarbeitung von Vorschlägen zum Abbau hemmend und zur Unterstützung fördernd wirkender Kontextfaktoren
- Information über die Erkrankung und deren Folgen sowie über die erforderlichen aktuellen und langfristigen Behandlungsmaßnahmen
- Förderung einer angemessenen Einstellung zur Erkrankung: Akzeptanz irreversibler Krankheitsfolgen, Motivation zu aktiver Krankheitsverarbeitung (Wandel vom Behandelten zum Handelnden) und Aufbau eines eigenverantwortlichen Gesundheitsbewusstseins
- Anleitung und Schulung zu eigenverantwortlichem Umgehen (Selbstmanagement) mit der Erkrankung
- Verhaltensmodifikation mit dem Ziel, eine krankheitsadäquate und gesundheitsförderliche Lebensweise zu fördern und gesundheitsschädliches Verhalten abzubauen

- Beratung und Anleitung von Bezugspersonen über den adäquaten Umgang mit der Rehabilitandin und den Folgen der Gesundheitsstörung
- Sozialmedizinische Beurteilung der Leistungsfähigkeit der Rehabilitanden
- Beratung im Hinblick auf die berufliche Tätigkeit und das Alltagsleben auf der Basis des erreichten Leistungsvermögens
- Planung und Anregung weiterer Maßnahmen (Nachsorge, Leistungen zur Teilhabe am Arbeitsleben, Indikationsstellung für weitere diagnostische und/oder therapeutische Maßnahmen) und Vorbereitung der Rehabilitanden darauf.

Charakteristisch für die medizinische Rehabilitation von Kindern ist also eine individuell geplante therapeutische Arbeit in Gruppen, an der die Rehabilitanden entwicklungsentsprechend aktiv mitwirken und die sich in der Regel über den ganzen Tag verteilen kann. Die Behandlung umfasst die Anleitung zu einer angemessenen Bewältigung der gesundheitlichen Schädigungen sowie der damit verbundenen Beeinträchtigungen auf der Ebene der Funktionen, Aktivitäten und Teilhabe. Bei bleibenden Folgeerscheinungen soll hierdurch auch die ggf. erforderliche Anpassung an ein verändertes Alltagsleben erleichtert werden. Das allgemeine Ziel besteht in einer „Hilfe zur Selbsthilfe", die für einen nachhaltigen Erfolg der Rehabilitation unerlässlich ist. Die Rehabilitanden sollen ein vertieftes Verständnis für die individuellen psychischen und sozialen Faktoren erlangen, die für die Entstehung und den Verlauf ihrer Gesundheitsstörung von Bedeutung sind. In diesem Zusammenhang sind die individuell relevanten Kontextfaktoren zu berücksichtigen mit geeigneter Unterstützung der Förderfaktoren und Abbau der hemmend wirkenden Faktoren (Barrieren). Leistungen aus anderen Hilfesystemen sollen mitüberlegt und ggf. angebahnt werden (SGB VIII familienunterstützende Maßnahmen)

Die Beurteilung der Qualität rehabilitativer Arbeit setzt eine angemessene **Basis- und Leistungsdokumentation** voraus (➤ Kap.40). Nach Beendigung bzw. Abschluss der medizinischen Rehabilitation erhalten daher – unter Wahrung der ärztlichen Schweigepflicht und des Datenschutzes – die behandelnden Ärzte und der zuständige Leistungsträger sowie ggf. der zuständige Medizinische Dienst einen Entlassungsbericht. Darin werden alle notwendigen Informationen über die abgeschlossene Rehabilitation dokumentiert: der Rehabilitationsverlauf unter Angabe der durchgeführten Therapien und die Ergebnisse der abschließenden Leistungsdiagnostik und einer sozialmedizinischen Beurteilung; diese umfassen z. B. die Stellungnahme

- zur Leistungsfähigkeit im Erwerbsleben unter Bezugnahme auf den beruflichen Kontext, bezogen auf den bisherigen Arbeitplatz bzw. den allgemeinen Arbeitsmarkt, bzw. bei Kindern und schulpflichtigen Jugendlichen auch zur Schulfähigkeit und erforderlichen schulischen Förderung
- zur Leistungsfähigkeit im Alltag bezogen auf die Selbstständigkeit bei den Verrichtungen des täglichen Lebens, insbesondere zur psycho-sozialen Situation und/oder
- zur Frage der Vermeidung oder Minderung von Pflegebedürftigkeit
- zur Krankheitsverarbeitung, zum Lebensstil einschließlich Risikofaktorenkonstellation und Motivation zur Lebensstiländerung
- Empfehlungen für weiterführende Leistungen zur Sicherung des Rehabilitationserfolgs (z. B. Leistungen zur Teilhabe am Arbeitsleben, stufenweise Wiedereingliederung, Rehabilitationssport oder Funktionstraining, Nachsorge)
- Empfehlungen zur Wiedereingliederung in das soziale Umfeld bzw. zur psychosozialen Betreuung
- ggf. Empfehlungen zu Helfern und Hilfesystemen (Eltern bzw. andere Bezugspersonen, Lehrer, Ärzte, Therapeuten, amtliche Hilfen).

39.13 Rehabilitationskette, Vernetzung

„Non-Profit"-Ansätze wie Kooperation mit Selbsthilfegruppen oder fach- bzw. diagnosebezogene Netzwerke (z. B. fachbezogene regionale Kooperationen von Rehabilitationseinrichtungen oder zu spezifischen Diagnosebereichen wie Asthma, Allergien, „Adipositasnetzwerke" in Rheinland-Pfalz, Saarland, Hessen und Schleswig Holstein, u. ä.) sind für die Patienten kurzfristig bereits von hohem Wert, aber sicher nicht auf Dauer ein prägendes Element im Gesundheitswesen. Hier sind eher politische Entwicklungen zu beachten, zu denen es auch schon Aktivitäten bzw. Stellungnahmen der Leistungsträger gibt: Die medizinische Rehabilitation ist ein Teil des gegliederten Systems der deutschen Gesundheitsversorgung. Gegenwärtig unterliegt das deutsche Gesundheitswesen einem dynamischen Wandel, der die medizinische Rehabilitation unmittelbar betrifft und zu einer zunehmenden Vernetzung der Behandlungssektoren bei Erwachsenen führt (z. B. ambulant – stationär, Akutmedizin – Rehabilitation) – damit ist für die Rehabilitation von Kindern, wo nicht mit Fachambulanzen und SPZs schon entsprechende Strukturen vorhanden sind, Ähnliches zu erwarten bzw. meistens noch eher zu fordern. Die Auswirkungen von gesundheitspolitischen Entscheidungen sind damit nicht mehr nur auf einen Versorgungssektor beschränkt. Dies gilt sowohl für die Einführung der DRG-Fallpauschalen für die Krankenhäuser und die damit verbundenen Abgrenzungsfragen als auch für die Disease-Management-Programme (DMP).
Seit 2004 wird die Qualitätssicherung der stationären medizinischen Rehabilitation von Kindern und Jugendlichen als Kooperationsprojekt der Renten- und der Krankenversicherung angesehen (Strukturerhebung und Strukturanalyse von Einrichtungen zur stationären medizinischen Rehabilitation von Kindern und Jugendlichen; Erfassung von Prozess- und Ergebnisqualität der Rehabilitationseinrichtungen). Anschlussprojekte betreffen die Entwicklung von Standards für die Kinder- und Jugendlichenrehabilitation sowie die Konzeption altersgerechter Befragungsinstrumente und kindbezogener Eltern- bzw. Angehörigenfragebögen. Die einmal angestoßenen Qualitätssicherungsaktivitäten im Bereich der stationären medizinischen Rehabilitation von Kindern und Jugendlichen sollen nicht abreißen, zumal die Akzeptanz in den Kliniken durchaus gegeben ist und entsprechende Kooperationsstrukturen mit pädiatrischen Klinikern und Kliniken, Fachgesellschaften und Fachverbänden geschaffen werden sollen. Andererseits sind hier wirtschaftliche Fragen oft noch ungeklärt und Manches kommt nur langsam voran.

39.14 Disease-Management-Programme

Disease-Management-Programme (DMP) sind strukturierte Behandlungsprogramme nach §137f. SGB V, mit denen Defizite in der Versorgung chronisch Kranker beseitigt und die Vernetzung der Versorgungsbereiche gestärkt werden sollen. Inhaltliche Anforderungen an die Ausgestaltung der DMP (§91 SGB V) werden durch den Gemeinsamen Bundesausschuss festgelegt und durch das zuständige Ministerium per Rechtsverordnung in Kraft gesetzt. DMP gibt es bislang im Erwachsenenbereich für Mammakarzinom, Diabetes mellitus Typ 1 und 2, koronare Herzerkrankung, Asthma bronchiale und COPD. In der Pädiatrie wurde mit dem DMP-Papier 2001 zum Asthma bronchiale eine vergleichbare Entwicklung eingeleitet.
Mit den DMP wird vor allem eine Optimierung in der akutmedizinischen Versorgung angestrebt, wobei die Steuerung und Koordination der verbindlich aufeinander abgestimmten Behandlungs- und Betreuungsprozesse meist durch niedergelassene Ärzte erfolgt. Grundsätzlich werden dabei auch Präventions- und Rehabilitationsleistungen der gesetzlichen Krankenversicherung berück-

sichtigt. Die formulierten Therapieziele sind am bio-medizinischen Krankheitsmodell der Akutmedizin ausgerichtet. Die medizinische Rehabilitation, die mit ihrem ganzheitlichen (bio-psychosozialen Krankheitsmodell) und verhaltensmedizinischen Behandlungskonzept eine wesentliche Rolle im Versorgungsprozess chronisch Kranker spielt, bleibt bisher nur unzureichend berücksichtigt. Aus Sicht der Rehabilitation stellen sich vor allem folgende Anforderungen an DMP:

- Die Ziele der medizinischen Rehabilitation müssen in die Behandlungsprogramme ausreichend integriert werden. Dazu gehören sowohl die berufliche Wiedereingliederung ins Erwerbsleben als auch die Teilhabe am Leben in der Gesellschaft.
- Die Schnittstellen zur Rehabilitation müssen eindeutig definiert sein und angeben, wann eine Reha-Leistung notwendig ist.
- Die Rehabilitation muss als gleichberechtigter Versorgungsbereich neben der Akutmedizin in den Behandlungsprogrammen verankert sein.

In einer gemeinsamen Erklärung vom 17.8.2004 haben Kranken- und Rentenversicherung festgestellt, dass grundsätzlich auch Rehabilitationsleistungen zu den im Rahmen der DMP vorzusehenden Behandlungsangeboten gehören. Gleichzeitig hielten sie fest, dass während der Durchführung von Leistungen zur medizinischen Rehabilitation alle medizinischen Grundlagen der strukturierten Behandlungsprogramme beachtet werden sollen. Das Antragsverfahren der gesetzlichen Rentenversicherung bleibt von dieser gemeinsamen Erklärung unberührt.

39.15 Integrierte Versorgung (§§ 140 a-d SGB V)

Integrierte Versorgung stellt neben Disease-Management-Programmen einen weiteren Reformansatz dar, mit dem die Kooperation und Koordination der Versorgungssektoren gestärkt werden soll. Der Gesetzgeber versteht unter integrierter Versorgung sowohl einen verschiedene Leistungssektoren übergreifenden Ansatz als auch eine interdisziplinär-fachübergreifende Versorgung. In integrierten Versorgungsmodellen können Krankenkassen und Leistungserbringer autonom Verträge (§75 Abs. 1 SGB V) über die Behandlung außerhalb des Sicherstellungsauftrags der Kassenärztlichen Vereinigungen abschließen. Die Versorgung kann damit auf einzelvertraglicher Grundlage durchgeführt werden. Benanntes Ziel ist, die integrierte Versorgung als neuen Bestandteil der Regelversorgung zu verankern und Qualität und Wirtschaftlichkeit der medizinischen Leistungen zu verbessern. Dazu sollen vor allem vertraglich vereinbarte Therapieleitlinien und Behandlungspfade dienen. Die Verträge zur integrierten Versorgung können grundsätzlich auch Rehabilitationsleistungen der Krankenversicherung umfassen und Rehabilitationseinrichtungen einbeziehen, soweit mit diesen ein Vertrag nach § 111 Abs. 2 SGB V besteht. Inwieweit dies für die pädiatrische vor allem die sozialpädiatrische Arbeit ein Fortschritt im Sinne der Kinder und Jugendlichen mit ihren Familien sein wird, muss sich auch nach langer Diskussion in den letzten Jahren in der Praxis noch erweisen.

Für die Einbindung der medizinischen Rehabilitation in integrierte Versorgungsmodelle ist es erforderlich, dass

- der Übergang von der Akutbehandlung in die Rehabilitation adäquat geregelt (z. B. über geeignete Kriterien, ggf. Assessment)
- die Reha-Fähigkeit der Versicherten Eingangskriterium für die Rehabilitation
- eine ausreichende Behandlungsdauer gewährleistet und
- bei einer Pauschalvergütung die Honorierung der Rehabilitationsleistung ausreichend hoch ist.

Ziel sollte sein, die Einbindung der Rehabilitation in die integrierte Versorgung zu erreichen, ohne dass dies zu einem Abbau von rehabilitativen Leistungen oder zu Qualitätseinbußen führt. Für die gesetzliche Rentenversicherung ist wesentlich, dass rentenversicherte Teilnehmer der integrierten Versorgung genauso gut rehabilitativ versorgt werden wie nicht rentenversicherte Teilnehmer. Mit der Anschlussheilbehandlung für Erwachsene hat die gesetzliche Rentenversicherung bereits ein bewährtes Verfahren etabliert, mit dem eine Nahtlosigkeit der Behandlung gegeben sein kann. In Rahmenvereinbarungen mit verschiedenen Krankenkassen haben die Deutsche Rentenversicherung Bund und mehrere Regionalträger bereits Einzelheiten zur Einbindung der medizinischen Rehabilitation in die integrierte Versorgung geregelt. Diese Rahmenvereinbarungen sollen vor allem die Zusammenarbeit der verschiedenen Leistungserbringer fördern.

39.16 Abgrenzung zwischen kurativer Medizin und medizinischer Rehabilitation

Der Wandel des Krankheitsspektrums in den zurückliegenden Jahrzehnten führte zu einem Überwiegen chronischer Krankheiten, die entsprechend ihrer Natur nur gebessert, aber nicht geheilt werden können und bei denen ein bio-medizinisch ausgerichteter Behandlungsansatz nur bedingt trägt (➤ Kap.35). Das Aufgabengebiet der medizinischen Rehabilitation in der Behandlung chronischer Erkrankungen mit einem umfassenden Behandlungsansatz nimmt im deutschen Gesundheitssystem eine wichtige Position ein. Ausgehend hiervon hat sich die medizinische Rehabilitation auch in der Pädiatrie so zu einem speziellen, relativ eigenständigen Teil der gesundheitlichen Versorgung mit einer spezifischen Zielsetzung und besonderen konzeptionellen Anforderungen entwickelt.

Während der akutmedizinische Ansatz tendenziell eher auf die Behandlung der Erkrankung selbst und die Behebung der gesundheitlichen Schädigung ausgerichtet ist bzw. im Schwerpunkt zunächst sein muss, liegt in der Rehabilitation demgegenüber der Schwerpunkt auf der Verbesserung und Wiederherstellung der gefährdeten oder bereits geminderten Leistungsfähigkeit in Alltag, in Familie, Freundeskreis, Schule und Berufsleben. Dazu gehört die Berücksichtigung der somatischen, psychischen und sozialen Dimensionen der Erkrankung und ihrer Folgen, einschließlich ihres Zusammenwirkens. Der Rehabilitation nutzt hierfür das bio-psycho-soziale Modell der ICF, das Gesundheit und Krankheit als Ineinandergreifen physiologischer, psychischer und sozialer Vorgänge beschreibt.

Trotz der durch Zielsetzung, konzeptionelle Anforderungen und institutionelle Differenzierung deutlich werdenden Eigenständigkeit der medizinischen Rehabilitation ist aus medizinischer Sicht eine strikte Trennung oder Abgrenzung von kurativer und rehabilitativer Medizin nicht möglich, auch weil die medizinisch-wissenschaftlichen Grundlagen beider Behandlungs-Ebenen übereinstimmen. Methoden und Verfahren der kurativen Medizin sind auch ein integraler Bestandteil der Rehabilitationsmedizin. Es gibt viele Behandlungsmaßnahmen, die sowohl in der Rehabilitation als auch in der kurativen Medizin gebräuchlich und erforderlich sind. Da die medizinische Rehabilitation neben der zur Erfüllung der rehabilitativen Aufgaben erforderlichen Diagnostik auch die Therapie von gesundheitlichen Schädigungen und in ihrer Folge die gezielte Verbesserung beeinträchtigter Funktionen und Fähigkeiten umfasst, ist der kurativ-medizinische Aspekt Bestandteil des rehabilitativen Gesamtkonzepts. Entscheidend für die Abgrenzung von Behandlungsphasen sind neben der für jeden der beiden Bereiche unterschiedlichen, charakteristischen Diagnostik und Therapie vor allem die jeweiligen spezifischen Behandlungsziele.

In § 27 SGB IX wird durch den Gesetzgeber verdeutlicht, dass auch die Leistungen der Krankenbehandlung grundsätzlich darauf ausgerichtet sein sollen, Behinderungen einschließlich chronischer Krankheiten abzuwenden oder zu beseitigen und eine Eingliederung von behinderten und chronisch kranken Menschen in seinen Kontext von Familie, Schule, Beruf und Gesellschaft zu gewährleisten. Damit wird auch die Krankenbehandlung auf eine rehabilitative Orientierung verpflichtet, die die Selbstbestimmung und Teilhabe am Leben in der Gesellschaft für behinderte und chronisch kranke Menschen stärken soll.
Ergänzend wurde mit dem SGB IX zudem grundlegend bestätigt, dass frührehabilitative Leistungen Teil der Krankenhausbehandlung sind (§39 Abs. 1 Satz 3 SGB V).
Die Übereinstimmung in übergeordneten Zielsetzungen und die gemeinsame Schnittmenge von therapeutischen Verfahren hebt jedoch die notwendige konzeptionelle Differenzierung zwischen kurativer medizinischer Versorgung und medizinischer Rehabilitation nicht auf. Die Schnittstelle zwischen den Behandlungssektoren wird durch das SGB IX inhaltlich nicht verändert. Die Erbringung von frührehabilitativen Leistungen im Krankenhaus kann nur solange erfolgen, wie Krankenhauspflegebedürftigkeit besteht.

Literatur

AWMF (2008) Leitlinien für Rehabilitation in der Kinder- und Jugendmedizin: Adipositas, Asthma bronchiale, Diabetes mellitus, Rehabilitation in der pädiatrischen Onkologie, Neurologische Rehabilitation, stationäre Rehabilitation in der Psychiatrie, Psychosomatik und Psychotherapie im Kindes- und Jugendalter, Rehabilitation bei angeborenen und erworbenen Herzerkrankungen im Kindes- und Jugendalter. Stand 3.11.2008; http://www.uni-duesseldorf.de/AWMF/ll/ll_070.htm#index

BAR (2008) Gemeinsames Rahmenkonzept der Gesetzlichen Krankenkassen und der Gesetzlichen Rentenversicherung für die Durchführung stationärer medizinischer Leistungen der Vorsorge und Rehabilitation für Kinder und Jugendliche. in Schriftenreihe der Bundesarbeitsgemeinschaft für Rehabilitation. Frankfurt: BAR.

BAR Frankfurt (Hrsg) (2007) Arbeitshilfe für die Rehabilitation und Teilhabe schädel-hirn-verletzter Kinder und Jugendlicher, Schriftenreihe der Bundesarbeitsgemeinschaft für Rehabilitation Heft 1, Bundesarbeitsgemeinschaft für Rehabilitation Frankfurt/Main http://infomed.mds-ev.de/sindbad.nsf/44eb5931ca44b6c9c12571e700442be5/3bc03b371c86ff5080256a50005bb59f/$FILE/BAR_AHi_Reha-SHT-KiJu_2007.pdf

Bauer Carl-Peter „Asthma bronchiale" in in BAR (Hrsg.) 2005 Rehabilitation und Teilhabe Wegweiser für Ärzte und andere Fachkräfte der Rehabilitation, 3. Aufl. DÄV ISBN 3-7691-0420-X; S 576–582

DGSPJ (2007) Allgemeine Rahmenempfehlungen zur ambulanten Rehabilitation von Kindern und Jugendlichen der Deutschen Gesellschaft für Sozialpädiatrie und Jugendmedizin und der Fachgesellschaft für Rehabilitation im Kindes- und Jugendalter. http://www.dgspj.de/pdfs/AmbReha.pdf http://www.dgspj.de/llambulantereha.php

DRV Bund (2007) Leitlinien zur sozialmedizinischen Beurteilung beim Rehabilitationszugang. http://www.deutsche-rentenversicherung-bund.de/nn_60542/SharedDocs/de/Inhalt/Zielgruppen/01__sozialmedizin__forschung/01__sozialmedizin/Begutachtung/leitlinien__sozmed__beurteilung__rehazugang.html

DRV Bund (2008) Anforderungsprofil für eine stationäre Einrichtung zur medizinischen Rehabilitation von Kindern und Jugendlichen mit 100 Rehabilitationsplätzen. Stand 6.10.2008 http://www.deutsche-rentenversicherung-bund.de/nn_39014/DRVB/de/Inhalt/Zielgruppen/Infos__f_C3_BCr__Rehaeinrichtungen/anforderungssprofile/profil__kinder__download,templateId=raw,property=publicationFile.doc/profil_kinder_download

DRV Bund (2008) Verzeichnis der bisher im gesamten Bundesgebiet eröffneten gemeinsamen Reha-Servicestellen für Rehabilitation. Stand 28.12.2008 http://www.reha-servicestellen.de/

Hartung K, Menger W, Schultze EG (1988) Kinderkuren und Kinderheilverfahren. Lübeck: Hansisches Verlagskontor.

Hollenweger J., Kraus de Camargo O. (Hrsg.) ICF-CY: Internationale Klassifikation der Funktionsfähigkeit, Behinderung und Gesundheit bei Kindern und Jugendlichen Verlag: Huber, Bern; Auflage: 1., Aufl. ISBN-10: 3456849214 ISBN-13: 978–3456849218

Hüller, E. „Neurodermitis" in in BAR (Hrsg.) 2005 Rehabilitation und Teilhabe Wegweiser für Ärzte und andere Fachkräfte der Rehabilitation, 3. Aufl. DÄV ISBN 3-7691-0420-X; S 582–587

Hüller E. „Adipositas" in in BAR (Hrsg.) 2005 Rehabilitation und Teilhabe Wegweiser für Ärzte und andere Fachkräfte der Rehabilitation, 3. Aufl. DÄV ISBN 3-7691-0420-X; S 588–593

Kinderkrebsinfo (2001) Positionspapier zur familienorientierten Rehabilitation bei krebskranken Kindern Mai 2001. http://www.kinderkrebsinfo.de/sites/migration_phase_2/kinderkrebsinfo/content/e2260/e5888/e5902/e5921/e5925/POSITP~2_ger.pdf

Kiosz D. „Mukoviszidose" in in BAR (Hrsg.) 2005 Rehabilitation und Teilhabe Wegweiser für Ärzte und andere Fachkräfte der Rehabilitation, 3. Aufl. DÄV ISBN 3-7691-0420-X; S. 569–576

Oepen, J. (05) „Allgemeine Lernbehinderungen und Entwicklungsstörungen" in in BAR (Hrsg.) 2005 Rehabilitation und Teilhabe Wegweiser für Ärzte und andere Fachkräfte der Rehabilitation, 3. Aufl. DÄV ISBN 3-7691-0420-X; S.216–224

Oepen, J. „Partielle (isolierte) Lernstörungen – Teilleistungsstörungen" in in BAR (Hrsg.) 2005 Rehabilitation und Teilhabe Wegweiser für Ärzte und andere Fachkräfte der Rehabilitation, 3. Aufl. DÄV ISBN 3-7691-0420-X; S 225–230

Oepen, J. „Hyperkinetische Störungen" in in BAR (Hrsg.) 2005 Rehabilitation und Teilhabe Wegweiser für Ärzte und andere Fachkräfte der Rehabilitation, 3. Aufl. DÄV ISBN 3-7691-0420-X; S 230–237

Oepen, J. „Intelligenzminderung und tiefgreifende Entwicklungsstörungen" in in BAR (Hrsg.) 2005 Rehabilitation und Teilhabe Wegweiser für Ärzte und andere Fachkräfte der Rehabilitation, 3. Aufl. DÄV ISBN 3-7691-0420-X;S 237–245

Schmidt-Ohlemann M. (2005) „Mehrfachbehinderungen bei Kindern und Jugendlichen" in BAR (Hrsg.) 2005 Rehabilitation und Teilhabe Wegweiser für Ärzte und andere Fachkräfte der Rehabilitation, 3. Aufl. DÄV ISBN 3-7691-0420-X; S 594–599

VDR (1998) Rahmenkonzept und indikationsspezifische Konzepte zur medizinischen Rehabilitation von Kindern und Jugendlichen. Frankfurt: VDR.

Verfahrensabsprache Familienorientierte Reha (2009)

http://www.aok-gesundheitspartner.de/imperia/md/gpp/bund/reha/leistungen/mutter_vater_kind/reha_verfahrensabsprache_fam

VIII Spezielle Themen der Sozialpädiatrie

VIII

40 Dokumentation in der Sozialpädiatrie

Christian Fricke

40.1 Einleitung

Die Dokumentation gehört zu den Bereichen der Medizin, die in den letzten Jahren eine boomende Entwicklung zu verzeichnen hatten. Diese Entwicklung stößt aufgrund der damit verbundenen Mehrbelastung im alltäglichen Arbeitsprozess teilweise auf massive Ablehnung durch das im Medizinbereich tätige Personal. Bei reflektierter Betrachtungsweise kann der Trend zur vermehrten Dokumentation aber in bestimmten Bereichen als positiv angesehen werden. Diese Aussage gilt insbesondere für die Sozialpädiatrie, da dieses Fachgebiet von interdisziplinärer Zusammenarbeit geprägt ist, die für alle Beteiligten nachvollziehbar sein muss und daher eine gründliche Datenerfassung erfordert. Zu hinterfragen ist die Sinnhaftigkeit der Dokumentation als Leistungsnachweis auf der Verwaltungsebene. Auch wenn das Forderung von Kostenträgern, einen Leistungsnachweis zu erhalten, nachvollziehbar ist, sollte diese Tätigkeit nur geringe Anteile der Personalressourcen und der eingesetzten finanziellen Ressourcen verbrauchen. Zu kritisieren sind Entwicklungen, bei denen ein Wettlauf in der Dokumentation zwischen Kostenträgern, Leistungserbringern, Behörden und anderen Beteiligten entsteht und als wesentlichen Effekt eine gegenseitige Ressourcenvernichtung nach sich zieht. Ein wichtiger Aspekt ist die Dokumentation zur Begegnung von Rechtsansprüchen.

Die Darstellung der Dokumentation in der Sozialpädiatrie erfolgt aus dem Grund in 4 Abschnitten. Grundlage ist die Publikation des Qualitätszirkels „Empfehlungen zur Dokumentation und schriftlichen Berichterstellung in Sozialpädiatrischen Zentren" (Scheying et al. 2007).

Formelle Anforderungen an ein schriftliches Dokument sind:

- Dokumentenechtheit (Bleistiftverbot)
- Lesbarkeit (Handschrift ist nur bedingt tauglich)
- Angabe von Ort und Datum
- Lesbare Unterschrift oder Unterschrift mit lesbarer Ergänzung zur Zurückverfolgbarkeit des Verfassers.

Sozialpädiatrie in der Praxis. http://dx.doi.org/10.1016/B978-3-437-31630-3.00040-7
Copyright © 2014 Elsevier GmbH. All rights reserved.

40.2 Dokumentation der inhaltlichen Arbeitsqualität sozialpädiatrischer Leistungen

Sie umfasst die juristisch nachvollziehbare Dokumentation folgender Bereiche:

- Dokumentation der biografischen Anamnese auch mit ergänzenden anamnestischen Aspekten
- Dokumentation von Untersuchungsergebnissen der Diagnostik
- Berichterstellung über die therapeutische Arbeit
- Verlaufsdokumentation diagnostischer und therapeutischer Maßnahmen, wiederholt alle 3–6 Monate
- Erfassung von diagnostischen und therapeutischen Erkenntnissen aller Beteiligten eines interdisziplinären sozialpädiatrischen Teams auch im Verlauf
- Dokumentation aller internen und externen technischen Untersuchungsergebnisse auch im Verlauf
- Dokumentation der interdisziplinären sozialpädiatrischen Arbeit, z. B. der Teamsitzungen mit patientenbezogener und zeitbezogener Zuordnungsmöglichkeit.

Nach § 43a SGB V ist die Erstellung eines **Behandlungsplans** verpflichtend. Nach den Kriterien des Altöttinger Papiers (Hollmann et al. 2009) wird das ärztliche zusammenfassende Gutachten in Form eines Arztbriefs erstellt. Der Arztbrief stellt somit gleichzeitig das Behandlungsgutachten mit dem daraus resultierenden Behandlungsplan dar. Wesentlich ist die allgemeine Verständlichkeit des Arztbriefs auch für den nicht unmittelbar entwicklungs- und sozialpädiatrisch tätigen Arzt. Der Inhalt des Arztbriefs wird vor der Fertigstellung mit den Eltern besprochen und abgestimmt. Im sozialpädiatrischen Verständnis sollte in der Regel der Arztbrief auch den Eltern zur Verfügung gestellt werden.

Eine Ausnahme kann dann erforderlich sein, wenn:

- Nicht zur Weitergabe bestimmte subjektive Einschätzungen enthalten sind
- Bei psychiatrischer Behandlung Angaben enthalten sind, die für eine Krankheitsbewältigung nachteilig sind
- Rechte anderer in die Behandlung einbezogener Personen berührt werden.

Das Weitergabeverbot sollte besonders kenntlich gemacht werden.

40.3 Dokumentation zum Leistungsnachweis

Die Leistungsdokumentation kann Basis einer leistungsgerechten Vergütung sein, aber auch eine Qualitätssteigerung der sozialpädiatrischen Arbeit nach sich ziehen. Es besteht häufig ein Konflikt zwischen Leistungserbringern und Kostenträgern hinsichtlich der Leistungsdokumentation. Die Leistungserbringer sind daran interessiert, möglichst wenig Zeit für Verwaltungsaufgaben und möglichst viel Zeit für die Patientenzuwendung zu haben. Die Kostenträger haben ein hohes Interesse an einer möglichst lückenlosen Leistungsdokumentation, um ihrerseits einen tragfähigen Vergütungsnachweis zu erhalten. Ein gangbarer Mittelweg besteht in einer zeitsparenden Dokumentation, in der Diagnosen, Leistungsdauer (in Minuten), Leistungsart (Diagnostik, Therapie, Beratung u. a.) und Leistungscharakter (Einzeltherapie, Gruppentherapie u. a.) patientenbezogen erfasst wird. Bei entsprechender EDV- gestützten Aufzeichnung ist dies in 2–4 Minuten pro Arbeitsstunde zu leisten.

In der **ambulanten Sozialpädiatrie** wird davon ausgegangen, dass zwei Drittel der Arbeitszeit direkt am Patienten erbracht werden. Für ärztliche wie nichtärztliche Therapeuten ist die Therapieeinheit mit 1 Stunde angesetzt, davon 45 Minuten Patientenkontakt und 15 Minuten für die Dokumentation, Vorbereitung und Teamarbeit. 10–15 % sind für Besprechungen (Teamzeit) sowie andere indirekte Tätigkeiten zur Patientenversorgung anzusetzen (Möller et al. 2007); auch die im Kinderbereich unvermeidlichen Therapieausfälle z. B. bei akuter Erkrankung sind hierin zu veranschlagen. Es ist davon auszugehen, dass sozialpädiatrische Arbeit weiterhin vorwiegend pauschal finanziert wird. Im Rahmen der Verhandlungen der Pauschale kann dann ohne Personenbezug und damit datenschutzgerecht durch neutrale Darstellung der Daten die erbrachte Leistung nachvollziehbar gemacht werden. Bei **stationärer sozialpädiatrischer Behandlung** ist eine den jeweiligen Fassungen der ICD-10 und des OPS-Katalogs entsprechende Dokumentation vorzunehmen. Eine themenzentrierte Synopse der beiden Verzeichnisse ist im „Kodierleitfaden Kinder- und Jugendmedizin" der GKinD (GKinD 2013) publiziert. Für die Sozialpädiatrie ist durch Einsatz der Mehrdimensionalen Bereichsdiagnostik (MBS) im EKPSA-Schema die Verschlüsselung von ICD-10-Ziffern in mehreren Bereichen erforderlich (Hollmann et al. 2009). Darüber hinaus sind die spezifischen, für die Sozialpädiatrie geeigneten OPS-Schlüssel einzusetzen. Speziell für die Sozialpädiatrie geschaffen wurde die OPS 9–403.1–8. Voraussetzung für die Anerkennung der Leistungen ist eine zeitgenaue Dokumentation der erbrachten Leistungen mit Rückvollziehbarkeit des Erbringers (Schmid et al. 2004).

40.3.1 Dokumentation zur Qualitätssicherung

Die im voranstehenden Abschnitt dargestellte Leistungserfassung dient zunächst dem Leistungsnachweis. Damit ist nicht automatisch eine ökonomische und qualitativ hochwertige Betriebsführung gegeben. Für eine qualitätsgesicherte Betriebsführung bietet sich die Installierung eines Qualitätsmanagementsystems (➤ Kap.41.5).

40.3.2 Dokumentation zur juristischen Absicherung

Die Dokumentation der abgefragten Informationen und erstellten bzw. eingeholten Befunde sowie des Therapieverlaufs ist Voraussetzung, um bei juristischen Auseinandersetzungen Vorwürfen begegnen zu können. Das Arztrecht verpflichtet den Arzt allgemein zur schriftlichen Dokumentation seiner Arbeit. Bei Haftungsfragen ist der Arzt beweispflichtig und daher auf die schriftliche Dokumentation angewiesen. Das Gleiche gilt analog für die anderen in einem SPZ diagnostisch und therapeutisch sowie beratend tätigen nichtärztlichen Mitarbeiter.Zu beachten ist auch das Recht der Eltern auf Akteneinsichtnahme und damit das Recht, die anamnestischen, diagnostischen und therapeutischen Abläufe zu rekonstruieren und die Einschätzung zu überprüfen (➤ Kap. 42). Werden die Dokumentationskriterien nicht erfüllt, besteht konsekutiv die Gefahr der Beweislastumkehr und damit eine deutliche Verschlechterung der Ausgangsposition. Eine lückenhafte oder fehlende Dokumentation ist eine äußerst schlechte Basis für ein Zivil- oder Strafrechtsverfahren.

Literatur

GKinD (2013) Kodierleitfaden Kinder- und Jugendmedizin. Version 2013. www.GKinD.de

Hollmann H, Kretzschmar C, Schmid RG (2009) Das Altöttinger Papier. Qualität in der Sozialpädiatrie Band 1. Bundesarbeitsgemeinschaft Sozialpädiatrische Zentren. Altötting: RS-Verlag.

Möller P, Hollmann H, Bücken St, Stegelmann J (2007) „Ökonomie". Handreichung zur Planung von Ökonomie im SPZ. In: Fricke C, Kretzschmar C, Hollmann H, Schmid RG. Qualität in der Sozialpädiatrie. Band 2. Altötting: RS-Verlag: 261–276.

Scheying H, Kalz G, Mehl B (2007) Empfehlungen zur Dokumentation und schriftlichen Berichterstellung in Sozialpädiatrischen Zentren. In: Fricke C, Kretzschmar C, Hollmann H, Schmid RG. Qualität in der Sozialpädiatrie. Band 2. Altötting: RS-Verlag: 279–294.

Schmid RG, Kieslich M, Stauffenberg von M, Lutterbüse N (2004). DRG in der Pädiatrie bei sozialpädiatrischer, neuropädiatrischer und pädiatrisch-psychosomatischer Behandlung. Kinder- und Jugendarzt 35: 838–840.

41 Qualitätssicherung in der Sozialpädiatrie

Christian Fricke

41.1 Einleitung

Seit vielen Jahren arbeiten die Deutsche Gesellschaft für Sozialpädiatrie und Jugendmedizin und die Bundesarbeitsgemeinschaft Sozialpädiatrischer Zentren (BAG SPZ) an der Formulierung von Standards, um die Qualität sozialpädiatrischer Arbeit weiterzuentwickeln und zu sichern. Mit Hilfe von Leitlinien wurden evidenzbasierte Kriterien in Diagnostik und Therapie eingeführt ergänzt durch Stellungnahmen zu verschiedenen Therapieverfahren und die Qualitätspapiere der BAG SPZ.

41.2 Evidenzbasierte Medizin

Die Prinzipien der evidenzbasierten Medizin (EbM) gewinnen wie in anderen Bereichen der Medizin auch bei Entwicklungsstörungen im Kindesalter zunehmend an Bedeutung. Diese

Sozialpädiatrie in der Praxis. http://dx.doi.org/10.1016/B978-3-437-31630-3.00041-9
Copyright © 2014 Elsevier GmbH. All rights reserved.

sollen eine wissenschaftlich fundierte und dem aktuellen Therapiestandard entsprechende Versorgung der Patienten ermöglichen, frei von politischen, wirtschaftlichen und ideologischen Zwängen und zu mehr Qualität und Rationalität in der medizinischen Versorgung führen.

Evidenzbasierte Medizin stützt sich auf **drei Säulen:** auf die individuelle klinische Erfahrung, die Werte und Wünsche des Patienten sowie den aktuellen Stand der klinischen Forschung. Dieser integrative Zugang ist nicht grundsätzlich neu, das besondere Merkmal ist der strukturierte und systematische Ansatz, mit dem eine möglichst transparente, zeitnahe und unverzerrte Berücksichtigung von Studienergebnissen erreicht werden soll.

Klassischerweise gliedert sich das strukturierte Vorgehen in der EbM in **fünf Schritte:**

1. Fragestellung: Übersetzung des klinischen Problems in eine beantwortbare Frage
2. Literaturrecherche: Suche nach relevanter und zuverlässiger externer Evidenz
3. Evidenzbewertung: kritische Beurteilung der relevanten Literatur
4. Umsetzung: der gewonnenen Einsichten in Abwägung der konkreten Situation
5. Evaluation: Selbstkritik und ggf. Anpassung der bisherigen Vorgehensweise.

Auch für den Geübten ist das skizzierte Vorgehen oftmals im Alltag der Patientenversorgung schwer umzusetzen. Einzelne Schritte, insbesondere die Literaturrecherche und -bewertung, bedürfen des Trainings. Journal Clubs und EbM-orientierte Literaturdienste bieten entsprechende Hilfestellung, indem sie zu relevanten Fragen eine Vorauswahl treffen und um kritische Bewertungen ergänzen. Die 1993 gegründete Cochrane Collaboration (CC) erstellt und publiziert als internationale gemeinnützige Organisation systematische Übersichtsarbeiten (Cochrane Database of Systematic Reviews) und stellt aktuelle Informationen allgemein zur Verfügung. Publizierte Studien werden dabei regelhaft in **5 Evidenzstufen** eingeteilt:

- Ia: wenigstens ein systematischer Review auf der Basis methodisch hochwertiger kontrollierter, randomisierter Studien (RCTs)
- Ib wenigstens eine ausreichend große, methodisch hochwertige RCT
- IIa wenigstens eine hochwertige Studie ohne Randomisierung
- IIb wenigstens eine hochwertige Studie eines anderen Typs quasi-experimenteller Studien
- III mehr als eine methodisch hochwertige nichtexperimentelle Studie
- IV Meinungen und Überzeugungen von angesehenen Autoritäten (aus klinischer Erfahrung); Expertenkommissionen; beschreibende Studien.

Das in Deutschland im Jahr 2000 gegründete Deutsche Netzwerk Evidenzbasierte Medizin (DNEbM) e. V. versteht sich als interdisziplinäre Plattform für Aktivitäten im Bereich der evidenzbasierten Medizin. Die Internetseite bietet eine Vielzahl nützlicher Hintergrundinformationen und Links (www.ebm-netzwerk.de).

Eine wichtige Institution, die sich für eine Verbesserung der Gesundheitsversorgung auf der Grundlage der EbM einsetzt, ist das im Jahr 2004 gegründete unabhängige Institut für Qualität und Wirtschaftlichkeit im Gesundheitswesen (IQWiG). Es ist überwiegend im Auftrag des Gemeinsamen Bundesausschusses (GBA) und des Bundesgesundheitsministeriums tätig. Zu seinen Aufgaben gehören u. a. die Bewertung von Operations- und Diagnoseverfahren, Arzneimitteln sowie Behandlungsleitlinien und die Erarbeitung der Grundlagen für neue Disease-Management-Programme (DMP).

41.3 Leitlinien der Deutschen Gesellschaft für Sozialpädiatrie und Jugendmedizin

Leitlinien sind nach Definition der Arbeitsgemeinschaft der wissenschaftlichen medizinischen Fachgesellschaften e. V. (AWMF) systematisch entwickelte Hilfen für Ärzte zur Entscheidungsfindung in spezifischen Situationen. Sie beruhen auf aktuellen wissenschaftlichen Erkenntnissen und in der Praxis bewährten Verfahren und sorgen für mehr Sicherheit in der Medizin, sollen aber auch ökonomische Aspekte berücksichtigen. Nach Straßburg (2003) sollen Leitlinien einfach formuliert sein, aber umfassende Aussagen zu Diagnostik, Indikation und Gegenindikation therapeutischer Verfahren enthalten, insbesondere auch die Bedingungen nennen, unter denen eine Therapie zu empfehlen ist.

41.3.1 Bedeutung der Leitlinien im ärztlichen Alltag

Leitlinien dienen als Entscheidungshilfen und sind rechtlich nicht verbindlich. Sie entbinden den Arzt nicht von der Überprüfung individueller Anwendbarkeit im konkreten Fall. Dies unterscheidet sie von Richtlinien. Im medizinischen Alltag haben Leitlinien mittlerweile erhebliche Bedeutung erlangt nicht, nur bei der Beantwortung haftungsrechtlicher Fragen, sondern auch bei Verhandlungen mit Kostenträgern.

In Deutschland hat sich die AWMF als Forum für Leitlinien etabliert. Diese sind im Internet publiziert und in drei Entwicklungsstufen kategorisiert. Lediglich Leitlinien der Stufe 3 (S3) sind als ausreichend evidenzbasiert anzusehen, entscheidend für eine entsprechende Einstufung ist ein systematisches und transparentes Vorgehen bei allen Schritten der Entwicklung. Der erforderliche Konsens aller beteiligten Fachgesellschaften wird in einem aufwändigen Prozess (sog. DELBI-Verfahren) von der AWMF moderiert.

Leitlinienstufen der AWMF:

- Stufe 1: im informellen Konsens durch Expertengruppe erarbeitet (S1-Leitlinie)
- Stufe 2: Ergebnisse eines systematischen Konsensusprozesses (evidenz- oder konsensusbasierte Leitlinie, S2e oder S2k-Leitlinie)
- Stufe 3: Leitlinien mit allen Elementen der systematischen Entwicklung (evidenz- und konsensusbasierte Leitlinie, S3-Leitlinie).

41.3.2 Grundlegende Probleme der Leitlinienerstellung in der Sozialpädiatrie

Aufgrund des Charakters der Sozialpädiatrie als „Querschnittswissenschaft" (Schlack 2000) und der Orientierung auf eine möglichst umfassende ganzheitliche und weniger krankheitsspezifische Betrachtungsweise sind Leitlinien nach den von der AWMF geforderten Qualitätsmerkmalen nur mit großen Schwierigkeiten zu erstellen. Die bislang zu einzelnen Störungsbildern existierenden parallelen S1Leitlinien benachbarter Fachgesellschaften (z. B. Neuropädiatrie, Kinder- und Jugendpsychiatrie) werden von der AWMF nicht mehr akzeptiert. Nach Ablauf deren Gültigkeit wird künftig nur noch eine von allen beteiligten Fachgesellschaften konsentierte Leitlinie auf S2 oder S3-Niveau akzeptiert. Dabei wird die jeweilige Leitlinie primär der/den zuerst anmeldenden Fachgesellschaft/en zugeordnet. S1 Leitlinien werden nur noch in wenigen Ausnahmefällen akzeptiert.

41.3.3 AWMF-Leitlinien der DGSPJ

Folgende Leitlinien wurden unter Mitarbeitet der Deutschen Gesellschaft für Sozialpädiatrie und Jugendmedizin erstellt (in Klammeren Niveau und Ablaufdatum):
- Therapie der Adipositas im Kindes- und Jugendalter (S3, 2014)
- Diagnostik des fetalen Alkoholsyndroms (S3, 31.12.2017)
- Sprachentwicklungsstörungen (SES), Diagnostik von, unter Berücksichtigung umschriebener Sprachentwicklungsstörungen (USES) (S2k, 15.12.2016)
- Umschriebene Entwicklungsstörung motorischer Funktionen (UEMF) (S3, 31.5.2016)
- Umgang mit Patienten mit nicht spezifischen, funktionellen und somatoformen Beschwerden (S3, 31.3.2017)
- Folgende Leitlinien werden unter der Federführung der DGSPJ in 2013 fertig gestellt:
- ADHS bei Kindern, Jugendlichen und Erwachsenen (S3)
- Sozialpädiatrische Nachsorge extrem unreifer Frühgeborener mit einem Geburtsgewicht unter 1.000 g (S2k)
- Folgende Leitlinien werden unter Beteiligung der DGSPJ in 2013 fertig gestellt:
- Intelligenzminderung und Lernbehinderung (S3)
- Säuglings- und Kleinkindpsychiatrie (S3)
- Behandlung von depressiven Störungen bei Kindern- und Jugendlichen (S3)
- Diagnostik und Behandlung von Kindern mit Lese- und/oder Rechtschreibstörung (S3)
- Störung der Geschlechtsentwicklung (S2k)

In engem Zusammenhang mit den Leitlinien der Deutschen Gesellschaft für Sozialpädiatrie und Jugendmedizin stehen die Arbeitspapiere der Qualitätszirkel der BAG Sozialpädiatrischer Zentren, welche die Leitlinien ergänzen bzw. Teilaspekte näher konkretisieren.

41.4 Qualitätspapiere der Bundesarbeitsgemeinschaft Sozialpädiatrischer Zentren Deutschlands (BAG SPZ)

41.4.1 Altöttinger Papier

Nach 5-jährigem Diskussionsprozess in der BAG SPZ wurde im Jahr 2001 das „Altöttinger Papier" mit der Formulierung von Grundlagen der Strukturqualität und Behandlung in sozialpädiatrischen Zentren (SPZ) mit breiter Mehrheit verabschiedet. Eine Aktualisierung ist für das Jahr 2013 geplant. Konkrete Anforderungen an personelle, apparative und räumliche Ausstattung der SPZ sind dort klar definiert, Aufgaben und Ziele sowie strukturelle Qualitätsanforderungen, insbesondere auch Anforderungsprofile für sämtliche beteiligten Berufsgruppen verbindlich formuliert. Als Basis für weitere Ausarbeitungen zur Prozessqualität wurden die Grundlagen der Mehrdimensionalen Bereichsdiagnostik in der Sozialpädiatrie (MBS) und ihre therapeutischen Konsequenzen dargestellt.

41.4.2 Qualitätszirkel der BAG SPZ

Aufgaben der Qualitätszirkel

Als weiterer Schritt zur konsequenten praktischen Umsetzung wurde im Jahr 2003 ein komplexes System verschiedener interdisziplinärer Qualitätszirkel etabliert.
Koordiniert durch einen Zentralen Qualitätsarbeitskreis haben die einzelnen Zirkel den Auftrag, die besonderen Voraussetzungen der sozialpädiatrischen Arbeitsweise herauszustellen, durch die eine umfassende und effiziente Diagnostik und Therapie ermöglicht wird. Dabei geht es explizit nicht darum, Leitlinien zu modifizieren oder zu ersetzen.
Das Themenspektrum umfasst Fragestellungen zu Konzeptionen, Methoden, Qualitätsmanagement, aber auch diagnosespezifischen Behandlungsabläufen. Die dort festgelegten Kriterien sind nicht als juristisch einklagbar angesehen, sind aber evidenzbasiert, mindestens aber auf Konsensbasis der jeweiligen beteiligten Experten formuliert. Sie beschreiben strukturiert inhaltliche und formale Arbeitsabläufe in sozialpädiatrischen Zentren, insbesondere für häufig dort behandelte Krankheitsbilder im Sinne klinischer Behandlungspfade. Mehrere Papiere stellen Instrumente zur Verbesserung der Prozessqualität zur Verfügung. Die inhaltliche Arbeit in sozialpädiatrischen Zentren wird damit nach innen und außen transparent dargestellt. Für neue, aber auch langjährige Mitarbeiter in den Zentren stellen die ausformulierten Positionen eine große Hilfe dar.

Zusammensetzung der Qualitätszirkel

Die Qualitätszirkel sind entsprechend der Grundstruktur sozialpädiatrischer Zentren interdisziplinär zusammengesetzt, die Leitung wird, wo immer möglich, durch einen zum Thema fachlich besonders ausgewiesenen Arzt übernommen. Die einzelnen Papiere werden im zentralen Qualitätsarbeitskreis intensiv bearbeitet und anschließend nach breiter Diskussion in den Regionalkonferenzen von der Vollversammlung der BAG SPZ verabschiedet. Als Rahmenempfehlungen werden sie nach Freigabe durch den Vorstand der Deutschen Gesellschaft für Sozialpädiatrie und Jugendmedizin publiziert.

Ergebnisse der Qualitätszirkel

Bis Ende 2012 konnten 33 Qualitätszirkel mit über 150 Mitgliedern ihre Arbeit erfolgreich abschließen. Die Resultate sind sowohl auf der Homepage der DGSPJ (www.dgspj.de) als auch in gedruckter Form veröffentlicht.
Die **Qualitätspapiere** der Bundesarbeitsgemeinschaft Sozialpädiatrischer Zentren:
- Mitarbeiterbefragung/Fragebogen für SPZ
- Zuweiserbefragung/Fragebogen für SPZ
- Elternbefragung/Fragebogen für SPZ
- „Ökonomie" im SPZ
- Dokumentation und schriftliche Berichterstattung im SPZ/Empfehlungen
- Gesprächsführung im SPZ
- Standards der psychologischen Diagnostik in SPZ
- Ressourcenanalyse zur Erstellung eines Behandlungsplans im SPZ
- Psychotherapie in der Sozialpädiatrie (Bestandsaufnahme)
- Sozialarbeit im SPZ
- Behandlung infantiler Zerebralparesen und ähnlicher Bewegungsstörungen
- Epilepsien

- Hyperkinetische Störungen – Aufmerksamkeits-Defizit-Hyperaktivitäts-Störung (ADHS)
- Kopfschmerzen und Migräne
- Regulations- und Beziehungsstörungen im Säuglings- und Kleinkindalter
- Spina bifida
- Sozialpädiatrische Nachsorge ehemals kritisch kranker Früh- und Neugeborener mit ihren Eltern
- Umschriebene Entwicklungsstörungen
- Umschriebene Entwicklungsstörungen der Wahrnehmung
- Vernachlässigung und Gewalt
- Ergotherapie
- Logopädie
- Physiotherapie
- Psychotherapie
- Heilpädagogik in der Sozialpädiatrie
- Autismus
- Hilfsmittelversorgung im Kindes- und Jugendalter im SPZ
- Nachsorge von Kindern und Jugendlichen nach Schädel-Hirn-Trauma
- Kinder- und Jugendpsychosomatik im SPZ
- Testverfahren und ausgewählte Fragebögen, die im SPZ Verwendung finden
- Psychische Störungen bei Kindern und Jugendlichen mit Intelligenzminderung
- Behandlung nichtorganischer Ausscheidungsstörungen: Enkopresis – Inkontinenz – Enuresis
- Stationäre sozialpädiatrische Behandlung.

Neben der kontinuierlichen Fortsetzung der Arbeit zu weiteren Themenbereichen besteht für den Zentralen Qualitätsarbeitskreis eine wesentlich Aufgabe darin, sich verstärkt um eine möglichst breite Umsetzung der erarbeiteten Positionen in den SPZ zu kümmern. Aktuell verlagert sich der Schwerpunkt auf die Förderung der Versorgungsforschung (➤ Kap. 45).

41.5 Qualitätsmanagementsysteme

41.5.1 Gesetzliche Vorgaben

Seit 2006 sind durch gesetzliche Rahmenvorgaben (SGB V, insb. §§ 135–137, SGB IX § 20, gemeinsamer Bundesausschuss, Richtlinie vom 1.1.06) alle Einrichtungen im Gesundheitswesen verpflichtet, ein Qualitätsmanagement(QM)-System einzuführen. Da der Gesetzgeber keine konkreten Vorgaben definiert hat, existieren insbesondere für ärztliche Praxen viele verschiedene Systeme, deren Grundelemente sich aber nur geringfügig voneinander unterscheiden.

41.5.2 Modell für Excellence der European Foundation for Quality Management (EFQM)

Im Folgenden wird das Modell für Excellence der European Foundation for Quality Management (EFQM) näher beleuchtet, weil es als umfassendes System zur ganzheitlichen Betrachtung von Organisationen entwickelt wurde und mittlerweile auch in Deutschland breit anerkannt und eingeführt ist.

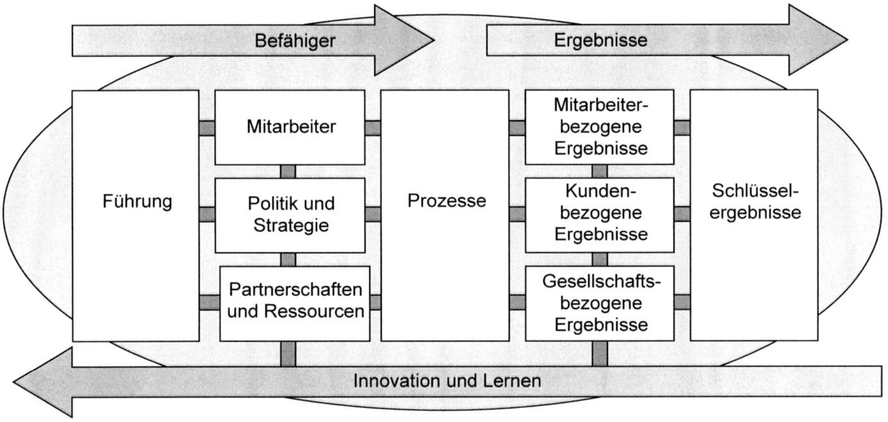

Abb. 41.1 EFQM-Modell© EFQM European Foundation for Quality Management

Die EFQM wurde 1989 auf Initiative der Europäischen Kommission von führenden Industrieunternehmen gegründet. In der Folgezeit hat sich die Mitgliederzahl deutlich erhöht, Organisationen aus dem Gesundheits- und Sozialbereich stellen mittlerweile 30 % der Gesamt-Mitgliederzahl.
Das EFQM-Modell stellt keine Liste von Qualitätsanforderungen dar, sondern betrachtet eine Organisation als Einheit. Wichtig ist die kontinuierliche Weiterentwicklung zu Excellence, definiert als überragende Vorgehensweisen. In der Überarbeitung des Modells in den Jahren 2003/2004 wurden besonders die Aspekte der Nachhaltigkeit und der sozialen Verantwortung hervorgehoben (➤ Abb. 41.1).

41.5.3 Kriterien des EFQM-Modells

Das Modell definiert 9 Kriterien und trennt diese in sogenannte „Befähiger" und „Ergebnisse". 5 Befähiger- und 4 Ergebniskriterien sind jeweils in Teilkriterien untergliedert. Zu allen insgesamt 32 Teilkriterien erläutern Ansatz- oder Orientierungspunkte jeweils wichtige Aspekte.
Die Befähiger-Kriterien beziehen sich auf die Aktivitäten und das Vorgehen einer Organisation (Abläufe und Prozesse). Dabei ist von besonderer Bedeutung, Strategie, Ziele und Maßnahmen angemessen zu verknüpfen und zu verdeutlichen, dass systematisch und präventiv vorgegangen wird. Außerdem wird bewertet, inwieweit intendiertes Vorgehen tatsächlich in die Praxis umgesetzt und gelebt wird und ob regelmäßige Bewertungen und Überprüfungen erfolgen, die sicherstellen, dass das Vorgehen veränderten Anforderungen entspricht.
Die Ergebniskriterien sind unterteilt in Mitarbeiter-, Kunden- und gesellschaftsbezogene sowie sog. Schlüsselergebnisse und beziehen sich auf das, was mit dem gewählten Vorgehen erreicht wurde. Es wird konkret überprüft, welchen Erfolg die in den Befähiger-Kriterien beschriebenen Ansätze zeigen.

41.5.4 Bewertungsmaßstäbe

Für die Bewertung des Reifegrades einer Organisation anhand des Modells wurde die sog. RADAR-Bewertungsmethodik entwickelt (Ergebnisse = **R**esults, dazu führende Vorgehensweisen =

Approach, Grad der Umsetzung = **D**eployment, Bewertung und Überprüfung = **A**ccessment und **R**eview). Dabei werden die Einzelbewertungen der insgesamt 32 Teilkriterien mit Punkten versehen und zu einer Gesamtbewertung zusammengefasst. Eine Bewertung erfolgt regelhaft zunächst als Selbstbewertung, die Aussagen zu Stärken und Verbesserungspotenzialen liefert, aus denen sich dann konkrete Projekte ableiten lassen. Eine erste realistische Selbstbewertung wird selten mit einem Ergebnis von mehr als 300 von 1.000 theoretisch möglichen Punkten enden. Bei hohem Reifegrad können externe Bewertungen mit objektivierten Vergleichsmöglichkeiten zu anderen Einrichtungen wichtige Impulse zur Weiterentwicklung geben.

41.5.5 EFQM und KTQ

Vergleicht man das EFQM-Modell mit dem in Deutschland eingeführten Zertifizierungsverfahren für Krankenhäuser „Kooperation für Transparenz und Qualität" (KTQ), nach dem auch viele Klinik-assoziierte SPZ zertifiziert werden, zeigen sich deutliche Unterschiede: KTQ basiert auf einem Bewertungskatalog, der die für die stationäre Patientenversorgung relevanten Aufgaben abdeckt und einen Konsens von Mindestanforderungen bzw. Mindeststandards darstellt. Dagegen orientiert sich das EFQM-Modell am Prinzip der kontinuierlichen Verbesserung und einer jeweils „besten Praxis", die als Vorbild für die Weiterentwicklung der eigenen Arbeit dient. Während sich im Bewertungskatalog von KTQ die im Krankenhaus geläufigen Begriffe finden, ist die Terminologie des EFQM-Modells durch die Nähe zu Management-Lehren und -Konzepten geprägt und muss regelhaft auf die Begriffe der jeweiligen Einrichtung übersetzt werden. Insgesamt ist aber bei entsprechender Adaptation durch die prinzipielle Offenheit des Systems, die Patienten- (Kunden), Mitarbeiter- und Außenorientierung („best practice"), das EFQM-Modell grundsätzlich gut geeignet, die Arbeit sozialpädiatrischer Einrichtungen abzubilden und zu bewerten.

41.5.6 DIN EN ISO 9001

Einige Sozialpädiatrische Zentren haben mittlerweile ein QM-System nach den Vorgaben des Deutschen Instituts für Normung e. V. (DIN) begonnen. Die europäische Norm EN DIN 9001ff. legt entsprechende Anforderungen fest, die 8 Hauptkapitel der Norm lauten:

- Kap. 1–3: Vorwort und Allgemeines
- Kap. 4: Qualitätsmanagementsystem (allgemeine Anforderungen, dokumentierte Anforderungen, QM-Handbuch, Lenkung von Dokumenten, Lenkung von Aufzeichnungen)
- Kap. 5: Verantwortung der Leitung
- Kap. 6: Management von Ressourcen
- Kap. 7: Produktrealisierung
- Kap. 8: Messung, Analyse und Verbesserung.

Überprüfungen erfolgen in Form „interner Audits" im weiteren Verlauf durch „externe Audits" von unabhängigen Organisationen. Eine Zertifizierung bescheinigt die Konformität mit den Anforderungen der Norm. Eine überarbeitete Fassung der DIN EN 9001 wird im Dezember 2008 veröffentlicht.

41.5.7 JCAHO

Die Joint Commission on Accreditation of Healthcare Organisations (JCAHO) in Trägerschaft der American Hospital Association (AHA) sowie amerikanischer und kanadischen Fachgesellschaften ist seit über 50 Jahren die führende Institution zur Überprüfung von Krankenhäusern in den USA. Es bestehen starke Bestrebungen, die JCAHO-Qualitätskriterien weltweit auszudehnen. Einige SPZ im universitären Bereich sind damit bereits konfrontiert.

41.5.8 Zertifizierung

Alle gängigen QM-Systeme bieten die Möglichkeiten einer Zertifizierung, deren Vielfalt dem Spektrum der angebotenen Verfahren entspricht. Unabhängig von der Wahl des Qualitätsmanagement-Systems sind Zertifizierungsprozesse für die jeweilige Einrichtung mit einem erheblichen Aufwand verbunden. In der Regel sind vor dem eigentlichen Zertifizierungsverfahren vorbereitende Arbeiten in der Einrichtung erforderlich, anteilige Personalkosten für einen QM-Beauftragten und Team-Mitglieder sind ebenso zu kalkulieren wie Aufwendungen für einen oder mehrere externe Berater sowie letztendlich für den eigentlichen Zertifizierungsprozess. Für die Startphase, für die im günstigsten Fall mindestens 1–2 Jahre veranschlagt werden müssen, ist mit Kosten deutlich über 10.000 € zu rechnen. Bedacht werden muss auch, dass jedes Zertifikat eine begrenzte Gültigkeit besitzt (EFQM 2 Jahre, DIN EN ISO und KTQ 3 Jahre) und eine regelmäßige Wiederholung (Rezertifizierung) erforderlich ist.
Der Wert eines solchen Prozesses ist daher umso größer, je mehr es gelingt, möglichst viele Mitarbeitende aktiv zu beteiligen und die wichtigsten Prozesse in einer für alle möglichst sinnvollen Weise zu strukturieren.

Literatur

Arbeitsgemeinschaft der Wissenschaftlichen Medizinischen Fachgesellschaften e. V. (2008) Leitlinien Sozialpädiatrie und Jugendmedizin. http://leitlinien.net/

Bäuerle T, Schubert HJ (2003) Das EFQM-Modell für Excellence. In: Schubert HJ. Management von Gesundheits- und Sozialeinrichtungen, Handlungsfelder – Methoden – Lösungen. Neuwied, Köln, München: Luchterhand.

Das Deutsche Cochrane Zentrum (2008) EbM in Deutschland. http://www.cochrane.de/de/ebhc.htm#ebmd

Deutsche Gesellschaft für Qualität (2004) Das EFQM-Modell für Excellence – wozu? http://www.deutsche-efqm.de/

Deutsches Netzwerk Evidenzbasierte Medizin
http://www.ebm-netzwerk.de

Fricke C, Hollmann H, Kretzschmar C, Schmid R (2007) Bundesarbeitsgemeinschaft Sozialpädiatrischer Zentren, Qualität in der Sozialpädiatrie Bd. 2. Altötting: RS-Verlag.

Gemeinsamer Bundesausschuss.
http://www.g-ba.de

Hildebrand R (1999) Das bessere Krankenhaus: Total quality planen, umsetzen, managen. Neuwied Kriftel: Luchterhand.

Hollmann H, Kretzschmar C, Schmid R (2008) Bundesarbeitsgemeinschaft Sozialpädiatrischer Zentren, Qualität in der Sozialpädiatrie Bd. 1. korrigierter Nachdruck. Altötting: RS-Verlag.

Institut für Qualität und Wirtschaftlichkeit im Gesundheitswesen.

http://www.iqwiq.de
Sackett DL et al. (1997) Was ist Evidenz-basierte Medizin und was nicht? Münch. Med. Wschr. 139 (44): 644–645.
Schlack HG (2000) Sozialpädiatrie. 2. Aufl. München Jena: Urban & Fischer.
Straßburg HM (2002) Die Leitlinien der Deutschen Gesellschaft für Sozialpädiatrie und Jugendmedizin. http://www.dgspj.de/llallgemeines.php
Straßburg HM (2003) Die Leitlinien der DGSPJ – aktueller Stand. Kinderärztliche Praxis: 264–265.

42 Rechtliche Grundlagen

Hans-Michael Straßburg

42.1 Grundgesetz

Für Kinder und Jugendliche in unserer Gesellschaft spielen, vor allem auch beim Vorliegen von Entwicklungsauffälligkeiten oder Behinderungen, eine Vielzahl von Gesetzen und Rechtsverordnungen eine wesentliche Rolle. Die grundsätzlichen Rechte und Pflichten eines jeden Menschen sind im **Grundgesetz (GG) für die Bundesrepublik Deutschland** von 1949 definiert. Hierin heißt es in Artikel 2, Absatz 2: „Jeder Mensch hat das Recht auf Leben und körperliche Unversehrtheit. Die Freiheit der Person ist unverletzlich. In diese Rechte darf nur aufgrund eines Gesetzes eingegriffen werden."
In Artikel 3, Absatz 3 steht: „Niemand darf wegen seines Geschlechts, seiner Abstammung, seiner Rasse, seiner Sprache, seiner Heimat und Herkunft, seines Glaubens, seiner religiösen oder politischen Anschauungen benachteiligt oder bevorzugt werden. Niemand darf wegen seiner Behinderung benachteiligt werden."
Ein Kind ist ab seiner Geburt, d. h. dem Durchtrennen der Nabelschnur, rechtsfähig und das Leben eines Neugeborenen und das eines Erwachsenen sind juristisch gleichwertig.
In Artikel 6 GG heißt es in Absatz 2: „Pflege und Erziehung der Kinder sind das natürliche Recht der Eltern und die zuvörderst ihnen obliegende Pflicht. Über ihre Betätigung wacht die staatliche Gemeinschaft."
Absatz 3: „Gegen den Willen der Erziehungsberechtigten dürfen Kinder nur aufgrund eines Gesetzes von der Familie getrennt werden, wenn die Erziehungsberechtigten versagen oder wenn die Kinder aus anderen Gründen zu verwahrlosen drohen."
Absatz 4: „Jede Mutter hat Anspruch auf den Schutz und die Fürsorge der Gemeinschaft."
Absatz 5: „Den unehelichen Kindern sind durch die Gesetzgebung die gleichen Bedingungen für ihre leibliche und seelische Entwicklung und ihre Stellung in der Gesellschaft zu schaffen wie den ehelichen Kindern."

Sozialpädiatrie in der Praxis. http://dx.doi.org/10.1016/B978-3-437-31630-3.00042-0
Copyright © 2014 Elsevier GmbH. All rights reserved.

Artikel 7 befasst sich mit dem Schulwesen. Hierin wird festgestellt, „dass das gesamte Schulwesen unter der Aufsicht des Staates steht". Die **Schulpflicht** ist die einzige Verpflichtung für Kinder. Als Ausdruck der Kulturhoheit der Länder werden die Einzelheiten im jeweiligen Landesrecht geregelt.
Aus Artikel 20, Absatz 1 GG ergibt sich die Verpflichtung des Staates, einen Mindeststandard menschenwürdigen Daseins, das sogenannte **soziokulturelle Existenzminimum,** sicherzustellen. Dieser Mindeststandard wird im Sozialgesetzbuch (SGB) konkretisiert, das Leitprinzip eines menschenwürdigen Daseins wird im Sozialgesetz XII definiert.

42.2 Sozialgesetzgebung

Bis zum Ende des 19. Jahrhunderts erfolgte die Unterstützung von armen und bedürftigen Menschen durch die Fürsorgeeinrichtungen der jeweiligen Gemeinden. Ab 1880 wurden sukzessive die Kranken-, Alters- und Invalidenversicherung eingeführt, die 1911 in der Reichsversicherungsordnung zusammengefasst wurden. Grundlage dieses Gesetzeswerks ist das sogenannte **Subsidiaritätsprinzip,** d.h. im Versicherungsfall können verschiedene Versicherungen in Anspruch genommen werden und, wenn keine Zuständigkeit besteht, ist letztendlich die **Sozialhilfe** nach Erschöpfung aller anderen Möglichkeiten für die Aufrechterhaltung des soziokulturellen Existenzminimums zuständig.
Das wichtigste Gesetzeswerk zur Regelung der sozialen Gerechtigkeit und Sicherheit in unserer Gesellschaft ist das **Sozialgesetzbuch (SGB).** Im ersten Buch steht in Artikel 1, dass es dazu beitragen soll, „ein menschenwürdiges Dasein zu sichern, gleiche Voraussetzungen für die freie Entfaltung der Persönlichkeit, insbesondere auch für junge Menschen, zu schaffen, die Familie zu schützen und zu fördern, den Erwerb des Lebensunterhaltes durch eine frei gewählte Tätigkeit zu ermöglichen und besondere Belastungen des Lebens, auch durch Hilfe zur Selbsthilfe, abzuwenden oder auszugleichen".
Das SGB gliedert sich derzeit in 12 Bücher (➤ Tab. 42.1); die für die Betreuung von Kindern und Jugendlichen wichtigsten sind das SGB V, VIII, IX, XI und XII.

Tab. 42.1 Sozialgesetzbücher

Sozialgesetzbuch (SGB)	Verabschiedung
SGB I – Allgemeiner Teil	1976
SGB II – Grundsicherung für Arbeitssuchende	2005
SGB III – Arbeitsförderung	1998
SGB IV – Sozialversicherung	1977
SGB V – Krankenversicherung	1989–2007
SGB VI – Rentenversicherung	1992
SGB VII – Unfallversicherung	1997
SGB VIII – Kinder- und Jugendhilfe	1991–1998
SGB IX – Rehabilitation	2001
SGB X – Datenschutz	1983
SGB XI – Pflegeversicherung	1995
SGB XII – Sozialhilfe	2005

42.2.1 SGB V – Krankenversicherung und medizinische Leistungen

Die **gesetzliche Krankenversicherung** ist in Deutschland im SGB V, § 1 definiert: Die Krankenversicherung als Solidargemeinschaft hat die Aufgabe, die Gesundheit der Versicherten zu erhalten, wiederherzustellen oder ihren Gesundheitszustand zu bessern. Die Versicherten sind für ihre Gesundheit mitverantwortlich; sie sollen durch eine gesundheitsbewusste Lebensführung, durch frühzeitige Beteiligung an gesundheitlichen Vorsorgemaßnahmen sowie durch aktive Mitwirkung an Krankenbehandlungen und Rehabilitationen dazu beitragen, den Eintritt von Krankheit und Behinderung zu vermeiden oder ihre Folgen zu überwinden. Die Krankenkassen haben den Versicherten dabei durch Aufklärung, Beratung und Leistungen zu helfen und auf gesunde Lebensverhältnisse hinzuwirken.

In SGB V, § 23 sind die **medizinischen Vorsorgeleistungen** definiert: Versicherte haben Anspruch auf ärztliche Behandlung und Versorgung mit Arznei-, Verband-, Heil- und Hilfsmitteln, wenn diese notwendig sind, um

1. eine Schwächung der Gesundheit, die in absehbarer Zeit voraussichtlich zu einer Krankheit führen würde, zu beseitigen,
2. einer Gefährdung der gesundheitlichen Entwicklung eines Kindes entgegenzuwirken,
3. Krankheiten zu verhüten oder deren Verschlimmerung zu vermeiden und
4. Pflegebedürftigkeit zu vermeiden.

Reichen bei Versicherten die obengenannten Leistungen nicht aus, kann die Krankenkasse aus medizinischen Gründen erforderliche ambulante Vorsorgeleistungen in anerkannten Kurorten erbringen. Die Satzung der Krankenkasse kann zu den übrigen Kosten, die den Versicherten im Zusammenhang mit dieser Leistung entstehen, einen täglichen Zuschuss (von derzeit bis zu 13,00 €) vorsehen. Bei ambulanten Vorsorgeleistungen für versicherte chronisch kranke Kleinkinder kann der Zuschuss (auf derzeit bis zu 21,00 €) erhöht werden.

Absatz 7: Medizinisch notwendige stationäre Vorsorgemaßnahmen für versicherte Kinder, die das 14. Lebensjahr noch nicht vollendet haben, sollen in der Regel für 4 bis 6 Wochen erbracht werden.

Im SGB V, § 26 sind die Kinderuntersuchungen **(Früherkennungsuntersuchungen, die sog. Vorsorgeuntersuchungen)** definiert: „Versicherte Kinder haben bis zur Vollendung des 6. Lebensjahres Anspruch auf Untersuchungen sowie nach Vollendung des 10. Lebensjahres auf eine Untersuchung zur Früherkennung von Krankheiten, die ihre körperliche oder geistige Entwicklung in nicht geringfügigem Maße gefährden. Zu den Früherkennungsuntersuchungen auf Zahn-, Mund- und Kieferkrankheiten gehören insbesondere die Inspektion der Mundhöhle, die Einschätzung und Bestimmung des Kariesrisikos, die Ernährungs- und Mundhygieneberatung sowie Maßnahmen zur Schmelzhärtung der Zähne und zur Keimzahlsenkung…"

In § 43 SGB V sind ergänzende **Leistungen zur Rehabilitation** aufgeführt: „Die Krankenkasse kann aus medizinischen Gründen in unmittelbarem Anschluss an eine Krankenhausbehandlung oder stationäre Rehabilitation erforderliche sozialmedizinische Nachsorgemaßnahmen für chronisch kranke und schwerstkranke Kinder, die das 12. Lebensjahr noch nicht vollendet haben, erbringen oder fördern, wenn die Nachsorge wegen der Art, Schwere und Dauer der Erkrankung notwendig ist, um den stationären Aufenthalt zu verkürzen oder die anschließende ambulante ärztliche Behandlung zu sichern".

In § 43a SGB V sind in der Fassung vom 21.7.2004 **nichtärztliche sozialpädiatrische Leistungen** festgeschrieben:

„Versicherte Kinder haben Anspruch auf nichtärztliche sozialpädiatrische Leistungen, insbesondere auf psychologische, heilpädagogische und psychosoziale Leistungen, wenn sie unter ärztlicher Verantwortung erbracht werden und erforderlich sind, um eine Krankheit zum frühestmöglichen Zeitpunkt zu erkennen und einen Behandlungsplan aufzustellen."
In § 43, Absatz 2 ist die sozialmedizinische Nachsorge als Leistung der gesetzlichen Krankenkassen unmittelbar im Anschluss an einen stationären Aufenthalt für chronisch kranke und schwerstkranke Kinder unter 12 Jahren definiert, wie es von dem Modellprojekt des Bunten Kreises eingeführt wurde.
In SGB V, § 119 in der Fassung vom 21.7.2004 werden Sozialpädiatrische Zentren definiert:

1. Sozialpädiatrische Zentren, die fachlich-medizinisch unter ständiger ärztlicher Leitung stehen und die Gewähr für eine leistungsfähige und wirtschaftliche sozialpädiatrische Behandlung bieten, können vom Zulassungsausschuss ... zur ambulanten sozialpädiatrischen Behandlung von Kindern ermächtigt werden. Die Ermächtigung ist zu erteilen, soweit und solange sie notwendig ist, um eine ausreichende sozialpädiatrische Behandlung sicher zu stellen.
2. Die Behandlung durch Sozialpädiatrische Zentren ist auf diejenigen Kinder auszurichten, die wegen der Art, Schwere und Dauer ihrer Krankheit oder einer drohenden Krankheit nicht von geeigneten Ärzten oder in geeigneten Frühförderstellen behandelt werden können. Die Zentren sollen mit den Ärzten und den Frühförderstellen eng zusammenarbeiten.

In § 119a wird die ambulante Behandlung in **Einrichtungen der Behindertenhilfe** beschrieben: Einrichtungen der Behindertenhilfe, die über eine ärztlich geleitete Abteilung verfügen, sind vom Zulassungsausschuss zur ambulanten ärztlichen Behandlung von Versicherten mit geistiger Behinderung zu ermächtigen, soweit und solange eine ausreichende ärztliche Versorgung dieser Versicherten ohne die besonderen Untersuchungs- und Behandlungsmethoden oder Kenntnisse der Ärzte in den Einrichtungen durch niedergelassene Ärzte nicht sichergestellt ist. Die Behandlung ist auf diejenigen Versicherten auszurichten, die wegen der Art oder Schwere ihrer Behinderung auf die ambulante Behandlung von diesen Einrichtungen angewiesen sind.

42.2.2 SGB VIII – Kinder- und Jugendhilfe

In diesem sehr ausführlichen Gesetzeswerk, das zuletzt 2007 geändert wurde, wird in § 1 das **Recht auf Erziehung, Elternverantwortung und Jugendhilfe** definiert:

1. Jeder junge Mensch hat ein Recht auf Förderung seiner Entwicklung und auf Erziehung zu einer eigenverantwortlichen und gemeinschaftsfähigen Persönlichkeit.
2. Pflege und Erziehung der Kinder sind das natürliche Recht der Eltern und die zuvörderst ihnen obliegende Pflicht. Über ihre Betätigung wacht die staatliche Gemeinschaft.
3. Jugendhilfe soll zur Verwirklichung des Rechts nach Absatz 1 insbesondere
   a. junge Menschen in ihrer individuellen und sozialen Entwicklung fördern und dazu beitragen, Benachteiligungen zu vermeiden oder abzubauen,
   b. Eltern und andere Erziehungsberechtigte bei der Erziehung beraten und
   c. Kinder und Jugendliche vor Gefahren für ihr Wohl schützen,
   d. dazu beitragen, positive Lebensbedingungen für junge Menschen und ihre Familien sowie eine kinder- und familienfreundliche Umwelt zu erhalten oder zu schaffen.

In § 2 sind die Aufgaben der Jugendhilfe zusammengestellt. Hierzu gehören u. a.

1. Angebote der Jugendarbeit, der Jugendsozialarbeit und des erzieherischen Kinder- und Jugendschutzes,
2. Angebote zur Förderung der Erziehung in der Familie,

3. Angebote zur Förderung von Kindern in Tageseinrichtungen und in Tagespflege,
4. Hilfe zur Erziehung und ergänzende Leistungen,
5. Hilfe für seelische behinderte Kinder und Jugendliche und ergänzende Leistungen,
6. Hilfe für junge Volljährige.

SGB VIII, § 8a befasst sich mit dem Schutzauftrag bei Kindeswohlgefährdung.

1. Werden dem Jugendamt gewichtige Anhaltspunkte für die Gefährdung des Wohls eines Kindes oder Jugendlichen bekannt, so hat es das Gefährdungsrisiko im Zusammenwirken mehrerer Fachkräfte abzuschätzen. Dabei sind die Personensorgeberechtigten sowie das Kind oder der Jugendliche einzubeziehen, soweit hierdurch der wirksame Schutz des Kindes oder des Jugendlichen nicht in Frage gestellt wird. Hält das Jugendamt zur Abwendung der Gefährdung die Gewährung von Hilfen für geeignet und notwendig, so hat es diese den Personensorgeberechtigten oder den Erziehungsberechtigten anzubieten.
2. In Vereinbarungen mit den Trägern von Einrichtungen und Diensten, die Leistungen nach diesem Buch erbringen, ist sicherzustellen, dass deren Fachkräfte den Schutzauftrag annehmen und bei der Abschätzung des Gefährdungsrisikos eine insoweit erfahrene Fachkraft hinzuziehen. Insbesondere ist die Verpflichtung aufzunehmen, dass die Fachkräfte bei den Personensorgeberechtigten oder den Erziehungsberechtigten auf die Inanspruchnahme von Hilfen hinwirken, wenn sie diese für erforderlich halten, und das Jugendamt informieren, falls die angenommenen Hilfen nicht ausreichend erscheinen, um die Gefährdung abzuwenden.
3. Hält das Jugendamt das Tätigwerden des Familiengerichts für erforderlich, so hat es das Gericht anzurufen; dies gilt auch, wenn die Personensorgeberechtigten oder die Erziehungsberechtigten nicht bereit oder in der Lage sind, bei der Abschätzung des Gefährdungsrisikos mitzuwirken. Besteht eine dringende Gefahr oder kann die Entscheidung des Gerichts nicht abgewartet werden, so ist das Jugendamt verpflichtet, das Kind oder den Jugendlichen in Obhut zu nehmen.

Demnach können bei Verdacht auf eine Gefährdung des Kindeswohls Kinder bis zu 48 h aus der Familie entfernt werden, um z. B. in einer Kinderklinik untersucht zu werden. Danach muss ein Beschluss des zuständigen Familiengerichts über das weitere Sorgerecht vorliegen.

§ 16 SGB VIII beschreibt die **allgemeine Förderung der Erziehung** in der Familie:

1. Müttern, Vätern, anderen Erziehungsberechtigten und jungen Menschen sollen Leistungen der allgemeinen Förderung der Erziehung in der Familie angeboten werden. Sie sollen dazu beitragen, dass Mütter, Väter und andere Erziehungsberechtigte ihre Erziehungsverantwortung besser wahrnehmen können. Sie sollen auch Wege aufzeigen, wie Konfliktsituationen in der Familie gewaltfrei gelöst werden können.
2. Leistungen zur Förderung der Erziehung in der Familie sind insbesondere
   a. Angebote der Familienbildung, die auf Bedürfnisse und Interessen sowie auf Erfahrung von Familien in unterschiedlichen Lebenslagen und Erziehungs-Situationen eingehen, die Familien zur Mitarbeit in Erziehungseinrichtungen und in Form der Selbst- und Nachbarschaftshilfe besser befähigen sowie junge Menschen auf Ehe, Partnerschaft und das Zusammenleben mit Kindern vorbereiten,
   b. Angebote der Beratung in allgemeinen Fragen der Erziehung und Entwicklung junger Menschen,
   c. Angebote der Familienfreizeit und der Familienerholung, insbesondere in belastenden Familiensituationen, die bei Bedarf die erzieherische Betreuung der Kinder einschließen.

Die Betreuung und Versorgung des Kindes in Notsituationen wird in § 20 SGB VIII beschrieben:

1. Fällt der Elternteil, der die überwiegende Betreuung des Kindes übernommen hat, für die Wahrnehmung dieser Aufgabe aus gesundheitlichen oder anderen zwingenden Gründen aus,

so soll der andere Elternteil bei der Betreuung und Versorgung des im Haushalt lebenden Kindes unterstützt werden, wenn
   a. wegen berufsbedingter Abwesenheit er nicht in der Lage ist, die Aufgabe wahrzunehmen,
   b. die Hilfe erforderlich ist, um das Wohl des Kindes zu gewährleisten,
   c. Angebote der Förderung des Kindes in Tageseinrichtungen oder in Tagespflege nicht ausreichen.
2. Fällt ein alleinerziehender Elternteil oder fallen beide Elternteile aus gesundheitlichen oder anderen zwingenden Gründen aus, so soll unter der obengenannten Voraussetzung das Kind im elterlichen Haushalt versorgt und betreut werden, wenn und solange es für sein Wohl erforderlich ist.

§ 22 SGB VIII beschreibt die Grundsätze der Förderung:

1. Tageseinrichtungen sind Einrichtungen, in denen sich Kinder für einen Teil des Tages oder ganztägig aufhalten und in Gruppen gefördert werden. Kindertagespflege wird von einer geeigneten Tagespflegeperson in ihrem Haushalt oder im Haushalt des Personensorgeberechtigten geleistet.
2. Tageseinrichtungen für Kinder und Kindertagespflege sollen
   a. die Entwicklung des Kindes zu einer eigenverantwortlichen und gemeinschaftsfähigen Persönlichkeit fördern,
   b. die Erziehung und Bildung in der Familie unterstützen und ergänzen,
   c. den Eltern dabei helfen, Erwerbstätigkeit und Kindererziehung besser miteinander vereinbaren zu können.

§ 27 SGB VIII beschreibt die **Hilfe zur Erziehung:**

1. Ein Personensorgeberechtigter hat bei der Erziehung eines Kindes oder eines Jugendlichen Anspruch auf Hilfe (Hilfe zur Erziehung), wenn eine dem Wohl des Kindes oder des Jugendlichen entsprechende Erziehung nicht gewährleistet ist und die Hilfe für seine Entwicklung geeignet und notwendig ist.
2. Hilfe zur Erziehung umfasst insbesondere die Gewährung pädagogischer und damit verbundener therapeutischer Leistungen.

Von besonderer Bedeutung ist die **Eingliederungshilfe für seelisch behinderte Kinder und Jugendliche** in § 35a SGB VIII:

1. Kinder oder Jugendliche haben Anspruch auf Eingliederungshilfe, wenn
   a. ihre seelische Gesundheit mit hoher Wahrscheinlichkeit länger als 6 Monate von dem für ihr Lebensalter typischen Zustand abweicht und
   b. daher ihre Teilhabe am Leben in der Gesellschaft beeinträchtigt ist oder eine solche Beeinträchtigung zu erwarten ist.

Von einer seelischen Behinderung bedroht im Sinne dieses Gesetzes sind Kinder und Jugendliche, bei denen eine **Beeinträchtigung ihrer Teilhabe am Leben** in der Gesellschaft nach fachlicher Erkenntnis mit hoher Wahrscheinlichkeit zu erwarten ist.

2. Hinsichtlich der Abweichung der seelischen Gesundheit … hat der Träger der öffentlichen Jugendhilfe die Stellungnahme
   a. eines Arztes für Kinder- und Jugendpsychiatrie und -psychotherapie,
   b. eines Kindes- und Jugendpsychotherapeuten und
   c. eines Arztes oder eines psychologischen Psychotherapeuten, der über besondere Erfahrungen auf dem Gebiet seelischer Störungen bei Kindern und Jugendlichen verfügt, einzuholen.

Die Stellungnahme ist auf der Grundlage der internationalen Klassifikation der Krankheiten (ICD-10) … zu erstellen. Dabei ist auch darzulegen, ob die Abweichung Krankheitswert hat oder auf einer Krankheit beruht.

3. Die Hilfe wird nach Bedarf im Einzelfall
   a. in ambulanter Form,
   b. in Tageseinrichtungen für Kinder oder in anderen teilstationären Einrichtungen,
   c. durch geeignete Pflegepersonen und
   d. in Einrichtungen über Tag und Nacht sowie sonstigen Wohnformen geleistet.

Ein Beispiel für eine durch das Jugendamt nach § 35 a SGB VIII zu finanzierende Erziehungshilfe ist die Betreuung von Kindern mit der Teilleistungsstörung „Legasthenie". Diese muss in einem Zusammenhang mit einer „drohenden seelischen Behinderung" stehen. Ein entsprechendes Gutachten kann von einem Arzt für Kinder- und Jugendpsychiatrie und Psychotherapie oder in einem sozialpädiatrischen Zentrum durch einen Arzt für Kinder- und Jugendmedizin in Zusammenarbeit mit einem klinischen Psychologen erstellt werden.

Die Finanzierung einer Legasthenie-Behandlung durch die Krankenkasse ist in der Regel nur über ein Gutachten zu erreichen, in dem der Begriff „Legasthenie" nicht erwähnt wird, da diese – bisher – von den Krankenkassen nicht als Krankheit im versicherungsrechtlichen Sinne anerkannt ist.

§ 42 SGB VIII befasst sich mit der **Inobhutnahme** von Kindern und Jugendlichen. Demnach ist das Jugendamt berechtigt und verpflichtet, ein Kind oder einen Jugendlichen in seine Obhut zu nehmen wenn

4. das Kind oder der Jugendliche um Obhut bittet oder
5. eine dringende Gefahr für das Wohl des Kindes und des Jugendlichen eine Inobhutnahme erfordert und
   a. die Personensorgeberechtigten nicht widersprechen oder
   b. eine familiengerichtliche Entscheidung nicht rechtzeitig eingeholt werden kann…
6. Freiheitsentziehende Maßnahmen im Rahmen der Inobhutnahme sind nur zulässig, wenn und soweit sie erforderlich sind, um eine Gefahr für Leib und Leben des Kindes oder des Jugendlichen oder Gefahr für Leib und Leben Dritter abzuwenden. Die Freiheitsentziehung ist ohne gerichtliche Entscheidung spätestens mit Ablauf des Tages nach ihrem Beginn zu beenden.

Die für die Kinder- und Jugendhilfe zuständige Institution ist das örtliche Jugendamt. Seine Aufgaben werden von einem Jugendhilfeausschuss koordiniert. Es ist zur Förderung der Jugendarbeit, für die Erziehungshilfe, für sozialpädagogische Familienhilfe, für das Erstellen eines Hilfeplans, für Eingriffe in das Sorgerecht bei Gefährdung des Kindswohls, Vormundschafts- und Pflegschaften, Sorge- und Umgangsrecht sowie Tageseinrichtungen und Tagespflege und verschiedenes andere zuständig. In akuten Situationen kann z. B. zur Klärung bei V. a. Kindesmisshandlung ohne richterlichen Beschluss für bis zu 48 h eine Beobachtung und Untersuchung des Kindes in einer Einrichtung veranlasst werden.

42.2.3 SGB IX – Rehabilitation und Teilhabe behinderter Menschen

In § 1 SGB IX wird zur Selbstbestimmung und Teilhabe am Leben in der Gesellschaft folgendes festgelegt: Behinderte oder von Behinderung bedrohte Menschen erhalten Leistungen nach diesem Buch und den für die Rehabilitationsträger geltenden Leistungsgesetzen, um ihre Selbstbestimmung und gleichberechtigte Teilhabe am Leben in der Gesellschaft zu fördern, Benachteiligungen zu vermeiden oder ihnen entgegenzuwirken. Dabei wird den besonderen Bedürfnissen behinderter und von Behinderung bedrohter Frauen und Kinder Rechnung getragen.

§ 2 SGB IX befasst sich mit der Definition von Behinderungen:

1. Menschen sind behindert, wenn ihre körperliche Funktion, geistige Fähigkeit oder seelische Gesundheit mit hoher Wahrscheinlichkeit länger als 6 Monate von dem für das Lebensalter typischen Zustand abweichen und daher ihre Teilhabe am Leben in der Gesellschaft beeinträchtigt ist. Sie sind von Behinderung bedroht, wenn die Beeinträchtigung zu erwarten ist.
2. Menschen sind ... schwerbehindert, wenn bei ihnen ein Grad der Behinderung von wenigstens 50 vorliegt.

Ab einem Grad der Behinderung von 50 wird ein **Schwerbehindertenausweis** ausgestellt, der unabhängig von Einkommen oder Besitz ist. Zusätzlich können Merkmale, die mit Großbuchstaben im Behindertenausweis eingetragen werden, anerkannt werden:

- H = Hilflosigkeit, mit der Notwendigkeit einer ständigen Beaufsichtigung
- B = Begleitperson erforderlich
- G = erhebliche Gehbeeinträchtigung
- aG = außergewöhnliche Gehbehinderung
- Bl = blind
- Gl = gehörlos
- RF = Befreiung von Rundfunkgebühren.

Im Einzelnen ermöglicht der Schwerbehindertenausweis u. a.:

- Steuerfreien Pauschalbetrag
- Steuererlass für das eigene Kraftfahrzeug
- Ansprüche auf Wohngeld und Wohnungsbauförderung
- Befreiung von Rundfunkgebühren
- Ermäßigung beim Fernsprechhauptanschluss
- Ggf. Parkerleichterung
- Unentgeltliche Beförderung im öffentlichen Nahverkehr (auch für Begleitpersonen)
- Vergünstigungen bei kulturellen Angeboten usw.
- Außerdem gibt es Besonderheiten beim Kündigungsschutz und Zusatzurlaub.

In § 4 SGB IX sind **Leistungen zur Teilhabe** festgeschrieben. Diese umfassen die notwendigen Sozialleistungen, um unabhängig von der Ursache der Behinderung

1. die Behinderung abzuwenden, zu beseitigen, zu mindern, ihre Verschlimmerung zu verhüten oder ihre Folgen zu mildern,
2. Einschränkungen der Erwerbsfähigkeit oder Pflegebedürftigkeit zu vermeiden, zu überwinden, zu mindern oder eine Verschlimmerung zu verhüten sowie den vorzeitigen Bezug anderer Sozialleistungen zu vermeiden oder laufende Sozialleistungen zu mindern,
3. die Teilhabe am Arbeitsleben entsprechend den Neigungen und Fähigkeiten dauerhaft zu sichern und
4. die persönliche Entwicklung ganzheitlich zu fördern und die Teilhabe am Leben in der Gesellschaft sowie eine möglichst selbstständige und selbstbestimmte Lebensführung zu ermöglichen und zu erleichtern.
5. Leistungen für behinderte oder von Behinderung bedrohte Kinder werden so geplant und gestaltet, dass nach Möglichkeit Kinder nicht von ihrem sozialen Umfeld getrennt und gemeinsam mit nichtbehinderten Kindern betreut werden können. Dabei werden behinderte Kinder alters- und entwicklungsentsprechend an der Planung und Ausgestaltung der einzelnen Hilfen beteiligt und ihre Sorgeberechtigten intensiv in Planung und Gestaltung der Hilfen einbezogen.

In § 6 werden die **Rehabilitationsträger** zusammengefasst. Diese können sein:

1. Gesetzlichen Krankenkassen
2. Bundesagentur für Arbeit
3. Träger der gesetzlichen Unfallversicherung
4. Träger der gesetzlichen Rentenversicherung
5. Träger der Kriegsopferversorgung und die Träger der Kriegsopferfürsorge
6. Träger der öffentlichen Jugendhilfe
7. Träger der Sozialhilfe.

In § 30 SGB IX sind **Früherkennung und Frühförderung** festgelegt:

1. Die medizinischen Leistungen zur Früherkennung und Frühförderung behinderter und von Behinderung bedrohter Kinder umfassen nach § 26, Absatz 2 auch:
   a. die medizinischen Leistungen der mit dieser Zielsetzung fachübergreifend arbeitenden Dienste und Einrichtungen,
   b. nicht-ärztliche, sozialpädiatrische, psychologische, heilpädagogische, psycho-soziale Leistungen und die Beratung der Erziehungsberechtigten, auch in fachübergreifend arbeitenden Diensten und Einrichtungen, wenn sie unter ärztlicher Verantwortung erbracht werden und erforderlich sind, um eine drohende oder bereits eingetretene Behinderung zum frühestmöglichen Zeitpunkt zu erkennen und einen individuellen Behandlungsplan aufzustellen. Leistungen nach Absatz 1 werden als **Komplexleistung** in Verbindung mit heilpädagogischen Leistungen (§ 56) erbracht.
2. Leistungen zur Früherkennung und Frühförderung behinderter und von Behinderung bedrohter Kinder umfassen des Weiteren nicht-ärztliche, therapeutische, psychologische, heilpädagogische, sonderpädagogische, psychosoziale Leistungen und die Beratung der Erziehungsberechtigten durch interdisziplinäre Frühförderstellen, wenn sie erforderlich sind, um eine drohende oder bereits eingetretene Behinderung zum frühestmöglichen Zeitpunkt zu erkennen oder die Behinderung durch gezielte Förder- und Behandlungsmaßnahmen auszugleichen oder zu mildern.

Nähere Verordnungen zur Komplexleistung werden im jeweiligen Landsrecht vorgenommen.

Heilpädagogische Leistungen sind nach § 56 SGB IX dann angebracht, wenn auf Grund fachlicher Erkenntnis zu erwarten ist, dass hierdurch

1. eine drohende Behinderung abgewendet oder der fortschreitende Verlauf einer Behinderung verlangsamt oder
2. die Folgen einer Behinderung beseitigt oder gemildert werden können.
3. Schwerbehinderte und schwerst mehrfachbehinderte Kinder, die noch nicht eingeschult sind, haben immer ein Recht auf heilpädagogische Leistungen.

In Verbindung mit Leistungen zur Früherkennung und Frühförderung (§ 30) und bei schulvorbereitenden Maßnahmen der Schulträger werden heilpädagogische Leistungen als Komplexleistungen erbracht.

42.2.4 SGB XI – Soziale Pflegeversicherung

Zur sozialen Absicherung des Risikos der Pflegebedürftigkeit ist als neuer eigenständiger Zweig der Sozialversicherung eine soziale Pflegeversicherung geschaffen worden. In den Schutz der sozialen Pflegeversicherung sind Kraft Gesetz alle einbezogen, die in der gesetzlichen Krankenversicherung versichert sind. Wer gegen Krankheit bei einem privaten Krankenversicherungsunter-

nehmen versichert ist, muss eine private Pflegeversicherung abschließen. Die Pflegeversicherung hat die Aufgabe, Pflegebedürftigen Hilfe zu leisten, die wegen der Schwere der Pflegebedürftigkeit auf solidarische Unterstützung angewiesen sind.
In den Richtlinien für Pflegeversicherung vom 7.11.1994 werden die Merkmale der Pflegebedürftigkeit, verschiedene Pflegestufen und Verfahren zur Feststellung der Pflegebedürftigkeit festgelegt. Sie sind für die Pflegekassen und den Medizinischen Dienst der Krankenkassen verbindlich.
Das Pflegeversicherungsgesetz gilt für alle pflegebedürftigen Menschen ab der Geburt. Es regelt überwiegend die Gewährung von Pflegesachleistungen und Pflegegeld, und dies unabhängig von Einkommen und Vermögen.
Pflegebedürftigkeit ist kein unveränderbarer Zustand, sondern ein Prozess, der durch präventive, therapeutische bzw. rehabilitative Maßnahmen und durch aktivierende Pflege beeinflussbar ist. Ziel ist es, trotz des Hilfsbedarfs eine möglichst weitgehende Selbstständigkeit des Pflegebedürftigen im täglichen Leben zu fördern, sie zu erhalten und wieder herzustellen. Dazu gehören auch Verbesserungen in der Kommunikation sowie ein besseres Zurechtfinden von geistig und seelisch behinderten, psychisch kranken und geistig verwirrten Menschen.
Nach § 14 SGB XI sind Personen pflegebedürftig, die wegen einer körperlichen, geistigen oder seelischen Krankheit oder Behinderung für die gewöhnlichen und regelmäßig wiederkehrenden Verrichtungen im Ablauf des täglichen Lebens auf Dauer, voraussichtlich aber für mindestens 6 Monate, in erheblichem Maße der Hilfe bedürfen. Die Krankheiten oder Behinderungen in diesem Sinne sind:

1. Verluste, Lähmungen und andere Funktionsstörungen am Stütz- und Bewegungsapparat,
2. Funktionsstörungen der inneren Organe und Sinnesorgane,
3. Störungen des zentralen Nervensystems sowie Antriebs-, Gedächtnis- und Orientierungsstörungen sowie endogene Psychosen, Neurosen und geistige Behinderungen.

Maßgeblich für die Beurteilung der Pflegebedürftigkeit ist ausschließlich die Einschränkung der Fähigkeit, bestimmte Verrichtungen des täglichen Lebens auszuüben und nicht Art und Schwere einer vorliegenden Erkrankung, z. B. die Diagnose einer bösartigen Erkrankung oder einer Behinderung wie Taubheit und Blindheit. Entscheidungen in einem anderen Sozialleistungsbereich über das Vorliegen einer Behinderung oder Gewährung einer Rente haben keine bindende Wirkung für die Pflegekasse und sagen nichts über das Vorliegen von Pflegebedürftigkeit aus. Ein Hilfebedarf kann nicht allein deshalb verneint werden, weil sich der Pflegebedürftige tagsüber außerhalb der Wohnung aufhält. Es werden **Grundpflege, Behandlungspflege und hauswirtschaftliche Versorgung** unterschieden. Pflegebedürftigkeit ist auch dann gegeben, wenn der Pflegebedürftige die Verrichtungen zwar motorisch ausüben, jedoch deren Notwendigkeit nicht erkennen bzw. diese Einsicht nicht in sinnvolles, zweckgerichtetes Handeln umsetzen kann, z. B. bei geistiger Behinderung, verminderter Orientierung, Nichterkennen von Personen usw.
Regelmäßig wiederkehrende Verrichtungen im Ablauf des täglichen Lebens sind z. B.:

- Im Bereich der Körperpflege Waschen, Duschen, Zahnpflege, Kämmen, Rasieren, Darm- und Blasenentleerung
- Im Bereich der Ernährung das mundgerechte Zubereiten und die Aufnahme der Nahrung
- Im Bereich der Mobilität das Aufstehen und Zubettgehen, das An- und Auskleiden, Gehen, Treppensteigen, Verlassen und Wiederaufsuchen der Wohnung
- Im Bereich der hauswirtschaftlichen Versorgung das Einkaufen, Kochen, Reinigen der Wohnung, Spülen, Wechseln und Waschen der Kleidung und das Heizen.

Ziel der Pflegehilfe ist die möglichst eigenständige Übernahme der Verrichtungen durch die pflegebedürftige Person, wobei das häusliche und soziale Umfeld zu berücksichtigen ist. Zur Unterstützung gehören auch die Anleitung der richtigen Nutzung von Hilfsmitteln sowie Pflege unter-

stützende Maßnahmen wie z. B. das Abklopfen bei an Mukoviszidose erkrankten Kindern. Maßnahmen der Krankenbehandlung und der medizinischen Rehabilitation oder der Behandlungspflege können bei der Feststellung des Pflegebedarfs nicht berücksichtigt werden.
Beaufsichtigung und Anleitung kommen insbesondere bei geistig und seelisch Behinderten in Betracht und richten sich darauf,

- körperliche, psychische und geistige Fähigkeiten zu fördern und zu erhalten,
- Eigen- und Fremdgefährdungen zu vermeiden z. B. durch unsachgemäßen Umgang mit Strom, Waffen und offenem Feuer sowie
- Ängste, Reizbarkeit und Aggressionen abzubauen.

Kriterien für die Zuordnung zu einer Pflegestufe sind neben den genannten Punkten die Häufigkeit des Hilfsbedarfs und ein zeitlicher Mindestaufwand. Folgende Pflegestufen werden unterschieden:
Pflegestufe I = erhebliche Pflegebedürftigkeit. Diese liegt vor, wenn mindestens einmal täglich Hilfebedarf bei mindestens zwei Verrichtungen aus einem oder mehreren Bereichen besteht. Der Zeitaufwand muss im Durchschnitt mindestens 1½ Stunden betragen.
Pflegestufe II = schwere Pflegebedürftigkeit. Hierbei müssen mindestens dreimal täglich zu verschiedenen Zeiten Pflegehilfe bei der Körperpflege, der Ernährung und der Mobilität benötigt werden. Der Aufwand muss im Tagesdurchschnitt mindestens 3 Stunden betragen.
Pflegestufe III = Schwerstpflegebedürftigkeit. Sie liegt vor, wenn jederzeit bei Tag und Nacht konkreter Hilfsbedarf anfallen kann und der Tagesdurchschnitt mindestens 5 Stunden Pflegeaufwand beträgt.
Seit dem 1.7.2008 besteht für Menschen mit erheblichen Einschränkungen ihrer Alltagskompetenz, die aber noch nicht die Bedingungen erfüllen, um in die Pflegestufe I aufgenommen zu werden, der Anspruch auf einen Betreuungsbetrag **(Pflegestufe 0).** Dies dürfte bei Kindern und Jugendlichen aber nur in Ausnahmefällen zutreffen.
Seit 2013 können Personen, die aufgrund demenzbedingter Fähigkeitsstörungen, geistigen Behinderungen oder psychischen Erkrankungen einen erheblichen allgemeinen Betreuungsbedarf haben, zusätzlich bis zu 200,- € monatlich für Betreuungsangebote in Anspruch nehmen, wenn sie sich nicht dauerhaft in einer stationären Einrichtung befinden.
Pflegebedürftige Kinder sind zur Feststellung des Hilfebedarfs mit einem gesunden Kind gleichen Alters zu vergleichen. Maßgebend für die Beurteilung des Hilfebedarfs bei einem Säugling oder Kleinkind ist nicht der natürliche, altersbedingte Pflegeaufwand, sondern nur der darüber hinausgehende Aufwand. Bei kranken und behinderten Kindern ist also der zusätzliche Hilfsbedarf zu berücksichtigen, der sich z. B. als Folge einer angeborenen Erkrankung, einer intensivmedizinischen Behandlung oder einer Operation im Bereich der Körperpflege, der Ernährung und der Mobilität ergibt und der u. a. in häufigeren Mahlzeiten oder zusätzlicher Körperpflege bzw. Lagerungsmaßnahmen bestehen kann.
Im ersten Lebensjahr liegt die Pflegebedürftigkeit im Sinne des Gesetzes nur ausnahmsweise vor, die Feststellung bedarf einer besonderen Begründung.
Die Leistungen bei Pflegebedürftigkeit sind bei der Pflegekasse zu beantragen. Diese trifft die Entscheidung unter maßgeblicher Berücksichtigung des Gutachtens des Medizinischen Dienstes der Krankenversicherung, aber auch der Aussagen von behandelnden Ärzten, insbesondere Hausärzten, und den Angaben verschiedener Personen, die den Betroffenen pflegen, ggf. auch externer Sachverständiger. In der Regel ist für die Überprüfung ein angekündigter Hausbesuch notwendig, dies gilt insbesondere auch im Widerspruchsverfahren.
Die Umsetzung des Pflegeversicherungsgesetzes stößt besonders bei der Beurteilung der Pflegebedürftigkeit von geistig behinderten Kindern und Jugendlichen, aber auch bei der Durchführung

spezieller Behandlungen, z. B. von Diäten oder einer PEG-Ernährung, auf Probleme, während die Beurteilung von Kindern mit Körper- oder Sinnesbehinderungen offensichtlich recht einheitlich möglich ist.
Es wird versucht, die Kriterien der altersbezogenen Pflegebedürftigkeit bzw. der Alltagsbewältigung in standardisierten Fragebögen, z. B. PEDI oder WeeFIM, zu erfassen. Darüber hinaus bemühen sich verschiedene Arbeitsgruppen, objektive Kriterien für die Beurteilung der Lebensqualität zu finden (z. B. KINDL-R und ICF-CY) (➤ Kap.16, ➤ Kap. 17).
Über die Pflegeversicherung hinaus müssen bei der Pflege und Förderung vor allem schwer Mehrfachbehinderter weiterhin auch die Sozialhilfeträger zur Finanzierung herangezogen werden.

42.2.5 SGB XII – Sozialhilfe

Aufgabe der **Sozialhilfe** ist es, dem Leistungsberechtigten die Führung eines Lebens zu ermöglichen, das der Würde des Menschen entspricht. Die Leistung soll sie so weit wie möglich befähigen, unabhängig von ihr zu leben; darauf haben auch die Leistungsberechtigten nach ihren Kräften hinzuarbeiten. Zur Erreichung dieser Ziele haben die Leistungsberechtigten und die Träger der Sozialhilfe im Rahmen ihrer Rechte und Pflichten zusammenzuwirken. Die Sozialhilfe kann als Geld- oder andere Dienstleistung erbracht werden. Alle Leistungen werden nur nach dem Maßstab der Bedürftigkeit erbracht.
Sozialhilfe erhält nicht, wer sich vor allem durch Einsatz seiner Arbeitskraft, seines Einkommens und seines Vermögens selbst helfen kann und wer die erforderliche Leistung von anderen, insbesondere von Angehörigen oder Träger anderer Sozialleistungen erhält (Subsidiaritätsprinzip).
§ 53 SGB XII regelt die **Eingliederungshilfe** als wesentliche Maßnahme der Sozialhilfe.

1. Personen, die durch eine Behinderung … wesentlich in ihrer Fähigkeit, an der Gesellschaft teilzuhaben, eingeschränkt oder von einer solchen wesentlichen Behinderung bedroht sind, erhalten Leistungen der Eingliederungshilfe, wenn und so lange nach der Besonderheit des Einzelfalls, insbesondere nach Art oder Schwere der Behinderung, Aussicht besteht, dass die Aufgabe der Eingliederungshilfe erfüllt werden kann. Personen mit einer anderen körperlichen, geistigen oder seelischen Behinderung können Leistungen der Eingliederungshilfe erhalten.
2. Von einer Behinderung bedroht sind Personen, bei denen der Eintritt der Behinderung nach fachlicher Erkenntnis mit hoher Wahrscheinlichkeit zu erwarten ist.
3. Besondere Aufgabe der Eingliederungshilfe ist es, eine drohende Behinderung zu verhüten oder eine Behinderung oder deren Folge zu beseitigen oder zu mildern und die behinderten Menschen in die Gesellschaft einzugliedern. Hierzu gehört insbesondere, den behinderten Menschen die Teilnahme am Leben in der Gemeinschaft zu ermöglichen oder zu erleichtern, ihnen die Ausübung eines angemessenen Berufs oder einer sonstigen angemessenen Tätigkeit zu ermöglichen oder sie so weit wie möglich unabhängig von Pflege zu machen.

Zu den Eingliederungshilfen gehören:

- Ambulante und stationäre ärztliche Behandlungsmaßnahmen
- Versorgung mit orthopädischen Hilfsmitteln
- Heilpädagogische Maßnahmen für noch nicht schulpflichtige Kinder
- Hilfen zur angemessenen Schulbildung
- Berufliche Ausbildung, Fortbildung und Umschulung
- Andere berufliche Eingliederungshilfen, z. B. beschützende Werkstätten
- Beihilfe.

Hierzu können auch Gewährungen spezieller Maßnahmen wie z. B. die Bereitstellung eines Schulbegleiters bei einem Kind mit Autismus gehören.
Eine zusätzliche Unterstützung kann in Form des Blindengeldes nach dem Landesrecht der Blindengesetze der Bundesländer in der Regel über die zuständigen Versorgungsämter (Amt für soziale Angelegenheiten) beantragt werden.
Bei besonderen familiären Belastungen kann u. a. auch die Landesstiftung „Mutter und Kind" unterstützen und z. B. durch die Familienkasse des Versorgungsamtes Zuschüsse gewähren, wenn frühzeitig auf die Notlage hingewiesen wurde.
Im Opferentschädigungsgesetz werden Regelungen für Personen getroffen, bei denen es ohne eigene Schuld zu bleibenden Gesundheitsschäden gekommen ist, z. B. bei Kindern, die von ihren Eltern schwer körperlich misshandelt wurden. Hier können unterschiedliche finanzielle Unterstützungen beim zuständigen Versorgungsamt beantragt werden.

42.3 Außerfamiliäre Betreuung von Kindern

Prinzipiell wird immer versucht, in den ersten drei Lebensjahren die Versorgung und Betreuung eines Kindes innerhalb der Familie, möglichst durch die Mutter, zu gewährleisten. Dies wird einerseits durch Verlängerung des **Mutterschutzurlaubs,** durch die Gewährung von **Elterngeld** und **Erziehungsurlaub** gewährleistet. In zunehmendem Maße wird vor allem nach Ende des ersten Lebensjahres die Betreuung der Kinder in **Kinderkrippen** staatlich gefördert, um einerseits gerade den Müttern den Wiedereinstieg in die Berufstätigkeit zu ermöglichen, um andererseits aber auch in den Fällen, in denen keine adäquate Betreuung der Kinder zu Hause gewährleistet ist, eine bessere Förderung zu ermöglichen. Ab dem 1.8.2013 haben alleKinder ab dem vollendeten 1. Lebensjahr einen **Rechtsanspruch auf einen Krippenplatz,** nach dem 3. Geburtstag besteht für alle Kinder ein Rechtsanspruch auf einen Kindergartenplatz (Kinderförderungsgesetz). 2013 wird für die Kinder, die nach dem 1. Lebensjahr zu Hause betreut werden, **Betreuungsgeld** bezahlt. Die genaue Regelung der Bereitstellung von Plätzen in Kindertagesstätten ist in den Bundesländern unterschiedlich, in einigen Bundesländern ist zumindest das letzte Kindergartenjahr vor der Einschulung kostenfrei. Die Diskussion innerhalb der Gesellschaft über einen weiteren Ausbau der Kindertagesstätten und insbesondere die Überwachung ihrer Qualität ist noch nicht beendet.
Bereits 1717 wurde in Preußen die **Schulpflicht** für alle Kinder eingeführt, in Bayern erst 1802. Diese beginnt derzeit für alle Kinder, die bis zum 30.9. eines Jahres 6 Jahre alt werden. Seit 2009 haben alle Kinder im Sinne der Inklusion einen Rechtsanspruch auf einen Platz in einer Regelschule. Die praktische Umsetzung dieses Rechtes wird in den Bundesländern unterschiedlich gehandhabt und führt zu Diskussionen über die Notwendigkeit des Fortbestandes von Förderschulen.

42.4 Weitere Rechtsverordnungen

Das **Bürgerliche Gesetzbuch** (BGB) regelt in §§ 1626–1698 b Fragen der elterlichen Sorge. In § 1626 wird festgelegt, dass Eltern das Recht und die Pflicht haben, für die Person und das Vermögen des Kindes zu sorgen. § 1626a regelt die elterliche Sorge nicht miteinander verheirateter Eltern, §§ 1671 und 1687 die heute übliche gemeinsame elterliche Sorge getrennt lebender Eltern. In

§ 1631 Abs. 2 ist das Recht der Kinder auf eine gewaltfreie Erziehung und das Verbot für körperliche Bestrafung, seelische Verletzungen und andere entwürdigende Maßnahmen durch Eltern bzw. Sorgeberechtigte festgelegt. In § 1666 sind die familiengerichtlichen Maßnahmen bei Kindeswohlgefährdung geregelt. § 1684 betrifft das Umgangsrecht des Kindes mit den Eltern, § 1685 das mit anderen Bezugspersonen. Die Pflichten eines Betreuers sind in § 1901, die Voraussetzungen für die Sterilisation eines nichteinwilligungsfähigen Betreuten im § 1905 festgelegt.

Eine Leihmutterschaft ist in Deutschland verboten, ebenso eine Eizellspende mit In-vitro-Fertilisation. Die Möglichkeit der Adoption durch homosexuelle Paare wird zur Zeit diskutiert.

Das **Strafgesetzbuch** (StGB) behandelt u. a. in den §§ 170 und 171 die Verletzung der Unterhaltspflicht sowie der Fürsorge- und Erziehungspflicht und in den §§ 176, 176 a und b den sexuellen Missbrauch an Kindern.

Das **Infektionsschutzgesetz** dient der Prävention und Eindämmung übertragbarer Erkrankungen. § 6 behandelt die meldepflichtigen Erkrankungen.

In den **Kindergarten- und Schulgesetzen** der Bundesländer sind u. a. die Rechte und Pflichten der Schüler und der Eltern (z. B. bei der Schulwahl) sowie die sonderpädagogische Förderung in Sonderschulen und allgemeinen Schulen (Integrationsmaßnahmen) geregelt.

Leistungsansprüche können u. a. aus den Steuergesetzen, dem Zuwanderungsgesetz, dem Asylbewerber-Leistungsgesetz, dem Wohnungsbauförderungsgesetz und der Kraftfahrzeughilfeverordnung abgeleitet werden.

Jungen Erwachsenen ab 18 Jahren, die aus medizinischen Gründen auf Dauer voll erwerbsgemindert sind, steht laut Bundesgesetz seit 2003 eine **bedarfsorientierte Grundsicherung** zu.

Landesspezifische Regelungen gelten für das **Blindengeld.** Außerdem gibt es besondere Regelungen für das **Kindergeld** bei behinderten Kindern.

Elterngeld ist eine Leistung des Bundes für ab 1.1.2007 geborene Kinder. Die zuständigen Stellen für die Beantragung sind in den Bundesländern festgelegt. Das Betreuungsgeld soll ab 1.8.2013 an die Eltern ausgezahlt werden, deren unter 3jähriges Kind keine öffentliche Kindertagesstätte besucht.

Das 2012 verabschiedete Patientenrechtegesetz soll vor allem die Rechte der Patienten in haftungsrechtlich relevanten Fragen klären.

Seit Anfang 2012 gilt auch das neue Bundeskinderschutzgesetz, das vor allem Kleinkinder vor Vernachlässigung, Verwahrlosung, Gewalt und Missbrauch schützen soll. In ihm werden die Frühen Hilfen, die Aufgaben der Familienhebammen und anderer ambulanter Dienste sowie die Rolle der Jugendämter klarer als bisher definiert (➤ Kap. 10.4; ➤ Kap. 36).

Die Beratungen zu einem bundeseinheitlichen **Präventionsgesetz** sind noch nicht abgeschlossen.

Viele weitere rechtliche Fragen müssen bei der Betreuung von Kindern, insbesondere bei Kindern mit Entwicklungsauffälligkeiten und Behinderungen, berücksichtigt werden. Beispiele hierfür sind:

- Kinderfreibetrag
- Wohngeld
- Erziehungsurlaub
- Pflegegesetze
- Adoptionsgesetze
- Kindschaftsrecht
- Mutterschutz
- Scheidungsrecht
- Opferentschädigungsrecht
- Jugendstrafrecht

- Schweigepflicht
- §34 StGB – rechtfertigender Notstand
- Sterbehilfe u. v. m.

42.5 UN-Kinderrechtskonvention

Am 20. November 1989 wurde von der Vollversammlung der Vereinten Nationen die Kinderkonvention als eine Zusammenstellung der Grundwerte im Umgang mit Kindern zu ihrem Schutz und ihrer gesellschaftlichen Beteiligung verabschiedet. In 54 Artikeln werde n vier Grundprinzipien abgehandelt. Hierauf wird im Kapitel 46 (➤ Kap. 46) näher eingegangen.

Literatur

Fegert JM (2000) Sozialpädiatrisch relevante gesetzliche Bestimmungen und Begutachtung. In: Schlack HG. Sozialpädiatrie. München: Urban & Fischer: 365–400.

Jung K, Preuss B (1992) Rechtsgrundlagen der Rehabilitation – Sammlung des gesamten Rehabilitationsrechts. Starnberg: Verlag RS Schultz.

Straßburg HM, Dacheneder W, Kreß W (2012) Eintwicklungsstörungen bei Kindern. Praxisleitfaden für die interdisziplinäre Betreuung. München: Elsevier

Internetlinks

www.bundesaerztekammer.de/downloads/BKiSchG_22.12.2011_final.pdf
www.ajfp.bs.ch/uno-kinderrechtskonvention.pdf
www.buergerliches-gesetzbuch.info
www.gesetze-im-internet.de/bgb/index
www.sozialgesetzbuch-bundessozialhilfegesetz.de

43 Ethik in der Sozialpädiatrie

Hans-Michael Straßburg

43.1 Einleitung

Eine besondere Verantwortung der Gesellschaft und speziell für alle, die im medizinischen System arbeiten, besteht bei ethischen Themen, insbesondere bei chronischer Krankheit und Behinderung. Ausgehend von den Nürnberger Prozessen nach dem Ende des nationalsozialistischen Terrorregimes und der **Deklaration von Helsinki** von 1968 wurden die ethischen Grundwerte in der Medizin wiederholt neu formuliert, zuletzt in der Deklaration von Seoul 2008.

Demnach werden als Grundlagen einer medizinischen Ethik folgende Prinzipien anerkannt:

- Respekt vor der Autonomie des Patienten
- Vermeidung von Schaden
- Hilfeleistung
- Gerechtigkeit.

Umstritten ist, inwieweit gesellschaftliche Interessen, z. B. sozio-ökonomischer Art, einen Stellenwert in einer medizinischen Ethik haben können.

Die Deutsche Gesellschaft für Muskelkranke hat Ende der 1990er Jahre einige **ethische Grundsätze** formuliert, die auch heute noch ihre Gültigkeit haben:

- Das Recht auf Leben für einen Menschen bedarf keiner Rechtfertigung.
- Eine Diskussion über einen „Lebensunwert" bedroht alle Behinderten.
- Eine Behinderung bzw. eine Krankheit darf nicht mit „Leid" gleichgesetzt werden.
- Kranke und Behinderte müssen ihren gleichberechtigten Platz in der Gesellschaft haben.
- Forschung und Beratung zu erblichen Erkrankungen ist mehr als eine Methode zur Vermeidung derselben.
- Jede Forschung zur Erklärung von Krankheiten und Behinderungen sollte zunehmend als Therapieforschung verstanden werden.
- Und schließlich: Die Erfahrung der Zeit des Nationalsozialismus darf nie vergessen werden.

Die heute gültigen ethischen Grundsätze sind auch in der Deklaration von Edinburgh und der Erklärung zur guten wissenschaftlichen Praxis (DCP) durch die Deutsche Forschungsgemeinschaft sowie in der Bioethikkonvention aus dem Jahr 2000 festgelegt.

Sozialpädiatrie in der Praxis. http://dx.doi.org/10.1016/B978-3-437-31630-3.00043-2
Copyright © 2014 Elsevier GmbH. All rights reserved.

43.2 Praktische Beispiele

Im Bereich der **Pränataldiagnostik** von angeborenen Behinderungen gibt es eine Vielzahl grundsätzlicher ethischer Probleme, z. B. bei der Frage, inwieweit Anlagestörungen des Gehirns oder des Rückenmarks ein ausreichender Grund für die Beendigung der Schwangerschaft sind. Ab welchem Schweregrad ist dies gerechtfertigt? Wer sollte die Beratung der Eltern hierzu durchführen? Sind evtl. schon leichte äußere Veränderungen wie eine Lippenspalte ein Grund, das Kind nicht zu akzeptieren, vielleicht sogar die Schwangerschaft zu beenden?
2007 wurde vom Obersten Gerichtshof in Österreich Eltern im Prozess gegen eine Frauenärztin Recht gegeben, weil diese in der 20. Schwangerschaftswoche Auffälligkeiten des Kopfes und der Wirbelsäule des Feten im Ultraschall nicht erkannt hatte. Nicht nur die Kosten des aus der Behinderung resultierenden Mehraufwands müssen lebenslang übernommen werden, sondern auch die Kosten für „sämtliche künftigen Aufwendungen, Pflegeleistungen und sonstige Vermögensnachteile, die mit der Obsorge, Pflege und Erziehung des Kindes im Zusammenhang stehen" (Kerbl 2008).
Wie bereits in früheren Prozessen wurde damit die Existenz eines Kindes mit einer Behinderung als „Schaden" eingestuft. Hiergegen haben die Gesellschaften für Kinder- und Jugendmedizin in Deutschland und Österreich mehrfach protestiert, unter anderem weil sie befürchten, dass so in Zukunft

- praktisch keine Pränataldiagnostik mehr durchgeführt werden kann,
- werdenden Eltern keine Befundinterpretationen mitgeteilt werden,
- stattdessen die Eltern nur die Mitteilung erhalten, dass eine Behinderung nicht ausgeschlossen werden kann und
- im Zweifelsfall immer zu einem Schwangerschaftsabbruch geraten werden muss.

Außerdem wird durch diese Rechtsprechung die Rolle von Menschen mit chronischen Krankheiten und Behinderungen in der Gesellschaft deutlich erschwert, wenn sich z. B. die Frage stellt, ob die Eltern für die nicht erfolgte Beendigung der Schwangerschaft entschädigt werden und der Mensch nur das lebende Resultat eines „ärztlichen Kunstfehlers" ist.
Nach heftigen Diskussionen in der Öffentlichkeit und den Fachgremien ist die **Präimplantationsdiagnostik** seit 2012 dann erlaubt, wenn mit einer hohen Wahrscheinlichkeit mit dem intrauterinen Tod oder einer schwersten Behinderung bei einem natürlich gezeugten Kind gerechnet werden muss. Noch ist aber offen, welche schwerwiegenden Diagnosen hierunter fallen (z. B. auch die Mukoviszidose?)
Einfache Geschlechtsbestimmungen oder prädiktive genetische Untersuchungen in der Frühphase der Schwangerschaft, die in Zukunft technisch immer einfacher im Blut der Mutter durchgeführt werden können, die künstliche Erzeugung von Geschwistern schwerstkranker Kinder zur Stammzelltransplantation oder Organentnahmen bei Neugeborenen mit Anenzephalie müssen auf Grund der aktuellen Bestimmungen in Deutschland abgelehnt werden. Eine operative Sterilisation darf nur mit Einwilligung darf nur mit glaubhafter Einwilligung des bzw. der Betroffenen nach Erreichung der Volljährigkeit vorgenommen werden. Auch die in den USA nach einem Gerichtsbeschluss vorgenommene operative Verstümmelung von schwer behinderten jungen Frauen zur Verhinderung der Geschlechtsreife bzw. zur „Pflegeerleichterung" wird in Deutschland abgelehnt..
Durch das **Gendiagnostikgesetz** von 2009 wurde in vielen Fällen Klarheit bei der genetischen Diagnostik geschaffen, die nur mit dem schriftlichen Einverständnis der Sorgeberechtigten im Rahmen einer differenzialdiagnostischen Untersuchung durchgeführt werden dürfen. Diese

Untersuchung kann nicht nur den Eltern Klarheit über die Ursache der Entwicklungsstörung geben, sondern auch eine differenzierte humangenetische Beratung anderer Familienmitglieder ermöglichen, z. B. bei weiterem Kinderwunsch der Mutter oder anderer weiblicher Familienmitglieder.

Nicht erlaubt sind hingegen bei nicht einwilligungsfähigen Kindern sog. **prädiktive Untersuchungen,** die zum aktuellen Zeitpunkt dem klinisch unauffälligen Untersuchten keinen Nutzen bringen, z. B. die molekulargenetische Untersuchung eines klinisch gesunden Kindes auf das Vorliegen einer sich erst im späteren Leben manifestierenden Erkrankung wie der Chorea Huntington, Morbus Alzheimer oder Brustkrebs – es sollte bis zur Mündigkeit das Recht auf Nichtwissen bestehen.

Aus diesem Grund ist auch die Mitteilung von Zusatzbefunden, die bei modernen molekulargenetischen Übersichtsverfahren festgestellt werden, nur in extremen Ausnahmen erlaubt.

Eine schwierige Frage gerade in der Kinder- und Jugendmedizin ist die **Einwilligung von Jugendlichen in medizinische Maßnahmen,** insbesondere bei folgenreichen und belastenden Eingriffen, z. B. in der Onkologie, aber auch bei komplexen Operationen. Zunehmend wird die Autonomie des minderjährigen Patienten auch bei der Behandlung chronischer Krankheiten und Behinderungen eine Rolle spielen, z. B. bei der Einleitung einer Langzeitbeatmung, der Therapie mit Stammzellen oder der Einführung neuer Therapieverfahren. Weitere äußerst schwierige ethische Probleme ergeben sich bei chronisch Kranken und Behinderten in der Intensivmedizin, z. B. bei Patienten im Koma oder bei der Entscheidung über **lebensverlängernde Maßnahmen** wie einer maschinellen Dauerbeatmung von Patienten mit Muskeldystrophie Duchenne oder der Anlage einer PEG-Sonde bei Patienten mit schwersten Hirnschädigungen.

Prinzipiell wird heute spätestens ab dem 14. Lebensjahr, wenn irgend möglich, die Einwilligung auch des betroffenen Jugendlichen, soweit er dazu in der Lage ist, neben der Einverständniserklärung der sorgeberechtigten Eltern eingefordert.

Ein weiteres schwieriges Problem besteht darin, dass viele Medikamente, gerade im Bereich der Intensivmedizin, der Onkologie und der Epileptologie für bestimmte Altersgruppen nicht bzw. noch nicht zugelassen sind. Hier muss eine Verordnung in Rahmen des **Off label use** erfolgen, wobei der verordnende Arzt das Risiko möglicher Nebenwirkungen mit den Sorgeberechtigten sehr differenziert im Vorfeld besprechen sollte. Es ist ein Anliegen der pädiatrischen Pharmakologie, die Verordnung gerade auch neuer wirksamer Medikamente bei Kindern und Jugendlichen unter strengen Auflagen mit der Verpflichtung zur genauen Dokumentation zu erleichtern.

43.3 Therapiestudien

Wissenschaftlich gesicherte, d. h. evidenzbasierte Aussagen zur Wirksamkeit therapeutischer Maßnahmen sind in aller Regel nur durch randomisierte, doppelblind durchgeführte Therapiestudien zu erhalten. Gerade auch hierbei muss neben der Einwilligung der Sorgeberechtigten möglichst frühzeitig auch die freiwillige Zustimmung der Kinder bzw. Jugendlichen angestrebt werden, wobei die gesetzlichen Bestimmungen des Datenschutzes, insbesondere auch das Recht auf Anonymisierung und die Zusicherung, durch Nichtteilnahme keine schlechtere Behandlung zu erhalten, gewährleistet sein müssen. In der Regel können solche Studien nur multizentrisch mit einem ausführlichen Studienprotokoll und einer Patientenversicherung durchgeführt werden. Einzelheiten hierzu sind im **Arzneimittelgesetz** vom August 2004 geregelt. Alle wissenschaftlichen Studien, insbesondere auch alle Therapiestudien müssen nach der Berufsordnung der deutschen Ärzte von einer unabhängigen **Ethikkommission,** die in der Regel an den Medizinischen Fakultäten und in den Landesärztekammern eingerichtet ist, geprüft werden. Dabei müssen auch die Strahlenschutzverordnungen, die Verwaltungs- und Verfahrensgesetze sowie andere rechtliche Bestimmungen berücksichtigt werden.

43.4 Ethische Grundsätze am Ende des Lebens

Gerade aufgrund der leidvollen Erfahrungen in der Zeit des Nationalsozialismus, während der ca. 200.000 behinderte, überwiegend psychisch kranke Menschen, darunter über 5.000 Kinder, systematisch ermordet wurden, werden ethische Fragen am Ende des Lebens in Deutschland besonders intensiv und empfindsam diskutiert. Dabei ist die große Bedeutung einer qualifizierten kinder- und jugendlichenspezifischen **Palliativmedizin** auch für Kinder mit schweren Mehrfachbehinderungen oder progredienten neurologischen Erkrankungen unbestritten und dieser Bereich wird zunehmend ausgebaut.
Die oft schwierige Entscheidung bei der Frage nach der Weiterbehandlung von schwerstbehinderten oder dauerhaft komatösen Patienten kann in einer Beratung durch ein unabhängiges **Ethikkomitee** zur Entlastung der behandelnden Pflegekräfte und Ärzte, aber auch der Angehörigen erörtert werden. Diagnosespezifische Lösungen in diesen Fragen gibt es nach der revidierten Fassung der **Einbecker Empfehlungen** nicht.

43.5 Prävention von Entwicklungsstörungen

Zur Vorbeugung bzw. Vermeidung von Entwicklungsstörungen kann eine Vielzahl völlig unterschiedlicher Maßnahmen eingesetzt werden, die immer aber auch unter ethischen Aspekten betrachtet werden müssen.
So spielt die **genetische Beratung** eines Paares mit Kinderwunsch dann eine wichtige Rolle, wenn ein Familienmitglied an einer speziellen erblichen Erkrankung leidet. Dabei können z. T. Prozentangaben über das bestehende Risiko gemacht werden, es können aber auch Chorionzotten- oder Amnionflüssigkeitsuntersuchungen bzw. frühe Ultraschalluntersuchungen durchgeführt werden. Seit 2012 ist auch in Deutschland unter bestimnmten Voraussetzungen die sogenannte **Präimplantationsdiagnostik,** die genetische Untersuchung auf eine definierte schwere Krankheit im Blastozystenstadium erlaubt.
In verschiedenen Kulturen hat eine genetische Beratung durchaus bindenden Charakter, z. B. bei Ashkenasi-Juden, in deren Familien Fälle von Tay-Sachs-Erkrankungen vorkommen, oder bei Familien, in denen eine Thalassaemia major bekannt ist.
Von der Ethik-Kommission der DAKJ wird gefordert, dass bei einer nach der 12. SSW festgestellten Fehlbildung des Feten vor der Durchführung eines Spätaborts eine obligatorische Beratung der Eltern sowohl durch einen Neonatologen als auch ein Mitglied eines Sozialpädiatrischen Zentrums stattzufinden hat.
Im Gegensatz zu anderen Ländern gibt es in Deutschland keine Substitution durch Nahrungsergänzungsstoffe in Grundnahrungsmitteln wie z. B. von Folsäure im Mehl (USA) zur Vermeidung eines Neuralrohrdefekts oder von Fluor im Trinkwasser (Schweiz).

Die Auswirkungen von Nikotin, Alkohol und Drogen auf die Entwicklung des Feten sind hinreichend bekannt, durch die Teilnahme an den Schwangerschaftsuntersuchungen, vor allem auch in der Endphase, können intrauterine Gedeihstörungen und Frühgeburtsbestrebungen oft so rechtzeitig erkannt werden, dass eine bleibende Schädigung des Kindes vermieden werden kann.

Auch wenn während der gesamten Schwangerschaft keinerlei Risikosymptome bestanden, können bei jeder **Geburt** akut schwerwiegende Komplikationen auftreten. Aus diesem Grund sollten Hausgeburten und Geburten in kleinen Entbindungseinrichtungen, in denen z. B. die Voraussetzungen für eine rasche Notsektion nicht gewährleistet sind, wenn irgend möglich nicht mehr stattfinden.

Nach der Geburt können durch die frühzeitige Erkennung und Behandlung von Erkrankungen, insbesondere bakteriellen Infektionen, bleibende Schäden am Gehirn vermieden werden. Die prophylaktische Gabe von Vitamin K und Vitamin D zur Vermeidung von Blutungen bzw. zur Rachitis-Prophylaxe ist allgemein anerkannt.

Unverändert werden selbst die allgemein empfohlenen Basisimpfungen von einem kleinen Teil der Bevölkerung überwiegend aus ideologischen Gründen abgelehnt. Wiederholte Forderungen nach einer Verpflichtung für Impfungen, u. a. für Mitarbeiter in Kindertagesstätten u. ä. haben sich bisher politisch nicht durchsetzen lassen.

Auch extrem Frühgeborene sollten die Impfungen mit ihrem chronologischen Alter, also beginnend mit der 9. Lebenswoche, erhalten.

Problematische Methoden zur Vermeidung von Entwicklungsstörungen sind u. a. die Embryonenselektion nach In-vitro-Fertilisation und der gezielte Fetozid bei Mehrlingsschwangerschaften.

Ethische Grundsätze bei der **Behandlung von extremen Frühgeborenen** sind in Grundsatzpapieren z. B. der Deutschen Akademie für Kinder- und Jugendmedizin zusammengestellt. Mit der Abnabelung ist das Kind eine juristisch selbstständige Person und damit ist der anwesende Arzt verpflichtet, primär alles, was zur Lebensrettung notwendig ist, durchzuführen. Eltern können einem Arzt bzw. einem Krankenhaus nicht verbieten, z. B. bei einer extremen Frühgeburtlichkeit keine lebenserhaltenen Maßnahmen für das Kind durchzuführen. Auch eine Leihmutterschaft bzw. eine Eizellspende sind in Deutschland nicht erlaubt.

43.6 Ausblick

Die im Jahr 2012 geführte öffentliche Diskussion zum Thema der Beschneidung bei Knaben aus religiösen Gründen hat die grundsätzliche Problematik zwischen den Persönlichkeitsrechten eines jeden Menschen ab der Geburt und traditionellen Maßnahmen, die diese nicht anerkennen, deutlich gemacht.

Insgesamt werden sich im Bereich der medizinischen Ethik in den nächsten Jahren viele zusätzliche, äußerst komplexe Probleme ergeben, die gesellschaftlich breit diskutiert und immer wieder neu überprüft werden müssen. Die Deutsche Akademie für Kinder- und Jugendmedizin stellt sich mit ihrer Kommission für ethische Fragen dieser Herausforderung.

Literatur

Hick C (2007) Klinische Ethik. Heidelberg: Springer

Hildt E (2012) Neuroethik. München: Reinhardt

Mitscherlich A, Mielke F (2004) Medizin ohne Menschlichkeit. Frankfurt: Fischer.

Straßburg HM, Dacheneder W, Kreß W (2012) Eintwicklungsstörungen bei Kindern. Praxisleitfaden für die interdisziplinäre Betreuung. München: Elsevier

Internetlinks

www.bundesaerztekammer.de/page.asp?his=1.100.1142

http://dakj.de/pages/aktivitaeten/stellungnahmen-und-empfehlungen/ethische-fragen.php

44 Begutachtung in der Sozialpädiatrie

Hans-Michael Straßburg

44.1 Einleitung

Gutachterliche Stellungnahmen sind in der Sozialpädiatrie von zunehmender Bedeutung. Verantwortlich hierfür ist die immer schwieriger werdende Differenzierung von juristischen und medizinischen Argumenten gerade auch bei Kindern und Jugendlichen mit chronischen Krankheiten und Behinderungen. In vielen Bereichen kann eine notwendige Festlegung nicht allein durch den Sachverstand einer einzelnen Person erreicht werden, oft werden sehr unterschiedliche Positionen von verschiedenen Interessensgruppen (z. B. spezialisierten Rechtsanwaltskanzleien) vertreten, deren Argumentation nur mit großem Sachverstand begegnet werden kann.
Sozialpädiatrische Gutachten können von vielen Institutionen in Auftrag gegeben werden, insbesondere von Zivil-, Sozial- und Strafgerichten, von Versorgungsämtern, dem Medizinischen Dienst der Krankenkassen, verschiedenen Versicherungen, insbesondere der Krankenversicherung, der Pflegeversicherung, der Rentenversicherung, Haftpflichtversicherungen, Behörden usw. Vor den gelegentlich von Privatpersonen oder Rechtsanwaltskanzleien angeforderten sogenannten „Parteiengutachten" wird in diesem Zusammenhang gewarnt, da diese in der Regel bei offiziellen Verfahren nur einen geringen Stellenwert haben.
Grundlage der Begutachtung können sehr unterschiedliche medizinische Fragestellungen und Rechtsprechungen sein. Hier sei u. a. auf das Grundgesetz Art. II, 2 hingewiesen: „Jeder hat das Recht auf Leben und körperliche Unversehrtheit", sowie auf das Strafgesetzbuch §§ 223 bis 231; hier steht u. a.: „Wer eine Körperverletzung mit Einwilligung der verletzten Person bzw. deren gesetzlichen Vertreters vornimmt, handelt nur dann rechtswidrig, wenn die Tat trotz der Einwilligung gegen die guten Sitten verstößt."
In kaum einem anderen Bereich der Medizin hat es eine stärkere Zunahme an Aufträgen gegeben wie im Bereich des medizinischen Gutachtenwesens. Man spricht auch von der „Verrechtlichung" der Gesellschaft. Es gibt hierfür zahlreiche Ursachen, u. a. ein höheres Rechtsbewusstsein der Bevölkerung im Sinne von mündigen Bürgern, eine Infragestellung hierarchischer Verwaltungsstrukturen, die Vereinfachung von Rechtsverfahren durch die Rechtschutzversicherungen und eine oft bestehende Erwartungshaltung des Patienten bzw. seiner gesetzlichen Vertreter. Darüber hinaus sind viele Bereiche von Gerichten und Versicherungen, vor allem aber beim Medizinischen Dienst der Krankenkassen in den vergangenen Jahren aufgebaut worden, die eine massive Zunahme von Begutachtungsfällen erklären.

Sozialpädiatrie in der Praxis. http://dx.doi.org/10.1016/B978-3-437-31630-3.00044-4
Copyright © 2014 Elsevier GmbH. All rights reserved.

Beispiele für überwiegend sozialpädiatrische Begutachtungen sind:
- Gutachtenaufträge von Gerichten in Straf- und Zivilverfahren, z. B. bei fraglichen Behandlungsfehlern
- Beurteilung der Einstufung für Pflegegeld bzw. der Festlegung von Pflegekosten, z. B. im Widerspruchsverfahren bei einem Gutachten des Medizinischen Dienstes
- Mitbegutachtung bei der Beantragung von Blindengeld oder Impffolgen
- Mitbegutachtung bei der Verordnung von Hilfsmitteln
- Begutachtung für Behörden, z. B. bei der Gewährung von Einzelintegration in Regeleinrichtungen (Krippe, Kindergarten, Schule) oder bei Fragen nach dem Jugendschutzgesetz
- Festlegung von Pflegestandards und ihre Kostenübernahme
- Beurteilung der Zuständigkeit einer Versicherung bei neuen Behandlungsverfahren usw.

Dazu kommen oft Aufträge zur Begutachtung bei der Ursachensuche einer Behinderung, wobei in der Regel geburtshilfliche, perinatologische, neuropädiatrische und neuroradiologische Gesichtspunkte ergänzt und ggf. zusammengefasst werden.

44.2 Durchführung eines Gutachtens

Unabhängig von der Fragestellung besteht die wesentliche Aufgabe bei einer Begutachtung darin, in einer auch für medizinische Laien klar verständlichen Sprache einen medizinischen Sachverhalt darzustellen. Wesentliche Voraussetzungen sind zum einen die strikte Neutralität des Gutachters und zum anderen die Fokussierung auf die Beantwortung der vom Auftraggeber gestellten Fragen. In der Regel werden zu Beginn eines freien Gutachtens die Fragestellungen nochmals wiederholt und dann die wesentlichen Dokumente bzw. unter Umständen auch die Ergebnisse von speziellen eigenen Untersuchungen am Patienten dargestellt. Die gesamten Befunde werden zusammenfassend bewertet und zum Abschluss des Gutachtens wird versucht, die gestellten Fragen eindeutig zu beantworten. Sofern sich der Gutachter bei seinen Aussagen sicher ist, sollte er den Begriff „mit an Sicherzeit grenzender Wahrscheinlichkeit" verwenden, ansonsten Begriffe wie „mit mehr als 50-prozentiger Wahrscheinlichkeit" oder „mit an Sicherheit grenzender Wahrscheinlichkeit auszuschließen". Bei haftungsrechtlichen Gutachten ist ein Behandlungsfehler, d. h. eine schuldhafte Verletzung der ärztlichen Sorgfaltspflicht, von einer mangelnden Sorgfaltspflicht zu unterscheiden.

Die Argumentation des Gutachters sollte sich an allgemein zugänglichen medizinischen Grundprinzipien, z. B. den Leitlinien der medizinischen Fachgesellschaften, orientieren, wobei unter Umständen auch der Kenntnisstand zum Zeitpunkt des im Gutachten zu bewertenden Geschehens berücksichtigt werden muss.

In der Regel ist die Erstellung eines Gutachtens eine Zusatzarbeit außerhalb der eigentlichen Dienstaufgaben und kann deshalb gesondert in Rechnung gestellt werden. Sofern keine Vorgaben vom Auftragsgeber gemacht werden, erfolgt dies nach dem Justizvergütungs- und -entschädigungsgesetz JVEG. Danach werden die Leistungen des Gutachters stundenweise abgerechnet, zusätzlich werden Schreibgebühren, Porto und Umsatzsteuer berücksichtigt. Bei Gerichtsgutachten muss der Gutachter häufig auf Anordnung zur mündlichen Verhandlung erscheinen, um seine schriftlich dargelegten Äußerungen nochmals zu erläutern. Es wird dringend geraten, dass jeder medizinische Gutacher eine private Berufshaftpflichtversicherung abgeschlossen hat.

44.3 Ärztliche Haftpflichtverfahren

Insgesamt wird davon ausgegangen, dass bei ca. 3 % aller Krankenhauspatienten sogenannte **„adverse events“** auftreten, das sind in Deutschland ca. 500.000 Fälle pro Jahr. Diese verteilen sich auf praktisch alle Fachgebiete, relativ am häufigsten auf chirurgische und geburtshilflich-gynäkologische Disziplinen. In Arzthaftpflichtverfahren wird in 70 % der Fälle von einem **Behandlungsfehler** ausgegangen, d. h. einer schuldhaften Verletzung der ärztlichen Sorgfaltspflicht, die schlechterdings nicht hätte passieren dürfen. In 20 % der Fälle bestehen Organisationsfehler, in 10 % Aufklärungsfehler. Sehr häufig lässt sich eine mangelnde Absprache bzw. Koordination zwischen ambulanter und stationärer Versorgung, ein Dokumentationsmangel, ein Übernahmeverschulden oder ein mangelhaftes Erkennen und Behandeln von Komplikationen nachweisen.
Beispiele für Behandlungsfehler seitens des Frauenarztes in der Perinatalzeit können sein:
- Mangelhafte Schwangerschaftsuntersuchung der Frau, z. B. fehlende Blutdruck- und Urinuntersuchungen
- Fehlerhafte Ableitung oder Interpretation eines Kardiotokogramms (CTG)
- Zu späte Verlegung der Schwangeren in ein Perinatalzentrum
- Fehlerhafte Behandlung von akuten Komplikationen unter der Geburt, z. B. Nabelschnurvorfall oder Schulterdystokie
- Zu lange Zeit zwischen Entschluss zur Notsektion und Entbindung (sogenannte EE-Zeit), die in jedem Fall kürzer als 25 Minuten sein muss.

Beispiele für Behandlungsfehler in der Neonatalzeit seitens des Kinderarztes können sein:
- Unzureichende Erstversorgung im Kreißsaal
- Unzureichende Aufrechterhaltung der Homöostase, z. B. der Körpertemperatur
- Unterlassung einer akuten Bluttransfusion
- Zu späte Erkennung von Komplikationen wie Sepsis, Meningitis, Schock, Pneumothorax oder zerebrale Anfälle
- Beatmungsfehler, z. B. Hyperventilation
- Vor allem mangelhafte Dokumentation!

Nicht nur das Verhalten der praktisch tätigen Ärzte oder Pflegerinnen und Pfleger, sondern auch die **Organisationsstruktur eines Krankenhauses** kann als Behandlungsfehler anerkannt werden, z. B.:
- Mangelhafte Räumlichkeiten, wenn der Fahrstuhl zwischen Kreißsaal und Sectio-OP defekt ist
- Wenn Erste-Hilfe-Geräte nicht bereitgestellt sind
- Wenn Hygienemaßnahmen nur unzureichend durchgeführt wurden
- Aber auch organisatorische Mängel bei der Erstellung des Dienstplans, z. B. fehlendes Ersatzpersonal oder zu wenig Personal, können als Ursache eines Behandlungsfehlers anerkannt werden
- Weitere wichtige Bereiche sind die mangelhafte Aufklärung vor einem medizinischen Eingriff, die fehlende schriftliche Einwilligung und die unzureichende Überwachung nach einem Eingriff, insbesondere wenn dieser ambulant durchgeführt wurde.

In Einzelfällen kann eine schier unübersehbare Fülle verschiedener Faktoren für die Entstehung von schweren, bleibenden Gesundheitsschäden gerade auch im Kindesalter verantwortlich sein, z. B. fehlerhafte Schutzgitter bei einem Kleinkindbett, Nichterfüllung von DIN-Normen an Gebäuden oder Mobiliar, Kissen, Schnüre und Decken in einem Bett, offene Fenster, mangelhafte Absicherung in einem Treppenhaus, fehlerhafte Fahrstuhlanlagen usw.
Gerade in einem Kinderkrankenhaus können durch Infektionen, z. B. Varizellen, Durchfallerreger, Herpes labialis, Staphylokokken, Aspergillen und vieles mehr, schwerwiegende Gesundheitsschäden bei Kindern verursacht werden. Ebenso kann es bei unzureichender Diagnostik von klini-

schen Symptomen zu Behandlungsfehlern kommen, wenn behandelbare Erkrankungen (Hirnabszess, Hirntumor, akute Appendizitis, Osteomyelitis, Schock und vieles mehr) nicht rechtzeitig erkannt werden.
Von besonderer Bedeutung ist die konsequente Analyse jedes akut verstorbenen Patienten, insbesondere auch im Kindesalter, wobei immer wieder an die Möglichkeit einer Kindesmisshandlung erinnert werden muss.

44.4 Zukünftige Entwicklung

Seit Jahren fordern verschiedene im Gesundheitswesen verantwortliche Institutionen mehr Offenheit im Umgang mit ärztlichen Behandlungsfehlern. Grundlagen hierfür sind vor allem:
- Ständig aktualisierte Aus- und Weiterbildung
- Ausreichend Zeit
- Rechtzeitiges Abgebenkönnen
- Sorgfältige, ehrliche Dokumentationen.

So sollten in jedem medizinischen Arbeitsbereich immer wieder auch Fehler offen diskutiert werden, wobei nicht nur nach Schuldigen gesucht werden darf, sondern vor allem eine gegenseitige Kontrolle, ein Qualitätsmanagement und die Stärkung der Eigenverantwortung aller Mitarbeiter angestrebt werden sollte.
Der **Medizinische Dienst der Krankenversicherung (MDK)** hat entsprechend § 275 des SGB V die Aufgabe, die gesetzlichen Kranken- und Pflegekassen so zu beraten, dass ihre Leistungen ausreichend, zweckmäßig und wirtschaftlich sind. Hierzu gehören u. a. Stellungnahmen zu Fragen von:
- Notwendigkeit, Art, Umfang und Dauer von Rehabilitationsleistungen
- Verordnung von Arznei-, Heil- und Hilfsmitteln
- Notwendigkeit und Dauer einer Krankenhausbehandlung
- Notwendigkeit und Dauer von häuslicher Krankenpflege.

So prüft der MDK das Vorliegen der Voraussetzungen für Pflegebedürftigkeit, gibt Empfehlungen für die Anerkennung einer Pflegestufe sowie Art und Umfang von Pflegeleistungen. Die Entscheidung über eine Leistung liegt aber stets bei den Kranken- und Pflegekassen.
Weiterhin beraten Mitarbeiter des MDK, die in der Regel Fachärzte mit einer Zusatzweiterbildung in Sozialmedizin sind, den Gemeinsamen Bundesausschuss (G-BA) bei der prinzipiellen Festlegung von Leistungen durch die Sozialversicherungen und erstellen vielfältige Gutachten, z. B. auch zu Fragen von Behandlungsfehlern.
Bei der sozialpädiatrischen Begutachtung, z. B. nach § 53 SGB XII, nach § 35 SGB VIII, bei Stellungnahmen zu Eingliederungshilfen, Beantragung von Reha-Maßnahmen, Hilfsmitteln, Blindengeld und Pflegebedürftigkeit usw. spielt die **Internationale Klassifikation der Funktionen im Kindes- und Jugendalter (ICF-CY)** eine immer größere Rolle (➤ Kap.16). Hierbei sind neben den medizinischen Diagnosen die Aktivität und Motivation des Betroffenen sowie seine Partizipation am Alltagsleben von entscheidender Bedeutung. Wesentliche Faktoren hierfür sind neben den objektiven, medizinisch relevanten Befunden auch die nicht immer einfach erfassbaren Ressourcen und seine Resilienz.
Diese individuellen Faktoren mögen im Einzelfall immer wieder Anlass geben, die Gesamtbeurteilung einer Begutachtung zu modifizieren. Hier ist neben der sozialmedizinischen die kinder- und jugendmedizinische Qualifikation in besonderem Maße gefordert und somit stellt die Begutachtung ein wesentliches Argument für die Einführung einer Zusatzweiterbildung Sozialpädiatrie dar.

Literatur

Fegert JM (2000) Sozialpädiatrisch relevante gesetzliche Bestimmungen und Begutachtung. In: Schlack HG (Hrsg) Sozialpädiatrie. München: Urban & Fischer: 365–400.

Straßburg HM (1999) Die Beurteilung perinataler Hirnschäden im neuropädiatrischen Gutachten. In: Bentele KHP, Kohlschütter A (Hrsg) (1999) Aktuelle Neuropädiatrie. Nürnberg: Novartis Pharma: 454–458.

Straßburg HM (2004) Schwere Schädigung durch mangelhafte Dokumentation? Postnatale Erstversorgung eines Frühgeborenen mit schwerer periventrikulärer Leukomalazie. Gynäkol Prax 28: 121–126.

Warnke A, Trott GE, Remschmidt H (Hrsg) (1997) Forensische Kinder- und Jugendpsychiatrie – ein Handbuch für Klinik und Praxis. Bern: Hans Huber.

45 Forschung in der Sozialpädiatrie

Peter Borusiak und Hans-Michael Straßburg

45.1 Einleitung

Unsere Gesellschaft steht vor einer Reihe Herausforderungen, die auch die Entwicklung von Kindern und Jugendlichen betreffen und entsprechende Fragen aufwerfen. Nur beispielhaft seien hier genannt:

- **Auswirkungen der neuen Medien auf die Entwicklung, das Lernen und das Verhalten der Kinder.** Unsere Gesellschaft hat sich in den letzten 10 Jahren drastisch durch die Einführung neuer Medien gewandelt. Das Kommunikationsverhalten und die Freizeitaktivitäten der Kinder und Jugendlichen ändern sich rapide. Welche Auswirkungen ergeben sich für die Entwicklung? Welche Konsequenzen müssen für die Bereiche Gesundheit, Lernen und Verhalten gezogen werden? Sind die bisherigen Beurteilungsmaßstäbe und Testverfahren noch zeitgemäß?
- **Zunehmende Kosten im Bereich der Heilmittelverordnungen.** Die Kosten für Ergotherapie, Logopädie und Physiotherapie im Kindesalter steigen an. Immer wieder wird eine „Medikalisierung der Pädagogik" vermutet bzw. unterstellt, d. h. es werde versucht, pädagogische Probleme mit medizinischen Therapien zu lösen. Stimmt das? Welche Therapieverfahren sind sinnvoll? Wie viel Therapie ist je nach Störungs- oder Krankheitsbild tatsächlich notwendig? Welche Ziele muss die Therapie verfolgen? Wann ist eine Therapie oder Behandlung gut?
- **Einfluss des kulturellen oder sog. Migrationshintergrundes.** Bestimmte Risikofaktoren konnten beim Studie Kinder- und Jugendlichen-Gesundheitssurvey (KiGGS) des Robert-Koch-Institutes Berlin klar herausgearbeitet werden. Gibt es weitere Faktoren und wie begegnen wir den Risikofaktoren? Erreichen wir mit unseren Angeboten alle Zielgruppen? Welche Hindernisfaktoren gibt es?
- **Frühe Hilfen.** Seit Jahrzehnten ist bekannt, dass ein möglichst früher Einsatz von unterstützenden Maßnahmen bei sozial schwachen Familien einen nachhaltigen positiven Effekt hat. Angesichts einer Reihe tragischer Fälle gibt es auch in Deutschland ein gestiegenes Bewusstsein und entsprechende Bemühungen der Politik, wobei sich vielerorts sehr unterschiedliche Pilotprojekte entwickelt haben. Welche Maßnahmen sind aber tatsächlich effektiv?
- **Lebensqualität.** Neben den rein medizinischen Ergebnissen steht zunehmend die Lebensqualität, gerade bei chronisch Kranken und behinderten Menschen im Vordergrund. Welche Maß-

Sozialpädiatrie in der Praxis. http://dx.doi.org/10.1016/B978-3-437-31630-3.00045-6
Copyright © 2014 Elsevier GmbH. All rights reserved.

nahmen wirken sich wie aus? Was beeinträchtigt die Lebensqualität tatsächlich, wie können wir diese verbessern?
- **Inklusion.** Wie wird diese umsetzbar sein? Wie wirkt sich gelebte Inklusion auf Menschen mit und ohne Krankheiten und Behinderungen aus? Welche Einschränkungen gibt es zu beachten? Welche zusätzlichen medizinischen Kenntnisse werden Einrichtungen, Behörden oder Pädagogen benötigen?

Die Sozialpädiatrie versteht sich als Querschnittswissenschaft, die sich mit der Entwicklung von Kindern und Jugendlichen und deren Störungen im Rahmen des familiären und sozialen Kontextes beschäftigt. In diesem Zusammenhang wird eine Fülle von Daten zur Kindesentwicklung erhoben – sei es in der kinder- und jugendärztlichen Praxis, im Sozialpädiatrischen Zentrum, in Rehabilitationskliniken oder im Bereich des öffentlichen Gesundheitswesens. Somit ist die Sozialpädiatrie prädestiniert, wertvolle Beiträge zu den oben genannten Fragen zu leisten. Allerdings ergibt sich eine Limitierung dadurch, dass die überwiegende Arbeit in der Sozialpädiatrie nicht an universitären oder anderen Forschungseinrichtungen erfolgt. Arztpraxen der Kinder- und Jugendmediziner sind für die Versorgung ihrer Patienten zuständig – eine universitäre Repräsentanz dieser „Basispädiatrie" gibt es kaum. Auch die Sozialpädiatrischen Zentren sind universitär mit bundesweit gerade zwei Lehrstühlen unterrepräsentiert. Ebenso gibt es im öffentlichen Gesundheitsdienst oder an den Rehabilitationskliniken wenige Ressourcen, die Vielzahl der erhobenen Daten auch auszuwerten.

45.2 Bisherige Aktivitäten und Ergebnisse

Eine Reihe wichtiger Informationen zur aktuellen Gesundheitssituation konnten aus der großen Studie des Robert-Koch-Institutes Berlin, dem Kinder- und Jugendlichen-Gesundheitssurvey (KiGGS) abgeleitet werden. In vielen weiteren Studien wurden darüber hinaus in den vergangenen Jahren wesentliche Grundlagen zur Beurteilung der Gesundheitssituation vor allem auch von Risikokindern in Deutschland gesammelt, z. B. in der oberbayerischen Perinatalstudie, der Hamburger Perinatalstudie, in Studien des Deutschen Jugendinstituts München usw.
- Um Forschungsprojekte im Bereich der Sozialpädiatrie in Deutschland bekannt zu machen und zu fördern, wurde 1997 von H. Bode eine jährlich stattfindende Tagung mit Workshopcharakter **„Forschung in der Sozialpädiatrie"** begründet. Hier wurden vor allem von Mitarbeitern Sozialpädiatrischer Zentren, Sozialpädiatrischen Arbeitsgruppen in Universitätskliniken und anderen aktuelle Forschungsthemen, die z. T. noch nicht abgeschlossen sind, vorgestellt. Beispiele für bisherige Themen sind die Evaluation diagnostischer Methoden im Bereich der verschiedenen Entwicklungsbereiche und Untersuchungen zu bestimmten definierten Krankheitsbildern wie Zerebralparesen, ADHS, Autismus, Epilepsie oder neuromuskulären Erkrankungen. Aber auch Forschungsprojekte zu Interaktionsstörungen, zur Auswirkung der Diagnosevermittlung oder bestimmter Untersuchungsmethoden auf die Krankheitsbewältigung werden vorgestellt.
- Die Deutsche Gesellschaft für Sozialpädiatrie und Jugendmedizin vergibt alle zwei Jahre den **Stefan-Engel-Preis** als höchste Auszeichnung für wissenschaftliche Studien. Außerdem wird bei jeder Jahrestagung seit 2007 ein Preis für die beste wissenschaftliche Arbeit im Bereich der **transkulturellen Sozialpädiatrie** vergeben. Die Deutsche Gesellschaft für Sozialpädiatrie und Jugendmedizin (DGSPJ) fördert ferner jährlich mit bis zu 5.000 € Forschungsprojekte aus allen Bereichen der Sozialpädiatrie.

Angesichts der Inhalte sozialpädiatrischer Arbeit ist eine Unterscheidung zwischen biomedizinischer und klinischer Forschung im klassischen Sinn und Versorgungsforschung sinnvoll.

45.3 Biomedizinische und klinische Forschung

Die biomedizinische und klinische Forschung beschäftigt sich überwiegend mit der **Wirksamkeit von diagnostischen und therapeutischen Maßnahmen unter Experimental- bzw. Idealbedingungen (=efficacy).**
Gerade im Bereich der Entwicklung von Kindern gibt es eine Reihe offener Fragen, was die Evaluation von Therapiemaßnahmen angeht. Spätestens seit der Metaanalyse von Dunst, Snyder und Mankinen von 1989 über die Wirksamkeit früher Interventionen ist die begrenzte Wirksamkeit der medizinisch-funktionellen Therapien evident. Dies darf nicht in einem therapeutischen Nihilismus enden, sondern bedarf umso mehr einer dezidierten Evaluation der eingesetzten Therapieverfahren. Die Devise „Viel hilft viel" kann heutzutage als überholt eingestuft werden. Neue Therapieansätze beziehen die Perspektive des betroffenen Kindes und der Familie mit ein und führen auch in Evaluationsstudien zu guten Ergebnissen. Als Beispiel hierfür sei das Canadian Occupational Performance Measure (COPM) genannt, was zunächst in der Ergotherapie evaluiert wurde, aber auch auf andere Ansätze angewandt werden kann.
Weitere Forschungsansätze in diesem Bereich überschneiden sich häufig mit anderen Fächern wie z. B. der Neuropädiatrie. Mehrere Studien, z. B. die Petö-Studie am Kinderzentrum München, haben sich mit der Evaluation von Therapiemaßnahmen bei Entwicklungsstörungen befasst. Auch therapeutische Interventionen wie das Heidelberger Elterntraining zur frühen Sprachförderung wurden entwickelt und evaluiert. Des Weiteren ist als aktuelles Projekt die deutsche Normierung des BOT-2 (Bruininks-Oseretsky Test of Motor Proficiency) unter Leitung des Kinderzentrums Maulbronn zu nennen.

45.4 Versorgungsforschung

Im Gegensatz zur klinischen Forschung fokussiert die Versorgungsforschung auf die wissenschaftliche Untersuchung der medizinischen Versorgung von Einzelnen und der Bevölkerung unter **Alltagsbedingungen.** Versorgungsforschung will Wissen erzeugen, analysieren und evaluieren

- über die Strukturen, Prozesse und Effekte der Gesundheitsversorgung
- über die Faktoren, die die Nachfrage, Produktion und Erbringung von Leistungen beeinflussen
- über Prozesse und Organisationen die gut laufen und solche, die verbessert werden müssen
- über Wege, die das Versorgungssystem effektiver und effizienter machen könnten

Es geht um **Wirkungen unter Routinebedingungen (=effectiveness),** was gleichzeitig einen Beitrag zur Sicherung bzw. Verbesserung der Qualität liefern kann und sollte. Bei den Forschungsansätzen steht die Ergebnisqualität im Zentrum der Überlegungen, wobei Struktur- und Prozessqualität der Versorgung als entscheidende Determinanten der Ergebnisqualität betrachtet werden. Wichtige Fragen der Qualitätsforschung sind, welche Prozesse und Organisationen gut laufen und welche verbessert werden müssen. Allerdings ist Qualität als Konstrukt nicht direkt messbar, weswegen eine Operationalisierung über Qualitätsindikatoren erfolgen muss.

An vielen Stellen in Deutschland werden Daten zur Prognose von Risikokindern, insbesondere sehr kleiner und extrem Frühgeborener, gesammelt, Die Studie zur Erfassung der Lebensqualität und Teilhabe von Kindern und Jugendlichen mit Zerebralparese wurde unter Federführung der Universität zu Lübeck im Rahmen der europaweiten **SPARCLE-Studie** durchgeführt und erbrachte wichtige Hinweise für die Behandlung von Kindern mit Zerebralparese. Diese seien nur beispielhaft angeführt. Auch eine Reihe von Untersuchungen zu verschiedenen Themen der Gesundheit inkl. Prävention, die am Institut für Soziale Pädiatrie und Jugendmedizin in München in den vergangenen Jahren durchgeführt wurden, sind in diesem Kontext zu sehen.
Ohne Zweifel können Forschungsaktivitäten gerade aus sozialpädiatrischen Zentren, aber auch aus Rehabilitationskliniken oder dem öffentlichen Gesundheitsdienst wesentliche Informationen liefern. Sie sind aus den Bedürfnissen der Praxis entstanden, verfolgen einen interdisziplinären Ansatz und haben einen hohen Informationswert für die Öffentlichkeit und die Fortbildung. In der Regel besteht keine Abhängigkeit von Industrieinteressen, andererseits haben Krankenversicherungen, der Medizinische Dienst der Krankenkassen und die öffentlichen Gesundheitsämter oft großes Interesse an den Ergebnissen. Es gibt aber gerade auch unter den aktuellen Bedingungen erhebliche Schwächen bei der Forschung in diesem Bereich. Dies ist besonders durch einen Mangel an Zeit und Personal, fehlende Ausbildung für wissenschaftliches Arbeiten, unzureichend ausgebildetes Personal, vor allem im Bereich von Epidemiologie und Public Health, sowie fehlende Drittmittel bedingt. Die DGSPJ sieht einen erheblichen Forschungsbedarf im Sinne der Evidence-based Health Care und hat im Jahr 2011 das Forschungsmemorandum Sozialpädiatrie verabschiedet. Hierin werden wichtige Forschungsinhalte für die kommenden Jahre, aber auch Forderungen an die staatlichen Träger der Forschungsförderung und andere forschungsunterstützende Stellen wie Stiftungen für eine verbesserte Unterstützung festgehalten.
Zusätzliche Aktivitäten zur Qualitätssicherung und Versorgungsforschung entstanden im **Zentralen Qualitätsarbeitskreis der Bundesarbeitsgemeinschaft Sozialpädiatrischer Zentren.** Nach Festlegung von strukturellen und personellen Qualitätsanforderungen der Arbeit in SPZ durch das **Altöttinger Papier** und einer Reihe von Qualitätspapieren zu Strukturen, Prozessen und verschiedenen Krankheitsbildern wird diese Arbeit aktuell mit einem Schwerpunkt im Sinne der Versorgungsforschung weiter fortgesetzt. In fünf Arbeitsgruppen werden die Themen Strukturen, Migration, Transition, Register für seltene Erkrankungen und Qualitätsindikatoren bearbeitet.

45.5 Ausblick

Die aktuellen Aktivitäten zur Qualitätssicherung und Versorgungsforschung müssen fortgesetzt werden. Auf der Ebene der individuellen Versorgung sind Indikatoren für die Qualität der Behandlung zu entwickeln. Der patientenbezogene Zielerreichungsgrad muss hierbei im Mittelpunkt stehen. Aspekte der Patientenpräferenzen als Ressource in der sozialpädiatrischen Behandlung entwicklungsgestörter und behinderter Kinder und Jugendlicher sind einzubeziehen. Entsprechende Ansätze patientenrelevanter Endpunkte müssen identifiziert, entwickelt, bewertet und evaluiert werden. Diese Aspekte müssen auch bei der Evaluation medizinisch-funktioneller Therapien berücksichtigt werden.
Diese Fülle von Aufgaben ist nicht aktueller Auftrag der bereits bestehenden Institutionen. Weitere Forschungsmöglichkeiten müssen geschaffen werden, um die reichlich vorhandenen Daten zu sichten, auszuwerten und in den Kontext einzuordnen.

Literatur

Dunst CJ, Snyder SW, Mankinen M (1989) Efficacy of early intervention in: Wang MC, Reynolds MC, Walberg HJ (Ed.) Handbook of Special Education, Vol.3: Low Incidence Conditions; Pergamon Press; Oxford

Pfaff H, Neugebauer EAM, Glaeske G, Schrappe M. (2011) Lehrbuch Versorgungsforschung. Schattauer; Stuttgart

Von Troschke J (2005) Versorgungsforschung in der Prävention und Gesundheitsförderung. Z Prävention 28: 68–72:

Internetlinks

www.bundesaerztekammer.de/30/versorgungsforschung
http://www.cp-studie.de/
www.dgspj.de/forschungdersp.php
www.dkvf.de/content/e36/e175/MemorandumzurVersorgungsforschung

46 Internationale Sozialpädiatrie

Harald Bode

46.1 Allgemeines

Sozialpädiatrie als familien- und gemeinwesenorientierte Gesundheitsförderung, Prävention, Früherkennung und Behandlung von Erkrankungen im Kindes- und Jugendalter hat international einen hohen Stellenwert. In allen Ländern spielt dabei die wechselseitige Beziehung von Armut, mangelnder Bildung und Krankheit eine herausragende Rolle. Weltweit besteht insbesondere für Kinder ein **sozioökonomischer Gradient von Gesundheit und Krankheit** trotz unterschiedlicher politischer Systeme und Strukturen der Gesundheitsversorgung. Damit hat die Sozialpädiatrie international die Gruppe der **sozial benachteiligten und der unterprivilegierten Kinder und Jugendlichen** als wesentliche Zielgruppe ihrer Aktivitäten.

International findet man verschiedene Begrifflichkeiten, die das Aufgabengebiet der Sozialpädiatrie umschreiben. Die Begriffe **„Social pediatrics"** und **„Child public health"** bezeichnen eher Forschung, Lehre und politikorientierte akademische Aktivitäten, während die Begriffe **„Community pediatrics"** und **„Community child health"** eher konkrete gesundheitsbezogene Interventionsmaßnahmen umschreiben.

Der Begriff **„Sozialpädiatrisches Zentrum"** ist in Deutschland geprägt worden. Inzwischen existieren sozialpädiatrische Zentren mit dieser oder ähnlicher Bezeichnung (z. B. „Kinderzentrum") und vergleichbarer Arbeitsweise auch in der Schweiz, Österreich und anderen europäischen und außereuropäischen Ländern.

„Child health and developmental centers" sind weltweit sehr verbreitet. Diese Einrichtungen sind je nach Struktur der Region und den Ressourcen des Landes unterschiedlich ausgestattete,

Sozialpädiatrie in der Praxis. http://dx.doi.org/10.1016/B978-3-437-31630-3.00046-8
Copyright © 2014 Elsevier GmbH. All rights reserved.

meist wohnortnahe Einrichtungen, in denen familienorientierte Prävention, Gesundheits- und Entwicklungsberatung, Förderung und teils auch medizinische Grundversorgung für Kinder, Jugendliche und ihre Familien angeboten werden.
Im Rahmen dieses Kapitels können nur einige internationale und nationale Organisationen/Fachgesellschaften mit sozialpädiatrischer Zielsetzung etwas vertieft dargestellt werden. Außerdem werden die sozialpädiatrischen Strukturen in einzelnen Ländern beispielhaft und orientierend beschrieben.

46.2 UN-Kinderrechte

Die **UN-Kinderrechtskonvention** ist eine Grundlage sozialpädiatrischen Denkens und Handelns. Sie wurde von der UN-Generalversammlung 1989 angenommen. Von der Bundesrepublik Deutschland wurde sie 1990 unterzeichnet und trat 1992 in Kraft. Mit Ausnahme der USA und Somalias haben alle Länder die Kinderrechtskonvention ratifiziert. Allerdings bedeutet die Ratifizierung nicht, dass es in den unterzeichnenden Staaten keine Verletzungen der Kinderrechte gäbe. Die Bundesregierung hat die Kinderrechtskonvention bislang nur unter dem Vorbehalt unterzeichnet, dass das deutsche Ausländerrecht Vorrang vor den Verpflichtungen der Konvention hat. Eine Abschiebung von Asyl suchenden bzw. Flüchtlingskindern und -jugendlichen aus Deutschland ist daher weiterhin möglich.
Die folgende Auflistung fasst die wichtigsten Kinderrechte (Grundrechte) zusammen:
- Recht auf Gleichbehandlung und Schutz vor Diskriminierung unabhängig von Religion, Herkunft und Geschlecht
- Recht auf einen Namen und eine Staatszugehörigkeit
- Recht auf Gesundheit
- Recht auf Bildung und Ausbildung
- Recht auf Freizeit, Spiel und Erholung
- Recht, sich zu informieren, sich mitzuteilen, gehört zu werden und sich zu versammeln
- Recht auf eine Privatsphäre und eine gewaltfreie Erziehung im Sinne der Gleichberechtigung und des Friedens
- Recht auf sofortige Hilfe in Katastrophen und Notlagen und auf Schutz vor Grausamkeit, Vernachlässigung, Ausnutzung und Verfolgung
- Recht auf eine Familie, elterliche Fürsorge und ein sicheres Zuhause
- Recht auf Betreuung bei Behinderung.

Der vollständige Text der Kinderrechtskonvention findet sich unter www.national-coalition.de/pdf/UN-Kinderrechtskonvention.pdf.
In der National Coalition (NC) haben sich Organisationen und Initiativen mit dem Ziel zusammengeschlossen, die UN-Kinderrechtskonvention in Deutschland bekanntzumachen und ihre Umsetzung voranzubringen. Die großen pädiatrischen Fachgesellschaften sind Mitglied der NC.

46.3 UN – Behindertenrechtskonvention

Weltweit gibt es etwa 650 Mio. Menschen mit Behinderungen. Die Gewährleistung der Menschenrechte für diese Menschen als staatliche Aufgabe und gesellschaftliche Verpflichtung bezweckt das Übereinkommen der Vereinten Nationen über die Rechte von Menschen mit Behinderungen **(UN-Behindertenrechtskonvention).** Die Konvention ist am 26. März 2009 in Deutschland in

Kraft getreten. Sie bestärkt sozialpädiatrisches Denken und Handeln in der Praxis nachhaltig. In ihrer Präambel wird zur Begriffsbestimmung ausgeführt: „Zu den Menschen mit Behinderungen zählen Menschen, die langfristige körperliche, seelische, geistige oder Sinnesbeeinträchtigungen haben, welche sie in Wechselwirkung mit verschiedenen Barrieren an der vollen, wirksamen und gleichberechtigten Teilhabe an der Gesellschaft hindern können."

Die offizielle deutsche Übersetzung der ursprünglich englischsprachigen Fassung erfolgte ohne Beteiligung behinderter Menschen und ihrer Verbände. Diese haben insbesondere kritisiert, dass der englische Begriff **„inclusion"** in der deutschen Fassung mit **„Integration"** übersetzt wurde. Inzwischen wird daher neben der amtlichen eine deutschsprachige „Schattenübersetzung" gezeigt, die den Bedenken der behinderten Menschen und ihren Verbänden Rechnung trägt. Weiterhin gibt es für eine Version in leichter Sprache.

Wesentliches Ziel der UN-Behindertenrechtskonvention ist es, die Menschenrechte auch behinderten Menschen zukommen zu lassen, sie in ihrer Selbstbestimmung und – möglichst vollständigen – gesellschaftlichen Teilhabe zu stärken. Die Konvention strebt eine inkl. Gesellschaft an, in der der Behindertenbegriff nicht mehr negativ gesehen wird. Behinderung wird vielmehr als Teil von Normalität und Verschiedenheit betrachtet. Als wesentlich wird diesbezügliche Aufklärung und Bewusstseinsbildung in der Gesellschaft angesehen.

Die wichtigsten Grundsätze der Konvention sind:

- Achtung der dem Menschen innewohnenden Würde, seiner individuellen Autonomie, der Freiheit, eigene Entscheidungen zu treffen, und seiner Unabhängigkeit
- Nichtdiskriminierung
- volle und wirksame Teilhabe an der Gesellschaft
- Achtung vor der Unterschiedlichkeit von Menschen mit Behinderungen und Akzeptanz dieser Menschen als Teil der menschlichen Vielfalt
- Chancengleichheit
- Zugänglichkeit
- Gleichberechtigung von Mann und Frau
- Achtung vor den sich entwickelnden Fähigkeiten von Kindern mit Behinderungen und die Achtung ihres Rechts auf Wahrung ihrer Identität.

Besondere Aufmerksamkeit hat in Deutschland das Recht auf **„inklusive Bildung"** erlangt. Ein gemeinsamer Schulbesuch behinderter und nicht behinderter Schüler („Schule für alle") ist Ziel der UN-Behindertenrechtskonvention. In Artikel 24 Abs. 2 heißt es:

„Bei der Verwirklichung dieses Rechts stellen die Vertragsstaaten sicher, dass

- **a.** Menschen mit Behinderungen nicht aufgrund von Behinderung vom allgemeinen Bildungssystem ausgeschlossen werden und dass Kinder mit Behinderung nicht aufgrund von Behinderung vom unentgeltlichen und obligatorischen Grundschulunterricht oder vom Besuch weiterführender Schulen ausgeschlossen werden;
- **b.** Menschen mit Behinderungen gleichberechtigt mit anderen in der Gemeinschaft, in der sie leben, Zugang zu einen integrativen (bzw. inklusiven), hochwertigen und unentgeltlichen Unterricht an Grundschulen und weiterführenden Schulen haben;
- **c.** angemessene Vorkehrungen für die Bedürfnisse des Einzelnen getroffen werden;
- **d.** Menschen mit Behinderungen innerhalb des allgemeinen Bildungssystems die notwendige Unterstützung geleistet wird, um ihre erfolgreiche Bildung zu erleichtern (bzw. zu ermöglichen);
- **e.** In Übereinstimmung mit dem Ziel der vollständigen Integration (bzw. Inklusion) wirksame individuell angepasste Unterstützungsmaßnahmen in einem Umfeld, das die bestmögliche schulische und soziale Entwicklung gestattet, angeboten werden."

Umfang und Ausmaß der schulischen Integration/Inklusion behinderter Kinder und Jugendlicher in Deutschland sind in vielen Bundesländern noch nicht gesetzlich geregelt. Sie werden in unterschiedlicher Form erprobt bzw. realisiert. Vielfältige Aktivitäten der Betroffen und anderer Interessengruppen haben einen mühsamen Veränderungsprozess angestoßen (➤ Kap. 4, ➤ Kap. 13).
Weitere konkrete Rechte bzw. Ziele der UN-Behindertenrechtskonvention sind u. a.
- barrierefreier Zugang zur physischen Umwelt, Information und Kommunikation (Art. 9)
- Angeborenes Recht auf Leben jedes Menschen (Art. 10)
- Gleichberechtigte rechtliche Stellung (Art. 12)
- Freiheit vor Ausbeutung, Gewalt und Missbrauch (Art. 16)
- Persönliche Mobilität/größtmögliche Unabhängigkeit (Art. 20)
- Achtung der Privatsphäre (Art. 22)
- Höchstmaß an Gesundheit ohne Diskriminierung (Art. 25)
- Frühestmögliche Habilitation und Rehabilitation mit multidisziplinärer Bewertung der individuellen Bedürfnisse und Stärken (Art. 26)

Jeder Vertragsstaat hat zunächst nach zwei Jahren, danach alle vier Jahre einen Bericht über die Umsetzung der UN-Behindertenrechtskonvention abzugeben. Deutschland hat seinen ersten Bericht im August 2011 vorgelegt.

46.4 Länderübergreifende Aktivitäten

Die **WHO** (Weltgesundheitsorganisation) führt seit Jahren vielfältige Programme und Projekte mit dem Ziel durch, Gesundheit und Entwicklung von Kindern und Jugendlichen zu verbessern. Die Fortschritte werden im Zwei-Jahres-Rhythmus publiziert. Die Programmprioritäten werden regelmäßig überprüft und an neue Gegebenheiten angepasst. Programmschwerpunkte sind Entwicklung von Kindern und Jugendlichen, gesundheitsbezogene Kinderrechte, Prävention und Behandlung von Erkrankungen. Programmdetails finden sich unter www.who.int/child_adolescent_health/enn.
UNICEF (United Nations International Childrens Emergency Fund) ist ein Organ der Vereinten Nationen. Sie unterstützt hauptsächlich in den Entwicklungsländern Kinder und Mütter in den Bereichen Gesundheit, Hygiene, Ernährung, Familienplanung und Bildung. In Notsituationen leistet sie humanitäre Hilfe. Das UNICEF Innocenti Research Center in Florenz trägt mit wissenschaftlichen Aktivitäten zur weltweiten Umsetzung der Kinderrechte bei.
Zahlreiche weitere internationale **Nichtregierungsorganisationen** leisten – meist durch Spenden finanziert – unverzichtbare konkrete Nothilfe, Gesundheitsförderung, Entwicklungshilfe sowie Hilfe zur Selbsthilfe für benachteiligte und notleidende Kinder, Jugendliche und ihre Familie in vielen Teilen der Welt.
Die **ISSOP** (International Society for Social Pediatrics and Child Health) entstand 2011 aus der **ESSOP** (European Society for Social Pediatrics and Child Health). Die ISSOP ist eine wissenschaftliche Gesellschaft, die mit einem nicht mehr nur europäischen, sondern einem globalen Profil allen mit Sozialpädiatrie und Kinder- und Jugendgesundheit befassten Berufsgruppen offensteht. Ziele der ISSOP sind Verbreitung des Wissens in Sozialpädiatrie, Stimulation der sozialpädiatrischen Forschung, Veranstaltungen von Tagungen und die Zusammenarbeit mit nationalen und internationalen Behörden, um lokale und globale Strategien zur Verbesserung der Gesundheit und des Wohlbefindens von Kindern und jungen Menschen zu entwickeln. Die ISSOP hat u. a. das Ziel formuliert, dass bis 2015 weltweit jedes Kind Zugang zu einem informierten Gesundheitsfürsorgeanbieter haben soll. Die offiziellen Sprachen der ISSOP sind Englisch, Französisch, Russisch

und Deutsch. Die ISSOP veranstaltet einmal jährlich an wechselnden Orten der Welt einen mehrtägigen Fachkongress mit unterschiedlichen Themen aus der Sozialpädiatrie. Sie hat in der Vergangenheit mehrfach in verschiedenen europäischen Ländern „Summer-Schools" zur Sozialpädiatrie veranstaltet.
Die Zeitschrift "Child: Care, Health and Development" ist das offizielle Journal der ISSOP.
Die **EACD** (European Academy of Childhood Disability) ist eine akademische Gesellschaft von Fachleuten, die mit behinderten Kindern arbeiten. Die mehr als 500 Mitglieder sind Pädiater, Neuropädiater, Physiotherapeuten, Ergotherapeuten, Sprachtherapeuten, Psychologen, Krankenschwestern, Sozialarbeiter und Sonderpädagogen aus über 35 europäischen Ländern. Ziele der EACD sind die Entwicklung von hochqualifizierter Forschung und Lehre bezogen auf Krankheiten mit Behinderungscharakter im Kindesalter, die Verbesserung der Behandlung dieser Kinder sowie die Etablierung europaweiter professioneller Standards. Es werden jährliche Kongresse an wechselnden Orten in Europa sowie Workshops zu verschiedenen Themen organisiert.
Der **CIPS** (Club International de Pédiatrie) stellt eine frankophone multidisziplinäre Gruppe von Praktikern aus acht verschiedenen Ländern dar, die an der Gesundheit von Kindern und Adoleszenten interessiert sind. Die sozialpädiatrischen Schwerpunkte der Arbeit sind u. a.: Kinder von Migranten, Schulgesundheit, Unfälle, Sprachentwicklung, schulische Integration, Gewalt, Kinder in der Familie, gemeinwesenorientierte Gesundheitsförderung. Der Club trifft sich zweimal jährlich an unterschiedlichen Orten, meist in Frankreich, aber auch in anderen Ländern.
Die **EUSUHM** (European Union for School and University Health and Medicine) hat die Entwicklung und Förderung von Gesundheitsangeboten und -förderung in Schulen und Universitäten zum Ziel. Deutschland ist in dieser Organisation über den Berufsverband der Ärzte im öffentlichen Gesundheitsdienst vertreten. Die EUSUHM veranstaltet in zweijährigen Intervallen internationale Tagungen an wechselnden Orten in Europa.
Die **Internationale Aktion Sonnenschein** hat seit den 1980er Jahren in über 20 Ländern der Welt, insbesondere in Osteuropa, Asien und Südamerika, bei der Gründung zahlreicher Kinderzentren mitgewirkt, die nach dem Vorbild des Kinderzentrums München, des ersten Sozialpädiatrischen Zentrums in Deutschland, und den Prinzipien der Entwicklungsrehabilitation arbeiten. Zu ihren Aufgaben gehört auch die Aus-, Fort- und Weiterbildung der dortigen Fachkräfte.

46.5 Sozialpädiatrie in einzelnen Ländern

46.5.1 Schweiz

Bevölkerungsbezogene Sozialpädiatrie wird durch die Arbeitsgruppe für Sozial- und Präventivpädiatrie der Foederatio Paedo-medicorum helveticorum (ärztliche Union für Kinder- und Jugendliche – fPmh) vertreten.
Eine Kommission für Sozial- und Präventivpädiatrie der schweizerischen Gesellschaft für Pädiatrie **(SGP)** beschäftigte sich 1978–1999 mit dem schweizerischen Gesundheitsheft, dem Manual der Vorsorgeuntersuchungen in der Praxis, mit verschiedenen Abhandlungen über Fragen der Volksgesundheit und dem gesellschaftlichen Strukturwandel sowie mit den Problemen der heranwachsender Kinder in ihrem Umfeld.
Eine neue Gruppe kümmert sich seit 2005 um die allgemeine Gesundheit der Kinder in ihrem physischen und sozialen Umfeld. Sie stützt sich bei ihren Empfehlungen für eine kindergerechte Pflege und Prävention während des Wachstums und zur Gesunderhaltung des Kindes auf epide-

miologische Grundlagen. **Ziele** der Gruppe sind die Förderung und Akzeptanz der Sozialpädiatrie in der fPmh, vor allem durch:
- Erforschung epidemiologischer Zusammenhänge und Bedürfnisabklärungen
- Ausarbeitungen von Empfehlungen zur Prävention von Krankheiten, zur Unfallverhütung während der Entwicklung des Kindes und zur Betreuung des Kindes und der Jugendlichen
- Einstehen für die Rechte des Kindes
- Eingehen auf die und Wahrnehmen der Sorgen der Bevölkerung.

Die sozial- und präventivpädiatrische Gruppe versteht sich als Experten- und Referenzgruppe für alle sozialpädiatrischen Fragen in der Schweiz:
- Sie will eine Plattform zur Diskussion von sozialpädiatrischen Ideen anbieten
- Sie registriert die sozialpädiatrischen Bedürfnisse in der Schweiz und in der SGP und leitet sie an die SGP und die fPmh weiter
- Sie unterstützt und begutachtet Projekte in Forschung, Ausbildung und Gesundheitsförderung
- Sie nimmt die sozialpädiatrischen Bedürfnisse der Kinder in der Schweiz wahr und versteht sich als Ansprechpartner für alle sozialpädiatrischen Probleme
- Sie vertritt die SGP und die fPmh bei Problemen der öffentlichen Gesundheitsförderung, die Kinder in der Schweiz betreffen
- Sie fördert den Informationsaustausch bezüglich Sozialpädiatrie innerhalb der Schweiz und im Ausland.

Themen der sozial- und präventivpädiatrische Gruppe waren zuletzt: Sonderpädagogik, Adipositas, Migranten, Stillen, Prävention.

Die Gruppe arbeitet auf 3 Ebenen:

politisch: Vorbereitung von Positionspapieren für die SGP-fPmh zu den Themen Prävention und öffentlicher Gesundheit, die Einfluss auf die Schweizer Politik und die Gesetze nehmen sollen
- strategisch: Repräsentation der SGP bei nationalen Präventionsprojekten (z. B. Impfungen)
- operational: Supervision bei den Revisionen des Schweizer Gesundheitsheftes (2006) und der Vorsorgeuntersuchungen (2011).

Andere Arbeitsgruppen befassen sich mit den Themen Schulgesundheit, Kinderschutz, SIDS und Ernährung. Die Gruppe veröffentlicht einen Jahresbericht in der Zeitschrift „Paediatrica“. Sie hat zu Fragen der Prävention Informationen für Eltern auf der Homepage der SGP und für Pädiater in der „Paediatrica“ veröffentlicht.

Die sozial- und präventivpädiatrische Gruppe besteht vor allem aus Kinderärzten, Kinderpsychiatern und Kinderchirurgen. Sie steht anderen interessierten Personen (vor allem Krankenschwestern und Sozialarbeitern) offen.

Die individualmedizinische Seite der Sozialpädiatrie wird in der Schweiz durch die Schweizerische Gesellschaft für Entwicklungspädiatrie (**SGEP**) vertreten. Sie wurde 2003 innerhalb der SGP gegründet und hat etwa 140 Mitglieder. Die SGEP ist eine Gesellschaft, die sich innerhalb der Schweizerischen Gesellschaft für Pädiatrie (SGP) für die Entwicklungspädiatrie einsetzt. **Entwicklungspädiatrie** befasst sich nach ihrem Verständnis mit der normalen Entwicklung von Kindern und Jugendlichen, ihren biologischen Voraussetzungen und den auf sie einwirkenden Umwelteinflüssen, sowie mit Entwicklungs- und Verhaltensstörungen im Kindes- und Jugendalter, ihrer Erforschung, Abklärung, Behandlung und Betreuung. Die SGEP setzt sich im Interesse der Klinik- und Praxispädiatrie in der Schweiz folgende Ziele:

a. Erarbeiten von Standards in Entwicklungspädiatrie.

b. Erarbeiten von entwicklungspädiatrischen Weiterbildungskonzepten und Anerkennungskriterien in Zusammenarbeit mit der SGP.

c. Anbieten und Unterstützen von Weiter- und Fortbildungsveranstaltungen für Pädiater und andere interessierte Fachgruppen.
d. Zusammenarbeit mit anderen Fachgruppen, die sich mit Kindern und Jugendlichen beschäftigen.
e. Initiieren und Unterstützen von Forschungsprojekten zur Erweiterung des Wissens über die kindliche Entwicklung und deren Störungen sowie zur Überprüfung von Abklärungs- und Therapiekonzepten.
f. Vertreten der Interessen der Kinder und Jugendlichen aus Sicht der Entwicklungspädiatrie gegenüber Behörden, Organisationen, Fachgruppen und Öffentlichkeit in Absprache mit den Gremien der Schweizerischen Gesellschaft für Pädiatrie SGP.

Wesentlicher Teil ihrer Aktivitäten war die Definition des Weiterbildungsgangs „Entwicklungspädiatrie". Dieser wurde als Schwerpunkt innerhalb der Pädiatrie im Herbst 2008 von der in der Schweiz für die Festlegung der Weiterbildungsordnungen zuständigen Foederatio Medicorum Helveticorum (FMH) anerkannt. Die Inhalte der Weiterbildung im Schwerpunkt Entwicklungspädiatrie entsprechen in vielem der in Deutschland angestrebten Zusatzweiterbildung Sozialpädiatrie. Die Gesamtdauer der Weiterbildung beträgt 2 Jahre, wovon 1 Jahr im Rahmen der Weiterbildung zum Facharzt für Kinder-und Jugendmedizin absolviert werden kann. Einzelheiten sind der Homepage der Gesellschaft zu entnehmen (http://www.entwicklungspaediatrie.ch).
Seit einigen Jahren existiert in Winterthur ein Sozialpädiatrisches Zentrum.

46.5.2 Österreich

Die AG Cerebrale Entwicklungsstörungen und die AG Sozialpädiatrie haben sich 2003 zur gemeinsamen **AG Entwicklungs- und Sozialpädiatrie** zusammengeschlossen.
Aktivitäten und Vorhaben zeigen sich in Form von Papieren, Beteiligungen an fach- bzw. gesundheitspolitischen Prozessen und an konkreten Arbeitsinhalten:
Im Sinne der Selbstdefinition des sozialpädiatrischen Anteils der AG wurde das Papier „Sozialpädiatrie in Österreich" verabschiedet und vom Vorstand der Österreichischen Gesellschaft für Kinder- und Jugendheilkunde (ÖGKJ) angenommen. Ein Organigramm zu Arbeitsfeldern der Sozialpädiatrie wurde entworfen. Es wurden Stellungnahmen zur flexiblen Schuleingangsphase, zur konduktiven Mehrfachförderung, zum Autismusscreening im Mutter-Kind-Pass, zum Musiktherapiegesetz und zur ambulanten Rehabilitation verfasst. Die Papiere „Standards zur Diagnostik und Therapie des ADHS" und „Flächendeckende Versorgung zur Habilitation beeinträchtigter Kinder und Jugendlicher am Beispiel Wiens" wurden erstellt.
Die Arbeitsgruppe war auch am „Gesundheitsplan Pädiatrie 2005" des Bundesministeriums für Gesundheit (BMG) mit dem Papier „Zur sozialpädiatrischen (und sozialkinderpsychiatrischen) Versorgung für entwicklungsbeeinträchtigte oder von Beeinträchtigung bedrohte Kinder und Jugendliche und deren Familien" beteiligt. Im Rahmen des vom BMG initiierten „Neurorehabilitationsplans für Kinder und Jugendliche in Österreich" wurde der Teil zur poststationären/ambulanten Betreuung beigetragen. Die AG beteiligte sich an Beratungen zur Pflegegesetznovelle, der Neuordnung der sozialpädiatrischen Versorgung und der Vorschuluntersuchung in Niederösterreich sowie an verschiedenen Fachtagungen.
Die AG war beteiligt an einer Kommission zur Vereinheitlichung der Durchführung der Pflegegeldbegutachtungen, wurde mehrfach zu Stellungnahmen zu Errichtungsanträgen von kinderpsychiatrischen Ambulatorien bzw. solcher zur Betreuung behinderter Kinder angefragt.

Neben den bereits früher etablierten „Entwicklungsdiagnostischen Ambulatorien" führen inzwischen 11 Einrichtungen den Titel „Sozialpädiatrie" in ihrem Namen.
Im Rahmen der Jahrestagungen der ÖGKJ wird eine eigene wissenschaftliche Sitzung abgehalten. Die AG hat etwa 100 Mitglieder und ist Mitglied der Deutschen Gesellschaft für Sozialpädiatrie und Jugendmedizin.
Aus der erlebten sozialpädiatrischen Unterversorgung und Defiziten im tertiären Versorgungssektor hat sich in den letzten Jahren eine Plattform **„Politische Kindermedizin"** gegründet. Sie befasst sich Organisationsstrukturen, die das Auftreten, Diagnostik und Verlauf von Krankheiten bei Kindern beeinflussen und will eine evidenzbasierte Gesundheitspolitik befördern. Es wird abgestrebt, in Anlehnung an das in angelsächsischen Ländern etablierte Prinzip der „child advocacy" fachkompetent Kinder und Eltern in gesundheitspolitischen Anliegen zu unterstützen, indem öffentlich Defizite in der österreichischen Kindermedizin thematisiert und notwendige Reformen und Verbessrungen angemahnt werden. Dies hat nicht unerwartet zu Konflikten mit vorgesetzten Behörden und der Politik geführt.
Die **„österreichische Liga für Kinder- und Jugendgesundheit"** wurde 2007 als gesundheits- und gesellschaftspolitisch orientierter Dachverband gegründet für Fachgesellschaften, Institutionen, Elternverbände und Einzelpersonen, die im Bereich der Kinder- und Jugendgesundheit tätig sind. Nach eigenen Angaben werden etwa 35.000 Professionelle und 1.8 Mio Eltern und betroffene über die Mitgliedsgesellschaften erreicht. Ziel ist die Verbesserung der präventiven, kurativen und rehabilitativen Gesundheitsversorgung von Kindern und Jugendlichen in Österreich. Jährlich wird ein Bericht zur „Lage der Kinder- und Jugendgesundheit in Österreich" veröffentlicht, der erhebliches politisches Echo und konkrete Folgen zeitigt: neu entwickelte Kindergesundheitsstrategien, Konzept „Frühe Hilfen", Österreichische Rahmengesundheitsziele mit Priorisierung der Kinder- und Jugendgesundheit, Kinder- und Jugendlichen-Versorgungsstudie, gesetzliche Vorgabe der verpflichtenden Prüfung von Auswirkungen jedes neuen Gesetzes auf die nächste Generation. Die Liga veranstaltet seit 2012 gemeinsam mit der ÖGKJ einen Lehrgang „Entwicklungs- und Sozialpädiatrie" mit 8 Modulen und insgesamt 108 Unterrichtseinheiten und fakultativen Vertiefungsseminaren mit derzeit etwa 70 Teilnehmern.

46.5.3 Großbritannien

Die **BACCH** (British Association for Community Child Health) ist die wichtigste Organisation, die sich mit gemeindenaher und sozialer Pädiatrie im Vereinigten Königreich befasst. Mitglieder sind Ärzte und andere Professionelle, die in der Pädiatrie oder für die Kindergesundheit im Gemeinwesen („Child Health in the Community") arbeiten. Die BACCH engagiert sich für die Kindergesundheit, stößt innovative gemeinwesenbezogene Praxisprojekte an und arbeitet für die Gleichheit in der Kindergesundheit. Einzelne Arbeitsschwerpunkte werden in speziellen Arbeitsgruppen behandelt:

- British Academy of Child Disability
- Child Protection Special Interest Group
- Child Public Health Interest Group.

Die BACCH ist dem RCPH (Royal College of Pediatrics and Child Health) angegliedert.
Die BACCH hat ein umfangreiches, gut strukturiertes Trainingsprogramm zum Erwerb der Kenntnisse, Fertigkeiten und Kompetenzen in der Ausbildung zum Spezialgebiet „Community Child Health" auf unterschiedlichen Levels der Spezialisierung ausgearbeitet, das landesweit zum

Einsatz kommt. Dieses Trainingsprogramm wurde bei der Erstellung des Themenkatalogs Sozialpädiatrie der DGSPJ berücksichtigt. Die folgende Auflistung zeigt die **Grundstruktur und die Weiterbildungsinhalte** für die Spezialisierung in „Community Child Health" des Royal College of Pediatrics and Child Health:

- Wissenschaftliche Grundlagen
- Allgemeine klinische Fertigkeiten
- Allgemeine technische Fertigkeiten
- Managementfertigkeiten
- Fertigkeiten in Forschung/Audit/Service-Evaluation
- Unterricht/Lernfähigkeiten
- Versorgungssysteme des Vereinigten Königreichs
  - Gesundheitsversorgung
  - Andere Versorgungssysteme
- Nationale/internationale Perspektiven
- Bevölkerungsbezogene Pädiatrie/Public Health
  - Screening und Surveillance
  - Impfungen und Kontrolle infektiöser Erkrankungen
  - Gesundheitsschutz, -förderung, -erziehung
  - Epidemiologie
  - Erfassung der Public-Health-Bedürfnisse
- Sozialpädiatrie
  - Benachteiligte Kinder
  - Kinderschutz
  - Adoption und Pflegefamilien
- Entwicklungspädiatrie und Pädiatrie für Behinderte
  - Allgemeine Prinzipien
  - Schwierigkeiten beim Lernen
  - Spezifische Lernstörungen
  - Probleme mit der Bewegung
  - Probleme mit der Koordination
  - Verlust von Fähigkeiten
  - Kommunikationsprobleme
  - Erworbene Behinderungen
  - Probleme mit dem Sehen
  - Probleme mit dem Hören
  - Kontinenzprobleme
- Verhaltenspädiatrie
- Unfälle und Verletzungen.

Eine mindestens zweijährige Weiterbildungszeit ist für diejenigen Ärzte vorgesehen, die später die Aufgaben eines „Consultant Community Pediatrician" übernehmen wollen. Es existieren Strukturvorgaben für die jeweiligen Ausbildungsstellen. Weiterbildungsmodule finden an regionalen und überregionalen akkreditierten Trainingszentren für „Community Child Health" statt. Man diskutiert in diesem Kontext auch die Etablierung eines „Master of Science" (MSc) in „Pediatrics and Child Health". Außerdem sind in den Dokumenten des Royal College of Pediatrics and Child Health (RCPCH) detailliert die Aufgaben und eine Arbeitsplatzbeschreibung eines „Consultant Pediatrician" im „Community Child Health" verzeichnet.

46.5.4 USA

Bereits 1970 wurde an der Kinderklinik in Monte Fiore/Albert Einstein College of Medicine in New York ein dreijähriges Trainingsprogramm in „Social Pediatrics" etabliert, das bis heute etwa 150 Pädiater absolviert haben. Das in der Bronx gelegene Ausbildungszentrum tritt für die Anliegen von benachteiligten Kindern und Familien in Armutsverhältnissen ein. Ein Großteil der Kursabsolventen ist weiterhin medizinisch in unterversorgten Gebieten tätig.

Seit 2002 ist die Entwicklungs- und Verhaltenspädiatrie (**„Developmental-Behavioral Pediatrics"**) als Subspezialität der Pädiatrie vom ABP (American Board of Pediatrics) anerkannt. Dazu ist nach Zertifizierung in Allgemeinpädiatrie (Dauer mindestens 3 Jahre Vollzeittätigkeit und Prüfung) und eine weitere 3-jährige Vollzeitweiterbildung mit anschließender Prüfung an einer zertifizierten Weiterbildungsstelle erforderlich. Ein Entwicklungs- und Verhaltenspädiater („Developmental-Behavioral Pediatrician") evaluiert, berät und behandelt Kinder, Jugendliche und ihre Familien bei einem weiten Spektrum von Entwicklungs- und Verhaltensproblemen. Die unten stehende Auflistung zeigt eine Zusammenfassung des umfangreichen Lernziel- und Ausbildungskatalogs. Dieser Katalog wurde bei der Erstellung des Themenkatalogs Sozialpädiatrie der DGSPJ berücksichtigt.

Entwicklungs- und Verhaltenspädiatrie in dem USA-Lernzielkatalog beinhaltet:

- Grundlagen der Entwicklungs- und Verhaltenspädiatrie
- Biologische Mechanismen in Entwicklung und Verhalten
- Familiäre und gesellschaftliche Faktoren
- Elemente der Beurteilung und Management
- Adaptation an allgemeine Gesundheitsprobleme und deren Behandlung
- Entwicklungs- und Verhaltensaspekte bei chronischen Erkrankungen und deren Behandlung
- Kognitive/adaptive Behinderungen
- Sprach- und Lernstörungen
- Motorische Behinderungen und multiple Handicaps
- Erkrankungen des Autismusspektrums
- Aufmerksamkeits-Defizit-Hyperaktivitäts-Störung
- Externalisierende Störungen
- Internalisierendes Verhalten und Störungen
- Substanzgebrauch und -missbrauch
- Kindesmisshandlung und -vernachlässigung
- Somatoforme Störungen und Schmerz
- Schlafprobleme
- Fütter- und Essstörungen
- Störungen der Elimination
- Sexualität
- Atypisches Verhalten
- Gesetze, Politik und Ethik
- Prinzipien der Forschung.

„Developmental-Behavioral Pediatrics" gehört in den USA zu den rasch wachsenden Subspezialitäten. Ende 2012 waren 651 Pädiater als „Developmental Behavior Pediatrician" in dieser Subspezialität vom ABP (American Board of Pediatrics) zertifiziert.

Die **SDBP** (Society of Developmental and Behavioral Pediatrics) ist eine internationale Organisation mit dem Ziel, die Gesundheit von Kindern und Adoleszenten zu verbessern, indem sie For-

schung, Lehre und klinische Praxis in „Developmental-Behavioral Pediatrics" befördert. Die Gesellschaft unterstützt das Verständnis der sozialen, erzieherischen und kulturellen Einflüsse auf Kinder. Die SDBP veranstaltet Jahrestreffen und lokale Veranstaltungen in den USA.
Die **AACPDM** (American Academy for Cerebral Palsy and Developmental Medicine) ist eine multidisziplinäre wissenschaftliche Gesellschaft, die sich die Aufgabe gestellt hat, die Zerebralparese und andere Erkrankungen mit Behinderungscharakter und Beginn in der Kindheit zu studieren, die Fortbildung von Fachleuten zur Behandlung dieser Erkrankungen zu fördern und die Lebensqualität von Menschen mit diesen Erkrankungen zu verbessern. Der Akademie gehören ähnliche Berufsgruppen an wie der EACD. Die AACPDM veranstaltet Jahrestagungen, diverse regionale und spezielle themenbezogene Meetings und Tutorien. Sie hat zu verschiedenen Problemen und Therapieverfahren bei Menschen mit Zerebralparesen umfangreich recherchierte Evidenzberichte publiziert.
Das **NICHD** (National Institute of Child Health and Human Development) ist der Kindergesundheit und Chancengleichheit verpflichtet. Das NICHD hat ein „Community Chart Health Research Network" (CCHN) etabliert, in dem mehrere Zentren wissenschaftlich kooperieren mit dem Ziel, die Interaktion von Gemeinwesen, Familien und Individuen mit biologischen Faktoren zu untersuchen und deren Einflüsse auf gesundheitliche Ungleichgewichte in Schwangerschaft und früher Kindheit zu erfassen.

46.5.5 Schweden

Schweden soll als Beispiel einer flächendeckenden staatlich gelenkten sozialpädiatrisch geprägten Gesundheitsstruktur dienen. Die Förderung und Überwachung der Kindergesundheit wird dort als prioritäre Aufgabe angesehen. Bereits vor über 60 Jahren wurden landesweit kostenlose **„Child Health Centers" (CHC)** gegründet. Ihre Aufgaben erweiterten sich mit den Jahren von Impfungen und Gesundheitsüberwachung zu Entwicklungskontrollen im Vorschulalter. Inzwischen gehören zusätzlich auch Elternerziehungsprogramme und Screeninguntersuchungen hinsichtlich Sprache, Seh-, Hörvermögen, Verhalten und psychomotorischer Entwicklung zu den Aufgaben der CHC. In ihnen werden nahezu 100 % der schwedischen Kinder von 0–6 Jahren jährlich untersucht.

46.5.6 Tschechien

Sozialpädiatrie in der Tschechischen Republik hat eine lange Tradition, aber keine einfache Geschichte. In kommunistischen Zeiten konnte man nicht über die Sozialpädiatrie sprechen. Es gaben damals keine Sozialprobleme, keine misshandelten Kinder. Dennoch haben Fachleute, vor allem Kinderärzte und Kinderpsychologen, dank ihrer internationalen Kontakte und Unterstützung besonders aus Deutschland die Sozialpädiatrie auf der Fachebene gut entwickelt. 1979 wurde das erste Lehrbuch der Sozialpädiatrie herausgegeben, das aber den Titel „Sozialpädiatrie" nicht tragen durfte, sondern „Sozialrechtliche Problematik im Kindesalter" hieß.
1992 wurde die Tschechische Gesellschaft für Sozialpädiatrie als fachliche und berufliche Vereinigung im Rahmen der Tschechischen Medizinischen Gesellschaft gegründet. Mitglieder sind nicht nur Mediziner, sondern auch diejenigen, die sich an der interdisziplinären Fürsorge für Kinder, insbesondere für bedrohte und behinderte Kinder, beteiligen. Die Gesellschaft hat etwa 150 Mitglieder.

Im Interessenbereich der Gesellschaft für Sozialpädiatrie stehen alle sozialpädiatrischen Tätigkeiten im erweiterten Sinne des Wortes, die sich aus der Beziehung zwischen Kind und Umgebung ergeben, die Bedürfnisse, Forderung und Rechte des Kindes in der Gesellschaft, vor allem die sozialpädiatrische Fürsorge für chronisch kranke, behinderte, vernachlässigte, missbrauchte und misshandelte Kinder sowie für Kinder in allen Formen der Ersatzfamilienfürsorge. Die Gesellschaft für Sozialpädiatrie will mit Gründung interdisziplinärer Teams beitragen zur Lösung von Problemen der Einrichtungen, die verschiedene Betreuungsformen für Kinder anbieten. Sie beteiligt sich an Forschungstätigkeiten und an der Fort- und Weiterbildung der Kinderärzte.
Die Tschechische Gesellschaft für Sozialpädiatrie hat gute Beziehungen zum Ministerium für Gesundheitswesen. Aktuell wird das Gesetz über die „Kinderzentren" (Sozialpädiatrische Zentren) beraten. Es handelt sich um ein ganz neues Modell der Fürsorge für Kinder nach dem deutschen Muster.
Nicht immer werden die Argumente der Tschechischen Gesellschaft für Sozialpädiatrie akzeptiert. Mehrere nicht staatliche Vereine übernehmen sehr oft die Fürsorge für behinderte und bedrohte Kinder, ohne fachliche Kenntnisse zu respektieren. Auch manche politischen Entscheidungen dienen nicht dem Wohl der Familie und ihrer Kinder. Die fachliche Kritik der Sozialpädiater findet leider kein großes Echo, sie ist für Medien und Öffentlichkeit nicht interessant.
International hat die tschechische Sozialpädiatrie sehr gute und enge Kontakte nach Deutschland, teilweise auch nach Großbritannien. Die Mitglieder beteiligen sich auch an internationalen Forschungsstudien.

46.5.7 Ukraine

In der Ukraine - Beispiel für ein Land mit einem sich mittlerweile differenzierenden Gesundheitssystem – war das Fachgebiet Sozialpädiatrie bis zum Jahr 1991 unbekannt. Sozialpädiatrie entwickelte sich seither dank einer Großspende des Betriebsrats des Volkswagenwerks im Rahmen der Wiedergutmachung im Zusammenhang mit der Beschäftigung von Fremdarbeitern während des Nationalsozialismus. Mit zusätzlicher Unterstützung der Bayerischen Staatsregierung wurde ein Ausbildungsprogramm für Ärzte, Psychologen, Therapeuten zum Aufbau eines ersten Sozialpädiatrischen Zentrums (SPZ) und Zentrums für Kinderneurologie an der Kinderklinik in Kiew realisiert. Ziel dieses Projekts ist die Verbesserung der gesundheitlichen Versorgung von Kindern mit Behinderungen und chronischen Krankheiten und ihren Familien. Rund 30 Fachleute wurden seit 1991 vornehmlich in München ausgebildet.
Ausgehend vom Zentrum in Kiew wurden rund 45 weitere Zentren in der Ukraine gegründet, davon 5 größere Zentren. Privatinstitutionen und Sponsoren finanzieren den Aufbau weiterer Zentren in privater Trägerschaft. Das Angebotsprofil ist unterschiedlich. In Kiew wurden rund ambulante 40.000 Konsultationen seit 1991 für Eltern mit Kindern und Jugendlichen mit Behinderung ermöglicht, 1.950 Kinder wurden stationär zu Diagnostik und Einleitung einer Therapie aufgenommen. Das SPZ in Kiew hat mittlerweile 98 Mitarbeiterinnen.
Seit 1991 wurden 18 ukrainisch-bayerische Symposien zur Sozialpädiatrie, Kinderneurologie/Neuropädiatrie und Neonatologie sowie Kinderchirurgie vornehmlich in der Ukraine organisiert. Diese Kongresse werden von Fachleuten aus der Ukraine, Weißrussland, Usbekistan, Turkmenistan und Russland besucht. In Kiew finden regelmäßige Seminare zu Themen der Sozialpädiatrie für Fachleute aus den GUS-Staaten statt. Derzeit läuft ein zertifiziertes Fortbildungscurriculum für Physiotherapeuten unter der Bezeichnung „Integrale Physiotherapie nach von Voss" als Ersatz für

nicht finanzierbare Bobath- oder Vojta-Kurse. Die SPZ in der Ukraine sind intensiv miteinander vernetzt. Seit 2012 existiert eine Zeitschrift „Soziale Pädiatrie und Rehabilitation“ unter deutscher Mitherausgeberschaft.
Als Folge des Aufbaus der Sozialpädiatrie in der Ukraine hat die Medizinische Akademie in Kiew beschlossen, einen Lehrstuhl für Sozialpädiatrie und Kinderneurologie im Jahr 2013 oder 2014 einzurichten. Der Aufbau des SPZ in Kiew veranlasste die Regierung der Ukraine, ein Gesetz zum Schutz von Kindern und Jugendlichen mit Behinderungen zu erlassen. Erste Elternselbsthilfegruppen haben sich gegründet.
Neben staatlich finanzierten Gesundheitseinrichtungen gewinnen privat finanzierte Institutionen (z. B. das Internationale Rehabilitationszentrum Kozijavkin) an Bedeutung, die z. T. internationale Kooperationspartner haben.
In der Ukraine existiert weiterhin in manchen Regionen eine hohe Morbidität für Zerebralparesen (3–4/1000). Mentale Retardierung, Syndrome, Verhaltensstörungen, Kinder mit Mehrfachbehinderungen, Kinder mit Behinderungen von alleinstehenden Müttern, Kinder aus sozial sehr benachteiligten Familien und Kinder mit Epilepsien stellen den Hauptanteil der Patienten der SPZ in der Ukraine. Es existieren noch große Versorgungs- und Aufklärungsdefizite, z. B. für Schwangere (landesweit organisierte und finanzierte Vorsorge – Untersuchungen fehlen), beim Aufbau eines effektiven perinatalen Versorgungskonzepts für gefährdete Früh- und Neugeborene, sowie bei lückenlosen Angeboten für Kinderimpfungen und zur Erfassung von Familien in Armut. Kinder mit mentalen Retardierungen werden ebenso unzureichend diagnostiziert, wie auch vor allem genetisch bedingte Syndrome, Verhaltensstörungen, Kinder mit Mehrfachbehinderungen, Kinder mit Behinderungen von alleinstehenden Müttern, Kinder aus sozial sehr benachteiligten Familien und Kinder mit Epilepsien etc. Der Status der Kinder mit mentalen Retardierungen ist sehr verbesserungswürdig. Gezielte und kompetente Diagnostik von erfahrenen Psychologen ist nur an wenigen Orten möglich. Ihre Förderung ist noch total unzureichend. Noch zu vielen von ihnen droht die Isolation. Bildungskonzepte für Kinder mit mentalen Entwicklungsstörungen existieren so gut wie noch gar nicht.
Aufgrund unzureichend finanzierter Gesundheitsversorgung in Folge eines immer noch mangelhaft funktionierenden staatlichen Gesundheitssystems und fehlender Krankenversicherungssysteme sind Familien mit Kindern mit Behinderung/Mehrfachbehinderung/chronischer Krankheit weiterhin auf ausländische Unterstützung angewiesen.

46.5.8 Indonesien

Indonesien ist in ein Entwicklungsland mit der viertgrößten Bevölkerung der Erde-. Ca. 50 % der Indonesier lebt unterhalb der Armutsgrenze (< 1,50 €/Tag). Die Gesundheitspolitik ist durch Überlebensstrategien gefordert. Hauptprobleme sind: 1. Infektionskrankheiten (TBC, Malaria etc.) 2. Untererernährung (1/4 der Kinder unter 5 Jahren) 3. hohe prä- und perinatale Mütter- und Kindersterblichkeit.
Die Regierung definiert Gesundheit als „ein Menschenrecht, das für […] das vom Land benötigte Arbeitspotenzial sorgt“: der Blick ist auf wirtschaftliches Wachstum gelenkt. Gesundheits- und Bildungsausgaben sind so niedrig wie in Indien. Die seit einem Jahrzehnt bestehenden 3 Krankenkassen nehmen chronisch kranke oder behinderte Kinder nicht auf. Es fehlt an ausgebildetem Gesundheitspersonal. Die Ratio Kinderarzt : Kinder ist in Indonesien 1 : 13.307 (in Deutschland 1 : 757). Wegen niedriger Gehälter arbeiten Ärzte in 2–3 Arbeitsstellen/Tag, ein Kinderarzt sieht ca.

50 Kinder/Tag; es gibt keine Ausbildung für Kindertherapien, eine erste Generation Sozialarbeiter sammelt Erfahrungen.
Ein Fundament der Gesundheitsversorgung für die arme Bevölkerung sind staatliche „Gesundheitszentren" samt Hilfs- und mobilen Zentren auf Rädern oder Booten in jedem Regierungsbezirk. Sie werden hauptsächlich von Krankenschwestern und Gemeindehelfern betreut, seltener von Ärzten. Sie sind jedoch ein wichtiger Ausgangspunkt für Gesundheitsfürsorge für Mutter und Kind. Sämtliche größere Gesundheitsprogramme wurden von internationalen Hilfsorganisationen wie UNICEFoder WHO finanziert und organisiert. Ihre diversen sozio-ökonomischen Hilfsmaßnahmen wurden im letzten Jahrzehnt von der UN-Kommission der Asia-Pazifikregion als „Millenniumsziele" gebündelt (MDG = Millennium Developmental Goals). Hauptziel: Halbierung der Armut bis 2015; Unterziele: Verminderung der Mangelernährung, Kinder- und Müttersterblichkeit. An den komplexen Ursachen wird gearbeitet.
Sozialpädiatrie wird als Subspezialisierung der Pädiatrie seit 1997 an einigen Universitäten gelehrt, ohne Curriculum. Auf Papier existieren alle 3 Interventionsebenen. In der Praxis konzentriert man sich, auch um die MDG zu erreichen, auf die primäre Praevention von Krankheiten. Kennenlernen der „Sozialen Umgebung des Kindes" heißt hier „Wie viele Bambusstückchen liegen herum?" (betreffend Tetanusinfektion durch Hautverletzungen) oder „Wie viele Personen leben in einem Raum? wie nah die Tiere?" Hinsichtlich kindlicher Entwicklung entwarf die Gesundheitsregierung Screeningprogramme, die in Erprobung sind.
Ganzheitliche Entwicklungstherapie in interdisziplinärer Teamarbeit samt Einbezug der Eltern wird bisher nur in Bandung, auf Java, realisiert, wo eine private Stiftung („YSK") seit 1987 mit deutscher und internationaler Hilfe ein erstes SPZ aufbaute. Der indonesische Kinderärzteverband ernannte es 1992 zum Modellzentrum, die EU finanzierte 2000 seinen Zentrumsbau, es behandelt jährlich ca. 12.000 Patienten, führt Weiterbildungen von Gesundheitszentrum bis Universität durch, aber überlebt nur mit Spenden, da politische Rahmenbedingungen für sekundäre und tertiäre Prävention fehlen.

Literatur

Blair M, Stewart-Brown S, Waterston T, Crowther R (2003) Child Public Health. Oxford. Oxford University Press.
Bode H (2004) Sozialpädiatrie – unverzichtbar in der pädiatrischen Fort- und Weiterbildung. Kinderärztliche Praxis 75: 301–309.
Kerbl R, Waldhauser F (2010) Politische Kindermedizin. Monatsschr Kinderheilkd 158: 324–325
Köhler L (1998) Child Public Health. Eur J Public Health 8: 253–255.
Kornfält R (2000) Survey of the pre-school child health surveillance programme in Sweden. Acta Paediatr Suppl 434: 2–7.
Largo R (2001) Weiterbildung in Entwicklungspädiatrie. Kinderärztliche Praxis 72: 316–317.
Levine M, Carey W, Crocker A (1999) Developmental Behavioral Pediatrics. 3. Aufl. Philadelphia: WB Saunders.
Lindstrom B, Spencer N (Hrsg) (1995): Social Pediatrics. Oxford: Oxford University Press.
Schlack H, von Kries R, Thyen U (Hrsg) (2009) Sozialpädiatrie. Gesundheitswissenschaft und pädiatrischer Alltag. Heidelberg: Springer
Waldhauser F, Harms E, Damm L, Kerbl L (2010). Politische Kindermedizin im internationalen Vergleich. Monatsschr Kinderheilkd 158: 326–337

Internetlinks

www.aacpdm.org
www.abp.org
www.bacch.org.uk
www.BACDis.org.uk
www.behindertenbeauftragter.de
www.cphig.org.uk
www.cpsig.org.uk
www.dgspj.de
www.docs4you.at
www.eacd.org
www.entwicklungspaediatrie.ch
www.eusuhm.org
www.issop.org
www.kinderjugendgesundheit.at
www.national-coalition.de
www.nichd.nih.gov
www.pediatrie-sociale.org
www.rcpch.ac.uk
www.sdbp.org
www.socialmedicine.org
www.swiss-paediatrics.org
www.theodor-hellbruegge-stiftung.de
www.unicef.org
www.who.int

Danksagung: N. Pellaud (Genf), K. Vavrik (Wien), N. Spencer (Warwick), F. Schneiberg (Prag) H. v. Voss (München) und S. v. Loh (Leipzig) für entscheidende Beiträge zum Kapitel 46.

IX Ausblick

IX

47 Ausblick

Harald Bode, Hans-Michael Straßburg und Helmut Hollmann

47.1 Die Zukunft der Sozialpädiatrie

Gesundheit und Krankheiten von Kindern und Jugendlichen müssen auch künftig im Kontext der gesellschaftlichen, wirtschaftlichen und politischen Entwicklungen gesehen werden, die nicht zuverlässig vorhersehbar sind. Allerdings ist aus den zurückliegenden Jahrzehnten ableitbar, dass die neuen Morbiditäten (z. B. chronische Erkrankungen, Adipositas, Entwicklungs- und Verhaltensauffälligkeiten, psychische Störungen, Störungen der sozialen Anpassung) einen hohen, evtl. weiter steigenden Anteil des medizinischen und therapeutischen Versorgungsbedarfs von Kindern und Jugendlichen ausmachen werden. Diese neuen Morbiditäten prägen in zunehmendem Maß auch die Lebens-, Lern- und Arbeitsbedingungen in Kindertagesstätten, Schulen und den Übergang in das Berufsleben.
Sozialpädiatrie befasst sich mit individuellen Kindern/Jugendlichen und mit bevölkerungsmedizinischen Themen. Sie bietet entsprechend ihrer Definition und ihrer Kompetenzen Hilfen und Wege zur Erhaltung oder Wiederherstellung von Gesundheit und normaler Entwicklung von Kindern und Jugendlichen an. Sie beinhaltet vertiefte Kenntnisse der entwicklungs-, familien- und umfeldorientierten interdisziplinären Behandlung von Störungen der Gesundheit und normalen Entwicklung und deren Auswirkungen. Die Zukunft des Querschnittsfachs Sozialpädiatrie ist daher auf verschiedenen Ebenen zu betrachten.

Sozialpädiatrie in der Praxis. http://dx.doi.org/10.1016/B978-3-437-31630-3.00047-X
Copyright © 2014 Elsevier GmbH. All rights reserved.

47.2 Inhaltliche Perspektiven

Wesentliche **Zukunftsaufgaben** der Sozialpädiatrie liegen in:
- Gesundheitsförderung und Krankheitsprävention
- Kinderschutz
- Stärkung sozial benachteiligter Kinder/Jugendlicher und ihrer Familien
- Prävention/Behandlung der neuen Morbiditäten im familiär-gesellschaftlichen Kontext
- Förderung der Entwicklung, Integration und Teilhabe von Kindern/Jugendlichen mit besonderen Bedürfnissen
- Stärkung der Elternkompetenz
- Transkulturelle Pädiatrie
- Qualitätssicherung, insbesondere bei:
  - Förder- und Integrationsmaßnahmen
  - Heil- und Hilfsmittelverordnungen
- Sozialpädiatrische Qualifizierung der in Praxis, Klinik und SPZ Tätigen durch Fort- und Weiterbildung
- Sicherung, Ausbau und Fortentwicklung sozialpädiatrischer Einrichtungen
- Forschung, vor allem Versorgungsforschung
- Vernetzung gesellschaftlicher Aktivitäten zugunsten des körperlichen, geistigen, seelischen und sozialen Wohls von Kindern/Jugendlichen
- Beratungsfunktion bei sozialpolitischen Themen wie Tagesstättenbetreuung, Schulen, Inklusion usw.
- Mitarbeit bei sozialrechtlichen und sozialethischen Themen
- Transitionskonzepte
- Interessenvertretung von Kindern/Jugendlichen und ihren Familien in Öffentlichkeit/Politik.

47.3 Zusatzweiterbildung (ZWB) Spezielle Sozialpädiatrie

Die ZWB Sozialpädiatrie wurde bereits 2008 als spezifische und zusätzliche fachärztliche Qualifikation bei der Bundesärztekammer (BÄK) beantragt worden. Diese Qualifikation geht über die von allen Kinder- und Jugendärzten zu fordernden Basiskompetenzen in Sozialpädiatrie hinaus. Nachdem der Antrag auf dem Deutschen Ärztetag 2010 überwiegend auf Grund von Bedenken niedergelassener Kinderärzte nicht behandelt wurde, liegt der Bundesärztekammer jetzt sowohl eine Definition der Sozialpädiatrie in der pädiatrischen Grundversorgung als auch der Antrag der DGSPJ zur Einführung einer Zusatzweiterbildung (ZWB) „Spezielle Sozialpädiatrie" vor. Auch wird es in der Gebührenordnung eine EBM-Abrechnungsziffer für sozialpädiatrische Leistungen von Kinderärzten geben. Wann nach einer Verabschiedung im Deutschen Ärztetag in den Landesärztekammern eine solche ZWB eingeführt werden kann, ist noch offen.
Die Weiterbildung in „Spezieller Sozialpädiatrie" wird vor allem im ambulanten Bereich (in kinder-/jugendärztlichen Praxen und SPZs) und im gewissen Umfang in Reha-Kliniken und dem öffentlichen Kinder- und Jugendgesundheitsdienst erfolgen können. Eine Kooperation von Kinder- und Jugendpsychiatrie, Sozialpädiatrie und Neuropädiatrie würde interessante fachliche und berufspolitische Perspektiven eröffnen und die Kinder- und Jugendmedizin vor allem im Bereich der

frühen Prävention und bei der Betreuung von Kindern mit Entwicklungsproblemen insgesamt stärken.

47.4 Sozialpädiatrische Zentren (SPZ)

Viele Eltern und Patienten erwarten spezielle sozialpädiatrische Kompetenz. Diese wird in den 145 SPZ in Deutschland weitgehend flächendeckend angeboten. Geschätzt 250.000 bis 300 000 Kinder und ihre Familien werden jährlich in SPZ betreut. Im Median kommt heute in Deutschland 1 SPZ auf 450.000 Einwohner. Die große Nachfrage nach Behandlungen in den SPZ und die langen Wartefristen sind ein Beleg für den hohen Bedarf. Präzise Richtgrößen für den künftigen Bedarf an SPZ existieren nicht. Insgesamt sollte man nur in sehr begrenztem Umfang weitere SPZ zulassen, sondern besser die vorhandenen SPZ durch Ausweitung der personellen und sachlichen Ausstattung an die bislang nicht befriedigte Nachfrage und neue Aufgabengebiete anpassen. Eine Anbindung an Kliniken für Kinder- und Jugendmedizin ist anzustreben.

Die Qualität der Arbeit in Sozialpädiatrischen Zentren wird auch künftig durch die Arbeit verschiedener interdisziplinär besetzter Qualitätszirkel kontinuierlich weiterentwickelt, die Ergebnisse werden in Form von **Qualitätspapieren** auf der Homepage der DGSPJ veröffentlicht. Für die Mitarbeiter müssen adäquate Fort- und Weiterbildungsmöglichkeiten angeboten werden. Eine wesentliche Aufgabe wird in den nächsten Jahren auch die Implementierung der Internationalen Klassifikation der Funktionsfähigkeit, Behinderung und Gesundheit (ICF) in einer auf Kinder mit sozialpädiatrischem Behandlungsbedarf zugeschnittenen Form sein. Hiermit soll erreicht werden, dass die Arbeit in SPZ, Frühförderstellen, Rehabilitationseinrichtungen u. a. vergleichbarer, transparent und alltagsrelevant abgebildet und gegenüber Kostenträgern noch klarer verdeutlicht werden kann.

Der Zugang zu einem SPZ wird nach §§ 119 und 43 a SGB V geregelt und nicht über Diagnosen. Deshalb können auch **chronisch kranke Kinder/Jugendliche** ein SPZ aufsuchen, wenn zusätzliche krankheitsübergreifende Störungen assoziiert sind oder als Folge der chronischen Erkrankung auftreten und ihre Behandlung ein interdisziplinäres Vorgehen erfordert. Dabei handelt es sich insbesondere um umschriebene Entwicklungsstörungen, Intelligenzminderungen, Verhaltens- und emotionale Störungen, somatoforme Störungen und schwierige psychosoziale Lebensumstände. Chronisch kranke Kinder sollten in SPZ betreut werden, die eng mit Kliniken für Kinder- und Jugendmedizin zusammenarbeiten, in denen die krankheitsspezifische Behandlung der Kinder erfolgt.

Bei chronischer Erkrankung oder Behinderung ist der Übergang vom Jugendlichen- ins Erwachsenenalter problematisch. Langjährige stützende und begleitende Strukturen wie SPZ oder pädiatrische Spezialambulanzen stehen nach dem 18. Lebensjahr nicht mehr oder nur noch eingeschränkt zur Verfügung. Gerade in SPZ müssen Konzepte zur **Transition** in die Erwachsenenwelt (Medizin, Beschäftigung, Sozialsystem) entwickelt und in Form von Transitionsplänen umgesetzt werden. Eine zentrale Forderung ist die Einrichtung sozialmedizinischer Zentren für Menschen mit geistiger und Mehrfach-Behinderung.

Um den Trägern der SPZs eine längerfristige Planung zu ermöglichen, sollte die Zulassungsdauer, die z. Zt. an vielen Stellen bei nur 2 Jahren liegt, auf mindestens 5, besser 10 Jahre verlängert werden.

47.5 Kinder- und jugendärztliche und therapeutische Praxen

Sozialpädiatrie wird verstärkt in kinder- und jugendärztlichen Praxen stattfinden müssen. Zur primären Prävention und Krankheitsfrüherkennung kommen neue sozialpädiatrische Aufgaben: Prävention vor und Kinderschutz bei Vernachlässigung/Misshandlung, transkulturelle Pädiatrie, Anwaltschaft für sozial benachteiligte Kinder/Jugendliche, Vermittlung gesundheitsfördernder Verhaltensweisen, Vermittlung von Elternkompetenzen, Begleitung und Stützung unsicherer oder überforderter Eltern, Behandlung von Kindern mit emotionalen und Verhaltensauffälligkeiten, Aufmerksamkeits- und Lernstörungen u. a. m. Die DGSPJ bietet zu diesen Themen seit Jahren meist ausgebuchte Fortbildungskurse an.

Ein „Qualitätslabel" ZWB Sozialpädiatrie muss sich positiv auf die Abrechnungsmöglichkeiten auswirken. Eine mit den Krankenkassen zu schließende „Sozialpädiatrie-Vereinbarung", die interdisziplinäre, z. T. nichtärztliche medizinische, psychologische und (heil-)pädagogische Maßnahmen als Zusatzleistung angemessen finanziert, könnte regionale Versorgungslücken schließen helfen. Eine solche Vereinbarung wäre von großem Interesse auch für die nichtärztlichen medizinisch-therapeutisch tätigen Berufsgruppen (Heilmittelerbringer) und für psychologisch und (heil-)pädagogisch tätige Fachleute. Sie entspräche dem heutigen sozialpädiatrischen Fachstandard einer vernetzten, interdisziplinären, entwicklungs- und familienorientierten Diagnostik und Therapie.

Kinderärztliche Einzelpraxen und damit eine flächendeckende hausärztliche pädiatrische Versorgung wird es in Zukunft immer weniger geben – umso mehr sollten Außensprechstunden o. ä. von sozialpädiatrisch versierten Kinderärzten gerade auch in ländlichen Regionen gefördert werden.

47.6 Kliniken für Kinder- und Jugendmedizin

In **Akutkliniken** mit ultrakurzen Liegezeiten und Patienten mit überwiegend schweren, z. T. sehr speziellen somatischen Erkrankungen gelingt es kaum, Ärzten in Weiterbildung alle Weiterbildungsinhalte der Kinder- und Jugendmedizin zu vermitteln. Man kann von ihnen nur verlangen, dass sie sozialpädiatrische Probleme identifizieren und die betreffenden Kinder, Jugendlichen und ihre Familien bei oder nach einer stationären Behandlung an sozialpädiatrisch kompetente Fachärzte weiterleiten. Sie müssen dazu sozialpädiatrische Basiskenntnisse besitzen. Zu deren Erwerb gibt es Fortbildungskurse.

Psychosomatische Stationen in Kliniken für Kinder- und Jugendmedizin können bei sinkender Bettenauslastung mit einem interdisziplinären bio-psycho-sozialen Behandlungskonzept den steigenden regionalen stationären sozialpädiatrischen Versorgungsbedarf innerhalb kinder- und jugendmedizinischer Einrichtungen mit abdecken. Auf diesen Stationen ist die Arbeit der Psychologen, Kinder- und Jugendlichenpsychotherapeuten, der Ergo-, Sprach- und Physiotherapeuten sowie (Heil-)Pädagogen sehr bedeutsam. Sie lässt sich bei genauer Dokumentation auch in entsprechenden OPS-Abrechnungsziffern abbilden.

Die 10 **sozialpädiatrischen Fachkliniken** in Deutschland haben eine überregionale Funktion in der spezialisierten Behandlung bei schwierigen Problemstellungen. Beispiele sind Mutter-Kind-Interaktions- und Verhaltensstörungen bei entwicklungsauffälligen und behinderten Kindern, schwere Ess-, Schluck- und Ernährungsstörungen, komplexe diagnostische und Rehabilitations-

maßnahmen, Hilfsmittelerprobungen, Traumatisierungsfolgen. Monatelange Wartezeiten zeigen, dass die Zahl stationärer sozialpädiatrischer Behandlungsplätze zu gering ist. Eine überregionale Absprache der Aufgabenbereiche mit dem Leistungsangebot der kinder- und jugendpsychiatrischen Kliniken und der stationären kinderneurologischen Rehabilitationseinrichtungen ist anzustreben und sachlich möglich.
Reha-Kliniken für Kinder/Jugendliche arbeiten heute überwiegend nach sozialpädiatrischen Prinzipien. Ihr Ausbau wird unterstützt durch das gesetzlich festgelegte Recht auf eine Mutter-Kind-Kur und deren Finanzierung durch die gesetzlichen Krankenkassen.
Reha-Kliniken werden ihre Behandlungskonzepte weiterentwickeln müssen, da anhaltende Effekte stationärer Reha-Behandlungen bislang nur begrenzt nachzuweisen sind. Begleitung im Lebensumfeld und lückenlose Reha-Ketten bieten dazu vielversprechende Ansätze. Ein wichtiger und weiter zu entwickelnder Bereich ist dabei die **ambulante Rehabilitation.**

47.7 Heil- und Hilfsmittel

Sozialpädiater werden als fachliche Instanz bei strittigen Verordnungsfragen von Heil- und Hilfsmitteln von Eltern, Vertragsärzten und Kostenträgern zu Rate gezogen und sehen hier eine ihrer Kernkompetenzen. Die Leitlinien zur Heilmittelverordnung der DGSPJ werden, möglichst auf Konsensbasis mit anderen beteiligten medizinischen und therapeutischen Berufsgruppen, regelmäßig aktualisiert. Standardisierte und alltagsrelevante Dokumentation von Befunden und Therapieverlauf sowie multizentrische Forschungsprojekte können dazu beitragen, einer zunehmenden undifferenzierten Heilmittelanwendung entgegenzuwirken zugunsten zielgerichteter, wirksamer und wirtschaftlicher Verordnungen. Es ist schwierig, dieses Ziel mit den Wünschen der Eltern und z. T. der Erzieher abzustimmen. Hier ist auch die Politik zu einer Positionierung aufgefordert. Der unübersichtliche Markt der Hilfsmittel erfordert eine kompetente und kontinuierliche sozialpädiatrische Bewertung auf dem Boden fundierter interdisziplinärer und multizentrischer Erfahrungen. Eine enge Kooperation mit dem Medizinischen Dienst der Krankenkassen würde interessante Synergiemöglichkeiten und fachliche Entwicklungen ermöglichen

47.8 Neue Versorgungsformen

Die in der **Komplexleistung Frühförderung** und der **sozialmedizinischen Nachsorge** verankerten Konzepte sind als ganzheitliche Ansätze zu begrüßen. Sie liegen im Interesse der Kinder/Jugendlichen und ihrer Familien. Politisch gut gemeint, ist ihre Umsetzung schwierig, die praktische Durchführung umständlich und bislang defizitär. Die in Deutschland sehr ausgeprägte „Versäulung“ der Leistungsbereiche und Träger und die föderale und kommunale Strukturpolitik machen eine effektive Arbeit und auskömmliche Finanzierung aus einer Hand über Zuständigkeitsgrenzen hinweg kaum möglich.
Es ist problematisch, wenn neue Kleinststrukturen geschaffen werden, die wirtschaftlich kaum überlebensfähig sind und unter administrativen Lasten leiden. Es wäre besser, wenn auch in Zukunft SPZ (§§ 43 a und 119 SGB V), hochspezialisierte ambulante Behandlung (§ 116b SGB V), interdisziplinäre Frühförderung (§§ 30 und 56 SGB IX) und sozialmedizinische Nachsorge (§ 4 3 Abs. 2 SGB V) von vernetzten Teams unter einem Dach bzw. einer administrativen

Zuständigkeit geführt werden könnten. Nachdem es den Kostenträgern trotz jahrelanger Gespräche und Verhandlungen nicht gelang, ist es die Aufgabe der Politik, durch Festlegung einer ausreichenden Finanzierung der Leistungen diese in ihrer fachlich notwendigen Qualität flächendeckend zu sichern. **Medizinische Versorgungszentren** (MVZ), in denen spezialisierte und hausärztlich tätige Kinder- und Jugendärzte, Ärzte anderer Berufsgruppen wie Orthopäden, Augenärzte oder Pädaudiologen, Psychologen und Therapeuten sowie Hilfsmittelanbieter räumlich und ggf. inhaltlich aufeinander abgestimmt arbeiten, werden parallel zu den SPZ entstehen. Die von der DGSPJ in Qualitätspapieren entwickelten Standards sollten für diese MVZ ein Maßstab sein.

47.9 Bevölkerungsmedizinische Aktivitäten

Sozialpädiater erkennen neue Gesundheitsgefährdungen und Morbiditäten früh. Sie sollten in Öffentlichkeit und Medien Stellung beziehen und fachlich beraten zu Themen wie soziale Ungleichheit, Medien- und Drogenkonsum, Schutz- und Risikofaktoren für Gesundheit und gute Entwicklung von Kindern und Jugendlichen in den heutigen Lebenswelten. Dem Öffentlichen Kinder- und Jugendgesundheitsdienst kommt dabei eine besonders wichtige Rolle zu. Seine Stellung und personelle Ausstattung muss gestärkt, die Qualifizierung der Mitarbeiter durch attraktive Stellenausstattung verbessert werden. Gesundheitswissenschaftlich begründete, **gemeinwesenbezogene Aktivitäten** zur Förderung der Kinder- und Jugendgesundheit müssen politisch und finanziell stärker unterstützt werden. Wichtige aktuelle Themen sind z. B. Kinderschutz, die spezifischen Anforderungen an Kinderkrippen, Entwicklungs- und Gesundheitsförderung im Vorschulalter, in Ganztagesschulen, Integrationseinrichtungen und anderen Formen institutioneller Betreuung, Einschulungsuntersuchungen, transkulturelle Pädiatrie und der Übergang (Transition) von Jugendlichen mit besonderen Bedürfnissen ins Erwachsenenalter.

Bevölkerungsbezogene epidemiologische Forschung hat überwiegend sozialpädiatrische Themen zum Inhalt und heute erfreulicherweise einen hohen gesellschaftlichen Stellenwert. Die Forschungsergebnisse werden von der Öffentlichkeit aufgegriffen und politisch umgesetzt. Beispiele sind der Kinder- und Jugendgesundheitssurvey, Berichte der Länder zur Kindergesundheit, Ergebnisse der Schuleingangsuntersuchungen. Erfreulicherweise werden diese Projekte inzwischen vielfach längerfristig fortgesetzt, sodass aus den longitudinal gewonnenen Ergebnissen wichtige repräsentative Erkenntnisse zum zeitlichen Verlauf von Gesundheit und Krankheiten von Kindern und Jugendlichen auf Bevölkerungsebene möglich werden. Daraus lassen sich Handlungsempfehlungen für die Gesundheits- und Sozialpolitik ableiten. Weitere bevölkerungsbezogene Themen der Sozialpädiatrie sind u. a. Präventions- und Früherkennungsmaßnahmen und die Versorgungsforschung.

47.10 Verankerung in Universitäten und Hochschulen

Eine deutlich stärkere Verankerung der Sozialpädiatrie in Universitäten und Fachhochschulen ist wünschenswert. Zurzeit findet außerhalb der medizinischen Fakultäten ein rascher Ausbau sozialpädiatrisch relevanter Bereiche insbesondere an **(Fach-)Hochschulen** statt. Studiengänge wie Sonderpädagogik, frühkindliche Pädagogik, integrative Frühförderung, interdisziplinäre Frühförderung, aber auch universitäre Studiengänge wie Entwicklungspsychologie, Public Health, Gesundheitswissenschaften und die Akademisierung der Therapeutenberufe mit neuen Bachelor- und Master-Studiengängen erhöht die Zahl qualifiziert und wissenschaftlich ausgebildeter Personen mit sozialpädiatrisch relevanten Kompetenzen.

Die **universitäre Sozialpädiatrie** sollte im Interesse der gesamten Kinder- und Jugendmedizin ebenfalls ausgebaut werden und mit diesen für die Zukunft der Kinder und Jugendlichen wichtigen anderen Ausbildungs- und Forschungseinrichtungen den interdisziplinären Kontakt suchen. Die universitäre Sozialpädiatrie wird sich mit individualmedizinischen und bevölkerungsbezogenen Themen der Gesundheitsförderung und der Prävention, Diagnostik und Therapie von Störungen der Gesundheit, Entwicklung und des Verhaltens vom Frühgeborenen bis zum Adoleszenten und jungen Erwachsenen befassen. Diese Themen haben einen hohen gesellschafts- und gesundheitspolitischen, aber auch einen ökonomischen Stellenwert.

Für die sozialpädiatrische Forschung ist eine angemessene **Finanzierung** erforderlich, um die komplexen Forschungsthemen – in Kooperation mit anderen nichtmedizinischen Wissenschaften – angehen zu können. Hier ist noch viel Überzeugungsarbeit bei den staatlichen und nichtstaatlichen Einrichtungen der Forschungsförderung nötig.

Literatur

AAP, AAFP et al. (2002) A Consensus Statement on Health Care Transitions for Young Adults with Special Health Care Needs. Pediatrics 111: 1.304–1.306.

Bode H (2004) Sozialpädiatrie – unverzichtbar in der pädiatrischen Fort- und Weiterbildung. Kinderärztliche Praxis 75: 301–309.

Bode H (2004) Welche Kinder brauchen ein SPZ? Kinderärztliche Praxis 75: 546–551.

Bode H (2007) Sozialpädiatrische Zentren – bei Entwicklungsstörungen im Kindesalter bewährt. Dtsch Arztebl 104: A1219-A1220.

Bode H (2007) Wenn Kinder erwachsen werden... Rückschritte statt Fortschritte. Medizin für Menschen mit geistiger oder mehrfacher Behinderung (Editorial) 4: 5–7.

Deutsche Gesellschaft für Sozialpädiatrie und Jugendmedizin. Qualität in der Sozialpädiatrie. http://www.dgspj.de/qspz.php

Deutsche Gesellschaft für Sozialpädiatrie und Jugendmedizin. Ambulante pädiatrische Rehabilitation. http://www.dgspj.de/llambulantereha.php

Ellsäßer G, Böhm A, Kuhn J et al. (2002) Soziale Ungleichheit und Gesundheit bei Kindern. Kinderärztliche Praxis 73: 248–257.

Hollmann H, Bode H (2007) Wie viele Sozialpädiatrische Zentren braucht das Land? Kinderärztliche Praxis 78: 276–279.

Kraus de Camargo O (2008) Übergang vom Jugendlichen- ins Erwachsenenalter: Neue Zukunftsoptionen auch für die Sozialpädiatrie. Kinderärztliche Praxis 79: 374–376.

Robert-Koch-Institut, Bundeszentrale für gesundheitliche Aufklärung (2008) Erkennen – Bewerten – Handeln: Zur Gesundheit von Kindern und Jugendlichen in Deutschland. Berlin Köln: Selbstverlag.

Index